Ecological Studies

Analysis and Synthesis

Volume 248

Ecological Studies is Springer's premier book series treating all aspects of ecology. These volumes, either authored or edited collections, appear several times each year. They are intended to analyze and synthesize our understanding of natural and managed ecosystems and their constituent organisms and resources at different scales from the biosphere to communities, populations, individual organisms and molecular interactions. Many volumes constitute case studies illustrating and synthesizing ecological principles for an intended audience of scientists, students, environmental managers and policy experts. Recent volumes address biodiversity, global change, landscape ecology, air pollution, ecosystem analysis, microbial ecology, ecophysiology and molecular ecology.

Graham P. von Maltitz • Guy F. Midgley •
Jennifer Veitch • Christian Brümmer •
Reimund P. Rötter • Finn A. Viehberg •
Maik Veste
Editors

Sustainability of Southern African Ecosystems under Global Change

Science for Management and Policy Interventions

Editors

Graham P. von Maltitz
Department of Botany and Zoology
Stellenbosch University
Stellenbosch, South Africa

Guy F. Midgley
Department of Botany and Zoology
Stellenbosch University
Stellenbosch, South Africa

Jennifer Veitch
South African Environmental Observation
Network (SAEON)
Cape Town, South Africa

Christian Brümmer
Thünen Institute of Climate-Smart
Agriculture
Braunschweig, Germany

Reimund P. Rötter
Tropical Plant Production and Agricultural
Systems Modelling (TROPAGS)
University of Göttingen
Göttingen, Germany

Finn A. Viehberg
Institute for Geography and Geology
University of Greifswald
Greifswald, Germany

Maik Veste
CEBra – Centrum für Energietechnologie
Brandenburg e.V.
Cottbus, Germany

ISSN 0070-8356   ISSN 2196-971X   (electronic)
Ecological Studies
ISBN 978-3-031-10950-8   ISBN 978-3-031-10948-5   (eBook)
https://doi.org/10.1007/978-3-031-10948-5

This work was supported by Johann Heinrich von Thünen-Institut

This Springer imprint is published by the registered company Springer Nature Switzerland AG
The registered company address is: Gewerbestrasse 11, 6330 Cham, Switzerland

Paper in this product is recyclable.

In memory off Mathieu Rouault

Mathieu Rouault passed away in early 2023. At the time he was the Director of the Nansen Tutu Centre for Marine Environmental Research and the South African National Research Foundation and SARChI Research Chairs for Ocean and Atmospheric Modelling, both in the Department of Oceanography at the University of Cape Town. He served three consecutive terms as President of the South African Society for Atmospheric Sciences (SASAS). The annual Stanley Jackson award for SASAS members who published an exceptional paper to enhance southern African atmospheric science and oceanography was won by Prof Rouault on several occasions. He was remembered for introducing the SASAS medal in recognition of an individual's outstanding contribution to research, education, or technical achievement in any of the SASAS fields. Prof Rouault's research encompassed a broad spectrum of ocean–atmosphere interactions including numerical modelling, experimental work at sea, meteorology, physical oceanography, climatology, and the impact of climate change and variability on marine ecosystems and water resources. His specific interest was in the impact and interplay of the El Niño-Southern Oscillation

(ENSO), Benguela Niño mechanisms, and the Agulhas Current on southern African weather and climate, in the context of climate variability and change. He participated in numerous multi-institutional research projects, both locally and internationally. Prof Rouault was a well-established and internationally recognised researcher and modeller as demonstrated by his international collaboration on numerous high impact publications, including Chap. 6 of this volume, co-authored with other internationally recognised scientists. His participation in the CLIVAR Atlantic and CLIVAR Africa panels was further witness to his international stature. Prof Rouault supervised numerous honours, master's, PhD, and postdoctoral students linked to a suite of research programmes in ocean and climate modelling that he initiated. In addition to his exemplary career, he had a passion for surfing, sailing, salsa, and his family—he really was so proud of his boys and loved taking them surfing at the weekends. Prof Rouault was popular among his peers and students, and he will be greatly missed.

Foreword

Dear reader,

Photo: Bundesregierung / Guido Bergmann

We will achieve more by working together. In the global pursuit of innovation to solve the grand challenges of our time, this applies particularly to science and research. Regions like southern Africa demonstrate the success of international research cooperation and the mutual benefit it is able to generate.

Ten years ago, the Federal Ministry of Education and Research initiated the "Science Partnerships for the Assessment of Adaptation to Complex Earth System Processes", SPACES for short, together with organisations in southern Africa. Cooperation between my Ministry, the South African Department of Science and Innovation, the Namibian Ministry for Higher Education, Technology and Innovation, and the Namibian National Commission on Research, Science and Technology provides the basis. The aim is to support joint projects conducted by German research institutions with partner institutions in South Africa, Namibia, and the neighbouring countries to improve our understanding of the region's sensitive ecosystems. The SPACES programme pools capacities and provides all those involved with access to a unique research infrastructure and key field sites.

As a result, major findings are now available about climate change and extreme weather processes as well as social-ecological aspects of food production that are vital for the region given the expected increase in heat waves, flooding, and droughts. The knowledge gained provides the basis for innovations, new technologies, and recommendations for action to promote the sustainable use of agricultural land as well as of coasts and seas. At the same time, it provides good orientation for political decision-making and transformation processes. The available results have already been included in publications of the IPCC and IPBES. This highlights their importance for sociopolitical debates and our actions as we move along the path set out by the Agenda 2030.

It is crucial for scientific findings to be quickly transferred to practical application. Such targeted transfer requires all stakeholders to be well-informed and well-trained. Besides sharing knowledge among researchers and promoting dialogue between science and policymakers, it is a matter of fostering young research talent. SPACES is also a pioneering programme in this respect: Young researchers from southern Africa and Germany have learned with and from each other by working on binationally supported research projects and participating in joint seminars, research cruises, and summer schools. They have gained an understanding of how terrestrial and marine landscapes in the region are changing and how their management can be improved. Furthermore, the cooperating partners developed an integrated training and knowledge sharing programme for both students and administrative staff. More than 100 people have already participated in the programme.

This book presents the latest findings for managing the valuable and diverse ecosystems in the temperate, subtropical, and tropical regions of southern Africa. These findings refer to processes in parts of another continent. Yet, they are also highly relevant for us in Europe because climate change does not stop at ocean margins or national borders. It influences marine currents moving from southern Africa to the North Atlantic. This also has an enormous impact on our temperate climate in northern Europe.

Today's world is growing closer together. International megatrends like sustainable management will shape future education, research, and innovation agendas. It is crucial that we cooperate, use the impetus provided by research, and translate its findings quickly into practical applications. I wish you interesting and inspiring reading. Above all, I hope it will give rise to many new ideas.

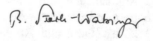

Bundesministerium
für Bildung
und Forschung

Member of the German Bundestag, Federal Minister   Bettina Stark-Watzinger
of Education and Research, Berlin, Germany

Over the last 50 years, many national governments have been led to invest more in counteractive and preemptive knowledge generation as growing human ambitions, numbers, and consumptive activities resulted in rapidly rising social-ecological impacts. Consequently, scientific evidence clarifying how society's future on Earth may be sustained convinced most countries to ratify the long list of international environmental commitments. The most prominent of those commitments is the United Nation's Sustainable Development Goals 2030.

Knowledge of Earth and social-ecological systems and our capacity to study those have arguably increased exponentially during the first two decades of this twenty-first century. This book produced by the SPACES II research programme is therefore a timely compendium and synthesis offering insights into current systems understanding complemented by southern African case studies pointing to potential management approaches.

The book aptly illustrates the immense complexity of environmental research and policymaking in southern Africa. The reader will come to appreciate how the challenges posed by the Sustainable Development Goals 2030 encompass a multitude of systemic interdependencies between the biophysical elements of the inherently variable regional Earth and social-ecological systems. Due to uncertainty within natural systems, discerning between natural variability and directional anthropogenic change requires long-term research by multi- and interdisciplinary research teams. Needless to say, these teams must include a broad range of social-ecologists to properly diagnose the role of humans as drivers of social-ecological processes and to codesign adaptive practices and policies where necessary.

The book's geographical coverage links the marine offshore systems with the inland savanna and clearly demonstrates that political boundaries are artificial constructs with little environmental relevance. Nationalist politicians and bureaucrats thus need to recognise that for a nation to prosper, regional and global cooperation towards Earth stewardship decision-making are essential elements. The SPACES II Programme as a multinational collaboration provides an example of the incremental benefits of international and regional science collaboration towards a sustainable future on Earth.

Of particular scientific interest is the mixed and complementary application of observational, experimental, and modelling approaches covered by the book.

By design, manipulative experiments can advance knowledge generation under simplified conditions to allow for the clarification of interdependencies between a small number of variables. So-called natural experiments can be useful to explore directional change over steep environmental gradients. Models are built on current and variable assumptions to be heuristic and predictive about relationships between key variables. Observation systems are designed to monitor those key variables over time and space. They are labour-intensive, time-consuming, and potentially costly, but in the long run can provide better answers to real-world ecosystem questions. All three approaches have recognised shortcomings but can be applied jointly to strengthen the total research output as illustrated by the relevant authors in this book.

Going forward it is inevitable that future research directions will increasingly be influenced by and addressed through highly instrumented national and global research infrastructures delivering higher volumes of data for improved vertical, horizontal, and temporal scope. These developments already require new cohorts of technologically literate researchers and data scientists to build on the experience and knowledge generated through the SPACES II and similar comprehensive environmental research programmes. Ultimately, as exemplified by this book, addressing society's local, regional, and global environmental grand challenges will be entirely dependent on open access to comprehensive long-term environmental data for analyses, predictive modelling, decision-making, and policy frameworks.

Past Managing and Founding Director: South African   Johan Christopher Pauw
Environmental Observation Network (SAEON),
National Research Foundation, Pretoria, South Africa

Acknowledgements

Today's interdisciplinary research on highly complex ecological systems can only be achieved through working in teams that bring together all relevant competences. In order to make the various results of the SPACES II programme including its publications possible, a large number of senior and junior staff members were involved in the implementation of the research activities and the analysis and interpretation of their results. We would like to thank the students, technicians, and ship crews but also the land users, landowners, farmers, and resource managers involved for their commitment to the research.

Special thanks go to Mari Bieri, who made the whole project and especially the summer schools possible and kept them on track with her tireless and always positive forward-looking work. Great thanks also to the regional coordinators and research managers in southern Africa for hosting and supporting the many scientists and students carrying out their "field work" at the study sites (on the land and the sea).

The editors gratefully acknowledge Lars Körner and his team at Springer Nature Publisher in Heidelberg for their help, patience, and technical support during the preparation of this Ecological Studies volume.

We would like to thank the Federal Ministry of Education and Research (BMBF) and the Project management DLR in Bonn and Project Management Jülich—PtJ in Rostock for their steady interest and support, especially Oliver Dilly, Olaf Pollmann, Dirk Schories, and Elisabeth Schulz. This book was only possible through the generous funding from BMBF's SPACES II programme which funded most of the background studies leading to the chapters in this book as well as funding the production of the book including ensuring it could be open access.

SPACES II
Science Partnerships for the
Adaptation to Complex Earth System
Processes in Southern Africa

Bundesministerium
für Bildung
und Forschung

The research projects listed below form the basis of this book and were carried out in the framework of the SPACES II collective research programme supported by the Federal Ministry for Education and Research (BMBF) based on a decision taken by the German Bundestag.

Acronym	Research project	Institution in Germany and partner in Southern Africa	Grant No.
ASAP	Agroforestry in Southern Africa—new pathways of innovative land use systems under a changing climate	Albert-Ludwigs-Universität Freiburg Leibniz-Zentrum für Agrarlandschaftsforschung (ZALF) e.V. Karlsruher Institut für Technologie—KIT CEBra—Centrum für Energietechnologie Brandenburg e.V., Cottbus Stellenbosch University Stellenbosch University of Pretoria Eduardo Mondlane University, Maputo, Mosambik The Copperbelt University (CBU) Kitwe, Sambia World Agroforestry Centre (ICRAF) Southern Africa Node Lilongwe, Malawi Southern African Science Service Centre for Climate Change and Adaptive Land Management (SASSCAL) Windhoek, Namibia	01LL1803A-D
EMSAfrica	Ecosystem Management Support for Climate Change in Southern Africa	Universität Bayreuth Johann Heinrich von Thünen-Institut: Institut für Agrarklimaschutz, Braunschweig Johann Wolfgang Goethe-Universität Frankfurt am Main Hochschule für Angewandte Wissenschaften Hamburg Friedrich-Schiller-Universität Jena Stellenbosch University University of the Witwatersrand University of Venda, Thohoyandou Rhodes Universit Grahamstown Council for Scientific and Industrial Research, Pretoria South African Environmental Observation Network, Pretoria, Grootfontein Agricultural Development Institute, Middelburg South African National Parks, Pretoria Southern African Science Service Centre for Climate Change and Adaptive Land Management, Windhoek	01LL1801A-E

(continued)

Acronym	Research project	Institution in Germany and partner in Southern Africa	Grant No.
ORYCS	Options for sustainable land use adaptations in savanna systems	Universität Potsdam Freie Universität Berlin Institut für sozial-ökologische Forschung (ISOE) GmbH Namibia University of Science and Technology, Windhoek University of Namibia Namibian Ministry of Environment, Forestry and Tourism	01LL1804A-C
SALDi	South African Land Degradation Monitor	Friedrich-Schiller-Universität Jena Universität Augsburg Eberhard Karls Universität Tübingen Deutsches Zentrum für Luft- und Raumfahrt e.V. (DLR) Agricultural Research Council, Pretoria South African National Parks, Pretoria Council for Scientific and Industrial Research, Pretoria South African National Space Agency, Pretoria Universities of Bloemfontein, Cape Town Pretoria, Stellenbosch, Witwatersrand Department of Agriculture, Land Reform and Rural Development, Pretoria Department of Forestry, Fisheries and the Environment, Pretoria Department of Water and Sanitation, Pretoria Eskom, GIS Centre of Excellence, Johannesburg South African Environmental Observation Network (SAEON): Expanded Freshwater and Terrestrial Environmental Observation Network (EFTEON), Pretoria White Waters Irrigation Board Mpumalanga White River Valley Conservation Board (WRVCB)	01LL1701A-D
SALLnet	South African Limpopo Landscapes Network	Georg-August-Universität Göttingen Senckenberg Gesellschaft für Naturforschung, Frankfurt Rheinische Friedrich-Wilhelms-Universität Bonn University of Limpopo, Sovenga University of Venda, Thohoyandou University of Witwatersrand, Johannesburg	01LL1802A-D

(continued)

Acronym	Research project	Institution in Germany and partner in Southern Africa	Grant No.
TRACES	Tracing Human and Climate impacts in South Africa	MARUM—Zentrum für Marine Umweltwissenschaften, Universität Bremen Universität Greifswald Friedrich-Schiller-Universität Jena Council for Geosciences (CGS) Pretoria University of KwaZulu Natal, Pietermaritzburg University of the Free State, Bloemfontein University of the Witwatersrand, Johannesburg	03F0798A-C
BANINO	Benguela Niños: Physical processes and long-term variability	GEOMAR Helmholtz-Zentrum für Ozeanforschung Kiel Leibniz-Institut für Ostseeforschung Warnemünde, Rostock Universität Hamburg National Fisheries Research Institute; INIP) Luanda and Namibe, Angola National Marine Information and Research Centre, Swakopmund, Namibia Gobabeb Research and Training Centre Gobabeb, Walvis Bay, Namib-Naukluft Park University of Cape Town	03F0795A-C
CASISAC	Changes in the Agulhas System and its Impact on Southern African Coasts	GEOMAR Helmholtz-Zentrum für Ozeanforschung Kiel Helmholtz-Zentrum Geesthacht GmbH (HZG) Geesthacht Universität Siegen Christian-Albrechts-Universität zu Kiel University of Cape Town Council for Scientific and Industrial Research, Stellenbosch University of Pretoria South African Environmental Observation Network, Pretoria	030796A-D
TRAFFIC	Trophic Transfer Efficiency in the Benguela Current	Leibniz-Zentrum für Marine Tropenforschung, Bremen Universität Bremen Universität Hamburg Johann Heinrich von Thünen-Institut: Institut für Seefischerei University of Cape Town Department of Forestry, Fisheries and the Environment, Cape Town Namibian Ministry of Fisheries and Marine Resources, Windhoek National Marine Information and Research Centre, Swakopmund University of Namibia	03F0797A-D
CaBuDe	Capacity Building Development	DAAD—Deutscher Akademischer Austauschdienst, Bonn	01LL1901A 03F0817A

Finally, the editors would like to thank all the contributing authors, without whose dedication this book would not have been possible. We are extremely grateful for all the reviewers who gave their time, sometimes against tight deadlines, to provide insightful reviews that greatly strengthened the quality of the final chapters: Augustine Ayantunde, Marion Bamford, Lisa Beal, Ermias Betemariam, Antonio Bombelli, Christian Brümmer, Manuel Chevalier, Richard Cowling, Anzel de Lange, Masih Eghdami, David Eldridge, Jennifer Fitchett, Ian Foster, Peter Gell, Graciela Gil-Romera, Joachim Hill, Andrew Hoell, Mario Hoppema, Klaus Kellner, Gregory Kiker, Nikolaus Kuhn, Jon Lovett, Beatriz Martinez, Simeon Materechera, Stephanie Midgley, Guy F. Midgley, Glenn Moncrieff, John Mupangwa, David Nash, Raphael Neukom, Tony Palmer, Johan Pauw, Laura Pereira, Romain Pirard, Ingo Richter, Ryan Rykaczewski, Andrew Skowno, Jasper Slingsby, Eric Smaling, Philip Stouffer, Rodolfo Cornejo Urbina, Brian van Wilgen, Jennifer Veitch, Maik Veste, Paul Vlek, Graham von Maltitz, Helen Wallace, Victor Wepener, and Corli Wigley-Coetsee. Finally to Willem Landman for the dedication.

Contents

Contributors

Issaka Abdulai Department of Crop Sciences, Georg-August-University of Göttingen, TROPAGS, Göttingen, Germany

Lamega Sala Alanda Division of Grassland Science, Department of Crop Sciences, University of Goettingen, Goettingen, Germany

Mina Anders Department of Crop Sciences, Functional Agrobiodiversity, Georg-August-University Göttingen, Göttingen, Germany

Ilse Aucamp Equispectives Research and Consulting Services, Pretoria, South Africa

Holger Auel BreMarE – Bremen Marine Ecology, Marine Zoology, University of Bremen, Bremen, Germany

Kingsley Kwabena Ayisi Risk and Vulnerability Science Center, University of Limpopo, Polokwane, South Africa

Jussi Baade Department for Physical Geography, Friedrich-Schiller-University, Jena, Germany
Department of Geography, Friedrich Schiller University Jena, Jena, Germany

Sara Yazdan Bakhsh Department of Agricultural Economics and Rural Development, University of Göttingen, Göttingen, Germany

Kai Behn Institute of Crop Science and Resource Conservation (INRES), University of Bonn, Bonn, Germany

Arne Biastoch GEOMAR Helmholtz Centre for Ocean Research Kiel, Kiel, Germany
Faculty of Mathematics and Natural Sciences, Kiel University, Kiel, Germany

Mari Bieri Thünen Institute of Climate-Smart Agriculture, Braunschweig, Germany

Niels Blaum Plant Ecology and Nature Conservation, University of Potsdam, Potsdam, Germany

Maya Bode-Dalby BreMarE – Bremen Marine Ecology, Marine Zoology, University of Bremen, Bremen, Germany

Bodo Bookhagen University of Potsdam, Potsdam, Germany

Mohammad Hadi Bordbar Leibniz Institute for Baltic Sea Research Warnemünde, Rostock, Germany

Lars Borrass Faculty of Environment and Natural Resources, University of Freiburg, Freiburg, Germany

Gennady Bracho-Mujica Tropical Plant Production and Agricultural Systems Modeling (TROPAGS), University of Göttingen, Göttingen, Germany

Peter Brandt GEOMAR Helmholtz Centre for Ocean Research Kiel, Kiel, Germany
Faculty of Mathematics and Natural Sciences, Kiel University, Kiel, Germany

Thomas Bringhenti Department of Crop Sciences, Georg-August-University Göttingen, TROPAGS, Göttingen, Germany

Bernhard Bruemmer Department of Agricultural Economics and Rural Development, University of Göttingen, Göttingen, Germany
Centre of Biodiversity and Sustainable Land Use (CBL), University of Göttingen, Göttingen, Germany

Christian Brümmer Thünen Institute of Climate-Smart Agriculture, Braunschweig, Germany

Nicole Burdanowitz Institute for Geology, Universität Hamburg, Hamburg, Germany

Hayley C. Cawthra Geophysics and Remote Sensing Unit, Council for Geoscience, Bellville, South Africa
African Centre for Coastal Palaeoscience, Nelson Mandela University, Port Elizabeth, South Africa

Paxie W. Chirwa Department of Plant and Soil Sciences, University of Pretoria, Pretoria, South Africa

Alicia Cimenti Institute for Socio-ecological Research, Frankfurt am Main, Germany

Paulo Coelho Instituto Nacional de Investigação Pesqueira e Marinha, Ministério da Agricultura e Pescas de Angola, Ilha de Luanda, Luanda, Angola

Anneliza Collett Directorate Land Use and Soil Management, Department of Agriculture, Land Reform and Rural Development, Pretoria, South Africa

Timo Conradi Plant Ecology, University of Bayreuth, Bayreuth, Germany

Bernhard Dalheimer Department of Agricultural Economics and Rural Development, University of Göttingen, Göttingen, Germany

Bastien Dieppois Centre for Agroecology, Water and Resilience, Coventry University, Coventry, UK

C. J. Dirk Plant Conservation Unit, Department of Biological Sciences, University of Cape Town, Cape Town, South Africa

Johannes Meyer zu Drewer University of Göttingen, Tropical Plant Production and Agricultural Systems Modelling (TROPAGS), Göttingen, Germany
Ithaka Institute for Carbon Strategies, Arbaz, Switzerland

Tim Dudeck ZMT, Leibniz Center for Tropical Marine Research, Bremen, Germany

Sabrina E. Duncan Thünen Institute of Sea Fisheries, Bremerhaven, Germany

Justin du Toit Grootfontein Agricultural Development Institute, South Africa

Frank Eckardt Environmental and Geographical Science, University of Cape Town, Cape Town, South Africa

Werner Ekau Leibniz Centre for Tropical Marine Research (ZMT), Bremen, Germany

Francois A. Engelbrecht Faculty of Science, Global Change Institute, University of the Witwatersrand, Witwatersrand, South Africa

Gregor Feig Department of Geography, Geoinformatics and Meteorology, University of Pretoria, Pretoria, South Africa
South African Environmental Observation Network, Colbyn, Pretoria, South Africa

Jan-Henning Feil Department of Agriculture, South Westphalia University of Applied Sciences, Soest, Germany
Department of Agricultural Economics and Rural Development, University of Göttingen, Göttingen, Germany

Nicole Costa Resende Ferreira Tropical Plant Production and Agricultural Systems Modeling (TROPAGS), University of Göttingen, Göttingen, Germany

Jemma Finch Discipline of Geography, School of Agricultural, Earth and Environmental Sciences, University of KwaZulu-Natal, Durban, South Africa

Heino O. Fock Thünen Institute of Sea Fisheries, Bremerhaven, Germany

Stefan Foord SARChI Chair on Biodiversity Value & Change, School of Mathematical & Natural Science, University of Venda, Thohoyandou, South Africa

Matthew Forrest Senckenberg Biodiversity and Climate Research Centre (SBiK-F), Frankfurt am Main, Germany

Tarryn Frankland Discipline of Geography, School of Agricultural, Earth and Environmental Sciences, University of KwaZulu-Natal, Pietermaritzburg, South Africa

Roger Funk Leibniz Centre for Agricultural Landscape Research (ZALF), Müncheberg, Germany

Katja Geißler University of Potsdam, Potsdam, Germany

Ursula Gessner German Aerospace Center (DLR), Wessling, Germany

L. Gillson Plant Conservation Unit, Department of Biological Sciences, University of Cape Town, Cape Town, South Africa

Christoph Glotzbach Department of Geosciences, University of Tübingen, Tübingen, Germany

Ingo Grass Department of Ecology of Tropical Agricultural Systems (490f), University of Hohenheim, Stuttgart, Germany

Andrew Green Discipline of Geology, University of KwaZulu-Natal, Durban, South Africa

Torsten Haberzettl Institute for Geography and Geology, University of Greifswald, Greifswald, Germany

Wilhelm Hagen BreMarE – Bremen Marine Ecology, Marine Zoology, University of Bremen, Bremen, Germany

Annette Hahn MARUM – Center for Marine Environmental Sciences, University of Bremen, Bremen, Germany

Eugene Hahndiek Nuwejaars Wetlands SMA, Bredasdorp, South Africa

Eliakim Hamunyelae University of Namibia, Windhoek, Namibia

Christiaan Harmse Department of Agriculture, Environmental Affairs, Land Reform and Rural Development, Northern Cape Province, Eiland Research Station, Upington, South Africa

Morgan Hauptfleisch Biodiversity Research Centre, Namibia University of Science and Technology, Windhoek, Namibia

Kai Heckel Department for Earth Observation, Friedrich Schiller University, Jena, Germany

Knut Heinatz Institute of Marine Ecosystem and Fishery Science, University of Hamburg, Hamburg, Germany

Robert Hering Plant Ecology and Nature Conservation, University of Potsdam, Potsdam, Germany

Tim Herkenrath University of Potsdam, Potsdam, Germany

Thomas Hickler Senckenberg Biodiversity and Climate Research Centre (SBiK-F), Frankfurt am Main, Germany
Institute of Physical Geography, Goethe University Frankfurt am Main, Frankfurt am Main, Germany

Steven I. Higgins Plant Ecology, University of Bayreuth, Bayreuth, Germany

Steven Hill Department of Remote Sensing, Institute of Geography and Geology, University of Wuerzburg, Wuerzburg, Germany

Martin Hipondoka University of Namibia, Windhoek, Namibia

Andreas Hirner German Aerospace Center (DLR), Wessling, Germany

Munir Hoffmann AGVOLUTION GmbH, Göttingen, Germany

Jenny A. Huggett Department of Forestry, Fisheries and the Environment (DFFE), Oceans and Coastal Research, Cape Town, South Africa
Department of Biological Sciences, University of Cape Town, Cape Town, South Africa

Birgit Hünicke Institute of Coastal Systems, Helmholtz-Centre hereon, Geesthacht, Germany

Rodrigue Anicet Imbol Koungue GEOMAR Helmholtz Centre for Ocean Research Kiel, Kiel, Germany

Katja Irob Freie Universität Berlin, Berlin, Germany

Ioana Ivanciu GEOMAR Helmholtz Centre for Ocean Research Kiel, Kiel, Germany

Isselstein Johannes Department of Crop Sciences, Division of Grassland Science, University of Goettingen, Goettingen, Germany
Centre of Biodiversity and Sustainable Land Use (CBL), University of Goettingen, Goettingen, Germany

Hans-Peter Kahle Faculty of Environment and Natural Resources, University of Freiburg, Freiburg, Germany

Kondwani Kapinga Biological Sciences Department, Mzuzu University, Mzuzu, Malawi

Manfred J. Kaufmann Marine Biology Station of Funchal, Faculty of Life Sciences, University of Madeira, Funchal, Portugal

Florian Kestel Leibniz Centre for Agricultural Landscape Research (ZALF), Müncheberg, Germany

Johann F. Kirsten Bureau for Economic Research, Stellenbosch University, Stellenbosch, South Africa

K. L. Kirsten Discipline of Geography, School of Agricultural, Earth and Environmental Sciences, University of KwaZulu-Natal, Durban, South Africa
Department of Geological Sciences, University of Cape Town, Cape Town, South Africa
Human Evolution Research Institute, University of Cape Town, Cape Town, South Africa

Jaap Knot Ladybrand, South Africa

Rolf Koppelmann Institute of Marine Ecosystem and Fishery Science, Universität Hamburg, Hamburg, Germany

Mareike Körner GEOMAR Helmholtz Centre for Ocean Research Kiel, Kiel, Germany
Faculty of Mathematics and Natural Sciences, Kiel University, Kiel, Germany

Ronja Kraus Institute for Social-Ecological Research (ISOE), Frankfurt, Germany
Stockholm University, Stockholm, Sweden

Harald Kunstmann Institute of Geography, University of Augsburg, Augsburg, Germany
Institute of Meteorology and Climate Research (IMK-IFU), Garmisch-Partenkirchen, Germany

Sunna Kupfer Kiel University, Kiel, Germany

Niko Lahajnar Institute of Geology, Universität Hamburg, Hamburg, Germany

Quang Dung Lam Tropical Plant Production and Agricultural Systems Modeling (TROPAGS), University of Göttingen, Göttingen, Germany

Tarron Lamont Oceans & Coasts Research Branch, Department of Forestry, Fisheries, and the Environment, Cape Town, South Africa
Department of Oceanography, University of Cape Town, Rondebosch, South Africa
Bayworld Centre for Research & Education, Cape Town, South Africa

Patrick Laux Institute of Geography, University of Augsburg, Augsburg, Germany
Institute of Meteorology and Climate Research (IMK-IFU), Garmisch-Partenkirchen, Germany

Jay J. Le Roux Afromontane Research Unit, Faculty of Natural and Agricultural Sciences, Bloemfontein, South Africa

Klinck Leonhard Department of Crop Sciences, Division of Grassland Science, University of Goettingen, Goettingen, Germany

Stefan Liehr Institute for Socio-ecological Research, Frankfurt am Main, Germany

Valerie M. G. Linden University of Venda, SARChI Chair on Biodiversity Value & Change, School of Mathematical & Natural Science, Thohoyandou, South Africa

Anja Linstädter Biodiversity Research/Systematic Botany, Institute of Biochemistry and Biology, University of Potsdam, Potsdam, Germany

Dirk Lohmann University of Potsdam, Potsdam, Germany

Antoinette Lombard Department of Social Work and Criminology, University of Pretoria, Hatfield, South Africa

Deon Louw NatMIRC, Ministry of Fisheries and Marine Resources-National Marine Information and Research Center, Swakopmund, Namibia

Michelle A. Louw Plant Ecology, University of Bayreuth, Bayreuth, Germany

Joke F. Lübbecke GEOMAR Helmholtz Centre for Ocean Research Kiel, Kiel, Germany
Faculty of Mathematics and Natural Sciences, Kiel University, Kiel, Germany

Deike Lüdtke Institute for Socio-ecological Research, Frankfurt am Main, Germany

Robert Luetkemeier Institute for Social-Ecological Research (ISOE), Frankfurt, Germany
Goethe University Frankfurt, Frankfurt, Germany

Trevor Lumsden Council for Scientific and Industrial Research (CSIR), Holistic Climate Change, Smart Places, South Africa

Amukelani Maluleke School for Climate Studies and Department of Botany and Zoology, Stellenbosch University, Stellenbosch, South Africa
Department of Geography, Geoinformatics and Meteorology, University of Pretoria, Pretoria, South Africa

Carola Martens Senckenberg Biodiversity and Climate Research Centre (SBiK-F), Frankfurt am Main, Germany
Institute of Physical Geography, Goethe University Frankfurt am Main, Frankfurt am Main, Germany

Bettina Martin Institute of Marine Ecosystem and Fishery Science, Universität Hamburg, Hamburg, Germany

Komainda Martin Department of Crop Sciences, Division of Grassland Science, University of Goettingen, Goettingen, Germany

Nkabeng Maruping-Mzileni Scientific Services, South African National Parks (SANParks), South Africa

Mohau Mateyisi Council for Scientific and Industrial Research (CSIR), Holistic Climate Change, Smart Places, South Africa

Meed Mbidzo Namibia University of Science and Technology, Windhoek, Namibia

Paul Mehlhorn Institute for Geography and Geology, University of Greifswald, Greifswald, Germany

Luisa Meiritz Institute of Geology, Universität Hamburg, Hamburg, Germany

Guy F. Midgley Department of Botany and Zoology, School for Climate Studies, Stellenbosch University, Stellenbosch, South Africa

Volker Mohrholz Leibniz Institute for Baltic Sea Research Warnemünde, Rostock, Germany

Tebatso Moloto CSIR Council for Scientific and Industrial Research, Pretoria, South Africa

Theunis Morgenthal Directorate Land Use and Soil Management, Department of Agriculture, Land Reform and Rural Development, Pretoria, South Africa

Christopher Morhart Faculty of Environment and Natural Resources, University of Freiburg, Freiburg, Germany

Hassane Moutahir Institute of Meteorology and Climate Research (IMK-IFU), Garmisch-Partenkirchen, Germany
The Mediterranean Center for Environmental Studies (CEAM), Valencia, Spain

Alex Msipa Department of Social Work and Criminology, University of Pretoria, Hatfield, South Africa

Edward Muhoko Plant Ecology, University of Bayreuth, Bayreuth, Germany

Shingirirai S. Mutanga Council for Scientific and Industrial Research (CSIR), Holistic Climate Change, Smart Places, South Africa

Saul E. Mwale Dag Hammarskjöld Institute for Peace and Conflict Studies, Copperbelt University, Kitwe, Zambia

Sasha Naidoo Council for Scientific and Industrial Research (CSIR), Holistic Climate Change, Smart Places, South Africa

Shingirai S. Nangombe Council for Scientific and Industrial Research (CSIR), Holistic Climate Change, Smart Places, South Africa

Nicholas P. Ndlovu Faculty of Environment and Natural Resources, University of Freiburg, Freiburg, Germany

William C. D. Nelson Tropical Plant Production and Agricultural Systems Modeling (TROPAGS), University of Göttingen, Göttingen, Germany

Frank H. Neumann Evolutionary Studies Institute, University of the Witwatersrand, Johannesburg, South Africa

Brent Newman Coastal Systems Research Group, CSIR, Durban, South Africa
Nelson Mandela University, Port Elizabeth, South Africa

Mandla Nkomo Solidaridad Southern Africa, Johannesburg, South Africa
CGIAR Excellence in Agronomy 2030 Initiative, IITA, Kasarani, Nairobi, Kenya

George Nyamadzaw Faculty of Environment and Natural Resources, University of Freiburg, Freiburg, Germany

Betserai I. Nyoka Department of Environmental Science, Bindura University of Science Education, Bindura, Zimbabwe
World Agroforestry Centre (ICRAF), Southern Africa Node, Lilongwe, Malawi

Jude Odhiambo School of Agriculture, University of Venda, Thohoyandou, South Africa

Insa Otte Department of Remote Sensing, Institute of Geography and Geology, University of Wuerzburg, Wuerzburg, Germany

Odhiambo Jude Julius Owuor Department of Plant and Soil Sciences, University of Venda, Thohoyandou, South Africa

Jonathan Padavatan Global Change Institute, Faculty of Science, University of the Witwatersrand, Witwatersrand, South Africa

Carsten Pathe Department for Earth Observation, Institute of Geography, Friedrich Schiller University Jena, Jena, Germany

Mirjam Pfeiffer Senckenberg Biodiversity and Climate Research Centre (SBiK-F), Frankfurt am Main, Germany

Anja K. van der Plas National Marine Information and Research Centre (NatMIRC), Ministry of Fisheries and Marine Resources (MFMR), Swakopmund, Namibia

Arthur Prigent GEOMAR Helmholtz Centre for Ocean Research Kiel, Kiel, Germany

Abel Ramoelo Centre for Environmental Studies (CFES), Department of Geography, Geoinformatics and Meteorology, University of Pretoria, Pretoria, South Africa

Markus Rauchecker Institute for Socio-ecological Research, Frankfurt am Main, Germany

Chris Reason University of Cape Town, Cape Town, South Africa

Alanna Rebelo Water Science Unit, Soil Climate and Water, Natural Resources and Engineering (ARC-NRE), Agricultural Research Council, Hilton, South Africa

Paul Renner Department for Earth Observation, Institute of Geography, Friedrich Schiller University Jena, Jena, Germany

Tim Rixen Leibniz Centre for Tropical Marine Research (ZMT), Bremen, Germany
Institute for Geology, Universität Hamburg, Hamburg, Germany

Marisa Roch GEOMAR Helmholtz Centre for Ocean Research Kiel, Kiel, Germany

Reimund Paul Rötter Tropical Plant Production and Agricultural Systems Modeling (TROPAGS), University of Göttingen, Göttingen, Germany

Centre of Biodiversity and Sustainable Land Use (CBL), University of Göttingen, Göttingen, Germany

Matthieu Rouault Nansen Tutu Center for Marine Environmental Research, Department of Oceanography, University of Cape Town, Cape Town, South Africa

Siren Ruhs GEOMAR Helmholtz Centre for Ocean Research Kiel, Kiel, Germany Institute for Marine and Atmospheric Research, Utrecht University, Utrecht, Netherlands

Oksana Rybchak Thünen Institute of Climate-Smart Agriculture, Braunschweig, Germany

Sara Santamaria-Aguilar Kiel University, Kiel, Germany

Enno Schefuß MARUM – Center for Marine Environmental Sciences, University of Bremen, Bremen, Germany

Simon Scheiter Senckenberg Biodiversity and Climate Research Centre (SBiK-F), Frankfurt am Main, Germany

Konstatin Schellenberg Department for Earth Observation, Institute of Geography, Friedrich Schiller University Jena, Jena, Germany

Martin Schmidt Leibniz Institute for Baltic Sea Research Warnemünde, Rostock, Germany

Christiane Schmullius Department for Earth Observation, Institute of Geography, Friedrich Schiller University Jena, Jena, Germany

Franziska U. Schwarzkopf GEOMAR Helmholtz Centre for Ocean Research Kiel, Kiel, Germany

Patricia Sebola Risk and Vulnerability Science Centre, University of Limpopo, Sovenga, South Africa

Thomas Seifert Faculty of Environment and Natural Resources, University of Freiburg, Freiburg, Germany
Department of Forestry and Wood Sciences, Stellenbosch University, Stellenbosch, South Africa

Anne F. Sell Thünen Institute of Sea Fisheries, Bremerhaven, Germany

Shanmugapriya Selvaraj Department for Earth Observation, Institute of Geography, Friedrich Schiller University Jena, Jena, Germany

Mmapatla P. Senyolo Department of Agricultural Economics and Animal Production, University of Limpopo, Polokwane, South Africa

Jonathan P. Sheppard Faculty of Environment and Natural Resources, University of Freiburg, Freiburg, Germany

Claire Siddiqui Leibniz Centre for Tropical Marine Research – ZMT, Bremen, Germany

Frances Siebert Unit for Environmental Sciences and Management, North-West University, Potchefstroom, South Africa

Gudeta W. Sileshi Addis Ababa University, College of Natural and Computational Sciences, Addis Ababa, Ethiopia
University of KwaZulu-Natal, Earth and Environmental Sciences, Pietermaritzburg, South Africa

Felix V. Skhosana Council for Scientific and Industrial Research (CSIR), Holistic Climate Change, Smart Places, South Africa
Council for Scientific and Industrial Research (CSIR), Pretoria, South Africa

Jasper A. Slingsby Department of Biological Sciences and Centre for Statistics in Ecology, Environment and Conservation, University of Cape Town, Cape Town, South Africa

Chris Smith Soil Conservation/LandCare, Department of Agriculture and Rural Development, Free State Province, Thaba Nchu, South Africa

Taylor Smith University of Potsdam, Potsdam, Germany

Felix Soltau University of Siegen, Siegen, Germany

Jessica Steinkopf Global Change Institute, Faculty of Science, University of the Witwatersrand, Witwatersrand, South Africa

Tercia Strydom Scientific Services, South African National Parks (SANParks), Skukuza, South Africa

Annette Swanepol Plant Production, Northern Cape Department of Agriculture, Environmental Affairs, Land Reform and Rural Development, Eiland Research Station, Upington, South Africa

Stephen Syampungani Copperbelt University, ORTARCHI – Oliver R Tambo African Research Initiative, Kitwe, Zambia

Manyana Tausendfruend Institute for Socio-ecological Research, Frankfurt am Main, Germany

Peter J. Taylor Department of Zoology & Entomology & Afromontane Research Unit, University of the Free State, Phuthaditjhaba, South Africa

Mulalo P. Thavhana Institute of Physical Geography, Goethe University Frankfurt am Main, Frankfurt am Main, Germany
Senckenberg Biodiversity and Climate Research Centre (SBiK-F), Frankfurt am Main, Germany

Humbelani Thenga Council for Scientific and Industrial Research (CSIR), Holistic Climate Change, Smart Places, South Africa

Sandy Thomalla CSIR Council for Scientific and Industrial Research, Pretoria, South Africa

Frank Thonfeld German Aerospace Center (DLR), Wessling, Germany

TBritta Tietjen Freie Universität Berlin, Berlin, Germany

Nele Tim Institute of Coastal Systems, Helmholtz-Centre hereon, Geesthacht, Germany

Jane Turpie Anchor Environmental Consultants, Tokai, South Africa
Environmental Policy Research Unit, School of Economics, University of Cape Town, Cape Town, South Africa

Kenneth Uiseb Ministry of Environment, Forestry and Tourism, Windhoek, Namibia

Marcel Urban Department for Earth Observation, Friedrich Schiller University, Jena, Germany
ESN (EnergiesystemeNord) GmbH, Jena, Germany

Shoopala Uugulu University of Namibia, Windhoek, Namibia

Athanasios T. Vafeidis Kiel University, Kiel, Germany

Carl D. van der Lingen Department of Biological Sciences, University of Cape Town, Cape Town, South Africa
Department of Forestry, Fisheries and the Environment (DFFE), Fisheries Management, Cape Town, South Africa

Jennifer Veitch Marine Offshore Node, South African Environmental Observation Network, Cape Town, South Africa

Hans M. Verheye Department of Biological Sciences, University of Cape Town, Cape Town, South Africa

Maik Veste CEBra – Centrum für Energietechnologie Brandenburg e.V., Cottbus, Germany
Institute for Environmental Sciences, Brandenburg University of Technology Cottbus-Senftenberg, Cottbus, Germany

Johanna von Holdt Environmental and Geographical Science, University of Cape Town, Cape Town, South Africa

Graham P. von Maltitz School for Climate Studies, Stellenbosch University, Stellenbosch, South Africa
South African National Biodiversity Institute, Cape Town, South Africa

Heike Wanke University of the West of England, Bristol, UK

Sina M. Weier Department of Zoology & Entomology & Afromontane Research Unit, University of the Free State, Phuthaditjhaba, South Africa

Joshua Weiss Anchor Environmental Consultants, Tokai, South Africa

Catrin Westphal Department of Crop Sciences, Functional Agrobiodiversity, Georg-August-University Göttingen, Göttingen, Germany

Ferdinand Wilhelm Senckenberg Biodiversity and Climate Research Centre (SBiK-F), Frankfurt am Main, Germany

Margit R. Wilhelm University of Namibia, Sam Nujoma Marine and Coastal Resources Research Centre (SANUMARC), and Department of Fisheries and Ocean Sciences, Henties Bay, Namibia

Matthias Zabel MARUM – Center for Marine Environmental Sciences, University of Bremen, Bremen, Germany

Zhenyu Zhang Institute of Geography, University of Augsburg, Augsburg, Germany
Institute of Meteorology and Climate Research (IMK-IFU), Karlsruhe Institute of Technology, Garmisch-Partenkirchen, Germany

Eduardo Zorita Institute of Coastal Systems, Helmholtz-Centre hereon, Geesthacht, Germany

Part I
Background

Coupled Earth System and Human Processes: An Introduction to SPACES and the Book

1

Graham P. von Maltitz ⓘ, Mari Bieri, Guy F. Midgley ⓘ, Jennifer Veitch ⓘ, Christian Brümmer ⓘ, Reimund P. Rötter ⓘ, and Maik Veste ⓘ

Abstract

Ecosystems in southern Africa are threatened by numerous global change forces, with climate change being a major threat to the region. Many climate change impacts and environmental-based mitigation and adaptation options remain poorly researched in this globally important biodiversity hotspot. This book is

G. P. von Maltitz (✉)
School for Climate Studies, Stellenbosch University, Stellenbosch, South Africa

South African National Biodiversity Institute, Cape Town, South Africa

M. Bieri · C. Brümmer
Thünen Institute of Climate-Smart Agriculture, Braunschweig, Germany
e-mail: christian.bruemmer@thuenen.de

G. F. Midgley
School for Climate Studies, Stellenbosch University, Stellenbosch, South Africa
e-mail: gfmidgley@sun.ac.za

J. Veitch
Marine Offshore Node, South African Environmental Observation Network, Cape Town, South Africa
e-mail: ja.veitch@saeon.nrf.ac.za

R. P. Rötter
Tropical Plant Production and Agricultural Systems Modelling, (TROPAGS) and Centre of Biodiversity and Sustainable Land Use (CBL), University of Göttingen, Göttingen, Germany
e-mail: reimund.roetter@uni-goettingen.de

M. Veste
CEBra – Centrum für Energietechnologie Brandenburg e.V., Cottbus, Germany

Brandenburg University of Technology Cottbus-Senftenberg, Institute for Environmental Sciences, Cottbus, Germany
e-mail: veste@cebra-cottbus.de

© The Author(s) 2024
G. P. von Maltitz et al. (eds.), *Sustainability of Southern African Ecosystems under Global Change*, Ecological Studies 248,
https://doi.org/10.1007/978-3-031-10948-5_1

a collection of chapters covering research undertaken in southern Africa by the German Federal Ministry of Education and Research's (BMBF) SPACES and SPACES II programs. SPACES II covered a wide range of global change-linked environmental issues ranging in scope from the impacts of ocean currents on global climate systems through to understanding how small-scale farmers may best adapt to the impacts of climate change. All the research has identified policy implications, and the book strives for a balance between presenting the detailed science underpinning the conclusions as well as providing clear and simple policy messages. To achieve this, many chapters in the book contextualize the issues through the provision of a mini-review and combine this with the latest science emulating out of the SPACES II program of research. The book therefore consolidated both past and the most current research findings in a way that will be of benefit to both academia and policy makers.

1.1   Introduction

This open-access Ecological Studies volume provides results and synthesis of key issues from the research program "Science Partnerships for the Adaptation to Complex Earth System Processes" (SPACES II), addressing the scientific, social and economic issues related to climate change impacts in southern Africa including terrestrial and marine ecosystems. It is written by 66 scientists from African nations together with 111 of their German and other European collaborators and summarizes, in 32 chapters, selected highlights from the latest research findings of SPACES II (2018–2022). These are of significant potential relevance for a better understanding of climate change impacts on marine and terrestrial ecosystems and may help to improve management options and guide environmental policy decisions. This is crucial considering projected African population increase in the context of very likely adverse impacts of climate change, including significant increases in aridity and warming, and the frequency of extreme weather events affecting both marine and terrestrial ecosystems and the human activities that depend upon them.

There is a particular value to such research in the southern African ecosystems, because their terrestrial ecosystems are among the last refugia in the world for their unique wildlife and rare plant diversity, and support a multiplicity of ecosystem services and human livelihoods. The oceans surrounding southern Africa furthermore comprise a critical bottleneck in the global thermohaline circulation, they act as a relatively poorly understood regulator of the global carbon balance, and they play an important role in sustaining marine biodiversity and the highly productive fisheries. These and other geographic advantages are further elucidated in Sect. 1.4, with reference to their coverage in relevant chapters.

In addition to key locational advantages for advancing research in global change science, southern Africa has a particular legacy of human scientific research capacity, some level of networking and extensive biophysical and social–ecological data available that can provide the basis for effective scientific advancement. However, the region lacks somewhat with regard to modern harmonized monitoring infrastructures and the networking required for quantitative and region-specific multisector impact assessments and applied efforts, such as development of evidence-based adaptation options. Although the understanding of the region's climate drivers has increased extensively over the past few years, there remain large gaps in knowledge and predictive skills of potential and actual impacts on the diverse ecosystems of the region. The development of the SPACES II suite of projects was positioned to address these with awareness of the facilitative environment described above.

1.2   Long-Term Southern African–German Scientific Cooperation and Background to SPACES

The scientific cooperation between German scientific institutions and southern Africa is based on decades of successful collaborations.

Joint research on biodiversity under climate change was the focus of the research program "BIOdiversity Monitoring Transect Analysis in Africa - southern Africa" (BIOTA South 2000–2006). Objectives of the program (Hoffman et al. 2010; Jürgens et al. 2010; Schmiedel and Jürgens 2010) were:

- Scientific support for sustainable use and conservation of biodiversity in Africa.
- A continental observation network in Africa, contributing to GEOSS (Global Earth Observation System of Systems).
- A network for observing land degradation and for developing measures to combat desertification in Africa.
- A network for capacity development and rural development in Africa.

Investigations of Earth system processes and their interactions at different spatial and temporal scales were the focus of the "Inkaba ye Africa" program in southern Africa (de Wit and Horsfield 2006, 2007). Furthermore, Germany also contributed to international marine research programs such as the "Benguela Current Large Marine Ecosystem" (BCLME) and "Benguela Environment Fisheries Interaction and Training" (BENEFIT) (Hampton and Sweijd 2008). Other bilateral and multilateral projects in marine research such as "NAMIBGAS" (Eruptions of methane and hydrogen sulfide from shelf sediments off Namibia, 2004–2007) and "GENUS" (Geochemistry and Ecology of the Namibian Upwelling System, 2009–2015) have put the current cooperation activities on a broad basis.

The goal of SPACES is to deepen both the thematic and geographical expansion of the research expertise acquired so far. With new research topics and with additional partners, SPACES has made a lasting contribution to the corresponding national programs and initiatives in the region. The promotion of young researchers is an essential component of SPACES. In addition to workshops and summer schools, an education and training component is an integral part of the program. To this end, the BMBF supports a German Academic Exchange Service (DAAD) scholarship program (master's and doctoral scholarships). In addition, training cruises for African students were offered on the German research vessels FS MERIAN, FS METEOR and FS SONNE.

Within its framework program "Research for Sustainable Development" (FONA), the German Federal Ministry of Education and Research (BMBF) funds research projects in key regions that are particularly affected by the impacts of climate change and to promote sustainable land-use and climate protection. The program was developed on the basis of multilateral discussions between Germany and South Africa in 2008, followed by two workshops in 2009 for a broad professional audience in Gobabeb and Henties Bay (Namibia) attended by scientists from Angola, Germany, Namibia and South Africa. The results of the workshop were then presented to potential partners in the South African and Namibian ministries and research institutions. The funding program "Science Partnerships for the Assessment of Complex Earth System Processes" (SPACES I) was initiated in 2012 under the FONA framework and followed by a second funding phase (SPACES II,) initiated in 2018 (PtJ 2021). Both SPACES I and SPACES II stemmed from national and international strategies and initiatives on global change and international partnerships. These include the German-South African Year of Science 2012, German Government's Africa Policy Guidelines of 2014, Strategy for the Internationalization of Science and Research (BMBF 2008), the BMBF's Africa Strategy (BMBF 2014a) and International Cooperation Action Plan of 2014 (BMBF 2014b), as well as the United Nations Sustainable Development Goals SDG13 "Urgent action to combat climate change and its impacts," SDG14 "Conserve and sustainably use the oceans, seas and marine resources for sustainable development," SDG15 "Protect, restore and promote sustainable use of terrestrial ecosystems, sustainably manage forests, combat desertification, and halt biodiversity loss," and SDG17 "Strengthen the means of implementation and revitalize the global partnership for sustainable development." Furthermore, the joint research contributes to the international programs of the United Nations Framework Convention on Climate Change (UNFCCC), UN Convention on Biodiversity (UNCBD) and UN Convention to Combat Desertification (UNCCD).

The core aim of SPACES I and SPACES II was to initiate collaborative research projects that contribute to the formulation of science-based recommendations on Earth system management, to ensure the sustainable use and conservation of the ecosystem services of the region. Both stress the provision of approaches tailored to the needs of end users.

The research focus of SPACES I (2014–2017) was defined as interactions between the geosphere, atmosphere and ocean as well as those between land and ocean, and biosphere and atmosphere. The main research themes of SPACES, with focus on assessment, were:

1. Coastal current systems in southern Africa, and their influence on land–ocean–atmosphere interactions, biogeochemical cycles and resource availability.

2. Quantification of the fluxes of carbon, water, nutrients and pollutants in rivers, estuaries and desert areas in terms of transport and transformation mechanisms, and implications for biodiversity and related ecosystem services.
3. Determinants of large-scale landscape evolution, hydrological changes and land-use change in southern Africa.
4. Describing, monitoring and conserving biodiversity in the face of habitat loss, and modeling potential changes based on predicted environmental and societal change.
5. Development and application of measures to restore and rehabilitate ecosystems damaged by human activities and natural processes.
6. Marine and terrestrial repositories of past climate and ecosystem change and their relevance to land–ocean–climate interactions.
7. Investigating the formation and evolution of the ecosystem.

SPACES I funded ten collaborative three-year research projects:

- AGULHAS—Regional and Global Relevance;
- ARS AfricaE—Adaptive Resilience of Southern African Ecosystems;
- IDESSA—Decision Support System for Rangeland Management;
- GENUS—Geochemistry and ecology of the Namibian upwelling system;
- GEOARCHIVES—Signals of Climate and Landscape Changes;
- GSI—Groundwater/seawater interaction along the South African south coast);
- LLL—Limpopo Living Landscapes;
- RAiN—Regional Archives for Integrated investigations;
- OPTIMAS—Sustainable Management of Savannah Ecosystems;
- SACUS—Southwest African coastal upwelling system and Benguela Niños;

The intention of the SPACES II was not the direct continuation of previous projects, but rather the thematic deepening and geographical expansion of the research competence acquired so far. With new research topics and with additional partners, SPACES II will sustainably contribute to the corresponding national programs and initiatives in southern Africa in light of new international challenges. Further, projects operating infrastructures for monitoring key environmental variables and land surface processes were asked to develop a transfer concept for potential implementation and long-term usage by southern African collaborators and partners after project termination. Capacity building was considered as a major action to achieve this goal and to avoid leaving many of the activities as temporary endeavors without sustainable benefits for local stakeholders.

The research focus was defined as interactions between the geosphere, atmosphere and ocean as well as those between land and ocean, and biosphere and atmosphere; however, the scope was widened to include interactions between the anthroposphere, geosphere, hydrosphere, biosphere and atmosphere. Important themes for both, SPACES I and SPACES II were defined as soil erosion, drought, local and regional shifts in plant species composition, the interaction of climate change and human impacts, such as land use and pollution, and changes in oceanic currents.

The second phase SPACES II (2018–2022) funded six terrestrial and three marine collaborative research projects (Table 1.1). The five main themes show the shift of focus from assessment to adaptations and include:

1. Seasonal and interannual variability and trends of coastal current systems, considering their influence on land–ocean–atmosphere interactions in southern Africa, and their implications for biogeochemical cycles and marine resources management,
2. The transport of carbon, water, nutrients and pollutants, considering their transformation mechanisms and dynamics in riverine, estuarine and coastal areas and in terms of their importance for population, biodiversity and ecosystem services,
3. The functioning of diverse landscapes in terms of sustainable land use, land-use change, carbon and water fluxes and their impacts on biodiversity, habitats and ecosystem services,
4. Management options for landscapes and their ecosystem components for societal resilience to environmental change,
5. Measures to restore and sustainably use degraded ecosystems for goals of resilience, adaptation and mitigation.

Networking and collaborations between research institutions were at the core of SPACES II. The program is carried out jointly with the South African Department of Science and Innovation (DSI) (at the start of the program, under its previous name Department of Science and Technology DST), and it is intended to contribute to the intensification of cooperation with the Ministry of Education at the Republic of Namibia. Premises of the cooperation are mutual added value through high-quality cooperation and focus on jointly defined areas, consideration of (country-)specific African and specifically German interests, partnership and ownership as well as continuity and reliability in the cooperation. Institutions of other neighboring sub-Saharan countries can be integrated into the projects accordingly.

1.3   SPACES II Training and Knowledge Exchange Program

In the long run, skills development through education and training is essential to economic development, security and stability in Africa, and BMBF sets the support of junior researchers and higher education as one of its central aims (BMBF 2008). As defined in the program call, all SPACES II projects were required to contribute to capacity development of the partnering institutions, promote young scientists, as well as facilitate scientific exchange and networking on thematic priority areas. Out of the ten target indicators that were set for the projects in the call, half were directly relevant to capacity development, namely (1) the number of jointly supervised student projects, (2) the development of joint training programs and utilization of research results in curriculum development, (3) training courses on intercultural

Table 1.1 Overview of the funded research projects in SPACES II (2018–2022) (PtJ 2021)

Acronym	Research project	Objectives	Study area
Landscapes in transition			
ASAP	Agroforestry in Southern Africa—new pathways of innovative land-use systems under a changing climate	Investigation of ecosystem services and environmental benefits of agroforestry systems (AFS) as an innovative, multipurpose land-use management practice in southern Africa.	Project partners in South Africa, Mozambique, Namibia, Zambia and Malawi
EMSAfrica	Ecosystem Management Support for Climate Change in Southern Africa	Dual impacts of climate change and land management in the savanna ecosystems of southern Africa. The interdisciplinary project aims to provide information to support decision-making concerning climate change adaptation and mitigation, and the sustainable management of ecosystems.	Three focal research areas are located along a precipitation gradient, from low to high precipitation in the summer rainfall area
ORYCS	Options for sustainable land-use adaptations in savanna systems	Impacts of wildlife-based land-use strategies in Namibian savannas on the feedbacks between wildlife movements and dynamics, vegetation and related ecosystem functions and services. Our study area is situated in the southwestern region of Etosha and includes communal conservancies, private game reserves and the Etosha National Park.	Semiarid Mopane savanna south-west of and including the western part of Etosha National Park, Namibia.
SALDi	South African Land Degradation Monitor	Development of a system for permanent observation for ecosystem changes and degradation based on satellite remote sensing. Improving procedures for assessing soil degradation with a focus on runoff-related soil erosion	Six study sites representing a major climate gradient from the semiarid winter-rainfall region in the southwest across the central semiarid year-round-rainfall region to the semihumid summer-rainfall region in the northeast.
SALLnet	South African Limpopo Landscapes Network	Functionality and resilience of multifunctional landscapes in southern Africa can be enhanced under climate change. The project focuses on three land-use types—rangelands, arable lands and orchards—that provide essential ecosystem services and are crucial for local livelihoods.	South African province of Limpopo

(continued)

Table 1.1 (continued)

Acronym	Research project	Objectives	Study area
Interaction land and sea			
TRACES	Tracing Human and Climate impacts in South Africa	Interdisciplinary investigations of the combined effects of climate change and anthropogenic impacts on aquatic and terrestrial ecosystems in eastern South Africa over the last 250 years. Existing data sets are completed by and compared with new information gained from estuary and terrestrial sediment archives.	I) Richards Bay with the catchments of the rivers Mhlathuze and Mfule and the Goedertrouw dam reservoir, II) Mkhuze swamps with the river catchments of Mkhuze and Pongola and the Pongolapoort dam reservoir, III) Olifants River catchment with the Loskop dam reservoir and two smaller reservoirs in the upstream.
Marine and coastal research			
BANINO	Benguela Niños: Physical processes and long-term variability	Improving the prediction of climate variability and impacts of climate change relevant for upwelling variability and its consequences for biological productivity.	South-East Atlantic Ocean encompassing the area off Angola, Namibia and South Africa
CASISAC	Changes in the Agulhas System and its Impact on Southern African Coasts	Changes in the oceanic conditions and regional sea level around southern Africa are explored through global ocean and coupled climate modeling. The highly variable Agulhas Current System is subject to changes in the hydrography and circulation in response to atmospheric variability and anthropogenic trends.	Agulhas current along the south African coastline
TRAFFIC	Trophic Transfer Efficiency in the Benguela Current	Fundamental research on the processes of the subsystems nBUS and sBUS (Benguela upwelling system) and their responses to climate change.	Benguela upwelling system off south-West Africa

competence; (4) capacity development of scientific personnel and (5) long-term development of joint German-African research and education capacity.

In line with these aims, capacity development in SPACES II was largely based on the following, BMBF-funded core activities:

1. Student fellowship program funded by the BMBF via the "Capacity Building and Development" (CaBuDe) scheme of the DAAD;
2. Short-term exchange grants for scientific visits under the CaBuDe;
3. Short courses and workshops in specialist research skills, as well as training on research vessels;

Capacity building for scientists and practitioners in general and specifically the training of early-career researchers were central aims of the SPACES II program. The integrated training program aimed at linking the capacities of the southern African and German research communities as well as strengthening the competencies in the key areas of SPACES II. These key competencies included for example various modeling approaches, ecosystem assessments, greenhouse gas measurements, generating earth observation products and field surveying methods. Some courses also focused on transferable skills, such as intercultural communication. Training formats ranged from summer/winter schools and workshops to joint conference sessions, tutorials and online materials.

Although the Covid-19 pandemic caused major disturbances to the program in terms of travel and meeting bans starting from early 2020, 11 courses and trainings were successfully completed by the beginning of the year 2022.

1.4   SPACES II Synthesis

This Ecological Studies volume summarizes new information and novel research approaches in the terrestrial and marine realms, and attempts where possible to integrate across these areas, especially with respect to their responses to global change phenomena. Although structured around projects of the SPACES II program, key chapters aim to give a coherent state-of-the-art summary of the dynamics of both ocean and terrestrial ecosystems as they are impacted by global change phenomena, by drawing on and synthesizing the deep legacy information and published work available for the region, including from the original 'Science Partnerships for the Assessment of Complex Earth System Processes' (SPACES I 2014–2017). It covers both the current and emerging understandings of the global change drivers of ecosystem dynamics, potential management options and an understanding of how future scenarios may be altered through policy and management interventions. Differing intensities of human use are considered where feasible, including levels of land management ranging from crop agriculture and agroforestry systems, through pasture and rangeland systems to natural vegetation as found in conservation areas for the terrestrial environment, and extractive industries in the oceanic realm. Common themes such as the drivers of net primary production and the dynamics of carbon fluxes are explored for both the terrestrial and as well as marine systems, highlighting both commonalities and particular differences between these different systems. In the marine realm, content ranges from fundamental biophysical aspects through to social–ecological issues such as food security and sustainability.

The book is divided into five main sections, with each section comprising a number of individual parts: (1) a policy relevant introduction to the region, (2) a discussion of the policy implications of large-scale earth system processes impacting on the region, (3) management and adaptation considerations, (4) monitoring options and (5) a synthesis with overall recommendations. In line with accepted practice in Intergovernmental Panel on Climate Change (IPCC) and Intergovernmental Science-Policy Platform on Biodiversity and Ecosystem Services (IPBES) assessments, the information aims to be policy relevant, not policy prescriptive. The tone of the book, while adhering to scientific writing rigor, aims to provide understandable information to an audience of policy makers, and those involved in the science–policy interface, who might not be domain specialists, but wish to understand the importance of the issues to guide policy and management interventions.

Climate variability and change is a central theme and is considered on its own (Chaps. 5, 6 and 7), or as it interacts with the terrestrial or marine realms (see Fig. 1.1). Where feasible, the interplay between components of all three realms that were the focus of specific research questions and activities is highlighted in chapters (Fig. 1.1). For example, in the marine section, the physical processes affecting currents and upwellings form the basis for considering impacts on and risks to ecological processes (Chaps. 8 and 9), and how these may translate to production and fisheries (Chaps. 2, 11 and 25). In the terrestrial environment, research ranges from the ecological processes impacted by climate change in the terrestrial system

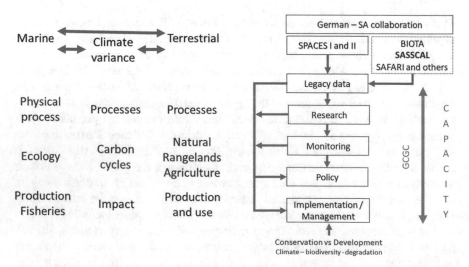

Fig. 1.1 Graphical abstract and conceptual overview of the book topics covered. GCGC = Global Change Grand Challenge

(Chaps. 14, 15, 16 and 17) to management and adaptation options for land uses ranging from natural systems (Chaps. 16 and 18) through extensively managed rangeland (Chaps. 16 and 19) systems and finally to intensively managed crop agricultural systems (Chaps. 20, 21, 22 and 23). Broader and integrative chapters consider physical processes driving the region's climate (Chaps. 8, 9 and 10), the region's role in regulating the global carbon balance (Chaps. 2 and 25) and feedbacks between climate and ecosystem productivity (Chaps. 2, 10, 11, 12, 14, 15, 16, 17, 18, 19, 20, 21, 22, 23, 24, 25, 26, 28 and 29).

1.5   Geographic Advantages for Global Change Research in Southern Africa

From an ecoregion and biodiversity perspective, southern Africa is globally unique, especially considering the high levels of endemism in both the terrestrial and marine ecosystems. Not only does the region have unique biodiversity including the plant diversity of the entire Cape Floristic Kingdom, but in addition it has extensive diversity of both terrestrial and marine habitats and biomes. A background to the southern African region, its unique ecosystems, the current threats to ecosystems as well as the macro-economics of the region is provided in the first four background chapters (Chaps. 2, 3 and 4).

Biomes in the southern African region are under increasing pressure, both from direct human activities and from the more indirect impacts of climate change (Chap. 3). Land transformation (Chap. 13), particularly for crop agriculture and plantations of exotic timber species has destroyed vast areas of natural habitat and spatially isolated areas of remaining natural habitat. Overharvesting, be it overfishing (Chaps. 2 and 25), overgrazing (Chaps. 15, 16, 17, 18 and 19), select use of individual natural species or deforestation, has degraded both the marine and terrestrial environment. A further major problem impacting on most ecosystems, but of particular concern to the Cape Floristic Region, is the extent of alien invasive organisms that have naturalized within natural habitats (Chap. 3). These organisms can displace indigenous biodiversity through direct competition for space, but also through altering the structural nature of the habitat or the disturbance regimes such as fire. Although these direct anthropomorphic threats are severe, it is the interplay between these and climate change that poses increased risk. For instance, agricultural fields may prevent plant and animal migration that would be required for the indigenous organisms to adapt to climate change impacts (Chaps. 2 and 3).

Climate change clearly poses a substantive threat to the unique biodiversity of the southern African subregion (Chaps. 2, 3 and 14). For instance a slight southward shift, possibly of as little as a few 10s of km, of the mid-latitude cyclones could have devastating impacts on the winter rainfall western cape region (Chaps. 6 and 7). Being at the tip of Africa, the region is impacted by numerous ocean currents that effectively meet. Global warming is anticipated to alter these currents, and this will have poorly understood impacts on local climate (Chap. 8). Equally, changes in the upwelling dynamics of the Benguela system could devastate the fishing industries along the west coast of South Africa, Namibia and Angola (Chaps. 9 and 11). Overall, climate change, including impacts from atmospheric CO_2 concentrations, may well cause entire biomes to shift in their spatial distribution (Chap. 14), adding additional pressures to, in some instances, already poorly conserved and vulnerable biodiversity (Chaps. 2 and 3). Similarly, climate induced changes to the movement of ocean currents and impacts from changed dissolved CO_2 concentrations, could cause major shifts to the marine biota (Chaps. 2, 8, 9 and 25).

From a research perspective, the region has a long history of well-maintained and well-documented research activities and data, as well as an extensive research infrastructure within the region, making it an ideal location for undertaking ongoing global change research within a developing world context (Chaps. 2 and 32). Germany, in particular, has had a long history of biodiversity and ecological research in Namibia and South Africa that has developed long-term and strong collaboration between German and southern African research institutes. Most chapters in this book are joint endeavors including both southern African and German research partners.

The southern African region is a globally important, but currently under-researched component of the global carbon cycles. The vast savanna and grassland regions of southern Africa represent a huge carbon store, much to this as soil organic carbon (Chaps. 2, 15–17). Anthropogenic impacts such as land cover change (Chap. 29), erosion (Chap. 13) and deforestation cause a large amount of terrestrial organic carbon to be emitted into the atmosphere as CO_2. However, these environments also represent potential carbon sinks, and through management may be important areas for global change mitigation (Chaps. 11, 17 and 30). Long-term trends in terrestrial carbon fluxes both from land cover change and from climate change, remain poorly understood, the direction of change in many instances not being certain (Chaps. 2, 12, 17, 24 and 30). While the terrestrial carbon fluxes are significant, they are dwarfed by the marine fluxes. The Benguela current upwellings along the west coast of South Africa, Namibia and Angola represents one of the highest primary production regions globally, though the long-term fate of this sequestered carbon is still poorly understood (Chaps. 2 and 25).

From a socioeconomic perspective, southern Africa is a developing region, with the development concerns and trajectory of southern Africa differing significantly from those of Europe. Wide-scale poverty, high population growth rates and major problems of unemployment, mean that national policies tend to favor short-term growth opportunities over environmental concerns. This leads to tensions between sustainability and the needs for short-term development opportunities. As such, this creates unique sets of development and environmental challenges that differ substantially from those of high-income countries (Chaps. 3 and 4). Identifying sustainable terrestrial land-use options and opportunities given both the developmental and climate change realities creates major policy challenges (Chaps. 16, 18–23).

1.5.1   Climate Change

The second part of the book is devoted to the large-scale biophysical processes and drivers impacting the climate in the subregion. This is introduced through descriptions of the past (Chap. 5), current (Chap. 6) and projected future climates (Chap. 7). In addition, it covers an emerging understanding of the impacts of southern ocean currents (Chaps. 8 and 9) as well as land–atmosphere feedbacks on climate (Chap. 10).

Southern Africa has unique sets of climate challenges. There is substantive evidence that the interior of the subcontinent is warming at a rate above the global norm with heat waves, droughts and severe storms all likely to become more common (Chaps. 6 and 7). In addition, the latest future prediction largely agrees that the already predominantly arid region is likely to become dryer in addition to becoming hotter (Chaps. 6 and 7). Implications of climate change suggest that major negative impacts to both natural and agricultural land-use systems are almost certain. Given the wide dependency of the region on natural resources, nature-based tourism and agricultural production, the implications for the region are severe, and adaptation will be paramount (Chaps. 3, 4, 15, 16, 17, 18, 19, 20, 21, 22 and 23). The marine environment is not immune to climate change, with major impacts on the fisheries industry being likely, and possibly already being felt (Chaps. 2, 8, 9, 11 and 25).

Understanding past climate is important for both understanding the conditions under which current biodiversity evolved, as well as understanding how biodiversity may respond to future climates. Detailed measures of past climate are also critically important for calibrating future climate models (Chaps. 5, 12 and 28).

The current climate plays an important role in determining current biodiversity patterns (Chap. 2). A feature of the current climate is the extent of interannual variability (Fig. 1.2). This is especially apparent regarding precipitation and periods of both droughts and above normal rainfall are normal (Chap. 6). Although there is a clear link between El Nino events and drought, this linkage is both complex and not absolute. Equally, La Nina events favor periods of high rainfall, but with only a weak correlation between the strength of the La Nina and rain (Chap. 6). Clearly, many other factors are also involved in determining local precipitation patterns, and understanding these drivers is critical for current and future climate predictions. In this regard, the roles of ocean currents and the possible ways they may change due to climate change are critical (Chaps. 8 and 9). In addition, land-use change on the terrestrial environment could have feedbacks into the predominantly convective precipitation patterns and this is explored in (Chap. 10).

The book provides an up-to-date overview of the current understanding of climate and its drivers in southern Africa. It starts by reviewing climate over the last millennium and the availability of long-term climate surrogates that can be used to calibrate climate models (Chap. 5). It then discusses current climate with a specific focus on the hydrological factors and the drivers of interannual variance (Chap. 6). This is followed by a consideration of future climate projections for the region, a

Fig. 1.2 The profound impacts of climate variance are illustrated in these repeat photos from Boesmanskop 2011 (grassy) and 2016 (no grass). This shows the transition from a wet period (early 2010s) to a period of summer droughts (mid to late 2010s). Photos J du Toit

factor important to most chapters within the book (Chap. 7). The two following chapters consider the roles of currents and upwellings in the climate system (Chaps. 8 and 9). The final chapter in this section considers the feedback between terrestrial land use, and specifically deforestation, and the climate system (Chap. 10).

1.5.2 Carbon Dynamics

Carbon dynamics, an important component of the global understanding of impacts and feedbacks of climate change, is a theme that runs through many of the SPACES II programs and book chapters pick up on this in all sections.

The southern African environment plays an important, but poorly understood role in global carbon dynamics. It is both a potential source and sink of carbon depending on a combination of climate change futures, management options and ecosystem responses. African savanna, because of its vast extent, is recognized as both an important carbon pool and representing a potentially important carbon sink (Chaps. 2 and 15). More arid environments, such as the Nama-Karoo biome, have smaller carbon stocks, but their carbon dynamics are very poorly understood (Chap. 17). Supporting carbon flux monitoring within the savanna has been an important component of the SPACES I and II programs (Chaps. 17 and 30).

Carbon dynamics in the marine environment of southern Africa are poorly understood, despite the west coast of southern Africa being an important upwelling system with exceptionally high net primary production. Understanding the fate of this biomass in terms of the role it plays as a major carbon sink is of global importance (Chap. 2 and 25).

1.6   Science in Support of Ecosystem Management

Part III of the book (Chaps. 11, 12, 13, 14, 15, 16, 17, 18, 19, 20, 21, 22, 23 24 and 25) focuses on science in support of ecosystem management. This section is focused on the policy relevant science to assist in the management of the southern African marine, freshwater and terrestrial ecosystems. It is grouped into four partly overlapping themes with each then developed through a number of interlinked chapters. Within the themes, chapters move from a general description of the sector and the threats that global change is placing on the sector to more focused discussions on potential policy and management interventions. The themes range from the marine and freshwater systems, through general terrestrial consideration, to rangelands and finally agricultural and agroforestry considerations. This represents a transition from management of relatively natural systems through partly transformed rangeland systems to fully transformed and intensely managed agricultural systems.

Human actions are having profound impacts on the natural ecosystems. Over-fishing and natural variability have fundamentally changed and reduced fish stocks from the production rich west coast of southern Africa. Understanding these changes and how climate change will impact on these already vulnerable resources is considered in (Chaps. 2, 11 and 25) (Fig. 1.3). To understand how these systems will change requires both an understanding of the drivers of upwell systems (Chaps. 8 and 9) as well as understanding the trophic ecology of the marine production system (Chaps. 2 and 25). This will have important implications for the management of the west coast fisheries industry, as well as potentially changing the so-called CO_2 pump and the degree to which this system sequesters carbon (Chap. 25). At the biome level, climate change may result in a shift in the spatial distribution of terrestrial biomes, favoring some and restricting others. These potential changes are explored in (Chap. 14).

Humans live on land and within the southern African region, and most human activity is related to the terrestrial landscape. The terrestrial landscape is also subdivided into areas with different levels of direct use and management. Understanding probable global change impacts on the terrestrial environment is therefore critical for its long-term sustainable management. Despite having been identified as an issue over a century ago, land degradation and more especially soil erosion, remains an important management consideration and this is instigated in (Chap. 13).

The predominant use of the natural or seminatural habitats of southern Africa is as rangeland, i.e., land used to support either or both livestock or indigenous game management. An overview of the savanna rangeland and its current threats is given in Chap. 15, and responses of savanna rangeland to drought are investigated in (Chap. 16). One strategy advocated for dealing with the management of arid savanna areas is to use these areas for wildlife management rather than livestock. This switch from livestock to wildlife has been observed to have taken place over the past few decades and a case study from the Etosha region of Namibia is investigated in (Chap. 18). Rates of primary production and how this may change given climate change are critical concerns for understanding climate change impacts. Long-term carbon flux measurements in the region are scarce, though there have been measurements in the savanna dating back to about the year 2000. Measurements in the vast Nama Karoo areas are fewer and this is explored in (Chap. 17).

Livestock management remains a key use of rangeland throughout the region and is of especial importance to Bantu tribes as in addition to the financial aspects, there are strong cultural aspects to livestock. The communal tenure of many traditional areas adds great complexity to achieving sustainable livestock management (Chap. 17). A feature of the savanna landscape is the long winter dry periods during which time grazing becomes scarce, especially during drought periods. Chapter 19 considers dealing with the feed gaps in such systems.

Although livestock management is important across most of the subregion, in areas with sufficient rain to support cropping, agriculture field crops become important as a livelihood strategy. Climate change is expected to have negative impacts on cropping due to general decline in precipitation over much of the region and longer and more frequent droughts. Chapter 20 provides an overview of the agricultural land-use systems of the region and the agricultural challenges that are being faced. The challenges are the development of integrated cultural landscapes, providing food security, ecosystem services and saving biodiversity.

Fig. 1.3 An important component of the spaces program was the extensive use of ship-based programs to sample biotic and abiotic component of the marine environment. Photos Zankl (1,3) and AF Sell (2)

In this context, agroforestry and other sustainable agricultural technologies are identified as potential adaptation opportunities (Chaps. 21 and 23). Chapter 22 considers macadamia, a crop growing in importance for both small-scale and large-scale farmers as a case study for impacts of climate change. In Chap. 23, overarching technology improvements and innovative strategies for small-scale farmers land management are considered.

1.7   Monitoring

Part IV of the book considers a number of emerging tools and processes for improved monitoring and modeling of the southern African environment (Chaps. 24, 25, 26, 27, 28, 29 and 30).

As discussed above, southern Africa has extensive legacy data that makes ongoing research in the region easier. However, maintaining and expanding monitoring networks is critical given the rapid changes that are being observed due to global change forces. New and cost-effective monitoring tools are also constantly emerging and evolving. Chapters 24 and 29 consider the emergence of new capability from remotely sensed products, while Chap. 30 considers South Africa's network of field-based data monitoring stations.

Marine monitoring for a better understanding of the primary productivity and carbon balances in the west coast upwelling systems is discussed in (Chap. 25), while (Chap. 26) considers the modeling of net primary production (NPP) in the terrestrial environment using both satellite-based and -modeled approaches.

The sediments in wetlands, estuaries and other coastal areas can provide long-term historical data based on pollutants and biological material that has been deposited. This can be interpreted both to know the state of the current environment (Chap. 27) and to reconstruct histories of based land cover (Chap. 28).

Chapter 29 considers how observational data can be used in support of policy, and Chap. 30 discusses research infrastructures for long-term environmental observation with a focus on greenhouse gas measurements.

1.8   Synthesis and Outlook

Section 1.5 of the book is a short synthesis of results. It contains results for a small study that reviewed lessons from north-south collaboration projects (Chap. 31). Overarching messages emerging from the research studies as well as suggestions for the future are given in Chap. 32.

References

BMBF - Bundesministerium für Bildung und Forschung (2008) Internationalisation of Education, Science and Research - Strategy of the Federal Government. Federal Ministry of Education and

Research
BMBF - Bundesministerium für Bildung und Forschung (2014a) Die Afrika-Strategie 2014-2018 - Afrika als Partner in Bildung und Forschung. Bundesministerium für Bildung und Forschung, Berlin, Germany
BMBF - Bundesministerium für Bildung und Forschung (2014b) Internationale Kooperation - Aktionsplan des Bundesministeriums für Bildung und Forschung. German Federal Ministry of Education and Research (BMBF), Bonn, Germany
de Wit M, Horsfield B (2006) Inkaba yeAfrica project surveys sector of earth from core to space. Eos Trans AGU 87(11):113. https://doi.org/10.1029/2006EO110002
de Wit M, Horsfield B (2007) Built on the shoulders of Alfred Wegener and Alex du Toit to apply German precision technology to the geological superlatives of South Africa. S Afr J Geol 110(2-3):165–174
Hampton I, Sweijd N (2008) Achievements and lessons learned from the Benguela Environment, Fisheries, Interaction and Training (BENEFIT) research programme. Afr J Mar Sci 30(3):541–564. https://doi.org/10.2989/AJMS.2008.30.3.9.643
Hoffman MT, Schmiedel U, Jürgens N (eds) (2010) Biodiversity in southern Africa. Volume 3: implications for landuse and management. – XII + 226 pp. + CDROM, Klaus Hess Publishers, Göttingen & Windhoek
Jürgens N, Haarmeyer DH, Luther-Mosebach J, Dengler J, Finckh M, Schmiedel U (eds) (2010) Biodiversity in southern Africa. Volume 1: patterns at local scale – the BIOTA observatories. – XX + 801 pp., Klaus Hess Publishers, Göttingen & Windhoek. Online version: https://www.biota-africa.org/PublPDF/Intern//BIOTA-Vol1.pdf, Online: https://www.biota-africa.org/PublPDF/Intern//BIOTA-Vol3.pdf
PtJ - Projektträger Jülich (ed.) (2021) SPACES II – science partnerships for the adaptation to complex earth system processes in southern Africa, Jülich, Germany
Schmiedel U, Jürgens N (eds) (2010) Biodiversity in southern Africa. Volume 2: patterns and processes at regional scale. – XII + 348 pp., Klaus Hess Publishers, Göttingen & Windhoek Online Version https://www.biota-africa.org/PublPDF/Intern//BIOTA-Vol2.pdf

Unique Southern African Terrestrial and Oceanic Biomes and Their Relation to Steep Environmental Gradients

2

Anne F. Sell, Graham P. von Maltitz ⓘ, Holger Auel, Arne Biastoch, Maya Bode-Dalby, Peter Brandt, Sabrina E. Duncan, Werner Ekau, Heino O. Fock, Wilhelm Hagen, Jenny A. Huggett, Rolf Koppelmann, Mareike Körner, Niko Lahajnar, Bettina Martin, Guy F. Midgley ⓘ, Tim Rixen, Carl D. van der Lingen, Hans M. Verheye, and Margit R. Wilhelm

Abstract

The southern African subcontinent and its surrounding oceans accommodate globally unique ecoregions, characterized by exceptional biodiversity and endemism. This diversity is shaped by extended and steep physical gradients or environmental discontinuities found in both ocean and terrestrial biomes. The region's biodiversity has historically been the basis of life for indigenous cultures and continues to support countless economic activities, many of them unsustainable, ranging from natural resource exploitation, an extensive fisheries industry and various forms of land use to nature-based tourism.

Anne F. Sell and Graham P. von Maltitz contributed equally as lead authors.

A. F. Sell (✉) · S. E. Duncan · H. O. Fock
Thünen Institute of Sea Fisheries, Bremerhaven, Germany
e-mail: anne.sell@thuenen.de

G. P. von Maltitz
School for Climate Studies, Stellenbosch University, Stellenbosch, South Africa

South African National Biodiversity Institute, Cape Town, South Africa

H. Auel · M. Bode-Dalby
University of Bremen, BreMarE - Bremen Marine Ecology, Marine Zoology, Bremen, Germany

A. Biastoch · P. Brandt
GEOMAR Helmholtz Centre for Ocean Research Kiel, Kiel, Germany

Kiel University, Faculty of Mathematics and Natural Sciences, Kiel, Germany

G. P. von Maltitz et al. (eds.), *Sustainability of Southern African Ecosystems under Global Change*, Ecological Studies 248,
https://doi.org/10.1007/978-3-031-10948-5_2

Being at the continent's southern tip, terrestrial species have limited opportunities for adaptive range shifts under climate change, while warming is occurring at an unprecedented rate. Marine climate change effects are complex, as warming may strengthen thermal stratification, while shifts in regional wind regimes influence ocean currents and the intensity of nutrient-enriching upwelling.

The flora and fauna of marine and terrestrial southern African biomes are of vital importance for global biodiversity conservation and carbon sequestration. They thus deserve special attention in further research on the impacts of anthropogenic pressures including climate change. Excellent preconditions exist in the form of long-term data sets of high quality to support scientific advice for future sustainable management of these vulnerable biomes.

W. Ekau
Leibniz Centre for Tropical Marine Research (ZMT), Bremen, Germany

W. Hagen
University of Bremen, BreMarE - Bremen Marine Ecology, Marine Zoology, Bremen, Germany

University of Bremen, Marum - Center for Marine Environmental Sciences, Bremen, Germany

J. A. Huggett
Department of Forestry, Fisheries and the Environment (DFFE), Oceans and Coastal Research, Cape Town, South Africa

University of Cape Town, Department of Biological Sciences, Cape Town, South Africa

R. Koppelmann · B. Martin
Universität Hamburg, Institute of Marine Ecosystem and Fishery Science, Hamburg, Germany

M. Körner
GEOMAR Helmholtz Centre for Ocean Research Kiel, Kiel, Germany

N. Lahajnar
Universität Hamburg, Institute for Geology, Hamburg, Germany

G. F. Midgley
School for Climate Studies, Stellenbosch University, Stellenbosch, South Africa

T. Rixen
Leibniz Centre for Tropical Marine Research (ZMT), Bremen, Germany

Universität Hamburg, Institute for Geology, Hamburg, Germany

C. D. van der Lingen
University of Cape Town, Department of Biological Sciences, Cape Town, South Africa

Department of Forestry, Fisheries and the Environment (DFFE), Fisheries Management, Cape Town, South Africa

H. M. Verheye
University of Cape Town, Department of Biological Sciences, Cape Town, South Africa

M. R. Wilhelm
University of Namibia, Sam Nujoma Marine and Coastal Resources Research Centre (SANUMARC), and Department of Fisheries and Ocean Sciences, Henties Bay, Namibia

2.1   Introduction

Southern Africa is a globally important hotspot of biodiversity and endemism. It hosts both individual taxa and entire ecological communities that are unique in the world. Its large marine biomes, while also containing many unique, endemic species, are known more for their exceptionally high biomass and productivity than for their overall species richness. For the purposes of this chapter, we have defined southern Africa as the area south of the Kunene River on the West Coast and Quelimane on the East Coast (approximately 17° S). As such, it incorporates all of Namibia, South Africa, Botswana, Eswatini (formerly Swaziland) and Lesotho, as well as most of Zimbabwe and the southern half of Mozambique. Aspects of the marine Angola Current biome between the Congo River (approximately 6°S) and the Kunene River are also included.

Identified as one of the world's 17 megadiverse nations, South Africa ranks in the top ten nations globally for plant species richness and is third for marine species endemism (Tolley et al. 2019). This is despite the fact that it is located mostly outside of tropical latitudes which host most of the world's species-rich ecosystems. With a landmass of 1.2 million km^2 and surrounding seas of 1.1 million km^2, South Africa (without its sub-Antarctic territories and waters) is among the smaller of the world's megadiverse countries—which together contain more than two-thirds of the world's biodiversity (Tolley et al. 2019). Tolley et al. (2019) reported that approximately 10% of the world's marine fish species, 7% of vascular plants, 5% of mammals, 7% of birds, 4% of reptiles, 2% of amphibians and 1% of freshwater fishes exist in South Africa. While they found limited information available on invertebrate groups overall, they stated that almost a quarter of global cephalopod species (octopus, squid and cuttlefish) are found in South African waters.

Southern African marine ecosystems offer habitats for a large variety of species. Even though species richness is lower than in the top-ranked ecosystems such as coral reefs, they host a high proportion of globally unique taxa. Griffiths et al. (2010) reported 12,715 marine eukaryotes in the ocean around South Africa, and the National Biodiversity Assessment in 2018 lists more than 13,000 marine species (Sink et al. 2019). However, these estimates constantly need updating owing to new species discoveries and taxonomic revisions. Many of the marine species in South African waters have so far been found nowhere else on the globe, and thus, the country is reported as having the third highest marine endemism rate after New Zealand (51%) and Antarctica (45%) (Sink et al. 2019). Accurate estimates of biodiversity and endemism are challenging to obtain, and records differ because studies vary in areal extent covered, methodology, data sources consulted, sampling effort and taxa investigated. Griffiths and Robinson (2016) concluded, based on then available comprehensive data sets for marine species, that around 28% to 33% of the marine taxa within the national boundaries of South Africa are endemic. It was, however, pointed out that sampling intensity varies greatly among habitats, and has been concentrated in the shallow near-shore areas. South African benthic invertebrates show peaks in species richness (regardless of a likely bias in sampling intensity), where species distribution ranges overlap when two biogeographic regions meet, particularly around the Cape Peninsula, where the influences of the Indian Ocean and the Atlantic Ocean converge (Awad et al. 2002). The region between Cape Point and Cape Agulhas, a transition zone between the cool-temperate West Coast and the warm-temperate South Coast biogeographic provinces, hosts distinct genetic lineages of several species which are unique to this zone (Teske et al. 2011). Maximum marine endemism has been found in the warm temperate Agulhas ecoregion along the South Coast of South Africa (Awad et al. 2002; Sink et al. 2019). However, several species like the African penguin, green sea turtle, abalone and many fish taxa are currently endangered due to habitat losses and other drivers.

The terrestrial ecosystems of southern Africa have a very rich biodiversity, particularly considering the subcontinent's low rainfall and subtropical-to-temperate climate. South Africa holds three of the world's 35 biodiversity hotspots (a measure of biological diversity combined with vulnerability to threats): the Cape Floristic region, Succulent Karoo Biome and Maputaland–Pondoland–Albany center of endemism (Tolley et al. 2019). There are an estimated 22,000 species of ferns, angiosperms and gymnosperms in southern Africa (Huntley 2003). Species diversity is not homogeneous and varies extensively throughout the region. In the southern African terrestrial habitats, floral speciation and endemism are particularly high within the Fynbos and Succulent Karoo Biomes. The succulent karoo vegetation represents the greatest floral species richness when compared globally to all areas with equivalent rainfall. Although the savanna has only slightly fewer plant species than the Fynbos Biome, the nature of the diversity is very different. Firstly, the southern African savanna is an order of magnitude greater in spatial extent than the fynbos. Secondly, there is a high alpha diversity, but low beta and gamma diversity in the savanna. This means there is a high diversity of species at any particular location, but these same species occur very widely (Huntley 2003). The fynbos, by contrast, has relatively few species at any specific location, that is a lower alpha diversity, but high differences in species between locations, thus high beta and gamma diversity (Cowling et al. 2003a; Scholes 2003). Plant speciation at the genus level is truly exceptional within the fynbos, with genera such as *Protea* and *Erica* showing immense and relatively recent speciation (Cowling et al. 2003b) along with Aizoaceae in the succulent karoo (Klak et al. 2004).

Faunal diversity is in part explained by the diverse flora (Proches and Cowling 2006). Terrestrial animal diversity, although less than plant diversity, is still high. About 960 species of birds, of which 98 are endemic, occur in the region (van Rensburg et al. 2002). Southern Africa has the richest reptile diversity in Africa, exceeding 490 species (Branch 2006) with 384 indigenous species in South Africa, Lesotho and Eswatini (Branch 2014), and including a diversity hotspot region for chameleons (Tolley et al. 2008). Insect diversity, though still not fully studied, is closely correlated with plant species diversity (Proches and Cowling 2006). Spider diversity is higher than in the African tropics and 71 spider families, 471 genera and 2170 species have been recorded from South Africa, of which 60% are endemic (Dippenaar-Schoeman et al. 2015).

Spatially extended and sometimes steep environmental gradients are thought to be largely responsible for the extensive biome-level diversity on land (Cowling et al. 2003b). These authors point out that high species richness is often related to dystrophic soils and disturbance, which prevents the establishment of shaded closed woodland and forest conditions even under high rainfall, and appears to promote high plant richness and endemism.

Deeper in time, the relative stability of major climatic and ocean features over the Quaternary and late Neogene periods has likely lowered extinction rates relative to many other parts of the world (Enquist et al. 2019). Over the Quaternary period, repeated glacial/interglacial cycles have caused pulsed equatorward/poleward shifts in the path of westerly rain-bearing frontal systems to the subcontinent and shifts in the position of trade winds (Stuut et al. 2004) and in sea level, exposing coastal shallows and platforms for periods of thousands to tens of thousands of years (Cowling et al. 2020).

Unlike on land, a certain degree of environmental stability appears to be a key factor promoting high biodiversity in the marine realm (Woodd-Walker et al. 2002; Robison 2004, 2009). Highly fluctuating coastal upwelling conditions on the southern African West Coast thus reduce biodiversity. In contrast, the interplay of the warm-water Agulhas Current with the cold-water Benguela Current at the southern coast sections as well as more stable conditions at the East Coast create high numbers of heterogeneous micro- and macrohabitats that enhance biodiversity and endemism.

Sharp spatial discontinuities shape the marine realm, namely via the topography, major ocean currents and wind-driven circulation affecting water properties—particularly temperature, salinity, oxygen and nutrient content—which determine the habitat structure and living conditions of organisms. One such major environmental discontinuity is generated by the convergence of the warm and saline Agulhas Current, which originates from the Indian Ocean and moves poleward (southward) along the East Coast of southern Africa, the South Atlantic Current and the cold eutrophic branches of the Antarctic Circumpolar Current from the south, feeding into the Benguela Current which flows equatorward (northward) along the West Coast of southern Africa (see Figs. 8.1 and 8.2; Chap. 8 and Fig. 9.5; Chap. 9).

Southern Africa may well be the area in which modern humans evolved (Fortes-Lima et al. 2022). The history of human impacts in this region is extensive. Small numbers of modern humans persisted in the southern Cape, in particular, during very adverse climatic conditions of Pleistocene stadials, supplementing their terrestrial diets with marine resources (Esteban et al. 2020; Wren et al. 2020), but only with likely local impacts. Coastal marine resources such as mussels and limpets were hand-harvested along the South African South Coast for food and other uses during the Middle and Late Stone Ages (Nelson-Viljoen and Kyriacou 2017). The San, a predominantly hunter-gatherer society, and the Khoekhoen, early pastoralists, have inhabited the region going back 150,000 to 260,000 years (Schlebusch et al. 2017). Bantu-speaking groups migrated into southern Africa by both a West African and an East African route, with settlements dating back approximately

1400 years (Vansina 1994; van Waarden 2002). These groups brought with them crop agriculture, metal smelting and cattle. Finally, there was European colonization of the area starting from the fifteenth century, including Portuguese settlements on the East Coast and Dutch settlements in the Cape region (Biggs and Scholes 2002). During the Holocene, human populations increased in size, and population growth of indigenous peoples and European colonization characterize the past millennium to century time scales. Tidal fish traps were in use from the late nineteenth century (Hine et al. 2010), before the advent of industrial-scale fishing for sole and Cape hakes in the early twentieth century (Durholtz et al. 2015). In the past few centuries, the introduction of modern technologies into the region has had sudden, massive impacts on both land and ocean biodiversity (Skowno et al. 2019). In the terrestrial environment, European settlers introduced guns to the area, which had devastating impacts on large mammals, especially elephants. In addition, many areas were destocked of their original wildlife, which was replaced with domestic livestock, often in fenced areas with artificial water points. Later there was further extensive transformation of natural vegetation to agricultural cropland.

Southern Africa has extensive and well-maintained legacy environmental data. For instance, the South African National Biodiversity Institute (SANBI) has the responsibility for coordinating the maintenance of national biodiversity information (www.sanbi.org/resources/infobases/). The South African Earth Observation Network (SAEON) maintains long-term environmental monitoring sites in South Africa (Chap. 30). Although at a slightly smaller scale, comprehensive research and data collections regarding biodiversity are also maintained in Namibia, Botswana, Zimbabwe and Mozambique. Much of the southern African historic data is well maintained and accessible in electronic format. Avian biodiversity is exceptionally well monitored through annual citizen science bird censuses (Hugo and Altwegg 2017).

2.2   Oceanic Biomes

The biogeography of the oceans is structured primarily through large-scale ocean circulation patterns, where frontal zones between water masses act as boundaries (Longhurst 2007). Biomes and provinces within the oceans have been defined using different classification methods (Hardman-Mountford et al. 2008; Oliver and Irwin 2008). Considering the oceans' biogeochemical properties, Fay and McKinley (2014) have distinguished 17 open-ocean biomes. Other authors based their biogeographic zonation of the oceans on a combination of physical and biogeochemical properties, as well as ecological communities (compare bioregions defined in Lombard et al. 2004; Sutton et al. 2017). Briggs and Bowen (2012) based their classification on fish distributions and defined the warm-temperate Benguela Province as a separate zone within the East Atlantic, and also the Agulhas Province, which is linked to the Indo-Pacific. In this chapter, we adopt a broad definition of 'oceanic biomes' for the marine realm, including not only abiotic criteria, but also ecological zonation patterns.

2.2.1   Oceanographic Gradients Shaping Southern African Marine Biomes

Satellite images and satellite-derived data products visualize some of the key characteristics of the marine biomes around southern Africa (Fig. 2.1; Good et al. 2020). On the West Coast, at about 17°S, the Angola-Benguela Frontal Zone (ABFZ) separates the warm oligotrophic waters off Angola from the Benguela Upwelling System (BUS) to the south, which is characterized by cold and nutrient-rich surface waters. In contrast, all along the East Coast of Africa, warm waters are found at the surface. The warm-water masses of the Angola Current in the north, and of the Agulhas Current in the south-east, form the oceanographic boundaries of the cold BUS (Hutchings et al. 2009; Kirkman et al. 2016; Chap. 9).

The circulation in Angolan waters is dominated by the Angola Current (AC) which transports warm tropical waters southward until reaching the ABFZ (Kopte

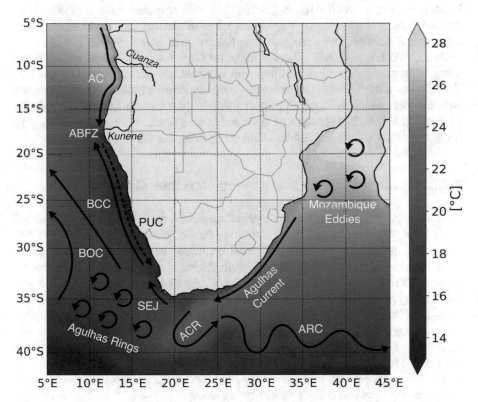

Fig. 2.1 Sea surface temperature (SST) around southern Africa; mean for the period 2003–2019. *AC* Angola Current, *ABFZ* Angola-Benguela Frontal Zone, *BCC* Benguela Coastal Current, *PUC* Poleward Undercurrent, *BOC* Benguela Offshore Current, *SEJ* Shelf Edge Jet, *ACR* Agulhas Current Retroflection, *ARC* Agulhas Return Current. Data: OSTIA product (https://resources. marine.copernicus.eu/product-detail/SST_GLO_SST_L4_REP_OBSERVATIONS_010_011/ INFORMATION), DOI: https://doi.org/10.48670/moi-00168

et al. 2018). South of the ABFZ, the subsurface Poleward Undercurrent (PUC) transports South Atlantic Central Water (SACW) poleward, providing warmer pelagic habitats in the otherwise cold Benguela environment. South-easterly trade winds of the South Atlantic Ocean drive the northward-flowing Benguela Current and the coastal upwelling, with surface waters being colder than the surrounding water masses. At the surface, the Benguela Current is a key dynamic feature, and branches into an oceanic part (Benguela Offshore Current, BOC) and a coastal part (Benguela Coastal Current, BCC). The BOC is part of the eastern limb of the Subtropical Gyre transporting subtropical Eastern South Atlantic Central Waters (ESACW) northward. The BCC can be seen as a coastal jet being part of the upwelling system (Siegfried et al. 2019).

The currents east and south of southern Africa provide an important part of the large-scale circulation in the Indian Ocean and establish a key link of the global conveyor belt circulation. As the western boundary current in this system, the Agulhas Current transports water of tropical and subtropical origin to the south. It flows along the South African South Coast, then overshoots the southern tip of the continent, before abruptly turning back into the Indian Ocean. At the Agulhas Current retroflection, large mesoscale eddies are shed which, together with small-scale currents like the Shelf Edge Jet, make the Cape Cauldron south-east of Africa a region of strong current interactions and vigorous air–sea exchange. As a result, the Agulhas waters provide a warm and saline component for the upper limb of the Atlantic Meridional Overturning Circulation (AMOC), and are important for global climate (Beal et al. 2011). Some of these waters also reach into the Benguela region (Fig. 2.1; for details on the Agulhas Current System, see Chap. 8).

2.2.2 Southern African Marine Biomes: A Brief Overview

Around southern Africa, different marine biomes or bioregions (*sensu* Lombard et al. 2004) can be distinguished, which are defined through their oceanographic properties, as well as by their biological communities. Five inshore bioregions have been described for the coast of South Africa alone (Lombard et al. 2004), namely (from east to west) the Delagoa, Natal, Agulhas, South-western Cape and the Namaqua bioregions (Fig. 2.2), the latter bordering the Namib bioregion off Namibia (Griffiths et al. 2010). As offshore bioregions, Lombard et al. (2004) defined the Atlantic Offshore, the Indo-Pacific, the West Indian Offshore and the South-west Indian Offshore Bioregions (see Fig. 2.2). In their spatial biodiversity assessment for South Africa, those authors further divided their five inshore and four offshore bioregions by depth strata into 'biozones,' units which may help to assess threat status and design protection measures.

Within the Benguela region or province, a description of ecologically meaningful spatial subunits has been summarized by Hutchings et al. (2009) and Kirkman et al. (2016). Recently, researchers have advised taking these boundaries into account to implement ecosystem-based management and effective conservation measures (Kirkman et al. 2016, 2019). The concept of marine biomes is also being integrated

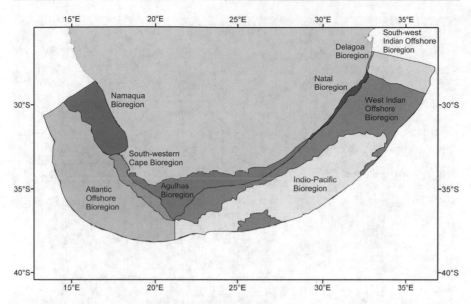

Fig. 2.2 South Africa's five inshore and four offshore bioregions, as defined by Lombard et al. (2004); from Griffiths et al. (2010), https://doi.org/10.1371/journal.pone.0012008.g004

into current efforts to implement Marine Spatial Planning (MSP) in the southern African countries (see Sect. 2.2.4 in this chapter).

The marine ecosystems around the southern African coasts cover the transition zone between Atlantic and Indio-Pacific biomes (Fig. 2.3). This gives rise to a remarkably rich biodiversity and a high amount of endemic species. For South Africa alone, more than 3500 species were classified as endemic during the Census of Marine Life, within a total of 12,715 reported eukaryotic species (Costello et al. 2010). The marine realm around the coasts of Angola, Namibia, South Africa and Mozambique comprises a large variety of habitats, including sandy and rocky shores, kelp forests, coral reefs and estuaries. Yet, an uneven sampling effort, in both geographical and taxonomic dimensions, as well as differential availability of region-specific identification guides and expertise, are expected to cause bias in the reported biodiversity (Griffiths 2005; Costello et al. 2010).

Depending on their distance from shore and their depth, the marine ecosystems can generally be grouped into three broad zones, namely coastal, benthic and pelagic ecosystems. For the West Coast of southern Africa, these zones have already been used in ecosystem mapping for South Africa's National Biodiversity Assessment (Sink et al. 2011), and in extending the approach across the national borders to cover the entire Benguela region (Kirkman et al. 2019). Alongshore, the southern African marine ecosystems are associated with the major biomes of the Angola Current region and the Benguela Upwelling System (BUS)—which together form the Benguela Current Large Marine Ecosystem (BCLME). The BCLME extends from the northern boundary of Angola southward along the West Coast

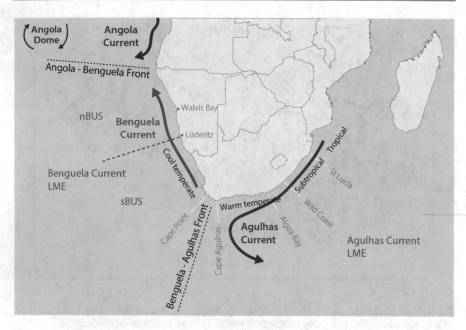

Fig. 2.3 Marine biomes and major current systems around southern Africa. The warm Angola and Agulhas Currents and the cold Benguela Current shape the ecosystems around the coasts of southern Africa. The four major biogeographic provinces (cool temperate, warm temperate, subtropical and tropical) each host characteristic assemblages of species. According to Teske et al. (2011), 'coastal phylogeographic breaks' (between Cape Point and Cape Agulhas, between Algoa Bay and the Wild Coast, and around St Lucia) characterize zones where distinct genetic lineages can be found in many coastal taxa. LME—Large Marine Ecosystem; sBUS and nBUS—southern and northern Benguela subsystem, respectively. (Figure adapted from Teske et al. 2011)

of southern Africa to the Cape of Good Hope in South Africa. The BUS hosts a cell of exceptionally intense upwelling off Lüderitz (around 27°S), which effectively divides the BUS into a northern and a southern part (Duncombe Rae 2005; Shannon 2009). Together with the adjacent Agulhas Current LME, these zones belong to the biogeochemical provinces BENG (Benguela Current coast) and EAFR (East African coast), respectively, according to Longhurst (2007).

Aside from their horizontal zonation with distance from shore, the southern African marine biomes are vertically structured and distinct: near-coastal biomes are clearly influenced by the contact with the seafloor and its benthic communities, and furthermore to varying degrees by river run-off or transport of water and minerals between land and sea (see Sect. 2.4 of this chapter). Further offshore, the steep slope of the continental shelf forms an important structure—shaping upwelling intensity, ocean currents and thereby also migration and dispersal routes for organisms. Far offshore, the open ocean is mainly structured by the diminishing intensity of light with increasing depth, leading to the typical oceanic zones: (1) the photic epipelagic zone reaching from the surface down to approximately 200 m (the zone with the largest productivity, based on photosynthesis), (2) the mesopelagic twilight zone from ~200 to 1000 m (inhabited also by vertically migrating species, which seek refuge here from predators during the day and ascend into the epipelagic at night to feed), (3) the bathypelagic zone (1000 to 4000 m) with no sunlight and no primary producers at all, and (4) the abyssopelagic zone (below 4000 m) with highest ambient pressure and temperatures close to 0°C. The ecology of the latter two zones of the global ocean is least studied, mainly due to the technical and logistical requirements to reach them with oceanographic instrumentation.

2.2.2.1 Angola Current Biome

The ecosystem properties north of the Angola-Benguela Frontal Zone (ABFZ), which is located at about 17°S, are driven by coastal trapped waves propagating southward from the Equatorial Atlantic, thereby resulting in seasons of dynamically driven upwelling and downwelling. Water is supplied from the open South Atlantic toward the eastern boundary by the South Equatorial Undercurrent and the South Equatorial Counter Current and transported southward along the coast by the Angola Current. Additionally, freshwater is supplied by river run-off from the Congo River and other rivers further south, such as the Cuanza River (Kirkman et al. 2016, and references therein, Siegfried et al. 2019). There is no significant wind-driven upwelling north of the ABFZ. Variable extents of low-oxygen waters occur and create an oxygen-depleted subsurface zone encompassing the Angola Dome region and extending toward the coast (Monteiro and van der Plas 2006). The main exploited resources of this subsystem include two *Sardinella* species and Kunene horse mackerel *Trachurus trecae* as pelagic fish, as well as demersal sparid fish such as *Dentex* spp., Angolan hake *Merluccius polli*, and deep-sea red crab *Chaceon* sp. (Kirkman et al. 2016). The biodiversity of demersal species in Angola is high, relative to the temperate Benguela ecosystem to the south (Huntley et al. 2019; Kirkman and Nsingi 2019).

In the steep environmental gradient at the ABFZ, high abundances of zooplankton (Verheye et al. 2001; Postel et al. 2007) and fish larvae (Ekau et al. 2001) have been observed. This front constitutes the northern boundary of the BCLME, which is divided into two oceanographically and ecologically distinct regions, the northern and the southern Benguela Upwelling subsystems.

2.2.2.2 The Northern Benguela Upwelling System (nBUS)

The northern Benguela Upwelling System (nBUS) extends from the Angola-Benguela Frontal Zone in the north (17°S) to the Lüderitz upwelling cell off Namibia in the south (27°S). Upwelling in the nBUS is perennial, with total cumulative upwelling being an order of magnitude greater than in the sBUS (Lamont et al. 2018). Extreme low-oxygen or even anoxic zones develop on the inner continental shelf off Namibia, due to a combination of factors: rather oxygen-poor SACW reaching the nBUS, and high oxygen consumption coupled with high sulfate reduction rates at the sea floor (microbial respiration and fermentation of the decaying organic matter deposited onto the shelf sediments). Hydrogen sulfide (H_2S) is maintained in large (1500 km^2) bacterial mats produced by sulfur-oxidizing, nitrate- and sulfur-storing bacteria, which prevent H_2S from constantly diffusing into the water column. However, gas (methane and H_2S) eruptions from this layer lead to seasonally occurring intrusions of hydrogen sulfide, with detrimental effects, including mass mortalities, on the surrounding marine organisms (Emeis et al. 2004; Brüchert et al. 2006; Currie et al. 2018).

Stock sizes of formerly important fisheries resources in the nBUS, specifically sardine (*Sardinops sagax*) and, to a lesser extent, anchovy (*Engraulis encrasicolus*), collapsed in the 1970s. Nowadays, abundant components of the food web include large scyphozoan jellyfishes (*Aequorea forskalea* and *Chrysaora fulgida*) (Roux et al. 2013), the bearded goby (*Sufflogobius bibarbatus*) (Utne-Palm et al. 2010) and mesopelagic fishes. Top predators include Cape fur seals (*Arctocephalus pusillus pusillus*), dolphins and seabirds, e.g., African penguins (*Spheniscus demersus*) and Cape gannets (*Morus capensis*) (Kirkman et al. 2016). Commercially exploited components include predatory demersal species—specifically the two hake species (*Merluccius paradoxus* and *M. capensis*) and monkfish (*Lophius vomerinus*)—and pelagic horse mackerel (*Trachurus capensis*) as well as deep-sea red crab (*Chaceon maritae*) (Kirkman et al. 2016).

2.2.2.3 The Southern Benguela Upwelling System (sBUS)

South of the permanent Lüderitz upwelling cell, in the southern Benguela Upwelling System (sBUS), low-oxygen water is less prevalent than in the nBUS and occurs in rather restricted locations. The sBUS is characterized by seasonal, wind-driven upwelling at discrete centers, with peaks occurring during austral spring and summer (Hutchings et al. 2009; Lamont et al. 2018). The densest communities of zooplankton have been reported downstream of the Namaqua and Cape Columbine upwelling cells, between the Orange River mouth and St Helena Bay (Pillar 1986; Huggett et al. 2009), in the major nursery grounds of commercially targeted fish species. The southern section of the sBUS is influenced by the Agulhas Current and Agulhas leakage into the South Atlantic, which occurs through shedding of mesoscale Agulhas rings and eddies (Beal et al. 2011). The landside boundary between the Benguela Current LME and the Agulhas Current LME is located at Cape Agulhas. The seaward transition, on the other hand, moves within certain limits as the variable mesoscale oceanographic features of the Agulhas leakage lead to different degrees of influence and exchange around the boundary between the two Large Marine Ecosystems.

The continental shelf of the Benguela region is particularly wide compared to the other eastern boundary upwelling systems. Thus, in conjunction with shelf-break fronts originating from the Shelf Edge Jet and inshore upwelling fronts, retention cells are formed, particularly in the sBUS, increasing the productivity (Flynn et al. 2020; Rixen et al. 2021). They facilitate nurseries for anchovies and other small pelagic fish species spawned west of Cape Agulhas as well as for mesopelagic pseudo-oceanic species, and provide rich feeding grounds for whales (Ragoasha et al. 2019; Dey et al. 2021).

Exploited marine resources in the sBUS include hakes, anchovy, sardine, a currently declining overexploited West Coast rock lobster (*Jasus lalandii*) population, linefish such as snoek (*Thyrsites atun*) and yellowtail (*Seriola lalandii*), and several tuna species (Kirkman et al. 2016). There is considerable exchange between the stocks of small pelagics of the sBUS and the Agulhas Current LME. Commercial and experimental fishing on mesopelagic resources has occurred in the sBUS where catches were dominated by the lanternfish *Lampanyctodes hectoris* (Tyler 2016), while this species as well as lightfishes (*Maurolicus walvisensis*, formerly reported as *M. muelleri*) were found to be common and abundant over the shelf of the entire BUS (Coetzee et al. 2009, 2018).

2.2.2.4 Agulhas Current LME

The Agulhas Current Large Marine Ecosystem stretches from Cape Agulhas to the northern end of the Mozambique Channel, and the ecosystem is driven by the swift southward-moving warm Agulhas Current. Intermittent upwelling occurs at the shelf edge and seasonal mixing takes place on the broad Agulhas Bank (Kirkman et al. 2016, and references therein). Strong winds from various directions are typical for the area. The Agulhas Bank provides spawning grounds for many commercially important species targeted by fisheries locally or after their migration to the West Coast. These include the hakes, sardine, anchovy, round herring (*Etrumeus whiteheadi*), horse mackerel, chokka-squid (*Loligo reynaudii*), Agulhas sole (*Austroglossus pectoralis*) and linefish, such as dusky kob (*Argyrosomus japonicus*) (Kirkman et al. 2016). Seabirds, particularly Cape gannets, are top predators of the region's pelagic fishes. The zooplankton community on the Agulhas Bank is dominated by a large copepod species, *Calanus agulhensis* (Verheye et al. 1994; Huggett et al. 2023), which is an important food item for pelagic fishes and squid. Highest copepod densities are often associated with a *quasi*-permanent ridge of cool upwelled water on the central and eastern parts of the bank that is thought to fuel local productivity as well as enhance retention (Huggett and Richardson 2000). Around the South Coast of South Africa in the Agulhas Current LME, biodiversity is particularly high due to the influence of the Indian Ocean biota (Gibbons and Hutchings 1996; Smit et al. 2017).

2.2.3   The Benguela Upwelling System: A Focus Region of SPACES Research

Research within the TRAFFIC [1] project of the SPACES II [2] program has set a focus on the Benguela Upwelling System (BUS), in order to increase our understanding of biodiversity, ecological functioning, carbon sequestration and particularly trophic transfer efficiency within its food web, and hence the mechanisms supporting the exceptionally high productivity of this ocean region.

---

[1] TRAFFIC - Trophic Transfer Efficiency in the Benguela Current
[2] SPACES II – Science Partnerships for the Adaptation to Complex Earth System Processes in Southern Africa

2.2.3.1 A Global Perspective on the Ecological Significance of the Benguela Region

Together with the Humboldt, Canary and California Current Systems, the Benguela Current System belongs to the world's most productive marine biomes and provides a significant portion of the global catch of wild fishes for human consumption. Common to the biomes in all four of these large Eastern Boundary Upwelling Systems (EBUS) is a rich supply of nutrients through coastal upwelling caused by the trade winds, which supports intense phytoplankton growth that forms the base for an exceptionally high marine productivity. At the same time, the Benguela Upwelling System (BUS) hosts communities with a high share of globally unique (i.e., endemic) marine taxa.

The Benguela Current Large Marine Ecosystem (BCLME) has been included in the Tentative List of UNESCO World Heritage Sites and has been classified as one of Outstanding Universal Value (OUV) based on three criteria: (1) the extremely high primary production sustaining a global hotspot of productivity and rich stocks of commercially targeted fish and crustaceans, as well as large populations of other fishes, seabirds and marine mammals; (2) the occurrence of many endemic or otherwise rare species; and (3) the massive genesis of seabird guano, used as natural fertilizer to enhance agriculture production.

2.2.3.2 Biome-Level Diversity

The communities within the marine biomes around southern Africa are shaped by both spatial boundaries and seasonal processes, which generate dynamic environmental gradients within the ocean.

The high productivity of the marine ecosystem in the Benguela region results from the process of wind-driven coastal upwelling, which defines the specific oceanographic conditions on the shelf and over the continental rise. Upwelling occurs seasonally in short (~10 day) cycles in the sBUS, while it is perennial in the nBUS (Hutchings et al. 2009; Lamont et al. 2018). During austral summer, hypoxic, nutrient-rich South Atlantic Central Water (SACW) from the Angola Gyre is transported into the northern Benguela, whereas during the winter season the oxygen-rich Eastern SACW (ESACW) spreads northward (Monteiro and van der Plas 2006; Monteiro et al. 2006; Mohrholz et al. 2008).

Because the Benguela Current Upwelling System has been a focus region of research in SPACES, its key elements are highlighted in the following. For more details see Chap. 11. Figures 2.4 and 2.5 below highlight selected taxa observed in the BUS during SPACES research cruises. Further taxa are depicted in Chap. 11.

Fig. 2.4 Selected species of mesopelagic invertebrates caught during SPACES cruise SO285 with the *Research Vessel Sonne* in 2021 (cruise report: https://doi.org/10.48433/cr_so285). Top left: cephalopod *Histioteuthis bonnellii*, top right: amphipod *Themisto gaudichaudii*, bottom left: a decapod shrimp, bottom right: a euphausiid (krill). Images not to same scale. © Solvin Zankl www.solvinzankl.com

2.2.3.2.1 Primary Production and Lower Trophic Levels

Upwelling brings nutrient-rich water from below the thermocline to the surface, providing nutrients for the growth of planktonic algae. As demonstrated for other regions (Ayón et al. 2008), plankton abundance and production increase under moderate upwelling conditions, both seasonally and locally (Grote et al. 2007; Bode et al. 2014). Strong turbulence during intense upwelling can hinder the primary production as phytoplankton cells are swirled out of the euphotic surface water or advected offshore. In addition, abundance and productivity are initially low in freshly upwelled water, because this water originates below the thermocline and its phytoplankton content is minimal. Therefore, the development of a diatom-dominated phytoplankton bloom in the nutrient-rich upwelling plume takes time to respond to upwelling conditions.

Microzooplankton, organisms within the size range from 20 to 200 μm, are distributed in a clear shelf-to-offshore zonation. Heterotrophic dinoflagellates prevail in cold, recently upwelled waters on the shelf, whereas in warmer waters at the shelf break, small copepod species dominate the microzooplankton. For protists, naked ciliates, small dinoflagellates and tintinnids, a clear preference has been shown for the warmer water masses surrounding an upwelling filament in the nBUS.

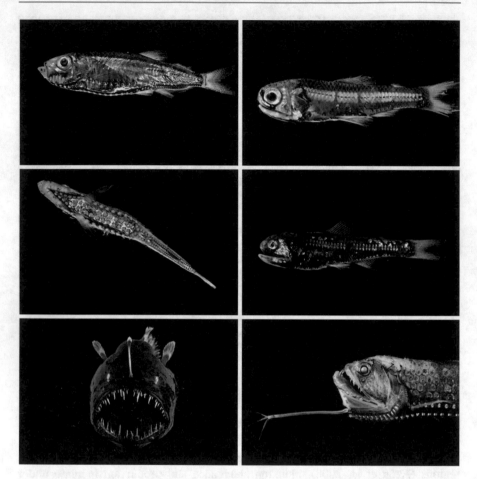

Fig. 2.5 Selected species of mesopelagic fishes caught during SPACES cruise SO285 with the *Research Vessel Sonne* in 2021 (cruise report: https://doi.org/10.48433/cr_so285). Top left: *Maurolicus walvisensis*, top right: *Diaphus hudsoni*, middle left: *M. walvisensis*—photophores as seen from below; middle right: *Diaphus dumerilii*; bottom left: meso- and bathypelagic angler fish *Melanocetus johnsonii*, bottom right: *Stomias boa*. Images not to same scale. © Solvin Zankl www.solvinzankl.com

In contrast, the copepods *Microsetella* spp., *Oithona* spp. and *Oncaea* spp. were associated with the cold water inside the filament (Bohata 2015).

2.2.3.2.2 Mesozooplankton

Zooplankton play a vital role in the functioning of marine ecosystems, providing the main energy pathway from primary producers to higher trophic levels, contributing significantly to carbon sequestration via the biological pump, and serving as sentinels of climate change (Richardson 2008; Batten et al. 2019). Organisms between 200 μm and 2 cm in size make up the mesozooplankton. As in other marine ecosystems, copepods dominate the mesozooplankton of the Benguela Upwelling System in terms of abundance, biomass and diversity. Consequently, they have been the most intensely studied zooplankton group compared to other taxa (Shannon and Pillar 1986; Verheye et al. 1992; Bode et al. 2014).

Zooplankton diversity generally decreases from the equator toward the poles and increases from the epipelagic toward the meso- and bathypelagic zones which reach from 200 to 1000 and from 1000 m to 4000 m depth, respectively (Woodd-Walker et al. 2002; Kosobokova et al. 2011; Bode et al. 2018). Since diversity in the ocean is usually connected to moderate levels of ecosystem disturbance (Woodd-Walker 2001; Woodd-Walker et al. 2002), zooplankton diversity in upwelling zones tends to be relatively low, increasing with distance from shore. This has also been shown for the BUS (Gibbons and Hutchings 1996). Typical copepod species of the Benguela upwelling community include the medium to large *Centropages brachiatus*, *Calanoides natalis* (previously *C. carinatus*), *Metridia lucens* and *Nannocalanus minor*, as well as several smaller members of the Paracalanidae and Clausocalanidae families (De Decker 1964, 1984). De Decker (1984) observed that copepod diversity in the nBUS and sBUS increased from less than 20 species over the shelf to more than 20 or 30 species typically found farther offshore in the South Atlantic central gyre. A patch of up to 47 species off the sBUS were likely transported within Agulhas rings originating from the warm Agulhas Current. An increase in zooplankton diversity from west to east around the coast of South Africa, i.e., toward the Indian Ocean, was also noted by Gibbons and Hutchings (1996), as species with temperate affinities give way to subtropical communities.

A major research task of the TRAFFIC project was to compare the two physically and biologically contrasting subsystems, the nBUS and sBUS, in order to better understand how zooplankton dynamics, community structure and functional traits influence food-web structures at higher trophic levels, and to disentangle biological from physicochemical effects. Predominantly herbivorous species such as the common copepod *C. natalis* become very abundant and play a key role in the food web during active upwelling events. The population size of this species is regulated through a complex mechanism involving offshore displacement of older life-cycle stages in maturing upwelled water plumes and temporary developmental arrest (diapause) of pre-adults at greater depth during prolonged periods of starvation. The latter coincides with the non-upwelling period and the diapausing individuals return shoreward into the productive surface zones with the onset of the next upwelling season (Verheye et al. 1991; Verheye et al. 2005). *C. natalis* clearly dominates the upwelling regions of the nBUS and the sBUS, while *Calanus agulhensis*, the dominant large copepod on the Agulhas Bank off the South African coast (De Decker et al. 1991; Verheye et al. 1994; Huggett et al. 2023), occurs at lower abundances in the sBUS (Huggett and Richardson 2000), and only more recently and sporadically in the nBUS (Rittinghaus 2021). Apart from spatial differences, there are also marked seasonal variations in the zooplankton of the BUS, which are closely coupled to the seasonality of the upwelling cycle. Besides the limited number of studies on non-copepod taxa, the distribution, dynamics and taxonomic composition of smaller calanoid and cyclopoid copepod species, which fit well into the prey-size spectra for larvae or juveniles of many fish species, are understudied (Verheye et al. 2016).

Since the 1950s to 1960s, substantial long-term changes occurred in the abundance, biomass and production of individual zooplankton species. Simultaneously, species and size composition of neritic (on the shelf) mesozooplankton communities have shifted in both the nBUS and sBUS subsystems (Huggett et al. 2009; Bode et al. 2014; Verheye et al. 2016). Abundances of neritic copepods have increased during recent decades by at least one order of magnitude in both subsystems, with turning points reached around the mid-1990s in the south and around the mid-2000s in the north, after which they declined. At the same time, there were marked changes in the copepod community structure, with a gradual shift in dominance from larger to smaller species in both subsystems. These major long-term changes in zooplankton communities are likely to have fundamental effects on biogeochemical processes, food-web structure and ecosystem functioning of the BUS as well as on the ecosystem services, such as fisheries, that ultimately rely on the zooplankton (Verheye et al. 2016). Researchers are currently investigating how this change in community size structure is related to climate change on the one hand, and to changes in the predation regime on the other, due to fluctuating planktivorous pelagic fish populations (e.g., sardine and anchovy).

2.2.3.2.3 Macrozooplankton and Jellyfishes

Macrozooplankton, organisms between 2 and 20 cm body length, provide an important link between higher and lower trophic levels and serve as the primary food for many species, particularly fish but also seabirds and marine mammals. They also play an important role in the export of carbon from the surface to the intermediate and deep ocean (Moriarty et al. 2013). In the BUS, the diversity of pelagic decapods is high, with 46 of 91 Atlantic species present (Schukat et al. 2013; Sutton et al. 2017). Euphausiids, especially *Euphausia hanseni*, represent a major portion of the macrozooplankton in the Benguela system, and their biomass peaks near the shelf edge; also large swarms of *Nyctiphanes capensis* may assemble above the shelf (Barange and Stuart 1991; Hutchings et al. 1991; Werner and Buchholz 2013).

Jellyfishes (e.g., *Chrysaora fulgida* and *Aequorea forskalea*) can be very abundant, particularly in the northern Benguela (Roux et al. 2013). Yet, jellyfishes and comb jellies (ctenophores) have been understudied (Brodeur et al. 2016; Gibbons et al. 2021), because of their poor quantitative representation in plankton nets. However, their role in the food web should receive further attention, especially since they have repeatedly increased in abundance under the influence of climate change and adverse fishing regimes (Lynam et al. 2006; Roux et al. 2013; Brodeur et al. 2016; Opdal et al. 2019). In the nBUS, jellyfishes appear to lead to dead-end food chains, since they have very limited nutritional value to top predators. An acceleration of this so-called jellification process has been hypothesized, should pronounced oxygen-minimum zones expand further under climate change (Ekau et al. 2018).

2.2.3.2.4 Cephalopods

Another characteristic of the pelagic food web is a high diversity of cephalopods (squids and cuttlefishes, specifically) which also distinguish the fauna of the Benguela region (and the Agulhas Current with its retroflection) from other regions of the Atlantic (Rosa et al. 2008). According to Tolley et al. (2019), South Africa hosts almost 25% of the world's cephalopod species. The ecoregions along the temperate coasts of southern Africa are inhabited by numerous species of cuttlefish, bobtails and squids, while they feature a rather low diversity of octopuses (Rosa et al. 2019). The role of cephalopods in the food webs around southern Africa has not yet been thoroughly assessed and requires further research, as already noted by Shannon et al. (2003). Therefore, within the TRAFFIC project, a combination of methods—microscopy, biochemical and genetic analyses—have been applied in order to investigate cephalopod feeding behaviour and their predatory interactions within the food webs of the Benguela upwelling subsystems.

2.2.3.2.5 Pelagic Fish Species

The Benguela ecosystem hosts a species spectrum typical of 'wasp-waist' upwelling systems, where a low number of so-called 'small pelagics' (the 'wasp-waist' with sardine, anchovy and round herring) control both the lower trophic levels (zooplankton) and the higher trophic levels such as tuna or hake and also seabirds and marine mammals (Cury et al. 2000). The small pelagic fishes are characterized by strong fluctuations in their stock sizes and the capacity to build up large stock sizes within a few years (Schwartzlose et al. 1999). Both the northern and the southern Benguela underwent multiple regime shifts since the 1950s caused by both environmental and anthropogenic factors leading to a significant reduction in demersal and pelagic catches in recent years (Jarre et al. 2013, 2015; Heymans and Tomczak 2016). In the northern Benguela, a particularly intense fishery for sardines collapsed around 50 years ago and stocks have not recovered since (Kainge et al. 2020). Sardine and particularly anchovy are still caught in the southern Benguela subsystem, but these forage fish species in the northern subsystem have been practically replaced by horse mackerel, bearded goby and shallow-water hake in the far north and by bearded goby, Hector's lanternfish (*Lampanyctodes hectoris*) and shallow-water hake in the southern part of the northern Benguela (Mwaala 2022).

2.2.3.2.6 Mesopelagic Fish Species

Mesopelagic fishes form an ecologically important component of the pelagic ecosystem, due to their biomass, their diversity, and their diel migrations, which greatly affect the vertical transport of carbon in the ocean (see Fig. 2.5). In the southern part of the sBUS, the proximity of the frontal system to the shelf break and prevailing retention cells, provide sufficient habitat for pseudo-oceanic species associated with continental shelf regions (Hulley and Lutjeharms 1989; Sutton et al. 2017). According to the latter authors, the landward extension of the distribution range of oceanic species and the seaward range extension of pseudo-oceanic species in the central and southern BUS (28–35°S) largely coincide with the 800-m isobath.

Hector's lanternfish *Lampanyctodes hectoris* (Myctophidae) and lightfish *Maurolicus walvisensis* (Sternoptychidae) have previously been shown to be the most abundant mesopelagic fishes in the region (Hulley and Prosch 1987). Because knowledge about the ecology and diversity of mesopelagic fishes was still limited, this group received particular attention in the TRAFFIC project. Spatially, communities differed between the nBUS and sBUS, as well as between on-the-shelf and offshore within each subsystem. These assemblages contained both tropical warm-water species and cold-water species as well as pseudo-oceanic species such as *L. hectoris*, which was prominent on the shelf of the sBUS (Duncan et al. 2022). During a research cruise in late austral summer (Feb-Mar 2019), those authors reported 88 mesopelagic fish species of 22 families in the two subsystems of the Benguela Upwelling System. The most diverse families were lanternfishes (Myctophidae) with 35 species, followed by Stomiidae (ten species) and Sternoptychidae (eight species). About half of all specimens caught were *Diaphus hudsoni* (Myctophidae), *M. walvisensis* (Sternoptychidae) and *Lampanyctus australis* (Myctophidae) (Duncan et al. 2022; Chaps. 3 and 11).

2.2.3.2.7 Demersal Fish Species

The most important nursery grounds for pelagic spawners, as well as for a wide variety of demersal and predatory fishes, are located on the shelf areas of northern-central Namibia, the West Coast of South Africa, the Agulhas Bank and in the small but significant KwaZulu-Natal Bight on the East Coast of South Africa (Hutchings et al. 2002, 2009). Eggs and larvae from spawning grounds on the western Agulhas Bank are transported alongshore to the West Coast in the strong Shelf Edge Jet, keeping them close to the shallower shelf regions, rather than dispersing them offshore (Grote et al. 2012).

Two species of hake co-occur in the Benguela, *Merluccius capensis* and *M. paradoxus*. The former is typically found in shallower zones, the latter at greater depths, leading to their common names of shallow-water and deep-water hake, respectively. As important fisheries resources, these species have been well studied. In general, hakes occupy high trophic levels in the food web. The diet of the shallow-water hake changes throughout their lifetime toward an almost exclusive fish prey and includes common cannibalistic feeding behavior as well as substantial predation on its con-generic *M. paradoxus*. In contrast, the diet of the deep-water hake consists of about one half of crustaceans and one half of fish prey, even for large adult specimens. The deep-water hake therefore usually appears at a lower trophic level than the shallow-water hake in the northern Benguela (Wilhelm et al. 2015), and also in the southern Benguela (van der Lingen and Miller 2014; Durholtz et al. 2015). Other commercially important demersal fish resources found in the Benguela are monkfish (*Lophius vomerinus*), which are especially important in the northern Benguela, kingklip (*Genypterus capensis*) and Agulhas sole (*Austroglossus pectoralis*) and West Coast sole (*A. microlepis*).

2.2.3.2.8 Functioning of Marine Food Webs in the Benguela Upwelling System

A classical food chain that develops from upwelling events is comprised of large-sized diatoms, followed by large-sized zooplankton (herbivorous euphausiids and copepods), which are consumed by small pelagic fishes. Long-term studies of the plankton communities off the West Coast of southern Africa revealed an increase in mesozooplankton abundance, accompanied by a shift to smaller-sized plankton taxa, especially copepods, since the 1950s–1960s and until the mid-1990s in the sBUS and the mid-2000s in the nBUS, respectively (Verheye et al. 2016). Those authors concluded that the observed changes in copepod abundance and size structure could be attributed to the complex interplay of local warming or cooling, increased primary production where upwelling intensified, combined with reduced predation pressure by pelagic fishes owing to increased fishing. Verheye et al. (2016) also emphasized that there is uncertainty about the relative importance of these bottom-up and top-down forcing mechanisms. Overall, the classic picture of a direct coupling between upwelling intensity and primary and secondary production with a short food chain cannot be sustained in the nBUS (Ekau et al. 2018). Secondary (and primary) production is highest either temporarily during moderate upwelling conditions (Cushing 1996) or spatially at some distance from the upwelling source. This confirms the 'optimal environmental window' hypothesis of Cury and Roy (1989), stating that larval fish survival and fish recruitment are dependent on upwelling intensity in a dome-shaped function in Ekman-type upwellings, where very low and very high wind speeds are detrimental. Furthermore, longer trophic pathways than previously thought are active in the food web of the nBUS, contributing to a higher complexity and thus lower transfer efficiency of biomass and energy from phytoplankton to fish (Schukat et al. 2014). This may partly explain the differences in fish production between individual Eastern Boundary Upwelling Systems, and specifically between the nBUS and sBUS despite similar primary production rates (Chavez and Messié 2009).

2.2.3.3 Productivity and Resource Utilization

Being an Eastern Boundary Upwelling System, its core characteristics provide the Benguela ecosystem with an exceptionally high productivity across the food chain from plankton to top predators, which lays the foundation for an effective provision of living marine resources and the development of large fisheries.

The Benguela region is inhabited by a variety of taxa, which are typical of upwelling systems and some of the fish species can attain high biomass levels. Several species are commercially important as fisheries resources including Cape hakes (*Merluccius capensis* and *M. paradoxus*), Cape and Cunene horse mackerels (*Trachurus capensis* and *T. trecae*, respectively), and small pelagics (sardine *Sardinops sagax*, anchovy *Engraulis encrasicolus* and round herring *Etrumeus whiteheadi*). Despite their smaller catches Crustaceans such as rock lobster (*Jasus lalandii*) and deep-water red crab (*Chaceon* sp.) are also commercially important (van der Lingen et al. 2006c; Kirkman et al. 2016; Kainge et al. 2020; Chap. 11).

In terms of their economic value, the most important fisheries in the nBUS subsystem are those for Cape hakes (most valuable) and horse mackerel (largest volume), whereas in the sBUS, Cape hakes (most valuable) and small pelagic species (largest volume) dominate the fisheries (Kainge et al. 2020; Chap. 11). Additionally, high biomasses of as yet not commercially targeted mesopelagic fishes occur in the Benguela region, and gobies as well as jellyfishes can also appear at high abundances, particularly in the nBUS (Lynam et al. 2006; Roux et al. 2013; Kirkman et al. 2016; Salvanes et al. 2018). The sBUS—from south of Lüderitz to Cape Agulhas—provides nursery grounds for most of that subsystem's ecologically and economically important fish species (Kirkman et al. 2016), including both hake species and small pelagics (Clupeiformes).

By the middle of the last century (1950 to late-1960s), sardine was the central fisheries resource in both subsystems, with peak catches beyond 1 million tons in the nBUS and close to half a million tons in the sBUS (for details see Chap. 11). Yet sardine catches declined rapidly due to overfishing, both off the Namibian and the South African coasts (van der Lingen et al. 2006c; Augustyn et al. 2018). In the sBUS, anchovy replaced sardine during the following three decades. In the nBUS, the collapse of the sardine population resulted in substantial reductions in the number of purse-seiners and labour-intensive canning factories and fish reduction plants, resulting in job losses for several thousands of people (Boyer and Oelofsen 2004), and the Namibian sardine fishery has essentially been replaced by the fishery for Cape horse mackerel.

The collapse of the sardine population has resulted in substantial ecosystem changes arising from changed trophic interactions, since sardine was the dominant forage fish species of a wide variety of predators, including other fishes, marine mammals and seabirds. In the nBUS, that change has been hypothesized to have promoted the proliferation of jellyfishes (Roux et al. 2013), and predators previously reliant on sardine as the main very fatty and nutritious food source, now utilize alternative food sources including bearded goby, lanternfish, horse mackerel, shallow-water hake and jellyfish. Goby appears to be a successful substitute when the other more nutritious species (horse mackerel in the north and lanternfish in the south) are not available, likely because of its high abundance in the system and tolerance to high temperature and low oxygen levels (Erasmus et al. 2021). In contrast, catches of the major South African fisheries have remained more or less stable over the past 70 years and pelagic fish are still abundant in the system. However, the most recent decline in sardine is concerning from both fisheries and ecosystem perspectives, particularly for endangered seabirds such as the African penguin, a species whose breeding success (Crawford et al. 2008) and mortality (Robinson et al. 2015) are correlated with the abundance and distribution of sardine.

An experimental fishery for mesopelagic fishes using midwater trawling was initiated as part of the South African small pelagic fishery in 2011, when a catch of 7000 tons was taken. That experiment was resumed in 2018 and 2019 with around 5000 tons taken each year, but the relatively high cost of fishing coupled with a general downturn in both the anchovy and sardine fisheries has resulted in its cessation (DEFF 2020).

2.2.3.4 Organizational Efforts Geared to Protect Marine Biodiversity

In 2007, the three coastal nations Angola, Namibia and South Africa founded the Benguela Current Commission (BCC), which in 2013 became a permanent intergovernmental institution through the Benguela Current Convention. Its purpose is to protect marine biodiversity and to promote sustainable use of the natural resources in the Benguela region through an ecosystem approach to ocean governance. All three nations defined national and common 'Ecologically and Biologically Significant Areas' (EBSAs), which have been and will be submitted to the Convention on Biological Diversity (CBD) and which form the basis for the designation of new Marine Protected Areas (Harris et al. 2022).

The International Union for Conservation of Nature, IUCN, has defined 'A Global Standard for the Identification of Key Biodiversity Areas' (IUCN 2016). Key Biodiversity Areas (KBAs) are 'Sites contributing significantly to the global persistence of biodiversity', in terrestrial, freshwater and marine ecosystems. Such KBAs are proposed and identified from the bottom up by local experts or private and government organizations through National Coordination Groups (NCGs). In order to promote an understanding of the spatial distribution and risk status of biodiversity, O'Hara et al. (2019) have linked the distribution ranges and conservation status (based on IUCN protected area categories) for thousands of marine species in order to present global maps of extinction risk of marine biodiversity.

The South African NCG is hosted by the South African National Biodiversity Institute (SANBI). South Africa hosts an exceptionally high number of endemic terrestrial and marine species (global rank 3 on the national level; SANBI). The National Biodiversity Assessment published by SANBI monitors and reports on the state of biological diversity in South Africa, in order to support political strategies for the conservation of biodiversity (Skowno et al. 2019). In Namibia, the National Biodiversity Assessments are hosted and published by the Ministry of Environment and Tourism (MET), Multi-lateral Environmental Agreements Division, supported by the National Biodiversity Strategies and Action Plan 2 (NBSAP2) steering committee (e.g., MET 2018).

2.2.4 Marine Spatial Planning in Southern Africa

Ongoing Marine Spatial Planning (MSP) in the Benguela Current Large Marine Ecosystem (BCLME) involves Angola, Namibia and South Africa, since part of the LME lies in each of their respective Exclusive Economic Zones (EEZs). Through the Benguela Current Convention, all three nations promote the vision of sustaining human and ecosystem well-being in the BCLME. With their MSP initiative, the countries in the Benguela region are among the first African coastal states to implement MSP, particularly since they pursue a transboundary perspective in their planning. The regional MSP strategy was adopted by the Benguela Current Convention (BCC 2018). The project MARISMA is a partnership between the BCC, its member states Angola, Namibia and South Africa and the government of Germany in pursuit of the sustainable development of the Benguela Current Large Marine Ecosystem. The project's goal is 'to maximize socio-economic benefits while ensuring the safeguarding of the marine ecosystem's health and maintenance of marine services provision' (www.benguelacc.org/marisma), leading to the very recent launch of the BCC GeoData Portal (https://geodata.benguelacc.org/). All three nations have divided their waters into several distinct planning zones, based on human uses and ecological boundaries (Finke et al. 2020; Fig. 2.6). Each nation developed a baseline report describing the actual status of their waters. The BCC countries prioritize—to varying degrees—the following objectives (see Finke et al. 2020 for details):

- Protection of biodiversity features of national, regional or global significance (focus on EBSAs);
- Providing access to fishing grounds, while protecting key fish habitats from adverse effects by human use, including fisheries;
- Securing mariculture locations;
- Enabling exploration and promoting sustainable use of geological resources;
- Guaranteeing maritime transport and disposal of dredge material;
- Allocating space for military training activities;
- Enabling responsible marine and coastal tourism;
- Protecting underwater infrastructure, e.g., cables; and
- Protecting maritime and underwater cultural heritage.

Finke et al. (2020) conclude that fostering ecosystem-based MSP in the context of strong economic growth agendas requires balanced and integrated governance and technical planning structures and processes. While the regional and national approaches taken are considered useful, their implementation will still need to overcome obstacles regarding funding, data needs, research data management, legislation and institutionalization (Finke et al. 2020).

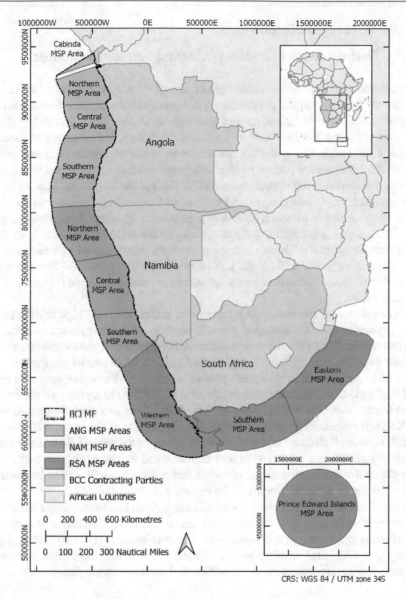

Fig. 2.6 The marine spatial planning (MSP) areas of Angola, Namibia and South Africa. Black dashed outline around the Benguela Current LME (from Finke et al. 2020)

2.3   Terrestrial Biomes

2.3.1   Environmental Gradients Shaping Terrestrial Biomes

Terrestrial biomes in southern Africa can be described simply in terms of the relative dominance by tree, shrub, grass and annual plant life forms (see Rutherford and Westfall 1986). The distribution and structure of these biomes are likely shaped by a mix of linked and independent gradients of climate, disturbance and substrate (geology and soils). Their distinct contemporary structure has evolved particularly during the Neogene, a period of declining atmospheric CO_2, increasing aridity, cooling and intensifying climate seasonality. During the Holocene, major climatic gradients include, broadly, a general tropical to temperature thermal regime from north (warm/humid) to south (cool/dry), an aridity gradient from west (dry) to east (wet), and a seasonality gradient from south-west (winter rainfall) to north east (summer rainfall). These climates provide the thermal and moisture regimes that underpin vegetation structural and related disturbance regimes that determine biome distributions, with some putative interactive role of substrate (soil texture and nutrient status).

The existence of flammable vegetation even under relatively high rainfall conditions appears to override assumed climatic controls of the dominance of trees, and thus the distribution of forests and woodlands. For this reason, flammable shrubland, grassland and savanna biomes may be found under rainfall conditions that have the potential to support taller and more closed vegetation with higher leaf area index over vast regions of the subcontinent (Bond et al. 2005). The dystrophic sandy soils of the south-western coastal plains and Cape Fold Mountains, and the deep sands of the Kalahari respectively support flammable shrublands of the Fynbos Biome, and the extensive well drained arid and mesic savannas of the central Kalahari. Heavier clay-rich soils are associated with arid and semiarid karoo shrublands that stretch from the arid western regions of Namibia and northern South Africa to the more mesic Thicket and Forest biomes of the eastern seaboard.

Paleoclimatic shifts associated with glacial periods have tended to increase the influence of cold and wet westerly rain-bearing frontal systems, and permitted a northern expansion of the associated winter-rainfall shrublands, while lower atmospheric CO_2 and aridity appears to have reduced the dominance of trees in savanna-dominated central southern Africa, and increased grass dominance.

2.3.2   Southern African Terrestrial Biomes

Southern Africa has a disproportionately high level of terrestrial diversity at the biome level. This is largely as a consequence of the climatic conditions and the intensity or frequency of disturbance. Within southern Africa nine uniquely different biome level vegetation structures have been identified (Figs. 2.7 and 2.8) (Olson et al. 2001; Rutherford et al. 2006). These are Forest, Fynbos, Succulent Karroo, Nama Karoo, Grassland, Savanna (including embedded halophytic pans and flooded

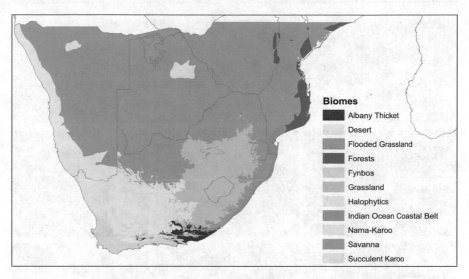

Fig. 2.7 Terrestrial biomes of southern Africa derived from Mucina and Rutherford (2006) for South Africa, merged with (Olson et al. 2001) for the rest of southern Africa. Olson's ecoregions were reclassified to match the South African biomes. Halophytics (salt pans) and flooded grasslands (e.g., the Okavango) are distinctive features in the Savanna Biome and not true biomes

grasslands), Thicket, Indian Ocean Tropical Belt (IOTB) and Desert. The Desert biome is the smallest biome in South Africa, but is better developed and extends northward along the Namibian coast to Angola. Despite this high biome diversity, it is savanna woodlands which dominate in southern African countries. These savannas can be divided into three main types, the more open arid woodland, moist woodlands which are referred to as the Miombo and only occur north of South Africa, and Mopani woodlands that dominate many of the low lying and hot river valley systems (Olson et al. 2001) (Fig. 2.7). The Fynbos Biome and Succulent Karoo Biome have exceptionally high plant species diversity and endemism. Key features of these biomes and the climatic and disturbance envelopes in which they occur are described below.

2.3.2.1 Savanna Biome

Savannas are by far the dominant biome of southern Africa covering 2.16 million km^2 (65%) of southern Africa and additionally extending north to as far as the Sahel, covering 60% of Africa (Scholes and Walker 1993). The Savanna Biome is distinctive in that there is a codominance of both a woody component including both perennial trees and shrubs (phanaerophytes) and herbaceous layer (hemicryptophytes) of predominantly C4 grasses, but also many forbs. Embedded in the savanna are also halophytics pans such as Etosha in Namibia and the Makgadikgadi Pan in Botswana, in total about 0.8% of the southern African region. In addition, there are flooded grasslands, the Okavango delta being the largest and most famous at a

Fig. 2.8 Typical appearances of southern African terrestrial biomes: (1) Desert, (2) Succulent-Karoo, (3) Indian-Ocean Coastal Belt, (4) Nama-Karoo, (5) Savanna, (6) Grassland, (7) Forest, (8) Albany Thicket, (9) Fynbos; 1–3 top, 4–6 middle, 7–9 bottom row. Photos 2, 5, 7–9 GvM. 1, 3, 4, 6 Dreamstime.com royalty free license

further 0.8% of the area. Both of these unique habitats add to the savanna habitat diversity and biodiversity.

Southern African savanna occurs over a rainfall gradient from less than 250 mm yr^{-1} to about 1800 mm yr^{-1} (Scholes and Walker 1993). It is, however, limited to areas that have wet summers and long dry winters. In southern Africa savanna is found in areas of unimodal rainfall, while in east Africa, it occurs in areas of bimodal rainfall. Tree cover in the savanna ranges widely from under 1% to about 80%, and is normally below the ecological limits for maximum tree canopy cover (Sankaran et al. 2005; Chap. 15). Effects of fire, herbivory and climatic variability play an important role in maintaining the dynamics of savanna systems, especially in the moist areas which in the absence of these disturbances could become forests (Sankaran et al. 2005; Chap. 14). Fire is a common occurrence ranging from occasional at the dry extreme to almost annual in moist areas (Archibald et al. 2010).

Herbivores, including the so-called mega herbivores and especially elephants, also appear to play a major role in the savanna dynamics (Bond et al. 2003; Stevens et al. 2016). The grass component of savanna is resistant to both grazing and fire impacts, mature trees tend to be fire resistant, but tree seedlings are often susceptible to fires. Crown fires are rare, and even if the crowns are destroyed, most species can resprout.

Southern African Savanna is part of the Sudano-Zambezian phytochorion (White 1983). A major distinction is between what is termed 'moist savanna' or Miombo in higher rainfall regions which has predominantly broad-leaf trees and a nutrient-poor substratum, as opposed to the 'arid savanna' or 'bushveld' in more arid regions with nutrient-rich substrates and predominantly fine-leafed trees, and finally the 'mopane savanna', which is similar to the arid savanna but dominated by the tree *Colocospermum mopane* (Huntley 1982; Chap. 15). Savannas have high alpha diversity (i.e., species richness at a local site) and contain 3–14 species per m^2 and 40–100 species per 0.1 ha, but with low local endemism (Scholes and Walker 1993).

2.3.2.2 Grassland Biome

The Grassland Biome is the second largest biome in southern Africa (360,000 km^2, 11%), and with the exception of small areas in the highlands of Zimbabwe, is almost exclusively limited to South Africa. A single stratum of hemicryptophytes is what characterizes the grasslands. Trees and shrubs are largely absent or limited to fire refugia. Five centers of plant endemism have been identified within the grasslands. The grasslands have a high flora diversity, especially of nongraminoid herbaceous plants with geophytes (perennial plants) being of particular importance. Grasslands have high alpha diversity and may contain 9–49 species per 100 m^2 plot, with a 1000 m^2 plot containing 55–100 species (Mucina and Rutherford 2006). C4 grasses dominate in the warmer and arid areas, while C3 grasses dominate at cooler high-altitude sites. The grasslands can be divided into two main classes, moist areas on leached and dystrophic soils that tend to be 'sour' with low palatability and the more arid grasses which are 'sweet', even when dormant, and palatable to livestock (Ellery 1992).

Grasslands occur over a wide range of rainfall from about 400 to 1200 mm annually, and over a wide range of soils, ranging from sea level to >3300 m altitude and located in areas that extend from frost free to snow covered in winter (O'Connor and Bredenkamp 2003). As in savannas, fire is an important driver of the ecology of grasslands, and is one of the main reasons that areas of high rainfall do not become forests. Small forest patches are often found within fire refugia in grasslands. There has been extensive debate as to why grasslands persist as grassland, despite the moister areas being able to support forest. Ellery (1992) suggested that fire and grazing regimes exclude woody plants. However, there appears to be a limit to either, as heavy grazing and altered fire regimes can change grass species composition and degrade the grassland. The grasslands are susceptible to invasions of alien plant species, including herbaceous species such as Pompom weed (*Campuloclinium macrocephalum*) and trees such as the Australian *Acacia mearnsii* (Black wattle), which may not only change the species composition but the entire structure. Invasion of indigenous woody species into grasslands is also an issue of concern, and this is a process similar to the bush encroachment experiences in savanna (Skowno et al. 2017; Chap. 15). The grasslands are one of the most anthropogenically transformed biomes as they are suitable for annual crops on the good soil flat areas and for timber plantations in the higher rainfall mountainous areas.

2.3.2.3 Nama-Karoo Biome

The Nama-Karoo Biome (260,000 km^2; 8% of the area) is found in South Africa and Namibia at altitudes between 550 and 1500 m. It is an arid environment dominated by perennial dwarf shrubs and annual grass. The vegetation furthermore includes leaf succulents, stem succulents, bulbous monocotyledons and annuals (Marloth 1908). Gibbs Russell (1987) classified the Nama-Karoo Biome as being 50% hemicryptophytes, and 25% each of chamaephytes and cryptophytes. Gradients across the biome go from strongly dwarf-shrub dominated in the west, to a higher grass ratio in the east where the biome meets the grassland and savanna biomes. Rainfall occurs predominantly in summer and ranges from about 60 to 400 mm/yr. Rainfall is highly variable, particularly in the most arid regions. Although the soils tend to be heavy and eutrophic, they are typically very shallow, with calcareous layers being common (Palmer and Hoffman 2003).

2.3.2.4 Desert Biome

Desert (112,000 km^2; 3% of the area) is found almost exclusively along the coast of Namibia, though a narrow band is found in South Africa in the Senqu (Orange) river valley. It is often separated into the sand deserts to the south and the gravel plains to north. It is dominated by ephemeral plants that respond rapidly to the rare and unpredictable rainfall. There are also some unique perennials such as the *Welwitschia mirabilis* as well as trees growing along river beds. The Desert Biome has extremely low rainfall and a mean of 20 mm yr^{-1} is misleading as to annual rainfall as there can be multiple years of zero rainfall followed by a single substantive rainfall event. Dense fog does, however, compensate in part for the low rainfall, with a number of species such as the *Stenocara gracilipes* beetle exhibiting unique adaptations to harvest the fog.

2.3.2.5 Succulent Karoo Biome

This biome (107,000 km^2) is found along the West Coast of South Africa and southern Namibia. It is characterized by winter rainfall of between 50 and 250 mm, supplemented by dewfall and fog. The biome has the highest biodiversity for any area globally with comparable rainfall, with very high levels of endemism. It is characterized by leaf-succulents and deciduous leafed dwarf shrubs dominated by the families Aizoaceae (Mesembryanthemaceae), Asteraceae, Crassulaceae and Euphorbiaceae (Desmet 2007). In addition, there are many annuals that form the main component of the mass spring floral display for which the biome is well known.

The Succulent Karoo Biome occupies a narrow niche between the Fynbos and the Nama-Karoo Biomes, where the former is characterized by winter rainfall, and the latter by summer rainfall. The Succulent Karoo vegetation is uniquely adapted to areas of low winter rainfall. Esler et al. (2015) have shown that it is fire that prevents the succulent karoo species from establishing within the higher rainfall fynbos. Any increase in precipitation that allows the fynbos to expand into the karoo would therefore be detrimental to the succulent karoo vegetation. Equally, any shift in seasonality toward summer rainfall dominance would likely allow for Nama-Karoo species to expand into the Succulent Karoo Biome.

2.3.2.6 Fynbos Biome

The Fynbos Biome is a small biome of about 85,000 km^2, only 2.6% of southern Africa and only occurs in South Africa, but has an extremely rich floral biodiversity of over 9000 vascular species (3% of the global vascular species) with 70% being endemic (Goldblatt and Manning 2002; Rebelo et al. 2006). It also has high faunal endemism with 55% of its 44 frog taxa, 84% of its 19 freshwater fish and 31% of its 234 butterflies being endemic (Critical Ecosystem Partnership Fund 2001). A distinguishing feature of the vegetation is codominance by perennial shrubs or small trees with sclerophyllous microphyllous (small-leaved) leaf characteristics, sedges and the grass like Restionaceae. True grasses (Poaceae) are very rare in the west, but slightly less rare in the east of the biome (Rutherford and Westfall 1986). The distinctive climatic feature of the region is the predominantly winterly precipitation, driven by the seasonal mid-latitude cyclonic systems that sweep across the region. During summer these fronts are mostly pushed further south and miss the Cape region. Precipitation varies greatly, partly due to the influence of the mountainous topography, with some areas being as low as 210 mm/yr, while the mountain tops are some of the wettest areas in southern Africa with Jonkershoek reaching about 3000 mm/yr. The biome can establish over a wide range of altitudes, soil and aspect features, all probably important in the supporting the vegetation (Cowling et al. 2003a).

The Fynbos Biome, together with the Succulent Karoo Biome, constitutes the Cape Floristic Kingdom, one of only six floristic kingdoms recognized globally, and the only one to be contained within a single country. The Fynbos Biome can be divided into two key vegetation complexes, the renosterveld vegetation found on the more fertile, clay and silt soils of the lowlands and fynbos which dominates the predominantly nutrient-poor mountainous regions of the Table Mountain Group. Atypical areas are also linked to heavier shale band soils and limestone inselbergs. Three families, the Proteaceae, Ericaceae and reed-like Restionaceae dominate the vegetation, with the Proteaceae and Ericaceae in particular showing an exceptionally high level of what appears to be relatively recent Neogene and Quaternary speciation (Cowling et al. 2003a).

The vegetation is fire dependent and requires regular burns for its persistence (van Wilgen et al. 1994; van Wilgen et al. 2010). Optimum fire frequency is considered to be between 10 and 15 years, with intensity of the fire effect largely dependent on the season in which the fire occurs. Although several species resprout after fire, the predominate regeneration strategy for most species is through reseeding, and a multitude of seeding strategies have evolved. Both pollination and in some cases, seeding is dependent upon complex insect plant interactions (Le Maitre and Midgley 1992). The Fynbos has proved to be exceptionally vulnerable to the invasion of alien plants and this is posing a major threat to the region. These aliens, in addition to displacing natural vegetation, can also impact on fire frequency and intensity, further impacting on indigenous biodiversity (Richardson et al. 1996).

2.3.2.7 Forest Biome

True closed canopy forests are uncommon in southern Africa. The Forest Biome is defined as dominated by a tree layer or stratum 'phanaerophytes' with 75% or more, overlapping crown cover and graminoids, if present, are rare (Bailey et al. 1999; von Maltitz et al. 2003). This differs substantially from the FAO definition of forests which would include much of southern Africa's savanna. An estimated 66,000 km^2—2% of the region—is forest, most of it occurring in Mozambique. This excludes a further 0.2% of the area being Mangrove forest (based on Olson et al. 2001; Mucina and Rutherford 2006). However, it is unclear how the Indian Ocean Coastal Belt Biome in South Africa (Mucina and Rutherford 2006) and the Forest Biome (Olson et al. 2001) should be differentiated within Mozambique. Forest within the region has two key origins, Afrotemperate forest occurring in the higher altitude mountains of Zimbabwe and South Africa, and extending to the southern Cape of South Africa where latitude compensates for altitude, and the more tropical forests extending down the coast from Mozambique. Unlike the other biomes where the biome forms a large block of consolidated land, the forests tend to be embedded in the other biomes, typically in fire refugia. They occur in areas of over 525 mm yr^{-1} in winter rainfall regions or over 725 mm yr^{-1} in summer rainfall regions.

The Forest Biome has high species diversity, with 1438 plant species recorded for just the South African component. On a per area basis forest species diversity (0.58 species per km^2) is second only to the fynbos (Geldenhuys 1992). Despite the low density of animals in forests, 14% of South Africa's bird and mammal taxa have been recorded from forests (Geldenhuys and MacDevette 1989).

2.3.2.8 Indian Ocean Coastal Belt Biome

The species rich 34,000 km^2 Indian Ocean Coastal Belt (IOCB) Biome is a narrow belt of vegetation stretching along the KwaZulu Natal coast in South Africa and along the coast of Mozambique in a vegetation type, which Olson et al. (2001) refer to as Southern Swahili coastal forests and woodlands. It contains both savanna and grassland elements and also includes embedded forest patches. The vegetation is a mosaic of areas with a factually and structurally savanna characteristics and areas of grassland characteristics. Although Moll and White (1978) recognized the uniqueness of the region, others such as Huntley (1982) have considered it part of the Savanna Biome and Low and Rebelo (1996) considered it a mix of grassland and savanna species. Rutherford and Westfall (1994) did not recognize it as an independent biome based of their classification systems, though Rutherford et al. (2006) finally gave it biome states, this based more on the tropical affinity of the biome, rather than its functional aspects, with the grasslands being either azonal or secondary in nature. The IOCB has unique species richness as it is part of the larger the Pondoland and Maputaland center of endemism hotspot, which is second only to the fynbos for species richness and endemism in South Africa, and contains over 8200 species of plants (23.5% endemic), 540 species of birds 200 mammal species, 200 reptiles (14.4% endemic) and 72 amphibians (15.3% endemic) (Conservation International 2014).

2.3.2.9 Albany Thicket Biome

Albany thicket is found exclusively within South Africa and is a 31,000 km^2 area representing 1% of southern Africa. It consists of a dense, almost impenetrable, tangle of short trees less than 3 m tall and with almost 100% canopy cover. An herbaceous understory may be present, though with minimal grass. It is in a transitional zone between winter and summer rainfall and is found predominantly on fertile soils. It is hot and dry, with unpredictable low rainfall of 200 to 950 mm yr^{-1} (Mucina and Rutherford 2006). The vegetation has affinities with the karoo, savanna and grassland vegetation, though Cowling et al. (2005) suggest it is an ancient biome which can be traced back to the Eocene. It has a high plant diversity of about 2000 species of which about 300 are endemic to the biome. The succulent nature of many of the species and the relative scarcity of grass means that fire is not a major feature of the natural biome, but becomes common when the biome is degraded and grass starts to dominate. The biome is exceptionally susceptible to degradation from overgrazing, and a large proportion of the biome is already partially or severely degraded.

2.3.3   Diversity of Southern African Terrestrial Biomes

2.3.3.1 Large-Scale Environmental Factors Shaping Terrestrial Biodiversity

The variability in mean annual precipitation (MAP), seasonality of rainfall and temperature gradients, coupled with fire and disturbances, are factors that have led to the exceptional natural diversity of faunal and floral species within the southern African environment (Cowling et al. 2003a). Africa's southernmost tip, the area of the Cape Floristic Kingdom, has evolved under a Mediterranean climate of winter rainfall and extreme summer water deficiency (Cowling et al. 2003b). As is discussed by Rouault et al. in Chap. 6 and shown in Fig. 2.9, the balance of the region has all year, or in most cases strongly summer-dominated precipitation with long arid winters, the perfect conditions for the development of grasslands and savanna vegetation (Huntley 1982; Scholes and Archer 1997). There are rainfall gradients going from less than 20 yr^{-1} precipitation in the West to over 3000 mm yr^{-1} in some of the mountainous areas (Fig. 2.9b). In addition, there is a high level of interseasonal variability of rainfall, linked mostly to the occurrence of El Niño Southern Oscillation (ENSO) cycles. This means that it is the norm, rather than the exception, to have periods of either above average rainfall, followed by similar periods with below average precipitation or even severe drought (Chaps. 6

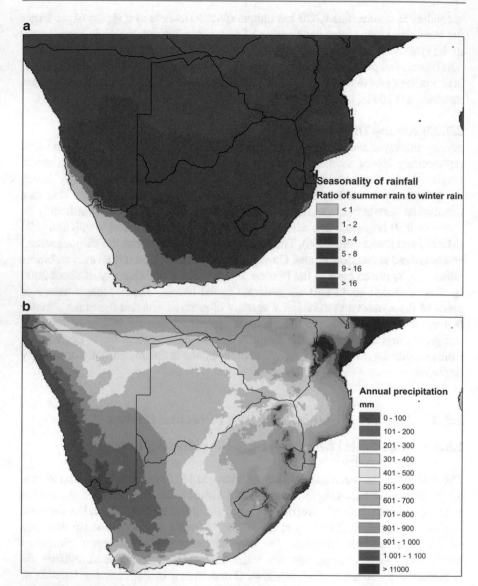

Fig. 2.9 (Top) Seasonality of rain expressed as the ratio of summer rain (Oct, Nov, Dec, Jan, Feb, Mar) to winter rain (Apr, May, Jun, Jul, Aug, Sep), and (Bottom) mean annual precipitation (based on WorldClim 1970 to 2000 data, https://www.worldclim.org/data/worldclim21.html)

and 29). The standard error of MAP is negatively correlated with rainfall, i.e., the most arid areas tend to have the most severe droughts but can also have exceptionally wet years. Not only has the region's unique fauna and flora evolved to deal with this variability, but it is probable that this variability is critical to maintaining the flora and fauna in its current state.

The oceans surrounding southern Africa play an important role in moderating coastal temperatures, the West Coast dominated by the cold Atlantic currents, while the East Coast is impacted by warm Indian Ocean tropical currents (Fig. 2.1; Chap. 6). The prevailing winds causing upwelling of cold water in the Benguela Current along the West Coast are largely responsible for the desert environments along the West Coast, but also contribute to the formation of fog banks that have led to unique adaptations in plants and animals who use this fog as their main source of moisture. The relatively high-altitude inland plateau is both cooler and experiences greater fluctuations in temperature. This area can experience severe frost in winter, an important determinant of the interface between grassland and savanna woodland (Ellery et al. 1991).

Fire has played a critical role in defining vegetation biomes, and the boundaries between biomes are in many instances defined by the occurrence or lack of fire (Bond et al. 2003; Esler et al. 2015). Forest patches within the grasslands are found in fire refugia (von Maltitz et al. 2003), the Fynbos ends and the Succulent Karoo starts when fire is no longer supported (Esler et al. 2015), frost kills the aboveground component of grass plants in the moist grasslands which leads to intense and frequent fires that prevent savanna expansion into the grasslands (Ellery et al. 1991). Within the grasslands, savanna and fynbos vegetation fire is critical for the biome maintenance and its frequency and intensity impacts on floristic structure. Other disturbances such as the above-mentioned rainfall variability, browsing and grazing pressures, including impacts from the mega herbivores are also gaining recognition for their role in biome maintenance.

2.3.3.2 Small-Scale Environmental Gradients Within Terrestrial Biomes

Within each biome there is a huge diversity of vegetation types driven by variation in local microclimatic conditions and soil properties. Within biomes, it is common to find areas with distinctly different geologies giving rise to very different soil nutrient status, and hence supporting completely different flora. For instance, much of the Fynbos is dominated by the nutrient-poor sandstone mountains of the Cape Folded Mountain belt. These bring about highly leached acidic soils. However, within the region, there are also limestone intrusions with base-rich soils. Similarly, much of the savanna is on eroded dystrophic granitic soils, sandstone or in the case of the extensive Kalahari, deep alluvial sands. These have distinctly different vegetation from the finer textured and nutrient-rich soils derived from basalt or shales.

Some areas of southern Africa have a great range of topography incorporating both relatively flat lowlands as well as mountains regions. Altitudinal gradients, varying aspects, differences in geologies and soils as well as the barriers to plant migration have contributed to extensive within biome diversity. For instance, in the Fynbos, the north facing and more harsh and arid mountain slopes are dominated by Restioids, while the Proteaceae are more prevalent on the cooler and moister southward oriented slopes at the same altitudes (Cowling et al. 2003b).

At a finer scale, there are typically unique plant communities associated with different catena positions. The savannas in particular have well-developed catenas where the ridge crest tends to have deep sandy soils, the midslope has shallow soils, often with a thin hydromorphic grass-belt along seep-lines in situations where moisture is forced to the surface by a clay plug. Clay soils with comparatively high nutrient content are found in the valley bottoms (Gertenbach 1983). On the granite-derived soils in Kruger National Park, the vegetation at the top of the catena has moist savanna features with trees being broad leafed and the grass largely unpalatable. By contrast the valley bottom has arid savanna characteristics of palatable grass and microphylls trees (see also Chap. 15). Unique habitats such a rivers, sodic areas and saltpans add additional habitat diversity within biomes.

Disturbance regimes can also play an important role in creating different habitats and diversity of the associated flora and fauna. In this regard the return periodicity of fire is of particular importance in all the fire dependent biomes. Frequent fires in the fynbos can prevent many of the more woody species, many of which are obligatory reseeders, from reaching reproductive maturity. In the savanna, frequent fire tends to favor grass over trees. A mosaic of different fire histories therefore assists in maintaining high diversity, both at the level of plant functional types as well as species (Parr and Brocket 1999). At an even finer scale, grazing lawns created by herbivores congregating, have higher nutrient states and a differing species composition than the surrounding savanna (Hempson et al. 2015).

Other disturbances such as grazing or browsing pressures, including the extreme impacts from mega herbivores such as elephants, the occurrence of droughts and floods, interact with fire in maintaining diversity. There is a growing acceptance that many of the southern African biomes are best understood based on disequilibrium theory (Nakanyala et al. 2018).

2.3.3.3 Primary and Secondary Productivity and Use

Modeled gross primary production (GPP) and mean annual net primary production (NPP) largely reflect mean annual rainfall patterns, except for areas with winter- and all-year rainfall, where NPP is substantially lower than the amount of rainfall would suggest. Individual biomes differ widely in their range of annual production (Table 2.1). Field-based studies on NPP are few (see Scholes et al. 2014), though there are numerous satellite-based estimates where trends in NPP are estimated (see Chap. 26).

Table 2.1 Estimates of mean GPP by biome based on Scholes et al. (2014)

Biome	Carbon production (mean) [gC m^{-2} yr^{-1}]	Carbon production (standard deviation) spatial [gC m^{-2} yr^{-1}]	Total biome GPP [TgC yr^{-1}]
Savanna	415	320	895
Grassland	645	304	232
Karoo	44	46	16
Fynbos	142	134	12
Thicket	381	264	12
Forest	977	281	64
Desert	1	0	1

For all biomes, except for Forest, the principle land use of untransformed vegetation can be described as rangeland, leading to secondary production in the form of livestock. However, the fynbos vegetation tends to have a very low carrying capacity for livestock or wildlife (over 30 hectares per large stock unit (LSU) (based on a 450 kg steer) for the renosterveld and over 100 ha LSU^{-1} for mountain fynbos) due to low levels of palatability, partly as a consequence of the key characteristics of the sclerophyllous plants and the low nutrient status of the soils. The Succulent Karoo Biome and Desert Biomes have very low carrying capacity for livestock due to the low rainfall, this despite them mostly having eutrophic soils. The remaining biomes are extensively used either for livestock production or increasingly for wildlife (see Chaps. 16 and 18). The grasslands (4–6 ha LSU^{-1}) and savanna (10–15 ha LSU^{-1}) in particular are highly suited to livestock and game. Although primary productivity increases with rainfall, soils tend to become more leached (Huntley 1982). There tends to be a critical point above about 600 mm yr^{-1} of precipitation where soils become leached and dystrophic, causing the grass and trees to switch from being predominantly palatable in winter, to being unpalatable when dry in winter (Huntley 1982). Soil nutrient status, texture and catena position can alter the precipitation level where this change takes place. This gives rise to what is termed 'sweetveld' which can be grazed all year in the arid areas and 'sourveld' which is only palatable during the summer in the moister areas. The interplay between NPP, palatability and grazing pressure means that there is high build-up of unpalatable grassy biomass in the high rainfall areas, and this is a driver for frequent fires in these areas (see Chap. 26). The Albany Thicket Biome is a remarkable outlier. In its natural state it has a higher NPP than might be expected for its low rainfall, and can support a high livestock of wildlife density. This is in large part due to the spekboom (*Portulacaria afra*) which has a joint CAM and C3 photosynthetic pathway and is both highly palatable and relatively fast-growing. Thicket is very sensitive to overgrazing, particularly by goats. When overutilized the thicket vegetation can 'collapse,' changing from a productive thicket (dense stands of small tree)-based system to a low-production annual grassland, with a concomitant loss of soil and biomass. There is evidence that the thicket is more resilient to wildlife and particularly elephant grazing, compared to grazing by livestock, and it has been identified as an area where rewilding might have extensive carbon sequestration benefits (Mills et al. 2005; Mills and Cowling 2010).

The Nama-Karoo Biome, despite its relatively low productivity, once supported vast herds of migratory herbivores. The vegetation has a carrying capacity of about 25–35 ha LSU^{-1} (Cowling and Roux, 1987). Human settlement has caused the area to be fenced into camps and it now supports an important sheep and goat industry (Masubelele et al. 2015). There are long-standing concerns over the impacts of overgrazing on the overall degradation of the karoo vegetation and in particular changes in the shrubs to grasses ratios. Acocks (1953) already raised concerns over the karoo shrubs encroaching into the grasslands. More recent data suggest that the opposite might in fact be true and that climate change is causing grasses to increase in the karoo (Masubelele et al. 2014). See Chap. 17 in this volume for recent work on the differences in production from highly grazed versus lightly grazed karoo areas.

Both the grasslands and the renosterveld are areas targeted for agricultural crop production. This has led to extensive transformation of these vegetation types in areas where good soils and adequate rainfall overlap. The grasslands and to a lesser extent the mountain Fynbos areas are also targeted for plantation forestry production, again causing transformation of the natural vegetation.

Forests have some of the highest rates of NPP in the region. However, southern Africa has very limited areas of true, closed-canopy forests, and they mostly lie in relatively cool climates which lower NPP rates. Higher production rates are observed in the more tropical coastal forests or on the East Coast of South Africa and Mozambique, rather than in the cooler mountain and southern Cape forests. Most of the South African forest is under conservation, though a limited amount of legal, sustainable timber and forest product harvesting takes place. In addition, an unsanctioned harvesting for traditional uses including traditional medicines also takes place, though there is a recent trend toward joint forest management to better regulate these traditional uses (Geldenhuys 2004). In Mozambique forests are extensively harvested, much of this illegally for export, but also for local charcoal production and the opening of new agricultural fields (Nielsen and Bunkenborg 2020; Woollen et al. 2016). Plantation forests of exotic pine, eucalyptus and wattle (*Acacias*) provide most of the region's timber, and pulp for paper, but represent a fully transformed, but productive habitat.

In addition to provision of grazing the natural vegetation provides humans with a multitude of what are termed 'veld products' or nontimber forest products. These products are especially important to the rural poor and vulnerable (Chap. 15). These products include medicinal plants for traditional remedies, especially from bark and geophytes, thatching grass, wild fruits, mushrooms and other foods, building material and craft material. Rural communities throughout southern Africa are also still dependent on fuelwood as the main source of domestic cooking fuel (Malimbwi et al. 2010). This dependency on wood fuel is partially reduced due to rural electrification, but even in electrified areas is still a common practice to use wood to spare costs. In Mozambique, charcoal is also the main urban domestic fuel (Mudombi et al. 2018) of the poor and there is an extensive charcoal trade between rural and urban centers (Smith et al. 2019). Within the Fynbos biome there is a large industry based on wild flower harvesting and the harvesting of rooibos tea (*Aspalathus linearis*), though both are now mostly grown in plantation format on transformed or semitransformed land (Louw 2006).

2.3.3.4 Organizational Efforts Geared to Protect Terrestrial Biodiversity

Southern Africa has a long and proud history of terrestrial biodiversity conservation. As a region it has, by global standards, both a high proportion of land under formal conservation and a high proportion of still largely natural environments. Historically, low human density, less destructive farming practices, unsuitability of land for year-round human habitation, and the occurrences of diseases such as malaria and trypanosomiasis (sleeping sickness) resulted in many areas remaining relatively natural. Western technology and medicine resulted in rapid population expansion in the twentieth century, coupled with opening of new land areas for agriculture, and this necessitated the need for formal conservation. Conservation effort is, however, not strategically allocated to conserving all biodiversity, and in many cases, it is not linked to the likely threats to the biodiversity of the biomes. The savanna is well conserved as in addition to large parts of the savanna biome being covered with formal conservation, there are also huge areas of private land under wildlife management, and extensive tracts of communal land in Namibia or the old hunting areas of Botswana which are managed by communities as part of Community-Based Natural Resource Management (CBNRM) tourism programs which adds to the conservation of the biome (see Chaps. 15 and 17).

Overall, the biodiversity-rich Fynbos Biome enjoys about 20% formal conservation; however, conservation differs significantly between the fynbos versus the renosterbos vegetation. Not only is the renosterbos poorly conserved at only 0.6%, but it has been extensively transformed for cropland (96% transformed). The conserved fynbos areas are predominantly in mountain regions, which are historically conserved as water catchments. Impacts from invasive alien vegetation occur throughout the biome and threaten the fynbos in both protected and unprotected areas. The Cape Action for People and the Environment (C.A.P.E.) (1998–2000), developed a 20-year strategy for conservation and sustainable development of the fynbos biodiversity hotspot. Funding from the Global Environment Facility (GEF), Critically Endanger Partnership Fund, National Government and numerous other donors has helped implement this plan.

Only about 3.5% of species and endemic-rich Succulent Karoo Biome is formally conserved. The Succulent Karoo Ecosystem Plan (SKEP) was a participative process to improve conservation of the endangered Succulent Karoo biome, and created a 20-year strategic plan (Driver et al. 2003). It used a strategic conservation planning approach to identify critical areas needing conservation. Further funded through the GEF, Critical Ecosystems Partnership Fund, in collaboration with other funders added more than 2.9 million hectares of biologically important land to the conservation estate via establishment of conservancies, signing of stewardship agreements, and the incorporation and designation of state land. This program included both South African and Namibia (Critical Ecosystem Partnership Fund 2001).

The Grassland Biome is extensively transformed into cropland and plantation forestry, and despite being one of the most threatened biomes, only has 3.2% formal conservation. The Grasslands Program was initiated by SANBI to strengthen conservation within the grasslands (DEA 2015).

The Nama-Karoo Biome has exceptionally low levels of formal conservation in South Africa, with only about 1% of the biome conserved. Much of the South African Nama-Karoo is considered degraded from overgrazing, but total land transformation is relatively limited (Hoffman 2014). Compared to South Africa, conservation levels are better in Namibia where about 17% of the Nama-Karoo Biome is conserved.

The Desert Biome is well conserved in both South Africa and Namibia. The Forests Biome within South Africa has an exceptionally high level of conservation as even forests outside of formal conservation areas have special protection under the National Forest Act. None of the Mozambique forest is within formal reserves, though a small proportion is in a hunting area. In South Africa both the IOCB and Albany Thicket have below the international 10% target for conservation, and both are vulnerable to degradation and transformation. Within Mozambique a small section of the IOCR is protected.

Initial conservation efforts in the region tended to focus on land with a low agricultural potential, rather than the strategic importance of the land from a biodiversity conservation perspective. South Africa has an active program to increase the extent of conservation, and in addition many areas throughout the region have been identified for management through studentship agreement. Several transboundary conservation initiatives have been implemented in the region and there is strong NGO support through organizations such as the Peace Park Foundation and WWF. Although conservation areas are relatively well managed and successful throughout most of the southern African region, funding for conservation remains problematic in all countries. In Mozambique conservation areas largely collapsed during the civil war and management capacity was lost. To help fill this gap, the private sector has become involved supporting formal conservation in initiatives such as the Carr foundations support to Gorongosa reserve in Mozambique, and the Frankfort Zoological Society's supports the Gonarezhou reserve in Zimbabwe. Ecotourism has also been used as a funding mechanism to support private and community conservation initiatives.

2.4   Connection Between Oceanic and Coastal Terrestrial Ecosystems

Land and ocean in the Benguela region of southern Africa are closely interlinked. The desert-dominated coastal strip along the West Coast is dependent on water and nutrient inputs via the sea fog, just as the coastal waters are influenced by nutrient inputs via desert dust and water inputs via submarine groundwater discharge. Climate change influences these material fluxes considerably (Bryant et al. 2007). The connections between these fluxes and ecological processes are not yet fully understood, and hence, the extent of this impact on pelagic and benthic ecosystems in the coastal waters off Namibia, on the Namib Desert and on the greenhouse-gas concentrations in the atmosphere has not yet been quantified.

The main transport vectors are winds, fog, river and groundwater discharges (Fig. 2.10). These vectors are driven by the complex interplay between the ocean and the atmospheric circulation over southern Africa and its teleconnections with climate anomalies reflected by indicators such as the El Niño Southern Oscillation (ENSO), Tropical Southern Atlantic Index (TSA) or St Helena Island Climate Index (HIX).

The relatively cold sea surface in the upwelling system of the Benguela Current in combination with subtropical subsidence lead to a rather stable lower atmosphere, which suppresses the formation of substantial precipitation. Therefore, other sources of moisture are of tremendous importance in the region. Due to the cold waters of

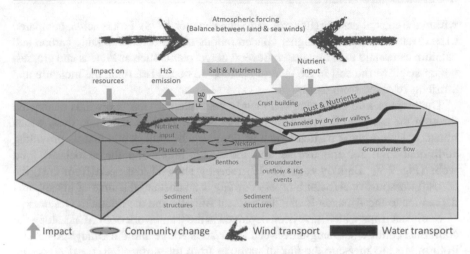

Fig. 2.10 Sketch of the main environmental factors, transport processes, and ecosystem components in the Namibian coastal upwelling-desert system (© Werner Ekau)

the Benguela Current, advected sea fog is a rather frequent feature (Schulze 1969; Lancaster et al. 1984) along the southern African coastal region and provides a major source of moisture for several local species (Hachfeld 2000; Hamilton III et al. 2003, Henschel and Seely 2008). Maximum monthly precipitation (dew) from fog can reach up to about 50 mm yr^{-1}, while annual mean rainfall amounts (excluding fog precipitation) for the region are only between ~20 and 40 mm yr^{-1} (Lancaster et al. 1984).

Fog thus is a main source of water for the Namib Desert ecosystem. It occurs mainly during the austral winter months when extremely cold and nutrient-/CO_2-rich water wells up along the Namibian coast (Eckardt and Schemenaur 1998). The generation of fog is enhanced when low atmospheric pressure systems develop in the south and upwelling is restricted to a narrow strip along the coast in the north. Under such circumstances warm moist air of the south and/or the open ocean is transported over upwelled water where it cools down. Sea surface temperatures have increased over the last 30 years and upwelling is assumed to strengthen in the course of global warming, raising the question of how this might affect the fog formation over the Namib Desert.

In the land-sea direction, another connecting water transport exists through submarine groundwater discharge (SGD). In Namibia for instance, the two most important coastal aquifers, the Omaruru Delta and the Kuiseb Dune, are closely linked to paleoriver channels filled with sands which provide the permeability for subsurface flow toward the sea. Submarine groundwater can have significant impact on shallow-water benthic communities (Starke et al. 2019).

Besides water, fog contains salt and nutrients which indicate that sea water is not its only source. It appears to act as a barrier scavenging all sorts of dust and other atmospheric compounds. When the water evaporates, salt crusts are formed within the coastal deserts (Eckardt and Schemenauer 1998). Fog is acidic compared to sea water, but contains higher concentrations of dissolved inorganic carbon and calcium carbonate than rain water. Several desert plants such as lichens and grasses, which stabilize the soil in the central Namib, are dependent upon the moisture and nutrients released from the fog.

Dust inputs into the ocean are mainly controlled by the bergwinds channeled through the ephemeral river beds (see Fig. 2.11). Besides inorganic dust, winds transport also organic particles such as desert-grass, lichens and insects providing additional food for the marine and the desert fauna and the associated food webs (Fig. 2.9). Barkley et al. (2019) recently showed that apart from dust, the aerosol transport of African biomass burning is a substantial source of phosphorus deposition to the Amazon Basin, the tropical Atlantic and the Southern Ocean.

Sediment traps which have been deployed in the framework of SPACES along the Namibian continental margin between the Walvis Ridge and the Orange River are, in turn, tools to measure the flux of particles from the surface into the deep ocean. Sediment trap results show that the majority of the exported material is residue from the pelagic food web but contains on average also ~9% to 17% of inorganic dust particles referred to as lithogenic matter (Vorrath et al. 2018). A long-term sediment

Fig. 2.11 Dust plumes along the Namibian coast (https://eoimages.gsfc.nasa.gov)

trap deployment close to Walvis Bay on the Namibian shelf (see Chap. 25 for more detailed information) showed furthermore that the contribution of lithogenic dust varies significantly on time scales of weeks but also seasonally and interannually. For instance, at the end of 2010, lithogenic dust particles contributed more than 40% to the trapped material whereas at the end of the year 2013 the contribution of lithogenic matter to total flux was less than 10%. These dust inputs are known to act as ballast in sinking particles which lowers the offshore advection of plankton blooms, and as essential sources of micronutrients such as iron which is assumed to limit plankton blooms in many other ocean regions (Rixen et al. 2019, and references therein). To what extent the varying dust inputs affect the development of plankton blooms still needs to be studied in more detail to better understand impacts of climate and local land use changes on the marine environment.

2.5   Southern African Biomes: Carbon Sources or Sinks?

2.5.1   Oceanic Biomes

Numerical models suggest that upwelling systems at lower latitudes such as the Benguela Upwelling System (BUS) in the south-eastern Atlantic Ocean act as net carbon dioxide (CO_2) sources to the atmosphere (Laruelle et al. 2014; Brady et al. 2019; Roobaert et al. 2019). In contrast, estimates based on data obtained during the SPACES program suggest that the nBUS acts as a CO_2 source while the sBUS acts as a CO_2 sink (Emeis et al. 2018; Siddiqui et al. 2023). These opposing functions are assumed to be a consequence of the combined effects of the marine solubility pump and the biological carbon pump on the partial pressure of CO_2 (pCO_2) in surface waters: if the pCO_2 in the ocean falls below pCO_2 in the atmosphere, the ocean takes up CO_2, whereas it emits CO_2 if the pCO_2 in surface waters exceeds that in the atmosphere. The solubility pump increases pCO_2 in seawater and favors CO_2 emission when upwelled water warms at the surface. Conversely, the biological carbon pump reduces the pCO_2 as it is driven by the photosynthesis of phytoplankton converting CO_2 into biomass which is subsequently exported into subsurface waters. TRAFFIC research has supported the conclusion that the sBUS operates as a regional CO_2 sink where the CO_2 uptake by the biological carbon pump exceeds the CO_2 release by the solubility pump, whereas the nBUS acts as a CO_2 source to the atmosphere (Siddiqui et al. 2023).

However, effects of these pumps in the BUS have to be linked to their opposing function in the Southern Ocean, since the water that upwells in the BUS originates from there. According to our current estimate, CO_2 uptake by the biological carbon pump in the BUS compensates up to 38% of its CO_2 loss in the Atlantic sector of the Southern Ocean (Chap. 25). Hence, changes in CO_2 uptake by the biological carbon pump in the BUS could significantly affect the marine carbon cycle and its role as a sink of anthropogenic CO_2 (Siddiqui et al. 2023).

Even though it is widely accepted that global change (Riebesell et al. 2007) and human impacts via, e.g., fisheries (Bianchi et al. 2021), affect the biological carbon pump, associated changes of its CO_2 uptake are still difficult to predict due to the low accuracy with which it can be determined. Accordingly, the development of new methods and observing strategies remain a great challenge in marine sciences, and efforts to protect the oceans and combat climate change need to be strengthened (Chap. 25).

2.5.2 Terrestrial Biomes

A 2014 South African assessment of terrestrial carbon stocks, repeated again in 2020 found that to a large extent standing carbon stocks follow rainfall patterns. Grasslands have one of the highest carbon stocks per unit area, in some cases approaching those of forests (Scholes et al. 2014; von Maltitz et al. 2020). This un-intuitive outcome is because of the extensive soil carbon stocks that accumulate particularly in the moist high-altitude grasslands. The savanna, despite having a slightly lower carbon stock per unit area, is a globally important carbon stock due to the vastness of the savanna regions.

Carbon fluxes remain poorly understood for the region and are covered in greater detail in subsequent chapters (Chaps. 17 and 30). A number of factors make understanding of the carbon fluxes complicated. Firstly, there are the overarching impacts from global climate change and the global fertilization impacts from raised CO_2 on standing plant biomass. According to Knowles et al. (2014), this should have a positive impact on most, but not all, biomes. Secondly, for most of the biomes, there is a gradual uptake of carbon over multiple years. This follows seasonal patterns of uptake during the wet periods, with emissions during the dry periods. However, in the fire-prone biomes, there are also major emissions during fire periods that might be every year in some grasslands to once in 20 years or more in some of the arid savanna and fynbos habitats. Especially in the fynbos, this leads to an almost total mortality of standing woody biomass, followed by reseeding and a gradual build-up of new biomass. Long-term trends are therefore difficult to ascertain, especially since fire intervals can vary greatly on the same parcel of land over time. Thirdly, the region has cycles of wet and arid periods that can span many years (Chaps. 3 and 6). This can mean that short-term rainfall-induced trends are very different from longer-term climate change and CO_2 fertilization trends. Further, most satellite-derived biomass products are only starting to give a sufficient time span of data to overrule the short-term rainfall fluctuations. Finally, human induced land degradation and restoration need to be considered (von Maltitz et al. 2018). Even using long-term data from National Parks can give misleading results. For instance, in the Kruger National Park elephants were almost absent when the park was proclaimed in 1902. Elephant numbers increased and were artificially held constant from 1967 to 1994 at about 7000 to 8000 animals; however, since 1994, their numbers have been allowed to increase and now stand at about 18,000. This has had profound impacts on standing tree biomass, as can be seen at the flux tower near Skukuza which has changed from a relatively dense savanna (30% tree cover) when initiated to a more open savanna at present (Scholes et al. 2001).

2.6   Impacts of Climate Change

Anthropogenic climate change is projected to affect individual terrestrial and marine biomes to different degrees. In marine ecosystems, regional atmospheric warming interacts with altered ocean currents. In terrestrial systems, aside from direct effects of warming, altered rainfall causes the main effects on biomes under climate change. Shifts in tropical cyclone behavior are not yet projected with reasonable confidence, although they may fundamentally change the extent and steepness of environmental gradients that have existed for millennia.

2.6.1   Impact of Climate Change on Marine Biomes

2.6.1.1 Climate Change Affecting Coastal Upwelling Systems

All Eastern Boundary Upwelling Systems (EBUSs) are dependent on the global circulation patterns of wind and ocean currents and thus are susceptible to alterations caused under climate change. Climate change affects EBUSs in different ways with partly opposing consequences. Global warming of the ocean surface will increase stratification and, hence, potentially reduce upwelling intensity. In contrast, increasing greenhouse gas concentrations in the atmosphere could force intensification of upwelling-favorable winds in EBUSs (stronger trade winds) and, therefore, strengthen upwelling intensity (Bakun et al. 2015). Thus, both opposing trends, the intensification and the relaxation of upwelling intensity, are possible under climate change scenarios. It is not yet clear which of the two processes will dominate where, but latitudinal differences in EBUSs' response to global climate change appear likely. Recent studies suggest that a poleward shift in subtropical high-pressure cells will lead to an intensification, and possibly seasonal expansion, of upwelling in poleward parts of EBUSs, whereas closer to the equator, upwelling intensity will be reduced (García-Reyes et al. 2015; Rykaczewski et al. 2015; Wang et al. 2015). Some evidence for this phenomenon has been observed in the Benguela System, including a significant recent decrease in upwelling in the nBUS, and a significant increase on the Agulhas Bank off South Africa (Lamont et al. 2018).

Sweijd and Smit (2020) showed that 99% of the area of all seven African Large Marine Ecosystems (LMEs) has warmed with rates of between 0.11°C per decade (Agulhas LME) and 0.58°C per decade (Canary LME) since the 1980s. They found that only 1% of the LME area was associated with cooling due to upwelling. The most intense warming hotspot in the nBUS was at the Angola-Benguela Frontal Zone, while the strongest cooling was at the boundary between the nBUS and sBUS, in the Lüderitz upwelling cell.

Sydeman et al. (2014) evaluated the evidence for an intensification of upwelling-favorable winds through a meta-analysis of existing studies and found support for the hypothesis that climate change is associated with stronger coastal winds and enhanced upwelling in EBUSs, including the Benguela. This in turn could lead to adverse effects such as stronger offshore advection, more frequent hypoxic events, and increased ocean acidification (Bakun et al. 2015). Recent studies within the

SPACES framework point out the extreme complexity of oceanographic processes on a smaller scale (e.g., eddies), which may also have modulating effects on the wind-driven upwelling off Lüderitz and Walvis Bay (Bordbar et al. 2021).

The Benguela Current Ecosystem is vulnerable to processes leading to deoxygenation. Increasing temperatures and decreasing oxygen levels have been measured—especially in the nBUS—for several decades (Stramma et al. 2008) with severe potential impact on the species there (Ekau et al. 2010). Deoxygenation is especially pronounced in the nBUS, as this subsystem is already defined by perennial low-oxygen water and hydrogen sulfide eruptions (Kirkman et al. 2016).

Whereas a variety of studies have shown that the impacts of global warming (and other anthropogenic stressors, e.g., fishing pressure or plastic pollution) on the marine food web and ecosystem health vary spatially, even within the Benguela Upwelling System, the region-specific effects in many areas of southern Africa have not been well studied (Hutchings et al. 2012; Kirkman et al. 2016; Verheye et al. 2016).

2.6.1.2 Climate Change Impacts on Marine Species in the Benguela

Climate change is expected to have complex effects with potentially far-reaching consequences in the vulnerable Benguela Upwelling System (Bakun et al. 2015; Bordbar et al. 2021). The contrasting local warming and cooling trends may have opposing effects on individual taxa and will cause complex ecosystem responses to climate change.

Phytoplankton production would principally be favored by enhanced coastal upwelling leading to elevated nutrient concentrations in surface waters, but it could simultaneously be reduced by greater turbulence and thus deeper mixed layers and light limitation (Bakun et al. 2015). Increased turbulence and offshore export may also cause a decoupling between primary producers and higher trophic levels (spatial mismatch). In contrast, new primary production further offshore may be enhanced. Alterations in phytoplankton and zooplankton communities will likely be the consequence, but are difficult to predict at the current state of knowledge. Certainly, such changes will propagate up the food web, e.g., impacting larval fish recruitment.

The expansion of oxygen-minimum zones (Stramma et al. 2008) affects the vertical distribution and migration of zoo- and ichthyoplankton (Ekau and Verheye 2005; Auel and Ekau 2009) that respond to the shift in oxyclines. Here, as in the Humboldt and California Current EBUSs (Bograd et al. 2008; Bertrand et al. 2010), the position of the oxycline is a determining factor in the structuring of the pelagic community as it limits the habitable space for pelagic organisms (Howard et al. 2020). In the zooplankton, certain species e.g., copepods of the families Eucalanidae, Rhincalanidae and Subeucalanidae, can better cope with oxygen minimum zones (OMZs) than others (Teuber et al. 2013; Teuber et al. 2019). The shifting of the oxycline in the nBUS could have severe impacts on fish recruitment due to the high sensitivity of many fish species to low-oxygen concentrations, and also via changing availability of zooplankton prey (Kreiner et al. 2009; Ekau et al. 2018). Overall, deoxygenation may have multiple ecological consequences, including range shifts in species distribution or alteration of vertical migration behavior, which in turn affect species interactions, as well as the local abundance and availability of commercially relevant living marine resources (Chan et al. 2019).

Potts et al. (2015) reviewed projected impacts of climate change (sea temperature, upwelling intensity, current strength, rainfall, pH, and sea level) on coastal fishes in southern Africa, distinguishing the effects on different groups of fishes (migratory, resident, estuarine-dependent or catadromous fish guilds) in separate biogeographical zones. The authors concluded that impacts will be diverse due to the different life strategies of coastal fishes, but process understanding would still need to be improved in order to derive sound predictions for the different zones of the southern African coasts.

In the northern Benguela, long-term trends in growth of fish stocks such as the deep-water hake were shown to be influenced by upwelling intensity, as well as by fisheries-induced evolution, which works in opposing directions for small and large fish (Wilhelm et al. 2020). Therefore, for each fished and non-fished taxon, understanding of their specific responses to the combined climate change and fishing impacts still needs to be improved.

For key species in the southern Benguela, Ortega-Cisneros et al. (2018a) performed a trait-based sensitivity assessment in order to evaluate their potential relative sensitivity to climate change. Several of the species included in their assessment are endemic to southern Africa, including white steenbras *Lithognathus lithognathus*, which was classified as particularly sensitive. The benthic invertebrate abalone (*Haliotis midae*) also has high expected vulnerability. Cape horse mackerel, shallow-water hake and sardine scored as having lower sensitivity to climate change than the other assessed species. Nevertheless, sardine, in particular, appears to suffer from the combined effects of climate change and fishing.

The South African fishery on small pelagics targets sardine, anchovy and (to a lesser degree) round herring, and is economically the country's second most valuable fishery (after the deep-sea trawl fisheries) and supports many employees. These characteristics, together with the fact that small pelagic fishes respond rapidly to environmental change in terms of their abundance and distribution patterns, have led to it being considered one of the fisheries most vulnerable to climate change (van der Lingen 2021). The sardine resource has been low for more than a decade, and is presently depleted, while both anchovy and sardine have shown changes in their distributions. Various adaptation measures have been suggested and/or implemented to mitigate climate change effects and to nurse the pelagic stocks. These range from sardine stock rebuilding programs and the importation of frozen sardines to reduce local fishing pressure while maintaining the processing infrastructure, to developing experimental fisheries on alternative resources, specifically the mesopelagic Hector's lanternfish (*Lampanyctodes hectoris*). Augustyn et al. (2018) present various rebuilding measures for the sardine fishery and point out that monitoring and scientific data collection are vital to continue recording the most important environmental variables and species to enable adaptive management of that fishery.

The complexity of interactions of partly antagonistic climate-driven processes makes overall predictions difficult, and strongly suggests that climate impact will differ spatially. The frequency of occurrence of harmful algal blooms has been suggested to increase under climate change, and negative effects on sardine (but not on anchovy or round herring) have been described (van der Lingen et al. 2016). In contrast to the other small pelagic species, sardine possesses finely meshed gill rakers which enable them to feed directly on small unicellular phytoplankton, but also making them vulnerable to the possible ingestion of toxic dinoflagellates such as *Gonyaulax polygramma* directly (van der Lingen et al. 2016). As sardine and anchovy differ in feeding behavior and diet, changes in local zooplankton community composition and size structure can also be expected to affect each fish species differently (van der Lingen et al. 2006a, b). Where larger zooplankton become more abundant, anchovy and round herring may be favored, whereas sardine may gain a competitive advantage where smaller zooplankton will dominate particularly under conditions of weak upwelling (van der Lingen et al. 2006b; Augustyn et al. 2018).

Physical mesoscale processes which directly impact the habitat of small pelagic fishes cause considerable 'noise' in oceanographic data, making detection of climate change signals difficult (van der Lingen and Hampton 2018). This complicates efforts to disentangle effects of climate change and fishing activities. Therefore, an integrated research response needs to be developed in order to improve forecasting of likely climate change impacts on vulnerable fish species and the related fishing sectors (van der Lingen 2021).

In a modeling study investigating the effects of several drivers on the southern Benguela upwelling system, Ortega-Cisneros et al. (2018b) found that warming had the greatest effect on species biomass, with mainly negative effects reported. Cephalopods are an exception to this rule, as there have been various records of their thriving under climate change (Golikov et al. 2013; Doubleday et al. 2016; Oesterwind et al. 2022). Mesopelagic fishes and large pelagic fish species were predicted to be the least and the most negatively affected groups, respectively, also when accounting for the effects of potential new fisheries on mesopelagics and round herring (Ortega-Cisneros et al. 2018b). Van der Sleen et al. (2022) incorporated sea surface temperature (SST) fluctuations in models of marine fish population dynamics at different trophic levels in ten LMEs (including the Agulhas Current LME). Using observed landings and SST data, they constructed their model to disentangle influences of climate from those of fishing pressure with some predictive power. Yet, in upwelling systems, responses to SST are partly opposing and generally more complex.

2.6.2 Impact of Climate Change on Terrestrial Biomes

Climate change is anticipated to have major impacts on the terrestrial biomes. A number of research initiatives starting from simple bioclimatic envelope approaches (Midgley et al. 2008) and progressing to complex Dynamic Vegetation Models

(Chap. 14). Outcomes from the modeling processes differ based on the different approaches, and the different climate futures used to drive the models, though shrinkage of many of the biomes appears inevitable. Climate change can be expected to affect the individual terrestrial biomes differently.

Savannas are a relatively resilient vegetation type as they have evolved under a multitude of disturbances; however, it is possible that climate change may move savannas beyond critical tipping points. Direct impacts from higher temperature are unlikely to affect savannas, though the increased temperatures may make the area more arid (DEA 2015). However, climate change poses a number of potential new threats to the biome. The C4 grasses evolved relatively recently under low global CO_2 concentrations (Bouchenak-Khelladi et al. 2009). As climate change is associated with elevated levels of atmospheric CO_2, this may lead to a shift in the competitive advantage of entire metabolic classes of plants. The C4 photosynthetic pathway currently gives the C4 grasses a water use efficiency advantage. Rising CO_2 levels appear, however, to favor C3 species, and indications are that this will favor trees over the grasses, possibly shifting the biome to higher tree density or even forest (Bond et al. 2003; Kgope et al. 2010; Chap. 14). Fire regimes under climate change remain a key unknown. Increased tree density might reduce fire risks; however, there is also a possibility that prolonged droughts and heat waves under climate change may lead to unprecedented 'fire storms' with unknown consequence to the tree-grass ratios (Staver et al. 2011).

Grasslands are considered as especially vulnerable to climate change, with models suggesting they will be encroached by other biomes and 'squeezed' to a fraction of their current extent (e.g., Midgley et al. 2008; DEA 2015; Chap. 14). Tree species are expanding into grasslands on the savanna grassland interface, effecting millions of ha of grasslands (Skowno et al. 2017). Grasslands may, however, expand into the Karoo in some circumstances (Chap. 15), and their greater prevalence in some Nama-Karoo shrublands may already be driving a novel wildfire regime.

The **Nama-Karoo Biome**, while being typified by the occurrence of dwarf shrubs of the Asteraceae, Poaceae, Aizoaceae, Mesembryanthemacaea, Liliaceae and Scrophulariaceae, may also have a large amount of grass, especially following periods of high rainfall (Palmer and Hoffman 2003). The aridity index of the biome has changed over the past 100 years and there is an increase in grass cover since the 1960s (Hoffman 2014). Fire is rare in the Nama-Karoo Biome, and although most of the Nama-Karoo shrubs are able to resprout after fire, fire could shift the vegetation to a more grassy state (Kraaij et al. 2017). Morgan et al. (2011) suggest that increased CO_2 concentrations may favor C4 grasses, potentially increasing the likelihood of fire which will further benefit grass more than shrubs (du Toit 2019). Projected climate change impacts on the Nama-Karoo vegetation are unclear, especially when utilizing projections of both climate and geology, but may reduce the Nama-Karoo Biome in South Africa, though cause its expansion in Namibia and Botswana (Guo et al. 2017).

The fauna and flora of the **Desert Biome** are critically dependent on the moisture provided through fog. This biome is considered exceptionally vulnerable to changes in fog occurrence, which is poorly understood, but considered to be declining under climate change (Mitchell et al. 2020).

For the **Succulent Karoo Biome**, any increase in precipitation that allows the fynbos species to expand would be detrimental. At the same time, any shift in seasonality toward stronger summer rainfall would likely allow for Nama-Karoo species expansion at the expense of the succulent karoo species. As such the Succulent Karoo Biome is considered as extremely vulnerable to small shifts in climate. Early biome niche models predict that the biome might shrink considerably as a result of a southward shift in the rain-bearing mid latitude cyclones (MacKellar et al. 2007; Midgley and Thuiller 2007). A model by Driver et al. (2012) based on more recent climate predictions found lesser evidence of extensive biome shrinkage.

The renosterveld element of the **Fynbos Biome** in particular has been highly fragmented through wide-scale transformation. The region is considered as very vulnerable to climate change impacts, due to the intrinsic nature of its high, but localized and isolated biodiversity. Increased temperature regimes could change fire frequency and intensity, factors that are known to impact fynbos biodiversity. Being at the southern tip of Africa, only altitude, and not southward migration, is available as an option for a range shift to avoid increased heat. A southerly shift of only a few 10s of km of the rainfall-bearing frontal systems that provide the unique winter rainfall could be devastating to the region.

Forests are sensitive to fire, and there are records of forests burning during intense hot and dry periods when the resultant damage can be catastrophic (Geldenhuys 1994). This makes them potentially vulnerable to climate change impacts where droughts and heat waves may become both more common and more intense.

Being close to the warm Indian Ocean Current, climate is subtropical in the **Indian Ocean Coastal Belt Biome**, but extremes are moderated by the coastal influence. Rainfall tends to be above 900 mm yr^{-1}. Rainfall occurs predominantly during summer in the south, but throughout the year toward the north. Climate change will have a lesser impact on temperature in the IOCB than in the inland biomes, and the east of the country where it occurs also appears to be slightly less vulnerable to an overall decrease in rainfall. Changes in fire regimes due to climate change might be one of the greatest climate change induced threats (DEA 2015).

Climate change impacts on the **Albany Thicket Biome** are poorly understood, but a predicted increase in rainfall might favor an incursion of savanna species including more grass into the biome. The Crassulacean Acid Metabolism (CAM) dominance of many of the species is less likely to benefit from raised global CO_2 which will favor C3 trees and to a lesser degree also C4 grasses. Increased occurrence of fire in this biome would be devastating to some species (DEA 2015).

The Desert Biome is projected to expand from its current distribution in coastal and southern Namibia and western southern Africa, toward the south and east, displacing Nama Karoo and savanna in western Botswana and northern South Africa. The iconic desert tree *Aloidendron dichotomum* may already be showing signs of population mortality in warmer and drier areas of its range in these regions (Foden et al. 2007).

2.7   Conclusions and Implications

The southern African subcontinent and its surrounding oceanic regions accommodate globally unique ecoregions, which are outstanding owing to their exceptional biodiversity, with the occurrence of a multitude of endemic species and, in the marine realm, a very high productivity. Southern African terrestrial and marine ecosystems are shaped by steep and extended environmental gradients; they are adapted to and dependent on these special conditions. The biota in terrestrial biomes are limited in potential adaptive range extensions toward the south. Therefore, the effects of climate change on specific biomes will call for changes in human use of resources on land and at sea, and for protective measures to safeguard the exceptionally high biodiversity of the subcontinent and its waters. In marine ecosystems, the effects of climate change are complex, especially since the underlying mechanisms are partly counteractive, particularly in the Benguela Upwelling System: global warming leads to stronger thermal stratification of the water column, impacting many biological processes (e.g., time of spawning, distribution) and biogeochemical vertical exchange. Simultaneously, higher gradients in air pressure and temperature changes over land and ocean also alter the regional wind fields. This may increase turbulence in the pelagic realm and intensify the upwelling of nutrient-rich waters, providing resources for pelagic food webs in some regions, or reduce upwelling in other regions. At the same time, various additional anthropogenic stressors including high fishing pressure are bound to threaten the ecological balance in marine ecosystems and their services such as the sequestration of CO_2. These processes may exacerbate the effects of climate change currently predicted.

Southern Africa, especially South Africa, has accumulated a long history of extensive biological research, unsurpassed by most areas outside of Europe and going back to the seventeenth century (Huntley 2003). These efforts have generated unique terrestrial and oceanic data sets of high value both for Africa and for most of the rest of the developing world. Data sets are stored in both European and local databases (e.g., PANGAEA, ODINAFRICA, ocean.gov.za) and involve numerous world-class national universities and research institutes. Many of these databases are constantly being expanded, becoming even more valuable. Their availability in extensive publications and ongoing studies on biodiversity and ecosystem functioning provide one of the most comprehensive settings for research investment both for regional and global benefit.

Current regional and international research regarding the biomes of southern Africa does not focus on the biology alone, but also addresses earth system processes and particularly climate change. Southern Africa has also generated extensive cross-disciplinary research interests beyond the continent, leading to numerous long-term collaborative research initiatives with Europe and other high-income nations. This includes programs such as Biodiversity Monitoring Transect Analysis (BIOTA) Africa, the Southern African Science Service Centre for Climate Change and Adaptive Land Management (SASSCAL), and Science Partnerships for the Adaptation to Complex Earth System Processes in Southern Africa (SPACES) I and

II programs focusing on the region. South Africa is also a driving partner in the All Atlantic Ocean Research and Innovation Alliance, so far including Atlantic coastal countries Brazil, South Africa, Argentina, Cabo Verde, Morocco, United States, Canada and several European states, and supported by the European Commission (relating to Galway Statement, Belém Statement and Washington Declaration; https://allatlanticocean.org/). These large initiatives increase the knowledge base needed to enable decisions toward sustainable management of human activities.

Their outstanding value, and at the same time their great vulnerability through climate change, habitat loss and various other human pressures, should make southern African biodiversity the focus of future research activities aimed at providing scientific support for ecosystem conservation measures and their management. The present digital age provides an excellent opportunity to create databases, making data sets readily available to simplify scientific and political collaboration and to exchange and thereby formulate socially, environmentally and economically acceptable management aims. However, various prerequisites need to be considered: the rich historic data sets need to be kept relevant by ensuring ongoing monitoring across all biota. This is a basic requirement for the analyses, meta-analyses and evaluations performed by scientists of different disciplines who need to formulate regular status reports on ecosystem health (e.g., status report on fisheries in South Africa, DEFF 2020) either for each country, or preferably through transnational organizations and institutional collaborations. These status reports could be evaluated by interdisciplinary committees to formulate up-to-date and clearly defined socio-economic management aims and guidelines, which at the same time protect the southern African biodiversity as well as ecosystem health and services. These steps advance society toward the global Sustainable Development Goals of finding socio-economic balance hand in hand with ecological balances.

Acknowledgments We appreciate the valuable comments of an anonymous reviewer and the editor, Christian Brümmer, on a previous version of this chapter. We thank Solvin Zankl for taking and providing the photographs of marine organism for Figs. 2.4 and 2.5 (www.solvinzankl.com). The SPACES II projects TRAFFIC, CASISAC and BANINO were funded by the German Federal Ministry of Education and Research (BMBF), FKZ 03F0797, 03F0796 and 03F0795.

References

Acocks JPH (1953) Veld types of South Africa. Memoirs of the Botanical Surveys South Africa, vol 28, Government Printer, South Africa, 192 p

Archibald S, Nickless A, Govender N, Scholes RJ, Lehsten V (2010) Climate and the inter-annual variability of fire in southern Africa: a meta-analysis using long-term field data and satellite-derived burnt area data. Glob Ecol Biogeogr 19:794–809

Auel H, Ekau W (2009) Distribution and respiration of the high-latitude pelagic amphipod *Themisto gaudichaudii* in the Benguela Current in relation to upwelling intensity. Prog Oceanogr 83:237–241

Augustyn CJ, Cockcroft A, Coetzee J, Durholtz D, van der Lingen CD (2018) Rebuilding South African fisheries: three case-studies. In: Garcia SM, Ye Y (eds) Rebuilding of Marine Fisheries part 2: case studies, Fisheries and Aquaculture Technical Paper No. 630/2. FAO, Rome

Awad AA, Griffiths CL, Turpie JK (2002) Distribution of South African marine benthic inverte-brates applied to the selection of priority conservation areas. Divers Distrib 8:129–145

Ayón P, Swartzman G, Bertrand A, Gutiérrez M, Bertrand S (2008) Zooplankton and forage fish species off Peru: large-scale bottom-up forcing and local-scale depletion. Prog Oceanogr 79:208–214

Bailey C, Shackleton C, Geldenhuys C, Moshe D, Flemming G, Vink E, Rathogwa N, Cawe S (1999) Guide to and summary of the meta-database pertaining to selected attributes of South African indigenous forests and woodlands. In: Division of Water EaFT, CSIR (ed) Report ENV-P-C 99027, Pretoria

Bakun A, Black BA, Bograd SJ, Garcia-Reyes M, Miller AJ, Rykaczewski RR, Sydeman WJ (2015) Anticipated effects of climate change on coastal upwelling ecosystems. Curr Clim Change Rep 1:85–93

Barange M, Stuart V (1991) Distribution patterns, abundance and population dynamics of the euphausiids *Nyctiphanes capensis* and *Euphausia hanseni* in the northern Benguela upwelling system. Mar Biol 109:93–101

Barkley AE, Prospero JM, Mahowald N, Hamilton DS, Popendorf KJ, Oehlert AM, Pourmand A, Gatineau A, Panechou-Pulcherie K, Blackwelder P, Gaston CJ (2019) African biomass burning is a substantial source of phosphorus deposition to the Amazon, tropical Atlantic Ocean, and Southern Ocean. Proc Natl Acad Sci 116:16216–16221

Batten SD, Abu-Alhaija R, Chiba S, Edwards M, Graham G, Jyothibabu R, Kitchener JA, Koubbi P, McQuatters-Gollop A, Muxagata E, Ostle C, Richardson A, Robinson K, Takahashi K, Verheye H, Wilson W (2019) A global plankton diversity monitoring program. Front Mar Sci 6:321

BCC (2018) Regional strategy for marine spatial planning in the Benguela Current Large Marine Ecosystem. Benguela Current Convention, Swakopmund

Beal LM, De Ruijter WP, Biastoch A, Zahn R, SCOR/WCRP/IAPSO Working Group 136 (2011) On the role of the Agulhas System in ocean circulation and climate. Nature 472:429–436

Bertrand A, Ballon M, Chaigneau A (2010) Acoustic observation of living organisms reveals the upper limit of the oxygen minimum zone. PLoS One 5:e10330

Bianchi D, Carozza David A, Galbraith Eric D, Guiet J, DeVries T (2021) Estimating global biomass and biogeochemical cycling of marine fish with and without fishing. Sci Adv 7:eabd7554

Biggs R, Scholes R (2002) Land-cover changes in South Africa 1911-1993: research in action. S Afr J Sci 98:420–424

Bode M, Kreiner A, van der Plas AK, Louw DC, Horaeb R, Auel H, Hagen W (2014) Spatio-temporal variability of copepod abundance along the 20°S monitoring transect in the northern Benguela upwelling system from 2005 to 2011. PLoS One 9:e97738

Bode M, Hagen W, Cornils A, Kaiser P, Auel H (2018) Copepod distribution and biodiversity patterns from the surface to the deep sea along a latitudinal transect in the eastern Atlantic Ocean (24°N to 21°S). Prog Oceanogr 161:66–77

Bograd SJ, Castro CG, Di Lorenzo E, Palacios DM, Bailey H, Gilly W, Chavez FP (2008) Oxygen declines and the shoaling of the hypoxic boundary in the California Current. Geophys Res Lett 35:1–6

Bohata K (2015) Microzooplankton of the northern Benguela upwelling system. PhD Thesis, Staats- und Universitätsbibliothek Hamburg, 163 pp

Bond W, Midgley G, Woodward F (2003) What controls South African vegetation - climate or fire? S Afr J Bot 69:79–91

Bond WJ, Woodward FI, Midgley GF (2005) The global distribution of ecosystems in a world without fire. New Phytol 165:525–538

Bordbar MH, Mohrholz V, Schmidt M (2021) The relation of wind-driven coastal and offshore upwelling in the Benguela Upwelling System. J Phys Oceanogr 51:3117–3133

Bouchenak-Khelladi Y, Anthony Verboom G, Hodkinson TR, Salamin N, Francois O, Ni Chong-haile G, Savolainen V (2009) The origins and diversification of C4 grasses and savanna-adapted ungulates. Glob Chang Biol 15:2397–2417

Boyer D, Oelofsen B (2004) Co-management: Namibia's experience with two large-scale industrial fisheries – sardine and orange roughy. In: Sumaila UR, Boyer D, Skogen MD, Steinshamn SI (eds) Namibia's fisheries: Ecological, Economic and Social Aspects. Eburon Academic Publishers, pp 333–356

Brady RX, Lovenduski NS, Alexander MA, Jacox M, Gruber N (2019) On the role of climate modes in modulating the air–sea CO_2 fluxes in eastern boundary upwelling systems. Biogeosciences 16:329–346

Branch WR (2006) A plan for phylogenetic studies of southern African reptiles. In: Branch WR, Tolley KA, Cunningham M, Bauer AM, Alexander G, Harrison JA, Turner AA, Bates MF (eds) SANBI biodiversity series. A plan for phylogenetic studies of southern African reptiles: proceedings of a workshop held at Kirstenbosch, February 2006. South African National Biodiversity Institute (SANBI)

Branch WR (2014) Conservation status, diversity, endemism, hotspots and threats. In: Bates MF, Branch WR, Bauer AM, Burger M, Marais J, Alexander GJ, de Villiers MS (eds) Atlas and red list of the reptiles of South Africa, Lesotho and Swaziland. Suricata Series, vol 1. South African National Biodiversity Institute (SANBI), Pretoria

Briggs JC, Bowen BW (2012) A realignment of marine biogeographic provinces with particular reference to fish distributions. J Biogeogr 39:12–30

Brodeur RD, Link JS, Smith BE, Ford M, Kobayashi D, Jones TT (2016) Ecological and economic consequences of ignoring jellyfish: a plea for increased monitoring of ecosystems. Fisheries 41:630–637

Brüchert V, Currie B, Peard KR, Lass U, Endler R, Dübecke A, Julies E, Leipe T, Zitzmann S (2006) Biogeochemical and physical control on shelf anoxia and water column hydrogen sulphide in the Benguela coastal upwelling system off Namibia. In: Neretin LN (ed) Past and present water column anoxia. Springer, Dordrecht

Bryant RG, Bigg GR, Mahowald NM, Eckardt FD, Ross SG (2007) Dust emission response to climate in southern Africa. J Geophys Res-Atmos 112:D09207, https://doi:10.1029/2005JD007025

Chan F, Barth JA, Kroeker KJ, Lubchenco J, Menge BA (2019) The dynamics and impact of ocean acidification and hypoxia - insights from sustained investigations in the northern California Current Large Marine Ecosystem. Oceanography 32:62–71

Chavez FP, Messié M (2009) A comparison of eastern boundary upwelling ecosystems. Prog Oceanogr 83:80–96

Coetzee J, Staby A, Krakstad J-O, Stenevik E (2009) Abundance and distribution of mesopelagic fish in the Benguela ecosystem, pp 44–47. In: Hampton I, Barange M, Sweijd N (eds). Benguela Environment Fisheries Interaction and Training Programme (BENEFIT) research projects. GLOBEC Report 25: ii, 126 p

Coetzee J, Merkle D, Shabangu F, Geja Y, Petersen J (2018) Results of the 2018 pelagic biomass survey. Fisheries/2018/DEC/SWG-PEL, vol 38. Department of Agriculture, Forestry and Fisheries, Cape Town

Conservation International (2014) Maputaland-Pondoland-Albany. Copy viewed on 17 March 2014. http://www.conservation.org/where/priority_areas/hotspots/africa/Maputaland-Pondoland-Albany/Pages/default.aspx

Costello MJ, Coll M, Danovaro R, Halpin P, Ojaveer H, Miloslavich P (2010) A census of marine biodiversity knowledge, resources, and future challenges. PLoS One 5:e12110

Cowling R, Roux P (eds) (1987) Karoo biome: a preliminary synthesis. Part 2-vegetation and history, vol 142. National Scientific Programmes Unit: CSIR

Cowling RM, Richardson DM, Mustart P (2003a) Fynbos. In: Cowling RM, Richardson DM, Pierce SM (eds) Vegetation of southern Africa. Cambridge University Press, Cambridge

Cowling RM, Richardson DM, Schulze RE, Hoffman MT, Midgley JJ, Hilton-Taylor C (2003b) Species diversity at the regional scale. In: Cowling RM, Richardson DM, Pierce SM (eds) Vegetation of southern Africa. Cambridge University Press, Cambridge

Cowling R, Procheş Ş, Vlok J, Van Staden J (2005) On the origin of southern African subtropical thicket vegetation. S Afr J Bot 71:1–23

Cowling RM, Potts AJ, Franklin J, Midgley GF, Engelbrecht F, Marean CW (2020) Describing a drowned Pleistocene ecosystem: last glacial maximum vegetation reconstruction of the Palaeo-Agulhas plain. Quat Sci Rev 235:105866

Crawford RJM, Underhill LG, Coetzee JC, Fairweather T, Shannon LJ, Wolfaardt AC (2008) Influences of the abundance and distribution of prey on African penguins *Spheniscus demersus* off western South Africa. Afr J Mar Sci 30:167–175

Critical Ecosystem Partnership Fund (2001) Ecosystem profile. The Cape floristic region, South Africa. https://www.cepf.net/resources/ecosystem-profile-documents/cape-floristic-region-ecosystem-profile-2001

Currie B, Utne-Palm AC, Salvanes AGV (2018) Winning ways with hydrogen sulphide on the Namibian shelf. Front Mar Sci 5:341

Cury P, Roy C (1989) Optimal environmental window and pelagic fish recruitment success in upwelling areas. Can J Fish Aquat Sci 46:670–680

Cury P, Bakun A, Crawford RJ, Jarre A, Quinones RA, Shannon LJ, Verheye HM (2000) Small pelagics in upwelling systems: patterns of interaction and structural changes in 'wasp-waist' ecosystems. ICES J Mar Sci 57:603–618

Cushing DH (1996) Towards a science of recruitment in fish populations, vol 7. Ecology Institute Oldendorf/Luhe Nordbünte

De Decker AHB (1964) Observations on the ecology and distribution of Copepoda in the marine plankton of South Africa. In: Investigational report of the Division of Sea Fisheries, vol 49. Department of Commerce and Industries, South Africa

De Decker A (1984) Near-surface copepod distribution in the south-western Indian and south-eastern Atlantic Ocean. Ann S Afr Mus 93:303–370

De Decker A, Kaczmaruk B, Marska G (1991) A new species of *Calanus* (Copepoda, Calanoida) from South African waters. Ann S Afr Mus 101:27–44

DEA (2015) Climate change adaptation plans for South African biomes. In: Kharika JRM, Mkhize NCS, Munyai T, Khavhagali VP, Davis C, Dziba D, Scholes R, van Garderen E, von Maltitz G, Le Maitre D, Archibald S, Lotter D, van Deventer H, Midgely G, Hoffman T (eds) Department of Environmental Affairs. Pretoria, South Africa

DEFF (2020) Status of the South African marine fisheries resources 2020. Department of Environment, Forestry and Fisheries, Cape Town, South Africa

Desmet P (2007) Namaqualand—a brief overview of the physical and floristic environment. J Arid Environ 70:570–587

Dey SP, Vichi M, Fearon G, Seyboth E, Findlay KP, Meynecke J-O, de Bie J, Lee SB, Samanta S, Barraqueta JLM, Roychoudhury AN, Mackey B (2021) Oceanographic anomalies coinciding with humpback whale super-group occurrences in the southern Benguela. Sci Rep 11:1–13

Dippenaar-Schoeman AS, Lyle R, Marais P, Haddad C, Foord S, Lotz L (2015) South African National Survey of Arachnida (SANSA): review of current knowledge, constraints and future needs for documenting spider diversity (Arachnida: Araneae). Trans R Soc South Africa 70:245–275

Doubleday ZA, Prowse TA, Arkhipkin A, Pierce GJ, Semmens J, Steer M, Leporati SC, Lourenço S, Quetglas A, Sauer W (2016) Global proliferation of cephalopods. Curr Biol 26:R406–R407

Driver A, Desmet P, Rouget M, Cowling R, Maze K (2003) Succulent Karoo ecosystem plan: biodiversity component. Technical Report CCU 1/03. Cape Conservation Unit, Botanical Society of South Africa

Driver A, Sink KJ, Nel JL, Holness S, Van Niekerk L, Daniels F, Jonas Z, Majiedt PA, Harris L, Maze K (2012) National Biodiversity Assessment 2011: an assessment of South Africa's biodiversity and ecosystems. Synthesis report. South African National Biodiversity Institute (SANBI) and Department of Environmental Affairs, Pretoria, South Africa

du Toit JCO (2019) Drivers of vegetation change in the eastern Karoo. PhD thesis, University of KwaZulu-Natal, Pietermaritzburg, South Africa, pp 183

Duncan SE, Sell AF, Hagen W, Fock HO (2022) Environmental drivers of upper mesopelagic fish assemblages in the Benguela Upwelling Systems. Mar Ecol Prog Ser 688:133–152

Duncombe Rae CM (2005) A demonstration of the hydrographic partition of the Benguela upwelling ecosystem at 26°40'S. Afr J Mar Sci 27:617–628

Durholtz M, Singh L, Fairweather T, Leslie R, van der Lingen C, Bross C, Hutchings L, Rademeyer R, Butterworth D, Payne A (2015) Fisheries, ecology and markets of South African hake. In: Arancibia H (ed) Hakes: biology and exploitation. John Wiley & Sons, Chichester

Eckardt FD, Schemenauer RS (1998) Fog water chemistry in the Namib Desert, Namibia. Atmos Environ 32:2595–2599

Ekau W, Verheye HM (2005) Influence of oceanographic fronts and low oxygen on the distribution of ichthyoplankton in the Benguela and southern Angola currents. Afr J Mar Sci 27:629–639

Ekau W, Hendricks A, Kadler S, Koch V, Loick N (2001) Winter ichthyoplankton in the northern Benguela upwelling and Angola-Benguela front regions: BENEFIT marine science. S Afr J Sci 97:259–265

Ekau W, Auel H, Pörtner H-O, Gilbert D (2010) Impacts of hypoxia on the structure and processes in pelagic communities (zooplankton, macro-invertebrates and fish). Biogeosciences 7:1669–1699

Ekau W, Auel H, Hagen W, Koppelmann R, Wasmund N, Bohata K, Buchholz F, Geist S, Martin B, Schukat A, Verheye HM, Werner T (2018) Pelagic key species and mechanisms driving energy flows in the northern Benguela upwelling ecosystem and their feedback into biogeochemical cycles. J Mar Syst 188:49–62

Ellery WN (1992) Classification of vegetation of the South African grassland biome. PhD thesis, University of the Witwatersrand, Johannesburg, South Africa, pp 211

Ellery W, Scholes R, Mentis M (1991) An initial approach to predicting the sensitivity of the South African grassland biome to climate change. S Afr J Sci 87:499–503

Emeis K-C, Brüchert V, Currie B, Endler R, Ferdelman T, Kiessling A, Leipe T, Noli-Peard K, Struck U, Vogt T (2004) Shallow gas in shelf sediments of the Namibian coastal upwelling ecosystem. Cont Shelf Res 24:627–642

Emeis K, Eggert A, Flohr A, Lahajnar N, Nausch G, Neumann A, Rixen T, Schmidt M, van der Plas A, Wasmund N (2018) Biogeochemical processes and turnover rates in the northern Benguela upwelling system. J Mar Syst 188:63–80

Enquist BJ, Feng X, Boyle B, Maitner B, Newman EA, Jørgensen PM, Roehrdanz PR, Thiers BM, Burger JR, Corlett RT, Couvreur TLP, Dauby G, Donoghue JC, Foden W, Lovett JC, Marquet PA, Merow C, Midgley G, Morueta-Holme N, Neves DM, Oliveira-Filho AT, Kraft NJB, Park DS, Peet RK, Pillet M, Serra-Diaz JM, Sandel B, Schildhauer M, Simova I, Violle C, Wieringa JJ, Wiser SK, Hannah L, Svenning JC, McGill BJ (2019) The commonness of rarity: global and future distribution of rarity across land plants. Sci. Adv. 5:eaaz0414

Erasmus VN, Currie B, Roux J-P, Elwen SH, Kalola M, Tjizoo B, Kathena JN, Iitembu JA (2021) Predatory species left stranded following the collapse of the sardine Sardinops sagax (Pappe, 1854) stock off the northern Benguela upwelling system: a review. J Mar Syst 224:103623

Esler K, Von Staden L, Midgley G (2015) Determinants of the fynbos/succulent Karoo biome boundary: insights from a reciprocal transplant experiment. S Afr J Bot 101:120–128

Esteban I, Marean CW, Cowling RM, Fisher EC, Cabanes D, Albert RM (2020) Palaeoenvironments and plant availability during MIS 6 to MIS 3 on the edge of the Palaeo-Agulhas plain (South Coast, South Africa) as indicated by phytolith analysis at Pinnacle Point. Quat Sci Rev 235:105667

Fay A, McKinley G (2014) Global open-ocean biomes: mean and temporal variability. Earth System Sci Data 6:273–284

Finke G, Gee K, Gxaba T, Sorgenfrei R, Russo V, Pinto D, Nsiangango SE, Sousa LN, Braby R, Alves FL, Heinrichs B, Kreiner A, Amunyela M, Popose G, Ramakulukusha M, Naidoo A, Mausolf E, Nsingi KK (2020) Marine spatial planning in the Benguela Current Large Marine Ecosystem. Environ Dev 36:100569

Flynn RF, Granger J, Veitch JA, Siedlecki S, Burger JM, Pillay K, Fawcett SE (2020) On-shelf nutrient trapping enhances the fertility of the southern Benguela upwelling system. J Geophys Res Oceans 125:e2019JC015948

Foden W, Midgley GF, Hughes G, Bond WJ, Thuiller W, Hoffman MT, Kaleme P, Underhill LG, Rebelo A, Hannah L (2007) A changing climate is eroding the geographical range of the Namib Desert tree aloe through population declines and dispersal lags. Divers Distrib 13:645–653

Fortes-Lima C, Mtetwa E, Schlebusch C (eds) (2022) Africa, the cradle of human diversity: cultural and biological approaches to uncover African diversity. Brill, Leiden

García-Reyes M, Sydeman WJ, Schoeman DS, Rykaczewski RR, Black BA, Smit AJ and Bograd SJ (2015) Under Pressure: Climate Change, Upwelling, and Eastern Boundary Upwelling Ecosystems. Front. Mar. Sci. 2:109. https://doi.org/10.3389/fmars.2015.00109

Geldenhuys CJ (1992) Richness, composition and relationships of the floras of selected forests in southern Africa. Bothalia 22: 205–233 https://doi.org/10.4102/abc.v22i2.847

Geldenhuys CJ (1994) Bergwind fires and the location pattern of forest patches in the southern Cape landscape, South Africa. J Biogeogr 21:49–62

Geldenhuys CJ (2004) Bark harvesting for traditional medicine: from illegal resource degradation to participatory management. Scand J For Res 19:103–115

Geldenhuys C, MacDevette D (1989) Conservation status of coastal and montane evergreen forest. In: Huntley BJ (ed) Biotic diversity in southern Africa: concepts and conservation. Oxford University Press, Cape Town

Gertenbach WD (1983) Landscapes of the Kruger National Park. Koedoe 26:9–121

Gibbons MJ, Hutchings L (1996) Zooplankton diversity and community structure around southern Africa, with special attention to the Benguela upwelling system. S Afr J Sci 92:63–75

Gibbons MJ, Haddock SH, Matsumoto GI, Foster C (2021) Records of ctenophores from South Africa. PeerJ 9:e10697

Gibbs Russell G (1987) Preliminary floristic analysis of the major biomes in southern Africa. Bothalia 17:213–227

Goldblatt P, Manning JC (2002) Plant diversity of the Cape region of southern Africa. Ann Mo Bot Gard 89:281–302

Golikov AV, Sabirov RM, Lubin PA, Jørgensen LL (2013) Changes in distribution and range structure of Arctic cephalopods due to climatic changes of the last decades. Biodiversity 14:28–35

Good S, Fiedler E, Mao C, Martin MJ, Maycock A, Reid R, Roberts-Jones J, Searle T, Waters J, While J, Worsfold M (2020) The current configuration of the OSTIA system for operational production of foundation sea surface temperature and ice concentration analyses. Remote Sens 12:720

Griffiths CL (2005) Coastal marine biodiversity in East Africa. Indian J Mar Sci 34:35–41

Griffiths CL, Robinson TB (2016) Use and usefulness of measures of marine endemicity in South Africa. S Afr J Sci 112:1–7

Griffiths CL, Robinson TB, Lange L, Mead A (2010) Marine biodiversity in South Africa: an evaluation of current states of knowledge. PLoS One 5:e12008

Grote B, Ekau W, Hagen W, Huggett J, Verheye H (2007) Early life-history strategy of Cape hake in the Benguela upwelling region. Fish Res 86:179–187

Grote B, Stenevik E, Ekau W, Verheye H, Lipiński M, Hagen W (2012) Spawning strategies and transport of early stages of the two Cape hake species, *Merluccius paradoxus* and *M. capensis,* in the southern Benguela upwelling system. Afr J Mar Sci 34:195–204

Guo D, Desmet PG, Powrie LW (2017) Impact of the future changing climate on the southern Africa biomes, and the importance of geology. J Geosci Environ Prot 5:1–9

Hachfeld B (2000) Rain, fog and species richness in the central Namib Desert in the exceptional rainy season of 1999/2000. Dinteria 26:113–146

Hamilton WJ III, Henschel JR, Seely MK (2003) Fog collection by Namib Desert beetles: correspondence. S Afr J Sci 99:181

Hardman-Mountford NJ, Hirata T, Richardson KA, Aiken J (2008) An objective methodology for the classification of ecological pattern into biomes and provinces for the pelagic ocean. Remote Sens Environ 112:3341–3352

Harris LR, Holness SD, Finke G, Amunyela M, Braby R, Coelho N, Gee K, Kirkman SP, Kreiner A, Mausolf E, Majiedt P, Maletzky E, Nsingi KK, Russo V, Sink KJ and Sorgenfrei R (2022)

Practical marine spatial management of ecologically or biologically significant marine areas: emerging lessons from evidence-based planning and implementation in a developing-world context. Front Mar Sci 9:831678. https://doi.org/10.3389/fmars.2022.831678

Hempson GP, Archibald S, Bond WJ, Ellis RP, Grant CC, Kruger FJ, Kruger LM, Moxley C, Owen-Smith N, Peel MJ, Smit IP, Vickers KJ (2015) Ecology of grazing lawns in Africa. Biol Rev 90:979–994

Henschel JR, Seely MK (2008) Ecophysiology of atmospheric moisture in the Namib Desert. Atmos Res 87:362–368

Heymans JJ, Tomczak MT (2016) Regime shifts in the northern Benguela ecosystem: challenges for management. Ecol Model 331:151–159

Hine P, Sealy J, Halkett D, Hart T (2010) Antiquity of stone-walled tidal fish traps on the Cape coast, South Africa. South Afr Archaeol Bull 65(191):35–44

Hoffman MT (2014) Changing patterns of rural land use and land cover in South Africa and their implications for land reform. J South Afr Stud 40:707–725

Howard EM, Penn JL, Frenzel H, Seibel BA, Bianchi D, Renault L, Kessouri F, Sutula MA, McWilliams JC, Deutsch C (2020) Climate-driven aerobic habitat loss in the California Current System. Sci Adv 6:eaay3188

Huggett J, Richardson A (2000) A review of the biology and ecology of *Calanus agulhensis* off South Africa. ICES J Mar Sci 57:1834–1849

Huggett J, Verheye H, Escribano R, Fairweather T (2009) Copepod biomass, size composition and production in the southern Benguela: Spatio–temporal patterns of variation, and comparison with other eastern boundary upwelling systems. Prog Oceanogr 83:197–207

Huggett JA, Noyon M, Carstensen J, Walker DR (2023) Patterns in the plankton – spatial distribution and long-term variability of copepods on the Agulhas Bank. Deep-Sea Res Part II 196:105265

Hugo S, Altwegg R (2017) The second southern African bird atlas project: causes and consequences of geographical sampling bias. Ecol Evol 7:6839–6849

Hulley PA, Lutjeharms JRE (1989) Lanternfishes of the southern Benguela region part 3. The pseudoceanic-oceanic interface. Ann S Afr Mus 98:1–10

Hulley P, Prosch R (1987) Mesopelagic fish derivatives in the southern Benguela upwelling region. S Afr J Mar Sci 5:597–611

Huntley B (1982) Southern African savannas. In: Huntley B, Walker B (eds) Ecology of tropical savannas. Springer, Berlin

Huntley B (2003) Introduction. In: Cowling RM, Richardson DM, Pierce SM (eds) Vegetation of southern Africa. Cambridge University Press, Cambridge

Huntley BJ, Russo V, Lages F, Ferrand N (2019) Biodiversity of Angola: science & conservation: a modern synthesis. Springer Nature, Cham

Hutchings L, Pillar SC, Verheye HM (1991) Estimates of standing stock, production and consumption of meso- and macrozooplankton in the Benguela ecosystem. S Afr J Mar Sci 11:499–512

Hutchings L, Beckley LE, Griffiths MH, Roberts MJ, Sundby S, van der Lingen C (2002) Spawning on the edge: spawning grounds and nursery areas around the southern African coastline. Mar Freshw Res 53:307–318

Hutchings L, van der Lingen CD, Shannon LJ, Crawford RJM, Verheye HMS, Bartholomae CH, van der Plas AK, Louw D, Kreiner A, Ostrowski M, Fidel Q, Barlow RG, Lamont T, Coetzee J, Shillington F, Veitch J, Currie JC, Monteiro PMS (2009) The Benguela Current: an ecosystem of four components. Prog Oceanogr 83:15–32

Hutchings L, Jarre A, Lamont T, Van den Berg M, Kirkman S (2012) St Helena Bay (southern Benguela) then and now: muted climate signals, large human impact. Afr J Mar Sci 34:559–583

IUCN (2016) A global standard for the identification of key biodiversity areas. Version 1.0. IUCN, Gland, Switzerland and Cambridge, UK

Jarre A, Ragaller SM, Hutchings L (2013) Long-term, ecosystem-scale changes in the southern Benguela marine pelagic social-ecological system: interaction of natural and human drivers. Ecol Soc 18(4):55

Jarre A, Hutchings L, Kirkman SP, Kreiner A, Tchipalanga PC, Kainge P, Uanivi U, van der Plas AK, Blamey LK, Coetzee JC, Lamont T, Samaai T, Verheye HM, Yemane DG, Axelsen BE, Ostrowski M, Stenevik EK, Loeng H (2015) Synthesis: climate effects on biodiversity, abundance and distribution of marine organisms in the Benguela. Fish Oceanogr 24:122–149

Kainge P, Kirkman SP, Estevão V, van der Lingen CD, Uanivi U, Kathena JN, van der Plas A, Githaiga-Mwicigi J, Makhado A, Nghimwatya L, Endjambi T, Paulus S, Kalola M, Antonio M, Tjizoo B, Shikongo T, Nsiangango S, Uahengo T, Bartholomae C, Mqoqi M, Hamukuaya H (2020) Fisheries yields, climate change, and ecosystem-based management of the Benguela Current Large Marine Ecosystem. Environ Dev 36:100567

Kgope BS, Bond WJ, Midgley GF (2010) Growth responses of African savanna trees implicate atmospheric [CO_2] as a driver of past and current changes in savanna tree cover. Austral Ecol 35:451–463

Kirkman SP, Nsingi KK (2019) Marine biodiversity of Angola: biogeography and conservation. In: Huntley BJ, Russo V, Lages F, Ferrand N (eds) Biodiversity of Angola. Springer Open, Cham 43–52

Kirkman SP, Blamey L, Lamont T, Field JG, Bianchi G, Huggett JA, Hutchings L, Jackson-Veitch J, Jarre A, Lett C, Lipiński MR, Mafwila SW, Pfaff MC, Samaai T, Shannon LJ, Shin Y-J, van der Lingen CD, Yemane D (2016) Spatial characterisation of the Benguela ecosystem for ecosystem-based management. Afr J Mar Sci 38:7–22

Kirkman SP, Holness S, Harris LR, Sink KJ, Lombard AT, Kainge P, Majiedt P, Nsiangango SE, Nsingi KK, Samaai T (2019) Using systematic conservation planning to support marine spatial planning and achieve marine protection targets in the transboundary Benguela ecosystem. Ocean Coast Manag 168:117–129

Klak C, Reeves G, Hedderson T (2004) Unmatched tempo of evolution in southern African semi-desert ice plants. Nature 427:63–65

Knowles T, Boardman P, Mugido W, Blignaut J (2014) Understanding potential climate change mitigation opportunities. National Terrestrial Carbon Sinks Assessment. Technical Report. Department of Environmental Affairs, Pretoria, South Africa

Kopte R, Brandt P, Claus M, Greatbatch RJ, Dengler M (2018) Role of equatorial basin-mode resonance for the seasonal variability of the Angola Current at 11°S. J Phys Oceanogr 48:261–281

Kosobokova KN, Hopcroft RR, Hirche H-J (2011) Patterns of zooplankton diversity through the depths of the Arctic's central basins. Mar Biodivers 41:29–50

Kraaij T, Young C, Bezuidenhout H (2017) Growth-form responses to fire in Nama-Karoo escarpment grassland, South Africa. Fire Ecol 13:85–94

Kreiner A, Stenevik EK, Ekau W (2009) Sardine *Sardinops sagax* and anchovy *Engraulis encrasicolus* larvae avoid regions with low dissolved oxygen concentration in the northern Benguela Current system J Fish Biol 74:270–277

Lamont T, García-Reyes M, Bograd S, van der Lingen C, Sydeman W (2018) Upwelling indices for comparative ecosystem studies: variability in the Benguela Upwelling System. J Mar Syst 188:3–16

Lancaster J, Lancaster N, Seely MK (1984) Climate of the central Namib Desert. Modoqua 14:5–61

Laruelle GG, Lauerwald R, Pfeil B, Regnier P (2014) Regionalized global budget of the CO_2 exchange at the air-water interface in continental shelf seas. Glob Biogeochem Cycles 28:1199–1214

Le Maitre D, Midgley J (1992) Plant reproductive ecology. In: Cowling R (ed) The ecology of fynbos: nutrients, fire and diversity. Oxford University Press, Oxford

Lombard AT, Strauss T, Harris J, Sink K, Attwood C, Hutchings L (2004) South African national spatial biodiversity assessment 2004. Technical Report, vol 4: Marine Component. South African National Biodiversity Institute, Pretoria

Longhurst AR (2007) Ecological Geography of the Sea. Academic Press, Amsterdam

Louw R (2006) Sustainable harvesting of wild rooibos (*Aspalathus linearis*) in the Suid Bokkeveld, Northern Cape. MSc thesis, University of Cape Town, pp 130

Low A, Rebelo A (eds) (1996) Vegetation of South Africa, Lesotho and Swaziland. Department of Environmental Affairs and Tourism (DEAT), Pretoria, South Africa

Lynam CP, Gibbons MJ, Axelsen BE, Sparks CAJ, Coetzee J, Heywood BG, Brierley AS (2006) Jellyfish overtake fish in a heavily fished ecosystem. Curr Biol 16:R492–R493

MacKellar N, Hewitson B, Tadross M (2007) Namaqualand's climate: recent historical changes and future scenarios. J Arid Environ 70:604–614

Malimbwi R, Chidumayo E, Zahabu E, Kingazi S, Misana S, Luoga E, Nduwamungu J (2010) Woodfuel. In: Chidumayo E, Gumbo D (eds) The dry forests and woodlands of Africa managing for products and services. Earthscan, London

Marloth R (1908) Das Kapland. Gustav Fischer, Jena

Masubelele ML, Hoffman M, Bond W, Gambiza J (2014) A 50 year study shows grass cover has increased in shrublands of semi-arid South Africa. J Arid Environ 104:43–51

Masubelele M, Hoffman M, Bond W (2015) Biome stability and long-term vegetation change in the semi-arid, south-eastern interior of South Africa: a synthesis of repeat photo-monitoring studies. S Afr J Bot 101:139–147

MET (2018) Sixth National Report to the convention on biological diversity (2014–2018). Ministry of Environment and Tourism, Republic of Namibia, Windhoek, Namibia

Midgley G, Thuiller W (2007) Potential vulnerability of Namaqualand plant diversity to anthropogenic climate change. J Arid Environ 70:615–628

Midgley G, Rutherford M, Bond WJ (2008) The heat is on: impacts of climate change on plant diversity in South Africa. South African National Biodiversity Institute (SANBI), Pretoria

Mills AJ, Cowling RM (2010) Below-ground carbon stocks in intact and transformed subtropical thicket landscapes in semi-arid South Africa. J Arid Environ 74:93–100

Mills A, Cowling R, Fey M, Kerley G, Donaldson J, Lechmere-Oertel R, Sigwela A, Skowno A, Rundel P (2005) Effects of goat pastoralism on ecosystem carbon storage in semiarid thicket, eastern Cape, South Africa. Austral Ecol 30:797–804

Mitchell D, Henschel JR, Hetem RS, Wassenaar TD, Strauss WM, Hanrahan SA, Seely MK (2020) Fog and fauna of the Namib Desert: past and future. Ecosphere 11:e02996

Mohrholz V, Bartholomae CH, van der Plas AK, Lass HU (2008) The seasonal variability of the northern Benguela undercurrent and its relation to the oxygen budget on the shelf. Cont Shelf Res 28:424–441

Moll E, White F (1978) The Indian Ocean Coastal Belt. In: Werger MJA (ed) Biogeography and ecology of southern Africa. Monographiae Biologicae, vol 31. Springer, Dordrecht

Monteiro PMS, van der Plas AK (2006) Low Oxygen Water (LOW) variability in the Benguela system: key processes and forcing scales relevant to forecasting. In: Shannon V, Hempel G, Malanotte-Rizzoli P, Moloney C, Woods J (eds) Benguela: predicting a large marine ecosystem. Large marine ecosystems, vol 14. Elsevier, Amsterdam

Monteiro PMS, van der Plas AK, Bailey GW, Malanotte-Rizzoli P, Duncombe Rae CM, Byrnes D, Pitcher G, Florenchie P, Penven P, Fitzpatrick J, Lass HU (2006) Low Oxygen Water (LOW) forcing scales amenable to forecasting in the Benguela ecosystem. In: Shannon V, Hempel G, Malanotte-Rizzoli P, Moloney C, Woods J (eds) Benguela: predicting a large marine ecosystem. Large marine ecosystems, vol 14. Elsevier, Amsterdam

Morgan JA, LeCain DR, Pendall E, et al. (2011) C4 grasses prosper as carbon dioxide eliminates desiccation in warmed semi-arid grassland. Nature 476:202–205 https://doi.org/10.1038/nature10274

Moriarty R, Buitenhuis ET, Le Quéré C, Gosselin MP (2013) Distribution of known macrozooplankton abundance and biomass in the global ocean. Earth Syst Sci Data 5:241–257

Mucina L, Rutherford MC (2006) The vegetation of South Africa, Lesotho and Swaziland, vol 19. South African National Biodiversity Institute (SANBI), Pretoria

Mudombi S, Nyambane A, von Maltitz GP, Gasparatos A, Johnson FX, Chenene ML, Attanassov B (2018) User perceptions about the adoption and use of ethanol fuel and cookstoves in Maputo, Mozambique. Energy Sustain Dev 44:97–108

Mwaala D (2022) Spatial, seasonal, interannual variability and long-term trends in the diet of Cape fur seals along the Namibian coast (1994–2018). MSc thesis, University of Namibia, Windhoek, Namibia, pp 127

Nakanyala J, Kosmas S, Hipondoka M (2018) The savannas: an integrated synthesis of three major competing paradigms. Int Sci Technol J Namibia 10:108–121

Nelson-Viljoen C, Kyriacou K (2017) Shellfish exploitation strategies at the Pinnacle Point Shell Midden Complex, South Africa, during the later stone age. J Island Coast Archaeol 12:540–557

Nielsen M, Bunkenborg M (2020) Natural resource extraction in the interior: scouts, spirits and Chinese loggers in the forests of northern Mozambique. J South Afr Stud 46:417–433

O'Connor T, Bredenkamp G (2003) Grassland. In: Cowling RM, Richardson DM, Pierce SM (eds) Vegetation of southern Africa. Cambridge University Press, Cambridge

Oesterwind D, Barrett CJ, Sell AF, Núñez-Riboni I, Kloppmann M, Piatkowski U, Wieland K, Laptikhovsky V (2022) Climate change-related changes in cephalopod biodiversity on the North East Atlantic Shelf. Biodivers Conserv 31:1491–1518

O'Hara CC, Villaseñor-Derbez JC, Ralph GM, Halpern BS (2019) Mapping status and conservation of global at-risk marine biodiversity. Conserv Lett 12:e12651

Oliver MJ, Irwin AJ (2008) Objective global ocean biogeographic provinces. Geophys Res Lett 35:L15601

Olson DM, Dinerstein E, Wikramanayake ED, Burgess ND, Powell GVN, Underwood EC, D'amico JA, Itoua I, Strand HE, Morrison JC, Loucks CJ, Allnutt TF, Ricketts TH, Kura Y, Lamoreux JF, Wettengel WW, Hedao P, Kassem KR (2001) Terrestrial ecoregions of the world: a new map of life on earth: a new global map of terrestrial ecoregions provides an innovative tool for conserving biodiversity. Bioscience 51:933–938

Opdal AF, Brodeur RD, Cieciel K, Daskalov GM, Mihneva V, Ruzicka JJ, Verheye HM, Aksnes DL (2019) Unclear associations between small pelagic fish and jellyfish in several major marine ecosystems. Sci Rep 9:2997

Ortega-Cisneros K, Yokwana S, Sauer W, Cochrane K, Cockcroft A, James N, Potts W, Singh L, Smale M, Wood A, Pecl G (2018a) Assessment of the likely sensitivity to climate change for the key marine species in the southern Benguela system. Afr J Mar Sci 40:279–292

Ortega-Cisneros K, Cochrane KL, Fulton EA, Gorton R, Popova E (2018b) Evaluating the effects of climate change in the southern Benguela upwelling system using the Atlantis modelling framework. Fish Oceanogr 27:489–503

Palmer A, Hoffman M (2003) Nama-Karoo. In: Cowling RM, Richardson DM, Pierce SM (eds) Vegetation of southern Africa. Cambridge University Press, Cambridge

Parr C, Brockett B (1999) Patch-mosaic burning: a new paradigm for savanna fire management in protected areas? Koedoe: African Protected Area Conservation and Science 42(2):117–130. https://doi.org/10.4102/koedoe.v42i2.237

Pillar S (1986) Temporal and spatial variations in copepod and euphausiid biomass off the southern and south-western coasts of South Africa in 1977/78. S Afr J Mar Sci 4:219–229

Postel L, da Silva AJ, Mohrholz V, Lass H-U (2007) Zooplankton biomass variability off Angola and Namibia investigated by a lowered ADCP and net sampling. J Mar Syst 68:143–166

Potts WM, Götz A, James N (2015) Review of the projected impacts of climate change on coastal fishes in southern Africa. Rev Fish Biol Fish 25:603–630

Procheş Ş, Cowling R (2006) Insect diversity in Cape Fynbos and neighbouring South African vegetation. Glob Ecol Biogeogr 15:445–451

Ragoasha N, Herbette S, Cambon G, Veitch J, Reason C, Roy C (2019) Lagrangian pathways in the southern Benguela upwelling system. J Mar Syst 195:50–66

Rebelo AG, Boucher C, Helme N, Mucina L, Rutherford MC (2006) Fynbos Biome. In: Mucina L, Rutherford MC (eds) The vegetation of South Africa, Lesotho and Swaziland. South African National Biodiversity Institute (SANBI), Pretoria

Richardson AJ (2008) In hot water: zooplankton and climate change. ICES J Mar Sci 65:279–295

Richardson DM, van Wilgen BW, Higgins SI, Trinder-Smith TH, Cowling RM, McKell DH (1996) Current and future threats to plant biodiversity on the Cape Peninsula, South Africa. Biodivers Conserv 5:607–647

Riebesell U, Schulz KG, Bellerby RGJ, Botros M, Fritsche P, Meyerhofer M, Neill C, Nondal G, Oschlies A, Wohlers J, Zollner E (2007) Enhanced biological carbon consumption in a high CO_2 ocean. Nature 450:545–548

Rittinghaus H (2021) On the distribution, population structure and ecological role of key species of the copepod family Calanidae in the Benguela upwelling system. BSc thesis, University of Bremen, Bremen, Germany, pp 67

Rixen T, Gaye B, Emeis K-C, Ramaswamy V (2019) The ballast effect of lithogenic matter and its influences on the carbon fluxes in the Indian Ocean. Biogeosciences 16:485–503

Rixen T, Lahajnar N, Lamont T, Koppelmann R, Martin B, van Beusekom JEE, Siddiqui C, Pillay K, Meiritz L (2021) Oxygen and nutrient trapping in the southern Benguela upwelling system. Front Mar Sci 8:1367

Robinson WM, Butterworth DS, Plagányi ÉE (2015) Quantifying the projected impact of the South African sardine fishery on the Robben Island penguin colony. ICES J Mar Sci 72:1822–1833

Robison BH (2004) Deep pelagic biology. J Exp Mar Biol Ecol 300:253–272

Robison BH (2009) Conservation of deep pelagic biodiversity. Conserv Biol 23:847–858

Roobaert A, Laruelle GG, Landschützer P, Gruber N, Chou L, Regnier P (2019) The spatiotemporal dynamics of the sources and sinks of CO_2 in the global coastal ocean. Glob Biogeochem Cycles 33:1693–1714

Rosa R, Dierssen HM, Gonzalez L, Seibel BA (2008) Large-scale diversity patterns of cephalopods in the Atlantic open ocean and deep sea. Ecology 89:3449–3461

Rosa R, Pissarra V, Borges FO, Xavier J, Gleadall IG, Golikov A, Bello G, Morais L, Lishchenko F, Roura Á (2019) Global patterns of species richness in coastal cephalopods. Front Mar Sci 6:469

Roux J-P, van der Lingen CD, Gibbons MJ, Moroff NE, Shannon LJ, Smith AD, Cury PM (2013) Jellyfication of marine ecosystems as a likely consequence of overfishing small pelagic fishes: lessons from the Benguela. Bull Mar Sci 89:249–284

Rutherford MC, Westfall RH (1986) Biomes of southern Africa - an objective categorization. Memoirs of the Botanical Survey of South Africa, No. 54. National Botanical Institute

Rutherford MC, Westfall RH (1994) Biomes of southern Africa - an objective categorization. Memoirs of the Botanical Survey of South Africa, No. 63. National Botanical Institute

Rutherford MC, Mucina L, Powrie LW (2006) Biomes and bioregions of southern Africa. In: Mucina L, Rutherford MC (eds) The vegetation of South Africa, Lesotho and Swaziland, vol 19. South African National Biodiversity Institute (SANBI), Pretoria

Rykaczewski RR, Dunne JP, Sydeman WJ, García-Reyes M, Black BA, Bograd SJ (2015) Poleward displacement of coastal upwelling-favourable winds in the ocean's eastern boundary currents through the 21st century. J Geophys Res Letters 42:6424–6431

Salvanes AGV, Christiansen H, Taha Y, Henseler C, Seivåg ML, Kjesbu OS, Folkvord A, Utne-Palm AC, Currie B, Ekau W (2018) Variation in growth, morphology and reproduction of the bearded goby (*Sufflogobius bibarbatus*) in varying oxygen environments of northern Benguela. J Mar Syst 188:81–97

Sankaran M, Hanan NP, Scholes RJ, Ratnam J, Augustine DJ, Cade BS, Gignoux J, Higgins SI, Le Roux X, Ludwig F, Ardo J, Banyikwa F, Bronn A, Bucini G, Caylor KK, Coughenour MB, Diouf A, Ekaya W, Feral CJ, February EC, Frost PG, Hiernaux P, Hrabar H, Metzger KL, Prins HH, Ringrose S, Sea W, Tews J, Worden J, Zambatis N (2005) Determinants of woody cover in African savannas. Nature 438:846–849

Schlebusch CM, Malmström H, Günther T, Sjödin P, Coutinho A, Edlund H, Munters AR, Vicente M, Steyn M, Soodyall H (2017) Southern African ancient genomes estimate modern human divergence to 350,000 to 260,000 years ago. Science 358:652–655

Scholes R (2003) Savanna. In: Cowling R, Richardson D, Pierce S (eds) Vegetation of southern Africa. Cambridge University Press, Cambridge

Scholes RJ, Archer SR (1997) Tree-grass interactions in savannas. Annu Rev Ecol Syst 28:517–544

Scholes R, Walker B (1993) An African savanna: synthesis of the Nylsvley study. Cambridge University Press, Cambridge

Scholes R, Gureja N, Giannecchinni M, Dovie D, Wilson B, Davidson N, Piggott K, McLoughlin C, Van der Velde K, Freeman A, Bradley S, Smart R, Ndala S (2001) The environment and vegetation of the flux measurement site near Skukuza, Kruger National Park. Koedoe 44:73–83

Scholes R, von Maltitz G, Archibald S, Wessels K, van Zyl T, Swanepoel D, Steenkamp K (2014) Terrestrial ecosystem carbon stocks in South Africa. South African National Carbon Sinks Assessment. Department of Environmental Affairs, Pretoria

Schukat A, Bode M, Auel H, Carballo R, Martin B, Koppelmann R, Hagen W (2013) Pelagic decapods in the northern Benguela upwelling system: distribution, ecophysiology and contribution to active carbon flux. Deep-Sea Res I Oceanogr Res Pap 75:146–156

Schukat A, Auel H, Teuber L, Lahajnar N, Hagen W (2014) Complex trophic interactions of calanoid copepods in the Benguela upwelling system. J Sea Res 85:186–196

Schulze BR (1969) The climate of Gobabeb. Scientific Papers of the Namib Desert Research Station 38:5–12

Schwartzlose R, Alheit J, Bakun A, Baumgartner T, Cloete R, Crawford J, Fletcher W, Green-Ruiz Y, Hagen E, Kawasaki T, Lluch-Belda D, Lluch-Cota S, Maccall A, Matsuura Y, Nevárez-Martínez M, Parrish R, Roy C, Serra R, Shust KV, Ward MN, Zuzunaga JZ (1999) Worldwide large-scale fluctuations of sardine and anchovy populations. S Afr J Mar Sci 21:289–347

Shannon L (2009) Benguela Current. In: Steele JH, Thorpe SA, Turekian KK (eds) Encyclopedia of Ocean Sciences, 2nd edn. Academic Press, London, pp 316–327

Shannon L, Pillar S (1986) The Benguela ecosystem. Part III. Plankton. Oceanogr Mar Biol Annu Rev 24:65–170

Shannon LJ, Moloney CL, Jarre A, Field JG (2003) Trophic flows in the southern Benguela during the 1980s and 1990s. J Mar Syst 39:83–116

Siddiqui C, Rixen R, Lahajnar N, van der Plas AK, Louw DC, Lamont T, Pillay K (2023) Regional and global impact of CO_2 uptake in the Benguela Upwelling System through preformed nutrients. Nat. Comm. 14:2582. https://doi.org/10.1038/s41467-023-38208-y

Siegfried L, Schmidt M, Mohrholz V, Pogrzeba H, Nardini P, Böttinger M, Scheuermann G (2019) The tropical-subtropical coupling in the Southeast Atlantic from the perspective of the northern Benguela upwelling system. PLoS One 14:e0210083

Sink K, Attwood C, Lombard A, Grantham H, Leslie R, Samaai T, Kerwath S, Majiedt P, Fairweather T, Hutchings L (2011) Spatial planning to identify focus areas for offshore biodiversity protection in South Africa. Final Report for the Offshore Marine Protected Area Project Cape Town: South African National Biodiversity Institute

Sink K, Van der Bank M, Majiedt P, Harris L, Atkinson L, Kirkman S, Karenyi N (2019) South African National Biodiversity Assessment 2018 Technical Report, Vol 4: Marine Realm. South African National Biodiversity Institute, Pretoria

Skowno AL, Thompson MW, Hiestermann J, Ripley B, West AG, Bond WJ (2017) Woodland expansion in South African grassy biomes based on satellite observations (1990–2013): general patterns and potential drivers. Glob Chang Biol 23:2358–2369

Skowno A, Poole C, Raimondo D, Sink K, Van Deventer H, Van Niekerk L, Harris L, Smith-Adao L, Tolley K, Zengeya T, Foden W, Midgley G, Driver A (2019) National Biodiversity Assessment 2018: the status of South Africa's ecosystems and biodiversity. Synthesis Report South African National Biodiversity Institute, an entity of the Department of Environment, Forestry and Fisheries, Pretoria

Smit AJ, Bolton JJ, Anderson RJ (2017) Seaweeds in two oceans: beta-diversity. Front Mar Sci 4:404

Smith HE, Jones D, Vollmer F, Baumert S, Ryan CM, Woollen E et al. (2019) Urban energy transitions and rural income generation: sustainable opportunities for rural development through charcoal production. World Dev 113:237–245

Starke C, Ekau W, Moosdorf N (2019) Enhanced productivity and fish abundance at a submarine spring in a coastal lagoon on Tahiti, French Polynesia. Front Mar Sci 6:809. https://doi.org/10.3389/fmars.2019.00809

Staver AC, Archibald S, Levin SA (2011) The global extent and determinants of savanna and forest as alternative biome states. Science 334:230–232

Stevens N, Erasmus BFN, Archibald S, Bond WJ (2016) Woody encroachment over 70 years in South African savannahs: overgrazing, global change or extinction aftershock? Philos Trans R Soc B: Biol Sci 371:20150437

Stramma L, Johnson GC, Sprintall J, Mohrholz V (2008) Expanding oxygen-minimum zones in the tropical oceans. Science 320:655–658

Stuut J-BW, Crosta X, Van der Borg K, Schneider R (2004) Relationship between Antarctic sea ice and southwest African climate during the late quaternary. Geology 32:909–912

Sutton TT, Clark MR, Dunn DC, Halpin PN, Rogers AD, Guinotte J, Bograd SJ, Angel MV, Perez JAA, Wishner K, Haedrich RL, Lindsay DJ, Drazen JC, Vereshchaka A, Piatkowski U, Morato T, Blachowiak-Samolyk K, Robison BH, Gjerde KM, Pierrot-Bults A, Bernal P, Reygondeau G, Heino M (2017) A global biogeographic classification of the mesopelagic zone. Deep-Sea Res Pt I 126:85–102

Sweijd N, Smit A (2020) Trends in sea surface temperature and chlorophyll-a in the seven African Large Marine Ecosystems. Environ Dev 36:100585

Sydeman WJ, García-Reyes M, Schoeman DS, Rykaczewski RR, Thompson SA, Black BA, Bograd SJ (2014) Climate change and wind intensification in coastal upwelling ecosystems. Science 345:77–80

Teske PR, Von der Heyden S, McQuaid CD, Barker NP (2011) A review of marine phylogeography in southern Africa. S Afr J Sci 107:1–11

Teuber L, Schukat A, Hagen W, Auel H (2013) Distribution and ecophysiology of calanoid copepods in relation to the oxygen minimum zone in the eastern tropical Atlantic. PLoS One 8:e77590

Teuber L, Hagen W, Bode M, Auel H (2019) Who is who in the tropical Atlantic? Functional traits, ecophysiological adaptations and life strategies in tropical calanoid copepods. Prog Oceanogr 171:128–135

Tolley KA, Chase BM, Forest F (2008) Speciation and radiations track climate transitions since the Miocene climatic optimum: a case study of southern African chameleons. J Biogeogr 35:1402–1414

Tolley K, da Silva J, Jansen van Vuuren B (2019) South African National Biodiversity Assessment 2018 Technical Report Vol 7: Genetic Diversity. South African National Biodiversity Institute, Pretoria

Tyler T (2016) Examining the feeding ecology of two mesopelagic fishes (*Lampanyctodes hectoris* & *Maurolicus walvisensis*) off the West Coast of South Africa using stable isotope and stomach content analyses. MSc thesis, University of Cape Town, 67 pp

Utne-Palm AC, Salvanes AG, Currie B, Kaartvedt S, Nilsson GE, Braithwaite VA, Stecyk JA, Hundt M, van der Bank M, Flynn B (2010) Trophic structure and community stability in an overfished ecosystem. Science 329:333–336

van der Lingen CD (2021) Adapting to climate change in the South African small pelagic fishery. In: Bahri T, Vasconcellos M, Welch DJ, Johnson J, Perry RI, Ma X, Sharma R (eds) Adaptive management of fisheries in response to climate change. FAO Fisheries and Aquaculture Technical Paper No 667. FAO, Rome

van der Lingen CD, Hampton I (2018) Climate change impacts, vulnerabilities and adaptations: Southeast Atlantic and Southwest Indian Ocean marine fisheries. In: Barange M, Bahri T, Beveridge MCM, Cochrane KL, Funge-Smith S, Poulain F (eds) Impacts of climate change on fisheries and aquaculture: synthesis of current knowledge, adaptation and mitigation options, FAO Fisheries and Aquaculture Technical Paper No 627. FAO, Rome

van der Lingen CD, Miller TW (2014) Spatial, ontogenetic and interspecific variability in stable isotope ratios of nitrogen and carbon of *Merluccius capensis* and *Merluccius paradoxus* off South Africa. J Fish Biol 85:456–472

van der Lingen CD, Fréon P, Hutchings L, Roy C, Bailey GW, Bartholomae C, Cockcroft AC, Field JG, Peard KR, van der Plas AK (2006a) Forecasting shelf processes of relevance to living marine resources in the BCLME. In: Shannon V, Hempel G, Malanotte-Rizzoli P, Moloney C, Woods J (eds) Large Marine Ecosystems. Elsevier, Amsterdam van der Lingen CD, Hutchings L, Field JG (2006b) Comparative trophodynamics of anchovy *Engraulis encrasicolus* and sardine *Sardinops sagax* in the southern Benguela: are species alternations between small pelagic fish trophodynamically mediated? S Afr J Mar Sci 28:465–477

van der Lingen CD, Shannon LJ, Cury P, Kreiner A, Moloney CL, Roux J-P, Vaz-Velho F (2006c) Resource and ecosystem variability, including regime shifts, in the Benguela Current System. In: Shannon V, Hempel G, Malanotte-Rizzoli P, Moloney C, Woods J (eds) Benguela: predicting a Large Marine Ecosystem, vol 14. Elsevier, Amsterdam

van der Lingen CD, Hutchings L, Lamont T, Pitcher GC (2016) Climate change, dinoflagellate blooms and sardine in the southern Benguela Current Large Marine Ecosystem. Environ Dev 17:230–243

van der Sleen P, Zuidema PA, Morrongiello J, Ong JLJ, Rykaczewski RR, Sydeman WJ, Di Lorenzo E, Black BA (2022) Interannual temperature variability is a principal driver of low-frequency fluctuations in marine fish populations. Commun Biol 5:1–8

van Rensburg B, Chown S, Gaston K (2002) Species richness, environmental correlates, and spatial scale: a test using South African birds. Am Nat 159:566–577

van Waarden C (2002) The early iron age of Botswana. In: Oxford Research Encyclopedia of Anthropology. Oxford University Press, Oxford

van Wilgen B, Richardson D, Seydack A (1994) Managing Fynbos for biodiversity: constraints and options in a fire-prone environment. S Afr J Sci 90:322–329

van Wilgen BW, Forsyth GG, De Klerk H, Das S, Khuluse S, Schmitz P (2010) Fire management in Mediterranean-climate shrublands: a case study from the Cape Fynbos, South Africa. J Appl Ecol 47:631–638

Vansina J (1994) A slow revolution: farming in subequatorial Africa. Azania: Archaeol Res Africa 29:15–26

Verheye H, Hagen W, Auel H, Ekau W, Loick N, Rheenen I, Wencke P, Jones S (2005) Life strategies, energetics and growth characteristics of *Calanoides carinatus* (Copepoda) in the Angola-Benguela frontal region. Afr J Mar Sci 27:641–651

Verheye H, Hutchings L, Huggett J, Painting S (1992) Mesozooplankton dynamics in the Benguela ecosystem, with emphasis on the herbivorous copepods. In: Payne A, Brink K, Mann K, Hilborn R (eds) Benguela Trophic Functioning. S Afr J Mar Sci, vol 12. pp 561–584

Verheye H, Hutchings L, Peterson W (1991) Life history and population maintenance strategies of *Calanoides carinatus* (Copepoda: Calanoida) in the southern Benguela ecosystem. S Afr J Mar Sci 11:179–191

Verheye HM, Hutchings L, Huggett JA, Carter RA, Peterson WT, Painting SJ (1994) Community structure, distribution and trophic ecology of zooplankton on the Agulhas Bank with special reference to copepods. S Afr J Sci 90:154–165

Verheye HM, Lamont T, Huggett JA, Kreiner A, Hampton I (2016) Plankton productivity of the Benguela Current Large Marine Ecosystem (BCLME). Environ Dev 17:75–92

Verheye HM, Rogers C, Maritz B, Hashoongo V, Arendse LM, Gianakouras D, Giddey CJ, Herbert V, Jones S, Kemp AD, Ruby C (2001) Variability of zooplankton in the region of the Angola-Benguela front during winter 1999: BENEFIT Marine Science. S Afr J Sci 97:257–258

von Maltitz G, Knowles T, Wiese L, Scholes R, Pienaar M, Davies C, Steenkamp K (2020) National Terrestrial Carbon Sinks Assessment: Technical Report (2020). Technical Report CSIR. Department of Environment, Forestry and Fisheries, Pretoria, South Africa

von Maltitz GP, Lindeque GHL, Kellner K (2018) A changing narrative on desertification and degradation in South Africa. Desertification: past, current and future trends. Nova Science Publishers, Inc, New York

von Maltitz GP, Mucina L, Geldenhuys C, Lawes M, Eeley H, Adie H, Vink D, Fleming G, Bailey C (2003) Classification system for South African indigenous forests: an objective classification for the Department of Water Affairs and Forestry. Environmental Report ENV-P-C 2003-017. CSIR, Pretoria, South Africa

Vorrath M-E, Lahajnar N, Fischer G, Libuku VM, Schmidt M, Emeis K-C (2018) Spatiotemporal variation of vertical particle fluxes and modelled chlorophyll a standing stocks in the Benguela upwelling system. J Mar Syst 180:59–75

Wang D, Gouhier TC, Menge BA, Ganguly AR (2015) Intensification and spatial homogenization of coastal upwelling under climate change. Nature 518:390–394

Werner T, Buchholz F (2013) Diel vertical migration behaviour in euphausiids of the northern Benguela Current: seasonal adaptations to food availability and strong gradients of temperature and oxygen. J Plankton Res 35:792–812

White F (1983) The vegetation of Africa, a descriptive memoir to accompany the UNESCO/AETFAT/UNSO vegetation map of Africa. Nat Resour Res 20:1–356

Wilhelm MR, Kirchner CH, Roux JP, Jarre A, Iitembu JA, Kathena JN, Kainge P (2015) Biology and fisheries of the shallow-water hake (*Merluccius capensis*) and the deep-water hake (*Merluccius paradoxus*) in Namibia. In: Arancibia H (ed) Hakes: biology and exploitation. Wiley Blackwell, Oxford

Wilhelm MR, Black BA, Lamont T, Paulus SC, Bartholomae C, Louw DC (2020) Northern Benguela *Merluccius paradoxus* annual growth from otolith chronologies used for age verification and as indicators of fisheries-induced and environmental changes. Front Mar Sci 7:315

Woodd-Walker RS (2001) Spatial distributions of copepod genera along the Atlantic meridional transect. Hydrobiologia 453:161–170

Woodd-Walker RS, Ward P, Clarke A (2002) Large-scale patterns in diversity and community structure of surface water copepods from the Atlantic Ocean. Mar Ecol Prog Ser 236:189–203

Woollen E, Ryan CM, Baumert S, Vollmer F, Grundy I, Fisher J, Fernando J, Luz A, Ribeiro N, Lisboa SN (2016) Charcoal production in the mopane woodlands of Mozambique: what are the trade-offs with other ecosystem services? Philos Trans R Soc B: Biol Sci 371:20150315

Wren CD, Botha S, De Vynck J, Janssen MA, Hill K, Shook E, Harris JA, Wood BM, Venter J, Cowling R (2020) The foraging potential of the Holocene Cape South Coast of South Africa without the Palaeo-Agulhas plain. Quat Sci Rev 235:105789

Environmental Challenges to Meeting Sustainable Development Goals in Southern Africa

3

Shingirirai S. Mutanga, Felix Skhosana, Mohau Mateyisi, Humbelani Thenga, Sasha Naidoo, Trevor Lumsden, Abel Ramoelo, and Shingirai S. Nangombe

Abstract

There is an inextricable link between ecosystem integrity and the potential for achieving sustainable development goals (SDG). This chapter highlights key ecosystem threats and their drivers within the southern African regional context to emphasize the role of earth system science in supporting the achievement of regional sustainable development goals. It describes how some major anthropogenic threats have unfolded in terrestrial, aquatic and marine ecosystems of the region. Earth system science is increasingly contributing to understanding how globally driven climate and environmental changes threaten these ecosystems, and in turn how these impact people's livelihoods. Long-term changes in rainfall variability, concomitant disruption of hydrological balances, impacts on ocean chemistry, together with more immediate impacts on the frequency and magnitude of extreme climate events are some of the critical global change drivers. While terrestrial ecosystems are already faced with encroachment by novel species, characterized by the proliferation of both invasive alien and endemic woody species, freshwater and marine ecosystems appear more immediately threatened by more local impacts, such as the accumulation of contaminants. Overall, predicted climate and environmental changes are projected to hamper

S. S. Mutanga (✉) · F. Skhosana · M. Mateyisi · H. Thenga · S. Naidoo · T. Lumsden · S. S. Nangombe
Council for Scientific and Industrial Research (CSIR), Holistic Climate Change, Smart Places, South Africa
e-mail: SMutanga@csir.co.za

A. Ramoelo
Centre for Environmental Studies (CFES), Department of Geography, Geoinformatics and Meteorology, University of Pretoria, Pretoria, South Africa

G. P. von Maltitz et al. (eds.), *Sustainability of Southern African Ecosystems under Global Change*, Ecological Studies 248,
https://doi.org/10.1007/978-3-031-10948-5_3

development trajectories and poverty reduction efforts, and possibly exacerbate adverse impacts on human livelihoods.

3.1   Introduction

Southern Africa's terrestrial, freshwater and marine ecosystems are highly diverse, unique in their biodiversity as described in Chap. 2 and of great regional significance to human livelihoods locally and regionally (IPBES 2019). This region also plays a role in global environmental sustainability, for example, through feedbacks via the carbon cycle, and therefore has global socioeconomic importance (Darwall et al. 2009; Davis-Reddy et al. 2017). Building an understanding of the trends in impacts and their effects in this region is of importance from the local to global level. Fundamental and applied science in this region, as conducted under the SPACES program, addresses information needs to support key environmental policy efforts, and in particular, the Sustainable Development Goals. The protection of life under water and on land remains one of the key strategic imperatives for most countries in the southern African region as expressed by SDG 14 (Life below water) and SDG 15 (Life on land), respectively.

SDG 14 seeks to conserve and sustainably use the oceans, seas and marine resources for sustainable development. SDG 15 focuses on protecting, restoring and promoting sustainable use of terrestrial ecosystems, sustainably manage forests, combat desertification, halting and reversing land degradation and reducing bio-diversity loss. Despite the critical role ecosystems play to sustain livelihoods and life on earth, these systems are increasingly threatened by the growing human population through habitat destruction or degradation, overharvesting and pollution (i.e., air, water and land) (Darwall et al. 2009; Galvani et al. 2016). These threats are superimposed on climate change and the associated extreme events (Dudgeon et al. 2006; IPCC 2007a, 2007b). For terrestrial ecosystem, climate change-related hazards include droughts, floods, heat waves and wildfires, occurring at a global scale, while for marine ecosystem of concern is ocean chemistry modification and sea-level rise which comes at a risk of coastal ecological infrastructure erosion (Darwall et al. 2009; McBean and Ajibade 2009; Kusangaya et al. 2014; Galvani et al. 2016). These risks are highlighted in SDG 13, focused on action to combat climate change and its impacts as well as to build resilience in responding to climate-related hazards and natural disasters. In southern Africa, these three SDG efforts (13, 14 and 15) are thus linked and mutually supportive.

This chapter considers how selected threats to terrestrial, freshwater and marine ecosystems may impede progress in attaining targets associated with these three SDGs. The chapter provides illustrative cases on the local relevance of these SDGs in some of the southern African countries (land areas south of 17° S) and discusses these briefly using empirical evidence on how ecosystems have been and are being modified, and the drivers behind these trends.

The chapter proposes that climate change and anthropogenic activities are the major long-term drivers threatening ecosystem function and the persistence of biodiversity in this region. A vast body of literature has documented the impacts of climate change on people's livelihoods (IPCC 2012) and illustrates how human-induced changes are creating conditions for unsustainable rapid changes in ecosystems despite some evidence for natural adaptive responses via biological evolution. Adverse trends include the worldwide deterioration of biodiversity, ecosystem functioning and ecosystem services (IPBES 2019). It is anticipated that climate and environmental change will hamper poverty reduction, or even exacerbate poverty in some if not all of its dimensions (IPBES 2019).

3.2   Ecosystems and Sustainable Development Goals Nexus

The year 2015 witnessed the convergence of world leaders adopting 17 Sustainable Development Goals (SDGs) that aim to "free humanity from poverty, secure a healthy planet for future generations, and build peaceful, inclusive societies as a foundation for ensuring lives of dignity for all" (Fig. 3.1, UN 2016). These goals are supported by 169 targets with over 200 indicators. All SDGs interact with one another, but since they are not by design an integrated set of global priorities and objectives (e.g., Nilsson et al. 2016), their interactions are complex and not always mutually supportive (Fonseca et al. 2020). The International Council for Science (ICS) explored the nature of interlinkages between the SDGs determining to what extent they reinforce or conflict with each other and found that SDG 2 (zero hunger), SDG 3 (good health and wellbeing), SDG 7 (affordable and clean energy), SDG 14 (life below water) and SDG 15 (life on land) were found to be the most synergistic with other goals (Griggs et al. 2017). This chapter focuses on some examples of how the "earth system science-focused" SDG14 and 15, whose scope covers issues of water and land-based systems respectively, provide a context for fundamental and applied science addressing the main ecosystem types covered in this book and the approaches through which these may be productively interrogated, i.e., terrestrial, aquatic ecosystems and ocean ecosystems and their inherent services. Southern African terrestrial biota has been classified variously into biomes and their marine biota into ecologically similar regions (see Chap. 2). In this chapter, we provide indicative examples of specific threats in the context of these ecological units.

Southern African ecosystems are found predominantly across tropical and subtropical climates, but with significant regions that fall within temperate, arid and hyperarid climate zones and even including an appreciably sized Mediterranean-type climatic zone in its southwestern reaches (Chap. 2, Midgley and Bond 2015). The subcontinent and its adjacent marine waters host a diversity of ecosystems, most of which are still relatively intact with reference to their preanthropogenic biodiversity (Chap. 2, Scholes and Biggs 2005). The evolutionary legacy of this region remains largely preserved, with unique elements of fauna and flora relatively well represented in an extensive conservation-based spatial network (Pio et al. 2014). This globally valuable resource represents a complex history of evolution,

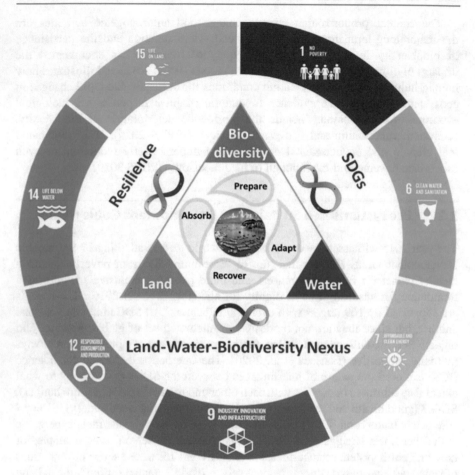

Fig. 3.1 The interaction between the SDGs, the land–water–biodiversity nexus and resilience (UN 2016)

for example, with arid and semiarid biomes retaining evidence of both ancient and recent diversification of desert groups (Klak et al. 2004), and fire-dependent C4 grasses that appear to have diversified extensively in association with changes in atmospheric CO_2 and climatic conditions (Spriggs et al. 2014). Freshwater biodiversity on land is also exceptional, with cichlids being an outstanding example (Zahradníčková et al. 2016). The southern African region is characterized by freshwater biomes including rivers such as the Zambezi, Orange River and Okavango Delta which coexist with azonal wetland ecosystems to support the associated biodiversity compositions. Along the coastal regions of southern Africa, marine biodiversity shows exceptional richness, also largely intact and with an expanding conservation effort showing some successes (Griffiths et al. 2000; Sowman et al. 2011). Biodiversity thus represents an extremely useful metric by which the successful achievement of goals in the nexus between SDGs 14 and 15, and sustainable development policy, can be assessed.

3.3   Drivers of Change, Typical Threats and SDG Implications

Southern African ecosystems are locally and regionally under threat from a range of climatic and nonclimatic drivers. While climate change trends and their impacts are already being detected in the region, local and direct impacts due to human use of ecosystems and biodiversity are well-known and extensive and represent critical short and medium-term pressures. These include land-use change due to deforestation and agricultural development, urbanization, environmental degradation (including freshwater quality and soil condition), illegal poaching, the informal and commercial overutilization of wild resources and invasions by alien species (IPCC 2021).

Terrestrial freshwater ecosystems are locally threatened by impacts such as canalization, inappropriate afforestation, deforestation and abstraction of water for agricultural, industrial and domestic activities (Meybeck 2003; Vörösmarty et al. 2004; Milly et al. 2008; Rahel and Olden 2008) that result in pollution and contamination of aquatic environments (Bashir et al. 2020). Marine systems are impacted largely by overharvesting and somewhat by pollution (Chaps. 2 and 8). Table 3.1 provides classic examples of ecosystem threats in the region.

Historical impacts over the past two centuries have mainly been due to land-use change and in the overutilization of ecosystems and wild species. Land-use change and climate change have been identified to be major drivers underlying the potential loss of biodiversity in southern Africa (MA 2005). While the region is one of the highly biodiverse regions of the world (WCMC 2000), biodiversity loss is a key concern given the impact on ecosystem functioning (Biggs et al. 2008). For example, Scholes and Biggs (2005) observed a 16% decline in wild species populations relative to the precolonial era, with a 95% confidence range of ±7%. These analyses averaged population sizes of the remaining plant and vertebrate groups in the major terrestrial biomes of southern Africa including the following countries (results brackets): Botswana (0.89), Lesotho (0.69), Mozambique (0.89), Namibia (0.91), South Africa (0.80), Swaziland (0.72) and Zimbabwe (0.76).

3.4   Natural Habitat Loss, Transformation and Degradation

3.4.1   Anthropogenic Land-Use Change

3.4.1.1 Transformation for Croplands and Commercial Timber Plantations

Population growth and economic growth are major drivers of land-use change, which in turn drive changes in the quantity and flow characteristics of water in lakes and river systems. In South Africa, a recent study showed 0.12% natural

Table 3.1 Select examples of studies of terrestrial and aquatic ecosystems threats within the Southern Africa context

Threats to ecosystems	Type of change to ecosystems	Ecosystem type	Country of the study	Case studies/Evidence
Climate change threats				
Droughts	Destruction of wetland ecosystem services. For example, the habitat and breeding areas of endangered crane bird species were perceived to be dwindling, affecting their reproduction.	Savanna	Zimbabwe	Severe 1968, 1973, 1982, 2004 1991/1992 season and 2012 seasons (Nangombe 2015)
Floods	More intense rainfall will cause soil capping, flash flooding, erosion and poor recharge.	Savanna Grasslands	Mozambique, Malawi, Zimbabwe	Cyclone Idai, 2019 and Tropical Storm Anna 2022.
Heat waves and Wildfires	Clearing of natural vegetation, destruction of natural habitats for wild species	Forests and Grasslands	South Africa	Western Cape, Central Karoo.
Terrestrial ecosystem threats: Anthropogenic land-use change				
Policy implementation	For example, Agrarian reform			
Deforestation	For example, impacts on biodiversity, impacts on ecosystem services, desertification	Forests and Woodlands	Southern Africa	In southern Africa, both the forests and woodlands cover 40% of the region and about 25,000 to 50,000 km^2 being cleared yearly at a rate exceeding regrowth (Scholes and Biggs 2004).
Bush encroachment	Increase of invasive or alien species outcompeting native plant species (e.g., *Seriphium plumosum* and *Acacia mearnsii*). Abandoned rainfed agriculture in rural areas are now hotspots for bush encroachment (e.g., Acacia species)	Grassland, Savanna	South Africa, Namibia	Invasive and alien plant species are reducing grazing resources in Southern Africa. While abandoned rainfed farms are hotspots for bush encroachment (Cho and Ramoelo 2019).

Alien invasive species on land	Abandoned land, degraded environment through overgrazing and high frequency, and inappropriate fire.	Grassland	Southern African savanna	Lantana camara invasion in Matabeleland South Province of Zimbabwe (Ncube et al. 2020).
Biodiversity loss on land	Clearing of natural vegetation for agriculture and human settlement development	All	Africa, SA	IPBES Africa Assessment report (Skowno et al. 2021)
Freshwater and marine ecosystem threats				
Flow alteration	Flow alteration in rivers and wetlands due to extraction of water, building of dams, interbasin transfers, urbanization, planting of higher water use crops (e.g., plantation forests, sugarcane), and climate change, resulting in changes in aquatic habitat availability.	Freshwater	Southern Africa	Impact of: • Extraction of water and building dams (Döll et al. 2009). • Snaddon et al. (1998). • Plantation forests in South Africa (Scott et al 2000). • Climate change across Southern Africa (Kusangaya et al. 2014; Banze et al. 2018).
Overharvesting of aquatic species	Decline in aquatic species.	Freshwater + Marine	South Africa	
Sea-level rise	Rising sea levels and (periodically) low river discharges are expected to increase freshwater salinity and soil salinity in coastal areas due to saltwater intrusion from the seaside.	Marine	Namibia	Boyer and Hampton (2001) observed a major decrease in Namibian sardine population and many other resources, which was associated with wide-scale advection of low-oxygen water into the northern Benguela from the Angola Dome in 1994, and the subsequent Benguela Niño of 1995.

(continued)

Table 3.1 (continued)

Threats to ecosystems	Type of change to ecosystems	Ecosystem type	Country of the study	Case studies/Evidence
Pollution	Pollution due to acid mine drainage, agricultural runoff, industrial effluent or poor management of sewage infrastructure.	Freshwater	South Africa	McCarthy (2011) described the impact of gold and coal mining on acid mine drainage in the Vaal River catchment. Oberholster et al. (2021) found that water quality in the Loskop Dam is degraded due to discharge from overloaded wastewater treatment plants and acid mine drainage, with levels of numerous chemical variables exceeding local and international guidelines NBA.
Biodiversity loss in aquatic systems	High variability of rainfall and droughts create diverse freshwater ecosystems. For example, inland wetlands are classified into 135 distinct types, while 222 distinct types were classified from rivers.	Freshwater	South Africa	

habitat were lost between 1990 and 2014 and accelerated to 0.24% between 2014 and 2018 (Skowno et al. 2021). The major drivers of natural habitat loss were settlement, agricultural and plantation forestry expansions (Skowno et al. 2021). The expansion of agriculture to feed a growing and developing population impacts water availability through the introduction of livestock, irrigation, with some rainfed crops and forest plantations having higher rates of transpiration and interception than the native vegetation (Jewitt 2006). In southern Africa, forests and woodlands cover 40% of the region and about 25,000 to 50,000 km^2 being cleared yearly at a rate exceeding regrowth (Scholes and Biggs 2004). Between 2001 and 2019, South Africa lost about 1.42 million hectares of tree cover, though most of this was from plantation forests, with only 11.9 Kha of indigenous forest loss. Forest loss in Mozambique was 3.8 Mha and 224 Kha in Zimbabwe over this same period (GFW 2022).

Wetlands are frequently drained to facilitate the planting of crops or other uses of land, thus significantly altering the hydrological characteristics of these unique environments (Vörösmarty et al. 2004; Darwall et al. 2009). The bidirectional interactions between the pursuit of food goals (SDG 2) and land and ecosystems (SDG 15) are nuanced and asymmetrical. For example, while land and its ecosystems clearly sustain food systems (with mostly positive interactions), food production often generates important land-related trade-offs which can have both positive and negative interactions (Pham-Truffert et al. 2020).

3.4.1.2 Transformation for Infrastructure (e.g., Hydropower, Dams and Urbanization)

The development of water resources infrastructure, such as dams, canals, wastewater treatment works and interbasin transfers, is necessary to sustain growing populations and economies. Perhaps, the greatest threat to imperiled freshwater species exists in the form of massive hydropower development projects currently underway in much of the developing world (Dudgeon 1999, 2000). This infrastructure results in the regulation of flows in streams and rivers, and the alteration of aquatic habitats (Meybeck 2003; Rolls et al. 2012). Environmental needs are often not prioritized, leading to overallocation of water resources (Poff et al. 2003). These problems are mitigated to some extent by the introduction of environmental flow requirements.

Urbanization results in the development of impervious areas and accompanying storm water infrastructure which rapidly discharges rainwater into streams and rivers, thus altering flow regimes (Walsh et al. 2005). Impervious areas and rapid discharge of water also hinder groundwater recharge, leading to less sustained base flows in river systems (Rolls et al. 2012). Overabstraction of groundwater, whether in urban areas or for agricultural use, also results in reduced base flows (Mccallum et al. 2013).

3.4.1.3 Policy Implementation Including Agrarian Reform

Land reform is a significant process throughout southern Africa that is unfolding rapidly with significant implications for food security (SDG6) but also manifold impacts relevant to SDG 15. The evolution of land reform programs and rangeland policy evolves over time in countries such as Zimbabwe and South Africa. Sibanda and Dube (2015) assessed the aftermath of land reform in Zimbabwe and found significant changes in land use and land cover between 2000 and 2010 with an increase in agricultural areas and a decrease in woodlands, specifically in newly resettled areas. In the same study, tree species diversity was found to be higher in unsettled areas relative to the post-redistribution resettled areas.

3.4.2 Woody Plant Proliferation

The direct effects of land-use change have also been coupled with climate and atmospheric CO_2 changes that appear to have increased rates of bush encroachment and the proliferation of invasive alien species (Chaps. 14, 15 and 16; Sibanda and Dube 2015). Over time, there has been an increase in woody plant cover across the terrestrial systems of Southern Africa causing a challenge in meeting the sustainable development goals. This proliferation in woody cover is both by bush encroachment and a rapid expansion of alien invasion plant species. The extent of woody proliferation in southern Africa is widespread as seen in Fig. 3.2, where Venter et al. (2018) showed a net greening in all southern African countries except for Madagascar where deforestation is extensive.

Driven by a combination of land use (overgrazing, fire suppression), climatic change regimes and rise in CO_2, bush encroachment is one of the most complex degradation phenomena in southern Africa (Bond et al. 2003; Kgope et al. 2010; Bond and Midgley 2012; Rohde and Hoffman 2012) and is discussed in greater detail in Chap. 15. It is defined as the directional increase in the cover of indigenous woody species in savanna and the invasion of the formerly grassland

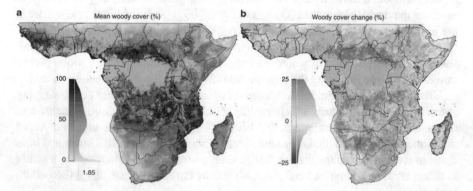

Fig. 3.2 (Venter et al. 2018): Woody plant cover dynamics over sub-Saharan Africa. Satellite observations of 30 years of fractional woody plant cover (**a**) reveal a dominant increasing trend (derived from the slope of the linear trend line between 1986 and 2016) (**b**). Histograms alongside color scales indicate data distributions. Gray areas were masked from the analysis and represent urban surfaces, wetland, cropland and forest (areas >40% cover by trees >5 m). Maps were constructed in Google Earth Engine (Figure and caption taken from Venter et al. (2018)

biome (O'Connor et al. 2014). The largely documented encroacher species in southern Africa are mostly nitrogen fixing legumes such as *Vachellia* and *Senegalia* (formerly *Acacia*) species, *Dichrostachys cinerea* and *Colophospermum mopane*, as well as nonlegumes (among others) such *as Terminalia sericea* and *Seriphium plumosum* (Stevens et al. 2017; Graham et al. 2020; Shikangalah and Mapani 2020; Lewis et al. 2021). Moreover, bush encroachment in South Africa and Namibia is estimated to be 10–20 and 26–30 million hectares, respectively (Bester 1999; Kraaij and Ward 2006; Daryanto et al. 2013; Eldridge et al. 2013). As much as bush encroachment comes with benefits such as the provisioning of woody fuels and woody material for multipurposes as well as the regulation of CO_2 through carbon sequestration, its impact on loss of biodiversity, water and grazing capacity leading to major reductions in meat, milk and other animal products is something to be not overlooked (Tallis and Kareiva 2007; Wigley et al. 2009; Trede and Patt 2015).

3.4.3   Alien Invasive Species

Alien invasive species (AIS) are any nonnative species introduced by humans into the new environment deliberately or nondeliberately. In this chapter, we focus on alien invasive plants AIPs such as aquatic weeds, arable weeds and woody weeds among other AIS such as viruses, fungi, insects and animals (Eschen et al. 2021). The main reasons for the introduction of invasive alien plants are to increase ecosystem services through rehabilitation, agroforestry and some as ornamental plants (e.g., *Lantana camara*) while others get introduced by accident (e.g., *Chromolaena odorata*) (Shackleton et al. 2019). *Prosopis*, one of the most widespread invasive species in southern Africa, was introduced into countries such as South Africa, Namibia and Botswana to provide ecosystem services including fuel wood and fodder. Similarly, the invasive Acacia species in Madagascar and South Africa have been introduced to provide timber, pulp for paper, bark for tannins and fuel wood (de Neergaard et al. 2005; Kull et al. 2007). *Opuntia ficus-indica* in South Africa is mainly used as fodder and food source (Henderson 2007; Shackleton et al. 2011).

As with many alien species around the world, these introduced IAPs become problematic by outcompeting the native species, spreading quickly (due to lack of biocontrol agents) and negatively affecting ecosystem services and livelihoods (van Wilgen et al. 2018; Shackleton et al. 2019). IAPs such as *Prosopis*, *Chromolaena* and *Lantana* have been documented to have severe impacts on the terrestrial system by reducing grazing capacity, biodiversity and moisture availability in these dry systems (Shackleton et al. 2014; Shackleton et al. 2019; Byabasaija et al. 2020; Kashe et al. 2020). *Prosopis*, which has long tap root that can utilize deep water sources, has spread rapidly in dry systems forming impenetrable thorny thickets that hinder maneuverability and injure animals (Hussain et al. 2020). IAPs in grasslands and Fynbos biomes have detrimental effects on stream flow due to high transpiration and hence reduce the country's mean annual runoff (Enright 2000). Le Maitre et al. (2016) estimated an annual water loss of 1444 million m^3 in South Africa largely by invasive *Acacias* (*Acacia mearnsii, A. dealbata* and *A. decurrens*) (34%) followed by *Pinus* species (19.3%) and *Eucalyptus* species (15.8%) among others.

Economical loses associated with invasive species in South Africa are estimated at US$1400 million in water resources, US$14 million y^{-1} for tourism and recreational and about US$ 52 ha^{-1} in pollination services (Pejchar and Mooney 2009; Nampala 2020). The increased biomass of the invasive species also increases the intensity of wildfires, which increases the damage from fire and the ability to control the fires. These negative effects, therefore, undermine the efforts to meet the Sustainable Development Goals (SDGs)-. Many programs have been put in place to eradicate invasive species before they become more widespread (Nampala 2020). Using biocontrol, such as *Dactylopius opuntiae* and *Opuntia ficus-indica,* in South Africa is now regarded as stable and desirable for food and fodder (Shackleton et al. 2011; Brown et al. 1985; Zimmermann and Moran 1991). The three phases of control that are effective are initial control (e.g., using biocontrol), follow-up (controlling seedlings, root suckers and regrowth) and maintenance (sustaining low and decreasing IAP numbers with annual control) (Martens et al. 2021).

3.5 Threats to Freshwater and Marine Ecosystem

Many aquatic species and ecosystems face an uncertain future. As human populations continue to expand their influence into the Earth's aquatic frontiers, conservation biologists are increasingly concerned about the implications for aquatic systems. In addition to the ongoing persistence of historically important threats facing aquatic environments, new stressors, including emerging diseases, the increasing taxonomic scope and ecological influence of invasive species, new industries and the accelerating trajectory of climate change, have the potential to radically alter the biological composition and ecological functioning of aquatic systems. More specifically, these alterations may manifest themselves in changes in aquatic biodiversity, individual life history patterns, communities, species distribution and range, and the extinction of vulnerable species (Dallas and Rivers-Moore 2014).

3.5.1 Overharvesting of Aquatic Species

Overharvesting of aquatic food sources is a major driver of the region's freshwater and marine ecosystem degradation, with the rich fish diversity of the region a significant target of increasingly effective extractive harvesting efforts mainly in the ocean. One of the most significant results has been the so-called "fishing-down" process illustrated by a historical decline in mean body size of the main harvested resources (Chaps. 11 and 25).

3.5.2   Coastal Impacts

Coastal regions will experience degradation from sea-level rise (SLR) combined with storm swells. Coral reefs will experience bleaching attributed to warmer oceans. Rising sea levels and (periodically) low river discharges are expected to increase soil salinity in coastal areas due to salt-water intrusion from the seaside. Boyer and Hampton (2001) observed a major decrease in Namibian sardine population and many other resources, which was associated with wide-scale advection of low-oxygen water into the northern Benguela from the Angola Dome in 1994 and the subsequent Benguela Niño of 1995. Some South African sea bird species have moved farther south over recent decades, partly due to climate change, though land-use change may also have contributed to this migration (Hockey and Midgley 2009; Hockey et al. 2011). It is considered that South African seabirds could be a valuable signal for climate change, particularly given the changes induced on prey species related to changes in physical oceanography. However deeper understanding of the ecology is needed so as to separate the influences of climate parameters from other environmental drivers (Crawford and Altwegg 2009).

3.5.3   Pollution

Aquatic ecosystems are the ultimate sinks for contaminants in the landscape (Bashir et al. 2020). Water pollution is the outcome of human activities such as urbanization, industrialization, mining and agriculture (Chap. 27). Excess pesticides and fertilizers, and sewage from residential and industrial areas, ultimately find their way to the aquatic environment, leading to eutrophication of freshwater ecosystems. Most of the great lakes of Southern Africa are in danger, with the extinction of as many as 200 fish species being recorded in Lake Victoria (Ryan 2020). The Vaal River catchment in South Africa is a hotspot for pollution due to acid mine drainage (McCarthy 2011), agricultural runoff and sewage leaks resulting from poor maintenance of infrastructure. Higher water temperatures, increased precipitation intensity and longer periods of low flows under climate change are expected to aggravate many forms of water pollution. These may include sediments, nutrients, dissolved organic carbon, pathogens, pesticides, salt and thermal pollution (Bates et al. 2008).

Although marine systems are generally larger and less constrained than riverine habitats, the industrialization of offshore waters through oil and gas platforms in some areas has been on a remarkable scale, while the increasing oil spills have posed a huge threat to ecosystems. A case in point is the devastation of the Mauritius oil spill with an estimated 1000 tons of oil leaking into the Indian Ocean, severely contaminating Mauritius's shoreline and lagoons (Shaama et al. 2020).

3.6 Climate Change, a Threat to Biodiversity and Ecosystem Functioning

Global climate change is likely to lead to significant changes across the southern African biomes, and seasonal rainfall will have implication on the ecosystems services they provide through the alteration of existing habitats, organism extinctions, water scarcity, and biodiversity and vegetation loss (Chaps. 13, 14, 15, 16, 17, 18, 19, 20, 21, 22 and 23). Southern Africa is anticipated to become hotter and dryer and discussed in detail in (Chaps. 6 and 7).

Extreme weather events together with anthropogenic activities might threaten the sustainability of Southern African ecosystem affecting species distribution through shifting habitat, changing the migration patterns, geographic range, emerging alien species and changing organisms' seasonal activity by altering life cycles of many terrestrial and marine species (Chaps. 14 and 26; UNEP 2012).

Projected increases in the likelihood of floods suggest a possibility of changes to the flow regimes in rivers (Dudgeon et al. 2006). Groundwater, which is critical to maintaining "low flows" and aquatic habitats during the drier periods, is likely to be impacted by changes in recharge rates due to increases in floods and droughts. Changes in flow regimes may affect channel geomorphology, longitudinal and lateral connectivity, aquatic habitat and biotic composition. Systematic quantification of loss and damage to ecological and coastal infrastructure around the riverbanks and coastal regions in Southern Africa could significantly inform estimation of adaptation needs and hence bolster attainment of SDG 15.

Regardless of which Shared Socioeconomic Pathway is used for predicting climate futures, current best estimates are that the southern Africa regions are likely to experience increased drought relative to 1850–1900 (IPCC 2021). Hydrological impacts from increased drought include reduced stream flow, resulting in degreased water storage in dams (Forzieri et al. 2014; Trambauer et al. 2014), long-term declines in rainfall (Rahman et al. 2015; Kruger and Nxumalo 2017), increased evaporation from water bodies and increased plant transpiration (Meybeck 2003). The reductions in stream flow and storage will be accompanied by climatically induced increases in the demand for water in environments such as agriculture (especially in terms of irrigation) and, to a lesser extent, power generation (Brown et al. 2013).

The Southern African region is considered to be water stressed, with South Africa and Namibia being the worst affected. In South Africa, more than half of the country's water management areas are in deficit (Alcamo and Henrichs 2002). Southern Africa has also been identified as a region characterized by a relatively high degree of flow alteration caused by the construction of dams (Döll et al. 2009). A review of studies on the impacts of climate change on stream flow found that reductions are projected for many basins in southern Africa including the Zambezi, Pungwe, Limpopo, Thukela, Okavango, Ruvhuma, Orange, Gwayi, Odzi and Sebakwe (Kusangaya et al. 2014). Given these patterns and the growing populations and economies in the region, the ability of aquatic ecosystems to provide ecosystem services is considered under threat. A challenge to the management of water in the region is the transboundary nature of many of the river basins, with 12 such basins existing across the Southern African Development Community countries (Kusangaya et al. 2014).

In the southern African region, drought occurred simultaneously with heat waves and the combined contributes to crop losses (Engelbrecht and Scholes 2021), stresses on regional water supplies and to widespread livestock mortality (Sivakumar 2007). In addition, reduced rainfall and increased drought frequency could result in a reduction in forage quality and quantity, hence affecting the dynamics and ecosystem function for wildlife and the vegetation (Nangombe 2015).

Heat waves are defined as warm extreme temperature events or excessively hot weather (Nairn and Fawcett 2013) that have socioeconomic and ecological impacts. Extreme temperatures are a threat to development in Southern Africa. The region experiences increased frequency of fires due to drastic increases in heat waves events (Engelbrecht et al. 2015; Garland et al. 2015; Mbokodo et al. 2020). Garland et al. (2015) reported warming over southern African region using the Conformal Cubic Atmospheric Model (CCAM) forced with the A2 emission scenario. Their model results suggested that extreme apparent temperature days in Africa are projected to increase in the future climate. This was in accord with Engelbrecht et al. (2015) who also projected substantial increases in the annual number of heat waves days over southern Africa. Moreover, a case study by Mbokodo et al. (2020) projected that there will be an increase in the number of hot extreme events in the most parts of the interior of South Africa throughout the year 2070–2099, while the number of cold events is decreasing. As a developing region (southern Africa), the substantial changes in the number of extreme temperature and heat wave events are a threat to a number of sectors including ecosystems, agriculture, water resources, energy demand and human health (Zuo et al. 2015).

For tropical and subtropical biomes, studies suggest that large landscapes in sub-Saharan Africa are prone to relatively fast shifts in vegetation structure and biodiversity due, in part, to shifts in fire regimes (Chap. 14, Lehmann et al. 2011; Bond and Midgley 2012; Moncrieff et al. 2014).

Foden et al. (2007) conducted a study on the distribution of *Aloe dichotoma* and observed changes in species distributions based on ~100-year (1904–2002) observational records. This study provides evidence that the range of a Namib Desert tree is shifting poleward, with extinction along trailing edge exceeding colonization along leading edge. Similar impacts are anticipated for countless other species.

Decision by African policy makers and stakeholders toward the attainment of SDG 15 goals must consider making urgent choices relating to trade-offs between biodiversity, carbon sequestration capacity of biomes and their direct ecosystems service. Such a decision could benefit from mechanistic studies that consider important plant functional types, herbivory and climate feedbacks (Huntley et al.

2014). Investment in early warning systems could help curb some of the major losses on ecosystems, especially around ecosystems that host critically endangered species.

3.7   An Analysis of SDGs and Ecosystem Threats

Changes in the biophysical environment, including droughts, floods, water quantity and quality, and degrading ecosystems, are expected to affect opportunities for people to generate income thus altering the synergistic nature of the SDGs. Interactions between targets for SDG 14 and SDG 15 with other SDGs show that generally, there are more synergies between goals and targets than there are trade-offs. Cumulative impacts from direct and indirect (via climate change) human pressures on marine and coastal ecosystems are potentially large and require concerted action in attaining both SDG 14 and SDG 13 (Griggs et al. 2017). The pursuit of food (SDG 2) and energy (SDG 7) goals can cause significant trade-offs with other SDGs, especially water (SDG 6) and ecosystems (SDG 15) (Pham-Truffert et al. 2020). Examples of interactions between SDG 14 and SDG 15 with other SDGs are outlined in Table 3.2.

3.7.1   Policy Implications for Ecosystem Protection and Restoration

Implementation of the adaptation component of the global climate policy is intractably linked to progress the SDG 14 and 15. This chapter posits that adaptation to the impacts of climate change-induced threats, in the least and developing countries in Southern Africa would be constrained without improved access to global climate finance by both private and public institution actors. Adaptation support in a form of technologies and means of implementation, of the global climate policy, would therefore accelerate the two SDGs through: fast tracking of ecosystem restoration and rehabilitation, e.g., nature-based solutions, intensify "working for" projects; protection of coastal settlements, ecological infrastructure and other uses of natural or seminatural ecosystems and landscapes for the delivery of ecosystem services; implementation of climate risk informed land-use planning. Reversal of some of the loss and damage and climate-proofing of infrastructure for all sectors of development (e.g., through improved design of dams, flood drainage and water reservoirs); and tailoring of climate services informed by researchers, service providers and fellow users' communities should not be negotiable.

Table 3.2 Examples of interactions between SDG 14 and SDG 15 with other SDGs

Interactions between SDG 14 and other SDGs	Interactions between SDG 15 and other SDGs
Sustainable management of fisheries in terms of supporting food security. Globally, fisheries play an important role in food security (SDG 2). SDG 14 includes the target to end overfishing and illegal, unreported and unregulated fishing and destructive fishing practices and implement science-based management plan to support restoring fish stocks in the shorted possible time and to produce the maximum sustainable yield as determined by their biological characteristics (WWF 2017).	Life on land (SDG 15) is impacted by the availability and quality of water (SDG 6), as such, SDG 15 sets a two-fold target of protecting inland freshwater ecosystems and the services they provide, and to reduce the impact of invasive alien species on water ecosystems (WWF 2017). While land and its ecosystem services clearly sustain food systems (mostly positive interactions), food production often generates important land-related trade-offs (both positive and negative interactions) (Pham-Truffert et al. 2020).
Water use (SDG 6) can impact the oceans, seas and marine resources referred to in Goal 14. Unregulated sewage disposals into these water bodies, as well as fossil-fuel mining and agricultural activities can have adverse impacts on the marine water resource, including the flora and fauna within it. Oceans and coastal ecosystems both affect and are affected by climate change, and this results in strong synergistic and bidirectional links between SDG 13 and SDG 14 (Griggs et al. 2017). An example of a synergy in achieving SDG 14 and SDG 13 is through conservation of coastal ecosystems acting as blue carbon sinks. A trade-off between SDG 13 and 14 is, for example, based on risks of coastal squeeze when trying to protect coasts from sea-level rise. Climate adaptation and coastal and marine protection measures need to be carefully managed to ensure that they do not conflict.	Agricultural intensification rarely leads to positive ecosystem impacts, for example, in some parts of sub-Saharan Africa, promoting food production can also constrain renewable-energy production (SDG 7) and terrestrial ecosystem protection (SDG 15) by competing for water and land (Nilsson et al. 2016). Agriculture's extensive land use also drives biodiversity loss (Lanz et al. 2018), as well as land degradation (Nowak and Schneider 2017). Conversely, limited land availability constrains agricultural production (Nilsson et al. 2016) An example of a positive interaction between SDG 7 and SDG 15 is that of renewable energy which can help decrease the role of firewood as an energy source in southern Africa, and so reduce the dangers of deforestation and help to protect habitats and ecosystems (WWF 2017)

3.8   Conclusion

There is an inextricable link between ecosystem function and the various SDG goals and targets. Invariably global climate change and the projected outlook threaten both the terrestrial and aquatic ecosystems impacting people's livelihoods. It is anticipated that climate and environmental change will hamper poverty reduction, or even exacerbate poverty in some or all of its dimensions. These changes, together with a shortage of adequate coping mechanisms and innovations to adapt to climate change, are to result in a surge in economic and social vulnerability of communities, particularly among the poor. While pollution could be regarded as a historical threat to aquatic biota, and thus outside the realm of a review focused on emerging threats, the scale of pollution impacts is accelerating in parallel with exponential human population growth and demographic population shifts to nearby surface water sources and coastal cities.

References

Alcamo J, Henrichs T (2002) Critical regions: A model-based estimation of world water resources sensitive to global changes. Aquat Sci 64:352–362. https://doi.org/10.1007/PL00012591

Banze F, Guo J, Xiaotao S (2018) Impact of climate change on precipitation in Zambeze river basin in Southern Africa Nature Environment and Pollution Technology An International Quarterly Scientific Journal Open Access 17, 1093–1103. Available at: www.neptjournal.com Accessed Feb 14, 2022

Bashir, I., Lone, F A, et al. (2020) "Concerns and threats of contamination on aquatic ecosystems," in Bioremediat Biotechnol: Sustainable Approaches to Pollution Degradation, pp. 1–26. Springer, Cham:https://doi.org/10.1007/978-3-030-35691-0_1

Bates BC et al (2008) Climate change and water. In: Technical paper of the Intergovernmental Panel on Climate Change. IPCC Secretariat, Geneva

Bester F (1999) Major problem, bush species and densities in Namibia. Agricola 10:1–3

Biggs R et al (2008) Scenarios of biodiversity loss in southern Africa in the 21st century. Glob Environ Chang 18(2):296–309. https://doi.org/10.1016/J.GLOENVCHA.2008.02.001

Bond WJ, Midgley GF (2012) Carbon dioxide and the uneasy interactions of trees and savannah grasses. Philos Trans R Soc B: Biol Sci 367(1588):601–612. https://doi.org/10.1098/RSTB.2011.0182

Bond WJ, Midgley GF, Woodward FI (2003) The importance of low atmospheric CO_2 and fire in promoting the spread of grasslands and savannas. Glob Chang Biol 9(7):973–982. https://doi.org/10.1046/j.1365-2486.2003.00577.x

Boyer DC, Hampton I (2001) An overview of the living marine resources of Namibia. Afr J Mar Sci 23:5–35

Brown C, Macdonald I, Brown S (1985) Invasive alien organisms in South West Africa/Namibia. Foundation for Research Development: CSIR, Pretoria, South Africa

Brown TC, Foti R, Ramirez JA (2013) Projected freshwater withdrawals in the United States under a changing climate. Water Resour Res 49(3):1259–1276. https://doi.org/10.1002/wrcr.20076

Byabasaija S et al (2020) Abundance, distribution and ecological impacts of invasive plant species in Maputo Special Reserve, Mozambique. Int J Biodivers Conserv 12(4):305–315. https://doi.org/10.5897/ijbc2020.1428

Cho MA, Ramoelo A (2019) Optimal dates for assessing long-term changes in tree-cover in the semi-arid biomes of South Africa using MODIS NDVI time series (2001–2018). Int J Appl Earth Obs Geoinf 81:27–36

Crawford RJM, Altwegg R (2009) Seabirds and climate change in southern Africa: some considerations. In Harebottle DM, et al. (eds) Proceedings of the 12th Pan-African Ornithological Congress. Animal Demography Unit, Cape Town, South Africa, pp. 1–5. Available at: http://opus.sanbi.org:80/jspui/handle/20.500.12143/3142 Accessed Sept 20, 2021

Dallas HF, Rivers-Moore N (2014) Ecological consequences of global climate change for freshwater ecosystems in South Africa. S Afr J Sci 110(5–6):1–11. https://doi.org/10.1590/sajs.2014/20130274

Darwall WRT et al (2009) The status and distribution of freshwater biodiversity in Southern Africa. IUCN and Grahamstown, South Africa, Gland

Daryanto S, Eldridge DJ, Wang L (2013) Spatial patterns of infiltration vary with disturbance in a shrub-encroached woodland. Geomorphology 194:57–64. https://doi.org/10.1016/J.GEOMORPH.2013.04.012

Davis-Reddy CL, Vincent K, Mambo J (2017) Socio-economic impacts of extreme weather events in Southern Africa. Climate risk and vulnerability: a handbook for southern Africa, pp 1–18. Available at: https://researchspace.csir.co.za/dspace/handle/10204/10148 Accessed Feb 14 2022

de Neergaard A et al (2005) Australian wattle species in the Drakensberg region of South Africa - an invasive alien or a natural resource? Agric Syst 85(3 SPEC. ISS):216–233. https://doi.org/10.1016/j.agsy.2005.06.009

Döll P, Fiedler K, Zhang J (2009) Global-scale analysis of river flow alterations due to water withdrawals and reservoirs. Hydrol Earth Syst Sci 13(12):2413–2432. https://doi.org/10.5194/HESS-13-2413-2009

Dudgeon D (1999) Tropical Asian streams: Zoobenthos, ecology and conservation. Hong Kong University Press, Hong Kong, p 830

Dudgeon D (2000) Conservation of freshwater biodiversity in oriental Asia: constraints, conflicts, and challenges to science and sustainability. Limnology 1(3):237–243. https://doi.org/10.1007/S102010070012

Dudgeon D, Arthington AH et al (2006) Freshwater biodiversity: importance, threats, status and conservation challenges. Biol Rev Camb Philos Soc 81(2):163–182. https://doi.org/10.1017/S1464793105006950

Eldridge DJ et al (2013) Impacts of shrub encroachment on ecosystem structure and functioning : towards a global synthesis. Ecol Lett 14(7):709–722. https://doi.org/10.1111/j.1461-0248.2011.01630.x.Impacts

Engelbrecht FA, Scholes RJ (2021) Test for Covid-19 seasonality and the risk of second waves. One Health 12:100202. https://doi.org/10.1016/J.ONEHLT.2020.100202

Engelbrecht F et al (2015) Projections of rapidly rising surface temperatures over Africa under low mitigation. Environ Res Lett 10(8):085004. https://doi.org/10.1088/1748-9326/10/8/085004

Enright WD (2000) The effect of terrestrial invasive alien plants on water scarcity in South Africa. Phys Chem Earth, Part B: Hydrol Oceans Atmos 25(3):237–242. https://doi.org/10.1016/S1464-1909(00)00010-1

Eschen R et al (2021) *Prosopis juliflora* management and grassland restoration in Baringo County, Kenya: Opportunities for soil carbon sequestration and local livelihoods. J Appl Ecol 58(6):1302–1313. https://doi.org/10.1111/1365-2664.13854

Foden W et al (2007) A changing climate is eroding the geographical range of the Namib Desert tree Aloe through population declines and dispersal lags. Divers Distrib 13:645–653. https://doi.org/10.1111/J.1472-4642.2007.00391.X

Fonseca LM, Domingues JP, Dima AM (2020) Mapping the sustainable development goals relationships. Sustainability 12(8):3359

Forzieri G et al (2014) Ensemble projections of future streamflow droughts in Europe. Hydrol Earth Syst Sci 18(1):85–108. https://doi.org/10.5194/hess-18-85-2014

Galvani AP et al (2016) Human–environment interactions in population and ecosystem health. Proc Natl Acad Sci 113(51):14502–14506. https://doi.org/10.1073/PNAS.1618138113

Garland RM et al (2015) Regional projections of extreme apparent temperature days in Africa and the related potential risk to human health. Int J Environ Res Public Health 12(10):12577–12604. https://doi.org/10.3390/IJERPH121012577

GFW (2022) South Africa Deforestation Rates & Statistics | GFW. Available at: https://www.globalforestwatch.org Accessed Feb 14, 2022

Graham SC, Barrett AS, Brown LR (2020) Impact of *Seriphium plumosum* densification on Mesic Highveld Grassland biodiversity in South Africa. R Soc Open Sci 7:192025. https://doi.org/10.1098/rsos.192025

Griffiths CL et al (2000) Functional ecosystems: rocky shores. In: Durham BD, Pauw JC (eds) Summary marine biodiversity status report for South Africa. National Research Foundation, Pretoria

Griggs DJ, et al. (2017) A guide to SDG interactions: from science to implementation. Paris: International Council for Science. Available at: http://pure.iiasa.ac.at/id/eprint/14591/ Accessed September 30, 2021

Henderson L (2007) Invasive, naturalized and casual alien plants in southern Africa: A summary based on the Southern African Plant Invaders Atlas (SAPIA). Bothalia 37(2):215–248. https://doi.org/10.4102/abc.v37i2.322

Hockey PAR, Midgley GF (2009) Avian range changes and climate change: a cautionary tale from the Cape Peninsula. Ostrich 80(1):29–34. https://doi.org/10.2989/OSTRICH.2009.80.1.4.762

Hockey PAR et al (2011) Interrogating recent range changes in South African birds: confounding signals from land use and climate change present a challenge for attribution. Divers Distrib 17(2):254–261. https://doi.org/10.1111/J.1472-4642.2010.00741.X

Huntley B et al (2014) Suborbital climatic variability and centres of biological diversity in the Cape region of southern Africa. J Biogeogr 41(7):1338–1351. https://doi.org/10.1111/JBI.12288

Hussain MI et al (2020) Invasive Mesquite (*Prosopis juliflora*), an allergy and health challenge. Plants 9(2):141. https://doi.org/10.3390/PLANTS9020141

IPBES (2019) Summary for policymakers of the global assessment report on biodiversity and ecosystem services of the Intergovernmental Science-Policy Platform on Biodiversity and Ecosystem Services. Available at: www.ipbes.net. Accessed: February 9, 2022

IPCC (2007a) In: Miller HL, Solomon SD, Qin M, Manning Z, Marquis CM, Averyt KB (eds) Climate change 2007: the physical science basis. Contribution of Working Group I to the fourth assessment report of the intergovernmental panel on climate change. Cambridge University Press, Cambridge and New York, NY

IPCC (2007b) Summary for policymakers. In: Miller HL, Solomon SD, Qin M, Manning Z, Marquis CM, Averyt KB (eds) Climate Change 2007: The Physical Science Basis. Contribution of Working Group I to the Fourth Assessment Report of the Intergovernmental Panel on Climate Change. Cambridge University Press, Cambridge and New York, NY

IPCC (2012) Managing the risks of extreme events and disasters to advance climate change adaptation. — European Environment Agency (no date). Available at: https://www.eea.europa.eu/data-and-maps/indicators/direct-losses-from-weather-disasters-1/ipcc-2012-managing-the-risks. Accessed Sept 30, 2021

IPCC (2021) Summary for policymakers. In: Masson-Delmotte V, Zhai P, Pirani A, Connors SL, Péan C, Berger S, Caud N, Chen Y, Goldfarb L, Gomis MI, Huang M, Leitzell K, Lonnoy E, Matthews JBR, Maycock TK, Waterfield T, Yelekçi O, Yu R, Zhou B (eds) Climate Change 2021: the physical science basis. Contribution of Working Group I to the sixth assessment report of the intergovernmental panel on climate change. Cambridge University Press. In Press. Available at: https://www.ipcc.ch/report/sixth-assessment-report-working-group-i/. Accessed Sept 20, 2021

Jewitt G (2006) Integrating blue and green water flows for water resources management and planning. Phys Chem Earth 31(15–16):753–762. https://doi.org/10.1016/j.pce.2006.08.033

Kashe K et al (2020) Potential impact of alien invasive plant species on ecosystem services in Botswana: A review on *Prosopis juliflora* and *Salvinia molesta*. In: Sustainability in developing countries. Springer, Cham, pp 11–31. https://doi.org/10.1007/978-3-030-48351-7_2

Kgope BS, Bond WJ, Midgley GF (2010) Growth responses of African savanna trees implicate atmospheric [CO2] as a driver of past and current changes in savanna tree cover. Austral Ecol 35(4):451–463. https://doi.org/10.1111/j.1442-9993.2009.02046.x

Klak C, Reeves G, Hedderson T (2004) Unmatched tempo of evolution in Southern African semi-desert ice plants. Nature 427(6969):63–65. https://doi.org/10.1038/nature02243

Kraaij T, Ward D (2006) Effects of rain, nitrogen, fire and grazing on tree recruitment and early survival in bush-encroached savanna, South Africa. Plant Ecol 186(2):235–246. https://doi.org/10.1007/s11258-006-9125-4

Kruger AC, Nxumalo MP (2017) Historical rainfall trends in South Africa: 1921–2015. Water SA 43(2):285–297. https://doi.org/10.4314/wsa.v43i2.12

Kull C et al (2007) Multifunctional, scrubby, and invasive forests? Mt Res Dev 27(3):224–231

Kusangaya S, Warburton ML et al (2014) Impacts of climate change on water resources in southern Africa: a review. Phys Chem Earth, Parts A/B/C 67–69:47–54. https://doi.org/10.1016/J.PCE.2013.09.014

Lanz B, Dietz S, Swanson T (2018) The expansion of modern agriculture and global biodiversity decline: an integrated assessment. Ecol Econ 144:260–277. https://doi.org/10.1016/J.ECOLECON.2017.07.018

Le Maitre DC et al (2016) Estimates of the impacts of invasive alien plants on water flows in South Africa. Water SA 42(4):659–672. https://doi.org/10.4314/wsa.v42i4.17

Lehmann CER et al (2011) Deciphering the distribution of the savanna biome. New Phytol 191(1):197–209. https://doi.org/10.1111/j.1469-8137.2011.03689.x

Lewis JR, Verboom GA, February EC (2021) Coexistence and bush encroachment in African savannas: the role of the regeneration niche. Funct Ecol 35(3):764–773. https://doi.org/10.1111/1365-2435.13759

MA (2005) Millenium ecosystem assessment synthesis report: ecosystems and human well-being

Martens C, et al. (2021) A practical guide to managing invasive alien plants: a concise handbook for land users in the Cape Floral Region South Africa. Capetown, South Africa. Available at: www.wwf.org.za/report/. Accessed Feb 14 2022

Mbokodo I et al (2020) Heatwaves in the future warmer climate of South Africa. Atmosphere 11(7):712. https://doi.org/10.3390/ATMOS11070712

McBean G, Ajibade I (2009) Climate change, related hazards and human settlements. Curr Opin Environ Sustain 1(2):179–186. https://doi.org/10.1016/j.cosust.2009.10.006

Mccallum AM et al (2013) River-aquifer interactions in a semi-arid environment stressed by groundwater abstraction. Hydrol Process 27(7):1072–1085. https://doi.org/10.1002/hyp.9229

McCarthy TS (2011) The impact of acid mine drainage in South Africa. S Afr J Sci 107(5/6):1–7. https://doi.org/10.4102/sajs.v107i5/6.712

Meybeck M (2003) Global analysis of river systems: from Earth system controls to Anthropocene syndromes. Philos Trans R Soc Lond B Biol Sci Philos T R Soc B 358(1440):1935–1955. https://doi.org/10.1098/rstb.2003.1379

Midgley GF, Bond WJ (2015) Future of African terrestrial biodiversity and ecosystems under anthropogenic climate change. Nat Clim Change 5(9):823–829. https://doi.org/10.1038/nclimate2753

Milly PCD et al (2008) Climate change: Stationarity is dead: Whither water management? Science 319(5863):573–574. https://doi.org/10.1126/science.1151915

Moncrieff GR et al (2014) Increasing atmospheric CO2 overrides the historical legacy of multiple stable biome states in Africa. New Phytol 201(3):908–915. https://doi.org/10.1111/NPH.12551

Nairn J, Fawcett R (2013) Defining heatwaves: heatwave defined as a heat-impact event servicing all community and business sectors in Australia

Nampala P (2020) Strategy for Managing Invasive Species in Africa 2021–2030

Nangombe SS (2015) Drought conditions and management strategies in Zimbabwe. In: Proceedings of the Regional Workshops on Capacity Development to Support National Drought Management Policies for Eastern and Southern Africa and the Near East and North Africa Regions, Addis Ababa, Ethiopia, pp. 5–8

Ncube B et al (2020) Spatial modelling the effects of climate change on the distribution of *Lantana camara* in Southern Zimbabwe. Appl Geogr 117:102172

Nilsson M, Griggs D, Visbeck M (2016) Policy: map the interactions between sustainable development goals. Nature 534:320–322. https://doi.org/10.1038/534320a

Nowak A, Schneider C (2017) Environmental characteristics, agricultural land use, and vulnerability to degradation in Malopolska Province (Poland). Sci Total Environ 590–591:620–632. https://doi.org/10.1016/J.SCITOTENV.2017.03.006

O'Connor TG, Puttick JR, Hoffman MT (2014) Bush encroachment in southern Africa: changes and causes. Afr J Range Forage Sci 31(2):67–88. https://doi.org/10.2989/10220119.2014.939996

Oberholster PF et al (2021) Assessing the adverse effects of a mixture of AMD and sewage effluent on a sub-tropical dam situated in a nature conservation area using a modified pollution index. Int J Environ Res 15(2):321–333. https://doi.org/10.1007/S41742-021-00315-3/TABLES/6

Pejchar L, Mooney HA (2009) Invasive species, ecosystem services and human well-being. Trends Ecol Evol 24(9):497–504. https://doi.org/10.1016/J.TREE.2009.03.016

Pham-Truffert M et al (2020) Interactions among Sustainable Development Goals: Knowledge for identifying multipliers and virtuous cycles. Sustain Dev 28(5):1236–1250. https://doi.org/10.1002/SD.2073

Pio DV et al (2014) Climate change effects on animal and plant phylogenetic diversity in southern Africa. Glob Chang Biol 20(5):1538–1549

Poff NL et al (2003) River flows and water wars: Emerging science for environmental decision making. Front Ecol Environ 1(6):298–306. https://doi.org/10.1890/1540-9295(2003)001[0298:RFAWWE]2.0.CO;2

Rahel FJ, Olden JD (2008) Assessing the effects of climate change on aquatic invasive species. Conserv Biol 22(3):521–533. https://doi.org/10.1111/j.1523-1739.2008.00950.x

Rahman K et al (2015) Declining rainfall and regional variability changes in Jordan. Water Resour Res 51:3828–3835. https://doi.org/10.1002/2015WR017153

Rohde RF, Hoffman TM (2012) The historical ecology of Namibian rangelands: vegetation change since 1876 in response to local and global drivers. Sci Total Environ 416:276–288. https://doi.org/10.1016/j.scitotenv.2011.10.067

Rolls RJ, Leigh C, Sheldon F (2012) Mechanistic effects of low-flow hydrology on riverine ecosystems: Ecological principles and consequences of alteration. Freshw Sci 31(4):1163–1186. https://doi.org/10.1899/12-002.1

Ryan JC (2020) Africa's Great Lakes in peril. World Watch; (United States) [Preprint]

Scholes RJ, Biggs R (2004) Ecosystem services in Southern Africa: a contribution to the Millennium Ecosystem Assessment, prepared by the regional-scale team of the Southern African Millennium Ecosystem Assessment, Pretoria, South Africa

Scholes RJ, Biggs R (2005) A biodiversity intactness index. Nature 434(7029):45–49. https://doi.org/10.1038/nature03289

Scott DF et al (2000) A re-analysis of the south African catchment afforestation experimental data. Water Research Commission

Shaama S et al (2020) Variations in the density and diversity of micro-phytoplankton and micro-zooplankton in summer months at two coral reef sites around Mauritius Island. J Sustain Sci Manag 15(4):18–33. https://doi.org/10.46754/jssm.2020.06.003

Shackleton S, Kirby D, Gambiza J (2011) Invasive plants – friends or foes? Contribution of prickly pear (Opuntia ficus-indica) to livelihoods in Makana Municipality, Eastern Cape, South Africa. Dev South Afr 28(2):177–193. https://doi.org/10.1080/0376835X.2011.570065

Shackleton RT et al (2014) Prosopis: a global assessment of the biogeography, benefits, impacts and management of one of the world's worst woody invasive plant taxa. AoB PLANTS 6. https://doi.org/10.1093/aobpla/plu027

Shackleton RT, Shackleton CM, Kull CA (2019) The role of invasive alien species in shaping local livelihoods and human well-being: a review. J Environ Manag 229:145–157. https://doi.org/10.1016/J.JENVMAN.2018.05.007

Shikangalah RN, Mapani BS (2020) A review of bush encroachment in Namibia: from a problem to an opportunity? J Rangel Sci 10(3)

Sibanda M, Dube T (2015) Assessing the aftermath of the fast track land reform programme in Zimbabwe on land-use and land-cover changes. Trans R Soc S Afr 70(2):181–186. https://doi.org/10.1080/0035919X.2015.1017865

Sivakumar MVK (2007) Interactions between climate and desertification. Agric For Meteorol 142(2–4):143–155. https://doi.org/10.1016/j.agrformet.2006.03.025

Skowno AL, Jewitt D, Slingsby JA (2021) Rates and patterns of habitat loss across South Africa's vegetation biomes. S Afr J Sci 117(1–2):1–5

Snaddon CD, Wishart MJ, Davies BR (1998) Some implications of inter-basin water transfers for river ecosystem functioning and water resources management in southern Africa. Aquat Ecosyst Health Manag 1(2):159–182. https://doi.org/10.1016/S1463-4988(98)00021-9

Sowman M et al (2011) Marine protected area management in South Africa: new policies, old paradigms. Environ Manag 47:573–583. https://doi.org/10.1007/s00267-010-9499-x

Spriggs EL, Christinb PA, Edwards EJ (2014) C4 photosynthesis promoted species diversification during the miocene grassland expansion. PLoS One 9(5):e97722. https://doi.org/10.1371/JOURNAL.PONE.0097722

Stevens N et al (2017) Savanna woody encroachment is widespread across three continents. Glob Chang Biol 23(1):235. https://doi.org/10.1111/gcb.13409

Tallis H, Kareiva P (2007) Ecosystem services. Curr Biol 15:R746–R748

Trambauer P et al (2014) Identification and simulation of space-time variability of past hydrological drought events in the Limpopo River basin, southern Africa. Hydrol Earth Syst Sci 18(8):2925–2942. https://doi.org/10.5194/hess-18-2925-2014

Trede R, Patt R (2015) Value added end-use opportunities for Namibian encroacher bush. Windhoek, Namibia

UN (2016) Transforming our world: The 2030 agenda for sustainable development. https://doi.org/10.1201/b20466-7

UNEP (2012) Climate Change Challenges for Africa: Evidence from selected Eu-Funded Research Projects. Nairobi, Kenya. https://doi.org/10.1016/B978-0-12-809665-9.09754-8

van Wilgen BW, et al (2018) The status of biological invasions and their management in South Africa in 2017. Stellenbosch

Venter Z, Cramer MD, Hawkins H (2018) Drivers of woody plant encroachment over Africa. Nat Commun 2018:1–7. https://doi.org/10.1038/s41467-018-04616-8

Vörösmarty C et al (2004) Humans transforming the global water system. Eos 85(48):1–6. https://doi.org/10.1029/2004EO480001

Walsh CJ et al (2005) The urban stream syndrome: current knowledge and the search for a cure. J N Am Benthol Soc 24(3):706–723. https://doi.org/10.1899/04-028.1

WCMC (2000) Earth's living resources in the 21st century. In: Goombridge B, Jenkins MD (eds) Global biodiversity: earth's living resources in the 21st century. World Conservation Press, Cambridge

Wigley BJ, Bond WJ, Hoffman MT (2009) Bush encroachment under three contrasting land use practices in a mesic South African savanna. Afr J Ecol 47(Suppl. 1):62–70. https://doi.org/10.1111/j.1365-2028.2008.01051.x

WWF (2017) The food-energy-water nexus as a lens for delivering the UN's Sustainable Development Goals in southern Africa. "Edited by T. von Bormann, M. Berchner, and G. Manisha. Cape Town, South Africa: WWF-SA (World Wide Fund for Nature). Available at: www.wwf.org.za/report/few_lens_for_SDGs

Zahradníčková P et al (2016) Species of Gyrodactylus von Nordmann, 1832 (Platyhelminthes: Monogenea) from cichlids from Zambezi and Limpopo river basins in Zimbabwe and South Africa: evidence for unexplored species richness. Syst Parasitol 93(7):679–700

Zimmermann HG, Moran VC (1991) Biological control of prickly pear, Opuntia ficusindica (Cactaceae), in South Africa. Agric Ecosyst Environ 37(1–3):29–35. https://doi.org/10.1016/0167-8809(91)90137-M

Zuo J et al (2015) Impacts of heat waves and corresponding measures: a review. J Clean Prod 92:1–12. https://doi.org/10.1016/J.JCLEPRO.2014.12.078

Overview of the Macroeconomic Drivers of the Region

4

Johann F. Kirsten, Bernhard Dalheimer, and Bernhard Brümmer

Abstract

The ecosystems in Southern Africa are impacted by economic activity and population growth and pressure. There are several macro-economic drivers shaping these economic and population pressures and it is for this reason that this chapter unpacks the macro-economic drivers in the region. With the economy of South Africa dominating the regional economy (90% of Gross value added) it makes sense to discuss to the macroeconomic situation in Southern Africa by referring to policy and macro indicators in South Africa as a proxy of the regional situation. We also focus on the Limpopo province which shares boundaries and an ecosystem with three other countries in Southern Africa. Starting from the general macropolicy situation, major macro indicators for the region, the country and Limpopo are presented, jointly with the specific challenges, regulatory frameworks and policies that govern the development processes in the region. We focus on environmental, agricultural and trade policy measures, including their interlinkages, and illustrate that they provide a volatile and uncertain environment for structural development of the agricultural sector.

J. F. Kirsten (✉)
Bureau for Economic Research, Stellenbosch University, Stellenbosch, South Africa
e-mail: jkirsten@sun.ac.za

B. Dalheimer · B. Brümmer
Department of Agricultural Economics and Rural Development & Centre for Biodiversity and Sustainable Land Use, University of Göttingen, Göttingen, Germany
e-mail: bernhard.dalheimer@uni-goettingen.de

© The Author(s) 2024
G. P. von Maltitz et al. (eds.), *Sustainability of Southern African Ecosystems under Global Change*, Ecological Studies 248,
https://doi.org/10.1007/978-3-031-10948-5_4

4.1   Introduction

The economy of the Southern Africa is dominated by the sheer size, diversity, and magnitude of the South African economy. South Africa is responsible for 90% of all gross value added in the region and its markets, policies, infrastructure have an important impact and effect on all the other economies and people of the region. The study region of Limpopo is imposed in the region of Southern Africa, which is a point in case of the heterogeneous economic conditions that can be found in Africa. Most predominantly, the stark contrast of the value addition in South Africa compared with those of neighboring countries, but also the different importance of sectors in other economies of the region is striking. However, these macroeconomic conditions critically determine economic, social and environmental value creation in the study region of Limpopo. Moreover, as Limpopo borders with Botswana, Zimbabwe and Mozambique, the macroeconomic developments of these regions—as much as potential differentials therein—are important to consider.

This chapter sets the stage by providing an overview of the macroeconomic drivers in the region using an analysis of South Africa indicators and trends as the reference point. We focus on environmental, agricultural and trade policy measures, including their interlinkages, and illustrate that they to provide a volatile and uncertain environment for structural development of the agricultural sector. Given the geographical location of the Limpopo region, the chapter adopts a hierarchical structure and provides insights at the supranational, the national and the regional level. In the rest of the chapter, we describe the macroeconomic trends starting from the broader region and subsequently narrowing down to the national and finally the regional level. Similarly, we provide key agricultural and trade policy insights again on these three levels. Finally, we conclude the chapter with recommendations for policy reforms.

4.2   Macroeconomic Trends in Southern Africa

Southern Africa[1] is responsible for about one third of Africa's GDP. Average per capita incomes are higher than the average per capita income of Africa. In the more recent past, growth in the region was slower than in most other parts of

[1] Southern Africa may refer to three different regional entities, each with a different composition of countries. First, in a geographical sense, it comprises countries that are south of the Congo river basin, but usually excluding the Democratic Republic of the Congo (DRC). Second, in a political sense, Southern Africa is often used to refer to members of the Southern African Development Community (SADC)—the regional economic and trade community—headquartered in Gaborone, Botswana. Third, Southern Africa also groups the countries of Botswana, Eswatini, Lesotho, Namibia and South Africa into a subgroup of the United nations (UN) geoscheme for Africa. In this chapter, we adopt a slightly modified denition of Southern Africa where we extend the UN subgroup with Zimbabwe and Mozambique to more adequately reflect the surrounding economies of our focus region, the province of Limpopo in the Republic of South Africa (RSA).

Africa, due to the already relatively large economies compared with others on the continent. In further comparison with the continent at large, the economies of southern Africa are characterized by homogeneously strong mining sectors and comparably large contributions from the secondary and tertiary sectors. Table 4.1 shows the relative contributions to total value added by sector in Southern Africa and selected southern African countries. With the exception of Eswatini, mining and utilities are responsible for between 10% and 15%, and in all countries, the secondary and tertiary sectors mostly make up for more than 70% of GDP. The importance of agriculture to national income on the other hand is the most diverse indicator, ranging from 2.1% in South Africa to almost one third in Mozambique (UNCTADstat 2022).

In the past, the development of manufacturing, predominantly in South Africa and Zimbabwe, benefited from larger investments in infrastructure, education and healthcare and led to improvements in related development indicators. Both regions were classified as lower-middle income countries since the 1980s. The region as a whole developed better than other countries in other parts of Africa in terms of income and other developmental outcomes. However, since the 1990s, increased competition with East Asia and South East Asia led to staggering manufacturing sectors and even deindustrialization in Zimbabwe. Today, poverty, inequality, corruption, HIV/AIDS and skilled workers emigration are the predominant impediments to GDP growth in the region (Nshimbi and Fioramonti 2014; Moyo et al. 2014).

As evidenced in Table 4.1, South Africa is by far the most important economy in the region with about 90% of the GDP of the political Southern Africa. Also compared with other countries in the region its GDP exceeds others by a factor of 10 or more in some cases. Given the predominant economic role of South Africa in the region, as well as the fact that Limpopo is a region within South Africa, the following section focuses on the macroeconomic situation in RSA in more detail.

4.2.1   The Situation in South Africa

Statistics South Africa (StatsSA) has in August 2021 released a new set of GDP numbers for South Africa. Real gross domestic product (GDP) is now measured at constant 2015 prices instead of 2010 prices. The revised estimate of GDP in 2020 is R5 521 billion, an increase of 11% compared with the previous estimate of R4 973 billion. The annual growth rate for 2020 was revised from −7.0% to −6.4%. With this GDP revised numbers, the South African GDP per capita is now R79 913 per person.

Based on the latest estimates by StatsSA, shown in Table 4.1, the tertiary sector makes up 65% of the South African economy—while the primary sector—mining and agriculture—which is dominant in the Limpopo region, is responsible for only 8.9% of gross value added in the South African economy.

The biggest concern for the South African economy is the alarming unemployment statistics. According to the most recent Quarterly Labour Force Survey

Table 4.1 Composition of GDP in Southern Africa and selected southern African economies (2019)

Sector	Botswana	Eswatini	Mozambique	Namibia	South Africa	Zimbabwe	Southern Africa
Gross value added	16.6	4.3	13.6	11.4	313.1	19.4	347.6
Agriculture, hunting, forestry, fishing	2.2	9.1	29.3	7.1	2.1	9.0	2.4
Construction	7.5	3.4	1.4	2.3	3.8	2.5	3.9
Mining and utilities	18.0	1.3	14.1	13.7	12.2	8.9	12.3
Manufacturing	5.8	30.6	10.2	12.7	13.2	12.3	13.1
Transport, storage and communications	6.8	4.8	9.9	4.8	9.8	11.4	9.4
Wholesale, retail trade, restaurants and hotels	21.43	15.6	12.4	13.4	15.1	22.5	15.4
Other services	38.3	35.1	22.7	45.9	43.8	33.4	43.5

Source: UNCTADstat (2022)

(QLFS), 7.8 million people were unemployed in the second quarter of 2021. The reality is even worse: besides the almost eight million officially classified as unemployed, we should add people who are not seen as part of the labor market because they have given up looking for work. These discouraged work seekers totaled 3.3 million in the second quarter. So, in broad terms there were 11.1 million people of working age not employed during the second quarter. Given an estimated labor force of 22.8 million, this means that the expanded unemployment rate in South Africa is now heading toward a staggering 50% (48.9% in the second quarter).

For some perspective, during the first quarter of 2008, which was the first time that the current version of the QLFS was published, 4.2 million people were unemployed. At the time, discouraged work seekers were 1.2 million. Therefore, 5.4 million were unemployed if the broader measure is used. In terms of the expanded definition of unemployment, roughly 5.5 million more people did not have a job in the second quarter of 2021 compared to early 2008. In the first quarter of 2008, the expanded unemployment rate was at an already worrisome 30%. For the expanded unemployment rate to deteriorate by roughly another 19 percentage points to almost 50% over the following decade suggests a major crisis.

A multitude of factors, including two very severe external shocks in the form of the global financial crisis (GFC) in 2007/8 (and its aftermath) and the COVID-19 pandemic since 2020, more than a decade of periodic Eskom (the South African power utility) load-shedding (switching of electricity to municipalities on a rotational basis to prevent total power grid collapse), and the devastating period of state capture help to explain why the labor market has deteriorated so significantly over the last decade and more. Not long after the release of the first iteration of the current QLFS in early 2008, the aftermath of the GFC resulted in a severe global economic downturn in 2009. As a result, real GDP in South Africa contracted by 1.5% in the same year. At the time, this was by far the worst GDP performance of the democratic era since 1994. As a result, 1.2 million private sector jobs were lost between 2008Q4 and 2010Q3, before a recovery followed.

The South African economy was already shedding jobs before the Covid-19 hard lockdown and the associated massive private sector job losses of 2.2 million in the second quarter of 2020. The weak pre-Covid labor market followed a sustained period where domestic real GDP growth was unable to keep up with the rate of population growth. Even after incorporating StatsSA's GDP rebasing and historical revisions that were released this week, real GDP growth averaged only 1% in the six years between 2014 and 2019. The weak growth meant that the economy was unable to absorb most of the new entrants into the job market. Compared to this period of weak output expansion, real GDP growth averaged 4.5% between 2003 and 2008, translating into average total annual private sector employment growth of 3.4% during these years. During a period of sustained robust GDP growth, the economy was able to generate jobs. One therefore needs to ask the critical questions such as what in government policy, political processes and general governance contributed to this weak growth performance. If policy levers, policy incentives and clean and effective government were in place (and which ones), would South Africa would have had better growth performance?

It is important to appreciate that the paltry GDP growth between 2014 and 2019 and the sharp rise in unemployment since 2008 was despite significantly more accommodate macropolicy settings than in the preceding period. Between 1994 and 2007, the SA Reserve Bank's repo policy interest rate averaged 12%. On average, the policy rate was almost halved to an average of 6.7% between 2008 and 2019. The policy rate was subsequently reduced dramatically further, amid the Covid shock last year and remains accommodate. Indeed, after adjusting for forward-looking inflation, the policy rate is negative in real terms.

In terms of fiscal policy, expressed as a share of the economy's size, main budget noninterest expenditure rose from just below 23% of GDP during the 2006/7 fiscal year to 26.5% in 2019/20. Expenditure rose further to 28.5% of GDP in 2020/21. The quoted numbers already incorporate the higher denominator (nominal GDP) after the recent GDP revisions. The key here is that despite a much lower policy interest rate and more government spending, real GDP growth severely underperformed in the years before Covid-19. This again flies against the argument that more government spending would bring about higher economic growth. While the pandemic has been a major blow to GDP and the local job market, weak output and employment growth precede it. Importantly, the sustained poor growth performance of recent years should not be laid at the door of excessively restrictive macroeconomic policy settings.

Except for gold, the prices of SA's major export commodities continued to rise in 2021Q1. Relative to 2020Q1, all South Africa's major export commodity prices were notably higher. **Rhodium** remained the star performer, but there were strong annual gains across the board. In some cases, the price gains continued in the early part of 2021Q2, with **palladium** reaching an all-time (nominal) high above $2900/oz. in the week ending 23 April. The platinum group metals (platinum, palladium and rhodium) continue to be supported by stricter vehicle emission standards in places like Europe and China. This increases the demand for these metals, which are used in catalytic converters.

The price movements have had a major impact on the contribution of the individual commodities to SA's total mineral export sales. In a nutshell, while the contribution of gold (long-term trend, 2020 being an exception), platinum (since 2015) and coal has declined, the opposite is true for iron ore, palladium and rhodium.

Regarding SA's import bill, the **Brent crude oil** price has increased at a faster tempo than anticipated. The sharp price gains in 2021Q1 were again driven by an improved outlook for global GDP growth and oil demand. Developments on the supply side also supported the price. In early March, the OPEC+ grouping of major oil producers agreed to extend oil output curbs. In addition, one-off events affected the oil price through the quarter. These included disruptive weather in Texas, which shut-in oil production totaling more than 10% of US oil supply, drone attacks on Saudi Arabia oil facilities and delays caused by the blockage in the Suez Canal. This increased shipping costs, as well as the oil price.

Focusing on the agricultural sector amidst all of the negative impacts and projections as a result of the COVID-19 pandemic, the South African agricultural sector has emerged as a shining light, growing by 13% in 2020 (StatsSA 2022). This represents a sharp turnaround and an illustration of the sector's ability to recover from extremely tough conditions over the past five years where the agricultural real growth rate averaged negative 1.3% per annum.

Compared to other economic sectors, the agriculture and food sector has been relatively insulated from the effects of the COVID-19 crisis because as an essential service, operations were allowed to continue, with the exception of alcoholic beverages and tobacco and initially also wool, mohair and cotton. Overall, agriculture was mainly affected in the short-run by a decline in sales due to the closure of hospitality, take-away-food outlets and informal trading and the ban on alcohol and tobacco sales greatly impacted the liquor and tobacco value chains. Moreover, the devaluation of the South Africa Rand during lockdown affected the cost of imported inputs like agrochemicals but also benefited exporting industries.

4.2.2 The Situation in Limpopo

The economy of Limpopo province has developed along similar trends as South Africa at large. Some differences emerge, however, because of the differences in the sectoral composition of GDP (Fig. 4.1).

As Fig. 4.1 illustrates, the primary sector has a much larger role in Limpopo than in South Africa at large. Mining and quarrying, as the largest industry, contributes about 28% of the provincial level GDP while agriculture, forestry and fishing add another 4%, totaling to about a third of provincial level GDP from the primary

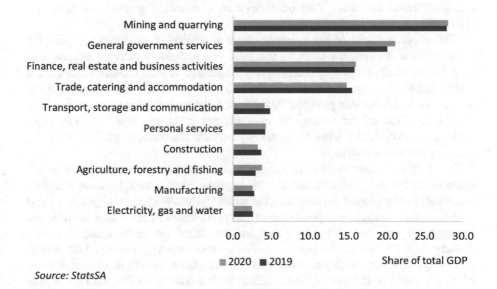

Source: StatsSA

Fig. 4.1 Composition of Limpopo province's GDP (2019 and 2020). Source: StatsSA

Employment

Source: before 2008 - Quantec, since 2008 - StatsSA QLFS

Fig. 4.2 Number of employed persons over the past 15 years in the Limpopo province. Source: before 2008—Quantec, since 2008—StatsSA QLFS

sector. Nationwide, the contribution of the primary sector is below 10%. The tertiary sector contributes to the GDP of Limpopo province to a similar extent as in national GDP but the secondary sector, with only 2.6% in 2020, is very underdeveloped compared to the rest of the country. This situation persisted already before the Covid-19 crisis and must thus be viewed as a structural property of Limpopo province.

The important role of the primary sector is visible in the trade statistics for Limpopo, too. Limpopo's share in the total value of South African exports was at 4.5% in 2020 and almost exclusively consisted of raw materials. Since the population of Limpopo province amounts to 5.8 million people, equivalent to 9.8% of the total South African population, the per capital export value from the province is less than half of the average number for South Africa. The most important agricultural exportables were fruit and nuts, which alone contributed 5.6% of the export value of the province.

The lackluster development of Limpopo province can at least to a large extent again be explained by the difficult conditions on the labor market. Figure 4.2 depicts the number of employed persons over the past 15 years. While a mild positive trend in employment numbers is present over the years between 2005 and 2018, when employment grew on average by 3.3% per year, this development came to an end already in 2019, before the Covid-19 shock hit the world economy. This shock, however, had a very strong negative effect on the labor market in Limpopo, even when compared to the rest of South Africa. With a drop of 10.7%, the loss of jobs was most marked among all provinces in South Africa.

The agricultural sector in Limpopo is dominated by horticultural production, followed by animal production activities, and field crop production. The contribution of horticulture to gross farming income amounts to about 60%, while the contributions of livestock and crop production are about 20% and 15%, respectively. Limpopo is well known for its substantive production of tomatoes, potatoes, onions, mangoes, avocadoes, citrus fruits and various nuts, confirming the dominance of the horticultural industry, which is water intensive and labor intensive. For tomatoes, pumpkins and subtropical fruits, Limpopo has the largest planted area in South Africa. In particular, Mopani district in Limpopo, where many of these crops are grown, is ranked sixth among all districts in SA with the highest agricultural employment numbers (29,000 workers).

The farm structure in Limpopo is characterized by a mixture of commercial, emerging and smallholder farms. There are about 3000 commercial farming units (540 of these are owned /operated by black farmers) in the province employing 97,400 farm workers, (63,000 are full-time employees).

4.3   The Policy Arena

At the Southern African level, the economic, trade and agricultural policies in RSA are embedded within the regional integration framework of the Southern Africa Development Community (SADC). The organization is one of the eight Regional Economic Communities (RECs) in Africa. Its current form emerged in the early 1990s with the aim of an intensified socioeconomic, political and security cooperation among its members. Between 1996 and 2012 a total of 27 protocols were ratified, which define the cooperation activities in various subfields, including agriculture and trade. In 2008, an additional free trade agreement (FTA) was implemented with the long-term targets to implement a customs union, a common market, a monetary union and eventually a single currency. However, besides the FTA, none of these milestones have been achieved to date (SADC 2022).

The main challenge of further integration within to SADC is overlapping memberships of its economies. For example, RSA and Botswana are also members in the Southern Africa Customs Union, Zambia joined the Common Market for Eastern and Southern Africa, and Tanzania is also part of the East African Community. Since all of these organizations target integration in overlapping or even identical areas, a number of regulatory, organizational and political conflicts of aims have arisen as impediments to negotiations and integration processes as a whole. In particular with regards to trade policy, where for instance tariff rates, quotas and rules of origin are regulated differently across these different economic integration endeavors, inducing contradicting trade governance that are not solved generically, but rather on a case by case basis.

However, efforts with regards to regional integration in SADC and neighboring RECs in the more recent past are likely to be outpaced by developments on the continental level. Since the foundation of the African Union (AU) in 2002, a number of socioeconomic, justice, political and security cooperation agreements have been negotiated at the continental level. On the one hand, this added a further layer of complexity to the already cumbersome regional integration processes. On the other hand, these AU-based initiatives allowed for some overarching, continent wide regulations. For example, in 2021, the African Continental Free Trade Area (AfCFTA) was launched, which aims at the implementation of free trade relationships across all of the African continent. Even though many of the details of the agreement remain to be negotiated, it is expected that in the medium term, the AfCFTA subsumes most of the regionally regulated trade relationships, such as those of SADC, and eventually forms a viable umbrella for most of intra-African trade.

In South Africa, the National Development Plan (NDP) identifies the key challenges facing South Africa as a country but argues that the country can eliminate poverty and reduce inequality by 2030. It emphasizes the importance of hard work, leadership and unity. It furthermore identifies Infrastructure Development, Job Creation, Health, Education, Governance, Inclusive Planning and the Fight against Corruption as key focus areas and spells out specific projects for each.

The NDP was developed after a detailed diagnostic assessment of the issues constraining economic development and improvement in social wellbeing. It identifies the critical interventions needed to improve education and health outcomes as well as grow the economy and reduce unemployment and alleviate poverty.

As the primary economic activity in rural provinces such as Limpopo, the NDP identifies agriculture as having the potential to create close to one million new jobs by 2030, a significant contribution to the overall employment target. To achieve this, the NDP proposes the following policy imperatives:

- Expand irrigated agriculture. Evidence shows that the 1.5 million ha under irrigation (which produce virtually all South Africa's horticultural harvest and some field crops) can be expanded by at least 500,000 ha through the better use of existing water resources and developing new water schemes.
- Use some underused land in communal areas and land-reform projects for commercial production.
- Pick and support commercial agricultural sectors and regions that have the highest potential for growth and employment.
- Support job creation in the upstream and downstream industries. Potential employment will come from the growth in output resulting from the first three strategies.
- Find creative combinations between opportunities. For example, emphasis should be placed on land that has the potential to benefit from irrigation infrastructure; priority should be given to successful farmers in communal areas, which would support further improvement of the area; and industries and areas with high potential to create jobs should receive the most support. All these will increase collaboration between existing farmers and the beneficiaries of land reform.
- Develop strategies that give new entrants access to product value chains and support from better-resourced players.

The NDP makes the following detailed recommendations, in achieving the goal of 1 million new jobs by 2030:

- Substantially increase investment in water resources and irrigation infrastructure where the natural resource base allows and improves the efficiency of existing irrigation to make more water available.
- Invest substantially in providing innovative market linkages for small-scale farmers in the communal and land reform areas, with provisions to link these farmers to markets in South Africa and further afield in the subcontinent.
- A substantial proportion of the agricultural output is consumed in the "food-away–from-home" market in South Africa. While this includes restaurants and take-away outlets, which are hardly relevant in most rural areas, it also includes school feeding schemes and other forms of institutionalized catering, such as food service in hospitals, correctional facilities and emergency food packages where the state is the main purchaser. As part of comprehensive support packages for farmers, preferential procurement mechanisms should be put in place to ensure that new entrants into agriculture can also access these markets.
- Create tenure security for communal farmers. Tenure security is vital to secure incomes from all existing farmers and for new entrants. Investigate the possibility of flexible systems of land use for different kinds of farming on communal lands.
- Investigate different forms of financing and vesting of private property rights to land reform beneficiaries that does not hamper beneficiaries with a high debt burden.
- There should be greater support for innovative public–private partnerships. South Africa's commercial farming sector is full of examples of major investments that have resulted in new growth and new job opportunities.
- Increase and refocus investment in research and development for the agricultural sector.

Although the NDP is very ambitious with the 1 million jobs in agriculture target, the plan and vision for the agricultural sector is much more inspiring and innovative than many of government's current action programs.

From our review of all the relevant policy documents and frameworks relevant to the agricultural and food sector, it is evident that the main focuses of these policies and plans are job creation in agriculture and dealing with process of empowerment and redistribution of resources and opportunities in the sector. Although food security and nutrition issues are mentioned and discussed as one of the main challenges it is merely seen as an important outcome if the agricultural and food sector performs optimally.

The National Development Plan is the only policy framework reviewed here that presents specific plans and suggestions on how to tackle the problems of food insecurity and malnutrition. To reduce the acute effects of poverty on millions of South Africans over the short term, the plan proposes to:

- Introduce a nutrition program for pregnant women and young children and extend early childhood development services for children under five.
- Ensure household food and nutrition security.
- Urgent action is required on several fronts: Households and communities. Proper nutrition and diet, especially for children under three, are essential for sound physical and mental development. The Commission makes recommendations on child nutrition, helping parents and families to break the cycle of poverty, and providing the best preparation for young children—including a proposal that every child should have at least two years of preschool education.

It has to be stated that these are just recommendations and suggestions and have yet to be translated into real policies with action programs and budgets. This point applies equally to most of the policy frameworks discussed above. The translation of strategic plans and policy frameworks into well-funded and efficiently executed government programs remain a major omission in the South African government. Implicitly most of the policies to promote increase agricultural production are listed and well argued—but never executed. South Africa has always the best plans and the most modern legislation but political rhetoric and nondelivery and general political insecurity flies against all these noble plans resulting in further job shedding and decrease in production.

4.3.1 Agricultural Policies

At the Southern African level, the key role of agricultural policies has been recognized in various of the regional organizations. SADC has recognized the importance of agriculture to food security in the region as the incomes of about 70% of the population in SADC countries depends on agriculture. Moreover, as agriculture in the region is relatively labor intensive, it is also heavily affected by the HIV/AIDS pandemic, which SADC identified as a major challenge for agricultural development. Given the mainly smallholder-based agricultural sectors of most member states, another pertinent issue is the access of smallholders to functioning product and factor markets, exacerbated by problems of road and marketing infrastructure. On the other hand, agricultural products account for about 13% of the total export value SADC, highlighting the key role of agricultural development from yet another perspective (SADC 2022).

The Dar-Es-Salaam declaration of agriculture and food security in the SADC region of 2004 addressed both market access as well as labor shortage problems. The declaration furthermore aimed at streamlining regional agricultural and food security policy with the Millennium Development Goals (MDG) of the United Nations (UN). In the short term, members agreed to facilitate smallholder access to improved seed varieties, fertilizers, agrochemicals, tillage services, farm implements and the construction of irrigation systems. For the medium and long terms, the SADC member countries also mandated the allocation of at least 10% of national budgets to agriculture and decided to implement national food reserves programs to buffer negative food supply shocks that are also pertinent in the region (SADC 2004).

In 2008, the SADC Multi-country Agricultural Productivity Programme implemented further measures to achieve the targets of the Dar-Es-Salaam declaration. The most overarching concern was low agricultural productivity in the region. The program stretches over a 15 year period and targets to enhance (1) food (crop, livestock and fishery) production and productivity through improved access to and sustainable use of agricultural productive assets, (2) diversification and value added through the establishment of a supportive policy and an adequate institutional environment for the development of efficient agroindustrial commodity chains, (3) disaster prevention, preparedness and mitigation through the implementation of a comprehensive strategy involving the development of drought and pest tolerant crop and livestock varieties, and (4) institutional collaboration in the region. These targets are mandated to be pursued and implemented by the member states.

The Government of National Unity (GNU) that was formed in April 1994 set in motion radical changes in the political economy of South Africa, but in agriculture policy changes had to wait until 1996 when the GNU was replaced by the ANC government whose policies over the following 27 years focused on the delivery of basic services, reducing poverty and expanding the payment of social grants to poor communities, all of which have implications for the demand for agricultural products.

In the agricultural sector, the most important policy initiatives in the post 1994 years included land reform; institutional restructuring in the public sector; the promulgation of new legislation, including the Marketing of Agricultural Products Act (No 47 of 1996) and trade policy and water and labor policies and laws within the framework of wider macroeconomic policy reform. At the same time, the political decision was made to dismantle the support services that favored white commercial farmers as part of the restructuring of government services to the population at large and reprioritization of government expenditure. The combined effect was that South African agriculture was exposed to all the volatilities of international commodity markets. The reduced levels of subsidization were supposed to reduce land values and hence to support land reform and the transformation of the agrarian economy. In reality, the impact was different: it made the process of integrating new and previously disadvantaged farmer communities into commercial agricultural value chains very difficult and was one of the main reasons for the failure of agrarian transformation in general and of land reform in particular (e.g., Hebinck and Cousins 2013). In an effort to correct these weaknesses, the government initiated various support programs to help land reform beneficiaries, including the Comprehensive Agricultural Support Programme (CASP), the Recapitalization and Development Programme (RECAP) and the Micro Finance Scheme for Agriculture (MAFISA), but these have met with little success (Kirsten et al. 2019).

The withdrawal of this support to white farmers had two consequences. First, it allowed the growth of very large-scale ("mega") farming operations (about 2600 (or 6.5%) of them), especially (but not exclusively) in intensive irrigated horticulture production (StatsSA 2020). Second, it was accompanied by the abolition of support measures, from direct subsidies to indirect market interventions, from funding of research and extension to the withdrawal of subsidies on conservation works (Vink 2000). The result was that black farmers were bereft of the support services that they had been denied under the previous regimes.

Unfortunately, the many attempts to remedy this situation (e.g., through the Comprehensive Agricultural Support Programme CASP and other programs) have been ex post, piecemeal and unsuccessful. Unless this is changed, commercial agriculture will remain white dominated, only slightly less racially segregated than in the 1980s, but with a strong bias against those smaller scale family farming operations whose development can do much to initiate growth and employment opportunities throughout the country, in the manner envisaged in the National Development Plan.

In the last three decades, the value of South African agricultural output more than doubled in real terms (DALRRD 2020). This growth has largely been driven by increased productivity, which has been underpinned by technological innovation, as well as growth in traditional export markets as well as access to new ones and has spanned across all subsectors of agriculture (livestock, horticulture and field crops).

Black farmers were largely excluded from the benefits of this agricultural growth while the various programs and plans were not sufficiently broad based to foster an inclusive and prosperous sector and, combined with the slow pace of land reform, contributed to frustrations among black farmers.

Despite remarkable growth during the last three decades, South Africa's agricultural sector remained plagued by dualism, mistrust and suboptimal performance. On the one hand, South Africa agriculture has surpassed the NDP targets in expanding a number of high-value commodities (citrus, macadamias, apples, table grapes, avocados, dairy and pork), but on the other hand the country has not fully achieved the jobs target and expansion of agriculture in the former homelands. The dualistic nature of the sector remains therefore entrenched. The only way this can change is through a capable and effective state (provincial and national), stable and conducive policy and investment environment; infrastructure development and services including electricity and water; and effective farmer support programs, among other support measures.

The expansion and growth of the industry over the last three decades was driven by new technology (irrigation, cultivation techniques, genetic material, etc.) and by the 2607 large farm enterprises that have capital and systems in place to invest, to expand and to export. Although these large enterprises are responsible for 67% of total farm output, cornerstone of the agricultural sector remains the many small family farms. More than 90% of all commercial farming units in South Africa are small family-based operations at different levels of commercial activity.

A number of prevailing and perpetuating cross-cutting factors are hindering inclusive growth and investment in the agriculture and agroprocessing value chains to a greater degree. These factors include continued policy ambiguities, mainly related to access to land and ownership as well as water rights, diminishing/unreliable infrastructural capacity (electricity, water, rail, roads, fresh produce markets and ports), major safety concerns for farmers and farm workers living in rural areas and a sharp increase in theft of stock and farming equipment, limited drive to opening new export markets and the inability to comply with stringent market access protocols, deteriorating biosecurity management, rising concentration and market power at food production, processing and distribution levels partly due to growing barriers to entry that limit access to key routes to market, and skills shortages and decaying research capacity in the country.

Moreover, coupled with consistently low and ineffective farmer support, in particular to subsistence and emerging farmers and high barriers to entry, these constraints contribute to an uncertain and unstable investment environment, resulting in limited growth and job creation in the sector, and thereby perpetuating inequality and exclusion of historically disadvantaged farmers and agripreneurs in agriculture and food value chains in the country.

The constraints and factors hindering the agricultural sector to grow and to transform to be more inclusive, are not new and have been identified in 1995 (White Paper on Agriculture), affirmed in 2001 (Agricultural Strategic Plan) and again in the diagnostics leading up to the National Development Plan (NDP) in 2011 and more recently by the High Level Panel Report led by former president Motlanthe and the Presidential Panel Report on Agriculture Land led by the late Vuyo Mahlati. They have not changed, and while some progress in some areas has been achieved since the adoption of the NDP, there are still areas of substantial underperformance. These include:

- Investment in agriculture and agroprocessing of the former homeland regions and state acquired land to drive overall productivity and commercial production to boost food security and alleviate poverty.
- Effective farmer support services and financing to increase black farmers' share of total agricultural output.
- Redistribution of agricultural land for sustainable agricultural production.
- Increased state capacity to open new export markets for a broader range of agricultural produce.
- Maintenance and upgrading of key infrastructure (roads, rail, ports, electricity.)
- Greater inclusion, participation and competition in agroprocessing.
- Transformation of agricultural and agroprocessing value chains, including the removal of barriers to entry.

4.3.2 Trade Policies

Trade policies constitute the key policy framework for the development of the South African economy. This holds in particular for Limpopo province, with its relatively strong reliance on the primary sector. Unleashing the full potential of agriculture in particular requires improved market access to high-value export chains. The development in this regard has been quite remarkable over the past decades, starting from the free trade agreement with the European Union (1999, fully implemented since 2004) and the regional free trade agreement within the

Southern African Development Community (fully implemented since 2012) up to the African Continental Free Trade Area (2020). Nevertheless, macroeconomic instability and policy incoherence remain substantial risk factors for agricultural value chains (SADC 2022).

Trade policy in South Africa has always had a strong regional focus. From the beginning of the Southern Africa Customs Union (SACU) more than 50 years ago, South Africa has been the driving and dominating force in regional integration. Currently, regional trade relations are mainly governed by the SADC free trade area. The loss of national decision power over tariffs and other trade-related policies that usually goes hand in hand with the formation of a free trade area or a customs union, has been not a major challenge for South Africa since the country accounts for about half of the current GDP of the SADC FTA. This retained national autonomy might explain why the regional integration within SADC FTA has remained somewhat subdued, with only 10% of the bloc's total trade being intra-FTA trade. While lack of physical infrastructure and the strong role of nontariff measures are common to many, in particular smaller, developing countries (Fiankor et al. 2021), the role of the policy process seems to be specific to SADC (Sikuko 2018), where a dominating influence seems to be exerted by South Africa.

The meager intraregional trade development within SADC has been present in many of the continents regional trade agreements. In the early 2000s, SADC and two more regional blocs (COMESA and EAC) aimed at forming a larger regional trade agreement, the Tripartite Free Trade Area. However, this initiative was never ratified but was replaced by a continent-wide initiative for boosting intra-African trade, the African Continental Free Trade Area (AfCFTA). This agreement is effective since 2019, and the expected effects on intra-African trade are huge. For instance, the World Bank (2020) estimates that intra-AfCFTA imports will double relative to the baseline by 2035. For total exports in agricultural products, the study expects an increase of more than 30% relative to the baseline. While the main effects for South Africa will materialize in increasing trade opportunities in manufacturing and services, the important role of agriculture in Limpopo suggests that for this province, new opportunities for intra-African exports might open up.

Overall, the trade policy of South Africa can be viewed as relatively liberalized, with open import and export markets. However, access to high-value export markets remains hampered by sanitary and phytosanitary measures (e.g., Kalaba et al. 2016) and the increasing role of private standards (e.g., Fiankor et al. 2020) that are increasingly used by developed countries' food importers and retailers.

4.4   Recommendations for Reform

In terms of **macroeconomic policies**, the hindrances to economic development are related to labor markets, monetary stability and external trade. The basic lessons to pursue these objectives can be learnt from other countries' experience and have frequently been put on the policy agenda. The willingness of policy makers to commit to these objectives, however, seems to be much less of a clear issue.

In light of these structural realities in the field of **agricultural policies**, the necessary guidelines to facilitate the growth and financial sustainability of the agricultural sector at large should be obvious and clear. What has been lacking over the past two decades is the practical implementation of the government policy frameworks and legislation, which, in turn, reinforced the lack of access among black farmers, and few opportunities within the input supply, agroprocessing and food retail sectors. The underlying factors behind this lack of implementation can be categorized into four broad streams. First, the limited government capacity to execute government programs together with a misalignment of functions and priorities between the three spheres of government. Second, the misallocation of the budget by the national and provincial governments. Third, the poor and uncoordinated transformation programs between government, private sector and civil society. Fourth, the abolishment of crucial institutions such as the Agricultural Credit Board and corporatization of farmer cooperates that were essential in coordinating and providing financial and nonfinancial support to farmers prior to 1998.

Biosecurity for plants and animals are becoming a critical issue as South Africa expands its production of exportable products. Government systems have time and again failed the agricultural sector in this regard with them unable to deal with animal diseases and plant diseases. At the same time, these impact South Africa—and especially the Limpopo province to trade with other countries in livestock products such as beef, game meat, and chicken and eggs. Stronger emphasis on developing and implementing standards and technical regulations in the field of biosecurity is therefore warranted, including cooperation with trading partners like the EU to both ensure the mutual acceptance of standards and technical regulations and to develop solutions for managing isolated outbreaks of pests or diseases in a better way. Promising ideas, e.g., "green corridor" zones with stricter regulation that could remain in export business even in case of phytosanitary problems elsewhere in the country, should be explored by bringing policy makers and private actors in the value chain from both South Africa and important destination markets together.

References

DALRRD (2020) Abstract of Agricultural Statistics, Pretoria: Department of Agriculture, Land Reform and Rural Development (DARLRRD)

Fiankor D-DD, Flachsbarth I, Masood A, Brümmer B (2020) Does GlobalGAP certification promote agrifood exports? Eur Rev Agric Econ 47(1):247–272. https://doi.org/10.1093/erae/jbz023

Fiankor D-DD, Haase O-K, Brümmer B (2021) The heterogeneous effects of standards on agricultural trade flows. J Agric Econ 72(1):25–46. https://doi.org/10.1111/1477-9552.12405

Hebinck P, Cousins B (2013) In the Shadow of Policy: Everyday Practices in South Africa's Land and Agrarian Reform, WITS University Press, Johannesburg

Kalaba M, Kirsten J, Sacolo T (2016) Non-tariff measures affecting agricultural trade in SADC. Agrekon 55(4):377–410

Kirsten JF, Machethe C, Ndlovu T, Lubambo P (2016) Performance of land reform projects in the North-West province of South Africa: Changes over time and possible causes. Development Southern Africa: Volume 33(4):442–458

Moyo S, Sill M, O'Keefe P (2014) The southern African environment: profiles of the SADC countries. Routledge

Nshimbi CC, Fioramonti L (2014) The will to integrate: South Africa's responses to regional migration from the SADC region. Afr Dev Rev 26:52–63

SADC (2004) Dar-es-salaam declaration on agriculture and food security in the SADC region

SADC (2022) SADC website. Retrieved in March 2022 at https://sadc.int.

Sikuko K (2018) Policy making in SADC: the missing link to advancing integration. In: Centre TL (ed) Monitoring regional integration in southern Africa, yearbook 2017/18. RSA, Tralac, Stellenbosch

StatsSA (2020) Census of Commercial Agriculture 2017, Pretoria: Statistics South Africa

StatsSA (2022) Statsitics South Africa. Retrieved in February 2022 at http://www.statssa.gov.za/

UNCTADStat (2022) Data center. Retrieved in March 2022 at https://unctadstat.unctad.org/EN/

Vink N (2000) Agricultural Policy Research in South Africa: Challenges for the Future. Agrekon 39:432–470

World Bank (2020) The African continental free trade area: economic and distributional effects. World Bank, Washington, DC. https://doi.org/10.1596/978-1-4648-1559-1

Part II

Drivers of Climatic Variability and Change in Southern Africa

Past Climate Variability in the Last Millennium 5

Eduardo Zorita, Birgit Hünicke, Nele Tim, and Matthieu Rouault

Abstract

We review our knowledge of the climate variability in southern Africa over the past millennium, based on information provided by proxy data and by climate simulations. Since proxy data almost exclusively record past temperature and/or precipitation, the review is focused on those two variables. Proxy data identify three thermal phases in the region: a medieval warm period around year 1000 CE (common era), a Little Ice Age until about the eighteenth century, and a clear warming phase since that temperature minimum until the present period. Variations of precipitation are different in the summer-rainfall and winter-rainfall regions. In the former, precipitation tends to accompany the temperature, with warm/humid and cold/dry phases. In the winter-rainfall zone, the variations are opposite to temperature. Thus, past precipitation variations display a see-saw pattern between the summer- and winter-rainfall zones. However, climate simulations do not display these three different hydroclimatic periods. Instead, the simulations show a clearly warm twentieth century and punctuated cooling due to volcanic eruptions, with otherwise little variations during the pre-industrial period. Also, the simulations do not indicate an anticorrelation between precipitation in the summer- and winter-rainfall zones. Possible reasons for these discrepancies are discussed.

E. Zorita (✉) · B. Hünicke · N. Tim
Institute of Coastal Systems, Helmholtz-Centre Hereon, Geesthacht, Germany
e-mail: eduardo.zorita@hereon.de

M. Rouault
Nansen Tutu Center for Marine Environmental Research, Department of Oceanography,
University of Cape Town, Cape Town, South Africa

© The Author(s) 2024
G. P. von Maltitz et al. (eds.), *Sustainability of Southern African Ecosystems under Global Change*, Ecological Studies 248,
https://doi.org/10.1007/978-3-031-10948-5_5

133

5.1   Introduction

Climatology is a branch of science where experiments are very difficult, if not impossible, to perform. Therefore, we are bound to use as much information from observations as possible, to try to understand the mechanisms of climate change and its possible impacts on society and ecosystems. Although climate models and basic physical considerations unambiguously agree that an increase in atmospheric greenhouse gases must lead to generally warmer temperatures, regional climate changes are more uncertain, in particular concerning other climate variables. For instance, questions such as whether warmer climates will lead to increased or decreased precipitation over a particular region are much more difficult to answer. The climate of the past can help find an answer.

The observational record is, however, usually too short, spanning at most the last 200 hundred years and more usually only the last few decades. Many relevant questions about the present climate, such as the unprecedented character of current temperatures, climate trends, and climate extremes, are better addressed by looking beyond the period covered by observations. This goal can be partially achieved by analysing indirect climate information from the so-called proxy data—tree rings, lake sediments, etc.—that are natural archives, sensitive to past environmental conditions. The period covered by these natural archives can be vast, but the past few centuries span climate conditions that are, from the geological perspective, not very much different from the present and future climate, so that the lessons learnt there may find applications for the understanding of present climate trends.

Simulations with Earth System Models that cover the past few hundred years can, together with proxy data, provide useful insights about the relevant climate mechanisms. Here, each of these two sources of information serve as independent confirmation (or rebuttal) of the other. Both are inherently uncertain, displaying different sources of error, and their combination leads to more robust conclusions than each of them taken in isolation would be able to provide.

In the following sections in this chapter, we will review the existing literature on climate variability in southern Africa during approximately the past millennium. We start with a summary of evidence available from proxies, both at the global scale and more specifically for southern Africa. These sections are followed by selected results obtained from climate simulations. Finally, we discuss their agreements and inconsistencies and conclude with the main take-home implications for future climate changes in this region.

5.2   The Climate of the Past Millennium: Global Background

Our knowledge of the climate of the past millennium is derived from indirect indicators that archive information about past environmental conditions. Trees tend to form thicker annual growth rings or produce wood of higher density in years with more suitable environmental conditions, usually warmer and/or wetter. These biological characteristics can be calibrated to reconstruct past physical magnitudes such as temperature or precipitation variations by statistically comparing recent tree rings to meteorological observations. Apart from dendroclimatological data, other proxy records also contain information from past environmental conditions: carbon and oxygen stable isotopes in old wood, oxygen isotopes in stalagmites, pollen assemblages in lake sediments, historical documents, etc. In this fashion, global networks of proxy data can be translated by means of complex statistical methods to annually resolved climate patterns in past periods (Li et al. 2010). These climate reconstructions, however, rely on some general assumptions that may not be always fulfilled. For instance, dendrochronological proxy records reflect the environmental conditions during the growing season, i.e., are seasonally biased. Also, other non-climatic factors may affect the growth of trees, such as availability of nutrients, fires, etc. Other types of records suffer from other corresponding caveats, so that it is not totally surprising that discrepancies between reconstructions derived from different proxy records arise. This highlights the need to combine different sources of information to reach robust conclusions.

In addition, depending on the method applied to translate the proxy information and on the network of proxy data, the reconstructions of past climate may differ on the amplitude of past climate variations and on the specific regional details. However, most of the temperature reconstructions published so far indicate that the Earth's climate of the past millennium can be described by relatively warm centuries around year 1000 CE (common era)—the Medieval Climate Anomaly or Medieval Warm Period, followed by colder centuries between around 1500 CE and 1850 CE—usually denoted as the Little Ice Age—which in turn were followed by a warming trend that has strongly intensified from around 1980 CE onwards until present (Crowley 2000).

These centennial climate fluctuations have been attributed to different *external climate forcings* (Schmidt et al. 2012). One is *volcanic activity* that tends to cool the global climate due to the volcanic aerosols ejected to the stratosphere, dimming the incoming solar radiation. *Solar output* itself is also variable through time. Land-use and forest cover can modulate the regional climate due to the implied changes in the surface reflectivity. Concerning *land use*, major changes have occurred in some regions such as Europe and East Asia over the past few centuries. Finally, *greenhouse gases* since the industrial revolution have contributed to the warming trend since the end of the Little Ice Age and are the single most important factor for the warming since the mid-twentieth century.

The variations of these external factors can be reconstructed by the analysis of polar ice cores. They archive the composition of the past atmosphere. Ice acidity records show sharp peaks due to the deposition of volcanic aerosols, allowing for an accurate dating and estimation of the strength of eruptions. The concentrations of cosmogenic isotopes such as ^{10}Be are indicative of past solar activity. These records show that around the Medieval Climate Anomaly volcanic activity was sparse and the Sun was stronger than in the ensuing centuries. By contrast, the Little Ice Age witnessed an intense and frequent volcanic activity and a weaker Sun. The recent decades are characterised by an almost constant solar output, relatively weak volcanism, and a very strong forcing due to greenhouse gases.

In addition, internal climate variations, not caused by any particular external factor but due to the slow variations of ocean currents and the inter-play between ocean and atmosphere, could have also contributed to some of these past variations. For instance, it is unclear yet as to whether the external factors could have been solely responsible for the Medieval Climate Anomaly or whether some sort of slowly varying internal mechanism might have contributed to generally warmer temperatures. This is perhaps more relevant at regional scales, for which internal climate patterns such as the El Niño-Southern Oscillation (ENSO) may have a stronger immediate influence than global external forcings.

5.3   The Climate of the Past Millennium: Southern Africa

Earlier literature reviews on the paleoclimate of southern Africa during the last 2000 years (Tyson and Lindesay 1992; Hannaford and Nash 2016) have identified the warm and cold climate phases in the past millennium previously mentioned in Sect. 5.2. A long 3000-year-long stalagmite record from Cold Air Cave in the Makapansgat Valley, which displays colour banding that is correlated to local temperature, also confirms the sequence of warm–cold–warm periods over the past millennium (Holmgren et al. 2001). According to this record, the Little Ice Age would have been about 1 °C colder than present. This is approximately confirmed by a stalagmite-based ^{18}O isotope record from the same site spanning the past 350 years, which has been interpreted as indicators of a sharp and well-defined cold multi-decadal period centred around 1720 CE (Sundqvist et al. 2013). The cooling may have amounted to 1.4 °C colder than present. The rise and fall in temperature in these different thermal phases would have been roughly homogeneous over the whole region.

The picture derived for precipitation is more nuanced. As explained in Chap. 6, southern Africa is characterised by two regions with different annual precipitation regimes: a (mostly) winter-precipitation region around Cape Town and a summer-precipitation region located further to the east and northeast (Reason 2017). During the Little Ice Age, the winter-precipitation region may have received more precipitation due to a northward displacement of the belt of westerly winds (Dunwiddie and LaMarche 1980), whereas the summer-precipitation zone of southern Africa would have faced a generally drier climate due to diminished evaporation from the Indian ocean leading to lower air humidity (Woodborne et al. 2015). The reversed precipitation pattern is found in the prior warmer centuries during the Medieval Climate Anomaly. In this period, proxy indicators of precipitation in the summer-rainfall region based on stable carbon isotopes in baobab trees show increased rainfall (Woodborne et al. 2015).

Thus, an important feature of the pattern of the variability of annual precipitation totals in southern Africa, as derived from proxy information, appears to be a see-saw pattern between the winter-precipitation zone in the southwest and the summer-

Fig. 5.1 Reconstructions of southern African precipitation over the past 500 years in the winter-rainfall zone (blue, left axis, based on cedar tree-ring widths) and summer-rainfall zone (red, right axis, based on concentrations of ^{13}C in baobab trees). Figure copied from Woodborne et al. (2015) (freely available)

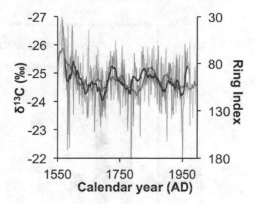

precipitation zone in the northeast (Fig. 5.1). Whether or not this see-saw type of variability recurs over time no matter which is the main external climate driver is indeed an interesting question, relevant for future climate changes.

The nineteenth century was relatively colder than the twentieth century. This period, with a more dense network of direct observations, offers the opportunity to test that working hypothesis suggested in the previous paragraph. Unfortunately, the available studies that provide more detailed analysis for the nineteenth century precipitation variations reach contradicting conclusions, indicating either drier (Nicholson et al. 2012) or wetter conditions (Neukom et al. 2014; Nash et al. 2016; Nash 2017) during the nineteenth century in the summer-precipitation zone. By contrast, precipitation in the winter-rainfall zone has likely remained temporally stable over the last two centuries (Nash 2017). An explanation for this discrepancy may lie in the different nature of the records analysed. Whereas Nicholson et al. analysed long instrumental and documentary records, the conclusion reached by the other studies is derived from a more comprehensive set of data, including, in addition to instrumental and documentary records and indirect proxies (dendroclimatological, corals). This highlights the difficulty of inferring past climates and the need to combine all available sources of information.

5.4 Paleoclimate Simulations with Earth System Models

Comprehensive climate models, very similar to those used to project the impact of greenhouse gases on future climate (Edwards 2011), have also been used to retrospectively simulate the climate of the past millennium (Fernández-Donado et al. 2013). These models, akin to weather prediction models, incorporate our knowledge of the main climate processes. They contain a representation not only of the atmosphere, but also of the ocean, of sea-ice, of soils and some of them also of the terrestrial and oceanic biosphere. All in all, they are one of the most complex software packages actually in use. Nevertheless, the climate system is very complex, with processes that typically occur over a vast range of spatial and temporal scales, from seconds to millennia and from millimetres to thousands of kilometres. Due

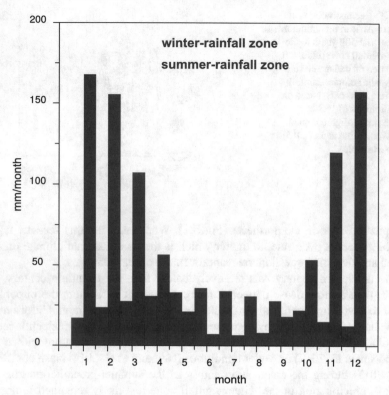

Fig. 5.2 An example of the annual cycle of precipitation in the two southern African precipitation zones simulated by the climate model MPI-ESM-P in the pre-industrial centuries (850–1800 CE). The magnitude of precipitation depends on the location of the selected model grid cells

to computing limitations, some of these processes need to be represented in a more simplified form. Typically, a climate model has a spatial resolution of about 100 km, and all smaller-scale processes are represented in an averaged fashion. Climate models, for instance, do not directly simulate the formation of clouds, but only the average effect of clouds over an area of typically Ī00 km long and wide. This leads to inaccuracies and uncertainties in the simulation of precipitation and to differences in climate projections obtained with different models. In spite of mentioned model uncertainties, climate models are indeed able to reasonably replicate the two precipitation regimes observed in southern Africa. An example is shown in Fig. 5.2, which can be compared with the corresponding figure in Chap. 6. This means that the main mechanisms behind these two precipitation regimes, namely extratropical cyclones for the winter-precipitation zone and convective precipitation within the South Indian Convergence Zone for the summer-precipitation zone (Cook 2000; Reason 2017), are reasonably well simulated by global climate models.

To simulate the climate of the past, climate models need to be driven by the external factors that provide or modulate the energy that reaches the Earth.

As mentioned in Sect. 5.2, in the past millennium, these external factors were volcanism, solar output, and atmospheric greenhouse gases (Schmidt et al. 2012). The magnitude of these factors in the past millennium can be approximately reconstructed from chemical analysis of polar ice cores, including the air bubbles trapped in them, and then used for climate simulations. There exists a relatively large set of climate models that has been used to estimate future climate change, as included in the different reports by the Intergovernmental Panel on Climate Change (IPCC). Some of these models have also been used to simulate the climate of the past centuries, more precisely the period 850–2005 CE. These simulations are part of the Coupled Model Intercomparison Project CMIP5 (Taylor et al. 2012). As indicated in the Introduction, the results of these simulations may differ from model to model, so that it is necessary to consider several models not only to identify the robust results but also to be aware of the uncertainties inherent in these simulations.

Figure 5.3 displays the near-surface air temperature in southern Africa simulated by a suite of climate models in the period 850–2005 CE. Table 5.1 lists these

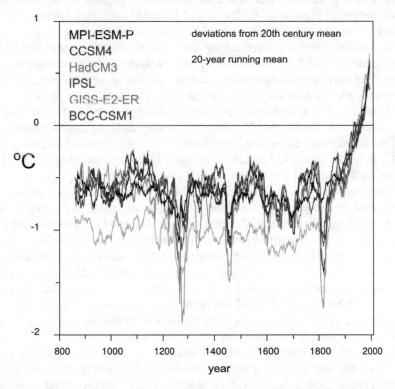

Fig. 5.3 Near-surface annual mean air temperature averaged in southern Africa (land areas between 10S-40S and 10E-45W) as simulated by a suite of climate models from the Climate Model Intercomparison Project CMIP5 for the period 850–2005 CE (Taylor et al. 2012). The time series represent deviations from the twentieth century mean temperature and have been smoothed with a 20-year running-mean filter

Table 5.1 List of global climate models and their spatial resolution (atmospheric model only) in geographical degrees used in this study

Model	Institution	latitude	longitude
MPI-ESM-P	Max-Planck-Institute for Meteorology (Germany)	1.875	1.875
CCSM4	National Center for Atmospheric Research (USA)	0.9	1.25
GISS-2-ER	Goddard Institute for Space Sciences (USA)	2	2.5
IPSL-CMA5-LR	Institute Pierre & Simon Laplace (France)	1.89	3.75
BCC-CSM1	Beijing Climate Centre (China)	2.81	2.81
HadCM3	Hadley Centre (UK)	1.875	3.75

climate models and their spatial resolution. All simulated temperatures display clear similarities but also differences. Most models, with the only exception of HadCM3, estimate colder pre-industrial temperatures of the order of 0.5 °C relative to the twentieth century mean. There seems to exist little doubt, from the models' perspective, that temperatures in the twentieth century have clearly been above the pre-industrial average level. The model HadCM3 estimates twice as cold pre-industrial temperatures with respect to the twentieth century mean.

The warming trend simulated during the twentieth century stands out compared to the trends in all other centuries, with the possible exception of the nineteenth century. The climate of the nineteenth century is, however, strongly impacted by the two first decades of intensive volcanism (see next paragraph), and by its associated strong cooling. The models that estimate the strongest impact of these eruptions (e.g., GISS-2-ER) are the ones for which the ensuing warming trend is also stronger. Even so, considering all models, the simulated twentieth century warming, about 1.2 °C, is larger than the bracket of nineteenth century warming of 0.5–0.8 °C spanned by the majority of models.

Further back in time, superposed to colder pre-industrial mean temperature, the models produce temperature variations that can be attributed to the impact of the external forcings. For instance, the clear cooling simulated after the mid-thirteenth century is due to the very strong eruption in Samalas (Indonesia) in 1258 CE (Guillet et al. 2017). Also clear in the figure are the series of eruptions in the early nineteenth century, one of them the famous Tambora eruption in 1815, which caused in Europe "the year without summer" (Raible et al. 2016) and serious societal disruptions there in 1816 CE. After around 1800 AD, the current warming trends set in, mainly caused by an increase of the solar output during the nineteenth century and by the increase of greenhouse gases in the second half of the twentieth century.

The succession of warm–cold–warm centennial or even multi-centennial periods (MCA-LIA-present) previously identified in the proxy-based reconstructions is not so clearly recognised in the simulations. Although the centuries around 1000 CE are slightly warmer in the simulations, the LIA is barely recognisable in the simulated series as a clearly cold differentiated centennial or multi-centennial period. As a result, the simulated temperature evolution during the pre-industrial period appears rather stable, with the interruptions caused by volcanism. In this respect, all models

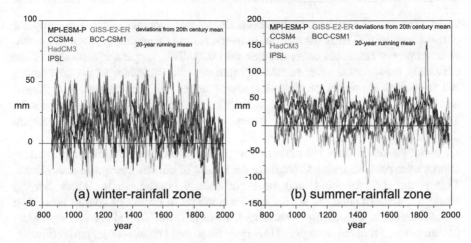

Fig. 5.4 Near-surface annual precipitation in southern Africa (land areas between 10S-40S and 10E-45W) as simulated by a suite of climate models from the Climate Model Intercomparison Project CMIP5 over the period 850–2005 CE, (**a**) for the winter-rainfall zone and (**b**) summer-rainfall zone (see Chap. 6). The time series represent deviations from the twentieth century mean precipitation and are smoothed with a 20-year running-mean filter

seem to agree. The possible reasons for the discrepancies between models and reconstructions are later discussed in Sect. 5.5.

Concerning past changes in precipitation, we again need to differentiate between the winter-rainfall zone and the summer-rainfall zone (see Chap. 2, Fig. 2.2). Since the mechanisms behind precipitation in these two zones are different, their long-term evolution could also diverge. The evolutions of simulated precipitation in these two zones in the past millennium are displayed in Fig. 5.4. Precipitation in the winter-rainfall zone (around Cape Town and along southern coast in the pre-industrial period) is in all simulations generally larger than in the twentieth century. The decadal variations of the pre-industrial mean precipitation are, however, much larger than for temperature, and a few decades in the simulations are indeed as dry as some decades of the twentieth century. Generally, the models do seem to indicate a tendency towards a drier climate in this region in the recent decades, in accordance with the analysis of observational data shown in Chap. 6. Precipitation in the summer-rainfall zone (continental northeast during the pre-industrial period, Fig. 5.4b) also shows a slight tendency to be larger than during the twentieth century. However, the suite of models displays a larger spread than for the winter-rainfall zone, so that the uncertainty is here also larger. One model, again the model HadCM3, behaves rather differently from the others and shows the opposite result.

Are the variations of the precipitation in the summer-rainfall and winter-rainfall zones mutually related? This is an interesting question regarding future climate change. If, for instance, the see-saw pattern suggested by proxy data (Sect. 5.3) is temporally stable, it is plausible that it will also be present in future climate, with enhanced rainfall in the summer zone and weaker in the winter zone, or vice versa. The model simulations, however, do not support this result. The temporal correlations, within each simulation, between annual mean precipitation in the two zones are very small (all cases smaller than 0.2). This happens irrespective of the timescale, for instance, after decadal or multi-decadal smoothing of the time series. All in all, the simulated precipitation shows rather wide interannual (not shown) and decadal variations, compared to any centennial or multi-centennial trends. The possible reasons for the disagreement between models and reconstructions are discussed in Sect. 5.5.

Variables other than temperature or precipitation may also be important to characterise past climates and estimate the impact of climate change on ecosystems. This is the case for wind, and more particularly coastal winds. Winds flowing along the western southern Africa coast are the main drivers of coastal upwelling in the Benguela Upwelling System and, therefore, are critical for the biological productivity of that ocean region. How upwelling could change under anthropogenic climate change was the focus of the hypothesis put forward by Bakun (1990); Bakun et al. (2015). According to this hypothesis, the intensity of upwelling-favourable winds should increase in the future due to the widening temperature difference between continental and oceanic surface caused by global warming. The studies that have analysed this hypothesis using observations or climate simulations are, however, not univocally conclusive (Sydeman et al. 2014). A confirmation by possible paleoclimate data from oceanic sediment cores, which may record past ocean productivity, and paleoclimate simulations could in theory shed light on this question. However, the analysis of paleoclimate simulations over the past millennium does not indicate that the intensity of upwelling-favourable winds had in the past varied hand-in-hand with global temperatures (Tim et al. 2016). Thus, these simulations do not in principle support Bakun's hypothesis. This can happen because the effect predicted by Bakun's hypothesis has been in the past too small compared to the natural wind variations, or because the modelled variations of past temperatures are unrealistically too narrow (Sect. 5.5). Other possible explanation assumes that state-of-the-art global climate models cannot realistically represent the reaction of coastal wind to variations in external forcing yet, due to their too coarse spatial resolution (Small et al. 2015).

In summary, the results from the global climate simulations regarding southern Africa can be interpreted as follows. The past evolution of temperatures clearly shows the impact of volcanic eruptions and, in the twentieth century, the impact of anthropogenic greenhouse gases. The twentieth century would have been the warmest of the past millennium. By contrast, precipitation would have varied little along the whole millennium, with a tendency of the twentieth century to be somewhat drier than the previous centuries but only in the winter-rainfall zone. The simulations do not show clear, well-defined, warm, cold, wet, or dry centuries in the pre-industrial centuries. Thus, the pre-industrial period appears in the simulations as a rather stable climatic background against which the 20th temperatures, but not so much precipitation, clearly stand out.

5.5   Comparison Between Proxy Data and Model Simulations

The combined analysis of climate simulations and climate reconstructions aims at a better understanding of the processes of climate variability and change and helps identify robust model results but also deficiencies that need to be addressed to increase our confidence in spatially detailed future climate projections.

The simulated temperatures over the past millennium show agreements with proxy-based reconstructions but also clear discrepancies. Both clearly suggest that the twentieth century has been a very warm period against the backdrop of the pre-industrial centuries. However, looking further back into the past, several climate reconstructions using independent data do show a well-defined cold period—the Little Ice Age, around 1700 CE—differentiated from a prior warmer period during medieval times. The simulations do not show these thermally differentiated periods. Instead, the simulations yield rather stable pre-industrial temperatures, punctuated by decadal cooling episodes caused by volcanic eruptions, which do not extend over several centuries.

One reason for this discrepancy may lie in the external forcing used to drive the climate models. The forcing used in the CMIP5 model suite of past millennium simulations follows a commonly agreed protocol that incorporates the knowledge of past variations of solar output and volcanism. These estimations are to some extent uncertain. Whereas volcanic forcing is relatively short lived, the variations of solar output display longer timescales, with cycles of several hundred years. The assumed amplitude of those solar variations of past solar output in the CMIP5 protocol represents a temporally narrower version of previous solar output reconstructions, which tended to have much wider, even tenfold, amplitude of variations (Schmidt et al. 2012). It is conceivable that this assumed reduction of the amplitude of past solar output variability turns out to be somewhat unrealistic. Stronger solar output variability in the past could have led to wider temperature variability at centennial timescales.

Concerning past precipitation, the proxy-based reconstructions in southern Africa indicate two important characteristics. One is the link between past precipitation and past temperature. Precipitation in the summer-rainfall zone varies, according to the reconstructions, in accordance with temperature, with lower precipitation during the Little Ice Age and higher precipitation during the Medieval Climate Anomaly. In the winter-rainfall zone, by contrast, the link between precipitation and temperature is opposite. Thus a second characteristic is the opposite variations in the summer-rainfall and winter-rainfall zones.

The simulations do not show this behaviour. Since the simulations do not present clearly defined thermal sub-periods, they cannot show a link between temperature and precipitation variations. Also, the simulated precipitation in the summer- and winter-rainfall zones does not appear anti-correlated in any of the model simulations. Should the see-saw character of precipitation variability be confirmed by further analysis of additional proxy records, it would raise the question as to why the models fail in this respect, and whether this deficiency is critical for more robust projections of future precipitation.

It is known that the simulation of precipitation can be a challenge for global climate models. Due to their coarse resolution (Table 5.1), the much smaller-scale formation of clouds and the condensation of water vapour to liquid water cannot be explicitly represented in the models. Instead, they use heuristic equations that translate large-scale humidity convergence and atmosphere stability to average precipitation over a whole grid cell. These heuristic, empirically derived, equations are prone to errors. One possible explanation is, therefore, that precipitation in the summer-rainfall zone, which is caused by convective systems associated to the South Indian Convergence Zone (Cook 2000), might not be perfectly simulated. However, other studies raise the question as to whether climate models realistically represent the connection between large-scale patterns of climate variability, such as ENSO, and southern African rainfall. It has been found that the CMIP5 models do not replicate the observed link between southern African rainfall and ENSO (Dieppois et al. 2015, 2019). Therefore, a targeted investigation on whether this deficiency is solved in the next generation of climate models is needed to increase the confidence of future precipitation projections in this region.

Concerning climate reconstructions based on proxy data, we have to bear in mind that the amount of those types of data in the southern African region is sparse compared to other regions, such as Europe or North America. On the other hand, there exist a few sources of historical data that can be further exploited. One example is the collection of log books from the British Navy, which contain valuable information about wind intensity and duration at daily timescales. These data can be used to reconstruct atmospheric variables in the region over the few past centuries (Hannaford et al. 2015). A more dense network of proxy data could allow to construct spatially resolved gridded *climate field reconstructions* (Tingley et al. 2012) covering the whole region, instead of a collection of climate reconstructions that refer to the relatively smaller subregions where the proxy records are located. A gridded reconstruction covering the whole southern African region would allow a much better comparison with the output of climate models. This also remains a research focus for the future.

5.6 Conclusions and Outlook

The main conclusions that can be derived from this brief review of the climate of southern Africa over the past millennium are as follows:

- Proxy records, such as dendroclimatological data, indicate a succession of a relatively warm medieval period around 1000 CE, following by colder centuries denoted as the Little Ice Age, and a warm twentieth century.

- Precipitation in the two relevant precipitation zones—winter-rainfall and summer-rainfall zones—was linked to the average regional temperatures, with precipitation in the summer-rainfall positively correlated to temperature. Precipitation in the winter-rainfall zone was negatively correlated to temperature.
- Climate simulations with state-of-the-art climate models do indicate a particularly warm twentieth century but show no thermally differentiated sub-periods nor correlations between temperature and precipitation. They do show, however, that the twentieth century might have been dry in the context of the past millennium.

These conclusions already indicate a possible way forward for future research. The disagreement between climate reconstructions and model simulations should be clarified in order to buttress the future climate projections obtained with those same climate models. Thereby, a particularly important aspect is the correlation between temperature and precipitation changes.

Acknowledgments This review was compiled in the frame of the project CASISAC (Changes in the Agulhas System and its impacts on southern African Coasts), funded by the German Federal Ministry of Education and Research (BMBF).

References

Bakun A (1990) Global climate change and intensification of coastal ocean upwelling. Science 247(4939):198–201

Bakun A, Black BA, Bograd SJ, Garcia-Reyes M, Miller AJ, Rykaczewski RR, Sydeman WJ (2015) Anticipated effects of climate change on coastal upwelling ecosystems. Current Climate Change Rep 1(2):85–93

Cook KH (2000) The South Indian convergence zone and interannual rainfall variability over Southern Africa. J Climate 13(21):3789–3804

Crowley TJ (2000) Causes of climate change over the past 1000 years. Science 289(5477):270–277

Dieppois B, Rouault M, New M (2015) The impact of ENSO on Southern African rainfall in CMIP5 ocean atmosphere coupled climate models. Climate Dyn 45(9):2425–2442

Dieppois B, Pohl B, Crétat J, Eden J, Sidibe M, New M, Rouault M, Lawler D (2019) Southern African summer-rainfall variability, and its teleconnections, on interannual to interdecadal timescales in CMIP5 models. Climate Dyn 53(5):3505–3527

Dunwiddie PW, LaMarche VC (1980) A climatically responsive tree-ring record from Widdringtonia cedarbergensis, Cape province. South Africa. Nature 286(5775):796–797

Edwards PN (2011) History of climate modeling. Wiley Interdiscip Rev Climate Change 2(1):128–139

Fernández-Donado L, González-Rouco J, Raible C, Ammann C, Barriopedro D, García-Bustamante E, Jungclaus JH, Lorenz S, Luterbacher J, Phipps S et al (2013) Large-scale temperature response to external forcing in simulations and reconstructions of the last millennium. Climate Past 9(1):393–421

Guillet S, Corona C, Stoffel M, Khodri M, Lavigne F, Ortega P, Eckert N, Sielenou PD, Daux V, Churakova OV et al (2017) Climate response to the Samalas volcanic eruption in 1257 revealed by proxy records. Nat Geosci 10(2):123–128

Hannaford MJ, Nash DJ (2016) Climate, history, society over the last millennium in Southeast Africa. Wiley Interdiscip Rev Climate Change 7(3):370–392

Hannaford MJ, Jones JM, Bigg GR (2015) Early-nineteenth-century southern african precipitation reconstructions from ships' logbooks. Holocene 25(2):379–390

Holmgren K, Tyson P, Moberg A, Svanered O (2001) A preliminary 3000-year regional temperature reconstruction for South Africa. South African J Sci 97(1):49–51

Li B, Nychka DW, Ammann CM (2010) The value of multiproxy reconstruction of past climate. J Amer Statist Assoc 105(491):883–895

Nash D (2017) Changes in precipitation over Southern Africa during recent centuries. In: Oxford research encyclopedia of climate science. Oxford University Press, Oxford. https://doi.org/10.1093/acrefore/9780190228620.013.539

Nash DJ, Pribyl K, Klein J, Neukom R, Endfield GH, Adamson GC, Kniveton DR (2016) Seasonal rainfall variability in Southeast Africa during the nineteenth century reconstructed from documentary sources. Climatic Change 134(4):605–619

Neukom R, Nash DJ, Endfield GH, Grab SW, Grove CA, Kelso C, Vogel CH, Zinke J (2014) Multi-proxy summer and winter precipitation reconstruction for southern Africa over the last 200 years. Climate Dyn 42(9):2713–2726

Nicholson SE, Klotter D, Dezfuli AK (2012) Spatial reconstruction of semi-quantitative precipitation fields over Africa during the nineteenth century from documentary evidence and gauge data. Quat Res 78(1):13–23

Raible CC, Brönnimann S, Auchmann R, Brohan P, Frölicher TL, Graf HF, Jones P, Luterbacher J, Muthers S, Neukom R et al (2016) Tambora 1815 as a test case for high impact volcanic eruptions: earth system effects. Wiley Interdiscip Rev Climate Change 7(4):569–589

Reason C (2017) Climate of Southern Africa. In: Oxford research encyclopedia of climate science. Oxford University Press, Oxford. https://doi.org/10.1093/acrefore/9780190228620.013.513

Schmidt G, Jungclaus JH, Ammann C, Bard E, Braconnot P, Crowley T, Delaygue G, Joos F, Krivova N, Muscheler R, et al (2012) Climate forcing reconstructions for use in PMIP simulations of the Last Millennium (v1. 1). Geosci Model Develop 5(1):185–191

Small RJ, Curchitser E, Hedstrom K, Kauffman B, Large WG (2015) The Benguela upwelling system: quantifying the sensitivity to resolution and coastal wind representation in a global climate model. J Climate 28(23):9409–9432

Sundqvist HS, Holmgren K, Fohlmeister J, Zhang Q, Matthews MB, Spötl C, Körnich H (2013) Evidence of a large cooling between 1690 and 1740 ad in Southern Africa. Sci Rep 3(1):1–6

Sydeman W, García-Reyes M, Schoeman D, Rykaczewski R, Thompson S, Black B, Bograd S (2014) Climate change and wind intensification in coastal upwelling ecosystems. Science 345(6192):77–80

Taylor KE, Stouffer RJ, Meehl GA (2012) An overview of CMIP5 and the experiment design. Bull Amer Meteorol Soc 93(4):485–498

Tim N, Zorita E, Hünicke B, Yi X, Emeis KC (2016) The importance of external climate forcing for the variability and trends of coastal upwelling in past and future climate. Ocean Sci 12(3):807–823

Tingley MP, Craigmile PF, Haran M, Li B, Mannshardt E, Rajaratnam B (2012) Piecing together the past: statistical insights into paleoclimatic reconstructions. Quat Sci Rev 35:1–22

Tyson PD, Lindesay JA (1992) The climate of the last 2000 years in Southern Africa. Holocene 2(3):271–278

Woodborne S, Hall G, Robertson I, Patrut A, Rouault M, Loader NJ, Hofmeyr M (2015) A 1000-year carbon isotope rainfall proxy record from South African baobab trees (Adansonia digitata l.). PLoS One 10(5):e0124202

Southern Africa Climate Over the Recent Decades: Description, Variability and Trends

6

Mathieu Rouault, Bastien Dieppois, Nele Tim, Birgit Hünicke, and Eduardo Zorita

Abstract

South of 15°S, southern Africa has a subtropical climate, which is affected by temperate and tropical weather systems and comes under the influence of the Southern Hemisphere high-pressure systems. Most rainfall occurs in austral summer, but the southwest experiences winter rainfall. Much of the precipitation in summer is of convective origin forced by large-scale dynamics. There is a marked diurnal cycle in rainfall in summer. The El Niño Southern Oscillation (ENSO) influences interannual rainfall variability. In austral summer, drought tends to occur during El Niño, while above-normal rainfall conditions tend to follow La Niña. During El Niño, higher than normal atmospheric pressure anomalies, detrimental to rainfall, occur due to changes in the global atmospheric circulation. This also weakens the moisture transport from the Indian Ocean to the continent. The opposite mechanisms happen during La Niña. On top of the variability related to ENSO, the Pacific Ocean also influences the decadal variability of rainfall. Additionally, the Angola Current, the Agulhas Current, the Mozambique Channel and the southwest Indian Ocean affect rainfall variability. Over the last 40 to 60 years, near-surface temperatures have increased over almost the whole region, summer precipitation has increased south of 10°S, and winter precipitation has mostly decreased in South Africa. Meanwhile, the

M. Rouault
Nansen Tutu Center for Marine Environmental Research, Department of Oceanography, University of Cape Town, Cape Town, South Africa

B. Dieppois (✉)
Centre for Agroecology, Water and Resilience, Coventry University, Coventry, UK

N. Tim · B. Hünicke · E. Zorita
Helmholtz-Zentrum Hereon, Institute for Coastal Systems – Analysis and Modeling, Geesthacht, Germany

© The Author(s) 2024
G. P. von Maltitz et al. (eds.), *Sustainability of Southern African Ecosystems under Global Change*, Ecological Studies 248,
https://doi.org/10.1007/978-3-031-10948-5_6

149

Agulhas Current and the Angola Current have warmed, and the Benguela Current has cooled.

6.1   Annual Cycle of Rainfall in Southern Africa

Climatologically speaking, sub-Saharan Africa can be divided into four regions: (i) West Africa, a classic monsoon climate; (ii) East Africa, fed by an intense flux of moisture coming from the Indian Ocean; (iii) Central Africa, which has a tropical climate, hot and wet with rainfall maximum following an annual latitudinal pattern and (iv) southern Africa, a relatively dry and hot climate that can be found south of 15°S to 20°S depending on longitude (Fig. 6.1). Therefore, in this chapter, we will define southern Africa as the continental areas south of 15°S. While there is a high range in annual rainfall and timing at the regional scale, southern Africa is characterized by distinct summer and winter rainfall regions (Figs. 6.2 and 6.3). There is also a region where rainfall is almost equal all year long to the far south. The longitudinal rainfall variation is due to a warm ocean to the east composed of the Agulhas Current, the Mozambique Channel and the southwest Indian Ocean. The longitudinal variation is also due to the South Atlantic high-pressure system's influence on the west, creating subsidence and preventing rainfall. The subsidence created to the east by the South Indian high-pressure system is offset by east-to-west moisture flux from the ocean to the continent and the presence of a convergence zone leading to air uplift and rain. Figure 6.2 shows the spatial distribution of the wettest month of the year in southern Africa. The red and blue domains are the area used to calculate the summer and winter rainfall total presented in Fig. 6.3 by averaging all grid points of each domain. It is also used to calculate the summer Standardized Precipitation Index (SPI) and calculate the Summer Rainfall Index (SRI), as was done by Dieppois et al. (2016, 2019).

Fig. 6.1 Mean 1950–2016 Global Precipitation Climatology Centre (GPCC) January to December annual total rainfall (mm) for southern Africa

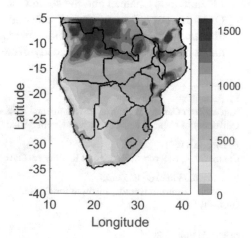

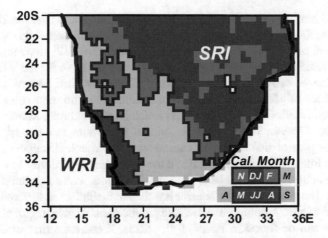

Fig. 6.2 Spatial distribution of southern Africa's wettest month of the year (calendar month: red, summer: November to March; blue winter: April to September). Lines and colors delineate the area used to calculate summer and winter rainfall averages and indices (SRI and WRI)

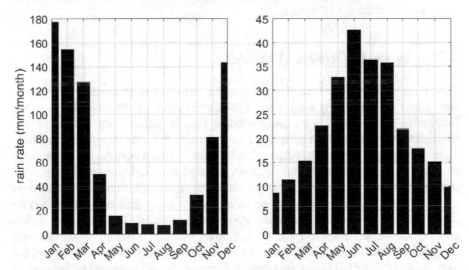

Fig. 6.3 Mean 1950–2016 annual cycle of rainfall in mm/month for southern Africa's southwest winter rainfall region (left) and summer rainfall region (right) as defined in Fig. 6.2

The greatest precipitation rate and maximum timing range are especially marked in South Africa. South Africa is divided into eight climatic regions by the South African Weather Service (Rouault and Richard 2003). The North-Western Cape has very little rainfall, with a winter maximum in June of 30 mm. The Southwestern Cape also has a maximum in June but differs with a maximum of 70 mm. Both regions constitute the winter rainfall region. The South Coast, which encompasses most of the Garden Route, experiences regular rainfall all year long, from about

30 mm to 40 mm per month, reaching about 100 mm in places over the Tsitsikamma region during October and November. There are even substantial spatial variations in rainfall within a region. For instance, Philippon et al. (2012) show that maximum monthly rainfall can vary from 240 mm to 30 mm per month for the Western Cape, primarily due to orography. The following five regions are part of the southern Africa summer rainfall region and have a maximum at different summer months. The Southern Interior has its maximum rainfall in late summer, March, with about 60 mm. The Western Interior has its rainfall maximum in January, with about 130 mm, but part of that region has a maximum in March (Dieppois et al. 2016). The Central Interior, KwaZulu-Natal and the North-Eastern Interior have maximum rainfall in January. KwaZulu-Natal is the wettest region, with a maximum in January of 130 mm. The North-Eastern Interior has the most significant difference between summer and winter. A comprehensive study of the annual cycle, regional differences and timings can be found in Favre et al. (2015). It indicates that while a rainfall deficit of 10 mm for a given month could be considered a drought in one region, it is not in another. Furthermore, the onset and demise of the rainy season are different for the eight areas making water management and mitigation of agricultural or hydrological drought difficult.

6.2   Synoptic Drivers of Rainfall

In southern Africa, much of the precipitation received in summer is of convective origin and forced by large-scale dynamics (Tyson and Preston-White 2000). Southern Africa has a subtropical climate and is affected by temperate and tropical weather systems. The country comes under the influence of the Southern Hemisphere high-pressure systems, but a heat low is found over the subcontinent in summer. This helps break the subsidence associated with high-pressure systems that prevent rain from occurring, allowing a diurnal cycle of rainfall to exist (Rouault et al. 2013). Most of the interior lies on an elevated plateau, and orography plays an important role in rainfall (Tyson and Preston-White 2000). The southwest region and the west coast receive most of their precipitation in austral winter through temperate systems such as cold fronts and cut-off lows, while the rest of the country gets most of its rainfall in austral summer. Summer rainfall is caused by large-scale synoptic systems leading to convection, but the diurnal cycle of rain has a substantial effect in the interior and along the East Coast. Such large rain-bearing systems include Tropical Temperate Troughs (TTTs; Chikoore and Jury 2010; Vigaud et al. 2012; Hart et al. 2012; Macron et al. 2014), Cut Off Lows (Favre et al. 2013), Mesoscale Convective Cloud systems (Blamey and Reason 2013) and occasionally tropical cyclones or tropical low-pressure systems associated with easterly waves (Malherbe et al. 2012). A substantial amount of rainfall in the summer rainfall region of southern Africa is due to TTTs and attendant cloud bands. The relative contribution of TTTs to rainfall during the summer ranges from 30% to 60%. While the annual cycle is dominant over southern Africa, intraseasonal rainfall oscillations occur with spectral peaks around 20 and 40 days associated with TTTs

(Chikoore and Jury 2010). The Madden-Julian Oscillation also affects rainfall on a 40–60 days' timescale (Pohl et al. 2007). Over the South Coast, nearly half of the annual rainfall is due to ridging anticyclones (Engelbrecht et al. 2015). The South Indian Anticyclone over the southwest Indian Ocean, Mozambique Channel and the Agulhas Current also plays an essential role in the source of moisture transported onto the interior plateau (Rapolaki et al. 2020; Imbol et al. 2021). Severe weather and flooding can occur due to cut-off lows, especially when the South Indian Anticyclone is quasi-stationary or blocking. Cut-off lows contribute significantly to rainfall over South Africa, especially along the south and east coasts and the Karoo regions (Favre et al. 2013). With some regional exceptions, spring is the season of the most substantial contribution by cut-off lows to annual rainfall. Over the Kalahari region, the most considerable contribution by cut-off lows to rainfall occurs during late summer. In contrast, the coastal regions, particularly the south coast, receive the most considerable cut-off low contribution to rainfall during the winter season (Favre et al. 2013). Over the south coast, cut-off lows have been associated with the autumn rainfall peak (Engelbrecht et al. 2015). For South Africa, the standardized amplitudes, indicative of the strength of the diurnal cycle across a region, are most substantial over the interior and along the East Coast, with up to 70% explained variance of hourly rainfall associated with the diurnal cycle (Rouault et al. 2013). The time of maximum precipitation is late afternoon to early evening in the interior and midnight to early morning along the Agulhas Current and inland in the northeast of the country (Rouault et al. 2013). In general, annual mean rainfall is overestimated by the regional climate model in South Africa (Favre et al. 2015), and also ocean–atmosphere global climate models (Dieppois et al. 2015, 2019). However, despite these general overestimations of rainfall totals, regional climate models have been demonstrated to simulate the main features of the annual cycle in circulation patterns and rainfall realistically across the southern African region (Engelbrecht et al. 2015).

6.3   Interannual Variability

The climate of southern Africa is highly variable and vulnerable to extremes such as droughts and floods. During the last decades, much has been gained on how the oceans can influence the climate of southern Africa on interannual scale (Richard et al. 2000, 2001; Fauchereau et al. 2003, 2009; Pohl et al. 2010; Crétat et al. 2012).

ENSO is the primary driver of southern Africa's rainfall interannual variability. Figure 6.4a shows the average normalized sea surface temperature and rainfall anomalies during the mature phase of El Niño in austral summer since 1982. This would be the average difference from normal during a typical El Niño event in austral summer. The normalized anomaly, which is presented in Fig. 6.4a, is a departure from the mean of November to March divided by the corresponding climatological standard deviation for 11 El Niño events. For instance, for values superior to one mean the anomaly is above one standard deviation. Roughly 66% of the values are between −1 and −1 standard if the data is normally distributed.

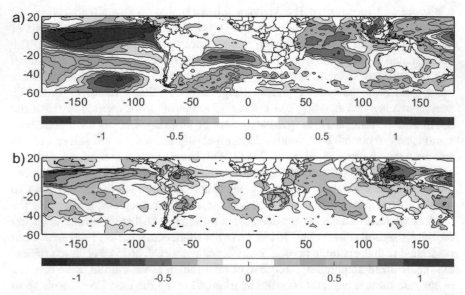

Fig. 6.4 (a), top: Average November–March sea surface temperature seasonal standardized anomalies during the mature phase of El Niño in austral summer (anomaly from the mean of November to March divided by the climatological corresponding standard deviation for 11 El Niño events). Blue/green is colder than normal; yellow/red is warmer than normal. (b), bottom: Average November–March rainfall seasonal standardized anomalies during the mature phase of El Niño in austral summer. Blue/green is wetter than normal; yellow/red is dryer than normal

As shown in Fig. 6.4a, most sea surface temperature anomalies are due to changes in global atmospheric and regional oceanic circulation, which are associated with El Niño. The impact of ENSO on sea surface temperature, land temperature, wind speed and rainfall offers predictability at the seasonal scale. This is because ENSO events' starting months precede the southern African rainy season, and atmospheric models relatively well reproduce this lagged relationship.

Note that we use the Oceanic Niño Index (ONI) provided by the Climate Prediction Center to detect El Niño and La Niña years on Figs. 6.4 and 6.5. In addition, in Fig. 6.4b, the precipitation rate is estimated from satellite remote sensing using the 2.5-by-2.5 degrees resolution Global Precipitation Climate Project dataset available since 1979. For sea surface temperature, we use the 1-by-1-degree resolution Reynolds SST Optimally Interpolated, available only since 1982. For that reason, we have selected the following 11 El Niños, 1982/1983, 1986/1987, 1987/1988, 1991/1992, 1994/1995, 1997/1998, 2002/2003, 2004/2005, 2006/2007, 2009/2010 and 2014/2015.

Although early works suggested that there were no strong relation between the strength of El Niño and the intensity and spatial extension of the southern African drought (Rouault and Richard 2003, 2005), Ullah et al. (2023) highlight significant statistical relationships between El Niño and large-scale extreme rainfall events. Indeed, seven of the ten strongest droughts since the 1950s happened during the

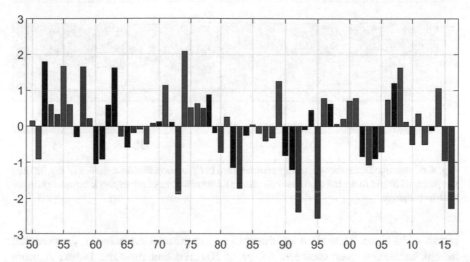

Fig. 6.5 November to March southern Africa summer rainfall Standardized Precipitation Index from 1950 to 2016. Red bars denote El Niño, and blue bars denote La Niña austral summers according to the ONI

mature phase of El Niño (Fig. 6.5), this provides ample early warning for drought monitoring. Droughts lasting two seasons always involve El Niño (Rouault and Richard 2003, 2005). The relationship between El Nino and drought occurrence has been particularly strong since the late 1970s (Fauchereau et al. 2003; Richard et al. 2000, 2001). However, this statistical relationship is not linear. For instance, the 1997/1998 strong El Niño did not lead to widespread drought (Lyon and Mason 2007), and severe drought can occur during a weak El Niño (Rouault and Richard 2005). Such nonlinear behavior in the El Niño-rainfall teleconnection has a significant impact on seasonal forecasts built on linear statistical models or in coupled or noncoupled general atmospheric circulation models, which, therefore, tends to be overconfident in predicting drought when El Niño occurs (Landman and Beraki 2012). Nevertheless, it may be noted that seasonal forecasts are skillful over the summer rainfall region of southern Africa during periods exhibiting significant ENSO forcing, even more so during La Niña, while forecasts are generally not skillful during neutral years (Landman and Beraki 2012).

The mechanisms linking the Pacific and southern Africa climate are relatively well understood and reasonably well simulated in state-of-the-art climate models (Dieppois et al. 2016, 2019). During El Niño events, near-surface divergence, detrimental to rainfall, is observed over southern Africa due to changes in the Walker and Hadley circulations. Along with the anomalous near-surface divergence, high-pressure anomalies (Fig. 6.6) over the landmass during El Niño also reduce the maritime moisture transport from the Indian Ocean to southern Africa.

A mid-tropospheric anticyclone becomes established and persistent during El Niño leading (not shown) to the anomalies presented in Fig. 6.6. Large-scale mid-troposphere disturbances (around 5000 m) are present globally during the mature

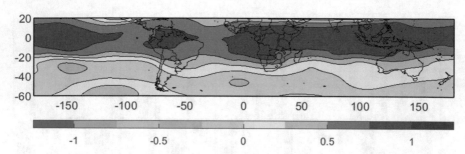

Fig. 6.6 Average December–February standardized El Niño normalized anomalies of the 500 hPa geopotential height in austral summer indicating global anomalies of atmospheric circulation in the mid-troposphere

phase of El Niño. There is a significant positive correlation between geopotential height anomalies over southern Africa at 500 hPa and the ONI Index. A more robust than normal Botswana high-pressure system causes enhanced subsidence for prolonged periods over much of the subcontinent. Moreover, El Niño is associated with extratropical atmospheric Rossby waves propagating poleward, influencing the southern hemisphere mid-latitude storm track (Fig. 6.6), and that could be responsible for an eastward shift of the South Indian Convergence Zone (Cook 2000, 2001). The eastward shift of the South Indian Convergence Zone is visible in Fig. 6.4b (bottom), where most of the large-scale synoptic-scale rain-bearing systems that affect southern Africa, such as Tropical temperate Troughs, preferably develop. The Indian Ocean warming during El Niño (Fig. 6.4a) also shifts atmospheric convection and rainfall eastward above the ocean, increasing subsidence above southern Africa and reducing the maritime moisture flux to the continent. During La Niña, a low-pressure anomaly is found over southern Africa, which favors the transport of moisture into southern Africa from the Indian Ocean, and consequently, rainfall is enhanced. Cut-off lows occur more frequently in a latitudinal band situated at lower latitudes during the La Niña years compared to El Niño years, during which these systems are located too far south of the country to contribute to seasonal rainfall amounts (Favre et al. 2012). El Niño and La Niña also change the wind strength along the coast and influence the ocean. In the upwelling system of the West Coast of South Africa, the Southern Benguela Current, El Niño often triggers lower than normal wind, weaker upwelling and warmer sea surface temperature due to the mid-latitude low-pressure anomalies shown in Fig. 6.6 (Rouault et al. 2010; Dufois and Rouault 2012; Tim et al. 2015; Blamey et al. 2015) and the opposite effect happens in the Northern Benguela (Rouault and Tomety 2022). During La Niña, the opposite occurs. Refer to Chap. 9 in this volume for more information on the northern and southern Benguela systems. El Niño also impacts the Western Cape rainfall and wind patterns in winter (Philippon et al. 2012).

Other modes of climate variability have been correlated to southern African rainfall, such as the Antarctic Annular Oscillation in winter, which is also called Southern Annular Mode (Reason and Rouault 2005; Malherbe et al. 2016) and the

Table 6.1 NDJFM synchronous cross-correlation between Summer Rainfall Index Standardized Precipitation Index (SRI SPI), ONI, South Indian Ocean Subtropical Dipole (SIOD) and Antarctic Annular Oscillation (AAO). In bold are correlations significant at the 95% confidence level

	SRI SPI	ONI	SIOD	AAO
SRI SPI	1	−0.67	**0.42**	−0.03
ONI		1	−0.45	**−0.1**
SIOD			1	0.16
AAO				1

South Indian Ocean Subtropical Dipole in summer (Behera and Yamagata 2001; Hoell et al. 2017).

Table 6.1 shows the correlation between the southern rainfall summer Standardized Precipitation Index, ONI, SIOD and AAO for austral summer (November to March). The strongest correlation, −0.67, is between ONI and SRI. There is a weaker correlation, 0.42, between SIOD and SRI but there is also a correlation of the same magnitude, −0.45, between ONI and SIOD, which means that the SIOD does not necessarily impact southern African rainfall directly but could be a symptom of the effect of ENSO on both the Indian Ocean and southern African rainfall although Hoell et al. (2017) proposed that the phase of the SIOD can disrupt or augment the southern Africa precipitation response to ENSO.

Next, we investigate the impact of ENSO on Water Management Areas and stream flow of South Africa, focusing on the seasonal average measured flow of water in all rivers of a specific Water Management Area. Observed stream flow monthly time series data were summed in summer from November to March per Water Management Areas from 1969 to 2004. The most striking result is that for Water Management Areas of South Africa situated in the summer rainfall region (Fig. 6.2), flows are between 1.6 and 2.8 times higher during La Niña years than during El Niño years (Rouault 2014). This confirms and extends the findings of Landman et al. (2001). For instance, from the 1970s, for WMA 6, Usutu Mhlathuze Swazi WMA (including the following major rivers: the Usutu River, Pongola River, Mhlathuze River, Mfolozi River and Mkuze River and covers the Oedertrouw dam and the am Mhlathuze River dam; Fig. 6.7), the five driest years are all El Niño years

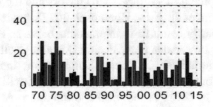

Fig. 6.7 Total summer stream flow volume averaged for all stations of the Water Management Area 6 (Usutu Mhlathuze Swazi). El Niño years are in red, and La Niña years are in blue. Stream flow is summed from November to March

(2015/2016, 1982/1983, 1994/1995, 2014/2015 and 1991/1992), and four out of five wettest years are La Niña years (1983/1984, 1987/1988, 2000/2001, 1988/1989 and 2010/2011). Four of the five lowest stream flow volume summer seasons occurred during El Niño (1982/1983, 2015/2016, 1994/1995, 1984/1985 and 1991/1992), although one occurred during la Niña. Three out of five of the highest stream flow volume years happened during La Niña.

Regional oceanographic and atmospheric features that have been reported to influence the southern African climate are the Agulhas Current (Jury et al. 1993; Nkwinkwa Njouodo et al. 2018; Imbol Nkwinkwa et al. 2021), the Angola Current (Rouault et al. 2003; Desbiolles et al. 2020; Koseki and Imbol Koungue 2021), the Mozambique Channel (Barimalala et al. 2020), the Angola low-pressure system (Crétat et al. 2019; Desbiolles et al. 2020) and the Botswana high-pressure system (Driver and Reason 2017). The Southern Annular Mode (Reason and Rouault 2005; Malherbe et al. 2016), the South Indian Ocean Dipole (Behera and Yamagata 2001; Hoell et al. 2017; Dieppois et al. 2016, 2019; Ullah et al. 2023) have also been linked to climate variability in southern Africa.

6.4   Decadal Variability of Southern Africa's Climate

As illustrated in Fig. 6.8, while interannual variability (2–8 years) mainly related to ENSO explains approximately 70% of the total rainfall variance on average over southern Africa, a third of rainfall variability is explained by variations on quasi-decadal (8–13 years) and interdecadal timescales (15–23 years; Dieppois et al. 2016, 2019). In addition, we note that the amplitudes of those decadal variations strengthen and weakened over the twentieth century in summer and winter rainfall areas (Dieppois et al. 2016). Teleconnections with global sea surface temperature, in particular, the Pacific Ocean and atmospheric circulation anomalies, reminiscent of the impact of ENSO, were shown to impact southern Africa's summer and winter rainfall at the interdecadal and quasi-decadal timescales (Dieppois et al. 2016, 2019). This could help understanding why, during some periods, ENSO does not strongly influence southern Africa as a cold or warm sea surface temperature background in the Pacific could make the impact of ENSO on the atmosphere weaker or stronger (Fauchereau et al. 2009; Pohl et al. 2018). This might also help understand the development of various ENSO flavors and their contrasted effects on southern Africa (Ratnam et al. 2014; Hoell et al. 2017). Indeed, the development of ENSO flavors might partly result from interactions between Pacific SST anomalies occurring on different timescales (e.g., Pacific Decadal Oscillation SST anomalies influencing ENSO anomalies). In austral summer, on interdecadal timescales, decadal ENSO-like forcing of the Pacific Decadal Oscillation (PDO) decadal variance indeed leads to shifts in the Walker circulation. At the regional scale, it contributes to ocean-atmosphere anomalies in the South Indian Ocean to shift the South Indian Convergence Zone toward the continent. According to Dieppois et al. (2016, 2019), combinations of ENSO forcing and ocean–atmosphere anomalies in the South Indian Ocean are thus needed in the relationship between Pacific Decadal

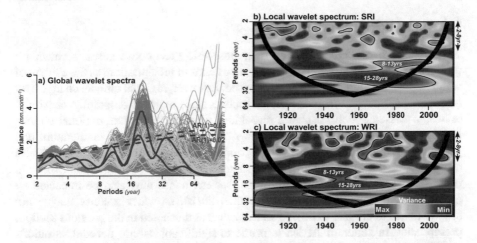

Fig. 6.8 Timescale patterns of variability for summer (SRI) and winter (WRI) southern African rainfall index (**a**) Global wavelet spectra of the SRI (red) and the WRI (blue), and of every grid-point used for their calculations (light red and blue). The dashed blue and red lines indicate the red noise spectra (**b–c**) Continuous wavelet power spectrum of the SRI and the WRI. Bold lines delineate the area under which power can be underestimated due to edge effects, wraparound effects and zero padding; thin contour lines show the 95% confidence limits. Note that SRI and WRI variability could be poorly represented from the mid-1990s due to a deficient number of rain gauges available in the Climatic Research Unit (CRU TS version 3.23) precipitation and temperature dataset that could substantially reduce interannual to interdecadal variability

Oscillation and southern African rainfall. The Interdecadal Pacific Oscillation (IPO) also drives such anomalies at the quasi-decadal timescale. At each timescale, colder or warmer Pacific SSTs result in changes to the Walker circulation: these interact with ocean-atmospheric modifications in the South Indian Ocean, which act together with varying degrees of importance to alter the intensity and longitudinal location of the South Indian Convergence Zone and, thus, to modulate the Tropical Trough developments and deep-convection over southern Africa (Dieppois et al. 2016, 2019; Pohl et al. 2018). The influence of tropical Pacific Ocean climate variability on austral winter southern African rainfall variability is almost negligible over the twentieth century. However, according to Philippon et al. (2012), the influence of tropical Pacific climate variability may have become more important since the 1970s. Decadal winter rainfall variability is strongly related to regional changes in the Southern hemisphere's subtropical high-pressure system and, thus, mid-latitude westerly low-pressure activity (Dieppois et al. 2016). This corroborates previous findings by Reason and Rouault (2005), highlighting that the Southern Annual Mode (SAM) influence is strong in austral winter.

6.5 Current Climate Trends

The increase in the concentration of atmospheric greenhouse gases is causing a global rise in temperatures and possibly also trends in precipitation, as the content of water vapor in the atmosphere has increased. However, regional climate change may be somewhat different from the global picture, since regional orography, proximity to water masses and atmospheric circulation may also modulate regional climate trends. In the following paragraphs, we present a summary of temperature and precipitation trends in southern Africa over the last decades, as these two variables are probably the most important for stakeholders. However, one should bear in mind that trends in water availability for societal needs may also be impacted by nonclimatic management decisions (Muller 2018), so water scarcity maybe not always linked to climate trends. In addition, and as discussed in the previous section, precipitation in southern Africa is prone to significant natural decadal variations (Dieppois et al. 2016, 2019; Mahlalela et al. 2019), so trends derived over the recent decades may result from the impact of greenhouse gas forcing but may also reflect shorter natural swings. Disentangling both, i.e., unambiguously attributing recent observed trends to anthropogenic climate forcing, requires a detailed analysis of observations and model simulations. In that sense, only a few studies highlighted an increase in the likelihood of extreme climate conditions, notably drought and fire-prone weather, in the southwestern regions of southern Africa (Otto et al. 2018; Zscheischler and Lehner 2022; Liu et al. 2023). In addition, according to climate future projections, Pohl et al. (2017) highlighted that southern Africa could expect fewer rainy days and more extreme rainfall by the end of the twenty-first century. This is also consistent with recent analyses identifying an increased severity of drought in South Africa from the mid-1980s (Phaduli 2018; Jury 2018). Therefore, an ongoing analysis of all available data sets to identify long-term trends in societal climate impacts due to climate change or other human impacts is paramount.

Beyond South Africa, southern Africa is not as densely covered by a net of meteorological stations as western Europe or North America, so estimating long-term climate trends from point observations may provide only a partial illustration of regional climate trends. Previous studies focusing on long-term trends were mainly based on meteorological stations' data (Jury 2018; Kruger and Nxumalo 2017; Ullah et al. 2021). Here, we use climate reanalysis and observations.

The reanalysis is essentially a retrospective hindcast using weather prediction models that incorporate the information from available meteorological observations. The trends in precipitation and air temperature over the last decades are shown here with the global reanalysis data sets, ERA5, of the ECMWF (European Centre for Medium-Range Weather Forecasts, Hersbach et al. 2020) and the observational data set CRU (Climate Research Units gridded time-series data set; Harris et al. 2020). Both are analyzed over the period 1979–2020. In Fig. 6.9, the trends in precipitation of the two data sets are shown for the two rainfall seasons, winter rainfall in June–August (JJA) and summer rainfall in December–February (DJF). The summer rainfall zone (SRZ) covers most of southern Africa, and the winter

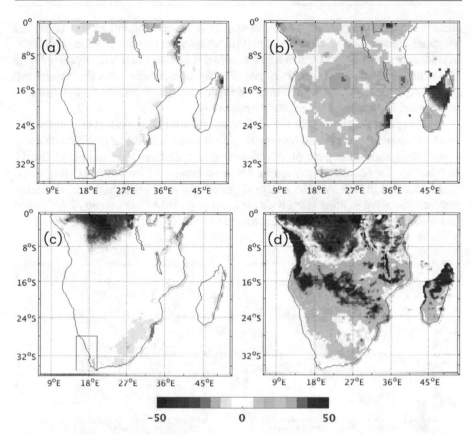

Fig. 6.9 Precipitation trends of CRU (**a**, **b**) and ERA5 reanalysis (**c**, **d**). Subfigures **a**, and **c**, left, are for the winter rainfall season (June–July–August), and subfigures **b** and **d**, right, are for the summer rainfall season (December–January–February). The winter rainfall zone is marked with black rectangular. Trends are given in mm/decade for the period 1979–2020. Based on typical standard deviation, trends of (**a**) 0.77, (**b**) 15.76, (**c**) 2.14 and (**d**) 31.8 mm/decade and larger are significant at a 95% level

rainfall zone (WRZ, black box in Figure 6.9a and c) covers the cape region in the southwest of South Africa.

For the winter rainfall zone, CRU and ERA5 provide somewhat different results. CRU shows a slight decrease in precipitation (Fig. 6.9a), while ERA5 shows an increase in the southwest of the winter rainfall zone (Fig. 6.9c), which has been found previously by Onyutha (2018) and MacKellar et al. (2014a, 2014b). In DJF in the summer rainfall zone, the precipitation has increased in most parts of southern Africa (Fig. 6.9b, d, Onyutha 2018; MacKellar et al. 2014a, 2014b; Kruger and Nxumalo 2017). Ullah et al. (2021) found a significant trend in the intensity of wet spells from 1965 to 2015 using daily observation and ERA5. Both CRU and ERA5 data sets show a drying at parts of the South African coast, which has also been found at the southeast coast by Jury (2018). Thus, the precipitation has mainly increased in the SRZ and decreased in the winter rainfall zone over the last 40 years. It must also be considered that the assessment provided by the sixth IPCC assessment report does not differentiate between summer and winter precipitation (Ranasinghe et al. 2021), so it is not directly comparable with the assessment provided here. Also, the available precipitation data sets do not always agree in this region, illustrating their inherent uncertainty.

Air temperature trends near the surface are shown in Fig. 6.10. The temperature has risen along the coast in both seasons. This has also been found in station data sets (Jury 2013) and in the pace of record-setting temperatures (McBride et al. 2021). In addition, ERA5 shows an area in central southern Africa where cooling or no trend occurred in DJF over the analyzed period. An even more pronounced cooling has been detected in other reanalysis data sets (e.g., ERA-Interim and JRA-55) and has also been found in annual mean trends by Jury (2013). Further analysis of the

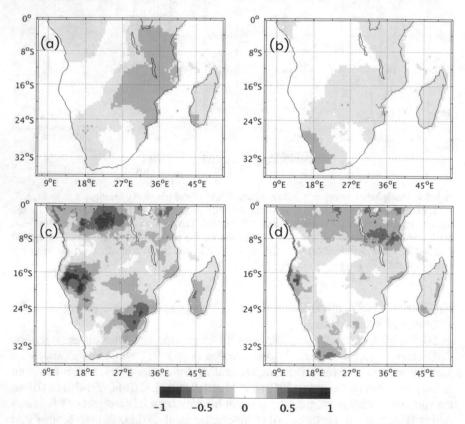

Fig. 6.10 2 m air temperature trends of CRU (**a**, **b**) and ERA5 reanalysis (**c**, **d**). Subfigures **a**, and **c**, left, are for winter (June–July–August), and subfigures **b** and **d**, right, are for summer (December–January–February). Trends are given in °C/decade for the period 1979–2020. Based on typical standard deviation, trends of (**a**) 0.09, (**b**) 0.11, (**c**) 0.19 and (**d**) 0.16 °C/decade and larger are significant at a 95% level

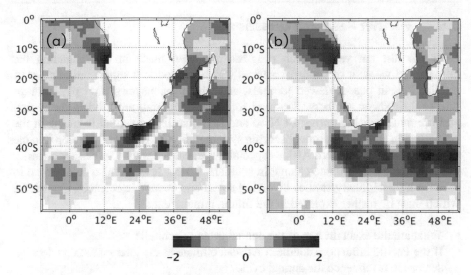

Fig. 6.11 Sea surface temperature (SST) trends of HadISST1, (**a**, left) for winter (June–July–August), (**b**, right) for summer (December–January–February). Trends are given in °C/period, which is 1958–2019. Based on typical standard deviation, trends of (**a**) 0.17 and (**b**) 0.35 degree/period and larger are significant at a 95% level

JRA-55 data set has shown that this cooling trend is likely an artifact of the data assimilation procedure. Fewer observational data can be assimilated in earlier years, and the introduction of increasingly more data can lead to this false-negative trend. The warming along the coast is accompanied by the warming of the adjacent oceans. The trends of sea surface temperature, as shown in Fig. 6.11, have been calculated using the interpolated observational data set, HadISST1 (Rayner et al. 2003). SSTs in the Agulhas Current and in the retroflexion, the region where the Agulhas Current turns eastward, have increased (Fig. 6.11 and Rouault et al. 2009), while SSTs in the Benguela Current have decreased (Fig. 6.11).

This warming in the Agulhas Current along the African coast of the Indian Ocean is upwind of the African continent. It may contribute to the positive trend in precipitation as the sea surface temperature of the Agulhas Current system influences southern Africa's rainfall (Nkwinkwa Njouodo et al. 2018; Imbol Nkwinkwa et al. 2021). The warming of the ocean off Angola could also have contributed to increased inland rainfall due to more significant atmospheric water content and subsequently increased moisture flux from the tropical Atlantic (Rouault et al. 2003).

To sum up, as found in basically all data sets analyzed here, the Agulhas Current, the Angola Current, the tropical Atlantic and the Indian Ocean have warmed, near-surface air temperatures along the coasts have risen, and rainfall has increased over large parts of the summer rainfall zone of southern Africa during austral summer while it has decreased over coastal areas of South Africa in both rainfall seasons of the last few decades.

6.6   Further Research Questions

Over the last ten years, much progress has been made in understanding the
climate and weather of southern Africa, which has led to a significant number of
publications in peer-reviewed journals, and which our paper cannot pay homage
due to their sheer numbers. Seasonal forecasts of rainfall are used in the region
and are more reliable during El Niño and La Niña. It is essential to maintain the
precipitation and temperature observing system in South Africa and develop it in the
other countries of southern Africa especially because we want to precisely quantify
the effect of global climate change in southern Africa. It would also be essential to
foster transdisciplinary studies involving social and economic sciences. Questions,
which must be further explored in the future, remain, e.g.,

- What are the exact drivers of the annual cycle of rainfall?
- If the Pacific influences southern African climate at the interannual and decadal
  scale, can it influence the annual cycle?
- How do the annual cycle and interannual variability of climate impact regional
  processes relate to agriculture, water resources, management of cities and
  harbors, or any other relevant societal matter?
- Why do the maxima of summer rainfall happen in different months across the
  SRZ?
- Is there a climate fluctuation in the probability of extreme events?
- What are the drivers of the Hadley circulation, Walker circulation and the
  temperate circulation, and what are their roles in the annual cycle of rainfall and
  temperature?
- What is the role of the oceans and land feedback on the annual cycle of
  precipitation and temperature?
- Does the extratropical ocean affect the atmospheric circulation of South Africa?
- What is the contribution of the different synoptic systems and the diurnal cycle
  of rainfall to the annual cycle of precipitation and temperature?
- Why is the timing of the maximum of the diurnal cycle at a different time of the
  day in southern Africa?
- Why are the onset and demise of the rainy season occurring at different times of
  the year?
- Why is there no robust relationship between the strength of El Niño or La Niña
  and their impact on the intensity and spatial extent of southern African rainfall
  and temperature?
- Why is La Niña better correlated with southern African rainfall than El Niño?
- What exact mechanisms link El Niño and La Niña to southern African rainfall?
- Why is ENSO not related to the southern African climate for some decades?
- What is the role of the adjacent ocean, namely the Agulhas Current, the Mozam-
  bique Channel and the Benguela Current, on the southern African Climate?
- What is the role of the Indian Ocean and the tropical Atlantic Ocean?
- Are all relevant processes included in the models used for seasonal rainfall and
  temperatures forecasts or for future climate projections for southern Africa?

Acknowledgments This work was supported by the NRF SARCHI chair on "modeling ocean-atmosphere-land interactions," the SPACE 2 project, the Nansen Tutu Centre, the EU H2020 TRIATLAS project under the grant agreement 817578, the Belmont Forum project EXEBUS and the PECO2 project. We like to thank NOAA for the NOAA OI V2 SST *data provided by the NOAA/OAR/ESRL PSL, Boulder, Colorado, USA.* We also thank Copernicus and ECMWF for making the ERA5 dataset easily accessible.

References

Barimalala R, Blamey RC, Desbiolles F, Reason CJ (2020) Variability in the Mozambique Channel trough and impacts on southeast African rainfall. J Clim 33(2):749–765. https://doi.org/10.1175/JCLI-D-19-0267.1

Behera SK, Yamagata Y (2001) Subtropical SST dipole events in the southern Indian Ocean. Geophys Res Lett 28:327–330. https://doi.org/10.1029/2000GL011451

Blamey RC, Reason CJC (2013) The role of mesoscale convective complexes in southern Africa summer rainfall. J Clim 26(5):1654–1668

Blamey LK, Shannon LJ, Bolton JJ, Crawford RJM, Dufois F, EversKing H, Griffiths CL, Hutchings L, Jarre A, Rouault M, Watermeyer K, Winker H (2015) Ecosystem change in the southern Benguela and the underlying processes. J Mar Syst 144:9–29

Chikoore H, Jury MR (2010) Intra-seasonal variability of satellite-derived rainfall and vegetation over southern Africa. Earth Interact 14:1–26

Cook KH (2000) The south Indian convergence zone and interannual rainfall variability over southern Africa. J Clim 13:3789–3804

Cook KH (2001) A southern hemisphere wave response to ENSO with implications for southern Africa precipitation. J Atmos Sci 15:2146–2162

Crétat J, Richard Y, Pohl B, Rouault M, Reason C, Fauchereau N (2012) Recurrent daily rainfall patterns over South Africa and associated dynamics during the core of the austral summer. Int J Climatol 32:261. https://doi.org/10.1002/joc.2266

Crétat J, Pohl B, Dieppois B, Berthou S, Pergaud J (2019) The Angola Low: relationship with southern African rainfall and ENSO. Clim Dyn 52(3):1783–1803

Desbiolles F, Howard E, Blamey RC, Barimalala R, Hart NC, Reason CJ (2020) Role of ocean mesoscale structures in shaping the Angola-low pressure system and the southern Africa rainfall. Clim Dyn 54:1–20. https://doi.org/10.1007/s00382-020-05199-1

Dieppois B, Rouault M, New M (2015) The impact of ENSO on southern African rainfall in CMIP5 ocean atmosphere coupled climate models. Clim Dyn 45(9):2425–2442. https://doi.org/10.1007/s00382-015-2480-x

Dieppois B, Pohl B, Rouault M, New M, Lawler D, Keenlyside N (2016) Interannual to Inter-decadal variability of winter and summer southern African rainfall, and their teleconnections. J Geophys Res: Atmospheres. Published online 121:6215

Dieppois B, Pohl B, Crétat J, Eden J, Sidibe M, New M, Rouault M, Lawler D (2019) Southern African summer-rainfall variability, and its teleconnections, on interannual to interdecadal timescales in CMIP5 models. Clim Dyn 53(5):3505–3527

Driver P, Reason CJC (2017) Variability in the Botswana high and its relationships with rainfall and temperature characteristics over southern Africa. Int J Climatol 37(S1):570–581. https://doi.org/10.1002/joc.5022

Dufois F, Rouault M (2012) Sea surface temperature in False Bay (South Africa): towards a better understanding of its seasonal and interannual variability. Cont Shelf Res 43:24–35. https://doi.org/10.1016/j.csr.2012.04.009

Engelbrecht CJ, Landman WA, Engelbrecht FA, Malherbe J (2015) A synoptic decomposition of over the cape south coast of South Africa. Clim Dyn 44:2589–2607

Fauchereau N, Trzaska S, Rouault M, Y., and Richard, Y. (2003) Rainfall variability and changes in southern Africa during the 20th century in the global warming context. Nat Hazards 29(2):139–145

Fauchereau N, Pohl B, Reason CJR, Rouault M, Richard Y (2009) Recurrent daily OLR patterns in the southern Africa/Southwest Indian Ocean region, implications for south African rainfall and teleconnexions. Clim Dyn 32:575. https://doi.org/10.1007/s00382-008-0426-2

Favre A, Hewitson B, Tadross M, Lennard C, Cerezo-Mota R (2012) Relationships between cut-off lows and the semiannual and southern oscillations. Clim Dyn 38:1473–1487

Favre A, Hewitson B, Lennard C, Cerezo-Mota R, Tadross M (2013) Cut-off lows in the South Africa region and their contribution to precipitation. Clim Dyn 41:2331–2351

Favre A, Philippon N, Pohl B, Kalognomou EA, Lennard C, Hewitson B, Nikulin G, Dosio A, Panitz HJ, Cerezo-Mota R (2015) Spatial distribution of precipitation annual cycles over South Africa in 10 CORDEX regional climate model present-day simulations. Clim Dyn 46:1799–1818. https://doi.org/10.1007/s00382-015-2677-z

Harris I, Osborn TJ, Jones P et al (2020) Version 4 of the CRU TS monthly high-resolution gridded multivariate climate dataset. Sci Data 7:109. https://doi.org/10.1038/s41597-020-0453-3

Hart NCG, Reason CJC, Fauchereau N (2012) Cloud bands over southern Africa: seasonality, contribution to rainfall variability and modulation by the MJO. Climate Dyn 41:1199–1212. https://doi.org/10.1007/s00382-012-1589-4

Hersbach H, Bell B, Berrisford P et al (2020) The ERA5 global reanalysis. Q J R Meteorol Soc 146:1999–2049. https://doi.org/10.1002/qj.3803

Hoell A, Funk C, Zinke J, Harrison L (2017) Modulation of the southern Africa precipitation response to the El Niño southern oscillation by the subtropical Indian Ocean dipole. Clim Dyn 48:2529–2540

Imbol Nkwinkwa AS, Rouault M, Keenlyside N, Koseki S (2021) Impact of the Agulhas current on southern Africa precipitation: a modelling study. J Clim 34(24):9973–9988

Jury MR (2013) Climate trends in southern Africa. S Afr J Sci 109(1):1–11

Jury MR (2018) Climate trends across South Africa since 1980. Water SA 44(2):297–307

Jury MR, Valentine HR, Lutjeharms JR (1993) Influence of the Agulhas current on summer rainfall along the southeast coast of South Africa. J App Meteo and Clim 32(7):1282–1287. https://doi.org/10.1175/1520-0450(1993)032

Koseki S, Imbol Koungue RA (2021) Regional atmospheric response to the Benguela Niñas. Int J Climatol 41:E1483–E1497

Kruger AC, Nxumalo MP (2017) Historical rainfall trends in South Africa: 1921–2015. Water SA 43(2):285–297

Landman WA, Beraki A (2012) Multi-model forecast skill for mid-summer rainfall over southern Africa. Int J Climatol 32:303–314

Landman WA, Mason SJ, Tyson PD, Tennant WJ (2001) Statistical downscaling of GCM simulations to streamflow. J Hydrol 252(1):221–236

Liu Z, Eden JM, Dieppois B, Conradie WS, Blackett M (2023) The April 2021 Cape Town wildfire: has anthropogenic climate change altered the likelihood of extreme fire weather? Bull Am Meteorol Soc 104:E298–E304

Lyon B, Mason SJ (2007) The 1997–98 summer rainfall season in southern Africa. Part I: Observations. J Clim 20(20):5134–5148

MacKellar N, New M, Jack C (2014a) Observed and modelled trends in rainfall and temperature for South Africa: 1960-2010. S Afr J Sci 110:1–13. https://doi.org/10.1590/sajs.2014/20130353

MacKellar N, New M, Jack C (2014b) Observed and modelled trends in rainfall and temperature for South Africa: 1960–2010. S Afr J Sci. 2014;110(7/8), Art. #2013-0353, 13 p. https://doi.org/10.1590/sajs.2014/20130353

Macron C, Pohl B, Richard Y, Bessafi M (2014) How do tropical temperate troughs form and develop over southern Africa? J Clim 27(4):1633–1647. https://doi.org/10.1175/JCLI-D-13-00175.1

Mahlalela PT, Blamey RC, Reason CJC (2019) Mechanisms behind early winter rainfall variability in the southwestern cape, South Africa. Climate Dyn 3(1):21–39

Malherbe J, Engelbrecht FA, Landman WA, Engelbrecht CJ (2012) Tropical systems from the Southwest Indian Ocean making landfall over the Limpopo River basin, southern Africa: a historical perspective. Int J Climatol 32:1018–1032. https://doi.org/10.1002/joc.2320

Malherbe J, Dieppois B, Maluleke P, Van Staden M, Pilay DL (2016) South Africa droughts and decadal variability. Nat Hazards 80:657–681

McBride CM, Kruger AC, Dyson L (2021) Trends in probabilities of temperature Records in the non-Stationary Climate of South Africa. Int J Climatol 2021:1–14

Muller M (2018) Cape Town's drought: don't blame climate change. Nature 59:174–176

Nkwinkwa Njouodo AS, Koseki S, Keenlyside N, Rouault M (2018) Atmospheric signature of the Agulhas current. Geophys Res Lett 45:5185–5193. https://doi.org/10.1029/2018GL077042

Onyutha C (2018) Trends and variability in African long-term precipitation. Stoch Environ Res Risk Assess 32:2721–2739. https://doi.org/10.1007/s00477-018-1587-0

Otto FEL, Wolski P, Lehner F, Tebaldi C, van Oldenborgh GJ, Hogesteeger S, Singh R, Holden P, Fuckar NS, Odoulami RC, New M (2018) Anthropogenic influence on the drivers of the Western cape drought 2015-2017. Environ Res Lett 13:124010

Phaduli E (2018) Drought characterization in South Africa under changing climate (Doctoral dissertation, University of Pretoria)

Philippon N, Rouault M, Richard Y, Favre A (2012) On the impact of ENSO on South Africa winter rainfall. Int J Climatol 32(15):2333–2347. https://doi.org/10.1002/joc.3403

Pohl B, Richard Y, Fauchereau N (2007) Influence of the madden–Julian oscillation on southern African summer rainfall. J Clim 20(16):4227–4242

Pohl B, Fauchereau N, Reason CJC, Rouault M (2010) Relationships between the Antarctic oscillation, the madden–Julian oscillation, and ENSO, and consequences for rainfall analysis. J Clim 23:238–254

Pohl B, Macron C, Monerie P A (2017) Fewer rainy days and more extreme rainfall by the end of the century in southern Africa. Sci Rep 7:46466

Pohl B, Dieppois B, Crétat J, Lawler D, Rouault M (2018) From synoptic to interdecadal variability in southern African rainfall: toward a unified view across time scales. J Clim 31(15):5845–5872

Ranasinghe R, Ruane AC, Vautard R, Arnell N, Coppola E, Cruz FA, Dessai S, Islam AS, Rahimi M, Carrascal DR, Sillmann J, Sylla MB, Tebaldi C, Wang W, Zaaboul R (2021) Climate change information for regional impact and for risk assessment. In: Masson Delmotte V, Zhai P, Pirani A, Connors SL, Péan C, Berger S, Caud N, Chen Y, Goldfarb L, Gomis MI, Huang M, Leitzell K, Lonnoy E, Matthews JBR, Maycock TK, Waterfield T, Yelekçi O, Yu R, Zhou B (eds) Climate Change 2021: The Physical Science Basis. Contribution of Working Group I to the Sixth Assessment Report of the Intergovernmental Panel on Climate Change. Cambridge University Press, Cambridge and New York, NY, pp 1767–1926. https://doi.org/10.1017/9781009157896.014

Rapolaki RS, Blamey RC, Hermes JC, Reason CJC (2020) Moisture sources associated with heavy rainfall over the Limpopo River basin, southern Africa. Clim Dyn 55(5):1473–1487

Ratnam JV, Behera SK, Masumoto Y, Yamagata T (2014) Remote effects of El Niño and Modoki events on the austral summer precipitation of southern Africa. J Clim 27:3802–3815

Rayner NA, Parker DE, Horton EB, Folland CK, Alexander LV, Rowell DP, Kent EC, Kaplan A (2003) Global analyses of sea surface temperature, sea ice, and night marine air temperature since the late nineteenth century. J Geophys Res 108(D14):4407. https://doi.org/10.1029/2002JD002670

Reason CJC, Rouault M (2005) Links between the Antarctic oscillation and winter rainfall over southwestern South Africa. Geophys Res Lett 32:L07705. https://doi.org/10.1029/2005GL0022419

Richard Y, Trzaska S, Roucou P, Rouault M (2000) Modification of the southern African rainfall variability/El Niño southern oscillation relationship. Clim Dyn 16:886–895

Richard Y, Fauchereau N, Poccard I, Rouault M, Trzaska S (2001) XXth century droughts in southern Africa spatial and temporal variability, teleconnections with oceanic and atmospheric conditions. Int J Climatol 21:873–885

Rouault M (2014) Impact of ENSO on South African Water Management Areas. Proceedings of 30th Annual Conference of South African Society for Atmospheric Science, 01–02 October 2014, Potchefstroom, South Africa. isbn 978-0-620-62777-1

Rouault M, Richard Y (2003) Spatial extension and intensity of droughts since 1922 in South Africa. Water SA 19:489–500

Rouault M, Richard Y (2005) Intensity and spatial extent of droughts in southern Africa. Geophys Res Lett 32:L15702. https://doi.org/10.1029/2005GL022436

Rouault M, Tomety FS (2022) Impact of the El Niño southern oscillation on the Benguela upwelling. J Phys Oceanogr, in press 52:2573

Rouault M, Florenchie P, Fauchereau N, Reason CJC (2003) South East Atlantic warm events and dn rainfall. Geophys Res Lett 29:13. https://doi.org/10.1029/2002GL014663

Rouault M, Penven P, Pohl B (2009) Warming of the Agulhas current since the 1980's Geophys. Res Lett 36:L12602. https://doi.org/10.1029/2009GL037987

Rouault M, Pohl B, Penven P (2010) Coastal oceanic climate change and variability from 1982 to 2009 around South Africa. Afr J Mar Sci 32(2):237–246

Rouault M, Roy SS, Balling RC (2013) The diurnal cycle of rainfall in South Africa in the austral summer. Int J Climatol 33:770–777. https://doi.org/10.1002/joc.3451

Tim N, Zorita E, Hünicke B (2015) Decadal variability and trends of the Benguela upwelling system as simulated in a high-resolution ocean simulation. Ocean Sci 11(3):483–502

Tyson PD, Preston-White RA (2000) The weather and climate of southern Africa. Oxford University Press Southern Africa, Oxford

Ullah A, Pohl B, Pergaud J, Dieppois B, Rouault M (2021) Intra-seasonal descriptors and extremes in southern African rainfall. Part I: summer climatology and statistical characteristics. Int J Climatol 42(9):4538–4563

Ullah A, Pohl B, Pergaud J, Dieppois B, Rouault M (2023) Intra-seasonal descriptors and extremes in southern African rainfall. Part II: summer teleconnections across multiple timescales. Int J Climatol, in press 43:3799

Vigaud N, Pohl B, Cretat J (2012) Tropical-temperate interactions over southern Africa simulated by a regional climate model. Climate Dyn 39:2895–2916. https://doi.org/10.1007/s00382-012-1314-3

Zscheischler J, Lehner F (2022) Attributing compound events to anthropogenic climate change. Bull Am Meteorol Soc 103(3):E936–E953

Projections of Future Climate Change in Southern Africa and the Potential for Regional Tipping Points

7

Francois A. Engelbrecht, Jessica Steinkopf, Jonathan Padavatan, and Guy F. Midgley ⓘ

Abstract

Southern Africa is a climate change hotspot with projected warming and drying trends amplifying stresses in a naturally warm, dry and water-stressed region. Despite model-projected uncertainty in rainfall change over the eastern escarpment of South Africa, strong model agreement in projections indicates that southern African is likely to become generally drier. Sharply increased regional warming and associated strong reductions in soil-moisture availability and increases in heat-waves and high fire-danger days are virtually certain under low mitigation futures. Changes are detectable in observed climate trends for the last few decades, including regional warming, drying in both the summer and winter rainfall regions, and increases in intense rainfall events. The southern African climate is at risk of tipping into a new regime, with unprecedented impacts, such as day-zero drought in the Gauteng province of South Africa, collapse of the maize and cattle industries, heat-waves of unprecedented intensity and southward shifts in intense tropical cyclone landfalls. Many of these adverse changes could be avoided if the Paris Accord's global goal were to be achieved, but research is urgently required to quantify the probabilities of such tipping points in relation to future levels of global warming. Adaptation planning is an urgent regional priority.

F. A. Engelbrecht (✉) · J. Steinkopf · J. Padavatan
Global Change Institute, Faculty of Science, University of the Witwatersrand, Witwatersrand, South Africa
e-mail: Francois.Engelbrecht@wits.ac.za

G. F. Midgley
School for Climate Studies, Stellenbosch University, Stellenbosch, South Africa

© The Author(s) 2024
G. P. von Maltitz et al. (eds.), *Sustainability of Southern African Ecosystems under Global Change*, Ecological Studies 248,
https://doi.org/10.1007/978-3-031-10948-5_7

169

7.1   Introduction

Southern Africa (here defined as Africa south of 10°S) was classified as a climate change hotspot by the Intergovernmental Panel on Climate Change (IPCC) Special Report on Global Warming of 1.5°C (SR1.5; Hoegh-Guldberg et al. 2018). It is a region with a warm climate and pronounced wet-dry seasonality, which is acknowledged to be water stressed in the context of naturally occurring droughts (Chap. 6), a growing population and the industrial ambitions of a developing economy (Chap. 4). Under low mitigation emission scenarios southern Africa is certain to become substantively warmer and likely drier (Engelbrecht et al. 2009, 2015b; Engelbrecht and Engelbrecht 2016; Hoegh-Guldberg et al. 2018; Lee et al. 2021), justifying its classification as a climate change hotspot, as under such scenarios, the options for adaptation will become limited. While drying is a general projection for the region, climate models also project likely subregional increases in intense rainfall events in eastern southern Africa, including the eastern escarpment region of South Africa and Mozambique, reflecting longer dry spells between more intense downpours (Ranasinghe et al. 2021).

The IPCC has recently assessed that general trends of drying and substantial warming can already be detected across this region (Ranasinghe et al. 2021). Moreover, the signal of increasing intense rainfall events in eastern southern Africa can also be detected in observed statistics over the last few decades (Ranasinghe et al. 2021). In Mozambique, increases either in the number of intense tropical cyclones (Fitchett 2018), or in the rainfall amount that they produce, likely contribute to the upward trend in the number of recorded intense rainfall events. A recent climate change attribution study has assessed that climate change has likely resulted in an increase in precipitation associated with the series of tropical cyclones that made landfall in Mozambique in 2022 (Otto et al. 2022).

Over the last six decades, average temperatures have been increasing at a surprising rate over southern African, at 2–4°C/century over large inland regions (Engelbrecht et al. 2015b; Kruger and Nxumalo 2016), with the highest warming rates recorded over northern Botswana and southern Zambia (Engelbrecht et al. 2015a). Extreme temperature events such as very hot days, heat-wave days and high fire-danger days have correspondingly increased sharply in frequency over the last several decades (Kruger and Sekele 2013). It is certain that further increases in oppressive temperature events will occur in the region for as long as global warming continues (Engelbrecht et al. 2015b; Garland et al. 2015; Seneviratne et al. 2021).

Generally, across most of southern Africa, intermodel agreement is strong across the ensembles of the Coupled Model Intercomparison Project Phase Six (CMIP6), Coordinated Regional Downscaling Experiment (CORDEX) and CORDEX-core ensembles, in terms of the general pattern of projected decreases in rainfall (Dosio et al. 2021). However, there is less agreement over the eastern escarpment areas of South Africa, where some models do not project general reductions in rainfall totals, but rather rainfall total increases (Lee et al. 2021).

Confidence in the projections of a general drying in southern Africa in a warmer world follows not only from model agreement, but also from the understanding of the dynamic circulation of the region in a warmer climate. The poleward displacement of the westerlies in a warmer world, one of the best document changes in circulation that can already be detected in the Southern Hemisphere, is associated with a reduction in frontal rainfall over South Africa's winter and all-year rainfall regions (the southwestern Cape, and the Cape south coast; Engelbrecht et al. 2009, 2015a). The southward displacement of the rain-bearing frontal systems of southern Africa occur in association with the strengthening of the subtropical high-pressure belt over southern Africa (Engelbrecht et al. 2009), a mechanism directly linked to the now infamous 2015–2017 Cape Town "day-zero" drought (Burls et al. 2019). A recent attribution study found that the likelihood of droughts of this magnitude occurring in South Africa's winter rainfall region has already increased by a factor of three as a consequence of anthropogenic climate change (Otto et al. 2018). During summer, the increase in the intensity and frequency of occurrence of the subtropical highs manifests in the mid-levels (Engelbrecht et al. 2009), including via the Kalahari high-pressure system. The more frequent occurrence of mid-level subsidence in mid-summer relate to longer dry spells, reduced precipitation and more sunlight reaching the surface, thereby contributing to sharply increased surface warming (Engelbrecht et al. 2009, 2015a).

In this chapter, our focus is on exploring in more detail the main climate change signal projected for southern Africa, namely that of a strongly warmer and generally drier climate, through the application of the CMIP6 ensemble (e.g., Fan et al. 2020)—the largest ensemble of global climate model (GCM) projections obtained to date. This includes an analysis of projected changes in weather extremes associated with a drastically warmer and generally drier climate. Increases in the number of heat-wave and high fire-danger days, for example, which may occur in conjunction with a general trend of drying, may have potentially devastating impacts on agriculture, water security, human and animal health, and biodiversity under low mitigation climate change futures.

7.2   Data and Methods

For projections in mean rainfall and temperature, the CMIP6 ensemble was used to derive average changes relative to a baseline period of 1850–1900 (i.e., a preindustrial baseline), under the SSP5–8.5 scenario (a largely unmitigated fossil fuel scenario). The projections were used to obtain changes in regional climate as a function of different levels of global warming, namely 1.5°C and 2°C (i.e., the end members of the global goal as defined by the Paris Accord), and 3°C and 4°C of global warming. Twenty-year moving averages are used to define periods representative of the above levels of global warming, separately for each GCM in the ensemble, as per the methodology of Lee et al. (2021).

For extreme event analysis, daily rainfall, average temperature, minimum and maximum temperature, relative humidity and surface wind speed data were obtained for the CMIP6 ensemble (Tebaldi et al. 2021) of GCM simulations. These six surface variables are essential for the calculation of the drought and fire indices that are key to the analysis undertaken here. Six CMIP6 models had available the mentioned six surface variables under the low mitigation scenario SSP5–8.5 (Socio-economic Pathway 5–8.5).

The relatively low resolution CMIP6 GCM data were interpolated to a common 1° latitude-longitude grid, toward a model-intercomparison of the projected climate change futures being undertaken. Since the extreme weather events of interest in this analysis, namely heat-waves in the presence of drought or reduced rainfall totals, are synoptic-scale features, their main characteristics are well-represented at 1° resolution. All changes are shown for a specific level of global warming relative to the preindustrial baseline period (1850–1990), which enables an assessment of the strengthening climate change signal as a function of the level of global warming.

Three extreme weather-event definitions were employed in the analysis. The first is the World Meteorological Organization (WMO) definition for heat-waves, as events when the maximum temperature at a specific location exceeds the average maximum temperature of the warmest month of the year by 5°C, for a period of at least 3 days (Engelbrecht et al. 2015b). The second is the Keetch-Byram drought index, D, which is defined in terms of a daily drought factor, dQ:

$$dQ = \frac{(203.2 - Q)\left[0.968 \times e^{(0.0875T + 1.5552)} - 8.30\right]}{\left[1 + 10.88 \times e^{(-0.001736R)}\right]} \tag{7.1}$$

Here R is the mean annual precipitation (mm) and Q (mm) is the soil-moisture deficiency that results from the interaction between rainfall and evaporation. Once Q has been updated by dQ, the drought index is calculated from the equation

$$D = \frac{10Q}{203.2} \tag{7.2}$$

Note that D ranges from 0 to 10, where $D = 10$ indicates completely dried out soil and vegetation (Keetch and Byram 1968; Engelbrecht et al. 2015b).

We employed the McArthur forest fire index (FFDI) (Dowdy et al. 2009) to quantify fire-danger risks:

$$\text{FFDI} = 2e^{(-0.45 + 0.987 \ln D + 0.0338T - 0.0345H + 0.0234U)} \tag{7.3}$$

Here T is maximum temperature (°C), H is relative humidity and U is average wind speed (measured at a height of 10 m in ms^{-1}).

7.3   Projected Changes in Rainfall and Temperature

7.3.1   Projected Changes in Annual Rainfall Totals

Changes are calculated with respect to the preindustrial baseline period of 1850–1900, with the pattern of change scaling in a remarkably stable way across increasingly higher levels of global warming. A generally drier future is projected for southern Africa by the CMIP6 SSP5–8.5 ensemble average, a signal that is expected to manifest even under 1.5°C of global warming but strengthening in amplitude at higher levels of global warming (Fig. 7.1). Overall, this analysis provides a clear picture of the potential avoided adverse impacts projected under a wide range of potential future warming scenarios that encompass the global goal according to the Paris Accord (top two panels), and beyond (bottom two panels), representing failure to meet the global goal.

There are three regions where increases in rainfall are consistently projected by the ensemble average: South Africa's KwaZulu Natal Province, east of the country's eastern escarpment; the most northern parts of the southern African domain and

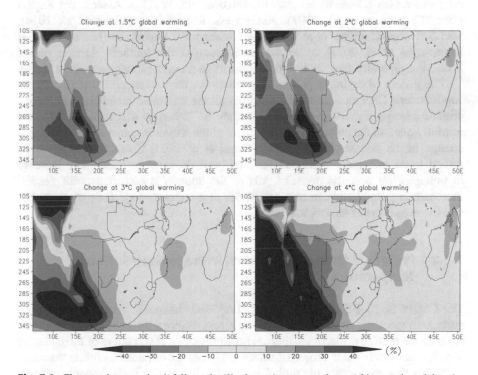

Fig. 7.1 Changes in annual rainfall totals (% change) over southern Africa projected by the CMIP6 SSP5–8.5 ensemble-average, across various levels of global warming reached with respect to the 1850–1900 baseline period, with the upper two panels directly representing the end members of the global mitigation goal expressed in the Paris Accord

the subtropical Atlantic Ocean along the coast of Angola. There is some variation across the ensemble members in terms of the projected pattern of rainfall change across southern Africa (see the next section, also see Dosio et al. 2021), but only two regions where model agreement is weak (that is, where conflicting signals of change are projected across the ensemble members). These regions are the KwaZulu-Natal Province of South Africa and the subtropical Atlantic Ocean west of Angola (Lee et al. 2021). Since model disagreement over these regions persists at high levels of global warming, the uncertainty is likely structural, rather than being caused by model internal variability. In the case of the KwaZulu-Natal province, this structural uncertainty may relate to the parameterization of convection over South Africa's steep eastern escarpment, and area long known to be associated with substantial model rainfall biases (Engelbrecht et al. 2002; Dedekind et al. 2016).

The pattern of general drying projected across the southern African domain has previously been linked to general increases in subtropical subsidence over southern Africa and the poleward displacement of frontal systems in winter (Engelbrecht et al. 2009, 2015b). This pattern of change is remarkably robust (in terms of the ensemble average, at least) across the CMIP6, CMIP5, CORDEX and CORDEX-core ensembles (Dosio et al. 2021). Moreover, the IPCC in Assessment Report Four (Christensen et al. 2007), Assessment Report Five (Niang et al. 2014), SR1.5 (Hoegh-Guldberg et al. 2018) and Assessment Report Six (Lee et al. 2021; Ranasinghe et al. 2021) made the assessment of the southern African region becoming generally drier, and/or to become more drought-prone in a warmer world.

The pattern of drying is particularly strong for the winter rainfall region of the southwestern Cape in South Africa, across all the ensemble members. The strong climate change signal over this region may be linked to a reduction in frontal rainfall linked to the poleward displacement of the westerlies, an already detectable change in the Southern Hemisphere (Goyal et al. 2021, Chap. 6) that has been linked to an increased likelihood for multiyear droughts to occur. The increase in precipitation over the northern part of the domain is consistent with general increases in precipitation in tropical Africa in a warmer world (Lee et al. 2021), and the expansion of the tropical belt.

7.3.2 Projected Changes in Annual Average Near-Surface Temperature

The CMIP6 SSP5–8.5 ensemble average projected changes in annual average near-surface temperature are shown in Fig. 7.2, across different levels of global warming. As with the rainfall projections, this analysis provides a clear picture of the potential avoided adverse impacts projected under a wide range of potential future warming scenarios that encompass the global goal according to the Paris Accord (top two panels), and beyond (bottom two panels), representing failure to meet the global goal.

Consistent with trends that can already be detected (Engelbrecht et al. 2015b), the strongest warming is centered over Botswana, extending across the western and

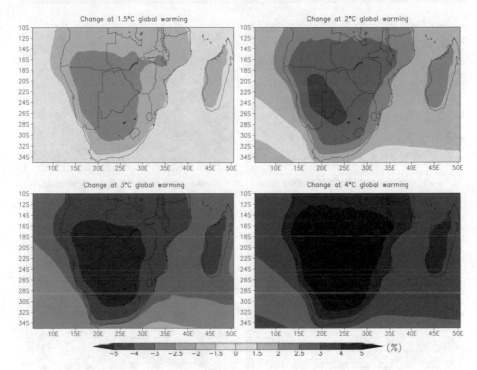

Fig. 7.2 Changes in near-surface mean annual temperature (°C) over southern Africa projected by the CMIP6 SSP5–8.5 ensemble-average, across various levels of global warming reached with respect to the 1850–1900 baseline period, with the upper two panels directly representing the end members of the global mitigation goal expressed in the Paris Accord

central interior regions of southern Africa. The interior regions of southern Africa are projected to warm at a higher rate than tropical Africa, while the moderating effect of the ocean also tempers the rate of warming over coastal areas. The relatively high rate of warming over subtropical interior southern Africa has been attributed to a strengthening of mid-level anticyclonic circulation and subsidence, which suppresses cloud formation and rainfall, resulting in more solar radiation reaching the surface, thereby driving the relatively high rate of temperature increase (Engelbrecht et al. 2009, 2015b).

7.3.3   Projected Changes in Extremes

With a view to gaining insight into weather extremes in southern Africa in a warmer world, projections of six CMIP6 GCMs are considered for which daily data are available for the variables of precipitation, minimum and maximum temperature, relative humidity and surface wind speed (allowing for the calculation of fire-danger indices). To facilitate comparison with Figs. 7.1 and 7.2, projected changes in

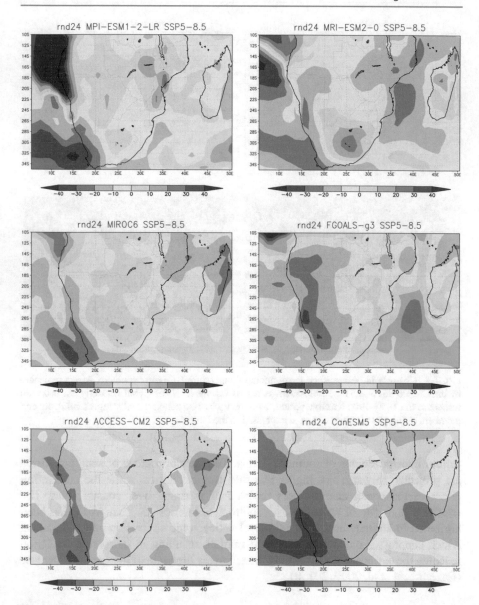

Fig. 7.3 Projected changes in annual rainfall totals (% change) over southern Africa under 3°C of global warming relative to preindustrial climate, as per an ensemble of six CMIP6 GCMs

annual rainfall totals (Fig. 7.3) and annual average surface temperature (Fig. 7.4) are plotted for each of the six GCMs under 3°C of global warming. This level of global warming is selected for extreme event analysis since it is likely to be associated with a clear climate change signal (as opposed to the relatively larger role that climate/internal variability may play under 1.5°C, and possibly 2°C, of global

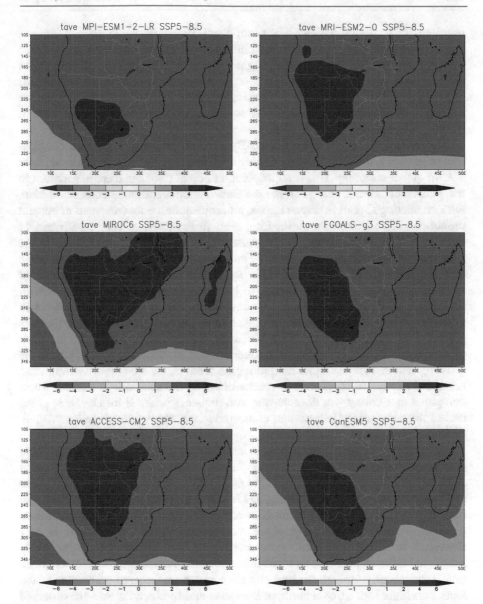

Fig. 7.4 Projected changes in annual average temperature (°C) over southern Africa under 3°C of global warming relative to preindustrial climate, as per an ensemble of six CMIP6 GCMs

warming). Moreover, under current international commitments to greenhouse gas reductions, the exceedance of even the 2°C threshold of global warming remains entirely possible.

The rainfall projections are indicated variation in rainfall patterns across the ensemble (Fig. 7.3). For example, three of the six projections indicate slight rainfall increases over South Africa's KwaZulu-Natal Province, while all six indicate pronounced drying over the winter rainfall region of the southwestern Cape. Three of the six projections also indicate slightly wetter conditions over all or parts of western Botswana (the larger CMIP6 ensemble shows model agreement in this region, in terms of a general signal of drying, see Sect. 7.3.1). All the simulations are indicative of pronounced warming over southern Africa, peaking over the western interior at levels of 4–6°C (Fig. 7.4).

The projected changes in the Keetch-Byram drought index (Fig. 7.5) are indicative of general reductions in soil-moisture availability across southern Africa under 3°C of global warming. This is an important finding: although there is variation in the pattern of rainfall change in the 6-member model ensemble considered here, with conflicting signals in some regions, all projections are in agreement of general reductions in soil-moisture availability, even in the areas of projected increases in rainfall. These reductions are the consequence of enhanced evaporation in the substantially warmer regional world. Thus, it is possible to conclude with some certainty that most of southern Africa is *likely* to become generally drier in terms of rainfall totals, but is *virtually certain* to become generally drier in terms of soil-moisture availability. This finding is consistent with strong model agreement in terms of projected decreases in soil-moisture as parameterized in CMIP6 GCMs (Wang et al. 2022; Zhai et al. 2020). Moreover, earlier work has indicated that such general reductions in soil-moisture in southern Africa translate to a shortening in the growing season in the summer rainfall region. That is, the amount of soil-moisture needed for crops to be planted is reached later in the season in a warmer world compared to a cooler world; moreover, soil-moisture peaks at lower values at the end of the rainy season in a warmer compared to a cooler world (Engelbrecht et al. 2015b).

Consistent with the sharp increases in temperature, the six GCMs considered here also project substantial increases in the number of heat-wave days over southern Africa. These increases range from 20 to 60 days per year over much of the western and central interior regions, implying that heat-waves, compared to the preindustrial threshold, will become a common and in some regions a semipermanent feature of summer climate. The geographical "center" of heat-wave increases is over Botswana in all of projections considered here. This pattern of change likely relates to increases in mid-level highs and associated subsidence (Engelbrecht et al. 2009), specifically through the intensification and more frequent occurrence of the Botswana high. Indeed, over northern Botswana and southern Zambia, the observed rate of increase in average temperature in decades has been about 4°C per century (Engelbrecht et al. 2015b) (Fig. 7.4).

In a generally drier regional world (Fig. 7.3), that is also warming at an almost unprecedented rapid rate, meteorological fire-danger may also be expected to increase. This is illustrated by Fig. 7.6, which shows the projected change in the number of high fire-danger days, as defined by the McArthur Fire Danger Index (Eq. 7.3). Substantial increases in the number of high fire-danger days, of between 20 and 80 days per year, are projected for extensive parts of the western and central interior, in some projections extending into the Limpopo River Valley. In relation

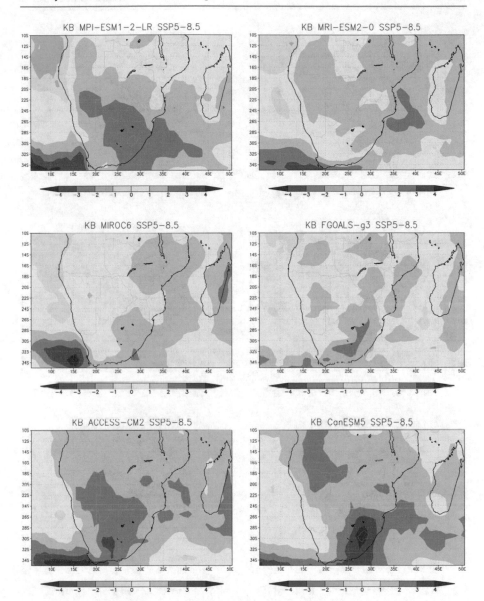

Fig. 7.5 Projected changes in the Keetch-Byram drought index over southern Africa under 3°C of global warming relative to preindustrial climate, as per an ensemble of six CMIP6 GCMs

to generally drier conditions, these changes also translate to a lengthening of the fire season in southern Africa (Engelbrecht et al. 2015b). It may be noted that one of the ensemble members analyzed (MPI-ESM1–2) is indicative of pronounced decreases in high fire-danger days over Zambia. This change is underpinned by

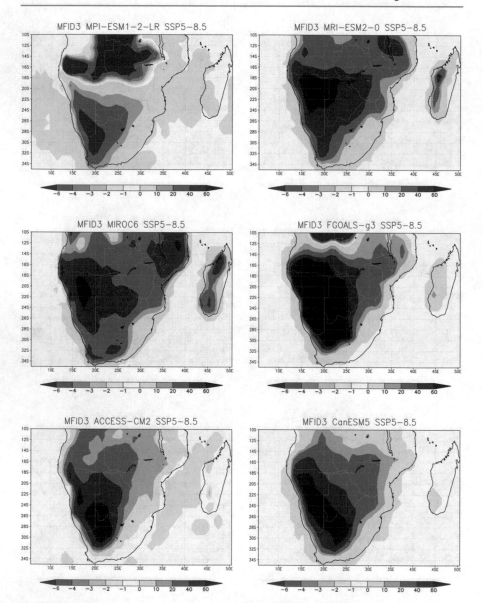

Fig. 7.6 Projected changes in the number of high fire-danger days per year over southern Africa under 3°C of global warming relative to preindustrial climate, as per an ensemble of six CMIP6 GCMs

rainfall increases (Fig. 7.3) and an increase in the number of rainfall days in this particular model projection.

Despite the likely wide-scale decreases in rainfall, including virtually certain decreases in rainfall in the west, and virtually certain increases in average temperatures, heat-wave days and high fire-danger days, general increases in intense rainfall

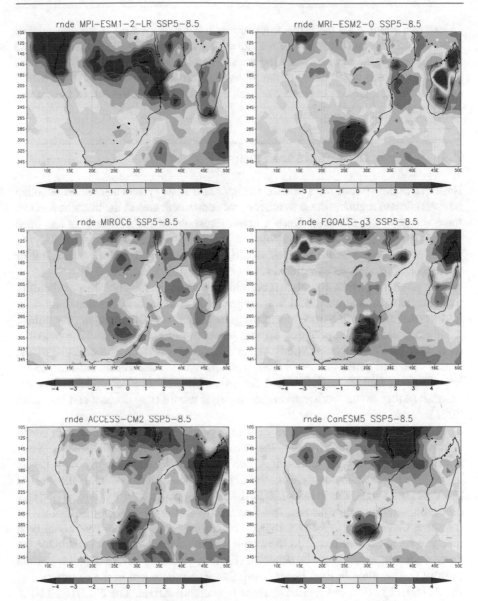

Fig. 7.7 Projected changes in the number of intense rainfall events (more than 20 mm of rain falling over an area of 10,000 km^2) over southern Africa under 3°C of global warming relative to preindustrial climate, as per an ensemble of six CMIP6 GCMs

events are likely in a warmer world in eastern southern Africa, including eastern South Africa and Mozambique (Ranasinghe et al. 2021). Such a trend can already be observed (Ranasinghe et al. 2021). The ensemble of six projections considered in Fig. 7.7 are indicative of the spatial variability in the projections of changes in extreme events. All the projections are indicative of increase in intense rainfall events across the northern part of the domain, a more consistent change compared to the model projections of changing annual rainfall totals. Northern Mozambique is included in this zone. Some of the projections are also indicative of the potential of pronounced increases in intense rainfall events over and to the east of the eastern escarpment.

7.4   The Risk of Regional Tipping Points

Regional tipping points refer to shifts in regional climate system that would establish a novel climate regime, where weather events unprecedented in the historical record have the potential to occur. Once a given threshold of global warming has been reached, these shifts are irreversible on the scale of human lifetimes. Tipping points in regional climate systems would, in all likelihood, induce ecological or socioeconomic changes at regional scales that are similarly irreversible.

For southern Africa, the almost certain reductions in soil-moisture availability and increases in heat-wave and fire-danger days, combined with a generally drier and warmer climate, hold the risks of triggering a number of regional tipping points. Four examples of such potential tipping points are discussed below: the potential of a "day-zero" drought in South Africa's Gauteng Province, the collapse of the maize crop and cattle industry across southern Africa, unprecedented heat-waves impacting on human mortality, and the risk of intense tropical cyclones making landfall further to the south than in the historical record (Engelbrecht and Monteiro 2021).

In September 2016, at the end of four consecutive years of drought in South Africa's summer rainfall region, the level of the Vaal Dam fell to 25%. Water restrictions were in place in South Africa's Gauteng Province, which depends on about 50% of its water supply from the integrated Vaal River system. If the level of the Vaal Dam should fall to below 20%, the Gauteng water supply would be severely compromised, for two reasons. The first relates to the engineering limitations of pumping water uphill to Johannesburg. The second relates to poor water quality at a dam level at 20% or lower, to the extent that the water would not be suitable for human consumption. It may be noted that the four-year-long period of below normal rainfall that resulted in dam levels being this low culminated in the occurrence of the 2015/16 El Niño and related drought in southern Africa. The 2015/16 El Niño was the strongest in recorded history, at least in terms of the magnitude of anomalies in the Niño 3.4 region, and there is evidence that climate change strengthened the event. Moreover, the IPCC in AR6 did not make high confidence statements about changes in El Niño and La Niña amplitudes and frequencies in a warmer world, it did assess that impacts are likely to strengthen in amplitude in most regions of the world (Lee et al. 2021). This finding, in conjunction with projections of generally drier conditions in southern Africa, reduced soil-moisture availability and increased temperatures and evaporation, suggest that the possibility exists that multiyear droughts in South Africa's eastern mega-dam region may occur more frequently, last longer and be more intense under higher levels of global warming. This in turn, suggests that the likelihood of the Vaal Dam's level falling below the critical threshold of 20% will increase in a warmer world, creating the possibility of a "day-zero" drought in the Gauteng Province. Such a drought is probably the largest climate change risk South Africa faces in the context of socioeconomic impacts (the Gauteng Province is South Africa's industrial heartland, where 15 million people live). A drought of duration and intensity to severely compromise Gauteng's water supply from the integrated Vaal River system has never occurred in the historical record, and if materialized would represent a tipping point in the regional climate system. The four-year drought including the 2015/16 El Nino was broken by good falls of rain in October 2016, but represents a near-miss. Quantifying the probability of a Gauteng day-zero drought under different levels of global warming should thus be a research priority. Given the potentially severe impacts of such a drought, a disaster risk reduction plan needs to be in place for such an event, even if it is low-probability event.

The terminology of "day-zero" droughts had its origin in the 2015–2017 Cape Town drought, during which the city came close to running out of water. This multiyear drought brought substantially reduced rainfall totals in the Theewater-skloof catchment, which is key to the City's water security. Associated with the substantially reduced rainfall totals projected for the southwestern Cape (Fig. 7.1) is the more frequent occurrence of multiyear droughts. A tipping point may be reached where multiyear droughts in the southwestern Cape will occur so frequently that it will impact on the City of Cape Town's sustainable growth, to the extent that it will require a new water resource. Desalination plants are often proposed as a solution in this regard, although the large electricity needs and excessive costs of the associate technologies render its implementation nontrivial. The mechanism underpinning day-zero droughts in the southwestern Cape is the poleward displacement of the Southern Hemisphere westerlies, a fingerprint of climate change that can already be detected (Goyal et al. 2021, Chap 6). Observed trends in rainfall in the winter rainfall region over the last thirty years are consistently negative (Wolski et al. 2021). Moreover, an attribution study concluded that the risks of day-zero-type droughts occurring in the southwestern Cape is already three times as large as in preindustrial times (Otto et al. 2018). The risk of day-zero-type droughts extends into the all-year rainfall region and South Africa's Eastern Cape Province (Archer et al. 2022).

Multiyear droughts occurring in association with intense and heat-waves pose risks to the agricultural sector, including the maize-crop (southern Africa's staple food) and the high commodity cattle industry. The 2015/2016 summer, experienced intense El Niño induced drought and heat-waves and was the driest in recorded history across South Africa's Free State and North West Provinces, which together produce more than 60% of South Africa's maize crop. The South Africa the maize crop was reduced by about 40% compared to yield of the previous summer. Botswana lost 40% of its cattle. IPCC warned that the collapse of both the maize crop and cattle industry are likely in southern Africa under 3°C of global warming (IPCC 2017). This assessment is based purely on the biophysical effects of heat-stress on the maize plant and cattle in a southern African climate that is warming drastically compared to the global rate of temperature increase, and which is likely to also become generally drier (Hoegh-Guldberg et al. 2018). However, if one also takes into account the socioeconomics of farming, including the ability of subsistence farmers and small commercial farmers to absorb the shocks of multiyear droughts becoming more intense, and occurring more frequently, the possibility exists that such droughts may occur more frequently. Generating probabilistic assessments of tipping points in these key commodities in southern Africa thus require a combined approach that is informed by both the physical science base and the socioeconomics of farming. To this end, more reliable seasonal forecasts would provide critical adaptation support as they could facilitate important decisions such as risk assessments relating to planting timing to minimize costly drought related crop failure. Cultivar or crop species selection matched to the projected climate conditions would also comprise a valuable potential adaptation option.

The 2015/16 El Niño brought heat-waves of unprecedented frequency and intensity to southern Africa, and there are clearly detectable upward trends in the frequency of occurrence of extreme (warm) temperature events in the region. Climate model projections indicate potentially devastating increases in heat-wave occurrences across southern Africa under high levels of global warming (Fig. 7.8; Seneviratne et al. 2021). It is also clear though, that heat-waves of unprecedented intensity will already occur in southern Africa in the near-term (the next twenty years). Millions of people live in informal housing in southern Africa, without air conditioning, and often without easy access to cool water. The elderly are particularly vulnerable to such heat-related stresses. The possibility exists of regional climate change in the near-term reaching a tipping point where heat-waves of unprecedented intensity and duration may kill thousands of people and livestock across southern Africa. Heat-adaptation plans in southern Africa will benefit from an enhanced understanding of the risk of heat-waves associated with high mortality occurring.

The fourth example of a tipping point worthy of highlighting is quite different from those related to oppressive temperatures and drought. It involves the potential landfall of intense tropical cyclones (that is, a category 4 or 5 hurricanes) at latitudes further to the south than ever recorded before in southern Africa. Global tropical cyclone statistics are indicative of the more frequent occurrence of intense systems as well as of landfall at more poleward locations. Warmer sea-surface temperatures in the southwest Indian Ocean, and in particular in the Mozambique Channel, may similarly allow for the more southward landfall of intense tropical cyclones in a warmer world. Indeed, recent decades has brought an increase in the number of category 4 and 5 hurricanes in the southwest Indian Ocean (Fitchett 2018), although actual landfall of a category 5 hurricane has never been recorded in Mozambique.

Intense tropical cyclone Idai reached category 4 status in the Mozambique Channel, before making landfall as category 3 hurricane at Beira around midnight on 14 March 2019 (Engelbrecht and Vogel 2021). In the destructive winds, storm surge and pluvial and fluvial flooding that followed, hundreds of people lost their lives. The total death toll in tropical cyclone Idai's path across Malawi, Mozambique and Malawi is estimated to have been more than 1000. This makes Idai the worst

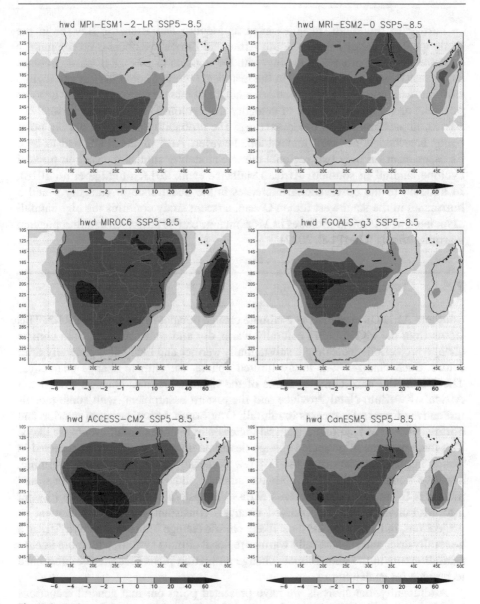

Fig. 7.8 Projected changes in the number of heat-wave days over southern Africa under 3°C of global warming relative to preindustrial climate, as per an ensemble of six CMIP6 GCMs

flood disaster in the history of Africa south of the equator. In Beira, there is some experience in local populations and disaster management agencies in terms of dealing with the impacts of tropical lows and cyclones. Tropical cyclone Idai serves as a stark reminder of how severe the impacts of category 3 to 5 hurricanes in southern Africa can be.

Further to the south, in cities such as Maputo and Richards Bay, or in the Limpopo River valley between South Africa and Zimbabwe, there is no community or governance experience in coping with the impacts of intense tropical cyclones as no such events have occurred in the historical record. Should the climate regime shift into a regime where such southern landfalls of intense tropical cyclones become possible, or where the landfall of category five cyclones start to occur in southern Africa, impacts may well be devastating. The probability of such a tipping point being breached is not well understood; however, AR6 of the IPCC had to base its assessment on only two regional climate modeling studies focused on tropical cyclone landfall in southern Africa (Malherbe et al. 2013: Muthige et al. 2018). In addition to the observational increases in the occurrence of category 4 and 5 hurricanes in the southwest Indian Ocean, a recent study confirms the high rainfall associated with tropical cyclones in Mozambique can be attributed to anthropogenic warming effects (Otto et al. 2022).

7.5   Conclusions

Southern Africa is classified as a climate change hotspot in the IPCC's SR1.5. This stems from the region being naturally warm, dry and water stressed, with climate change projections indicating a substantially warmer and likely also generally drier future. Such changes will imply limited options for climate change adaptation. There is model uncertainty in terms of the signal of rainfall changes over South Africa's KwaZulu-Natal Province and the eastern escarpment, with some models indicative of general increases in rainfall. Over South Africa's eastern interior, and northward over Mozambique, general increases in intense rainfall events are likely. The main patterns of projected change described above can already be detected in trends in observed data over the last few decades: substantial regional warming, negative trends in rainfall in both the summer and winter rainfall regions, and increases in intense rainfall events across the eastern escarpment and northward into Mozambique. These observed changes, consistent with the assessment of projections, indicate to us the most likely climate change future of southern Africa: a generally drier and substantially warmer regional climate system, with more intense rainfall events in the east. Climate change adaptation plans first and foremost need to prepare for such a future.

Additionally, the analysis we have presented point out that general reductions in soil-moisture availability are virtually certain to occur across the region, even in regions where the model ensemble average, or individual models, are indicative of increases in rainfall. This is a consequence of enhanced evaporation in the warmer regional climate, and implies a generally shorter growing season and longer wildfire season across the region. Moreover, substantial increases in the number of heat-wave days and high fire-danger days are virtually certain to occur under high levels of global warming.

Against this background, it is clear that the potential exists for the regional climate system to tip into a new regime, where unprecedented climate change impacts may start to occur in southern Africa. Examples include the possibility of a day-zero drought in Gauteng, the collapse of the maize crop and cattle industry, the occurrence of heat-waves of unprecedented intensity, and the landfall of tropical cyclones as far south as Maputo, the Limpopo River Valley, or Richards Bay. More research is urgently required to quantify the probabilities of these and additional tipping points being reached. This information is critical for the identification and implementation of climate change adaptation plans and actions, especially as any of these adaptation options are likely to be extensive, expensive and requiring of long lead time for implementation. It is clear though, that the risk exists of the southern African region to become less habitable under high levels of global warming. Imagine for example, a future southern Africa without its staple food maize, without a cattle industry, with frequent intense heat-waves impacting on human health and mortality, and with frequent long-lasting droughts hampering industrial development and the sustainable growth of cities. It is clear that southern Africa as a region needs to advocate strongly for climate change mitigation, and contribute its fair share to this mitigation effort, with the aim of avoiding these tipping points being reached in the first place.

Acknowledgements The analysis undertaken here was supported by the Carnegie Foundation of New York through the Global Change Institute at the University of the Witwatersrand and the FOCUS-AFRICA project, which has received funding from the European H2020 Research and Innovation program (Grant Agreement 869575). The tipping point assessment was underpinned by research grants from the National Research Foundation (Project Longdryspell, Grant 136480) in South Africa and the Southern African Science Services Centre for Climate Change and Adaptive Land Management (SASSCAL), via its Research Grant TIPPECC (Climate change information for adapting to regional tipping Points). The content of this document reflects only the authors' views. The European Commission is not responsible for any use that may be made of the information the paper contains. We thank the climate modeling groups that produced and made available their CMIP6 model output, and acknowledge the World Climate Research Programme's Working Group on Coupled Modelling for their coordination of CMIP6.

References

Archer E, Du Toit J, Timm EC, Hoffman T, Landman W, Malherbe J, Sterng M (2022) The 2015-19 multi year drought in the Eastern Cape, South Africa: it's evolution and impacts on agriculture. J Arid Environ 196:104630

Burls NJ, Blamey RC, Cash BA, Swenson ET, Al FA, Bopape M-JM, Straus DM, Reason CJC (2019) The Cape Town "day zero" drought and Hadley cell expansion. Npj Clim Atmos Sci 2:27. https://doi.org/10.1038/s41612-019-0084-6

Christensen JH et al (2007) In: Solomon S, Dqin M, Chen Z, Mmarquis AB, Averyt M, Miller HL (eds) Regional climate projections climate change 2007: the physical science basis. Contribution of Working Group I to the fourth assessment report of the intergovernmental panel on climate change. Cambridge University Press, Cambridge

Dedekind Z, Engelbrecht FA, Van der Merwe J (2016) Model simulations of rainfall over southern Africa and its eastern escarpment. Water SA 42:129–143. https://doi.org/10.4314/wsa.v42i1.13

Dosio A, Jury MW, Almazroui M, Ashfaq M, Diallo I, Engelbrecht FA, Klutse NAB, Lennard C, Pinto I, Sylla MB, Tamoffo AT (2021) Projected future daily characteristics of African precipitation based on global (CMIP5, CMIP6) and regional (CORDEX, CORDEX-CORE) climate models. Clim Dyn 57:3135–3158. https://doi.org/10.1007/s00382-021-05859-w

Dowdy AJ, Mills GA, Finkele K, De Groot W (2009) Australian fire weather as represented by the McArthur Forest fire danger index and the Canadian Forest fire weather index. CAWCR Technical Report number 10, Centre for Australian Weather and Climate Research. Melbourne, Vic., Australia

Engelbrecht CJ, Engelbrecht FA (2016) Shifts in Köppen-Geiger climate zones over southern Africa in relation to key global temperature goals. Theor Appl Climatol 123:247–261. https://doi.org/10.1007/s00704-014-1354-1

Engelbrecht FA, Monteiro PMS (2021) The IPCC assessment report six working group I report and southern Africa: reasons to take action. S Afr J Sci 117(9/10):10.17159/sajs.2021/12679

Engelbrecht FA, Vogel CH (2021) When early warning is not enough. One Earth 4:1055–1058. https://doi.org/10.1016/j.oneear.2021.07.016

Engelbrecht FA, Rautenbach CJ d W, McGregor JL, Katzfey JJ (2002) January and July climate simulations over the SADC region using the limited-area model DARLAM. Water SA 28:361–374

Engelbrecht FA, McGregor JL, Engelbrecht CJ (2009) Dynamics of the conformal-cubic atmospheric model projected climate-change signal over southern Africa. Int J Climatol 29:1013–1033. https://doi.org/10.1002/joc.1742

Engelbrecht CJ, Landman WA, Engelbrecht FA, Malherbe J (2015a) A synoptic decomposition of rainfall over the cape south coast of South Africa. Clim Dyn 44:2589–2607. https://doi.org/10.1007/s00382-014-2230-5

Engelbrecht FA, Adegoke J, Bopape M-J, Naidoo M, Garland R, Thatcher M, McGregor J, Katzfey J, Werner M, Ichoku C, Gatebe C (2015b) Projections of rapidly rising surface temperatures over Africa under low mitigation. Env Res Lett 10:085004. https://doi.org/10.1088/1748-9326/10/8/085004

Fan X, Miao C, Duan Q, Shen C, Wu Y (2020) The performance of CMIP6 versus CMIP5 in simulating temperature extremes over the global land surface. J Geophys Res Atmos 125(18):e2020JD033031

Fitchett JM (2018) Recent emergence of CAT5 tropical cyclones in the South Indian Ocean. S Afr J Sci 114:11–12. https://doi.org/10.17159/sajs.2018/4426

Garland R.M., Matooane M., Engelbrecht F.A., Bopape M-JM, Landman W.A., Naidoo M., Van der Merwe J. And Wright C.Y. (2015). Regional projections of extreme apparent temperature days in Africa and the related potential risk to human health. Int J Environ Res Public Health 12 12577-12604. https://doi.org/10.3390/ijerph121012577

Goyal R, Sen Gupta A, Jucker M, England MH (2021) Historical and projected changes in the southern hemisphere surface westerlies. Geophys Res Lett 48:e2020GL090849. https://doi.org/10.1029/2020GL090849

Hoegh-Guldberg O, Jacob D, Taylor M, Bindi M, Brown S, Camilloni I, Diedhiou A, Djalante R, Ebi KL, Engelbrecht F, Guiot J, Hijioka Y, Mehrotra S, Payne A, Seneviratne SI, Thomas A, Warren R, Zhou G (2018) Impacts of 1.5°C global warming on natural and human systems. In: Pörtner H-O, Roberts D, Skea J, Shukla PR, Pirani A, Moufouma-Okia W, Péan C, Pidcock R, Connors S, Matthews JBR, Chen Y, Zhou X, Gomis MI, Lonnoy E, Maycock T, Tignor M, and Waterfield T (eds) Global warming of 1.5°C. An IPCC special report on the impacts of global warming of 1.5°C above pre-industrial levels and related global greenhouse gas emission pathways, in the context of strengthening the global response to the threat of climate change, sustainable development, and efforts to eradicate poverty. Masson-Delmotte V, Zhai greenhouse gas emission pathways. In Press

Keetch JJ, Byram GM (1968) A drought index for fire control. Res. Pap. SE-38 (Asheville, NC: US Department of Agriculture, Forest Service, southeastern Forest Experiment Station) p 32 (revised November 1988)

Kruger AC, Nxumalo M (2016) Surface temperature trends from homogenized time series in South Africa: 1931–2015. Int J Climatol 37:2364. https://doi.org/10.1002/joc.4851

Kruger AC, Sekele SS (2013) Trends in extreme temperature indices in South Africa: 1962–2009. Int J Climatol 33:661–676

Lee JY, Marotzke J, Bala G, Cao L, Corti S, Dunne JP, Engelbrecht F, Fischer E, Fyfe JC, Jones C, Maycock A, Mutemi J, Ndiaye O, Panickal S, Zhou T (2021) Future global climate: scenario-based projections and near-term information. In: Masson-Delmotte V, Zhai P, Pirani A, Connors SL, Péan C, Berger S, Caud N, Chen Y, Goldfarb L, Gomis MI, Huang M, Leitzell K, Lonnoy E, Matthews JBR, Maycock TK, Waterfield T, Yelekçi RY, Zhou B (eds) Climate change 2021: the physical science basis. contribution of working group I to the sixth assessment report of the intergovernmental panel on climate change. Cambridge University Press, Cambridge

Malherbe J, Engelbrecht FA, Landman WA (2013) Projected changes in tropical cyclone climatology and landfall in the Southwest Indian Ocean region under enhanced anthropogenic forcing. Clim Dyn 40:1267–1286. https://doi.org/10.1007/s00382-012-1635-2

Muthige M, Malherbe J, Engelbrecht F, Grab S, Beraki A, Maisha R, Van Der Merwe J (2018) Projected changes in tropical cyclones over the south West Indian Ocean under different extents of global warming. Env Res Lett 13:065019. https://doi.org/10.1088/1748-9326/aabc60

Niang I, Ruppel OC, Abdrabo MA, Essel A, Lennard C, Padgham J, Urquhart P (2014) Africa. In: Barros VR, Field CB, Dokken DJ, Mastrandrea MD, Mach KJ, Bilir TE, Chatterjee M, Ebi KL, Estrada YO, Genova RC, Girma B, Kissel ES, Levy AN, MacCracken S, Mastrandrea PR, White LL (eds) Climate change 2014: impacts, adaptation, and vulnerability. Part B: regional aspects. Contribution of working group II to the fifth assessment report of the intergovernmental panel on climate change. Cambridge University Press, Cambridge, and New York, NY, pp 1199–1265

Otto EL, Wolski P, Lehner F, Tebaldi C, Van Oldenborgh GJ, Hogesteeger S, Singh R, Holden P, Fučkar NS, Odoulami RC, New M (2018) Environ Res Lett 13:124010

Otto FEL, Zachariah M, Wolski P, Pinto I, Barimalala R, Nhamtumbo B, Bonnet R, Vautard R, Philip S, Kew S, Luu LN, Heinrich D, Vahlberg M, Singh R, Arrighi J, Thalheimer L, Van Aalst M, Li S, Sun J, Vecchi G, Harrington LJ (2022) Climate change increased rainfall associated with tropical cyclones hitting highly vulnerable communities in Madagascar, Mozambique & Malawi. Word Weather Attribution Service. https://www.worldweatherattribution.org/climate-change-increased-rainfall-associated-with-tropical-cyclones-hitting-highly-vulnerable-communities-in-madagascar-mozambique-malawi/

Ranasinghe R, Ruane AC, Vautard R, Arnell N, Coppola E, Cruz FA, Dessai S, Islam AS, Rahimi M, Carrascal DR, Sillmann J, Sylla MB, Tebaldi C, Wang W, Zaaboul R (2021) Climate change information for regional impact and for risk assessment. In: Masson-Delmotte V, Zhai P, Pirani A, Connors SL, Péan C, Berger S, Caud N, Chen Y, Goldfarb L, Gomis MI, Huang M, Leitzell K, Lonnoy E, Matthews JBR, Maycock TK, Waterfield T, Yelekçi O, Yu R, Zhou B (eds) Climate change 2021: the physical science basis. contribution of working group I to the sixth assessment report of the intergovernmental panel on climate change. Cambridge University Press, Cambridge, New York, NY, pp 1767–1926. https://doi.org/10.1017/9781009157896.014

Seneviratne SI, Zhang X, Adnan M, Badi W, Dereczynski C, Di Luca A, Ghosh S, Iskandar I, Kossin J, Lewis S, Otto F, Pinto I, Satoh M, Vicente-Serrano SM, Wehner M, Zhou B (2021) Weather and climate extreme events in a changing climate. In: Masson-Delmotte V, Zhai P, Pirani A, Connors SL, Péan C, Berger S, Caud N, Chen Y, Goldfarb L, Gomis MI, Huang M, Leitzell K, Lonnoy E, Matthews JBR, Maycock TK, Waterfield T, Yelekçi O, Yu R, Zhou B (eds) Climate change 2021: the physical science basis. contribution of working group i to the sixth assessment report of the intergovernmental panel on climate change. Cambridge University Press, Cambridge, and New York, NY, pp 1513–1766. https://doi.org/10.1017/9781009157896.013

Tebaldi C, Debeire K, Eyring V, Fischer E, Fyfe J, Friedlingstein P, Knutti R, Lowe J, O'Neill B, Sanderson B, van Vuuren D, Riahi K, Meinshausen M, Nicholls Z, Tokarska KB, Hurtt G, Kriegler E, Lamarque J-F, Meehl G et al (2021) Climate model projections from the scenario model intercomparison project (Scenariomip) of cmip6. Earth Syst Dyn 12:253–293. https://doi.org/10.5194/esd-12-253-2021

Wang A, Kong X, Chen Y, Ma X (2022) Evaluation of soil moisture in CMIP6 multimodel simulations over conterminous China. J Geophys Res Atmospheres 127:e2022JD037072. https://doi.org/10.1029/2022JD037072

Wolski P, Conradie S, Jack C, Tadross M (2021) Spatio-temporal patterns of rainfall trends and the 2015–2017 drought over the winter rainfall region of South Africa. Int J Climatol 44:E1303–E1319. https://doi.org/10.1002/joc.6768

Zhai J, Mondal SK, Fischer T et al (2020) Future drought characteristics through a multi-model ensemble from CMIP6 over South Asia. Atmospheric Res 246:105111. https://doi.org/10.1016/j.atmosres.2020.105111

The Agulhas Current System as an Important Driver for Oceanic and Terrestrial Climate

8

Arne Biastoch, Siren Rühs, Ioana Ivanciu, Franziska U. Schwarzkopf, Jennifer Veitch (ID), Chris Reason, Eduardo Zorita, Nele Tim, Birgit Hünicke, Athanasios T. Vafeidis, Sara Santamaria-Aguilar, Sunna Kupfer, and Felix Soltau

Abstract

The Agulhas Current system around South Africa combines the dynamics of strong ocean currents in the Indian Ocean with eddy–mean flow interactions. The system includes an associated interoceanic transport towards the Atlantic, Agulhas leakage, which varies on both interannual and decadal timescales. Agulhas leakage is subject to a general increase under increasing greenhouse gases, with higher leakage causing a warming and salinification of the upper ocean in the South Atlantic. The far-field consequences include the impact of the

A. Biastoch (✉)
GEOMAR Helmholtz Centre for Ocean Research Kiel and Kiel University, Kiel, Germany
e-mail: abiastoch@geomar.de

S. Rühs
GEOMAR Helmholtz Centre for Ocean Research Kiel, Kiel, Germany

Institute for Marine and Atmospheric Research, Utrecht University, Utrecht, Netherlands

I. Ivanciu · F. U. Schwarzkopf
GEOMAR Helmholtz Centre for Ocean Research Kiel, Kiel, Germany

J. Veitch
The South African Environmental Observation Network, Pretoria, South Africa

C. Reason
University of Cape Town, Cape Town, South Africa

E. Zorita · N. Tim · B. Hünicke
Helmholtz-Zentrum Hereon, Geesthacht, Germany

A. T. Vafeidis · S. Santamaria-Aguilar · S. Kupfer
Kiel University, Kiel, Germany

F. Soltau
University of Siegen, Siegen, Germany

© The Author(s) 2024
G. P. von Maltitz et al. (eds.), *Sustainability of Southern African Ecosystems under Global Change*, Ecological Studies 248,
https://doi.org/10.1007/978-3-031-10948-5_8

Agulhas Current on the Benguela Upwelling system, a major eastern boundary upwelling system that supports a lucrative fishing industry. Through sea surface temperatures and associated air–sea fluxes, the Agulhas Current system also influences regional climate in southern Africa, leading to a heterogeneous pattern of rainfall over southern Africa and to a reduction of precipitation in most areas under global warming conditions. Changes in the Agulhas Current system and the regional climate also cause changes in regional sea-level and wind-induced waves that deviate from global trends. Combining these oceanic changes with extreme precipitation events, global warming can considerably amplify flood impacts along the coast of South Africa if no adaptation measures are implemented.

8.1   Introduction

The waters around South Africa host a unique setting of ocean currents and water masses that is mainly determined by the Agulhas Current system, one of the world's most intense and vigorous western boundary current systems. Owing to the southern termination of the African continent being located further north than the other continents in the Southern Hemisphere, circulation and hydrography of the Agulhas Current system are impacted by both the Atlantic and Indian Ocean dynamics. In turn, the Agulhas Current system also is an important driver of global oceanic climate, in particular, of the Atlantic Meridional Overturning Circulation (AMOC).

We start our chapter with an outline of the characteristics and dynamics of the Agulhas Current flowing southward along the South African coast (Sect. 8.2).

A significant portion of the Agulhas Current provides an interoceanic transport of mass, heat, and salt into the South Atlantic. This "Agulhas leakage" becomes part of the global overturning circulation and has far-field implications for the Atlantic circulation and hydrography. This includes a direct impact on one of the world's four highly productive eastern boundary upwelling systems: the Benguela Current upwelling system. Dynamics and oceanic impacts of Agulhas leakage are analysed in Sect. 8.3.

Both the Agulhas Current east of Africa and Agulhas leakage (south)west of Africa influence the regional climate through its impact on sea surface temperature and associated air–sea flux patterns. The result is a heterogeneous distribution of rainfall over southern Africa with drying in the future, but also potential risks for extreme events such as droughts and floodings, amplified under a warming climate (Sect. 8.4).

Changes in the Agulhas Current and leakage can affect the regional oceanic heat budget, leading, in combination with eustatic sea-level rise, to regionally variable changes in coastal sea levels. Additionally, changes in climatic drivers (e.g., pressure, wind) can potentially affect the characteristics of waves, which constitute a primary driver of coastal flooding in the region. Together, those changes in the oceanic drivers, increased precipitation, and river discharge can exacerbate coastal impacts through compound flooding along the South African coasts (Sect. 8.5).

Here we provide an introduction to the individual components of the Agulhas Current system, their temporal evolution during ongoing anthropogenic climate change, and their impact on oceanic and terrestrial climate, as well as flood vulnerability of African coastlines. The individual sections contain new insights from research within the project "Changes in the Agulhas System and its Impact on Southern African Coasts" (CASISAC, BMBF grant 03F0796). An important aspect of this chapter is the thread evolving from basic understanding to predictions and to impact studies, while specific details are published in separate topical studies. Technically, this chapter follows a chain of high-resolution ocean models under past atmospheric conditions, climate and wave models under future conditions, towards impact studies with direct socio-economic consequences in a changing climate.

8.2   The Agulhas Current

The Agulhas Current is the western boundary current of the South Indian Ocean (Lutjeharms 2006). Its structure and transport are shaped by a combination of wind-related and thermohaline drivers from the atmosphere. While the trade winds over the Indian Ocean set up a horizontal basin-scale circulation in the Indian Ocean (Biastoch et al. 2009; Loveday et al. 2014), an additional global component arrives from the Pacific Ocean (Le Bars et al. 2013; Durgadoo et al. 2017). As a result, the Agulhas Current flows southward along the South African coast in the Indian Ocean and transports warm and saline waters from equatorial to higher latitudes. The combination of its own inertia and the westerly winds in the Southern Hemisphere cause the Agulhas Current to overshoot the southern tip of Africa and to retroflect back into the Indian Ocean. Since this retroflection is not complete, a significant part of the Agulhas Current finds its way into the South Atlantic. This "Agulhas leakage" combines the subtropical gyres in the Indian Ocean and the South Atlantic towards one "supergyre" (Speich et al. 2007; Biastoch et al. 2009). Agulhas leakage is not only a direct inflow, but also happens in the form of mesoscale Agulhas rings and filaments, which make the region around southern Africa one of the most eddying regions in the world ocean (Chelton et al. 2011).

Fully constituted at $\sim$27° S, the Agulhas Current flows with surface velocities of 1.5 m s^{-1} and above (Lutjeharms 2006). At 32° S, it reaches down to below 2000 m, in a typical v-shaped profile that hugs the African continental slope. With a width of 200 km, it transports more than 75 Sv (1 Sv $= 10^6$ m^3 s^{-1}) (Bryden et al. 2005). Owing to recirculations in the Southwest Indian Ocean Subgyre, the transport increases and reaches 84 Sv at $\sim$34° S (Beal et al. 2015). On its further way towards the south, the shelf widens, leaving more room for evolving instabilities. South of Cape Agulhas, the Agulhas Current flows into the South Atlantic, before it abruptly turns back towards the east into the Indian Ocean as the Agulhas Return Current (Lutjeharms 2006). The dynamics of the retroflection are still not fully understood but built on a combination of the inertia of the current, the bathymetry, and the strength and position of the westerlies (Beal et al. 2011). At the retroflection, large mesoscale meanders are generated by loop exclusion as well as barotropic and baroclinic instabilities, leaving Agulhas rings with diameters of several 100 km and velocities reaching below 2000 m depth as a prominent characteristic (Van Aken et al. 2003b). Agulhas rings transport warm and saline waters from the Indian Ocean into the Atlantic Ocean and act as contributors to the surface branch of the Atlantic Meridional Overturning Circulation (AMOC), although the major portion recirculates in the subtropical gyre within the South Atlantic (Speich et al. 2001; Beal 2009; Rühs et al. 2019).

Velocities and transport of the Agulhas Current are subject to temporal variability at a range of different timescales. The seasonal cycle with a minimum in austral winter and a maximum in austral summer is related to the wind field driving the subtropical gyre in the Indian Ocean (McMonigal et al. 2018). Although parts of the interannual variability have been linked to El Niño/Southern Oscillation (ENSO), the dominant mode of interannual variability in the Pacific Ocean (Putrasahan et al. 2016; Elipot and Beal 2018), it was also shown in a model study that the Agulhas Current can generate a year-to-year variability in the range of the observed variability through internal instabilities (Biastoch et al. 2009).

A prominent feature of the Agulhas Current is the intermittent perturbations that impact the Agulhas Current. These "Natal Pulses" occur ∼1.6 times per year (Elipot and Beal 2015) but strongly vary from year to year (Yamagami et al. 2019). Caused by eddy–mean flow interaction of arriving mesoscale eddies from the Mozambique Channel and from the South-East Madagascar Current (Biastoch and Krauss 1999), the Natal Pulses rapidly travel downstream and have the potential to generate the shedding of Agulhas rings (Schouten et al. 2002).

Figure 8.1 shows the circulation around South Africa as simulated in the eddy-rich ocean general circulation model INALT20 (Schwarzkopf et al. 2019) driven by the recent JRA55-do atmospheric forcing dataset (Tsujino et al. 2018) over the past six decades from 1958 to 2019. The Agulhas Current is fed by flow through the Mozambique Channel and the South-East Madagascar Current and reaches highest mean south-westward velocities of $1.6 \, \text{m s}^{-1}$ at ∼33.5° S. On its way further south, the Agulhas Current crosses the location of the Agulhas Current Timeseries (ACT) array (Beal and Elipot 2016, indicated by the black line in Fig. 8.1a) before it detaches from the coast. At the ACT transect, the Agulhas Current is represented by a ∼350-km-wide coastal, surface-intensified current reaching down to ∼1500 m depth (Fig. 8.1c). Further downstream, it changes its direction towards the west, where it retroflects between 15° E and 20° E into the meandering Agulhas Return Current back into the Indian Ocean. Although the 20-year mean surface velocity shows a straight path of the Agulhas Current and a robust meandering Agulhas Return Current, the velocity field is highly variable and eddying, not only along its main path but especially where the retroflection takes place (Fig. 8.1b). Another area of high variability is a corridor towards the Atlantic Ocean where Agulhas rings

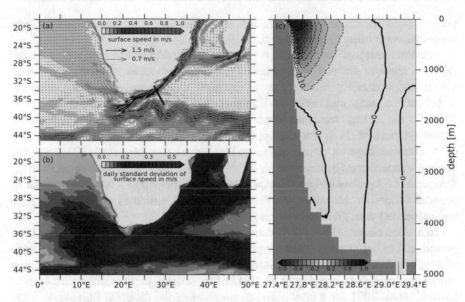

Fig. 8.1 Circulation in the Agulhas Current system around South Africa as simulated by a 1/20° ocean model (Schwarzkopf et al. 2019): (**a**) mean and (**b**) standard deviation of surface speed in the period 2000–2019 (in (**a**) vectors shown every 15th grid point). (**c**) Mean section across the Agulhas Current at ~34° S (location indicated in (**a**))

transport Indian Ocean waters into the neighbouring basin, contributing to Agulhas leakage.

8.3   Agulhas Leakage and Its Impact on the South Atlantic and the Benguela Upwelling System

Agulhas leakage is defined as the transfer of relatively warm and salty water from the Agulhas Current in the Indian Ocean to the South Atlantic Ocean (Lutjeharms 2006; Beal et al. 2011). It constitutes a key process of the global overturning circulation (Broecker 1991) and has been suggested to impact regional to global climate variability through various processes on a vast range of timescales (Beal et al. 2011).

Agulhas leakage occurs at the retroflection of the Agulhas Current. It is mediated in large parts through anticyclonic Agulhas rings (Schouten et al. 2000) as well as cyclonic mesoscale eddies and filaments that are shed at the retroflection and travel north-westward into the South Atlantic. The generation of these mesoscale features in the retroflection region has been linked to barotropic instabilities of the Southern Hemisphere supergyre (that is, the interconnected subtropical gyres of the Atlantic, Pacific, and Indian Ocean) (Elipot and Beal 2015; Weijer et al. 2013), but individual ring shedding events can further be impacted by mesoscale upstream perturbations (Biastoch et al. 2008; Schouten et al. 2002) as well as regional sub-mesoscale variability (Schubert et al. 2019, 2021).

The turbulent and intermittent (sub-)mesoscale eddy-driven component makes Agulhas leakage difficult to measure. Different approaches to estimate its magnitude and variability from ocean observations were introduced. The evaluation of subsurface floats and surface drifter pathways yielded the canonical number of 15 Sv for the leakage transport in the upper 1000 m (Richardson 2007) and a more recent estimate of 21 Sv for the upper 2000 m (Daher et al. 2020). The imprints of Agulhas leakage on observed sea surface height (Le Bars et al. 2014) and sea surface temperature (Biastoch et al. 2015) patterns have been used to reconstruct timeseries of Agulhas leakage transport anomalies, revealing a large interannual to decadal variability. However, the observation-based estimates are limited in time and associated with great uncertainties.

To test the observation-based estimates, to retrieve timeseries over a longer time period, and to better understand the processes that determine Agulhas leakage variability, ocean and climate models have been employed. In this context, Lagrangian model analysis (van Sebille et al. 2018) has proven particularly valuable (see Schmidt et al. 2021 for a review and comparison of the different Lagrangian tools and experiment designs used to estimate Agulhas leakage). Model-based studies suggest that interannual to decadal variability in Agulhas leakage can be related to larger-scale changes in the Southern Hemisphere winds (Biastoch et al. 2009; Durgadoo et al. 2013; Cheng et al. 2018). Moreover, they indicate that Agulhas leakage has increased since the 1960s (Biastoch et al. 2009; Rouault et al. 2009), due to a strengthening in the westerlies caused by increasing anthropogenic greenhouse gases and ozone depletion (Biastoch and Böning 2013; Ivanciu et al. 2021). A continuation of Agulhas leakage strengthening during future climate change and a resulting enhanced transport of salt into the South Atlantic may stabilise the Atlantic Meridional Overturning Circulation (Weijer et al. 2002; Biastoch and Böning 2013; Biastoch et al. 2015), while warming and ice sheet melting are projected to weaken it (Beal et al. 2011).

Nevertheless, due to, e.g., (i) outstanding challenges in correctly representing Agulhas leakage in ocean and climate models (at minimum mesoscale resolving resolution is needed), (ii) the dependence of ocean-only model simulations on the availability and accuracy of the atmospheric forcing, as well as (iii) uncertainty regarding the future emission of anthropogenic greenhouse gases (GHGs) and the potential recovery of the ozone depletion, there are still many open questions regarding past and future changes of Agulhas leakage and their impact on climate (change).

The exact temporal evolution of Agulhas leakage, in particular over the last decades, is still ambiguous. While some ocean model simulations (for example, a simulation in INALT20 under CORE forcing Schwarzkopf et al. 2019) indicate a nearly continuous increase since the mid-1960s that further accelerates in the 1990s and 2000s, other model simulations (for example, a simulation in INALT20 under JRA55-do forcing Schmidt et al. 2021) and observation-based reconstructions (for example, a timeseries reconstructed from HadISST Biastoch et al. 2015) do not

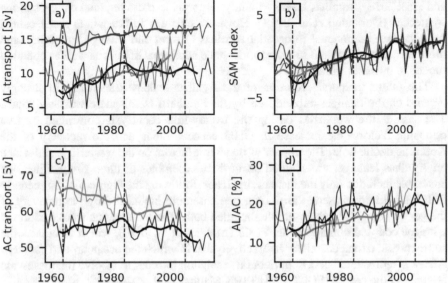

Fig. 8.2 Temporal evolution of (**a**) Agulhas leakage (AL), (**b**) annual Southern Annular Mode (SAM) index calculated following Marshall (2003), (**c**) Agulhas current (AC), and (**d**) ratio AL / AC within simulations with the eddy-rich ocean model configuration INALT20 under JRA55-do forcing (Schmidt et al. 2021, black) and CORE forcing (Schwarzkopf et al. 2019, gray). An AL timeseries reconstructed from HadISST (Biastoch et al. 2015, pink line in (**a**)) and the station-based annual SAM index (pink line in (**b**)) are also shown

exhibit a significant trend over the full time period and show a levelling since the 1990s (Fig. 8.2a). Previous studies suggest that the different temporal evolutions of Agulhas leakage could be due to a different representation of the wind fields in the different atmospheric forcing datasets. However, such a relationship is not trivial and cannot be represented by simple integrative parameters such as the proposed Southern Annular Mode index (SAM, a measure of the strength of the westerly winds that shows only minor differences between the different simulations, Fig. 8.2b). It should also be emphasised that not only the temporal evolution of Agulhas leakage differs between the different simulations, but also that of other components of the Agulhas Current system. In particular, in contrast to Van Sebille et al. (2009) but consistent with Loveday et al. (2014), Durgadoo et al. (2013), Cheng et al. (2018), there is no clear relationship between the temporal evolution of the strength of the Agulhas Current and Agulhas leakage (Fig. 8.2c). Rather, interannual and longer-term fluctuations in Agulhas leakage represent a changing proportion of the transport of the Agulhas Current flowing into the South Atlantic (Fig. 8.2d). Hence, Agulhas leakage and Agulhas Current variability are driven by distinct processes and show different responses to changes in Southern Hemisphere winds. While the Agulhas Current responds to larger-scale changes in subtropical wind curl, including westerlies and trades and their modulation via ENSO (Elipot and Beal 2018), Agulhas leakage mainly responds to more regional changes in the westerlies (Durgadoo et al. 2013). However, changes in the winds alone cannot fully explain interannual to decadal Agulhas Current and leakage variability, and more research is required to better understand additional drivers as well as potential modes of intrinsic variability.

The future temporal evolution of the Agulhas leakage transport will strongly depend on the changes experienced by the Southern Hemisphere westerly winds. The fate of the westerlies during the twenty-first century is controlled by two opposing factors: the increase in GHG concentrations and the recovery of the Antarctic ozone hole. The impact of these two factors on the westerlies, and hence on Agulhas leakage, was studied using three ensembles of three simulations: one ensemble included only the increase in GHGs, following the high-emission scenario SSP5-8.5 (Meinshausen et al. 2020), one ensemble included only the recovery of the ozone hole, and one ensemble included both. The simulations were performed with the coupled climate model FOCI (Matthes et al. 2020), which calculates the stratospheric ozone chemistry interactively and in which the ocean around southern Africa is represented at 0.1° horizontal resolution in order to resolve the mesoscale features of the region (FOCI_INALT10X Matthes et al. 2020).

Timeseries of the westerly winds and of Agulhas leakage in the three ensembles are depicted in Fig. 8.3e and f, respectively. The increase in GHGs leads to a pronounced poleward intensification of the westerlies and, as a result, to a positive Agulhas leakage trend of 0.36 ± 0.12 Sv per decade until the end of the century. This translates in about 28% more Agulhas leakage entering the Atlantic Ocean at the end of the twenty-first century compared to the current day. In contrast, the recovery of the ozone hole leads to a weakening of the westerly winds and to a weak but significant decrease in Agulhas leakage of -0.13 ± 0.12 Sv per decade. This implies that, in the absence of an increase in GHGs, Agulhas leakage would be 7% weaker at the end of the century compared to today. When the impacts of ozone recovery and increasing GHGs are considered together, the impact of the increasing GHGs dominates. Agulhas leakage exhibits a trend of 0.18 ± 0.12 Sv per decade, which implies an increase of 13% at the end of the century compared to today. These results are dependent on the high-GHG-emission scenario used.

The future increase in Agulhas leakage has implications for the Atlantic Ocean. The waters contained in Agulhas rings are warmer and more saline compared to the surrounding waters (Van Aken et al. 2003b). Observations of Agulhas rings (Van Aken et al. 2003b; Giulivi and Gordon 2006) revealed that below a well-mixed surface layer, the rings carry subtropical mode water formed in the southwestern Indian Ocean, South Indian Ocean Central Water, and Sub-Antarctic Mode Water, while in the underlying intermediate layer the Antarctic Intermediate Water dominates, but the Red Sea Water is also present. As parts of Agulhas leakage feed into the upper limb of the AMOC, changes in its transport are linked to changes in the thermohaline properties of the Atlantic Ocean. This is revealed by a composite analysis, whereby low-pass filtered (5-year cut-off period) and detrended temperature and salinity anomalies were selected for the years when Agulhas leakage exceeded the 90th percentile of its distribution (Fig. 8.3). Periods

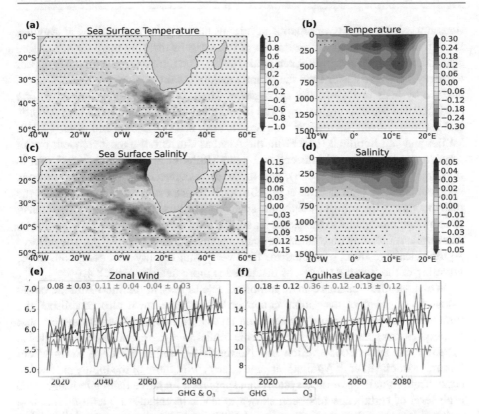

Fig. 8.3 Composites of (**a**) sea surface temperature anomalies, (**b**) temperature anomalies aver aged over 20° S–40° S, in °C, (**c**) sea surface salinity anomalies, and (**d**) salinity anomalies averaged over 20° S–40° S for the years when Agulhas leakage exceeds the 90th percentile of its distribution. The output from the simulations with fixed GHGs was used. The anomalies, as well as Agulhas leakage, were low-pass filtered to retain variations with periods above 5 years and a linear trend was removed. The stippling masks anomalies that are not significantly different from the time mean according to the Monte Carlo method. Timeseries of (**e**) zonal mean zonal wind averaged between 45° S–60° S, in m s⁻¹, (**f**) Agulhas leakage in Sv for the ensemble (which may average out some of the individual ensemble members in this highly stochastic process) that includes only ozone recovery (blue), only the increase in GHGs (orange), and both forcings (black). The dashed lines depict the corresponding linear trends, and the numbers at the top of the panels give the values of the trends per decade

of high Agulhas leakage are associated with positive sea surface temperature (SST) and surface salinity anomalies, which propagate north-westward from the Agulhas retroflection region into the South Atlantic (Fig. 8.3a, c). These temperature and salinity anomalies extend below the surface, as seen in Fig. 8.3b and d, which shows the vertical profile of the anomalies averaged over the latitudinal band 20° S–40° S. The temperature anomalies extend down to 1000 m, while significant salinity anomalies can be found down to about 750 m. Therefore, an increase in Agulhas leakage leads to a warming and salinification of the South Atlantic. While the model does not exhibit salinity anomalies at intermediate depth, observations of Agulhas rings found positive salinity anomalies at these depths marking the presence of Red Sea Intermediate Water (van Aken et al. 2003a). The temperature anomalies appear to decay westward faster than the salinity anomalies do, as they are damped at the surface by heat release to the atmosphere. There is observational and modelling evidence that waters originating from the Agulhas region reach the North Atlantic and its deep convection regions (van Sebille et al. 2011; Biastoch and Böning 2013; Weijer and van Sebille 2014). From the Agulhas Current, the most frequent transit time to the North Brazil Current is 7 years (Rühs et al. 2019), to 26° N one to two decades (Rühs et al. 2013), and to the deep convection regions between one and four decades (van Sebille et al. 2011). These are estimates for the peak in the distributions of the transit times to the specific locations. The fastest reported transit time of Agulhas waters to the North Atlantic is only 4 years (van Sebille et al. 2011). The Agulhas thermohaline anomalies can potentially affect the AMOC. The positive Agulhas leakage trend predicted for the twenty-first century will contribute to the warming of the South Atlantic, as discussed in more detail in Sect. 8.4.

While Agulhas leakage has an important role to play in the Indo-Atlantic ocean exchange segment of the global conveyor belt circulation, it also establishes the direct interaction between a western and an eastern boundary current system that is unique among the world's oceans. This results in a region of intense turbulence (Matano and Beier 2003; Veitch and Penven 2017) within the Cape Basin where features associated with Agulhas leakage interact with those of the highly productive southern Benguela upwelling system that supports a lucrative fishing industry. This high level of turbulence has been shown to result in enhanced lateral mixing and reduced surface chlorophyll (Rossi et al. 2008) and, therefore, productivity within the Cape Basin. More direct impacts of the Agulhas on the Benguela upwelling system include its contribution to the development of a shelf-edge jet current (Veitch et al. 2017) that transports fish eggs and larvae from their spawning ground on the Agulhas Bank to their nursery area within St Helena Bay (Shelton and Hutchings 1982; Fowler and Boyd 1998). Furthermore, this jet current presents a barrier to cross-shelf exchanges (Barange et al. 1992; Pitcher and Nelson 2006) on the southern Benguela shelf, which helps to promote both the concentration of upwelled nutrients and nearshore retention, thereby enhancing productivity. Additionally, Agulhas Rings have been observed to have a role to play in the generation of large upwelling filaments that have the potential to cause the offshore advection of large quantities of nutrient-rich waters (Duncombe-Rae et al. 1992). The characteristics of the upwelling source waters of the southern Benguela are a key component of the productive marine ecosystem.

They enter the system directly from the south (Tim et al. 2018) via Agulhas rings at the continental slope, causing cross-shelf intrusions of water (and its properties) from the Agulhas leakage into the upwelling region (Baker-Yeboah et al. 2010).

8.4   Impact on Climate in Southern Africa

Southern African climate is strongly modified by the high-altitude interior plateau and the termination of the relatively narrow landmass in the mid-ocean subtropics that allows the Agulhas to flow in close proximity to the Benguela upwelling system. As a result of these moderating oceanic and topographic factors, surface land temperatures are typically less extreme in southern Africa than would be expected, and there are important implications for weather system development and rainfall patterns.

As a simple example, Durban (29° 53'S) on the east coast adjacent to the Agulhas has an annual mean rainfall of 1019 mm compared to about 50 mm for Alexander Bay (28° 35'S) on the west coast in the central Benguela upwelling system. Annual mean temperatures at Durban are almost 5 °C warmer than those at Alexander Bay. The influence of the broader Agulhas Current region on South African climate in particular has long been recognised. Large surface heat fluxes were associated with the southern Agulhas Current (Walker and Mey 1988). Latent heat fluxes in the central Agulhas Current (south of Port Alfred) have been found to be about 75% greater in the core of the current than further seawards and up to about 7 times greater than those measured inshore of the current, leading to modifications of the marine boundary layer and large increases in precipitable water content as air advected across the current (Lee-Thorp et al. 1999). These high fluxes associated with the core of the current are difficult to represent in operational models since it is typically only ~70–80 km wide (Rouault et al. 2003). Statistical relationships between interannual variability of SST in the Agulhas Current and summer rainfall over central and eastern South Africa were found (Walker 1990; Mason 1995). Evidence that the relative proximity of the Agulhas Current core to the coastline helps to account for the large increase in average rainfall along the east coast was given (Jury et al. 1993). For example, Port Elizabeth (~34° S), where the continental shelf is wide and the current far from the coast, has an annual average rainfall of 624 mm, whereas Durban (~30° S) with its narrow shelf is much wetter on average (1019 mm). Given that South Africa is semi-arid but with generally mild temperatures, most research on the influences of the Agulhas Current on regional climate has focused on rainfall or on the development of rain-producing weather systems. Emphasis has also typically been placed on the summer half of the year, since this is by far the dominant rainfall season over almost all of southern Africa.

Various model studies have explored relationships between SST variability in the Agulhas Current region and southern African rainfall together with the potential mechanisms involved. In the simplest case, these studies have imposed idealised SST anomalies in the South-West Indian Ocean on the climatological SST forcing fields applied to coarse resolution atmospheric general circulation models (AGCMs). Warming in the Agulhas region resulted in statistically significant increased rainfall over southeastern Africa via enhanced latent fluxes over the SST anomaly, and advection of the anomalously moist unstable air towards the landmass (Reason and Mulenga 1999). While the greatest model response was found in summer, there were also large rainfall increases in autumn and particularly spring. In a similar idealised AGCM experiment, it was found that smoothing out the current so that the observed SST in the broader Agulhas Current region was replaced by zonally averaged SST and hence cooled led to a southward shift and weakening of midlatitude cyclonic weather systems tracking south of South Africa and reduced rainfall over southern South Africa (Reason 2001). The same type of experiment was performed about two decades later with a regional climate model (Nkwinkwa Njouodo et al. 2018). The results confirmed earlier work that the core of the Agulhas Current is associated with sharp gradients in SST and sea-level pressure, together with a band of convective cloud, and sometimes rainfall. Under favourable synoptic conditions, rainfall can then also occur over the nearby coastal landmass. There is evidence that SST gradients associated with Agulhas Current eddies and meanders affect the vertical air column up to the tropopause (Desbiolles et al. 2018).

Other experiments have provided evidence that latent heat fluxes from the Agulhas Current can significantly impact the rainfall over coastal South Africa of high-impact weather events such as cut-off lows (Singleton and Reason 2006, 2007) and mesoscale convective systems (Blamey and Reason 2009). In all cases, the presence of a low-level wind jet blowing across the current towards the land was important in transporting moist, unstable air that, when forced to rise by the coastal mountains, led to heavy rainfall. By removing the effect of the current in the model, it was shown that most of the moisture transported by the jet evaporated off the current.

Over eastern South Africa, long-lived mesoscale convective systems, which are often associated with heavy rainfall in summer, tend to occur downstream of the Drakensberg mountains and over the northern Agulhas Current where convective available potential energy (CAPE) and wind shear environments are favourable (Morake et al. 2021). More generally, the northern Agulhas Current and adjacent southeastern Africa have been identified as one of the convective hotspots in the global atmosphere (Brooks et al. 2003; Zipser et al. 2006). Lagrangian trajectory analyses have confirmed that the Agulhas Current is one of the important moisture sources for summer rainfall over a large interior region in subtropical southern Africa (the Limpopo River Basin) on both seasonal and synoptic scales (Rapolaki et al. 2020, 2021). However, its role seems to be one of enhancing moisture uptake along the trajectory path (which often originates in the midlatitude South Atlantic), associated with large-scale weather systems such as ridging anticyclones and cloud bands, rather than as a source region in its own right. Nevertheless, summer dry spell frequencies (the number of wet days) have decreasing (increasing) trends between 1981–2019 over the Eastern Cape/central South Africa related to changes in moisture fluxes / winds over the southern Agulhas Current region (Thoithi et al. 2021).

In terms of the southern Agulhas Current region, there are as yet no continuous measurements of the strength of the Agulhas leakage here. Thus, analysis of the impact of this leakage on regional climate has usually been based on their inferred fingerprints on oceanic sea surface temperatures. For instance, a stronger advection of warm water masses from the Indian Ocean by a stronger leakage leads to warmer sea surface temperatures in the southeast Atlantic. Due to this data limitation, the simulations with ocean–atmosphere models have been conducted to provide more direct analysis of the regional climate anomalies that, in the simulations, may be correlated with the intensity of the Agulhas Current and leakage. These simulations also allow for a quantification of their contribution to the long-term trends of regional precipitation, and this is both in the current and in projected future climate. In the following, we summarise some of these analysis obtained for both periods with the global coupled ocean–atmosphere model (FOCI_INALT10X Matthes et al. 2020), with interactive ozone chemistry, used in Sect. 8.3) for the past (1951–2013) and future (2014–2099, SSP5-8.5 scenario). These simulations were used to drive a high-resolution regional atmospheric model (CCLM, COSMO model in CLimate Mode, https://www.clm-community.eu/) that can better represent regional precipitation (Tim et al. 2023). Agulhas leakage is defined here as the amount of water crossing the Good Hope Line within a 5-year window, thus leaving the Agulhas system and entering the South Atlantic (Tim et al. 2018). A similar study based on a different atmosphere–ocean global model CCSM3.5 had previously been conducted for the current period (Cheng et al. 2018), and thus we can assess the robustness of the results when a different model is used.

In the simulations of the current and future climate, a stronger Agulhas leakage leads, as expected, to warmer SST southwest of the Western Cape region as mentioned in the previous section (Fig. 8.3a). This results in a relatively high-positive temporal correlation ($r = \sim 0.5$) between Agulhas leakage and SST in the Agulhas retroflection region, southwest of it, and in the corridor where Agulhas rings transport warm Indian Ocean water into the South Atlantic. The increase in Agulhas leakage over the last decades (as described in the previous section) and the warming of the Agulhas Current system (Rouault et al. 2010, 2009) both result in warming of the SSTs southwest of the Western Cape region, up to 2 °C (historical period) and 4 °C (in the scenario period). A linear regression analysis of the simulated temperature and leakage timeseries indicates that around 1/6 of both warming rates is due to the increase of Agulhas leakage.

In the historical period, Agulhas leakage and precipitation along the southeast coast of South Africa are found to be positively correlated (Fig. 8.4a). A more intense Agulhas leakage imprints the SSTs patterns in the region, with warmer SSTs in the retroflection region and colder SSTs in the southwest Indian Ocean. Warmer SSTs are linked with higher convective precipitation at the southeast coast of South Africa during summer (Walker 1990), a result also found in the simulations with the model CCSM3.5 (Cheng et al. 2018). By contrast, the correlations of Agulhas leakage with precipitation in the winter rainfall zone around the Western Cape region are negative (Fig. 8.4a), as also found by Cheng et al. (2018). This may be due to the modification of cyclonic activity by the SSTs in this season, as discussed before. Concerning the long-term trends, the simulated precipitation displays a weak negative trend in most of southern Africa, with some precipitation intensification along the southeast coast (Fig. 8.4c). Around 1/10 of this trend can be statistically explained by Agulhas leakage (Fig. 8.4e). The areas with a positive

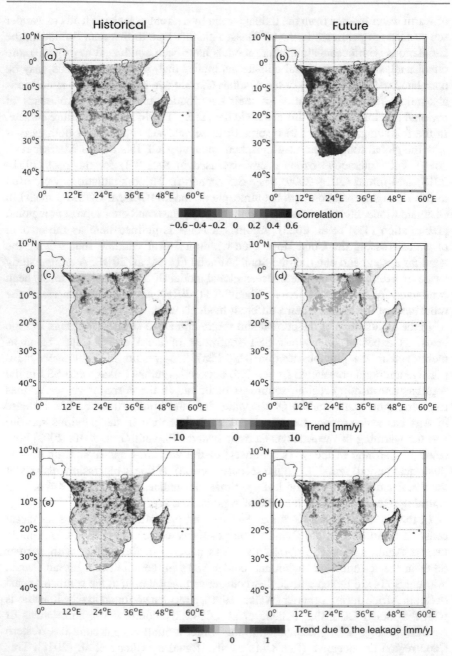

Fig. 8.4 The correlation of (**a**, **b**) Agulhas leakage (FOCI simulation) and the precipitation (CCLM simulation), (**c**, **d**) the precipitation trend over the simulation period, and (**e**, **f**) the contribution of Agulhas Leakage to the precipitation trend, in the (left column) historical simulation covering the period 1951–2013 and (right column) the scenario simulation covering the period 2014–2099. At a significance level of 95%, correlations of (**a**) 0.26 and (**b**) 0.22 and larger are significant. Based on typical standard deviation, precipitation trends of (**c**) 1.2 mm/y and (**d**) 0.76 mm/y and larger are significant

precipitation trend are also the ones that are positively correlated to the leakage. In the Western Cape region (Fig. 8.4d), precipitation displays a long-term decline. Both trends reflect the long-term impact of the intensifying leakage on precipitation at the southeast coast of South Africa.

Regarding changes in precipitation in the future scenario simulation, Agulhas leakage and precipitation in South Africa appear negatively correlated (Fig. 8.4b), contrary to the historical period. Thus, the simulation indicates both an intensification of leakage and a diminishing precipitation along the whole coast and the southern inland, as also found by, e.g., Dosio et al. (2019) and Rojas et al. (2019). Again, around 1/10 of the trend in precipitation is due to Agulhas leakage (Fig. 8.4f). The change in the dependency of Agulhas leakage strength and precipitation comparing the current and future periods requires further investigation. One explanation could be related to the southward shift of the Agulhas Current in the future scenario. According to this explanation, Agulhas leakage intensifies, while the Agulhas Current weakens (Ivanciu et al. 2022) and is displaced poleward away from the coast (Yang et al. 2016). As found by Jury et al. (1993), the precipitation at the southeast coast is stronger when the core of the Agulhas Current is located closer to the coast. This leads us to the conclusion that the future trend in simulated precipitation in the winter rainfall zone may be directly linked to the strength of the Agulhas leakage, whereas the trend in precipitation at the southeast coast may be more strongly linked to the position and intensity of whole Agulhas Current system. It should be, however, kept in mind that other remote influences, for instance that of ENSO, can also modulate precipitation trends in this region (see Chap. 6).

8.5   Impact on Coasts

With more than 2500 km of coastline, South Africa is largely exposed to flooding and erosion from extreme water levels and waves. The majority of the largest South African cities such as Cape Town and Durban are located at the coastline, comprising around 40% of the South African population and 60% of the country's economy (CSIR 2019). Exposure to coastal hazards is further exacerbated by socio-economic development. Coastal cities are growing and developing at a rapid rate (ASCLME/SWIOFP 2012), with some areas of the coast having experienced a building boom over the past two decades (Smith et al. 2007). The main ports of South Africa such as Durban, Cape Town, and Port Elizabeth/Gqeberha are located in big cities along this coastline and provide the largest trades for the region (Mather and Stretch 2012). In addition, the tourism sector has a large contribution to the economy of the country (Fitchett et al. 2016). However, the risk of flooding has not been taken into account in coastal development planning, and several coastal cities, such as those previously mentioned as well as resorts, are built in low-lying areas, estuaries, and beachfront close to the high-water mark (Theron and Rossouw 2008; Mather and Stretch 2012), where the probability of flooding is often the highest.

As a result, coastal flooding and erosion from extreme events can potentially cause large damages along the South African coastline. This was the case in March

2007, when a combined spring tide and extreme swells produced up to a billion Rand (about 70 Mill €) of damages along the region of KwaZulu-Natal, South Africa (Smith et al. 2007). Despite these large damages, the event of March 2007 is estimated of having an average return period of only 10–12 years (Palmer et al. 2011). As was the case in the event of March 2007, flooding can arise from a combination of different drivers such as tides, waves, and storm surges but also in combination with precipitation and river discharge. These so-called compound events are particularly important for estuarine areas, where the interaction between the flood drivers can exacerbate flood impacts (Zscheischler and Seneviratne 2017).

Changes in the climate of the region due to changes in the circulation and atmospheric patterns discussed in former Sects. 8.2, 8.3, and 8.4 may lead to changes in the flood drivers and sea level. The IPCC AR6 (IPCC 2021) estimates a likely global mean sea-level rise (SLR) by 2100 up to 1.01 m under a very high emissions scenario (SSP5-8.5, upper likely range). However, higher sea-level rise scenarios above the likely range cannot be excluded due to large uncertainties in the ice sheet processes. In addition, regional deviations from the global estimates can be expected along the South African coastline as the west coast is dominated by variations in offshore buoyancy by the Benguela system (such as those caused by wind-driven coastal Kelvin waves) (Schumann and Brink 1990), while water levels along the east coast are influenced by fluctuations in the Agulhas Current system (van Sebille et al. 2010). However, the effects of the Agulhas System on coastal sea levels are not fully understood due to the variability of several characteristics of the Agulhas current system (e.g., position of the core, eddies) and the lack of in situ measurements (Nhantumbo et al. 2020).

In addition, future warming of the ocean will not be spatially uniform, so regional differences in the thermal expansion of the water column are also expected (Yin et al. 2010). For instance, the sea-level trend in the period 1992–2018 monitored by satellite altimetry shows that along the southern African coastline sea level has risen (Hamlington et al. 2020) faster than the global rate of 3.1 mm/year over the same period (Cazenave et al. 2018). However, the data from tidal-monitoring stations near Cape Town in the period 1957–2017 display temporal data gaps, especially in the last decades of the twentieth century, a situation that is common in Africa (Woodworth, Philip L. et al. 2007) and the linear trend derived from those data points to a rate of sea level of about 2 mm/year (Dube et al. 2021a). This is in agreement with the results of the analysis of the Durban tide gauge (Mather 2007), and of other Namibian and South African tide gauges (Mather et al. 2009). Nevertheless, even along the southern African coast, regional differences in the relative sea-level trends are apparent, with sea level along the east coast having risen more rapidly (2.7 mm/year) than along the south (1.5 mm/year) and the west coast (1.9 mm/year). It is therefore difficult to conclude whether sea-level rise in this area has recently accelerated, since the measurements of tidal gauges and of satellite altimetry have different characteristics and very few studies exist so far (Brundrit 1984; Mather et al. 2009). Following state-of-the-art methods and using tide gauge and altimetry records, Allison et al. (2022) suggest 7–14% higher SLR projections compared to the global IPCC estimates along the South African coastline, driven by larger contributions of all components of the sea-level budget. However, further research is needed for establishing the effects of the Agulhas system on coastal sea levels along the South African coast and to assess the impacts of the future potential changes in the Agulhas system (discussed in Sects. 8.2 and 8.3) on coastal sea-level rise.

Mean sea-level rise and the possible intensification of sea-level extremes are expected to impact coastal regions of southern Africa, as has been observed in recent years for the Cape Town area (Dube et al. 2021b,a). Among the various flood drivers, wind waves have the relative largest contribution to total extreme water levels along the South African coastlines (Theron et al. 2010). A potential intensification of waves in this region due to a projected intensification of winds under high-GHG-emission scenarios may exacerbate present flood risks. Therefore, predicting and analysing possible changes in wave climate during this century is essential for coastal risk and impact assessments for the South African coast.

Although global wave projections for the twenty-first century under different climate change scenarios already exist (e. g. Semedo et al. 2013; Hemer et al. 2015; Casas-Prat et al. 2018; Morim et al. 2020), regional effects can be omitted due to the coarse resolution of atmospheric forcing used at global scales. In CASISAC, we generated a coherent wave hindcast (1958–2018) and wave projections (2014–2098) for a high-end scenario (SSP5-8.5), using Time Delay Neural Networks (TDNN), and downscaled atmospheric forcing from the CCLM model discussed in Sect. 8.4. To train the TDNN, we used mean sea-level pressure (MSLP) data (Kanamitsu et al. 2002) as a forcing predictor and an existing global wave hindcast (Durrant et al. 2013) based on the same reanalysis dataset as response. It is important to mention that the TDNN is able to account for the processes included in the modelling of the training data, but it is not able to capture future changes of these processes or other processes not included such as, e.g., the interaction with the Agulhas current (Grundlingh and Rossouw 1995). This interaction can produce large increases in offshore wave heights, which can be a hazard for the shipping industry, but little effects are observed in coastal waves (Barnes and Rautenbach 2020).

Results show an average increase of 0.2 m in the mean H_S (2.7 m; Fig. 8.5) and 0.6 m in the 1% highest H_S (5.3 m; Fig. 8.6). However, along the west and south coasts, the increase of the mean H_S is of 0.3 m, while a smaller increase of 0.1 m is observed at the east coast. The 1% highest H_S show a similar pattern with increases of 0.8 and 0.9 m at the west and south coast, respectively, and 0.3 m at the east coast. Distinguishing between seasons, the largest increase of 0.3 m of mean H_S appears

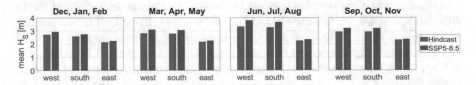

Fig. 8.5 Changes in mean significant wave height (Hs) between the hindcast and SSP5-8.5 scenario

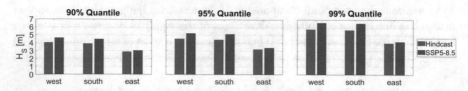

Fig. 8.6 Changes in the 10%, 5%, and 1% highest waves between the hindcast and SSP5-8.5 scenario

in winter, opposing 0.2 m in the other seasons. The 1% highest waves are predicted to increase 0.7 m in winter and spring and 0.5 m in summer and autumn. These changes in the wave climate are also affecting the extreme waves, increasing the 1-in-100 year H_S up to 1.5 m at some locations of the west coast and up to 0.5 m at the east coast.

Due to the high flood exposure of population and assets of the South African coastline, the Council for Scientific and Industrial Research (CSIR) of South Africa in cooperation with universities and government departments produced a national coastal flood assessment in order to support climate-resilient development (Lückvogel 2019). This assessment was performed for the majority of the coastline of South Africa providing a qualitative flood hazard index based on elevation and extreme events. We refined and extended this analysis by employing the simplified hydrodynamic model LISFLOOD-FP (Bates et al. 2005) to simulate flooding and provide quantitative estimates of flood characteristics for the 100-year return water level along the entire South African coast, at 90-m resolution, using the MERIT Digital Elevation Model (Yamazaki et al. 2017). We used extreme events derived from tidal levels from the global FES2014 (produced by Noveltis, Legos, and CLS and distributed by Aviso+, with support from CNES, https://www.aviso.altimetry. fr Carrère et al. 2015); non-tidal residuals for the present day analysis from the ocean model INALT20 (Schwarzkopf et al. 2019) under JRA55-do forcing (Schmidt et al. 2021); for the future SSP5-8.5 scenario outputs from the climate model FOCI-INALT10X (Matthes et al. 2020) (both described in Sects. 8.2 and 8.3; the latter from an experiment with prescribed atmospheric chemistry); and waves from the newly developed wave hindcast discussed above. The wave contribution was included as the wave set-up, which relates the average increase in coastal water levels by offshore waves, typically 20% of the offshore H_S.

The results of the flood simulations show that low-lying regions along estuaries are the areas most exposed to flooding from the current 1-in-100 year event of oceanic drivers (tides, storm surges, and waves). The main coastal cities of South Africa (Cape Town, Port Elizabeth, and Durban) are also affected, albeit to a smaller extent (Fig. 8.7). In these three cities, the largest flooded areas are also observed along the low-lying riverine and harbour regions. The largest flood impacts occur in Cape Town where several settlements, resorts, and a natural reserve are flooded. Under the SSP5-8.5 scenario, we find an average increase of more than 100% in flood extent along the South African coastline from the 1-in-100 year event, which is mostly driven by SLR. However, the potential changes in the wave climate of

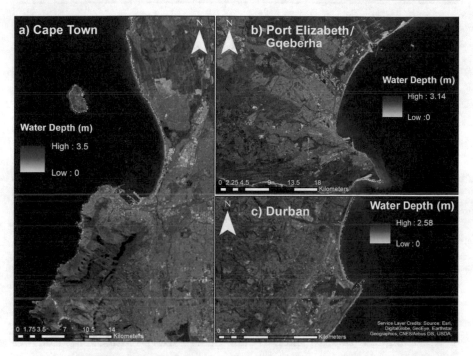

Fig. 8.7 Flood estimation for the present 1-in-100 years extreme event for: (**a**) Cape Town, (**b**) Port Elizabeth/Gqeberha, and (**c**) Durban

South Africa predicted under the SSP5-8.5 scenario and discussed before cause an increase in both flood extent and depth for the 1-in-100 years event (Fig. 8.8). The increase in extreme wave height leads to an increase up to 10% in flood depth as well as an increase of 19% in flood extent compared to present estimates (Fig. 8.8). These increases vary spatially along the coast due to the variability of the waves but also due to changes in the coastal characteristics. For example, there is an increase of 0.74 m in the 1-in-100 year H_S at Cape Town and of 0.51 m at Durban, but we find a larger increase in flood depths at the latter due to the characteristics of the floodplain. For Durban, coastal flooding can be further exacerbated in the future from the changes in precipitation discussed in Sect. 8.4, which can amplify flooding by compounding effects with the oceanic drivers as discussed later in this section.

Future coastal flooding estimates are primarily driven by SLR, but changes in wave climate solely will increase flooding and thus need to be considered in coastal planning and management in this region. Both SLR and changes in wave climate may also increase coastal erosion, which in turn can amplify coastal flooding by reducing the protection offered by natural systems such as dunes and wetlands.

In estuarine areas, coastal flooding can also occur from a combination of ocean drivers (i.e., water levels and waves) and river discharge. The South African coastline is particularly prone to compound flooding from waves and river discharge as cold fronts, cut-off lows, and cyclones produce large swells but also heavy rainfalls

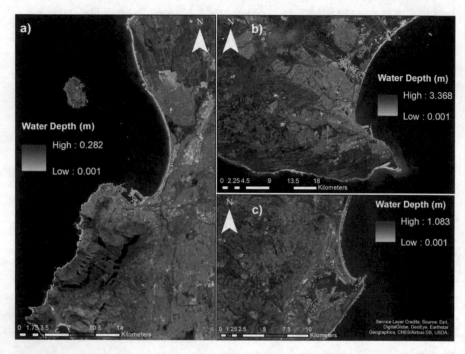

Fig. 8.8 Flood differences between the present 1-in-100 years event and the 1-in-100 years under SSP5-8.5 scenario (not accounting for SLR) for: (**a**) Cape Town, (**b**) Port Elizabeth/Gqeberha, and (**c**) Durban. Flood differences are caused due to changes in wave climate under SSP5-8.5

that can cause fluvial flash floods (Pyle and Jacobs 2016). However, compound flooding in South Africa has not been addressed yet, despite the large number of estuaries along the South African coastline that can be prone to compound flooding (van Niekerk et al. 2020). To explore the interactions between flood drivers in South African estuaries, we investigated compound flooding from river discharge and waves through the case study of the Breede estuary (Kupfer et al. 2021). This estuary is one of the largest permanently open estuaries (van Niekerk et al. 2020) and has the fourth largest annual runoff in South Africa (Taljaard 2003). Using the hydrodynamic modelling suite Delft3D (Lesser et al. 2004), we simulated river discharge, tides, waves, and their interactions to analyse the effects of each driver on the resulting flooding. For detailed description of the study, see Kupfer et al. (2021).

Our results showed differences in flood depth and extent when comparing the compound scenarios (Fig. 8.9). Co-occurring extreme river discharge and extreme waves (S_{TWQ}) result in an increase of 45% in flood extent in the upper part of the domain, compared to the low river discharge scenario (S_{TW}). However, when the waves are omitted, only a reduction of 10% of the flood extent is observed in regions of the estuary mouth and centre, where populated areas are located. An increase in the magnitude of the waves causes an increase of 12% in flood extent and up to 40 cm in flood depth, mostly in the lower part of the estuary. Therefore, waves

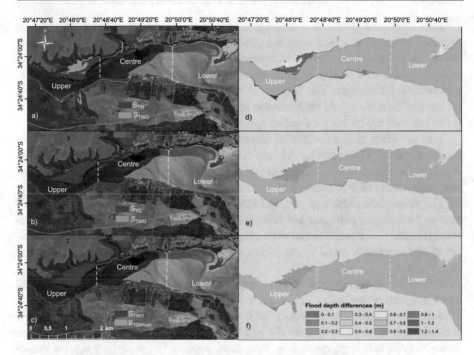

Fig. 8.9 Comparison of flood extents of the compound and excluding driver scenarios at the Breede estuary (South Africa) (left panel, (**a**), (**c**) and (**e**)) and differences in flood depths (right panel). Panel (**b**) shows the flood depths of S_{TWQ}–S_{TW}, (**d**) shows S_{TWQ}–S_{TQ}, and (**f**) $S_{TWQextr}$ (Kupfer et al. 2021)

can exacerbate flooding when combined with an extreme river discharge event by blocking river flow to the sea. Compound flooding can be further amplified in these regions where an increase in precipitation under a high GHG scenario is expected (Sect. 8.4).

8.6   Summary

The proper simulation of the Agulhas Current system dynamics requires sufficient ocean resolution, such that the mesoscale processes dominating the region can be simulated. At the same time, the variability of the Agulhas system in ocean-only simulations largely depends on the applied atmospheric forcing. The Agulhas leakage has been subject to decadal changes in the past and is likely to increase substantially towards the end of the century owing to rising GHGs, as predicted by our coupled climate model. The increase in Agulhas leakage is expected to result in a warming and salinification of the southeastern South Atlantic, impacting further downstream water masses in the Benguela upwelling system and in the pan-Atlantic circulation including the AMOC.

The increased offshore warming associated with the projected increase in Agulhas leakage, coupled with a hypothesised increase in upwelling favourable winds in the southern Benguela upwelling system and therefore cooler coastal waters, would lead to the enhancement of the already intense shelf-edge frontal system and jet current. The impact of the Agulhas, both directly and indirectly, on the southern Benguela region needs to be continually monitored to support good governance of the highly productive system that supports a lucrative fishing industry.

The warming also provides an important boundary condition through SST to the atmosphere. The southern African climate is therefore directly impacted by the Agulhas Current system. An increase of Agulhas leakage under global warming is, although not exclusively, linked to decreasing precipitation in South Africa.

The atmospheric changes under a high GHGs scenario cause an intensification of extreme waves along the coast of South Africa, which combined with (regional) sea-level rise and increased precipitation events has the potential to increase compound flooding along the coast of South Africa, considerably amplifying impacts to coastal communities.

The different disciplines in this chapter have outlined the need for inter- and transdisciplinary collaboration, a core attribute of CASISAC. When a regional climate model relies on a proper SST distribution under current-day and future conditions, it is, in particular for the Agulhas Current system, important to simulate those at sufficient high resolution. The same applies to the impact studies examining the vulnerability of the African coastlines. Detailed precipitation maps and distributions of regional level in response to ocean currents and spatial warming are required. Even in a collaborative project, this is not fulfilled up to the last instance because of the large range of scales and the different importance of mechanisms for different disciplines. It is important to keep in mind that our results are based on numerical models. In particular for the oceanic fields, currents and hydrography, but also sea level, waves and flooding, observations are required to ground-truth the models and to correct model deficiencies originating from a poor representation of unresolved processes.

Our studies outline the strong impact of the surrounding ocean on the land climate in Southern Africa. Changes through global warming arising from large-scale climate models need not only to be downscaled but may also be regionalised, in our case to correctly take into account the specific effect of the Agulhas Current system.

References

Allison LC, Palmer MD, Haigh ID (2022) Projections of 21st century sea level rise for the coast of South Africa. Environ Res Commun 4(2). https://doi.org/10.1088/2515-7620/ac4a90

ASCLME/SWIOFP (2012) Transboundary Diagnostic Analysis of the Large marine Ecosystems of the Western Indian Ocean. Volume 1: Baseline, vol 1

Baker-Yeboah S, Flierl G, Sutyrin G, Zhang Y (2010) Transformation of an Agulhas eddy near the continental slope. Ocean Sci 6(1):143–159. https://doi.org/10.5194/os-6-143-2010

Barange M, Pillar SC, Hutchings L (1992) Major pelagic borders of the Benguela upwelling system according to euphausiid species distribution. South African J Marine Sci 12(1):3–17. https://doi. org/10.2989/02577619209504686

Barnes MA, Rautenbach C (2020) Toward operational wave-current interactions over the Agulhas current system. J Geophys Res Oceans 125(7):1–21. https://doi.org/10.1029/2020JC016321

Bates PD, Dawson RJ, Hall JW, Horritt MS, Nicholls RJ, Wicks J (2005) Simplified two-dimensional numerical modelling of coastal flooding and example applications. Coastal Eng 52(9):793–810. https://doi.org/10.1016/j.coastaleng.2005.06.001

Beal LM (2009) A time series of Agulhas undercurrent transport. J Phys Oceanogr 39:2436–2450. https://doi.org/10.1175/2009JPO4195.1

Beal LM, Elipot S (2016) Broadening not strengthening of the Agulhas current since the early 1990s. Nature 540(7634):570–573. https://doi.org/10.1038/nature19853

Beal LM, De Ruijter WPM, Biastoch A, Zahn R, Members of SCOR/WCRP/IAPSO Working Group 136 (2011) On the role of the Agulhas system in ocean circulation and climate. Nature 472(7344):429–436. https://doi.org/10.1038/nature09983

Beal LM, Elipot S, Houk A, Leber GM (2015) Capturing the transport variability of a western boundary jet: results from the Agulhas current time-series experiment (ACT). J Phys Oceanogr 45(5):1302–1324. https://doi.org/10.1175/JPO-D-14-0119.1. http://journals. ametsoc.org/doi/10.1175/JPO-D-14-0119.1

Biastoch A, Böning CW (2013) Anthropogenic impact on Agulhas leakage. Geophys Res Lett 40:1138–1143. https://doi.org/10.1002/grl.50243

Biastoch A, Krauss W (1999) The role of Mesoscale Eddies in the source regions of the Agulhas current. J Phys Oceanogr 29(9):2303–2317. https://doi.org/10.1175/1520-0485(1999)029<2303:TROMEI>2.0.CO;2

Biastoch A, Lutjeharms JRE, Böning CW, Scheinert M (2008) Mesoscale perturbations control inter-ocean exchange south of Africa. Geophys Res Lett 35(20):L20602. https://doi.org/10. 1029/2008GL035132

Biastoch A, Böning CW, Schwarzkopf FU, Lutjeharms JRE (2009) Increase in Agulhas leakage due to poleward shift of Southern Hemisphere westerlies. Nature 462(7272):495–498. https:// doi.org/10.1038/nature08519

Biastoch A, Durgadoo JV, Morrison AK, Van Sebille E, Weijer W, Griffies SM (2015) Atlantic multi-decadal oscillation covaries with Agulhas leakage. Nat Commun 6:10082. https://doi.org/ 10.1038/ncomms10082

Blamey R, Reason C (2009) Numerical simulation of a mesoscale convective system over the east coast of South Africa. Tellus A Dyn Meteorol Oceanogra 61(1):17–34. https://doi.org/10.1111/ j.1600-0870.2008.00366.x

Broecker WS (1991) The great ocean conveyor. Oceanogr 4:79–89. http://www.jstor.org/stable/ 43924572

Brooks HE, Lee JW, Craven JP (2003) The spatial distribution of severe thunderstorm and tornado environments from global reanalysis data. Atmospheric Res 67:73–94. https://doi.org/10.1016/ S0169-8095(03)00045-0

Brundrit GB (1984) Monthly mean sea level variability along the west coast of Southern Africa. South African J Marine Sci 2(1):195–203. https://doi.org/10.2989/02577618409504368

Bryden HL, Beal LM, Duncan LM (2005) Structure and transport of the Agulhas current and its temporal variability. J Oceanogr 61:479–492. https://doi.org/10.1007/s10872-005-0057-8

Carrère L, Lyard F, Cancet M, Guillot A (2015) FES 2014, a new tidal model on the global ocean with enhanced accuracy in shallow seas and in the Arctic region. In: EGU General Assembly, Vienna

Casas-Prat M, Wang XL, Swart N (2018) CMIP5-based global wave climate projections including the entire Arctic Ocean. Ocean Modell 123:66–85. https://doi.org/10.1016/j.ocemod.2017.12. 003

Cazenave A, Meyssignac B, Ablain M, Balmaseda M, Bamber J, Barletta V, Beckley B, Benveniste J, Berthier E, Blazquez A, Boyer T, Caceres D, Chambers D, Champollion N, Chao B, Chen J, Cheng L, Church JA, Chuter S, Cogley JG, Dangendorf S, Desbruyères D, Döll P, Domingues C,

Falk U, Famiglietti J, Fenoglio-Marc L, Forsberg R, Galassi G, Gardner A, Groh A, Hamlington B, Hogg A, Horwath M, Humphrey V, Husson L, Ishii M, Jaeggi A, Jevrejeva S, Johnson G, Kolodziejczyk N, Kusche J, Lambeck K, Landerer F, Leclercq P, Legresy B, Leuliette E, Llovel W, Longuevergne L, Loomis BD, Luthcke SB, Marcos M, Marzeion B, Merchant C, Merrifield M, Milne G, Mitchum G, Mohajerani Y, Monier M, Monselesan D, Nerem S, Palanisamy H, Paul F, Perez B, Piecuch CG, Ponte RM, Purkey SG, Reager JT, Rietbroek R, Rignot E, Riva R, Roemmich DH, Sørensen LS, Sasgen I, Schrama EJ, Seneviratne SI, Shum CK, Spada G, Stammer D, van de Wal R, Velicogna I, von Schuckmann K, Wada Y, Wang Y, Watson C, Wiese D, Wijffels S, Westaway R, Woppelmann G, Wouters B (2018) Global sea-level budget 1993-present. Earth Syst Sci Data 10(3):1551–1590. https://doi.org/10.5194/essd-10-1551-2018

Chelton DB, Schlax MG, Samelson RM (2011) Global observations of nonlinear mesoscale eddies. Prog Oceanogr 91(2):167–216. https://doi.org/10.1016/j.pocean.2011.01.002

Cheng Y, Beal LM, Kirtman BP, Putrasahan D (2018) Interannual Agulhas leakage variability and its regional climate imprints. J Climate 31(24):10105–10121. https://doi.org/10.1175/JCLI-D-17-0647.1

CSIR (2019) Green Book: Adapting South African settlements to climate change. www.greenbook.co.za

Daher H, Beal LM, Schwarzkopf FU (2020) A new improved estimation of Agulhas leakage using observations and simulations of lagrangian floats and drifters. J Geophys Res Oceans 125(4). https://doi.org/10.1029/2019JC015753

Desbiolles F, Blamey R, Illig S, James R, Barimalala R, Renault L, Reason C (2018) Upscaling impact of wind/sea surface temperature mesoscale interactions on Southern Africa austral summer climate. Int J Climatol 38(12):4651–4660. https://doi.org/10.1002/joc.5726

Dosio A, Jones RG, Jack C, Lennard C, Nikulin G, Hewitson B (2019) What can we know about future precipitation in Africa? Robustness, significance and added value of projections from a large ensemble of regional climate models. Climate Dyn 53(9):5833–5858. https://doi.org/10.1007/s00382-019-04900-3

Dube K, Nhamo G, Chikodzi D (2021a) Flooding trends and their impacts on coastal communities of Western Cape Province, South Africa. GeoJournal 0123456789. https://doi.org/10.1007/s10708-021-10460-z

Dube K, Nhamo G, Chikodzi D (2021b) Rising sea level and its implications on coastal tourism development in Cape Town, South Africa. J Outdoor Recreat Tour 33:100346. https://doi.org/10.1016/j.jort.2020.100346

Duncombe-Rae C, Shillington F, Agenbag J, Taunton-Clark J, Grundlingh M (1992) An Agulhas ring in the South Atlantic Ocean and its interaction with the Benguela upwelling frontal system. Deep-Sea Res 39(11/12):2009–2027

Durgadoo J, Loveday B, Reason C, Penven P, Biastoch A (2013) Agulhas leakage predominantly responds to the southern hemisphere westerlies. J Phys Oceanogra 43(10). https://doi.org/10.1175/JPO-D-13-047.1

Durgadoo JV, Rühs S, Biastoch A, Böning CW (2017) Indian ocean sources of Agulhas leakage. J Geophys Res Ocean 122(4):3481–3499. https://doi.org/10.1002/2016JC012676

Durrant T, Hemer M, Trenham C, Greenslade D (2013) CAWCR wave hindcast 1979-2010. v10. https://doi.org/10.4225/08/523168703DCC5

Elipot S, Beal LM (2015) Characteristics, energetics, and origins of agulhas current meanders and their limited influence on ring shedding. J Phys Oceanogr 45(9):2294–2314. https://doi.org/10.1175/JPO-D-14-0254.1

Elipot S, Beal LM (2018) Observed Agulhas current sensitivity to interannual and long-term trend atmospheric forcings. J Climate 31(8):3077–3098. https://doi.org/10.1175/JCLI-D-17-0597.1

Fitchett JM, Grant B, Hoogendoorn G (2016) Climate change threats to two low-lying South African coastal towns: risks and perceptions. South African J Sci 112(5/6):1–9. https://doi.org/10.17159/sajs.2016/20150262

Fowler JL, Boyd AJ (1998) Transport of anchovy and sardine eggs and larvae from the western Agulhas bank to the west coast during the 1993/94 and 1994/95 spawning seasons. South African J Marine Sci 19(1):181–195. https://doi.org/10.2989/025776198784127006

Giulivi CF, Gordon AL (2006) Isopycnal displacements within the Cape Basin thermocline as revealed by the hydrographic data archive. Deep Sea Res Part I Oceanograph Res Papers 53(8):1285–1300. https://doi.org/10.1016/j.dsr.2006.05.011. https://www.sciencedirect.com/science/article/pii/S0967063706001397

Grundlingh M, Rossouw M (1995) Wave attenuation in the Agulhas current. South African J Sci 91(7):357–359

Hamlington BD, Gardner AS, Ivins E, Lenaerts JT, Reager JT, Trossman DS, Zaron ED, Adhikari S, Arendt A, Aschwanden A, Beckley BD, Bekaert DP, Blewitt G, Caron L, Chambers DP, Chandanpurkar HA, Christianson K, Csatho B, Cullather RI, DeConto RM, Fasullo JT, Frederikse T, Freymueller JT, Gilford DM, Girotto M, Hammond WC, Hock R, Holschuh N, Kopp RE, Landerer F, Larour E, Menemenlis D, Merrifield M, Mitrovica JX, Nerem RS, Nias IJ, Nieves V, Nowicki S, Pangaluru K, Piecuch CG, Ray RD, Rounce DR, Schlegel NJ, Seroussi H, Shirzaei M, Sweet WV, Velicogna I, Vinogradova N, Wahl T, Wiese DN, Willis MJ (2020) Understanding of contemporary regional sea-level change and the implications for the future. Rev Geophys 58(3):1–39. https://doi.org/10.1029/2019RG000672

Hemer M, Trenham C, Durrant T, Greenslade D (2015) CAWCR Global wind-wave 21st century climate projections. v2. https://doi.org/10.4225/08/55C991CC3F0E8. https://data.csiro.au/collection/csiro:13500v2

IPCC (2021) Climate Change 2021: The Physical Science Basis. Contribution of Working Group I to the Sixth Assessment Report of the Intergovernmental Panel on Climate Change. Technical Report In Press. https://www.ipcc.ch/report/ar6/wg1/downloads/report/IPCC_AR6_WGI_Full_Report.pdf

Ivanciu I, Matthes K, Wahl S, Harlaß J, Biastoch A (2021) Effects of prescribed CMIP6 ozone on simulating the Southern Hemisphere atmospheric circulation response to ozone depletion. Atmos Chem Phys 21(8):5777–5806. https://doi.org/10.5194/acp-21-5777-2021

Ivanciu I, Matthes K, Biastoch A, Wahl S, Harlaß J (2022) Twenty-first-century southern hemisphere impacts of ozone recovery and climate change from the stratosphere to the ocean. Weather Climate Dyn 3(1):139–171. https://doi.org/10.5194/wcd-3-139-2022. https://wcd.copernicus.org/articles/3/139/2022/

Jury MR, Valentine HR, Lutjeharms JR (1993) Influence of the Agulhas current on summer rainfall along the southeast coast of South Africa. J Appl Meteorol Climatol 32(7):1282–1287. https://doi.org/10.1175/1520-0450(1993)032<1282:IOTACO>2.0.CO;2

Kanamitsu M, Ebisuzaki W, Woollen J, Yang SK, Hnilo JJ, Fiorino M, Potter GL (2002) NCEP–DOE AMIP-II reanalysis (R-2). Bull Amer Meteorol Soc 83(11):1631–1644. https://doi.org/10.1175/BAMS-83-11-1631. https://journals.ametsoc.org/downloadpdf/journals/bams/83/11/bams-83-11-1631.xml

Kupfer S, Santamaria-Aguilar S, van Niekerk L, Lück-Vogel M, Vafeidis A (2021) Investigating the interaction of waves and river discharge during compound flooding at Breede Estuary, South Africa. Natl Hazards Earth Syst Sci Discuss 22:1–27. https://doi.org/10.5194/nhess-2021-220

Le Bars D, Dijkstra HA, De Ruijter WPM (2013) Impact of the Indonesian throughflow on Agulhas leakage. Ocean Sci 9(5):773–785. https://doi.org/10.5194/os-9-773-2013

Le Bars D, Durgadoo JV, Dijkstra HA, Biastoch A, De Ruijter WPM (2014) An observed 20-year time series of Agulhas leakage. Ocean Sci 10(4):601–609. https://doi.org/10.5194/os-10-601-2014

Lee-Thorp A, Rouault M, Lutjeharms J (1999) Moisture uptake in the boundary layer above the Agulhas current: a case study. J Geophys Res Oceans 104(C1):1423–1430. https://doi.org/10.1029/98JC02375

Lesser GR, Roelvink JA, van Kester JA, Stelling GS (2004) Development and validation of a three-dimensional morphological model. Coastal Eng 51(8–9):883–915. https://doi.org/10.1016/j.coastaleng.2004.07.014

Loveday BR, Durgadoo JV, Reason CJC, Biastoch A, Penven P (2014) Decoupling of the Agulhas leakage from the Agulhas current. J Phys Oceanogr 44(7):1776–1797. https://doi.org/10.1175/JPO-D-13-093.1

Lück-vogel M (2019) Green Book- Coastal Flooding Hazard Assessment. Technical Report, CSIR, Pretoria

Lutjeharms JR (2006) The Agulhas Current. Springer, Berlin. https://doi.org/10.1007/3-540-37212-1

Marshall GJ (2003) Trends in the Southern Annular mode from observations and reanalyses. J Climate 16(24):4134–4143. https://doi.org/10.1175/1520-0442(2003)016<4134:TITSAM>2.0.CO;2

Mason SJ (1995) Sea-surface temperature–South African rainfall associations, 1910–1989. Int J Climatol 15(2):119–135. https://doi.org/10.1002/joc.3370150202

Matano R, Beier E (2003) A kinematic analysis of the Indian/Atlantic interocean exchange. Deep-Sea Res 50:229–249. https://doi.org/10.1016/S0967-0645(02)00395-8

Mather AA (2007) Linear and nonlinear sea-level changes at Durban, South Africa. South African J Sci 103(11–12):509–512

Mather AA, Stretch DD (2012) A perspective on sea level rise and coastal storm surge from Southern and Eastern Africa: a case study near Durban, South Africa. Water 4(4):237–259. https://doi.org/10.3390/w4010237

Mather AA, Garland GG, Stretch DD (2009) Southern African sea levels: corrections, influences and trends. African J Marine Sci 31(2):145–156. https://doi.org/10.2989/AJMS.2009.31.2.3.875

Matthes K, Biastoch A, Wahl S, Harlaß J, Martin T, Brücher T, Drews A, Ehlert D, Getzlaff K, Krüger F, Rath W, Scheinert M, Schwarzkopf FU, Bayr T, Schmidt H, Park W (2020) The flexible ocean and climate infrastructure version 1 (FOCI1): mean state and variability. Geosci Model Develop 13(6):2533–2568. https://doi.org/10.5194/gmd-13-2533-2020

McMonigal K, Beal LM, Willis JK (2018) The seasonal cycle of the South Indian ocean subtropical gyre circulation as revealed by Argo and Satellite data. Geophys Res Lett 45(17):9034–9041. https://doi.org/10.1029/2018GL078420

Meinshausen M, Nicholls ZRJ, Lewis J, Gidden MJ, Vogel E, Freund M, Beyerle U, Gessner C, Nauels A, Bauer N, Canadell JG, Daniel JS, John A, Krummel PB, Luderer G, Meinshausen N, Montzka SA, Rayner PJ, Reimann S, Smith SJ, van den Berg M, Velders GJM, Vollmer MK, Wang RHJ (2020) The shared socio-economic pathway (SSP) greenhouse gas concentrations and their extensions to 2500. Geosci Model Develop 13(8):3571–3605. https://doi.org/10.5194/gmd-13-3571-2020

Morake D, Blamey R, Reason C (2021) Long-lived mesoscale convective systems over eastern South Africa. J Climate 34(15):6421–6439. https://doi.org/10.1175/JCLI-D-20-0851.1

Morim J, Trenham C, Hemer M, Wang XL, Mori N, Casas-Prat M, Semedo A, Shimura T, Timmermans B, Camus P, Bricheno L, Mentaschi L, Dobrynin M, Feng Y, Erikson L (2020) A global ensemble of ocean wave climate projections from CMIP5-driven models. Sci Data 7(1):1–10. https://doi.org/10.1038/s41597-020-0446-2

Nhantumbo BJ, Nilsen JE, Backeberg BC, Reason CJ (2020) The relationship between coastal sea level variability in South Africa and the Agulhas Current. J Marine Syst 211:103422. https://doi.org/10.1016/j.jmarsys.2020.103422

Nkwinkwa Njouodo AS, Koseki S, Keenlyside N, Rouault M (2018) Atmospheric signature of the Agulhas current. Geophys Res Lett 45(10):5185–5193. https://doi.org/10.1029/2018GL077042

Palmer BJ, van der Elst R, Mackay F, Mather AA, Smith AM, Bundy SC, Thackeray Z, Leuci R, Parak O (2011) Preliminary coastal vulnerability assessment for KwaZulu-Natal, South Africa. J Coastal Res (64):1390–1395. https://www.jstor.org/stable/e26482118

Pitcher GC, Nelson G (2006) Characteristics of the surface boundary layer important to the development of red tide on the southern Namaqua shelf of the Benguela upwelling system. Limnol Oceanogr 51(6):2660–2674. https://doi.org/10.4319/lo.2006.51.6.2660. https://aslopubs.onlinelibrary.wiley.com/doi/pdf/10.4319/lo.2006.51.6.2660

Putrasahan D, Kirtman BP, Beal LM (2016) Modulation of SST interannual variability in the Agulhas leakage region associated with ENSO. J Clim 29(19):7089–7102. https://doi.org/10.1175/JCLI-D-15-0172.1

Pyle DM, Jacobs TL (2016) The port alfred floods of 17–23 October 2012: A case of disaster (mis)management? Jàmbá J Disaster Risk Stud 8(1):1–8. https://doi.org/10.4102/jamba.v8i1.207

Rapolaki R, Blamey R, Hermes J, Reason C (2020) Moisture sources associated with heavy rainfall over the Limpopo River Basin, Southern Africa. Climate Dyn 55(5):1473–1487. https://doi.org/10.1007/s00382-020-05336-w

Rapolaki R, Blamey R, Hermes J, Reason C (2021) Moisture sources and transport during an extreme rainfall event over the Limpopo River Basin, Southern Africa. Atmospher Res 105849. https://doi.org/10.1016/j.atmosres.2021.105849

Reason C (2001) Evidence for the influence of the Agulhas current on regional atmospheric circulation patterns. J Climate 14(12):2769–2778. https://doi.org/10.1175/1520-0442(2001)014<2769:EFTIOT>2.0.CO;2

Reason C, Mulenga H (1999) Relationships between South African rainfall and SST anomalies in the southwest Indian ocean. Int J Climatol J R Meteorol Soc 19(15):1651–1673. https://doi.org/10.1002/(SICI)1097-0088(199912)19:15<1651::AID-JOC439>3.0.CO;2-U

Richardson PL (2007) Agulhas leakage into the Atlantic estimated with subsurface floats and surface drifters. Deep-Sea Res I 54:1361–1389. https://doi.org/10.1016/j.dsr.2007.04.010

Rojas M, Lambert F, Ramirez-Villegas J, Challinor AJ (2019) Emergence of robust precipitation changes across crop production areas in the 21st century. Proc Natl Acad Sci 116(14):6673–6678. https://doi.org/10.1073/pnas.1811463116. https://www.pnas.org/content/116/14/6673.full.pdf

Rossi V, Löpez C, Sudre J, Hernández-García E, Garçon V (2008) Comparative study of mixing and biological activity of the Benguela and canary upwelling systems. Geophy Res Lett 35(11). https://doi.org/10.1029/2008GL033610. https://agupubs.onlinelibrary.wiley.com/doi/pdf/10.1029/2008GL033610

Rouault M, Reason C, Lutjeharms J, Beljaars A (2003) Underestimation of latent and sensible heat fluxes above the Agulhas current in NCEP and ECMWF analyses. J Climate 16(4):776–782. https://doi.org/10.1175/1520-0442(2003)016<0776:UOLASH>2.0.CO;2

Rouault M, Penven P, Pohl B (2009) Warming in the Agulhas current system since the 1980's. Geophys Res Lett 36(12). https://doi.org/10.1029/2009GL037987. https://agupubs.onlinelibrary.wiley.com/doi/pdf/10.1029/2009GL037987

Rouault M, Pohl B, Penven P (2010) Coastal oceanic climate change and variability from 1982 to 2009 around South Africa. African J Marine Sci 32(2):237–246. https://doi.org/10.2989/1814232X.2010.501563

Rühs S, Durgadoo JV, Behrens E, Biastoch A (2013) Advective timescales and pathways of Agulhas leakage. Geophys Res Lett 40(15):3997–4000. https://doi.org/10.1002/grl.50782

Rühs S, Schwarzkopf F, Speich S, Biastoch A (2019) Cold vs. warm water route-sources for the upper limb of the Atlantic meridional overturning circulation revisited in a high-resolution ocean model. Ocean Sci 15(3). https://doi.org/10.5194/os-15-489-2019

Schmidt C, Schwarzkopf FU, Rühs S, Biastoch A (2021) Characteristics and robustness of Agulhas leakage estimates: an inter-comparison study of Lagrangian methods. Ocean Sci 17:1067–1080. https://doi.org/10.5194/os-17-1067-2021

Schouten MW, de Ruijter WPM, van Leeuwen PJ, Lutjeharms JRE (2000) Translation, decay and splitting of Agulhas rings in the southeastern Atlantic Ocean. J Geophys Res 105(C9):21913–21925. https://doi.org/10.1029/1999JC000046

Schouten MW, de Ruijter WPM, van Leeuwen PJ (2002) Upstream control of Agulhas ring shedding. J Geophys Res 107(10.1029). https://doi.org/10.1029/2001JC000804

Schubert R, Schwarzkopf FU, Baschek B, Biastoch A (2019) Submesoscale impacts on mesoscale Agulhas dynamics. J Adv Model Earth Syst 11(8):2745–2767. https://doi.org/10.1029/2019MS001724

Schumann EH, Brink KH (1990) Coastal-trapped waves off the coast of South Africa: generation, propagation and current structures. J Phys Oceanogr 20:148–162. https://doi.org/10.1175/1520-0485(1990)020<1206:CTWOTC>2.0.CO;2

Schwarzkopf FU, Biastoch A, Böning CW, Chanut J, Durgadoo JV, Getzlaff K, Harlaß J, Rieck JK, Roth C, Scheinert MM, Schubert R (2019) The INALT family - A set of high-resolution nests for the Agulhas current system within global NEMO ocean/sea-ice configurations. Geosci Model Dev 12(7):3329–3355. https://doi.org/10.5194/gmd-12-3329-2019

Schubert R, Gula A, Biastoch A (2021) Submesoscale impacts on Agulhas leakage and Agulhas cyclone formation. Nat Commun Earth Environ 2(197). https://doi.org/10.1038/s43247-021-00271-y

Semedo A, Weisse R, Behrens A, Sterl A, Bengtsson L, Günther H (2013) Projection of global wave climate change toward the end of the twenty-first century. J Climate 26(21):8269–8288. https://doi.org/10.1175/JCLI-D-12-00658.1. http://journals.ametsoc.org/doi/abs/10.1175/JCLI-D-12-00658.1

Shelton P, Hutchings L (1982) Transport of anchovy, engraulis capensis gilchrist, eggs and early larvae by a frontal jet current. JConsintExplorMer 40:185–198. https://doi.org/10.1093/icesjms/40.2.185

Singleton A, Reason C (2006) Numerical simulations of a severe rainfall event over the Eastern Cape Coast of South Africa: sensitivity to sea surface temperature and topography. Tellus A Dyn Meteorol Oceanogr 58(3):335–367. https://doi.org/10.1111/j.1600-0870.2006.00180.x

Singleton A, Reason C (2007) A numerical model study of an intense cutoff low pressure system over South Africa. Monthly Weather Rev 135(3):1128–1150. https://doi.org/10.1175/MWR3311.1

Smith AM, Guastella LA, Bundy SC, Mather AA (2007) Combined marine storm and SAROS spring high tide erosion events along the KwaZulu-Natal coast in March 2007. South African J Sci 103(7–8):274–276. URL https://hdl.handle.net/10520/EJC96710

Speich S, Blanke B, Madec G (2001) Warm and cold water routes of an OGCM thermohaline conveyor belt. Geophys Res Lett 28(2):311–314. https://doi.org/10.1029/2000GL011748

Speich S, Blanke B, Cai W (2007) Atlantic meridional overturning circulation and the Southern Hemisphere supergyre. Geophys Res Lett 34:1–5. https://doi.org/10.1029/2007GL031583

Taljaard S (2003) Intermediate determination of the resource directed measures for the Breede River Estuary. Technical Report, Department of Water Affairs and Forestry, Stellenbosch, South Africa

Theron BAK, Rossouw M (2008) Analysis of potential coastal zone climate change impacts and possible response options in the Southern African region. In: Coastal Climate Change Impacts Southern Africa, pp 1–10

Theron A, Rossouw M, Barwell L, Maherry A, Diedericks G, de Wet P (2010) Quantification of risks to coastal areas and development: wave run-up and erosion. In: CSIR 3rd Biennial Conference 2010. http://researchspace.csir.co.za/dspace/handle/10204/4261

Thoithi W, Blamey RC, Reason CJ (2021) Dry spells, wet days, and their trends across Southern Africa during the summer rainy season. Geophys Res Lett 48(5). https://doi.org/10.1029/2020GL091041

Tim N, Zorita E, Schwarzkopf FU, Rühs S, Emeis KC, Biastoch A (2018) The impact of Agulhas leakage on the central water masses in the Benguela upwelling system from a high-resolution ocean simulation. J Geophys Res Oceans 123(12):9416–9428. https://doi.org/10.1029/2018JC014218

Tim N, Zorita E, Hünicke B, Birgit, Ivanciu I (2023) The impact of the Agulhas current system on precipitation in southern Africa in regional climate simulations covering the recent past and future. Weather Clim Dyn 4:381–397. https://doi.org/10.5194/wcd-4-381-2023 (open access)

Tsujino H, Urakawa S, Nakano H, Small RJ, Kim WM, Yeager SG, Danabasoglu G, Suzuki T, Bamber JL, Bentsen M et al (2018) JRA-55 based surface dataset for driving ocean–sea-ice models (JRA55-do). Ocean Modell 130. https://doi.org/10.1016/j.ocemod.2018.07.002

van Aken H, van Veldhoven A, Veth C, de Ruijter W, van Leeuwen P, Drijfhout S, Whittle C, Rouault M (2003a) Observations of a young Agulhas ring, Astrid, during MARE in March 2000. Deep Sea Res Part II Top Stud Oceanogr 50(1):167–195. https://doi.org/10.1016/S0967-0645(02)00383-1. https://www.sciencedirect.com/science/article/pii/S0967064502003831. Inter-ocean exchange around Southern Africa

Van Aken HM, Van Veldhoven AK, Veth C, De Ruijter WP, Van Leeuwen PJ, Drijfhout SS, Whittle CP, Rouault M (2003b) Observations of a young Agulhas ring, Astrid, during MARE in March 2000. Deep Res Part II Top Stud Oceanogr 50(1):167–195. https://doi.org/10.1016/S0967-0645(02)00383-1

van Niekerk L, Adams JB, James NC, Lamberth SJ, MacKay CF, Turpie JK, Rajkaran A, Weerts SP, Whitfield AK (2020) An Estuary Ecosystem Classification that encompasses biogeography and a high diversity of types in support of protection and management. Afr J Aquat Sci 45(1–2):199–216. https://doi.org/10.2989/16085914.2019.1685934

Van Sebille E, Biastoch A, van Leeuwen PJ, de Ruijter WPM (2009) A weaker Agulhas current leads to more Agulhas leakage. Geophys Res Lett 36(3):L03601. https://doi.org/10.1029/2008GL036614

van Sebille E, Beal LM, Biastoch A (2010) Sea surface slope as a proxy for Agulhas current strength. Geophys Res Lett 37(9):L09610. https://doi.org/10.1029/2010GL042847

van Sebille E, Beal LM, Johns WE (2011) Advective time scales of Agulhas leakage to the north Atlantic in surface drifter observations and the 3D OFES model. J Phys Oceanogr 41(5):1026–1034. https://doi.org/10.1175/2011JPO4602.1. https://journals.ametsoc.org/view/journals/phoc/41/5/2011jpo4602.1.xml

van Sebille E, Griffies SM, Abernathey R, Adams TP, Berlof P, Biastoc A, Blanke B, Chassignet EP, Cheng Y, Cotter CJ, Deleersnijder E, Döös K, Drake H, Drijfhout S, Gar SF, Heemink AW, Kjellsson J, Koszalka IM, Lange M, Lique C, MacGilchrist GA, Marsh R, Adame GCM, McAdam R, Nencioli F, Paris CB, Piggott MD, Polton JA, Rühs S, Shah SH, Thomas MD, Wang J, Wolfram PJ, Zanna L, Zika JD (2018) Lagrangian ocean analysis: fundamentals and practices. Ocean Modell 121:49–75. https://doi.org/10.1016/j.ocemod.2017.11.008

Veitch JA, Penven P (2017) The role of the Agulhas in the Benguela current system: a numerical modeling approach. J Geophys Res Oceans 122(4):3375–3393. https://doi.org/10.1002/2016JC012247, https://agupubs.onlinelibrary.wiley.com/doi/pdf/10.1002/2016JC012247

Veitch J, Hermes J, Lamont T, Penven P, Dufois F (2017) Shelf edge jet currents in the southern Benguela: a modelling approach. J Marine Syst 188:27–38. https://doi.org/10.1016/j.jmarsys.2017.09.003

Walker ND (1990) Links between South African summer rainfall and temperature variability of the Agulhas and Benguela current systems. J Geophys Res Oceans 95(C3):3297–3319. https://doi.org/10.1029/JC095iC03p03297. https://agupubs.onlinelibrary.wiley.com/doi/pdf/10.1029/JC095iC03p03297

Walker ND, Mey RD (1988) Ocean/atmosphere heat fluxes within the Agulhas retroflection region. J Geophys Res Oceans 93(C12):15473–15483. https://doi.org/10.1029/JC093iC12p15473

Weijer W, van Sebille E (2014) Impact of Agulhas leakage on the Atlantic overturning circulation in the CCSM4. J Climate 27(1):101–110. https://doi.org/10.1175/JCLI-D-12-00714.1

Weijer W, de Ruijter WP, Sterl A, Drijfhout SS (2002) Response of the Atlantic overturning circulation to South Atlantic sources of buoyancy. Global Planetary Change 34:293–311. https://doi.org/10.1016/S0921-8181(02)00121-2

Weijer W, Zharkov V, Nof D, Dijkstra HA, De Ruijter WP, Van Scheltinga AT, Wubs F (2013) Agulhas ring formation as a barotropic instability of the retroflection. Geophys Res Lett 40(20):5435–5438. https://doi.org/10.1002/2013GL057751

Woodworth, Philip L, Aman A, Aarup T (2007) Sea-level monitoring in Africa. African J Marine Sci 29(3):321–330. https://doi.org/10.2989/AJMS.2007.29.3.2.332

Yamagami Y, Tozuka T, Qiu B (2019) Interannual variability of the natal pulse. J Geophys Res Ocean 124(12):9258–9276. https://doi.org/10.1029/2019JC015525

Yamazaki D, Ikeshima D, Tawatari R, Yamaguchi T, O'Loughlin F, Neal JC, Sampson CC, Kanae S, Bates PD (2017) A high-accuracy map of global terrain elevations. Geophys Res Lett 44(11):5844–5853. https://doi.org/10.1002/2017GL072874

Yang H, Lohmann G, Wei W, Dima M, Ionita M, Liu J (2016) Intensification and poleward shift of subtropical western boundary currents in a warming climate. J Geophys Res Oceans 121(7):4928–4945. https://doi.org/10.1002/2015JC011513

Yin J, Griffies SM, Stouffer RJ (2010) Spatial variability of sea level rise in twenty-first century projections. J Climate 23(17):4585–4607. https://doi.org/10.1175/2010JCLI3533.1

Zipser EJ, Cecil DJ, Liu C, Nesbitt SW, Yorty DP (2006) Where are the most intense thunderstorms on Earth? Bull Amer Meteorol Soc 87(8):1057–1072. https://doi.org/10.1175/BAMS-87-8-1057

Zscheischler J, Seneviratne SI (2017) Dependence of drivers affects risks associated with compound events. Sci Adv 3(6):1–11. https://doi.org/10.1126/sciadv.1700263

Physical Drivers of Southwest African Coastal Upwelling and Its Response to Climate Variability and Change

9

Peter Brandt, Mohammad Hadi Bordbar, Paulo Coelho, Rodrigue Anicet Imbol Koungue, Mareike Körner, Tarron Lamont, Joke F. Lübbecke, Volker Mohrholz, Arthur Prigent, Marisa Roch, Martin Schmidt, Anja K. van der Plas, and Jennifer Veitch ⓘ

P. Brandt (✉)
GEOMAR Helmholtz Centre for Ocean Research Kiel, Kiel, Germany

Faculty of Mathematics and Natural Sciences, Kiel University, Kiel, Germany
e-mail: pbrandt@geomar.de

M. H. Bordbar · V. Mohrholz · M. Schmidt
Leibniz Institute for Baltic Sea Research Warnemünde, Rostock, Germany
e-mail: hadi.bordbar@io-warnemuende.de; volker.mohrholz@io-warnemuende.de;
martin.schmidt@io-warnemuende.de

P. Coelho
Instituto Nacional de Investigação Pesqueira e Marinha, Ministério da Agricultura e Pescas de
Angola, Ilha de Luanda, Luanda, Angola

R. A. Imbol Koungue · M. Körner · J. F. Lübbecke · A. Prigent · M. Roch
GEOMAR Helmholtz Centre for Ocean Research Kiel, Kiel, Germany
e-mail: rimbol@geomar.de; mkoerner@geomar.de; jluebbecke@geomar.de;
aprigent@geomar.de; mroch@geomar.de

T. Lamont
Oceans & Coasts Research Branch, Department of Forestry, Fisheries, and the Environment, Cape
Town, South Africa

Department of Oceanography, University of Cape Town, Rondebosch, South Africa

Bayworld Centre for Research & Education, Cape Town, South Africa
e-mail: tlamont@dffe.gov.za

A. K. van der Plas
National Marine Information and Research Centre, Ministry of Fisheries and Marine Resources,
Swakopmund, Namibia; Anja.VanDerPlas@mfmr.gov.na

J. Veitch
The South African Environmental Observation Network, Cape Town, South Africa
e-mail: ja.veitch@saeon.nrf.ac.za

© The Author(s) 2024
G. P. von Maltitz et al. (eds.), *Sustainability of Southern African Ecosystems
under Global Change*, Ecological Studies 248,
https://doi.org/10.1007/978-3-031-10948-5_9

Abstract

The southeastern tropical Atlantic hosts a coastal upwelling system characterized by high biological productivity. Three subregions can be distinguished based on differences in the physical climate: the tropical Angolan and the northern and southern Benguela upwelling systems (tAUS, nBUS, sBUS). The tAUS, which is remotely forced via equatorial and coastal trapped waves, can be characterized as a mixing-driven system, where the wind forcing plays only a secondary role. The nBUS and sBUS are both forced by alongshore winds and offshore cyclonic wind stress curl. While the nBUS is a permanent upwelling system, the sBUS is impacted by the seasonal cycle of alongshore winds. Interannual variability in the region is dominated by Benguela Niños and Niñas that are warm and cold events observed every few years in the tAUS and nBUS. Decadal and multidecadal variations are reported for sea surface temperature and salinity, stratification and subsurface oxygen. Future climate warming is likely associated with a southward shift of the South Atlantic wind system. While the mixing-driven tAUS will most likely be affected by warming and increasing stratification, the nBUS and sBUS will be mostly affected by wind changes with increasing winds in the sBUS and weakening winds in the northern nBUS.

Abbreviations

ABA	Angola-Benguela area
ABFZ	Angola-Benguela Frontal Zone
AC	Angola Current
BC	Benguela Current
BCC	Benguela Coastal Current
BLLCJ	Benguela Low Level Coastal Jet
BOC	Benguela Offshore Current
CMIP6	Coupled Model Intercomparison Project phase 6
ENSO	El Niño-Southern Oscillation
ESACW	Eastern South Atlantic Central Water
EUC	Equatorial Undercurrent
GC	Guinea Current
GCUC	Gabon-Congo Undercurrent
GUC	Guinea Undercurrent
LOW	Low-Oxygen Water
nBUS, sBUS	Northern, southern Benguela Upwelling System
OMZ	Oxygen Minimum Zone
PUC	Poleward Undercurrent
SAA	South Atlantic Anticyclone
SACW	South Atlantic Central Water
SEC, nSEC, cSEC, sSEC	South Equatorial Current and its northern, central, and southern branches
SEJ	Shelf-edge jet
SEUC	South Equatorial Undercurrent

SECC	South Equatorial Countercurrent
SST	Sea Surface Temperature
tAUS	Tropical Angolan Upwelling System

9.1   Introduction

At the eastern boundaries of the tropical and subtropical Atlantic and Pacific oceans, four highly productive ecosystems are located. In the southeastern Atlantic, the Benguela Current Large Marine Ecosystem sustains important fisheries for the three coastal countries Angola, Namibia, and South Africa (Jarre et al. 2015b; Kainge et al. 2020). This region undergoes important climate variability such as extreme warm and cold events off Angola and Namibia, which are termed Benguela Niños and Niñas, respectively (Shannon et al. 1986). They do not only have drastic consequences for the marine ecosystem (Gammelsrød et al. 1998) but also influence the climate over large parts of southern Africa (Reason and Smart 2015; Rouault et al. 2003). Ongoing climate change will affect the eastern boundary regions among others by its effect on the wind field and the resulting wind-driven upwelling, by enhanced warming, increased stratification and ocean deoxygenation (Gruber 2011). Climate predictions for this region using climate models are mostly hampered by long-standing biases in the climate mean-state and variability (Li et al. 2020; Richter 2015; Richter and Tokinaga 2020). Improving predictions of climate and associated impacts on biogeochemistry and ecosystems strongly relies on the representation of the oceanic and coastal upwelling and related processes (Zuidema et al. 2016).

Following Jarre et al. (2015b), the eastern boundary upwelling system of the South Atlantic can be divided into three subsystems: the tropical Angolan upwelling system (tAUS, ~6–17°S), the northern Benguela upwelling system (nBUS, ~17–27°S) and the southern Benguela upwelling system (sBUS, ~27–35°S) that can roughly be associated with the three coastal countries, Angola, Namibia and South Africa, respectively (Fig. 9.1). Typical characteristics of the physical forcing of the marine ecosystems and their temporal variability varies among these subsystems. Southeasterly trade winds prevail in the southeastern Atlantic connecting the subtropical and tropical atmosphere (Fig. 9.1b). The strongest winds are found in the nBUS off the Namibian coast and are associated with the atmospheric Benguela Low Level Coastal Jet (BLLCJ) (Patricola and Chang 2017). The BLLCJ marks the boundary between the Angolan Low and the South Atlantic Anticyclone (SAA). In the tAUS, north of about 17°S, winds are substantially weaker with marginal seasonal variations marked by slightly enhanced southerly winds in austral spring and calm winds in austral winter. The seasonal variations in the position of the SAA with a northwestward shift in austral autumn and a southeastward shift in austral spring result in westerly winds during winter in the sBUS, while southerly winds in the nBUS remain relatively steady throughout the year (Veitch et al. 2009).

The wind-driven ocean circulation is characterized by vigorous zonal currents in the equatorial region and mostly meridional currents along the eastern boundary of the South Atlantic (Fig. 9.1a). Among the most energetic currents is the Equatorial

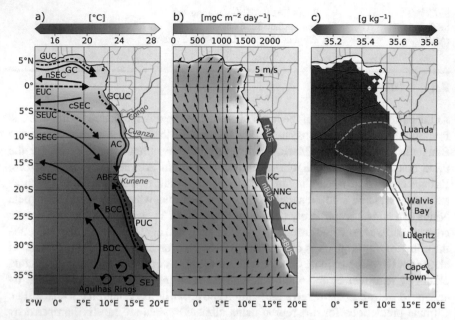

Fig. 9.1 Mean background conditions and circulation schematic for the eastern boundary upwelling system of the South Atlantic. (**a**) Sea surface temperature in the eastern tropical and subtropical South Atlantic obtained from OSTIA-SST with circulation schematic superimposed, (**b**) net primary production as deduced from satellite observations from Ocean Productivity site with surface wind vectors (ASCAT) superimposed and (**c**) absolute salinity on the potential density surface $\sigma_\theta = 26.3$ kg m^{-3} (Argo float data averaged for 2006–2020). Data in (**a**) and (**b**) are averaged for 2002–2019. In (**a**), surface (solid arrows) and thermocline (dashed arrows) current branches shown are the Guinea Undercurrent (GUC), the Guinea Current (GC), the Equatorial Undercurrent (EUC), the northern, central and southern branches of the South Equatorial Current (nSEC, cSEC and sSEC), the South Equatorial Undercurrent (SEUC), the South Equatorial Countercurrent (SECC), the Gabun-Congo Undercurrent (GCUC), the Angola Current (AC), the Poleward Undercurrent (PUC), the Benguela Offshore and Coastal Currents (BOC and BCC), and the shelf-edge jet (SEJ). Also marked in (**a**) is the Angola-Benguela Frontal Zone (ABFZ) at about 17°S and the three rivers Congo, Cuanza and Kunene. In (**b**) the latitude range of the three subregions, the tropical Angolan and the northern and southern Benguela upwelling systems (tAUS, nBUS, sBUS) as well as mean latitudes of the dominant upwelling cells, the Kunene cell, the Northern Namibian cell, the Central Namibian cell and the Lüderitz cell (KC, NNC, CNC, LC) are marked. In (**c**), the 70 µmol kg^{-1} oxygen concentration contours at 130 m depth (light blue dashed line) and at 250 m depth (black line) are included. Oxygen data is from Schmidtko et al. (2017)

Undercurrent (EUC) transporting South Atlantic Central Water (SACW) from the western boundary eastward (Johns et al. 2014). When approaching Africa, most of the water recirculates into the South Equatorial Current (SEC), forming a northern branch (nSEC) slightly north of the equator and a central branch (cSEC) south of the equator (Kolodziejczyk et al. 2014). Along the northern coast of the Gulf of Guinea, the slightly offshore located Guinea Current (GC) and the Guinea Undercurrent (GUC) attached to the continental slope flow toward the eastern boundary (Fig.

9.1a) (Djakouré et al. 2017; Herbert et al. 2016). The equatorward extension of the GC and the GUC, together with the remnants of the EUC, supply the southward Gabon-Congo Undercurrent (GCUC) (Wacongne and Piton 1992), which continues as the Angola Current (AC) in the tAUS (Kopte et al. 2017), and eventually as the Poleward Undercurrent (PUC) through the nBUS into the sBUS (Mohrholz et al. 2008; Nelson 1989). Additional transport of SACW toward the east and the south originating in the near-equatorial belt is carried by the South Equatorial Undercurrent (SEUC) mainly at the thermocline level and by the South Equatorial Countercurrent (SECC) (Siegfried et al. 2019). Along the eastern boundary, the generally southward flowing SACW meets the colder and fresher Eastern SACW (ESACW) flowing northward within the Benguela Current (BC). The confluence of these two currents at about 17°S causes a strong meridional sea surface temperature (SST) gradient that is termed the Angola-Benguela Frontal Zone (ABFZ) (Fig. 9.1). The BC can be described as being composed of an offshore branch (Benguela Offshore Current, BOC) and a coastal branch (Benguela Coastal Current, BCC) (Siegfried et al. 2019). After converging in the ABFZ, the eastern boundary flow turns westward, forming—together with the BOC—the southern branch of the SEC (sSEC) that constitutes the main westward branch of the South Atlantic subtropical gyre.

The surface ocean at the eastern boundary is characterized by the presence of warm and fresh tropical surface water in the northern tAUS, a salinity maximum in the southern tAUS and colder and slightly fresher surface waters south of the ABFZ (Fig. 9.1). South of the ABFZ the surface shows a particular temperature minimum along the coast that is characteristic for the permanent wind-driven coastal upwelling in the nBUS. The sBUS is connected to the Agulhas Current that transports warmer waters from the Indian Ocean around the southern tip of Africa. The two upper-ocean water masses in the region, the SACW and the ESACW, are quite distinct on a specific density surface with SACW being characterized by higher salinities compared to ESACW (Fig. 9.1c). SACW is the water mass of the southern hemisphere subtropical gyre, which is transported by the North Brazil Current toward the equatorial current system and eventually reaches the eastern boundary via the different eastward current branches mentioned above. When arriving at the eastern boundary, it is already low in oxygen with oxygen being further reduced due to high consumption near the highly productive eastern boundary upwelling system. In the area of the ABFZ, it meets the equatorward-flowing oxygen-rich ESACW representing a mixture of SACW and Indian Ocean central water formed in the Cape Basin (Duncombe Rae 2005; Mohrholz et al. 2008). Below the surface layer, the SACW arriving from the north and flowing further poleward within the PUC is thereby continuously mixed with the ESACW and eventually upwells in the northern Benguela (Mohrholz et al. 2008).

The oxygen distribution at the eastern boundary of the South Atlantic is characterized by an open ocean oxygen minimum zone (OMZ) approximately located between the equator and 20°S with a core depth of about 400 m (Karstensen et al. 2008; Monteiro et al. 2008). The OMZ is a result of weak ventilation in a region bounded by the BC and sSEC in the south and the well-ventilated equatorial region in the north (Brandt et al. 2015). Theoretically, the existence of the OMZ is explained by the Ekman pumping (i.e., the wind stress curl-driven vertical transport out of the oceanic mixed layer into the stratified ocean below) in the area of the SAA resulting in a geostrophic thermocline flow toward the west and the equator thereby forming the shadow zone of the ventilated thermocline (Luyten et al. 1983). Closer to the surface, local oxygen consumption due to high primary production plays a more important role and low-oxygen regions with partly anoxic conditions are found on the shelf as far south as 26°S (Bartholomae and van der Plas 2007) or even in the sBUS in shallow, near-coastal regions (Jarre et al. 2015a).

Although the primary productivity is high along the whole eastern boundary from the Gulf of Guinea to the southern tip of Africa (Fig. 9.1b), the processes relevant for the nutrient supply required to support the primary productivity might vary substantially across the different subregions. While there is high productivity at about 6°S reaching far offshore and driven by the enhanced nutrient supply associated with the river run-off of the Congo (Hopkins et al. 2013; Sena Martins and Stammer 2022), the enhanced near-coastal productivity along the eastern boundary is instead driven by an upward supply of nutrients to the euphotic zone. Physical drivers of the upward nutrient supply are (1) alongshore winds resulting in a near-surface offshore Ekman transport (i.e., the wind-forced horizontal transport perpendicular to the wind direction) supplied by near-coastal upwelling, (2) the cyclonic wind stress curl driving Ekman suction near the coast (i.e., the wind stress curl-driven vertical transport into the oceanic mixed layer out of the stratified ocean below) and (3) upwelling associated with the passage of coastal trapped waves and vertical mixing (Fig. 9.2) (Bordbar et al. 2021; Zeng et al. 2021).

Main topics that will be addressed in this chapter are the mean state and the seasonal cycle of the eastern boundary circulation, upwelling processes, associated biological productivity and low-oxygen regions that are expected to expand under warming climate conditions. We will present main characteristics of the observed interannual to decadal variability, including extreme warm and cold events, decadal temperature and wind changes and evidence of ongoing oxygen changes. Possible future changes will be discussed by using model projections for warming climate scenarios. To capture a comprehensive picture of climate variability and change, a recommendation for the development of the observing system will be provided.

9.2   Eastern Boundary Upwelling System of the South Atlantic

The different physical drivers of the Benguela Current Large Marine Ecosystem that extends from the Congo River mouth to the southern tip of Africa are the basis of three different subsystems: the tAUS, the nBUS and the sBUS. Main differences are (1) warm tropical surface waters in the tAUS compared to cooler subtropical surface waters in the nBUS and sBUS (Fig. 9.3b), (2) strong wind forcing and corresponding wind-driven upwelling resulting in high primary productivity in the nBUS and sBUS compared to the weak wind forcing in the tAUS (Fig. 9.3a and c) and (3) permanent upwelling in the nBUS compared to strong seasonal variability in the tAUS and

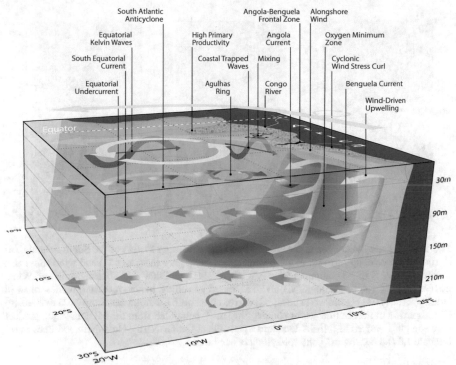

Fig. 9.2 Schematic view of the upper-ocean dynamics and the wind forcing in the southeastern Atlantic. The anticyclonic wind stress curl in the subtropics of the South Atlantic is associated with southerly winds along the African coast. The wind maximum is shifted slightly offshore resulting in a cyclonic wind stress curl in near-coastal areas. Both alongshore winds and cyclonic wind stress curl drive the upwelling in the sBUS and nBUS, which supplies the nutrients for the high primary productivity. In the tAUS, north of the Angola-Benguela Frontal Zone, winds are weak and the upwelling is mostly associated with the propagation of coastal trapped waves mainly forced via equatorial Kelvin waves impinging at the eastern boundary. Mixing during the upwelling wave phases plays the major role in supplying nutrients to the euphotic zone. Further to the north, south of the equator, high primary productivity is associated with nutrient input from the Congo River. Low-oxygen regions that extend far into the open ocean are found mainly north of the Angola-Benguela Frontal Zone and south of the equator. Further to the south, mostly associated with high productivity and enhanced oxygen consumption, low-oxygen regions (partly even anoxic regions) can be found on the shelf in the nBUS and locally in the sBUS

sBUS (Fig. 9.4). Oceanic oxygen conditions reveal the existence of a deep OMZ in the tAUS, extremely low-oxygen conditions on the shelf in the nBUS and only locally low-oxygen conditions in shallow waters on the shelf in the sBUS (Fig. 9.5). The three subsystems are described in the following subsections.

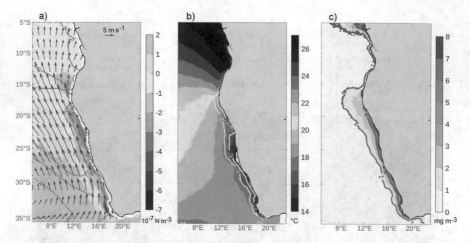

Fig. 9.3 Climate mean-state in the eastern boundary upwelling system of the South Atlantic. (**a**) Surface wind velocity (arrows; m s^{-1}), wind speed (solid grey lines; m s^{-1}) and wind stress curl (color shading; N m^{-3}) computed from ASCAT records over 2008–2020 (Ricciardulli and Wentz 2016). (**b**) SST (color-shading; °C) and SST-based upwelling index calculated as described in Bordbar et al. (2021) (solid white lines; °C) derived from MODIS-Aqua data from 2003 to 2020. (**c**) Chlorophyll-a concentration (color-shading; mg m^{-3}) computed from the MODIS-Aqua product between 2003 and 2020 (NASA Goddard Space Flight Center 2018). Note that solid lines in (**c**) indicate 1.0 and 4.0 mg m^{-3} chlorophyll-a concentration

9.2.1 The Tropical Angolan Upwelling System

The near-coastal area between the Congo River mouths and the ABFZ hosts the tAUS. It is characterized by a tropical stratification with fresh and warm waters at the surface located above saltier waters below. Freshwater input is due to precipitation and the main rivers, the Congo with the river mouth at about 6°S and the Cuanza at about 9°S. Along the coast, a weak southward current, the AC, transports SACW from the equatorial region toward the ABFZ (Tchipalanga et al. 2018). The mean core velocity of the AC at 11°S was found to be only 5–8 cm s^{-1} at a depth of about 50 m superimposed by substantially stronger intraseasonal, seasonal and interannual variability (Kopte et al. 2017). Alongshore winds and near-coastal wind stress curl are generally weak and are not suitable to produce a mean wind-driven upwelling system as it is found further south in the nBUS (Ostrowski et al. 2009; Zeng et al. 2021).

The oxygen distribution off Angola is characterized by the presence of an eastern boundary OMZ that was first described using data from the German Atlantic Meteor Expedition by Wattenberg (1929). It is located between the oxygen-richer regions at the equator and the paths of the BC and sSEC (Fig. 9.1c). The OMZ is a consequence of reduced oxygen supply in the shadow zones of the ventilated thermocline as theoretically predicted by Luyten et al. (1983) in combination with enhanced oxygen consumption due to high primary productivity. Dissolved oxygen concentration in the OMZ off Angola was measured regularly below 35 μmol kg^{-1} with lowest

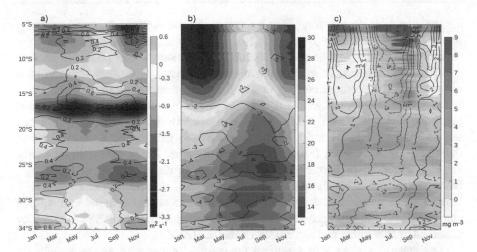

Fig. 9.4 Climatological seasonal cycles in the eastern boundary upwelling system of the South Atlantic. (**a**) Ekman offshore transport (color shading; $m^2 \, s^{-1}$) and wind stress curl-driven upwelling velocity (contours; $m \, d^{-1}$), (**b**) SST (color shading; °C) and the upwelling index based on SST (contours; °C), and (**c**) chlorophyll-a concentration (color shading; $mg \, m^{-3}$) and sea level anomaly (contour; cm). In (**a**) Ekman transport is the zonal transport in the nearest grid point to the coast, whereas the wind stress curl-driven upwelling, SST, SST-based index, chlorophyll-a concentration and sea level anomaly were averaged over a 150 km band along the coast. The reference period for (**a**) and (**b**) is 2008–2020 and 2003–2020, respectively. The climatological monthly mean for chlorophyll-a concentration and sea level anomaly in (**c**) were computed over 2003–2020 and 1993–2018, respectively

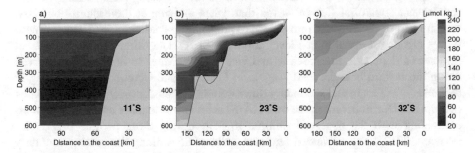

Fig. 9.5 Mean oxygen distribution across the continental slope and shelf. (**a**) Dissolved oxygen concentration along 11°S in the tAUS, (**b**) 23°S in the nBUS and (**c**) 32°S in the sBUS. Mean oxygen values were derived from repeat shipboard hydrographic sections

values close to the eastern boundary at about 400 m depth (Fig. 9.5a). The OMZ off Angola is thus more pronounced than its northern hemisphere counterpart, but substantially more oxygenated than the OMZs at the eastern boundary of the Pacific Ocean (Karstensen et al. 2008).

The seasonal cycle of SST is characterized by low temperatures during austral winter (JAS) and generally warmer waters from November to May (Fig. 9.4b). The seasonal cycle is mostly controlled by that of the atmospheric fluxes, specifically the solar radiation and the latent heat flux. Upper-ocean warming due to shortwave radiation experiences its annual minimum during austral winter primarily because of a seasonal maximum in solar zenith angle and the expansion of the stratocumulus cloud deck. The seasonal cooling due to the latent heat flux peaks in May which is associated with weak seasonal changes of relative humidity and wind speed (Scannell and McPhaden 2018). Sea surface salinity (SSS) is strongly reduced after maximum rainfall and river discharge of the Congo and Cuanza rivers at the beginning of the year (Sena Martins and Stammer 2022). Low-salinity waters are first observed in January off northern Angola directly south of the Congo River mouth and a few months later further to the south (Lübbecke et al. 2019). During some years the reduction in SSS during late austral summer and autumn can be observed to approach the ABFZ. A secondary minimum in SSS is regularly observed in November and December (Kopte et al. 2017).

The generally weak winds undergo a seasonal cycle with weakest southerly (upwelling-favorable) winds during the main upwelling season from July to September corresponding to a reduced Ekman offshore transport (Fig. 9.4a). Similarly, the wind stress curl-driven upwelling is particularly weak during austral winter north of the ABFZ (Fig. 9.4a). Thus, the winds are suggested not to be responsible for the high primary productivity occurring off Angola during this period (Fig. 9.4c) (Ostrowski et al. 2009; Zeng et al. 2021). The seasonal upwelling and downwelling phases in the tAUS associated with the upward and downward movement of the thermocline (Kopte et al. 2017) are instead predominantly remotely forced from the equatorial Atlantic. Semiannual wind forcing along the equator generates semiannual equatorial Kelvin waves that, when arriving at the eastern boundary, transfer part of their energy into poleward propagating coastal trapped waves. These waves can be observed in sea level anomaly (SLA) data, where elevated sea level indicates a depressed thermocline and vice-a-versa. Maxima in SLA are observed along the eastern boundary off Angola in February and October corresponding to the onset of the main and secondary downwelling seasons, respectively. The months of June and December mark the onset of the main and secondary upwelling seasons, coinciding with the observed minima in SLA during these periods (Fig. 9.4c) (Rouault 2012). The enhanced seasonal cycle of SLA along the eastern boundary could be related to a resonance of the equatorial basin established by eastward and westward propagating equatorial Kelvin and Rossby waves, respectively. At the eastern boundary, this resonance results in an enhancement of the semiannual and annual cycles, however, of different vertical structure (second baroclinic mode for the semiannual cycle compared to a higher baroclinic mode of the annual cycle) (Brandt et al. 2016; Kopte et al. 2018).

The stratification of the upper ocean in the tAUS is related to the vertical movement of the thermocline and further impacted by atmospheric heat and freshwater fluxes and river run-off. Strong upper-ocean stratification is observed during the downwelling seasons (February to April and October, November), while the stratification is weak during July to September (Kopte et al. 2017; Zeng et al. 2021). In the absence of upwelling-favorable winds, vertical mixing might be the main process responsible for the upward transport of nutrients into the euphotic zone. Zeng et al. (2021) could indeed show by using a tidal model that mixing induced by internal tides generated at the continental slope and propagating toward the coast contribute to near-coastal mixing (i.e., water shallower than 50 m) and associated local reduction of SST. Higher primary productivity during the main upwelling season could be explained as a result of weaker stratification: while the tidal energy available for mixing is almost constant throughout the year, the reduced stratification during austral winter allows stronger mixing and results in enhanced surface cooling and upward nutrient supply near the coast.

Superimposed on the seasonal cycle are intraseasonal or subseasonal fluctuations most prominently visible in SLA data and subsurface moored velocity records (Kopte et al. 2017; Polo et al. 2008). They are associated with coastal trapped waves either locally generated by local wind fluctuation or remotely generated along the equator or even further upstream in the tropical North Atlantic (Illig et al. 2018; Imbol Koungue and Brandt 2021). Contrary to the eastern boundary in the South Pacific, where such waves can coherently be observed in SLA data from the equator to about 27°S, in the South Atlantic the signal fades out already at 12°S. The difference could be explained by the higher baroclinic modes in the Atlantic compared to the Pacific, where the first baroclinic mode dominates, resulting in a stronger dissipation of wave energy along its path in the Atlantic (Illig et al. 2018). Intraseasonal coastal trapped waves off Angola show particular spectral peaks at about 90 and 120 days. These waves are mostly associated with remote equatorial forcing either by zonal wind forcing in the eastern equatorial Atlantic (90-day waves) or by the establishment of a resonant equatorial basin mode (120-day waves), respectively. The superposition of the intraseasonal waves with seasonal or interannual waves was found to be able to enhance or reduce the seasonal cycle as well as to impact Benguela Niños and Niñas (Imbol Koungue and Brandt 2021).

9.2.2   The Northern Benguela Upwelling System

The nBUS stretches between the ABFZ in the north and the Lüderitz upwelling cell in the south and is the transition zone between two source water masses, tropical SACW and subtropical ESACW (Junker et al. 2017; Mohrholz et al. 2008). The SACW, that dominates the region north of about 17°S, is warmer and more saline than the ESACW (Fig. 9.1c). It carries nutrients and is oxygen depleted. In contrast, the ESACW is well oxygenated. Over the nBUS, the fraction of SACW is decreasing poleward between unity near the ABFZ and almost zero near the Lüderitz upwelling cell. Hence, temperature and salinity but also the macronutrient and oxygen concentrations vary throughout the nBUS.

Figure 9.6, which represents a long-term average (2002 to 2016) produced from output of an ocean circulation model (Bordbar et al. 2021; Siegfried et al. 2019), shows the general circulation pattern guiding SACW and ESACW toward the nBUS. The simulated flowlines shown here follow the isopycnal $\sigma_\theta=26.3$ kg m^{-3}. Since potential density of a water mass is conserved below the surface layer, flowlines on a certain density level depict the spreading of the water mass with this specific

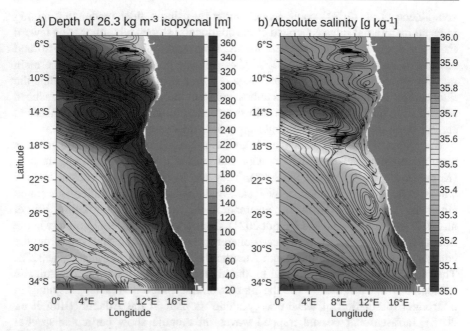

Fig. 9.6 Circulation and characteristics of the core water mass that upwells along the Southwest African coast. The water mass is defined by the potential density $\sigma_\theta = 26.3$ kg m^{-3}. (**a**) Depth of the isopycnal $\sigma_\theta = 26.3$ kg m^{-3} overlaid with flowlines. (**b**) Salinity on the isopycnal $\sigma_\theta = 26.3$ kg m^{-3} overlaid with flowlines. The isopycnal $\sigma_\theta = 26.3$ kg m^{-3} outcrops in the nBUS and the sBUS with water upwelling into the surface layer. The upwelled water originates from different source regions. The salinity distribution reveals the role of the nBUS as a mixing hotspot between water of tropical and subtropical origin. Results are derived from an ocean circulation model

density. Figure 9.6a shows the depth of the $\sigma_\theta = 26.3$ kg m^{-3} isopycnal. Water with this density outcrops and is upwelled in the nBUS and the sBUS, indicating it as a core water mass of the whole BUS. Also, in the north at around 8°E, 14°S, the $\sigma_\theta = 26.3$ kg m^{-3} isopycnal is elevated and associated with a cyclonic flow pattern. Some flowlines are continuing from the northern rim of this pattern into the nBUS suggesting the transport of warm and high-saline SACW from the tropical Atlantic into the nBUS (Lass and Mohrholz 2008). Similarly, flowlines starting in the south, e.g., at 35°S near the eastern boundary (Fig. 9.6), ending up in the nBUS revealing the northward transport of cold and fresh ESACW.

The long-term averaged flow is essentially in Sverdrup balance (Mohrholz et al. 2008; Siegfried et al. 2019; Small et al. 2015) by which the wind stress curl (Fig. 9.3a) determines the vertically integrated circulation. The simulated circulation shown in Fig. 9.6 is more general and also considers baroclinic details. However, its large-scale flow features can be understood from the Sverdrup balance: The northwestward flow off Namibia and South Africa (cf. BOC in Fig. 9.1) is related to the large-scale anticyclonic wind stress curl (Fig. 9.3a). In turn, the broad band of the southeastward directed flow off Angola corresponds to the area with weak but extended cyclonic wind stress curl. The Sverdrup balance also explains the sudden westward turn of the flow at about 16°S as related to a permanently present patch of strongly enhanced cyclonic wind stress curl in the Kunene upwelling cell (Fig. 9.3a). South of the ABFZ the numerical model results show a strong poleward flow that leaves the shelf with another westward turn near the Lüderitz upwelling cell at about 26°S.

The almost alongshore wind drives the coastal upwelling in the entire nBUS (Bordbar et al. 2021; Junker et al. 2015; Lass and Mohrholz 2008). The upwelled water is SACW and ESACW originating from just below the surface mixed layer down to approximately 200 m depth (Siegfried et al. 2019). It forms a narrow coastal band of nutrient-enriched surface water with relatively low temperature and enhanced chlorophyll-a concentration, which is visible in satellite-derived data (Fig. 9.3). In the nBUS the SST contrast between coast and open ocean locally amounts to more than 3°C (Fig. 9.3b) and is particularly large in the area between Walvis Bay and Lüderitz (Fig. 9.1). Here, features like fronts and filaments are observed that govern the exchange of heat and matter between the coastal and the open ocean (Bettencourt et al. 2012; Hösen et al. 2016; Mohrholz et al. 2008; Muller et al. 2013; Veitch and Penven 2017). Strength and seaward extension of upwelling in the nBUS vary along the coast and features several upwelling cells with characteristically low coastal SST (Bordbar et al. 2021; Chen et al. 2012; Shannon 1985). The most prominent cells are the Kunene cell at 17°S and the Lüderitz cell at 27°S (Fig. 9.4). Other upwelling cells are the Northern Namibian cell at 20°S and the Central Namibian or Walvis Bay cell at 23°S. The typical "coastal drop" of the alongshore wind off Namibia, i.e., coastal winds are weaker than the winds offshore, implies a cyclonic wind stress curl. The associated Ekman suction causes upwelling and plays an important role in the offshore region of the entire BUS (Figs. 9.3a and 9.4a) (Fennel 1999; Fennel et al. 2012). In general, the spatial pattern of the cyclonic wind stress curl narrows poleward, which matches the pattern of chlorophyll-a concentration fairly well (Fig. 9.3) (Bordbar et al. 2021; Fennel 1999). The Ekman suction shows seasonal maxima and minima appearing during different times of the year in the different cells. In the Kunene cell it is enhanced during March–October, whereas it peaks between September and April in the Walvis Bay cell at 23°S (Fig. 9.4a). The SST-based upwelling index is intensified between March and June everywhere across the nBUS (Fig. 9.4b), which follows the pattern of neither the coastal upwelling nor the Ekman suction.

The wind-driven upwelling is accompanied by swift geostrophic alongshore currents composed of the equatorward coastal jet at the surface and the PUC, an offshore countercurrent at subsurface. The structure and the relative strengths of the different eastern boundary current branches are governed by the separation of the alongshore wind maximum from the coast and the magnitude of the associated onshore wind stress curl (Fennel 1999; Fennel et al. 2012). The coastal jet amounts up to about 0.4 m s^{-1} near the ABFZ at about 17°S with velocities decreasing southward (Junker et al. 2019). In large-scale circulation maps it is often marked as the coastal branch of the Benguela Current (Fig. 9.1a). Associated with the reduced wind north of the Kunene cell, the coastal jet almost disappears there. Here, the

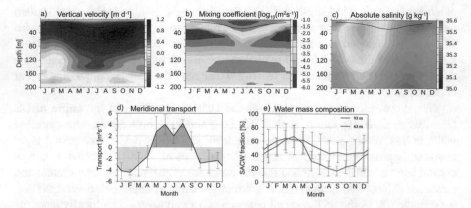

Fig. 9.7 The Walvis Bay upwelling cell at 23°S. Climatological seasonal cycles off Walvis Bay of (**a**) vertical velocity, (**b**) vertical mixing coefficient, (**c**) salinity. The quantities are the areal average over a 1° × 1° rectangle around 14°E, 23°S. The black line indicates the mixed layer depth. The monthly climatology (2002–2016) is derived from the results of an ocean circulation model forced by observed momentum, heat and mass fluxes (Bordbar et al. 2021; Siegfried et al. 2019). Climatological seasonal cycles of (**d**) meridional transport and (**e**) water mass composition (SACW vs. ESACW) at 63 m (red line) and 93 m (blue line) based on data from a mooring operated from 2003 to 2016 at 14°E, 23°S (Junker et al. 2017). Note that positive (negative) values in (**d**) indicate equatorward (poleward) transport. Whiskers represent the monthly climatological standard deviation

AC arriving from the north merges with PUC and continues southward through the ABFZ into the nBUS (Mohrholz et al. 2008).

In the following, we will describe the physical processes within the Walvis Bay upwelling cell at 23°S based on results from an ocean model simulation (Bordbar et al. 2021; Siegfried et al. 2019) and measurements taken by a long-term mooring on the Namibian shelf (14°E, 23°S) at a water depth of about 130 m (Junker et al. 2017). The vertical velocity (Fig. 9.7a) is primarily driven by the alongshore wind and peaks near the mixed layer depth (Fennel et al. 2012). Its seasonal cycle follows that of the alongshore wind with maxima in April and October. The mixed layer deepens from June to September concurrent with the lowest seasonal SST (Fig. 9.4b). Vertical mixing (Fig. 9.7b) forces entrainment of subsurface water into the surface mixed layer and is controlled by both, the seasonal cycle of the wind and the net surface heat flux. The maximum vertical mixing occurs in July and is concurrent with surface cooling (Fig. 9.4b) and deepening of the mixed layer. The seasonal variation of the salinity off Walvis Bay is visible over the entire upper 200 m depth (Fig. 9.7c). Salinity exhibits a seasonal maximum in April and a minimum in September corresponding to a dominance of SACW and ESACW, respectively. The meridional transport on the shelf as observed at the mooring position (Fig. 9.7d) (Junker et al. 2017; Mohrholz et al. 2008) shows a seasonal flow reversal that can be attributed to the seasonal cycle of the local wind stress curl (Junker et al. 2015). The observed flow is directed poleward from October–April and directed equatorward from March–September, which leads to the alternating influences of SACW and

ESACW on the Namibian shelf visible in the water mass fractions as obtained from moored hydrographic measurements (Fig. 9.7e). Mohrholz et al. (2008) showed that this alternation regulates the oxygen concentration on the Namibian shelf as well. Local oxygen consumption further modifies the oxygen conditions on the shelf and shapes the very specific ecosystem (Schmidt and Eggert 2016).

9.2.3 The Southern Benguela Upwelling System

The northern and southern Benguela upwelling systems are separated by the perennial Lüderitz upwelling cell, where the continental shelf also narrows abruptly toward the north. The sBUS extends southward from the Lüderitz cell along the entire west coast of South Africa and eastward to Cape Agulhas along the south coast (Boyd and Nelson 1998). Whereas the large-scale, depth integrated flow regime of the nBUS, from the ABFZ to Lüderitz, is driven by the alongshore winds and the wind stress curl, the large-scale dynamics of the sBUS is dominated by nonlinearities via turbulence associated with the shedding of Agulhas Rings, eddies and filaments at the Agulhas retroflection south of the African continent (Veitch et al. 2010). This intense offshore turbulence within the Cape Basin is unique among eastern boundary upwelling systems and presents itself as a distinct juxtaposition to the relatively quiescent shelf region of the sBUS (Veitch et al. 2010). Intense submesoscale variability in the Cape Basin leads to extreme vertical motions that have been shown to reduce productivity in coastal upwelling regions (Gruber et al. 2011) as well as in the offshore domain of the sBUS (Rossi et al. 2008).

The transition between the relatively quiescent shelf and the turbulent offshore region is marked by an intensified shelf-edge jet (SEJ) (Fig. 9.1a). It arises from the geostrophic adjustment of the particularly intense temperature front between the strongly seasonal cold upwelling regime at the coast and offshore waters that are warmed and modulated by highly variable influx of Agulhas waters (Veitch et al. 2018). This jet not only has an important role in transporting fish eggs and larvae from their spawning ground on the Agulhas Bank to their nursery area in St Helena Bay (Hutchings et al. 2009), it also limits cross-shelf exchanges (Barange and Pillar 1992; Pitcher and Nelson 2006). Despite seasonal and higher frequency modulations, the jet is a permanent feature (Nelson and Hutchings 1983) that tends to be situated over the 200–500 m isobaths and has been described as a convergent north-west oriented system on the western Agulhas Bank that funnels into the west coast (Shannon and Nelson 1996), bifurcating at Cape Columbine (32°50′S) where its inshore section veers into St Helena Bay (Lamont et al. 2015).

Satellite imagery (Demarcq et al. 2007; Lutjeharms and Stockton 1987) and model studies (Veitch and Penven 2017) have revealed that upwelling filaments in the southern Benguela do not extend as far offshore as their northern Benguela counterparts. The mean position of the upwelling front is approximately coincident with the location of the shelf-edge (Shannon 1985) and is commensurate with the frontal system that drives the intense SEJ. The limited offshore extent of upwelling filaments is therefore related both to the turbulent influx of warm Agulhas waters beyond the shelf-edge as well as to the SEJ that serves to inhibit their offshore expansion. While this frontal system helps to maintain the retentive nature of the southern Benguela shelf, it also contributes to the generation of cyclonic eddies that have been observed to form throughout the year in the vicinity of the upwelling front along the 200 m isobath (Lutjeharms and Stockton 1987) and to migrate predominantly in a west-southwestward direction (Hall and Lutjeharms 2011). The modeling results of Rubio et al. (2009) confirmed their generation at the upwelling front followed by their offshore migration, but also quantified the huge volume of water they carry, that could potentially be highly productive coastal waters transporting fish eggs and larvae from the Agulhas Bank to St Helena Bay as well (Hutchings et al. 1998).

Upwelling is strongly seasonal in the sBUS, being strongest during austral spring and summer months (Fig. 9.4a) and modulated with a period of 5–6 days due to the passage of cyclones and continental lows (Nelson and Hutchings 1983). Discrete upwelling cells are located in regions of enhanced wind stress curl and primarily where there are changes in coastline orientation (Lamont et al. 2018; Shannon and Nelson 1996). The two southernmost cells on the west coast, Cape Columbine and Cape Peninsula, are associated with a narrowing of the southern Benguela shelf and the only two prominent embayments of the Benguela system, namely St. Helena Bay and Table Bay. During active upwelling periods a cold plume develops at the distinct promontory of Cape Columbine (Taunton-Clark 1985), while a cyclonic eddy develops in its lee (Penven et al. 2000). These features produce a dynamic boundary between the nearshore and offshore regimes and create a highly retentive region within St. Helena Bay that is crucial for primary production, fish recruitment, but also has implications for low-oxygen water (LOW) on the southern Benguela shelf (Fig. 9.5c).

Aside from the shelf-edge frontal system, the existence of multiple fronts on the broad southern Benguela shelf was conceptualized by Barange and Pillar (1992) and observed by Lamont et al. (2015) to develop and merge in accordance with upwelling-favorable wind conditions. The proposed mechanism of Andrews and Hutchings (1980) that upon reaching a front, offshore advecting particles follow isopycnals and are subducted has been identified as crucial for nutrient-trapping and oxygen dynamics on the southern Benguela shelf (Flynn et al. 2020). This mechanism arises from secondary, ageostrophic circulations associated with frontogenesis and include upward (downward) velocities on the warm (cold) side of fronts (Capet et al. 2008).

The boundaries of the sBUS are dominated by well-ventilated ESACW. The varying presence of southern Benguela LOW is therefore limited to nearshore regions, developing in response to local dynamics (upwelling, retention, stratification and advection) and biogeochemical processes that are strongly dependent on seasonal wind fluctuations (Monteiro and van der Plas 2006). The formation of LOW has been observed throughout the sBUS (Jarre et al. 2015a), but occurs less frequently off the Namaqua shelf than within the St Helena Bay region where hypoxia persists throughout the year in the bottom waters (Fig. 9.5c), suggesting a permanent reservoir of LOW (Lamont et al. 2015). The formation of these low-oxygen bottom waters is driven by the decay of large phytoplankton blooms in the inner and mid-shelf regions during the upwelling season when retention and stratification is high (Pitcher and Probyn 2017). While wind mixing during winter months causes reoxygenation of the entire water column at nearshore stations, it is unable to erode the LOW reservoir at the bottom at depths greater than 50 m (Lamont et al. 2015). By the beginning of the upwelling season, in early austral spring, the shelf bottom waters are most oxygenated due to the on-shelf entrainment of the well-ventilated ESACW. For the duration of the upwelling season, the bottom oxygen concentrations within St Helena Bay progressively decrease due to continual draw-down of decaying organic matter. Flynn et al. (2020) demonstrated that nutrients available to be upwelled are augmented by regenerated nutrients from the previous summer and early winter that are trapped on the shelf by dynamics associated with the frontal system. This gives rise to enhanced primary production, decay and oxygen consumption. During periods of upwelling-favorable winds, the bottom pool of LOW is advected shoreward, priming the conditions for episodic anoxic events in St Helena Bay that occur toward the end of the upwelling season in the shallower nearshore environments, where high biomass dinoflagellate blooms (or "red-tides") are retained due to persistent downwelling or to the relaxation of upwelling-favorable winds. Their decay causes extreme oxygen consumption and oxygen depletion throughout the water column, leading to rock lobster walk-outs (Cockcroft 2001) as well as to major fish mortalities (Matthews and Pitcher 1996). New high temporal resolution dissolved oxygen measurements between February 2019 and October 2020, off Hondeklip Bay along the Namaqua shelf, revealed minimum oxygen concentrations during austral winter at a location offshore of the upwelling front suggesting the importance of lateral fluxes for the oxygen seasonality. This oxygen seasonality appears to be driven by the breakdown of the upwelling front during winter followed by periods of enhanced lateral mixing, which allows offshore advection of oxygen-depleted water from the nearshore environment (Rixen et al. 2021).

9.3   Interannual Variability

Around the ABFZ, SST varies from year to year with extreme warm and cold events occurring irregularly every few years. These events have been termed Benguela Niños and Niñas, respectively, to highlight their similarity to the Pacific El Niño phenomenon (Shannon et al. 1986). During these events, SST in the Angola-Benguela area (ABA, 8°E-coast; 10–20°S) can exceed the climatological value by more than 2°C (Fig. 9.8).

Both warm and cold events typically start with an SST anomaly off the Angolan coast in austral fall, which then spreads into the eastern equatorial Atlantic in the following months (Fig. 9.9). Since they have been first described in the 1980s, a number of different processes that contribute to the generation of these events have been determined. They can be broadly classified into remote and local forcing mechanisms (Richter et al. 2010). Remote forcing from the equatorial Atlantic

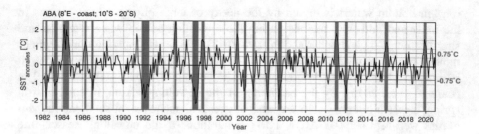

Fig. 9.8 Monthly detrended interannual SST anomalies (NOAA OISST) averaged in the Angola-Benguela Area (ABA, 8°E-coast; 10–20°S) from January 1982 to December 2020. The red and blue rectangles highlight the extreme warm and cold events in the region, respectively. The horizontal red and blue lines show the standard deviation of the interannual SST anomalies in the ABA. The criterion for an extreme event is defined by an SST anomaly exceeding the threshold of ±1 standard deviation for at least three consecutive months

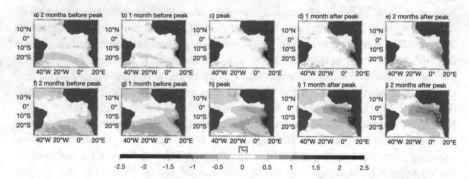

Fig. 9.9 Temporal evolution of Benguela Niños and Niñas. Composite maps of monthly detrended SST anomalies derived from observations (NOAA OISST) and computed from four extreme warm events (1984; 1986; 1995 and 2001) for (**a**) 2 months before the peak, (**b**) 1 month before the peak, (**c**) peak, (**d**) 1 month after the peak, (**e**) 2 months after the peak. Bottom panels (**f–j**) show the same as (**a–e**) but computed from six extreme cold events (1982, 1983, 1992, 1997, 2004 and 2005). Note that we included only extreme events that occurred during the period 1982–2020 and peaked between March and May

happens via the propagation of equatorial and coastal trapped waves. Variations of the zonal wind stress in the western equatorial Atlantic excite equatorial Kelvin waves that propagate eastward along the equatorial waveguide. Once they reach the eastern boundary, part of their energy is transmitted poleward along the African coast as coastal trapped waves. These waves are associated with a deflection of the thermocline that is linked to subsurface and subsequently surface temperature anomalies. A weakening (strengthening) of the easterly trade winds in the western to central equatorial Atlantic excites a downwelling (upwelling) Kelvin wave that is associated with a warm (cold) subsurface temperature anomaly and a strengthening (weakening) of the poleward flow of tropical warm waters, leading to a positive (negative) SST anomaly in the ABA. Many studies have shown the importance of this remote forcing mechanism, both for observed individual events (Florenchie et al. 2003; Rouault et al. 2018) and as a dominant mechanism in generating SST anomalies off Angola in ocean model simulations (Bachèlery et al. 2016, 2020; Florenchie et al. 2004; Imbol Koungue et al. 2017; Lübbecke et al. 2010). The impact of interannual coastal trapped waves on near-coastal temperatures is found to be strongest in the tAUS and northern nBUS. Based on model simulations, Bachèlery et al. (2020) showed that they are dominantly associated with the second and third coastal trapped wave modes. Coastal trapped waves of the less-dissipative first mode can be traced further south eventually reaching the sBUS, but their impact on near-coastal temperatures is weak.

In addition to the remote forcing, local processes can cause SST anomalies and lead to the generation of Benguela Niño and Niña events (Lübbecke et al. 2019; Polo et al. 2008; Richter et al. 2010). Variations in the local winds result in changes in the wind-driven upwelling of cold subsurface waters, anomalies in the latent heat flux from the ocean to the atmosphere and modulations of the meridional currents. Anomalous freshwater input from precipitation and river run-off might additionally impact SSTs via changes in stratification and the creation of barrier layers that inhibit the upward mixing of cold subsurface waters (Lübbecke et al. 2019). Illig et al. (2020) suggested that local wind forcing is crucial to explain the timing and spatial evolution of Benguela Niños and the subsequent warming in the eastern equatorial Atlantic. Moreover, Hu and Huang (2007) showed using reanalysis data that locally forced warming over the Angola-Benguela upwelling region is likely to generate westerly wind anomalies along the equatorial Atlantic one to two months later. An example of such a connection is the occurrence of the 2019 Benguela Niño (Fig. 9.8) recently studied by Imbol Koungue et al. (2021). This extreme warm event that developed along the coasts of Angola and Namibia between October 2019 and January 2020, was found to be forced by a combination of local and remote forcing with local forcing leading the remote forcing by one month. Remote and local forcing generally can be connected via changes of the SAA (Illig et al. 2020; Imbol Koungue et al. 2019; Lübbecke et al. 2010; Richter et al. 2010).

Benguela Niños and Niñas can have large impacts on the marine ecosystem as well as on the precipitation over Southwest Africa. The extreme warm event of 1995 has been associated with observed mortalities in sardine, horse mackerel and kob (Gammelsrød et al. 1998) as well as a southward shift of the sardine population (Boyer and Hampton 2001). Benguela Niño events are linked to above average rainfall over western Angola and Namibia via enhanced evaporation and moisture flux (Reason and Smart 2015; Rouault et al. 2003) while Benguela Niña events are associated with reduced precipitation along the Angolan coast (Koseki and Imbol Koungue 2021). For the benefits of the southern African countries and the coastal communities, it is thus desirable to predict such warm and cold events. A promising approach toward the prediction of SST anomalies off Angola and Namibia is based on the time it takes the equatorial and coastal waves to cross the basin and propagate poleward along the coast. Imbol Koungue et al. (2017) combined data from real time PIRATA buoys, altimetry and outputs from an ocean linear model to define an index of equatorial Kelvin wave activity. In agreement with the remote forcing mechanism described above, they found a high correlation for a second-mode equatorial Kelvin wave leading SST anomalies in the ABA by one month.

The sBUS varies primarily due to fluctuations associated with the seasonal shift of the SAA (Shannon and Nelson 1996). Interannual fluctuations in the sBUS, in terms of both upwelling strength and related LOW variability (Johnson and Nelson 1999; Monteiro et al. 2006), are strongly tied to variations in the SAA, which are related to both the phase of the Southern Annular Mode and El Niño-Southern Oscillation (ENSO) (Sun et al. 2019). For instance, Dufois and Rouault (2012) and Rouault et al. (2010) demonstrated that Pacific El Niño events tend to be associated with a northward shift of the SAA and a concomitant reduction of upwelling-favorable winds and enhanced coastal SSTs in the sBUS, with the opposite occurring during Pacific La Niña events. While confirming the influence of ENSO on the position of the SAA, Sun et al. (2019) found, however, no clear relationship between ENSO and the upwelling strength in the sBUS, especially for austral winter.

Located southwest of South Africa, the Agulhas retroflection is an important region of interbasin exchange of heat, salt and energy (Beal et al. 2011) and can also have a direct influence on the nutrient availability and biological responses observed on the adjacent shelf ecosystems (Roy et al. 2007; Roy et al. 2001). Interannual fluctuations of Agulhas leakage therefore have important implications for the sBUS. Using an improved method to identify the location of the core and edges of the Agulhas Current, a recent study by Russo et al. (2021) examined the variability of the Agulhas retroflection between 1993 and 2019. The Agulhas retroflection was located generally between 40.5–38.4°S and 15.0–20.0°E. Although seasonal variations were not statistically significant, the Agulhas retroflection extended further west during summer than in winter. During the 1993–2019 period, a total of seven events with Agulhas retroflections occurring further eastward were identified, with five events (1999, 2000–2001, 2008, 2013, 2019) being classified as early retroflections (east of 22.5°E), and two events (2014 and 2018) associated with the shedding of large Agulhas Rings. Their study described that early retroflection events during the first half of the 1993–2019 period tended to be more extreme but less frequent, while the latter part of study period showed less extreme but more frequent events (Russo et al. 2021). Changes in the mean position of the Agulhas retroflection, as well as the frequency and length of early retroflection events, can substantially impact the influx of Agulhas Current waters. This likely influences the stability and maintenance of the SEJ responsible for transport of fish eggs and larvae from the Agulhas Bank to the west coast of South Africa (Veitch et al. 2018).

9.4   Decadal Variations and Multidecadal Trends

Compared to the global average, the southeastern tropical Atlantic experienced moderate sea surface warming during the last 40 years (Bulgin et al. 2020). For the Southwest African coast, Sweijd and Smit (2020) reported the strongest warming trend near the ABFZ of more than 0.4°C decade^{-1} from 1981 to 2019. Warming

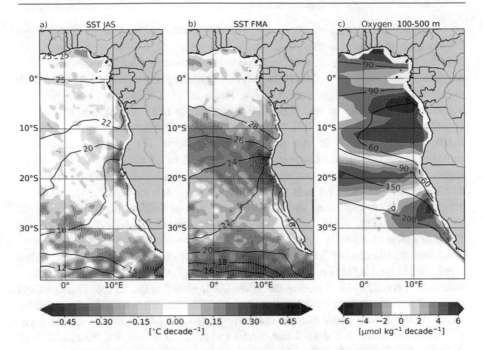

Fig. 9.10 Observed decadal trends in SST and oxygen. SST trends during (**a**) austral winter and (**b**) late austral summer evaluated between 1982 and 2017 using ERA5 reanalysis data (Hersbach et al. 2020). Black contours indicate seasonal mean SST. Grey dots indicate that the linear regressions are significant at the 95% level according to the Student's t-test. (**c**) Dissolved oxygen trend in 100–500 m (μmol kg^{-1} decade^{-1}) for the period 1960–2010. Black contours indicate mean 100–500 m oxygen concentration. Oxygen data is from Schmidtko et al. (2017)

was particularly enhanced in the tAUS and northern nBUS in comparison to the southern nBUS and the sBUS partly even showing cooling trends (Fig. 9.10). Seasonally warming patterns differ substantially with strongest warming during the satellite period observed during late austral summer with more than 0.3°C decade^{-1} (Fig. 9.10b). Vizy and Cook (2016), using high-resolution observations and reanalysis datasets over the period 1982–2013, showed from the analysis of the ocean surface heat balance that the austral summer SST warming trend along the Angolan/Namibian coast is associated with an increase in net surface heat flux. In addition, they showed a decrease in coastal upwelling due to atmospheric circulation changes related to a poleward shift of the SAA and an intensification of the Southwest African thermal low. The observed warming that varied substantially among different SST datasets was found to be associated with a slight southward shift of the ABFZ only (Prigent et al. 2020a; Vizy et al. 2018). However, only weak changes in the latitude of the front are expected as the position is largely determined by the shoreline orientation, bathymetry and associated wind stress (Shannon et al. 1987).

Upper-ocean decadal changes are well-detectable using data from the Argo observation program. Since 2000 the oceans have been filled with Argo floats continuously measuring temperature and salinity from the surface down to 2000 m of the water column within 10-day cycles (Desbruyères et al. 2017; Roemmich et al. 2015). From Argo observations, Roch et al. (2021) suggested that the southeastern Atlantic around the ABFZ is shifting from subtropical to tropical upper-ocean conditions. In this region, the vertical stratification maximum of the thermocline/pycnocline is located only slightly below the mixed layer. Hence, upper-ocean stratification changes can be assumed to be closely linked to mixed layer changes. The trend pattern of the vertical stratification maximum in the southeastern Atlantic that was derived following Roch et al. (2021), reveals a stratification increase of up to 30% decade^{-1} within 7–25°S for the time period of 2006–2020 (Fig. 9.11a). This area is located around the ABFZ. North of this region, which is actually the area with highest stratification values in the mean field, the stratification is decreasing. South of 30°S stratification changes are rather weak. The mixed layer depth shows a strong shoaling trend of around 8 m decade^{-1} within 5°E-coast, 2–18°S during the 2006–2020 period (Fig. 9.11b). This overlaps regionally with the area of the largest stratification increase. Clearly, the shoaling trend of the mixed layer cannot continue in the long-term as the mixed layer thickness has to reach a lower limit. North and south of this region the mixed layer is deepening. Within 30–40°S and west of the 0°-meridian, the deepening is largest with almost 10 m decade^{-1}.

The pronounced stratification enhancement in the area around the ABFZ is associated with a warming and freshening of the mixed layer (up to 2°C decade^{-1} and around 0.3 g kg^{-1} decade^{-1}, respectively, Fig. 9.11c and d). North and south of this region the mixed layer temperature depicts a cooling trend (Fig. 9.11c). However, along the coast from 30°S to Cape of Good Hope, a relatively weak warming trend (up to 0.5°C decade^{-1}) can be observed. In contrast, the freshening of the mixed layer can be found all the way south to 40°S. Yet, east of the 35.8 g kg^{-1} isohaline from the mean field, no trend is found (Fig. 9.11d).

To conclude, the observed enhancement of stratification around the ABFZ during the Argo observation period (2006–2020) together with the warming, freshening and shoaling of the mixed layer resulted in a southward expansion of background conditions associated with tropical upwelling systems (Roch et al. 2021). Besides in the equatorial region, the decrease in stratification can be associated with a cooling and salinification of the mixed layer. In contrast, off the coast of South Africa and southern Namibia weak trends can be observed including a slight warming and a deepening of the mixed layer.

Long-term changes have also been observed in the occurrence of Benguela Niño and Niña events. Prigent et al. (2020a) reported, relative to the period 1982–1999, an about 30% reduction of the March–April–May interannual SST variability around the ABFZ during 2000–2017 (Fig. 9.12). The weakened interannual ABA SST variability goes along with a reduced influence of the remote forcing by equatorial wind stress variability (cf. Sect. 9.3). Indeed, the lower zonal wind stress variability in the western equatorial Atlantic reported by Prigent et al. (2020b) tends to reduce

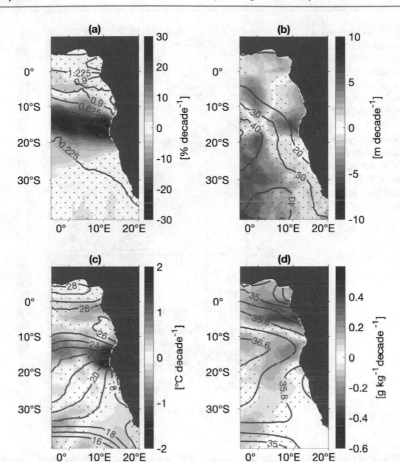

Fig. 9.11 Observed decadal trends of vertical stratification maximum and mixed layer properties. Trends inferred from Argo data for the period 2006–2020 for (**a**) vertical stratification maximum, (**b**) mixed layer depth, (**c**) mixed layer temperature and (**d**) mixed layer salinity. Trends are computed from the anomalies relative to the climatological seasonal cycle. Grey contour lines show the mean fields for the period of 2006–2020, respectively. Labels of the contour lines in (**a**) need to be multiplied by 10^{-3} to receive the squared Brunt-Väisälä frequency, N^2 [s^{-2}]. A positive trend of MLD refers to mixed layer deepening. Areas where the trend is not of 95% significance are stippled. Adapted from Roch et al. (2021)

the equatorial Kelvin wave activity that is an important driver of SST variability in the ABA (Imbol Koungue et al. 2017). In addition, the strong link between the fluctuations in the SAA strength and ABA SST (Lübbecke et al. 2010) has diminished since 2000. The Angola-Benguela region and the equatorial Atlantic are known to be strongly connected (Hu and Huang 2007; Illig et al. 2020; Lübbecke et al. 2010; Reason et al. 2006) and interannual SST variability has indeed also weakened along the equator. In particular, a 31% reduction of the interannual May–June–July SST variability was found in the eastern equatorial Atlantic (20°W-

Fig. 9.12 Weakening of interannual SST variability in the southeastern Atlantic. Difference of March–April–May SST anomalies standard deviation between 2000–2017 and 1982–1999 from ERA5. The blue shading depicts a reduction of the SST variability during 2000–2017 relative to 1982–1999. Adapted from Prigent et al. (2020a)

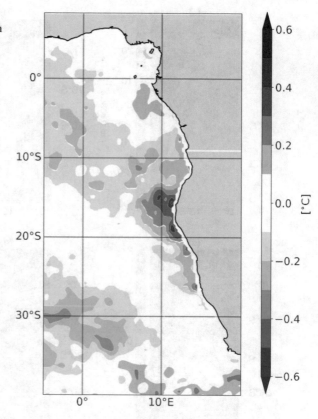

0°; 3°S-3°N) since 2000. The weakened SST variability in the eastern equatorial Atlantic was attributed to the reduced positive ocean-atmosphere feedback, the so-called Bjerknes feedback and increased thermal damping (Prigent et al. 2020b; Silva et al. 2021). Previously, Tokinaga and Xie (2011) reported a reduction of the eastern equatorial Atlantic variability over the period 1950–2009. They attributed the reduction in SST variability to a basin-wide warming, which is most pronounced in austral winter, reducing the annual cycle through positive ocean-atmosphere feedback.

The warming of the upper ocean during the recent decades is suggested to be the main reason for the ongoing deoxygenation of the ocean (Schmidtko et al. 2017). It is mainly attributed to a reduction of the oxygen solubility in warmer waters, but other mechanisms such as reduced subduction due to enhanced stratification or enhanced biological productivity might play a role as well (Oschlies et al. 2018). Downward oxygen trends and an expansion of the OMZ in the southeastern tropical Atlantic were first reported by Stramma et al. (2008). For the 100–500 m layer that include the core depth of the southeastern tropical Atlantic OMZ, available data for the period 1960 to 2010 predominantly suggests a deoxygenation in the southeastern Atlantic (Schmidtko et al. 2017). Oxygen reduction seems to be more confined to the eastern boundary in the south. Near the ABFZ, oxygen concentrations might even increase (Fig. 9.10c). While the long-term oxygen trends (1960–2010) presented by Schmidtko et al. (2017) largely represent open-ocean conditions and are based on relatively sparse data coverage in the South Atlantic, Pitcher et al. (2021) found in 20-year oxygen time series taken at the shelf and continental slope off Walvis Bay, 23°S, no discernible deoxygenation trend. Along the southern Benguela coast Pitcher et al. (2014) reported no significant deoxygenation trend over the past 50 years as well, but an increase in the frequency of episodic anoxic events during recent years. The lack of long-term trends in dissolved oxygen concentrations on the shelf may be associated with the lack of significant phytoplankton biomass changes observed in this region (Lamont et al. 2019).

The general expectation is that long-term changes in upwelling-favorable winds, and hence the frequency and intensity of upwelling events, will result in variations in phytoplankton production and biomass on similar scales (Verheye et al. 2016), with related consequences for LOW formation and zooplankton biomass. However, to date, *in situ* and satellite observations have been unable to demonstrate clear link-ages between physical forcing and subsequent ecosystem responses. For the African Large Marine Ecosystems, Sweijd and Smit (2020) found a general warming around Africa varying between 0.1 and 0.4°C decade^{-1}, while trends in productivity are much more heterogeneous with a tendency of enhanced productivity in the nBUS and sBUS and reduced productivity in the ABFZ. Besides the long-term warming in the nBUS, decreases in upwelling-favorable winds have been observed (Jarre et al. 2015b). Additionally, Lamont et al. (2018) reported substantially less upwelling-favorable wind in the nBUS since 2009, with a tendency toward fewer upwelling days and an increase in the number of events as upwelling has become less continuous. In contrast, the sBUS has shown a tendency toward increased upwelling and hence, long-term cooling (Lamont et al. 2018; Rouault et al. 2010). Annual to multidecadal variations in upwelling have been associated with shifts in the magnitude and position of the SAA, which drives upwelling-favorable winds in the region. Southward shifts of the SAA (Jarre et al. 2015b) appear to be responsible for the increased upwelling in the sBUS, with concomitantly less upwelling in the northern nBUS. Generally increasing phytoplankton biomass levels appear to be associated with the overall reduction in upwelling, as well as reduced grazing pressure from zooplankton communities, on the northern Benguela shelf (Lamont et al. 2019). In contrast, in the sBUS, increases in phytoplankton biomass levels on the Agulhas Bank during summer seem to be related to elevated nutrient levels arising from the observed increase in upwelling-favorable winds. Generally, the correspondence between long-term changes in upwelling-favorable winds and phytoplankton biomass on the west coast of South Africa has been more variable, with clear seasonal differences (Lamont et al. 2019).

9.5 Summary, Discussion and Recommendation for the Future Observing System

In this chapter we have presented a general view of our current understanding of the physical drivers of the eastern boundary upwelling system of the South Atlantic. The first part presents a description of the mean and the seasonal cycle of the eastern boundary upwelling system with a focus on local wind and remote equatorial forcing, tidal mixing and eddies. The description of the climatological situation is followed by observational evidence of the interannual variability including extreme warm and cold events, i.e., Benguela Niños and Niñas, as well as fluctuations in the Agulhas leakage. Decadal and multidecadal changes that are evident in the available observational datasets might be associated with internal variability of the climate system or with ongoing climate change due to global warming. Here, we showed evidence of long-term warming particularly enhanced near the ABFZ, of ocean deoxygenation particularly in the region of the deep OMZ, and also a reduction in the interannual variability associated with long-term thermocline deepening and strengthening stratification.

The upwelling system is separated into three subsystems: (1) the seasonally varying, mixing-driven tAUS, (2) the permanently wind-driven nBUS and (3) the seasonally varying, wind-driven sBUS. Seasonal variations in the tAUS are dynamically driven via the propagation of equatorial and coastal trapped waves that are remotely forced in the equatorial Atlantic. When arriving in the tAUS, the dominantly semiannual, coastal trapped waves are associated with upward and downward movements of the thermocline. Primary and secondary upwelling seasons are then established during phases of elevated thermocline in July/August and December/January, respectively. The upward nutrient supply during phases of elevated thermocline and reduced upper-ocean stratification is induced by vertical mixing predominantly forced by internal tides generated at the continental slope and shelf edge. The resulting productivity maximum in August is found to be delayed by about a month relative to the upwelling wave phases (Fig. 9.4c).

The nBUS and sBUS are wind-driven upwelling systems with localized enhanced upwelling cells, i.e., the Kunene, Northern Namibian, Walvis Bay and Lüderitz cells. Both regions, the nBUS and the sBUS, can be differentiated by their seasonality: while the sBUS is characterized by a strong seasonal cycle with enhanced wind forcing in austral spring and summer, the nBUS is a permanent upwelling system. Both the alongshore winds and the cyclonic wind stress curl that is established by the offshore displacement of the BLLCJ and the resulting onshore weakening of the wind are important mechanisms of near-coastal upwelling. Besides the wind forcing, the sBUS is strongly impacted by the turbulence associated with the shedding of Agulhas rings, eddies and filaments.

The observational record generally shows a warming trend of the sea surface since the 1980s, which is strongest in the ABFZ. While there is mostly warming in the nBUS, temperature data from the sBUS also indicate local cooling trends (Fig. 9.10a, b). A differential behavior of near-coastal and open ocean SST under

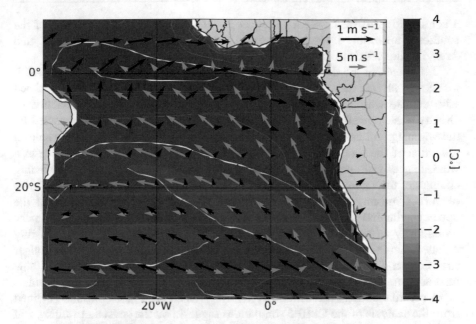

Fig. 9.13 Projected centennial sea surface temperature and wind changes in the South Atlantic. Difference of mean SST between 2070–2099 and 1970–1999 (shading) and difference of mean surface wind velocity between 2070–2099 and 1970–1999 (black arrows) from CMIP6 ensemble. CMIP6 ensemble mean surface wind velocity during 1970–1999 are shown as grey arrows. The CMIP6 ensemble is composed of 21 models. Prior to analysis the CMIP6 model data were interpolated on a common $1° \times 1°$ grid

climate warming was suggested by Bakun (1990) as a consequence of intensifying, upwelling-favorable winds due to increased air-temperature and sea-level pressure gradients between ocean and continent. To test this hypothesis, Sydeman et al. (2014) analyzed the available literature and found for the BUS evidence for a wind intensification only south of about 20°S. Such pattern, however, was found to be in general agreement with a southward shift of the SAA (Jarre et al. 2015b).

To further examine the mean SST response to sustained anthropogenic global warming, the worst-case scenario of the Shared Socioeconomic Pathway 5-8.5 (SSP5-8.5) applied to 21 General Circulation Models participating to the sixth phase of the Coupled Model Intercomparison Project (CMIP6) (Eyring et al. 2016) is used. Together with the equatorial Atlantic around 20°W, the Southwest African coast from 5°S to 30°S depicts the strongest warming (Fig. 9.13). Relative to 1970–1999, during 2070–2099 the SST in the region 0–10°S, 5–15°E (10–20°S; 5–15°E) is projected to increase by 3.2°C (3.3°C). In addition to the SST changes, the surface wind velocity field is also projected to change. Relative to 1970–1999, during 2070–2099, two major changes appear (Fig. 9.13): (1) the equatorial Atlantic easterly winds are found to decrease, (2) the southerly winds off Southwest Africa increase (decrease) south (north) of 20°S. The wind changes are in general agreement with a southward shift of the SAA (Yang et al. 2020) leading to a weakening of the southerly, upwelling-favorable winds in the tAUS and the northern nBUS and strengthening winds in the southern nBUS and sBUS (Lima et al. 2019).

Differences in the driving mechanisms of the different regions, i.e., the tAUS, nBUS and sBUS, may result in different responses to future changes associated with climate warming. Projections for the tAUS indicate a further reduction of already weak upwelling-favorable winds. This region is likely more impacted by the warming and associated increase in stratification. A likely scenario includes a reduction of the thermocline response to remote equatorial forcing resulting in a weakening of seasonal and interannual variations. The prediction of corresponding changes of the marine ecosystem would require the understanding of the response of vertical mixing to temporal and spatial changes in the stratifications at the upper continental slope and shelves. The nutrient supply that fuels the primary productivity in the tAUS was suggested to be associated with vertical mixing induced dominantly by internal tides. It was found to be weaker for stronger stratification (Zeng et al. 2021), thus representing a possible mechanism how increased stratification in a warming climate might impact the marine ecosystem.

In accordance with a recent study by Wang et al. (2015), our results obtained from the analysis of the CMIP6 simulations suggest that the coastal upwelling will become more intense in the sBUS, while in the northern nBUS, similarly to the tAUS, upwelling-favorable winds are weakening. Wang et al. (2015) additionally suggested a change in the timing of upwelling with the upwelling season starting earlier and ending later at high latitudes (sBUS) and only weak changes at low latitudes (nBUS). This is also in agreement with a general poleward expansion of the tropics. Changes in the stratification suggest that a tropical stratification with fresh and warm surface waters expands southward into regions where surface waters are characterized by a salinity maximum (Roch et al. 2021).

While ocean warming and associated physical processes such as reduced oxygen solubility in seawater and reduced vertical exchange will be a dominant factor in expected future oxygen changes, low-oxygen regions in shallow waters on the shelf largely depend on the biological productivity and the related oxygen consumption. Here, strong interactions between physical and biological processes are expected but are not well understood for the present upwelling systems (Lamont et al. 2019). In general, current coupled physical–biogeochemistry–biological models do not reproduce observed patterns for oxygen changes in the ocean's thermocline making regional projections of future oxygen changes highly uncertain (Oschlies et al. 2018).

While CMIP6 climate models are improved compared to results from the previous phase, they still show substantial biases in the tropical Atlantic indicating the need for model improvements (Li et al. 2020; Richter and Tokinaga 2020). This is particularly the case when coupling physical, biogeochemical and ecosystem models. Improved understanding of physical processes, their interaction with the ecosystem and possible changes under climate warming relies on the availability of high-quality observations. Following the recommendations by Foltz et al. (2019)

made for the tropical Atlantic observing system, we suggest the following key points for the observing system at the eastern boundary of the South Atlantic:

1. The southeastern Atlantic particularly lacks a sustained observing system for the eastern boundary circulation and its variability associated with local and remote forcing. Subsurface moorings (shielded against fishing activity) and other advanced *in situ* platforms should be operated to obtain continuous time series at high temporal resolution to capture long-term changes in key parameters and to identify long-term changes in physical processes with enhanced short-term variability.
2. Observations of upper-ocean mixing using autonomous platforms such as moorings or gliders are required for estimates of vertical heat fluxes and upward nutrient supply. Turbulence observations must include simultaneous observations of stratification and velocity variability on short temporal scales. As vertical mixing will likely change under the impact of warming and increasing stratification, studying processes such as internal tides and waves, wind-generated near inertial waves, surface waves and other mixed layer processes on the shelf and at the continental slope will be of critical importance for the understanding of the future development of primary productivity.
3. As satellite data is often unavailable or uncertain in near-coastal regions the use of the full suite of ocean observations is recommended to fill the gaps. Particularly installing additional sensors for atmospheric and upper-ocean measurements on existing autonomous platforms such as drifters, sail-drones, wave-gliders, floats and gliders together with multidisciplinary research cruises is recommended.
4. Of critical importance will be the improvement of hindcast and forecast models as well as reanalysis products still showing large biases (Tchipalanga et al. 2018; Zuidema et al. 2016). This can be achieved by dedicated process studies to improve parameterizations of or to identify new processes in upwelling regions. Freely providing physical and biogeochemical data to open data centers to be used in assimilation systems and for the initialization of forecast models would be an important step to enhance the understanding of the mechanisms of upwelling variability and changes and to improve ocean predictions, which is crucially needed for the management of fisheries and the marine environment in the coastal areas of Southwest Africa.
5. Enhancing research capabilities in the coastal countries as well as international cooperation is a necessary condition to tackle challenges of ongoing climate change and climate variability.

Acknowledgments This work was supported by the German Federal Ministry of Education and Research in the frame of the BANINO (03F0795) and EVAR (03F0814) projects and by the EU H2020 under grant agreement 817578 TRIATLAS project. This work was further funded by the German Research Foundation through grant 511812462 (IM 218/1-1). We thank M. Müller (GEOMAR) for help with the graphics.

References

Andrews WRH, Hutchings L (1980) Upwelling in the southern Benguela current. Prog Oceanogr 9:1–81. https://doi.org/10.1016/0079-6611(80)90015-4

Bachèlery ML, Illig S, Dadou I (2016) Interannual variability in the south-East Atlantic Ocean, focusing on the Benguela upwelling system: remote versus local forcing. J Geophys Res Oceans 121:284–310. https://doi.org/10.1002/2015jc011168

Bachèlery ML, Illig S, Rouault M (2020) Interannual coastal trapped waves in the Angola-Benguela upwelling system and Benguela Niño and Niña events. J Mar Syst 203:103262. https://doi.org/10.1016/j.jmarsys.2019.103262

Bakun A (1990) Global climate change and intensification of Coastal Ocean upwelling. Science 247:198–201. https://doi.org/10.1126/science.247.4939.198

Barange M, Pillar SC (1992) Cross-shelf circulation, zonation and maintenance mechanisms of Nyctiphanes-Capensis and Euphausia-Hanseni (Euphausiacea) in the northern Benguela upwelling system. Cont Shelf Res 12:1027–1042. https://doi.org/10.1016/0278-4343(92)90014-B

Bartholomae CH, van der Plas AK (2007) Towards the development of environmental indices for the Namibian shelf, with particular reference to fisheries management. Afr J Mar Sci 29:25–35. https://doi.org/10.2989/Ajms.2007.29.1.2.67

Beal LM, De Ruijter WPM, Biastoch A et al (2011) On the role of the Agulhas system in ocean circulation and climate. Nature 472:429–436. https://doi.org/10.1038/nature09983

Bettencourt JH, Lopez C, Hernandez-Garcia E (2012) Oceanic three-dimensional Lagrangian coherent structures: a study of a mesoscale eddy in the Benguela upwelling region. Ocean Model 51:73–83. https://doi.org/10.1016/j.ocemod.2012.04.004

Bordbar MH, Mohrholz V, Schmidt M (2021) The relation of wind-driven coastal and offshore upwelling in the Benguela upwelling system. J Phys Oceanogr 51:3117–3133. https://doi.org/10.1175/jpo-d-20-0297.1

Boyd AJ, Nelson G (1998) Variability of the Benguela current off the cape peninsula, South Africa. S Afr J Mar Sci Suid-Afrikaanse Tydskrif Vir Seewetenskap 19:27–39. https://doi.org/10.2989/025776198784126665

Boyer DC, Hampton I (2001) An overview of the living marine resources of Namibia. S Afr J Mar Sci Suid-Afrikaanse Tydskrif Vir Seewetenskap 23:5–35. https://doi.org/10.2989/025776101784528953

Brandt P, Bange HW, Banyte D et al (2015) On the role of circulation and mixing in the ventilation of oxygen minimum zones with a focus on the eastern tropical North Atlantic. Biogeosciences 12:489–512. https://doi.org/10.5194/bg-12-489-2015

Brandt P, Claus M, Greatbatch RJ et al (2016) Annual and semiannual cycle of equatorial Atlantic circulation associated with basin-mode resonance. J Phys Oceanogr 46:3011–3029. https://doi.org/10.1175/Jpo-D-15-0248.1

Bulgin CE, Merchant CJ, Ferreira D (2020) Tendencies, variability and persistence of sea surface temperature anomalies. Sci Rep 10:7986. https://doi.org/10.1038/s41598-020-64785-9

Capet X, McWilliams JC, Molemaker MJ et al (2008) Mesoscale to submesoscale transition in the California current system. Part II: frontal processes. J Phys Oceanogr 38:44–64. https://doi.org/10.1175/2007jpo3672.1

Chen ZY, Yan XH, Jo YH et al (2012) A study of Benguela upwelling system using different upwelling indices derived from remotely sensed data. Cont Shelf Res 45:27–33. https://doi.org/10.1016/j.csr.2012.05.013

Cockcroft AC (2001) Jasus lalandii 'walkouts' or mass strandings in South Africa during the 1990s: an overview. Mar Freshw Res 52:1085–1094. https://doi.org/10.1071/Mf01100

Demarcq H, Barlow R, Hutchings L (2007) Application of a chlorophyll index derived from satellite data to investigate the variability of phytoplankton in the Benguela ecosystem. Afr J Mar Sci 29:271–282. https://doi.org/10.2989/Ajms.2007.29.2.11.194

Desbruyères D, McDonagh EL, King BA et al (2017) Global and full-depth ocean temperature trends during the early twenty-first century from Argo and repeat hydrography (vol 30, pg 1985, 2017). J Clim 30:7577–7577. https://doi.org/10.1175/Jcli-D-17-0181.1

Djakouré S, Penven P, Bourlès B et al (2017) Respective roles of the Guinea current and local winds on the coastal upwelling in the northern gulf of Guinea. J Phys Oceanogr 47:1367–1387. https://doi.org/10.1175/Jpo-D-16-0126.1

Dufois F, Rouault M (2012) Sea surface temperature in False Bay (South Africa): towards a better understanding of its seasonal and inter-annual variability. Cont Shelf Res 43:24–35. https://doi.org/10.1016/j.csr.2012.04.009

Duncombe Rae CM (2005) A demonstration of the hydrographic partition of the Benguela upwelling ecosystem at 26°40'S. Afr J Mar Sci 27:617–628. https://doi.org/10.2989/18142320509504122

Eyring V, Bony S, Meehl GA et al (2016) Overview of the coupled model Intercomparison project phase 6 (CMIP6) experimental design and organization. Geosci Model Dev 9:1937–1958. https://doi.org/10.5194/gmd-9-1937-2016

Fennel W (1999) Theory of the Benguela upwelling system. J Phys Oceanogr 29:177–190. https://doi.org/10.1175/1520-0485(1999)029<0177:TOTBUS>2.0.CO;2

Fennel W, Junker T, Schmidt M et al (2012) Response of the Benguela upwelling systems to spatial variations in the wind stress. Cont Shelf Res 45:65–77. https://doi.org/10.1016/j.csr.2012.06.004

Florenchie P, Lutjeharms JRE, Reason CJC et al (2003) The source of Benguela Niños in the South Atlantic Ocean. Geophys Res Lett 30:1505. https://doi.org/10.1029/2003gl017172

Florenchie P, Reason CJC, Lutjeharms JRE et al (2004) Evolution of interannual warm and cold events in the Southeast Atlantic Ocean. J Clim 17:2318–2334. https://doi.org/10.1175/1520-0442(2004)017<2318:EOIWAC>2.0.CO;2

Flynn RF, Granger J, Veitch JA et al (2020) On-shelf nutrient trapping enhances the fertility of the southern Benguela upwelling system. J Geophys Res Oceans 125:e2019JC015948. https://doi.org/10.1029/2019JC015948

Foltz GR, Brandt P, Richter I et al (2019) The tropical Atlantic observing system. Front Mar Sci 6:206. https://doi.org/10.3389/fmars.2019.00206

Gammelsrød T, Bartholomae CH, Boyer DC et al (1998) Intrusion of warm surface water along the Angolan-Namibian coast in February-march 1995: the 1995 Benguela Niño. S Afr J Mar Sci Suid-Afrikaanse Tydskrif Vir Seewetenskap 19:41–56. https://doi.org/10.2989/025776198784126719

Gruber N (2011) Warming up, turning sour, losing breath: ocean biogeochemistry under global change. Philos Trans Royal Soc A-Math Phys Eng Sci 369:1980–1996. https://doi.org/10.1098/rsta.2011.0003

Gruber N, Lachkar Z, Frenzel H et al (2011) Eddy-induced reduction of biological production in eastern boundary upwelling systems. Nat Geosci 4:787–792. https://doi.org/10.1038/Ngeo1273

Hall C, Lutjeharms JRE (2011) Cyclonic eddies identified in the Cape Basin of the South Atlantic Ocean. J Mar Syst 85:1–10. https://doi.org/10.1016/j.jmarsys.2010.10.003

Herbert G, Bourlès B, Penven P et al (2016) New insights on the upper layer circulation north of the Gulf of Guinea. J Geophys Res Oceans 121:6793–6815. https://doi.org/10.1002/2016jc011959

Hersbach H, Bell B, Berrisford P, Hirahara S et al (2020) The ERA5 global reanalysis. Q J R Meteorol Soc 146:1999–2049. https://doi.org/10.1002/qj.3803

Hopkins J, Lucas M, Dufau C et al (2013) Detection and variability of The Congo River plume from satellite derived sea surface temperature, salinity, ocean colour and sea level. Remote Sens Environ 139:365–385. https://doi.org/10.1016/j.rse.2013.08.015

Hösen E, Möller J, Jochumsen K et al (2016) Scales and properties of cold filaments in the Benguela upwelling system off Luderitz. J Geophys Res Oceans 121:1896–1913. https://doi.org/10.1002/2015jc011411

Hu ZZ, Huang B (2007) Physical processes associated with the tropical Atlantic SST gradient during the anomalous evolution in the southeastern ocean. J Clim 20:3366–3378. https://doi.org/10.1175/Jcli4189.1

Hutchings L, Barange M, Bloomer SF et al (1998) Multiple factors affecting south African anchovy recruitment in the spawning, transport and nursery areas. S Afr J Mar Sci Suid-Afrikaanse Tydskrif Vir Seewetenskap 19:211–225. https://doi.org/10.2989/025776198784126908

Hutchings L, van der Lingen CD, Shannon LJ et al (2009) The Benguela current: an ecosystem of four components. Prog Oceanogr 83:15–32. https://doi.org/10.1016/j.pocean.2009.07.046

Illig S, Bachèlery ML, Cadier E (2018) Subseasonal coastal-trapped wave propagations in the southeastern Pacific and Atlantic oceans: 2. Wave characteristics and connection with the equatorial variability. J Geophys Res Oceans 123:3942–3961. https://doi.org/10.1029/2017jc013540

Illig S, Bachèlery ML, Lübbecke JF (2020) Why Do Benguela Niños Lead Atlantic Niños? J Geophys Res Oceans 125:e2019JC016003. https://doi.org/10.1029/2019JC016003

Imbol Koungue RA, Brandt P (2021) Impact of Intraseasonal waves on Angolan warm and cold events. J Geophys Res Oceans 126:e2020JC017088. https://doi.org/10.1029/2020JC017088

Imbol Koungue RA, Illig S, Rouault M (2017) Role of interannual kelvin wave propagations in the equatorial Atlantic on the Angola Benguela current system. J Geophys Res Oceans 122:4685–4703. https://doi.org/10.1002/2016jc012463

Imbol Koungue RA, Rouault M, Illig S et al (2019) Benguela Niños and Benguela Niñas in Forced Ocean simulation from 1958 to 2015. J Geophys Res Oceans 124:5923–5951. https://doi.org/10.1029/2019jc015013

Imbol Koungue RA, Brandt P, Lübbecke J et al (2021) The 2019 Benguela Niño. Front Mar Sci 8:800103. https://doi.org/10.3389/fmars.2021.800103

Jarre A, Hutchings L, Crichton M et al (2015a) Oxygen-depleted bottom waters along the west coast of South Africa, 1950-2011. Fish Oceanogr 24:56–73. https://doi.org/10.1111/fog.12076

Jarre A, Hutchings L, Kirkman SP et al (2015b) Synthesis: climate effects on biodiversity, abundance and distribution of marine organisms in the Benguela. Fish Oceanogr 24:122–149. https://doi.org/10.1111/fog.12086

Johns WE, Brandt P, Bourlès B et al (2014) Zonal structure and seasonal variability of the Atlantic Equatorial Undercurrent. Clim Dyn 43:3047–3069. https://doi.org/10.1007/s00382-014-2136-2

Johnson AS, Nelson G (1999) Ekman estimates of upwelling at cape columbine based on measurements of longshore wind from a 35-year time-series. S Afr J Mar Sci Suid-Afrikaanse Tydskrif Vir Seewetenskap 21:433–436. https://doi.org/10.2989/025776199784125971

Junker T, Schmidt M, Mohrholz V (2015) The relation of wind stress curl and meridional transport in the Benguela upwelling system. J Mar Syst 143:1–6. https://doi.org/10.1016/j.jmarsys.2014.10.006

Junker T, Mohrholz V, Siegfried L et al (2017) Seasonal to interannual variability of water mass characteristics and currents on the Namibian shelf. J Mar Syst 165:36–46. https://doi.org/10.1016/j.jmarsys.2016.09.003

Junker T, Mohrholz V, Schmidt M et al (2019) Coastal trapped wave propagation along the southwest African shelf as revealed by moored observations. J Phys Oceanogr 49:851–866. https://doi.org/10.1175/Jpo-D-18-0046.1

Kainge P, Kirkman SP, Estevao V et al (2020) Fisheries yields, climate change, and ecosystem-based management of the Benguela current large marine ecosystem. Environ Dev 36:100567. https://doi.org/10.1016/j.envdev.2020.100567

Karstensen J, Stramma L, Visbeck M (2008) Oxygen minimum zones in the eastern tropical Atlantic and Pacific oceans. Prog Oceanogr 77:331–350. https://doi.org/10.1016/J.Pocean.2007.05.009

Kolodziejczyk N, Marin F, Bourlès B et al (2014) Seasonal variability of the equatorial undercurrent termination and associated salinity maximum in the Gulf of Guinea. Clim Dyn 43:3025–3046. https://doi.org/10.1007/s00382-014-2107-7

Kopte R, Brandt P, Dengler M et al (2017) The Angola current: flow and hydrographic characteristics as observed at 11°S. J Geophys Res Oceans 122:1177–1189. https://doi.org/10.1002/2016jc012374

Kopte R, Brandt P, Claus M et al (2018) Role of Equatorial Basin-mode resonance for the seasonal variability of the Angola current at 11°S. J Phys Oceanogr 48:261–281. https://doi.org/10.1175/Jpo-D-17-0111.1

Koseki S, Imbol Koungue RA (2021) Regional atmospheric response to the Benguela Niñas. Int J Climatol 41:E1483–E1497. https://doi.org/10.1002/joc.6782

Lamont T, Hutchings L, van den Berg MA et al (2015) Hydrographic variability in the St. Helena Bay region of the southern Benguela ecosystem. J Geophys Res Oceans 120:2920–2944. https://doi.org/10.1002/2014jc010619

Lamont T, Garcia-Reyes M, Bograd SJ et al (2018) Upwelling indices for comparative ecosystem studies: variability in the Benguela upwelling system. J Mar Syst 188:3–16. https://doi.org/10.1016/j.jmarsys.2017.05.007

Lamont T, Barlow RG, Brewin RJW (2019) Long-term trends in phytoplankton chlorophyll a and size structure in the Benguela upwelling system. J Geophys Res Oceans 124:1170–1195. https://doi.org/10.1029/2018jc014334

Lass HU, Mohrholz V (2008) On the interaction between the subtropical gyre and the subtropical cell on the shelf of the SE Atlantic. J Mar Syst 74:1–43. https://doi.org/10.1016/j.jmarsys.2007.09.008

Li XW, Bordbar MH, Latif M et al (2020) Monthly to seasonal prediction of tropical Atlantic Sea surface temperature with statistical models constructed from observations and data from the Kiel climate model. Clim Dyn 54:1829–1850. https://doi.org/10.1007/s00382-020-05140-6

Lima DCA, Soares PMM, Semedo A et al (2019) How will a warming climate affect the Benguela coastal low-level wind jet? J Geophys Res-Atmos 124:5010–5028. https://doi.org/10.1029/2018JD029574

Lübbecke JF, Böning CW, Keenlyside NS et al (2010) On the connection between Benguela and equatorial Atlantic Niños and the role of the South Atlantic anticyclone. J Geophys Res Oceans 115:C09015. https://doi.org/10.1029/2009jc005964

Lübbecke JF, Brandt P, Dengler M et al (2019) Causes and evolution of the southeastern tropical Atlantic warm event in early 2016. Clim Dyn 53:261–274. https://doi.org/10.1007/s00382-018-4582-8

Lutjeharms JRE, Stockton PL (1987) Kinematics of the upwelling front off southern Africa. S Afr J Mar Sci Suid-Afrikaanse Tydskrif Vir Seewetenskap 5:35–49. https://doi.org/10.2989/025776187784522612

Luyten JR, Pedlosky J, Stommel H (1983) The ventilated thermocline. J Phys Oceanogr 13:292–309. https://doi.org/10.1175/1520-0485(1983)013<0292:TVT>2.0.CO;2

Matthews SG, Pitcher GC (1996) Worst recorded marine mortality on the south African coast. In: Yasumoto T, Oshima Y, Fukuyo Y (eds) Harmful and toxic algal blooms. Intergovernmental Oceanographic Commission of UNESCO, Sendai, Japan, pp 89–91

Mohrholz V, Bartholomae CH, van der Plas AK et al (2008) The seasonal variability of the northern Benguela undercurrent and its relation to the oxygen budget on the shelf. Cont Shelf Res 28:424–441. https://doi.org/10.1016/j.csr.2007.10.001

Monteiro PMS, van der Plas AK (2006) Low Oxygen Water (LOW) variability in the Benguela system: key processes and forcing scales relevant to forecasting. In: Shannon V, Hempel G, Malanotte-Rizzoli P et al (eds) Large marine ecosystems. Elsevier, Amsterdam, pp 71–90. https://doi.org/10.1016/S1570-0461(06)80010-8

Monteiro PMS, van der Plas A, Mohrholz V et al (2006) Variability of natural hypoxia and methane in a coastal upwelling system: oceanic physics or shelf biology? Geophys Res Lett 33:L16614. https://doi.org/10.1029/2006gl026234

Monteiro PMS, van der Plas AK, Melice JL et al (2008) Interannual hypoxia variability in a coastal upwelling system: ocean-shelf exchange, climate and ecosystem-state implications. Deep-Sea Res I Ocean Res Pap 55:435–450. https://doi.org/10.1016/j.dsr.2007.12.010

Muller AA, Mohrholz V, Schmidt M (2013) The circulation dynamics associated with a northern Benguela upwelling filament during October 2010. Cont Shelf Res 63:59–68. https://doi.org/10.1016/j.csr.2013.04.037

NASA Goddard Space Flight Center (2018) Moderate-resolution Imaging Spectroradiometer (MODIS) Aqua Chlorophyll Data, 2018: Reprocessing. NASA OB.DAAC. https://doi.org/10.5067/AQUA/MODIS/L3B/CHL/2018. Accessed 20 Feb 2019

Nelson G (1989) Poleward motion in the Benguela area. In: Neshyba SJ, Mooers CNK, Smith RL et al (eds) Poleward flows along Eastern Ocean boundaries. Springer, New York, NY, pp 110–130. https://doi.org/10.1007/978-1-4613-8963-7_10

Nelson G, Hutchings L (1983) The Benguela upwelling area. Prog Oceanogr 12:333–356. https://doi.org/10.1016/0079-6611(83)90013-7

Oschlies A, Brandt P, Stramma L et al (2018) Drivers and mechanisms of ocean deoxygenation. Nat Geosci 11:467–473. https://doi.org/10.1038/s41561-018-0152-2

Ostrowski M, da Silva JCB, Bazik-Sangolay B (2009) The response of sound scatterers to El Niño- and La Niña-like oceanographic regimes in the southeastern Atlantic. Ices J Mar Sci 66:1063–1072. https://doi.org/10.1093/icesjms/fsp102

Patricola CM, Chang P (2017) Structure and dynamics of the Benguela low-level coastal jet. Clim Dyn 49:2765–2788. https://doi.org/10.1007/s00382-016-3479-7

Penven P, Roy C, de Verdiere AC et al (2000) Simulation of a coastal jet retention process using a barotropic model. Oceanologica Acta 23:615–634. https://doi.org/10.1016/S0399-1784(00)01106-3

Pitcher GC, Nelson G (2006) Characteristics of the surface boundary layer important to the development of red tide on the southern Namaqua shelf of the Benguela upwelling system. Limnol Oceanogr 51:2660–2674. https://doi.org/10.4319/lo.2006.51.6.2660

Pitcher GC, Probyn TA (2017) Seasonal and sub-seasonal oxygen and nutrient fluctuations in an embayment of an eastern boundary upwelling system: St Helena bay. Afr J Mar Sci 39:95–110. https://doi.org/10.2989/1814232x.2017.1305989

Pitcher GC, Probyn TA, du Randt A et al (2014) Dynamics of oxygen depletion in the nearshore of a coastal embayment of the southern Benguela upwelling system. J Geophys Res Oceans 119:2183–2200. https://doi.org/10.1002/2013jc009443

Pitcher GC, Aguirre-Velarde A, Breitburg D et al (2021) System controls of coastal and open ocean oxygen depletion. Prog Oceanogr 197:102613. https://doi.org/10.1016/j.pocean.2021.102613

Polo I, Lazar A, Rodriguez-Fonseca B et al (2008) Oceanic kelvin waves and tropical Atlantic intraseasonal variability: 1. Kelvin wave characterization. J Geophys Res Oceans 113:C07009. https://doi.org/10.1029/2007jc004495

Prigent A, Imbol Koungue RA, Lübbecke JF et al (2020a) Origin of weakened Interannual Sea surface temperature variability in the southeastern tropical Atlantic Ocean. Geophys Res Lett 47:e2020GL089348. https://doi.org/10.1029/2020GL089348

Prigent A, Lübbecke JF, Bayr T et al (2020b) Weakened SST variability in the tropical Atlantic Ocean since 2000. Clim Dyn 54:2731–2744. https://doi.org/10.1007/s00382-020-05138-0

Reason CJC, Smart S (2015) Tropical south East Atlantic warm events and associated rainfall anomalies over southern Africa. Front Environ Sci 3:24. https://doi.org/10.3389/fenvs.2015.00024

Reason CJC, Florenchie P, Rouault M et al (2006) 10 Influences of large scale climate modes and agulhas system variability on the BCLME region. In: Shannon V, Hempel G, Malanotte-Rizzoli P et al (eds) Large marine ecosystems. Elsevier, Amsterdam, pp 223–238. https://doi.org/10.1016/S1570-0461(06)80015-7

Ricciardulli L, Wentz FJ (2016) Remote sensing systems ASCAT C-2015 daily ocean vector winds on 0.25 deg grid, version 02.1. Remote Sensing Systems. http://www.remss.com/missions/ascat. Accessed 1 June 2019

Richter I (2015) Climate model biases in the eastern tropical oceans: causes, impacts and ways forward. Wiley Interdiscip Rev Climate Change 6:345–358. https://doi.org/10.1002/wcc.338

Richter I, Tokinaga H (2020) An overview of the performance of CMIP6 models in the tropical Atlantic: mean state, variability, and remote impacts. Clim Dyn 55:2579–2601. https://doi.org/10.1007/s00382-020-05409-w

Richter I, Behera SK, Masumoto Y et al (2010) On the triggering of Benguela Niños: remote equatorial versus local influences. Geophys Res Lett 37:L20604. https://doi.org/10.1029/2010gl044461

Rixen T, Lahajnar N, Lamont T et al (2021) Oxygen and nutrient trapping in the southern Benguela upwelling system. Front Mar Sci 8:730591. https://doi.org/10.3389/fmars.2021.730591

Roch M, Brandt P, Schmidtko S et al (2021) Southeastern tropical Atlantic changing from subtropical to tropical conditions. Front Mar Sci 8:748383. https://doi.org/10.3389/fmars.2021.748383

Roemmich D, Church J, Gilson J et al (2015) Unabated planetary warming and its ocean structure since 2006. Nat Clim Chang 5:240–245. https://doi.org/10.1038/Nclimate2513

Rossi V, Lopez C, Sudre J et al (2008) Comparative study of mixing and biological activity of the Benguela and canary upwelling systems. Geophys Res Lett 35:L11602. https://doi.org/10.1029/2008gl033610

Rouault M (2012) Bi-annual intrusion of tropical water in the northern Benguela upwelling. Geophys Res Lett 39:L12606. https://doi.org/10.1029/2012gl052099

Rouault M, Florenchie P, Fauchereau N et al (2003) South East tropical Atlantic warm events and southern African rainfall. Geophys Res Lett 30:8009. https://doi.org/10.1029/2002gl014840

Rouault M, Pohl B, Penven P (2010) Coastal oceanic climate change and variability from 1982 to 2009 around South Africa. Afr J Mar Sci 32:237–246. https://doi.org/10.2989/1814232x.2010.501563

Rouault M, Illig S, Lübbecke J et al (2018) Origin, development and demise of the 2010-2011 Benguela Niño. J Mar Syst 188:39–48. https://doi.org/10.1016/j.jmarsys.2017.07.007

Roy C, Weeks S, Rouault M et al (2001) Extreme oceanographic events recorded in the southern Benguela during the 1999-2000 summer season. S Afr J Sci 97:465–471

Roy C, van der Lingen CD, Coetzee JC et al (2007) Abrupt environmental shift associated with changes in the distribution of cape anchovy Engraulis encrasicolus spawners in the southern Benguela. Afr J Mar Sci 29:309–319. https://doi.org/10.2989/Ajms.2007.29.3.1.331

Rubio A, Blanke B, Speich S et al (2009) Mesoscale eddy activity in the southern Benguela upwelling system from satellite altimetry and model data. Prog Oceanogr 83:288–295. https://doi.org/10.1016/j.pocean.2009.07.029

Russo CS, Lamont T, Krug M (2021) Spatial and temporal variability of the Agulhas retroflection: observations from a new objective detection method. Remote Sens Environ 253:112239. https://doi.org/10.1016/j.rse.2020.112239

Scannell HA, McPhaden MJ (2018) Seasonal mixed layer temperature balance in the southeastern tropical Atlantic. J Geophys Res Oceans 123:5557–5570. https://doi.org/10.1029/2018jc014099

Schmidt M, Eggert A (2016) Oxygen cycling in the northern Benguela upwelling system: modelling oxygen sources and sinks. Prog Oceanogr 149:145–173. https://doi.org/10.1016/j.pocean.2016.09.004

Schmidtko S, Stramma L, Visbeck M (2017) Decline in global oceanic oxygen content during the past five decades. Nature 542:335–339. https://doi.org/10.1038/nature21399

Sena Martins M, Stammer D (2022) Interannual variability of The Congo River plume-Induced Sea surface salinity. Remote Sens 14:1013. https://doi.org/10.3390/rs14041013

Shannon LV (1985) The Benguela ecosystem. Part I. evolution of the Benguela, physical features and processes. Oceanogr Mar Biol 23:105–182

Shannon LV, Nelson G (1996) The Benguela: large scale features and processes and system variability. In: Wefer G, Berger WH, Siedler G et al (eds) The South Atlantic: present and past circulation. Springer, Berlin, Heidelberg, pp 163–210. https://doi.org/10.1007/978-3-642-80353-6_9

Shannon LV, Boyd AJ, Brundrit GB et al (1986) On the existence of an El-Niño-type phenomenon in the Benguela system. J Mar Res 44:495–520. https://doi.org/10.1357/002224086788403105

Shannon LV, Agenbag JJ, Buys MEL (1987) Large-scale and mesoscale features of the Angola-Benguela front. S Afr J Mar Sci Suid-Afrikaanse Tydskrif Vir Seewetenskap 5:11–34

Siegfried L, Schmidt M, Mohrholz V et al (2019) The tropical-subtropical coupling in the Southeast Atlantic from the perspective of the northern Benguela upwelling system. PLoS One 14:e0210083. https://doi.org/10.1371/journal.pone.0210083

Silva P, Wainer I, Khodri M (2021) Changes in the equatorial mode of the tropical Atlantic in terms of the Bjerknes feedback index. Clim Dyn 56:3005–3024. https://doi.org/10.1007/s00382-021-05627-w

Small RJ, Curchitser E, Hedstrom K et al (2015) The Benguela upwelling system: quantifying the sensitivity to resolution and coastal wind representation in a global climate model. J Clim 28:9409–9432. https://doi.org/10.1175/Jcli-D-15-0192.1

Stramma L, Johnson GC, Sprintall J et al (2008) Expanding oxygen-minimum zones in the tropical oceans. Science 320:655–658. https://doi.org/10.1126/Science.1153847

Sun XM, Vizy EK, Cook KH (2019) Land-atmosphere-ocean interactions in the southeastern Atlantic: interannual variability. Clim Dyn 52:539–561. https://doi.org/10.1007/s00382-018-4155-x

Sweijd NA, Smit AJ (2020) Trends in sea surface temperature and chlorophyll-a in the seven African large marine ecosystems. Environ Dev 36:100585. https://doi.org/10.1016/j.envdev.2020.100585

Sydeman WJ, Garcia-Reyes M, Schoeman DS et al (2014) Climate change and wind intensification in coastal upwelling ecosystems. Science 345:77–80. https://doi.org/10.1126/science.1251635

Taunton-Clark J (1985) The formation, growth and decay of upwelling tongues in response to the mesoscale wind field during summer. In: Shannon LV (ed) South African Ocean colour and upwelling experiment. Sea Fisheries Research Institute, Cape Town (South Africa), pp 47–61

Tchipalanga P, Dengler M, Brandt P et al (2018) Eastern boundary circulation and hydrography off Angola: building Angolan oceanographic capacities. Bull Am Meteorol Soc 99:1589–1605. https://doi.org/10.1175/Bams-D-17-0197.1

Tokinaga H, Xie SP (2011) Weakening of the equatorial Atlantic cold tongue over the past six decades. Nat Geosci 4:222–226. https://doi.org/10.1038/Ngeo1078

Veitch JA, Penven P (2017) The role of the Agulhas in the Benguela current system: a numerical modeling approach. J Geophys Res Oceans 122:3375–3393. https://doi.org/10.1002/2016jc012247

Veitch J, Penven P, Shillington F (2009) The Benguela: a laboratory for comparative modeling studies. Prog Oceanogr 83:296–302. https://doi.org/10.1016/j.pocean.2009.07.008

Veitch J, Penven P, Shillington F (2010) Modeling equilibrium dynamics of the Benguela current system. J Phys Oceanogr 40:1942–1964. https://doi.org/10.1175/2010jpo4382.1

Veitch J, Hermes J, Lamont T et al (2018) Shelf-edge jet currents in the southern Benguela: a modelling approach. J Mar Syst 188:27–38. https://doi.org/10.1016/j.jmarsys.2017.09.003

Verheye HM, Lamont T, Huggett JA et al (2016) Plankton productivity of the Benguela current large marine ecosystem (BCLME). Environ Dev 17:75–92. https://doi.org/10.1016/j.envdev.2015.07.011

Vizy EK, Cook KH (2016) Understanding long-term (1982-2013) multi-decadal change in the equatorial and subtropical South Atlantic climate. Clim Dyn 46:2087–2113. https://doi.org/10.1007/s00382-015-2691-1

Vizy EK, Cook KH, Sun XM (2018) Decadal change of the South Atlantic Ocean Angola-Benguela frontal zone since 1980. Clim Dyn 51:3251–3273. https://doi.org/10.1007/s00382-018-4077-7

Wacongne S, Piton B (1992) The near-surface circulation in the northeastern corner of the South-Atlantic Ocean. Deep-Sea Res Part a-Oceanographic Research Papers 39:1273–1298. https://doi.org/10.1016/0198-0149(92)90069-6

Wang DW, Gouhier TC, Menge BA et al (2015) Intensification and spatial homogenization of coastal upwelling under climate change. Nature 518:390–394. https://doi.org/10.1038/nature14235

Wattenberg H (1929) Die Durchlüftung des Atlantischen Ozeans: (Vorläufige Mitteilung aus den Ergebnissen der Deutschen Atlantischen Expedition). ICES J Mar Sci 4:68–79. https://doi.org/10.1093/icesjms/4.1.68

Yang H, Lohmann G, Krebs-Kanzow U et al (2020) Poleward shift of the Major Ocean gyres detected in a warming climate. Geophys Res Lett 47:e2019GL085868. https://doi.org/10.1029/2019GL085868

Zeng Z, Brandt P, Lamb KG et al (2021) Three-dimensional numerical simulations of internal tides in the Angolan upwelling region. J Geophys Res Oceans 126:e2020JC016460. https://doi.org/10.1029/2020JC016460

Zuidema P, Chang P, Medeiros B et al (2016) Challenges and prospects for reducing coupled climate model SST biases in the eastern tropical Atlantic and Pacific oceans: the U.S. CLIVAR eastern tropical oceans synthesis working group. Bull Am Meteorol Soc 97:2305–2327. https://doi.org/10.1175/Bams-D-15-00274.1

Regional Land–Atmosphere Interactions in Southern Africa: Potential Impact and Sensitivity of Forest and Plantation Change

10

Zhenyu Zhang [iD], Patrick Laux [iD], Jussi Baade [iD], Hassane Moutahir [iD], and Harald Kunstmann [iD]

Abstract

Southern Africa is experiencing increasing land transformation and natural vegetation losses. Deforestation is one type of this land degradation where there are indigenous forests present, and afforestation of other nature ecosystems with timber plantations. This study performs regional coupled land–atmosphere model simulations using the Weather Research and Forecast (WRF) model with a resolution of 12 km, to assess the impact of forest and plantation cover change on regional climate in southern Africa. Three WRF simulations were designed for different land covers: (i) MODIS-derived land cover for the year 2000 (baseline), (ii) Landsat-based forest and plantation change map during 2000–2015 overlain on the baseline and (iii) theoretical forest and plantations removal relative to the baseline. Modeling results suggest that conversion of forest and plantations

Z. Zhang (✉) · P. Laux · H. Kunstmann
Institute of Geography, University of Augsburg, Augsburg, Germany

Institute of Meteorology and Climate Research (IMK-IFU), Karlsruhe Institute of Technology, Campus Alpin, Garmisch-Partenkirchen, Germany
e-mail: zhenyu.zhang@kit.edu; zhenyu.zhang@partner.kit.edu; patrick.laux@kit.edu; harald.kunstmann@kit.edu

J. Baade
Department for Physical Geography, Friedrich-Schiller-University, Jena, Germany
e-mail: jussi.baade@uni-jena.de

H. Moutahir
Institute of Meteorology and Climate Research (IMK-IFU), Karlsruhe Institute of Technology, Campus Alpin, Garmisch-Partenkirchen, Germany

The Mediterranean Center for Environmental Studies (CEAM), Valencia, Spain
e-mail: hassane.moutahir@kit.edu

© The Author(s) 2024
G. P. von Maltitz et al. (eds.), *Sustainability of Southern African Ecosystems under Global Change*, Ecological Studies 248,
https://doi.org/10.1007/978-3-031-10948-5_10

landscape to croplands and sparse vegetated land may result in a warmer and drier local climate, increasing daytime temperature by up to 0.6°C during the austral summer, and regulation of energy exchanges by decreasing the latent heat flux. In addition, results suggest that the removal of forest cover in northern part of southern Africa may decrease local precipitation recycling by around 1.2%. While the benefits of conserving native forests are obvious from an ecological perspective, afforestation considerations still require more detailed and local-scale treatments along the soil–vegetation–atmosphere continuum.

10.1 Introduction

Forests cover more than 30% of the global land area and play an important role in the Earth's system (FAO 2020; Hansen et al. 2013; IPBES 2019). Forests can absorb anthropogenic CO_2 emissions, store large carbon pools and modulate energy and water exchanges at the land–atmosphere interface (Alkama and Cescatti 2016). They are therefore considered an important mechanism for climate change mitigation (Shukla et al. 2019). However, it is reported that the forest area worldwide has declined by about 178 million ha since 1990, while the net forest loss in Africa (3.9 million ha/year) has been increasing progressively during the last three decades (FAO 2020). Such large-scale loss of forest cover reduces carbon sequestration and modifies surface energy budget and cloud formation, therefore impacts on the climate, both at regional and global scales (Bonan 2008; Wees et al. 2021).

The impact of the decline of forest cover on the regional climate is complex as the nonlinear vegetation-climate feedbacks and biogeophysical mechanisms vary in space and time. Large-scale forest cover loss generally leads to an increase of surface albedo, resulting cooling of the surface by reflecting more radiation into the atmosphere. This radiative effect is found to be dominant in temperate and boreal regions (e.g., Brovkin et al. 2006; Lee et al. 2011). On the other hand, the corresponding reduction in leaf area and stomatal resistance reduces transpiration rates, which leads to a locally warmer and drier climate, particularly in tropical rain forests (e.g., Bonan 2008; Zeng et al. 2021), but may result in more groundwater available in temperate regions (Rebelo et al. 2022). The compound effect of large-scale forest loss at different geographic locations may vary depending on which of these processes dominate (Davin and de Noblet-Ducoudré 2010; Jach et al. 2020).

Localized loss in forest cover, such as the conversion of forest to croplands and pasture, affects the energy balance and moisture cycling, and therefore the pattern and amount of precipitation (Sheil and Murdiyarso 2009; Leite-Filho et al. 2020). For example, Zeng et al. (2021) found that the recent deforestation between 2000 and 2014 over the Albertine Rift Mountains of Central Africa increased the regional warming by around 0.05°C during the dry season. Eghdami and Barros (2020) examined the implication of tropical forest loss on orographic precipitation in the eastern Andes, and found an increase in light rainfall and a decrease in moderate rainfall over the mountains. In the Amazon rainforest, Lejeune et al. (2015) performed climate simulations under projected deforestation and complete deforestation scenarios. They found an increase of annual mean temperature of 0.5°C and a decrease of precipitation of 0.17 mm/day, and further suggested the changes reach 0.8°C and 0.22 mm/day under total deforestation. Laux et al. (2017) applied a theoretical deforestation scenario over a coastal river basin in Central Vietnam. They found that deforestation caused only marginal differences in surface energy partitioning and did not clearly affect surface air temperature and precipitation. To date, most of the modeling studies have focused on the climate impact of boreal forest loss at high latitudes and deforestation in the moist tropics, with few studies addressing regional climate change in response to forest loss in southern temperate regions, such as southern Africa.

Southern Africa is not considered to be a rich forested region, and its native forests are generally distributed in the northernmost countries located in the tropics. The Food and Agriculture Organization (FAO) reported that losses of native forest cover in southern Africa are associated with deforestation, land-use change, land degradation, forest industry activities and an increase in wildfires due to climate change (FAO 2015, 2020; Wees et al. 2021). The case of South Africa is slightly different, where tree cover has been increasing due to expanding forestry plantations, alien tree invasion and bush encroachment. Intensive commercial forestry practices, such as harvesting and planting timber plantation, have been mentioned to result in high rates of tree cover change (Curtis et al. 2018).

This study tries to fill the gap by investigating the impact of change in forest and plantation cover on local climate in southern Africa. This work is conducted within the framework of the SPACE2 joint project: South Africa Land Degradation Monitor (SALDi). Within the SALDi project, our focus is on the use of a coupled regional climate modeling approach to assess land–atmosphere interactions in the context of land degradation in southern Africa on a subcontinental scale. Land-cover changes associated with native forest loss and forestry plantations are important components of land degradation in southern Africa; therefore, its impact on regional climate is investigated in this project.

10.2   Specific Objectives

Using the WRF regional coupled land–atmosphere model, the specific objectives of this study are:

1. To evaluate the effects of two forest and plantations modified experiments on land–atmosphere interactions based on the recent stage of their cover change (2000–2015) and an extreme forest and plantation removal case in southern Africa, and
2. To quantify the general impacts of forest and plantation cover on the local and regional climate of southern Africa.

10.3 Data and Methods

10.3.1 Regional Coupled Land–Atmosphere Model

The Advanced Research version of the Weather Research and Forecast (WRF) model version 4.1 coupled with the Noah Land Surface model (LSM) is used in this case study. The WRF model (Skamarock et al. 2012) is a nonhydrostatic, fully compressible and terrain-following coordinate model developed at National Center for Atmospheric Research (NCAR). The model has been widely used for climate dynamic downscaling and regional climate impact applications. By coupling with Noah LSM, WRF is able to accurately represent interactions between the land surface and the lower atmosphere due to its physically-based model processes. To explore the land-cover changes on subgrid scales, the Noah LSM using the mosaic approach of land cover (Noah mosaic) is adopted in the study. The Noah mosaic approach (Li et al. 2013) takes into account heterogeneous land surfaces by specifying land cover with N main categories (here $N = 3$) in subgrid scale, with weightings based on the fractional coverage, instead of using the dominant land cover over each grid as in unified Noah LSM. The subgrid heterogeneity of land cover is necessary to be considered, as forest and plantation cover change is rather sporadic across the regions considered and may not modify the dominant land cover type in the whole grid cell. The Noah mosaic approach has been used in assessing the climate impact of localized deforestation in previous studies (e.g., Wang et al. 2021; Zeng et al. 2021). In the WRF model, the different effects of land cover types on land–atmosphere interactions are represented by various predetermined biogeophysical properties, specified in lookup tables, such as leaf area index, albedo, emissivity and roughness length. These will be applied in the calculation of the radiation and energy balance, as well as in the vertical transmission of moisture, heat and momentum, which further affect the temperature and moisture fields near the surface and atmospheric evolution. It should be noted that the ecological impact processes are not able to be represented in the regional coupled land–atmosphere model, and forestry plantations and native forests are treated equally in land-cover categories and in the lookup tables.

All simulations carried out in this study share the same model domain setups and physical options. The model domain has a spatial resolution of 12 km, covering the area of southern Africa (Fig. 10.1a). The model domain has 35 vertical levels with the upper boundary set to 50 hPa. The atmospheric lateral boundary conditions are provided by the Reanalysis version 5 of the European Centre for Medium-Range Weather Forecasts at a 3-hourly interval. Based on a literature review of climate dynamic downscaling applications over southern Africa (Crétat et al. 2012; Ratna et al. 2014; Ratnam et al. 2013; Zhang et al. 2023), the following model physics schemes were selected. The Betts-Miller-Janjic cumulus scheme (Janjić 1994) is used to parameterize the subgrid-scale processes of convective clouds. The WRF single-moment 6-class microphysics scheme (Hong and Lim 2006) was used to simulate the water phase exchanges in the atmosphere, and the Rapid Radiative Transfer Model for General Circulation Models (Iacono et al. 2008) is used to

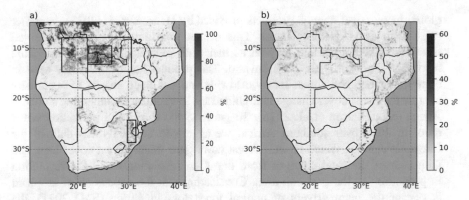

Fig. 10.1 (a) Forest and plantation cover over southern Africa from MODIS-derived land cover for the year 2000, and (b) satellite-observed forest and plantation cover loss (Hansen et al. 2013) over southern Africa between 2000 and 2015. All the datasets are aggregated to the WRF 12-km grid. The black rectangles in subplot (a) shows the location of the three areas for analysis (labeled A1, A2 and A3, respectively)

simulate the long-wave and short-wave radiation. The land surface turbulent fluxes are simulated by a revised Monin-Obukhov surface layer scheme (Jiménez et al. 2012), and the Yonsei University planetary boundary layer scheme (Hong et al. 2006) is used to parameterize the vertical transport concerning mass, moisture and energy fluxes transport in the planetary boundary layer.

10.3.2 Experiment Design

In order to evaluate the impact of forest and plantation cover modification on land–atmosphere interactions over southern Africa, three experiments with contrasting land-cover maps were modeled:

1. For the baseline experiment (CTL), the land-use and land-cover map from the Moderate Resolution Imaging Spectroradiometer (MODIS) 30-s product (Friedl et al. 2010) was used for representing the landscape features of the year 2000. The MODIS land-cover product provides 20 categories based on the International Geosphere-Biosphere Programme (IGBP) land-cover classifications. Those include evergreen needleleaf forest, evergreen broadleaf forest, deciduous needleleaf forest, deciduous broadleaf forest, mixed forests, closed shrublands, open shrublands, woody savannas, savannas, grasslands, permanent wetlands, croplands, urban and built up, cropland/natural vegetation mosaic, snow and ice, barren or sparsely vegetated, water, wooded tundra, mixed tundra and barren tundra.
2. The first land-cover change experiment (EXP1) represents the changed forest and plantation cover from 2000 to 2015 according to Landsat satellite observations. The forest and plantation cover change is generated from the high-resolution

global forest cover change products provided by Hansen et al. (2013), hereafter referred to as HANSEN-dataset. This dataset provides the satellite-observed forest and plantations loss as well as their gain information at a 30-m spatial resolution from the year 2000 onward. This product has been widely used in quantifying regional forest loss worldwide and has been shown to accurately capture the forest cover change in highland and lowland areas, including southern Africa (e.g., Hansen et al. 2016). To generate the forest and plantation cover map for 2015 over southern Africa, we aggregate the HANSEN-data of the forest net change, which is their forest cover gain minus forest cover loss, and then convert the total changes from the 30-m resolution to 12-km resolution in proportion to the grid cell area. Considering that the expansion of cropland is one of the main drivers of natural forest loss in Africa (FAO 2021), the conversion of forest to cropland is taken into account in the forest and plantation cover-modified land-cover map. The total change is superimposed on the forest categories in the original land-cover map of the year 2000, and an elevation threshold of 300 m a.s.l. is used to conceptually distinguish the lowland cropland and bare (sparse vegetated) ground in higher elevation area, which is similar to the approach followed in Wang et al. (2021) and Zeng et al. (2021). Cultivated land in the lowlands could be artificially irrigated and intensified throughout the year. However, crops at higher elevations are mostly rain-fed and usually sparsely vegetated during the dry season. If there is a net decrease in forest and plantation cover on a grid cell, we decrease the percentage of forest-type categories proportionately, and we treat the area as croplands if the elevation is lower than 300 m a.s.l., or we treat the area as barren or sparsely vegetated category if the elevation is above 300 m a.s.l. If there is an increase in forest and plantation cover in a grid cell, we increase the percentages of forest-type categories and decrease other land-cover categories proportionately.

3. The second forest and plantation cover change experiment (EXP2) represents an extreme scenario with a general forest and plantation cover removed, relative to the baseline (CTL). This would represent a forest loss of 100% for the tropical forest case study, and the removal of any indigenous forest and forestry plantations for the South African case study. The percentages of forest-type categories from the original MODIS land cover were converted into croplands or barren ground for all grid cells based on the above-described processes.

All three experiments (CTL, EXP1 and EXP2) use the same model configurations as described above, along with the soil texture map provided by Harmonized World Soil Database version 1.2. To isolate the impact of forest and plantation cover change-driven climate impacts, the simulations of each of the three experiments were run for the period from September 2014 to March 2015. Regarding the model spin-up period, previous studies found evidence that a one-month model spin-up period is sufficient to reach an equilibrium surface variable condition over southern Africa (e.g., Crétat et al. 2012; Ratnam et al. 2013, 2016). In this study case, we use two months as the model spin-up period, which is sufficient for the investigation. The austral summer months from November to March are chosen for analysis.

We addressed the two research objectives for three analysis areas with extensive forest and plantation cover, namely sites A1, A2 and A3 (Fig. 10.1a). Area A1 covers the northernmost part of southern Africa and has the highest native forest cover and Area A2 considers the same region but at a larger scale (and incorporates A1). Therefore, the impact of forest cover change can to some extent be explored at different scales. Forest and woodland losses have been confirmed to have occurred in sites A1 and A2 in the last decades (Mendelsohn 2019; Phiri et al. 2022). These areas are dominated by all-year-round rainfall with a hot climate. The vegetation is mainly rain-fed woody savanna/woodlands and deciduous broadleaf forests. Study site A3 is in a different climate region and biome from the last two and is located along the northern part of the Drakensberg Mountain Range, with plantation cover changes related to extensive forestry activities in this area. This area has a temperate climate and receives most of its rainfall from the Indian Ocean during the austral summer months. The cover of the plantations varies remarkably in this area and is mainly influenced by timber plantation footprint and by harvesting and planting of new rotations.

10.4   Results and Discussion

10.4.1 Validation of Model Performance

This study aims at investigating the impact of forest cover change and removal based on experimental simulation comparisons, rather than a comprehensive evaluation of the simulation results. Therefore, we validate the simulation results of the CTL experiment with observation-based high-resolution gridded datasets. The air temperature from the Climatic Research Unit (CRU) dataset (Harris et al. 2020) and the precipitation from Climate Hazards group Infrared Precipitation with Stations (CHIRPS) dataset (Funk et al. 2015) are used for validation. In order to facilitate direct comparison, the simulated temperature and precipitation are interpolated into the grids of the corresponding reference datasets. For southern Africa in the region south of 5° S, we calculate the spatial correlation from the spatial maps of values between the interpolated model results and the reference dataset.

The comparison of the spatial pattern of the averaged air temperature at 2 m above the ground is shown in Fig. 10.2. The simulation yields high spatial correlations of air temperature with the CRU dataset ($R = 0.98$, $P < 0.01$). For the monthly averaged temperature, the simulation also represents the temporal variations quite well (not shown), with slight deviations ranging from $-1.2°C$ to $0.9°C$. The results of the comparison illustrate that WRF, in general, represents the spatiotemporal variations of air temperature over southern Africa reasonably well.

In terms of precipitation, the simulation was able to replicate similar patterns compared to the reference dataset, showing the precipitation bands over the vegetated area of the northern most part of southern Africa as well as in the high mountains in the east of South Africa (Fig. 10.3). The spatial correlation coefficient between the simulation and the reference dataset is 0.76 ($P < 0.01$). However, the

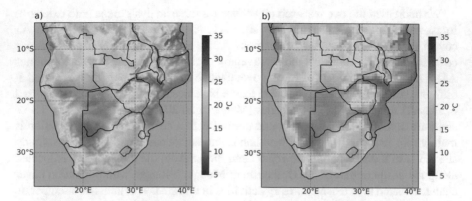

Fig. 10.2 Comparison of mean near-surface temperature between (**a**) WRF CTL simulation and (**b**) CRU dataset for the period Nov–Mar 2014/2015

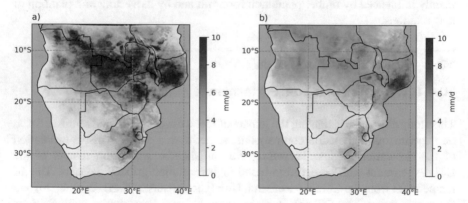

Fig. 10.3 Comparison of daily precipitation between (**a**) WRF CTL simulation and (**b**) CHRIPS dataset for the period Nov–Mar 2014/2015

simulated precipitation is generally overestimated (wet bias), which may be related to the fact that grided precipitation data usually underestimate precipitation over complex terrain area. Specific to the three analysis areas, the overestimation of precipitation reaches up to 40%. Nevertheless, such wet biases are often reported in precipitation simulation by dynamic downscaling models (e.g., Crétat et al. 2012; Ratna et al. 2014; Ratnam et al. 2012).

10.4.2 Impacts of Current Forest and Plantation Cover Change on Regional Climate and Land–Atmosphere Interactions

In EXP1, the change in forest and plantation cover is mainly reflected by a slight modification of percentage value in forest-type in the land-cover map. This forest and plantation cover change was found to modify the local precipitation

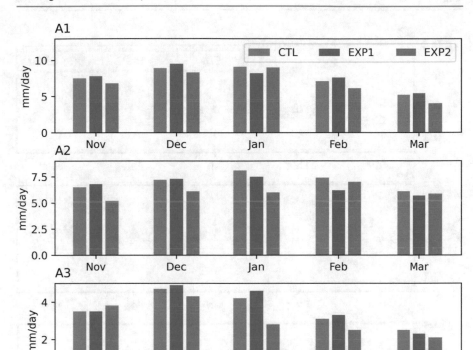

Fig. 10.4 Simulated monthly precipitation (mm/d) for the three experiments (CTL, EXP1 and EXP2) averaged over the analysis area shown in Fig. 10.1a, in southern Africa

only slightly. The monthly precipitation in the three analysis areas has changed moderately by about 1 mm/day (Fig. 10.4). The surface soil moisture content in EXP1 only marginally differs from the CTL (Fig. 10.5). Moreover, the standard deviations have similar values, indicating a comparable daily variation in soil moisture between CTL and EXP1.

Figure 10.6 illustrates the derived diurnal cycle of air temperature, sensible heat flux, latent heat flux and specific humidity for three analysis areas. All these surface variables show similar diurnal variations for CTL and EXP1 (Fig. 10.6). Some differences can be identified, including a slight increase in daytime air temperature and sensible heat flux, and a decrease in latent heat and specific humidity. These effects are comparatively small, because the changes in land cover are resolved at the subgrid scale in the WRF simulations, small percentage change in land-cover types very slightly influence the land–atmosphere coupling strength. Nevertheless, using the WRF Noah Mosaic approach, the forest and plantation cover change during the period 2000–2015 is simulated to be able to produce potential climate impacts in southern Africa.

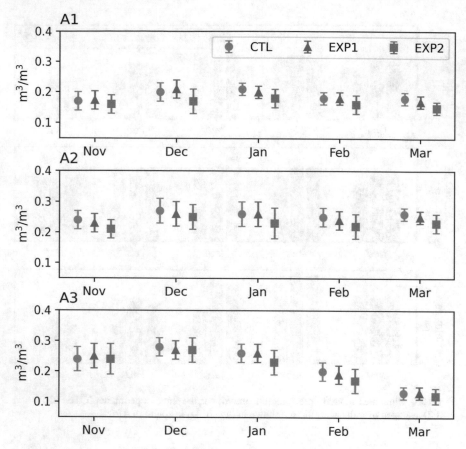

Fig. 10.5 Simulated mean surface soil moisture content for the three experiments (CTL, EXP1 and EXP2) for the three analysis areas shown in Fig. 10.1a, in southern Africa. The whiskers depict the standard deviation

10.4.3 Potential Impacts of Forest and Plantation Removal on Land–Atmosphere Interactions

The simulated total removal of forest and plantation removal experiment (EXP2) was found to have a remarkable impact on surface variables. In most cases, EXP2 exhibits a lower precipitation than CTL and EXP1 (Fig. 10.4). The surface soil moisture in EXP2 is generally lower than CTL in all three analysis areas due to the loss of the forest and plantation cover (Fig. 10.5), despite the variations of precipitation. It is evident that the standard deviation of soil moisture in EXP2 is larger than that of CTL and EXP1, indicating that the canopy interception of precipitation by forest and plantations reduces the soil moisture variations relative to bare ground and crop cover.

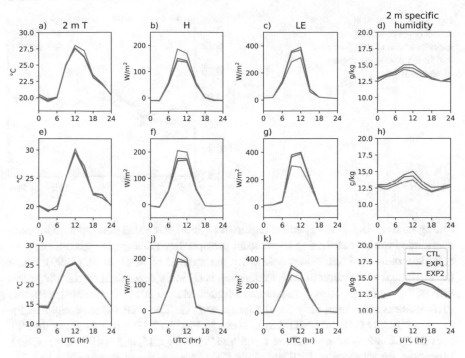

Fig. 10.6 Diurnal cycle of simulated 2 m temperature (first column), sensible heat flux H (second column), latent heat flux LE (third column), and 2 m specific humidity (fourth column) for the three experiments (CTL, EXP1 and EXP2) averaged over the three analysis areas A1 (top row), A2 (middle row) and A3 (bottom row) in southern Africa, respectively. The x-axis represents the hour in UTC time

As shown in Fig. 10.6, the daytime air temperature in EXP2 is marginally higher than that in CTL and EXP1. The removal of forest and plantations increases the temperature around 0.3 to 0.6°C at the diurnal peaks for three study sites. This higher air temperature is associated with increased sensible heat flux, decreased latent heat flux and decreased specific humidity during the daytime. This is attributed to the fact that transpiration and canopy evaporation are greatly reduced following the entire transfer of forest and plantation cover to croplands and barren ground. Moreover, even though the removal of forest and plantation cover can enhance the evaporation capacity of the soil, the overall evapotranspiration is still reduced, due to the fact that the water availability over southern Africa is mostly insufficient and depends mainly on rainfall. Overall, our modeling results that the total loss of forest and plantation cover could result in a warmer and drier local climate. It is worth noting that, for the three study regions with their different sizes, the above impacts on land–atmosphere interactions are consistent to a large degree.

The precipitation recycling ratio (PRR), defined as the contribution of the evaporated water to the precipitation within the same region, is a commonly used measure for quantifying land–atmosphere interactions (i.e., Arnault et al. 2016;

Fig. 10.7 Calculated
monthly precipitation
recycling ratio (PRR) for area
A2, in the northmost region
of southern Africa

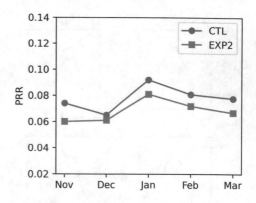

Zhang et al. 2019). For a specific region, a higher PRR value corresponds to more evaporated water contributing to the local precipitation (Zhang et al. 2022). Here we calculate monthly PRR values following the method of Schär et al. (1999). Since the analytical calculation of water recycling is strongly dependent on the shape and size of the area (e.g., Rios-Entenza and Miguez-Macho 2014; Trenberth 1999), this PRR value is calculated only for the large area A2. The PRR values computed for the experiments CTL and EXP2 are shown in Fig. 10.7. The results show that the overall PRR values of area A2 are 6.9% and 7.8% for CTL and EXP2, respectively. Moreover, for all months, PRR values in EXP2 are smaller than the values in CTL, which indicates that the simulation of the removal of forest and plantation cover reduces the local water recycling. This result is supportive to the knowledge that large extent of forest cover contributes to positive feedbacks to the precipitation, i.e., more forest cover increases the evapotranspiration and the local precipitation.

Although studies have shown that increased surface albedo associated with deforestation cools the surface and may mitigate climate warming in some regions (Lee et al. 2011; Williams et al. 2021), our findings suggest that the decrease in forest and plantation cover in southern Africa may lead to a warmer and drier local climate during the rainy austral summer. This may be related to the very strong evaporative capacity (e.g., the potential evapotranspiration) in southern Africa (Trabucco and Zomer 2019). It is worth noting that the drier and warmer effect may accumulate over the years, due to the long-term memory of soil moisture and the identified positive feedback on precipitation.

10.5   Summary and Outlook

In this study, we addressed the potential impact of forest and plantation cover change on regional climate and land–atmosphere interactions over southern Africa. Regional coupled land–atmosphere WRF model simulations were performed, with the different extent of forest-type categories prescribed as the land surface boundary. Based on the results presented above, it is concluded that the extreme experiment of the removal of forest cover in southern Africa may result in a warmer and drier climate during the austral summer months. The local precipitation recycling ratio was simulated have notably decreased due to the removal of forest cover, suggesting that forest and plantation positively contribute to local precipitation. Although the model results are affected by noise (internal model variabilities), our results for different sized areas and with different extents of forest and plantation cover change give sufficient evidence for conclusions.

Our results suggest that decision- and policymakers should focus attention on sustainable land management strategies to reduce the loss of native forests and properly manage forestry plantations for climate change mitigation. As southern Africa is not rich in forest resources, preserving what remains of the original and native forests is particularly important. Using established nature conservation areas to protect forest resources should be one of the most effective method. Since the major drivers of local deforestation in Africa include agriculture and cutting trees for firewood, preventative measures could include sustainable and appropriate population growth, the development of sustainable agriculture, as well as the use of renewable energy. Of course, robust policy interventions need to joint consideration of human and ecological sustainability and development and the related trade-offs. Governments and nongovernmental organizations should take the initiative to develop vigorous ecosystem restoration plans to re-establish forests and put long-term sustainability ahead of short-term concerns, which can be most effective in adapting to climate change.

This study provides a first step toward analyzing how forest and plantation modification affects the land–atmosphere interactions over southern Africa. A limitation of this study is that the aggregation of the change in forest cover into the coupled land–atmosphere model at mesoscale grid prevents many land surface characteristics from being properly represented. The imperfect classification of land-cover categories and the lookup table values assigned for land surface modeling lead to additional uncertainties in WRF simulations. Indeed, the nondistinction between native forests and forestry plantations, and the lack of consideration of ecological processes limit the model ability to simulate the potential role of forest and plantations in climate change in more detail. A fine spatial model resolution (< 4 km) would be beneficial for further investigation of the climate impact of forest and plantation cover change at convective scale and for extreme rainfall events. In particular, when combining in-situ measurements with very high-resolution modeling (< 1 km) for specific hotspot areas, the microclimate impact by forest and plantations can be investigated in more depth. Additionally, long-term model simulations and considering model internal uncertainties would allow an even deeper analysis.

Acknowledgments The work on this chapter benefited from funding by the Federal Ministry of Education and Research (BMBF) for the SPACES II Joint Project: South Africa Land Degradation Monitor (SALDi) (BMBF grant 01LL1701 A–D) within the framework of the Strategy "Research for Sustainability" (FONA) www.fona.de/en. HM is supported by the Generalitat Valenciana and the European Social Fund (APOSTD2020).

References

Alkama R, Cescatti A (2016) Biophysical climate impacts of recent changes in global forest cover. Science 351(6273):600–604. https://doi.org/10.1126/science.aac8083

Arnault J, Knoche R, Wei J, Kunstmann H (2016) Evaporation tagging and atmospheric water budget analysis with WRF: A regional precipitation recycling study for West Africa. Water Resour Res 52(3):1544–1567. https://doi.org/10.1002/2015WR017704

Bonan GB (2008) Forests and climate change: Forcings, feedbacks, and the climate benefits of forests. Science 320(5882):1444–1449. https://doi.org/10.1126/science.1155121

Brovkin V, Claussen M, Driesschaert E, Fichefet T, Kicklighter D, Loutre MF, Matthews HD, Ramankutty N, Schaeffer M, Sokolov A (2006) Biogeophysical effects of historical land cover changes simulated by six Earth system models of intermediate complexity. Clim Dyn 26(6):587–600. https://doi.org/10.1007/s00382-005-0092-6

Crétat J, Pohl B, Richard Y, Drobinski P (2012) Uncertainties in simulating regional climate of Southern Africa: sensitivity to physical parameterizations using WRF. Clim Dyn 38(3–4):613–634. https://doi.org/10.1007/s00382-011-1055-8

Curtis PG, Slay CM, Harris NL, Tyukavina A, Hansen MC (2018) Classifying drivers of global forest loss. Science 361(6407):1108–1111. https://doi.org/10.1126/science.aau3445

Davin EL, de Noblet-Ducoudré N (2010) Climatic impact of global-scale deforestation: radiative versus nonradiative processes. J Clim 23(1):97–112. https://doi.org/10.1175/2009JCLI3102.1

Eghdami M, Barros AP (2020) Deforestation impacts on orographic precipitation in the tropical Andes. Front Environ Sci 8(November):1–14. https://doi.org/10.3389/fenvs.2020.580159

FAO (2015) Southern Africa's forests and people. FAO, Rome

FAO (2020) Global forest resources assessment 2020 – key findings. FAO, Rome. https://doi.org/10.4060/ca8753en

FAO (2021) Global forest resources assessment 2020 – Remote sensing survey. FAO, Rome. https://www.fao.org/forest-resources-assessment/remote-sensing/fra-2020-remote-sensing-survey/en/

Friedl MA, Sulla-Menashe D, Tan B, Schneider A, Ramankutty N, Sibley A, Huang X (2010) MODIS collection 5 global land cover: algorithm refinements and characterization of new datasets. Remote Sens Environ 114(1):168–182. https://doi.org/10.1016/j.rse.2009.08.016

Funk C, Peterson P, Landsfeld M, Pedreros D, Verdin J, Shukla S, Husak G, Rowland J, Harrison L, Hoell A, Michaelsen J (2015) The climate hazards infrared precipitation with stations—a new environmental record for monitoring extremes. Sci Data 2(1):150066. https://doi.org/10.1038/sdata.2015.66

Hansen MC, Potapov PV, Moore R, Hancher M, Turubanova SA, Tyukavina A, Thau D, Stehman SV, Goetz SJ, Loveland TR, Kommareddy A, Egorov A, Chini L, Justice CO, Townshend JRG (2013) High-resolution global maps of 21st-century forest cover change. Science 342(6160):850–853. https://doi.org/10.1126/science.1244693

Hansen MC, Potapov PV, Goetz SJ, Turubanova S, Tyukavina A, Krylov A, Kommareddy A, Egorov A (2016) Mapping tree height distributions in Sub-Saharan Africa using Landsat 7 and 8 data. Remote Sens Environ 185:221–232. https://doi.org/10.1016/j.rse.2016.02.023

Harris I, Osborn TJ, Jones P, Lister D (2020) Version 4 of the CRU TS monthly high-resolution gridded multivariate climate dataset. Sci Data 7(1):109. https://doi.org/10.1038/s41597-020-0453-3

Hong S-Y, Lim JJ (2006) The WRF single-moment 6-class microphysics scheme (WSM6). Asia-Pac J Atmos Sci 42:129–151

Hong S-Y, Noh Y, Dudhia J (2006) A new vertical diffusion package with an explicit treatment of entrainment processes. Mon Weather Rev 134(9):2318–2341. https://doi.org/10.1175/MWR3199.1

Iacono MJ, Delamere JS, Mlawer EJ, Shephard MW, Clough SA, Collins WD (2008) Radiative forcing by long-lived greenhouse gases: calculations with the AER radiative transfer models. J Geophys Res 113(D13):D13103. https://doi.org/10.1029/2008JD009944

IPBES. (2019). Global assessment report on biodiversity and ecosystem services of the inter-governmental science-policy platform on biodiversity and ecosystem services. https://doi.org/10.5281/ZENODO.5517154

Jach L, Warrach-Sagi K, Ingwersen J, Kaas E, Wulfmeyer V (2020) Land cover impacts on land-atmosphere coupling strength in climate simulations with WRF over Europe. J Geophys Res Atmos 125(18). https://doi.org/10.1029/2019JD031989

Janjić ZI (1994) The step-mountain eta coordinate model: further developments of the convection, viscous sublayer, and turbulence closure schemes. Mon Weather Rev 122(5):927–945. https://doi.org/10.1175/1520-0493(1994)122<0927:TSMECM>2.0.CO;2

Jiménez PA, Dudhia J, González-Rouco JF, Navarro J, Montávez JP, García-Bustamante E (2012) A revised scheme for the WRF surface layer formulation. Mon Weather Rev 140(3):898–918. https://doi.org/10.1175/MWR-D-11-00056.1

Laux P, Nguyen PNB, Cullmann J, Van TP, Kunstmann H (2017) How many RCM ensemble members provide confidence in the impact of land-use land cover change? Int J Climatol 37:2080–2100. https://doi.org/10.1002/joc.4836

Lee X, Goulden ML, Hollinger DY, Barr A, Black TA, Bohrer G, Bracho R, Drake B, Goldstein A, Gu L, Katul G, Kolb T, Law BE, Margolis H, Meyers T, Monson R, Munger W, Oren R, Paw UKT, Richardson AD, Schmid HP, Staebler R, Wofsy S, Zhao L (2011) Observed increase in local cooling effect of deforestation at higher latitudes. Nature 479(7373):384–387. https://doi.org/10.1038/nature10588

Leite-Filho AT, Costa MH, Fu R (2020) The southern Amazon rainy season: the role of deforestation and its interactions with large-scale mechanisms. Int J Climatol 40(4):2328–2341. https://doi.org/10.1002/joc.6335

Lejeune Q, Davin EL, Guillod BP, Seneviratne SI (2015) Influence of Amazonian deforestation on the future evolution of regional surface fluxes, circulation, surface temperature and precipitation. Clim Dyn 44(9–10):2769–2786. https://doi.org/10.1007/s00382-014-2203-8

Li D, Bou-Zeid E, Barlage M, Chen F, Smith JA (2013) Development and evaluation of a mosaic approach in the WRF-Noah framework. J Geophys Res Atmos 118(21):11,918–11,935. https://doi.org/10.1002/2013JD020657

Mendelsohn JM (2019) Landscape changes in Angola. In: Huntley B, Russo V, Lages F, Ferrand N (eds) Biodiversity of Angola. Springer, Cham. https://doi.org/10.1007/978-3-030-03083-4_8

Phiri D, Chanda C, Nyirenda VR, Lwali CA (2022) An assessment of forest loss and its drivers in protected areas on the Copperbelt province of Zambia: 1972–2016. Geomat Nat Haz Risk 13(1):148–166. https://doi.org/10.1080/19475705.2021.2017021

Ratna SB, Ratnam JV, Behera SK, Rautenbach CJ d W, Ndarana T, Takahashi K, Yamagata T (2014) Performance assessment of three convective parameterization schemes in WRF for downscaling summer rainfall over South Africa. Clim Dyn 42(11–12):2931–2953. https://doi.org/10.1007/s00382-013-1918-2

Ratnam JV, Behera SK, Masumoto Y, Takahashi K, Yamagata T (2012) A simple regional coupled model experiment for summer-time climate simulation over southern Africa. Clim Dyn 39(9–10):2207–2217. https://doi.org/10.1007/s00382-011-1190-2

Ratnam JV, Behera SK, Ratna SB, de Rautenbach CJW, Lennard C, Luo JJ, Masumoto Y, Takahashi K, Yamagata T (2013) Dynamical downscaling of austral summer climate forecasts over Southern Africa using a regional coupled model. J Clim 26(16):6015–6032. https://doi.org/10.1175/JCLI-D-12-00645.1

Ratnam JV, Behera SK, Doi T, Ratna SB, Landman WA (2016) Improvements to the WRF seasonal hindcasts over South Africa by bias correcting the driving sintex-F2v CGCM fields. J Clim 29(8):2815–2829. https://doi.org/10.1175/JCLI-D-15-0435.1

Rebelo AJ, Holden PB, Hallowes J, Eady B, Cullis J, Esler KJ, New MG (2022) The hydrological impacts of restoration: A modelling study of alien tree clearing in four mountain catchments in South Africa. J Hydrol 127771:127771

Rios-Entenza A, Miguez-Macho G (2014) Moisture recycling and the maximum of precipitation in spring in the Iberian Peninsula. Clim Dyn 42(11–12):3207–3231. https://doi.org/10.1007/s00382-013-1971-x

Schär C, Lüthi D, Beyerle U, Heise E (1999) The soil–precipitation feedback: a process study with a regional climate model. J Clim 12(3):722–741. https://doi.org/10.1175/1520-0442(1999)012<0722:TSPFAP>2.0.CO;2

Sheil D, Murdiyarso D (2009) How forests attract rain: an examination of a new hypothesis. Bioscience 59(4):341–347. https://doi.org/10.1525/bio.2009.59.4.12

Shukla PR, Skea J, Calvo Buendia E, Masson-Delmotte V, Pörtner HO, Roberts DC, Zhai P, Slade R, Connors S, van Diemen R (2019) IPCC, 2019: climate change and land: an IPCC special report on climate change, desertification, land degradation, sustainable land management, food security, and greenhouse gas fluxes in terrestrial ecosystems. IPCC, Geneva

Skamarock WC, Klemp JB, Duda MG, Fowler LD, Park S-H, Ringler TD (2012) A multiscale nonhydrostatic atmospheric model using Centroidal Voronoi Tesselations and C-grid staggering. Mon Weather Rev 140(9):3090–3105. https://doi.org/10.1175/MWR-D-11-00215.1

Trabucco A, Zomer R (2019) Global aridity index and potential evapotranspiration (ET0) climate database v2. Figshare Dataset. https://doi.org/10.6084/m9.figshare.7504448.v3

Trenberth KE (1999) Atmospheric moisture recycling: role of advection and local evaporation. J Clim 12(5 II):1368–1381. https://doi.org/10.1175/1520-0442(1999)012<1368:amrroa>2.0.co;2

Wang D, Wu J, Huang M, Li LZX, Wang D, Lin T, Dong L, Li Q, Yang L, Zeng Z (2021) The critical effect of subgrid-scale scheme on simulating the climate impacts of deforestation. J Geophys Res Atmos 126:1–12. https://doi.org/10.1029/2021jd035133

Wees D, Werf GR, Randerson JT, Andela N, Chen Y, Morton DC (2021) The role of fire in global forest loss dynamics. Glob Chang Biol 27(11):2377–2391. https://doi.org/10.1111/gcb.15591

Williams CA, Gu H, Jiao T (2021) Climate impacts of U.S. forest loss span net warming to net cooling. Sci Adv 7(7). https://doi.org/10.1126/sciadv.aax8859

Zeng Z, Wang D, Yang L, Wu J, Ziegler AD, Liu M, Ciais P, Searchinger TD, Yang Z, Chen D, Chen A, Li LZX, Piao S, Taylor D, Cai X, Pan M, Peng L, Lin P, Gower D et al (2021) Deforestation-induced warming over tropical mountain regions regulated by elevation. Nat Geosci 14(1):23–29. https://doi.org/10.1038/s41561-020-00666-0

Zhang Z, Arnault J, Wagner S, Laux P, Kunstmann H (2019) Impact of lateral terrestrial water flow on land-atmosphere interactions in the Heihe River basin in China: fully coupled modeling and precipitation recycling analysis. J Geophys Res Atmos 124(15):8401–8423. https://doi.org/10.1029/2018JD030174

Zhang Z, Arnault J, Laux P, Ma N, Wei J, Shang S, Kunstmann H (2022) Convection-permitting fully coupled WRF-Hydro ensemble simulations in high mountain environment: impact of boundary layer- and lateral flow parameterizations on land-atmosphere interactions. Clim Dyn 59(5-6):1355–1376. https://doi.org/10.1007/s00382-021-06044-9

Zhang Z, Laux P, Baade J, Arnault J, Wei J, Wang X, Liu Y, Schmullius C, Kunstmann H (2023) Impact of alternative soil data sources on the uncertainties in simulated land-atmosphere interactions. Agri For Meteorol 339(March):109565. https://doi.org/10.1016/j.agrformet.2023.109565

Part III

Science in Support of Ecosystem Management

Studies of the Ecology of the Benguela Current Upwelling System: The TRAFFIC Approach

11

Bettina Martin, Holger Auel, Maya Bode-Dalby, Tim Dudeck, Sabrina Duncan, Werner Ekau, Heino O. Fock, Wilhelm Hagen, Knut Heinatz, Manfred J. Kaufmann, Rolf Koppelmann, Tarron Lamont, Deon Louw, Tebatso Moloto, Anne F. Sell, Sandy Thomalla, and Carl D. van der Lingen

B. Martin (✉) · K. Heinatz · R. Koppelmann
Institute of Marine Ecosystem and Fishery Science, University of Hamburg, Hamburg, Germany
e-mail: bmartin@uni-hamburg.de

H. Auel · M. Bode-Dalby · W. Hagen
BreMarE - Bremen Marine Ecology, Marine Zoology, University of Bremen, Bremen, Germany

T. Dudeck · W. Ekau
ZMT, Leibniz Center for Tropical Marine Research, Bremen, Germany

S. Duncan · H. O. Fock · A. F. Sell
Thünen Institute of Sea Fisheries, Bremerhaven, Germany

M. J. Kaufmann
Marine Biology Station of Funchal, Faculty of Life Sciences, University of Madeira, Funchal, Portugal

T. Lamont
Oceans & Coasts Research Branch, Department of Forestry, Fisheries, and the Environment, Cape Town, South Africa

Department of Oceanography, University of Cape Town, Rondebosch, South Africa

Bayworld Centre for Research & Education, Cape Town, South Africa

D. Louw
NatMIRC, Ministry of Fisheries and Marine Resources-National Marine Information and Research Center, Swakopmund, Namibia

Debmarine Namibia (DBMN), Windhoek, Namibia

T. Moloto · S. Thomalla
Southern Ocean Carbon-Climate Observatory (SOCCO), Council for Scientific and Industrial Research (CSIR), Cape Town, South Africa

© The Author(s) 2024
G. P. von Maltitz et al. (eds.), *Sustainability of Southern African Ecosystems under Global Change*, Ecological Studies 248,
https://doi.org/10.1007/978-3-031-10948-5_11

Abstract

Under the umbrella of SPACES (Science Partnerships for the Adaptation to Complex Earth System Processes in Southern Africa), several marine projects have been conducted to study the coastal upwelling area off southwestern Africa, the Benguela Upwelling System (BUS). The BUS is economically important for the bordering countries due to its large fish stocks. We present results from the projects GENUS and TRAFFIC, which focused on the biogeochemistry and biology of this marine area. The physical drivers, the nutrient distributions, and the different ecosystem components were studied on numerous expeditions using different methods. The important aspects of the ecosystem, such as key species and food web complexity were studied for a later evaluation of trophic transfer efficiency and to forecast possible changes in this highly productive marine area. This chapter provides a literature review and analyses of own data of the main biological trophic components in the Benguela Upwelling System gathered during two cruises in February/March 2019 and October 2021.

11.1 Introduction

The Benguela Upwelling System (BUS) is one of four major Eastern Boundary Upwelling Systems (the others are the California, Canary and Humboldt Upwelling Systems) that are among the most productive marine ecosystems and account for up to 20% of global fish catches (Bonino et al. 2019). The BUS extends about 2000 km along the eastern margin of the South Atlantic between Cape Agulhas (35°S) and the Angola-Benguela Front at ~17°S (Sakko 1998). It is bounded by the Agulhas Current in the south and the confluence of the Benguela Current with the Angola Current at the Angola-Benguela Front in the north (Fig. 11.1), which are the two largest warm-water bodies in the area (Carter 2011). The BUS is divided by the intense upwelling cell off Lüderitz (26°–27°S) into a southern (sBUS) and a northern (nBUS) subsystem (Bakun 1996). Characteristic of the BUS, in relation to the other three eastern boundary upwelling systems, is the width and depth of the coastal shelf that often extends up to 250 km offshore (Bordbar et al. 2021), providing space for the establishment of both shelf break/oceanic fronts at the shelf edge through coastal jets at about 350 to 500 m depth (Ragoasha et al. 2019), and upwelling fronts closer to the coast (Mann and Lazier 1991), with subsequent effects

C. D. van der Lingen
Fisheries Management, Department of Forestry, Fisheries and the Environment, DFFE, Cape Town, South Africa

Department of Biological Sciences, University of Cape Town, Cape Town, South Africa

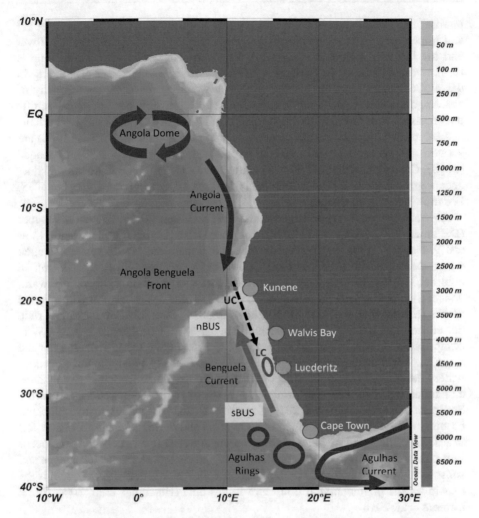

Fig. 11.1 Map of the Benguela Current Region, *LC* Luederitz Upwelling Cell, *UC* seasonally-varying poleward undercurrent, *nBUS* northern Benguela Upwelling System, *sBUS* Benguela Upwelling System

on nutrient supply (Flynn et al. 2020; Rixen et al. 2021) and possible larval trapping on the shelf (see Tiedemann and Brehmer 2017, for the Central East Atlantic).

Upwelling is the movement of surface water away from the coast and is caused by wind drag and Coriolis force. These water masses are replaced by cold and nutrient-rich water from deeper layers by coastal upwelling and Ekman transport (Ekman 1905). A second type of upwelling is the offshore upwelling facilitated through Ekman pumping (Rykaczewski and Checkley 2008; Bordbar et al. 2021). Whereas Ekman transport is dependent on a constant wind field, Ekman pumping responds to gradients in wind strength. This so-called wind stress curl is more intense in the sBUS than in the nBUS (Fennel and Lass 2007; Bordbar et al. 2021) and important offshore upwelling is indicated for the Cape Columbine/Cape Town and the Lüderitz regions up to Walvis Bay. Wind stress curl-driven upwelling is further responsible for the surfacing of the poleward undercurrent originating in the nBUS, where coastally trapped Kelvin waves may modulate the intensity of the Angola Current. With increased wind stress curl, offshore upwelling uplifts undercurrent water masses to the surface creating highly variable surface currents and intensified upwelling. Important so-called upwelling cells (Shannon 1985) are located at Cape Frio (19°S), Walvis Bay (23°S), Lüderitz (25–26°S), Namaqua (30°S), Cape Columbine (32°S), and Cape Town (34°S). These regionally distinct oceanographic conditions form the basis for separate ecoregions within the BUS. For more details, see Chap. 9.

Upwelling events vary seasonally and locally (Carter 2011). In austral summer (December to March), warm oligotrophic water from Angola (South Atlantic Central Water, SACW) is transported southward and therefore can dominate upwelled water masses in the nBUS at that time of the year (Hutchings et al. 2006). In contrast, Eastern South Atlantic Central Water (ESACW) is transported northward, constituting the upwelled water mass in the sBUS, and also in austral winter in the nBUS (Monteiro et al. 2008; Mohrholz et al. 2014). The ESACW is oxygen-rich, but carries comparatively less nutrients than the SACW further north. ESACW is comprised of a mixture of central waters from the Indian Ocean which enter the Cape Basin as intrusions from the Agulhas Current retroflection region, with central waters transported across the South Atlantic from the Brazil-Malvinas Confluence, and mode waters which are formed just north of the sub-Antarctic Front in the Southern Ocean (Kersalé et al. 2018; Lamont et al. 2015). Maxima in phytoplankton biomass emerge in spring (September–October) and during late summer/early fall in nBUS, while south of South Africa, on the Agulhas Bank shelf, maxima occur in fall (March–April) (Lamont et al. 2018). It is likely that excessive turbulence and substantial offshore advection in the Lüderitz cell, which is active year-round, is the reason for the phytoplankton minimum in this area (Hutchings et al. 2006, 2012; Lamont et al. 2018).

The high phytoplankton productivity in the BUS, however, is in surprising contrast to the relatively low productivity of the higher trophic levels as compared to the Humboldt Current System (Messié and Chavez 2015). Nutrients reaching the surface during upwelling are expected to be rapidly re-exported, given the substantial offshore transport of surface waters driven by upwelling-favorable winds, so that organisms of the higher trophic levels cannot effectively utilize the available production. The result of this inefficient retention of nutrients is, that, despite short trophic pathways, fisheries yields are relatively low. However, management effectiveness and industrial capitalization cannot be ruled out as a cause for the different rates in fisheries production.

From a human perspective, and regarding the opportunities of harvesting living marine resources, the productivity of an ecosystem's upper trophic levels is of particular interest. Many of the commercially targeted marine species are predatory fish at the upper end of the food chain. Inherently, upwelling systems are characterized by exceptionally high productivity, the degree of which can be assessed as the ecosystem's trophic transfer efficiency (TTE), alternatively called energy transfer efficiency. In this chapter, we will provide literature and new data on the main biological trophic components in the Benguela Upwelling System. Two research cruises were conducted to cover seasonal variations. A summary of the results is in progress and will be provided elsewhere.

11.2   Previous Research and Hypotheses

The TRAFFIC (Trophic TRAnsfer eFFICency in the Benguela Current) project is part of the SPACES II program and was conducted between 2018 and 2022. The members of the TRAFFIC consortium have had close collaborations with regional scientific organizations in southern Africa (NATMIRC*, BCC*, UNAM*, UCT*, and DFFE*) or with national projects (BIOTA Africa*, NAMIBGAS*, BENEFIT* (for abbreviations see end of chapter) for many years. These—mostly disciplin-specific—precursory works have laid the scientific foundation for the predecessor SPACES I project GENUS (Geochemistry and Ecology of the Namibian Upwelling System). GENUS was a contribution to the international IMBER (Integrated Marine Biosphere Research) initiative of the IGBP (International Geosphere-Biosphere Programme) and was built on the established regional research collaborations BENEFIT (Benguela Environment Fisheries Interaction and Training, 1997–2007) and BCLME (Benguela Current Large Marine Ecosystem, since 2002). Many crucial data sources used as the knowledge base for TRAFFIC were developed in GENUS and other predecessor projects. In addition to providing valuable scientific results for understanding climate-induced changes in upwelling areas and the ecosystem services associated with them, TRAFFIC has deepened the collaboration with scientists in the partner countries Namibia and South Africa.

The GENUS project (see Ekau et al. 2018) has shown that, regarding the overall net flux of carbon dioxide, the northern Benguela subsystem releases CO_2 into the atmosphere, while the southern subsystem takes up CO_2 (Emeis et al. 2018). Previously, this difference was attributed solely to different oceanographic conditions. The nutrient- and CO_2-rich South Atlantic Central Water (SACW) is a main water supply to the northern Benguela, whereas the comparatively nutrient- and CO_2-poor Eastern South Atlantic Central Water (ESACW) forms the main upwelling water in the southern subsystem (Fig. 11.1). Consequently, upwelling in the nBUS promotes the emission of CO_2 to the atmosphere on the Namibian shelf and the export of carbon from the euphotic zone to the deep ocean, readable from the formation of the carbon-rich silt layers at the seabed. In contrast, upwelling in the sBUS leads to a net uptake of CO_2 and is accompanied by lower carbon sedimentation rates (Mollenhauer et al. 2004). Differences in fishery yields and results of biological studies show that in contrast to the relatively short food chain in the sBUS, which has been considered typical of upwelling systems, the food web is more complex in the nBUS. Because primary production is very similar in the northern and southern subsystems (Barlow et al. 2009), these relationships indicate more efficient utilization within the food chain in the southern Benguela. The aim of the TRAFFIC project has been to unravel the biological processes leading to the differences in the TTE between the food webs of the two subsystems of the BUS.

TRAFFIC also relates studies of the food web to ongoing climate-related changes in the BUS. Recent findings from modeling studies have shown a poleward shift in subtropical high-pressure areas due to global climate change (Garcia-Reyes et al. 2015; Rykaczewski et al. 2015; Wang et al. 2015). As a result, the trade winds in the sBUS will likely intensify, whereas the wind speeds and upwelling intensities in the nBUS will likely weaken. Our investigation will unravel the effects of ongoing changes in physical forcing on the overall productivity and the food web structure of the Benguela Current subsystems.

Historically, it has been assumed that around 90% of material and energy is lost by metabolic activity from one trophic level to the next higher trophic level and that only 10% reaches the next level (Lindeman 1942). This value, however, can be very variable. Eddy et al. (2020) compiled data from several studies based on Ecopath with Ecosim models and calculated that the TTE ranges from 0.3% to 52.0% between trophic levels 2–3 and 3–4 with means of 12.0% in polar/subarctic-boreal regions, 9.6% in temperate regions, 8.6% in tropical/subtropical regions, and 8.0% in upwelling regions. In a warmer world, i.e., under climate change, the TTE may decrease due to higher metabolic losses. Freshwater plankton in artificial ponds that have been exposed for seven years to 4 °C warming relative to ambient conditions showed a decrease of TTE by up to 56% (Barneche et al. 2021). Projections by du Pontavice et al. (2020) also assume a decrease of TTE until 2100 under the RCP 8.5 global warming scenario, which would be associated with an increase of about 4.8 °C in global mean temperatures.

Different physical conditions affect primary production (PP) and subsequent consumers, thereby determining the efficiency with which the produced biomass is carried through the food web. Low-latitude stratified ecosystems are dominated by small phytoplankton and carbon is routed through many trophic levels (TLs) before reaching pelagic fish. The overall TTE (the transfer of energy from primary to secondary producers and higher trophic levels) is furthermore driven by the complexity of the food web (see also Armengol et al. 2019). The mean number of TLs between primary producers and fish is around 6 in oceanic, 4 in coastal, but only 2.5 in upwelling regions (Ryther 1969; Eddy et al. 2020). A short food chain generally results in a high trophic transfer efficiency, for example when large chain-forming diatoms are consumed directly by sardines, without an intermediate level of zooplankton consumers (Moloney et al. 1991; van der Lingen et al. 2006a). The timing of the development of the different components in the food web, i.e., the temporal match and mismatch of TLs (Cushing 1990), is crucial for an efficient overall TTE. However, alternative dead-end scenarios for carbon transport may exist when stochastic blooms of salps consume the entire primary production and carbon does not reach higher TLs but sinks into deeper water layers at an increased rate (Martin et al. 2017). Consequently, zooplankton composition as well as the food web structure determine the amount of carbon and energy reaching upper trophic levels such as fish, seabirds, and marine mammals, and ultimately fisheries.

From this point of view, ecosystems dominated by the zooplanktivorous (krill-dominated diet) but partly piscivorous horse mackerel (Pillar and Barange 1998; Kadila et al. 2020) will be less productive in comparison to those dominated by anchovy and/or sardines that feed on smaller (copepods) zooplankton but also phytoplankton (van der Lingen et al. 2006a). Results of the GENUS project revealed that trophic interactions and the community structure of trophic levels are not as simple and straight-forward as previously thought with Schukat et al. (2014) showing that trophic roles of calanoid copepods in the nBUS were far more complex than merely linking phytoplankton to pelagic fish.

Recent work has suggested that so-called "dead end" species (e.g., jellyfish and salps that feed on primary and/or secondary producers/consumers and were previously considered to be rarely consumed by predators) can be trophically important (Hays et al. 2018; Gibbons et al. 2021). Specifically, they can outcompete planktivorous fishes by forming intensive blooms when conditions are favorable and increase the export of organic matter to deeper layers by producing fast-sinking fecal pellets and mass mortality events. This reduces the energy available for higher trophic levels and the recycling of nutrients within the epipelagic realm.

The TRAFFIC project set out to closely investigate and compare the nBUS and the sBUS ecosystems in relation to the underlying oceanographic and biogeochemical processes. Based on the concept of three alternative structures of the food chain (Fig. 11.2) and on recent climate models that suggest an intensification of winds and upwelling in the sBUS, in contrast to a weakening of the upwelling intensity in the nBUS (Garcia-Reyes et al. 2015; Rykaczewski et al. 2015; Wang et al. 2015), TRAFFIC investigated how current conditions influence productivity, carbon export and food chain structure in the two subsystems, and hence their trophic transfer efficiency and potential to support top predators and fisheries.

11.3 Major Biological Components of the Benguela Upwelling System

For the comparison of the TTE of the northern and southern BUS it is not only crucial to get a picture of the community structure and food web complexity, but it is also necessary to identify the starting conditions such as the efficiency of primary production and to follow the energy from the base of the food web to top predator level. For this purpose, two research cruises in the TRAFFIC project have been undertaken: a first one with RV METEOR in austral summer (M153, February/March 2019) and a second one with RV SONNE at the end of austral winter (SO285, September/October 2021). During these cruises, samples were collected with different gears in order to measure physical drivers, and to quantify biomass, standing stocks and plankton and fish composition of the two ecosystems. Additionally, experiments were set up onboard the ships to measure vital rates such as primary production, respiration, growth, metabolism and grazing of various planktonic organisms.

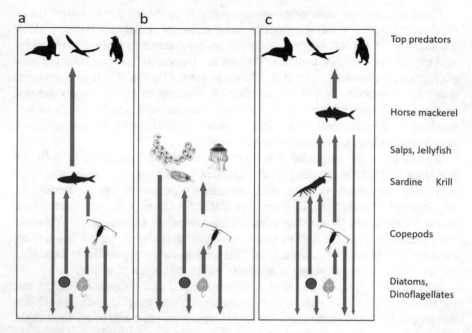

Fig. 11.2 Three alternative simplified food web structures which may establish in the Benguela Upwelling System and lead to different trophic transfer efficiencies: (**a**) Typical short food chain of an Eastern Boundary Upwelling System; highly efficient transfer of carbon to upper trophic layers (common in sBUS); (**b**) food chain structure during mass occurrences of gelatinous zooplankton (salps and jellyfish), decoupling from higher trophic levels (occasionally in nBUS and sBUS); (**c**) long food chain with less efficient overall trophic transfer, common in nBUS. Blue arrows: upward transport of energy and matter through the food chain; brown arrows: export flux of energy and matter

11.3.1 Abiotic Parameters and Chlorophyll Measurements

A high-speed remotely operated towed vehicle (ROTV, TRIAXUS) was used during cruise M153 in 2019 (see also Rixen et al. 2021) to measure temperature, salinity, oxygen content, nitrate, chlorophyll *a* (Chl *a*) and other pigments, turbidity, photosynthetic active radiation (PAR) and hydroacoustics on several transects in the nBUS and sBUS (Fig. 11.3). Zooplankton was analyzed using a mounted Video Plankton Recorder (Möller et al. 2012). The vehicle was towed at a speed of 8 knots with a horizontal offset out of the vessel's wake, undulating vertically between 5 and 180 m, depending on the water depth.

Vertical profiles of conductivity, temperature, pressure, oxygen, fluorescence, turbidity and photosynthetically active radiance (PAR) were obtained using a CTD in-situ (Fig. 11.4, left). These data were compared with satellite images of temperature and chlorophyll at the surface (Fig. 11.4, right) to provide background information for future evaluations.

Fig. 11.3 TRIAXUS on deck of RV METEOR

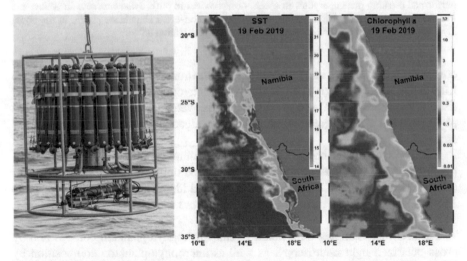

Fig. 11.4 Left: CTD with rosette; Right: Satellite data of sea surface temperature (SST) and Chl *a* during the expedition M153 in February 2019

11.3.2 Phytoplankton and Microzooplankton

The base of the food web, i.e., the primary producers and the microzooplankton were investigated to determine the quality and quantity of food available for higher trophic levels. For this purpose, water was taken from different depths using a Niskin bottle rosette attached to the CTD and filtered to determine the Chl *a*

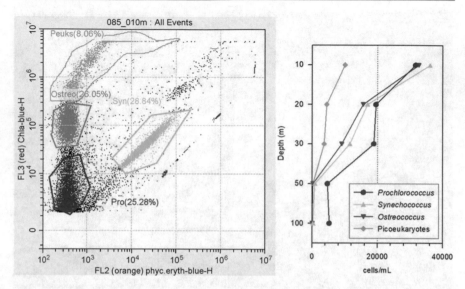

Fig. 11.5 Left: Bivariate plot of red fluorescence (FL3) vs. orange fluorescence (FL2) for acquired flow cytometry at 10 m depth, station 85 (30°S, 16°30′E). Preliminary gating was performed marking picoeukaryotes in green, *Ostreococcus* in pink, *Synechococcus* in yellow and *Prochlorococcus* in blue. Right: From flow cytometry converted abundance profile in the upper layer (10–100 m) of the main picophytoplankton groups at station 85

content by spectrometry and the pigment composition by HPLC (High-Performance Liquid Chromatography). The results will be intercompared with remotely sensed data. A fluoroprobe was used in addition to the water samples to analyze different phytoplankton groups (green-algae, blue-green algae, diatoms, and cryptophyta) *in situ*.

In order to assess the contribution and the community composition of picophytoplankton (0.2–2 µm) water samples were analyzed onboard using a flow cytometer (CytoFLEX, Beckman Coulter) for counting and identifying the main groups like *Prochlorococcus*, *Synechococcus,* and *Ostreococcus* as well as picoeukaryotes (Fig. 11.5).

Further water samples were taken to study nanophytoplankton composition by cross polarized light microscopy, as well as microphytoplankton composition by inverse microscopy in the home laboratories and to assess the photosynthetic fitness of the phytoplankton onboard the ship using the Fast Repetition Rate Fluorometry (FRRF, Fasttracka II, Chelsea Technology, UK). FRRF is a noninvasive method to measure the activity of primary producers using Chl *a* fluorescence (Oxborough et al. 2012). Small plankton were caught with an Apstein net (20 µm mesh size) to study the trophic positions and nutritional quality of phyto- and microplankton applying stable isotope and fatty acid analyses. The taxonomical composition was determined using fluid imaging (FlowCam). The FlowCam takes pictures of the organisms found in a sample (Fig. 11.6), which can be analyzed subsequently by

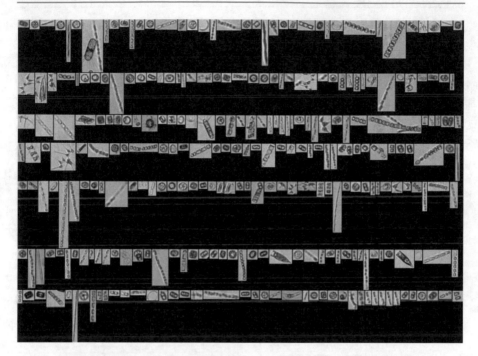

Fig. 11.6 FlowCam images from a diatom bloom in the sBUS during cruise SO285

a deep learning program which has been trained using plankton samples from the Benguela ecosystem.

Marine life does not react immediately following upwelling events but rather with certain time lags. Initially, abundance and productivity are low in recently upwelled and highly turbulent waters (Ayon et al. 2008; Ekau et al. 2018), because upwelling water originates from the central water layer below the thermocline, where the phytoplankton stock is low. Furthermore, strong turbulence inhibits phytoplankton growth by transporting the organisms out of the range of photosynthetic active radiance. Production peaks in moderate upwelling and in quiescent phases after upwelling events (Grote et al. 2007; Bode et al. 2014), forming an optimal environmental window (Cury and Roy 1989). The development of a diatom-dominated phytoplankton bloom in the nutrient-rich upwelling plume and a community succession from diatoms to flagellates requires time to respond to upwelling conditions. The increase of phytoplankton biomass by an order of magnitude takes approximately two weeks (Hansen et al. 2014). A mixed population of dinoflagellates, coccolithophores, and microflagellates was detected on cross-shelf transects off Walvis Bay in newly upwelled waters (<13 days old) close to the coast. In contrast, diatoms dominated maturing waters (13–55 days old) 40 to 250 km off the coast, whereas dinoflagellates prevailed in waters older than 55 days after the upwelling event.

Fig. 11.7 Multinet Midi used to sample microzooplankton

Primary production and respiration rates were measured by incubation of water from the most productive layer (Deep Chlorophyll Maximum) at in-situ temperatures under different light conditions in a plankton-wheel.

In marine ecosystems, generally 60% to 70% of primary production is consumed by microzooplankton and 10% to 40% by mesozooplankton (Calbet 2001; Calbet and Landry 2004), with microzooplankton also being an important dietary component of mesozooplankton (Bollens and Landry 2000; Calbet and Saiz 2005). Phytoplankton growth rates and microzooplankton grazing were studied using the dilution method after Landry (1993). Landry (1993) postulated higher algae growth in water with less microzooplankton predators and undisturbed growth in the highest dilution. The experiment concomitantly gives information about the grazing activity of the microplankton.

Microzooplankton for taxonomic and stable isotopic analyses were collected with a HydroBios Midi (mouth area 0.25 m²) Multinet (multiple opening/closing net) system equipped with five nets (55 μm mesh size) in discrete depth intervals (Fig. 11.7). Vertical hauls were conducted with hauling speed of 0.2 m/s from 100 m depth up to the surface.

Microzooplankton (< 200 μm, sampled with 55 μm meshed nets) of the nBUS is often dominated by mixotrophic and heterotrophic dinoflagellates, tintinnids and small copepods (Bohata 2016). Figure 11.8 shows some of the microzooplankton organisms sampled during cruise M153.

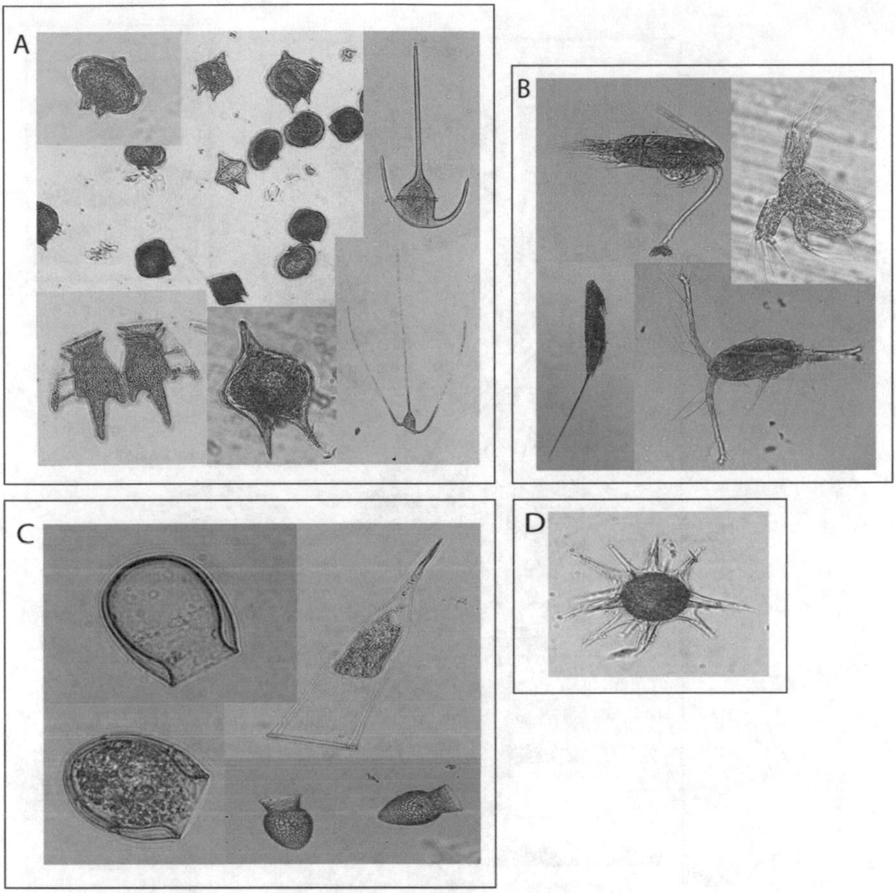

Fig. 11.8 Four important microplankton groups: (**a**) Mixo- und heterotrophic Dinoflagellata, (**b**) small Copepoda and nauplia, (**c**) Tintinnida, (**d**) Radiolaria

The microplankton distribution patterns revealed a shelf—offshore zonation and clear temperature associations (Fig. 11.9). Heterotrophic dinoflagellates such as *Protoperidinium* and *Noctiluca scintillans* prevailed in <15 °C cold, recently upwelled water on the shelf, whereas subsequent succession stages in 15–20°C warm surface water on the shelf were dominated by small copepods such as *Oncaea*, *Oithona* and *Microsetella*. *Protoperidinium*, Tintinnidae and the mixotrophic dinoflagellate *Ceratium* were abundant in decreasing order in >20 °C warm surface water at the shelf break. Tintinnidae contributed >37% to microzooplankton at the medium-warm shelf break, followed by *Oncaea*, *Microsetella* and *Protoperidinium*. The cold water and shelf break areas were dominated by *Oncaea*, followed by *Protoperidinium* and *Ceratium*. The warm offshore region was dominated by Tintinnidae comprising >30% of total abundance. Mixotrophic (*Ceratium*) and heterotrophic (*Protoperidinium*) dinoflagellates were also very abundant here,

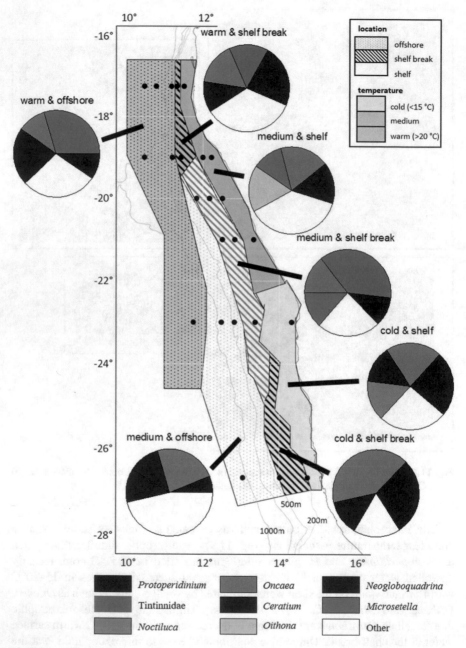

Fig. 11.9 Dominant microplankton taxa in different habitat zones of the northern Benguela Upwelling System during September/October 2011 (modified after Ekau et al. 2018)

collectively making up >30% of total abundance. The medium-warm offshore area was characterized by high abundances of *Oncaea* (> 23%) and the Foraminifera *Neogloboquadrina* (>14%) (see also Bohata 2016).

11.3.3 Mesozooplankton

Mesozooplankton plays a key role in the energy transfer from primary production and microzooplankton to higher trophic levels. Due to their short generation times and direct coupling to physical drivers, zooplankton reacts sensitively to climatic change and can be used as indicators of environmental change (Hays et al. 2005). Cyclopoid and calanoid copepods usually dominate the mesozooplankton communities (on average 70%–85%) in the nBUS and sBUS, playing a key role in sustaining marine fish stocks as a principal food source for larvae, juveniles and adults and sometimes all three stages (Hansen et al. 2005; Bode et al. 2014; Verheye et al. 2016). Furthermore, Bivalvia larvae can be sporadically dominant at near-shore regions, while Appendicularia (*Oikopleura*), Thaliacea (Doliolida, Salpida), Amphipoda and Euphausiacea can contribute substantially to abundance and/or biomass further offshore. Along the continental slope in the nBUS, the krill species *Euphausia hanseni* plays an important role in the active carbon flux from the productive shelf to the adjacent open ocean and into the deep sea because of its pronounced diel vertical migration (Werner and Buchholz 2013).

Mesozooplankton sampling consisted of vertical hauls with a multiple opening/closing net (HydroBios Multinet Midi, five nets, 0.25 m² mouth opening, 200 μm mesh size) at 0.5 m/s hauling speed. Samples were taken from discrete depth layers down to ~10 m above sea floor (minimum bottom depth 55 m at inshore stations, maximum sampling depth 1500 m offshore). Additional material of larger and more mobile species (krill, decapods) were collected from double oblique hauls of a Multinet Maxi (0.5 m² opening area, five nets of 300 μm mesh size; HydroBios). All samples were presorted onboard and potential key zooplankton and other species of the food web were deep-frozen at −80 °C for trophic biomarker analyses (stable isotopes and fatty acids). Key copepod species (Fig. 11.10) were selected from the net samples for in-situ experiments to measure respiration, egg production and fecal pellet production.

During periods of active upwelling, the BUS zooplankton communities on the shelf are dominated by the biomass-rich herbivorous-omnivorous copepod *Calanoides natalis* (ex *C. carinatus*; Bode et al. 2014) and small calanoid (esp. *Acartia, Clausocalanus, Ctenocalanus, Paracalanus, Calocalanus* spp.) and cyclopoid (esp. *Oithona* spp.) copepods (Verheye et al. 2016; Bode-Dalby et al. 2022). With increasing bottom depth closer to the continental slope, *Centropages brachiatus* and *Metridia lucens* occur at higher abundances. Further offshore, the copepod community is more diverse, due to mixing of cold- and warm-water species as well as deeper-dwelling diel vertical migrants such as *Pleuromamma* spp. When upwelling ceases and warmer water masses intrude/expand onto the shelf, the shelf community is replaced by medium-sized copepod species such as *Nannocalanus*

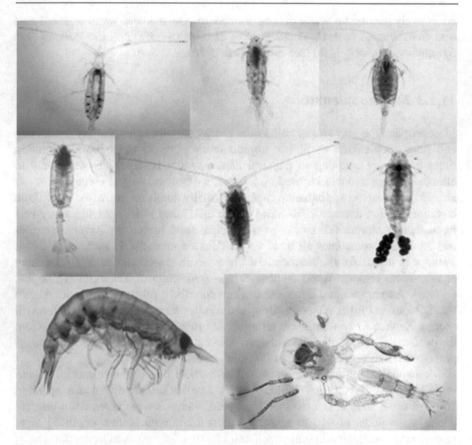

Fig. 11.10 Copepods and amphipods in the Benguela Upwelling System. Top row: *Pareucalanus sewelli, Pleuromamma quadrungulata, Candacia* sp.; second row: *Euchaeta* sp., *Gaetanus pileatus, Euchirella similis* female with two egg sacs; bottom row: *Vibilia armata, Phronima* sp.

minor (Schukat et al. 2013, 2014; Bode et al. 2014). A striking difference between the northern and southern Benguela copepod community is the absence of *Calanus agulhensis* in the nBUS, as it is apparently advected offshore and away from the nBUS by the strong and permanent Lüderitz upwelling cell. So far, *C. agulhensis* has not been recorded in the nBUS. *C. agulhensis* is the most abundant calanoid copepod on the Agulhas Bank, a major spawning ground for sardines and anchovies in the sBUS (Peterson et al. 1992; Huggett and Richardson 2000; Richardson et al. 2003). Later/older copepodid stages and adults of *C. agulhensis* are advected to the South African west coast, where its abundance is associated with warmer offshore waters (Huggett and Richardson 2000; Huggett et al. 2007).

Phytoplankton growth after an upwelling event is followed by increasing copepod abundance about 20–23 days after initial upwelling (Postel et al. 1995; Hutchings et al. 2006). In contrast to the sBUS, the seasonal signal in the nBUS is often diffuse with high interannual variability. Upwelling events, as well as zooplankton abundances, can be strongly pulsed with huge local and interannual variability and multiple interacting factors. Long-term data series derived from the Namibian monitoring program emphasize the complex interannual variability in the nBUS, where years with intense upwelling in spring can be followed by strong warm-water intrusions of tropical Angola Current water masses in late summer (Bode et al. 2014; Martin et al. 2015). Such years with strong seasonal temperature gradients were characterized by high copepod abundances suggesting a strong link between zooplankton distribution and physical forcing (Bode et al. 2014).

In the sBUS, zooplankton abundance was positively correlated to upwelling intensity (Verheye et al. 1998), although seasonal cycles of mesozooplankton differ depending on the subregion (Verheye et al. 2016). For instance, around St. Helena Bay (32–33°S), mesozooplankton populations usually peak during late summer and show a distinct decline in autumn. On the western Agulhas Bank (35°S), on the other hand, maximum mesozooplankton abundance usually occurs during late autumn and spring. Differences in mesozooplankton abundances and community structure are not only caused by bottom-up mechanisms such as upwelling intensities and phyto-plankton availability, but also by the distribution patterns of different planktivorous fish ("small pelagics") and their life-history stages along the southwestern African coast (Verheye et al. 1998; Hutchings et al. 2006; van der Lingen et al. 2006b). Furthermore, stochastic mass occurrences ("blooms") of gelatinous zooplankton such as salps or jellyfish can eliminate other plankton and reset the succession of the entire pelagic community (Martin et al. 2017).

Besides predator-prey interactions and food web structure, the availability of dissolved oxygen and the vertical extent of the oxygen minimum zone (OMZ) strongly determines the distribution of zooplankton in the BUS, especially in the nBUS (Auel and Verheye 2007; Ekau et al. 2010). Such OMZs occur regularly on the Namibian shelf in the nBUS (Schmidt and Eggert 2016) and around St. Helena Bay in the sBUS (Pitcher et al. 2014), yet, at different extents. The specific conditions on the Namibian shelf are favorable for benthic sulfur bacteria, which may form thick mats and, during occasional anoxic conditions, cause hydrogen sulfide eruptions (Schmidt and Eggert 2016). Many zooplankton species can cope with the upwelling-driven, highly pulsed productivity regime, strong advective processes and the regionally pronounced OMZs. The dominant copepod *C. natalis* is well adapted to the highly dynamic upwelling regime with its reproductive strategy, lipid accumulation, ontogenetic vertical migration, and dormant phase (diapause) of C5 copepodids at depth (Auel et al. 2005; Verheye et al. 2005; Auel and Verheye 2007; Schukat et al. 2013, 2014; Bode et al. 2015). Females of *C. natalis* and other species such as *M. lucens* or *C. agulhensis* maintain their populations in the productive shelf region through diel vertical migration (DVM) between surface currents and subsurface counter-currents (Timonin 1997; Huggett and Richardson 2000; Loick et al. 2005). Species that can retreat into OMZs for at least part of the day have various advantages, e.g., finding refuge from predators (Loick et al. 2005). The extent of DVM can also be adapted to upwelling intensities and food availability: *C. natalis* and *C. agulhensis* showed very pronounced DVM during periods of increased advection and high Chlorophyll *a* concentrations, whereas DVM was reduced during quiescence of upwelling and low phytoplankton concentrations (Verheye and Field 1992; Huggett and Richardson 2000).

Since the 1950s, long-term changes of zooplankton abundance and biomass have been observed in the entire BUS (Verheye and Richardson 1998; Verheye 2000; Huggett et al. 2009; Hutchings et al. 2009; Verheye et al. 2016). Around Walvis Bay (23°S), copepod abundances increased six-fold from the 1980s to the early 2000s, followed by a decline after 2005 (Hutchings et al. 2009). In contrast to the nBUS, the sBUS has been studied more regularly in terms of zooplankton abundance and community structure making long-term assessments more reliable (reviewed by Verheye et al. 2016). A 100-fold increase in total copepod abundance (cyclopoids and calanoids) was reported for the sBUS between 1950 and 1995 (Verheye et al. 1998). Between 1988 and 2003 copepod biomass and production along the entire sBUS coast were around one order of magnitude higher than in the late 1970s (Huggett et al. 2009). This long-term increase in copepod abundance was accompanied by increasing wind stress and upwelling intensities (Shannon et al. 1992; Verheye 2000), and it also coincided with the onset of commercial fishing since the 1950s (Verheye et al. 2016). Since the mid-1990s, copepod abundance decreased again slightly; thus, the decline in copepod abundance started one decade earlier than in the nBUS (Verheye et al. 2016).

In both subsystems, there has also been a size shift in the mesozooplankton communities from larger to smaller species during the last decades (Verheye and Richardson 1998; Verheye et al. 2016). In the 1950s, euphausiids (esp. *Euphausia lucens* and *Nyctiphanes capensis*) and large to medium-sized copepods such as *C. natalis*, *R. nasutus,* and *C. brachiatus* prevailed in the species composition of St. Helena Bay. From the late 1980s onward, smaller copepod species such as "small calanoids" (mostly Clausocalanidae and Paracalanidae) and the cyclopoid *Oithona* spp. became clearly dominant (Hutchings et al. 2012; Verheye et al. 2016). The shift from larger to smaller species can be an indicator of ocean warming, whereas a cooling trend by up to 0.5 °C per decade has been evident from the 1980s onward due to intensification of upwelling in this region (Rouault et al. 2010; Verheye et al. 2016). Both oceanographic and biological processes (bottom-up control) together with predation effects (top-down control), particularly size-selective feeding of sardines and anchovies, seemed to cause these changes in the zooplankton communities (Verheye et al. 1998; Verheye and Richardson 1998; Hutchings et al. 2012). The decline of larger copepods in the St. Helena Bay region since the mid-1990s coincided with a marked increase in biomass of small pelagic fish such as anchovy, which potentially prey on these copepods (Hutchings et al. 2012; Verheye et al. 2016). In the nBUS, no clear predator-prey relationships between zooplankton and fish have been identified to date. After the decline of anchovies in the mid-1990s no increase of larger copepods was detected (Verheye et al. 2016). Hence, the relative importance of bottom-up vs. top-down effects remains uncertain, but it is clear that such changes in zooplankton have fundamental effects on biogeochemical processes, food web structure and thus ecosystem functioning and services.

Due to considerable interannual variability and different patterns in various sub-regions of the BUS, trends in one region and season cannot be extrapolated to other regions in the BUS, emphasizing the need for high spatial, seasonal, and taxonomic coverage of continuous monitoring programs (Huggett et al. 2009; Kirkman et al. 2016; Verheye et al. 2016). Thus far, the discontinuous and heterogeneous nature and the relatively poor data of the BUS compared to time series from other systems do not allow far-reaching conclusions about the synchronicity of fluctuations of zooplankton biomass and abundance at spatial scales similar to those found for fish species (Batchelder et al. 2012). This emphasizes the need for appropriate and concrete actions proposed by the Benguela Current Commission to advance sustainable development of the BUS goods and services (Verheye et al. 2016). Characterizing zooplankton communities by functional types and not only focusing on large species will help improving predictive biogeochemical and ecosystem models. The community structure of the small calanoid copepods in the BUS has not been well distinguished so far (Bode-Dalby et al. 2022). There are contradictory and uncertain mentions of *Microcalanus* (=*Clausocalanus*?) and *Pseudocalanus* (=*Ctenocalanus*?) spp. (Verheye et al. 2016); thus, it is not known how diversity of small copepods and their functional role has changed over the last decades.

11.3.4 Macrozooplankton and Micronekton

To investigate the trophic transfer efficiency in midwater ecosystem components, biomass size spectra comprising all major taxa encountered, i.e., fish, crustacean and gelatinous plankton (see Fock and Czudaj 2019) were analyzed, as well as diurnal feeding patterns and food composition of key fish species in combination with stable isotope ratios for selected fish species, medusa and crustacean plankton and other micronekton.

The vertical distributions of macrozooplankton and micronekton were ana-lyzed using depth-stratified net catches (Multinet-maxi, Multinet-midi, Rectangular Midwater Trawl RMT, Figs. 11.7, 11.11) as well as horizontal surface sampling (Neuston catamaran). These were used to assess abundance, biomass, and species composition, and also to gain information about the behavior of fish larvae in relation to hydrography on meso-spatial scales and at high vertical resolution. Fishes were captured mostly at night in double oblique hauls down to 500 m depth.

In order to be able to construct normalized biomass size spectra (NBSS), samples were analyzed using digital imaging tools such as the ZooScan (Fig. 11.12). Similar to the FlowCam method for phytoplankton and microzooplankton, organisms like krill, chaetognaths, and fish larvae were scanned and measured digitally. Thus, size and volume could be calculated in addition to taxonomic classification and abundance. The age, RNA/DNA ratios, fatty acid composition and C/N isotope content of commercially important fish larvae (mostly *Trachurus capensis*, *Sardinops sagax*, and *Engraulis encrasicolus*) were analyzed to indicate their fitness and condition.

Fig. 11.11 Mesopelagic fishes, large crustaceans and gelatinous plankton were collected using a Rectangular Midwater Trawl (RMT 8)

A Kongsberg EK80 hydroacoustic system was used on the first cruise (M153) to detect fish and biomass aggregations of smaller nekton. A configuration using a frequency of 38 kHz and a long pulse duration of 1.024 ms allowed the detection of biomass down to 750 m depth (Fig. 11.13). Smaller particles could be detected using a 200 kHz transducer, but only to a depth of about 150 m. By continuous activation of the echosounder, the hydroacoustic systems were able to document the vertical distribution of biomass, diel vertical migration and behavioral changes of the spatial distributions of organisms like zooplankton and fish. In this way, the mesopelagic zone could be monitored continuously, which led to the first documentation of deep-scattering layers (DSL) in the Benguela ecosystem between 300 m and 600 m depth. The intense diel vertical migration between the DSL and the

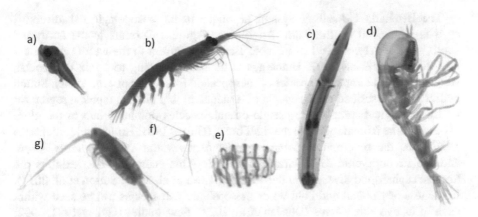

Fig. 11.12 Zooscan pictures of (**a**) *Trachurus capensis* larvae, (**b**) Euphausiacea, (**c**) Chaeto-gnatha, (**d**) Amphipoda: *Paraphronima* sp., (**e**) Salpidae, (**f**) Pluteus larvae, (**g**) Copepoda

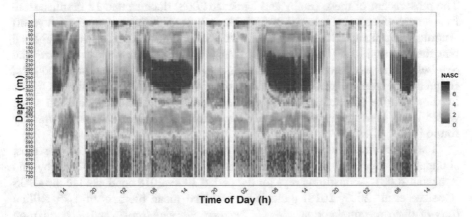

Fig. 11.13 Diel vertical migration from the Deep-Scattering Layer over a 72 h cycle at Station 18, sBUS, recorded by the EK80 during M153. Nautical Area Scattering Coefficient (NASC, as log m^2 nm^{-2}) was calculated over 10 m depth bins and 10 min intervals and serves as a proxy for biomass

surface layers, dominated by certain euphausiids and mesopelagic fishes, shows the strong connection between the deep sea and the productive euphotic zone (see Fig. 11.13). During the cruise SO285, a more powerful hydroacoustic system was used. This EK60 configuration with four frequencies (18, 38, 120 and 200 kHz) allows for the distinction between acoustic response curves, or echoes, of different taxonomic groups and can help to distinguish between acoustic biomass of jellyfish, krill and different species of fishes. To further identify the origin of the acoustic biomass, rectangular midwater trawls were carried out in specific layers of high biomass. These reference hauls can then be compared to the acoustic signals of these layers to associate biomass with certain species or groups.

The Benguela Upwelling System is unique in its mesopelagic fish diversity, as it is influenced by the warm Angola and Agulhas Currents to the north and south, respectively, as well as the cold Benguela Current to the west (Duncombe-Rae 2004; Lett et al. 2007; Hutchings et al. 2009), leading to a mix of tropical, subtropical, and temperate species of mesopelagic fish (Duncan et al. 2022). Sutton et al. (2017) described the Benguela as a unique global biogeographical region for its mesopelagic fauna, where pseudo-oceanic species dominate due to the close distance of the frontal region to the shelf break (Hulley 1981; Hulley and Lutjeharms 1989), i.e., the myctophids *Lampanyctodes hectoris* and *Symbolophorus boops,* and the sternoptychid *Maurolicus walvisensis*. This ecoregion also exhibits the highest cephalopod diversity in the Atlantic (Rosa et al. 2008; Sutton et al. 2017). Influences of tropical and cold-water mesopelagic fish species can be seen within each of its two subsystems (Duncan et al. 2022). Few studies (Hulley 1981, 1992; Rubiés 1985; Hulley and Prosch 1987; Hulley and Lutjeharms 1989; Armstrong and Prosch 1991) have investigated the mesopelagic fish community in the Benguela. The most recent of these (Staby and Krakstad 2008) documented 18 families from five orders of mesopelagic fish using data collected during research surveys off Angola, Namibia and South Africa over the period 1985–2005. This study reported that the Myctophidae (*Diaphus* spp. off Angola and *Lampanyctodes hectoris* and *Symbolophorus boops* off Namibia and South Africa) prevailed and occurred most frequently in research trawls particularly over the shelf and slope, followed by the Sternoptychidae (predominantly *Maurolicus walvisensis*; also in shelf and slope waters), and then the Phosichthyidae (*Phosichthys argenteus*) that were typically found further offshore.

Lampanyctodes hectoris and *Maurolicus walvisensis* appear to be the most abundant mesopelagic fishes in the region (Hulley and Prosch 1987; Staby and Krakstad 2008). Acoustic surveys for the period 2006–2018 conducted in the sBUS (Coetzee et al. 2009, 2018) indicate a combined mean biomass of 1.25 million tons of three mesopelagic species (*L. hectoris*, *M. walvisensis* and *S. boops*), on average split between *M. walvisensis* and the myctophids 2:1. Dense aggregations were mostly found between 31°S and 35°S. As compared to mesopelagic biomass estimates for the Humboldt Current System, a biomass of 2–11 million tons for one single species (*Vinciguerria lucetia*) was estimated beyond the shelf break along the whole coast off Peru (see Cornejo and Koppelmann 2006). Similar to the biomass of small pelagics, the biomass of mesopelagic fish appears to be low in relation to system primary production.

Most studies have focused on *L. hectoris* and *M. walvisensis* (Hulley 1981; 1992; Rubiés 1985; Hulley and Prosch 1987; Hulley and Lutjeharms 1989), and also on mesopelagic fish larvae (Olivar 1987; Olivar and Beckley 1994; Ekau and Verheye 2005). However, comparative studies of mesopelagic fish communities between these two dynamic upwelling systems as well as biological and ecological studies are lacking, especially including species of mesopelagic families such as Gonostomatidae, Stomiidae, Phosichthyidae, Bathylagidae, and Melamphaidae, among others (Staby and Krakstad 2008).

Data collected during the TRAFFIC research surveys show that the overall abundance of mesopelagic fish did not differ between the northern (north of Walvis Bay) and southern Benguela subsystems, but species accumulation curves indicate that the nBUS has a higher mesopelagic fish richness (Duncan et al. 2022). However, there is high heterogeneity within each subsystem, which demonstrates the need for increased sampling of these organisms at appropriate spatial scales. Species counts revealed 88 mesopelagic species for the TRAFFIC campaign so far, as compared to 131 species listed in the South African *Africana* data base and 98 listed in the BENEFIT project (Staby and Krakstad 2008). Seven mesopelagic fish communities have been identified on the shelf and slope of the Benguela system. These include one shelf group in each of the nBUS and sBUS, as well as several offshore groups (Fig. 11.14). The shelf group in the sBUS has low diversity and is dominated by *Maurolicus walvisensis,* which is a shelf/slope-associated species (Hulley and Prosch 1987; Prosch 1991) and corroborates the findings of Coetzee et al. (2009). In contrast, the shelf of the nBUS had very low abundance of mesopelagic fishes and is dominated by gobies and jellyfishes (Roux et al. 2013; unpublished data). The only species that defined the shelf ecosystem assemblage in the nBUS was the myctophid

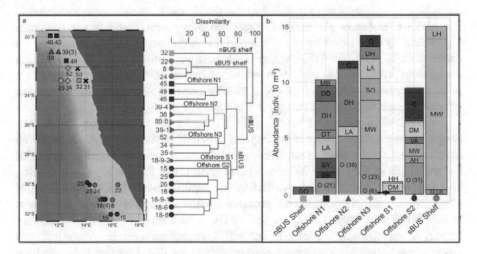

Fig. 11.14 Modified figure from Duncan et al. (2022) showing (**a**) Station map (left) with clusters of mesopelagic fish communities based on hierarchical cluster analysis (right) using the average linkage method on Bray-Curtis dissimilarity matrix for Hellinger-transformed abundance data. Panel (**b**) reflects the abundance of dominant species in each cluster, corrected by the number of stations representative of each cluster. Those species listed contribute at least 5% to the fraction of mesopelagic fishes and those contributing less have been combined to the category "other" with the total number of species contributing to "other" in parenthesis. Abbreviations stand for *DD Diaphus dumerilii, DH Diaphus hudsoni, DM Diaphus meadi, DT Diaphus taaningi, HH Hygophum hanseni, LH Lampanyctodes hectoris, LA Lampanyctus australis, SY Symbolophorus barnardi, SO Symbolophorus boops* (all Myctophidae), *MB Melanolagus bericoides* (Bathylagidae), *SB Stomias boa* (Stomiidae), *C Cyclothone* spp. (Gonostomatidae), *AH Argyropelecus hemigymnus, MW Maurolicus walvisensis* (both Sternoptychidae), *VA Vinciguerria attenuata* (Phosichthyidae), and *O* other unspecified fishes

Diaphus dumerilii. This species has a tropical distribution with populations along the equatorial Atlantic, both off the coast of Brazil and in the eastern Atlantic (Hulley 1981; Czudaj et al. 2021). Offshore stations in the sBUS were dominated by gonostomatid *Cyclothone* spp. as well as the myctophid *Diaphus meadi*, with the exception of one community, which was dominated by the myctophid *Hygophum hanseni*, one of the most abundant myctophids circumglobally, in the area of the Subtropical Convergence (Hulley 1981 and references cited therein). In the nBUS, offshore stations consist of three communities, two of which are dominated by the myctophid *Diaphus hudsoni*, and one where *M. walvisensis* prevails.

Environmental factors that drive mesopelagic fish species composition in the Benguela are water masses, surface chlorophyll *a* and oxygen concentrations. As the nBUS is more influenced by SACW and the Angola Current, species that are classified as having tropical patterns can be found, such as *Diaphus dumerilii* and *Diaphus taaningi* (Hulley 1981), while the sBUS has a higher abundance of fishes with temperate and convergence patterns (Hulley 1981), such as *D. meadi*. Hulley (1992) also found that the fishing depth and temperature influenced the downslope distribution of species. As tropical ecosystems tend to have a higher diversity of organisms than temperate regions, this is also reflected in the order Stomiiformes, for which 15 species were identified in the sBUS, while species were identified in the nBUS (Duncan et al. 2022). Overall, the subsystems found in the Benguela Upwelling System show mesopelagic fish diversity, that is highly influenced by differences in water masses, oxygen concentrations and currents. However, data on seasonality of mesopelagic fish communities are still lacking and there is a need for further investigation in order to fully assess the diversity and abundance of mesopelagic fishes in the Benguela Current.

Normalized biomass size spectra (Fig. 11.15) for the micronekton and macro-zooplankton offshore components beyond the shelf break front, both indicate an exchange across the shelf break front, given the differences in slopes in particular for the nBUS as well as the influence of the shelf break Benguela jet, transporting anchovy larvae downstream spawned in the Agulhas Bank region. The latter is supported by the significant fish group normalized biomass with a body mass below 0.1 g wet mass (WM) in the sBUS, which mostly comprises anchovy larvae, *E. encrasicolus*. In turn, in the nBUS, euphausiids are dominating in this biomass range, mainly *Euphausia hanseni*, which are dependent on the upwelling regime. Given that TTE model parameters are the same for the two closely related subsystems, TTE is lower for the nBUS, i.e., the NBSS slope is steeper (-1.37) as compared to the sBUS (-1.07, Fig. 11.15).

11.3.5 Higher Trophic Levels

The Benguela ecosystem hosts the species spectrum typical of upwelling systems and supports multiple species in higher trophic levels, including crustaceans, cephalopods, fishes, marine mammals and seabirds. A total of 133 fish species from 40 families are listed for the Benguela Current LME (www.Seararoundus.com).

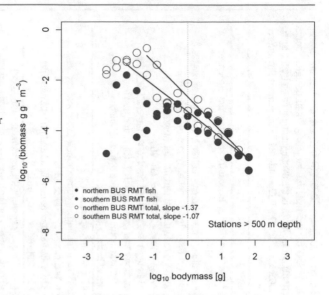

Fig. 11.15 Normalized biomass size spectra for the northern and southern Benguela subsystems offshore micronekton and macrozooplankton from RMT8 samples obtained during the M153 late summer cruise, March 2019. "Total" refers to total community biomass, i.e. fish plus macroplankton

Some of these species (e.g., small pelagic fishes) can attain high biomass levels, and several are commercially important as fisheries resources including the Cape hakes (*Merluccius capensis* and *M. paradoxus*), Cape and Cunene horse mackerels (*Trachurus capensis* and *T. trecae*), and small pelagics (sardine *Sardinops sagax*, anchovy *Engraulis encrasicolus* and round herring *Etrumeus whiteheadi*). Substantially smaller catches of crustaceans such as rock lobster (*Jasus lalandi*) and deep-water crab (*Chaceon* sp.) are also commercially important (van der Lingen et al. 2006b; Kirkman et al. 2016; Kainge et al. 2020). Additionally, high biomasses of as yet not commercially targeted mesopelagic fishes, gobies and jellyfish occur in the Benguela, the latter two in the nBUS in particular (Lynam et al. 2006; Roux et al. 2013; Kirkman et al. 2016; Salvanes et al. 2018). The southern part of the Benguela—from Luederitz to Cape Agulhas—provides nursery grounds for most of that subsystem's ecologically and economically important fish species (Kirkman et al. 2016), including both hake species, the small pelagics, and horse mackerel.

11.3.6 Commercial Fishery

Important fisheries (in terms of economic value) in the Benguela are those for Cape hakes (most valuable) and horse mackerels (largest volume) in the nBUS, and Cape hakes (most valuable) and small pelagics (largest volume) in the sBUS (Kainge et al. 2020). The fishing gear, product utilization and markets, average catches, and management strategies of these fisheries are summarized in Table 11.1, and catch time-series shown in Fig. 11.16.

Historically, sardine dominated landings from 1950 to the late-1960s in both subsystems, with peak catches of >1 million tons in the nBUS and close to ½

Table 11.1 The fishing gear, species, product utilization, average annual catches (rounded to the nearest 100 tons; 2000–2016; ± std. dev.), management strategies, and present stock status (Kainge et al. (2020) for Namibia and DEFF (2020) for South Africa) for the fisheries targeting small pelagics, horse mackerel and Cape hakes off Namibia (the nBUS) and South Africa (the sBUS and Agulhas Bank system). Also shown is the average annual catch by the three fisheries combined, and the % contribution by the marine fisheries sector to Namibia's and South Africa's GDP. *OMP* Operational Management Procedure, *PUCL* Precautionary Upper Catch Limit, *TAC* Total Allowable Catch

Fishery	nBUS	sBUS
Small pelagics (purse-seine; by-catch in midwater trawl fishery)	• *Sardinops sagax, Engraulis encrasicolus, Etrumeus whiteheadi* • Sardine frozen and canned for human consumption (export and local); anchovy and round herring reduced to fish meal and fish oil (export) • Average catch: 22,300 ± 12,200 • Management = assessment and annual TAC (only for sardine) • Present status = sardine heavily depleted; fishery closed (under moratorium) from 2018 onward;	• *Sardinops sagax, Engraulis encrasicolus, Etrumeus whiteheadi* • Sardine frozen and canned for human consumption and pet food (local); anchovy and round herring reduced to fish meal and fish oil (export) • Avg. catch: 431,900 ± 110,900 • Management = OMP used to set annual TACs for anchovy and sardine; PUCL for round herring • Present status = sardine depleted; anchovy optimal; round herring abundant
Horse mackerel (midwater trawl and demersal trawl; juveniles taken by the small pelagic fishery)	• *Trachurus capensis* (dominant). *T. trecae* • Frozen for human consumption (local and export) • Avg. catch: 279,800 ± 55,000 • Management = assessment and annual TAC • Present status = biologically sustainable	• *Trachurus capensis* • Frozen for human consumption (local and export) • Avg. catch: 29,800 ± 8300 • Management = annual TAC and effort limitation • Present status = optimal
Cape hakes (demersal trawl; long-line)	• *Merluccius capensis, M. paradoxus* • Fresh and frozen for human consumption (mostly export to EU markets) • Avg. catch: 144,100 ± 18,000 • Management = assessment and annual TAC • Present status = overfished	• *Merluccius capensis, M. paradoxus* (dominant) • Fresh and frozen for human consumption (local and export) • Avg. catch: 138,500 ± 13,700 • Management = OMP used to set annual TAC (both species combined) • Present status = *M. capensis* abundant; *M. paradoxus* optimal
Combined average catch 2000–2016	446,200 ± 69,800	600,200 ± 120,300
Fisheries GDP contribution	3.6%	<1.0%

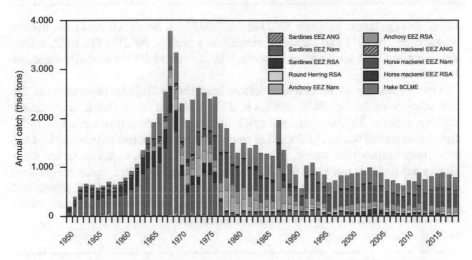

Fig. 11.16 Time series of catches of sardine, round herring, anchovy, horse mackerel and Cape hakes in the EEZs of Angola, Namibia and South Africa (RSA), 1950–2019. Catches for Cape hakes and horse mackerels off Namibia updated from Kainge et al. (2020) and for small pelagics from FAO (Namibian catches were included in South African catches from 1950 to 1989 and only specifically recorded post-independence (1990). Earlier catches are estimated by subtracting the South African catches (see below) from FAO catches. Catches for Cape hakes, horse mackerels and small pelagics off South Africa updated from DFFE (2020) and from FAO

million in the sBUS (Fig. 11.16), but catches of this species then declined rapidly off both Namibia and South Africa due to overfishing (van der Lingen et al. 2006b; Augustyn et al. 2018). Anchovy replaced sardine in South African catches for the next three decades, and sardine catches increased briefly before declining again to recent depleted levels (DEFF 2020) arising from prolonged poor recruitment and increased fishing mortality (Augustyn et al. 2018). However, despite changes in species dominance, and including a small contribution made by round herring, catches by the South African small pelagic fishery have been relatively stable (Fig. 11.16).

The Namibian sardine fishery has essentially been replaced by the fishery for horse mackerels (*Trachurus capensis* and *T. trecae*), with catches of the latter increasing rapidly from the early 1970s to a peak of almost 660,000 t in 1983. Since then this has been Namibia's major fishery in terms of volume despite a drop from peak catches to around 300,000 t annually since the mid-1990s (excluding 1997 when a catch of only 125,000 t was taken). Substantially smaller (by almost an order of magnitude) quantities of horse mackerel (*T. capensis* only) are caught off South Africa, mostly from the Agulhas Bank off the south coast. The Namibian fishery for hake started in the early 1960s with catches increasing rapidly to 820,000 t in 1972 owing to poor control and increased fishing by foreign vessels (Kainge et al. 2020). Improved control has resulted in sharply reduced catches, which then increased again during the 1980s before declining sharply again in the early 1990s (Fig. 11.16). Since the turn of the century however, catches have been relatively stable and fluctuated between 100,000 and 200,000 t. South Africa's hake fishery was initiated in 1917 and increased steadily to a peak of 300,000 t in 1972, before declining and then leveling out between 100,000 and 150,000 t annually since the late-1970s.

Overall, catches by these major sectors have shown different trajectories in the two subsystems. In the nBUS, catches initially increased rapidly but then decreased rapidly, with the highest combined catch (just over 2 million t) occurring in 1969. Combined annual catches in the nBUS since 2000 have been low (around 20% of the maximum attained) but steady. In the sBUS, catches in each sector increased more slowly and declines were less or did not occur because of species replacements (i.e., anchovy for sardine), and the highest combined catch occurred in 1988. Combined annual catches in the sBUS since 2000 have averaged 70% of maximum catch, albeit with more variability (42%–93%) than in the nBUS.

11.4 Conclusion

Like every upwelling system, the Benguela system reacts to changes in physical forcing and is expected to be responsive to present and future climate changes, as it is extremely sensitive to global, regional, and local fluctuations in atmospheric circulation patterns (Bakun 1990; Bakun et al. 2010; Demarcq 2009). Various coastal upwelling systems have undergone dramatic changes (called ecosystem regime shifts) in their ecosystem structure and fishery productivity in the past (Alheit and Bakun 2010; Cury and Shannon 2004; Finney et al. 2010). These changes are not thought to have been triggered exclusively by human actions, but also by global or regional physical drivers (Overland et al. 2010; Rykaczewski and Checkley 2008). Although some models differ in their implications to projected climate change (Wang et al. 2010), most calculations and assessments that address the consequences of potential climate change in coastal upwelling areas postulate an intensification of physical forcing (wind) that results in stronger, more persistent, or more widespread upwelling, particularly in the poleward regions of these systems (Bakun 1990; Bakun et al. 2010).

We summarized TRAFFIC results together with data from the literature and previous projects in a synoptic presentation to shed light on trophic transfer efficiencies in the nBus and sBUS. First results indicate that trophic transfer efficiency can be modified by metabolic processes and behavior of the animals, for instance in terms of predator-prey interactions. Schukat et al. (2021) compared life history traits of dominant calanoid copepods in the Humboldt Current System (HCS) with those in the nBUS to infer effects of behavior on trophic transfer efficiency. The authors concluded that higher transfer efficiency within the HCS was correlated with the lack of ontogenetic vertical migration of *Calanus chilensis,* making it easily accessible to epipelagic predators during all life stages. In contrast, in the nBUS, the large copepod *Calanoides natalis* tends to perform vertical migration through the oxygen minimum zone, taking it out of reach for hypoxia-sensitive predators and hence preventing efficient transfer toward higher trophic levels (Schukat et al.

2021). To establish a holistic view of the carbon transfer processes information from other ecological components have to be added. For example, the role of bacteria and DOC fluxes is still not understood. DOC export is estimated to account for around 20% of the global passive export production (Roshan and DeVries 2017). Model estimates of DeVries and Weber (2017) indicate that large plankton produce primarily labile DOC, which is rapidly remineralized within several days, whereas small plankton produce more non-labile DOC that persists for years and contributes to carbon export and sequestration. Investigations on small-sized plankton are in progress. At the other side of the size spectrum, higher trophic levels other than fish seabirds and marine mammals should be given more attention to fully assess trophic transfer efficiency.

Acknowledgments We thank the captains and crews of the research vessels RV Sonne and RV Meteor for their excellent support. We would also like to thank all colleagues from Namibia, South Africa, and Germany for their collaboration during the field campaigns. We are grateful indepted to the responsible authorities in Namibia and South Africa for making the research in the region possible. We also thank two anonymous reviewers and the editor, Jennifer Veitch, for valuable comments. The project TRAFFIC was funded by the Bundesministerium für Bildung und Forschung (BMBF), FKZ 03F0797. MJK received partial support by the project iFADO (EAPA_165/2016), INTERREG Atlantic Area Programme.

Glossary of Organizations and Projects

[The below list is limited to those organizations and projects cited in this chapter, and is not intended to represent a comprehensive overview of all marine initiatives in southern Africa]

Organizations

NATMIRC	National Marine information and Research Center, Swakopmund, Namibia
BCC	Benguela Current Convention, Swakopmund, Namibia
UNAM	University of Namibia, Windhoek and Henties Bay, Namibia
UCT	University of Cape Town, Cape Town, South Africa
DFFE	Department of Forestry, Fisheries and the Environment, Cape Town, South Africa

Projects

BIOTA Africa	BIOdiversity Monitoring Transect Analysis in Africa. The German Federal Ministry of Education and Research (BMBF) was open to fund the initiative, meanwhile

several African countries and partner institutions added funding

NAMIBGAS	funded by BMBF
BENEFIT	Benguela Environment Fisheries Interaction and Training Programme, funded by a number of local, regional and international research and development sources
GENUS (2009–2015)	Geochemistry and Ecology in the Namibian Upwelling System, funded by BMBF
TRAFFIC (2019–2022)	Trophic Transfer Efficiency in the Benguela Current, funded by BMBF

References

Alheit J, Bakun A (2010) Population synchronies within and between ocean basins: apparent teleconnections and implications as to physical–biological linkage mechanisms. J Mar Syst 79:267–285

Armengol L, Calbet A, Franchy G et al (2019) Planktonic food web structure and trophic transfer efficiency along a productivity gradient in the tropical and subtropical Atlantic Ocean. Sci Rep 9(1):1–19. https://doi.org/10.1038/s41598-019-38507-9

Armstrong MJ, Prosch RM (1991) Abundance and distribution of the mesopelagic fish *Maurolicus muelleri* in the Southern Benguela system. South Afr J Mar Sci 10:13–28. https://doi.org/10.2989/02577619109504615

Auel H, Verheye HM (2007) Hypoxia tolerance in the copepod *Calanoides carinatus* and the effect of the intermediate oxygen minimum layer on copepod vertical distribution in the northern Benguela Upwelling System and Angola-Benguela front. J Exp Mar Biol Ecol 352:234–243

Auel H, Hagen W, Ekau W, Verheye HM (2005) Metabolic adaptations and reduced respiration of the copepod *Calanoides carinatus* during diapause at depth in the Angola-Benguela front and northern Benguela upwelling regions. Afr J Mar Sci 27(3):653–657

Augustyn CJ, Cockcroft A, Coetzee J et al (2018) Rebuilding South African fisheries: three case-studies. In: Garcia SM, Ye Y (eds) Rebuilding of marine fisheries Part 2, case studies. FAO fisheries and aquaculture technical paper 630/2. FAO, Rome, pp 107–143

Ayon P, Schwartzmann G, Bertrand A et al (2008) Zooplankton and forage fish species off Peru: large-scale bottom-up forcing and local-scale depletion. Prog Oceanogr 79:208–214

Bakun A (1990) Global climate change and intensification of coastal ocean upwelling. Science 247(4939):198–201

Bakun A (1996) Patterns in the ocean: ocean processes and marine population dynamics. University of California Sea Grant, in cooperation with Centro de Investigaciones Biológicas del Noroeste, La Paz, p 323

Bakun A, Field DB, Redondo-Rodriguez A, Weeks SJ (2010) Greenhouse gas, upwelling-favourable winds, and the future of coastal ocean upwelling ecosystems. Glob Change Biol 16:1213–1228

Barlow R, Lamont T, Mitchell-Innes B et al (2009) Primary production in the Benguela ecosystem, 1999–2002. Afr J Mar Sci 31(1):97–101

Barneche DR, Hulatt CJ, Dossena M et al (2021) Warming impairs trophic transfer efficiency in a long-term field experiment. Nature 592:76–79

Batchelder HP, Mackas DL, O'Brien TD (2012) Spatial-temporal scales of synchrony in marine zooplankton biomass and abundance patterns: a world-wide comparison. Prog Oceanogr 97:15–30

Bode M, Kreiner A, van der Plas AK et al (2014) Spatio-temporal variability of copepod abundance along the 20°S monitoring transect in the northern Benguela Upwelling System from 2005 to 2011. PLoS One 9:e97738. https://doi.org/10.1371/journal.pone.0097738

Bode M, Hagen W, Schukat A et al (2015) Feeding strategies of tropical and subtropical calanoid copepods throughout the eastern Atlantic Ocean - latitudinal and bathymetric aspects. Prog Oceanogr 138:268–282

Bode-Dalby M, Würth R, Oliveira LDF et al (2022) Small is beautiful: the important role of small copepods in carbon budgets of the southern Benguela upwelling system. J Plankton Res 45:110–128. https://doi.org/10.1093/plant/fbac061

Bohata K (2016) Microzooplankton of the northern Benguela Upwelling System. PhD thesis, University of Hamburg, Hamburg

Bollens GCR, Landry MR (2000) Biological response to iron fertilization in the eastern equatorial Pacific (IronEx II). II. Mesozooplankton abundance, biomass, depth distribution and grazing. Mar Ecol Prog Ser 201:43–56

Bonino G, Di Lorenzo E, Masina S, Iovino D (2019) Interannual to decadal variability within and across the major eastern boundary upwelling systems. Sci Rep 9:19949. https://doi.org/10.1038/s41598-019-56514-8

Bordbar MH, Mohrholz V, Schmidt M (2021) The relation of wind-driven coastal and offshore upwelling in the Benguela Upwelling System. J Phys Oceanogr 51(10):3117–3133

Calbet A (2001) Mesozooplankton grazing effect on primary production: a global comparative analysis in marine ecosystems. Limnol Oceanogr 46(7):1824–1830

Calbet A, Landry MR (2004) Phytoplankton growth, microzooplankton grazing, and carbon cycling in marine systems. Limnol Oceanogr 40:51–57

Calbet A, Saiz E (2005) The ciliate-copepod link in marine ecosystems. Aquat Microb Ecol 38:157–167

Carter R (2011) Appendix 1B: water column. Draft Report. Namibian Marine Phosphate (Pty) Ltd

Coetzee J, Staby A, Krakstad J-O, Stenevik E (2009) Abundance and distribution of mesopelagic fish in the Benguela ecosystem. In: Hampton I, Barange M, Sweijd N (eds) Benguela environment fisheries interaction and training program (BENEFIT) research projects, GLOBEC Rep, vol 25, pp 44–47

Coetzee J, Merkle D, Shabangu F et al (2018) Results of the 2018 pelagic biomass survey Fisheries, SWG-PEL 38:1–14

Cornejo R, Koppelmann R (2006) Distribution patterns of mesopelagic fishes with special reference to *Vinciguerria lucetia* Garman 1899 (Phosichthyidae: Pisces) in the Humboldt current region off Peru. Mar Biol 146:15719–11537

Cury P, Roy C (1989) Optimal environmental window and pelagic fish recruitment success in upwelling areas. Can J Fish Aquat Sci 46:670. https://doi.org/10.1139/f89-086

Cury P, Shannon L (2004) Regime shifts in upwelling ecosystems: observed changes and possible mechanisms in the northern and southern Benguela. Prog Oceanogr 60:223–243

Cushing DH (1990) Plankton production and year-class strength in fish populations: an update of the match/mismatch hypothesis. Adv Mar Biol 26:249–293

Czudaj S, Koppelmann R, Möllmann C et al (2021) Community structure of mesopelagic fishes constituting sound scattering layers in the eastern tropical North Atlantic. J Marine Syst 224:103635

DEFF (Department of Environment, Forestry and Fisheries) (2020) Status of the South African marine fisheries resources 2020. DEFF, Cape Town. 112 + vi pp

Demarcq H (2009) Trends in primary production, sea surface temperature and wind in upwelling systems (1998–2007). Prog Oceanogr 83:376–385

DeVries T, Weber T (2017) The export and fate of organic matter in the ocean: new constraints from combining satellite and oceanographic tracer observations. Global Biogeochem Cy 31:535–555

du Pontavice H, Gascuel D, Reygondeau G et al (2020) Climate change undermines the global functioning of marine food webs. Glob Chang Biol 26:1306–1318

Duncan SE, Sell AF, Hagen W et al (2022) Environmenral drivers of upper mesopelagic fish assemblages in the Benguela Upwelling System. Mar Ecol Prog Ser 688:133–152. https://doi.org/10.3354/meps14017

Duncombe-Rae CMD (2004) A demonstration of the hydrographic partition of the Benguela upwelling ecosystem at 26°40'S. Afr J Mar Sci 27(3):617–628. https://doi.org/10.2989/18142320509504122

Eddy TD, Bernhardt JR, Blanchard JL et al (2020) Energy flow through marine ecosystems: confronting transfer efficiency. Trends Ecol Evol 36:76–86

Ekau W, Verheye HM (2005) Influence of oceanographic fronts and low oxygen on the distribution of ichthyoplankton in the Benquela and southern Angola Currents. Afr J Mar Sci 27:629–639

Ekau W, Auel H, Pörtner H-O, Gilbert D (2010) Impacts of hypoxia on the structure and processes in pelagic communities (zooplankton, macro-invertebrates and fish). Biogeosciences 7:1669–1699

Ekau W, Auel H, Hagen W et al (2018) Pelagic key species and mechanisms driving energy flows in the northern Benguela upwelling ecosystem and their feedback into biogeochemical cycles. J Mar Syst 188:49–62

Ekman VW (1905) On the influence of the earth's rotation on ocean-currents. Ark Mat Astro Fysik 2(11):1–53

Emeis KC, Anja Eggert A, Flohr A et al (2018) Biogeochemical processes and turnover rates in the Northern Benguela Upwelling System. J Mar Syst 188:63–80

Fennel W, Lass U (2007) On the impact of wind curls on coastal currents. J Mar Syst 68:128–142

Finney BP, Alheit J, Emeis KC et al (2010) Paleoecological studies on variability in marine fish populations: a long-term perspective on the impacts of climatic change on marine ecosystems. J Mar Syst 79:316–326

Flynn RF, Granger J, Veitch JA et al (2020) On-shelf nutrient trapping enhances the fertility of the Southern Benguela Upwelling System. J Geoph Res 125:1–24. https://doi.org/10.1029/2019JC015948

Fock HO, Czudaj S (2019) Size structure changes of mesopelagic fishes and community biomass size spectra along a transect from the equator to the Bay of Biscay collected in 1966–1979 and 2014–2015. ICES J Mar Sci 76:755–770

Garcia-Reyes M, Sydeman WJ, Schoeman DS et al (2015) Under pressure: climate change, upwelling, and eastern boundary upwelling ecosystems. Front Mar Sci 2. https://doi.org/10.3389/fmars.2015.00109

Gibbons MJ, Skrypzeck H, Brodeur RD et al (2021) A comparative review of macromedusae in eastern boundary currents. Oceanogr Mar Biol: Annl Rev 59:371–482

Grote B, Ekau W, Hagen W et al (2007) Early life-history strategy of Cape hake in the Benguela upwelling region. Fish Res 86:179–187

Hansen FC, Cloete RR, Verheye HM (2005) Seasonal and spatial variability of dominant copepods along a transect off Walvis Bay (23°S), Namibia. Afr J Mar Sci 27(1):55–63

Hansen A, Ohde T, Wasmund N (2014) Succession of micro-and nanoplankton groups in ageing upwelled waters off Namibia. J Mar Syst 140:130–137

Hays GC, Richardson AJ, Robinson C (2005) Climate change and marine plankton. Trends Ecol Evol 20:337–344

Hays GC, Doyle TK, Houghton JDR (2018) A paradigm shift in the trophic importance of jellyfish? Trends Ecol Evol 33(11):874–884

Huggett JA, Richardson AJ (2000) A review of the biology and ecology of Calanus agulhensis off South Africa. ICES J Mar Sci 57:1834–1849

Huggett JA, Richardson AJ, Field JG (2007) Comparative ecology of the copepods Calanoides carinatus and Calanus agulhensis – the influence of temperature and food. Afr J Mar Sci 29(3):473–490

Huggett J, Verheye HM, Escribano R, Fairweather T (2009) Copepod biomass, size composition and production in the Southern Benguela: Spatio-temporal patterns of variation, and comparison with other eastern boundary upwelling systems. Prog Oceanogr 83:197–207

Hulley PA (1981) Results of the research cruises of the FRV 'Walther Herwig' to South America. LVII. Family Myctophidae. Arch Fischereiwiss. Heenemann Verlagsgesellschaft mbH, Berlin

Hulley PA (1992) Upper-slope distributions of oceanic lanternfishes (family: Myctophidae). Mar Biol 114:365–383

Hulley PA, Lutjeharms JRE (1989) Lanternfishes of Southern Benguela region. Part 3: the pseudoceanic-oceanic interface. Ann S Afr Museum 98:409–435

Hulley PA, Prosch RM (1987) Mesopelagic fish derivatives in the Southern Benguela upwelling region. S Afr J Mar Sci 5:597–611

Hutchings L, Verheye HM, Hugget JA et al (2006) Variability of plankton with reference to fish variability in the Benguela current large marine ecosystem - an overview. In: Shannon V, Hempel G, Malanotte-Rizzoli P, Moloney CL, Woods J (eds) Benguela: predicting a large marine ecosystem, Large marine ecosystems, vol 14. Elsevier, Amsterdam, pp 91–124

Hutchings L, van der Lingen CD, Shannon LJ et al (2009) The Benguela current: an ecosystem of four components. Prog Oceanogr 83:15–32

Hutchings L, Jarre A, Lamont T et al (2012) St Helena Bay (southern Benguela) then and now: muted climate signals, large human impact. Afr J Mar Sci 34:559–583

Kadila HK, Nakwaya DN, Butler M, Iitembu JA (2020) Insights into feeding interactions of shallow water Cape hake (*Merluccius capensis*) and Cape horse mackerel (*Trachurus capensis*) from the Northern Benguela (Namibia). Reg Stud Mar Sci 34:101071

Kainge P, Kirkman SP, Estevao V et al (2020) Fisheries yields, climate change, and ecosystem-based management of the Benguela Current Large Marine Ecosystem. Environ Develop 36:100567

Kersalé M, Lamont T, Speich S et al (2018) Moored observations of mesoscale features in the Cape Basin: characteristics and local impacts on water mass distributions. Ocean Sci 14:923–945

Kirkman SP, Blamey L, Lamont T et al (2016) Spatial characterisation of the Benguela ecosystem for ecosystem-based management. Afr J Mar Sci 38:1–16

Lamont T, Hutchings L, van den Berg MA et al (2015) Hydrographic variability in the St Helena Bay region of the southern Benguela ecosystem. J Geophys Res Oceans 120:2920–2944

Lamont T, García-Reyes M, Bograd SJ et al (2018) Upwelling indices for comparative ecosystem studies: variability in the Benguela Upwelling System. J Mar Syst 188:3–16

Landry MR (1993) Estimating rates of growth and grazing mortality of phytoplankton by the dilution method. In: Kemp PF, Sherr BF, Sherr EB, Cole JJ (eds) Handbook of methods in aquatic microbial ecology. Lewis Publishers, Boca Raton, FL, pp 715–722

Lett C, Veitch J, van der Lingen CD, Hutchings L (2007) Assessment of an environmental barrier to transport of ichthyoplankton from the southern to the northern Benguela ecosystems. Mar Ecol Prog Ser 347:247–259

Lindeman RL (1942) The trophic dynamic aspects of ecology. Ecology 23:399–418

Loick N, Ekau W, Verheye HM (2005) Water-body preferences of dominant calanoid copepod species in the Angola-Benguela frontal zone. Afr J Mar Sci 27:597–608

Lynam CP, Gibbons MJ, Axelsen BE et al (2006) Jellyfish overtake fish in a heavily fished ecosystem. Curr Biol 16. https://doi.org/10.1016/j.cub.2006.06.018

Mann KH, Lazier JRN (1991) Dynamics of marine ecosystems. Biological/physical interactions in the oceans. Blackwell Publishing

Martin B, Eggert A, Koppelmann R et al (2015) Spatio-temporal variability of zooplankton biomass and environmental control in the Northern Benguela Upwelling System: field investigations and model simulation. Mar Ecol 36:637–658

Martin B, Koppelmann R, Kassatov P (2017) Ecological relevance of salps and doliolids in the northern Benguela Upwelling System. J Plankton Res 39:290–304

Messié M, Chavez FP (2015) Seasonal regulations of primary production in eastern boundary upwelling systems. Prog Oceanogr 134:1–18

Mohrholz V, Eggert A, Junker T et al (2014) Cross shelf hydrographic and hydrochemical conditions and their short term variability at the northern Benguela during a normal upwelling season. J Mar Syst 140:92–110

Mollenhauer G, Schneider R, Jennerjahn T et al (2004) Organic carbon accumulation in the South Atlantic Ocean: its modern, mid-Holocene and last glacial distribution. Glob Planet Chang 40:249–266

Möller KO, John MSt, Temming A et al (2012) Marine snow, zooplankton and thin layers: indications of a trophic link from small-scale sampling with the video plankton recorder. Mar Ecol Prog Ser 468:57–69

Moloney CL, Field JG, Lucas MI (1991) The size-based dynamics of plankton food-webs. II. Simulations of three contrasting southern Benguela food webs. J Plankton Res 13:1039–1092

Monteiro PMS, van der Plas AK, Mélice J-L, Florenchie P (2008) Interannual hypoxia variability in a coastal upwelling system: ocean–shelf exchange, climate and ecosystem-state implications. Deep Sea Res I 55:435–450

Olivar MP (1987) Larval development and spawning of *Diaphus hudsoni* in the Benguela Current region. Mar Biol 94:605–611

Olivar MP, Beckley LE (1994) Influence of the Agulhas Current on the distribution of lanternfish larvae off the southeast coast of Africa. J Plankton Res 16:1759–1780

Overland JE, Alheit J, Bakun A et al (2010) Climate controls on marine ecosystems and fish populations. J Mar Syst 79:305–315

Oxborough K, Moore CM, Suggett DJ et al (2012) Direct estimation of functional PSII reaction center concentration and PSII electron flux on a volume basis: a new approach to the analysis of Fast Repetition Rate Fluorometry (FRRf) data. Limnol Oceanogr Methods 10:142–154

Peterson WT, Hutchings L, Huggett JA, Largier JL (1992) Anchovy spawning in relation to the biomass and the replenishment rate of their copepod prey on the western Agulhas Bank. S Afr J Mar Sci 12:487–500

Pillar SC, Barange M (1998) Feeding habits, daily ration and vertical migration of the Cape horse mackerel off South Africa. S Afr J Mar Sci 19:263–274

Pitcher GC, Probyn TA, du Randt A et al (2014) Dynamics of oxygen depletion in the nearshore of a coastal embayment of the southern Benguela Upwelling System. J Geophys Res-Oceans 119:2183–2200

Postel L, Arndt EA, Brenning U (1995) Rostock zooplankton studies off West Africa. Helgoland Marine Res 49:829–847

Prosch RM (1991) Reproductive biology and spawning of the myctophid *Lampanyctodes hectoris* and the sternoptychid *Maurolicus muelleri* in the southern Benguela Ecosystem. S Afr J Mar Sci 10:241–252

Ragoasha N, Herbette S, Cambon G et al (2019) Lagrangian pathways in the southern Benguela Upwelling System. J Mar Syst 195:50–56

Richardson AJ, Verheye HM, Mitchell-Innes BA et al (2003) Seasonal and event-scale variation in growth of *Calanus agulhensis* (Copepoda) in the Benguela Upwelling System and implications for spawning of sardine *Sardinops sagax*. Mar Ecol Prog Ser 254:239–251

Rixen T, Lahajnar N, Lamont T et al (2021) Oxygen and nutrient trapping in the southern Benguela Upwelling System. Front Mar Sci 8. https://doi.org/10.3389/fmars.2021.730591

Rosa R, Dierssen HM, Gonzalez L, Seibel BA (2008) Large-scale diversity patterns of cephalopods in the Atlantic open ocean and deep sea. Ecology 89:3449–3461

Roshan S, DeVries T (2017) Efficient dissolved organic carbon production and export in the oligotrophic ocean. Nature Comm 8:1–8

Rouault M, Pohl B, Penven P (2010) Coastal oceanic climate change and variability from 1982 to 2009 around South Africa. Afr J Mar Sci 32:237–246

Roux JP, van der Lingen CD, Gibbons MJ et al (2013) Jellyfication of marine ecosystems as a likely consequence of overfishing small pelagic fishes: lessons from the Benguela. Bull Mar Sci 89:249–284

Rubiés P (1985) Zoogeography of the Lanternfishes (Osteichthyes, Myctophidae) of Southwest Africa. In: International symposium for upwelling of West Africa, p 573–586

Rykaczewski RR, Checkley DM (2008) Influence of ocean winds on the pelagic ecosystem in upwelling regions. PNAS 105(6):1965–1970

Rykaczewski RR, William JP, Sydeman WJ et al (2015) Poleward displacement of coastal upwelling-favourable winds in the ocean's eastern boundary currents through the 21st century. Geophys Res Lett 42:6424–6431

Ryther JH (1969) Photosynthesis and fish production in the sea. Science 166(3901):72–76

Sakko AL (1998) The influence of the Benguela Upwelling System on Namibia's marine biodiversity. Biodivers Conserv 7:419–433

Salvanes AGV, Christiansen H, Taha Y et al (2018) Variation in growth, morphology and reproduction of the bearded goby (*Sufflogobius bibarbatus*) in varying oxygen environments of northern Benguela. J Mar Syst 188:81–97

Schmidt M, Eggert A (2016) Oxygen cycling in the northern Benguela Upwelling System: modelling oxygen sources and sinks. Prog Oceanogr 149:145–173

Schukat A, Teuber L, Hagen W et al (2013) Energetics and carbon budgets of dominant calanoid copepods in the northern Benguela Upwelling System. J Exp Mar Biol Ecol 442:1–9

Schukat A, Auel H, Teuber L et al (2014) Complex trophic interactions of calanoid copepods in the Benguela Upwelling System. J Sea Res 85:186–196

Schukat A, Hagen W, Dorschner S et al (2021) Zooplankton ecological traits maximize the trophic transfer efficiency of the Humboldt Current upwelling system. Prog Oceanogr 193:102551

Shannon LV (1985) The Benguela ecosystem. Part I. Evolution of the Benguela, physical features and processes. Oceanogr Mar Biol Ann Rev 23:105–182

Shannon LV, Crawford RJM, Pollock DE et al (1992) The 1980s - a decade of change in the Benguela ecosystem. S Afr J Mar Sci 12:271–276

Staby A, Krakstad J-O (2008) Review of the state of knowledge, research (past and present) of the distribution, biology, ecology and abundance of non-exploited mesopelagic fish (Order Anguilliformes, Argentiniformes, Stomiiformes, Myctophiformes, Aulopiformes) and the bearded goby (*Sufflogobius bibarbatus*) in the Benguela ecosystem. Report on BCLME project LMR/CF/03/08, p 84

Sutton T, Clark MR, Dunn DC et al (2017) A global biogeographic classification of the mesopelagic zone. Deep Sea Res I 126:85–102

Tiedemann M, Brehmer P (2017) Larval fish assemblages across an upwelling front: indication for active and passive retention. Est Coast Shelf Sci 187:118–133

Timonin AG (1997) Diel vertical migrations of *Calanoides carinatus* and *Metridia lucens* (Copepoda: Calanoida) in the northern Benguela upwelling area. Okeanologiya 37:868–873

van der Lingen CD, Hutchings L, Field JG (2006a) Comparative trophodynamics of anchovy *Engraulis encrasicolus* and sardine *Sardinops sagax* in the southern Benguela: are species alternations between small pelagic fish trophodynamically mediated? Afr J Mar Sci 28:465–477

van der Lingen CD, Shannon LJ, Cury P, et al (2006b) Resource and ecosystem variability, including regime shifts, in the Benguela current system. In: Shannon V, Hempel G, Malanotte-Rizzoli P, Moloney C, Woods J (eds) Benguela: predicting a large marine ecosystem, Large marine ecosystems, 14:147–184

Verheye HM (2000) Decadal-scale trends across several marine trophic levels in the Southern Benguela Upwelling System off South Africa. Ambio 29:30–34

Verheye HM, Field JG (1992) Vertical distribution and diel vertical migration of *Calanoides carinatus* (Krøyer, 1849) developmental stages in the southern Benguela upwelling region. J Exp Mar Biol Ecol 158:123–140

Verheye HM, Richardson AJ (1998) Long-term increase in crustacean zooplankton abundance in the southern Benguela upwelling region (1951-1996): bottom-up or top-down control? ICES J Mar Sci 55:803–807

Verheye HM, Richardson AJ, Hutchings L et al (1998) Long-term trends in the abundance and community structure of coastal zooplankton in the southern Benguela system, 1951-1996. S Afr J Mar Sci 19:317–332

Verheye HM, Hagen W, Auel H et al (2005) Life strategies, energetics and growth characteristics of *Calanoides carinatus* (Copepoda) in the Angola-Benguela front region. Afr J Mar Sci 27(3):641–652

Verheye HM, Lamont T, Huggett JA et al (2016) Plankton productivity of the Benguela Current Large Marine Ecosystem (BCLME). Environ Dev 17:75–92

Wang M, Overland JE, Bond NA (2010) Climate projections for selected large marine ecosystems. J Mar Syst 79:258–266

Wang D, Gouhier TC, Menge BA, Ganguly AR (2015) Intensification and spatial homogenization of coastal upwelling under climate change. Nature 518:390–394

Werner T, Buchholz F (2013) Diel vertical migration behaviour in Euphausiids of the northern Benguela current: seasonal adaptations to food availability and strong gradients of temperature and oxygen. J Plankton Res 35:792–812

The Application of Paleoenvironmental Research in Supporting Land Management Approaches and Conservation in South Africa

12

K. L. Kirsten, C. J. Forbes, J. M. Finch, and L. Gillson

Abstract

Research into past environments and climates of South Africa has significantly grown in recent decades, owing to its rich archeological heritage and high biodiversity. The paleoscience community has worked toward an improved understanding of long-term climate and environmental dynamics, yet the application and dissemination of such information into the realm of conservation and land-use management have remained limited. In this chapter, we briefly explore the current state of paleoenvironmental research in South Africa, recent methodological advancements and potential applications of paleoresearch for natural resource management and conservation. We advocate for a more integrated research approach, bringing together the fields of ecology, ecosystem restoration, conservation biology and paleoecology, as an avenue toward tackling

Supplementary Information The online version contains supplementary material available at https://doi.org/10.1007/978-3-031-10948-5_12.

K. L. Kirsten (✉)
Discipline of Geography, School of Agricultural, Earth and Environmental Sciences, University of KwaZulu-Natal, Pietermaritzburg, South Africa

Department of Geological Sciences, University of Cape Town, Cape Town, South Africa

Human Evolution Research Institute, University of Cape Town, Cape Town, South Africa

C. J. Forbes · L. Gillson
Plant Conservation Unit, Department of Biological Sciences, University of Cape Town, Cape Town, South Africa

J. M. Finch
Discipline of Geography, School of Agricultural, Earth and Environmental Sciences, University of KwaZulu-Natal, Pietermaritzburg, South Africa

G. P. von Maltitz et al. (eds.), *Sustainability of Southern African Ecosystems under Global Change*, Ecological Studies 248,
https://doi.org/10.1007/978-3-031-10948-5_12

uncertainties in conservation and land-use management practices. We use a case study from the Kruger National Park, to demonstrate the benefits of incorporating a long-term perspective in understanding the natural variability and thresholds of an ecological system, and thereby inform more sound natural resource management strategies and conservation planning.

12.1 Introduction

The state of a landscape is the product of current and past environmental factors, including climatic, biotic and anthropogenic factors. Growing anthropogenic impacts on the climate and the biosphere has brought into focus the dynamic nature of our planet and highlighted the need to understand long-term climate and ecological processes, both retrospectively in terms of historical patterns and processes, and as a basis for informing future projections (Willis et al. 2010). Under current climate change scenarios, environmental systems are expected to undergo adaptive responses leading to novel ecosystems with no modern analogue (Lovejoy 2007; Williams and Jackson 2007). Long-term ecological observatories, monitoring programs and advancements in remote sensing techniques have allowed for ecosystem dynamics to be mapped and monitored at a large spatial scale; however, these are temporally restricted to the last few decades. Among other applications, long-term datasets form the basis for model building and validation for future climate projections. For more recently established observatories, it is impractical to wait for decades to generate long-term monitoring data (Rull 2014). Given the limited temporal range of historical data, alternative approaches are required to gain insight as to how climates and ecosystems operate over ecologically meaningful timescales.

To provide a deeper understanding of environmental variability over longer timescales, researchers have turned to natural archives of environmental change, such as marine, lake and wetland sediments. These archives offer an indirect record of ecological variability, and the underlying mechanisms influencing ecosystem change (Jackson and Hobbs 2009). For instance, through the identification of fossil pollen preserved in sedimentary archives, paleoresearch attempts to reconstruct past vegetation composition. This is achieved by transferring a species, environmental envelope from the extant ecosystem processes within which it occurs to the fossilized occurrences within the paleorecord (Williams and Jackson 2007; Willis et al. 2010). This approach offers a window into environmental change over the course of deposition and contributes to understanding environmental factors such as climate and hydrology, and biological factors such as species niche parameters and geographical range (Seddon et al. 2014). In many cases, baseline conditions are perceived to be those which occurred prior to intensive human involvement and modification. Yet, ecosystems are anything but static in nature, but rather dynamic with a range of natural variability (Froyd and Willis 2008). Fundamentally, systems are in continual flux, responding to external forces within a defined boundary range and baselines can shift depending on the timescale of observation. This is a key consideration when developing a management strategy and identifying ecologically realistic management targets (Forbes et al. 2018). Here, paleoenvironmental research can present unique insights in determining a system's threshold and resilience to change by revealing former multiple interacting processes (Willis et al. 2010; Gillson and Marchant 2014; Gillson 2015).

Paleoecology as a discipline has progressed from a largely qualitative base typified by the descriptive reconstruction of past environments to a considerably more quantitative science. This is mainly due to statistical innovations that have allowed for cross-validation of sites and proxies and the detection of regime shifts and tipping points (e.g., Line et al. 1994; Birks et al. 2012; Seddon et al. 2014; Blaauw et al. 2020). This, coupled with the development of more robust chronologies (see Blaauw et al. 2007; Bronk Ramsey 2008; Aquino-López et al. 2018) has provided direct correlations between sites, ultimately creating a regional perspective. The application and perceived relevance of the research outputs from quantitative, temporally constrained environmental reconstructions has increased. For example, paleoclimatic data now forms a critical component of the Intergovernmental Panel on Climate Change (IPCC) Physical Science basis (IPCC 2021), and the reassessment of the Ramsar Convention on Wetlands, on an individual site basis, to include paleoenvironmental insights into the natural ecological character of wetland systems for better management practices (Finlayson et al. 2016; Gell et al. 2016; Gell 2017). Paleoenvironmental perspectives also increase our understanding of climate versus human-driven fire regimes, through charcoal-based paleofire records, assisting in fire management, conservation and restoration efforts (McWethy et al. 2013; Iglesias et al. 2015; Maezumi et al. 2021).

There have been numerous calls for the closer integration of ecology, ecosystem restoration, conservation biology and paleoecology (e.g., Willis and Birks 2006; Dearing 2008; Froyd and Willis 2008; Willis and Bhagwat 2010; Birks 2012; Gillson and Marchant 2014; Gillson 2015; Davidson et al. 2018) leading to a greater application of paleoenvironmental and paleoecological datasets to answer questions of conservation and management relevance. Yet, concrete applications of paleoecology in conservation and management remain scarce. Rull (2014, p. 2) criticizes this lack of synergy and warns of "delaying the advancement of ecological knowledge and the potential impact of its applications on ... nature conservation and the sustainable use of ecological services." High resolution, multiproxy paleoenvironmental studies reveal the importance of long-term data in broadening our comprehension of alterations in ecosystem services (ES), ecosystem resilience and variability and incorporating knowledge into ecosystem assessments and management (Dearing et al. 2012b; Gillson 2015; Jeffers et al. 2015). In the South African context, there are numerous Holocene-age paleoecological studies (Table S12.1); however, few adequately demonstrate the impactful application of paleoecological data for land management (Fig. 12.1). Such applied paleoecology is a major development in the field, which is beginning to gain traction in South Africa, working to address the "fundamental disconnect" between the disciplines of ecology and paleoecology (Gillson and Duffin 2007; Forbes et al. 2018).

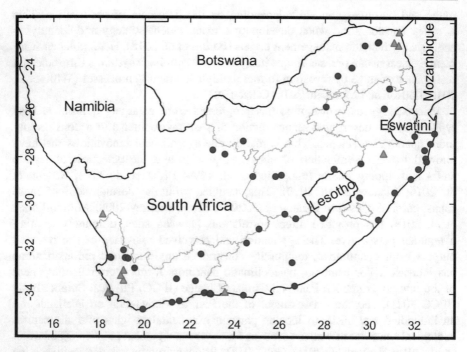

Fig. 12.1 Spatial distribution of selected Holocene-age paleosites across South Africa, noting those which included applied paleoecological aspects (triangles) (Full details and associated references for these sites are in the Supplementary Material)

12.2 Evolution of South African Paleoenvironmental Research

The paleoresearch community in South Africa has emerged from relatively slow beginnings, compounded by a range of region-specific limitations. These include the often cited lack of organic sedimentary archives due to the arid and semiarid climatic setting (Chase and Meadows 2007), and, more generally, a lack of capacity, funding and isolated research environment, at least up until the mid-1990s. There are now several active research groups and organizations across the country, with strong evidence of ongoing international collaboration (e.g., Haberzettl et al. 2014). In recent years, growing interest, and investment, in the story of human evolution has propelled the paleosciences forward, as a means of providing climate and environmental context to the development of early modern humans (Meadows 2015). Through this inherent geographical advantage, South Africa is broadly recognized as a priority research area for the paleosciences and has benefitted from a dedicated national funding instrument, the National Research Foundation African Origins Platform.

Paleoenvironmental research in South Africa has traditionally focused on pollen-based vegetation reconstructions, using wetland sediments, as a means to infer past climatic conditions (e.g., Martin 1956; Coetzee 1967). Innovative strides were taken to combat the effects of dry and strongly seasonal climate, which hampers accumulation of long, continuous sedimentary deposits. Thus, an expansion into a broader range of unconventional archives began (Meadows 2015), including pan sediments (e.g., Scott 1988), cave sediments (e.g., Thackeray 1992), rock hyrax middens (e.g., Scott et al. 2004) and coprolites (e.g., Carrion et al. 2000). Fitchett et al. (2017) noted a rapid increase in both the number of studies published and the number of proxies used in recent years, as researchers moved from a pollen-dominated narrative to include additional physical, chemical and biological proxies (Meadows 2014). In particular, studies employing isotopes (e.g., Smith et al. 2002; Esterhuysen and Smith 2003), geochemistry (e.g., Wündsch et al. 2016, 2018; Strobel et al. 2019) and diatoms (e.g., Kirsten et al. 2018, 2020) have become more prevalent (Fitchett et al. 2017). This shift toward a more multiproxy approach follows international trends, with the use of multiple independent lines of evidence as a means to strengthen interpretations and identify inconsistencies.

Although local paleoenvironmental research has remained largely qualitative, recent shifts toward a more quantitative analytical science have begun. The advent of transfer functions and various modeling approaches to paleodata have provided measurable comparisons to modern ecosystems (Anderson 1995; Birks et al. 2012). The incorporation of paleodata for use in probability density functions (PDF), such as that employed by the software CREST (Chevalier 2021), has assisted in reconstructing several climatic variables, including winter and summer temperature and precipitation, mean annual aridity and rainfall seasonality, by assigning modern climatic envelopes to fossil pollen assemblages across southern Africa (e.g., Chevalier and Chase 2016). Additionally, a transfer function was developed to quantitatively reconstruct relative sea level along the east coast of South Africa through the analysis of the modern elevation preferences of intertidal salt-marsh foraminifera (Strachan et al. 2014, 2015). Such quantitative approaches are underpinned by their modern data coverage, following which a recent move to address this modern data deficit is apparent (see, for example, Sobol et al. 2019; Strobel et al. 2020), still more work is needed in this area. These quantitative applications can deliver a direct source of information to support climate projections, with a scope to further develop southern African training databases to better constrain local predictions.

Nevertheless, the application of paleoenvironmental data implicitly relies on robust and well-constrained chronologies. Early studies developed age models based on linear interpolation of often uncalibrated radiocarbon age determinations; however, age-depth modeling of sediment sequences has developed into a complex, multisample, statistical approach. This transition has assisted in cross-validating environmental trends on a local, regional and global scale. For greater refinement, extensive work was undertaken in determining regional marine reservoir effects for the eastern to southeastern coast (Maboya et al. 2018) and the west coast (Dewar et al. 2012) of South Africa, where previously the marine carbon component was overlooked. Beyond the accelerator mass spectrometry (AMS) radiocarbon and optically stimulated luminescence dating methods, the incorporation of chronological markers (Neumann et al. 2011), lead-210 (see Forbes et al. 2018) and even paleomagnetic secular variations (see Haberzettl et al. 2019) have assisted in refining age-depth models. These analytical developments have greatly benefitted the interpretation of data and temporally constrained environmental events from sites across South Africa.

Despite the paleocommunity being active in advancing the field, there are several gaps in the knowledge of the role and application of paleoecology in informing sustainable land-use management, restoration and conservation. By focusing on the last 5000 years, key reference conditions for restoration and conservation efforts can be ascertained due to notable climatic deviations, including the mid-Holocene Altithermal, arrival of pastoralism in southern Africa, Medieval Climate Anomaly, Little Ice Age, European settlement and twentieth century global change drivers. However, even with global efforts heeding the call to put applied paleoecology into practice, there is limited implementation, bar a few studies (e.g., Forbes et al. 2018; Cramer et al. 2019; MacPherson et al. 2019; Dirk and Gillson 2020; Gillson et al. 2020; Dabengwa et al. 2021) (Fig. 12.1).

12.3 Shifting Mindsets: The Combination of Paleoecology and Restoration Ecology

Defining restoration and management targets requires a nuanced understanding of landscape history and an acknowledgement that variability and disturbance are normal. For example, an increase in tree cover in savannas might be considered undesirable if it is caused by the unprecedented disruptions of the twentieth century, including CO_2 enrichment and fire suppression, but might be considered tolerable or advantageous, if representing a return to former tree cover following land abandonment (Fig. 12.2). Furthermore, in the late twentieth century, ecological paradigms shifted, with the recognition that change rather than balance was the norm for most ecosystems (Pickett et al. 1997). Early conservation tactics aimed at preventing change, such as fire suppression and culling of animals, largely failed to stabilize ecosystems and ecologists realized that a new ecology of flux was needed (Gillson 2015).

In restoration ecology, managers aim to restore a degraded ecosystem whose structure, composition and function has been compromised, usually as a result of anthropogenic factors, for example deforestation, agriculture, urbanization, pollution or the invasion of nonnative species (Falk 2017). To define restoration goals, a reference condition is needed. In the case of a once-off disturbance event, the reference condition can be the state of the ecosystem prior to the disturbance, or the state of a neighboring ecosystem that has not experienced the disturbance. However, in most cases, defining reference conditions is much more complicated. Disturbance and variability are normal in most ecosystems; climate and fire regime changes over time can be an integral process in system functionality and many landscapes have been managed by people for centuries or even millennia. These factors leave a lasting legacy on landscapes that must be considered in restoration plans (Higgs et al. 2014; Johnstone et al. 2016; Manzano et al. 2020). What then should be the reference conditions that inform restoration management?

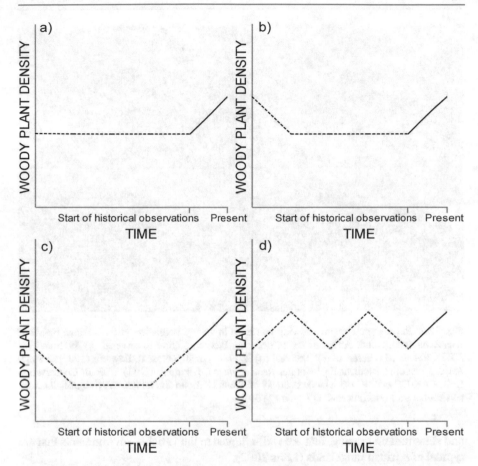

Fig. 12.2 Hypothetical graphs indicating observed increase in woody vegetation cover in the recent past (solid line) and different possible late Holocene landscape histories (dotted lines) (**a**) the recent increase in tree cover is unprecedented within the timeframe examined (**b**) the recent increase represents an recovery from past woody vegetation clearance (**c**) the recent increase represents an recovery from past woody vegetation clearance, but woody cover is now higher than it was before clearance (**d**) woody vegetation cover varies cyclically over time

In Sub-Saharan Africa, there is a case to be made that at least 500 years of paleoecological data are needed to define an appropriate range of variability, this captures the status of ecosystems and social-ecological systems prior to European settlement, and well before the onset of postindustrial anthropogenic climate change (Gillson and Marchant 2014; Gillson 2015). Data from fossil pollen, together with historical documents and archeological records, can be used to reconstruct the influence of anthropogenic and environmental change on vegetation cover and form a basis for informing restoration of degraded landscapes. These data can also be useful in restoring traditional management techniques, such as patch mosaic burning

Fig. 12.3 Repeat photographs showing increases in woody vegetation in the savanna biome of KwaZulu-Natal, South Africa in recent decades. Weenen Nature Reserve in (**a**) 1955 and (**b**) 2011. Weenen Middlerest in (**c**) 1955 and (**d**) 2011. Original photos: D. Edwards (1955) © South African National Biodiversity Institute. *Repeat photos*: J. Puttick (2011) © Plant Conservation Unit, UCT. CC BY-NC 4.0. Photos courtesy of Timm Hoffman and rePhotoSA [http://ibali.uct.ac.za/s/rephotosa/] (Hoffman and O'Connor 1999)

and transhumant grazing, that are well-adapted to the variable environments that are typical of African rangelands (Laris 2002).

Here, savannas are used as an example case study to illustrate the importance of long-term data in defining reference conditions and informing restoration targets. In recent decades, woody vegetation has increased in many savannas (Wigley et al. 2010; O'Connor et al. 2014; Hoffman et al. 2019) (Fig. 12.3), with several factors interacting to determine tree density. Tree abundance varies in response to climate (especially rainfall), fire, herbivory, soil nutrients and CO_2 (Bond and Midgley 2012). Furthermore, savannas are important rangelands and humans have manipulated both fire and herbivory to sustain grazing for domestic animals and retain a diverse array of ES. This manipulation of savannas through fire and herbivory was disrupted due to the eighteenth-century European colonization and settlement. Hunting of elephants for ivory and extermination of predators, for example, would have affected herbivory and therefore tree recruitment and vegetation structure (Hempson et al. 2015; Venter et al. 2017). In the late nineteenth century, Rinderpest wiped out significant proportions of herbivores, probably driving pulses of tree and shrub recruitment and leading to unusually high tree densities (Ofcansky 1981; Holdo et al. 2009). In addition, many colonial governments instigated policies of fire suppression and later prescribed burning that disrupted natural fire regimes and traditional patterns of fire management, again facilitating tree growth (van Wilgen et al. 2014; Humphrey et al. 2020).

Such fire management policies were often continued beyond independence and in many cases are only recently being reviewed. At the same time, rising levels of CO_2 further enhanced tree recruitment (Scheiter and Higgins 2012), while land-use and settlement patterns changed, with increasing sedentarization of previously transhumant populations. With increasing urbanization and government grants, destocking of formerly heavily grazed areas has occurred in recent decades (Blair et al. 2018). Many savannas were unusually heavily wooded by the opening of the twentieth century, due to the unusual and rapidly changing conditions of the past two centuries. Therefore, when many national parks were founded in the early decades of the twentieth century, they were likely in a state of atypically high tree density compared with preceding centuries and millennia. Using protected areas as reference conditions for other, more heavily degraded areas is therefore often inappropriate and highlights the potential role of paleoecology in establishing reference conditions for savanna restoration.

Where paleoecology shows that tree abundance is unprecedented or outside of the historical range of variability (Fig. 12.3a and c), appropriate restoration might include attempts to reduce tree cover, for example through larger herbivore populations or more intense burns. Such interventions might be deemed unnecessary, where tree cover is recovering from past clearance or undergoing cyclical change (Fig. 12.3b and d). With so many variables at play, it is not surprising that the structure of savannas is highly variable over both space and time. Therefore, the appropriate response to increasing woody cover depends on understanding the history of the landscape, requiring long-term data that extends beyond the timescale of intensive human impact in the past few centuries (Gillson and Marchant 2014; Gillson 2015).

Sediment cores retrieved from the Kruger National Park, and spanning the last 5000 years, assisted in identifying a series of alternate stable states in savanna vegetation, from studying interactions between local hydrology, climate, fire and herbivory (Gillson and Ekblom 2009, 2020; Ekblom and Gillson 2010). These states are largely determined by the interplay between rainfall and fire, but transitions between states can be facilitated or discouraged by management actions that alter herbivory and fire. In this way, the impacts of global change, can be ameliorated at least to some extent by management actions at landscape scales (Midgley and Bond 2015). Fossil pollen data assisted managers and ecologists in the Kruger National Park in developing suites of monitoring endpoints, known as Thresholds of Potential Concern (TPCs) (Rogers 2003; Gillson and Duffin 2007). These thresholds define upper and lower limits of acceptable change in key environmental parameters, for example tree cover. When the measured parameters approach the thresholds, management interventions are triggered that bring the variable back into the accepted range of variability, or alternatively the TPC is re-evaluated. Managers chose limits of variability to tree cover, which if crossed would trigger management responses such as relocation of elephants or changes in fire management (Gillson and Duffin 2007; van Wilgen and Biggs 2011; Gillson 2015). Thus, long-term records can assist in defining the historical range of variability and identifying ecological thresholds, which are significant in restoration and ecosystem management due to their impacts on ecological process and biodiversity. The application of these insights into past variability can assist in the management and restoration of terrestrial ecosystems by determining whether to control increases in woody plants (scenarios a and b in Fig. 12.2) or where paleoecology shows ancient open grassland or heathland systems.

Paleoecology can similarly be used to describe the "natural ecological character" of ecosystems (Finlayson et al. 2005; Gell et al. 2013, 2018; Davidson 2016; Gell 2017), a requirement under the Ramsar Convention which was ratified in 1971. An example from the Murray-Darling Basin (MDB), Australia, shows how a lack of long-term data can lead to incorrect assumptions about "baselines" leading to the perpetuation of degraded states (Finlayson et al. 2017). Decadal–centennial timescales allowed the previous centuries of human impact to be understood, providing a more realistic target for wetland restoration and management of the surrounding landscape policies. Long-term data contextualizing recent vegetation changes and insight into the history of landscapes, environmental change and land use is particularly vital at the current time, an example would be management pressure to afforest open landscapes that are perceived as degraded or denuded forests (Bond et al. 2019). In fact, many open landscapes such as savannas, grasslands and heathlands are valuable in terms of biodiversity and ES. Therefore, it is essential that ancient ecosystems are distinguished from degraded systems if misguided restoration plans on ancient open and biodiverse ecosystems are to be avoided. As shown in the examples of the Kruger National Park and Murray-Darling Basin, this can only be achieved if the history of landscapes is properly understood.

12.4   A Look into the Future: Applied Paleoecology

With the increase and advancements of paleostudies within South Africa, much can be done to operationalize and mainstream paleoecology for sustainable development in the region. Recent applied paleoecological studies seek to combine methodologies and promote interdisciplinary and transdisciplinary (TD) research, thus encouraging a past-present-future continuum (Dawson et al. 2011; Birks 2012; Gillson and Marchant 2014; Marchant and Lane 2014; Gillson 2015) to frame methodological approaches. This continuum is multifaceted, encompassing a wide variety of methodologies that must be taken into account at a past, present and future level to ensure a comprehensive research approach (Table 12.1).

Table 12.1 Examples of mixed methods that can be used to explore the past–present–future continuum in interdisciplinary and TD research

Past	Paleoecology and paleoclimate science, analysis of orthorectified historical aerial photographs, repeat terrestrial photographs, GIS mapping, remote sensing, documentation analysis of historical records and long-term monitoring
Present	Stakeholder engagement via interviews and workshops, vegetation surveys, analysis of modern pollen trap data and experiments
Future	Scenario planning and various modeling techniques such as adaptive cycle, Bayesian network, systems- and agent-based simulation models

12.4.1   What Operational Approaches Are Needed to Implement Ecosystem-Based Management Actions Based on Applied Paleoecology?

Paleoecology is in a unique position to offer insights into historical variability, thereby providing clues regarding climatic and anthropogenic-induced impacts over time. Policymakers and land managers can use the paleoresults in a strategic way within existing systemic structures at multiple governance scales (Monat and Gannon 2015). Coherence and coordination with governance strategies that focus on ES and human well-being values, stewardship, environment-friendly technology, innovation and investment would therefore be an ideal leverage point for mainstreaming paleoecology into the sustainability development dialogue, which contains both environmental and social dimensions (Dearing 2008; Bond and Morrison-Saunders 2011; Grace 2015; Morrison-Saunders et al. 2015; Díaz et al. 2018; Folke et al. 2021). In this regard, the analysis of multiple paleoproxies can contribute to understanding environmental change at global to local scales, from the interaction between regional biophysical drivers such as climate and fire to changes in vegetation and grazing patterns. However, it is of utmost importance that the (applied) paleoecological community coordinates research efforts to align with governance mechanisms at multiple governance scales.

South Africa's climate change and biodiversity policy supports and promotes coordinated and cross-sectoral implementation of Ecosystem-based Adaptation (EbA) (DEA and SANBI 2016; DEFF 2019). EbA is a globally-recognized approach that advocates for a climate change response that also has socio-economic and biodiversity/ecological cobenefits, thus contributing to sustainable development (Secretariat of the Convention on Biological Diversity 2009; Vignola et al. 2009; Pasquini and Cowling 2015; Aronson et al. 2019). Therefore, ecosystem-based approaches are a potential leverage point to incorporate applied paleoecology into land management and decision-making. Paleoresearch should be mainstreamed into governance structures and planning processes (including adaptive management and policy frameworks, e.g., Dearing et al. 2012a, b; Gillson and Marchant 2014), bearing in mind possible contextual factors that would either constrain or enable environmental mainstreaming efforts (Dalal-Clayton and Bass 2009; Bass et al. 2011; Pasquini and Cowling 2015; Food and Agriculture Organization (FAO) of the

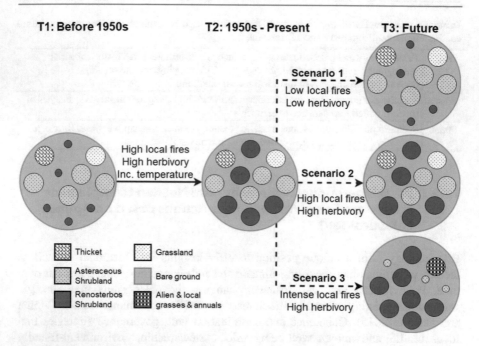

Fig. 12.4 Conceptual model showing a transition from T1 which is ca. AD 750–1950s to T2 which is ca. AD 1950s to present, and then to a potential future state (T3). Scenarios 1–3 represent potential transitions and the drivers of change associated with them. Herbivory is by livestock between ca. AD 1705–1973 and by reintroduced large indigenous herbivores since AD 1973–present. Scenario 1 shows a reversal to a pre-1950s state. Scenario 2 shows if the current state remains as is. It is hypothesized that, should an environmental threshold be crossed in the future, an alternative stable state of degraded Renosterveld may be attained (Scenario 3). This occurs as the intensity of local fires continues to increase, and the grazing of reintroduced large indigenous herbivores persists

United Nations 2016), and therefore impact the likelihood of sustained resilience and social dynamics on South African landscapes.

An interesting interface exists between paleoecology and modeling which further develops the holistic past-present-future continuum and could also improve the utilization of end-products. Despite the scarcity of research that combines long-term paleoecological data with system dynamics (e.g., United Kingdom agro-ecosystem study by McKay et al. 2019), a recent study at a lowland conservation site (Elandsberg Private Nature Reserve) in the Cape Floristic Region of South Africa showed that this can be achieved (Dirk 2021). The study noted the development of the site pre- and post-1950s to present and proposed potential future shifts to one of three alternative stable states (Fig. 12.4). It is indeterminate as to when a forthcoming regime shift will occur and the shift could be due to inappropriate levels of land-use disturbance (fire and overgrazing) and/or climate change. It has been observed that Degraded Renosterveld would consist of more than 60% bare ground, and would be homogenous at the landscape level, with Elytropappus rhinocerotis and alien plant species dominating the area (Forbes et al. 2018). The TD study utilized high temporal resolution, multiproxy paleodata (fossil pollen, spores and charcoal) to infer the alterations of a provisioning Ecosystem Service (plant biodiversity) and two land-use drivers of change (fire and herbivory) to define the historical range of variability. Participatory system dynamics—including a multistakeholder engagement workshop and semistructured interviews with commercial farmers, conservation practitioners and government officials—was used to unravel the temporal complexity of the area. This approach was used to identify feedbacks in the dynamic SES structure and analyze potential scenarios in response to grazing and fire policy, management practices and climate change. The end-product included a simulation model interface (Story Interface in Stella® Architect, 2019. Isee Systems Inc.), which facilitates engagement with interactive paleodata visualizations enabled by the system dynamics model. The end-product could then be used as a participatory tool or boundary object (Star and Griesemer 1989; Fischer and Riechers 2019) to encourage dialogue regarding unexpected simulation results and future scenarios to provide information to aid in land-use management and the promotion of resilience in the region.

When amalgamating techniques from a variety of disciplines, it is imperative to prioritize the politics and social dynamics associated with not only the formation of knowledge, but also its application outside of academia (Roux et al. 2017; Biermann et al. 2020). To achieve sustainable ecosystem management, the manner in which data is obtained and utilized should incorporate stakeholder participation and mutual learning (Knight et al. 2008). Moreover, the applied paleoecological Community of Practice (CoP) needs to consider the benefits of reflection, and the compilation and dissemination of insights on how and why we engage with stakeholders. A common practice in climate change development programs is a process of monitoring and evaluation (M&E) to assess desired outcomes and impacts, and capturing lessons learned during project implementation (Spearman 2011; Bours et al. 2014). Such principles of transparency and replicability are equally important in applied research. Applied paleoecology is "reflexive" by nature because it uses techniques to gather data that describes the past and develops insights for the present and future. By harnessing best practices from the reflexive nature of paleoecology, the applied paleoecology CoP should be reflecting and reporting on the operational research processes they employ as innovative approaches and methodologies emerge. Documenting and sharing the novel process steps and methodological adaptations are essential for case-specific, multisector and geographically diverse contexts in southern African is an essential knowledge management practice for advancing this field (e.g., Dirk 2021).

Community buy-in and stakeholder ownership is essential for addressing knowledge gaps, integrating diverse knowledge streams and bringing about meaningful change. Therefore, novel approaches whereby researchers and stakeholders iteratively and collaboratively formulate research questions based on real-world problems, use mixed methods and reflect on the applicable evidence and end-products is the fundamental nature of sustainability science (Kates et al. 2009). In addition to using participatory approaches for multistakeholder collaboration at research conception and implementation, long-term data needs to be relevant and effective for communities that require the information. Thus, (applied) paleoecologists need to assume the duty of disseminating paleooutputs which are converted and presented in accessible formats for all relevant stakeholder groups (be they are from the public, private, or civil society sector). Applied paleoecological outputs packaged into useful management decision-support tools will hopefully empower and motivate land-users to practice biodiversity conservation and use natural resources judiciously (Gelderblom et al. 2003; Jackson et al. 2009).

12.5 Conclusions

The South African paleocommunity, in collaboration with international partners and funders, has actively sought to understand and decipher past environmental change. Historically, this has often taken on the form of relatively low temporal resolution studies emphasizing environmental reconstruction over a variety of timescales (decadal–centennial–millennial timescales), with the presentation of the generated data often being inaccessible to other disciplines and end-users, due to the specialist format or the lack of dissemination.

However, over the last decade or so, the development of the science has been recognized as a valuable instrument in informing on the natural variability and thresholds of an ecological system. Globally, paleoenvironmental research has transitioned to encompass quantitative approaches including modeling, experimentation and observation to directly inform natural resource management practices.

Thus, to take timely action in a world that is facing global problems (such as climate change, poverty and pandemics such as COVID-19), the paleocommunity together with other stakeholders (as multiactor levers) should consider using a past-present-future continuum. Through this lens, innovation and advancement is propelled with an emphasis on recognizing and initiating points of leverage to incorporate paleoecological understandings of long-term change into numerous institutionalized governance levels.

The inclusion of applied paleoecology by the South African paleocommunity, by actively incorporating reflexivity into their research outcomes and providing opportunities for knowledge sharing, will enable them to tailor their work for effective context-based interventions by other researchers and practitioners. It is recommended that reflexive outputs such as lessons learnt, conceptual frameworks, historical range of variability and limits of acceptable change could be compiled and disseminated in the format of policy briefs, grey literature and shared with other knowledge holders (e.g., restoration ecology) Gillson et al. 2021.

Lastly, the next iteration of paleoecological outputs must take into account the end-user and the practicality of the data to effectively guide sustainable land management through the incorporation of policy-relevant concepts such as ES, ecosystem function and social-ecological systems resilience (Dirk 2021).

Acknowledgments We acknowledge the German Federal Ministry of Education and Research (BMBF) project TRACES (Tracing Human and Climate impacts in South Africa) project number: 03F0798A. We further acknowledge the collaborative project "Regional Archives for Integrated Investigations" (RAiN), which is embedded in the international research program SPACES (Science Partnership for the Assessment of Complex Earth System Processes). KLK is supported by DAAD within the framework of the Climate Research for Alumni and Postdocs in Africa (climapAfrica) program (Reference no. 57576494) with funds of the German Federal Ministry of Education and Research, the DSI-NRF Centre of Excellence in Palaeosciences (Grant Ref#: COE2021NGP-KD), NRF CPRR, 120806, "2020–2022 Novel isotopic analysis and U-series dating of terrestrial carbonates to understand the last 1 million years of southern African hydroclimate variability" (PI: Pickering) and the UCT VC 2030 Leadership Program Operational Support, 2019–2023 (PI: Pickering). LG would like to acknowledge National Research Foundation (NRF) Competitive Program for Rated Researchers (Grant Number 118538), NRF/African Origins Platform (Grant Number 117666), NRF/Global Change Grand Challenge/SASSCAL (Grant number 118589), and the University of Cape Town Vice Chancellor's Future Leaders Fund. The publisher is fully responsible for the content.

References

Anderson NJ (1995) Using the past to predict the future: lake sediments and the modelling of limnological disturbance. Ecol Model 78:149–172. https://doi.org/10.1016/0304-3800(94)00124-Z

Aquino-López MA, Blaauw M, Christen JA, Sanderson NK (2018) Bayesian analysis of 210 Pb dating. J Agric Biol Environ Stat 23:317–333. https://doi.org/10.1007/s13253-018-0328-7

Aronson J, Shackleton S, Sikutshwa L (2019) Joining the puzzle pieces: reconceptualising ecosystem- based adaptation in South Africa within the current natural resource management and adaptation context. Cape Town, South Africa

Bass S, Banda JLL, Chiotha S, et al. (2011) Mainstreaming the environment in Malawi's development: Experience and next steps.

Biermann C, Kelley LC, Lave R (2020) Putting the anthropocene into practice: methodological implications. Ann Assoc Am Geogr 111:808–818

Birks HJB (2012) Ecological palaeoecology and conservation biology: Controversies, challenges, and compromises. Int J Biodivers Sci Ecosyst Serv Manag 8:292–304. https://doi.org/10.1080/21513732.2012.701667

Birks HJB, Lotter AF, Juggins S, Smol JP (2012) Tracking environmental Change using lakes sediment: volume 5 data handling and numerical techniques. Springer, London

Blaauw M, Bakker R, Christen JA et al (2007) A Bayesian framework for age modeling of radiocarbon-dated peat deposits: case studies from the Netherlands. Radiocarbon 49:357–367

Blaauw M, Christen JA, Aquino-López MA (2020) A Review of Statistics in Palaeoenvironmental Research. J Agric Biol Environ Stat 25:17–31. https://doi.org/10.1007/s13253-019-00374-2

Blair D, Shackleton CM, Mograbi PJ (2018) Cropland abandonment in South African smallholder communal lands: Land cover change (1950–2010) and farmer perceptions of contributing factors. Land 7:1–20. https://doi.org/10.3390/land7040121

Bond WJ, Midgley GF (2012) Carbon dioxide and the uneasy interactions of trees and savannah grasses. Philos Trans R Soc Lond Ser B Biol Sci 367:601–612. https://doi.org/10.1098/rstb.2011.0182

Bond AJ, Morrison-Saunders A (2011) Re-evaluating sustainability assessment: aligning the vision and the practice. Environmental Impact Assessment 31:1–7. https://doi.org/10.1016/j.eiar.2010.01.007

Bond WJ, Stevens N, Midgley GF, Lehmann CER (2019) The Trouble with Trees: Afforestation Plans for Africa. Trends Ecol Evol 34:963–965. https://doi.org/10.1016/j.tree.2019.08.003

Bours D, McGinn C, Pringle P (2014) Guidance note 3: Theory of Change approach to climate change adaptation programming

Bronk Ramsey C (2008) Deposition models for chronological records. Quat Sci Rev 27:42–60

Carrion JS, Brink JS, Scott L, Binneman JNF (2000) Palynology and palaeo-environment of Pleistocene hyaena coprolites from an open-air site at Oyster Bay, Eastern Cape coast, South Africa. S Afr J Sci 96:449–453

Chase BM, Meadows ME (2007) Late Quaternary dynamics of southern Africa's winter rainfall zone. Earth Sci Rev 84:103–138. https://doi.org/10.1016/j.earscirev.2007.06.002

Chevalier M (2021) crestr an R package to perform probabilistic climate reconstructions using fossil proxies. Clim Past Discuss 18:1–35. https://doi.org/10.5194/cp-2021-153

Chevalier M, Chase BM (2016) Determining the drivers of long-term aridity variability: a southern African case study. J Quat Sci 31:143–151. https://doi.org/10.1002/jqs.2850

Coetzee JA (1967) Pollen analytical studies in East and Southern Africa. Palaeoecology of Africa 3:125–147

Cramer MD, Power SC, Belev A et al (2019) Are forest-shrubland mosaics of the Cape Floristic Region an example of alternate stable states? Ecography 42:717–729. https://doi.org/10.1111/ecog.03860

Dabengwa AN, Gillson L, Bond WJ (2021) Resilience modes of an ancient valley grassland in South Africa indicated by palaeoenvironmental methods. Environ Res Lett 16(5):1–31. https://doi.org/10.1080/14484846.2018.1432089

Dalal-Clayton B, Bass S (2009) The challenges of environmental mainstreaming: experience of integrating environment into development institutions and decisions. IIED, London

Davidson NC (2016) Editorial: understanding change in the ecological character of internationally important wetlands. Mar Freshw Res 67:685–686. https://doi.org/10.1071/MF16081

Davidson TA, Bennion H, Reid M et al (2018) Towards better integration of ecology in palaeoecology: from proxies to indicators, from inference to understanding. J Paleolimnol 60:109–116. https://doi.org/10.1007/s10933-018-0032-1

Dawson TP, Jackson ST, House JI et al (2011) Beyond predictions: biodiversity conservation in a changing climate. Science 332:53–58

DEA, SANBI (2016) Strategic framework and overarching implementation plan for ecosystem-based adaptation (EbA) in South Africa: 2016–2021. DEA, SANBI, Pretoria

Dearing JA (2008) Landscape change and resilience theory: A palaeoenvironmental assessment from Yunnan, SW China. The Holocene 18:117–127. https://doi.org/10.1177/0959683607085601

Dearing JA, Bullock S, Costanza R et al (2012a) Navigating the perfect storm: research strategies for social ecological systems in a rapidly evolving world. Environ Manag 49:767–775

Dearing JA, Yang X, Dong X et al (2012b) Extending the timescale and range of ecosystem services through paleoenvironmental analyses, exemplified in the lower Yangtze basin. Proc Natl Acad Sci 109:E1111–E1120. https://doi.org/10.1073/pnas.1118263109

DEFF (2019) Ecosystem based adaptation Action Plan and Priority Areas Mapping report. DEFF, Pretoria

Dewar G, Reimer PJ, Sealy J, Woodborne S (2012) Late-Holocene marine radiocarbon reservoir correction (ΔR) for the west coast of South Africa. The Holocene 22:1481–1489. https://doi.org/10.1177/0959683612449755

Díaz S, Pascual U, Stenseke M et al (2018) Assessing nature's contributions to people. Science 359:270–272

Dirk CJ (2021) Using applied palaeoecology and participatory system dynamics modelling to investigate changes in ecosystem services in response to climate and social-ecological drivers within the Middle Berg River Catchment. University of Cape Town, Cape Town

Dirk CJ, Gillson L (2020) Using paleoecology to inform restoration and conservation of endangered heathlands. Past Global Changes Magazine 28:20–21. https://doi.org/10.22498/pages.28.1.3

Ekblom A, Gillson L (2010) Hierarchy and scale: Testing the long term role of water, grazing and nitrogen in the savanna landscape of Limpopo National Park (Mozambique). Landsc Ecol 25:1529–1546. https://doi.org/10.1007/s10980-010-9522-x

Esterhuysen AB, Smith JM (2003) A comparison of charcoal and stable carbon isotope results for the Caledon River Valley, Southern Africa, for the period 13,500–5000 yr BP. South African Archaeol Bull 58:1–5. https://doi.org/10.2307/3889151

Falk DA (2017) Restoration ecology, resilience, and the axes of change. Ann Mo Bot Gard 102:201–216. https://doi.org/10.3417/2017006

Finlayson CM, Bellio MG, Lowry JB (2005) A conceptual basis for the wise use of wetlands in northern Australia - linking information needs, integrated analyses, drivers of change and human well-being. Mar Freshw Res 56:269–277. https://doi.org/10.1071/MF04077

Finlayson CM, Clarke SJ, Davidson NC, Gell P (2016) Role of palaeoecology in describing the ecological character of wetlands. Mar Freshw Res 67:687–694. https://doi.org/10.1071/MF15293

Finlayson C, Baumgartner LJ, Gell P (2017) We need more than just extra water to save the Murray-Darling Basin. In: The Conversation. http://theconversation.com/we-need-more-than-just-extra-water-to-save-the-murray-darling-basin-80188

Fischer J, Riechers M (2019) A leverage points perspective on sustainability. People and Nature 1:115–120

Fitchett JM, Grab SW, Bamford MK, Mackay AW (2017) Late quaternary research in southern Africa: progress, challenges and future trajectories. Trans R Soc S Afr 72:280–293

Folke C, Polasky S, Rockstrom J et al (2021) Our future in the Anthropocene biosphere. Ambio 50:834–869. https://doi.org/10.1007/s13280-021-01544-8

Food and Agriculture Organization (FAO) of the United Nations (2016) Mainstreaming ecosystem services and biodiversity into agricultural production and management in East Africa

Forbes CJ, Gillson L, Hoffman MT (2018) Shifting baselines in a changing world: Identifying management targets in endangered heathlands of the Cape Floristic Region, South Africa. Anthropocene 22:81–93. https://doi.org/10.1016/j.ancene.2018.05.001

Froyd CA, Willis KJ (2008) Emerging issues in biodiversity & conservation management: The need for a palaeoecological perspective. Quat Sci Rev 27:1723–1732. https://doi.org/10.1016/j.quascirev.2008.06.006

Gelderblom CM, van Wilgen BW, Nel JL et al (2003) Turning strategy into action: Implementing a conservation action plan in the Cape Floristic Region. Biol Conserv 112:291–297

Gell PA (2017) Using paleoecology to understand natural ecological character in Ramsar wetlands. Past Global Changes Magazine 25:86–87. https://doi.org/10.22498/pages.25.2.86

Gell P, Mills K, Grundell R (2013) A legacy of climate and catchment change: The real challenge for wetland management. Hydrobiologia 708:133–144. https://doi.org/10.1007/s10750-012-1163-4

Gell PA, Finlayson CM, Davidson NC (2016) Understanding change in the ecological character of Ramsar wetlands: Perspectives from a deeper time - Synthesis. Mar Freshw Res 67:869–879. https://doi.org/10.1071/MF16075

Gell PA, Perga M-E, Finlayson CM (2018) Changes over time. In: Hughes J (ed) Freshwater ecology and conservation: approaches and techniques. Oxford University Press, Oxford, pp 283–305

Gillson L (2015) Biodiversity conservation and environmental change: using palaeoecology to manage dynamic landscapes in the Anthropocene. Oxford University Press, Oxford

Gillson L, Duffin KI (2007) Thresholds of potential concern as benchmarks in the management of African savannahs. Philos Trans R Soc B Biol Sci 362:309–319. https://doi.org/10.1098/rstb.2006.1988

Gillson L, Ekblom A (2009) Resilience and thresholds in Savannas: Nitrogen and fire as drivers and responders of vegetation transition. Ecosystems 12:1189–1203. https://doi.org/10.1007/s10021-009-9284-y

Gillson L, Ekblom A (2020) Using palaeoecology to explore the resilience of southern African savannas. Koedoe 62:1–12. https://doi.org/10.4102/koedoe.v62i1.1576

Gillson L, Dirk C, Gell P (2021) Using long-term data to inform a decision pathway for restoration of ecosystem resilience. Anthropocene, 36, p.100315

Gillson L, Marchant R (2014) From myopia to clarity: Sharpening the focus of ecosystem management through the lens of palaeoecology. Trends Ecol Evol 29:317–325. https://doi.org/10.1016/j.tree.2014.03.010

Gillson L, MacPherson AJ, Hoffman MT (2020) Contrasting mechanisms of resilience at mesic and semi-arid boundaries of fynbos, a mega-diverse heathland of South Africa. Ecol Complex 42:100827. https://doi.org/10.1016/j.ecocom.2020.100827

Grace WR (2015) Sustainability and the SDGs: a systems perspective. University of Queensland, St Lucia

Haberzettl T, Baade J, Compton J et al (2014) Paleoenvironmental investigations using a combination of terrestrial and marine sediments from South Africa - The RAIN (Regional Archives for Integrated iNvestigations) approach. Zentralblatt für Geologie und Paläontologie, Teil I:55–73. https://doi.org/10.1127/zgpI/2014/0055-0073

Haberzettl T, Kirsten KL, Kasper T et al (2019) Using 210 Pb-data and paleomagnetic secular variations to date anthropogenic impact on a lake system in the Western Cape, South Africa. Quat Geochronol 51:53–63. https://doi.org/10.1016/j.quageo.2018.12.004

Hempson GP, Archibald S, Bond WJ (2015) A continent-wide assessment of the form and intensity of large mammal herbivory in Africa. Science 350:1056–1061

Higgs E, Falk DA, Guerrini A et al (2014) The changing role of history in restoration ecology. Front Ecol Environ 12:499–506. https://doi.org/10.1890/110267

Hoffman MT, O'Connor TG (1999) Vegetation change over 40 years in the Weenen/ Muden area, KwaZulu-Natal: evidence from photo-panoramas. Afr J Range Forage Sci 16:71–88

Hoffman MT, Rohde RF, Gillson L (2019) Rethinking catastrophe? Anthropocene 25:100189

Holdo RM, Sinclair ARE, Dobson AP et al (2009) A Disease-Mediated Trophic Cascade in the Serengeti and its Implications for Ecosystem C. PLoS Biol 7:e1000210. https://doi.org/10.1371/journal.pbio.1000210

Humphrey GJ, Gillson L, Ziervogel G (2020) How changing fire management policies affect fire seasonality and livelihoods. Ambio 50:475–491. https://doi.org/10.1007/s13280-020-01351-7

Iglesias V, Yospin GI, Whitlock C (2015) Reconstruction of fire regimes through integrated paleoecological proxy data and ecological modeling. Front Plant Sci 5:1–12. https://doi.org/10.3389/fpls.2014.00785

IPCC (2021) Climate change 2021: the physical science basis. Contribution of working group I to the sixth assessment report of the intergovernmental panel on climate change. Cambridge University Press, Cambridge

Jackson ST, Hobbs RJ (2009) Ecological restoration in the light of ecological history. Science 325:567–569. https://doi.org/10.1126/science.1172977

Jackson ST, Gray ST, Shuman B (2009) Paleoecology and resource management in a dynamic landscape: case studies from the Rocky Mountain headwaters. Conserv Paleobiol Sci Pract 15:61–80

Jeffers ES, Nogué S, Willis KJ (2015) The role of palaeoecological records in assessing ecosystem services. Quat Sci Rev 112:17–32. https://doi.org/10.1016/j.quascirev.2014.12.018

Johnstone JF, Allen CD, Franklin JF et al (2016) Changing disturbance regimes, ecological memory, and forest resilience. Front Ecol Evol 14:369–378. https://doi.org/10.1002/fee.1311

Kates RW, Clark WC, Corell R et al (2009) Sustainability science. Science 292:641–642

Kirsten KL, Haberzettl T, Wündsch M et al (2018) A multiproxy study of the ocean-atmospheric forcing and the impact of sea-level changes on the southern Cape coast, South Africa during the Holocene. Palaeogeogr Palaeoclimatol Palaeoecol 496:282–291. https://doi.org/10.1016/j.palaeo.2018.01.045

Kirsten KL, Kasper T, Cawthra HC et al (2020) Holocene variability in climate and oceanic conditions in the winter rainfall zone of South Africa—inferred from a high resolution diatom record from Verlorenvlei. J Quat Sci 35:572–581. https://doi.org/10.1002/jqs.3200

Knight AT, Cowling RM, Rouget M, Balmford A, Lombard AT, Campbell BM (2008) Knowing but not doing: Selecting priority conservation areas and the research-implementation gap. Conserv Biol 22:610–617

Laris P (2002) Burning the seasonal Mosaic: preventative burning strategies in the wooded Savanna of Southern Mali. Hum Ecol 30:155–186

Line JM, ter Braak CJF, Birks HJB (1994) WACALIB version 3.3 - a computer program to reconstruct environmental variables from fossil assemblages by weighted averaging and to derive sample-specific errors of prediction. J Paleolimnol 10:147–152. https://doi.org/10.1007/BF00682511

Lovejoy TE (2007) Paleoecology and the path ahead. Front Ecol Environ 5:456. https://doi.org/10.1890/1540-9295(2007)5[456:PATPA]2.0.CO;2

Maboya ML, Meadows ME, Reimer PJ et al (2018) Late Holocene marine radiocarbon reservoir correction for the south and east coast of South Africa. Radiocarbon 60:571–582

MacPherson AJ, Gillson L, Hoffman MT (2019) Between- and within-biome resistance and resilience at the fynbos-forest ecotone, South Africa. The Holocene 29:1801–1816. https://doi.org/10.1177/0959683619862046

Maezumi SY, Gosling WD, Kirschner J et al (2021) A modern analogue matching approach to characterize fire temperatures and plant species from charcoal. Palaeogeogr Palaeoclimatol Palaeoecol 578:110580. https://doi.org/10.1016/j.palaeo.2021.110580

Manzano S, Julier ACM, Dirk CJ et al (2020) Using the past to manage the future: the role of palaeoecological and long-term data in ecological restoration. Restor Ecol 28:1335–1342. https://doi.org/10.1111/rec.13285

Marchant R, Lane P (2014) Past perspectives for the future: foundations for sustainable development in East Africa. J Archaeol Sci 51:12–21

Martin ARH (1956) The ecology and history of Groenvlei. S Afr J Sci 52:187–192

McKay DIA, Dearing JA, Dyke JG et al (2019) To what extent has sustainable intensification in England been achieved? Sci Total Environ 648:1560–1569

McWethy DB, Higuera PE, Whitlock C et al (2013) A conceptual framework for predicting temperate ecosystem sensitivity to human impacts on fire regimes. Glob Ecol Biogeogr 22:900–912. https://doi.org/10.1111/geb.12038

Meadows ME (2014) Recent methodological advances in Quaternary palaeoecological proxies. Prog Phys Geogr 38:807. https://doi.org/10.1177/0309133314540690

Meadows ME (2015) Seven decades of Quaternary palynological studies in southern Africa: A historical perspective. Trans R Soc S Afr 70:103–108. https://doi.org/10.1080/0035919X.2015.1004139

Midgley GF, Bond WJ (2015) Future of African terrestrial biodiversity and ecosystems under anthropogenic climate change. Nat Clim Chang 5:823–829. https://doi.org/10.1038/nclimate2753

Monat JP, Gannon TF (2015) What is systems thinking? A review of selected literature plus recommendations. Am J Syst Sci 4.11–26. https://doi.org/10.5923/j.ajss.20150401.02

Morrison-Saunders A, Pope J, Bond A (eds) (2015) Handbook of sustainability assessment. Edward Elgar Publishing, Cheltenham

Neumann FH, Scott L, Bamford MK (2011) Climate change and human disturbance of fynbos vegetation during the late Holocene at Princess Vlei, Western Cape, South Africa. The Holocene 21:1137–1149. https://doi.org/10.1177/0959683611400461

O'Connor TG, Puttick JR, Hoffman MT (2014) Bush encroachment in southern Africa: changes and causes. Afr J Range Forage Sci 31:67–88. https://doi.org/10.2989/10220119.2014.939996

Ofcansky TP (1981) The 1889–1897 Rinderpest epidemic and the rise of British and German Colonialism in Eastern and Southern Africa. J Afr Stud 8:31–38

Pasquini L, Cowling RM (2015) Opportunities and challenges for mainstreaming ecosystem-based adaptation in local government: evidence from the Western Cape, South Africa. Environ Dev Sustain 17:1121–1140

Pickett STA, Ostfeld RS, Shacha M, Likens GE (eds) (1997) The ecological basis of conservation; heterogeneity, ecosystems, and biodiversity. Springer Science & Business Media, New York

Rogers KH (2003) Adopting a heterogeneity paradigm: implications for management of protected savannas. In: du Toit JT, Rogers KH, Biggs HC (eds) The Kruger experience: ecology and management of Savanna heterogeneity. Island Press, Washington, DC, pp 41–58

Roux DJ, Nel JL, Cundill G, O'farrell P, Fabricius C (2017) Transdisciplinary research for systemic change: who to learn with, what to learn about and how to learn. Sustainability Science 12:711–726

Rull V (2014) Time continuum and true long-term ecology: from theory to practice. Front Ecol Evol 2:1–7. https://doi.org/10.3389/fevo.2014.00075

Scheiter S, Higgins SI (2012) How many elephants can you fit into a conservation area. Conserv Lett 5:176–185. https://doi.org/10.1111/j.1755-263X.2012.00225.x

Scott L (1988) Holocene environmental change at western Orange Free State pans, South Africa, inferred from pollen analysis. Palaeoecol Afr 19:109–119

Scott L, Marais E, Brook GA (2004) Fossil hyrax dung and evidence of Late Pleistocene and Holocene vegetation types in the Namib Desert. J Quat Sci 19:829–832. https://doi.org/10.1002/jqs.870

Secretariat of the Convention on Biological Diversity (2009) Connecting biodiversity and climate change mitigation and adaptation: report of the second ad hoc technical expert group on biodiversity and climate change. Secretariat of the Convention on Biological Diversity, Montreal

Seddon AWR, Mackay AW, Baker AG et al (2014) Looking forward through the past: identification of 50 priority research questions in palaeoecology. J Ecol 102:256–267. https://doi.org/10.1111/1365-2745.12195

Smith JM, Lee-Thorp JA, Sealy JC (2002) Stable carbon and oxygen isotopic evidence for late Pleistocene to middle Holocene climatic fluctuations in the interior of Southern Africa. J Quat Sci 17:683–695. https://doi.org/10.1002/jqs.687

Sobol MK, Scott L, Finkelstein SA (2019) Reconstructing past biomes states using machine learning and modern pollen assemblages: a case study from Southern Africa. Quat Sci Rev 212:1–17. https://doi.org/10.1016/j.quascirev.2019.03.027

Spearman M (2011) Making Adaptation Count: Concepts and options for monitoring and evaluation of climate change adaptation. World Resources Institute: Washington, D.C.

Star SL, Griesemer JR (1989) Institutional ecology, translations' and boundary objects: Amateurs and professionals in Berkeley's Museum of Vertebrate Zoology, 1907–39. Soc Stud Sci 19:387–420

Strachan K, Hinch JM, Hill T, Barnett RL (2014) A late Holocene sea-level curve for the east coast of South Africa. S Afr J Sci 110:1–9

Strachan KL, Hill TR, Finch JM, Barnett RL (2015) Vertical zonation of foraminifera assemblages in Galpins salt marsh, South Africa. J Foraminifer Res 45:29–41

Strobel P, Kasper T, Frenzel P et al (2019) Late Quaternary palaeoenvironmental change in the year-round rainfall zone of South Africa derived from peat sediments from Vankervelsvlei. Quat Sci Rev 218:200–214. https://doi.org/10.1016/j.quascirev.2019.06.014

Strobel P, Haberzettl T, Bliedtner M et al (2020) The potential of $\delta 2Hn$-alkanes and $\delta 18Osugar$ for paleoclimate reconstruction – a regional calibration study for South Africa. Sci Total Environ 716:137045. https://doi.org/10.1016/j.scitotenv.2020.137045

Thackeray AI (1992) The Middle Stone Age south of the Limpopo River. J World Prehist 6:385–440. https://doi.org/10.1007/BF00975633

van Wilgen BW, Biggs HC (2011) A critical assessment of adaptive ecosystem management in a large savanna protected area in South Africa. Biol Conserv 144:1179–1187. https://doi.org/10.1016/j.biocon.2010.05.006

van Wilgen BW, Govender N, Smit IPJ, Macfadyen S (2014) The ongoing development of a pragmatic and adaptive fire management policy in a large African savanna protected area. J Environ Manag 132:358–368. https://doi.org/10.1016/j.jenvman.2013.11.003

Venter ZS, Hawkins H-J, Cramer MD (2017) Implications of historical interactions between herbivory and fire for rangeland management in African savannas. Ecosphere 8:e01946. https://doi.org/10.1002/ecs2.1946

Vignola R, Locatelli B, Martinez C, Imbach P (2009) Ecosystem-based adaptation to climate change: what role for policy-makers, society and scientists? Mitig Adapt Strateg Glob Chang 14:691–696

Wigley BJ, Bond WJ, Hoffman MT (2010) Thicket expansion in a South African savanna under divergent land use: Local vs. global drivers? Glob Chang Biol 16:964–976. https://doi.org/10.1111/j.1365-2486.2009.02030.x

Williams JW, Jackson ST (2007) Novel climates, no-analog communities, and ecological surprises. Front Ecol Environ 5:475–482. https://doi.org/10.1890/070037

Willis KJ, Bhagwat SA (2010) Questions of importance to the conservation of biological diversity: Answers from the past. Clim Past 6:759–769. https://doi.org/10.5194/cp-6-759-2010

Willis KJ, Birks HJB (2006) What is natural? The importance of a long-term perspective in biodiversity conservation and management. Science 314:1261–1265

Willis KJ, Bailey RM, Bhagwat SA, Birks HJB (2010) Biodiversity baselines, thresholds and resilience: Testing predictions and assumptions using palaeoecological data. Trends Ecol Evol 25:583–591. https://doi.org/10.1016/j.tree.2010.07.006

Wündsch M, Haberzettl T, Kirsten KL et al (2016) Sea level and climate change at the southern Cape coast, South Africa, during the past 4.2 kyr. Palaeogeogr Palaeoclimatol Palaeoecol 446:295–307. https://doi.org/10.1016/j.palaeo.2016.01.027

Wündsch M, Haberzettl T, Cawthra HC et al (2018) Holocene environmental change along the southern Cape coast of South Africa – insights from the Eilandvlei sediment record spanning the last 8.9 kyr. Glob Planet Chang 163:51–66. https://doi.org/10.1016/j.gloplacha.2018.02.002

Soil Erosion Research and Soil Conservation Policy in South Africa

13

Jussi Baade ⓘ, Ilse Aucamp ⓘ, Anneliza Collett, Frank Eckardt ⓘ,
Roger Funk ⓘ, Christoph Glotzbach ⓘ, Johanna von Holdt ⓘ,
Florian Kestel ⓘ, Jaap Knot, Antoinette Lombard ⓘ,
Theunis Morgenthal ⓘ, Alex Msipa, and Jay J. Le Roux ⓘ

J. Baade (✉)
Department of Geography, Friedrich Schiller University Jena, Jena, Germany
e-mail: jussi.baade@uni-jena.de

I. Aucamp
Equispectives Research and Consulting Services, Pretoria, South Africa
e-mail: ilse@equispectives.co.za

A. Collett · T. Morgenthal
Directorate Land Use and Soil Management, Department of Agriculture, Land Reform and Rural
Development, Pretoria, South Africa
e-mail: AnnelizaC@dalrrd.gov.za; TheunisM@dalrrd.gov.za

F. Eckardt · J. von Holdt
University of Cape Town, Environmental and Geographical Science, Cape Town, South Africa
e-mail: frank.eckardt@uct.ac.za; johanna.vonholdt@uct.ac.za

R. Funk · F. Kestel
Leibniz Centre for Agricultural Landscape Research (ZALF), Müncheberg, Germany
e-mail: rfunk@zalf.de; Florian.Kestel@zalf.de

C. Glotzbach
Department of Geosciences, University of Tübingen, Tübingen, Germany
e-mail: christoph.glotzbach@uni-tuebingen.de

J. Knot
Independent Consultant, Ladybrand, South Africa

A. Lombard · A. Msipa
Department of Social Work and Criminology, University of Pretoria, Hatfield, South Africa
e-mail: Antoinette.Lombard@up.ac.za

J. J. Le Roux
Afromontane Research Unit, Faculty: Natural and Agricultural Sciences, University of the Free
State, Bloemfontein, South Africa
e-mail: LeRouxJJ@ufs.ac.za

© The Author(s) 2024
G. P. von Maltitz et al. (eds.), *Sustainability of Southern African Ecosystems
under Global Change*, Ecological Studies 248,
https://doi.org/10.1007/978-3-031-10948-5_13

335

Abstract

Soil erosion has been identified as an issue in South African farming for more than a century. Erosion of land surfaces by water or wind is a natural process which might be accelerated directly by human impact on land surface properties, e.g., vegetation and soils. An assessment of soil erosion risk indicates average soil loss rates two orders of magnitude larger than long-term soil formation rates. This challenging condition clearly underlines the need for continuous application of established policies and principles as well as emerging modes of conservation agriculture in farming activities in most parts of South Africa. In addition, conservation agriculture has been shown to have positive effects on the cost–value ratio, but diffusion and adoption of this innovative approach still meet resistance often founded in traditional faith and belief systems. However, to cope with challenges from global climate change, e.g., intensified extreme weather conditions (droughts and flooding), strengthened resilience of farming systems is required to i) meet increased domestic and global demand for food and ii) to put into practice sustainable management to diminish on-site and off-site damages from soil erosion on the way to reach sustainable development goals.

13.1   Introduction

Due to the interaction of internal (e.g., tectonic) and external (e.g., climate) processes the surface of the Earth has been constantly changing for millions of years. Plate tectonics might create mountain ranges by uplifting rocks while climate induced erosion, denudation and weathering decays rocks and minerals and lowers the mountain ranges (e.g., England and Molnar 1990). Thus, erosion and denudation are natural processes which can be accelerated by human impact resulting in what is then called soil erosion, either by water or wind or direct human action like ploughing and other means of removal of soils (Bennett 1939, Shakesby 2003, Baade 2006).

Many authors around the World (Oldeman et al. 1991, FAO 2019) and in southern Africa (Hoffman and Ashwell 2001, Boardman et al. 2012, FAO and ITPS 2015) consider soil erosion as the major land degradation process. In addition, biological (e.g., loss of soil organic matter), physical (e.g., soil compaction) and chemical degradation (e.g., nutrient loss, acidification and salinization) of soils have to be noted. In South Africa, early mentioning of soil erosion issues, initially often related to livestock farming and overgrazing date back to the eighteenth and nineteenth centuries (Hoffman and Ashwell 2001, Beinart 2003, Rowntree 2013). Widespread recognition of land degradation and soil erosion and the development of land degradation and land conservation policies started in the late nineteenth and early twentieth centuries (Cooper 1996, Beinart 2003) and provide the foundation for the status quo and future developments.

This contribution will focus on the soil erosion problem in South Africa, its extent, existing soil conservation measures and policies, socioeconomic dimensions and the challenges for future sustainable agricultural soil use preventing land degradation as much as possible. To set the scene, we first look at rates of late Quaternary and current geological erosion and soil formation. We will then briefly review the general effects of human induced soil erosion on the soils and the environment, such as the on- and off-site damages from soil erosion. Acknowledging, that the problem of soil erosion has already been noticed in the eighteenth century and soil conservation policies developed in the early twentieth century, we then review the policy development. The next two sections provide an overview of the current extent of soil erosion by both water and wind in South Africa. We subsequently examine the socioeconomic dimension of soil erosion and land degradation, focusing on stakeholder's and farmer's perspective. In conclusion, we identify some major challenges for soil conservation and soil conservation policies in South Africa. In general, the scope of this contribution is to provide a nation-wide overview and to bridge natural science-based findings and socioeconomic aspects of human induced soil erosion. Due to the rather complex physical, historical and socioeconomic causes of land degradation and soil erosion (e.g., Meadows and Hoffman 2002), it will not be possible to consider all aspects. Despite focusing on South Africa, we are convinced that many aspects discussed may be applicable to many parts of southern Africa.

13.2 Erosion and Denudation

The evaluation of the impact of human induced soil erosion on soil degradation requires us to determine the long-term geological background (kyr-scale) erosion and denudation rates which are mainly controlled by tectonic uplift modulated by climate variability (e.g., Molnar and England 1990, Raymo and Ruddiman 1992, Binnie and Summerfield 2013). Tectonic uplift results in physical erosion which produces fresh mineral surfaces, available for chemical weathering. Increased tectonic uplift and physical erosion takes place in active mountain ranges (e.g., Himalaya, Andes), but the majority of the Earth's surface, is characterized by low tectonic activity and less pronounced topography, including large parts of South Africa.

Cosmogenic nuclide methods are well suited to determine long-term denudation rates and relate them to observed current soil erosion rates (von Blanckenburg 2006). Production of cosmogenic nuclides is highest at the Earth surface and decreases exponentially, so that most nuclides are produced in the upper few meters of Earth's surface (Bierman 1994). The concentration of cosmogenic nuclides in a surface sample is inversely related to the long-term denudation rate (Lal 1991). Thus, cosmogenic nuclide-derived denudation rates average timescales of 10^2 years in active tectonic areas with several m kyr^{-1} erosion to 10^5 years in inactive areas with several mm kyr^{-1} erosion (von Blanckenburg 2006). Often the nuclide concentration is well mixed in the soil by physical and biological processes (Schaller et al. 2018, Glotzbach et al. 2016), and therefore, recent human-induced increase in current soil erosion usually does not impact the cosmogenic nuclide concentration measured at the surface.

South Africa is characterized by a wide variety of landforms, such as flat high plateaus, steep mountains and hilly to flat coastlines (Partridge et al. 2010). Despite these great landscape variability, long-term erosion rates derived from cosmogenic nuclides do not vary much and most rates are in the order of 1 to 10 mm kyr^{-1} with an average of 3.6 ± 3.1 mm kyr^{-1} (see Glotzbach et al. 2016 for a review and additional references). It is noteworthy that erosion rates show no simple dependency on either topography, climate or rock type. For example, the steep mountains of the Cape (slope of up to >30°) do erode as slowly as the lowlands in the Kruger National Park with a rate of only ~5 mm kyr^{-1} (cf. Scharf et al. 2013, Glotzbach et al. 2016). A feasible explanation is that differences in rock type and associated erodibility are offset by other parameters, such that steep regions with weathering resistant rocks (e.g., quartzite) do erode at similar rates like less resistant rocks (e.g., basalts) in flat terrain. Chemical weathering and soil production rates are in the same range as observed long-term denudation rates and suggest that the transformation from rock into soil is very likely in steady state over long timescales (e.g., Chadwick et al. 2013, Decker et al. 2011).

Short-term (decades) sediment yield rates have been determined throughout Africa using gauging stations and reservoirs establishing yields from 0.2 to 15,700 t km^2 yr^{-1} with a median at 160 t km^2 yr^{-1} (Vanmaercke et al. 2014). In the Kruger National Park (KNP), short-term (decades) and long-term (hundred thousand of years) sediment yields have been determined by sediment trapping in dams and cosmogenic nuclides (Reinwarth et al. 2019, Glotzbach et al. 2016). Long-term erosion rates are tightly clustered around a value of ~5 mm kyr^{-1}, whereas short-term rates range between 5 and 75 mm kyr^{-1} and are on average ~ 6-times higher than the long-term rates (Fig. 13.1). The same trend with even higher short-term rates was reported by Decker et al. (2011) in south-central parts of South Africa. The rather low long-term, geologic erosion rates in the KNP can be accounted for by very low tectonic activity, low relief and rather low precipitation and weathering rates in the Lowveld. Based on the fact that the study area located in the KNP was newer exposed to European style agriculture and ploughing of soils, the current sediment yield rates were expected to be low as well. As is evident from Fig. 13.1, some catchments yield current values as low as the long-term rates, but others yielded much higher values. A similar trend can be seen elsewhere in South Africa. Only a small fraction of the catchment sediment yield rates (Vanmaercke et al. 2014) are in the range of long-term denudation rates (~3.6 mm kyr^{-1} or ~ 10 t km^2 yr^{-1}) (Glotzbach et al. 2016).

Are these differences caused by human impact or can they be explained by climate change or methodological differences? A number of studies have shown, that erosion rates determined by sediment trapping can yield either higher or lower rates compared to long-term rates. In some cases, this was clearly attributable to the climate-induced variability in sediment transport and the magnitude of events occurring during the sediment trapping period. It is believed that inclusion or

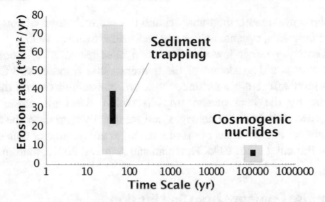

Fig. 13.1 Erosion rates [t km^{-2} yr^{-1}] of dam-locked river catchments in the Kruger National Park derived from sediment trapping averaged over a few decades and cosmogenic nuclides averaging of hundred thousand years. For conversion of erosion rates between volume and mass, we use the continental crust density of about 2.7 t m^{-3} (Compilation: C. Glotzbach)

exclusion of large infrequent flood events during the monitoring period is an explanation for the different results. Some studies observed an order of magnitude increase in denudation rates over the past decades and attributed them to human impact (Hewawasam (2003) in Sri Lanka and Raab et al. (2018) in Italy).

While the observed difference between long-term and short-term sediment yield and erosion rates in the pristine Kruger National Park environment is not easy to explain cosmogenic nuclide derived denudation rates provide a proxy for long-term soil formation rates—assuming a steady state—of about 5 mm kyr^{-1} equivalent to 13.5 t km^{-2} yr^{-1}. Current pristine erosion rates in the Lowveld are in the order of about 30 mm kyr^{-1} equivalent to 81 t km^{-2} yr^{-1}. Given the existing studies, it is suggested that sustainable agriculture should aim at limiting current soil loss in and sediment yield from catchments to the range of values presented here, i.e., to 10 to 100 t km^{-2} yr^{-1}.

13.3   Soil Erosion Due to Human Impact

Under steady-state conditions, natural, geologic erosion is governed by internal and external forces and is believed to be close to an equilibrium state between these two major forces. We can assume that, humans interfere with this balance by removing directly or indirectly (by livestock farming) the natural vegetation cover protecting the soil against the kinetic energy of rainfall or wind as well as disturbing the moisture conditions at the surface. Ploughing or other mechanical disturbance from working the soils or by livestock farming further alters the structure and stability of the soil facilitating its removal, i.e., soil erosion by wind or water (Bennett 1939, Shakesby 2003, Baade 2006). Considering the processes involved, soil erosion comprises the entrainment, the transport and the sedimentation of soil and related material like nutrients, organic matter and fertilizers (Morgan 1995).

Due to generally rather low soil formation rates (Sect. 13.2), human induced soil erosion causes a degradation of the nonrenewable resource soil (FAO 2019). Damages, which affect the area under use, are considered on-site damages and usually borne by the land owner. Impacts which affect the closer and wider surroundings are called off-site damages and are usually carried by the community. Estimates indicate that the costs of off-site damages are far greater than on-site ones (Clark 1985, Pimentel et al. 1995, Hoffman and Ashwell 2001, Boardman 2021).

13.3.1 On-Site Damages from Soil Erosion

The most important on-site damage from soil erosion is the removal of the soil and related material like seeds or plants, soil organic matter, fertilizer, and other nutrients from the fields eventually depleting the thickness and the fertility of the soil (Lal 2015). This removal might occur in the form of deflation by wind or sheet (or inter-rill) and rill erosion caused by rainfall impact (i.e., splash erosion) and overland flow. Where overland flow concentrates, (ephemeral) gully erosion can remove considerable amounts of soil in addition. These removal processes are selective leaving coarser lag deposits behind. The effect is an extensive gradual depletion and degradation of the fine soil material at the surface thinning the ecologically important upper soil horizon which provides the base for crop production and livestock farming, among other functions (du Preez et al. 2020). When it comes to erosion by water, occurring generally in hilly terrain, redeposition within a site is another aspect of concern, e.g., due to the accumulation of sediments, crusting, burial of plants and harvest losses. The same is true for wind erosion, which additionally might affect crops by abrasion (Funk and Reuter 2006).

Usually this type of extensive, subtle soil erosion progresses slowly and often unnoticed. Visible effects include a mottled development of vegetation cover, eventually spotty exposures of parent material, and variation in forage or crop yields lowering the farmers return and eventually the Gross Domestic Product of a country. Nonetheless, estimates for the costs of on-site damages by soil erosion have not yet been established for South Africa (Turpie et al. 2017), despite early requests to do so (Braune and Looser 1989).

Compared to this, the specific type of erosion caused by the concentration of overland flow in hilly terrain, i.e., gully erosion, results in often spectacular, well visible linear erosion features dissecting the land in a way that arable farming is precluded and livestock farming constricted. Often these areas are then called badlands (Fig. 13.2, Boardman and Foster 2008; Foster and Boardman 2020). A specific issue of gully erosion is the fact that this process is governed by the overland flow originating in the uphill catchment area of the gully. Given the generally steep slopes at the gully head, gully erosion and gully head retreat are often characterized by reinforcing feedback loops and very difficult to control or stabilize. Eventually,

Fig. 13.2 Example of a strongly dissected former cultivated field, north of Ladybrand, Eastern Free State Province. In the foreground a gully head is clearly visible and brush to keep off livestock (Photo: J. Baade 2018)

gully erosion might even impede or inhibit forestry activities or result in completely dissected and barren land (Bennett 1939, Boardman et al. 2012).

13.3.2 Off-Site Damage from Soil Erosion

The material removed by soil erosion from a site will be transported and eventually deposited in the closer or wider surroundings. Details of the processes and the effects are specific to the eroding and transporting agent: water or wind, and need to be examined separately.

The extent of off-site damage from soil erosion by water is controlled by the magnitude of the rainfall and runoff event causing erosion and surface runoff as well as the configuration, density and connectivity of the drainage system downstream of the eroding sites (Rowntree 2012, Msadala and Basson 2017). Often these events cause sediment rich waters to end up downstream. Higher magnitude events might further flood roads, houses and other properties causing impacts from little discomforts to considerable damage. Some of the eroded material will be deposited along the way when flooding of river banks occurs. But, in South Africa, a considerable amount of the material is deposited in the dams and reservoirs (Braune and Looser 1989, Rooseboom et al. 1992, Msadala and Basson 2017) which serve as important sources of water to humans and semiarid agriculture. The remaining sediments will be carried on by the main rivers to the oceans where they get deposited in estuaries, coastal lakes, harbors or on the shelf, e.g., the mudbelt off the Orange River mouth (Compton et al. 2010) (see Chaps. 27 and 28).

Using DWS (2017) data on reservoir siltation and the approach by Turpie et al. (2017) the monetary damage caused by the siltation of dams and reservoirs in South Africa sums up to about 2 billion ZAR for the period 1980 to 2000. This estimate is based on the costs to reestablish the lost water storage volume, only. It does not consider the costs for water treatment plants and other efforts to produce drinking water from heavily silted water (e.g., along the Caledon River). In addition, not included is the ecological damage caused by water enriched in silt, nutrient and possibly pesticides to the aquatic food resources in the receiving water courses including the oceans (Turpie et al. 2017) (cf. Chap. 31).

Off-site damage from wind erosion can be divided into those that occur immediately and those that show more long-term effects (Funk and Reuter 2006). Sand deposits are often found directly at the field boundaries after an event, filling ditches, developing fence-line dunes or covering traffic routes (Holmes et al. 2012). The immediate effects are further related to the dust emissions (Fig. 13.3). They cause reduction in visibility affecting traffic safety, and air pollution with particles of the fractions PM_{10} and $PM_{2.5}$, which are harmful to human health (Vos et al.

Fig. 13.3 Dust event near Bultfontein, Free State Province. Dust traps are collecting saltating and suspended sediments originating from a bare peanut field in August 2018. PM10 concentrations peaked at 2500 ppm during this two-hour event and resulted in the collection of 26 grams of sediment from the 4 traps (Photo: F. Eckardt 2018)

2021). Contamination of crops and fruits by dust deposits can also be considered as immediate off-site damage in the agricultural sector. Long-term damages are caused by repeated sand and dust input into adjacent sensitive areas, such as settlements or natural aquatic or terrestrial biotopes.

13.4   Soil Erosion and Conservation Policy in South Africa

13.4.1 Development of Soil Conservation Policy

According to Kanthack (1930, p. 516), the "devastation of large areas due to soil erosion" has been an issue for the general public and the farmers at least since the beginning of the twentieth century. Rowntree (2013) provides a recent review of early discussions of land degradation in the *Agricultural Journal of the Cape of Good Hope* published between the 1890s and 1910s. This concern stimulated the development and further refinement of soil conservation policies. A detail account of these developments in the twentieth century up to 1994 is provided by Cooper (1996) who presents a unique and in-depth perspective of soil conservation policies in South Africa focusing on the human dimension thereof. According to her, the first documented policies on soil conservation go back to the early Cape settlers who provided directives to land users in the form of "Placaaten" starting already in the seventeenth century (Cooper 1996, 83 f.). Most of these included indirect reference to soil conservation.

The first attempt by the Union of South Africa to formally investigate the effect of land degradation, especially soil erosion, was through the Drought Investigation Commission in 1923, followed by a Soil Erosion Conference in 1929 (Adler 1985). Initial governmental funding of erosion-control schemes to rehabilitate and prevent further erosion took place in the 1930s (Bennett 1939, Beinart 2003). The first formal legislation to address soil degradation was through the Soil Conservation Act, Act 45 of 1946 (SCA_1946, Table 13.1, Hoffman and Ashwell 2001).

The Union Government soon realized that without support, farmers will be unable to successfully rehabilitate eroded and degraded land and a number of schemes were developed, namely the Grass Ley Crop Scheme (1958–1972), Veld Reclamation Scheme (1966–1971) and the Stock Reduction Scheme (1969–1979).

Table 13.1 Overview of important acts related to soil degradation and soil conservation

Year	Act No	Name	Abbrev.
1946	45	Soil Conservation Act	SCA_1946
1949	6	Soil Conservation Amendment Act	SCAA_1949
1960	37	Soil Conservation Amendment Act	SCAA_1960
1967	15	Soil Conservation Amendment Act	SCAA_1967
1983	43	Conservation of Agricultural Resources Act	CARA
1998	107	National Environmental Management Act	NEMA
2004	10	National Environmental Management Biodiversity Act	NEM:BA

Based on these interventions, considerable subsidies and other investments to combat and rehabilitate land degradation were provided to, e.g., lower stock numbers per unit area to combat erosion due to overgrazing, terracing of arable land and the development of contour banks or runoff bunds (Cooper 1996, von Maltitz et al. 2019).

The amended Soil Conservation Act of 1967, repealed the SCA_1946 (Theron 1985) and in 1983 was replaced by the Conservation of Agricultural Resources Act (CARA) (South Africa 1983). It is important to note that CARA build on the previous act's success and shortcomings. Many of the current concepts were born from the preceding Soil Conservation Acts (Departement van Landbou-Tegniese Dienste 1966). CARA was at its inception applicable to white owned agricultural land, only, and excluded the African self-governing homeland areas. This exclusion of former homeland areas was repealed through the Abolition of Racially Based Land Measures Act, Act 108 of 1991. Within homeland areas the "Betterment" scheme was used to improve land use planning and land degradation. But these efforts were met with considerable resistance and also had unintended environmental consequences like the degradation of resources around the newly build villages (Hoffman and Ashwell 2001).

The scope of CARA includes control measures on soil cultivation, alien invasive plants, veld management, veld fires, management of wetland and soil conservation works. The act made provision for the establishment of a number of incentive schemes and subsidies for soil conservation, agricultural land rehabilitation, the reduction of stock numbers, the establishment of specific crops to enhance soil fertility and combating of alien plants. The schemes were governed by regulations (South Africa 1984) under CARA and a variety of manuals and guidelines were made available (e.g., Department of Agriculture 1984, 1997, Russell 1998).

Any soil conservation activities were based on farm plans drafted by soil conservation technicians and were kept on file at the local extension office. CARA further made provision for the expropriation of land for the restoration or reclamation of the farm. Based on this, a number of farms were expropriated due to the extent and severity of erosion on these farms. Examples from the Eastern Cape include a number of farms in the Molteno area as well as the Weenen Nature Reserve that was converted in 1975. Further examples exist where townlands (commonage) were so overgrazed that they were handed over (in 1973) to Agriculture Technical Services and converted to a research station, e.g., Adelaide Research Station, Eastern Cape (pers. Comm. Craig Trethewe 2021).

Although numerous attempts have been made within the National Ministerial Department responsible for Agriculture after 1994 to put in place a more inclusive act that conforms to the constitution, CARA remains the only agricultural legislation governing natural resource use and protection of agricultural land. Its role to oversee agricultural land management has in part been replaced by the National Environmental Management Act (NEMA) in 1998 and the National Environmental Management Biodiversity Act (NEM:BA) in 2004 (Table 13.1, De Villiers and Hill 2008). After 1994 large sections of the act became redundant and inactive. Funding for soil conservation committees ceased, although a few committees are still semiactive through the initiative of farmers. Financial provision for schemes stopped because the schemes before 1994 exclusively assisted white commercial farmers.

Post 1994 the role of funding rehabilitation degraded land is shared between the Department of Environment, Forestry and Fisheries (DAFF) Working for Water (WfW) and the Department of Agriculture Land Reform and Rural Development (DALRRD) LandCare Program. The LandCare program was established in 1997 in South Africa (Mulder and Brent 2006). Here, LandCare is a labor-intensive public works social program aimed at poverty alleviation and job creation while assisting rural communities to improve their livelihoods through soft interventions (e.g., clearing of alien invasive plants, fencing to rehabilitate cultivated fields, promoting better livestock production and conservation agriculture) (Kepe et al. 2004). According to Nabben and Nduli (2001), the new postapartheid policy direction was toward commitment to address the needs of people living in the former impoverished homelands, community empowerment and partnership with government. Kepe et al. (2004) stated that natural resources can ultimately contribute to poverty alleviation if key principles of LandCare are considered, i.e., that land degradation is addressed and sustainable natural resource utilization is achieved. A similar program, the Working for Wetlands, was initiated by the Department of Water Affairs in 1996 and later transferred to the Department of Environmental Affairs. Similar to LandCare, the program focuses on poverty alleviation by providing work to marginalized groups. Working for Wetland concentrates on the rehabilitation of wetland systems through, e.g., building erosion structures and sediment traps, plugging artificial drainage systems, revegetation and bioengineering. Part of the program is also to conclude contractual agreements with landowners where rehabilitation took place to secure the sustainability of the interventions and also to develop ecotourism opportunities through the establishment of bird hides and boardwalks (Dini and Bahadur 2016).

Recently, the DALRRD has developed a Conservation Agriculture (CA) Policy for South Africa (DAFF 2017). The policy aims at promoting sustainable management practices to increase soil cover, biological diversity and minimum soil disturbance. It is grounded on the principles of farmer empowerment, addressing social inequality, implementing sustainable agricultural practices and knowledge development and sharing. The CA principle of minimum or no tillage is increasingly adopted by (commercial) farmers as it reduces mechanization and input costs (see Box 13.1 for a personal view). Rotation of monocrops is preferred above intercropping in annual dryland cropping, although none is widely practiced (Van Antwerpen et al. 2021). Although CA promotes the increase of soil cover either through crop residues or cover crops, the preservation of crop residues is a challenge in mixed cropping/livestock systems, since many farmers rely on crop residues for overwintering of livestock (Thierfelder et al. 2015). As part of South Africa's UNCCD Land Degradation Neutrality (LDN) Targets (von Maltitz et al. 2019), 60,000 km^2 cultivated land needs to be converted to CA systems by 2030. There is considerable literature pointing to CA as a farming system of importance within a South African context (De Wit et al. 2015, Van Antwerpen et al. 2021). A recent study by Smith (2021) estimated that the total area of crop-livestock systems under CA in South Africa is 16,300 km^2 with the highest adoption in the Western Cape (51%). However, CA adoption under semicommercial and small holder systems is only 0.8%.

Box 13.1 On the Implementation of Conservation Agriculture (CA) in the Free State Province: A Farmer's and CA Consultant's Personal View by J. Knot

It is fulfilling to be a steward of the soil. Working with small and commercial farmers is exciting especially when they are zealous in applying all sorts of soil conservation and regenerative farming practices. What is a farmer's opinion about soil degradation and soil conservation? Here, I'll share some of my findings, perceptions and interpretations.

Farmers are not ignorant of soil degradation. They see the gullies (dongas) become deeper and wider and obviously note that their maize yield (mostly without synthetic inputs) dropped over time. The question is what they can do about it? What can individual farmers do inside a current collective system where action is defined by faith and belief systems?

At district level these farmers cannot do much as grazing laws and regulations, etc. are the responsibility of the government. But, success stories at farm and village level have been noted: farmers that adopted CA applied compost, manure and soil cover. Fields were not ploughed anymore and tillage reduced enormously. Cover crops have been adopted into the maize production systems especially as relay cropping. They managed to a certain extend to reverse the traditional uncontrolled "free for all" grazing on their crop residues and winter cover crops.

Commercial farmers have probably more advanced and better access to social media, internet research on reading about soil erosion and land degradation. Many conventional commercial farmers however are not convinced that tillage is necessarily degrading the soil. Many if not all (conventional) farmers measure farm performance in financial terms only. If the crop yields decline then many find resort in using improved seed cultivars and apply higher fertilizer rates. Simply said, the crop yields remain the same, but the soil quality decreases gradually. The journey of soil conservation along with keeping farming profitable is a far more difficult road to travel. It requires more hands-on management, but it is the MUST-road to travel. Look after the soil and it will look after you, your crops and the livestock.

It appears that combating soil degradation in southern Africa will best be via a bottom-up approach driven by devoted, visionary, regenerative lead farmers in a supportive innovative environment who successfully implement CA and stimulate farmer groups around them to also try, research and practice CA. Soil and water conservation is related to implementation of principles

(continued)

Box 13.1 (continued)
but the HOW TO DO remains farm-specific. Unfortunately, as agricultural extension and research support is lacking soil conservation is solely on the shoulders of the farmers.

Soil conservation committees provided an important means to promote conservation of agricultural resources (Theron 1983) and to the implementation of CARA. Although the success and efficiency of soil conservation committees varied, clear legislative and regulative guidelines existed (South Africa 1983, 1984). Based on this, farmers must be responsible for interventions and ideally share the cost of interventions. In addition, policies and legislation must be implemented by a capable and motivated staff component and knowledgeable farmers to ensure that land degradation can be quickly identified, and the most appropriate soil conservation interventions can be applied. In the past considerable emphasis existed on training a skilled workforce both on a technical but also academic and managerial level. For this purpose, curriculums within agricultural engineering and extension were developed at Technicon's to train soil conservation technicians and practitioners.

The recent Land Degradation Neutrality (LDN) document revolves around three strategies of avoiding degradation, reducing degradation and restoring degradation (Cowie et al. 2018, von Maltitz et al. 2019). Any policy needs to consider all three with the emphasis on the first. Farmers can avoid and reduce further degradation by adopting practices and behavior that conserve and improve ecological capital, e.g., the soils. However, it remains very difficult for farmers or land users to rehabilitate severely eroded land and any intervention requires long-term dedication and sacrifice.

13.4.2 Soil Erosion and Soil Conservation Research Development

The visible effects of land degradation and the desire to optimize production stimulated research interest and created the awareness that a better understanding of soils and vegetation in South Africa would be needed (Cooper 1996). Thus, the Land Type survey came forth from the need to have a nationwide map of soil information as good soil data is imperative for setting policy guidelines on soil conservation. From early 1970 up to 2000 various soil scientists contributed to define homogenous areas called land types according to terrain, climate and soil (Land Type Survey Staff 2012). The land type survey remains the only nationwide survey of soils in South Africa. A similar exercise was completed by Acocks (1988) to define vegetation types for livestock and agricultural production. In addition to this, considerable research on the most sustainable veld management strategies to prevent land degradation was undertaken.

Since 1994, research emphases have shifted away from innovative veld and cropping system research at Government owned research institutes. Greater emphasis is on on-site trials focusing on smallholder farming and communal farming systems testing traditional, low-input approaches. The focus of extension services has shifted toward project implementation aiming at poverty alleviation and small-scale farmer assistance. Within commercial livestock farming, holistic farming and high intensity grazing has been adopted because of the perceived benefits the adaptive management decision-making framework brings. Holistic farming or Holistic Management (Savory and Butterfield 2016) is however seen as highly controversial from an ecological point of view (e.g., Briske et al. 2011, Mann and Sherren 2018). Through intensification of livestock rotational systems (smaller camps and faster rotation) and higher camp stocking rates, grass utilization efficiency is increased but grass recovery becomes longer, stimulating cover, nutrient cycling and carbon sequestration (Gosnell et al. 2020).

13.4.3 A Successful Soil Conservation Policy

Strong legislation and regulations regarding soil conservation developed over a period of 40 years spanning three acts. The CARA act is still relevant, but it is an old legislation that needs review. It does not consider new philosophies of holistic management, carbon sequestration and conservation agriculture. Tools to assist with implementing the act have also progressed significantly. Without sound implementation any form of legislation is ineffective. A strong and capable personnel corps is needed with an enduring institutional memory, an issue discussed as well in Europe (Boardman and Vandaele 2010), for administrating the acts, regulations, policies and strategies on soil conservation.

There exists a wealth of data on methods to prevent soil degradation and to restore degraded soils. A good example is the World Overview of Conservation Approaches and Technologies (WOCAT) containing a database of various sustainable land management practices including a guideline book for southern Africa (Liniger and Studer 2019). Nonetheless, the need for constant data on the state of the resource and research on testing new technological advances in sustainable land management is essential. Although remote sensing provides a rather new dimension in natural resource monitoring (see Chaps. 24 and 29) the collection and monitoring of ground base measurements still remain essential. Although sufficient research is available on basic methods to prevent, reduce and restore soil degradation, continued training and research on newer technologies toward building human capacity are needed. Considering the time for rehabilitated land to recover, the benefit for current land users remains minimal. Policy decision makers should understand that benefits after land rehabilitation are only visible over the long-term and often beyond the lifespan of a farmer.

13.5   The Extent of Soil Erosion by Water in South Africa

While long-term, geological erosion and denudation rates seem to be quite low and uniform over South Africa, current rates of soil erosion show a strong spatial variation. Due to a still missing uniform soil erosion model which would assess sheet and rill erosion as well as gully erosion at the same time, these two forms need to be looked at separately. Together, they are good proxies for the on-site damages from soil erosion by water, while sediment yield assessments are a good proxy for the off-site damages.

In order to assess the actual risk of soil loss from sheet and rill erosion in South Africa, the Universal Soil Loss Equation (USLE, Wischmeier and Smith 1978) was interfaced in a Geographical Information System (GIS) and applied to the whole country. The (R)USLE represents the globally most frequently applied soil erosion by water model (Borrelli et al. 2021). It has sufficient simplicity for a risk assessment (not actual rates) on a national scale by incorporating the main factors causing soil erosion, i.e., rainfall erosivity, soil erodibility, topography and vegetation cover. Details on the method and the data used to account for the main factors are provided in Le Roux et al. (2008).

To additionally assess concentrated flow erosion a gully erosion location map for South Africa was created by visual interpretation and vectorization satellite imagery acquired in 2008 to 2012 (Mararakanye and Le Roux 2012). SPOT 5 satellite imagery was utilized because it provides high resolution air photo-like quality for erosion mapping and was acquired from government agencies for the whole country. As a result, the study successfully mapped over 150,000 gully erosion features ranging from just a few square meters to several hectares of surface area each. However, one has to note that this mapping exercise is a first step in providing an assessment on gully erosion in South Africa. In particular, it does not consider the status of a gully (active or inactive) and does not provide any estimates on the age or lifetime of the gullies and thus gully erosion rates.

Figure 13.4 illustrates the distribution of areas with a high erosion risk for sheet and rill erosion, basically under conventional tillage operations on arable land. Arable land is defined here as the interpretive groupings of Land Types (Land Type Survey Staff 2012) with cultivation potential ranging from least suitable (requires careful management) to most suitable (for safely and profitably cultivating crops) (Schoeman et al. 2002). Areas are classified as having a moderate to high erosion risk when the average annual soil loss rate exceeds $1200 \text{ t km}^{-2} \text{ yr}^{-1}$ (thereafter shortly called water erosion risk areas). This classification applies to about 25% of the land in South Africa. In total, approximately $96,350 \text{ km}^2$ and 30% of all potential arable land can be considered at risk of water erosion. The average predicted sheet and rill erosion soil loss rate for arable land in South Africa is $1260 \text{ t km}^{-2} \text{ yr}^{-1}$, a value recently confirmed by Borrelli et al. (2017, Fig. 4). Looking at the total area classified as water erosion risk areas (Table 13.2), the Eastern Cape ranks first, followed by the Free State and the Northern Cape provinces. But if one considers

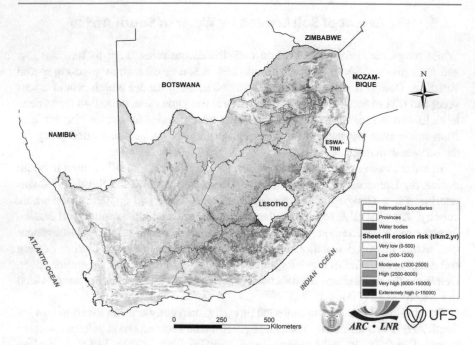

Fig. 13.4 Water erosion risk map of South Africa (emphasizing sheet-rill erosion) (modified from Le Roux et al. 2008, p. 310)

Table 13.2 Assessment of the extent of soil erosion by water (sheet and rill erosion and gully erosion) in South Africa

Province	Total Area [km²]	Area affected by sheet and rill erosion [km²]	Area affected by sheet and rill erosion [%]	Area affected by gully erosion [km²]	Area affected by gully erosion [%]
Western Cape	129,462	20,653	16.0	254	0.2
Northern Cape	372,889	54,071	14.5	1608	0.4
North West	104,882	14,853	14.2	108	0.1
Free State	129,825	55,039	42.4	647	0.5
Eastern Cape	168,966	61,886	36.6	1518	0.9
Limpopo	125,755	44,227	35.2	587	0.5
Gauteng	18,178	7757	42.7	1	0
Mpumalanga	76,495	25,784	33.7	174	0.2
KwaZulu Natal	94,361	21,380	22.7	875	0.9
Total	1,220,813	305,650	25.0	5772	0,5

the proportion of areas at risk compared to the total area, then Gauteng and the Free State show up as hotspots followed by Eastern Cape, Limpopo and Mpumalanga.

Figure 13.5 illustrates the distribution of gullies in the country and shows that i) all provinces are affected by gully erosion and that ii) there is a clear overlap with

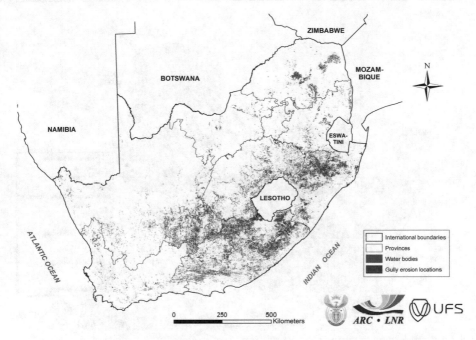

Fig. 13.5 Gully erosion location map of South Africa (from Mararakanye and Le Roux 2012, p. 213)

the water erosion risk areas (Fig. 13.4). Overall, about 0.5% of South Africa's land surface is dissected by gullies (Table 13.2). The largest areas covered by gullies are in the Northern and Eastern Cape provinces and the highest density of gullies are found in the Eastern Cape and KwaZulu Natal. These findings confirm earlier assessments by Hoffmann and Ashwell (2001, 151) about the "provinces most badly affected by soil degradation." In addition to the extent of gully erosion, Mararakanye and Le Roux (2012) determined that 1818 km^2 and 5.5% of arable land is affected by gully erosion. Here, poor farming practices as well as the trend toward agricultural intensification can be considered a major cause of gully erosion.

Finally, the assessment of catchment sediment yield provides a reasonable proxy for the off-site damages from soil erosion by water. Given the increasing threat of reservoir siltation, the Water Research Commission (WRC) identified the need to improve the original sediment yield map of South Africa (Rooseboom et al. 1992). The revised sediment yield map (Msadala et al. 2010) was produced using latest reservoir siltation data in probabilistic and empirical modeling. One of the improvements involved the identification of new regional boundaries based on above-mentioned USLE study of Le Roux et al. (2008). Furthermore, revised sediment yield confidence bands were developed using recent bathymetric survey data for 157 dams obtained from the Department of Water and Sanitation dam survey information book (DWS 2006).

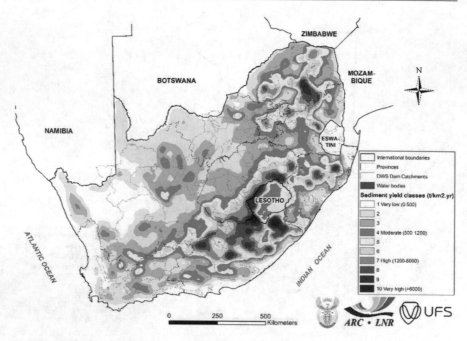

Fig. 13.6 Sediment yield map of South Africa (modified from Msadala et al. 2010, p. 243)

Figure 13.6 illustrates that the highest sediment yields are predicted in the eastern parts of the country. The spatial pattern is rather similar to the water erosion risk and gully erosion maps (Fig. 13.3, Fig. 13.4). Sediment yield ranges between 4 and 1510 t km^{-2} yr^{-1} with an average of 207 t km^{-2} yr^{-1}. It is noteworthy that differences between sediment yield and soil loss can be high (Walling 1983) and often sediment yield is lower than estimated soil loss or soil detachment in the corresponding catchment area. Nonetheless, comparing the three figures provides evidence of quite similar patterns of hotspots of sheet and rill erosion, gully erosion and sediment yield across the country and a strong correlation with the distribution of dispersive and duplex soils in South Africa (see Fey 2010).

Recently, Borelli et al. (2017) published a RUSLE-based global modeling approach to assess the potential rates of soil detachment by water in the years 2001 and 2012 considering land use change, but omitting climate change. In good agreement with earlier results of Le Roux et al. (2008) they report an average annual loss of about 1600 t km^{-2} yr^{-1} in cropland for 2012 (Borelli et al. 2017, Fig. 4). In addition to this, Borrelli et al. (2020) assessed future developments for the year 2070 based on Representative Concentration Pathways (RCP). According to this, sheet and rill erosion are supposed to remain on the current level or decrease slightly for RCP 2.6 and RCP 4.5 and increase slightly for RCP 8.5. However, due to a number of uncertainties, monitoring the development by, e.g., continuous monitoring of reservoir siltation is advised.

13.6   The Extent of Soil Erosion by Wind in South Africa

Soil erosion has been identified as one of South Africa's biggest environmental problem already decades ago, but the discussion has been clearly dominated by water erosion while the problems caused by wind erosion are often considered less severe and overlooked (Laker 2004). Wind erosion research related to southern Africa can be divided into two main topics. One is the consideration of wind erosion as a geomorphological or natural process in arid environments, forming landscapes and covering large areas and long timescales (e.g., Holmes 2015). The other one is the recent effects of wind erosion on agricultural land caused by human activities and contributing to soil degradation (Wiggs and Holmes 2011, Eckardt et al. 2020, Vos et al. 2021, Salawu-Rotimi et al. 2021).

As is the case all over the world, data available on wind erosion in South Africa is very sparse. In a review, Laker (2004) refers to just a few available quantifications with soil loss rates reported for wind erosion ranging from 1100 to 5900 $t\,km^{-2}\,a^{-1}$. In addition, he refers to findings by Schoeman et al. (2002) identifying hotspots of wind erosion in the coastal belts as well as parts of the Northern Cape, the Free State and the Northwest Provinces.

More recently, wind erosion research has increasingly addressed the problems of aeolian dust emissions on agricultural land. Wiggs and Holmes (2011) were the first to examine both issues on agricultural land in the Free State Province. Here, sandy soils, strong winds and impacts of cultivation promote the processes of wind erosion. The controlling factors vary in time and space, with saltation during high winds as a key factor. High wind speeds are not only associated with the passages of frontal systems, but also with the diurnal cycle of wind resulting in many erosion events during early daytime hours. When the fields are bare, the roughness and moisture are the crucial parameters to mitigate erosion. Perpendicular orientation of tillage induced roughness to the main wind direction is therefore a particularly effective measure to mitigate against wind erosion.

The Free State Province has been identified as a frequent dust source as well by Eckardt et al. (2020) evaluating satellite images for the decade 2006–2016 from the Spinning Enhanced Visible and Infrared Imager (SEVIRI) (Fig. 13.7). Particularly land used for rain fed agriculture lead to increased events during drought phases, which are accompanied by reduced ground cover by vegetation. A relation to the diurnal variation of the wind speed could also be shown. Most of the dust plumes were traveling in the direction of the Indian Ocean, and rather of minor extent compared to those in the northern hemisphere. Eckardt et al. (2020) also concludes that the individual field conditions are more important for the emissions than land use in general.

The close proximity of natural and agricultural sources of dust in semiarid environments makes it also difficult to distinguish between them. Thus, a clear identification of possible source areas and the factors favoring dust emissions are important prerequisites to improve models as well as to effectively apply possible mitigation measures. Ground-based investigations of natural and agricultural dust

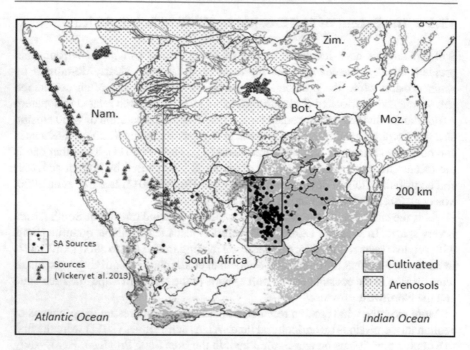

Fig. 13.7 Map of southern Africa's major dust sources based on satellite image surveys (2005 to 2008, Vickery et al. 2013) and (2006 to 2016, Eckardt et al. 2020) (modified from Eckardt et al. 2020)

emission hotspots were made by von Hóldt et al. (2019) and Vos et al. (2019, 2021) to identify the most relevant controlling factors. The PI-SWERL (Portable In-Situ Wind ERosion Lab) device used allows to derive functional relationships between the force of the wind, the friction velocity (u_*), and soil properties (particle size distribution, aggregation) and soil surface characteristics (roughness, crusts), expressed as dust emission potentials.

The Revised Wind Erosion Equation (RWEQ, Fryrear et al. 1998) is used in many parts of the World as basis to assess the wind erosion risk and to identify potential hotspots. The estimation is based on soil properties (erodibility and crusting), a meteorological parameter combining transport capacity of the wind (erosivity) and the ratio between precipitation and evaporation, and the land use to derive vegetation cover as well as landscape roughness (for details see Kestel et al. 2023). If only these most basic influencing factors are considered, the following picture emerges for South Africa (Fig. 13.8, Fig. 13.9). High sand contents and a low ability to form aggregates result in the highest susceptibilities, especially in the north-western parts of South Africa (Fig. 13.8). The climatic erosivity is highest along the coast line with decreasing tendency inland and from west to east (Fig. 13.9). The combination of all parameters results in highest susceptibilities to wind erosion in the northwestern parts of South Africa, amounting to 5.5% in the moderate to highest susceptibility level (Fig. 13.10). The soil losses assigned to the classes are in accordance with

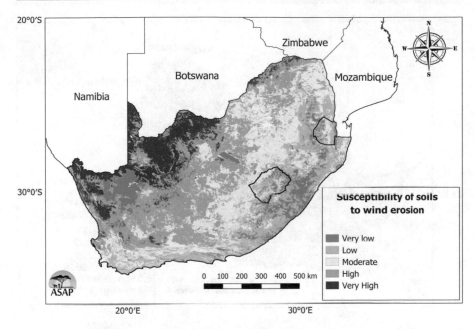

Fig. 13.8 Susceptibility of soils to wind erosion in South Africa (data sources: de Sousa et al. 2020, Fischer et al. 2008, compilation: F. Kestel)

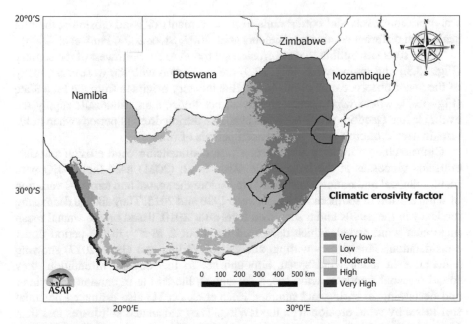

Fig. 13.9 Climatic erosivity factor in South Africa derived from mean monthly wind speed and the ratio of precipitation and evaporation (data source: Abatzoglou et al. 2018, compilation: F. Kestel)

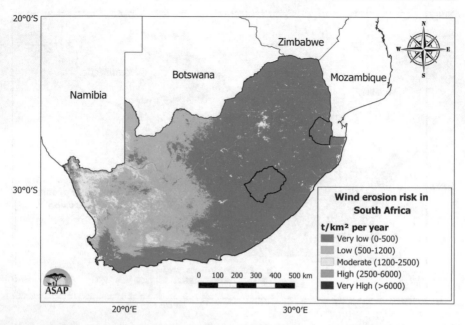

Fig. 13.10 Wind erosion risk in South Africa, 15-year average (2005–2019) (modified from Kestel et al. 2023)

water erosion levels and corresponds to measurements of wind erosion at the plot scale from different continents (Bielders et al. 2002, Sterk 2003, Funk et al. 2004).

While soils susceptible to wind erosion are found to the northwest of the country (Fig. 13.8.) and major winds (Fig. 13.9) are associated with the west coast, many of the major observed dust events in satellite imagery originate from the Free State (Fig. 13.7), where wind speeds are lower, but luvisols and arenosols supply the available fine fraction, which becomes available during drought periods when fields remain devoid of crop cover for extended periods of time.

Climate change impacts are a further factor influencing wind erosion and dust emission processes at the larger scale. Zhao et al. (2021) used the RWEQ with meteorological and remote sensing data to explore the spatial and temporal variation of wind erosion in southern Africa between 1990 and 2015. They show a decreasing tendency in the 1990s and a stabilized level after 2010, based on the annual mean maximum wind speed, which decreased by about 2 m s^{-1} in the period under consideration. This agrees with an analysis of Wright and Grab (2017) showing a decrease in mean wind speed, although not to this degree. In addition, they found seasonal trends in wind speed deviations, increasing in autumn and winter and decreasing in spring and summer. Zhao et al. (2021) also estimated potential soil losses by wind erosion with the RWEQ. They calculated soil losses less than 1000 t km^{-2} yr^{-1} for most parts for southern Africa, but also maximum values up to 17,000 t km^{-2} yr^{-1}. The values are not verified by any measure, but they show relative changes in the wind erosion risk over the regarded 25 years caused by climatic factors and associated vegetation cover changes.

13.7   The Socioeconomic Dimension of Soil Erosion

The African population is experiencing a number of socioeconomic problems such as poverty, food insecurity and increased mortality rates (UNEP 2015). These hardships are intensified by land degradation which further disadvantage people and leads to migration and conflict over arable land. Land degradation has an influence on productivity which may affect food security and the livelihoods of those who derive their well-being from practicing small scale farming or working on commercial farms (Hamdy and Aly 2014). Barbier and Hochard (2016) reason that the overall poverty in developing countries may be influenced by the concentration of communal populations on degrading, as opposed to improving, agricultural land. In South Africa for example, 68% of the country's land surface is occupied by range land and are mostly utilized for livestock agriculture. Large parts of these rangeland are already experiencing different levels of degradation. Gully erosion, scrub encroachment and a general decrease in vegetation cover pose a threat to the production of livestock, the farmer's livelihood, and the production of food in South Africa (Rabumbulu and Badenhorst 2017).

Land degradation is often identified as the consequence of existing social and economic conditions experienced by the land users and workers (Abu Hammad and Tumeizi 2012). These social and economic conditions include population growth, poverty, overgrazing, deforestation and access to agriculture extension, infrastructure, opportunities and constraints created by market access as well as policies and general government effectiveness (Jouanjean et al. 2014).

A substantial proportion of people in rural and urban areas depend on agricultural production as the main source of employment and livelihood. Therefore, population growth has an impact on land as an important economic sector in terms of food production, employment generation and improving the livelihoods of the poor to alleviate poverty (Kangalawe and Lyimo 2010). In South Africa, there has been a significant migration to urban areas. The push factors include poverty and unemployment, with the prospect of receiving high wages in urban areas being a major pull factor. The outmigration is a threat to the growth and productivity of rural agriculture due to the loss of human capital. However, studies also indicated that remittances from migrant workers are often used to purchase agricultural inputs (Mbata and Mofokeng 2021).

The socioeconomic development of smallholder farmers is dependent on access to profitable markets, thus markets where they obtain information, farm organizations and income (van Tilburg and van Schalkwyk 2012). This can contribute to profit incentives and empower farmers to upgrade yield production and mitigate land degradation through acquisition of sustainable land use information and practices, which in turn may contribute to household income and food security. In South Africa, for example, it is broadly understood that smallholder farmers experience difficulties to access profitable markets due to poor infrastructure, long distance to access output and input markets, expensive transport costs, absence of information (regarding markets, production and environmental issues such as land degradation), lack of technical assistance (training on sustainable land use) and inefficient record keeping practices (Ngqangweni et al. 2016). These issues have been also confirmed in a case study of the South Africa Land Degradation Monitor (SALDi) project in the Ladybrand area, Free State Province of South Africa (see Box 13.2).

Deep-rooted poverty leads to overdependence on natural resources for livelihoods which in some instances have undermined the capacity of the population to manage the resources sustainably (Kangalawe and Lyimo 2010). Kirui (2016) and Mbata and Mofokeng (2021) observe that the connection between land degradation and poverty is greater in rural areas of developing countries such as South Africa, where the livelihoods of the majority of the population are attached to agriculture. Poverty inhibits farmers to have access to equipment that enhances the rehabilitation of the land (Birungi 2007). Most small-scale farmers live barely on subsistence level and do not have the capacity to use purchased inputs or to pay for labor to use the labor-intensive conservation technologies (Birungi 2007). Thus, farmers are disadvantaged if they are unable to utilize effective land productivity enhancing inputs such as fertilizers which contributes to the degradation of natural resources (Kirui 2016). Poor small-scale farmers are often unable to compete for resources, including high quality and productive land and are therefore restricted to peripheral land that cannot sustain their practices which prolong land degradation and advance poverty (Birungi 2007). Even where small-scale farmers may have access to productive land and infrastructure in South Africa, the sociopolitical context may prevent them from using the land. This can play out through corruption, nepotism or imbalances in power relations.

Box 13.2 South Africa Land Degradation Monitor (SALDi) Case Study on Small-Scale Farmers and Land Care Workers' Perceptions of Land Degradation: Main Results

A case study in the Ladybrand area in the Free State Province that includes commercial, small scale and commonage farmers was conducted between 2019 and 2021. Information was gathered through observation, focus group discussions and face to face interviews (for details see Msipa 2022). The study investigates the community perceptions about land degradation and the impact of land degradation on the livelihoods of the communities. The study shows that the main driver for farming in the area is the provision of a livelihood. Soil quality and land degradation is seen as something that prevents commercial gains.

Commercial farmers have more access to resources such as fertilizer, equipment and manpower to prevent soil degradation. Although there are some conservation farmers in the area, it requires a different mindset. Doing

(continued)

Box 13.2 (continued)

things nature's way makes it more difficult to manage your farm and it appears "less neat." From a commercial farming perspective, the study area is a high input driven farming society.

Commonage farmers rent municipal land to farm on. About 60 farmers from the area use the commonage to graze cattle. Many people using the commonage are poor and don't have resources to conserve the soil, have no training, no equipment and limited options. It is also difficult to manage the numbers of the cattle and because of the bad condition of the land, yields are low and the contribution to their livelihood security is limited. Farmers are reluctant to invest in land that they do not own. Commonage farmers list their biggest challenges as lack of access to land, limited space, lack of equipment, no access to capital, bad management skills and no safety nets. They feel that the land is "sour" and needs something added to the soil to improve it.

Small scale farmers also have challenges with land degradation. The degradation is attributed to invasive plants, lack of fences, proximity to waste sites and urban areas and illegal grazing. They attempt to do things like rotational grazing to protect the soil, but due to illegal grazing this often does not work, as they are left with no grazing for their cattle. Due to the proximity to the urban area, their land is susceptible to veld fires started in the township. Land degradation causes lower yields, which with the other challenges, have a real impact on their livelihoods. Small scale farmers in this area tend not to grow commercial crops because they don't have the equipment to plant and harvest crops. When renting equipment timing becomes an issue, and sometimes they are late with planting or harvesting. Buying things like fertilizers and chemicals to control weeds are beyond their budgets and often the yield of the crops does not cover these costs. Thus, they report that crop farming is not profitable at all. There is a dire need for mentorship programs and skills development. Small-scale farmers feel that they cannot produce the same quality as the commercial farmers, and that the markets favor the products of commercial farmers because these are of a higher quality due to their access to resources.

The poor and food insecure households may contribute to land degradation because they are unable to set land aside for given periods of time (fallow), make investments in land improvements or use cost-effective external inputs (Birungi 2007). Given the over dependency on natural resources, stagnation or reduction in agricultural productivity due to land degradation imposes serious income and livelihood constraints for rural and urban households and therefore leading to poverty. Poverty contributes to land degradation and the latter contributes to poverty; it is a cyclic process, as the commonage and small-scale farmers in Ladybrand confirm.

13.8   Challenges for Soil Conservation in South Africa

Soil erosion is the most important soil degradation process in South Africa and
has been considered an issue for over a century. Already at the beginning of the
twentieth century, this problem was addressed in policy, and around the mid of
the century, extensive measures were taken to combat soil erosion on (commercial)
arable land and rangeland. Visible features include contour tillage and runoff bunds,
usually considered effective measures to diminish soil erosion by water as well
as wind breaks to control wind erosion. In addition, strict guidelines concerning
livestock numbers accompanied by subsidies to reduce stocks were implemented.

Nonetheless, sustainable soil use (in agriculture and beyond) remains a challenge. The urgency to control soil erosion by the application of established policies
and principles as well as emerging modes of conservation agriculture (CA) in
farming activities is best illustrated by the following comparison (Table 13.3): The
average geological denudation and erosion rate, indicating long term soil formation
rates varies in South Africa between 3 and 27 t km^{-2} yr^{-1} (Table 13.3). Compared
to this, the mean USLE-based predicted sheet and rill erosion soil loss rate for arable
land is assessed to range between 1260 and 1750 t km^{-2} yr^{-1} and the predicted soil
loss from wind erosion is about 750 t km^{-2} yr^{-1}. Thus, there is evidence that current
soil loss is two orders of magnitude higher than the long-term soil formation rates
and still more than one order of magnitude higher than values for tolerable soil loss
rates reported for Australia (20 to 85.5 t km^{-2} yr^{-1}, FAO 2019).

The United Nations Convention to Combat Desertification (UNCCD) recently
began endorsing a response hierarchy that invests resources to avoid future degradation, followed by a reduction in current degradation and lastly to restore degraded
land (Cowie et al. 2018). Specifically, in South Africa it will not be feasible to
rehabilitate erosion features with large and expensive structures at a broad or
catchment scale due to limited financial resources. Not only are large structures
costly, structures in dispersive soils enhance subsurface accumulation of water
and cause further erosion around structure walls (van Zijl et al. 2014). Thus, it is
important to prevent further erosion by protecting susceptible areas that are currently
not eroded (Le Roux and van der Waal 2020).

It is postulated that prevention measures cost far less than repairing the on- and
off-site damage caused by soil erosion (Boardman 2021). It is imperative to prevent

Table 13.3 Summary of soil formation, denudation and erosion rates in South Africa (for details
see Sects. 13.2 and 13.5)

Process	Time span [yr]	Rates [t km^{-2} yr^{-1}]	Remarks
Soil formation	10^4–10^5	3–27	South Africa (range)
Soil formation	10^4–10^5	13.5	In Kruger National Park, Lowveld
Sediment yield	10^1–10^2	81.0	In Kruger National Park, Lowveld
Sheet and rill erosion	10^1	1260–1750	USLE soil loss risk for arable land
Wind erosion	10^1	~ 750	RWEQ soil loss risk,

erosion with appropriate soil conservation measures and expansion of conservation agriculture (CA) in cultivated areas, as well as to protect (natural) vegetation from overgrazing through rotational grazing management systems. The potential of CA to protect and improve soil health has been well documented (De Wit et al. 2015, Van Antwerpen et al. 2021). Under CA, crop residues are retained on the soil surface to protect it from the erosive impact of rainfall, runoff, and wind. The use of cover crops can further increase the crop canopy- and ground cover on the soil, while the presence of permanent and strong living root systems in the soil greatly enhance the resistance of the soil against erosion. Ultimately, the increased organic matter level in the soil is the key factor stabilizing cultivated lands against the devastating effect of erosion. Soil erosion prevention by means of CA will not only prevent soil loss and sustain agricultural production, but will also prevent siltation of water resources and increase the life span of dams and reservoirs in South Africa.

In recent years, devastating droughts and local to regionalized flood events have severely affected South Africa. Agriculture as one of the main sectors of the economy in South Africa ranging from the intensive, large-scale, commercial agricultural to the low-intensity, small-scale, and subsistence farming, will in any case be strongly influenced by climate change in the coming decades. Thus, there is a need to strengthen the resilience of farming operations to i) meet increased domestic and global demand and ii) to put into practice sustainable management to diminish on site and off-site damages from soil erosion on the way to reach sustainable development goals.

Tangible and reciprocal partnerships between research institutes, farmers, consumers and government are needed to foster knowledge of sustainable land use practices. A broader view is needed to recover productivity and promote sustainable land management practices including good land use practices and efficient engineering design implementation. Land degradation also needs to be seen within the bigger picture of agricultural and nonagricultural land use and activities. Good soil conservation and sustainable agricultural practices intersect with the preservation of high potential agricultural land, conservation of natural ecosystems, including wetlands, good catchment management and the conservation of water resources.

Acknowledgments We like to thank all people contributing to the ideas developed in this chapter during field work or in numerous discussions. Finalizing this contribution benefited from funding by the Federal Ministry of Education and Research (BMBF) for the SPACES2 Joint Projects: South Africa Land Degradation Monitor (SALDi) (BMBF grant 01LL1701 A–D) and ASAP—Agroforestry in southern Africa—new Pathways of innovative land use systems under a changing climate (BMBF grant 01LL1803 A–D) within the framework of the Strategy "Research for Sustainability" (FONA) www.fona.de/en. We are thankful for the critical and helpful comments of the reviewers, J. Boardman and I. Foster, as well as the editor, G. von Maltitz, to an earlier draft of this manuscript.

References

Abatzoglou JT, Dobrowski SZ, Parks SA, Hegewisch KC (2018) Terraclimate, a high-resolution global dataset of monthly climate and climatic water balance from 1958-2015. Scientific Data 5:170191. https://doi.org/10.1038/sdata.2017.191(2018)

Abu Hammad A, Tumeizi A (2012) Land degradation: socioeconomic and environmental causes and consequences in the eastern Mediterranean. Land Degrad Dev 23(3):216–226. https://doi.org/10.1002/ldr.1069

Acocks JPH (1988) Veld types of South Africa. Memoirs of the Botanical Survey of South Africa 57(3)

Adler ED (1985) Soil conservation in South Africa. Bulletin 406. Department of Agriculture and Water Supply, Pretoria

Baade J (2006) Soil erosion. In: Geist H (ed) The Earth's changing land: an encyclopedia of land-use and land-cover change. Greenwood, Westport, pp 539–542

Barbier EB, Hochard JP (2016) Does land degradation increase poverty in developing countries? PLoS One 11(5):1–12

Beinart W (2003) The rise of conservation in South Africa. In: Settlers, livestock, and the environment 1770–1950. Oxford University Press, Oxford

Bennett HH (1939) Soil conservation. McGraw-Hill, New York

Bielders CL, Rajot J-L, Amadou M (2002) Transport of soil and nutrients by wind in bush fallow land and traditionally managed cultivated fields in the Sahel. Geoderma 109:19–39

Bierman PR (1994) Using in situ produced cosmogenic isotopes to estimate rates of landscape evolution: a review from the geomorphic perspective. J Geophys Res 99B:13885–13896

Binnie SA, Summerfield MA (2013) Rates of denudation. In: Shroder J, Marston RA, Stoffel M (eds) Treatise on geomorphology, Mountain and hillslope geomorphology, vol 7. Academic Press, San Diego, CA, pp 66–72

Birungi PB (2007) The linkages between land degradation, poverty and social capital in Uganda. University of Pretoria. Environmental Economics. PhD Thesis. Available http://hdl.handle.net/2263/25109. Accessed 22 June 2020

Boardman J (2021) How much is soil erosion costing us? Geography 106(1):32–38. https://doi.org/10.1080/00167487.2020.1862584

Boardman J, Foster I (2008) Badland and gully erosion in the Karoo, South Africa. J Soil Water Conserv 63(4):121A–125A. https://doi.org/10.2489/jswc.63.4.121A

Boardman J, Vandaele K (2010) Soil erosion, muddy floods and the need for institutional memory. Area 42(4):502–513. https://doi.org/10.1111/j.1475-4762.2010.00948.x

Boardman J, Hoffman MT, Holmes PJ, Wiggs GFS (2012) Soil erosion and land degradation. In: Holmes P, Meadows M (eds) Southern African geomorphology: recent trends and new directions. Sun Media, Bloemfontein, pp 307–328

Borrelli P, Robinson DA, Fleischer LR, Lugato E, Ballabio C, Alewell C, Meusburger K, Modugno S, Schütt B, Ferro V, Bagarello V, Oost KV, Montanarella L, Panagos P (2017) An assessment of the global impact of 21st century land use change on soil erosion. Nat Commun 8(1). https://doi.org/10.1038/s41467-017-02142-7

Borrelli P, Robinson DA, Panagos P, Lugato E, Yang JE, Alewell C, Wuepper D, Montanarella L, Ballabio C (2020) Land use and climate change impacts on global soil erosion by water (2015-2070). Proc Natl Acad Sci USA 117(36):21994. https://doi.org/10.1073/pnas.2001403117

Borrelli P, Alewell C, Alvarez P, Anache JAA, Baartman J, Ballabio C, Bezak N, Biddoccu M, Cerdà A, Chalise D, Chen S, Chen W, De Girolamo AM, Gessesse GD, Deumlich D, Diodato N, Efthimiou N, Erpul G, Fiener P, Freppaz M, Gentile F, Gericke A, Haregeweyn N, Hu B, Jeanneau A, Kaffas K, Kiani-Harchegani M, Villuendas IL, Li C, Lombardo L, López-Vicente M, Lucas-Borja ME, Märker M, Matthews F, Miao C, Mikoš M, Modugno S, Möller M, Naipal V, Nearing M, Owusu S, Panday D, Patault E, Patriche CV, Poggio L, Portes R, Quijano L, Rahdari MR, Renima M, Ricci GF, Rodrigo-Comino J, Saia S, Samani AN, Schillaci C, Syrris V, Kim HS, Spinola DN, Oliveira PT, Teng H, Thapa R, Vantas K, Vieira D, Yang JE, Yin S, Zema DA, Zhao G, Panagos P (2021) Soil erosion modelling: a global review and statistical analysis. Sci Total Environ 780:146494. https://doi.org/10.1016/j.scitotenv.2021.146494

Braune E, Looser U (1989) Cost impacts of sediments in South African rivers. In: Hadley RF, Ongley ED (eds) Sediment and the environment. IAHS Press/IAHS Publication, Wallingford

Briske DD, Sayre NF, Huntsinger L, Fernández-Giménez M, Budd B, Derner JD (2011) Origin, persistence, and resolution of the rotational grazing debate: integrating human dimensions into rangeland research. Rangel Ecol Manag 64(4):325–334. https://doi.org/10.2111/REM-D-10-00084.1

Chadwick OA, Roering JJ, Heimsath AM, Levick SR, Asner GP, Khomo L (2013) Shaping post-orogenic landscapes by climate and chemical weathering. Geology 41:1171–1174. https://doi.org/10.1130/G34721.1

Clark EHI (1985) The off-site costs of soil erosion. J Soil Water Conserv 40(1):19–22

Compton JS, Herbert CT, Hoffman MT, Schneider RR, Stuut J-B (2010) A tenfold increase in the Orange River mean Holocene mud flux: implications for soil erosion in South Africa. The Holocene 20(1):115–122. https://doi.org/10.1177/0959683609348860

Cooper A (1996) Soil conservation policy in South Africa, 1910–1992: the 'human dimension'. Doctoral dissertation, University of Natal, Durban. Online available: https://researchspace.ukzn.ac.za/bitstream/handle/10413/8664/Cooper_Amanda_1997.pdf

Cowie AL, Orr BJ, Castillo Sanchez VM, Chasek P, Crossman ND, Erlewein A, Louwagie G, Maron M, Metternicht GI, Minelli S, Tengberg AE, Walter S, Welton S (2018) Land in balance: the scientific conceptual framework for land degradation neutrality. Environ Sci Policy 79:25–35. https://doi.org/10.1016/j.envsci.2017.10.011

DAFF (Department of Agriculture, Forestry and Fisheries) (2017) Draft conservation agriculture policy. Director: Land and Soil Management, Department of Agriculture, Land Reform and Rural Development, Pretoria

de Sousa LM, Poggio L, Batjes NH, Heuvelink GBM, Kempen B, Riberio E, Rossiter D (2020) SoilGrids 2.0: producing quality-assessed soil information for the globe. Soil Discuss 7:1–37. https://doi.org/10.5194/soil-2020-65

De Villiers CC, Hill RC (2008) Environmental management frameworks as an alternative to farm-level EIA in a global biodiversity hotspot: a proposal from the cape floristic region, South Africa. JEAPM 10(04):333–360

De Wit MP, Blignaut JN, Knot J, Midgley S, Drimie S, Crookes DJ, Nkambule NP (2015) Sustainable farming as a viable option for enhanced food and nutritional security and a sustainable productive resource base. Synthesis report. Green Economy Research Report, Green Fund. Development Bank of Southern Africa, Midrand

Decker JE, Niedermann S, de Wit MJ (2011) Soil erosion rates in South Africa compared with cosmogenic ^{3}He based rates of soil production. S Afr J Geol 114:475–488

Departement van Landbou-Tegniese Dienste (1966) Handleiding oor die toepassing van die Grandbewaringswet (Afr.). Staatsdrukker, Pretoria

Department of Agriculture (1984) Manual for the subsidy scheme (Act 43 of 1983) division of soil protection. Department of Agriculture, Pretoria

Department of Agriculture (1997) National Soil Conservation Manual. Department of Agriculture, Pretoria

Dini J, Bahadur U (2016) South Africa's national wetland rehabilitation programme: working for wetlands. In: Finlayson CM, Everard M, Irvine K, McInnes R, Middleton B, van Dam A, Davidson NC (eds) The wetland book. Springer, Dordrecht, pp 1–7. https://doi.org/10.1007/978-94-007-6172-8_145-2

du Preez CC, Kotzé E, van Huysteen CW (2020) Southern African soils and their susceptibility to degradation. In: Holmes PJ, Boardman J (eds) Southern African landscapes and environmental change. Routledge, London, pp 29–52

DWS (Department of Water and Sanitation) (2006) Dam list. Ministry of Water and Environmental Affairs, Republic of South Africa, Pretoria

DWS (Department of Water and Sanitation) (2017) Hydrological services - Surface water: reservoir sites. online https://www.dwa.gov.za/Hydrology/Verified/HyCatalogue.aspx

Eckardt FD, Bekiswa S, Von Holdt J, Jack C, Kuhn NJ, Mogane F, Murray JE, Ndara N, Palmer AR (2020) South Africa's agricultural dust sources and events from MSG SEVIRI. Aeolian Res 47:100637. https://doi.org/10.1016/j.aeolia.2020.100637

England P, Molnar P (1990) Surface uplift, uplift of rocks, and exhumation of rocks. Geology 18:1173–1177

FAO (Food and Agriculture Organization of the United Nations) (2019) Soil erosion: the greatest challenge to sustainable soil management. FAO, Rome

FAO (Food and Agriculture Organization of the United Nations) and ITPS (Intergovernmental Technical Panel on Soils) (2015) Status of the World's soil resources (SWSR) – Main report. FAO, Rome

Fey M (2010) Soils of South Africa. Cambridge University Press, Cambridge

Fischer G, Nachtergaele F, Prieler S, van Velthuizen HT, Verelst L, Wiberg D (2008) Global agro-ecological zones assessment for agriculture (GAEZ 2008). IIASA, Laxenburg, Austria and FAO, Rome, Italy

Foster IDL, Boardman J (2020) Monitoring and assessing land degradation: new approaches. In: Holmes PJ, Boardman J (eds) Southern African landscapes and environmental change. Routledge, London, pp 249–274

Fryrear DW, Saleh A, Bilbro JD, Schromberg HM, Stout JE, Zobeck TM, Schomberg HM, Stout JE, Zobeck TM (1998) Revised wind erosion equation. USDA-ARS, Southern Plains Area Cropping Systems Research Laboratory, Technical Bulletin 1

Funk R, Reuter HI (2006) Wind erosion. In: Boardman J, Poesen J (eds) Soil erosion in Europe. Wiley, Chichester, pp 563–582

Funk R, Skidmore EL, Hagen LJ (2004) Comparison of wind ero-sion measurements in Germany with simulated soil losses by WEPS. Environ Model Softw 19:177–183

Glotzbach C, Paape A, Baade J, Reinwarth B, Rowntree KM, Miller JK (2016) Cenozoic landscape evolution of the Kruger National Park as derived from cosmogenic nuclide analyses. Terra Nova 28(5):316–322. https://doi.org/10.1111/ter.12223

Gosnell H, Grimm K, Goldstein BE (2020) A half century of holistic management: what does the evidence reveal? Agric Hum Values 37(3):849–867

Hamdy A, Aly A (2014) Land degradation, agriculture productivity and food security. In Proceedings, fifth international scientific agricultural symposium "Agrosym 2014", Jahorina, Bosnia and Herzegovina, October 23-26, 2014, pp 708–717

Hewawasam T (2003) Increase of human over natural erosion rates in tropical highlands constrained by cosmogenic nuclides. Geology 31(7):597–600. https://doi.org/10.1130/0091-7613(2003)031<0597:IOHONE>2.0.CO;2

Hoffman T, Ashwell A (2001) Nature divided: land degradation in South Africa. University of Cape Town Press, Cape Town

Holmes P (2015) The Western Freee state Panfield: a landscape of myriad pans and lunettes. In: Grab S, Knight J (eds) Landscape and landforms of South Africa. Springer, Cham, pp 139–145

Holmes PJ, Thomas DSG, Bateman MD, Wiggs GFS, Rabumbulu M (2012) Evidence for land degradation from Aeolian sediment in the west-Central Free State Province, South Africa. Land Degrad Dev 23(6):601–610. https://doi.org/10.1002/ldr.2177

Jouanjean M, Tucker J, te Velde DW (2014) Understanding the effects of resource degradation on socio-economic outcomes in developing countries. Overseas Development Institute, London. Available https://cdn.odi.org/media/documents/8830.pdf. Accessed 06 Aug 2020

Kangalawe RYM, Lyimo JG (2010) Population dynamics, rural livelihoods and environmental degradation: some experiences from Tanzania. Environ Dev Sustain 12:985–997

Kanthack FE (1930) The alleged desiccation of South Africa. Geogr J 76(6):516–521

Kepe T, Saruchera M, Whande W (2004) Poverty alleviation and biodiversity conservation: a South African perspective. Oryx 38(2):143–145

Kestel F, Wulf M, Funk R (2023) Spatiotemporal variability of the potential wind erosion risk in Southern Africa between 2005 and 2019. Land Degrad Dev 34:2945. https://doi.org/10.1002/ldr.4659

Kirui OK (2016) Economics of land degradation and improvement in Tanzania and Malawi. In: Nkonya E, Mirzabaev A, von Braun J (eds) Economics of land degradation and improvement–a global assessment for sustainable development. Springer, Cham, pp 609–649. https://doi.org/10.1007/978-3-319-19168-3_20

Laker MC (2004) Advances in soil erosion, soil conservation, land suitability evaluation and land use planning research in South Africa, 1978-2003. S Afr J Plant Soil 21(5):345–368. https://doi.org/10.1080/02571862.2004.10635069

Lal D (1991) Cosmic ray labeling of erosion surfaces: in situ nuclide production rates and erosion models. Earth Planet Sci Lett 104:424–439. https://doi.org/10.1016/0012-821X(91)90220-C

Lal R (2015) Soil Erosion. In: Nortcliff S (ed) Task force: soil matter. Solution under foot. GeoEcology essays. Catena, Reiskirchen, pp 39–48

Land Type Survey Staff (2012) Land types of South Africa: digital map (1:250,000 scale) and soil inventory databases. ARC-Institute for Soil, Climate and Water, Pretoria

Le Roux JJ, van der Waal B (2020) Gully erosion susceptibility modelling to support avoided degradation planning. S Afr Geogr J 102(3):406–420. https://doi.org/10.1080/03736245.2020.1786444

Le Roux JJ, Morgenthal TL, Malherbe J, Sumner PD, Pretorius DJ (2008) Water erosion prediction at a national scale for South Africa. Water SA 34(3):305–314

Liniger H, Studer RM (2019) Sustainable rangeland management in sub-Saharan Africa – guidelines to good practice. WOCAT, Washington, DC. Online available: https://www.wocat.net/library/media/174/

Mann C, Sherren K (2018, 1848) Holistic management and adaptive grazing: a trainers' view. Sustainability 10(6)

Mararakanye N, Le Roux JJ (2012) Gully erosion mapping at a national scale for South Africa. S Afr Geogr J 94(2):208–218. https://doi.org/10.1080/03736245.2012.742786

Mbatha MW, Mofokeng NR (2021) A Literature Review of the Impact and Implication of Out-Migration on Rural Agriculture in South Africa. J Soc Sci Human 18(19):1–12

Meadows ME, Hoffman MT (2002) The nature, extent and causes of land degradation in South Africa: legacy of the past, lessons for the future? Area 34(4):428–437

Molnar P, England P (1990) Late Cenozoic uplift of mountain ranges and global climate change: chicken or egg? Nature 346:29–34

Morgan RPC (1995) Soil erosion and conservation. Longman, Harlow

Msadala VC, Basson GR (2017) Revised regional sediment yield prediction methodology for ungauged catchments in South Africa. J South Afr Inst Civil Eng 59(2):28–36. https://doi.org/10.17159/2309-8775/2017/v59n2a4

Msadala V, Gibson L, Le Roux J, Rooseboom A, Basson GR (2010) Sediment yield prediction for South Africa: 2010th edition. Water Research Commission. Report 1765/1/10. WRC, Pretoria

Msipa AM (2022) Small-scale farmers and land care workers' perceptions of land degradation and how it influences their livelihoods: an explorative study in Ladybrand. Unpublished Master's Thesis, University of Pretoria, Pretoria

Mulder J, Brent AC (2006) Selection of sustainable rural agriculture projects in South Africa: case studies in the LandCare programme. J Sustain Agric 28(2):55–84

Nabben T, Nduli N (2001) Lessons learned in international partnerships the LandCare partnership between Western Australia and South Africa. Paper presented at Institute of Public Administration National Conference, Sydney, Australia 28–30 November

Ngqangweni S, Mmbengwa V, Myeki L, Sotsha K, Khoza T (2016) Measuring and tracking smallholder market access in South Africa. National Agricultural Marketing Council Working Paper (NAMC/WP/2016/03)

Oldeman LR, Hakkeling RTA, Sombroek WG (1991) World map of the status of human-induced soil degradation: an explanatory note. Wageningen, ISRIC

Partridge TC, Dollar ESJ, Moolman J, Dollar LH (2010) The geomorphic provinces of South Africa, Lesotho and Swaziland: a physiographic subdivision for earth and environmental scientists. Trans R Soc S Afr 65(1):1–47. https://doi.org/10.1080/00359191003652033

Pimentel D, Harvey C, Resosudarmo P, Sinclair K, Kurz D, McNair M, Crist S, Shpritz L, Fitton L, Saffouri R, Blair R (1995) Environmental and economic costs of soil erosion and conservation benefits. Science 267(5201):1117–1123. https://doi.org/10.1126/science.267.5201.1117

Raab G, Scarciglia F, Norton K, Dahms D, Brandová D, de Castro PR, Christl M, Ketterer ME, Ruppli A, Egli M (2018) Denudation variability of the Sila Massif upland (Italy) from decades to millennia using 10Be and 239+240Pu. Land Degrad Dev 29(10):3736–3752. https://doi.org/10.1002/ldr.3120

Rabumbulu M, Badenhorst M (2017) Land degradation in the West-Central Free State: human-induced or climate variability, the perceptions of Abrahamskraal–Boshof district farmers. S Afr Geogr J 99(3):217–234. https://doi.org/10.1080/03736245.2016.1231623

Raymo M, Ruddiman WF (1992) Tectonic forcing of late Cenozoic climate. Nature 359:117–122

Reinwarth B, Petersen R, Baade J (2019) Inferring mean rates of sediment yield and catchment erosion from reservoir siltation in the Kruger National Park, South Africa: an uncertainty assessment. Geomorphology 324:1–13. https://doi.org/10.1016/j.geomorph.2018.09.007

Rooseboom A, Verster E, Zietsman HL, Lotriet HH (1992) The development of the new sediment yield map of South Africa. Water Research Commission report no. 297/2/92. WRC, Pretoria

Rowntree KM (2012) Fluvial geomorphology. In: Holmes P, Meadows M (eds) Southern African geomorphology: recent trends and new directions. Sun Media, Bloemfontein, pp 97–140

Rowntree KM (2013) The evil of sluits: a re-assessment of soil erosion in the Karoo of South Africa as portrayed in century-old sources. J Environ Manag 130:98–105. https://doi.org/10.1016/j.jenvman.2013.08.041

Russell WB (1998) Conservation of farmland in KwaZulu-Natal. KZN Department of Agriculture, Pietermaritzburg

Salawu-Rotimi A, Lebre PH, Vos HC, Fister W, Kuhn N, Eckardt FD, Cowan DA (2021) Gone with the wind: microbial communities associated with dust from emissive farmlands. Microb Ecol 82:859. https://doi.org/10.1007/s00248-021-01717-8

Savory A, Butterfield J (2016) Holistic management: a commonsense revolution to restore our environment. Island Press, Washington, DC

Schaller M, Ehlers TA, Lang KAH, Schmid M, Fuentes-Espoz JP (2018) Addressing the contribution of climate and vegetation cover on hillslope denudation, Chilean coastal cordillera (26°–38°S). Earth Planet Sci Lett 489:111–122. https://doi.org/10.1016/j.epsl.2018.02.026

Scharf TE, Codilean AT, de Wit M, Janson JD, Kubik PW (2013) Strong rocks sustain ancient postorogenic topography in Southern Africa. Geology 41:331–334. https://doi.org/10.1130/G33806.1

Schoeman JL, van der Walt M, Monnik KA, Thackrah A, Malherbe J, Le Roux RE (2002) Development and application of a land capability classification system for South Africa. ARC-ISCW report no. GW/A/2000/57. Department of Agriculture, Pretoria

Shakesby RA (2003) Soil erosion. In: Matthews JA, Bridges EM, Caseldine CJ, Luckman AJ, Owen G, Perry AH, Shakesby RA, Walsh RPD, Whittaker RJ, Willis KJ (eds) The encyclopaedic dictionary of environmental change. Arnold, London, p 590

Smith HJ (2021) An assessment of the adoption of conservation agriculture in annual crop-livestock systems in South Africa. FAO, Rome

South Africa (1983) Conservation of agricultural resources act, Act 43 of 1983. Government Gazette 8673:883. (27 April 1983)

South Africa (1984) Regulations: conservation of agricultural resources act, Act 43 of 1983. Government Gazette 9238:R1048. (25 May 1984)

Sterk G (2003) Causes, consequences and control of wind erosion in Sahelian Africa: a review. Land Degrad Dev 14:95–108

Theron CHB (1983) Internal dynamics of soil conservation committees. Conservation Forum 2:12–14

Theron CHB (1985) Conservation chronicles-part two. Conservation Forum 3:22–25

Thierfelder C, Rusinamhodzi L, Ngwira AR, Mupangwa W, Nyagumbo I, Kassie GT, Cairns JE (2015) Conservation agriculture in Southern Africa: advances in knowledge. Renewable agriculture and food systems, 30. Conservation Chronicles-part two. Conservation Forum 3:22–25

Turpie JK, Forsythe KJ, Knowles A, Blignaut J, Letley G (2017) Mapping and valuation of South Africa's ecosystem services: a local perspective. Ecosyst Serv 27(B):179–192

UNEP (United Nations Environmental Programme) (2015) The economics of land degradation in Africa, ELD Initiative, Bonn. Available www.eld-initiative.org. Accessed 09 Nov 2019s

van Antwerpen R, Laker MC, Beukes DJ, Botha JJ, Collett A, du Plessis M (2021) Conservation agriculture farming systems in rainfed annual crop production in South Africa. S Afr J Plant Soil 2021:1–16. https://doi.org/10.1080/02571862.2020.1797195

van Tilburg A, van Schalkwyk HD (2012) Strategies to improve smallholders' market access. In: van Schalkwyk HD, Groenewald JA, Fraser GCG, Obi A, van Tilburg A (eds) Unlocking markets to smallholders: lessons from South Africa, Mansholt publication series, vol 10. Wageningen, Wageningen Academic Publishers, pp 35–58

van Zijl GM, Ellis F, Rozanov A (2014) Understanding the combined effect of soil properties on gully erosion using quantile regression. S Afr J Plant Soil 31(3):163–172. https://doi.org/10.1080/02571862.2014.944228

Vanmaercke M, Poesen J, Broeckx J, Nyssen J (2014) Sediment yield in Africa. Earth Sci Rev 136:350–368. https://doi.org/10.1016/j.earscirev.2014.06.004

Vickery KJ, Eckardt FD, Bryant RG (2013) A sub-basin scale dust plume source frequency inventory for southern Africa, 2005–2008. Geophys Res Lett 40(19):5274–5279

von Blanckenburg F (2006) The control mechanisms of erosion and weathering at basin scale from cosmogenic nuclides in river sediment. Earth Planet Sci Lett 242:224–239. https://doi.org/10.1016/j.epsl.2005.11.017

von Holdt JRC, Eckardt FD, Baddock MC, Wiggs GFS (2019) Assessing landscape dust emission potential using combined ground-based measurements and remote sensing data. J Geophys Res Earth 124:1080–1098. https://doi.org/10.1029/2018JF004713

von Maltitz GP, Gambiza J, Kellner K, Rambau T, Lindeque L, Kgope B (2019) Experiences from the South African land degradation neutrality target setting process. Environ Sci Pol 101:54–62

Vos HC, Fister W, Eckardt FD, Palmer AR, Kuhn NJ (2019) Physical crust formation on sandy soils and their potential to reduce dust emissions from croplands. Land 9:503. https://doi.org/10.3390/land9120503

Vos HC, Fister W, von Holdt JR, Eckardt FD, Palmer AR, Kuhn NJ (2021) Assessing the PM10 emission potential of sandy, dryland soils in South Africa using the PI-SWERL. Aeolian Res 53:10747. https://doi.org/10.1016/j.aeolia.2021.100747

Walling DE (1983) The sediment delivery problem. J Hydrol 65(1–3):209–237

Wiggs G, Holmes P (2011) Dynamic controls on wind erosion and dust generation on west-central free state agricultural land, South Africa. Earth Surf Process Landf 36:827–838. https://doi.org/10.1002/esp.2110

Wischmeier WH, Smith DD (1978) Predicting rainfall erosion losses, a guide to conservation planning. USDA agricultural handbook no 537. USDA, Washington, DC

Wright MA, Grab SW (2017) Wind speed characteristics and implications for wind power generation: cape regions, South Africa. S Afr J Sci 113(7/8):2016–0270. https://doi.org/10.17159/sajs.2017/20160270

Zhao C, Zhang H, Wang M, Jiang H (2021) Impacts of climate change on wind erosion in Southern Africa between 1991 and 2015. Land Degrad Dev 32:2169–2182

Biome Change in Southern Africa

14

Steven I. Higgins, Timo Conradi, Michelle A. Louw, Edward Muhoko, Simon Scheiter ⓘ, Carola Martens, Thomas Hickler, Ferdinand Wilhelm, Guy F. Midgley ⓘ, Jane Turpie, Joshua Weiss, and Jasper A. Slingsby

S. I. Higgins (✉) · T. Conradi · M. A. Louw · E. Muhoko
Plant Ecology, University of Bayreuth, Bayreuth, Germany
e-mail: Steven.Higgins@uni-bayreuth.de

S. Scheiter · F. Wilhelm
Senckenberg Biodiversity and Climate Research Centre (SBiK-F), Frankfurt am Main, Germany

C. Martens · T. Hickler
Senckenberg Biodiversity and Climate Research Centre (SBiK-F), Frankfurt am Main, Germany

Institute of Physical Geography, Goethe University Frankfurt am Main, Frankfurt am Main, Germany

G. F. Midgley
School for Climate Studies and Global Change Biology Group, Department of Botany and Zoology, Stellenbosch University, Stellenbosch, South Africa

J. Turpie
Anchor Environmental Consultants, Cape Town, South Africa

Environmental Policy Research Unit, School of Economics, University of Cape Town, Cape Town, South Africa

J. Weiss
Anchor Environmental Consultants, Cape Town, South Africa

J. A. Slingsby
Department of Biological Sciences and Centre for Statistics in Ecology, Environment and Conservation, University of Cape Town, Cape Town, South Africa

Fynbos Node, South African Environmental Observation Network (SAEON), Pretoria, South Africa

© The Author(s) 2024
G. P. von Maltitz et al. (eds.), *Sustainability of Southern African Ecosystems under Global Change*, Ecological Studies 248,
https://doi.org/10.1007/978-3-031-10948-5_14

Abstract

Biomes are regional to global vegetation formations characterised by their structure and functioning. These formations are thus valuable for both quantifying ecological status at sub-regional spatial scales and defining broad adaptive management strategies. Global changes are altering both the structure and the functioning of biomes globally, and while detecting, monitoring and predicting the outcomes of such changes is challenging in Southern Africa, it provides an opportunity to test biome theory with the goal of guiding management responses and evaluating their effectiveness. Here, we synthesise what is known about recent and expected future biome-level changes from Southern Africa by reviewing progress made using dynamic global vegetation modelling (based on archetypal plant functional types), phytoclime modelling (based on species-defined plant growth forms) and phenome monitoring (based on the seasonal timing of vegetation activity). We furthermore discuss how monitoring of indicator species and indicator plant growth forms could be used to detect and monitor biome-level change in the region. We find that all the analysis methods reviewed here indicate that biome-level change is likely to be underway and to continue, but that the analytical approaches and methods differed substantially in their projections. We conclude that the next phases of research on biome change in the region should focus on reconciling these differences by improving the empirical opportunities for model verification and validation.

14.1 Introduction

Biomes are conceptual constructs that categorise terrestrial ecosystems into structural and functional units and thereby help us organise our knowledge on how ecosystems work (Moncrieff et al. 2016). Despite their importance, there is surprisingly little consensus on how to define biomes (Moncrieff et al. 2016). Most biome schemes use a combination of plant growth forms, leaf phenology and sometimes climate to recognise formations that include Forest (evergreen, deciduous and mixed), Savannas (mixed tree and grass formations), Shrublands, Grasslands, Deserts and high elevation or high latitude Tundra. Whittaker (1975), refining ideas developed by Schimper (1903) made the case that biomes are strongly dependent on climate, while observing that it was not possible to predict the dominant biome in seasonally dry, subtropical climates, which happen to cover vast areas of the planet. Bond (2005) addressed this conundrum, hypothesising that fire and herbivores override climate forcing in the subtropics by preventing vegetation from attaining its "climatic potential", while (Walter 1973) drew attention to how soils and orographic processes may also cause vegetation to deviate from climatic potential.

The manifold impacts of anthropogenic climatic change make it extremely likely that biomes are changing in character and distribution, given that there is ample evidence for wide-scale redistribution of global biomes under the changing climates of the Pleistocene and earlier epochs (Huntley et al. 2021). Indeed, several studies have used satellite imagery to detect changes in vegetation cover and ecosystem functioning in recent decades that are large enough to qualify as biome shifts (Seddon et al. 2016; Higgins et al. 2016; Buitenwerf et al. 2015, Song et al. 2018; Zhu et al. 2016). Such shifts can have large impacts on biodiversity, for example, the number of endemic species has been shown to be higher in areas of relative past biome constancy (Huntley et al. 2016, 2021).

One way of forecasting biome shifts is to use Dynamic Global Vegetation Models (DGVMs). DGVMs are ambitious models that seek to simulate how resource assimilation, growth, competition and consumption (fire, herbivory) processes interact over ecological time scales and thereby how biomes might shift in response to changes in the climate system (Prentice et al. 2007). DGVMs also account for the impacts of changes in atmospheric chemistry, in particular the plant physiological impacts of enhanced CO_2 levels (which increases carbon uptake through photosynthesis and reduces water loss through transpiration Walker et al. 2021). The aDGVM is a dynamic global model that has been specifically developed to model the biome boundaries between forest, savanna and grassland and is therefore particularly suited to modelling the Southern African sub-region (Scheiter and Higgins 2009; Scheiter et al. 2012). How DGVMs define biomes however varies considerably. In the aDGVM, the relative cover of C3 grasses, C4 grasses, savanna trees and forest trees is used to define biomes.

An alternative method for forecasting biome shifts is to use a data-driven approach to represent plant growth forms. Conradi et al. (2020), for example, estimated the climatic suitability of geo-locations for plant growth forms typically used to define biomes. This was achieved by parameterising physiological growth models for 23,500 African plant species categorised into the growth forms, projecting the distribution of climatically suitable geo-locations for each species and then calculating the proportion of species of each growth form for which a geo-location was suitable. This proportion was used to characterise the climatic suitability of a geo-location for each growth form. This approach allows the researchers to forecast changes in the ability of a geo-location's climate to support different plant growth forms. The climate suitability of a geo-location for different plant growth forms describes its capacity to support different types of plants that ecologists use to define biomes. The vector of growth form suitability scores can then be used to classify different geo-locations into groups. The groups can be conceptualised as phytoclimes, where a phytoclime is defined as a geographic region where the climate's influence on growth form suitability is similar. Phytoclimes align conceptually with Holdridge (1947) and Whittaker (1975) who emphasised links between climate and vegetation formations even if phytoclimes establish these links differently. Using the term phytoclime emphasises that such analyses model the potential of a region's climate to support different plant growth forms; phytoclimes are not equivalent to biomes. Rather, the phytoclimate describes which plant types could potentially grow at a location, whereas a range of processes ignored by phytoclimes such as competition, facilitation, herbivory, predation, dispersal and historical contingencies determine how climate suitability translates into vegetation formations and biomes.

Southern Africa is a challenging arena for forecasting biome shifts because climate and consumption (fire and herbivory) processes interact to shape the regions' major vegetation formations. South Africa alone has 9 recognised biomes (Mucina and Rutherford 2006), and the biome concept has for decades shaped and structured both the practice of ecological science in the country and the national environmental policy. This is well illustrated by the biodiversity vulnerability assessment, which was part of the South African country study on climate change (Rutherford et al. 1999). The Rutherford et al. (1999) study developed the case that changes in the climate forces that regulate the distribution of the biomes of South Africa would lead to a large and dramatic re-organisation of the country's vegetation. A combination of effective science communication and the severity of the impacts predicted in the study served to stimulate an intensification of research on climate change impacts in the region. Perhaps more importantly, the report has had a sustained impact on South African climate change policy. Here, we review recent analyses of how climatic change may impact on the distribution of the vegetation of Southern Africa.

The forecasts of predictive models such as DGVMs or phytoclime models have uncertainties originating from multiple sources. These include uncertainty on the key processes represented in the models, the parameterisation of the processes, the Global Circulation Models (GCMs) used to forecast future climates that ecologists use to force their ecological models and the global emission scenarios that these GCMs assume (Thuiller et al. 2019). It follows that such models need to be critically evaluated using independent information. Indeed, much can be inferred about the trajectories of change that ecosystems are on by observing the recent past using remote sensing and on-the-ground monitoring. For this reason, we would suggest that modelling approaches endeavour to provide output in a form that can be compared to monitoring data. In particular, we highlight the potential of using the MODIS satellite record and field-based monitoring using Biome Shift Monitoring Phytometers (BISMOPs) and indicator species.

14.2   Phytoclimes

14.2.1 Phytoclime Methods

A phytoclime analysis of Southern Africa (Higgins et al. 2023) used species distribution data from the National Vegetation Database (Rutherford et al. 2012), ACKDAT (Rutherford et al. 2003) and BIEN version 4.1 (http://bien.nceas.uscsb. edu), thereby combining South Africa's excellent vegetation data legacy with a leading global data base on plant species distributions. These distribution data and the CHELSA 2.0 climate data (Karger et al. 2017), which scales CMIP6 climate projections to a 1 km resolution, were used to parameterise a process-based physiological plant growth model, the TTR-SDM (Thornley Transport Resistance Species Distribution Model Higgins et al. 2012), for 5006 species. The model considers how spatial and monthly variations in temperature, soil moisture, solar radiation and atmospheric CO_2 concentrations influence the growth of plant species. The model fitting procedure estimates the influence of these environmental factors on growth that are consistent with the observed distribution data. The fitted models were projected in geographical space to identify climatically suitable grid cells for each species. The growth form of each of these species was then used to group species by growth form. The analysis assigned species to the following growth forms, trees, shrubs, C3 grasses, C4 grasses, restioids, geophytes, annual forbs, other forbs, succulents and climbers using several data bases: C4 grass data base—Osborne et al. (2014), succulent plant data base—Eggli and Hartmann (2001), Eggli and Nyffeler (2020), POSA—http://newposa.sanbi.org, BIEN—http://bien.nceas. uscsb.edu and GIFT—Weigelt et al. (2020).

The average of the geographical projections of the climate suitability scores of each species belonging to a growth form was then used to estimate the suitability of the sub-region for each of the 10 growth forms (Fig. 14.1). The growth form suitability scores of grid cells were then classified, using unsupervised classification, to yield phytoclimes—geographical regions where the climate favours plant growth forms in similar ways.

Using CHELSA 2.0, climate projections derived from the CMIP6 (Eyring et al. 2016) programme future (225, 2055, 2085) shifts in phytoclimes were projected. These projections considered uncertainty in TTR-SDM model used (different TTR-SDM variants make different assumptions, 4 variants were considered), the GCM used (5 GCMs were considered) and the SSP assumed (3 SSPs were considered).

14.2.2 Phytoclime Findings

The suitability surfaces for each growth form (Fig. 14.1) revealed patterns that are broadly consistent with prior knowledge, and we highlight a few of these patterns here. The arid region associated with the Namib was generally unsuitable for all growth forms, as were the higher lying parts of Lesotho. In general, suitability for growth forms gradually increased from west to east (Fig. 14.1). C3 grasses had preferences for the south coast and eastern coast of South Africa, and this preference area extended over the escarpment and into the highveld areas including the Soutpansberg mountains and Inyanga mountains in Zimbabwe. C4 grasses by contrast showed a preference for the north eastern part of the sub-region. Both grass types revealed a low preference for the arid central and western parts of the sub-region, although this trend was stronger for C3 than for C4 grasses. Perennial forbs had a lower preference for an arid area starting at the Cunene River mouth and extending towards the south coast. Climatic suitability for annual forbs was often higher than for perennial forbs, and in particular they revealed less aversion to the aforementioned arid area. Geophytes had a high preference for the south coast and grassland regions, including the Soutpansberg mountains in South Africa and Inyanga mountains in Zimbabwe. Restioids exhibited the same preferences as geophytes, but a more distinct aversion to almost all other parts of the sub-region. Succulents exhibited preferences for the Cape provinces of South Africa, central

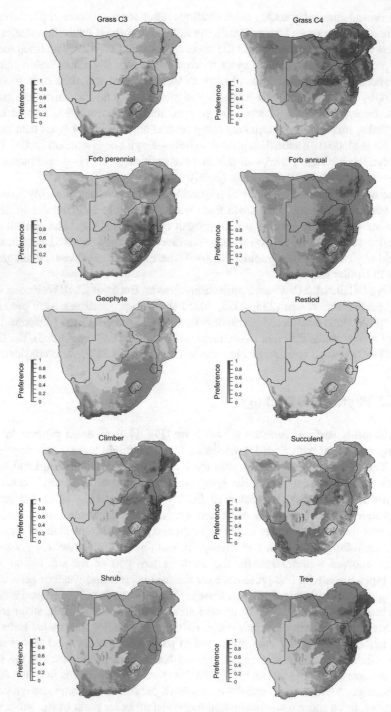

Fig. 14.1 Ambient suitability surfaces for growth forms derived from the TTR-RED-LD model. The preference scores are the averaged preference scores of the species belonging to each growth form, transformed to scale between 0 and 1

Namibia and a region centred on the intersection of the border between South Africa, Botswana and Zimbabwe. Shrubs showed preference for the southern coast regions and the central east of the sub-region, and they showed low preferences for the arid regions of Namibia and the Kalahari. Trees showed a preference for the north eastern part of the sub-region and had an aversion to an arid triangle that extends from the Cunene River mouth to the southern coast. Climbing plants had a similar preference surface to trees.

Additional insight can be gathered by examining how the growth form suitability values shown in Fig. 14.1 change over time (from the present to the end of the century). Figure 14.2 shows the rate of change (% change in suitability per 100 years) in growth form suitability averaged over the different TTR-SDM variants, SSPs and GCMs. The rates of change were frequently as high as 35% per 100 years, forecasting a fundamental change in the influence of climate on the region's vegetation. The most striking trend is that eastern South Africa is forecast to experience an increase in suitability for C4 grasses, trees and climbers and to a lesser extent shrubs, succulents and annual forbs (Fig. 14.2). Furthermore, Lesotho is forecast to be climatically more suitable for all growth forms in the future, suggesting that these high-lying parts of the Drakensberg mountains could serve as a climate refuge for many species and growth forms if they could migrate there. A similar trend can be seen in the Cape Fold mountains. A comparison of Figs. 14.1 and 14.2 suggests that C4 grasses, which under ambient conditions had an intermediate preference for the central plateau, will find this region more suitable in the future. These same areas will become less suitable for C3 grasses. C4 grasses were also forecast to find the fynbos, karoo and Namib more favourable in the future. C3 grasses were predicted to find the south coast, an area for which they have a high ambient preference, less suitable in the future. Perennial forbs, which under ambient conditions had a low preference for an arid triangle from the Cunene River mouth to southern coast, were predicted to find this area slightly more suitable in the future. Geophytes were largely predicted to face decreases in climate suitability except in Lesotho and surrounding areas. Restios were predicted to face a marked decrease in climatic suitability in the areas in which they currently have a high climatic suitability (the south coast). Projected loss and gain patterns in climbers and trees resembled those projected for C4 grasses. Succulents were projected to face losses in their ambient high-preference areas such as the Succulent Karoo and the area around the intersection of the borders between South Africa, Botswana and Zimbabwe). Shrubs were projected to find the Drakensberg and surrounding areas more suitable, but parts of the bushveld of South Africa, Botswana and Zimbabwe as well as the Albany Thicket biome of the Eastern Cape less suitable.

When the ambient growth form preferences (Fig. 14.1) are classified into phytoclimes, it is possible to produce a range of phytoclime maps. Figure 14.3 illustrates phytoclimes for the region when assuming 6–32 phytoclimes. There is no a priori reason why the region should have 6 or 32 phytoclimes. That is, this analysis explicitly acknowledges that phytoclimes are abstractions designed to help us understand climate's influence on a region's vegetation ecology, but they do not represent an absolute truth. In this analysis, 24 different phytoclime maps were generated (4

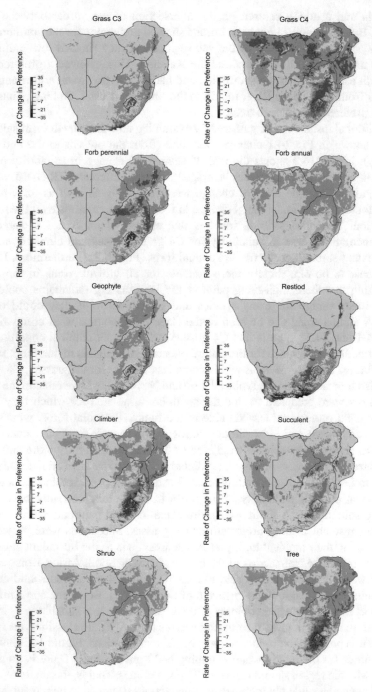

Fig. 14.2 The mean rate of predicted change in growth form suitability (% change in suitability per 100 years) for the 10 growth forms considered in this study. The rate of change is calculated assuming a linear change in preference over time (from the present to the end of the century) and using TTR-variant, GCM and SSP as covariates

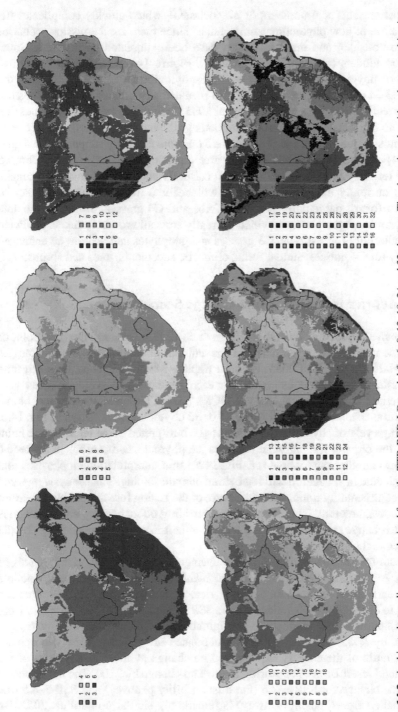

Fig. 14.3 Ambient phytocolime maps derived from the TTR-RED-LD model for 6, 9, 12, 18, 24 and 32 clusters. The clustering algorithm detects multivariate discontinuities in the growth form suitability surfaces displayed in Fig. 14.1 to delimit the phytoclimes

TTR-SDM variants × 6 numbers of phytoclimes), which quickly complicates the interpretation of how phytoclimes may change since there are 3 emission scenarios (SSPs) to consider, and these scenarios have been simulated by different global circulation models (we consider 5 GCMSs). Figure 14.4 provides an illustrative example of phytoclime change that uses a 6-phytoclime map generated using the TTR-RED-LD model variant, the SSP 585 scenario as predicted by the gfdl-esm-4 GCM for a climatology centred on the year 2085. This single scenario illustrates that phytoclime 6 (climate that currently supports grassland and fynbos) will decrease in area, mostly due to losses to phytoclime 5 (climate currently supporting savanna) (Fig. 14.4). Figure 14.5 reveals that the area currently occupied by phytoclime 5 will lose territory to phytoclime 4 (which currently supports more arid savannas). The area currently under phytoclime 5 will become increasingly unsuitable for all growth forms, particularly perennial forbs and C4 grasses. Similarly, the area currently under phytoclimate 6, which currently suits all growth forms, is predicted to lose suitability for restioids, C3 grasses and geophytes and acquire an enhanced suitability for C4 grasses, annual forbs, climbers, succulents, trees and shrubs.

14.2.3 Synthesis of Phytoclime Change Scenarios

To synthesise phytoclime change, Higgins et al. (2023) recorded the time point of phytoclime changes observed between the ambient climatology and climatologies centred at 2025, 2055 and 2085. Using a Kaplan–Meier estimator, the mean time to phytoclime change was estimated for each location. The averaged mean year of phytoclime change, averaged over GCM, SSP, TTR-SDM variant and biome classification scheme (i.e. 6, 9, 12, 18, 24 or 32 phytoclimes) is shown in Fig. 14.6. This average year of phytoclime change is to be interpreted as the relative time point at which the climate forcing which defines the phytoclimes is sufficient to force a change into another phytoclime (cf. Fig. 14.5); this interpretation emphasises that realised change in vegetation will lag behind climate forcing. This analysis revealed a strong continentality trend, with the centre of the region forecast to change earlier than the coastal regions. This suggests an overriding effect of temperature which is forecast to change more severely in the interior than in coastal regions (Engelbrecht and Engelbrecht 2016).

The patterns summarised in Fig. 14.6 average away some important sources of variation. Anchor Environmental, *under stakeholder review* explored the phytoclime change trends in more detail for South Africa. They found that very few areas are expected to change by 2025 under both the SSP 126 and the SSP 585, with relatively modest changes across phytoclime configurations by 2055 under both pathways. However, by 2085, the changes in phytoclimates are forecast to be widespread. Very few parts of the country experienced no change at all with only parts of the Fynbos and Desert Biomes appearing to have no change by 2100. The likelihood of phytoclime change for SSP 1–2.6 (the Sustainability pathway) and SSP 5–8.5 (the Fossil-Fuelled Development pathway) is remarkably similar up until ca. 2055. By 2025, the likelihood of phytoclime change appears to be mostly limited to a few

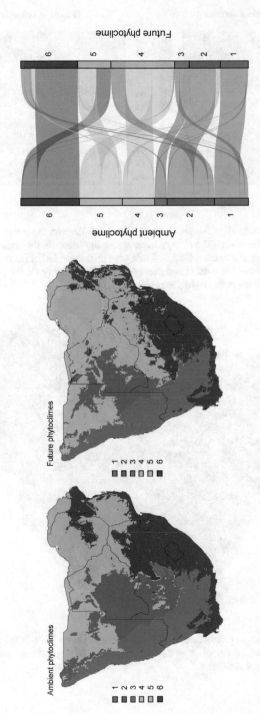

Fig. 14.4 Ambient and future (climatology centred on year 2085, SSP 585, GCM gfdl-esm4) phytoclime maps of Southern Africa derived from the TTR-RED-LD model for a 6-phytoclime classification. The right-hand panel summarises the phytoclime transitions between the two maps

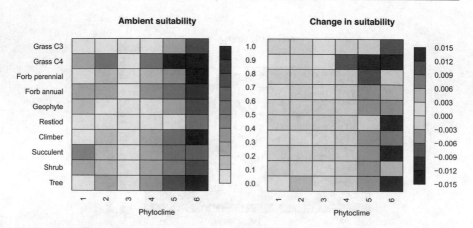

Fig. 14.5 Growth form suitability for the 6-phytoclime map derived from the TTR-RED-LD model under ambient conditions and the change in the suitability scores in the ambient phytoclime regions forecast for the year 2085 under SSP 585 using the gfdl-esm4 GCM (as in Fig. 14.4). The values in the left panel express the normalised average climatic suitability for the growth forms in the phytoclimes and the values in the right panel show their change

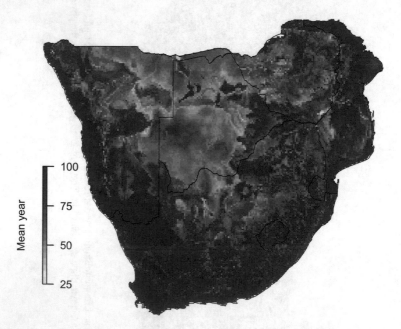

Fig. 14.6 The average time to phytoclime transition averaged across TTR-SDM model variants ($n = 4$), SSPs ($n = 3$), climate model ($n = 5$) and the number of phytoclimes assumed ($n = 6$) using a Kaplan–Meir survival estimator

isolated areas of the Kalahari Duneveld and Eastern Kalahari Bushveld Bioregions in the North West and Northern Cape Provinces of South Africa. The likelihood of phytoclime change increases notably by 2055. By 2055, a high change score region extends from the Kgalagadi to the south and east. Moderate to high change scores are also predicted for large areas of the Grassland Biome (particularly, the Dry and Mesic Grassland Bioregions) by 2055. Moderate to high change scores are also predicted for large areas of the Central Bushveld Bioregion spanning the Savanna Biome in the North West and Limpopo Province. Phytoclime change scores are higher under SSP 5–8.5. Change to South Africa's coastline occurs later, with projected phytoclime change scores remaining low at the 2055 time slice, which may be due to the cooling effect of the proximity to the ocean. The Succulent Karoo, Nama-Karoo, Albany Thicket and Fynbos biomes overall are likely to be least affected. The northern half of the Kruger National Park in Limpopo also appears to remain relatively unchanged in the 2055 time slice.

By the 2085 time slice, the phytoclime change scores are high throughout South Africa, only a few areas, such as patches of mountainous Fynbos in the Western Cape, and isolated areas of Namaqualand and the Desert Biome remain unchanged. Lower change scores were derived for many areas of the Eastern Cape and Southern KwaZulu-Natal (Sub-Escarpment and Drakensberg Grassland) and the Bushmanland Bioregion of the Nama-Karoo. The central interior of the country had very high phytoclime change scores. The Lowveld and Sub-escarpment Savanna areas of Mpumalanga, Limpopo and KwaZulu-Natal exhibited large increases in change scores between 2055 and 2085.

Overall, the Savanna and Grassland Biomes have the highest phytoclime change scores. This is perhaps partly due to their size, occupying 59% of South Africa's terrestrial extent. This matches findings by Midgley et al. (2011) who identified the Orange River Basin and Highveld plateau as being climate change hotspots.

For planning, it is useful to combine spatial information on both the likelihood of ecological change due to climate change and the level of protection against other anthropogenic impacts. For this purpose, Anchor (2022) calculated a vulnerability index that considered the level of ecosystem health (DEA, 2019) and the level of conservation protection (DFFE, 2021). The resulting vulnerability indices were summarised by the recognised biomes of South Africa and contrasted for SSP1-2.6 (Fig. 14.7) and SSP 5–8.5 (Fig. 14.8). Across both SSPs, the extent of high vulnerability areas increases substantially over time, particularly between the period centred on 2055 and 2085. There is almost no difference between the SSPs in the 2011 to 2040 period, since protected areas ensure vulnerability remains very low in this time window. The most noticeable changes in the following period centred on 2055 are the increase in the extent of areas with high vulnerability > 0.5 in the Kalahari, Bushveld and southern Kruger National Park (all in the Savanna Biome). There is also an overall increase in vulnerability in the Bushmanland Bioregion of the Nama-Karoo. In the period centred on 2085, the extent of very high vulnerability (>0.75) increases dramatically, particularly under SSP 5–8.5 (under SSP 1–2.6, various areas in the Grassland and Savanna regions in Gauteng, Free State, Mpumalanga, North West and Limpopo remain slightly more stable). There

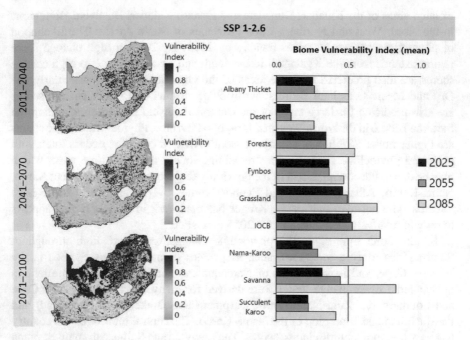

Fig. 14.7 The vulnerability index for change of vegetation under SSP 1–2.6 for the periods 2011–2040, 2041–2070 and 2071–2100 (left) and the biome vulnerability index (mean vulnerability index per biome) for each time period's median year. The vulnerability index considers phytoclimatic change, land use change and protected area status

are also substantial increases in vulnerability across KwaZulu-Natal and areas of the Northern Cape. The Desert Biome area, eastern parts of the Succulent Karoo, the mountainous areas of the Fynbos and eastern Fynbos Biome areas remain the most stable.

The increase in mean biome vulnerability index can also be seen between the SSPs with slight increases between 1–2.6 and 5–8.5. For SSP 5–8.5, in the earliest time period (2025), the Indian Ocean Coastal Belt (IOCB) has the highest biome vulnerability index with 0.52, followed by Grassland with 0.46 and Fynbos 0.36. The Desert and Albany Thicket have the lowest overall. These values largely reflect the existing extents of modified land in each of the biomes with the IOCB and Grassland having the highest values of all nine terrestrial biomes (Skowno et al. 2021). By 2055, Grassland, Savanna and IOCB have the highest biome vulnerability indices with values of approaching 0.55. The Desert and Albany Thicket biomes continue to have the lowest. By 2085, the biome vulnerability index of Savanna increases to 0.92, followed by Grassland (0.83) and IOCB (0.81). Desert remains the lowest but increases to 0.31 (still the only biome below 0.5) followed by Fynbos with 0.52. Similar between-biome differences are forecast under SSP 1–2.6 albeit with slightly lower scores.

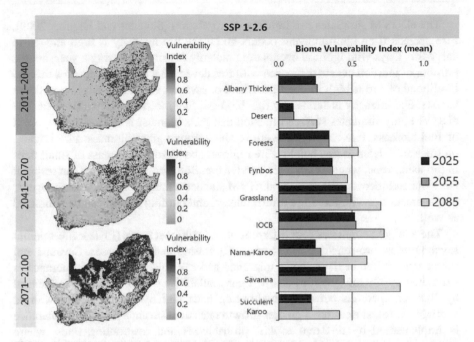

Fig. 14.8 The vulnerability index for change of vegetation under SSP 5–8.5 for the periods 2011–2040, 2041–2070 and 2071–2100 (left) and the biome vulnerability index (mean vulnerability index per biome) for each time period's median year. The vulnerability index considers phytoclimatic change, land use change and protected area status

14.3   Insights from DGVM Modelling of Biomes

14.3.1 DGVM Methods

Here, we analysed results from four different DGVMs, three global ones that were not adapted for the study region and the aDGVM originally developed for Africa (Scheiter and Higgins 2009; Scheiter et al. 2012). As the global models do not represent all major biome types in the study region, we first describe the methodology and results for aDGVM and provide more detail than for the global models. The aDGVM uses concepts and processes commonly used in other DGVMs such as leaf-level ecophysiology or the representation of the carbon cycle (Prentice et al. 2007). In addition, processes such as carbon allocation to different plant compartments or leaf phenology are adjusted based on environmental conditions. The aDGVM is individual-based and simulates growth, biomass, allometry, reproduction and mortality of individual trees. This approach enables simulations of the impacts of disturbances such as fire, herbivory or fuelwood collection on individual plants while relating these impacts to plant traits such as tree height or stem diameter. In contrast to trees, grasses are only simulated by super-individuals representing grasses between and under tree crowns.

The aDGVM simulates fire impacts on vegetation (Scheiter and Higgins 2009). Fire spreads if an ignition event occurs and if the fire intensity is high enough to carry fire. Days with ignition events are randomly distributed during a year, but the number of ignition events decreases with tree cover. This approach ensures that the likelihood of fire is high in open grassland or savanna vegetation but low in dense forests. Fire intensity is defined by fuel biomass, fuel moisture and wind speed. The aDGVM only simulates surface fire such that grass biomass is the main component of fuel biomass. Fire removes the entire aboveground grass biomass. Fire impacts on trees are related to tree height. Fire removes aboveground biomass of small trees in the flame zone, whereas tall trees survive fire. Both grasses and trees can resprout after fire and recover. By default, aDGVM simulates natural fire but anthropogenic management fire with fixed fire return interval and burning season can be simulated as well.

The aDGVM simulates four different plant functional types (PFTs): fire-tolerant savanna trees, fire-sensitive forest trees, C_4 grasses and C_3 grasses. Savanna and forest trees differ in their shade tolerance and fire tolerance. Shade tolerance is simulated by linking growth rates to light availability which is in turn influenced by light competition between neighboring trees and by shading. At low light availability, forest trees have greater growth rates than savanna trees. Fire tolerance is implemented by different topkill probabilities and resprouting rates, where savanna trees have lower topkill probabilities and higher resprouting rates than forest trees. Based on these assumptions, forest trees outcompete savanna trees in closed forest stands without fire, whereas savanna trees can survive in open and fire-driven environments. The difference between C_4 and C_3 grasses is simulated based on the physiological differences between C_4 and C_3 photosynthesis and shade tolerance. To simulate differences in shade tolerance, we used different light competition parameters for C_4 and C_3 grasses, which describe how shading effects by neighboring plants influence the light availability and photosynthetic rate of a target plant. The relative abundances of these four PFTs are influenced by the prevailing environmental conditions, competition between individual plants, and disturbance regimes.

Similar to suitability surfaces of different growth forms derived from the TTR-RED-LD model, relative abundances of the four PFTs derived from aDGVM simulations can be used to create suitability surfaces. In addition, vegetation simulated by aDGVM can be classified into different biome types using simulated model state variables. Following the classification scheme developed by Martens et al. (2021), vegetation is classified as desert if tree cover is below 10% and grass biomass is below 1.5 t/ha and as grassland if tree cover is below 10% and grass biomass is above 1.5 t/ha. Grassland is further separated into C_3 or C_4 grassland based on the relative abundance of C_3 and C_4 grass PFTs. If tree cover is between 10 and 80%, vegetation is classified as either woodland, if forest tree cover exceeds savanna tree cover, or as savanna, if savanna tree cover exceeds forest tree cover. This separation between woodland and savanna reflects the absence or presence of regular fire, which favours savanna trees. Savannas are further split into C_4 or C_3 savanna, depending on the abundance of grass PFTs. If the total tree cover (i.e.

forest and savanna tree cover) exceeds 80%, vegetation is classified as forest. Hence, biome shifts, simulated for example in response to climate change or changes in land use, occur if simulated biomass of different grass PFTs or cover of different tree PFTs exceeds or falls below the respective thresholds used for biome classification. While this scheme represents important biomes of Southern Africa, it ignores biomes including Fynbos or Karoo. In the current version, aDGVM lacks PFTs required to simulate these biome types, including CAM plants, succulent shrubs and flammable shrubs (Moncrieff et al. 2015).

Martens et al. (2021) used aDGVM to study climate change impacts on future vegetation in Africa for an ensemble of climate change scenarios. Specifically, simulations were conducted for the period between 1971 and 2099 for RCP4.5 and RCP8.5. For each scenario, down-scaled climate forcing from six different GCMs was available. Down-scaling to a spatial resolution of 0.5° was conducted with the variable-resolution conformal-cubic atmospheric model (CCAM, McGregor 2005; Engelbrecht et al. 2015; Engelbrecht and Engelbrecht 2016). All details of the modeling protocol are provided by Martens et al. (2021). Here, we evaluate the results provided by Martens et al. (2021) focusing on the Southern African study region.

Several DGVMs were run globally with harmonised past to future environmental forcing and land use data within the Intersectoral Impact Model Intercomparison Project (ISIMIP; https://www.isimip.org). The modelling protocol is described in Frieler et al. (2017). Only three models produced outputs that enabled us to analyse results per PFT and to transform the results with land use into simulations of the potential natural vegetation, which we considered more relevant here than grid cell averages with land use. Therefore, we analysed results for these three models: LPJ-GUESS (Smith et al. 2014), ORCHIDEE (Krinner et al. 2005) and CARAIB (Dury et al. 2011). As the global models have not been adapted for Africa and do not represent all major biomes in the study region well, we only analysed mean results across all three DGVMs and available climate scenarios (four for LPJ-GUESS and CARAIB and two for ORCHIDEE, whereby results for each DGVM were weighted equally), and we used a rather coarse biome classification. More details of the analyses are described by Wilhelm (2021).

14.3.2 DGVM Findings

aDGVM simulations showed that most of the study region is suitable to support both C_3 and C_4 grasses, except the Namib region and Lesotho (Fig. 14.9) with low precipitation and low temperature, respectively. Suitability is generally higher for C_4 grasses than for C_3 grasses. Hence, in our simulations C_4, grasses dominate the grass layer except a small region in the Karoo (Fig. 14.10). The model simulated increasing suitability for tree PFTs from west to east with low suitability in the Namib region and the Karoo, intermediate suitability in the savanna regions of Namibia, Botswana, South Africa and Zimbabwe and high suitability along the south and east coast of South Africa and Mozambique. Suitable regions for savanna

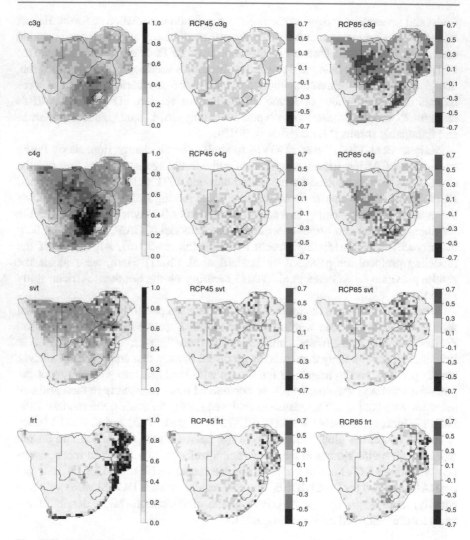

Fig. 14.9 Suitabilities/Niches in 2000–2019 and their changes until 2080–2099 under RCP4.5 and RCP8.5 for plant functional types (PFTs) as simulated by aDGVM. PFTs simulated by aDGVM are C$_3$ grasses (c3g), C$_4$ grasses (c4g), fire-tolerant savanna trees (svt) and shade-intolerant forest trees (frt). Suitability surfaces (left) are based on maximum values in 2000–2019. For grasses, 0.9 times the maximum value of both grass PFTs was used to scale grid cells. For savanna trees, 0.9 times the maximum savanna tree canopy cover was used because inherently the savanna tree PFT rarely has a closed canopy. Forest tree canopy cover was used as suitability surface for forest trees. Changes are derived from the differences between 2080–2099 and 2000–2019. The simulation setup is described in Martens et al. (2021)

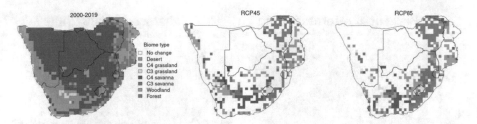

Fig. 14.10 Biomes in 2000–2019 and biome changes until 2080–2099 under RCP4.5 and RCP8.5 simulated by aDGVM. Grid cells were classified into biomes based on grass biomass, dominance of C_3 or C_4 grasses, tree cover, dominance of savanna or forest tree types. In the biome change subfigures, biomes that were simulated for 2080–2099 for grid cells where biome transitions were simulated are shown. This figure is based on Martens et al. (2021)

trees and forest trees are almost disjoint with little overlap in their distributions. While savanna trees dominate fire-driven savanna regions with intermediate tree cover, forest trees dominate forests in the East of the study region. This result is not surprising given that savanna trees in aDGVM are fire-tolerant and able to outcompete fire-intolerant forest trees in fire-driven regions, whereas forest trees are shade-tolerant and able to outcompete shade-intolerant savanna trees in dense forests.

Under climate change, aDGVM simulates changes in the suitability of different PFTs. Suitability of C_3 grasses was predicted to increase in almost the entire study region. Suitability of C_4 grasses was predicted to increase in Namibia, Botswana and Mozambique, whereas it was predicted to decrease in Zimbabwe and most of South Africa. These results can be explained by CO_2 fertilisation effects that, in aDGVM, enhance C_3 photosynthesis but not C_4 photosynthesis. Suitability of savanna trees was predicted to decrease primarily in Namibia, Botswana and the North of South Africa, whereas it was predicted to increase in Zimbabwe and most of South Africa. Suitability of forest trees was predicted to increase in South Africa, but decreases were simulated in the forest regions of Mozambique. Taken together, these changes cause woody encroachment and an increase of woody biomass in the entire study region until the end of the century (not shown, Martens et al. 2021). Patterns of change of suitability were similar for RCP4.5 and RCP8.5 for all PFTs, but on average, changes were higher for RCP8.5.

Predicted changes in suitability of PFTs imply changes in simulated biome patterns (Fig. 14.10). Increases in the suitability of tree PFTs implied transitions from desert to woodland (along the coast of Namibia), from grassland to savanna or woodland (e.g. in the Karoo) and from savanna to woodland or forest (e.g. East of South Africa, Botswana). In Mozambique, both forest dieback and transitions to woodland as well as transitions from woodland to forest were predicted.

The ensemble mean of the three global DGVMs roughly reproduced the main distribution of biomes across the study region as reconstructed in a global PNV (potential natural vegetation) map. Major shrub biomes, such as the Nama Karoo, however, were not distinguished (Fig. 14.11). The models also confirmed the

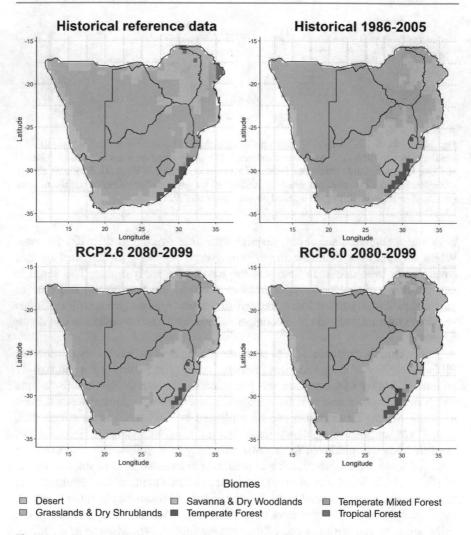

Fig. 14.11 Biomes according to expert reconstruction (reference data) based on Haxeltine and Prentice (1996); multi-model ensemble mean (3 DGVMs and 2–4 GCMs) for recent past (mean 1986–2005) and future (mean 2080–2099) for RCP2.6 and RCP6.0

increasing suitability for tree PFTs from west to east predicted by aDGVM. For the future, the DGVMs predicted woody encroachment and increasing tree cover in particular in eastern parts of the study area, and much more pronounced under RCP6.0, as a result of higher CO_2 fertilisation effects on woody plants under this scenario. Biome shifts were also much more pronounced under RCP6.0 and concentrated in the eastern part of the study region and, under RCP 6.0, the northwestern fringe (Fig. 14.12). These results were roughly in line with those by the aDGVM, but they clearly differ for large parts of Zimbabwe, where the aDGVM

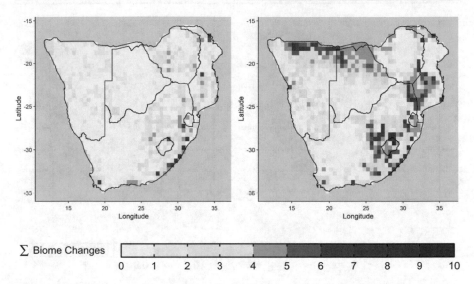

Fig. 14.12 Sum of simulations with biome change between historical (mean 1986–2005) and future (mean 2080–2099) under RCP2.6 (left) and RCP6.0 (right) according to a multi-model ensemble

predicted increasing climatic suitability for fire-tolerant savanna trees (Fig. 14.9), while the global DGVM ensemble predicted decreasing woody leaf area index (Fig. 14.13). These differences might have been caused by differences in vegetation process representations or by using different climate models as forcing. The general strong woody encroachment into savannas is consistent with results from other DGVM-based studies (Scholze et al. 2006; Gonzalez et al. 2010).

Regarding biome stability, simulations with LPJ-GUESS spanning the last 140,000 years suggest a remarkable biome constancy along the southern and eastern coast and in the southern Kalahari, which appears to be correlated with high species diversity (Huntley et al. 2016, 2021). Simulations for the future with climate scenario data input, however, suggest substantial biome shifts in these areas of previous biome constancy (Huntley et al. 2021).

14.4   Monitoring Biome Change

Process-based forecasting models such as the phytoclime models and aDGVM are influenced by process and parameter uncertainty, which can make them difficult to interpret. A complementary alternative is to use statistical monitoring of change that is already ongoing. Global change impacts are already manifesting, meaning that statistical detection and description of the recent trajectories of change should also be a priority for adaptation and mitigation science. That is, process-based models are essential for exploring future scenarios of change, but statistical analysis of recent

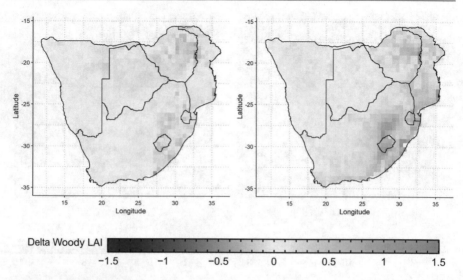

Fig. 14.13 Woody LAI change between recent past (mean 1986–2005) and future (mean 2080–2099) under RCP2.6 (left) and RCP6.0 (right), according to a multi-model ensemble mean (3 DGVMs and 2–4 GCMs)

trajectories of change can help us understand change that is already ongoing. In this section, we describe three promising change monitoring and detection activities.

14.4.1 Birds as Indicators of Biome Change

Indicator species are organisms which are easily monitored and whose status reflects the state of the environment in which they are found (Siddig et al., 2016). The use of indicator species as early warning systems in climate change science is still in its infancy. Anchor (2022) has recently suggested a range of bird species that could be used as indicators of biome change. Birds have the advantage that they are readily observed by a large population of hobby birdwatchers and citizen scientists and are also mobile enough to respond immediately to changes in climate and vegetation structure. The premise is that species that are strongly associated with a particular biome can serve as sensitive indicators for changes in biome structure and functioning that are not necessarily detectable using satellites or standard vegetation monitoring. A comparable approach has been used by BirdLife International for global analyses, and however the BirdLife protocol needs to be adapted to the sub-region. Anchor (2022) identified 16 bird species with range boundaries that correspond with biome boundaries and where the range boundaries have relatively hard edges. Birds as indicators have the potential to run as a citizen science project which would reduce field costs and enhance public engagement. However, the lure of volunteer field work should not distract from the fact that a substantial investment is nonetheless required to ensure that an efficient computational back-end exists that

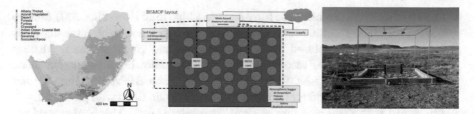

Fig. 14.14 Biome shift monitoring phytometers (BISMOPs) have been installed in the major biomes of South Africa (left panel). A BISMOP is a mini-common garden experiment that monitors the performance of the plant growth forms characteristic of the biomes of South Africa. Each BISMOP monitors 3 individuals of 11 plant species and records their NDVI over a diurnal cycle as well as associated climate system variables (middle panel). The Nieuwoudtville Hantam Botanical Garden BISMOP (right panel)

allows data to be uploaded, processed and visualised in real time. Another issue is that detailed research will be required to understand and anticipate the different ways in which bird species may respond to changes in climate versus vegetation structure (e.g. Seymour et al. 2015) or direct human influence on habitat structure and resource availability (e.g. Weideman et al. 2020).

14.4.2 Phytometers as Indicators of Biome Change

A phytometer is a special kind of indicator species: a phytometer is a transplanted plant whose growth, survival or reproduction are used to provide information on environmental conditions. Phytometers have the advantage that their performance can be interpreted as an integrative measure of the environment. Here, we provide an illustration of how phytometers could be used to monitor climatically induced biome changes. The system we describe is a biome shift monitoring phytometer (BISMOP), which we have implemented in South Africa (Fig. 14.14).

A BISMOP is a miniature common garden experiment where plants representative of the biomes of South Africa are grown in common garden settings in the different biomes. BISMOP plants are selected based on two criteria. First, their distributions should closely match the distribution of the biomes. Second, they should represent a plant growth form typical of the biome in question. The existing BISMOPS used 11 vascular plant species. Individuals were planted in 9 litre pots filled with perlite substrate. Perlite was used because it has good hydraulic properties (reasonable drainage and water retention) and is commercially available in various quality classes ensuring that the same substrate can be sourced in the future. Three replicate plants of each species are grown in each BISMOP. The planted pots and one control (34 in total) are arranged in a 250 × 200 cm area. An aluminum frame supports a camera system 250 cm above the pots. The cameras are based on the Raspberry Pi Camera Module 2, which uses a Sony IMX219 8-megapixel sensor and a wide angle and low distortion lens (MX219 Camera Module 160°FoV). The camera modules have no infra-red filter and use a blue filter (Roscolux 2007 Storaro Blue, Rosco, UK). The blue filter blocks most red and green light, allowing near-infrared (NIR) to be captured in the camera's red and green channels, whereas the camera's blue channel records mostly blue light allowing a vegetation index with similar properties to NDVI to be calculated. Each BISMOP station further records air temperature and air humidity. The control pot, filled with perlite only, is instrumented with a soil moisture and soil temperature probe. This bare ground pot is used to calculate a simple water balance that allows the current level of moisture stress at the BISMOP site to be estimated. Images from the cameras are captured every daylight hour and environmental variables are recorded every 10 minutes.

The data from the BISMOP systems produce time series (at daily or hourly resolution) that allow the analysis of how the physiological performance of individual plants are influenced by climate system variables. Since the data represent key South African plant growth forms, growing in environmental situations that represent the diversity of climate systems in the region, the data will additionally be valuable for testing the predictions of DGVM and phytoclime models.

14.4.3 Remote Sensing of Phenome Change

There is a long history of monitoring vegetation activity using satellite-based remote sensing. Buitenwerf et al. (2015), for example, developed a method for analysing change in a satellite-based vegetation activity using NDVI data (NDVI, normalised difference vegetation index) from the AVHRR satellite program. The primary advantage of the (Buitenwerf et al. 2015) method is that it stratifies the phenological metrics into groups, or phenomes, thereby allowing change in the phenological metrics to be analysed by group. This allows the change analysis to be stratified by phenological groups, which in turn aids change detection since phenologically divergent signals are not diluted by averaging.

Higgins et al. (2023) applied this method to the Southern African region; this analysis used MODIS EVI (enhanced vegetation index), which has a higher resolution (the $1 \times 1\,km^2$ MOD13 product spanning 2000–2019 was used) than the AVHRR NDVI used by Buitenwerf et al. (2015). EVI is additionally less prone to saturation at high vegetation density and less sensitive to atmospheric sources of error than NDVI. Higgins et al. (2023) further focused exclusively on protected areas, which removes the potential for confounding the effects of land use and climate on vegetation activity. The world database on protected areas was used (UNEP-WCMC and IUCN 2021) to select areas that do not contain land use effects (nature reserves, game parks, national parks and forest reserves, Fig. 14.15).

Following Buitenwerf et al. (2015), Higgins et al. (2023) estimated 21 phenological metrics for each phenological year for the time series in each $1\,km^2$ pixel (Fig. 14.16). The metrics describe the annual phenological curve and include metrics with time units (e.g. day of peak EVI) and metrics in EVI units (e.g. maximum EVI). To quantify change, the mean of each metric was calculated for the first and second half of the record. Figure 14.17 summarises by how many standard

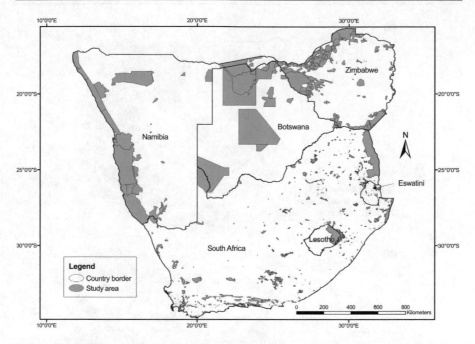

Fig. 14.15 To exclude land use effects, phenological change was analysed in the protected areas of Southern Africa. Protected areas were derived from the world database on protected areas (UNEP-WCMC and IUCN 2021)

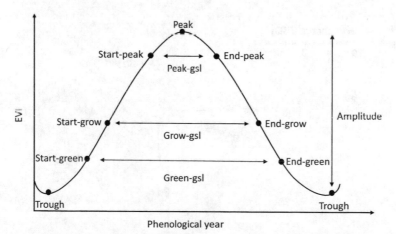

Fig. 14.16 A schematic representation of the EVI activity of a remotely sensed grid cell. At each labelled point, metrics in both EVI units and day of year units are recorded. The integral of the EVI between two subsequent trough days, although not labelled in this diagram, is also calculated. gsl: growing season length

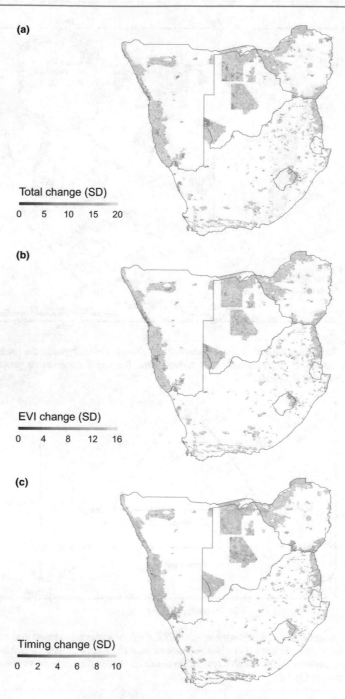

Fig. 14.17 Changes in vegetation phenological activity between 2000 and 2019 reported in standard deviations. (**a**) Aggregated change of all metrics. (**b**) Aggregated changes in metrics representing vegetation activity (measured in EVI units). (**c**) Aggregated changes in metrics that represent the timing of the vegetation activity (measured in days or day of the year)

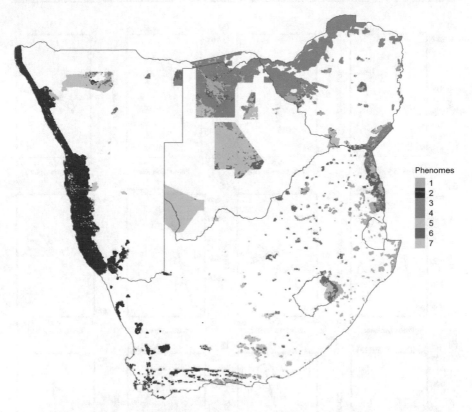

Fig. 14.18 Phenomes of the protected areas of Southern Africa. The phenomes were produced by performing an unsupervised clustering of the phenological metrics as defined in Fig. 14.16. The 7 phenomes represent zones with similar phenological behaviour

deviations these metrics have changed revealing widespread change hotspots in Namibia, Botswana and Zimbabwe (Fig. 14.17). These analyses revealed shifts in both the timing magnitude of vegetation activity.

The ecological interpretation of the pattern of change is aided by classifying pixels into groups (Buitenwerf et al. 2015 phenomes) with similar phenological behaviour. Higgins et al. (2023) used the phenological metrics (Fig. 14.16) to classify pixels into 7 phenomes (Fig. 14.18). The phenome classification identified regions superficially consistent with desert, arid savanna, bushveld, grassland, woodland savanna and fynbos.

For each phenome (Fig. 14.18), the mean phenological signature in the first (2000–2009) and second (2011–2019) parts of the time series is shown in Fig. 14.19. The vectors in Fig. 14.19 show the average change in phenological metrics in both the EVI (vegetation activity) and time (timing of activity) dimensions between the first and second parts of the time series. Although the phenological signature

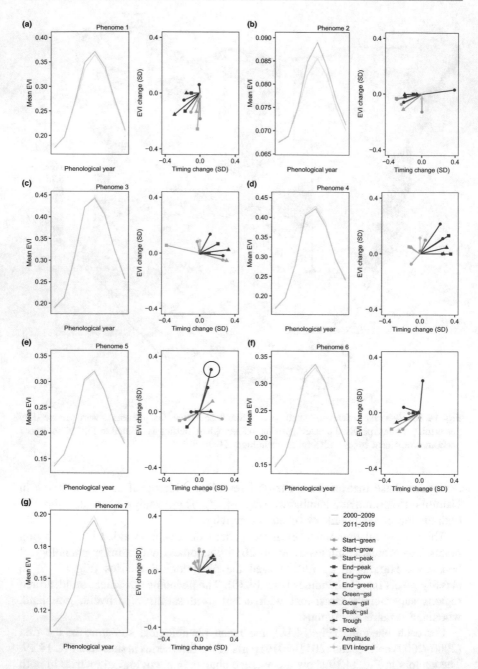

Fig. 14.19 Phenological change per phenome. The phenological signature of each phenome is plotted for the first (2000–2009, blue lines) and second (2011–2019, green lines) parts of the time series. The vector plots show change in all metrics (cf. Fig. 14.16) in both EVI and time dimensions. For example, the circle plotted phenome 5 shows change in the metric "Trough." In this example, the trough day has increased (it is delayed) and its EVI value has increased. Vectors parallel to the axes represent metrics that have only EVI or time dimensions

plots suggest minor absolute changes in each phenome's phenological signature, the vector plots reveal large relative changes.

Phenome 1 covered a diverse range of Cowling's phytogeographical regions (Cowling et al. 2004), from the Zambezian region, the Kalahari-Highveld Transition Zone, the Tongaland-Pondoland Region and Afromontane Region to the Cape region. It spread from the Moremi National Park in northwest Botswana through national parks in South Africa such as the Songimvelo in the northeast to the Maloti-Drakensberg in the southeast. The growing season length decreased in phenome 1, caused by both a delayed start and an earlier end of the growing season.

Phenome 2 was primarily distributed in the xeric Karoo-Namib and the Cape regions, spreading southwards from the Skeleton Coast in northwestern Namibia towards the Anysberg in South Africa. This phenome was characterised by a low vegetation activity and the amplitude and integral of the annual NDVI decreased over the study period. A decrease in the length of the growing season was also detected, caused by both a delayed start and an earlier end of the growing season.

Phenome 3 was mainly distributed in the Zambezian phytogeographical region, spreading from the Bwabwata in northeast Namibia through the Chobe in northern Botswana to the Hwange in northwest Zimbabwe. Further in Zimbabwe, it was distributed in the Charara in the north as well as the Gonarezhou in the south. This phenome revealed an increase in the amplitude and integral of the annual EVI signal. The length of the GSL did not change substantially although the timing metrics were delayed, with the exception of the onset of the peak and peak of the growing season which were earlier.

Phenome 4 was mainly distributed in the Zambezian and the Kalahari-Highveld Transition Zone phytogeographical regions. This phenome stretched from the Moremi in northern Botswana through the Chewore in northern Zimbabwe to the northern parts of Kruger in northeastern South Africa. The growing season length extended over the observation period, caused by both an earlier onset and a delayed end. As a consequence, the integral of the EVI increased and as did the peak EVI which was also delayed.

Phenome 5 occurred mainly in the Zambezian and the Kalahari-Highveld Transition Zone phytogeographical regions, spreading from eastern Etosha in northern Namibia towards the Kalahari in central Botswana to the eastern parts of Kruger in northeast South Africa and the Tsehlanyane in eastern Lesotho. In this phenome, the growing season length decreased mostly due to delayed start of the growing season. In addition, the peak, integral and amplitude of the EVI also decreased, combining to produce an overall decrease in vegetation activity.

Phenome 6 was primarily distributed in the Zambezian and the Kalahari-Highveld Transition Zone phytogeographical regions, spreading from the Khaudum in northeast Namibia through the Kalahari in central Botswana to the eastern parts of Kruger in South Africa and the western parts of Tsehlanyane in Lesotho. The growing season length in this phenome decreased primarily due to an earlier end of the growing season. Moreover, the peak, integral and amplitude of EVI decreased.

Phenome 7 was primarily distributed in the Kalahari-Highveld Transition Zone phytogeographical region, spreading from western Etosha in northern Namibia through the Kgalagadi in southwestern Botswana to Camdeboo in southern South Africa. In this phenome, the growing season length increased due to both an earlier start and a delayed end to the growing season, although the delayed end effect dominated. The peak, integral and amplitude of EVI all increased in this phenome.

14.5 Discussion

This review summarises recent analyses of vegetation change in Southern Africa. The focus is on predictive simulations of vegetation structure and function, as derived from both established mechanistically based dynamic global vegetation models (DGVMs) and novel species-based modelling of the climatic preferences of major plant growth forms (phytoclimatic modelling). The review additionally suggests options for monitoring vegetation and biome change that can be used to detect and understand change as well as to test the ability of the prediction models to provide information robust enough to support policy and management actions. Over the past two decades, the capacity to develop such models has evolved rapidly from the rudimentary yet insightful correlative modelling work of Rutherford et al. (1999) at both biome and species levels, to more sophisticated correlative modelling (Thuiller et al. 2006) and earlier dynamic vegetation modelling efforts (Scheiter and Higgins 2009; Moncrieff et al. 2014). The results of these early efforts raised awareness and concerns regarding the potential severity of climate change impacts, despite the level of uncertainty that was clearly communicated. However, reported observations of historically observed changes tempered these projections even further, showing trends that ran counter to projections in some cases (Masubelele, Stevens), or revealing changes consistent with projections (Stevens et al.) that may have occurred via unanticipated mechanisms. Two main causes of these divergences have been noted, namely the importance of disturbance via fire in over-riding climatic drivers of change, and the role of rising atmospheric CO_2 via increased water use efficiency and productivity that were not accounted for in correlative approaches. The mechanistically based approaches described here provide new insights that permit a deeper understanding that may be of more value for policy and management guidance.

We first discuss what light the results of the modelling approaches used here sheds on our understanding of the current representation of biomes and vegetation types in the region. The (Rutherford and Westfall 1986) treatment of biomes of Southern Africa, updated by Mucina and Rutherford (2006), remains one of the clearest in terms of its simple plant growth form definitions (that is, simple combinations of Raunkiaer plant life forms trees, shrubs, grasses and annual plants). Their mapping of combinations of dominant growth forms at the 100 km grid scale provides a useful template for comparison, and it remains the basis of fundamental regional ecological distinction for a wide range of applications, including the planning of adaptation responses. Despite being based purely on growth form dominance, their approach uncovered floristic distinctiveness between biomes (e.g.

Gibbs Russel et al.), which makes a comparison with the phytoclime approach extremely interesting.

It is most insightful to compare the phytoclimes identified by successive increases in the number of species clusters with the biomes as mapped by Mucina and Rutherford (2006). It must be noted that the phytoclime maps represent the potential of the climate and substrate to support different growth forms (Fig. 14.1), whereas ecological processes not considered by the phytoclimatic analysis, such as fire, herbivory, competition, fertility and dispersal, influence which combination of growth forms will actually grow at a given location.

At its most parsimonious 6-cluster level (Fig. 14.3), the arid and semi-arid phytoclime unit 1 closely matches the Desert, Nama-Karoo and Succulent Karoo biome borders with Fynbos Biome in the south and Savanna, Grassland and Albany thicket biomes in the east. A hyper-arid subdivision of desert indicates a level of sensitivity lacking in the Mucina and Rutherford (2006) definition, while three phytoclimes are discerned in what is traditionally seen as Savanna across Namibia, Botswana, Zimbabwe and Mocambique. Thus, a predominance of tropical assemblages overwhelms the purely plant life form approach of Mucina and Rutherford (2006). Even at the 9-cluster level (an equivalent number of biomes recognised by Mucina and Rutherford 2006), hyper-arid and subtropical assemblages dominate, with latitudinal divisions emerging to differentiate the 6-phytoclime classification's units 1, 2 and 6, more clearly separating grassland and savanna-type biomes in central and northern South Africa and differentiating further North/South divisions in arid and semi-arid shrubland and savanna types in the west. Phytoclime units comparable with Succulent karoo, Grassland, Thicket, Coastal Belt and Fynbos biomes emerge only when clustering 18 phytoclimes levels and higher, while the forest biome does not emerge at all.

Overall, unit 6 in the 6-phytoclime classification retained coherence at successive clustering levels while also showing low concordance with purely plant life form-defined biomes for this region. This is a novel revelation of interesting climatic control of vegetation, mediated by plant form and function, in this region. This could be further refined by expanding the phytoclime analysis to consider how soil fertility modulates the influence of climate. However, low confidence in currently available soil fertility products precludes consideration of soil fertility.

The representation of Mucina and Rutherford (2006) biomes by DGVM simulations is also interesting in how the arid- and semi-arid Succulent Karoo and Desert biomes are somewhat recognisable, but C4-dominated Grassland, C4-co-dominated Nama-Karoo and Savanna biomes occupy the central portion of the region, with increasing dominance of trees from the arid west to the more mesic eastern reaches, resulting in a less refined differentiation relative to both (Mucina and Rutherford 2006) and the phytoclime approach. The DGVMs only consider a limited set of plant functional types, and it is likely that consideration of additional functional types such as succulent plants and different types of shrubs (Gaillard et al. 2018) may enhance the ability of the DGVMs to represent the Mucina biome map more closely.

The phytoclime and DGVM analyses do make predictions of where we should expect vegetation change in the future, but these are quite different. The aDGVM, for example, forecasts that the core of the large C4 savanna biome in the central interior does not change with climate change (Fig. 14.10), whereas this is an area of high change in the phytoclime analysis (e.g. Fig. 14.4). This is confirmed when examining how the suitability of geographic space for aDGVM plant types changes across Southern Africa with changing climate. The aDGVM predicts, for example, that central Botswana will become increasingly suitable for C3 grasses and to a lesser extent for C4 grasses and that C4 grasses will decrease in central South Africa (Fig. 14.9). The phytoclimatic analysis by contrast suggests that C3 grass suitability will decrease almost everywhere except Lesotho and that C4 grass suitability will increase over the Namibian coast, the South African coast and the higher altitude areas of the region but will decrease over most of the central and northern parts of the region (Fig. 14.2). In effect, the phytoclimatic analysis is emphasising temperature effects which will favour C4 grasses, whereas the aDGVM is emphasising CO_2 effects which will favour C3 grasses.

The strong positive effects of increasing atmospheric CO_2 on the photosynthesis and the growth of C3 plants are a general feature of the DGVMs. However, DGVMs that do not simulate nutrient limitation most likely overestimate these effects (Hickler et al. 2015; Martens et al. 2021). Nonetheless, the woody encroachment observed in the eastern part of the study region was simulated by the LPJ-GUESS model (Wilhelm 2021), which does include a nitrogen cycle and nitrogen limitation of plant growth (Smith et al. 2014).

A further reason for these differences might be that the aDGVM model predicts the realised niche of the functional types, that is, the biomass that can accumulate under a given climate under the influence of fire and competition, whereas the phytoclimatic suitability surfaces represent estimates of the climate's potential to support different plant growth forms. In effect, the phytoclimatic suitabilities (Fig. 14.1) represent the climatic limits hard coded in earlier developed DGVM model versions. These global DGVMs include hard-coded limits that relate to cold temperature tolerance and are thus unlikely to be important in this study region. The aDGVM by contrast does not include hard-coded climatic limits for its different plant types. Thus, in aDGVM, biome borders in the study region are mainly determined by the competition between grasses and trees as mediated by the simulated frequency and intensity of fire.

One of the apparently important similarities between projections made by the models is that tree dominance will increase in parts of the savanna regions (Figs. 14.4 and 14.9). For example, the region currently occupied the phytoclime suitable for grasses (phytoclime 6, Fig. 14.4) will be increasingly occupied by a phytoclime suitable for trees (phytoclime 5, see also Fig. 14.5). It is now well established that these regions are centres of widespread thickening of woody plants and that this process may drive substantive changes in the functioning and diversity of these landscapes over vast areas (e.g. Stevens et al review).

The phenome analysis (Higgins et al. 2023) provides convincing evidence that the vegetation of all parts of the sub-region is responding to climate change. This analysis is unique in that it focuses on change in protected areas and thereby excludes the potential confounding effects of land use change. The phenome analysis reveals that plants in some regions are experiencing an increase in growing season length (phenomes 1, phenome 4, phenome 7), whereas decreases were observed in other regions (phenome 2, phenome 6). In addition, in some regions there was an overall increase in vegetation activity (phenome 3, phenome 4, phenome 7), whereas for others there was an overall decrease (phenome 5) in vegetation activity. These findings are in agreement with previous work. For example, in northeast South Africa, Masia et al. (2018) reported early green-up dates and early leaf drop dates among species typical of phenomes 1 and 6. Furthermore, in the Cape Floristic region of South Africa, herbarium specimen of species has been used to show that between 1901 and 2009, leaf flowering has advanced by 12 days (Williams et al. 2021). Phenomes 1 and 2 represent this region, and they both show an early start to the growing season. Higgins et al. (2023)'s findings further agree with the early greening patterns reported by Whitecross et al. (2017) who used MODIS NDVI data to detect early green-up dates along a latitudinal gradient from Zambia to South Africa. Phenomes 1 and 4 cover portions of their study region, and they both show an early start to the growing season. Overall the findings of the phenome study emphasise that different parts of the study region are responding in qualitatively different ways to changes in the climate system. Further work is needed to ground verify the changes detected by the Earth observation satellites used in the phenome analysis. The analysis reviewed in this study provides the data for guiding such field work. In parallel, additional work is needed to attribute these changes to particular climate system drivers.

This study illustrates that modelling the biomes of Southern Africa with DGVMs reveals priorities for model development. For example, simulating field experiments with a DGVM identified that improving the representation of demographic events associated with drought is a priority (Baudena et al. 2015). In a more general sense, DGVMs can be improved by calibrating and or testing them using the Earth observation system (EOS) data; the potential of this avenue is under-exploited, particularly in Southern Africa. Using multiple lines of EOS evidence (e.g. biome maps, NPP and GPP, fire activity, surface temperature) would improve the identifiability of DGVM parameterisations and the robustness of their predictions. Furthermore, more regionalised approaches might be necessary in order to capture, for example, the peculiarities of the Karoo which are poorly captured by the PFT's in existing DGVMs. Strategically, a balance between parameterisation and calibration needs to be found: short- and medium-term predictions can be enhanced by calibration, whereas understanding is best enhanced by adjusting parameterisations and assumptions.

Our premise when initiating the work reported on here was that using different methods to analyse biome-level change in vegetation in the Southern Africa would allow us to make robust inferences about the nature of change. However, we have learned that different inference methods represented by statistical inference of the capacity of the regions climate to support different plant growth forms (the phyto-climatic analysis), forward simulation of ecosystem dynamics (DGVM simulations) and monitoring of change using remote sensing time series of vegetation activity (the phenome analysis) have yielded different insights. The only unifying insight is that change is widespread. This unfortunately leaves us in the position where we have to advocate increased effort to refine the forecasting models. Satellite-based monitoring of vegetation change using the phenome method as described here and phytometer monitoring as represented by the BISMOP system are promising avenues for generating the data needed to test the models' predictions and the assumption upon which they are based. However, it is also clear that experimental work aimed at exploring the interactions between CO_2 fertilisation, temperature increases and moisture deficits is necessary to resolve the disparate predictions of the phytoclimatic and dynamic vegetation models. This means that atmospheric CO_2 experiments designed to quantify the interactive effects of CO_2 enrichment, temperature and moisture should be a regional research infrastructure priority.

Acknowledgments We thank the Jinfeng Change, Phillip Ciais and Louis Francois for providing results from the ORCHIDEE and CARAIB DGVMs and the ISIMIP team for providing harmonised DGVM forcing data. This research was supported by BMBF SPACES project EMSAfrica, grant number 01LL1801A. Edward Muhoko acknowledges a doctoral scholarship provided by the DAAD.

References

Anchor Environmental (under review) Biome level plans for adaptation strategy. Project reference number 2085, https://anchorenvironmental.co.za/resources/draft-biome-level-plans-biodiversity-adaptation-strategy

Baudena M, Dekker SC, van Bodegom PM, Cuesta B, Higgins SI, Lehsten V, Reick CH, Rietkerk M, Scheiter S, Yin Z, Zavala MA (2015) Forests, savannas, and grasslands: bridging the knowledge gap between ecology and dynamic global vegetation models. Biogeosciences 12(6):1833–1848

Bond WJ (2005) Large parts of the world are brown or black: a different view on the 'green world' hypothesis. J Veget Sci 16(3):261–266

Buitenwerf, R, Rose, L, Higgins, S.I., (2015). Three decades of multi-dimensional change in global leaf phenology. Nat Climate Change 5(4):364–368

Conradi T, Slingsby JA, Midgley GF, Nottebrock H, Schweiger AH, Higgins SI (2020) An operational definition of the biome for global change research. New Phytol 227:1294–1306

Cowling RM, Richardson DM, Pierce SM (2004) Vegetation of Southern Africa. Cambridge University Press, Cambridge

Dury M, Hambuckers A, Warnant P, Henrot A, Favre E, Ouberdous M, François L (2011) Responses of European forest ecosystems to 21st century climate: assessing changes in interannual variability and fire intensity. iForest Biogeosci Forestry 4:82–99.

Eggli U, Hartmann HEK (2001) Illustrated handbook of succulent plants I–VI. Illustrated Handbook of Succulent Plants, 1st edn. Springer, Berlin

Eggli U, Nyffeler R (eds) (2020). Monocotyledons. Illustrated Handbook of Succulent Plants, 2nd edn. Springer, Berlin.

Engelbrecht CJ, Engelbrecht FA (2016) Shifts in Koeppen-Geiger climate zones over Southern Africa in relation to key global temperature goals. Theoret Appl Climatol 123(1):247–261

Engelbrecht F, Adegoke J, Bopape M-J, Naidoo M, Garland R, Thatcher M, McGregor J, Katzfey J, Werner M, Ichoku C, Gatebe C (2015) Projections of rapidly rising surface temperatures over Africa under low mitigation. Environ Res Lett 10(8):085004

Eyring V, Bony S, Meehl GA, Senior CA, Stevens B, Stouffer RJ, Taylor KE (2016) Overview of the coupled model intercomparison project phase 6 (CMIP6) experimental design and organization. Geosci Model Devel 9(5):1937–1958

Frieler K, Lange S, Piontek F, Reyer CPO, Schewe J, Warszawski L, Zhao F, Chini L, Denvil S, Emanuel K, Geiger T, Halladay K, Hurtt G, Mengel M, Murakami D, Ostberg S, Popp A, Riva R, Stevanovic M, Suzuki T, Volkholz J, Burke E, Ciais P, Ebi K, Eddy TD, Elliott J, Galbraith E, Gosling SN, Hattermann F, Hickler T, Hinkel J, Hof C, Huber V, Jägermeyr J, Krysanova V, Marcé R, Schmied HM, Mouratiadou I, Pierson D, Tittensor DP, Vautard R, van Vliet M, Biber MF, Betts RA, Bodirsky BL, Deryng D, Frolking S, Jones CD, Lotze HK, Lotze-Campen H, Sahajpal R, Thonicke K, Tian H, Yamagata Y (2017) Assessing the impacts of 1.5 c global warming – simulation protocol of the inter-sectoral impact model intercomparison project (ISIMIP2b). Geosci Model Devel 10(12):4321–4345

Gaillard C, Langan L, Pfeiffer M, Kumar D, Martens C, Higgins SI, Scheiter S (2018) African shrub distribution emerges via a trade-off between height and sapwood conductivity. J Biogeogr 45(12):2815–2826

Gonzalez P, Neilson RP, Lenihan JM, Drapek RJ (2010) Global patterns in the vulnerity of ecosystems to vegetation shifts due to climate change. Global Ecol Biogeogr 19(6):755–768

Haxeltine A, Prentice IC (1996) Biome3: An equilibrium terrestrial biosphere model based on ecophysiological constraints, resource availability, and competition among plant functional types. Global Biogeochem Cycles 10(4):693–709

Hickler T, Rammig A, Werner C (2015) Modelling co2 impacts on forest productivity. Current Forestry Rep 1(2):69–80

Higgins SI, O'Hara RB, Bykova O, Cramer MD, Chuine I, Gerstner E-M, Hickler T, Morin X, Kearney MR, Midgley GF, Scheiter S (2012) A physiological analogy of the niche for projecting the potential distribution of plants. J Biogeogr 39(12):2132–2145

Higgins SI, Buitenwerf R, Moncrieff GR (2016) Defining functional biomes and monitoring their change globally. Global Change Biol 22(11):3583–3593

Higgins SI, Conradi T, Ongole S, Slingsby J., (2023). Projecting biome shifts from species distributions for Southern Africa. Manuscript in preparation 000:00–00

Holdridge LR (1947) Determination of world plant formations from simple climatic data. Science 105(2727):367–368

Huntley B, Collingham YC, Singarayer JS, Valdes PJ, Barnard P, Midgley GF, Altwegg R, Ohlemuller R (2016) Explaining patterns of avian diversity and endemicity: climate and biomes of southern africa over the last 140,000 years. J Biogeogr 43(5):874–886

Huntley B, Allen JRM, Forrest M, Hickler T, Ohlemuller R, Singarayer JS, Valdes PJ (2021) Projected climatic changes lead to biome changes in areas of previously constant biome. J Biogeogr 48:11

Karger DN, Conrad O, Böhner J, Kawohl T, Kreft H, Soria-Auza RW, Zimmermann NE, Linder HP, Kessler M (2017) Climatologies at high resolution for the earth's land surface areas. Sci Data 4(1):170122

Krinner G, Viovy N, de Noblet-Ducoudré N, Ogee J, Polcher J, Friedlingstein P, Ciais P, Sitch S, Prentice IC (2005) A dynamic global vegetation model for studies of the coupled atmosphere-biosphere system, Global Biogeochem Cycles 19:GB1015. https://doi.org/10.1029/2003GB002199

Martens C, Hickler T, Davis-Reddy C, Engelbrecht F, Higgins SI, von Maltitz GP, Midgley GF, Pfeiffer M, Scheiter S (2021) Large uncertainties in future biome changes in Africa call for le climate adaptation strategies. Global Change Biol 27(2):340–358

Masia ND, Stevens N, Archibald S (2018) Identifying phenological functional types in savanna trees. Afr J Range Forage Sci 35(2):81–88

McGregor JL (2005) C-CAM: geometric aspects and dynamical formulation. CSIRO Atmospheric Research technical paper 70. CSIRO Atmospheric Research. Aspendale, Victoria, Australia, 43 pp

Moncrieff GR, Scheiter S, Bond WJ, Higgins SI (2014) Increasing atmospheric CO2 overrides the historical legacy of multiple stable biome states in Africa. New Phytol 201(3):908–915

Moncrieff GR, Scheiter S, Slingsby JA, Higgins SI (2015) Understanding global change impacts on South African biomes using dynamic vegetation models. S Afr J Bot 101(SI):16–23

Moncrieff GR, Bond WJ, Higgins SI (2016) Revising the biome concept for understanding and predicting global change impacts. J Biogeogr 43(5):863–873

Mucina L, Rutherford MC (2006) The vegetation of South Africa, Lesotho and Swaziland. Strelitzia 19. South African National Biodiversity Institute, Pretoria

Higgins SI, Conradi T, Muhoko E (2023) Shifts in vegetation activity of terrestrial ecosystems attributable to climate trends. Nat Geosci 16(2):147–153

Osborne CP, Salomaa A, Kluyver TA, Visser V, Kellogg EA, Morrone O, Vorontsova MS, Clayton WD, Simpson DA (2014) A global database of c4 photosynthesis in grasses. New Phytol 204(3):441–446

Prentice IC, Bondeau A, Cramer W, Harrison SP, Hickler T, Lucht W, Sitch S, Smith B, Sykes MT (2007) Dynamic global vegetation modeling: Quantifying terrestrial ecosystem responses to large-scale environmental change, pp 175–192. Springer, Berlin

Rutherford M, Westfall R (1986) Biomes of Southern Africa: an objective categorization. Memoirs of the Botanical Survey of South Africa, Botanical Research Institute

Rutherford MC, Midgley GF, Bond WJ, Powrie LW, Roberts R, Allsopp J (1999) Climate change impacts in southern Africa. Report to the National Climate Change Committee, chapter Plant biodiversity: vulnerability and adaptation assessment. South African Country Study on Climate Change, pp 1–58. Department of Environment Affairs and Tourism, Pretoria

Rutherford M, Powrie L, Midgley G (2003) ACKDAT: a digital spatial database of distributions of South African plant species and species assemblages. S Afr J Bot 69(1):99–104. Special Issue — Acocks' Veld Types

Rutherford M, Mucina L, Powrie L (2012) The South African national vegetation database: history, development, applications, problems and future. S Afr J Sci 108(1/2):8

Scheiter S, Higgins SI (2009) Impacts of climate change on the vegetation of Africa: an adaptive dynamic vegetation modelling approach (aDGVM). Global Change Biol 15:2224–2246

Scheiter S, Higgins SI, Osborne CP, Bradshaw C, Lunt D, Ripley BS, Taylor LL, Beerling DJ (2012) Fire and fire-adapted vegetation promoted C4 expansion in the late Miocene. New Phytol 195(3):653–666

Schimper AFW (1903) Plant-geography upon a physiological basis. Oxford University Press, Oxford

Scholze M, Knorr W, Arnell NW, Prentice IC (2006) A climate-change risk analysis for world ecosystems. Proc Natl Acad Sci USA 103(35):13116–13120

Seddon AWR, Macias-Fauria M, Long PR, Benz D, Willis KJ (2016) Sensitivity of global terrestrial ecosystems to climate variability. Nature 531(7593):229–232

Seymour CL, Simmons RE, Joseph GS, Slingsby JA (2015) On bird functional diversity: species richness and functional differentiation show contrasting responses to rainfall and vegetation structure in an arid landscape. Ecosystems 18(6):971–984

Skowno AL, Jewitt D, Slingsby JA (2021). Rates and patterns of habitat loss across South Africa's vegetation biomes. South African J Sci 117(1/2):1–5

Smith B, Warlind D, Arneth A, Hickler T, Leadley P, Siltberg J, Zaehle S (2014) Implications of incorporating n cycling and n limitations on primary production in an individual-based dynamic vegetation model. Biogeosciences 11:2027–2054

Song X-P, Hansen MC, Stehman SV, Potapov PV, Tyukavina A, Vermote EF, Townshend JR (2018) Global land 766 change from 1982 to 2016. Nature 560(7720):639–643

Thuiller W, Midgley FG, Rougeti M, Cowling RM (2006) Predicting patterns of plant species richness in megadiverse South Africa. Ecography 29(5):733–744

Thuiller W, Guaguen M, Renaud J, Karger DN, Zimmermann NE (2019) Uncertainty in ensembles of global biodiversity scenarios. Nat Commun 10(1):1446

UNEP-WCMC and IUCN (2021). Protected Planet: The World Database on Protected Ar772 eas (WDPA) [Online], [July 2021], Cambridge, UK: UNEP-WCMC and IUCN. Available at: www.protectedplanet.net

Walker AP, De Kauwe MG, Bastos A, Belmecheri S, Georgiou K, Keeling RF, McMahon SM, Medlyn BE, Moore DJP, Norby RJ, Zaehle S, Anderson-Teixeira KJ, Battipaglia G, Brienen RJW, Cabugao KG, Cailleret M, Campbell E, Canadell JG, Ciais P, Craig ME, Ellsworth DS, Farquhar GD, Fatichi S, Fisher JB, Frank DC, Graven H, Gu LH, Haverd V, Heilman K, Heimann M, Hungate BA, Iversen CM, Joos F, Jiang MK, Keenan TF, Knauer J, Korner C, Leshyk VO, Leuzinger S, Liu Y, MacBean N, Malhi Y, McVicar TR, Penuelas J, Pongratz J, Powell AS, Riutta T, Sabot MEB, Schleucher J, Sitch S, Smith WK, Sulman B, Taylor B, Terrer C, Torn MS, Treseder KK, Trugman AT, Trumbore SE, van Mantgem PJ, Voelker SL, Whelan ME, Zuidema PA (2021) Integrating the evidence for a terrestrial carbon sink caused by increasing atmospheric co2. New Phytol 229(5):2413–2445

Walter H (1973) Vegetation of the earth in relation to climate and the eco-physiological conditions. Springer, New York

Weideman EA, Slingsby JA, Thomson RL, Coetzee BTW (2020) Land cover change homogenizes functional and phylogenetic diversity within and among African savanna bird assemblages. Landscape Ecol 35(1):145–157

Weigelt P, König C, Kreft H (2020) GIFT - a global inventory of oras and traits for macro ecology and biogeography. J Biogeogra 47(1):16–43

Whitecross MA, Witkowski ET, Archibald S (2017) Assessing the frequency and drivers of early-greening in broad-leaved woodlands along a latitudinal gradient in southern africa. Austral Ecol 42(3):341–353

Whittaker RH (1975) Communities and Ecosystems. Macmillan Publishing, New York.

Wilhelm F (2021). Multi-model analysis of climate-change-induced vegetation changes in the 21st century. Master's Thesis

Williams TM, Schlichting CD, Holsinger KE (2021) Herbarium records demonstrate changes in flowering phenology associated with climate change over the past century within the Cape Oristic Region, South Africa. Climate Change Ecol 1:100006

Zhu ZC, Piao SL, Myneni RB, Huang MT, Zeng ZZ, Canadell JG, Ciais P, Sitch S, Friedlingstein P, Arneth A, Cao CX, Cheng L, Kato E, Koven C, Li, Y, Lian X, Liu YW, Liu RG, Mao JF, Pan YZ, Peng SS, Penuelas J, Poulter B, Pugh TAM, Stocker BD, Viovy N, Wang XH, Wang YP, Xiao ZQ, Yang H, Zachle S, Zeng N (2016) Greening of the earth and its drivers. Nat Climate Change 6(8):791–795

Biodiversity and Ecosystem Functions in Southern African Savanna Rangelands: Threats, Impacts and Solutions

15

Katja Geißler, Niels Blaum, Graham P. von Maltitz [ID], Taylor Smith,
Bodo Bookhagen, Heike Wanke, Martin Hipondoka,
Eliakim Hamunyelae, Dirk Lohmann, Deike U. Lüdtke,
Meed Mbidzo, Markus Rauchecker, Robert Hering, Katja Irob,
Britta Tietjen, Arnim Marquart, Felix V. Skhosana, Tim Herkenrath,
and Shoopala Uugulu

K. Geißler (✉) · N. Blaum · T. Smith · B. Bookhagen · D. Lohmann · R. Hering · T. Herkenrath
University of Potsdam, Potsdam, Germany
e-mail: kgeissle@uni-potsdam.de

G. P. von Maltitz
Stellenbosch University, Stellenbosch, South Africa

South African National Biodiversity Institute, Cape Town, South Africa

H. Wanke
University of the West of England, Bristol, UK

M. Hipondoka · E. Hamunyelae · S. Uugulu
University of Namibia, Windhoek, Namibia

D. U. Lüdtke · M. Rauchecker
ISOE - Institute for Social Ecological Research & SBiK-F, Frankfurt, Germany

Senckenberg Biodiversity and Climate Research Centre, Frankfurt, Germany

M. Mbidzo
Namibia University of Science and Technology, Windhoek, Namibia

K. Irob · B. Tietjen
Freie Universität Berlin, Berlin, Germany

A. Marquart
North-West University, Potchefstroom, South Africa

F. V. Skhosana
Council for Scientific and Industrial Research (CSIR), Pretoria, South Africa

G. P. von Maltitz et al. (eds.), *Sustainability of Southern African Ecosystems under Global Change*, Ecological Studies 248,
https://doi.org/10.1007/978-3-031-10948-5_15

Abstract

Savanna rangelands provide diverse communities across southern Africa with livestock and wildlife-based livelihoods, as well as extensive ecosystem services. Historical usage patterns, however, are increasingly challenged by widespread degradation. While regional- and local-scale policy initiatives have attempted to minimize damage and increase the sustainability of savanna rangelands, poverty, land tenure and shifting climate conditions all exacerbate ongoing degradation. Here, we detail the environmental and political setting of southern African savanna rangelands, causes and implications of rangeland degradation, and discuss possible strategies toward improved regional ecosystem management. We present recent knowledge on how degradation by bush encroachment influences biodiversity and biodiversity-mediated ecosystem functioning of semiarid savanna rangelands with the aim of improving rangeland management strategies. Improved rangeland management requires a broad approach which integrates both socioeconomic and ecological frameworks, built upon improved understanding of the strong couplings between flora, fauna, water and land-management strategies.

15.1 Biophysical Features of African Savanna Rangelands

Southern Africa's vegetation is dominated by tropical savannas (Fig. 15.1) which can be distinguished from other vegetation biomes by their codominance of trees and C4 grasses (see Chap. 2). Within the southern African savanna, there is a vast variety of types, such as the different ecoregions as defined by Olson et al. (2001) (Fig. 15.1). They are home to diverse endemic floras and faunas including the earth's greatest diversity of ungulates. Most of the natural land area of southern Africa can be considered as rangeland (Ellis and Ramankutty 2008), i.e., natural ecosystem habitats managed for grazing livestock and wildlife (Allen et al. 2011).

Climate A climatic feature of all southern African savannas, which occur over a wide rainfall gradient from approximately 250 mm to 1800 mm mean annual precipitation (MAP), is almost exclusive summer rainfall with long dry winters (see Fig. 15.2, Huntley 1982; Scholes and Archer 1997). Interannual variability in rainfall, with the most arid areas having the greatest coefficient of variation in rainfall, means that years of both intense drought and near flooding are common (see Chap. 7). The drought and above average rainfall are linked to El Niño–Southern Oscillation cycles and as such periods of either above or below average rainfall are the norm. This has profound impacts on the carrying capacity for livestock and wildlife with animal numbers increasing during rainy periods but with die-offs occurring during droughts (Shackleton 1993).

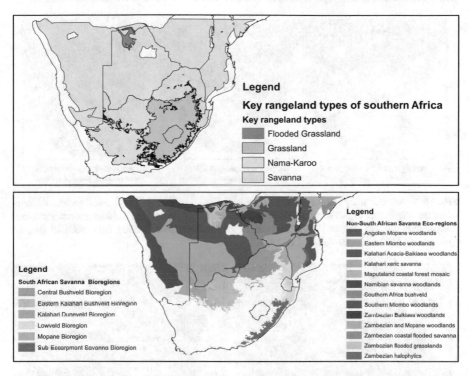

Fig. 15.1 Map of savanna and other rangeland vegetations of southern Africa (ecoregions based on Olson et al. 2001, bioregions on Mucina and Rutherford 2006)

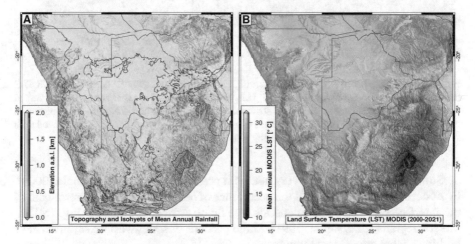

Fig. 15.2 Topographic Overview of southern Africa based on ETOPO1 data (**a**). Mean annual rainfall is derived from CHIRPS data and averaged from 1980 to 2020 (Funk et al. 2015). Red line indicates 200 mm/yr isohyet, and the blue line is 400 mm/yr. (**b**) Land surface temperature is derived from MODIS data 2000–2021 https://lpdaac.usgs.gov/products/mod11a1v006/

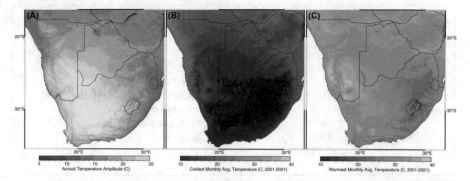

Fig. 15.3 Annual temperature seasonality. (**a**) Long-term (2001–2021) average annual temperature amplitude (e.g., warmest to coldest month). (**b**) long-term coldest monthly average temperature and (**c**) warmest monthly average temperature. Data derived from MODIS (https://lpdaac.usgs.gov/products/mod11a1v006/)

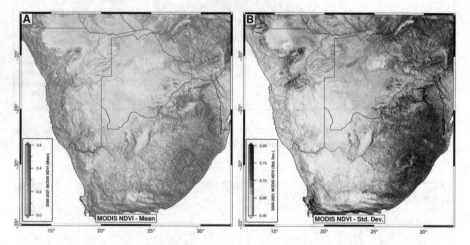

Fig. 15.4 Satellite-based Normalized Differential Vegetation Index (NDVI) derived from MODIS data (https://lpdaac.usgs.gov/products/mod11a1v006/). Sixteen-day measurements were averaged from 2000 to 2021 and their mean (**a**) and standard deviation (**b**) are shown. Note the low vegetation cover and low standard deviation in northern Namibia. Savanna regions are partly associated with slightly higher standard deviation, suggesting seasonal changes (light red colors in **b**)

Summers tend to be hot to very hot, with potentially cool nights (Scholes 2004). Frost occurs in some areas, but the occurrence of heavy frost is a key determinant of the interface between savanna and the true South African grasslands (Fig. 15.1) in moister regions (Ellery et al. 1991). Large portions of south-western savannas have highly variable annual temperature cycles (Fig. 15.3).

Vegetation and Soil Due to the high variation in climate and geology, there is a large degree of variation in vegetation types and vegetation amount over the climatic and geological range of the savannas (Figs. 15.1 and 15.4). Changes in geology can be clearly discerned at a regional scale; texture and nutrient status of soils have a profound influence at a local scale. The position on the slope, i.e., the catena position, leads to distinctive changes in vegetation characteristics (see Chap. 2). For instance, on the granite soils of the Kruger National Park, top of slopes are sandy, dystrophic, leached soils and are dominated by broadleaved trees and unpalatable grasses. In the valley at the bottom of slopes, the soil has a finer texture and higher nutrient status and is dominated by fine leaved trees and palatable grasses. A mid-slope hydromorphic grassland fringed by *Terminalia* species is also common (Scholes 2004).

Scientists differ on how the savannas should be subcategorized. Huntley (1982) points out a clear distinction between what he terms moist/dystrophic savannas (Miombo) and arid/eutrophic savannas (bushveld). In addition, the vast hot, low-lying areas are covered in near monospecific stands of the mopane tree (*Colophospermum mopane*), and though part of Huntley's arid savanna is sufficiently different for Olson et al. (2001) to include them as a unique bioregion. Key differences between these three savanna types are given in Table 15.1. Mukwada (2018) suggests that the arid savanna can be broken into five types, moist/mesic dystrophic savanna, dry eutrophic savannas, dry miombo woodlands, mopane woodlands, teak woodlands and *Terminalia-Combretum* and *Acacia* woodlands. He, however, makes no attempt to map the distribution of these woodlands. Olson et al. (2001) identify nine savanna types (see Fig. 15.1). Savanna vegetation is often

Table 15.1 Generalized key distinctions between arid and moist savanna after Huntley (1982) and Scholes (2004)

	Moist Savanna	Arid Savanna	Mopani Savanna
Rainfall per year	More than 600 mm	Less than 650 mm	Less than 650 mm
Soils	Mostly dystrophic	Mostly eutrophic	Mostly eutrophic
Key tree genera	*Brachystegia, Julbernardia, Burkea, Ochna*	*Vachellia, Senegalia* (previously *Acacia*), *Commiphora*	*Colophospermum*
Tree leaf morphology	Marginally deciduous large leaflets or broadleaf	Strongly deciduous fine leaf nanophyllous	Deciduous large broadleaf
Key tree defense and palatability	High tannins nonspinescent leaf N < 2.5%	Moderate tannins mostly spinescent leaf N > 2.5%	Moderate tannins nonspinescent leaf N > 2.5%
Key grass genera	*Andropogon, Schizacharium, Loudetia*	*Stipagrostis, Panicum, Enneapogon, Aristida*	*Anthephora, Aristida, Eragrostis, Schmidtia*
Grass palatability	coarse leafed and low palatability	fine leafed and high palatability	fine leafed and high palatability
Fire	very frequent every 1–3 years	less frequent every 10 years or more	less frequent every 10 years or more

mapped to an even finer detail within individual countries; for instance, in South Africa, Mucina and Rutherford (2006) identify 87 distinct savanna vegetation types grouped into 6 bioregions.

The woody cover of savanna varies from 0% in hydromorphic grasslands through to about 80% in areas of over 600 mm MAP. Above this, the canopy fully closes and the grass layer is lost. Savanna systems persist at tree densities below those that could be expected from rainfall (Sankaran et al. 2005). Many ecologists consider them to be disequilibrium systems where disturbances such as fire, herbivory, droughts and floods are key to their functioning (Scholes 2004). Therefore, any change in herbivory and fire can shift a savanna to either an increased or decreased tree or bush density (see Sect. 15.4.1).

15.2   Land Tenure and Grazing Systems

The management and use of the savanna rangeland is closely tied to the tenure arrangements, with three dominant tenure patterns being found in all the southern African countries, with country specific nuances. These are commercial tenure, either on freehold or leasehold land, communal tenure and conservation areas on state land (Table 15.2). Early European colonialists tended to establish large commercial farms/ranches that were initially managed almost exclusively for livestock production. These farms tended to be stocked at stocking rates considered to optimize economic return. Production of livestock was almost exclusively for marketing purposes and over time, complex management strategies were developed based on stocking rates and grazing rotation. Individual farmers used different stocking rates and rotations, this despite legislated maximum stocking rates in most countries. Initially wildlife was seen as competing with grazing for livestock and commercial farmers actively prevented wildlife from their land (ABSA 2003). A change in wildlife tenure legislation during the 1970s effectively changed wildlife from being royal game to a commodity which commercial farmers could own and financially exploit (Carruthers 2008). This resulted in a huge conversion of livestock land to mixed livestock and wildlife or simply wildlife. ABSA (2003) found that over 5000 game ranches and 4000 mixed game and livestock farms were developed in South Africa alone. Low value cattle land, such as that adjacent to the Kruger National Park in South Africa, is now some of the most valuable savanna rangeland and being run as exclusive private wildlife destinations. Similar trends to wildlife based land use are found in all the southern African countries.

Communal management of rangelands has historically been for subsistence use of livestock and is very common in much of South Africa. The rangeland is managed as a commonage and is typically stocked at ecological carrying capacity. In some cases, livestock is kept for cultural and investment purposes and offtake for sale is only when cash is desperately needed. In such cases, the driving force in management is to maximize numbers rather than turnover or quality. In Botswana, there is a higher degree of commercial cattle production on the communal

Table 15.2 Approximate land allocation to different tenure and land use (based on Potts 2012), as well as typical livestock and wildlife usage patterns

	Approximate percentage per country	Livestock	Mixed	Wildlife
Communal (1)	SA 13% Namibia 48% Botswana 48% Zimbabwe 47% eSwatini 42% Mozambique (5)	Predominant use of communal land and is typically open access. Stocked at ecological carrying capacity (CC)	Occasional e.g., Namibian conservancies, Botswana around Okavango	Very unique situations e.g., Botswana around Okavan-go. The basis of the Zimbabwe CAMPFIRE program. Namibian conservancies
Commercial Freehold and concessions (2)	SA 87% Namibia 52% Botswana 4.8% Zimbabwe 47% (3) eSwatini 44% (4) Mozambique (5)	The common traditional European settler farming model. Stocked at economic CC	Common trend made possible by changes in legislation to allow game ownership.	Common trend of private nature reserves, Tuli block Botswana—greater Kruger South Africa
State conservation land (4)	SA 8% Namibia 38% Botswana 29% Zimbabwe 27% eSwatini 20% Mozambique 22%			Formal conservation areas and state forests.

(1) Land reform is currently taking place in all southern African countries. The communal tenure refers to land historically allocated to traditional communities
(2) Land reform has led to some traditionally commercial or state land being redistributed to communities; this is most prevalent in Zimbabwe. Statistics refer to historic allocations, which change over time through tenure reform
(3) This relates to freehold land prior to the 2000 land reform process in Zimbabwe
(4) Conservation land is often included as a component of the commercial or communal land and is therefore in some cases double accounted in the statistics
(5) In Mozambique, land is allocated to commercial farming under a long term lease process. Currently it is difficult to obtain accurate statistics between communal and commercial land use due to informal land occupations and an evolving tenure regime

rangelands with individual farmers with large commercial herds drilling boreholes to open up new exclusive grazing areas.

There has been, however, a trend to what is termed Community Based Natural Resource Management (CBNRM) in all countries. This started with the CAMPFIRE program in Zimbabwe, but it is probably Botswana and Namibia where this is most developed (Bond 2001). In Botswana, areas that were previously termed Wildlife Management Areas have been converted into Community Management Areas. Many of these areas are now managed by communities for ecotourism or hunting, with wildlife rather than livestock. In Namibia, communal conservancies also opt for wildlife management to gain income from tourism to reduce poverty of communal households. Furthermore, wildlife tourism is promoted as a sustainable land use option and an alternative to livestock farming (Berzborn and Solich 2013). Beyond CBNRM programs, the governments try to correct colonial injustices with land reforms and resettlement programs but with different paths. On one side, the fast track and highly conflictive land reform in Zimbabwe and on the other side the slow resettlement program in Namibia are two extremes of land reform in southern Africa. Land reform is directed mainly to the redistribution of freehold land of white farmers, but also of state land (Breytenbach 2004). CBNRM and land reform contribute to more equity in land tenure and conservation, but until today, it was not possible to turn around colonial injustices. Furthermore, both development pathways generate new borders, conflicts and uncertainties.

Southern African savannas have exceptionally high levels of formal state protection as conservation areas, but the level of actual protection varies across the countries (Table 15.2). Moreover, management in wildlife conservation areas has evolved over time. In the past, there was sometimes extensive management intervention in terms of fire management, elephant density controls or changing the balance of predator and prey species and the provision of artificial watering points. The more recent trend, particularly in the larger reserves, is to now allow systems to regulate naturally with fewer management interventions and a reduction in artificial watering points.

15.3   Savanna Rangelands as Source of Food, Fodder and Valuable Ecosystem Services

As per their definition, one of the main uses of rangelands is to produce livestock and increasingly wildlife. In addition to this, the savanna rangelands of southern Africa provide many additional ecosystem services critical to maintaining both local livelihoods (Shackleton and Shackleton 2004) as well as being an important global carbon sink (Scholes 2004). The benefits derived from the rangelands differ substantially by the tenure system and management objectives (Table 15.2). Especially in communal areas, rangelands provide the greatest diversity in additional ecosystem services. Shackleton and Shackleton (2004) identify eight categories of direct benefits: fuelwood, construction material, wild foods, medicine, household implements, fodder, fibers and cash sales, as well as four indirect benefits: spiritual, cultural, indigenous knowledge and ecological services accruing to community's resident within the savanna rangelands. A multitude of studies suggests that the provisioning of fuelwood is often the single biggest value of the savanna rangelands to local communities (e.g., Dovie et al. 2004). Most rural communities are almost fully dependent on fuelwood for household cooking energy. The wood use varies widely, depending on household size and dead wood availability, with Dovie et al. (2004) estimating approximately 700 kg/capita/year usage. Use of charcoal is not that common in rural areas, but is a major fuel source in many cities, with the communal areas deriving substantive income from the sale of charcoal (Baumert et al. 2016). Charcoal trade differs substantially across the subregion. South Africa, Botswana, eSwatini and Zimbabwe having relatively low or no charcoal dependency, while Mozambique, Zambia, Malawi and Tanzania having urban centers almost totally dependent on charcoal as the fuel for low income households. For instance, the town of Maputo has shown an increase from most households preferring firewood (60%) to charcoal (17%) in 1990 to 87% of informal households using charcoal in 2004 and 96% using charcoal by 2016 (Mudombi et al. 2018). Estimates from Tanzania suggest that rural income from charcoal production exceeds the value of rural subsistence agriculture (Luoga et al. 2000), though harvesting for charcoal is often done by external contractors, with limited financial benefit accruing to the local residents (Baumert et al. 2016). Namibia is a major producer of charcoal from eradicating bush encroachment, much of this is exported (Shikangalah and Mapani 2020).

In communal areas, rangelands provide a critical safety net during times of need such as droughts (Shackleton and Shackleton 2004). Although fruits, nuts, insects and mushrooms provide an important nutritional supplement during droughts, these are indispensable to household wellbeing. In addition, the poor in the village often use woodland products for applications such as brooms, or even make and sell these and other crafts for a cash income (Shackleton et al. 2010).

Cattle and other livestock have important cultural value throughout the subregion (Shackleton and Shackleton 2004). In addition to provision of fodder, the savanna rangeland also provides material for the building of cattle kraals (corrals), and indeed for the construction of traditional homesteads which use both poles for the structure and grass for thatching (Dovie et al. 2004).

Medicinal plants from the rangelands are also very important and valuable from both a cultural, spiritual and medicinal perspective. Bulbs, bark and leaves are harvested for a wide variety of treatments (Mander 1998).

In commercial areas, the use of rangeland has traditionally focused on subsistence-based livestock production, with limited additional benefits to the land-owner. A trend in all southern countries has been a move from livestock toward mixed livestock and wildlife or to total wildlife.

In conservation areas, the key benefit from the rangeland is biodiversity preser-vation. However, in these and all other rangeland areas, there is also growing realization of the important role of the savanna rangelands in terms of being a globally important carbon sink (see Chap. 17).

15.4 Indicators and Drivers of Degradation in African Savanna Rangelands

Rangeland degradation is a growing threat to many savanna regions in southern Africa. Degradation is expressed through lowered groundwater tables and reduced water quality, soil erosion and decreased soil fertility, and both loss of—and changes in—plant species density, diversity and palatability as well as bush encroachment. Both climatic (e.g., changes in precipitation, CO_2 levels) and anthropogenic (e.g., overgrazing, fire suppression and the socioecological framework) drivers contribute to rangeland degradation through overlapping and often synergistic processes.

15.4.1 Climate, Atmospheric CO_2, Overgrazing and Fire Suppression

The consensus explanation for savanna degradation is that overgrazing reduces the grass layer, which in turn increases the opportunities for bush establishment by lower fuel loads for natural fires and reduced grass competition with emerging bush seedlings (e.g., Walter 1939; Archer et al. 1995; Higgins et al. 2000; Ward 2005). When grasses disappear, bare-soil patches become increasingly prevalent, and are highly vulnerable to soil erosion by wind and water.

Both browsing wildlife and natural fires play an important role in maintaining the coexistence of grasses and bushes in healthy savanna ecosystems. The replacement of indigenous browsers and grazers by domestic livestock and the artificial prevention of hot fires are key anthropogenic drivers of intensified rangeland degradation. Indigenous browsers in general suppress bush establishment (Trollope and Dondofema 2003); elephants even uproot mature trees (Dahlberg 2000). Frequent fires in moist savannas limit bush density by killing bush saplings and even large trees if their fire-resistant bark has been stripped by browsers (Yeaton 1988). In more arid savannas, where fires are less frequent, fires often coincide with a series of above-average rainfall seasons. High rainfall generally encourages abundant grass biomass that is then available as fuel for fires (Joubert et al. 2008). It is exactly those wet years where soil moisture conditions are also optimal for bush sapling establishment (Joubert et al. 2008; Lohmann et al. 2014).

Under changing climate conditions—particularly increased rainfall variability and extended droughts—the rate and scale of rangeland degradation is expected to increase; however, site-specific soil, vegetation and land use characteristics make generalization difficult (e.g., Ward 2005; Bond and Midgley 2012; Stevens et al. 2016; Archer et al. 2017). Evidence increasingly suggests that higher atmospheric CO_2 will play an additional role in driving rangeland degradation by reducing the advantage in water use efficiency of C4 grasses compared to C3 bushes and trees (Archer et al. 1995; Bond and Midgley 2012; Stevens et al. 2016).

15.4.2 Socioecological Framework and Policy

The relation between drivers of unsustainable land use (overgrazing and fire suppression) and their sociopolitical and socioeconomic context is rooted in colonialism and apartheid. Colonial expansion was tantamount with grabbing of indigenous lands by white settlers, mining- and farming companies (52% in Namibia in 1990, 87% in South Africa in 1936) (Potts 2012). The establishment of nature reserves for wildlife protection by the colonial authorities exacerbated the loss of land for indigenous communities (Berzborn and Solich 2013). The subsequent territorial encapsulation of indigenous people on remaining lands led to overpopulation in comparison to the white farming areas, which were characterized by large farm sizes (Schnegg et al. 2013). Furthermore, the communal lands in Namibia and South Africa often had worse climatic, geomorphological and soil conditions (Hoffman and Todd 2000; Menestrey Schwieger and Mbidzo 2020). These conditions together with overpopulation led to overstocking in communal areas, despite each communal farmer keeping only few livestock. The villagization, induced by colonial authorities and the apartheid regime, in combination with kraaling and cultivation near the homestead to be able to protect livestock and crops worsened the situation. It led to overgrazing and trampling within and near the settlements. Illegal fencing of communal land by those who can afford fencing materials increases grazing pressure in the remaining commonage by effectively excluding other residents from using the fenced off area. Although it is illegal, it has been observed in communal areas of Namibia. However, fencing by a community could be viewed as a strategy to deal with drought conditions by protecting grazing resources for the future (Kashululu and Paul 2020). In the region of today's Namibia, the combat against animal diseases led to further territorial encapsulation of black smallholders. The veterinary cordon fence (VCF) dividing the northern areas from the rest of the country was supposed to protect the farming area south of the VCF from animal diseases such as foot and mouth disease (FMD) and to secure export markets. Since the mid-1960s, outbreaks have been limited to the communal areas north of the VCF (see Chap. 18; Schneider 2012). While the farming area south of the VCF, where nearly all white farmers and also communal areas are located, has been well protected against FMD, the new borders and fences represented a conspicuous obstacle to pastoral mobility.

In what is now Namibia and South Africa, before the advent of colonialism, indigenous communities used pastoral grazing systems similar to modern rotational grazing and veld fires to control bush encroachment and to stimulate grass growth (Beinart 2003; Hoffman 2014; Rohde and Hoffman 2012; Menestrey Schwieger and Mbidzo 2020). Colonial authorities in South West African Hereroland and other reserves and the Native Trust in South Africa forcibly intervened into the practices of black smallholders restricting livestock movement, dictating maximum stocking numbers, fencing-off land and establishing grazing or stock fees. As a result, the communal farmers' mobility, on which their traditional farming practices were based upon, was further reduced, which subsequently intensified overgrazing and trampling. (Delius and Schirmer 2000; Menestrey Schwieger and Mbidzo 2020).

Furthermore, state authorities banned and discouraged the practice of veld fires for freehold as well as communal farmers in the late nineteenth and early twentieth centuries. They feared that fires would damage vegetation and get out of control, even though the beneficial properties were known (Beinart 2003; Rohde and Hoffman 2012; Hoffman 2014; Humphrey et al. 2021). By disrupting traditional pastoralist activities and intentionally giving scarce resources to indigenous communities, colonial authorities and the apartheid regime enforced the diversification of pastoralist activities of communal farmers, which are still practiced today. Absentee farmers work outside of communal areas and use their wages to buy cattle and to increase their herds, which are herded by family members or paid herders. Herd sizes of absentee farmers tend to depend more on their wages than on environmental considerations (Schnegg et al. 2013).

The colonial and apartheid state in today's Namibia and South Africa established several assistance programs for water supply infrastructure, livestock vaccination, new livestock breeds, farming techniques and the drought relief programs. The governmental assistance enabled livestock farming beyond natural limits and avoiding destocking in the face of drought especially for freehold farmers showing favorable treatment of white land users (Lange et al. 1998; Delius and Schirmer 2000; Bollig 2013; Kreike 2009; Menestrey Schwieger and Mbidzo 2020). Summing up, the disruption of pastoralist practices of communal farmers as well as the favorable government assistance for freehold farmers led to overstocking and in the end land degradation.

15.5   Degradation of African Savanna Rangelands

15.5.1 Bush Encroachment

Bush encroachment has been defined as the increase in the density of woody vegetation in grassland and savanna ecosystems (e.g., De Klerk 2004; Smit 2004; Ward 2005; O'Connor et al. 2014; Archer et al. 2017). It occurs across all southern African savannas (Fig. 15.5), with increasing severity along a moisture gradient (O'Connor et al. 2014).

Bush encroachment describes a change in savanna vegetation structure and function impacting ecosystem services and ecosystem disservices worldwide (e.g., Li et al. 2016). From a rangeland perspective, it is regarded as a major form of land degradation. It diminishes the availability and quality of forage for grazing animals, with impacts on meat, milk and leather (Shikangalah and Mapani 2020). Bush encroachment can also lead to an increase in certain animal diseases (e.g., Bollig and Osterle 2008). Bush thickets have been blamed for offering hiding spots for predators to ambush livestock (Vehrs and Heller 2017). The impenetrability of thorny thickets increases animal injuries and hinders the accessibility of limited grass forage (Mcleon 1995). A study in Zimbabwe showed that for every percent

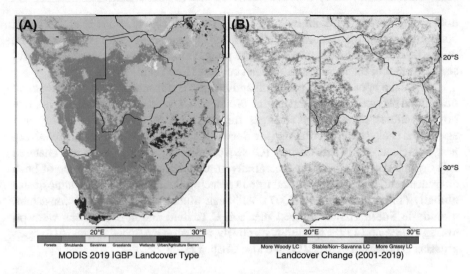

Forests Shrublands Savannas Grasslands Wetlands Urban/Agriculture Barren

MODIS 2019 IGBP Landcover Type

More Woody LC   Stable/Non-Savanna LC   More Grassy LC

Landcover Change (2001-2019)

Fig. 15.5 Changes in nonforest and nonanthropogenic landcovers. (**a**) MODIS landcover classification 2019 grouped by functional type. Note the preponderance of shrublands and grasslands in much of southern Africa. (**b**) Changes in landcover (2001–2019). Blue areas show a move toward more woody (e.g., woody savanna, shrublands) landcover, brown areas show a move toward more grassy (e.g., grasslands, open savanna) landcover. Grey areas show no change in landcover type, as well as anthropogenic or other natural (e.g., forest) landcovers. More land area has shifted toward grasslands; however, the spatial pattern of these changes is complex (https://lpdaac.usgs.gov/products/mod11a1v006/)

increase in bush cover grass biomass decreased by 26.5 kg/ha. In the Molopo region in South Africa veld productivity for cattle was suppressed by 970 kg/ha of dry matter resulting in a reduction in grazing capacity from 8.7 to 45 ha per large stock unit (LSU) (Moore and Odendaal 1987). Wigley et al. (2009) reported a 20% decline in grazing capacity in a mesic savanna rangeland in KwaZulu-Natal, South Africa over a 63-year period between 1937 and 2000. In Namibia, where the original carrying capacity (also known as grazing capacity) before bush encroachment was 10 ha/LSU it decreased to 30 ha/LSU (de Klerk 2004; NAU 2010). Due to the reductions in grazing capacity and related consequences, in 2012, the Namibian Meat Board estimated a decline of about 30% to 64% in cattle numbers in commercial farms when compared to 1959 (Demas et al. 2012). This decline was estimated to be equivalent to an economic loss in meat production of N$700–1.6 billion (US$ 94.8 million–US$ 217 million) per annum (NAU 2010; Trede and Patt 2015). For Zimbabwe, woody encroachment caused a loss of £16 million (US$74 million) in beef production already in 1948 (Hattingh 1952). This is all worrisome and can lead to significant declines in food security, such as in Namibia, where livestock production (mainly beef production) is estimated to contribute up to 75% to the total agricultural output (Shikangalah and Mapani 2020).

Dense bush also negatively impacts important cultural services especially tourism by obstructing game viewing (Demas et al. 2012) and by decreasing densities of certain wild animals (e.g., Wigley et al. 2009; Demas et al. 2012). In communal areas, the cultural standing as well as the royal value is compromised where bush encroachment prevents cattle production, because cattle are a powerful status symbol in many southern African cultures (Reed et al. 2015).

Despite the apparent and predominantly negative impacts of bush encroachment on cultural services, fodder and animal production, recent studies increasingly show positive effects of structural changes for, e.g., soil properties, biodiversity and associated ecosystem functioning (see Sects. 15.5.3 and 15.5.4). Also, the relatively new consciousness is that Carbon (C) sequestration by woody plants and changed soil organic carbon stocks are potentially an important regulating service of bush encroachment, and can play a vital role in reducing C emissions into the atmosphere globally (Tallis and Kareiva 2007). Although global patterns are not consistent, a study in South Africa showed that soil C content and soil C stocks were on average 148% and 117% greater, respectively, in bush-encroached compared to open grasslands (Dlamini et al. 2019, see also Chap. 17).

15.5.2 Soil Erosion, Soil Nutrient and Soil Moisture Decline

High stocking rates of livestock or wildlife can decrease soil water by two processes: (i) trampling can destroy soil porosity, which decreases infiltration and (ii) grazing decreases plant cover critical for soil organic matter, which is important for water storage (water holding capacity) (McNaughton et al. 1988). High stocking rates reducing the density of vegetation cover and size of vegetation patches (overgrazing) also increase soil erosion, thus leading to an overall loss of water and nutrient resources from the ecosystem (Rietkerk and van de Koppel 1997). Degradation of soil resources will largely determine primary production of vegetation (fodder) also in the long-term.

It has been argued, however, that the increase in thick dense bush and the spread of the unpalatable grasses, which is normally viewed as degradation of savanna rangelands, may actually help controlling soil erosion (e.g., Shikangalah and Mapani 2020). Aboveground plant parts covering the ground decrease the erosive impacts of raindrops and high winds. The root system anchoring the soil offers stability to the soil structure. Moreover, bush encroachment mainly by legumes that fix nitrogen increases nitrogen mineralization. Greater soil fertility with bush cover was observed in many African savanna rangelands (Belsky 1994; Smit 2004; Sitters et al. 2013; Dlamini et al. 2019; Sandhage-Hofmann et al. 2020; Mogashoa et al. 2021).

While nitrogen-fixing legumes may indeed increase nutrients in the soil and can improve the soil microclimate by shade (Metzger et al. 2014), they possibly deplete soil moisture by high evapotranspiration and lowered infiltration (Archer et al. 2017). Bush encroachment alters soil moisture availability (Geissler et al. 2019a) and will largely determine both the recovery potential of the remaining herbaceous vegetation and the establishment of new bush seedlings leading to further bush encroachment.

Fig. 15.6 Poorly managed water point in an area that is already vulnerable to nitrate contamination by its shallow soil and dominated by *Senegalia* species. Note: water is leaking from the trough and dissolves nitrate from animal feces that can contaminate groundwater

15.5.3 Decline in Water Quality and Groundwater Recharge

Although often less obvious, a decreased water quality (i.e., surface and subsurface water) is another important sign of rangeland degradation. A water quality parameter of concern in many catchments of this world is nitrate (e.g., Weitzman et al. 2021), since the very complex biogeochemical nitrogen cycle can be influenced in various ways. Regarding rangeland degradation, a shift in species composition and/or biomass production, particularly toward nitrogen fixing *Senegalia* species and other legumes, will alter the nitrogen cycle in the unsaturated zone and consequently influence the nitrate concentration in groundwater carried in with infiltration water.

Poor water point management (Fig. 15.6) can further the impacts and amplify groundwater pollution. Moreover, water points are areas of the highest animal excrement deposition. Soil erosion could free nitrate pools, but also lead to a loss of fungi and bacteria that are usually involved during ammonification in the soil zone, an intermediary step of the nitrogen cycle toward nitrate (Hiscock et al. 1991).

An example of dynamics in water quality in rangelands is drawn from an area in south-western Etosha National Park, including surrounding conservancies, livestock farms and private game reserves using water quality index (WQI). Among other easy measurable parameters, this index includes nitrate concentration. Comparison of historical (pre-2000) and recent (2020) data of 68 boreholes shows a reduction in water quality at 40% of the boreholes. The number of boreholes with deteriorated water quality in livestock farms and private game reserves was at least two times higher than in conservancies and national park. In contrast, the number of boreholes with improved water quality was at least two times lower. This shows that signs of rangeland degradation in terms of water quality are more pronounced in livestock and private game reserves than in conservancies and national parks.

The effect of bush encroachment on groundwater recharge is a topic of controversy since the differences in water consumption between grasses and bushes are still unclear (Scholes and Archer 1997; O'Connor et al. 2014). Nevertheless, water loss in Namibia through bush encroachment is estimated to be around 12 million m^3 on 10,000 ha (NAU 2010). Indeed, Groengroeft et al. (2018) found deep drainage about 3-fold lower in the area below the canopy than in intercanopy patches. They explain the decrease in the amount of deep percolating water by lower infiltration and higher evapotranspiration in bush dominated patches and conclude that bush encroachment is likely to reduce groundwater recharge.

15.5.4 Biodiversity

Bush encroachment is associated with negative, but also neutral and positive consequences for species richness of plants and animals, and biodiversity-associated ecosystem functioning as part of the degradation process in African savanna rangelands (e.g., Chown 2010; Eldridge et al. 2011). A decline in ecosystem functioning has serious consequences for ecosystem services and the regulatory processes of the systems, which can further exacerbate degradation. A higher biodiversity increases both, the multifunctionality and the resilience of an ecosystem. The so-called "Insurance Hypothesis" assumes that higher species richness increases the probability that functionally redundant species buffer a specific function against environmental fluctuation or future threats. (Buisson et al. 2019).

15.5.4.1 Plant Diversity

The impacts of bush encroachment on plant diversity in savanna rangelands depend on local conditions of climate and soils, in particular on the level of disturbance and resource stress (soil moisture and nutrients, see Sect.15.5.3), the heterogeneity of these resources and viable soil seed banks. Several studies have reported a decline in the diversity of understory vegetation with increasing bush cover (e.g., Scholes and Archer 1997; Angassa 2005; Mogashoa et al. 2021). Other studies found no effect or a decrease in diversity only above a certain bush cover and only for specific functional groups (perennial grasses, forbs) (Dreber et al. 2018). In contrast, Belay et al. (2013) showed that plant diversity increased with bush density.

Bush encroachment also supports the invasion of alien species (e.g., Chromolaena in South Africa, Wigley et al. 2009), although the susceptibility of rangelands varies between the vegetation types. By changing the structure and function of soils, invasive alien plants exacerbate the risk of losing native plant diversity. The Southern African Plant Invaders Atlas (SAPIA) database contains currently records for over 500 invasive plant species in South Africa, Lesotho and eSwatini (Zengeya and Wilson 2020).

15.5.4.2 Plant Diversity Mediates Soil Moisture, Groundwater Recharge and Primary Production

The hydrological cycle is in various ways interlinked with biodiversity and ecosystem functions. The availability of soil water obviously controls the primary production of vegetation. Less obvious are biodiversity-mediated feedbacks on the water cycle, controlling local climate and ultimately the availability of freshwater as an ecosystem service to the human beneficiaries.

Depending on plant size and morphology, rainfall is differently intercepted by the above-ground parts of plants and either directly returned to the atmosphere via evaporation from the canopy or funneled to the soil below via stemflow and throughfall (Yuan et al. 2016). Water reaching the soil can affect not only the amount of water available to plants but also the amount of groundwater recharge and runoff, controlling soil erosion. However, most dryland plants—particularly those in Africa—are poorly explored. Comparing the impacts of morphologically different African bush encroacher species such as *Terminalia sericea*, *Dichrostachys cinerea*, *Colophospermum mopane*, *Senegalia mellifera*, *Vachellia reficiens* and *Catophractes alexandri* on rainfall interception and subsequent soil moisture is key focus of the recent research program SPACES.

Once water has entered the soil, diverse rooting systems of different plant species can either improve local soil-water availability by increasing the preferential flow along the root systems into deeper soil layers or decrease local soil-water availability by a complete uptake of water via spatial niche partitioning (Lee et al. 2018). Modern stable isotopes analyses of water indicate a sharing of water resources between many species of bushes, trees and grasses in the upper soil of African savanna rangelands (Kulmatiski et al. 2010; Beyer et al. 2018; Geissler et al. 2019b; Uugulu 2022). These results indicate that diverse vegetation has a potentially higher water use efficiency (biomass produced per unit water used) compared to species-poor ecosystems. High water-use efficiency maximizes the relative amount of water kept in the ecosystem; however, groundwater recharge potentially decreases as fewer precipitation events reach below the effective root zone. The positive impact of plant diversity on water use efficiency was confirmed by an ecohydrological simulation study (Irob et al. 2022). The simulation results also showed a positive effect of functional diversity on vegetation cover, especially the cover of perennial herbaceous vegetation. As a result of these changes, the biomass production of sites with functionally diverse vegetation increased, leading to improved fodder production.

15.5.4.3 Animal Diversity

Bush encroachment will strongly impact animal diversity in African savanna rangelands through changes in the structural diversity of vegetation. Structural diversity not only determines the amount and quality of habitats (i.e., niche space), but also the heterogeneity of resources and the interactions between species (e.g., Hering et al. 2019). While an initial increase in bush cover increases structural diversity, severe bush encroachment has the opposite effect (e.g., Blaum et al. 2007a)—it creates homogeneous and structure-poor environments that lead to the loss of habitats and species niches. Thus, animal diversity often shows hump-shaped responses to increasing amounts of bush cover (e.g., Chown 2010). Studies along bush-cover gradients in the Kalahari in South Africa have shown similar responses for insects and spiders (Blaum et al. 2009; Hering et al. 2019) and for small- to medium-sized mammalian carnivores (Blaum et al. 2007a, b). Maximum diversity for all taxonomic groups was found at a mean bush cover of 15% (Fig. 15.7). Similar responses to altered structural diversity linked to bush encroachment have been reported for birds in the Rooipoort Nature Reserve in South Africa (Sirami et al. 2009) and lizards in central Namibia (Meik et al. 2002). Though hump-shaped distributions recur for some taxonomic groups, linear negative effects of bush encroachment were observed for reptiles (Wasiolka and Blaum 2011) and rodents (Blaum et al. 2007c; Fig. 15.7).

The loss of structural diversity due to degradation of savanna vegetation may even cascade up to large herbivores and top predators. In South Africa's Kruger National Park, the reduction of grass cover induced by an increasing bush layer led to declines in grazing herbivores such as zebras (*Equus burchellii*) and wildebeests (*Connochaetus taurinus*), which in turn were replaced by browsing herbivores such as kudus (*Tragelaphus strepsiceros*) and giraffes (*Giraffa camelopardalis*) (Smit and Prins 2015). Likely, the changes in herbivore communities caused by structural changes in vegetation negatively affect top predators. Lowered preferred prey densities, poor visibility and restricted movement spaces may reduce their hunting efficiency and alter hunting behavior (e.g., Muntifering et al. 2006).

Significant changes in hunting behavior due to bush encroachment have already been noted for different taxa. For example, the Spotted sand lizard (*Pedioplanis l. lineoocellata*) is classified as a "sit and wait forager." That is, it sits and hides in a grass tussock and waits for prey to pass by. In bush dominated habitats with low food availability, the Spotted sand lizard changes its foraging behavior to "actively foraging." Individuals actively search for prey and travel significantly longer distances, which increases their own predation risk. This behavioral plasticity acts as a buffer mechanism against the negative effects of bush encroachment and allows this species to still occur in heavily bush encroached savannas, albeit at significantly lower densities (Blumröder et al. 2012).

These examples illustrate that consequences of changes in structural diversity caused by bush encroachment may vary considerably across taxonomic groups and that multiscale feedback mechanisms can impact trophic cascades and complex interactions in food webs.

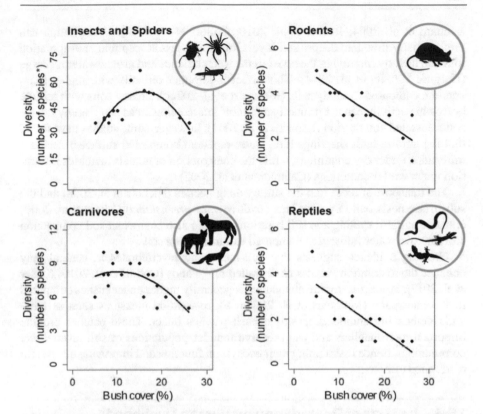

Fig. 15.7 Relationship between bush cover and animal diversity in Kalahari savanna rangelands in South Africa. From top to bottom: Insects and Spiders, Small- and medium-sized mammalian carnivores, rodents, reptiles. Redrawn after Blaum et al. 2007a, b, c, 2009; Wasiolka and Blaum 2011

15.5.4.4 Animal Diversity Mediates Soil Moisture and Soil Nutrient Dynamics

The awareness of animal biodiversity for ecosystem functioning has developed rapidly in the past decades. Early work has focused primarily on indigenous large mammals and has largely overlooked the role of small animals only until recently. Bioturbating animals such as termites, ants, beetles and other invertebrates can alter the micro-soil environment. They modify soil texture, build tunnels (macropores) and transport nutrients into the soil (e.g., termites, dung beetles) that affect a variety of soil functions, such as nutrient accessibility, water holding capacity and water sorption (Rückamp et al. 2012), which in turn impact vegetation (e.g., Traore et al. 2015). Termites for example, promote plant growth directly around their nests, thereby altering the vegetation composition and structure of entire landscapes (Bonachela et al. 2015).

Macropores of all sizes and forms created by termites, ants or beetles that open to the soil surface can strongly improve water infiltration into soils (e.g.,

Léonard et al. 2004; Brown et al. 2010; Colloff et al. 2010). Precipitation can preferentially flow into deeper soil layers, making it less susceptible to evaporation loss and thereby improving the ecosystems water balance and groundwater recharge (Bargués Tobella et al. 2014). This effect is stronger on soils with higher loam content, compared to sandy soils (Marquart et al. 2020c). Loamy soils with termite burrowing activity can act primarily as runoff interception areas and thereby reduce water loss and soil erosion (Léonard et al. 2004); however, some studies have found that the termite-built sheetings (i.e., water-repellent compacted surfaces) decrease infiltration under dry conditions, while the construction of tunnels increases infiltration under wetter conditions (Cammeraat et al. 2002).

The transport of feces into the soil by dung beetles (Nichols et al. 2008) and the subsurface nests build by meat ants (*Iridomyrmex greensladei*) (Nkem et al. 2000) enhance nutrient cycling and soil aeration, and can also counteract soil compaction and improve water infiltration compared to surrounding soils.

Growing evidence suggests that soil-burrowing invertebrates in general may enhance the restoration success of degraded rangelands (Colloff et al. 2010; Kaiser et al. 2017). Since, macropore abundance is generally higher under bushes compared to the interspace (Marquart et al. 2020a, b), restoration measures such as area-wide de-bushing should keep single bush patches intact. These patches provide important microhabitats and can preserve founder populations of soil invertebrate communities, hence maintaining their ecosystem function and improving ecosystem restoration efforts.

15.6   Impacts of Degradation on Human Livelihoods

With ecosystem degradation, the lives and livelihoods of those dependent on ecosystem services become vulnerable and in the case of the poorest, they might even get devastated. According to the Consortium on Ecosystems and Poverty in sub-Saharan Africa (CEPSA), the livelihood of most people living in arid and semiarid African rangelands directly depends on natural resources (Shackleton et al. 2008) through agriculture, fishing and hunting (Barbier and Hochard 2018). The equivalent income share of such products might constitute up to one third of total income of these households. Therefore, declining ecosystem services often (i) lead to a steady erosion of livelihood assets, (ii) increase vulnerability by making people less able to withstand external shocks, (iii) increase the risk of widespread disaster and (iv) might even exacerbate existing conflicts and give rise to new conflicts over access to ecosystem services (Shackleton et al. 2008). In other words, rangeland degradation is not only linked to increased poverty, but rather entails serious consequences for poverty alleviation in general.

One of the most obvious signs of land degradation is declining biomass production. It leads not only to reduced agricultural crop productivity but also to difficulties to support livestock numbers in rangeland systems (Tully et al. 2015) affecting the livelihoods of both pastoral communities and cattle farmers. To manage such lands and support livestock, an increased investment of management costs (i.e., supplementation or bush control) would be necessary (e.g., Lal 2015). In 2016 in Namibia, animal fodder was ranked as the eighth largest imported good, with total fodder imports valuing close to 4 billion Namibian Dollar (NAD), equivalent to the gross value addition of the entire agricultural sector (Honsbein et al. 2017). Such economic investments will, however, only be accessible to those who can afford them and may on top of that be coupled with indebtedness, especially in the case of subsistence and poor farmers (Gomiero 2016).

By the end of the last century, 75% of the population of sub-Saharan Africa depended on subsistence farming (Sanchez et al. 2007). In 2017, an estimate of 4 million households practiced subsistence farming in South Africa (RSA 2017) and in Namibia, today, approximately 48% of rural households depend on it (MET 2020). Most people living in rural areas and depend on agriculture have relatively low purchasing power and are thus rarely able to purchase the required supplementary fodder that livestock needs for a healthy growth and weight gain. Therefore, livestock is undernourished, which in turn implicates an insufficient nutrition and calorie provision for consumers (Schnegg et al. 2013). Moreover, with underweight livestock, potential sales are not as profitable and household capital is confined (Barrett and Bevis 2015). The consequence is a restriction to invest in more sustainable practices, let alone the ability to provide sustenance (i.e., food, health).

Degradation of rangelands can therefore force farmers to abandon their lands and either look for new ones to meet the surviving needs of current and future generations (Gomiero 2016), or look for new opportunities to earn a livelihood else-where. Such a situation might also lead to the diversification of pastoralist activities or migration movements. Rangeland degradation can therefore foster urbanization via rural-urban migration (Marchiori et al. 2012). This in turn entails far-reaching consequences, e.g., a loss of local knowledge in rural areas, disintegration of local communities, wage competition in urban areas and as a result migration decisions of urban households. However, given the lack of substantial assets, household members of the poorest seeking additional working opportunities (off-farm or outside of agriculture) tend to seek them locally or migrate only temporarily for short distances (Banerjee and Duflo 2007). This still forces many to exploit natural resources in their surrounding environments to supplement their consumption and income or to support their family members, giving the potential for a downward spiral if no sufficient pathways to improve livelihoods can be offered (see also Barbier and Hochard 2018; Suich et al. 2015).

Besides an increased vulnerability to poverty, land degradation can also dampen economic growth, especially in countries where agriculture and livestock farming is the engine for economic development. Many African economies are very dependent on climate sensitive sectors such as agriculture, forestry and fishery (Diao et al. 2010). Without proper sustainable land management practices to reduce rangeland degradation, the effects on the cumulative loss in agricultural Gross Domestic Product (GDP) and on overall poverty at a national level can therefore be enormous. For example, in South Africa, the success of industrial development depends partially on the agricultural sector and its improvements (Poonyth et al. 2001), and about 10% of formal employment is related to agriculture (RSA 2017). Also in Namibia, agriculture is a key sector of the economy. It is not only the largest employer but also critical to livelihoods and food security (MET 2020).

15.7   Strategies to Mitigate the Effects of Degradation

One of the most effective ways to reduce pressure on degraded rangelands is to reduce livestock numbers by selling animals. However, destocking has not been a common strategy to cope with limited grazing resources on communal land, mainly due to social prestige and wealth attached to having a large herd of cattle. Traditionally, one of the main strategies to deal with drought and declining grazing resources in communal areas has been to move livestock (particularly cattle) from degraded to nondegraded areas during the dry season (Twyman et al. 2001). This practice is still being largely done especially in the north-eastern parts of Namibia and involves herders moving herds of cattle over long distances in search for fodder. Since the twentieth century, many parts of Namibia have seen restricted movement of cattle, mainly due to limited land and policies leading to overpopulation and restricted red meat transfer between regions (see Sect. 15.4.2). After independence, government subsidies have also enabled people to keep livestock without the necessity to move them. This encourages many livestock farmers to keep animals during periods of fodder scarcity, resulting in further degradation of rangelands.

Yet, there have been some advances in successful mitigation of rangeland degradation: To detain bush encroachment and promote de-bushing, a growing industry of bush utilization can benefit farmers using it for economic gain; either as possible animal fodder in times with scarce resources or for charcoal production. The goal is mostly reducing rather than eliminating woody species completely (de Klerk 2004; Haussmann et al. 2016). With sustainable bush harvesting, the former threat of bush encroachment has become an important opportunity for economic income. In Namibia, more than 5000 jobs have been created in the biomass sector initiated by development projects (e.g., Bush Control and Biomass Utilization, BCBU) since 2015 (GIZ 11/10/2020). This has also led to an increase in wages and created a new opportunity for rural employment. With at least 2.6% of global charcoal export (USD 34.1 million in 2018), Namibia has reached the top rank in Africa and rank 12 word-wide (South Africa with 1.5%: 2nd in Africa, 18th world-wide) (Shikangalah and Mapani 2020). It is also estimated that in Namibia, the plain utilization of firewood (e.g., from bushes) currently ranges between 0.5 and 1 million tons per year and is projected to reach 1.2 million in 2025 (Stafford et al. 2017). The future use of these large-scale de-bushed areas, however, remains uncertain. An immediate return to livestock production is difficult and requires well planned restoration and aftercare measures, since viable grass seeds in the soil are lacking, the reintroduction of native savanna grasses is extremely difficult (van den Berg and Kellner 2005; Kinyua et al. 2010), and bushes often resprout (Bhattachan et al. 2014). Resaturation includes active and passive options such as furrowing, brush packing and revegetation (e.g., van den Berg and Kellner 2005; Harmse et al. 2016). Bush control techniques by prescribed fire, as well as manual and chemical methods (e.g., Trollope 2011; Joubert et al. 2012; Lohmann et al. 2014) are aftercare measurements to prevent resprouting and reestablishment of bushes. However, the potential impacts of these measures on biodiversity and ecosystem functioning remain topics of active debate and research.

Wildlife-based management as a strategy to approach rangeland degradation is increasingly applied but not well understood. Increased browsing of wildlife on seedlings and saplings has the potential to mitigate bush encroachment and to contribute toward conservation (McGranahan 2008; Irob et al. 2022). Additionally, wildlife has shown more resilience to local environmental change compared to livestock, which might make it a better management strategy in the face of degradation. The wide forage selection of wild herbivores allows them to cope with scarcity in palatable grass species in degraded lands (Taylor and Walker 1978, cited in McGranahan 2008). Growing evidence suggests that wildlife-based land uses offer higher socioeconomic benefits compared to livestock in semiarid rangelands in both Namibia and South Africa (Lindsey et al. 2013). There is consensus on the potential economic benefits that wildlife can provide through consumptive and nonconsumptive utilization (Lindsey et al. 2013; NACSO 2019).

As a response to changing environmental and socioeconomic conditions, a shift from cattle farming to wildlife farming has been observed in South African rangelands since the 1960 (Chaminuka et al. 2012) and more recently in Namibia. While this shift in land use can be profitable on commercial freehold farms, such a shift can be challenging in other communities and communal managed systems. Since wildlife is a common pool resource in communal areas, the allocation of direct benefits (e.g., income) is much harder to implement. In Namibia, the presence of Community Based Natural Resource Management (CBNRM) such as conservancies and community forests covering about 22% of the total land area (NACSO 2019) provides however an opportunity to develop wildlife management as a land use option to deal with rangeland degradation. Whereas Namibia has community-based conservation areas, where communities can manage and benefit from wildlife, more reforms in land tenure rights may be required in countries where such legal frameworks are still in progress.

Evidence from ecological studies suggests that individuals and communities can successfully reduce land degradation (Herrick et al. 2013), and restoration might be possible. However, mitigation measures can be very expensive, elaborate and local people might often not have the capacity to implement them sufficiently without support (Bourne et al. 2017). This is either because financial support is lacking (Inman 2020) or because the knowledge about these measures or how to implement them correctly is missing. Additionally, evidence suggests that complete recovery will hardly be achieved (Parkhurst et al. 2021) especially in systems where overgrazing is a continuous threat. Proper, institutionalized aid might promote restoration and could potentially advance resilience (Bourne et al. 2017).

15.8   Conclusions

1. Rangeland degradation is an ongoing threat to many savannas in Southern Africa, which is not only expressed through loss of fodder and soil. Symptoms comprise bush encroachment, lowered groundwater tables, reduced water quality, soil fertility, invasive species and both loss of—and changes in—plant and animal diversity. Neither ecological nor socioeconomic knowledge alone is sufficient to understand complex socioecological relationships and effectively manage savanna rangeland systems.
2. Concerns and actions related to the loss of forage production should be broadened to include biodiversity and its multiple functional relationships with the water and nutrient cycle, primary production and trophic interactions.
3. It is important to strengthen alternative strategies for conserving and restoring ecosystems and improving human well-being, for example, by implementing local solutions through community-based grazing management (e.g., communal conservancies) and land use diversification.
4. A shift from cattle farming to wildlife farming has the potential to contribute toward biodiversity conservation, the control of species causing bush-encroachment and increased economic benefits for local communities. We note, however, that more scientific evidence, reforms in land tenure rights and improved restoration practices are needed.

References

ABSA (2003) Game ranch profitability in Southern Africa. Rivonia, the SA. Financial Sector Forum: 73 pp

Allen VG et al (2011) An international terminology for grazing lands and grazing animals. Grass Forage Sci 66(1):2–28. https://doi.org/10.1111/j.1365-2494.2010.00780.x

Angassa A (2005) The ecological impact of bush encroachment on the yield of grasses in Borana rangeland ecosystem. Afr J Ecol 43:14–20. https://doi.org/10.1111/j.1365-2028.2005.00429.x

Archer S, Schimel DS, Holland EA (1995) Mechanisms of shrub land expansion: land use, climate or CO2. Clim Chang 29:91–99

Archer SR, Andersen EM, Predick KI, Schwinning S, Steidl RJ, Woods SR (2017) Woody plant encroachment: causes and consequences. In: Briske DD (ed) Rangeland systems. Springer International Publishing, pp 25–84. https://doi.org/10.1007/978-3-319-46709-2_2

Banerjee AV, Duflo E (2007) The economic lives of the poor. J Econ Perspect 21(1):141–167. https://doi.org/10.1257/jep.21.1.141

Barbier EB, Hochard JP (2018) Land degradation and poverty. Nat Sustain 1(11):623–631. https://doi.org/10.1038/s41893-018-0155-4

Bargués Tobella A, Reese H, Almaw A, Bayala J, Malmer A, Laudon H, Ilstedt U (2014) The effect of trees on preferential flow and soil infiltrability in an agroforestry parkland in semiarid Burkina Faso. Water Resour Res 50(4):3342–3354. https://doi.org/10.1002/2013WR015197

Barrett CB, Bevis LEM (2015) The self-reinforcing feedback between low soil fertility and chronic poverty. Nat Geosci 8(12):907–912. https://doi.org/10.1038/ngeo2591

Baumert S et al (2016) Charcoal supply chains from Mabalane to Maputo: who benefits? Energy Sustain Dev 33:129–138. https://doi.org/10.1016/j.esd.2016.06.003

Beinart W (2003) The rise of conservation in South Africa. Settlers, livestock, and the environment 1770–1950. Oxford University Press, Oxford. checked on 8/3/2021

Belay TA, Totland Ø, Moe SR (2013) Ecosystem responses to woody plant encroachment in a semiarid savanna rangeland. Plant Ecol 214:1211–1222. https://doi.org/10.1007/s11258-013-0245-3

Belsky AJ (1994) Influences of trees on savanna productivity: tests of shade, nutrients, and treegrass competition. Ecology 75:922–932

Berzborn S, Solich M (2013) Pastoralism and nature conservation in Southern Africa. In: Bollig M, Schnegg M, Wotzka H-P (eds) Pastoralism in Africa. Past, present, and future. Berghahn, New York, pp 440–472

Beyer M, Hamutoko JT, Wanke H, Gaj M, Koeniger P (2018) Examination of deep root water uptake using anomalies of soil water stable isotopes, depth-controlled isotopic labeling and mixing models. J Hydrol 566:122–136

Bhattachan A, D'Odorico P, Dintwe K, Okin GS, Collins SL (2014) Resilience and recovery potential of duneland vegetation in the southern Kalahari. Ecosphere 5:2–14

Blaum N, Rossmanith E, Popp A, Jeltsch F (2007a) Shrub encroachment affects mammalian carnivore abundance and species richness in semiarid rangelands. Acta Oecol 31:86–92

Blaum N, Rossmanith E, Schwager M, Jeltsch F (2007b) Responses of mammalian carnivores to land use in arid Kalahari rangelands. Basic Appl Ecol 8:552–564

Blaum N, Rossmanith E, Jeltsch F (2007c) Land use affects rodent communities in Kalahari savannah rangelands. Afr J Ecol 45:189–195

Blaum N, Seymour C, Rossmanith E, Schwager M, Jeltsch F (2009) Changes in arthropod diversity along a land use driven gradient of shrub cover in savanna rangelands: identification of suitable indicators. Biodivers Conserv 18(5):1187–1199. https://doi.org/10.1007/s10531-008-9498-x

Blumröder J, Eccard J, Blaum N (2012) Behavioural flexibility in foraging mode of the spotted sand lizard (Pedioplanis l. lineoocellata) seems to buffer negative impacts of savanna degradation. J Arid Environ 77.149–152

Bollig M (2013) Chapter 10: Social-ecological change and institutional development in a pastoral Community in North-western Namibia. In: Bollig M, Schnegg M, Wotzka H-P (eds) Pastoralism in Africa. Past, present, and future. Berghahn, New York, pp 316–340

Bollig M, Osterle M (2008) Changing communal land tenure in an East African pastoral system: institutions and socio-economic transformations among the Pokot of NW Kenya. Z Ethnol 133(2):301–322

Bonachela JA, Pringle RM, Sheffer E, Coverdale TC, Guyton JA, Caylor KK, Levin SA, Tarnita CE (2015) Termite mounds can increase the robustness of dryland ecosystems to climatic change. Science 347(6222):651–655. https://doi.org/10.1126/science.1261487

Bond I (2001) CAMPFIRE and the incentives for institutional change, in African wildlife and livlihoods. The promise and performance of community conservation. James Currey, Oxford

Bond WJ, Midgley GF (2012) Carbon dioxide and the uneasy interaction of trees and savannah grasses. Philos Trans R Soc Lond B Biol Sci 367:601–612

Bourne A, Muller H, De Villiers A, Alam M, Hole D (2017) Assessing the efficiency and effectiveness of rangeland restoration in Namaqualand, South Africa. Plant Ecol 218(1):7–22. https://doi.org/10.1007/s11258-016-0644-3

Breytenbach W (2004) Land reform in Southern Africa. In: Hunter J (ed) Who should own the land? Analyses and views on land reform and the land question in Namibia and southern Africa. Konrad-Adenauer-Stiftung; Namibia Institute for Democracy, Windhoek Namibia, pp 46–63

Brown J, Scholtz CH, Janeau J-L, Grellier S, Podwojewski P (2010) Dung beetles (Coleoptera: Scarabaeidae) can improve soil hydrological properties. Appl Soil Ecol 46(1):9–16. https://doi.org/10.1016/j.apsoil.2010.05.010

Buisson E, Stradic SL, Silveira FAO, Durigan G, Overbeck GE, Fidelis A, Fernandes GW, Bond WJ, Hermann J-M, Mahy G, Alvarado ST, Zaloumis NP, Veldman JW (2019) Resilience and restoration of tropical and subtropical grasslands, savannas, and grassy woodlands. Biol Rev 94(2):590–609. https://doi.org/10.1111/brv.12470

Cammeraat LH, Willott SJ, Compton SG, Incoll LD (2002) The effects of ants' nests on the physical, chemical and hydrological properties of a rangeland soil in semi-arid Spain. Geoderma 105(1):1–20. https://doi.org/10.1016/S0016-7061(01)00085-4

Carruthers J (2008) "Wilding the farm or farming the wild"? The evolution of scientific game ranching in South Africa from the 1960s to the present. Trans R Soc South Africa 63(2):160–181. https://doi.org/10.1080/00359190809519220

Chaminuka P, McCrindle CME, Udo HMJ (2012) Cattle farming at the wildlife/livestock Interface: assessment of costs and benefits adjacent to Kruger National Park, South Africa. Soc Nat Resour 25(3):235–250. https://doi.org/10.1080/08941920.2011.580417

Chown SL (2010) Temporal biodiversity change in transformed landscapes: a southern African perspective. Philos Trans R Soc Lond B Biol Sci 365:3729–3742

Colloff MJ, Pullen KR, Cunningham SA (2010) Restoration of an ecosystem function to revegetation communities: the role of invertebrate macropores in enhancing soil water infiltration. Restor Ecol 18:65–72. https://doi.org/10.1111/j.1526-100X.2010.00667.x

Dahlberg AC (2000) Landscape(s) in transition: an environmental history of a village in north-East Botswana. J South Afr Stud 26(4):759–782. https://doi.org/10.1080/03057070020008260

de Klerk JN (2004) Bush encroachment in Namibia. Report on phase 1 of the bush encroachment research, monitoring, and management project. Edited by Ministry of Environment and Tourism, Government of the Republic of Namibia. Published. John Meinert Printers Windhoek

Delius P, Schirmer S (2000) Soil conservation in a racially ordered society: South Africa 1930–1970. J South Afr Stud 26(4):719–742. https://doi.org/10.1080/713683610

Demas D, Sylvia D, Nancy M, Forman S, Jaap A, Stephane F, Philip S (2012) Livestock competitiveness, economic growth and opportunities for job creation in Namibia

Diao X, Hazell P, Thurlow J (2010) The role of agriculture in African development. World Dev 38(10):1375–1383. https://doi.org/10.1016/j.worlddev.2009.06.011

Dlamini P, Mbanjwa V, Gxasheka M, Tyasi L, Sekhohola-Dlamini L (2019) Chemical stabilisation of carbon stocks by polyvalent cations in plinthic soil of a shrub-encroached savanna grassland, South Africa. Catena 181:104088. https://doi.org/10.1016/j.catena.2019.104088

Dovie DBK, Witkowski ETF, Shackleton CM (2004) The fuelwood crisis in Southern Africa – relating fuelwood use to livelihoods in a Rural Village. GeoJournal 60:123–133. https://doi.org/10.1023/B:GEJO.0000033597.34013.9f

Dreber N, van Rooyen SE, Kellner K (2018) Relationship of plant diversity and bush cover in rangelands of a semi-arid Kalahari savannah, South Africa. Afr J Ecol 56:132–135. https://doi.org/10.1111/aje.12425

Eldridge DJ, Bowker MA, Maestre FT, Roger E et al (2011) Impacts of shrub encroachment on ecosystem structure and functioning: towards a global synthesis. Ecol Lett 14:709–722

Ellery WN, Scholes RJ, Mentis MT (1991) An initial approach to predicting the sensitivity of the South African grassland biome to climate change. S Afr J Sci 87:499–503

Ellis EC, Ramankutty N (2008) Putting people in the map: anthropogenic biomes of the world. Front Ecol Environ 6:439–447. https://doi.org/10.1890/070062

Funk C, Peterson P, Landsfeld M, Pedreros D, Verdin J, Shukla S et al (2015) The climate hazards infrared precipitation with stations—a new environmental record for monitoring extremes. Sci Data 2(1):1–21

Geissler K, Hahn C, Joubert D, Blaum N (2019a) Functional responses of the herbaceous plant community explain ecohydrological feedbacks of savanna shrub encroachment. Perspect Plant Ecol Evol Syst 39:125458. https://doi.org/10.1016/j.ppees.2019.125458

Geissler K, Heblack J, Uugulu S, Wanke H, Blaum N (2019b) Partitioning of water between differently sized shrubs and potential groundwater recharge in a semiarid savanna in Namibia. Front Plant Sci 10. https://doi.org/10.3389/fpls.2019.01411

GIZ (11/10/2020): Namibia: Bush waste boosts the economy. Laufs, Johannes. Deutsche Gesellschaft für Internationale Zusammenarbeit. Available online at https://www.giz.de/en/workingwithgiz/91230.html, checked on 7/26/2021

Gomiero T (2016) Soil degradation, land scarcity and food security: reviewing a complex challenge. Sustainability 8(3):281. https://doi.org/10.3390/su8030281

Groengroeft A, de Blécourt M, Classen N, Landschreiber L, Eschenbach A (2018) Acacia trees modify soil water dynamics and the potential groundwater recharge in savanna ecosystems. In: Revermann R, Krewenka KM, Schmiedel U, Olwoch JM, Helmschrot J, Jürgens N (eds)

Climate change and adaptive land management in southern Africa – assessments, changes, challenges, and solutions, Biodiversity & Ecology, vol 6. Klaus Hess Publishers, Göttingen & Windhoek, pp 177–186. https://doi.org/10.7809/b-e.00321

Harmse CJ, Kellner K, Dreber N (2016) Restoring productive rangelands: a comparative assessment of selective and non-selective chemical bush control in a semi-arid Kalahari savanna. J Arid Environ 135:39–49

Hattingh ER (1952) Comments on thorn scrub control with herbicides in Africa. Weeds 1(4):372–373

Haussmann NS, Kalwij JM, Bezuidenhout S (2016) Some ecological side-effects of chemical and physical bush clearing in a southern African rangeland ecosystem. S Afr J Bot 102:234–239. https://doi.org/10.1016/j.sajb.2015.07.012

Hering R, Hauptfleisch M, Geißler K, Marquart A, Schoenen M, Blaum N (2019) Shrub encroachment is not always land degradation: insights from ground-dwelling beetle species niches along a shrub cover gradient in a semi-arid Namibian savanna. Land Degrad Dev 30(1):14–24. https://doi.org/10.1002/ldr.3197

Herrick JE, Sala OE, Karl JW (2013) Land degradation and climate change: a sin of omission? Front Ecol Environ 11(6):283. https://doi.org/10.1890/1540-9295-11.6.283

Higgins SI, Bond WJ, Trollope WSW (2000) Fire, resprouting and variability: a recipe for grass-tree coexistence in savanna. J Ecol 88(2):213–229. https://doi.org/10.1046/j.1365-2745.2000.00435.x

Hiscock KM, Lloyd JW, Lerner DN (1991) Review of natural and artificial denitrification of groundwater. Water Res 25(9):1099–1111

Hoffman MT (2014) Changing patterns of rural land use and land cover in South Africa and their implications for land reform. J South Afr Stud 40(4):707–725. https://doi.org/10.1080/03057070.2014.943525

Hoffman MT, Todd S (2000) A national review of land degradation in South Africa: the influence of biophysical and socio-economic factors. J South Afr Stud 26(4):743–758. https://doi.org/10.1080/713683611

Honsbein D, Shiningavamwe K, Iikela J, de la Puerta Fernandez ML (2017) Animal feed from Namibian encroacher bush. Available online at https://www.dasnamibia.org/download/brochures/GIZ-UNDP-MAWF-Animal-Feed-Manual-2017.pdf

Humphrey GL, Gillson L, Ziervogel G (2021) How changing fire management policies affect fire seasonality and livelihoods. Ambio 50:475–491

Huntley BJ (1982) Southern African savannas. In: Huntley BJ, Walker BH (eds) Ecology of tropical savannas. Springer-Verlag, Berlin Heidelberg. ISBN 978-3-642-68786-0

Inman EN (2020) Community conservation and restoration of degraded land in semi-arid Namibia in the context of climate change. Dissertation, University of Western Australia. School of Biological Sciences, checked on 8/3/2021

Irob K, Blaum N, Baldauf S, Kerger L, Strohbach B, Kanduvarisa A, Lohmann D, Tietjen B (2022). Browsing herbivores improve the state and functioning of savannas: a model assessment of alternative land use strategies. https://doi.org/10.22541/au.162682707.72443437/v1

Joubert DF, Rothauge A, Smit GN (2008) A conceptual model of vegetation dynamics in the semiarid Highland savanna of Namibia, with particular reference to bush thickening by Acacia mellifera. J Arid Environ 72:2201–2210. https://doi.org/10.1016/j.jaridenv.2008.07.004

Joubert DF, Smit GN, Hoffman MT (2012) The role of fire in preventing transitions from a grass dominated state to a bushthickened state in arid savannas. J Arid Environ 87:1e7

Kaiser D, Lepage M, Konaté S, Linsenmair KE (2017) Ecosystem services of termites (Blattoidea: Termitoidae) in the traditional soil restoration and cropping system Zaï in Northern Burkina Faso (West Africa). Agric Ecosyst Environ 236(Supplement C):198–211. https://doi.org/10.1016/j.agee.2016.11.023

Kashululu R-MP, Paul H (2020) Chapter 8: The fencing question in Namibia: a case study in Omusati region. In: Odendaal W, Werner W (eds) "Neither here nor there". Indigeneity, marginalisation and land rights in post-independence Namibia. Legal Assistance Centre, Windhoek Namibia, pp 163–182. Available online at https://library.wur.nl/webquery/wurpubs/fulltext/521219

Kinyua D, McGeoch LE, Georgiadis N, Young TP (2010) Short-term and long-term effects of soil ripping, seeding and fertilization on the restoration of a tropical rangeland. Restor Ecol 18:226–233

Kreike E (2009) De-globalisation and deforestation in colonial Africa: closed markets, the cattle complex, and environmental change in North-Central Namibia, 1890–1990. J South Afr Stud 35(1):81–98. https://doi.org/10.1080/03057070802685585

Kulmatiski A, Beard KH, Verweij RJT, February EC (2010) A depth-controlled tracer technique measures vertical, horizontal and temporal patterns of water use by trees and grasses in a sub-tropical savanna. New Phytol 188:199–209. https://doi.org/10.1111/j.1469-8137.2010.03338.x

Lal R (2015) Restoring soil quality to mitigate soil degradation. Sustainability 7(5):5875–5895. https://doi.org/10.3390/su7055875

Lange G-M, Barnes JI, Motinga DJ (1998) Cattle numbers, biomass, productivity and land degradation in the commercial farming sector of Namibia, 1915-95. Dev South Afr 15(4):555–572. https://doi.org/10.1080/03768359808440031

Lee E, Kumar P, Barron G, Greg A, Hendryx SM (2018) Impact of hydraulic redistribution on multispecies vegetation water use in a semiarid savanna ecosystem: an experimental and modeling synthesis. Water Resour Res 54(6):4009–4027. https://doi.org/10.1029/2017WR021006

Léonard J, Perrier E, Rajot JL (2004) Biological macropores effect on runoff and infiltration: a combined experimental and modelling approach. Agric Ecosyst Environ 104(2):277–285. https://doi.org/10.1016/j.agee.2003.11.015

Li H, Shen H, Chen L, Liu T, Hu H, Zhao X, Zhou L, Zhang P, Fang J (2016) Effects of shrub encroachment on soil organic carbon in global grasslands. Sci Rep 6(1):28974. https://doi.org/10.1038/srep28974

Lindsey PA, Havemann CP, Lines RM, Price AE, Retief TA, Rhebergen T, Van der Waal C, Romañach SS (2013) Benefits of wildlife-based land uses on private lands in Namibia and limitations affecting their development. Oryx 47(01):41–53. https://doi.org/10.1017/S0030605311001049

Lohmann D, Tietjen B, Blaum N, Joubert DF, Jeltsch F (2014) Prescribed fire as a tool for managing shrub encroachment in semi-arid savanna rangelands. J Arid Environ 107:49–56. https://doi.org/10.1016/j.jaridenv.2014.04.003

Luoga EJ, Witkowski ETF, Balkwill K (2000) Economics of charcoal production in miombo woodlands of eastern Tanzania: some hidden costs associated with commercialization of the resources. Ecol Econ 35(2):243–257. https://doi.org/10.1016/S0921-8009(00)00196-8

Mander M (1998) Marketing of indigenous medicinal plants in South Africa: a case study in Kwazulu-Natal. Food and Agriculture Organization of the United Nations, Rome. https://doi.org/10.13140/2.1.1073.4084

Marchiori L, Maystadt J-F, Schumacher I (2012) The impact of weather anomalies on migration in sub-Saharan Africa. J Environ Econ Manage 63(3):355–374. https://doi.org/10.1016/j.jeem.2012.02.001

Marquart A, Eldridge DJ, Geissler K, Lobas C, Blaum N (2020a) Interconnected effects of shrubs, invertebrate-derived macropores and soil texture on water infiltration in a semi-arid savanna rangeland. Land Degrad Dev 31(16):2307–2318. https://doi.org/10.1002/ldr.3598

Marquart A, Geissler K, Heblack J, Lobas C, Münch E, Blaum N (2020b) Individual shrubs, large scale grass cover and seasonal rainfall explain invertebrate-derived macropore density in a semi-arid Namibian savanna. J Arid Environ 176:104101. https://doi.org/10.1016/j.jaridenv.2020.104101

Marquart A, Goldbach L, Blaum N (2020c) Soil-texture affects the influence of termite macropores on soil water infiltration in a semi-arid savanna. Ecohydrology 13(8):e2249. https://doi.org/10.1002/eco.2249

McGranahan DA (2008) Managing private, commercial rangelands for agricultural production and wildlife diversity in Namibia and Zambia. Biodivers Conserv 17(8):1965–1977. https://doi.org/10.1007/s10531-008-9339-y

Mcleon G (1995) Environmental change at Letlhakeng in the Kweneng District. Botsw Notes Rec 27:81–298

McNaughton SJ, Ruess RW, Seagle SW (1988) Large mammals and process dynamics in African ecosystems: herbivorous mammals affect primary productivity and regulate recycling balances. Bioscience 38:794–800. https://doi.org/10.2307/1310789

Meik JM, Jeo RM, Mendelson JR, Jenks KE (2002) Effects of bush encroachment on an assemblage of diurnal lizard species in central Namibia. Biol Conserv 106:29–36

Menestrey Schwieger DA, Mbidzo M (2020) Socio-historical and structural factors linked to land degradation and desertification in Namibia's former Herero 'homelands'. J Arid Environ 178:1–7. https://doi.org/10.1016/j.jaridenv.2020.104151

MET (2020) Fourth National Communication to the United Nations Framework Convention on Climate Change. Edited by Ministry of Environment and Tourism. Government of the Republic of Namibia, Windhoek

Metzger JC, Landschreiber L, Grongroft A, Eschenbach A (2014) Soil evaporation under different types of land use in southern African savanna ecosystems. J Plant Nutr Soil Sci 177:468–475. https://doi.org/10.1002/jpln.201300257

Mogashoa R, Dlamini P, Gxasheka M (2021) Grass species richness decreases along a woody plant encroachment gradient in a semi-arid savanna grassland, South Africa. Landsc Ecol 36:617–636. https://doi.org/10.1007/s10980-020-01150-1

Moore A, Odendaal A (1987) The economic implications of bush encroachment and bush control in a weaner calf production system in the Thorny Bushveld of the Molopo area. Afr J Range Forage Sci 4(4)

Mucina L, Rutherford MC (eds) (2006) The vegetation of South Africa, Lesotho and Swaziland. South African National Biodiversity Institute (Strelitzia), Pretoria, p 19

Mudombi S et al (2018) User perceptions about the adoption and use of ethanol fuel and cookstoves in Maputo, Mozambique. Energy Sustain Dev 44:97–108. https://doi.org/10.1016/j.esd.2018.03.004

Mukwada G (2018) Savanna ecosystems of southern Africa. In: Holmes PJ, Boardman J (eds) Southern landscapes and environmental change. Taylor and Francis. https://doi.org/10.4324/9781315537979-10

Muntifering JR, Dickman AJ, Perlow LM et al (2006) Managing the matrix for large carnivores: a novel approach and perspective from cheetah (*Acinonyx jubatus*) habitat suitability modelling. Anim Conserv 9:103–112

NACSO (2019) State of community conservation in Namibia: a review of communal conservancies, community forests and other CBNRM initiatives (p. 88) [Annual report]. http://www.nacso.org.na/resources/state-of-community-conservation

NAU (Namibia Agricultural Union) (2010) The effect of bush encroachment on groundwater resources in Namibia: a desk top study (Issue December)

Nichols E, Spector S, Louzada J, Larsen T, Amezquita S, Favila ME (2008) Ecological functions and ecosystem services provided by Scarabaeinae dung beetles. Biol Conserv 141(6):1461–1474. https://doi.org/10.1016/j.biocon.2008.04.011

Nkem JN, Lobry de Bruyn LA, Grant CD, Hulugalle NR (2000) The impact of ant bioturbation and foraging activities on surrounding soil properties. Pedobiologia 44(5):609–621. https://doi.org/10.1078/S0031-4056(04)70075-X

O'Connor TG, Puttick JR, Timm Hoffman TM (2014) Bush encroachment in southern Africa: changes and causes. Afr J Range Forage Sci 31(2):67–88. https://doi.org/10.2989/10220119.2014.939996

Olson DM et al (2001) Terrestrial ecoregions of the world: a new map of life on earth. Bioscience 51(11):933. https://doi.org/10.1641/0006-3568(2001)051[0933:TEOTWA]2.0.CO;2

Parkhurst T, Prober SM, Hobbs RJ, Standish RJ (2021) Global meta-analysis reveals incomplete recovery of soil conditions and invertebrate assemblages after ecological restoration in agricultural landscapes. J Appl Ecol. https://doi.org/10.1111/1365-2664.13852

Poonyth D, Hassan R, Kirsten JF, Calcaterra M (2001) Is agricultural sector growth a precondition for economic growth? The case of South Africa. Agrekon 40(2):269–279. https://doi.org/10.1080/03031853.2001.9524950

Potts D (2012) Land alienation under colonial and white settler governments in southern Africa historical land 'grabbing'. In: Allan T, Keulertz M, Sojamo S, Warner J (eds) Handbook of land and water grabs in Africa. Foreign direct investment and food and water security, 1st edn. Routledge, Abingdon, pp 24–42. checked on 8/3/2021

Reed MS, Stringer LC, Dougill AJ, Perkins JS, Atlhopheng JR, Mulale K, Favretto N (2015) Reorienting land degradation towards sustainable land management: linking sustainable livelihoods with ecosystem services in rangeland systems. J Environ Manag 151:472–485. https://doi.org/10.1016/j.jenvman.2014.11.010

Rietkerk M, van de Koppel J (1997) Alternate stable states and threshold effects in semi-arid grazing systems. Oikos 79:69–76

Rohde RF, Hoffman MT (2012) The historical ecology of Namibian rangelands: vegetation change since 1876 in response to local and global drivers. Sci Total Environ 416:276–288. https://doi.org/10.1016/j.scitotenv.2011.10.067

RSA (ed) (2017) Agriculture and land. National Treasury, Republic of South Africa, Pretoria. Chapter 9; (Provincial Budgets and Expenditure Review 2010/11–2016/17)

Rückamp D, Martius C, Bornemann L, Kurzatkowski D, Naval LP, Amelung W (2012) Soil genesis and heterogeneity of phosphorus forms and carbon below mounds inhabited by primary and secondary termites. Geoderma 170:239–250. https://doi.org/10.1016/j.geoderma.2011.10.004

Sanchez P, Palm C, Sachs J, Denning G, Flor R, Harawa R et al (2007) The African millennium villages. Proc Natl Acad Sci USA 104(43):16775–16780. https://doi.org/10.1073/pnas.0700423104

Sandhage-Hofmann A, Löffler J, Kotze E, Weijers S, Wingate V, Wundram D, Weihermüller L, Pape R, du Preez CC, Amelung W (2020) Woody encroachment and related soil properties in different tenure-based management systems of semiarid rangelands. Geoderma 372:114399

Sankaran M, Hanan NP, Scholes RJ, Ratnam J, Augustine DJ, Cade BS, Gignoux J et al (2005) Determinants of Woody Cover in African Savannas. Nature 438. (December 2005):846–849. https://doi.org/10.1038/nature04070

Schnegg M, Pauli J, Greiner C (2013) Chapter 11: Pastoral belonging: causes and consequences of part-time pastoralism in North-Western Namibia. In: Bollig M, Schnegg M, Wotzka H-P (eds) Pastoralism in Africa. Past, present, and future. Berghahn, New York, pp 341–364

Schneider HP (2012) The history of veterinary medicine in Namibia. J S Afr Vet Assoc 83(1):1–11. https://doi.org/10.4102/jsava.v83i1.4

Scholes RJ (2004) Savanna. In: Cowling RM, Richardson DM, Pierce SM (eds) Vegitation of Southern Africa. Cambridge University Press, Cambridge

Scholes RJ, Archer SR (1997) Tree-grass interactions in savannas. Annu Rev Ecol Syst 28:545–570

Shackleton CM (1993) Are the communal grazing lands in need of saving? Dev South Afr 10(1):65–78. https://doi.org/10.1080/03768359308439667

Shackleton C, Shackleton S (2004) The importance of non-timber forest products in rural livelihood security and as safety nets: a review of evidence from South Africa. S Afr J Sci 100:558–664

Shackleton C, Shackleton S, Gambiza J, Nel E, Rowntree K, Urquhart P (2008) Links between ecosystem services and poverty alleviation. Situation analysis for arid and semi-arid lands in southern Africa. Edited by Consortium on Ecosystems and Poverty in Sub-Saharan Africa (CEPSA)

Shackleton S, Cocks M, Dold T, Kaschula S, Mbata K, Mickels-Kokwe G, von Maltitz GP (2010) Non-wood forest products: description, use and management. In: Chidumayo EN, Gumbo DJ (eds) The dry forests and woodlands of Africa: managing for products and services. Earthscan, London

Shikangalah RN, Mapani BS (2020) A review of bush encroachment in Namibia: from a problem to an opportunity? J Rangel Sci 10(3):251–266

Sirami C, Seymour C, Midgley G, Barnard P (2009) The impact of shrub encroachment on savanna bird diversity from local to regional scale. Divers Distrib 15:948–957

Sitters J, Edwards PJ, Venterink HO (2013) Increases of soil C, N, and P pools along an acacia tree density gradient and their effects on trees and grasses. Ecosystems 16:347–357

Smit GN (2004) An approach to tree thinning to structure southern African savannas' for long-term restoration from bush encroachment. J Environ Manag 71:179–191. https://doi.org/10.1016/j.jenvman.2004.02.005

Smit IPJ, Prins HHT (2015) Predicting the effects of woody encroachment on mammal communities, grazing biomass and fire frequency in African savannas. PLoS One 10:e0137857

Stafford W, Birch C, Etter H, Blanchard R, Mudavanhu S, Angelstam P, Blignaut J, Ferreira L, Marais C (2017) The economics of landscape restoration: benefits of controlling bush encroachment and invasive plant species in South Africa and Namibia. Ecosyst Serv 27:1–10. https://doi.org/10.1016/j.ecoser.2016.11.021

Stevens N, Erasmus BFN, Archibald S, Bond WJ (2016) Woody encroachment over 70 years in South African savannahs: overgrazing, global change or extinction aftershock? Philos Trans R Soc B 371(1703):20150437. https://doi.org/10.1098/rstb.2015.0437

Suich H, Howe C, Mace G (2015) Ecosystem services and poverty alleviation: a review of the empirical links. Ecosyst Serv 12:137–147. https://doi.org/10.1016/j.ecoser.2015.02.005

Tallis H, Kareiva P (2007) Ecosystem services. Curr Biol 15(18):R746–R748

Traore S, Tigabu M, Jouquet P, Ouedraogo SJ, Guinko S, Lepage M (2015) Long-term effects of Macroternies termites, herbivores and annual early fire on woody undergrowth community in Sudanian woodland, Burkina Faso. Flora 211:40–50. https://doi.org/10.1016/j.flora.2014.12.004

Trede R, Patt R (2015) Value added end-use opportunities for Namibian encroacher bush

Trollope WSW, Dondofema F (2003) Role of fire, continuous browsing and grazing in controlling bush encroachment in the arid savannas of the Eastern Cape province in South Africa. In: Allsopp N, Palmer AR, Milton SJ, Kerley GIH, Kirkman KP, Hurt R, Brown C (eds) Rangelands in the new millennium: Proceedings of the 7th international rangeland conference, Durban, South Africa. International Rangeland Congress, Durban, pp 408–411

Tully K, Sullivan C, Weil R, Sanchez P (2015) The state of soil degradation in sub-Saharan Africa: baselines, trajectories, and solutions. Sustainability 7(6):6523–6552. https://doi.org/10.3390/su7066523

Twyman C, Dougill A, Sporton D, Thomas D (2001) Community fencing in open rangelands: self-empowerment in Eastern Namibia. Rev Afr Polit Econ 28(87):9–26. https://doi.org/10.1080/03056240108704500

Uugulu, S (2022) Estimation of groundwater recharge along a precipitation gradient for Savanna aquifers in Namibia. PhD Thesis, University of Namibia, 180 pages

van den Berg L. and Kellner K. (2005) Restoring degraded patches in a semi-arid rangeland of South Africa. J. Arid Environ 6::497–511. https://doi.org/10.1016/j.jaridenv.2004.09.024

Vehrs HP, Heller GR (2017) Fauna, fire, and farming: landscape formation over the past 200 years in pastoral East Pokot, Kenya. In: Gale Group (ed) Human ecology. Springer New York LLC, New York. https://doi.org/10.1007/s10745-017-9926-1

Walter H (1939) Grassland, Savanne und Busch der arideren Teile Afrikas in ihrer ökologischen Bedingtheit. Jahrbücher für wissenschaftliche Botanik 87:750–860

Ward D (2005) Do we understand the causes of bush encroachment in African savannas? Afr J Range Forage Sci 22(2):101–105

Wasiolka B, Blaum N (2011) Comparing biodiversity between protected savanna and adjacent non-protected farmland in the southern Kalahari. J Arid Environ 75:836–841

Weitzman JN, Brooks JR, Mayer PM, Rugh WD, Compton JE (2021) Coupling the dual isotopes of water ($\delta 2 H$ and $\delta 18 O$) and nitrate ($\delta 15 N$ and $\delta 18 O$): a new framework for classifying current and legacy groundwater pollution. Environ Res Lett 16(4):1

Wigley BJ, Bond WJ, Hoffman MT (2009) Bush encroachment under three contrasting land-use practices in a Mesic South African savanna. Afr J Ecol 47(SUPPL. 1):62–70. https://doi.org/10.1111/j.1365-2028.2008.01051.x

Yeaton RI (1988) Porcupines, fires and the dynamics of the tree layer of Burkea africana savanna. J Ecol 76:1017–1029

Yuan C, Gao G, Fu B (2016) Stemflow of a xerophytic shrub (Salix psammophila) in northern China: implication for beneficial branch architecture to produce stemflow. J Hydrol 539:577–588

Zengeya TA, Wilson JR (2020) The status of biological invasions and their management in South Africa in 2019. South African National Biodiversity Institute, Kirstenbosch and DSI-NRF Centre of Excellence for Invasion Biology, Stellenbosch, p 71. https://doi.org/10.5281/zenodo.3947613

Managing Southern African Rangeland Systems in the Face of Drought: A Synthesis of Observation, Experimentation and Modeling for Policy and Decision Support

16

Simon Scheiter (ID), Mirjam Pfeiffer (ID), Kai Behn (ID), Kingsley Ayisi (ID), Frances Siebert (ID), and Anja Linstädter (ID)

Abstract

Savanna rangelands cover large areas of southern Africa. They provide ecosystem functions and services that are essential for the livelihoods of people. However, intense land use and climate change, particularly drought, threaten biodiversity and ecosystem functions of savanna rangelands. Understanding how these factors interact is essential to inform policymakers and to develop sustainable land-use strategies. We applied three different approaches to understand the impacts of drought and grazing on rangeland vegetation: observations, experimentation and modeling. Here, we summarize and compare the main results from these approaches. Specifically, we demonstrate that all approaches consistently show declines in biomass and productivity in response to drought periods, as well as changes in community composition toward annual grasses and forbs. Vegetation recovered after drought periods, indicating vegetation resilience.

S. Scheiter (✉) · M. Pfeiffer
Senckenberg Biodiversity and Climate Research Centre (SBiK-F), Frankfurt am Main, Germany
e-mail: simon.scheiter@senckenberg.de

K. Behn
Institute of Crop Science and Resource Conservation (INRES), University of Bonn, Bonn, Germany

K. Ayisi
Risk and Vulnerability Science Centre, University of Limpopo, Sovenga, South Africa

F. Siebert
Unit for Environmental Sciences and Management, North-West University, Potchefstroom, South Africa

A. Linstädter
Biodiversity Research/Systematic Botany, Institute of Biochemistry and Biology, University of Potsdam, Potsdam, Germany

© The Author(s) 2024
G. P. von Maltitz et al. (eds.), *Sustainability of Southern African Ecosystems under Global Change*, Ecological Studies 248,
https://doi.org/10.1007/978-3-031-10948-5_16

439

However, model extrapolation until 2030 showed that vegetation attributes such as biomass and community composition did not recover to values simulated under no-drought conditions during a ten-year period following the drought. We provide policy-relevant recommendations for rangeland management derived from the three approaches. Most importantly, vegetation has a high potential to regenerate and recover during resting periods after disturbance.

16.1 Introduction

Semiarid savannas cover approximately 20% of the global land area (Sankaran et al. 2005) and occupy extensive areas in the global subtropics and tropics of Africa, South America, Asia and northern Australia. Savannas are characterized by a continuous layer of grasses and forbs, interspersed with woody vegetation consisting of trees and shrubs. Fire, mammalian herbivores and pronounced seasonality with distinct dry and wet seasons have shaped the vegetation structure in savannas throughout their evolution. The coexistence of various plant life forms (Linstädter et al. 2014; Siebert and Dreber 2019), their structural heterogeneity and the unique floristic and faunistic elements (Du Toit and Cumming 1999) convey high biodiversity to savannas. Although savanna vegetation has evolved resilience strategies to cope with disturbances such as drought and herbivory (Charles-Dominique et al. 2017; Wigley et al. 2018), intense disturbances, climate change and land-use change may lead to transitions into alternative vegetation states once tipping points are exceeded (Higgins and Scheiter 2012; Pausas and Bond 2020; Staver et al. 2011). Alternative vegetation states can be wood-dominated states (Staver et al. 2011) or degraded states with low vegetation cover and little forage availability for herbivores (Oomen et al. 2016).

Savanna vegetation provides essential ecosystem services to people and supports their livelihood (Ferner et al. 2018; Matsika et al. 2013; Shackleton et al. 2005; Chap. 15). Direct benefits from communal rangelands (Fig. 16.1, "Ecosystem Services") include products ranging from cattle grazing, fuelwood for cooking and heating, collection of edible plants, fruits and nuts, to harvesting of medicinal plants (Matsika et al. 2013). Tourism and recreation in national parks and game reserves may create additional revenue (Kalvelage et al. 2020). Yet, nonsustainable land-use practices combined with climate change (Fig. 16.1, "Environmental drivers") are considerable threats to biodiversity and ecosystem services in savannas (Fig. 16.1, "Threats"). Increasing human population density (Fig. 16.1, "Socioeconomic drivers") in most cases entails land-use intensification and rising pressure on natural resources that can reduce ecosystem resilience (Buisson et al. 2019). For instance, overstocking of rangelands with domestic livestock can lead to a decrease in rangeland productivity and cause shifts in vegetation composition toward nonpalatable species or woody encroachment (Stevens et al. 2017). Such land-use impacts may result in a degraded state not suitable for livestock (Fig. 16.1, "Adverse Effects"). Resting times as implemented in rotational grazing systems

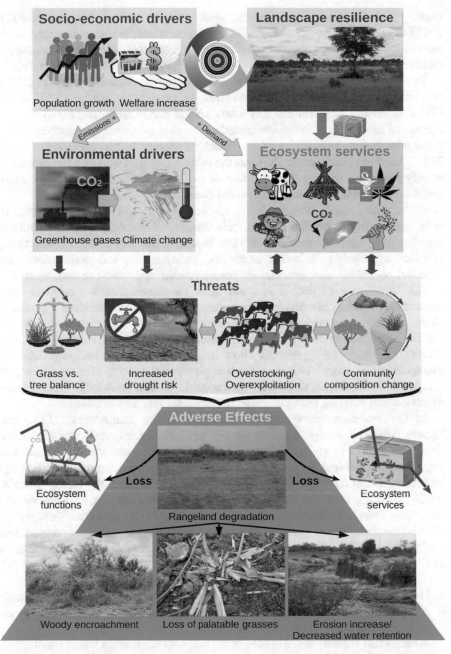

Fig. 16.1 Savanna rangelands are influenced by complex interactions between environmental drivers, socioeconomic drivers and vegetation. They experience multiple threats through climate change and land-use change that can lead to a loss of important ecosystem functions and ecosystem services as well as to rangeland degradation

(Savory and Butterfield 2016), or the presence of browsers (Venter et al. 2017, 2018) can counteract such undesirable vegetation states. Whether degraded states are permanent, or vegetation is resilient and can recover (Fig. 16.1, "Landscape resilience") determines if rangelands will remain available for livestock grazing.

In addition to direct land-use impacts, anthropogenic climate change (Fig. 16.1, "Environmental drivers") exhibits pressure on savanna vegetation (IPCC 2021; Chap. 7). Rising temperatures and lower annual precipitation combined with an increasing frequency of extensive drought events are projected to negatively affect plant growth and the capacity of rangelands to support herbivores (Ruppert et al. 2015). Elevated atmospheric CO_2, on the other hand, may enhance the growth of woody plants and thereby facilitate woody encroachment and transitions to wood-dominated vegetation states (Midgley and Bond 2015).

Given that savanna rangelands are highly dynamic social-ecological systems, it is challenging to predict how climate change, biodiversity loss and land-use pressure may influence future ecosystem functions and services. Interdisciplinary research approaches including scientists, policymakers and stakeholders are required to derive management recommendations and land-use strategies that ensure livelihoods of people in rural areas (Marchant 2010). Complementary research approaches can derive such understanding: (1) Observational approaches allow assessing vegetation dynamics under natural environmental conditions. (2) Experimental approaches allow manipulation of selected biotic and abiotic drivers. (3) Modeling approaches allow integration of system understanding and extrapolation of system behavior. Each approach has specific aims and strengths (Table 16.1). Combining them facilitates acquiring robust knowledge on ecosystem dynamics and deriving policy-relevant management recommendations.

In the two consecutive projects "Limpopo Living Landscapes" (LLL) and "South African Limpopo Landscapes network" (SALLnet), we combined these three approaches to understand the impacts of drought and grazing on savanna vegetation in the Limpopo province, South Africa. Here, we compiled key findings and management recommendations for their sustainable use in the face of drought, while benefiting from the complementarity of observations, experimentation and modeling. Then, we parameterized a vegetation model using data from DroughtAct, a long-term drought and grazing experiment conducted during the two projects, and projected future rangeland vegetation dynamics until 2030. Our research questions are:

1. How does the vegetation in Limpopo's semiarid rangelands respond to drought and grazing? Do the responses of our three complementary approaches (observations, experimentation and modeling) agree?
2. Is rangeland vegetation resilient to drought events of different lengths, and which role does resting play in this context?
3. What are recommendations for policy and decision-making to improve the resilience of savanna rangelands in the face of drought?

Table 16.1 Comparison of different research approaches (field observations, field experimentation and modeling) applied to understand rangeland dynamics. The table focuses on relevant aspects within the LLL and SALLnet projects but does not represent a general comparison of these approaches

	Observation (short-term or long-term)	Experimentation	Modeling
Realism and capturing of all interactions	• Observations of real world • Capture economic reality and actual land use • Can be biased by unknown factors, missing or nonrecorded features, site selection	• Experiments conducted in real world • Control scenarios often similar to the situation in field observation, plus additional treatments • Can be biased by unknown factors, missing or nonrecorded features, site selection	
Historical evidence	• Long-term observations particularly suitable to derive historical evidence	• Long-term experiments can provide historical evidence	
Creation of empirical data	• Empirical data directly derived by observations	• Empirical data directly derived by experiments	
Disentangle joint effects and interactions of drivers	• Possible when all factor combinations of relevant drivers are present and recorded • Crossed space-time substitution enables quantification of interactions	• Full-factorial experiment can assess interactions between a limited number of factors and drivers	• Full-factorial experiment can assess interactions between all factors and drivers represented by model • Model runtime constrains number of factors
Assess sensitivity to drivers and treatments	• Possible when all factor combinations and intensity levels of relevant drivers are present and recorded • Crossed space-time substitution enables sensitivity analysis	• Possible when all factor combinations and intensity levels of relevant drivers are present and recorded • Sensitivity can be assessed for a small number of factors and intensity levels, as every level needs its own treatment	• Gradual changes of intensity of all factors and drivers represented by model are possible • Model runtime can be a constraint

(continued)

Table 16.1 (continued)

	Observation (short-term or long-term)	Experimentation	Modeling
Detect tipping points	• Possible particularly in long-term observations if regular monitoring data are available	• Possible if the intensity and duration of treatments in experiment are sufficient	• Intensity and duration of all drivers and treatments in model can be varied until tipping point is reached • Tipping points and associated threshold can be detected if prevalent in the model
Extrapolation	• Possible for the range of environmental gradients captured in observations	• Possible for the range of treatments captured in experiment • Response patterns of experiment may be applied to similar ecosystems	• Generally possible to extrapolate for new environmental conditions, treatments or disturbance regimes • Uncertainty beyond data used for parameterization
Long term predictions			• Modeling approaches can provide predictions for any time frame (past, present, future) when adequate environmental forcings are available
Uncertainty/Accuracy	• Uncertainty due to unknown factors or sampling issues	• Uncertainty due to unknown factors, sampling issues, or choice of treatments in factorial design	• Models represent current knowledge and mechanisms considered important for specific study questions • Uncertainty due to mechanisms not considered in models, and data used for parameterization and benchmarking • Uncertainty due to extrapolation beyond parameterization

16.2   Study Area

The Limpopo province occupies the Northeast of South Africa (Fig. 16.2). It exhibits considerable variation in soils, topography and climatic conditions. Mean annual precipitation (MAP) ranges between less than 200 mm and more than 1000 mm, and mean annual temperature ranges between 18°C and 28°C (New et al. 2002). The Limpopo province provides rich natural resources that allow the abundant production of agricultural goods such as livestock, vegetables, cereals, fruits and tea. While commercial farms produce these products on a large scale, the Limpopo province also hosts some of the most underprivileged rural areas of South Africa (Lehohla 2012). There, smallholder and subsistence farming rely heavily on available natural resources. Rural communities are particularly vulnerable to environmental change, disruptions such as drought and changes in the social-economic conditions (Gbetibouo 2009; Twine et al. 2003). Overall, approximately 90% of the area of the Limpopo province is utilized by rural and commercial farming, with around 10.5% of the area used for agriculture and 81% used for livestock and game (Maluleke et al. 2016). The Limpopo province also hosts conservation areas and national parks, including parts of the Kruger National Park and the Vhembe Biosphere Reserve (Pool-Stanvliet 2013), creating income through tourism. The current population in the Limpopo province is around 5.9 million people (Stats SA 2021) and growing.

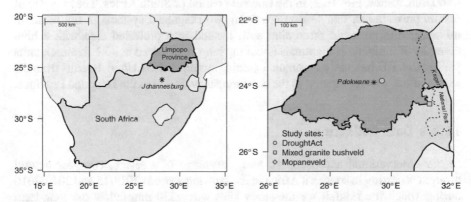

Fig. 16.2 Maps showing the Limpopo Province in South Africa, and the study sites of observational studies (mixed granite bushveld, mopaneveld, Sect. 16.3), and experimental and modeling studies (DroughtAct, Sects. 16.4 and 16.5). Note that the sites of observational studies represent clusters of several study sites

16.3 Observational Approach

16.3.1 Background: Observational Approaches to Study Combined Effects of Drought and Grazing

The design of observational studies commonly relies on recording temporal and spatial variation in ecosystem characteristics along local to continent-wide environmental gradients (Table 16.1). For example, study sites can be arranged along steep regional climate gradients to assess the impact of future climatic conditions via a space-for-time substitution (Blois et al. 2013). Moreover, long-term monitoring allows the detection of early warning signals for rapid ecosystem state changes (Arena et al. 2018; Buitenwerf et al. 2011). Observational studies are particularly suitable to understand drought effects on ecosystem function and services for landscapes with several land-use types. In multiple observational studies, species- and trait-based methods have helped assess the extent of taxonomic and functional responses to drought (Ruppert et al. 2015; Wigley-Coetsee and Staver 2020). However, understanding of patterns and drivers of forb communities in African rangeland systems is still limited (Siebert and Dreber 2019). Therefore, our research aimed to assess how forbs, compared to grasses, respond to drought and land-use change or rangeland intensification. We conducted observational studies across ecosystem types with varying annual precipitation. Sites were located in a semiarid mixed granite bushveld (MAP: ~630 mm, 2 sites, Fig. 16.2) and mopaneveld (MAP: ~460 mm, 3 sites, Fig. 16.2) in the Lowveld region of South Africa. The majority of results presented in this section are from two rangeland systems in the semiarid mixed granite bushveld on similar soil, located in a protected area with a high diversity of indigenous mammals (stocking rates maintained at ~0.1 livestock units per hectare, LU/ha), and a communal grazing system (>1.0 LU/ha). Results from the semiarid site were compared to the drier mopaneveld site in the Limpopo Province.

16.3.2 Data Collection

In our observational approach, we took advantage of a severe, two-year natural drought occurring in southern Africa in the growing seasons 2014/15 and 2015/2016 during which the rainfall for the study sites was ~330 mm below the long-term average (52% and 72% reduction relative to MAP for the two study regions). Complete floristic surveys (i.e., plant individual counts per species) were conducted in October 2016 (representing the "in-drought" survey) within permanent 1 m^2 plots situated in a protected area and on communal rangeland, which was repeated in the peak growth season (i.e., January) of the postdrought year 2017. Herbaceous plants were divided into four plant functional types (PFTs): perennial grasses, annual grasses, perennial forbs, annual forbs. Directly adjacent to the permanent plots in both rangeland systems, standing plant biomass was clipped and hand-sorted into forbs and grasses. Biomass was oven-dried (70 ° C, > 48 h) and weighed (Siebert et al. 2020). We derived aboveground net primary production (ANPP) per rangeland type and year from the clipped aboveground standing biomass using the "peak standing biomass" method (Ruppert and Linstädter 2014). We assessed the relative contribution of forbs and grasses to ANPP.

16.3.3 Data Analysis

Herbaceous species composition across different rangeland types and rainfall was explored using Nonmetric Multidimensional Scaling (NMDS) analyses compiled in PRIMER 6 software. To assess the difference between drought and post-drought herbaceous biomass per life form, a repeated measures analysis of variance (ANOVA) combined with the Bonferroni *post-hoc* significance test was performed. A two-way ANOVA type Hierarchical Linear Model (HLM) was used to test the effects of rainfall year and rangeland type on variation in grass and forb diversity indices. Significant differences in response to rangeland type and rainfall year were tested using effect sizes (Cohen's *d*, Ellis and Steyn 2003).

16.3.4 Key Results and Discussion

The two-year natural drought had significant effects on ANPP and community composition, irrespective of study sites' MAP. Although a previous meta-analysis has shown that protected areas with a high proportion of perennial grasses are usually more resistant to drought than those dominated by annual grasses (Ruppert et al. 2015), protected areas and sites with intense grazing pressure were equally affected by the drought. However, protected areas showed a more pronounced postdrought recovery (Fig. 16.3), especially in comparison to sites with a long disturbance history (Klem 2018; Minnaar 2020).

Grass and forb biomass were equally low in both rangeland systems during drought. However, postdrought recovery differed between the two rangeland systems and between the two life forms (Fig. 16.3). Forb biomass increased significantly after the drought at both sites, although the increase was much higher in the protected area. We recorded a five-fold increase in grass biomass postdrought in the protected area, opposed to a much weaker one in the communal rangeland. The proportion of grass to forb biomass was also significantly higher in the protected area (Siebert et al. 2020).

Compared to the communal rangeland, grass species richness and diversity were lower in the protected area during the drought but increased significantly after the drought (Klem 2018). Forb richness and diversity were equally low at both sites and increased significantly postdrought, but in the protected area only. These results suggest that long-term exposure to intensive livestock grazing may deplete the herbaceous seedbank (O'Connor 1991). In the drier mopaneveld study, drought had a similar negative effect on both grass and forb species richness and diversity in a moderately grazed protected area (Minnaar 2020).

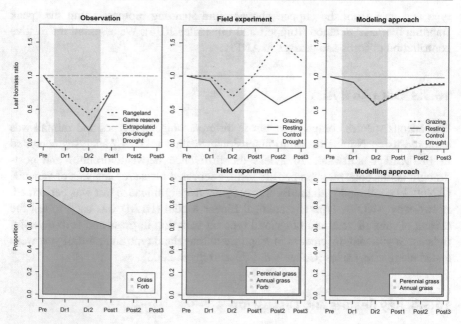

Fig. 16.3 Responses of leaf biomass ratio and the proportion of different PFTs to a 2-year drought in observations, field experiment and modeling. Leaf biomass ratio represents ANPP in observations, ANPP in field experiments and leaf biomass in the modeling approach. Time slices considered are "Pre": year predrought; "Dr1" and "Dr2": first and second drought year; "Post1," "Post2," "Post3": three years after drought allowing vegetation recovery. For the observations, drought and postdrought data is based on (Siebert et al. 2020). Predrought biomass ratio was extrapolated using post- and predrought ratios from Van Staden (2016) and Minnaar (2020), due to a lack of predrought data for the study sites

During the drought, perennial PFTs dominated the herbaceous layer in both rangeland systems. In the communal rangeland, perennial grasses had a more pros-trate, lawn-like growth form, which can be connected to an improved tolerance to combined effects of drought and intensive livestock grazing (Hempson et al. 2015b). Grass and forb PFTs that persisted during the drought were often characterized by clonal growth and bud position close to the soil surface. Both rangeland systems had few annual grasses during the drought. Annual forbs increased in abundance after the drought, but only in the protected area (Klem 2018). This higher postdrought recovery of annuals is in line with the results of a global meta-analysis (Ruppert et al. 2015).

In conclusion, poor rangeland conditions resulting from long-term intensive livestock grazing in the communal rangeland could explain the observed weak positive effects of postdrought rainfall on species and functional composition of grasses (Fynn and O'Connor 2000). Both life forms displayed persistence of generalist tolerator species. Plants, such as *Digitaria eriantha* and various perennial forb species have a life-history strategy that tolerates defoliation through the ability to resprout from stored resources and a bud bank at or below ground level (Archibald et al. 2019), which can be ascribed to a long evolutionary history of grazing (O'Connor 1995). Although the drought in the growing seasons 2014/15 and 2015/2016 was severe, both rangeland types displayed high ecosystem resilience. However, the duration of this drought event may have been short enough for vegetation recovery after a substantial rain. Possibly, a longer period of below-average rainfall and high temperatures may have led to more pronounced effects on plant community composition (as confirmed by the experiments in Sect. 16.4).

Forb cover and biomass increased in response to postdrought rainfall. Increases in forb cover are often regarded as indicators of land degradation (Camp and Hardy 1999). In the current study, the dominance of perennial forbs in the communal rangeland system during and after drought illustrates their functional importance. Forbs are particularly important for securing forage when grass biomass is low.

16.3.5 Recommendations Derived from the Results and Outlook

Our results on rangeland vegetation responses to a 2-year natural drought event provided valuable insights into drought resistance of African semiarid savanna ecosystems. This ecosystem type experienced long evolutionary adaptation to grazing by large herbivores. Therefore, African grassland species have developed traits that convey tolerance to heavy browsing, grazing and defoliation by fire and allow rapid postdisturbance recovery. These patterns are similar to what we observed through their postdrought response, irrespective of the rangeland type. Furthermore, we found that the response of different herbaceous PFTs is not necessarily aligned. Changes in the species and functional composition of both PFTs are good indicators of long-term effects on the resilience of these rangeland ecosystems.

Therefore, studying the patterns of vegetation change from predrought, during drought, to postdrought conditions in an uncontrolled experimental setting could contribute to an improved understanding of the prolonged effects of drought combined with heavy grazing. Such observational approaches are even more informative when comparing results with those from experiments and model simulations.

16.4 Experimental Approach

16.4.1 Background: Experimental Approaches to Study Combined Effects of Drought and Grazing

Field experiments offer valuable opportunities to study ecological processes and the impact of climatic or land-use factors under controlled conditions (Table 16.1). Many studies have either evaluated grazing (Díaz et al. 2007; Linstädter et al. 2014) or drought (Cherwin and Knapp 2012; Tielbörger et al. 2014) as drivers for functional changes. However, field experiments focusing on combined and potentially interactive effects of rangeland management and centennial-scale drought on South

Africa's semiarid rangeland vegetation are lacking. Hence, "DroughtAct" has been specifically designed to address this research gap. The experiment also aimed at comparing the effects of drought events of different lengths (two versus six years), and to evaluate suitable rangeland management options during and after a drought. DroughtAct also contributes to the large, international and coordinated drought experiment DroughtNet (https://drought-net.colostate.edu/) which aims to assess ecosystem responses to centennial-scale drought.

16.4.2 Experimental Setup

The DroughtAct experiment started with a pretreatment year in the rainy season 2013/14, followed by six treatment years. It was implemented on a grazing camp of the Syferkuil experimental farm (Fig. 16.2) belonging to the University of Limpopo and was maintained as a collaborative effort of researchers from South Africa and Germany (Munjonji et al. 2020). The experiment has a full-factorial design with four-block repetitions of eight treatment plots. At the core of the experiment are four treatment combinations that were maintained for the whole duration of the experiment (six years): Ambient rainfall and grazing (D−G+, control), ambient rainfall and resting (D−G−, Fig. 16.4), drought and grazing (D+ G+, Fig. 16.4) and drought and resting (D+ G−). To directly compare the effects of a 6-year drought and a 2-year drought, we terminated the experimental drought on four plots per block after 2 years. We observed rangeland characteristics on plots with a drought history (H+) or without (H−), both under grazed and rested conditions.

For the drought treatments, passive rainout-shelters were used (Yahdjian and Sala 2002). The shelters had a size of 6 x 6 m and consisted of metal constructions with transparent polycarbonate roof sheets. Sheets covered two-thirds of the area

Fig. 16.4 Two treatments of the "DroughtAct" experiment. The left picture shows a plot under ambient rainfall and resting (D−G−) fenced with strain wire to prevent cattle grazing and chicken wire to also restrict entry of smaller mammals. The right picture shows a drought treatment in combination with grazing (D+ G+). The plot is covered by a rainout shelter and contains three movable cages to temporarily exclude grazing. Pictures by K. Behn

and thus reduced ambient rainfall by ca. 66%. We chose this level of reduction to reflect the site-specific rainfall history by turning a year with average rainfall into a year of centennial-scale drought (Knapp et al. 2017), following the guidelines of DroughtNet (https://drought-net.colostate.edu/). The perimeters of drought plots were trenched with plastic foil down to 1 m depth to inhibit lateral water flow that would counteract rainfall reduction (Mudongo et al. 2020). The camp was subject to moderate rotational grazing with cattle. Rested plots were fenced with strain wire to prevent grazing of cattle and with chicken wire to restrict entry of smaller mammals. We extensively checked for experimental artifacts and found that rainout shelters did not significantly alter light interception and microclimate (Mudongo et al., unpublished). No evidence of changed grazing preferences resulting from the rainout shelters was observed. With a height of at least 2 m, cattle could easily walk and graze below the roofs. The transparency of the polycarbonate shields did not provide enough shade to make the shelters a preferred resting place.

16.4.3 Data Collection

We assessed treatment effects on vegetation structure and composition, and on vegetation-mediated ecosystem functions and services. Here, we only report data collection for aboveground net primary production (ANPP) and functional vegetation composition, i.e., proxies that were considered in the observational study (see Sect. 16.3). At the time of peak standing biomass, we estimated ANPP of each plot. The sampling procedure differed between grazed and fenced plots (see Munjonji et al. 2020 for details). In brief, we harvested biomass on three 1 m^2 harvesting quadrats on grazed plots (G+), where grazing was excluded for a given vegetation period with the aid of moveable cages (Fig. 16.4). In adjacent grazed quadrats, we assessed residual biomass with a nondestructive approach. In nongrazed plots (G−), the nondestructive approach was used, and in some years, biomass was additionally harvested in three smaller squares with a size of 0.25 m^2. On each square plot, we recorded all occurring species, visually assessed their cover and measured their average height. We assigned each species to a PFT with the same a-priori approach used in the observational study. We distinguished three PFTs: perennial grasses, annual grasses and forbs. Facultative perennials and biennials were classified as perennials, following Linstädter et al. (2014).

16.4.4 Data Analysis

For all squares that were sampled using the nondestructive approach, we used data on species' cover and height to estimate their relative contribution to ANPP. To this end, we applied a biomass–biovolume calibration developed for the experiment. To assess the drought and postdrought impacts, we calculated response ratios (Mackie et al. 2019) by dividing the mean value of the leaf biomass in all sampling quadrats belonging to the (post)drought treatments by the respective values of the control treatment. We put response ratios in relation to the experimental reduction of precipitation (Behn et al. 2022). The significance of the (post)drought effect was tested using ANOVA and TukeyHSD *post-hoc* test. For the PFTs, we calculated their relative contribution to the overall ANPP for each treatment and year.

16.4.5 Key Results

Our analyses of ANPP in predrought years, within-drought years and postdrought years across grazed and rested plots, and across plots with different drought duration showed marked differences in drought resistance and resilience (Behn et al. 2022). Similar as for the natural drought captured in the observational study (Sect. 16.3), the two-year experimental drought had significant negative effects on ANPP (Fig. 16.3). Compared to nondrought conditions, ANPP was on average reduced by 30% under grazed conditions in the second year of drought. Under rested conditions, the effect was even stronger with up to 50% reduction.

We found a fast recovery after the 2-year drought particularly under grazed conditions (Fig. 16.3), where already the first postdrought year did not show a significant deviation from the control. Under rested conditions, postdrought ecosystem performance remained on average below control level, even though differences were not significant from the second postdrought year on. However, the 6-year drought had devastating effects on ecosystem performance with an ANPP reduction of up to 80% and thus even exceeding the 66% reduction of precipitation. Grazing had an ambivalent role because its impacts were beneficial in the initial two years but became detrimental under ongoing drought.

Regarding the relative abundance of PFTs, there were notable differences concerning the drought duration. Perennial grasses tended to increase their relative abundance during the first years of drought and in the recovery phase after short drought at the expense of annual grasses (Fig. 16.3). With prolonged drought however, perennial grasses showed a strong decrease in relative abundance while forbs increased. Thus, species such as *Chamaecrista mimosoides* (L.) Greene and *Monsonia angustifolia* E.Mey. ex A.Rich. could be identified as relative winners of prolonged droughts. In absolute numbers, however, there were no winners when looking at biomass production per area.

16.4.6 Recommendations Derived from the Results and Outlook

Our results give valuable indications on both research and rangeland management practices in the face of drought. They stress the importance of a detailed understanding of the effects. Aridity and grazing management both altered ANPP and caused changes in vegetation structure and species composition. In combination, these effects impacted forage quantity, quality and palatability. Therefore, the changes caused by drought and grazing management can serve as warning signals for degradation. Management needs to be adapted and consider the ambivalent role that grazing and resting can have under different drought durations and intensities.

The DroughtAct experiment in Limpopo, South Africa, was the blueprint for a similar experiment in the Waterberg region of Namibia (Namtip project, https://www.namtip.uni-bonn.de/). The Tippex experiment of Namtip further improved the concept of DroughtAct and added a more detailed sampling approach, including the prominent role of the soil seed bank and different grazing intensities. With climate change, decreasing rainfall and changing rainfall patterns, drought experiments are crucial to understand the effects on vegetation to predict and mitigate degradation. Therefore, experiments like DroughtAct are essential due to their direct results, the data supply for vegetation modeling and to help scientists implement and improve drought experiments across the globe.

16.5 Modeling Approach

16.5.1 Background: Modeling Savanna Rangelands

Models represent properties, processes and functions of real-world systems in quantitative ways to improve system understanding. Thereby, they represent the current system understanding, but do not account for the full complexity of an ecosystem. Models are usually developed for specific research questions and only represent mechanisms considered relevant for those questions. As system understanding advances, models evolve through the inclusion of new knowledge, data and processes. Their main advantage is the applicability for a wide range of scenarios, such as past or future climate change scenarios, management scenarios, or factor combinations not considered in experiments (Table 16.1).

For savanna rangelands, several modeling approaches have been applied, including heuristic differential equation models (Baudena et al. 2010; Scheiter and Higgins 2007; van Langevelde et al. 2003), agent-based models (Fust and Schlecht 2018; Kuckertz et al. 2011) and process-based dynamic vegetation models (DVMs) (Pfeiffer et al. 2019; Scheiter and Higgins 2009). Within SALLnet, we improved and applied the individual-based dynamic vegetation models aDGVM (Scheiter and Higgins 2009) and aDGVM2 (Scheiter et al. 2013). Both models were originally developed to simulate grass-tree dynamics in savannas, but they differ in their representation of plant diversity. While aDGVM simulates dynamics of four PFTs (forest and savanna trees, C_3 and C_4 grasses), aDGVM2 simulates community assembly processes to create plant communities adapted to biotic and abiotic drivers.

16.5.2 Improving aDGVM and aDGVM2

We identified and resolved several limitations of aDGVM and aDGVM2. First, the herbaceous layer is commonly poorly represented in DVMs (Pfeiffer et al.

2019). While DVMs typically represent several woody PFTs, they aggregate grasses and forbs in C_3 herbaceous and C_4 herbaceous PFTs. However, when studying grazing impacts on productivity, diversity and forage quality in savanna rangelands, a representation of different grass PFTs and forbs is essential. Therefore, we included annual and perennial grass types into aDGVM2 by adjusting reproduction, carbon allocation and mortality (Pfeiffer et al. 2019). Annual grasses in aDGVM2 preferentially allocate carbon to rapid leaf growth and high seed production and die after one growing season. Perennial grasses preferentially invest carbon into root and storage compartments to enhance survival. Forbs have not yet been included in aDGVM2.

Second, in previous model versions, land-use activities such as grazing and fuel-wood harvesting were poorly represented. Therefore, we included a cattle grazing model into aDGVM2 (Pfeiffer et al. 2019). In this model, cattle owners prescribe the grazing regime. Animals graze selectively with a preference for grass patches with high quantities of living leaf biomass, low ratios of dead-to-live grass biomass and high palatability. Palatability is assumed to increase with specific leaf area (SLA) and leaf nitrogen content. Thereby, grazing directly impacts the abundance of annual and perennial grasses. We further coupled aDGVM with routines to simulate grazing and fuelwood harvesting (Scheiter et al. 2019). Grazing nonselectively removes a prescribed amount of grass biomass, and fuelwood harvesting removes prescribed amounts of woody biomass. Harvesting was related to tree stem diameter, preferring trees with a stem diameter between 5 and 10 cm (Twine and Holdo 2016).

Third, aDGVM and aDGVM2 did not consider shrubs. We included shrubs into aDGVM2 (Gaillard et al. 2018) and assumed that differences between trees and shrubs are related to trade-offs between water availability, light availability and height. Trees generally invest more into height growth, which is an advantage in dense and light-limited environments. Contrastingly, shrubs with a multistemmed architecture have a higher sapwood area and improved water transport capacity. Having multiple stems entails lower height growth and a competitive disadvantage in light-limited ecosystems (Gaillard et al. 2018).

16.5.3 Key Results

When considering natural vegetation dynamics without land use, aDGVM results showed increases in woody biomass and biome transitions to wood-dominated vegetation states until 2099 in response to the representative concentration pathway (RCP) scenarios RCP4.5 and RCP8.5 (Martens et al. 2021; Scheiter et al. 2018). RCP4.5 is a modest-high climate mitigation scenario where carbon emissions peak toward the middle of the century, whereas RCP8.5 is a low climate mitigation scenario with high carbon emissions and energy consumption (van Vuuren et al. 2011). Biomass increases resulted from CO_2 fertilization effects on tree growth. Grassland and savanna areas were most susceptible to biome transitions (Martens et al. 2021; Scheiter et al. 2018; Chap. 14). Simulations showed that vegetation dynamics lag behind environmental forcing and that observed vegetation states can deviate from those expected under given environmental conditions (Pfeiffer et al. 2020; Scheiter et al. 2020). Lagged responses result from the different velocities at which processes such as ecophysiology, population dynamics or succession operate.

Increasing grazing pressure under varying annual rainfall regimes reduced grass productivity and grass biomass at study sites in South Africa and altered the composition of the grass layer (Pfeiffer et al. 2019). Specifically, community composition shifted toward a higher abundance of annuals with increasing grazing intensity. Annual grasses became dominant once the grazing demand exceeded a critical value between 1.5 and 3 LU/ha for a rainfall gradient between 253 and 926 mm MAP, assuming a daily dry matter demand of 12.5 kg/LU/day. Such changes occurred when grazing intensity exceeded the carrying capacity of perennial grasses, i.e., when regrowth of perennial grass biomass was insufficient to cover animal demand. Consequently, the grass type preferred by grazers switched from perennials to annuals. Resting periods without animals were necessary for biomass recovery and regeneration of perennial grasses (Pfeiffer et al. 2019). Recovery periods were site-specific and up to 8 and 17 years at arid and humid sites respectively. Recovery was faster at more arid sites because these sites had a lower biomass without grazers that they had to reach, and because annual grasses that are more abundant at the arid sites recover faster than perennial grasses that are more abundant at the humid sites. Drought impacts have not been considered explicitly in these simulations, but see Sect. 16.6.1 for aDGVM2 simulations of the DroughtAct experiment.

We used optimization techniques to identify grazing and fuelwood harvesting intensities that were well-sustained by vegetation at Bushbuckridge, South Africa, and maximized the economic value of the land-use system (Scheiter et al. 2019). The economic value included, for example, milk, meat, dung and the cultural status conveyed by owning cattle. Simulations indicated that the optimal animal number was only 0.076 LU/ha whereas observed animal numbers are, for example, up to 0.75 LU/ha on the farm containing the DroughtAct experiment, 0.88 LU/ha in communal grazing lands north of Acornhoek (Parsons et al. 1997) and > 1 LU/h at a site used for the observations (Sect. 16.3). Similarly, optimal fuelwood harvesting intensities were lower than observed intensities.

16.5.4 Recommendations Derived from the Results and Outlook

Simulation results indicate that in the absence of land use and herbivory, grasslands and savannas are highly susceptible to woody encroachment and transition to woody vegetation states due to climate change and elevated CO_2 (Martens et al. 2021; Scheiter et al. 2018). Land use (fuelwood harvesting and grazing) can alter the velocity of such transitions (Scheiter and Savadogo 2016). Vegetation changes are considerably slower than changes in environmental drivers. Therefore, vegetation will continue changing even if humanity reduces greenhouse gas emissions and manages to stabilize the climate system (Pfeiffer et al. 2020; Scheiter et al. 2020). Such delayed responses of vegetation need consideration when developing management plans for savanna rangelands.

Intense grazing strongly decreases grass biomass and productivity and shifts grass communities toward a higher abundance of annual grasses (Pfeiffer et al. 2019). According to the simulation results, vegetation is resilient and recovery is possible during resting periods without grazing. To ensure full recovery of heavily grazed vegetation (more than ca. 2.5 LU/ha) to the productivity levels expected in the absence of grazing, resting times of 5–17 years were simulated, with shorter resting times (5–8 years) for more arid sites and longer resting times (14–17 years) for more humid sites. After low-intensity grazing (less than ca. 1.2 LU/ha), resting times of less than 6 years were sufficient. Yet, to ensure sustained forage for animals, full recovery might not be necessary and sorter resting times are sufficient. Downsizing of herds, provision of additional fodder, as well as rotational herding strategies that include rangeland and cropland are possible methods to achieve resting (Pfeiffer et al. 2019, 2022).

16.6   Integrating Observations, Experiments and Modeling

Observation, experimentation and modeling ideally complement each other, and data and information are shared between these approaches (Fig. 16.5). Field-based observations provide insights into system functionality, identify relevant system components and are necessary for a qualitative system conceptualization. The resulting conceptual system understanding is the basis for hypothesis formulation that can be evaluated in specifically tailored experiments. The strength of observations and experiments (Sects. 16.3 and 16.4) is that they directly measure system dynamics and responses to natural environmental conditions or different experimental treatments (Table 16.1). Results from experiments provide insights allowing focus re-evaluation for further observations. They also yield system-specific quantitative data for model parameterization, calibration and benchmarking, but are often resource-limited regarding treatment numbers, factor combinations, study sites, replicates and duration. Quantitative experimental data are the basis for the deduction of system dependencies that allow the development of quantitative process models. The steps from observation to experimentation and model development signify an increasing degree of system generalization and abstraction, with a narrowing focus on processes and components deemed most relevant for the targeted questions. Process-based models (Sect. 16.5) enable scenario testing, spatiotemporal extrapolation and testing of system sensitivities and allow exploration of possible trajectories for future climate change, management scenarios or regional upscaling. Knowledge gained from models can then inform both experimentation and observation to re-evaluate the focus of further research, help generate new hypotheses and improve system understanding.

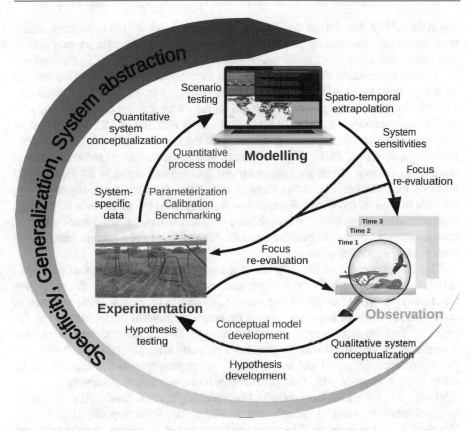

Fig. 16.5 Linking observations, experiments and modeling—how different research approaches inform each other

16.6.1 Applying aDGVM2 to DroughtAct

Observations and experiments showed clear responses of vegetation to drought. Therefore, we replicated the DroughtAct experiment in aDGVM2 simulations (1) to test if aDGVM2 can simulate observed vegetation responses to drought and (2) to test if vegetation can recover between the end of the 6-year experimental drought in 2020 and 2030. We simulated all factor combinations described in Sect. 16.4 but used 40 instead of 4 replicates per treatment and continued simulations until 2030. Meteorological data were not available for the DroughtAct site and for future conditions to allow postdrought simulations. Data for a neighboring meteorological station had data gaps (https://www.weathersa.co.za/) and were therefore unsuitable to conduct simulations. Hence, we selected climate forcing data from climate model simulations to find the model and RCP scenario that agreed best with precipitation from station data and the EWEMBI climate product for 2008 to 2017 (Lange 2019). We found the best fit for precipitation from the CMIP5 simulations conducted with the IPSL-CM5A-LR model for RCP8.5 (Dufresne et al. 2013) combined with a transformation. Specifically, we shifted the simulated climate data by one year to the future and multiplied daily precipitation with a factor of 0.73. For drought treatments, we reduced daily rainfall by 66% for a 2-year and 6-year period starting in the growing season 2014/2015 following DroughtAct. For grazing treatments, we assumed an average number of 30 LU on the 40 ha grazing camp that includes the DroughtAct experiment, with a per-capita daily demand of 12.5 kg dry matter. We simulated a rotational grazing system with animals present during four 30-day periods per year (i.e., 120 days per year) and resting times without animals between the grazing periods. Timing of presence and absence was equal in all years. Daily biomass removal per simulated hectare was determined by randomly assigning animals to each of the 40 ha. Simulations were repeated in the absence of grazing. However, differences between simulations with and without grazing were minor as drought impacts overrode grazing impacts. We therefore only present simulations with grazing in the following.

The first two years of drought led to a steep decline in mean annual plot-level leaf biomass compared to the control scenario (Figs. 16.3, 16.6a). Biomass was reduced by ca. 40% after the second year of drought. In the 6-year drought, biomass stabilized at values of approximately 40% below the control during the 3rd to 6th years of drought. Postdrought recovery was rapid during the first two years after drought before slowing down (Fig. 16.6a). Five years after the end of the 2-year and the 6-year drought, plot-level leaf biomass was approximately 5% and 8% below the control, respectively. In 2030, plot-level biomass was still approximately 3% and 6% below the control for the 2-year and 6-year drought, respectively. At grass patch-level, biomass response to drought was similar to plot-level response (Fig. 16.6b). However, at patch-level leaf biomass fully recovered to values simulated in the control. This result indicates that surviving or newly established grass patches were resilient to drought, whereas slow recruitment and recolonization of bare ground inhibited recovery at the plot level.

Plot-level biomass loss during drought was caused by increased mortality of perennial grasses and lower productivity of surviving grasses. While mortality was 1–3 times higher than control during the first two drought years, it was almost 20 times higher during the 3rd and 4th drought years. Remarkably, the higher mortality persisted for five years after drought treatments and then reached values simulated for the control. Mortality of perennial grasses during drought allowed annual grasses to colonize the study site and their fraction in the population was 1.5 times higher during the first two years and 2.5 times higher from the 3rd drought year onward (Fig. 16.6c). Their abundance decreased during the recovery period but was still 1.4 and 1.8 times higher in the 2-year and 6-year drought scenarios at the end of the simulation period.

Overall, simulations indicated resilience of all considered model variables. We found no tipping point behavior or transition into an alternative vegetation state that persisted after drought. However, the velocity of recovery differed between the 2-year and the 6-year drought and between considered variables. For instance, while patch-level grass biomass recovered within 2 or 3 years after the drought treatment,

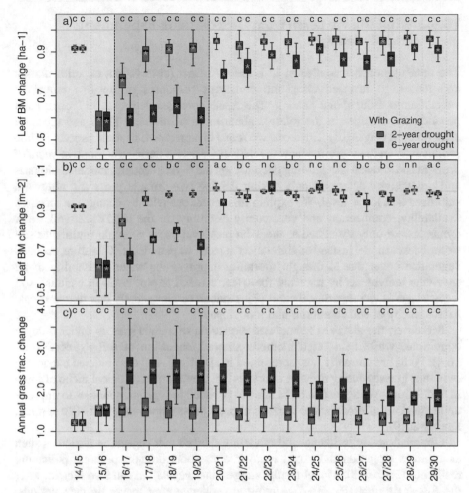

Fig. 16.6 Impacts of a 2-year and a 6-year drought on simulated grass biomass and PFTs, and postdrought recovery relative to nondrought control scenario. Panels show (**a**) biomass at the site scale, (**b**) biomass at the grass patch scale and (c) fraction of annual grasses in the grass population. The yellow shading indicates the 2-year drought period, the yellow and red shading the 6-year drought period and the green shading the recovery period until 2030 after the termination of drought treatments. Letters above boxes indicate the significance of the difference in the mean based on the t-test between treatment and control (n: not significant; a: $p < 0.05$; b: $p < 0.01$; c: $p < 0.001$)

attributes of community composition did not fully recover during the simulation period. This indicates that resting or at least reduced grazing intensity becomes increasingly important with increasing drought duration.

16.6.2 What Did We Learn from Observations, Experiments and Modeling?

The three approaches applied in LLL and SALLnet (observation, experimentation and modeling) revealed clear and consistent responses of savanna rangeland vegetation to drought and grazing. The results included decreases in biomass and productivity, changes in the relative abundances of different PFTs and recovery after drought (Fig. 16.3). However, we found disagreement in some aspects. While productivity in the experiments was relatively resistant during the first two years of drought and showed accelerating decreases in the 6-year drought, modeled biomass and productivity decreased only during the first two drought years and afterward stabilized at a lower level. We explain this model behavior by changes in resource availability, competition and community assembly. In the aDGVM2, soil water dynamics are only simulated at the 1-ha plot-level using a simple multilayer soil water bucket model instead of simulation at patch or plant level. Therefore, reduced vegetation cover due to drought mortality improves the water availability of all surviving individuals by lowering the water demand at plot level. In contrast, soil water status in the experiments and observations varies spatially due to microscale heterogeneity of soil conditions and topography.

Moreover, the aDGVM2 simulated expansion of annual grasses under drought conditions, while field-based methods showed expansion of forbs (Sects. 16.3, 16.4). Forbs are currently not represented in aDGVM2, and the modeled expansion of annual grasses during drought indicates that they fill the ecological niche of forbs in simulations. We argue that forbs need to be added to vegetation models to improve understanding of drought and grazing impacts on savanna rangelands (Siebert and Dreber 2019).

Observations, experiments and modeling showed that vegetation attributes such as biomass or productivity can recover after short drought periods, indicating resilience (Fig. 16.3). Model results suggest that vegetation can also recover after the 6-year drought (Fig. 16.6), hinting at resilience after longer drought periods. However, the rate and duration of recovery differ between vegetation attributes. In particular, community-related attributes show considerable delays when recovering after grazing or drought, primarily due to delayed community assembly and recolonization. How these model results agree with observed vegetation dynamics remains open because results related to recovery from the 6-year drought experiments are not yet available. Regular monitoring of the field sites during the following years would provide valuable information on vegetation recovery and the predictive power of the aDGVM2.

Within LLL and SALLnet, observations, experiments and modeling independently have provided valuable insights into ecosystem resistance and resilience to drought and grazing. However, added value emerged when all three activities were combined and informed each other. DroughtAct experiments allowed in-depth investigation of specific aspects that have been identified in observations, for example, the role of forbs in mitigating drought effects. Experiments allowed direct control and manipulation of environmental drivers, for example, precipitation reduction via rainout shelters, soil moisture control via trenching of experimental plots, or regulation of soil nutrient levels via fertilizer application. In observational studies, such insights are only possible by space-for-time substitution (Blois et al. 2013) or if all treatment combinations are observable. Observational studies usually lack predrought data, as drought events are not planned or easily predictable. Such predrought data are available in controlled experiments.

Field-based methods informed model development. Comparing model results with observational data revealed key aspects not captured by aDGVM2 and led to stepwise model improvement. For example, such comparisons motivated the implementation of annual and perennial grasses as distinct PFTs (Pfeiffer et al. 2019), adjustments of perennial-grass mortality during drought, consideration of buffering effects due to storage reserves during the first drought year and adjustments of postdrought colonization of bare ground (this study, Behn et al. 2022). We identified the representation of water sharing among plant individuals as a model limitation that may lead to overestimates in drought resistance once the number of plant individuals has declined due to drought mortality. A refined soil water scheme that explicitly simulates plant water availability at grass patch level may be required to reproduce the progressive decline of biomass and productivity during the 6-year drought observed in the experiment. We also found lacking forbs as functional types in the model likely influences modeled community response to drought. Using knowledge on the function and ecology of forbs gained in observations and experiments is therefore essential for future model development.

In return, the model allowed prognostic extrapolation of postdrought vegetation recovery during the 2020s. Such model-based extrapolations contribute toward synergies between observations, experiments and models, and they can provide added value to the conclusions drawn from experimental results (Behn et al. 2022). Model results can generate new hypotheses for future experimental studies. For example, we used the aDGVM2 to simulate all factorial treatment combinations realized in the DroughtAct experiment. We could conduct further simulations for factorial combinations not considered in the experiment, test additional management options, rotational grazing systems and climate change scenarios, or investigate how repeated drought affects rangelands with drought history.

16.6.3 Recommendations for Decision-Makers

Based on evidence from three different approaches, we derived the following conclusions and recommendations:

1. *Grasslands and savanna rangelands are susceptible to climate change, particularly to drought, and overgrazing.* Impacts include woody encroachment and transitions to woody vegetation states, or, if heavily utilized, to declining forage quality and availability, increased erosion, and transitions to degraded

states. Contrastingly, dead biomass accumulation resulting from underutilization reduces rangeland productivity.

2. *Recovery of productivity, biomass and community composition is possible after drought and grazing, but related to drought duration, grazing intensity and precipitation.* We found rangeland resilience to short drought or low grazing, and quick recovery after such disturbances. In contrast, long drought or intense grazing requires recovery periods of at least two years, particularly if plant community composition has changed.

3. *Resting times without grazing or very low grazing intensity are necessary to allow the recovery of perennial grasses.* Resting times of 1–2 years with sufficient precipitation are a "window of opportunity of regeneration" (Linstädter et al. 2014), and ensure long-term forage quality for grazers. Resting times of more than two years may be required in grazing systems affected by longer droughts. Rotational grazing systems or forage supply during and after drought can be applied to manage resting times. The holistic management approach (Savory and Butterfield 2016) could provide a template for grazing and resting schemes.

4. *Forbs have high value to secure essential ecosystem functions and services within rangeland systems and need consideration in management.*

5. *Capacity building is crucial to make results available to relevant stakeholders.* It should include training courses on field methods or modeling, stakeholder workshops and contributions to regional or global initiatives such as the South African Risk and Vulnerability Atlas, IPBES or IPCC.

16.7   Outlook

By replicating the DroughtAct experiments with aDGVM2 and comparing model results to the experiment, we identified several opportunities for improving the aDGVM2 for rangelands. As highlighted in Sect. 16.6, improvements should focus on the soil water model. A fine-scale representation of the soil water status allows more precise simulations of plant water availability and drought response. Additionally, aDGVM2 should be improved to capture the diversity of vegetation. Future model development could include more detailed PFTs in the herbaceous layer, such as increasers or decreasers, grass types propagating by stolons or rhizomes, geophytes, xerophytes, succulents or perennial forbs. Adaptations to disturbances such as a bud bank, water- and carbon-storage organs or chemical defenses could be included (Archibald et al. 2019; Pausas et al. 2018). The representation of the woody layer could be improved by including protection against herbivory such as thorns, spines or a cage architecture (Charles-Dominique et al. 2017; Wigley et al. 2018).

In addition, modeling animal behavior and plant-animal interactions in more detail is desirable. This includes, for example, utilization of observed timing and duration of animal presence and absence at a study site instead of using a probabilistic approach. Model development could improve the representation of herbivore selection of particular plants, the feeding duration and amount of biomass removed per plant, and the average number of plants affected per day based on observations. Rangelands are often not exclusively utilized by domesticated animals but also by wild animals. Therefore, aDGVM2 could include different herbivore functional types (Hempson et al. 2015a) with distinctive forage preferences. Finally, aDGVM2 currently does not simulate population dynamics and movement of animals. These processes are particularly relevant when modeling wild animals. Reproduction or mortality of animals could be included and related to their nutritional status using previous work (Pachzelt et al. 2013). DVMs could be coupled with agent-based models to simulate animal movement (Clemen et al. 2021; Fust and Schlecht 2018; Tang and Bennett 2010) and to simulate behavior and decision-making of pastoralists. For instance, the DECUMA household model has previously been coupled to the SAVANNA ecosystem model to study drought impacts on socioeconomic systems in Kenya (Boone et al. 2011; Boone and Galvin 2014).

Modeling drought impacts requires consideration of vegetation resistance to drought, recovery and drought-induced mortality. Modeling plant mortality is still challenging and a significant source of uncertainty in vegetation models (Hartmann et al. 2018). Ecophysiological processes, plant traits, or plant trait syndromes affect drought resistance and mortality (Sankaran 2019), and detailed knowledge of these aspects is required. A cascade of processes describes the establishment niche after disturbance (Holt 2009), including recovery of individual plants, colonization of bare ground, establishment and succession, and drives postdrought recovery. Different simulated recovery rates of grass biomass at plot- and patch-level and changes in the fractions of annual and perennial grasses after drought indicate that such processes are, to a certain degree, represented by aDGVM2. However, DVMs typically operate at large spatial scales and thus ignore small-scale processes such as seed dispersal, dynamics of the seed bank and colonization. Drought conditions can influence seed dynamics and decouple species represented in the seed bank from species in the aboveground community (Basto et al. 2018), and seed mortality increased under drought in a watering experiment (Harrison and LaForgia 2019). While those studies were not conducted in savanna rangeland systems, observations and experiments such as DroughtAct can provide information on recovery and the establishment niche. This information is valuable for parameterizing recovery and successional effects in models. In addition to increased drought risk, vegetation will experience higher atmospheric CO_2 under future conditions. Elevated CO_2 may fertilize plants, particularly growth of woody vegetation, and thereby modify competitive interactions between grasses, forbs and woody vegetation in savanna rangelands (Midgley and Bond 2015). Elevated CO_2 might mitigate drought impacts via increased water use efficiency and enhance recovery after drought.

Our analyses focused on grazing and drought effects at a single study site. We argue that more holistic study approaches at the landscape or regional level are required to account for the multifunctionality of southern African landscapes (Rötter et al. 2021). Rural areas in southern Africa are typically used for cattle grazing, fuelwood collection and crop production. Cattle connect rangeland and cropland when feeding on crop residues or in rotational grazing systems (Pfeiffer et al. 2022). On larger scales, landscapes can be mosaics of rural areas, commercial farms, plantations, conservation areas and game reserves. Management policies need to consider these diverse land-use forms and the socioeconomic interests of stakeholders utilizing the natural resources. Combining rangeland models, crop models, economic models and agent-based models that simulate animal behavior and the decision-making of stakeholders can provide a valuable tool for decision support and the development of regional-level management policies.

The high environmental variability in savanna rangelands and the increasing likelihood of droughts make it challenging to derive management policies that ensure the sustained availability of invaluable ecosystem functions and services. Droughts can amplify the effects of grazing on vegetation and increase the risk that vegetation shifts into degraded states with low forage quantity and quality. More model simulations are required to better assess the impacts of repeated drought and different grazing strategies. Such simulations can provide insights on minimum or optimal resting times between grazing treatments or optimal resting times after drought periods of different length and therefore account for the fact that resting times of several years as assumed in our simulations might not be applicable in reality. Combining observations, experiments and models can help develop early warning signals that indicate the risk of degradation. Such indicators can then inform the developers of management intervention strategies to keep the socioecological system in a safe operating space. A model well parameterized and tested with observational and experimental data can systematically simulate a large ensemble of different management and drought scenarios. These model results can then help estimate the risk of undesirable vegetation change and degradation under different scenarios.

16.8   Conclusions

In the LLL and SALLnet projects, we combined observations, experiments and models to understand the response of rangeland vegetation to drought and grazing. The approaches consistently showed that drought causes substantial losses in biomass and productivity of grasses and shifts in community composition from the dominance of perennials to annuals or forbs. However, we did not find tipping point behavior and irreversible transitions to alternative vegetation states that persisted after drought. Rangelands were able to recover from drought and grazing impacts, in model results even after a 6-year drought period. However, in the model results, some vegetation attributes did not fully recover to the vegetation state simulated under no-drought conditions until 2030. We highlight the following conclusions:

1. Resting times are necessary to allow vegetation recovery after grazing and drought and to ensure continued provision of essential ecosystem services to people. Recovery rates and resting times depend on the length and intensity of grazing and drought.
2. Combining observations, experiments and models is essential to understand rangeland ecology and to forecast impacts of future climate change and land-

use on rangeland vegetation. Using such an integrated approach, we were able to derive policy-relevant recommendations based on evidence from three approaches (see Sect. 16.6).

3. Further research is required, in particular, to understand drought impacts on the complex interactions between woody vegetation, grasses and forbs, and to understand how plant traits influence resilience to grazing and drought.
4. Based on our methods, we can develop early warning signals that indicate potential undesired vegetation shifts and the necessity of management intervention.

Acknowledgments The "Limpopo Living Landscapes (LLL)" and "South African Limpopo Landscapes Network (SALLnet)" projects are part of the SPACES/SPACES II programs (Science Partnerships for the Adaptation to Complex Earth System Processes in Southern Africa), funded by the BMBF (German Federal Ministry of Education and Research, grants 01LL1304B, 01LL1304D, 01LL1802B and 01LL1802C) under FONA (Research for Sustainable Development).

References

Archibald S, Hempson GP, Lehmann C (2019) A unified framework for plant life-history strategies shaped by fire and herbivory. New Phytol 224:1490–1503. https://doi.org/10.1111/nph.15986

Arena G, van der Merwe H, Todd SW, Pauw MJ, Milton SJ, Dean WRJ, Henschel JR (2018) Reflections, applications and future directions of long-term ecological research at Tierberg. Afr J Range Forage Sci 35:257–265. https://doi.org/10.2989/10220119.2018.1513072

Basto S, Thompson K, Grime JP, Fridley JD, Calhim S, Askew AP, Rees M (2018) Severe effects of long-term drought on calcareous grassland seed banks. Npj Clim Atmospheric Sci 1:1–7. https://doi.org/10.1038/s41612-017-0007-3

Baudena M, D'Andrea F, Provenzale A (2010) An idealized model for tree-grass coexistence in savannas: the role of life stage structure and fire disturbances. J Ecol 98:74–80

Behn K, Pfeiffer M, Mokoka V, Mudongo, E, Ruppert J, Scheiter S, Ayisi K, Linstädter A (2022): Assessing the relevance of drought duration on dryland rangelands: an experimental and modelling study. Tropentag 2022, Prague, Czech Republic, 14–16 September 2022 (Poster).https://doi.org/10.13140/RG.2.2.24307.09768

Blois JL, Williams JW, Fitzpatrick MC, Jackson ST, Ferrier S (2013) Space can substitute for time in predicting climate-change effects on biodiversity. Proc Natl Acad Sci 110:9374–9379. https://doi.org/10.1073/pnas.1220228110

Boone RB, Galvin KA (2014) Simulation as an approach to social-ecological integration, with an emphasis on agent-based modeling. In: Manfredo MJ, Vaske JJ, Rechkemmer A, Duke EA (eds) Understanding society and natural resources: forging new strands of integration across the social sciences. Springer Netherlands, Dordrecht, pp 179–202. https://doi.org/10.1007/978-94-017-8959-2_9

Boone R, Galvin K, Burn Silver S, Thornton P, Ojima D, Jawson J (2011) Using coupled simulation models to link pastoral decision making and ecosystem services. Ecol Soc 16. https://doi.org/10.5751/ES-04035-160206

Buisson E, Le Stradic S, Silveira FAO, Durigan G, Overbeck GE, Fidelis A, Fernandes GW, Bond WJ, Hermann J-M, Mahy G, Alvarado ST, Zaloumis NP, Veldman JW (2019) Resilience and restoration of tropical and subtropical grasslands, savannas, and grassy woodlands. Biol Rev 94:590–609. https://doi.org/10.1111/brv.12470

Buitenwerf R, Swemmer AM, Peel MJS (2011) Long-term dynamics of herbaceous vegetation structure and composition in two African savanna reserves. J Appl Ecol 48:238–246. https://doi.org/10.1111/j.1365-2664.2010.01895.x

Camp KGT, Hardy MB (1999) Veld condition assessment. In: Hardy MB, Hurt CR, Camp KGT, Smith JMB, Tainton NM (eds) Veld in KwaZulu-Natal. KwaZulu-Natal. Department of Agriculture and Environmental Affairs, Cedara, pp 18–31

Charles-Dominique T, Barczi J-F, Roux EL, Chamaille-Jammes S, McArthur C (2017) The architectural design of trees protects them against large herbivores. Funct Ecol 31:1710–1717. https://doi.org/10.1111/1365-2435.12876

Cherwin K, Knapp A (2012) Unexpected patterns of sensitivity to drought in three semi-arid grasslands. Oecologia 169:845–852. https://doi.org/10.1007/s00442-011-2235-2

Clemen T, Lenfers UA, Dybulla J, Ferreira SM, Kiker GA, Martens C, Scheiter S (2021) A cross-scale modeling framework for decision support on elephant management in Kruger National Park, South Africa. Ecol Inform 62:101266. https://doi.org/10.1016/j.ecoinf.2021.101266

Díaz S, Lavorel S, McINTYRE S, Falczuk V, Casanoves F, Milchunas DG, Skarpe C, Rusch G, Sternberg M, Noy-Meir I, Landsberg J, Zhang W, Clark H, Campbell BD (2007) Plant trait responses to grazing – a global synthesis. Glob Chang Biol 13:313–341. https://doi.org/10.1111/j.1365-2486.2006.01288.x

Du Toit JT, Cumming DHM (1999) Functional significance of ungulate diversity in African savannas and the ecological implications of the spread of pastoralism. Biodivers Conserv 8:1643–1661. https://doi.org/10.1023/A:1008959721342

Dufresne J-L, Foujols M-A, Denvil S, Caubel A, Marti O, Aumont O, Balkanski Y, Bekki S, Bellenger H, Benshila R, Bony S, Bopp L, Braconnot P, Brockmann P, Cadule P, Cheruy F, Codron F, Cozic A, Cugnet D, de Noblet N, Duvel J-P, Ethé C, Fairhead L, Fichefet T, Flavoni S, Friedlingstein P, Grandpeix J-Y, Guez L, Guilyardi E, Hauglustaine D, Hourdin F, Idelkadi A, Ghattas J, Joussaume S, Kageyama M, Krinner G, Labetoulle S, Lahellec A, Lefebvre M-P, Lefevre F, Levy C, Li ZX, Lloyd J, Lott F, Madec G, Mancip M, Marchand M, Masson S, Meurdesoif Y, Mignot J, Musat I, Parouty S, Polcher J, Rio C, Schulz M, Swingedouw D, Szopa S, Talandier C, Terray P, Viovy N, Vuichard N (2013) Climate change projections using the IPSL-CM5 earth system model: from CMIP3 to CMIP5. Clim Dyn 40:2123–2165. https://doi.org/10.1007/s00382-012-1636-1

Ellis SM, Steyn HS (2003) Practical significance (effect sizes) versus or in combination with statistical significance (p-values): research note. Manag Dyn J South Afr Inst Manag Sci 12:51–53. https://doi.org/10.10520/EJC69666

Ferner J, Schmidtlein S, Guuroh RT, Lopatin J, Linstädter A (2018) Disentangling effects of climate and land-use change on West African drylands' forage supply. Glob Environ Chang 53:24–38. https://doi.org/10.1016/j.gloenvcha.2018.08.007

Fust P, Schlecht E (2018) Integrating spatio-temporal variation in resource availability and herbivore movements into rangeland management: RaMDry—an agent-based model on livestock feeding ecology in a dynamic, heterogeneous, semi-arid environment. Ecol Model 369:13–41. https://doi.org/10.1016/j.ecolmodel.2017.10.017

Fynn RWS, O'Connor TG (2000) Effect of stocking rate and rainfall on rangeland dynamics and cattle performance in a semi-arid savanna, South Africa. J Appl Ecol 37:491–507. https://doi.org/10.1046/j.1365-2664.2000.00513.x

Gaillard C, Langan L, Pfeiffer M, Kumar D, Martens C, Higgins SI, Scheiter S (2018) African shrub distribution emerges via height - sapwood conductivity trade-off. J Biogeogr 45:2815–2826

Gbetibouo GA (2009) Understanding farmers perceptions and adaptations to climate change and variability: the case of the Limpopo Basin farmers South Africa. IFPRI Discussion Paper 849. IFPRI, Washington, DC

Harrison S, LaForgia M (2019) Seedling traits predict drought-induced mortality linked to diversity loss. Proc Natl Acad Sci 116:5576–5581. https://doi.org/10.1073/pnas.1818543116

Hartmann H, Moura CF, Anderegg WRL, Ruehr NK, Salmon Y, Allen CD, Arndt SK, Breshears DD, Davi H, Galbraith D, Ruthrof KX, Wunder J, Adams HD, Bloemen J, Cailleret M, Cobb R, Gessler A, Grams TEE, Jansen S, Kautz M, Lloret F, O'Brien M (2018) Research frontiers for improving our understanding of drought-induced tree and forest mortality. New Phytol 218:15–28. https://doi.org/10.1111/nph.15048

Hempson GP, Archibald S, Bond WJ (2015a) A continent-wide assessment of the form and intensity of large mammal herbivory in Africa, vol 350. Science, p 1056

Hempson GP, Archibald S, Bond WJ, Ellis RP, Grant CC, Kruger FJ, Kruger LM, Moxley C, Owen-Smith N, Peel MJS, Smit IPJ, Vickers KJ (2015b) Ecology of grazing lawns in Africa. Biol Rev 90:979–994. https://doi.org/10.1111/brv.12145

Higgins SI, Scheiter S (2012) Atmospheric CO_2 forces abrupt vegetation shifts locally, but not globally. Nature 488:209–212

Holt RD (2009) Bringing the Hutchinsonian niche into the 21st century: ecological and evolutionary perspectives. Proc Natl Acad Sci 106:19659–19665. https://doi.org/10.1073/pnas.0905137106

IPCC (2021) Climate change 2021: the physical science basis. Contribution of working group I to the sixth assessment report of the Intergovernmental Panel on Climate Change. Cambridge University Press, Cambridge

Kalvelage L, Revilla Diez J, Bollig M (2020) How much remains? Local value capture from tourism in Zambezi, Namibia. Tour Geogr 1:1–22. https://doi.org/10.1080/14616688.2020.1786154

Klem J (2018) Drought responses of forb and grass communities in communal and protected rangelands. MSc dissertation, North-West University, South Africa

Knapp AK, Ciais P, Smith MD (2017) Reconciling inconsistencies in precipitation–productivity relationships: implications for climate change. New Phytol 214:41–47. https://doi.org/10.1111/nph.14381

Kuckertz P, Ullrich O, Linstädter A, Speckenmeyer E (2011) Agent-based modeling and simulation of a Pastoralnomadic land use system. Simul Notes Eur 21:147–152

Lange, S., 2019. EartH2Observe, WFDEI and ERA-interim data merged and bias-corrected for ISIMIP (EWEMBI). V. 1.1. GFZ Data Services

Lehohla P (2012) Census 2011 Census in brief. Statistics South Africa

Linstädter A, Schellberg J, Brüser K, García CAM, Oomen RJ, du Preez CC, Ruppert JC, Ewert F (2014) Are there consistent grazing indicators in drylands? Testing plant functional types of various complexity in South Africa's grassland and savanna biomes. PLoS One 9:e104672. https://doi.org/10.1371/journal.pone.0104672

Mackie KA, Zeiter M, Bloor JMG, Stampfli A (2019) Plant functional groups mediate drought resistance and recovery in a multisite grassland experiment. J Ecol 107:937–949. https://doi.org/10.1111/1365-2745.13102

Maluleke, M., Malungani, T., Steenkamp, K. (Eds.), 2016. Limpopo environment outlook report 2016. Limpopo Provincial Government, Department of Economic Development, Environment and Tourism

Marchant R (2010) Understanding complexity in savannas: climate, biodiversity and people. Curr Opin Environ Sustain 2:101–108. https://doi.org/10.1016/j.cosust.2010.03.001

Martens C, Hickler T, Davis-Reddy C, Engelbrecht F, Higgins SI, von Maltitz GP, Midgley GF, Pfeiffer M, Scheiter S (2021) Large uncertainties in future biome changes in Africa call for flexible climate adaptation strategies. Glob Chang Biol 27:340–358. https://doi.org/10.1111/gcb.15390

Matsika R, Erasmus BFN, Twine WC (2013) Double jeopardy: the dichotomy of fuelwood use in rural South Africa. Energy Policy 52:716–725. https://doi.org/10.1016/j.enpol.2012.10.030

Midgley GF, Bond WJ (2015) Future of African terrestrial biodiversity and ecosystems under anthropogenic climate change. Nat Clim Chang 5:823–829

Minnaar, C., 2020. Drought effects on the herbaceous community structure of transformed Mopaneveld. MSc dissertation, North-West University

Munjonji L, Ayisi KK, Mudongo EI, Mafeo TP, Behn K, Mokoka MV, Linstädter A (2020) Disentangling drought and grazing effects on soil carbon stocks and CO2 fluxes in a semi-arid African savanna. Front Environ Sci 8:207. https://doi.org/10.3389/fenvs.2020.590665

New M, Lister D, Hulme M, Makin I (2002) A high-resolution data set of surface climate over global land areas. Clim Res 21:1–25

O'Connor TG (1991) Patch colonisation in a savanna grassland. J Veg Sci 2:245–254. https://doi.org/10.2307/3235957

O'Connor TG (1995) Transformation of a savanna grassland by drought and grazing. Afr J Range Forage Sci 12:53–60. https://doi.org/10.1080/10220119.1995.9647864

Oomen RJ, Ewert F, Snyman HA (2016) Modelling rangeland productivity in response to degradation in a semi-arid climate. Ecol Model 322:54–70. https://doi.org/10.1016/j.ecolmodel.2015.11.001

Pachzelt A, Rammig A, Higgins SI, Hickler T (2013) Coupling a physiological grazer population model with a generalized model for vegetation dynamics. Ecol Model 263:92–102. https://doi.org/10.1016/j.ecolmodel.2013.04.025

Parsons DAB, Shackleton CM, Scholes RJ (1997) Changes in herbaceous layer condition under contrasting land use systems in the semi-arid lowveld, South Africa. J Arid Environ 37:319–329. https://doi.org/10.1006/jare.1997.0283

Pausas JG, Bond WJ (2020) Alternative biome states in terrestrial ecosystems. Trends Plant Sci 25:250–263. https://doi.org/10.1016/j.tplants.2019.11.003

Pausas JG, Lamont BB, Paula S, Appezzato-da-Glória B, Fidelis A (2018) Unearthing below-ground bud banks in fire-prone ecosystems. New Phytol 217:1435–1448. https://doi.org/10.1111/nph.14982

Pfeiffer M, Langan L, Linstädter A, Martens C, Gaillard C, Ruppert JC, Higgins SI, Mudongo EI, Scheiter S (2019) Grazing and aridity reduce perennial grass abundance in semi-arid rangelands – insights from a trait-based dynamic vegetation model. Ecol Model 395:11–22. https://doi.org/10.1016/j.ecolmodel.2018.12.013

Pfeiffer M, Kumar D, Martens C, Scheiter S (2020) Climate change will cause non-analogue vegetation states in Africa and commit vegetation to long-term change. Biogeosciences 17:5829–5847

Pfeiffer M, Hoffmann MP, Scheiter S, Nelson W, Isselstein J, Ayisi K, Odhiambo JJ, Rötter R (2022) Modeling the effects of alternative crop-livestock management scenarios on important ecosystem services for smallholder farming from a landscape perspective, Biogeosciences, 19:3935–3958. https://doi.org/10.5194/bg-19-3935-2022

Pool-Stanvliet R (2013) A history of the UNESCO man and the biosphere Programme in South Africa. South Afr J Sci 109. https://doi.org/10.1590/sajs.2013/a0035

Rötter RP, Scheiter S, Hoffmann MP, Pfeiffer M, Nelson WCD, Ayisi K, Taylor P, Feil J-H, Bakhsh SY, Isselstein J, Lindstädter A, Behn K, Westphal C, Odhiambo J, Twine W, Grass I, Merante P, Bracho-Mujica G, Bringhenti T, Lamega S, Abdulai I, Lam QD, Anders M, Linden V, Weier S, Foord S, Erasmus B (2021) Modeling the multi-functionality of African savanna landscapes under global change. Land Degrad Dev 32:2077–2081. https://doi.org/10.1002/ldr.3925

Ruppert JC, Linstädter A (2014) Convergence between ANPP estimation methods in grasslands — a practical solution to the comparability dilemma. Ecol Indic 36:524–531. https://doi.org/10.1016/j.ecolind.2013.09.008

Ruppert JC, Harmoney K, Henkin Z, Snyman HA, Sternberg M, Willms W, Linstädter A (2015) Quantifying drylands' drought resistance and recovery: the importance of drought intensity, dominant life history and grazing regime. Glob Chang Biol 21:1258–1270. https://doi.org/10.1111/gcb.12777

Sankaran M (2019) Droughts and the ecological future of tropical savanna vegetation. J Ecol 107:1531–1549. https://doi.org/10.1111/1365-2745.13195

Sankaran M, Hanan NP, Scholes RJ, Ratnam J, Augustine DJ, Cade BS, Gignoux J, Higgins SI, Roux XL, Ludwig F, Ardo J, Banyikwa F, Bronn A, Bucini G, Caylor KK, Coughenour MB, Diouf A, Ekaya W, Feral CJ, February EC, Frost PGH, Hiernaux P, Hrabar H, Metzger KL, Prins HHT, Ringrose S, Sea W, Tews J, Worden J, Zambatis N (2005) Determinants of woody cover in African savannas. Nature 438:846–849. https://doi.org/10.1038/nature04070

Savory A, Butterfield J (2016) Holistic management: a commonsense revolution to restore our environment, 3rd edn. Island Press, Washington, DC

Scheiter S, Higgins SI (2007) Partitioning of root and shoot competition and the stability of savannas. Am Nat 170:587–601

Scheiter S, Higgins SI (2009) Impacts of climate change on the vegetation of Africa: an adaptive dynamic vegetation modelling approach (aDGVM). Glob Chang Biol 15:2224–2246

Scheiter S, Savadogo P (2016) Ecosystem management can mitigate vegetation shifts induced by climate change in West Africa. Ecol Model 332:19–27. https://doi.org/10.1016/j.ecolmodel.2016.03.022

Scheiter S, Langan L, Higgins SI (2013) Next generation dynamic global vegetation models: learning from community ecology. New Phytol 198:957–969. https://doi.org/10.1111/nph.12210

Scheiter S, Gaillard C, Martens C, Erasmus BFN, Pfeiffer M (2018) How vulnerable are ecosystems in the Limpopo province to climate change? South Afr J Bot 116:86–95. https://doi.org/10.1016/j.sajb.2018.02.394

Scheiter S, Schulte J, Pfeiffer M, Martens C, Erasmus BFN, Twine WC (2019) How does climate change influence the economic value of ecosystem Services in Savanna Rangelands? Ecol Econ 157:342–356. https://doi.org/10.1016/j.ecolecon.2018.11.015

Scheiter S, Moncrieff GR, Pfeiffer M, Higgins SI (2020) African biomes are most sensitive to changes in CO_2 under recent and near-future CO_2 conditions. Biogeosciences 17:1147–1167

Shackleton CM, Shackleton SE, Netshiluvhi TR, Mathabela FR (2005) The contribution and direct-use value of livestock to rural livelihoods in the Sand River catchment, South Africa. Afr J Range Forage Sci 22:127–140

Siebert F, Dreber N (2019) Forb ecology research in dry African savannas: knowledge, gaps, and future perspectives. Ecol Evol 9:7875–7891. https://doi.org/10.1002/ece3.5307

Siebert F, Klem J, Van Coller H (2020) Forb community responses to an extensive drought in two contrasting land-use types of a semi-arid Lowveld savanna. Afr J Range Forage Sci 37:53–64. https://doi.org/10.2989/10220119.2020.1726464

Stats SA (2021) Statistical release P0302: mid-year population estimates 2021

Staver AC, Archibald S, Levin S (2011) Tree cover in sub-Saharan Africa: rainfall and fire constrain forest and savanna as alternative stable states. Ecology 92:1063–1072

Stevens N, Lehmann CER, Murphy BP, Durigan G (2017) Savanna woody encroachment is widespread across three continents. Glob Chang Biol 23:235–244. https://doi.org/10.1111/gcb.13409

Tang W, Bennett DA (2010) Agent-based modeling of animal movement: a review. Geogr Compass 4:682–700. https://doi.org/10.1111/j.1749-8198.2010.00337.x

Tielbörger K, Bilton MC, Metz J, Kigel J, Holzapfel C, Lebrija-Trejos E, Konsens I, Parag HA, Sternberg M (2014) Middle-eastern plant communities tolerate 9 years of drought in a multi-site climate manipulation experiment. Nat Commun 5:5102. https://doi.org/10.1038/ncomms6102

Twine WC, Holdo RM (2016) Fuelwood sustainability revisited: integrating size structure and resprouting into a spatially realistic fuelshed model. J Appl Ecol 53:1766–1776. https://doi.org/10.1111/1365-2664.12713

Twine W, Moshe D, Netshiluvhi T, Siphugu V (2003) Consumption and direct-use values of savanna bio-resources used by rural households in Mametja, a semi-arid area of Limpopo province, South Africa. South Afr J Sci 99:467–473

van Langevelde F, van de Vijver CADM, Kumar L, van de Koppel J, de Ridder N, van Andel J, Skidmore AK, Hearne JW, Stroosnijder L, Bond WJ, Prins HH, Rietkerk MM (2003) Effects of fire and herbivory on the stability of savanna ecosystems. Ecology 84:337–350

Van Staden N (2016) Herbaceous species diversity, redundancy and resilience of Mopaneveld across different land-uses. MSc dissertation, North-West University

van Vuuren DP, Edmonds J, Kainuma M, Riahi K, Thomson A, Hibbard K, Hurtt GC, Kram T, Krey V, Lamarque J-F, Masui T, Meinshausen M, Nakicenovic N, Smith SJ, Rose SK (2011) The representative concentration pathways: an overview. Clim Chang 109:5–31. https://doi.org/10.1007/s10584-011-0148-z

Venter ZS, Hawkins H-J, Cramer MD (2017) Implications of historical interactions between herbivory and fire for rangeland management in African savannas. Ecosphere 8:e01946. https://doi.org/10.1002/ecs2.1946

Venter ZS, Cramer MD, Hawkins H-J (2018) Drivers of woody plant encroachment over Africa. Nat Commun 9:2272. https://doi.org/10.1038/s41467-018-04616-8

Wigley BJ, Fritz H, Coetsee C (2018) Defence strategies in African savanna trees. Oecologia 187:797–809. https://doi.org/10.1007/s00442-018-4165-8

Wigley-Coetsee C, Staver A (2020) Grass community responses to drought in an African savanna. Afr J Range Forage Sci 37:43–52. https://doi.org/10.2989/10220119.2020.1716072

Yahdjian L, Sala OE (2002) A rainout shelter design for intercepting different amounts of rainfall. Oecologia 133:95–101. https://doi.org/10.1007/s00442-002-1024-3

A Fine Line Between Carbon Source and Sink: Potential CO_2 Sequestration through Sustainable Grazing Management in the Nama-Karoo

17

Oksana Rybchak, Justin duToit, Amukelani Maluleke, Mari Bieri, Guy F. Midgley (ID), Gregor Feig, and Christian Brümmer (ID)

Abstract

Semiarid South African ecosystems are managed for livestock production with different practices and intensities. Many studies have found grazing to be an important driver of vegetation change; however, its impacts on carbon fluxes remain poorly studied. Unsustainable management over the past 200 years has led to an increase of degraded areas and a reduction in species diversity, but destocking trends in the past three decades may be facilitating a recovery of net primary productivity and vegetation cover in some areas. This chapter provides a brief historical overview on livestock management practices and their likely impact on carbon exchange in the Nama-Karoo Biome. We present a case study based on five years of eddy covariance measurements, in which effects of past and current livestock grazing on CO_2 exchange were studied. Two sites with different livestock management but similar climatic conditions formed the basis

O. Rybchak (✉) · M. Bieri · C. Brümmer
Thünen Institute of Climate-Smart Agriculture, Braunschweig, Germany
e-mail: oksana.rybchak@thuenen.de; mari.bieri@thuenen.de; christian.bruemmer@thuenen.de

J. du Toit
Grootfontein Agricultural Development Institute, Middelburg, South Africa

A. Maluleke · G. F. Midgley
School for Climate Studies and Department of Botany and Zoology, Stellenbosch University, Stellenbosch, South Africa
e-mail: amukelani.maluleke@thuenen.de; gfmidgley@sun.ac.za

G. Feig
Department of Geography, Geoinformatics and Meteorology, University of Pretoria, Pretoria, South Africa

South African Environmental Observation Network, Colbyn, Pretoria, South Africa
e-mail: gregor@saeon.ac.za

© The Author(s) 2024
G. P. von Maltitz et al. (eds.), *Sustainability of Southern African Ecosystems under Global Change*, Ecological Studies 248,
https://doi.org/10.1007/978-3-031-10948-5_17

471

for this preliminary effort to improve the understanding of carbon exchange and its drivers under contrasting management regimes. The case study revealed that net CO_2 exchange is near-neutral over an annual scale, with precipitation distribution emerging as the main controlling factor of subannual variance. Although CO_2 release at the lenient grazing site was slightly higher than at the experimental grazing site, longer time series are likely needed in such variable ecosystems to make a pronouncement regarding long-term net fluxes. Given their vast extent, livestock rangelands may have an important effect on regional carbon balance.

17.1 Livestock Grazing Systems in the Nama-Karoo

The Nama-Karoo is the third largest biome in South Africa, characterized by vast, flat plains in the lower Karoo ($\sim$ 800 m.a.s.l.) and rugged reliefs in the upper Karoo ($\sim$ 1300 m.a.s.l.) with shallow, weakly developed lime-rich soils (Dean and Milton 1999; Mucina et al. 2006). It is dominated by shrubs, grasses, herbs and geophytes. The major disturbance to vegetation in the Nama-Karoo has been extensive livestock grazing (mainly sheep) restrained within paddock boundaries since the last decades of the nineteenth century (Skead 1982; Archer 2000). Approximately 60% of the Nama-Karoo Biome is characterized by moderately to severely degraded soil and vegetation cover (change in species assemblage) (Mucina et al. 2006). Many research efforts have been focused on understanding the causes of land and vegetation degradation in the Nama-Karoo Biome, and inappropriate grazing management is considered one of the major drivers (Hoffman and Ashwell 2001). Regardless of extensive theoretical background, a fully predictive understanding of the effects of livestock grazing on biodiversity and ecosystem processes in semiarid Nama-Karoo ecosystems remains elusive (Bekele 2001; Tilman et al. 2012; Rutherford and Powrie 2013).

17.1.1 Historical Overview

Springbok (*Antidorcas marsupialis*) was the most numerous native grazer in the Nama-Karoo prior to large-scale human livestock management (Dean and Milton 1999). Namaqua Afrikaner sheep grazing (transhumance [1] or nomadic [2] ranching) has been present in the Karoo for approximately 1800 years among the Khoekhoen people (Truter et al. 2015). In the eighteenth century, the abundance of ungulates in the Karoo was largely determined by surface water availability (Milton 1993;

---

[1] Transhumance livestock ranching is a mobile method of land management based on regular (generally predictable) seasonal movements of livestock along the same paths and rangelands.

[2] Nomadic ranging is characterised by a continuous, irregular and unpredictable movement pattern to find new rangelands in which to graze.

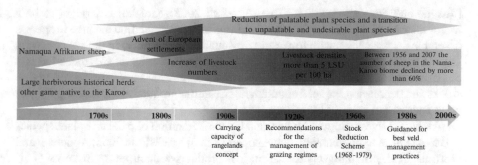

Fig. 17.1 Historical overview of livestock grazing in the Karoo (adapted from Milton 1993; Archer 2000; Truter et al. 2015; Du Toit and O'Connor 2020, Shaw 1873; Acocks 1966, 1979; Roux and Vorster 1983; McLeod 1997; Hoffman et al. 1999)

Owen-Smith and Danckwerts 1997; Archer 2000; Du Toit and O'Connor 2020). In the 1800s, with the advent of European settlers, the number of livestock increased to much higher stocking rates, which negatively affected the vegetation cover in the Nama-Karoo (Fig. 17.1) (Shaw 1873; Archer 2000).

In the 1880s, the concept of rangelands carrying capacity, which determines the number of livestock that can be sustainably kept per unit area, was being implemented as grazing systems in South Africa (Sayre 2008). Farmers used the carrying capacity concept to determine the sustainable stocking rates. The carrying capacity concept was seen as an imposition of foreign land management practices over essentially incompatible indigenous ecosystems (McLeod 1997). By the end of the nineteenth century, large herbivorous historical herds of Springbok had been substantially reduced due to the direct displacement by domestic livestock and reduced grass production (Fig. 17.1) (Hoffman et al. 1999, 2018). By the mid-nineteenth century, the number of sheep had reached about 5 million, and by the 1930s, it had reached 45 million (Van den Berg et al. 2019). By 1904, the density of livestock had increased to the highest recorded level (more than 5 large stock unit (LSU [3]) per 100 ha) and remained high until the late 1960s (Hoffman et al. 1999; Van den Berg et al. 2019). In the early 1920s, most of Nama-Karoo had been transformed into fenced rangelands for livestock grazing (mainly sheep and mohair goats, which favor higher-quality plant species) (Hoffman et al. 1999).

The historical trends in land use and alteration of the grazing regime have been responsible for changes in vegetation cover and species composition (Rutherford and Powrie 2013; Du Toit and O'Connor 2020), which are likely to translate into effects on ecosystem functioning (Susiluoto et al. 2008; Pérez-Hernández and Gavilán 2021). Many farms in the Nama-Karoo Biome have undergone a variety of historical land management practices, regimes and stocking at unsustainable rates

---

[3] Large Stock Unit is a reference unit that facilitates the aggregation of livestock of different species and ages (1 LSU is equivalent to 450 kg beef steer).

(Teague and Dowhower 2003; O'Farrell et al. 2008). Grazing is thought to be a major factor responsible for the perceived vegetation degradation (change in species assemblages to species with lower grazing quality/palatability) in the area (Acocks 1966; Owen-Smith and Danckwerts 1997). Even when animals are stocked at low densities, continuous selective grazing benefits unpalatable species by depleting root stocks and decreasing the seed abundance, size and reproductive success of palatable plant species (Acocks 1979; Milton 1993).

Continued livestock grazing has also led to a reduction of palatable plant species and a transition to unpalatable plant species (Kraaij and Milton 2006; Anderson and Hoffman 2007), making Nama-Karoo less productive than before in terms of its grazing capacity (Tidmarsh 1951; Roux and Theron 1987), with effects amplified during times of below long-term average precipitation (du Toit and O'Connor 2014).

Following a drought period (1900–1915) with mean annual precipitation of 288 mm (significantly (20%) lower than the long-term average of 373 mm), a commission of inquiry disseminated recommendations for the management of livestock grazing, separating it into paddocks (du Toit et al. 1923; du Toit and O'Connor 2014). These recommendations helped to optimize livestock distribution on rangelands with rest periods for paddocks as needed. The final report of the Drought Investigation Commission indicated how livestock grazing could degrade Karoo ecosystems (du Toit et al. 1923). One of the main recommendations was to stop stockading animals at night in the same place and limit livestock to paddocks, where livestock remains for a period of a few weeks to months before moving to the next paddock. Equilibrium-based models of vegetation response to livestock grazing and precipitation provided guidance to landowners for best veld management practices (Roux and Vorster 1983; Moll and Gubb 1989). These models are based on recommendations on the control of livestock grazing regimes (paddocks division) (du Toit et al. 1923) and grazing trials near Middelburg in the eastern Karoo (Tidmarsh 1951). Discussions on the benefits of lenient, heavy, or intermediate grazing resulted in the development of rotational grazing systems.

17.1.2 Current Grazing Practices in the Nama-Karoo

Of the total land area of South Africa, 84% is currently used for extensive grazing, and the majority of the Nama-Karoo ($\sim$ 95%) is commercial rangeland under freehold tenure, mainly used for extensive livestock production (sheep and goats) (Roux et al. 1981; Hoffman et al. 1999). As discussed under the previous section, heavy historical grazing has led to widespread degradation of vegetation in the Nama-Karoo, often to the extent beyond which vegetation composition cannot be naturally restored despite livestock removal (Curtin 2002; Snyman 2003; Anderson and Hoffman 2007). However, this trend of heavy grazing has reversed in many regions in the last few decades (Timm Hoffman et al. 2018). It has recently been concluded that since the significant reduction in the numbers of domestic livestock, more than 95% of the greater Karoo region can be viewed as "natural."

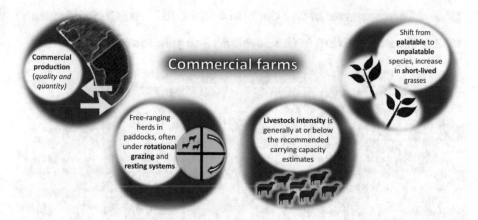

Fig. 17.2 Main characteristics of commercial rangelands system in South Africa (adapted from Hoffman et al. 1999; Todd and Hoffman 1999)

There are two main land tenure systems in operation in South Africa, which differ in land management practices, namely: communal and commercial (more than 80% of land in South Africa) (Hoffman et al. 1999; Todd and Hoffman 1999); in the Nama-Karoo Biome almost the entire area is under commercial livestock production. Commercial rangelands have clear boundaries and exclusive property rights (Fig. 17.2) (Palmer and Ainslie 2005). Management systems can be divided into the following categories: continuous grazing, rotational grazing, short-duration grazing and nonselective grazing system (McCabe 1987). In the continuous grazing system, livestock are constantly kept on one rangeland which is not divided into paddocks, so the stock density should be low, but with selective livestock grazing (for example, small ruminants may be pickier and spend more time looking for high quality palatable plants, while unpalatable species are often discarded) (McCabe 1987; O'Connor et al. 2010). The rotational grazing system (multiple paddocks) was developed and recommended by the Commission of Inquiry to improve rangeland conditions and enhance wildlife habitat through appropriate resting periods (du Toit et al. 1923). The short duration grazing system was advocated based on the untested hypotheses of Acocks (1966) to improve sustained forage and livestock productivity and prevent land degradation (Acocks 1966; Savory 1978; Holechek et al. 2000). The main characteristic of this system is that livestock is rotated through many (≥8) paddocks for a relatively short period of time. Theoretically, it would reduce the competitive advantage of unpalatable species. While discussions regarding the benefits of the different grazing management practices are ongoing (Roberts 1969; Parsons et al. 1983; Westoby et al. 1989), most commercial farms in the Karoo practice some form of rotational grazing with rest periods from several months to more than a year (Hoffman 1988). Outside commercial farms, in, e.g., National Parks, where animals cannot be fenced and rotated, continuous grazing is practiced.

17.2   Components of the Carbon Cycle and Their Quantification

17.2.1 Carbon Cycling in the Semiarid Rangelands of South Africa

Box 17.1 Carbon in Terrestrial Ecosystems
Terrestrial ecosystems contain carbon in the form of plants, animals, soils and microorganisms (bacteria and fungi) (Schimel 1995). Of total carbon, plants and soils account for the largest share. Most of the carbon in terrestrial ecosystems is in organic forms (compounds formed by living organisms such as leaves, roots, dead plant and the organic residues from the decomposition of previous living tissues). The terrestrial carbon cycle includes photosynthesis (carbon sequestration), ecosystem respiration (carbon release) and storage (biomass and soil storage).

This section summarizes the current understanding of the terrestrial carbon cycle (Box 17.1) (with a focus on the factors (such as climatic variables, biodiversity change and livestock grazing) that affect the uptake or release of CO_2 in semiarid rangelands.

The carbon cycle is governed by net ecosystem exchange (NEE), which is the balance of gross primary production (GPP) (photosynthesis) and ecosystem respiration ($R_{eco} = R_a{}^4 + R_h{}^5$) (Box 17.2) (Ciais et al. 2011). Carbon exchange between the plants and the atmosphere occurs through photosynthesis and respiration (Fig. 17.3). During photosynthesis, plants use solar energy to absorb CO_2 from the atmosphere (by diffusion through the stomata) and water from the soil to produce carbohydrates. Most of the absorbed CO_2 is reemitted. The amount of CO_2 converted to carbohydrates during photosynthesis is called gross primary production (GPP). Plants also release CO_2 into the atmosphere through respiration (corresponding to the exhalation of plants); plant cells use carbohydrates produced during photosynthesis as energy. An ecosystem usually absorbs and releases carbon due to these processes taking place simultaneously.

Box 17.2 Ecosystem Respiration (R_{eco})
Ecosystem respiration is the sum of all respiration processes of living organisms in an ecosystem. Ecosystem respiration can be divided into autotrophic respiration (R_a), which comes from organisms with the sun as their main source of energy (i.e., plants), and heterotrophic respiration (R_h) from heterotrophic organisms (e.g., microbial decomposition of residues and soil organic matter), whose main source of energy are other organisms.

---

[4] Autotrophic respiration.

[5] Heterotrophic respiration.

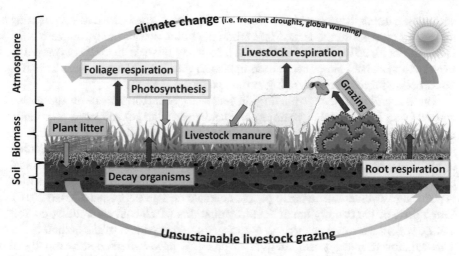

Fig. 17.3 The schematic diagram describes the ecosystem–atmospheric carbon exchange in the semiarid grazing ecosystems (adapted from Schimel 1995; Asner and Archer 2010)

Some of the aboveground biomass is consumed by livestock, and carbon compounds are transferred through the food chain (Spangler et al. 2021). Most of the carbon they consume is exhaled in the form of CO_2 produced by aerobic respiration. Eventually, some of the carbon returns to the soil as manure, which can also increase rangeland yields, thereby improving soil carbon storage (Gross and Glaser 2021), or into the atmosphere through intestinal fermentation (Lee et al. 2017).

Most of the carbon is ultimately re-emitted to the atmospheric carbon pool via heterotrophic and autotrophic respiration. Most of the carbon in the soils comes in the form of dead plants, which is decayed by microorganisms (consumed by bacteria and fungi) during the decomposition process (Gougoulias et al. 2014). Over time, the metabolic processes of microorganisms decompose most of the organic matter into CO_2. The decomposition process releases carbon into the atmosphere at a rate that depends on the chemical composition of the dead tissues and environmental conditions (dry conditions, flooding and low temperatures slow down decomposition) (Bardgett et al. 2008; Paz-Ferreiro et al. 2012). Plant material can take years to decades (large trees) to decompose, and carbon is temporarily stored in soil organic matter.

The effectiveness of the terrestrial absorption of CO_2 depends on the transition of carbon to long-lived forms (i.e., trees and woody shrubs). Management practices could increase the carbon sink potential due to the inertia of these "slow" carbon pools. The difference between R_{eco} and GPP determines how much carbon is released or absorbed by the ecosystem without interferences that removes carbon from the ecosystem, such as harvest or fire (Prentice et al. 2001). This carbon balance can be estimated from changes in carbon stocks or by measuring CO_2 fluxes between the ecosystem and the atmosphere (Sect. 17.2.2). In the absence of disturbance, R_{eco} should balance GPP, and NEE would be zero. Anthropogenic activities, natural disturbances and climate variability change GPP and R_{eco} causing temporary changes in the terrestrial carbon pool and, consequently, nonzero NEE (Williams et al. 2014; Hamidi et al. 2021). A steady increase in GPP is expected to result in a sustained larger net carbon uptake so that increased terrestrial carbon is not processed through the respiring carbon pools.

The African continent is playing an increasingly important role in the global carbon cycle (due to global warming and land cover change) and has a potentially significant impact on climate change (Epple et al. 2016; Simpkin et al. 2020). In general, it is estimated that the African continent is a small carbon sink (Valentini et al. 2014), but due to the lack of long-term CO_2 measurements in many critical ecosystems on the continent, the uncertainty of these estimates is high. South Africa is particularly vulnerable to the impacts of climate change (DEA and SANBI 2016). Large parts of the country are affected by droughts (Archibald et al. 2009; du Toit 2017), which can cause changes in the magnitude and pattern of the carbon cycle. The duration, frequency and strength of droughts have increased substantially in recent decades, particularly in semiarid ecosystems. Droughts could interact with other biotic and abiotic changes (global warming, grazing intensities) and thus could fundamentally alter the function, structure and vegetation composition of ecosystems (Abdulai et al. 2020; Malik et al. 2020). Moreover, droughts are the main cause of interannual variability in the carbon balance of semiarid ecosystems and are closely linked to terrestrial carbon cycling at various scales (Archibald et al. 2009; Merbold et al. 2009). The distribution of sequestered carbon in the semiarid ecosystems of South Africa, as well as their potential capacity to store and sequester carbon in soils, is still uncertain (Brent et al. 2011; Von Maltitz et al. 2020). Compared to other ecosystems, semiarid shrub/grassland ecosystems are more susceptible to subtle environmental and land management changes in terms of function and structure (Milton and Dean 2021). Semiarid grassland ecosystems with high root productivity store most of their carbon in soils, the turnover of which is relatively slow (Leifeld et al. 2015; Mureva et al. 2018). Viewed together, these traits of semiarid shrub/grasslands would emphasize their importance for sustainable land management.

17.2.2 Methods for Quantifying Carbon Uptake and Release

The scarce observation networks in and around the African continent mean that Africa is one of the weakest links in our understanding of the global carbon cycle. CO_2 can be measured over a range of scales, from a small landscape to a global extent (Fig. 17.4). There are several sites that monitor terrestrial CO_2 in South Africa (Feig et al. 2017 and the references therein):

- Stations equipped with high-precision spectroscopic analyzers for atmospheric concentration: the Cape Point Global Atmosphere Watch station; the Eskom Ambient Air Quality Monitoring network of stations mainly Elandsfontein and

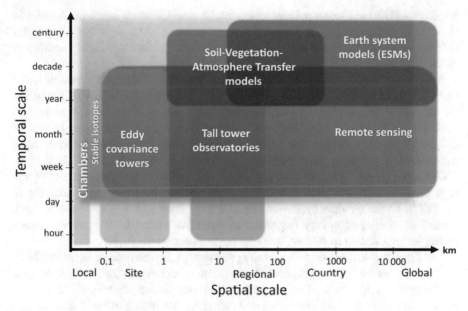

Fig. 17.4 The schematic representation of the methods for quantifying carbon uptake and release integrated across time and space scales (adapted from Ciais et al. 2010)

Medupi located stations in the vicinity of major coal power stations in the country.

- The network of eddy covariance (EC) flux towers: Skukuza, Malopeni, Agincourt, Vuwani; Middelburg Karoo (study sites in this chapter), Cathedral Peak, Welgegund.

Eddy covariance (EC) has been proven to be the most effective method of measuring carbon exchange between the ecosystem and atmosphere on a landscape scale (Swinbank 1951; Verma 1990; Moncrieff et al. 1996; Twine et al. 2000; Burba 2013). In the last two decades, the technology has become reliable enough (due to the continuous development of computer acquisition, sensors such as high-performance CO$_2$ analyzers, ultrasonic anemometers and data processing capacity) to continuously measure fluxes over several years, allowing the study of seasonal and annual variability of CO$_2$. EC has the advantages of a strong theoretical basis, few theoretical assumptions and wide application range, but also imposes high hardware requirements, such as fast response sensors and rapid data acquisition. It is widely used to study the ecosystem–atmosphere carbon exchange over crops, forests and grassland sites, but is difficult to apply over sloping areas and heterogeneous canopies. To understand the responses of NEE to climatic and anthropogenic changes, it is important to separate and analyze GPP and R$_{eco}$ flux components, which can be derived using two common partitioning methods: (1) the night-time-based flux partitioning method is based on the modeling of the night ecosystem respiration (GPP = 0) extrapolated to daytime as a temperature function (Reichstein et al. 2005); (2) the day-time-based flux partitioning method is based on light-response curves that are fitted to the measurements of daytime NEE (Lasslop et al. 2010).

Chamber measurements were the dominant technique for monitoring field-scale soil CO_2 flux for almost a century (Lundegårdh 1927; Acosta et al. 2013) until the EC technique became the standard method (Aubinet et al. 2012). There are several methods available to implement an automated technique for the measurement of CO_2: an open dynamic chamber (CO_2 exchange is measured using the increase in concentration inside the chamber) (Livingston and Hutchinson 1995; Liang et al. 2003) and a closed dynamic chamber (air is directed from the chamber headspace to a portable CO_2 analyzer and returned to the chamber) methods (Rochette et al. 1997). Chamber measurements are relatively easy to use and adaptable to a variety of studies, especially important in situations where the EC method cannot be implemented such as on smaller plots under different treatment.

Leaf gas exchange occupies a central position in the analysis of photosynthetic processes. This has resulted in rapid development of techniques used for commercial and research purposes to directly measure the net rate of photosynthetic carbon assimilation of individual leaves or plant canopies, combining infrared gas analysis and chlorophyll fluorescence capabilities (Farquhar et al. 1980; Von Caemmerer and von Farquhar 1981; Field and Mooney 1990). Systems use a closed transparent chamber that measures the change in the concentration of CO_2 of the air flowing across the chamber.

Measurements of greenhouse gas concentrations, typically from tall towers (at elevations high enough to be representative of greenhouse gas concentrations in the planetary boundary layer), are used to refine preliminary estimates of greenhouse gas release and uptake (Tans 1993; Haszpra et al. 2015). Measurements of atmospheric CO_2 concentrations and transport model simulations are used to constrain surface fluxes using inverse modeling (the atmospheric transport model is linearized and the transport operator is inverted to relate emissions to a measured concentration at low and medium resolutions) (Enting and Mansbridge 1989; Masarie et al. 2011).

Stable isotopes have been shown to be a conversion tool for distinguishing between two different sources of an element (e.g., soil vs. plant C) (Peterson and Fry 1987; Whitman and Lehmann 2015). This is an important method for high-precision, small-volume, automated and relatively rapid measurements for multiscale geochemical cycle studies using the traditional dual inlet Isotope Ratio Mass Spectrometry (IRMS) with cryogenic extraction (large air samples) or Gas Chromatography-Isotope Ratio Mass Spectrometry (GC-IRMS) (smaller air samples, less accurate) (Reinnicke et al. 2012).

Traditional field measurements of CO_2 using the above-mentioned methods are the most accurate approach for obtaining reliable data, but they are difficult to carry out over large areas. Remote sensing methods can measure the spectral reflectance of vegetation and analyze carbon-stock dynamics using spectral band imaging (Situmorang et al. 2016). Satellite data can be easily collected and used to estimate the aboveground carbon stocks by generating regression models that focus on the relationship between observations and satellite image vegetation indices. Remote sensing, however, is sensitive to the vegetation structure, texture, shadow and does not provide a direct estimate of the above-ground biomass (Gerber 2000; Ramankutty et al. 2007).

Physical models of Soil-Vegetation-Atmosphere Transfer (SVAT) are fundamental mathematical representation of the ecosystem–atmosphere interactions and prediction of the surface fluxes (energy fluxes, carbon flux, evapotranspiration) (Breil et al. 2017; Bigeard et al. 2019). In general, SVAT models are considered overly parameterized due to the limited data availability for calibration, which affects the robustness of the parameters. Local SVAT models are based on accurate descriptions of the energy balance of the ecosystem canopy, while larger-scale models use simplified assumptions that are based on typical deposition rates instead of site-specific parameters.

Earth system models (ESMs) differ in their representation of many key processes (e.g., vegetation dynamics, carbon–nitrogen interactions, physiological effects of CO$_2$ increase, climate sensitivity, etc.) (Friedlingstein et al. 1999; Kolby Smith et al. 2016). The modern fully coupled carbon-climate ESMs have been triggered by studies on the feedback between climate change and the carbon cycle, which can be classified into three categories depending on their complexity (Hajima et al. 2014): conceptual models, intermediate complexity ESMs (EMICs) and ESMs based on general circulation models (GCMs).

17.3   Multiyear CO$_2$ Budgets Under Different Grazing Intensities: A Case Study from the Nama-Karoo

17.3.1   Site Description and Measurement Setup

In October 2015, two (EC) towers were installed 1.5 km apart at Middelburg, Eastern Cape, to measure the ecosystem–atmosphere exchange of carbon, water and heat flux (Fig. 17.5). To be able to study the impacts of grazing, the towers were placed at lenient grazing (LG) and experimental grazing (EG) sites. A three-dimensional sonic anemometer (CSAT3, Campbell Scientific Inc., Logan, UT, USA), mounted 3 m above the ground, was used to measure three orthogonal wind components in conjunction with an enclosed path fast-response Infra-Red Gas Analyzer (IRGA) Li-7200 (IRGA, Li-Cor, Lincoln, NE, USA) for CO$_2$ and H$_2$O measurements and the extended weather station (temperature/humidity probe to record relative humidity and air temperature, tipping bucket rain gauge for precipitation, net radiometer for radiation components, two heat flux plates, soil moisture and soil temperature probes).

The studied ecosystems are located in the Nama-Karoo Biome at an altitude of 1310 m.a.s.l. (Mucina et al. 2006) (Fig. 17.5). The vegetation is codominated by dwarf shrubs (perennial, both succulent and nonsucculent) and grasses (short-lived and perennial), including shrubs, sedges, geophytes and herbs (O'Connor and Roux 1995; du Toit et al. 2018). The warm season benefits the growth of grasses, while

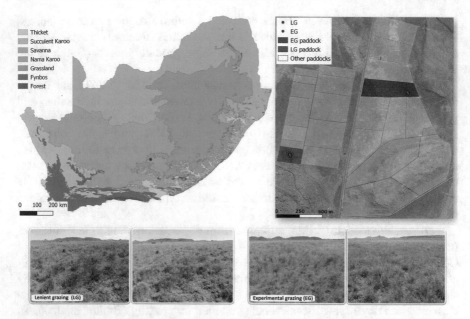

Fig. 17.5 South African biomes with location of the study sites marked as a red circle and location of paddocks on the top right side, and pictures in the dry and growing seasons for the lenient grazing (LG) (bottom left) and the experimental grazing (EG) sites (bottom right) (South African Environmental GIS Data 2013) (Base map: Satellite, Map data ©2015 Google)

the growth of dwarf shrubs is promoted in the cool season (du Toit and O'Connor 2014). The soils are loamy at both research sites (Roux, 1993). Four main seasons can be identified: cold and dry winter (June–August), warm and relatively dry spring (September–November), hot and wet summer (December–February) and warm and relatively wet autumn (March–May). In summer, it is usually hot during the day (30–40°C) and moderately warm at night (10–16°C), while in winter, days are moderate to warm (14–25°C) and nights are cold (−4–4°C). The long-term mean annual temperature is 15°C. Precipitation and droughts are unpredictable (Booysen and Rowswell 1983). Annual precipitation in the years 1889–2013 ranged from 163 mm to 749 mm, with a mean annual precipitation of 374 mm (du Toit and O'Connor 2014). Precipitation mainly occurs in the summer and autumn, with March being the rainiest month. Seasonality and amount of precipitation, including long-term droughts and wet periods, are important drivers of ecosystem processes in the area, especially for the vegetation dynamics, composition, structure and functioning (Anderson and Hoffman 2007; Du Toit and O'Connor 2020). Droughts are common in the Nama-Karoo region with severe droughts occurring approximately every 20 years (du Toit 2017).

The LG site (31°25′20.97″ S, 25°1′46.38″ E) has been grazed by sheep and cattle (Fig. 17.5) using a rotational grazing system (about 2 weeks of grazing followed by 24–26 weeks of rest) at recommended stocking rates of 1/16 animal units per hectare ($AU\ ha^{-1}$) since the 1970s. In terms of botanical composition, the site is considered an excellent "reference" site, with high species diversity and co-dominance of grasses (*Digitaria eriantha* (palatable perennial grass), *Sporobolus fimbriatus* (palatable perennial grass) and dwarf shrubs (*Pentzia globosa* (palatable nonsucculent dwarf-shrub) and *Eriocephalus ericoides* (palatable shrub)) (Fig. 17.5) (du Toit and Nengwenani 2019). The EG site (31°25′48.69″ S, 25°0′57.70″ E) was grazed by Dorper sheep using a two-paddock rotational system (120 days grazing followed by 120 days rest) at stocking rates approximately double the recommended rate ($2/16\ AU\ ha^{-1}$) as part of an experimental trial from 1988 to 2007 (Fig. 17.5). The Dorper breed is described as a hardy sheep that prefers shrubs to grasses (Brand 2000). This intensive grazing treatment has eradicated almost all palatable species and nearly all dwarf shrubs. Therefore, the site is dominated by *Aristida diffusa* (unpalatable perennial grass), *Aristida congesta* (short-lived unpalatable grass) and *Tragus koelerioides* (creeping unproductive grass) (van Lingen 2018). As a result, the EG site is degraded from an agricultural perspective, having transformed from a diverse grassy shrubland to unpalatable semiarid grassland. The site was ungrazed from 2008 to 2017, but did not recover in terms of species composition, where the bulk of fodder available to animals comprised grasses of relatively low palatability. In July 2017, Dorpers were reintroduced at a slightly higher stocking rate ($1/5\ AU\ ha^{-1}$) (Fig. 17.5), and the paddock was continuously grazed unless food capacity was too low (for short periods). Vegetation biomass has never been completely removed by grazers (Du Toit and O'Connor 2020). In addition, nongrowing plants retain their quality well (almost like standing hay) and remain palatable to animals. Compared with the 10 plant species at the EG site, the LG site shows a clearly higher species richness, with 32 species. Climatic conditions of the two sites are similar.

We analyzed and defined the following periods as hydrological years: Year I (Nov 2015–Oct 2016), Year II (Nov 2016–Oct 2017), Year III (Nov 2017–Oct 2018), Year IV (Nov 2018–Oct 2019) and Year V (Nov 2019–Oct 2020).

17.3.2 Ecosystem–Atmosphere CO₂ Exchange

17.3.2.1 Diurnal Variations of Carbon Fluxes

The mean diurnal variations of carbon fluxes were compared between the EG and LG sites and past (years I–II) and current (years III–V) livestock grazing due to the reintroduced livestock grazing at the EG site in the end of the year II (Fig. 17.6). The patterns of carbon fluxes were grouped by dry (June–December) and growing (January–May) seasons. In our study, positive numbers indicate net CO_2 release to the atmosphere and negative numbers net CO_2 sequestration by the ecosystem through higher photosynthetic uptake by plants than respiratory losses. The positive nighttime values of CO_2 indicate that CO_2 is released into the atmosphere by ecosystem respiration processes (Sect. 17.2.1). During the daytime (between 6:00 and 18:00 LT), however, CO_2 uptake by photosynthesis is higher than CO_2 release by ecosystem respiratory processes, and thus, the CO_2 values are negative. As shown

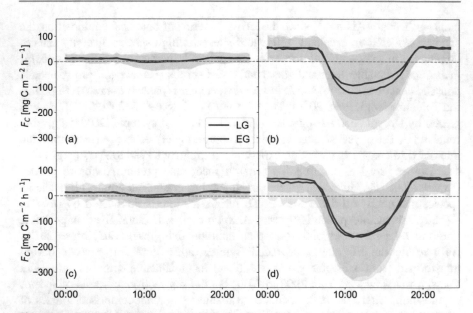

Fig. 17.6 Mean diurnal carbon fluxes in the dry (June–December) (**a**, **c**) and growing (January–May) (**b**, **d**) seasons for years I–II (top) and III–V (bottom) in the (blue) lenient grazing (LG) and (red) experimental grazing (EG) sites. Shaded area indicates ±1 standard deviation. Positive numbers refer to net CO_2 release to the atmosphere, while negative numbers indicate net CO_2 sequestration by the ecosystem

in Fig. 17.6, diurnal CO_2 fluctuations at the study sites show stable positive CO_2 release during the night and negative CO_2 fluxes during daytime, peaking around noon.

Averaging all dry seasons, the mean daily CO_2 fluxes were 14 mg C m^{-2} h^{-1} and 13 mg C m^{-2} h^{-1} for the LG and EG sites, respectively. These values are quite low, implying that net carbon uptake was limited due to the lack of water availability and the inactivity of vegetation.

During the growing seasons, diurnal cycles of carbon fluxes for both sites showed predominance of carbon uptake during the day and only respiration at night (Fig. 17.6b, d). However, there were small differences between the CO_2 fluxes of the study ecosystems. Peak of the mean diurnal CO_2 uptake (higher negative NEE) was higher in years I–II at the EG site with unpalatable grass species as dominant vegetation cover (124 mg C m^{-2} h^{-1}) than at the LG site (94 mg C m^{-2} h^{-1}) (Fig. 17.6a, c). However, the situation changed when heavy livestock grazing was reintroduced at the EG site (July 2017) and the sites showed similar peak of the mean diurnal CO_2 uptake in years III–V (160 mg C m^{-2} h^{-1} for both sites).

In general, during the measurement period, the highest hourly carbon uptake rates were observed in year V with the highest amount of precipitation during the growing season. This was also the year when the highest carbon release was measured after the longest dry period in the previous year.

17.3.2.2   Seasonal NEE, GPP and R$_{eco}$ Variations

The daily sums of NEE, R$_{eco}$ and GPP showed clear seasonal variability (Fig. 17.7) that followed the patterns of precipitation. The length of the wet season (NEE < 0 on a daily basis) varied from year to year depending on the distribution of precipitation throughout the year (from 60 (year I) to 150 wet days (year III)) (Fig. 17.8).

At the study sites, the magnitude of daily NEE ranged from -4.2 g C m^{-2} d^{-1} to 5.4 g C m^{-2} d^{-1} during the measurement period. At both sites, carbon sequestration was always highest during the growing season and decreased during the dry season, with the lowest values observed in June–October (Fig. 17.8). The occurrence of the maximum daily NEE uptake during the growing seasons over the five years measurement period demonstrated the important role that precipitation and soil moisture play in the CO$_2$ uptake and release rates in the semiarid water-limited ecosystems (Fig. 17.7).

The dry periods were mainly characterized by low precipitation (<10 mm month^{-1}), low soil water content (<15%), and almost inactive (low NEE, GPP and R$_{eco}$) carbon cycling between the ecosystem and the atmosphere (Fig. 17.7). Mean daily CO$_2$ rates during the dry seasons of the measurement period were similar ($\sim$130 mg C m^{-2} d^{-1}) with maximum R$_{eco}$ and GPP of 2.1 g C m^{-2} d^{-1} and 1.9 g C m^{-2} d^{-1} for both sites.

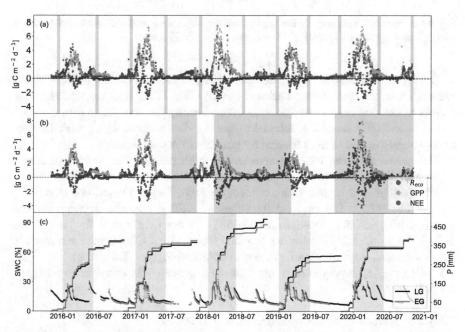

Fig. 17.7 Daily cumulative measured net ecosystem exchange (NEE) and partitioned component fluxes (i.e., gross primary production (GPP), ecosystem respiration (R$_{eco}$)) across different grazing intensities for (**a**) lenient grazing (LG) and (**b**) experimental grazing (EG) sites, (**c**) daily means of soil water content (SWC, left) and cumulative precipitation (P, right). The blue and red patterns represent livestock periods in the LG and EG sites, respectively

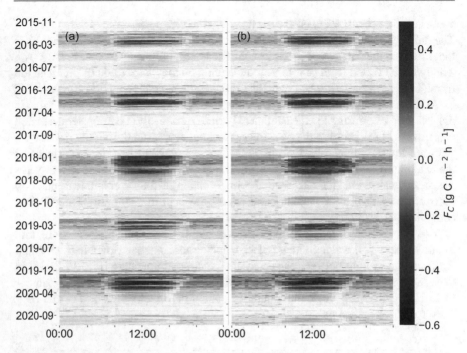

Fig. 17.8 Temporal dynamics of the hourly carbon fluxes for the entire measurement period in the (a) lenient grazing (LG) and (b) experimental grazing (EG) sites

During the November–December transition period from the dry to the growing season, there was an increased release of CO_2 (Fig. 17.8). R_{eco} responded immediately to the first major precipitation events turning the ecosystem into a carbon source, while GPP showed a delayed response (~1–4 weeks, during which carbon uptake begins to rise as assimilation by grasses and herbs increases).

The growing season was characterized by high precipitation (>65% of annual precipitation) and high soil water content (15%–35%) compared to the dry season, which resulted in a higher water content available for plants and enhanced CO_2 uptake (Fig. 17.7). The mean daily values of NEE during the growing seasons (January–May) of the measurement period (-74 mg C m^{-2} d^{-1} for the LG site and $-$ 152 mg C m^{-2} d^{-1} for the EG site) showed high CO_2 uptake compared to the dry seasons with maximum values of GPP and R_{eco} of 7.6 g C m^{-2} d^{-1} and 7.7 g C m^{-2} d^{-1}, respectively, in the peak of the growing season (February–March) for both sites.

17.3.2.3 Seasonal and Annual Carbon Balances

Seasonal and annual carbon balances were estimated to illustrate the carbon source/sink potential of the studied ecosystems (Fig. 17.9). The seasonal and annual ecosystem–atmosphere CO_2 fluxes show how climatic (water availability and its distribution) and land management (grazing intensity) factors affect carbon budgets.

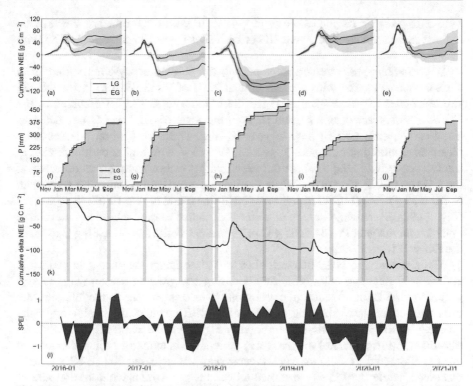

Fig. 17.9 Annual cumulative net ecosystem exchange (NEE) (**a–e**), annual cumulative precipitation (P) (**f–j**) for the years I–V, (**k**) cumulative delta NEE (EG NEE–LG NEE) with shaded areas that represent livestock grazing periods at the LG (green) and EG (gray) sites and (**l**) Standardized Precipitation Evapotranspiration Index (SPEI) (Beguería et al. 2014) monthly basis

Similar mean seasonal carbon releases were observed (50 g with a seasonal mean uncertainty [6] of 24 g C m^{-2} for the LG site and 45 g with a seasonal mean uncertainty of 24 g C m^{-2} for the EG site) during the dry seasons (June–December), with the highest carbon release in the dry season of year V (Fig. 17.8). Although the studied ecosystems temporarily acted as carbon sinks during high precipitation events (e.g., July–August 2016), this was not reflected in seasonal budgets.

While only minor differences in NEE occurred during the dry seasons at the study sites between years I–V, significant variations were observed during the growing seasons (January–May). The length of the wet season and strength of carbon uptake varied significantly from 2 to 6 months during the measurement period. Net carbon release was observed on a seasonal basis during the growing seasons of years I and IV, while those of years II, III and V showed enhanced carbon sequestration. Mean seasonal carbon uptake during the growing seasons of the measurement periods was

---

[6] Uncertainty was estimated as random uncertainty after (Finkelstein and Sims 2001) plus uncertainty that originates from gap filling after (Lucas-Moffat et al. 2018).

-24 ± 15 g C m^{-2} for the LG site and -49 ± 19 g C m^{-2} for the EG site with the highest carbon uptake in the year III for both sites ($\sim -120 \pm 26$ g C m^{-2}) (Fig. 17.7).

Over the five-year measurement period, the studied ecosystems varied from carbon sink to source with a mean annual NEE of 26 ± 39 g m^{-2} for the LG site compared to -5 ± 42 g m^{-2} for the EG site (Fig. 17.9). Compared to the EG site, which acted as a carbon sink during years II–III, the LG site acted as a net carbon sink (-NEE) only in year III (Fig. 17.9b, c). Carbon releases were observed annually at both sites in years I, IV and V, with higher carbon release at the LG site (Fig. 17.9a, d, e). In the year II, a net carbon release was observed at the LG site (24 ± 36 g m^{-2}), while the EG site acted as a carbon sink ecosystem (-31 ± 38 g m^{-2}) (Fig. 17.9b). Similar NEE was found in year III (92 ± 49 g m^{-2}) (Fig. 17.9c). Meanwhile, a lack of water contributed to the highest carbon release rates during the year IV (84 ± 35 g C m^{-2} for the LG site and 60 ± 38 g C m^{-2} for the EG site) (Fig. 17.9d).

The difference in NEE between sites was statistically significant in years I-II and showed negative delta NEE trends during the growing season (Fig. 17.9k). Following the reintroduction of continuous livestock grazing to the EG site, the differences decreased, showing positive delta NEE trends at the beginning of the growing season in years III–V (temporarily higher uptake at LG) (Fig. 17.9k). As a result of the five years of measurement period, the cumulative NEE indicates that the LG site acted as a carbon source ecosystem and released 131 ± 39 g C m^{-2} of carbon, while the EG site was a small carbon sink with a sequestration rate of -22 ± 42 g C m^{-2}.

17.3.3 Carbon Flux Drivers

17.3.3.1 Historical and Current Grazing

Understanding the effects of livestock grazing on carbon exchange is indispensable for predicting and assessing the feedbacks between global change and carbon cycles (Okach et al. 2019; Ondier et al. 2021). Many studies suggest that livestock grazing reduces productivity and CO_2 exchange by decreasing photosynthetic biomass and altering soil water capacity, increasing soil compaction, often resulting in losses of soil organic matter necessary for biomass development. A wide range of factors contribute to the recovery of the grazing area, such as the intensity and duration of grazing and the availability of soil moisture and nutrients (Leriche et al. 2003; Seymour et al. 2010).

In the Karoo study, a comparison of the carbon exchange during years I and II allowed us to compare a site showing the impacts of past overgrazing (EG) to a leniently grazed site with near-natural species composition (LG). After grazing was reintroduced to the EG site, we further observed the impacts of current heavy grazing at the EG site, compared to parallel lenient grazing at the LG site (years III–V).

Based on the results of the first two years of measurements, we found the LG site to act as a net carbon source (84 g C m^{-2} with an annual mean uncertainty of 42 g C m^{-2}), while the EG site was carbon neutral (-4 g C m^{-2} with an annual mean uncertainty of 41 g C m^{-2}) with temporarily higher carbon sequestration rates (Fig. 17.7). This indicates that the change in species composition (due to past overgrazing) and the recovery period (2007–2017) from long-term disturbance (regrowth of biomass) resulted in an increase of the carbon sink strength at the EG site. Differences in grasses and shrubs and their response to pulse rain and drought events in the studied ecosystems, in turn, may explain the differences in carbon sink strength between the sites. For example, *Aristida diffusa* is a drought-tolerant dominant grass at the EG site (Du Toit and O'Connor 2020). In a study conducted by Zhou et al. (2012), the fine root biomass of perennial grasses was more abundant in the top layer of the soil than those of shrubs. This implies that the grass cover in the EG ecosystem can respond more quickly to precipitation events by using discontinuous and erratic water sources in the upper soil layers, unlike shrubs (with deep-root systems) that use water in deeper soil layers (Canadell et al. 1996; Hipondoka et al. 2003; Zhou et al. 2012). Furthermore, the higher soil organic carbon inputs of perennial grasses and their slower decomposition allow them to store more soil organic carbon than shrubs. After continuous heavy grazing was reintroduced at the EG site in July 2017, the difference in NEE between sites was reduced in years III–V (Fig. 17.9). Despite this, differences in grazing intensity were not reflected as differences in carbon fluxes, and the EG site still indicated slightly higher carbon sink strength than the LG site. Furthermore, our results support the assumption that in highly seasonal systems, grazing pressure may be enhanced when coupled with drought stress. After a long drought in year IV (with just 5 mm of precipitation in June–November), the highest respiration peak was observed at the beginning of the growing season in year V. Despite the fact that continuous heavy grazing (with stocking rate and duration many times higher than recommended) was reintroduced at the EG site, we found great NEE resilience of the EG site that was dominated by grasses and which had a low species richness. Similar annual NEE ranging from -98 g C m^{-2} yr^{-1} to 21 g C m^{-2} yr^{-1} have been observed in the semiarid Kendall grassland, USA with precipitation of 345 mm (63% in summer) (Scott et al. 2010). During 2011–2013, Räsänen et al. (2017) reported annual carbon budgets of -85 C m^{-2} yr^{-1}, 67 C m^{-2} yr^{-1} and 139 C m^{-2} yr^{-1} in the Welgegund atmospheric measurement station grassland ecosystem, South Africa (540 mm). On a broader perspective, Valentini et al. (2014) reported a small carbon sink of 0.61 ± 0.58 Pg C yr^{-1} (-20 g C m^{-2} yr^{-1}) for the entire African continent. In our study, we found an annual mean of 20 g C m^{-2} yr^{-1} for the LG site and -9 g C m^{-2} yr^{-1} for the EG site, illustrating the fine line between net carbon releases and sequestration due to differing land use and management.

17.3.3.2 Water Availability

Semiarid ecosystems, which are also referred to as water-limited ecosystems, usually receive between 200 and 700 mm precipitation annually (Gallart et al. 2002). For periods within this range, the mean annual precipitation was 319 mm in 1899–1937, 401 mm in 1938–1952, 353 mm in 1953–1984, 426 mm in 1985–2010, 481 mm in 2011–2015 and 384 mm in 2016–2020 (du Toit and O'Connor 2014). In these environments, it is typically warm enough for physiological activity, and thus, temperature is not considered a limiting factor to ecosystem production (Archibald et al. 2009). The pulsed input of water in semiarid ecosystems is understood to drive ecosystem responses such as ecosystem water use, productivity and respiration (Archibald et al. 2009; Williams et al. 2009) with water availability in these environments characterized by a strong seasonality component (Merbold et al. 2009). In the Karoo ecosystems, as the growing seasons progressed, negative carbon fluxes (correlated with daytime light intensity) occurred during the daytime, whereas positive fluxes (carbon release) were observed at night. Plant germination and small increase in GPP was observed at the onset of summer precipitation events (beginning of the growing season), while R_{eco} responded rapidly to precipitation, so ecosystems acted as a small source of carbon. In contrast, the ecosystems turned into a carbon sink when GPP exceeded R_{eco} at the peak of the growing season. Limited water availability slowed ecosystem activity during dry seasons, mainly indicating an inactive (carbon-neutral) ecosystem. This seasonality is described by the widely applied pulse-reserve paradigm to explain ecosystem functional responses to inputs of rain or soil water (Kutsch et al. 2008; Yepez and Williams 2009). This paradigm brings to attention the thresholds and lags in ecosystem responses to water availability. The timing, degree and duration to which these responses are controlled and sustained however remains poorly understood.

While the pulse-paradigm attempts to describe ecosystem responses to water inputs, it has rarely been tested with high frequency medium term observation approaches. In addition, the nonlinearity of ecosystem responses to water inputs highlights the observed rapid responses (respiration pulses) to even small wetting events in these ecosystems (i.e., R_{eco} began to increase rapidly with precipitation of 9 mm during the transition months from the dry to the growing season in the water-limited Karoo ecosystems). To further complicate matters, these responses are seemingly not only dependent on water inputs and availability but also on the physiological recovery of the ecosystem from preceding dry periods which may modify the sensitivity of responses. In the Karoo ecosystems, for example, high carbon releases were observed after a long drought period (June–November 2019) with only 5 mm of precipitation at the beginning of the growing season in year V. Therefore, it appears that the status of antecedent vegetation state, soil condition and the timing of rainfall significantly influence ecosystem responses.

17.4 Potential Adjustments and Recommendations for C Sequestration

Considering the wide coverage of rangelands in southern Africa and their likely ongoing recovery from previously unsustainable stocking rates, they may play an important role in carbon sequestration (Dean et al. 2015; Wigley et al. 2020). In semiarid ecosystems, the rate of carbon sequestration may be relatively slow, but appears to be substantial when scaled up under optimal management regimes. Sustainable livestock farming should take into account the needs and aims of both public (carbon sequestration and food security) and private stakeholders (income and sustainability of production) (Eisler et al. 2014; Hasselerharm et al. 2021). While climate-smart management practices appear to offer multiple benefits, it will be critical to support their development based on good evidence to better inform policies that could enhance such multiple benefits.

As shown by O'Reagain and Turner (1992), key decisions determining long-term influence in vegetation dynamics in southern African rangelands systems include stocking rate, management system and livestock type. The majority of Nama-Karoo is commercial rangeland, and the adoption of climate-smart management practices in these systems could therefore have a large potential impact on climate mitigation. A wide range of management proposals has been made, from destocking to intensive "nonselective grazing" as extreme ends of a management continuum. Some practitioners have made remarkable claims of large sequestration potential from intensive grazing that is applied with some temporal precision (Frith 2020). The results that we present here suggest that a site that is subjected to intensive grazing and then rested has a higher carbon sequestration potential than a leniently grazed site, based on results for 2016 and 2017, and that this sequestration appears somehow resilient to the reintroduction of grazing even at higher than recommended rates. Whether this resilience would persist over time is unknown, as is the influence of other factors such as drought. It is anticipated that species diversity would remain low as the continued presence of grazers would prevent ingress of palatable shrubs and grasses. In the case of a drought, it would be predicted that grasses may die out (Du Toit and O'Connor 2020), and the agricultural potential would collapse in the absence of drought-tolerant dwarf shrubs. There is also a strong interaction with rainfall amount, showing a greater sequestration capacity in leniently grazed vegetation when water availability is higher and well distributed throughout the growing season. Taken together, this suggests that the ecosystem composition change may have facilitated a greater resilience of carbon sequestration to grazing and drought, perhaps due to greater dominance of unpalatable species that continue to sequester carbon irrespective of grazing intensity or drought stress. Such a result requires confirmation in a wider range of sites, as this would be important in designing optimal management strategies based on composition and rainfall variability. In highly seasonal systems with large interannual variability, such as the Nama-Karoo, long time series are particularly important in understanding the patterns of land-atmosphere carbon exchange and ecosystem carbon balance. Furthermore, an important axis of future investigation would be to explore different rates of grazing in "good quality" vegetation. Scientifically based evidence derived from techniques such as described here could be pivotal in supporting policy application and cost-benefit analysis (Snyman 1998; Schurch et al. 2021).

17.5 Conclusions

Large uncertainties remain in the understanding of carbon dynamics of southern African semiarid rangelands. In this chapter, we presented a case study on the impacts of livestock grazing on carbon exchange in the Nama-Karoo biome in South Africa. Historically, the Nama-Karoo biome was over-stocked with livestock; however, a recent reduction in stocking rates has facilitated recovery of primary productivity and vegetation cover in many areas. We found that a leniently grazed study site was a slight net source of carbon during the five-year measurement period (wet year III and extremely dry year IV compared to the long-term mean annual precipitation (373 mm), while a previously overgrazed, recovering site, characterized by unpalatable species, had higher carbon sequestration potential and was nearly carbon neutral. Furthermore, the previously overgrazed site seemed to show better resilience to drought. Taking the vast areal importance of livestock rangelands in South Africa, we recommend exploring the impact of different intensities of livestock grazing on carbon balance on natural ecosystems.

References

Abdulai I, Hoffmann MP, Jassogne L et al (2020) Variations in yield gaps of smallholder cocoa systems and the main determining factors along a climate gradient in Ghana. Agric Syst 181:102812. https://doi.org/10.1016/j.agsy.2020.102812

Acocks JPH (1966) Non-selective grazing as a means of veld reclamation. Proc Annu Congr Grassl Soc South Africa 1:33–39. https://doi.org/10.1080/00725560.1966.9648517

Acocks JPH (1979) The flora that matched the fauna. Bothalia 12:673–709. https://doi.org/10.4102/abc.v12i4.1442

Acosta M, Pavelka M, Montagnani L et al (2013) Soil surface CO2 efflux measurements in Norway spruce forests: comparison between four different sites across Europe—from boreal to alpine forest. Geoderma 192:295–303

Anderson PML, Hoffman MT (2007) The impacts of sustained heavy grazing on plant diversity and composition in lowland and upland habitats across the Kamiesberg mountain range in the Succulent Karoo, South Africa. J Arid Environ 70:686–700. https://doi.org/10.1016/j.jaridenv.2006.05.017

Archer S (2000) Technology and ecology in the Karoo: a century of windmills, wire and changing farming practice. J South Afr Stud 26:675–696. https://doi.org/10.1080/03057070020008224

Archibald SA, Kirton A, Van Der Merwe MR et al (2009) Drivers of inter-annual variability in net ecosystem exchange in a semi-arid savanna ecosystem, South Africa. Biogeosciences 6:251–266. https://doi.org/10.5194/bg-6-251-2009

Asner GP, Archer SR (2010) Livestock and the global carbon cycle. In: Steinfeld, H., Mooney, HA, Schneider, F., Neville, LE (ed) Livestock in a changing landscape: drivers, consequences and responses. pp. 69–82, Island Press, Washington, DC

Aubinet M, Vesala T, Papale D (2012) Eddy covariance: a practical guide to measurement and data analysis. Springer Science & Business Media, Dordrecht

Bardgett RD, Freeman C, Ostle NJ (2008) Microbial contributions to climate change through carbon cycle feedbacks. ISME J 2:805–814. https://doi.org/10.1038/ismej.2008.58

Beguería S, Vicente-Serrano SM, Reig F, Latorre B (2014) Standardized precipitation evapotranspiration index (SPEI) revisited: parameter fitting, evapotranspiration models, tools, datasets and drought monitoring. Int J Climatol 34:3001–3023. https://doi.org/10.1002/joc.3887

Bekele SG (2001) Grasshopper ecology and conservation in the Nama-Karoo. 1–226

Bigeard G, Coudert B, Chirouze J et al (2019) Ability of a soil–vegetation–atmosphere transfer model and a two-source energy balance model to predict evapotranspiration for several crops and climate conditions. Hydrol Earth Syst Sci 23:5033–5058. https://doi.org/10.5194/hess-23-5033-2019

Booysen J, Rowswell DI (1983) The drought problem in the Karoo areas. Proc Annu Congr Grassl Soc South Africa 18:40–45. https://doi.org/10.1080/00725560.1983.9648979

Brand TS (2000) Grazing behaviour and diet selection by Dorper sheep. Small Rumin Res 36:147–158. https://doi.org/10.1016/S0921-4488(99)00158-3

Breil M, Panitz H-J, Schädler G (2017) Impact of soil-vegetation-atmosphere interactions on the spatial rainfall distribution in the Central Sahel. Meteorol Zeitschrift 26:379–389. https://doi.org/10.1127/metz/2017/0819

Brent A, Hietkamp S, Wise R, O'Kennedy K (2011) Estimating the carbon emissions balance for South Africa. South African J Econ Manag Sci 12:263–279. https://doi.org/10.4102/sajems.v12i3.216

Burba G (2013) Eddy covariance method-for scientific, industrial, agricultural, and regulatory applications. LI-COR Biosciences, Lincoln, Nebraska Copyright

Canadell J, Jackson RB, Ehleringer JR et al (1996) Maximum rooting depth of vegetation types at the global scale. Oecologia 108:538–595

Ciais P, Dolman H, Dargaville R, et al (2010) GEO carbon strategy

Ciais P, Bombelli A, Williams M et al (2011) The carbon balance of Africa: synthesis of recent research studies. Philos Trans R Soc A 369:1–20. https://doi.org/10.1098/rsta.2010.0328

Curtin CG (2002) Livestock grazing, rest, and restoration in arid landscapes. Conserv Biol 16:840–842

DEA, SANBI (2016) Strategic framework and overarching implementation plan for ecosystem based adaptation (EbA) in South Africa: 2016–2021. DEA/SANBI, Pretoria

Dean WRJ, Milton SJ (1999) The Karoo, ecological patterns and processes, (eds). Cambridge University Press, Cambridge

Dean C, Kirkpatrick JB, Harper RJ, Eldridge DJ (2015) Optimising carbon sequestration in arid and semiarid rangelands. Ecol Eng 74:148–163

du Toit JCO (2017) Droughts and the quasi-20-year rainfall cycle at Grootfontein in the eastern Karoo, South Africa. Grootfontein Agric 17:36–42

du Toit JCO, Nengwenani TP (2019) Boesmanskop compositional data 2007–2019. Unpublished data

du Toit JCO, O'Connor TG (2014) Changes in rainfall pattern in the eastern Karoo, South Africa, over the past 123 years. Water SA 40:453–460. https://doi.org/10.4314/wsa.v40i3.8

Du Toit JCO, O'Connor TG (2020) Long-term influence of season of grazing and rainfall on vegetation in the eastern Karoo, South Africa. African J Range Forage Sci 37:159–171. https://doi.org/10.2989/10220119.2020.1725122

du Toit HSD, Gadd SM, Kolbe GA et al (1923) Final report of the drought investigation commission. Cape Times Limited. Gov Printers, Cape Town

du Toit JCO, Ramaswiela T, Pauw MJ, O'Connor TG (2018) Interactions of grazing and rainfall on vegetation at Grootfontein in the eastern Karoo. African J Range Forage Sci 35:267–276. https://doi.org/10.2989/10220119.2018.1508072

Eisler MC, Lee MRF, Tarlton JF et al (2014) Agriculture: steps to sustainable livestock. Nat News 507:32

Enting IG, Mansbridge JV (1989) Seasonal sources and sinks of atmospheric CO2 direct inversion of filtered data. Tellus Ser B Chem Phys Meteorol 41:111–126

Epple C, García Rangel S, Jenkins M, Guth M (2016) Managing ecosystems in the context of climate change mitigation: a review of current knowledge and recommendations to support ecosystem-based mitigation actions that look beyond terrestrial forests. Montreal

Farquhar GD, von Caemmerer S, von Berry JA (1980) A biochemical model of photosynthetic CO2 assimilation in leaves of C3 species. Planta 149:78–90

Feig GT, Joubert WR, Mudau AE, Monteiro PMS (2017) South African carbon observations: CO_2 measurements for land, atmosphere and ocean. S Afr J Sci 113:2–5. https://doi.org/10.17159/sajs.2017/a0237

Field CB, Mooney HA (1990) Leaf chamber methods for measuring photosynthesis under field conditions. Remote Sens Rev 5:117–139. https://doi.org/10.1080/02757259009532125

Finkelstein PL, Sims PF (2001) Sampling error in eddy correlation flux measurements. J Geophys Res Atmos 106:3503–3509. https://doi.org/10.1029/2000JD900731

Friedlingstein P, Joel G, Field CB, Fung IY (1999) Toward an allocation scheme for global terrestrial carbon models. Glob Chang Biol 5:755–770. https://doi.org/10.1046/j.1365-2486.1999.00269.x

Frith S (2020) The evidence for holistic planned grazing. In: Green meat. pp 89–106

Gallart F, Sole A, Puigdefabregas J, Lazaro R (2002) Badland systems in the mediterranean. In: Bull LJ, Kirkby MJ (eds) Dryland rivers: hydrology and geomorphology of semi-arid channels. Wiley, Chichester, pp 299–326

Gerber L (2000) Development of a ground truthing method for determination of rangeland biomass using canopy reflectance properties. Afr J Range Forage Sci 17:93–100. https://doi.org/10.2989/10220110009485744

Gougoulias C, Clark JM, Shaw LJ (2014) The role of soil microbes in the global carbon cycle: tracking the below-ground microbial processing of plant-derived carbon for manipulating carbon dynamics in agricultural systems. J Sci Food Agric 94:2362–2371

Gross A, Glaser B (2021) Meta-analysis on how manure application changes soil organic carbon storage. Sci Rep 11:5516. https://doi.org/10.1038/s41598-021-82739-7

Hajima T, Kawamiya M, Watanabe M et al (2014) Modeling in Earth system science up to and beyond IPCC AR5. Prog Earth Planet Sci 1:29. https://doi.org/10.1186/s40645-014-0029-y

Hamidi D, Komainda M, Tonn B et al (2021) The effect of grazing intensity and sward heterogeneity on the movement behavior of Suckler cows on semi-natural grassland. Front Vet Sci 8:639096. https://doi.org/10.3389/fvets.2021.639096

Hasselerharm CD, Yanco E, McManus JS et al (2021) Wildlife-friendly farming recouples grazing regimes to stimulate recovery in semi-arid rangelands. Sci Total Environ 788:147602

Haszpra L, Barcza Z, Haszpra T et al (2015) How well do tall-tower measurements characterize the CO2 mole fraction distribution in the planetary boundary layer? Atmos Meas Tech 8:1657–1671. https://doi.org/10.5194/amt-8-1657-2015

Hipondoka MHT, Aranibar JN, Chirara C et al (2003) Vertical distribution of grass and tree roots in arid ecosystems of Southern Africa: niche differentiation or competition? J Arid Environ 54:319–325. https://doi.org/10.1006/jare.2002.1093

Hoffman M (1988) The rationale for karoo grazing systems: criticisms and research implications. S Afr J Sci/S-Afr TYDSKR Wet 84:556–559

Hoffman T, Ashwell A (2001) Nature divided: land degradation in South Africa. University of Cape Town Press, Cape Town

Hoffman M, Cousins B, Meyer T, et al (1999) The karoo: Historical and contemporary land use and the desertification of the Karoo. In The Karoo: Ecological patterns and processes. Cambridge University Press, Cambridge, pp. 257–273

Hoffman M, Skowno A, Bell W, Mashele S (2018) Long-term changes in land use, land cover and vegetation in the Karoo drylands of South Africa: implications for degradation monitoring§. Afr J Range Forage Sci 35:209–221. https://doi.org/10.2989/10220119.2018.1516237

Holechek JL, de Souza Gomes H, Molinar F, Society for Range Management (2000) Short-duration grazing: the facts in 1999. Rangelands 22. https://doi.org/10.2458/azu_rangelands_v22i1_holechek

Kolby Smith W, Reed SC, Cleveland CC et al (2016) Large divergence of satellite and Earth system model estimates of global terrestrial CO2 fertilization. Nat Clim Chang 6:306–310. https://doi.org/10.1038/nclimate2879

Kraaij T, Milton SJ (2006) Vegetation changes (1995-2004) in semi-arid Karoo shrubland, South Africa: effects of rainfall, wild herbivores and change in land use. J Arid Environ 64:174–192. https://doi.org/10.1016/j.jaridenv.2005.04.009

Kutsch WL, Hanan N, Scholes B, et al (2008) Response of carbon fluxes to water relations in a savanna ecosystem in South Africa. Biogeosciences 5(6):1797–1808

Lasslop G, Reichstein M, Papale D et al (2010) Separation of net ecosystem exchange into assimilation and respiration using a light response curve approach: critical issues and global evaluation. Glob Chang Biol 16:187–208. https://doi.org/10.1111/j.1365-2486.2009.02041.x

Lee MA, Todd A, Sutton MA et al (2017) A time-series of methane and carbon dioxide production from dairy cows during a period of dietary transition. Cogent Environ Sci 3:1385693. https://doi.org/10.1080/23311843.2017.1385693

Leifeld J, Meyer S, Budge K et al (2015) Turnover of grassland roots in mountain ecosystems revealed by their radiocarbon signature: role of temperature and management. PLoS One 10:e0119184. https://doi.org/10.1371/journal.pone.0119184

Leriche H, Le Roux X, Desnoyers F et al (2003) Grass response to clipping in an African savanna: testing the grazing optimization hypothesis. Ecol Appl 13:1346–1354. https://doi.org/10.1890/02-5199

Liang N, Inoue G, Fujinuma Y (2003) A multichannel automated chamber system for continuous measurement of forest soil CO$_2$ efflux. Tree Physiol 23:825–832. https://doi.org/10.1093/treephys/23.12.825

Livingston GP, Hutchinson GL (1995) Enclosure-based measurement of trace gas exchange: applications and sources of error. In: Matson PA, Harriss RC (eds) Biogenic trace gases : measuring emissions from soil and water, vol 51. Blackwell Science, Oxford, pp 14–51

Lucas-Moffat AM, Huth V, Augustin J et al (2018) Towards pairing plot and field scale measurements in managed ecosystems: using eddy covariance to cross-validate CO$_2$ fluxes modeled from manual chamber campaigns. Agric For Meteorol 256–257:362–378. https://doi.org/10.1016/j.agrformet.2018.01.023

Lundegårdh H (1927) Carbon dioxide evolution of soil and crop growth. Soil Sci 23:417–453

Malik AA, Swenson T, Weihe C et al (2020) Drought and plant litter chemistry alter microbial gene expression and metabolite production. ISME J 14:2236–2247. https://doi.org/10.1038/s41396-020-0683-6

Masarie KA, Pétron G, Andrews A et al (2011) Impact of CO2 measurement bias on carbon tracker surface flux estimates. J Geophys Res 116:D17305. https://doi.org/10.1029/2011JD016270

McCabe K (1987) Veld management in the Karoo. The Naturalist 31:8–15

McLeod SR (1997) Is the concept of carrying capacity useful in variable environments? Oikos 79:529. https://doi.org/10.2307/3546897

Merbold L, Ardö J, Arneth A et al (2009) Precipitation as driver of carbon fluxes in 11 African ecosystems. Biogeosciences 6:1027–1041. https://doi.org/10.5194/bg-6-1027-2009

Milton SJ (1993) Studies of herbivory and vegetation change in Karoo shrublands

Milton SJ, Dean WRJ (2021) Anthropogenic impacts and implications for ecological restoration in the Karoo, South Africa. Anthropocene 36:100307

Moll EJ, Gubb AA (1989) Southern African shrublands. Academic Press, New York, pp. 145–175

Moncrieff JB, Malhi Y, Leuning R (1996) The propagation of errors in long-term measurements of land-atmosphere fluxes of carbon and water. Glob Chang Biol 2:231–240

Mucina L, Rutherford MC, Palmer AR et al (2006) Nama-Karoo Biome. In: Mucina L, Rutherford MC (eds) The vegetation of South Africa, Lesotho and Swaziland. SANBI, Pretoria, pp 324–347

Mureva A, Ward D, Pillay T et al (2018) Soil organic carbon increases in semi-arid regions while it decreases in humid regions due to Woody-Plant encroachment of grasslands in South Africa. Sci Rep 8:15506. https://doi.org/10.1038/s41598-018-33701-7

O'Connor TG, Roux PW (1995) Vegetation changes (1949-71) in a semi-arid, grassy dwarf shrubland in the Karoo, South Africa: influence of rainfall variability and grazing by sheep. J Appl Ecol 32:612–626

O'Connor TG, Kuyler P, Kirkman KP, Corcoran B (2010) Which grazing management practices are most appropriate for maintaining biodiversity in South African grassland? Afr J Range Forage Sci 27:67–76. https://doi.org/10.2989/10220119.2010.502646

O'Farrell PJ, Le Maitre DC, Gelderblom C et al (2008) Applying a resilience framework in pursuit of sustainable land-use development in the Little Karoo, South Africa. In: Burns M, Weaver A (eds) Advancing sustainability science in South Africa. SUN, Stellenbosch, pp 383–432

O'Reagain PJ, Turner JR (1992) An evaluation of the empirical basis for grazing management recommendations for rangeland in southern Africa. J Grassl Soc South Africa 9:38–49

Okach DO, Ondier JO, Kumar A et al (2019) Livestock grazing and rainfall manipulation alter the patterning of CO_2 fluxes and biomass development of the herbaceous community in a humid savanna. Plant Ecol 220:1085–1100. https://doi.org/10.1007/s11258-019-00977-2

Ondier JO, Okach DO, Onyango JC, Otieno DO (2021) Ecosystem productivity and CO_2 exchange response to the interaction of livestock grazing and rainfall manipulation in a Kenyan savanna. Environ Sustain Indic 9:100095. https://doi.org/10.1016/j.indic.2020.100095

Owen-Smith N, Danckwerts JE (1997) Herbivory. In: Cowling RM, Richardson DM, Pierce SM (eds) Vegetation of Southern Africa. Cambridge University Press, Cambridge, pp 397–420

Palmer AR, Ainslie AM (2005) Grasslands of South Africa. In: Grasslands of the world. p 77

Parsons AJ, Leafe EL, Collett B et al (1983) The physiology of grass production under grazing. II. Photosynthesis, crop growth and animal intake of continuously-grazed swards. J Appl Ecol 20:127–139

Paz-Ferreiro J, Medina-Roldán E, Ostle NJ et al (2012) Grazing increases the temperature sensitivity of soil organic matter decomposition in a temperate grassland. Environ Res Lett 7:14027. https://doi.org/10.1088/1748-9326/7/1/014027

Pérez-Hernández J, Gavilán RG (2021) Impacts of land use changes on vegetation and ecosystem functioning : old Field secondary succession. Plan Theory 10:990. https://doi.org/10.3390/plants10050990

Peterson BJ, Fry B (1987) Stable isotopes in ecosystem studies. Annu Rev Ecol Syst 18:293–320

Prentice IC, Farquhar GD, Fasham MJR, et al (2001) The carbon cycle and atmospheric carbon dioxide

Ramankutty N, Gibbs HK, Achard F et al (2007) Challenges to estimating carbon emissions from tropical deforestation. Glob Chang Biol 13:51–66

Räsänen M, Aurela M, Vakkari V et al (2017) Carbon balance of a grazed savanna grassland ecosystem in South Africa. Biogeosciences 14:1039–1054. https://doi.org/10.5194/bg-14-1039-2017

Reichstein M, Falge E, Baldocchi D et al (2005) On the separation of net ecosystem exchange into assimilation and ecosystem respiration: review and improved algorithm. Glob Chang Biol 11:1424–1439. https://doi.org/10.1111/j.1365-2486.2005.001002.x

Reinnicke S, Juchelka D, Steinbeiss S et al (2012) Gas chromatography/isotope ratio mass spectrometry of recalcitrant target compounds: performance of different combustion reactors and strategies for standardization: GC/IRMS of recalcitrant target compounds. Rapid Commun Mass Spectrom 26:1053–1060. https://doi.org/10.1002/rcm.6199

Roberts BR (1969) The multicamp controversy: a search for evidence. In: Proceedings of the veld management conference, Bulawayo, Rhodesie Management Conference. pp. 41–57

Rochette P, Ellert B, Gregorich EG et al (1997) Description of a dynamic closed chamber for measuring soil respiration and its comparison with other techniques. Can J Soil Sci 77:195–203

Roux PW, Theron GK (1987) Vegetation change in the Karoo biome. Karoo biome a Prelim Synth Part:50–69

Roux PW, Vorster M (1983) Vegetation change in the Karoo. Proc Annu Congr Grassl Soc South Africa 18:25–29. https://doi.org/10.1080/00725560.1983.9648976

Roux PW, Vorster M, Zeeman PJL, Wentzel D (1981) Stock production in the Karoo region. Proc Annu Congr Grassl Soc South Africa 16:29–35

Rutherford MC, Powrie LW (2013) Impacts of heavy grazing on plant species richness: a comparison across rangeland biomes of South Africa. South African J Bot 87:146–156. https://doi.org/10.1016/j.sajb.2013.03.020

Savory A (1978) A holistic approach to ranch management using short duration grazing. In: Proceedings of the first international rangeland congress. Denver, Colorado, pp 555–557

Sayre NF (2008) The genesis, history, and limits of carrying capacity. Ann Assoc Am Geogr 98:120–134. https://doi.org/10.1080/00045600701734356

Schimel DS (1995) Terrestrial ecosystems and the carbon cycle. Glob Chang Biol 1:77–91. https://doi.org/10.1111/j.1365-2486.1995.tb00008.x

Schurch MPE, McManus J, Goets S et al (2021) Wildlife-friendly livestock management promotes mammalian biodiversity recovery on a semi-arid Karoo farm in South Africa. Front Conserv Sci 2:6

Scott RL, Hamerlynck EP, Jenerette GD et al (2010) Carbon dioxide exchange in a semidesert grassland through drought-induced vegetation change. J Geophys Res 115:G03026. https://doi.org/10.1029/2010JG001348

Seymour CL, Milton SJ, Joseph GS et al (2010) Twenty years of rest returns grazing potential, but not palatable plant diversity, to Karoo rangeland, South Africa. J Appl Ecol 47:859–867. https://doi.org/10.1111/j.1365-2664.2010.01833.x

Shaw J (1873) On the changes going on in the vegetation of South Africa through the introduction of the merino sheep. J Linn Soc London Bot 14:202–208

Simpkin P, Cramer L, Ericksen PJ, Thornton PK (2020) Current situation and plausible future scenarios for livestock management systems under climate change in Africa. CCAFS Work Pap

Situmorang JP, Sugianto S, Darusman D (2016) Estimation of carbon stock stands using EVI and NDVI vegetation index in production forest of lembah Seulawah sub-district, Aceh Indonesia. Aceh Int J Sci Technol 5:126–139

Skead C (1982) Historical mammal incidence in the Cape Province. Department of Nature and Environmental Conservation, Cape Town

Snyman HA (1998) Dynamics and sustainable utilization of rangeland ecosystems in arid and semi-arid climates of southern Africa. J Arid Environ 39:645–666

Snyman HA (2003) Revegetation of bare patches in a semi-arid rangeland of South Africa: an evaluation of various techniques. J Arid Environ 55:417–432. https://doi.org/10.1016/S0140-1963(02)00286-0

South African Environmental GIS Data (2013) South Africa – Biomes. In: OpenAfrica. https://africaopendata.org/dataset/e2ef64dd-6577-4a73-9eff-871677b2dd92/resource/620721d9-2605-44dc-81f8-cfb5af9169da/download/rsabiomc4xkhi.zip%0A%0A

Spangler D, Tyler A, McCalley C (2021) Effects of grazer exclusion on carbon cycling in created freshwater wetlands. Land 10:805. https://doi.org/10.3390/land10080805

Susiluoto S, Rasilo T, Pumpanen J, Berninger F (2008) Effects of grazing on the vegetation structure and carbon dioxide exchange of a Fennoscandian fell ecosystem. Arctic, Antarct Alp Res 40:422–431. https://doi.org/10.1657/1523-0430(07-035)[SUSILUOTO]2.0.CO;2

Swinbank WC (1951) The measurement of vertical transfer of heat and water vapor by eddies in the lower atmosphere. J Atmos Sci 8:135–145

Tans PP (1993) Observational strategy for assessing the role of terrestrial ecosystems in the global carbon cycle: scaling down to regional levels. In: Scaling physiological processes. Elsevier, pp 179–190

Teague WR, Dowhower SL (2003) Patch dynamics under rotational and continuous grazing management in large, heterogeneous paddocks. J Arid Environ 53:211–229

Tidmarsh C (1951) Pasture research in South Africa. Progress report No. 3. Veld management studies: 1934–1950. Middelburg, Cape Province

Tilman D, Reich PB, Isbell F (2012) Biodiversity impacts ecosystem productivity as much as resources, disturbance, or herbivory. Proc Natl Acad Sci U S A 109:10394–10397. https://doi.org/10.1073/pnas.1208240109

Todd S, Hoffman MT (1999) A fence-line contrast reveals effects of heavy grazing on plant diversity and community composition in Namaqualand, South Africa. Plant Ecol 142:169–178. https://doi.org/10.1023/A:1009810008982

Truter WF, Botha PR, Dannhauser CS et al (2015) Southern African pasture and forage science entering the 21st century: past to present. Afr J Range Forage Sci 32:73–89. https://doi.org/10.2989/10220119.2015.1054429

Twine TE, Kustas WP, Norman JM et al (2000) Correcting eddy-covariance flux underestimates over a grassland. Agric For Meteorol 103:279–300. https://doi.org/10.1016/S0168-1923(00)00123-4

Valentini R, Arneth A, Bombelli A et al (2014) A full greenhouse gases budget of Africa: synthesis, uncertainties, and vulnerabilities. Biogeosciences 11:381–407. https://doi.org/10.5194/bg-11-381-2014

Van den Berg L, Du Toit JCO, Van Lingen M, Van der Merwe H (2019) The effect of sheep farming on the long-term diversity of the vegetation of the Karoo – a review. Grootfontein Agric 19:66–72

van Lingen M (2018) Afrikaner/Hereford compositional data 2018. Unpublished data

Verma SB (1990) Micrometeorological methods for measuring surface fluxes of mass and energy. Remote Sens Rev 5:99–115

Von Caemmerer S, von Farquhar GD (1981) Some relationships between the biochemistry of photosynthesis and the gas exchange of leaves. Planta 153:376–387

Von Maltitz G, Scholes B, Pienaar M, et al (2020) National terrestrial carbon sinks assessment 2020. Technical Report. Pretoria, South Africa

Westoby M, Walker B, Noy-Meir I (1989) Opportunistic management for rangelands not at equilibrium. J Range Manag 42:266. https://doi.org/10.2307/3899492

Whitman T, Lehmann J (2015) A dual-isotope approach to allow conclusive partitioning between three sources. Nat Commun 6:8708. https://doi.org/10.1038/ncomms9708

Wigley BJ, Augustine DJ, Coetsee C et al (2020) Grasses continue to trump trees at soil carbon sequestration following herbivore exclusion in a semiarid African savanna. Ecology 101:e03008

Williams CA, Hanan N, Scholes RJ, Kutsch W (2009) Complexity in water and carbon dioxide fluxes following rain pulses in an African savanna. Oecologia 161:469–480. https://doi.org/10.1007/s00442-009-1405-y

Williams IN, Torn MS, Riley WJ, Wehner MF (2014) Impacts of climate extremes on gross primary production under global warming. Environ Res Lett 9:94011. https://doi.org/10.1088/1748-9326/9/9/094011

Yepez EA, Williams DG (2009) Precipitation pulses and ecosystem carbon and water exchange in arid and semi-arid environments. In: De la Barrera E, Smith WK (eds) Perspectives in biophysical plant ecophysiology: a tribute to park S. Nobel. Universidad Nacional Autónoma de México, México, p 27

Zhou Y, Pei Z, Su J et al (2012) Comparing soil organic carbon dynamics in perennial grasses and shrubs in a saline-alkaline arid region, northwestern China. PLoS One 7. https://doi.org/10.1371/journal.pone.0042927

Trends and Barriers to Wildlife-Based Options for Sustainable Management of Savanna Resources: The Namibian Case

Morgan Hauptfleisch, Niels Blaum, Stefan Liehr, Robert Hering, Ronja Kraus, Manyana Tausendfruend, Alicia Cimenti, Deike Lüdtke, Markus Rauchecker, and Kenneth Uiseb

Abstract

Use of wildlife as an alternative or complimentary rural livelihood option to traditional farming has become popular throughout southern Africa. In Namibia, it is considered a climate change adaptation measure since livestock productivity has declined across much of the country in the past few decades. In contrast with neighboring South Africa, Namibian landowners and custodian often avail large open areas to this purpose, such as in the communal conservancies where fences are prohibited. The SPACES II ORYCS project considered wildlife management in a multiple land-use and tenure study area in Namibia's arid Kunene region. The aim was to investigate positive and negative impacts of the inclusion of wildlife on livelihoods and ecosystem services. Movement is recognized as an important survival strategy for wildlife in arid landscapes such as Namibia's north-west, and this study found that movement barriers within and between the land uses could present a challenge to wildlife survival and productivity. Notwithstanding, wildlife persisted in crossing many of these barriers, including the national veterinary cordon fence to satisfy their requirements. This often led to human–wildlife conflict, especially with elephants and predators. Interviews

M. Hauptfleisch (✉)
Biodiversity Research Centre, Namibia University of Science and Technology, Windhoek, Namibia
e-mail: mhauptfleisch@nust.na

N. Blaum · R. Hering
Plant Ecology and Nature Conservation, University of Potsdam, Potsdam, Germany

S. Liehr · R. Kraus · M. Tausendfruend · A. Cimenti · D. Lüdtke · M. Rauchecker
Institute for Socio-ecological Research, Frankfurt am Main, Germany

K. Uiseb
Ministry of Environment, Forestry and Tourism, Windhoek, Namibia

© The Author(s) 2024
G. P. von Maltitz et al. (eds.), *Sustainability of Southern African Ecosystems under Global Change*, Ecological Studies 248,
https://doi.org/10.1007/978-3-031-10948-5_18

found that despite this conflict, an understanding of the need for wildlife and general biodiversity provided complimentary livelihood opportunities and improved land productivity.

18.1 Wildlife as a Complementary or Alternative Livelihood Strategy in Arid Environments

Free-range beef production has been the cornerstone of Namibia's rural livelihood strategy since colonial times (Lange et al. 1998). For freehold farmers south of the veterinary cordon fence (see Fig. 18.1) cattle farming is a commercial venture on individually owned freehold units, while north of the fence it was, and remains, mostly for cultural, ploughing and homestead use (Groves and Tjiseua 2020; Mendelsohn 2006). Crop farming is limited to the far north and north-east of the country where rainfall is above 600 mm per annum. Most of Namibia however is considered marginal for livestock farming as a result of its aridity (Mendelsohn 2010). Cattle numbers peaked in the 1950s with around two and a half million head and has since gradually declined. Reasons for the decline include land degradation as a result of overgrazing and deteriorating grazing production caused by bush thickening and encroachment (De Klerk 2004).

Another factor affecting rangeland productivity for cattle, linked to abovementioned land degradation and bush encroachment, is purported to be climate change (Midgley et al. 2005; Reid et al. 2008). Much of Namibia is expected to receive

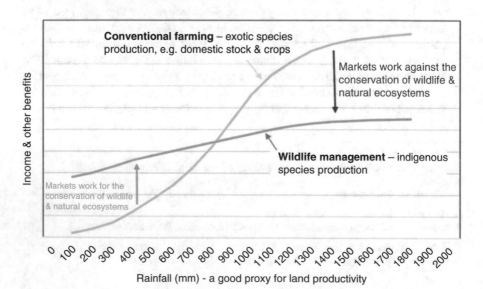

Fig. 18.1 Revenue and benefits from wildlife and commercial farming in relation to annual rainfall (Brown 1996)

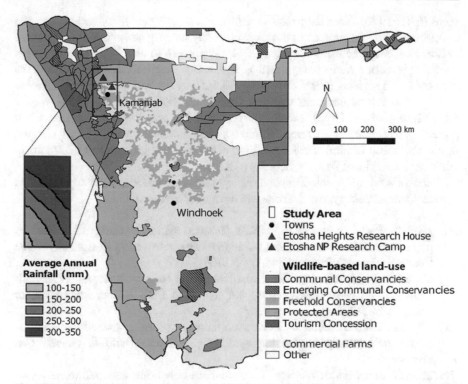

Fig. 18.2 Wildlife land use in Namibia, highlighting the ORYCS project study area (ORYCS 2018)

lower and more erratic rainfall (Barnes et al. 2012; Midgley et al. 2005) leading to increased pressure on rangelands and farm revenue. Wildlife and tourism-related revenue streams are considered effective strategies to either replace or compliment livestock production on communal and commercial farmland (Crawford and Terton 2016). Brown (1996) postulated that livestock and other agricultural enterprises are marginal at best considering Namibia's poor and erratic rainfall. As illustrated in Fig. 18.1, wildlife revenues and benefits seem to outperform conventional farming in arid environments while commercial agriculture is more viable at higher rainfall. Since the SPACES II ORYCS study site (blue box in Fig. 18.2) receives annual rainfall of between 250 and 400 mm per year, this area is ideal to investigate in terms of the value wildlife provides, and the attitudes of persisting livestock farmers toward wildlife. More detail of the project is provided later in this subchapter.

Since the promulgation of the Nature Conservation Ordinance in the 1970s, commercial farmers on freehold land were given right of use and ownership of wildlife on their properties, driven by the economic benefit that adding this income stream on their farms could bring (GRN 1975; Lindsey et al. 2013; Schalkwyk et al. 2010). These rights were only afforded to communal residents after Namibia's independence with the gazetting of the Nature Conservation Amendment Act of

1996 (GRN 1996). This led to the establishment of communal conservancies, able to utilize wildlife through meat production, trophy hunting and eco-tourism (Ashley and Barnes 2020; Murphy and Roe 2004; Schalkwyk et al. 2010). Consequently, wildlife protection and management is now being practiced across 45.6% of the country's land surface. (GRN 2018) (Fig. 18.2). This has brought with it an increase in large indigenous mammal numbers from under 500,000 in the 1960s to over 3.5 million (Brown 1996), mostly through wildlife breeding, translocations and protection as economic resource. Within communal conservancies alone, it resulted in a contribution of almost N$ 9.8 billion to net national income and 5200 jobs in addition to traditional farming benefits (GRN 2018).

Namibia's wildlife protection and management systems vary similarly across its different land-tenure systems. These can be categorized as follows:

(i) National Parks (green in Fig. 18.2): Gazetted national conservation areas as part of Namibia's commitment to biodiversity conservation as signatory to the Convention on Biological Diversity (CBD);
(ii) Communal conservancies: Registered under the Nature Conservation Amendment Act of 1996 (GRN 1996) not separated from human settlements or livestock by fencing;
(iii) Freehold Conservancies: A loose association of privately owned game farms, mostly individually fenced but agreeing to common wildlife conservation objectives; and
(iv) Tourism concessions relying on eco-tourism benefiting national conservation and sustainable use objectives.

A unique element to the Namibian wildlife management sector is the large variation in spatial scale of wildlife management units. Three of Namibia's national parks are in excess of 2 million hectares, and since communal conservancies are unfenced, there are clusters of conservancies which allow for free movement of wildlife across vast areas. As example, the north-western conservancy cluster stretching from the veterinary cordon fence in the south to the Kunene river in the north spans over 4.2 million hectares (Bollig and Menestrey Schwieger 2014). By contrast a private game farm can legally be as small as 1000 hectares and is fenced (Nature Conservation Ordinance 4 of 1975). The social-ecological feedbacks, diversity and productivity of wildlife versus livestock land use have been studied on individually fenced neighboring commercial farms during the study, but ecological differences caused by spatial size variations in wildlife management units have not been assessed. Against the background of climate change, the research project aimed to close the knowledge gap in the design of adapted land use management options for semiarid regions. This includes differing perspectives of stakeholders on the interactions between wildlife management and ecosystem services in fenced areas compared to areas where wildlife movements are unrestricted, and to consider them in shaping adequate solutions (Hauptfleisch 2018).

18.2 Wildlife Management in the Etosha South-West Landscape Case Study (The SPACES II ORYCS Project supported by the Biodiversity Economy in Landscapes Project)

The SPACES II project ORYCS (Options for sustainable land use adaptations in savanna systems: Chances and risks of emerging wildlife-based management strategies under regional and global change) investigated ecological and social factors associated with wildlife-based land-use strategies at vastly varying spatial scale in the western part of Etosha National Park and surrounding commercial and communal areas (blue box in Fig. 18.2). This multiple land use and management landscape covering more than 30,000 km^2 provides an opportunity to compare the ecological and social linkages within and between each management type containing wildlife. This includes Ehirovipuka and #Khoadi−//Hôas Communal Conservancies, Etosha National Park (over 2.2 million hectares), Etosha Heights Private Reserve (a larger fenced area of 50,000 hectares) and individually fenced cattle and wildlife farms with an average size of +/−5000 hectares each. Table 18.1 illustrates the different wildlife-based land uses, and specific benefits they derive from wildlife consumptive and nonconsumptive use.

18.2.1 The Importance of Long-Distance Mobility for Wild Herbivore Survival in the Arid Savanna

It is well known in ecological science that the main survival strategy of large mammals in arid landscapes is movement (Bailey 2004; Wato et al. 2018). Herbivores disperse to find growing vegetation, followed by carnivores seeking the migrating protein. Impressive "treks" of large springbok herds were regularly documented in the nineteenth and twentieth centuries (Scully 1913), and John Skinner was convinced that fences would not have stopped the movement of these large herds (Skinner 1993). Rinderpest and breach-load firearms largely laid waste to the herds and their migration. Humans have used a similar strategy of regular movement in arid areas, and even today the Ovahimba pastoralists outwit the unpredictable droughts of northern Namibia and southern Angola through following scattered rain-fed grasslands (Gibson 1977). In contrast, fencing is seen to be a key drought buffer strategy of the modern western-trained pastoralist (Barnes and Denny 1991). Controlling the movement of livestock through camping, it allows for micro-level rotational grazing and the protection of grazing camps for drought conditions (Tainton 1999). Ironically, fencing allowed arid-savanna farmers to apply Savory's high intensity/long rest method to pastures which mimic the large herd migrations by trampling and heavily grazing a rangeland before moving to the next. (Savory 1983; Savory and Parsons 1980) This has led to most livestock range, no-matter how marginal to be checkerboarded by fencing.

Table 18.1 Characteristics of different land-use units which derive benefits from wildlife within the Etosha South-West Landscape study area

Land-use type	Tenure type	Livelihood commodities	Management of wildlife	Benefits from wildlife
National Park	Government owned, proclaimed a National Park under Ordinance 31 or 1967	Wildlife only	Extensive passive management	National reputational benefits in protecting critically endangered species, tourism, source of breeding stocks for other parks and communal conservancies, wildlife research
Private Nature Reserve	Private ownership of multiple farms amalgamated with no internal fencing. Private concessions for tourism and/or hunting with tourism or hunting operators. No formal proclamation.	Ecotourism and hunting (wildlife only)	Privately managed, consumptive use of wildlife is regulated by government through the Nature Conservation Ordinance 4 of 1975 (GRN 1975)	Exclusive "high-end" eco-tourism, occasional hunting for rations
Individual hunting farm	Private ownership either by an individual or Closed Corporation	Mostly mixed wildlife and livestock	Privately managed, consumptive use of wildlife is regulated by government through the Nature Conservation Ordinance 4 of 1975 (GRN 1975)	Trophy hunting, commercial meat hunting, products from wildlife skins
Communal conservancy	Owned by government but managed by an elected conservancy committee as mandated through the Nature Conservation Amendment Act of 1996 (Act 5 of 1996) (GRN 1996)	Mixed wildlife and livestock, and occasionally crop production for own use	Managed by the conservancy committee with support from Namibian and international nongovernmental organizations (NGOs). Tourism ventures are mostly operated through concessions by recognized tourism specialist companies	Eco-tourism, trophy hunting, meat hunting, products from wildlife skins

18.2.2 Private Wildlife Farming and Fencing

Many Namibian farmers have over the past 30 years gradually added wildlife to their production activities, and indigenous wild herbivores have been reintroduced to the micro-fenced land as personally owned assets. Namibian legislation currently requires game-proof fencing as a condition of private wildlife ownership on farms (GRN 1975). Fences are known to be effective in limiting and containing some animal diseases. Namibia's commercial beef production is protected from Foot-and-Mouth (FMD) disease by a through the veterinary cordon fence which traverses the country for 1300 km. The double-fence intends to prevent possibly FMD infected cloven-hoofed mammals (wild and domestic) from carrying the disease to the commercial farmland south of the fence.

There has been scant research on the effect of fencing on wildlife movement in Namibia, although alarming mortalities of wildlife along fences have been recorded, mostly in neighboring Botswana (Mbaiwa and Mbaiwa 2006; Williamson 1981). Anecdotally, farmers immediately south of the Namibian veterinary cordon fence (VCF) have become increasingly aware of regular breaches of the fence, and wildlife moving into their farmland, especially those bordering Etosha National Park ("Close to N$500 million needed to fix damaged Etosha fence" n.d.). This is discussed in more detail below.

18.2.3 Wildlife Movement and the Veterinary Cordon Fence (VCF)

The complexity of the situation increases when not only local interests are considered, but also international stakeholders who influence local fence construction. A VCF stretches from west to east through the country (red line in Fig. 18.3) and runs along the southern boundary of Etosha National Park. It was constructed in the 1960s after the occurrence of FMD outbreaks in the northern part of the country (Schneider 2012). According to the Animal Health Act Regulations (GRN 2018), today's area south of the cordon fence is declared free from the contagious bovine pleuropneumonia (CBPP) and FMD, the latter being a highly infectious disease that may be transmitted from wildlife to livestock (Gadd 2012). Only beef originating from this area south of the fence is allowed to be imported to the European Union (Kreike 2009; Gadd 2012).

Veterinary fences and strict controls enable commercial farmers to participate in international markets (McGahey 2011). In 2019, beef with the value of around 44.8 million euros was imported from Namibia to the EU (European Commission's Agricultural and Rural Development Department 2021). However, the benefits and the long-term maintenance of this fence is increasingly under scrutiny among different interest groups: On the one hand, negative effects on wildlife ecology are increasingly observed (Martin 2005); On the other hand, the trade regulations set by the World Organization for Animal Health (OIE) implore government to maintain the fence which is frequently damaged by elephants and other wildlife

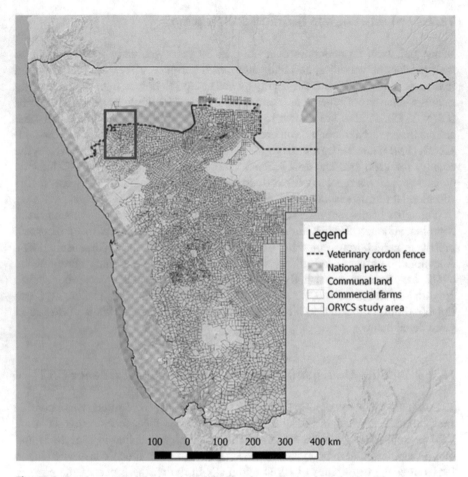

Fig. 18.3 Land tenure and the location of the Veterinary Cordon fence in Namibia

(Grant et al. 2008; Smit 2018). Interviews with local farmers in 2019 revealed their concern regarding dependency on the EU, as their business heavily relies on the fence's integrity. Considering the costs of the VCF, Scoones et al. (2010) suggest detachment from the fence-dependent strategy and instead to apply other mechanism like commodity-based trade: setting the focus on an acceptable risk coming with the product to be traded. This may require seeking other international markets or lobbying the EU to amend mechanisms of disease control. In this regard, certification was suggested to adapt to additional markets and allow more farmers to benefit. This strategy might be interesting for Namibia whose vision is to also develop the livestock sector north of the VCF in order to integrate farmers into international markets (Meat Board of Namibia 2015).

The presence of the VCF across the Etosha South-West Landscape is a conundrum for livelihoods in the area. This is detailed in the next section which

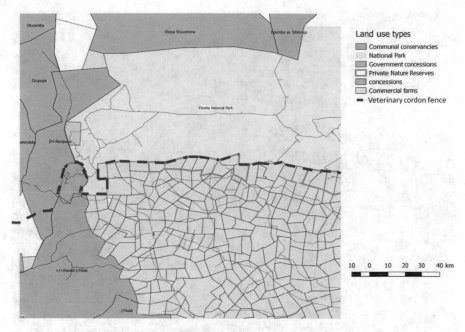

Fig. 18.4 Land use in the Etosha South-West Landscape

documented livestock farmers' attitudes toward wildlife within the landscape. The barrier to improve cattle commercial value in the southern portion of the landscape clearly reduces the ability of wildlife to move and impedes peripheral benefits to communities surrounding the Park. When scrutinizing land use along the southern boundary of Etosha (Fig. 18.4), three wildlife sanctuaries have been established on rewilded commercial livestock land. Ongava Private Wildlife Reserve, Etosha Heights Private Reserve and the Karros hunting farms derive their revenues entirely from consumptive and nonconsumptive wildlife utilization. These areas are comprised of combined rewilded livestock farms and benefit from resident wildlife and, in some cases from wildlife movement across the VCF. Despite regular maintenance of the fence the need for wildlife to move in relation to resource availability and territorial expansion causes regular breaching of the fence by large wildlife.

This study used camera trap monitoring of three regularly breached sites (e.g., Plate 18.1) along the fence over a four week period in October 2019. Over a combined 81 trap days, 10,362 wildlife-fence interactions of 20 different species were recorded at the VCF with only 1058 individuals turning back at the fence-line the rest moving through the fence-line. This preference to traverse the fence with little hesitation indicates that the cross-fence movement is regular and habitual. The highest recorded fence-crossing species was springbok (*Antidorcas marsupialis*) with 8454 separate interactions with the fence followed by gemsbok (*Oryx gazella*) with 1058 and common eland (*Taurotragus oryx*) with 486 observations. During the study period 38.54% of the movements were from the National Park

EH003   20/01/2022 11:44:53   PIR Trigger   18 30°C 86°F   10

Plate 18.1 Camera trap image of a fence breach used by wildlife to cross the veterinary cordon fence. The elephant in the center of the photo is between the farm fence (left) and the national park fence (right). The breaches are mostly created by elephants

into the southern commercial sanctuaries and 61.46% into Etosha National Park. Movements are detailed in Table 18.2.

Two key determinants likely drove the extensive movement of wildlife across the VCF in the study area. (i) surface water resources—the study area is almost exclusively supplied with surface water by pumping groundwater into artificial reservoirs. As in most arid savannas the distribution of these reservoirs is a key determinant of the distribution patterns of wildlife throughout the landscape (Crosmary et al. 2012). Matson et al. (2006) found that black-faced impala (*Aepyceros melampus petersi*) seldom moved further than 1 km from water in Etosha, while elephant movements were also found to be highly influenced by available surface water (De Beer et al. 2006; Wato et al. 2018) but would move up to 70 km from water in search of sufficient forage (Leggett 2006). Water is scarce in the Etosha National Park portion of the landscape and mostly limited to waterholes along the 19^0 latitude (Fig. 18.5). This was partly to reduce the movement of wildlife, and hence human–wildlife conflict, in farms south of the Park. Evidence in Table 18.2 disproves the effectiveness of this strategy. The high density of artificial surface water in the commercial farmland south of the Park is likely a significant attractant for wildlife.

Table 18.2 Movement of wildlife across the Veterinary Cordon Fence between Etosha National Park and neighboring Private Nature Reserve farms in October 2019

Species	Number of animals			
	Southward	Northward	Turning at the fence	Moving along the fence
Springbok (Antidorcas marsupialis)	2968	5011	284	191
Gemsbok (Oryx gazella)	389	587	39	43
Common eland (Taurotragus oryx)	230	245	4	7
Giraffe (Giraffa giraffa angolensis)	47	39	0	10
African elephant (Loxodonta africana)	38	10	0	6
Greater kudu (Tragelaphus strepsiceros)	14	2	0	6
Blue wildebeest (Connochaetes taurinus)	9	16	5	4
Steenbok (Raphicerus campestris)	8	16	1	22
Cheetah (Acinonyx jubatus)	6	2	0	0
Black-backed jackal (Canis mesomelas)	5	4	1	5
Burchell's plains zebra (Equus quagga burchellii)	5	5	14	1
Red hartebeest (Alcelaphus buselaphus caama)	4	3	1	2
Common warthog (Phacochoerus africanus)	3	1	0	0
Hartmann's mountain zebra (Equus zebra hartmannae)	3	0	0	0
Leopard (Panthera pardus)	2	0	0	1
Caracal (Caracal caracal)	1	0	0	0
Brown hyena (Hyaena brunnea)	0	6	0	1
Scrub hare (Lepus saxatilis)	0	1	0	0
Total:	**3727**	**5943**	**335**	**298**

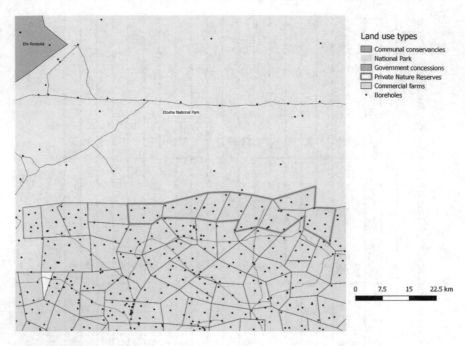

Fig. 18.5 Water-point distribution in the west of Etosha National Park and surrounding land use

Telemetric monitoring of giraffes in the Etosha South-West Landscape provided an apt example of this (Fig. 18.6). Giraffe female 3620 was fitted with a satellite tag in Etosha National Park (19.0147^{0}S, 14.8196°E) 18.77 km due north of the southern boundary of the Park on 25 July 2020. In the late dry season of 2021, Fig. 18.6 clearly illustrates the reliance of this animal and assumed associated herd on a waterhole outside the Park while continuing foraging within the Park.

Seasonal foraging preferences were also determined to affect movements. The Namibia University of Science and Technology's Biodiversity Research Centre (NUST-BRC) and Potsdam University satellite collared a total of 57 springbok, eland, kudu and giraffe to investigate their movements in the Etosha South-West Landscape. Over 1.8 million GPS-localizations at a high temporal resolution (every 5–15 min) were recorded considering vegetation, management structures such as water points and fences, and land use type. The species that did not challenge fences to a great degree were sedentary browsing kudu which occupied very restricted areas of a few dozens of square kilometers and spend most of their time in woody vegetation with only slight seasonal variation (Fig. 18.7 left). Nomadic, mixed feeding elands used vast areas of hundreds of square kilometers, constantly moved around and shifted vegetation use from shrublands during dry season (May to October) to more grasslands during the rainy season (November to April) and crossed the VCF multiple times (Fig. 18.7 center). Mixed feeding springboks formed temporal home ranges, which shifted in the course of the year and used a

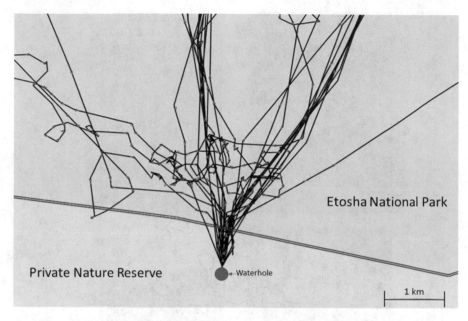

Fig. 18.6 Tracks of four-hourly movements of giraffe 3260 between September 1 and October 15, 2021. Light green shading represents Etosha National Park, light yellow area an adjacent private nature reserve and the double red line the VCF (Earthranger 2021)

few hundreds of square kilometers in total. Springboks mostly used open grasslands but shifted to more shrubby vegetation in late dry and early rain season (Fig. 18.7 right), mostly crossing the VCF to do so. The longest distances travelled away from the mostly visited waterholes was 44.8 km (mean: 5.4 ± 10.2 SD) for kudu, 73.4 km (mean: 10.2 ± 10.2 SD) for eland and 86.9 km (mean: 8.4 ± 12.2 SD) for springbok.

Thousands of animal–fence interactions occurred all year long and reached highest numbers right before the rain season started. Despite porosity of the VCF in some areas, fences limited movements on a large scale (Fig. 18.8). For instance, eland home-ranges differed vastly between areas without fences, compared to fenced areas (Etosha National Park: 22895 km^2, Etosha Heights Private Reserve: 480 km^2, private farm with livestock and wildlife: 99 km^2).

18.3   North-West Namibian Farmers' Perception on Coexistence with Wildlife

The results of Sect. 18.2 suggest that rural communities in the study area live in close proximity with wildlife, some without choosing so. This is not unique with wildlife and human coexistence being prevalent in many African countries and is expected to increase globally (Lamarque et al. 2009; Carter and Linnell 2016). Interaction with wildlife can generate ecological and societal challenges and various conflicts

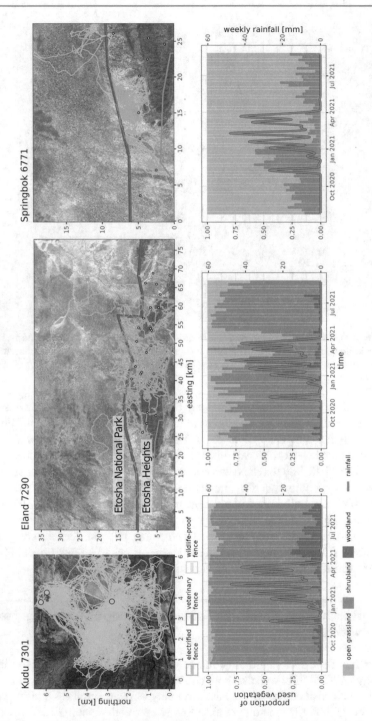

Fig. 18.7 Examples of antelope movements and used vegetation type in the Etosha South-West Landscape. Shown are one year movement tracks (first row) of three antelope species (left—kudu, center—eland, right—springbok) and proportion of weekly used vegetation (second row) with general vegetation types (see legend) and amount of weekly rainfall (blue line). Note that the scales of the maps differ

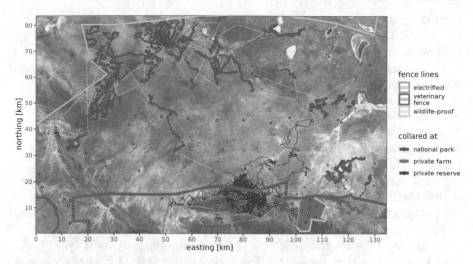

Fig. 18.8 Tracks of three eland females from August 2020 to August 2021 collared at three places of different land use. Fence lines of different types are indicated. Water sources are shown as blue points (note: points north of the red line not necessarily represent actual water points). Note: data of eland with green points contains data recording gaps due to transmission schedule, one day sections of consecutive points are connected by gray lines, gaps last three days each

(Martin et al. 2020). This introduces a range of risks for both (Carter et al. 2020). On the one hand, humans can degrade and fragment habitats of wildlife or even directly kill certain wildlife species. On the other hand, wildlife can be a reason for damages and threats to human well-being (Ceauşu et al. 2018). The intention of successful coexistence is often to maximize human benefits; nevertheless, the interaction can also result in nonbeneficial outcomes and can induce effects that are harmful to human interests (Dorresteijn et al. 2017; Campagne et al. 2018). Yet, wildlife populations are essential for a functional ecosystem and provide both benefits and costs for human society (Ceauşu et al. 2018).

In recent years there has been more scientific interest on topics exploring human–wildlife conflict and coexistence (e.g., Jordan et al. 2020; Lamarque et al. 2009; Marker and Boast 2015; Matinca 2018; Seoraj-Pillai and Pillay 2017). In Africa, for example, Namibia is one of the countries that includes conservation and sustainable use of natural resources in its constitution, resulting in a noticeable conservation success and increased wildlife numbers (MET and NACSO 2018; Jones and Barnes 2006). However, according to the Namibian Ministry of Environment, Forestry and Tourism (MEFT) (2020), human–wildlife conflicts remain an ongoing challenge.

To learn more about the driving forces of human–wildlife conflicts and farmer attitudes toward certain wildlife species, structured interviews (surveys) were conducted within the Etosha South-West Landscape study area. Interviews were conducted with 20 commercial and ten communal farmers. Specific activities on the commercial freehold farms include commercial livestock (cattle, sheep, goat), wildlife (conservation hunting, photo tourism, meat hunting/sales), a mixture of both or other activities (e.g., charcoal production). Communal farmers, on the other hand, live and manage state land that is allocated to a traditional authority; they do not have private ownership. They mostly earn their livelihood with small-stock farming but can also benefit indirectly from utilization of wildlife through the conservancy (MEFT 2020). This can be either through tourist activities or, within the allocated limits of a government controlled wildlife utilization quota, meat and trophy hunting. The interviews with commercial farmers were conducted in English or German and the interviews with communal farmers were held in their indigenous language with the help of a local translator. The survey included questions about general attitudes toward specific wildlife species. The survey also included questions about cost and benefits of wildlife as well as opinions about wildlife heritage value. In addition, the same survey was also used to investigate the perception of biodiversity (see below) and incorporated another set of questions about the farmers' knowledge background and policies about biodiversity.

A key aspect of human–wildlife coexistence is attitudes toward species. All respondents felt *happy*, *excited*, or *fine* about sharing land with giraffes, oryxes and springboks (Fig. 18.9). In contrast, leopards, hyenas, and elephants evoked contrasting feelings among the farmers. The majority of respondents expressed negative feelings toward hyenas and elephants: 64% had negative feelings (*afraid*, *angry*, *disgusted*) toward hyenas and 56% toward elephants. Overall, leopards were perceived slightly more positively, with 56% of the respondents feeling *happy*, *excited* or *fine* about them. However, the attitudes toward these animals strongly

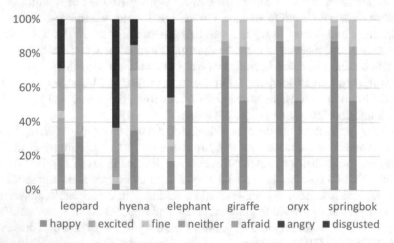

Fig. 18.9 Livestock vs. wildlife-based farmers' general feelings toward wildlife species; red colors represent negative feelings, green colors represent positive feelings; each left bar per category indicates the feelings of the livestock farmers, the right of the wildlife-based farmers; Answers from structured interviews: $N = 30$, with $N_{livestock} = 24$, $N_{wildlife-based} = 6$

depended on the farm income-generating activity (livestock vs. wildlife-based [1]; Fig. 18.9).

Wildlife-based farmers felt exclusively positive (*happy, excited*) toward elephants and leopards, while livestock farmers have a variety of feeling toward these species. For leopards, in each case 25% of the livestock farmers responded they felt *afraid*, or *angry*, and 4% of them were even *disgusted* by them. Notwithstanding, 46% of the livestock farmers felt positive toward them (*happy*: 21%; *excited*: 21%; *fine*: 4%). Elephants evoked slightly more negative feelings among the livestock farmers (*afraid*: 17%; *angry*: 33%; *disgusted*: 13%). However, the species most negatively regarded by livestock farmers was spotted hyena. In each case, only 4% felt *happy* or *fine*, whereas 13% felt *afraid*, 29% felt *angry* and 33% *disgusted*.

The study highlights that nondestructive species, such as antelope species or giraffes are viewed exclusively as positive (Fig. 18.9). One reason might be that this group has a benefit-cost ratio that is strongly shifted toward benefits. Without generating high maintenance or costs, these animals can contribute to livelihoods as an additional source of income through live and meat sales or conservation hunting practices and tourism (MEFT 2020). In addition, farmers who do not directly benefit from them, also do not have problems, instead, they refer to these animals as "beautiful to look at" (Interviewee no. 18, commercial livestock farmer) or as a "nice experience for the children" (Interviewee no. 25, communal livestock farmer). In contrast, especially predators such as leopards and hyenas as well as elephants evoke negative feelings among many livestock farmers. These species are predominantly associated with costs as they depredate livestock or damage crops/infrastructure and can be a threat to human well-being. However, the feelings toward these animals were highly dependent on the main farm income-generating activity (i.e., management type) (Fig. 18.9). For example, a farmer with wildlife-based activities (e.g., conservation hunting, photo tourism) might perceive hyenas, leopards, and elephants as financial assets. In this case, the benefits outweigh the possible conflicts, and therefore, wildlife roaming around locally is viewed positively. In contrast, for many livestock farmers, predators and elephants create conflicts by killing livestock or damaging infrastructure without generating a direct benefit.

In conclusion, the attitudes of communities sharing a landscape with nondestructive species were mostly positive across all land-use types, while perceptions toward predators and elephants were highly dependent on the farm management type.

---

[1] Within the study, livestock refers to farms with only livestock (18) and the ones with a mixture of livestock and wildlife (6). Wildlife-based refers to only wildlife-based (i.e. farms that offer conservation hunting and/or photo tourism; 5) or other activities (e.g. charcoal production; 1). Communal farmers (10) were only included in the "livestock only" group, as they only indirectly benefit from wildlife through the Conservancy. Commercial farmers (20) however practice a greater variety of main farm income-generating activities.

18.4   Namibian Farmers' Perceptions of Biodiversity

Attitudes toward wildlife, and tolerance thereof, often depend on a person's understanding of being part of ecosystems biodiversity, which has been found to differ significantly between communal and commercial farmers. Hence, it is important to examine why people tolerate or do not tolerate co-occurring with wildlife, as well as to enhance, and develop strategies to improve coexistence (Carter et al. 2020; Slagle and Bruskotter 2019).

Biodiversity has been defined by the Convention on Biological Diversity (CBD 1992) as "diversity within species, between species, genes and of ecosystems" (CBD 1992: Art. 2) and has received extensive global research. However, how the different local population groups in the study area, including both communal and commercial farmers, relates to biodiversity concepts is unclear (Bischofberger et al. 2016). Farmers have a daily experience of the natural environment, which brings them to have a more functional view on biodiversity that might differ from the scientists' view (Kelemen et al. 2013).

In addition to perceptions of wildlife, farmers' perceptions of their natural environment were examined, which could influence their land management decisions (Warren 1996). Farmers were asked for their definition of biodiversity. Furthermore, the farmers' evaluation of the temporal change (i.e., now vs. 10 years ago) of the number of different species of large mammals (heavier than 1 kg), small mammals, reptiles, insects and plants gave us an insight of the farmers' perception of biodiversity in practice. To connect the farmers' perception of biodiversity to their farming activities, we asked if specific animal groups (i.e., carnivores, grazing herbivores, browsing herbivores, big herbivores, small mammals, reptiles and insects) have a positive or negative impact on the land productivity. Here, land productivity refers to broad indicators of land health such as soil health, biomass productivity and livestock and wildlife well-being.

Figure 18.10 summarizes that 37% of all farmers did not know or did not have a direct translation for the term "biodiversity" (Fig. 18.10). Consequently, a short explanation was given; in a few cases it was still unclear and the question has been reformulated as "what is nature for you?". Most farmers' own description of biodiversity referred to natural elements such as plants, animals and soil (Fig. 18.10), of which animals are mentioned most often. Similar numbers of commercial and communal farmers included concepts of interrelationships (e.g., diversity, balance) and beneficial elements (e.g., resources, food and farming) in their understanding of biodiversity. Apart from that, some distinct differences between commercial and communal farmers appeared in the descriptions of biodiversity. Commercial farmers used terms closer to scientific concepts (e.g., ecosystems, species) more frequently. Furthermore, only commercial farmers linked sustainability and religious aspects to it. In contrast, for communal farmers "biodiversity is humans, wildlife and trees" (Interviewee no. 26, communal livestock farmer). In other words, they included humans significantly more often as part of biodiversity. In addition, time-related aspects of biodiversity, like *future* and *past*, held more importance for them. This

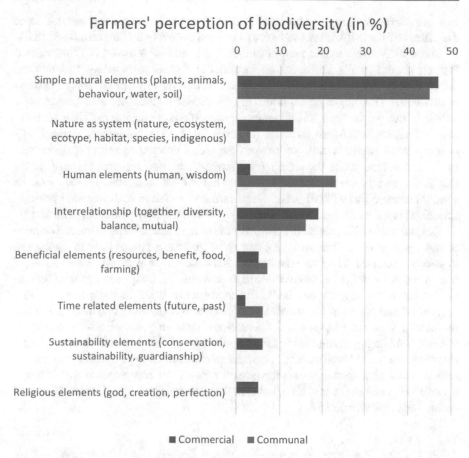

Fig. 18.10 Comparison between commercial and communal farmers' keywords to define biodiversity

seems to indicate a much more inclusive view of biodiversity and nature regarding humans being a part of it in addition to animals and plants, an understanding common to indigenous communities on different continents (Descola 2012). In comparison, commercial farmers understood humans outside of biodiversity and rather saw the linkages between humans and nature, with mutual influence, but also with clear separation: "we have to manage nature in order to preserve it" (Interviewee no. 9, commercial wildlife-based farmer).

Regarding the temporal change in numbers of different species, the majority of communal farmers observed a stable state, while the majority of commercial farmers noted changes (increase or decrease) in species composition except for small mammals and reptiles. Moreover, most communal farmers only saw a positive effect of big browsing and grazing herbivores especially for bush control but did not mention grazing competition with their livestock. They also did not see an effect on land productivity of the other animal groups (carnivores, small mammals, reptiles).

In contrast, the majority of commercial farmers expressed a positive perception of the effects of the different animal groups on land productivity (see also Sect. 18.3).

These differences in the answers might have various reasons. All interviewed farmers shared family sentiment on farming, and here, some ethnic and cultural differences could explain the discrepancies in the views of biodiversity (Olbrich et al. 2009). This might be explained by the closer relation between humans and nature in the worldviews of indigenous communities, opposing the clear human–nature division in modern societies (Descola 2012). The communal farmers' all-encompassing understanding of nature might lead to a perception of a static state in the ecosystem, while the commercial farmers' use and resource view of nature might induce a perception of change. Only half of the communal farmers went to school compared to 95% of commercial farmers, which at least attended primary school. This may also lead to different views of biodiversity (Kelemen et al. 2013).

Despite some differences, the majority of farmers acknowledged the importance of most groups of animals and the place value that these groups have in the system to keep the balance. They also mentioned that the number of animals plays a role, if one group is too large the system would be unbalanced. Despite the general feeling of communal farmers that number of different species is stable, many farmers, both communal and commercial, are aware of ecosystem complexity and its cycles. They mentioned seasonal changes and fluctuations between years of aridity and good rainfall influencing animal and plant populations to shrink or thrive accordingly. However, the term "biodiversity" remains an abstract concept for everyone but its understanding interestingly encompasses a richness of aspects from the different dimensions, such as the scientific, but also the economic, social and ethical (Gayford 2000; Wals and Weelie 1997).

18.4.1 Case Study on Human–Elephant Conflict in the Etosha South-West Landscape

African elephants (*Loxodonta africana*) are a keystone and charismatic species, however, reports on conflicts between elephants and humans have increased throughout southern Africa (Cumming and Jones 2005; Twine and Magome 2007). Specifically in Namibia elephant populations increased as elephants result of conservation efforts and its economic value to the tourism and hunting sectors (MET 2007b; CITES 2016; Matinca 2018). While from the nature conservation perspective this trend is seen as a success, land users are increasingly challenged by elephants dispersing to communal and commercial farmland, reoccupying areas from which they have been absent for many years (MET 2018).

To complement the investigation of wildlife movements and perceptions on different land uses, farmer–elephant interactions were explored within the Etosha South-West Landscape. Many farmers live adjacent or in close proximity to Etosha National Park and, despite the previously described benefits derived from wildlife (Fig. 18.2), are prone to experience conflicts with wildlife (MET 2007a; Mackenzie and Ahabyona 2012). As part of this case study, a semistructured qualitative survey was conducted with 11 communal and commercial farmers and nine wildlife experts familiar with or experiencing human–elephant conflicts in the study area. Furthermore, observations of group discussions at a local meeting on human–elephant conflicts (governmental representatives, commercial and resettlement farmers) revealed insights into the local interactions and challenges associated with elephants.

The interviews and observations uncovered a rather conflicting interrelationship between farmers and elephants. In agreement with a study by Shaffer et al. (2019), the characteristics of these human–elephant conflicts were influenced by elephant and human variables and framed by press and pulse disturbances in the landscape. Land users perceived an increase in the local elephant population, as well as an influx from other areas, leading to destruction and over-use of vegetation resources, and a high demand on surface water. Elephants were further found to destroy fences which farmers use for livestock and wildlife management. Most of the survey respondents had a negative attitude toward elephants influencing their livelihoods, triggering fence construction as a barrier for elephant movement. These new fences likely affect wildlife movement and migration and could result in a displacement of conflict to different areas, and not a reduction.

Coexistence of humans and elephants is improved by the reduction of associated costs and the increase of benefits from the elephants (Thouless 1994). This is crucial when choosing well-adapted conflict mitigation and prevention measures in the face of future challenges such as climate and land use change that are expected to exacerbate competition over shrinking resources (Otiang'a-Owiti et al. 2011; cited by Shaffer et al. 2019).

There seemed to be a considerable influence of sentiments of the international community on this very localized conflict in the study area. Stakeholders proposed that consumptive use of elephants within the area is critical to be included in options for conflict mitigation. Trophy hunting, or as recently termed conservation hunting, was found to be an attractive option for affected stakeholders. In addition to reducing the local elephant population, it can provide considerable income which will increase tolerance from farmers, and also provide a means to repair damage to water infrastructure and fences. In 2020 elephants were offered as trophies for on average US$ 43,000 in neighboring Botswana (Sguazzin n.d.). In contrast, international animal welfare organizations reject any use and especially trophy hunting of elephants without having an understanding of local contexts ("South Africa and Namibia hit out hunting trophy ban" n.d.). As a response to the conflict in the study area, 22 elephants were sold and translocated from the study site immediately following this case study. A future study on the effect of this action on tolerances of the farming community toward elephants compared to results found during the study has been initiated.

18.5 Chapter Conclusion

Namibia has a unique blend of multiple land use and tenure activities which have seen benefits from wildlife. Within the arid study area of north-west Namibia it was found that almost all of these include livelihood benefits from wildlife to some extent through consumptive (hunting and meat) and/or nonconsumptive (eco-tourism) means. Depending on the strategies of wildlife use, there are however negative impacts, particularly from human–carnivore and human–elephant conflict.

Interviews with stakeholders in the study area clearly indicated that there has been an increasing trend of land-use practices toward extensive inclusion of wildlife into livelihood strategies for commercial as well and communal land users around Etosha National Park. This has resulted in an overall increase of wildlife distribution across the study area, as well as and an increase in overall abundance. A critical distinction from neighboring South Africa's increase in wildlife numbers on private land is that wildlife in Namibia had the ability to naturally occupy communal farmland in the extensive and unfenced communal conservancies. Uniquely this has also been evident in commercial farmland south of Etosha National Park where some landowners benefit through tourism and sustainable hunting from the movement of wildlife across boundaries and even across and through fences to reach grazing and water resources in the greater area. The arid landscape undoubtedly drives wildlife movements for survival, even across substantial barriers such as the Veterinary Cordon Fence.

Complexity in individual rural livelihood choices has an interesting impact on the commercial value of wildlife resources. Using the case study of elephants while within Etosha National Park, it is argued to be an exclusively beneficial and valuable asset. It not only attracts tourism, but also plays a role in improved ecological function in the Park since it reduces bush encroachment and makes browsing resources available to other herbivores through its wasteful browsing methods (Bothma and Du Toit 2016). As soon as that same animal leaves the Park onto commercial livestock farmland, its value immediately disintegrates, and it becomes a significant cost to the farmers since it damages fences and water infrastructure and competes for grazing and browse resources with livestock. The interplay between economics, ecology, conservation, social cohesion and sentiment is complex, but teasing out the dynamics will be an important guiding tool for decision-making and livelihood strategies in this dynamic environment.

There was significant governmental and nongovernmental assistance for communal farmers to develop communal conservancies and include wildlife use since the promulgation of the Nature Conservation Amendment Act 5 of 1996. This, together with the incentive of benefiting from wildlife while still being able to practice livestock farming has led to more acceptance of wildlife. For commercial farmers the inclusion of wildlife was based solely on their socio-economic needs.

Wildlife survival in this arid area with erratic rainfall patterns is dependent on their ability to move in order to find water and forage. The study investigated a large sample set of herbivore movements across the landscape, and with fencing being the major impediment to movement, its effects on disrupting movements and loss in required energy for the ungulates are clear indications of the negative effect fencing may have on extensive wildlife productivity.

Acknowledgments We thank the Namibian Ministry of Environment Forestry and Tourism for assistance and access to study areas. We thank the Giraffe Conservation Foundation, UNAM School of Veterinary Medicine for field collaboration, Etosha Heights Private Reserve and Etosha National Park for providing and facilitating use of the study area, members of the Ehirovipuka and #Khoadi-//Hôas Conservancies and farmers of the Kamanjab and Outjo districts for their responses to surveys. The research was conducted as part of the ORYCS project (01LL1804) that is funded by the German Federal Ministry of Education and Research (BMBF) in the context of its funding platform "Research for sustainable development" (FONA) and the research program "SPACES II – Science Partnerships for the Adaptation/Adjustment to Complex Earth System Processes." The Biodiversity Economy in Landscapes project through support from the Deutsche Gesellschaft für Internationale Zusammenarbeit (GIZ) provided additional support to the NUST Biodiversity Research Centre's work.

References

Ashley C, Barnes J (2020) Wildlife use for economic gain: the potential for wildlife to contribute to development in Namibia. CRC Press, Windhoek

Bailey DW (2004) Management strategies for optimal grazing distribution and use of arid rangelands. J Anim Sci 82:E147–E153

Barnes DL, Denny RP (1991) A comparison of continuous and rotational grazing on veld at two stocking rates. J Grassl Soc South Afr 8:168–173. https://doi.org/10.1080/02566702.1991.9648285

Barnes JI, MacGregor J, Alberts M (2012) Expected climate change impacts on land and natural resource use in Namibia: exploring economically efficient responses. Pastor Res Policy Pract 2:1–23

Bischofberger J, Reutter C, Liehr S, Schulz O (2016) The integration of stakeholder knowledge – how do Namibian farmers perceive natural resources and their benefits? In: Universität für Bodenkultur Wien (BOKU) (Hg.): solidarity in a competing world. Tropentag 2016, Vienna

Bollig M, Menestrey Schwieger DA (2014) Fragmentation, cooperation and power: institutional dynamics in natural resource governance in North-Western Namibia. Hum Ecol 42:167–181. https://doi.org/10.1007/s10745-014-9647-7

Bothma J d P, Du Toit JG (2016) Game ranch management. Van Schaik, Pretoria

Brown CJ (1996) The outlook for the future. Namib Environ 1:15–20

Campagne C, Roche P, Salles J (2018) Looking into Pandora's box: ecosystem disservices assessment and correlations with ecosystem services. Ecosyst Serv 30:126–136. https://doi.org/10.1016/j.ecoser.2018.02.005

Carter N, Baeza A, Magliocca N (2020) Emergent conservation outcomes of shared risk perception in human-wildlife systems. Conserv Biol 34(4):903–914. https://doi.org/10.1111/cobi.13473

Carter N, Linnell JD (2016) Co-adaptation is key to coexisting with large carnivores. Trends Ecol Evol 31(8):575–578. https://doi.org/10.1016/j.tree.2016.05.006

Ceauşu S, Graves R, Killion A, Svenning J, Carter N (2018) Governing trade-offs in ecosystem services and disservices to achieve human-wildlife coexistence. Conserv Biol 33:1–11. https://doi.org/10.1111/cobi.13241

Convention on biological diversity. (1992) Rio De Janeiro. Retrieved 8 Jan 2015, from http://www.cbd.int/doc/legal/cbd-en.pdf

Convention on International Trade in Endangered Species of Wild Fauna and Flora (CITES) (2016) CoP17 Prop. 14. CITES Secretariat, Geneva. Proposal to CITES. Available at https://cites.org/sites/default/files/eng/cop/17/E-CoP17-Prop-14.pdf

Crawford A, Terton A (2016) Review of current and planned adaptation action in Namibia. International Institute for Sustainable Development, Winnipeg

Crosmary W, Valeix M, Fritz H, Madzikanda H, Côté SD (2012) African ungulates and their drinking problems: hunting and predation risks constrain access to water. Anim Behav 83:145–153

Cumming D, Jones B (2005) Elephants in southern Africa: management issues and options. WWF-Sarpo occasional paper number 11, Harare. Available at http:// d2ouvy59p0dg6k.cloudfront.net/downloads/cumming___jones__ 2005__ele-phants_in_sthn__africa___mgmt_options___issues__wwf_sarpo_occ_.pdf

Descola P (2012) Beyond nature and culture. HAU J Ethnogr Theory 2:473. 10.14318/hau2.1.021

De Beer Y, Kilian W, Versfeld W, Van Aarde RJ (2006) Elephants and low rainfall alter woody vegetation in Etosha National Park, Namibia. J Arid Environ 64:412–421

De Klerk J (2004) Bush encroachment in Namibia. Ministry of Environment and Tourism, Windhoek

Dorresteijn I, Schultner J, Collier N, Hylander K, Senbeta F, Fischer J (2017) Disaggregating ecosystem services and disservices in the cultural landscapes of southwestern Ethiopia: a study of rural perceptions. Landsc Ecol 32(11):2151–2165. https://doi.org/10.1007/s10980-017-0552-5

Earthranger (2021) Protecting wildlife with real-time data. Available at https://nust.pamdas.org/

European Commission's Agricultural and Rural Development Department (2021) Beef Trade Data, last update 13/08/2021. Available at https://agridata.ec.europa.eu/extensions/DashboardBeef/BeefTrade.html

Gadd ME (2012) Barriers, the beef industry and unnatural selection: a review of the impact of veterinary fencing on mammals in Southern Africa. In: Somers MJ, Hayward M (eds) Fencing for conservation: restriction of evolutionary potential or a riposte to threatening processes? Springer New York, New York, NY, pp 152–186. https://doi.org/10.1007/978-1-4614-0902-1

Gayford C (2000) Biodiversity education: a teacher's perspective. Environ Educ Res 6(4):347–361

Gibson GD (1977) Himba epochs. Hist Afr 4:67–121

Government of the Republic of Namibia (GRN) (1975) Nature conservation ordinance 4 of 1975, last amendment in 2003. Namibia

Government of the Republic of Namibia (GRN) (1996) Nature Conservation Amendment Act 5 of 1996

Government of the Republic of Namibia (GRN) (2018) Animal Health Regulations: Animal Health Act, Animal health regulations: animal health act. Available at: https://gazettes.africa/archive/na/2018/na-government-gazette-dated-2018-12-28-no-6803.pdf

Grant C, Bengis R, Balfour D, Peel M, Davies-Mostert W, Kilian H, Little R, Smit I, Garaï M, Henley M, Anthony B, Hartley P (2008) Controlling the distribution of elephants. In: Scholes RJ, Mennell KG (eds) Elephant management: a scientific assessment for South Africa. Wits University Press, Johannesburg. https://doi.org/10.18772/22008034792

Groves D, Tjiseua V (2020) The mismeasurement of cattle ownership in Namibia's northern communal areas. Nomadic Peoples 2:255–271

Hauptfleisch M (2018) The challenges with managing rangelands for wildlife. In: OPTIMASS - a joint Namibian-German research project. Plant ecology and nature conservation. Potsdam University, Potsdam, pp 60–61

Jones BT, Barnes JI (2006) Human wildlife conflict study: Namibian case study. WWF Glob Species Program Namibia 264:1–102

Jordan N, Smith B, Appleby R, van Eeden L, Webster H (2020) Addressing inequality and intolerance in human-wildlife coexistence. Conserv Biol 34(4):803–810. https://doi.org/10.1111/cobi.13471

Kelemen E, Nguyen G, Gomiero G, Kovács E, Choisis J, Choisis N, Paoletti M, Podmaniczky L, Ryschawy J, Sarthou J, Herzog F, Dennis P, Balázs K (2013) Farmers' perceptions of biodiversity: lessons from a discourse-based deliberative valuation study. Land Use Policy 35:318–328. https://doi.org/10.1016/j.landusepol.2013.06.005

Kreike E (2009) De-globalisation and deforestation in colonial Africa: closed markets, the cattle complex, and environmental change in north-Central Namibia, 1890-1990. J South Afr Stud 35(1):81–98. https://doi.org/10.1080/03057070802685585

Lamarque F, Anderson J, Fergusson R, Lagrange M, Osei-Owusu Y, Bakker L (2009) Human-wildlife conflict in Africa. Causes, consequences and management strategies. Ed. Food and Agriculture Organization of the United Nations (FAO) Food and Agriculture Organization of the United Nations (FAO), Rome. http://www.fao.org/3/i1048e/i1048e00.pdf. Accessed 3 Nov 2020

Lange G, Barnes JI, Motinga DJ (1998) Cattle numbers, biomass, productivity and land degradation in the commercial farming sector of Namibia, 1915-95. Dev South Afr 15:555–572

Leggett KE (2006) Home range and seasonal movement of elephants in the Kunene Region, northwestern Namibia. Afr Zool 41:17–36

Lindsey PA, Havemann CP, Lines RM, Price AE, Retief TA, Rhebergen T, Van der Waal C, Romañach SS (2013) Benefits of wildlife-based land uses on private lands in Namibia and limitations affecting their development. Oryx 47:41–53

Mackenzie CA, Ahabyona P (2012) Elephants in the garden: financial and social costs of crop raiding. Ecol Econ 75:72–82. https://doi.org/10.1016/j.ecolecon.2011.12.018

Marker L, Boast L (2015) Human–Wildlife conflict 10 years later: Lessons learned and their application to Cheetah conservation. Hum Dimens Wildl 20(4):302–309. https://doi.org/10.1080/10871209.2015.1004144

Martin J, Chamaillé-Jammes S, Waller DM (2020) Deer, wolves, and people: costs, benefits and challenges of living together. Biol Rev Camb Philos Soc 95(3):782–801. https://doi.org/10.1111/brv.12587

Martin RB (2005) The influence of veterinary control fences on certain wild large mammal species in the Caprivi, Namibia. In: Osofsky SA (ed) Conservation and development interventions at the wildlife/livestock interface: implications for wildlife, livestock and human health. Proceedings of the Southern and East African experts panel on designing successful conservation. AHEAD (Animal Health fo. Gland: IUCN (Occasional papers of the IUCN Species Survival Commission), pp 27–39. Available at http://www.wcs-ahead.org/book/AHEADbook27MB.pdf. Accessed 19 Sept 2019

Matinca A (2018) Human-wildlife conflict in Northeastern Namibia: CITES, elephant conservation and local livelihoods, Culture and Environment in Africa Series, vol 12. Cologne African Studies Centre, Cologne

Matson TK, Goldizen AW, Jarman PJ, Pople AR (2006) Dispersal and seasonal distributions of black-faced impala in the Etosha National Park, Namibia. Afr J Ecol 44:247–255

Mbaiwa JE, Mbaiwa OI (2006) The effects of veterinary fences on wildlife populations in Okavango Delta, Botswana

McGahey DJ (2011) Livestock mobility and animal health policy in southern Africa: the impact of veterinary cordon fences on pastoralists. Pastoralism 1(14):1–29

Meat Board of Namibia (2015) Common vision of the livestock & meat industry of Namibia. Strenghtening Cooperation towards a Shared Plan for Economic Growth

Mendelsohn J (2006) Farming systems in Namibia. Research & Information Services of Namibia, Windhoek

Mendelsohn J (2010) Atlas of Namibia–a portrait of the land and its people, 3rd edn. Sunbird Publishers, Cape Town

Midgley G, Hughes G, Thuiller W, Drew G, Foden W (2005) Assessment of potential climate change impacts on Namibia's floristic diversity, ecosystem structure and function

Ministry of Environment and Tourism (MET) (2007a) Policy on tourism and wildlife concessions on state land

Ministry of Environment and Tourism (MET) (2007b) Species Management Plan: Elephants. Available at https://www.iucn.org/sites/dev/files/import/downloads/namibia_elephant_management_plan_dec__2007.pdf

Ministry of Environment and Tourism (MET) (2018) Revised national policy on human wildlife conflict management 2018–2027

Ministry of Environment and Tourism (MET); Namibian Association of CBNRM Support Organisations (NACSO) (2018) The state of community conservation in Namibia. A review of communal conservancies, community forests and other CBNRM initiatives (Annual Report 2017)

Ministry of Environment, Forestry and Tourism (MEFT) (2020) Annual progress report 2019–2020. Ed. Ministry of Environment, Forestry and Tourism (MEFT)

Murphy C., Roe D. (2004) Livelihoods and tourism in communal area conservancies. Livelihoods CBNRM Namib. Find. WILD Proj. Final Tech. Rep. Wildl. Integr. Livelihood Diversif. Proj

Olbrich R, Quaas MF, Baumgärtner S (2009) Sustainable use of ecosystem services under multiple risks – a survey of commercial cattle farmers in semi-arid rangelands in Namibia

ORYCS (2018) Options for sustainable land use adaptations in savanna systems: Chances and risks of emerging wildlife-based management strategies under regional and global change. Project Document. German Federal Ministry of Education and Research. Berlin.

Otiang'a-Owiti GE, Nyamasyo S, Emalel E, Onyuro R (2011) Impact of climate change on human-wildlife conflicts in East Africa. Kenya Veterinarian 35:103–110

Reid H, Sahlén L, Stage J, MacGregor J (2008) Climate change impacts on Namibia's natural resources and economy. Clim Policy 8:452–466

Savory A (1983) The Savory grazing method or holistic resource management. Rangel Arch 5:155–159

Savory A, Parsons SD (1980) The Savory grazing method. Rangel Arch 2:234–237

Schalkwyk DL van, McMillin KW, Witthuhn RC, Hoffman LC (2010) The contribution of wildlife to sustainable natural resource utilization in Namibia: a review. Sustainability 2(11): 3479–3499

Schneider HP (2012) The history of veterinary medicine in Namibia. J S Afr Vet Assoc 83(1):1–11. https://doi.org/10.4102/jsava.v83i1.4

Scoones I, Bishi A, Mapitse N, Moerane R, Penrith ML, Sibanda R, Thomson G, Wolmer W, Wolmer W (2010) Foot-and-mouth disease and market access: challenges for the beef industry in southern Africa. Pastoralism 1(2). https://doi.org/10.3362/2041-7136.2010.010

Scully WC (1913) Further reminiscences of a south African pioneer. TF Unwin, London

Seoraj-Pillai N, Pillay N (2017) A meta-analysis of human–wildlife conflict. South African and global perspectives. Sustainability 9(1):34. https://doi.org/10.3390/su9010034

Sguazzin A (n.d.) Botswana sells rights to kill elephants for $43,000 per head [WWW Document]

Shaffer LJ, Khadka KK, Van Den Hoek J, Naithani KJ (2019) Human-elephant conflict: a review of current management strategies and future directions. Front Ecol Evol 6:1–12. https://doi.org/10.3389/fevo.2018.00235

Skinner JD (1993) Springbok (Antidorcas marsupialis) treks. Trans R Soc South Afr 48:291–305

Slagle K, Bruskotter J (2019) Tolerance for wildlife. In: Frank B, Glikman JA, Marchini S (eds) Human–wildlife interactions, Bd. 28. Cambridge University Press, Cambridge, pp 85–106

Smit E (2018) Ministry opts for cheaper Etosha fence the environment ministry is working on a simplified design for a boundary fence for Namibia's flagship park. Namibia Sun, 12 December. Available at https://www.namibiansun.com/news/ministry-opts-for-cheaper-etosha-fence2018-12-12

South Africa and Namibia hit out hunting trophy ban [WWW Document], (n.d.). URL https://www.scotsman.com/news/world/south-africa-and-namibia-hit-out-hunting-trophy-ban-1498111. Accessed 30 Dec 21

Tainton N (1999) Veld management in South Africa. University of Natal Press, Pietermaritzburg

Thouless CR (1994) Conflict between humans and elephants on private land in northern Kenya. Oryx 28(2):119–127. https://doi.org/10.1017/S0030605300028428

Twine W, Magome H (2007) Interactions between elephants and people. In: Scholes RJ, Mennel KG (eds) Elephant management. A scientific assessment of South Africa. Witwatersrand University Press, Johannesburg, pp 206–240. https://doi.org/10.18772/22008034792.15

Wals A, van Weelie D (1997) Environmental education and the learning of ill-defined concepts: the case of biodiversity. S Afr J Environ Educ 17:4–11

Warren D (1996) Indigenous knowledge, biodiversity conservation, and development. In: James VU (ed) Sustainable development in third world countries: applied and theoretical perspectives. Praeger, Westport, pp 81–88

Wato YA, Prins HH, Heitkönig I, Wahungu GM, Ngene SM, Njumbi S, Van Langevelde F (2018) Movement patterns of African elephants (Loxodonta africana) in a semi-arid savanna suggest that they have information on the location of dispersed water sources. Front Ecol Evol 6:167

Williamson JE (1981) An assessment of the impact of fences on large herbivore biomass in the Kalahari. Botsw Notes Rec 13:107–110

Feed Gaps Among Cattle Keepers in Semiarid and Arid Southern African Regions: A Case Study in the Limpopo Province, South Africa

19

Sala Alanda Lamega, Leonhard Klinck, Martin Komainda, Jude Julius Owuor Odhiambo, Kingsley Kwabena Ayisi [iD], and Johannes Isselstein

Abstract

Rural livestock farmers in the semiarid and arid areas of Southern Africa face large uncertainties due to a high intraseasonal and year-to-year variability in rainfall patterns which affect forage resources. Creating resilient communal livestock farming systems will require the understanding of feed gaps as perceived by livestock farmers as well as an assessment of available feed resources. In this chapter, we estimated the annual feed balance (i.e., forage supply minus forage demand) based on statistical data and described the perception of feed gaps across 122 livestock farmers in Limpopo province, South Africa. In addition, we analyzed available feed and soil resources during the dry season across land use types. We found a negative feed balance, an indication of feed gaps for livestock farms, mainly during the winter and spring seasons. Farmers perceived a combination of factors such as drought, infrastructure, capital, and access to land as the major causes of feed gaps. Furthermore, our analyses of feed and soil

S. A. Lamega (✉) · L. Klinck · M. Komainda
Department of Crop Sciences, Division of Grassland Science, University of Goettingen, Goettingen, Germany
e-mail: sala.lamega@uni-goettingen.de

J. J. O. Odhiambo
Department of Plant and Soil Sciences, University of Venda, Thohoyandou, South Africa

K. K. Ayisi
Risk and Vulnerability Science Centre, University of Limpopo, Polokwane, South Africa

J. Isselstein
Department of Crop Sciences, Division of Grassland Science, University of Goettingen, Goettingen, Germany

Centre of Biodiversity and Sustainable Land Use (CBL), University of Goettingen, Goettingen, Germany

© The Author(s) 2024
G. P. von Maltitz et al. (eds.), *Sustainability of Southern African Ecosystems under Global Change*, Ecological Studies 248,
https://doi.org/10.1007/978-3-031-10948-5_19

resources point to low crude protein (e.g., ~5% in rangeland biomass) and poor soil nutrient contents (e.g., $\%N < 0.1$). To support rural policies and improve the performance of communal livestock systems, there is a need to combine the most appropriate site-specific options in optimizing the feed supply.

19.1   Introduction

... And also the effects of global warming, we are feeling it here. This drought, it might take long, it can be here for a very long time. We experience it almost every year and every year it's a little bit harsher than in the previous year. (farmer from Maruleng Municipality, Limpopo Province)

In many parts of southern Africa, livestock plays a very important role in the livelihood of rural dwellers (Nyamushamba et al. 2017). According to a report by Köhler-Rollefson (2004), livestock contributes, in cash only, up to 38% to the agricultural Gross Domestic Product in the region, and about 90% of the livestock keepers can be classified as smallholders. A smallholder is often characterized as a resource-constrained farmer that operates livestock primarily for subsistence purposes but also as a major risk-alleviating activity (Köhler-Rollefson 2004). Keeping livestock has been reported to improve household income through sales of animals, milk, and dairy products (Maleko et al. 2018). Smallholders also depend on cattle production for household consumption and, in a mixed-crop livestock system, the integration of cattle also provides benefits such as dung for manure, and draught power for tillage cropping and transport (Thornton and Herrero 2015). In the Limpopo province of South Africa, keeping livestock in the smallholder context remains a cultural-based strategy important for financial security (Marandure et al. 2020). With respect to the smallholder livestock farming sector in the province, Stroebel et al. (2011) reported small herd size (for instance, less than 10 head of cattle) with low or no-input management and poor breeding objectives. Hence, the sector is generally characterized by low productivity (Mapiye et al. 2019). Despite the already challenged livestock production systems, climate change and variability pose an additional threat, representing a major concern (Nardone et al. 2010). Throughout the southern African region, there is evidence of negative effects of lower rainfall, increased temperature, and prolonged droughts (Archer et al. 2019; Makuvaro et al. 2018; Simelton et al. 2013; Ziervogel et al. 2014) with adverse effects on livestock and the livelihoods of smallholder farmers in the arid and semiarid areas (Batisani et al. 2021; Descheemaeker et al. 2016). In South Africa, Ziervogel et al. (2014) and Archer et al. (2019) have explicitly demonstrated climate anomalies such as exacerbated weather events (e.g., prolonged drought, extended heat waves, change in the distribution and frequency of rainfall, drying up of water bodies). Such changes have significant negative impacts, particularly for smallholder livestock and mixed-crop livestock systems that are associated with natural grazing on communal rangelands and rain-fed agriculture (Thornton and Herrero 2015). Prolonged drought, as a result of annual or seasonal variation in the rainfall patterns, is reported to be the most challenging or damaging by its effect on rangelands (Godde et al. 2020; Vetter et al. 2020) and on rain-fed agricultural systems (Meza et al. 2021). It is now widely accepted that alterations in forage provision will increase with climatic variability (Godde et al. 2021) leading to feed gaps.

For livestock, a feed or a forage gap generally addresses a period during which the animal's feed/forage demand is higher than the feed/forage supply. As explained by Moore et al. (2009), a feed gap is a consequence of the combination of bio-economic factors such as seasonal forage growth, livestock feed intake, farmers' objectives, and financial capacities. In the communal smallholder livestock context, a feed gap is also dependent on additional factors such as herd size, structure, and management, or natural resource governance (Vetter et al. 2020). A feed balance may undergo considerable seasonal variation within one year or vary considerably from one year to another due to environmental factors (e.g., high interannual rainfall variability) that govern rangelands' biomass productivity. Therefore, two types of feed gaps occur which can be referred to as a "regular" feed gap and an "irregular" feed gap. A regular feed gap occurs every year on account of the seasonal changes in forage growth (e.g., autumn to winter, winter to spring, or summer to autumn), while an irregular feed gap typically occurs once every few years due to a year-to-year variability (e.g., years of severe drought in 2015–2016 and recently 2018–2020). In livestock production systems, feed gaps are important phenomena setting the potential for farm productivity. As argued by Bell (2009) and Moore et al. (2009), the capacity of a livestock-keeping enterprise to maintain or sustain animals during periods of feed gaps is regarded as the safe carrying capacity of the enterprise that could improve profitability. This is because feed gaps, whether regular or irregular, may affect the livestock directly or indirectly, consequently affecting productivity.

A direct effect of a feed gap according to Moore et al. (2009) reduces the forage intake by livestock, forcing the animals to lose weight. According to Schlecht et al. (1999), the variation of the forage availability from the rainy to dry seasons not only leads to a decline in feed quantity but also in its nutritive quality. For instance, during a feed gap, the energy provided to cattle from the dry and fibrous (i.e., less nutritious) pasture is not sufficient leading to a catabolism of their body tissue. Therefore, a feed gap, when it occurs, does not only contribute to the decline in the maintenance of the cattle energy status but also has economic implications for the farmer (return on sales).

Moore et al. (2009) further argued that feed gaps may affect livestock indirectly through decreased and poor sperm production, and ovulation rates all of which have significant effects on breeding performance. For instance, beef bull calves that are fed below their maintenance requirements (in terms of energy and protein) may encounter sexual immaturity with decreased sperm production (Thundathil et al. 2016). Therefore, nutrition deficiency caused by feed scarcity during the dry season would first affect the livestock's residual feed intake. This would cause a decline in feed efficiency in relation to cattle growth rate, consequently affecting morphological development. Additionally, nutrition deficiency is also known to impact lactation and embryo survival affecting the reproductive capacity of the livestock systems (Thundathil et al. 2016).

A very recent integrated drought risk assessment by Meza et al. (2021) revealed that the Limpopo province of South Africa is one of the most exposed provinces to extreme drought, resulting in decreased rangeland productivity and crop yields. Thus, the frequent and major drought periods facing cattle keepers could be considered extended feed gap periods. A sound assessment of the seasonal livestock feed gaps through the perceptions of vulnerable livestock farmers, and data on available feed resources during the dry period (quality and utilization) may be crucial for the development of adequate recommendations. Providing adequate supplementary nutrients to nutritionally-challenged livestock in periods of feed gaps will be crucial in improving livestock production and increasing profitability (Bell et al. 2017). For this, we assessed the contribution of crop residues to the feeding regime of cattle, to clearly identify periods where feed is unavailable to meet animal's demand.

One of the urgent priorities is to find a proper way to deal with the seasonal feed gaps for rural livestock farmers to facilitate resilience towards improved livestock systems. The principal goal of this chapter is to inform the general public and policy makers on climate-induced feed gaps that represent a threat during periods of feed scarcity, particularly to communal livestock production.

19.2   Materials and Methods

19.2.1 Study Area

The study was conducted in Limpopo, the northernmost province of South Africa which is characterized by semiarid climatic conditions with low and variable precipitation (Mpandeli et al. 2015). The province receives about 600 mm of rainfall per annum, most of which occurs between October and April. The summer season (December–February) is hot and wet with an average maximum temperature of about 27°C while the winter (June–August) is cool and dry with an average minimum temperature of 15°C. Soils in the study area are predominantly reddish-brown loamy sand soils of low nutrient content (Munjonji et al. 2020). The typical natural vegetation is an open bush savanna woodland and natural grasslands, i.e., rangelands, dominated by C_4 grass species. Based on a recent survey, the population increased from 5.4 million to nearly 6 million by 2016 with 38.2% of all households involved in agricultural activities and 36% in livestock production (Stats SA 2018). However, livestock keeping is mostly integrated with cropping activities where maize (*Zea mays L.*), cowpea (*Vigna unguiculata*), groundnut (*Arachis hypogaea*), butternut (*Cucurbita moschata*), spinach (*Spinacia oleracea*), and water melon (*Citrullus lanatus*) were the most frequently and simultaneously cultivated crops. The vast majority of cattle farmers (95%) are users of communal lands with variable herd sizes (5–80) due to resource endowment. Moreover, the several government-owned natural reserves (e.g., rangelands) in the province remained a constraint as it reduces the availability of agricultural and grazing areas for livestock farming (Rootman et al. 2015). The most widespread breed is a cross-breed between Nguni and Brahman cattle and the respective pure breeds. Other popular breeds include Bonsmara and Afrikaner.

19.2.2 Data Collection and Analysis

Data used for this chapter were collected from two sets of surveys and a focus group discussion conducted at different stages of a research project. Firstly, the preliminary survey was conducted from September to November 2018 across 32 cattle farms in the arid and semiarid areas of the Limpopo province on the basis of communal livestock keeping (Klinck et al. 2022). A follow-up survey was carried out from June to September 2019 across 90 cattle farms (see more details in Lamega et al. 2021) (Fig. 19.1). The surveys were conducted using a semistructured questionnaire instrument (KoBoToolbox) (Deniau et al. 2017) which was delivered on a basis of a personal interview with the farmers. The questionnaire mainly assessed farmers' perception of (i) months of feed unavailability; (ii) feeding regimes and strategies; (iii) weight losses during feed gaps and (iv) adaptation responses/constraints to adaptation. Additionally, open-ended interviews with selected farmers were conducted to further explore the perceived feed gap challenges. The responses were recorded, transcribed, and reported based on Miles et al. (2014). In 2020, a one-day online feedback workshop was conducted with a few key farmers to discuss research results and identify management options. Selected results are averaged and reported in this chapter.

Secondly, aside from the perceptions of farmers on the seasonality of feed gaps and their effects on livestock production, the likelihood of winter feed gaps was further evaluated through the assessment of grazed rangeland biomass, crop residues, feed supplements, and selected soil nutrient levels. For instance, on communal rangelands and cropping lands, rangeland biomass and crop residues were sampled respectively by cutting from inside a 50 cm by 50 cm quadrat along a longitudinal transect (5 m apart). At the farm-level, we collected whenever possible (i.e., if the farmer had access), supplemental feed residues that may be used to feed cattle during that period. Collected feed samples were oven-dried at 60°C, ground, and then analyzed for the relative abundance of stable isotopes of nitrogen using an elemental analyzer (NA 1110; Carlo Erba, Milan) interfaced (ConFlo III; Finnigan MAT, Bremen) to an isotope ratio mass spectrometer (Delta Plus; Finnigan MAT). The nitrogen content in the feed samples is given as mass ratio in dry matter (%N) which was then multiplied by 6.25 to obtain crude protein concentration in the respective feed sample. In addition, soil samples (0–10 cm, diameter 2 cm) were taken after the removal of biomass on rangelands or cropping lands. Per quadrat, three samples were taken, which consist of 15 subsamples from one transect at a particular site. The soil was homogenized, cleared of any foreign materials, dried at 105°C, sieved (2 mm), and analyzed using the Calcium Acetate Lactate (CAL) extractable method (Schüller 1969). Soil pH was determined in water

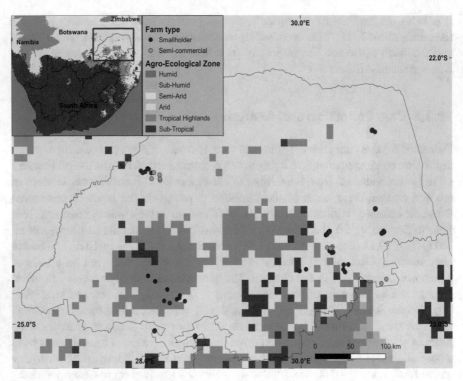

Fig. 19.1 Locations of sampled farms across semiarid and arid zones in Limpopo. In total, $N = 122$ livestock farms including 11 semicommercially oriented livestock farms (dotted light blue). Classification data for agroecological zones obtained from (HarvestChoice and IFPRI 2015)

while the concentrations of P and K were determined in continuous flow analysis coupled to a UV/VIS spectrophotometer (San System, Skalar, the Netherlands). The remaining nutrient concentrations were determined using atomic absorption spectrometry (AAnalyst 400, Perkin Elmer Inc., Waltham, USA).

Finally, we calculated feed balances based on statistical data. However, an uncertain number of young and old livestock is kept in the smallholder sector of Limpopo. According to (DAFF 2021), a total of 860,000 heads of cattle were kept in the Limpopo province in 2020. We assumed an average live weight of 450 kg cattle to obtain an estimate of tropical livestock units (TLU = 250 kg live weight) with every TLU consuming 10 kg dry matter daily. These values consequently represent the cattle livestock forage demand. We further derived an estimate of crop residue yields from maize production as based on Kutu (2012) who reports a stover proportion of 0.41 for maize production in Limpopo. The so-calculated maize residue amount was added to an estimate of rangeland biomass, as extracted from Martens et al. (2020) and Avenant (2019), to obtain an estimate of the forage supply. The survey data was analyzed in R using descriptive statistics to report on the perception of feed gaps across farmers and characterize the quality of feed and soil resources across sites.

19.3   Results

19.3.1   Estimation of Feed Balance in the Limpopo Province

The severity of feed deficit in the cattle livestock sector of Limpopo was derived by calculating feed balances. According to our calculation, about 1,484,753 TLU are kept in Limpopo per year. With a daily forage demand of 10 kg DM per day and TLU, an estimated annual forage demand of about 5.7 million tons for cattle is expected (see Table 19.1).

In many parts of southern Africa, major forage resources for cattle livestock may constitute rangeland biomass and maize residues from cropping lands (Homann-Kee Tui et al. 2015; Masikati et al. 2015). On the supply side we, consequently, used maize production and rangeland biomass production to estimate forage supply. According to Avenant (2019), approximately 7.4 million ha of rangeland is available for grazing in the Limpopo province. Maize is the most commonly grown crop, especially on smallholder farms. Statista (2021) estimated a total volume of 231.000 t maize in 2020 (Table 19.1). According to Kutu (2012), who has analyzed maize production systems in two locations in the Limpopo province, a stover proportion of 0.41 of total aboveground maize biomass can be assumed. Using this proportion, we estimated a total of 160,525 t of maize stover biomass that is potentially available to be used as forage when maize is harvested which usually takes place in March (autumn) at the end of the wet season. A reliable calculation for the productivity of rangeland is far more complex. We used the results of

Table 19.1 Annual forage balance for the Limpopo region as derived from official maize production amounts, an estimate for the stover production, and an estimate for the whole rangeland dry matter accumulation on an annual basis (tons) and compared to the forage demand of all cattle livestock as expressed in tropical livestock units (TLU). *DM* dry matter

Year	What	Value	Reference
2020	Maize grain yield (*t*)	231,000	(Statista 2021)
	Stover % (total aboveground maize)	0.41	(Kutu 2012)
	Total maize biomass total (*t*)	391,525	Calculated
	Stover biomass total (*t*)	160,525	Calculated
Mean 2011–2019	Rangeland biomass supply (*t* DM/ha)	0.54	(Martens et al. 2020)
	Rangeland available for grazing (ha)	7,400,000	(Avenant 2019)
	Rangeland biomass supply (*t*)	4,015,968	Calculated
	Annual feed demand cattle (*t*)	470,850	Calculated
	Feed supply total (*t*)	4,176,494	Calculated
	Feed demand total (*t*)	5,650,200	Calculated
	Balance: supply − demand (*t*)	**−1,473,706**	Calculated

modeled rangeland productivity for the province and for our study sites (Martens et al. 2020) to calculate the seasonal rangeland productivity across the arid and semiarid zones which gave an annual estimate of 0.228 t C/ha per year. Assuming that dry matter (DM) biomass contains 42% C giving an annual value of 0.54 t DM/ha of rangeland which was applied to a rangeland area of 7.4 million ha (76% of the total rangeland area Table 19.1). Not all of the rangeland area in Limpopo is considered suitable for grazing, because of shrub and tree cover, area protection, or urbanization. Consequently, we found an annual feed supply of 4,176,494 t that is unable to sustain the demand of cattle (5,650,200 t), resulting in a negative feed balance (Table 19.1).

Avenant (2019) used a different approach to calculate the carrying capacity of rangeland in the study area. Using the estimated values for rangeland production in that study, 0.488 t DM/ha is very close to the value used in our approach (0.54 t DM/ha). According to our estimation, we found a shortage in feed supply on an annual basis, taking into account that there are two major constraints underlying our calculations. Firstly, we only used predominantly statistical data, and we did not consider livestock species other than cattle although small ruminants are also important forage consumers in the region. In addition, we did not account for forage quality which is likely limiting the utilization capacity of maize residues and rangeland biomass during a large part of the year. According to Descheemaeker et al. (2018), the requirements of metabolizable energy (ME) range from 45 to 65 MJ ME/day per animal. As known from other studies, maize residues never reach values >6 MJ ME/kg DM when harvested at physiological maturity (Terler et al. 2019). In addition, grass ME concentration ranges usually between 6.5 and 10.3 MJ/kg DM in the dry and the wet season respectively, which points to a shortage of forage with sufficient quality in the dry season. But not only quality is likely limiting in the dry season. When using the annual forage balance data for monthly calculations, we found strong support for a serious shortage in feed supply during winter and spring (Table 19.2). Forage quantity and likely quality are, consequently, critical issues for the livestock sector.

Moreover, to check the assumptions made for the calculation of the feed balance, a sensitivity analysis was carried out where, under a constant average live weight of 450 kg, the daily forage DM intake was varied from 10 to 5 kg (Fig. 19.2a) or, under a constant average forage intake of 10 kg per day, the live weight varied from 450 to 300 kg (Fig. 19.2b). These calculations have an effect on the annual feed requirement. The result show that already at about 7 kg DM intake per day a negative balance is no longer to be expected (Fig. 19.2a). On the other hand, a positive balance can only be expected at an average herd weight of 300 kg DM which is unusually low. The assumption made about live weight, consequently, underestimates the problem of the feed gap evaluation. For the exact forage requirement, however, it would be good to generate accurate information on the variation of forage intake of the cattle in Limpopo which is the prerequisite to understanding the contribution of other potential forage sources.

Table 19.2 Derived seasonal feed balance as monthly feed supply from rangeland and maize stover (*t*) against the seasonal feed demand by cattle livestock (*t*). *TLU* Tropical livestock units, *DM* dry matter

Season	Month	DM demand cattle TLU (*t*)	Maize stover (*t*)	Rangelands (*t*)	Feed balance
Summer	Jan	470,850	0	1,866,444	1,395,594
Summer	Feb	470,850	0	1,866,444	1,395,594
Autumn	Mar	470,850	160,525	1,571,619	1,261,294
Autumn	Apr	470,850	53,508	1,571,619	1,154,277
Autumn	May	470,850	17,836	1,571,619	1,118,605
Winter	Jun	470,850	0	108,063	**−362,787**
Winter	Jul	470,850	0	108,063	**−362,787**
Winter	Aug	470,850	0	108,063	**−362,787**
Spring	Sep	470,850	0	469,254	**−1596**
Spring	Oct	470,850	0	469,254	**−1596**
Spring	Nov	470,850	0	469,254	**−1596**
Summer	Dec	470,850	0	1,866,444	1,395,594

19.3.2 Feed Gap as Perceived by Livestock Farmers

Across the arid and semiarid zones, winter and spring are the seasons of feed deficit according to the farmers. While feed shortages are perceived to be most severe during September and October (spring), the duration of experienced shortages was generally one month longer for some farmers (3.4 vs. 2.4 months) (see Lamega et al. (2021)). The heterogeneity between farms plays an important role in the perceptions of the seasonal patterns of feed gaps. For instance, farmers' perceptions of feed gaps did not differ significantly during winter as both mixed-crop livestock and specialist livestock-only farmers were equally affected. However, the perceptions of feed gaps in autumn and spring differed between both farming systems irrespective of their locations.

Cattle livestock farmers did not follow a controlled mating schedule for selective breeding but allowed for natural breeding instead. Animals from farmers that are little endowed (> 50 cattle head) were reported to be weaned at around 7 months, whereas typical smallholder farmers (< 20 cattle head) reported a weaning age of about 11 months. Calves commonly wean later when they receive milk of poorer nutritive value from their dams. During drought, pregnant and lactating cows suffer from nutrient deficiency which is likely mirrored in the lower reproductive performance of the offspring. Furthermore, limited flexibility in securing water availability is a limiting factor in the feed-drought nexus. Access to water sources is tightly linked to access to land and thus taps, boreholes, dams, or streams.

Smallholder farmers in our study area perceived the phenomenon of "drought" particularly manifesting in its biophysical dimension, that is, the perception in the decline in water availability and rangeland productivity. Thus, livestock husbandry under (semi)arid conditions requires a form of adaptive capacity that allows farmers and herders to respond flexibly. For example, by producing their own feed or seeking

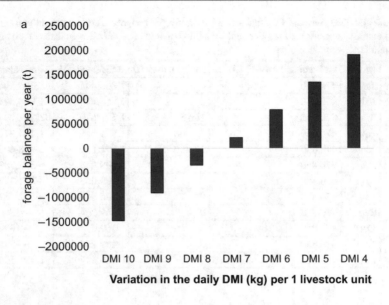

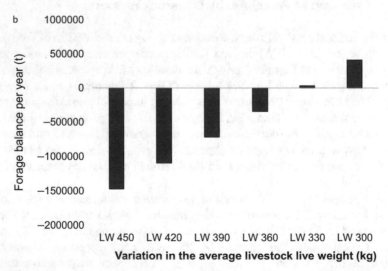

Fig. 19.2 Sensitivity analysis of forage balance calculation as affected by a) the variation in daily dry matter intake (DMI) (4–10 kg) per 1 livestock unit (450 kg) and b) the variation in cattle live weight (300–450 kg) consuming a daily dry matter of 10 kg

out extensive grazing lands, they could face the harsh climatic condition. Access to and utilization of extensive rangelands is crucial when animals (and herders) are required to cover greater distances to water sources, during prolonged droughts when dams and communal watering holes dry up. Farmers would then move their animals to alternative water sources further away or fetch water with motorized vehicles. If feed in the dry period is already critically limited, the additional caloric costs, i.e., animals covering extra distances for pasture and water, may translate into poor livestock health (Ouédraogo et al. 2021).

A rough on-farm assessment of the body condition score (BCS) demonstrated that animals relying solely on communal rangelands are indeed on average closer to drought-induced starvation (BCS of ~2.01 with 0 = emaciated and 5 = over-fat). In many cases, in communal livestock systems, livestock farmers or managers do not look into maximizing operating profit, instead maximizing or maintaining herd size, remains the priority (Stroebel et al. 2011; Tavirimirwa et al. 2019). Therefore, the risk of feed gaps may not only be associated with the unproductivity of rangeland during the dry season but it may also be related to the high costs of producing/purchasing feed, concentrates, and or conservation of forage. It is likely that farmers that have access to capital are more flexible in their modes of feed provision (Chikowo et al. 2014). Such farmers may draw from a variety of on-farm produced crops, forage, silage, and commercial supplements. To some extent, these livestock farmers that are more endowed may dispose of private boreholes and wells to alleviate the impacts of feed gaps. Moreover, in areas with sufficient annual rainfall, ground water may be important in maintaining the productivity of rangeland biomass, hence reducing feed gap risks significantly. In the arid zones of Western Limpopo, some farms even employed water-intensive fodder crops like sugar cane (*Saccharum officinarum*) or Blue Buffalo Grass (*Cenchrus ciliaris*). Also, in the arid zones of Eastern Limpopo, livestock may graze on Mopane tree leaves (*Colosphospermum mopane*), which are available on the rangelands but become scarce with the extended dry period.

Furthermore, farmers perceived feed shortages not directly as a result of bio-physical drought, but rather linked to low overall farm profitability and low returns in investments (Fig. 19.3). Aside from the obvious climate-induced drought, farmers mentioned a variety of limitations including insufficient technical extension support, poor local beef demand, poor access to external markets, and contract farming. These limitations were all perceived as impediments to profitability and business growth.

One farmer related the exclusive nature of contracts in the retail sector to favoring commercially-oriented farmers only:

We [small – semi farmers] don't get access to Spar [supermarket] . . . direct straight. We are under someone else, it's a middleman. We can't grow. From 1914 to today, no successful farming in here, we just do farming for pleasure or whatever, to make a living.

Commercially-oriented cattle production, on the other hand, requires high-caloric and nutritious feed throughout the year for regular off-take to auctions and abattoirs. Supplements thus play a crucial role, whether produced on-farm or bought off-farm and it requires a certain financial margin for investments in feedstuffs (Fig. 19.3). In contrast, in the communal setup, a feed gap is essentially linked to the availability of grazing areas that accommodate community-level stocking density. Additional feed is rather linked to farm types (if the farmer engages in cropping) or capital (if the farmer can purchase feed). Since smallholders are mostly financially

Fig. 19.3 Concept map summarizing perceived root causes (gray) and feed gap mitigation strategies (white) for livestock farmers during feed gaps. Relational arrows indicate enhancing (+) or reducing (−) effects on feed supply

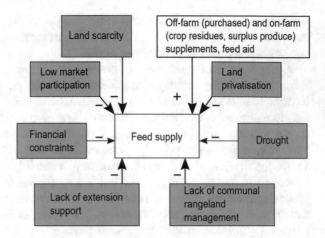

constrained, they tend to be low adopters of feed gap strategies. The most common strategy is the use of readily available crop residues during autumn (Table 19.2), which serves as an additional feed input for livestock farms at no cost. Under severe drought conditions, where crop residues alone are not enough, farmers may reduce their livestock number to balance feed requirements. These strategies are associated with the socioeconomic challenges of the smallholder livestock sector that render it vulnerable to feed gaps (Lamega et al. 2021; Mapiye et al. 2009; Marandure et al. 2020).

19.3.3 Results of Available Feed and Soil Resources

19.3.3.1 Feeding Resources

Cattle rely heavily on the productivity of rangelands. In the study area in particular, rainfall patterns have created a vegetation gradient that may differ from the arid to the semiarid zones. According to Mpofu et al. (2017), the veld type (an indigenous grazing and or browsing vegetation composed of any sort of plant species capable to reproduce itself undecidedly under existing environmental conditions) varies from sweet and mixed in arid areas to sourveld in semiarid areas with prevailing grass species such as *Panicum maximum, Aristida transvaalenesis, Eragrostis curvula,* and *Themeda triandra.* A sweet veld according to Trollope et al. (1990) is a veld that retains acceptable nutritive values of its forage plants after maturity, utilizable throughout the year by livestock while a sourveld shows sharp declines in forage quality with ongoing maturation. A mixed veld is an intermediate veld between the sour and sweetveld with an acceptable quality supply of forage to the livestock.

Our analysis in terms of crude protein (CP) concentration of the dry rangeland biomass in winter showed low herbage quality across the studied sites in the Limpopo province with a maximum of 5.3% (Table 19.3). Hence, the quality of the fibrous and dead herbage is poor. Even lower values of 2.7% CP were reported in a previous study by Moyo et al. (2012) in the winter period due to low growth

Table 19.3 Cattle feeding resources and their crude protein concentration (%) during the winter dry season: Site regroups about two to three villages where rangeland biomass is collected on communal rangelands, crop residues, and supplements are collected from selected farms

Feed resource	Site	Number of samples	Crude protein (%)
Rangeland biomass	Site 1	10	5.3 ± 1.7
Rangeland biomass	Site 2	10	4.6 ± 1.8
Rangeland biomass	Site 3	26	4.2 ± 1.4
Crop residues	Site 1	5	9.7 ± 1.3
Crop residues	Site 2	10	4.5 ± 0.9
Crop residues	Site 3	–	–
Feed supplements	Site 1	7	11.3 ± 3.3
Feed supplements	Site 2[a]	1	10.7
Feed supplements	Site 3	12	12.2 ± 9.4
Tree leaves	Site 1	10	9.1 ± 1.5

[a]Only one farmer at Site 2 had access to feed supplements

and senescence. Nevertheless, in situations where there is hardly any herbage to consume, mineral nutrients may help livestock to cover some of its elemental demand irrespective of low protein or energy concentrations. The mineral nutrient concentration is likely insufficient to meet the livestock's nutritional demand (Lamega et al. 2021). In response to the dry and fibrous pasture during the dry season with low CP concentration, cattle may increase the selective retention time for feed particles in the rumen, hence improving fiber digestion. However, this response to the feed gap is hardly adequate to avoid the loss in body tissue which is associated with reduced nutrient supply and metabolic processes (Moore et al. 2009; Schlecht et al. 1999). The scarcity of grazing resources in terms of quality (Table 19.3) and quantity (Tables 19.1 and 19.2, Fig. 19.2) along with increasing bush encroachments on the grazing rangelands (Mogashoa et al. 2021) is, therefore, a call for supplementary feeding.

Crop residues are the first source of additional feed across the study sites. In mixed-crop livestock systems in particular, crop residues represent supplementary feed for livestock in the dry season (Masikati et al. 2015). Therefore, the management of these residues may differ significantly in relation to the utilization as feed (Rusinamhodzi et al. 2016). Generally, the availability of crop residues coincides temporally with times when rangeland productivity declines (in terms of quantity and quality, see Fig. 19.4), making them a valuable feed resource. Crops such as maize, pumpkin, groundnut, and cabbage are found in the fields and the straw and stover left at harvest are used for livestock feed. In line with this, Mapiye et al. (2009), explored the cattle keeping system among 218 smallholder farmers in the study province and showed that about 70% of the total farmers used crop residues to cope with the feed shortages during the dry season. The importance of crop residues is further demonstrated in Fig. 19.5 as a farmer collects and stores for use in periods of feed gaps.

Fig. 19.4 The communal grazing resource during the dry seasons in the arid and semiarid areas of Limpopo. Left picture shows cattle browsing on shrubs, and right picture shows the scarcity of rangeland biomass in the dry period during winter

Fig. 19.5 A farmer's supplemental feed made up of dry crop residues and tree leaves. An option for feed gap mitigation

The crop residues in the present study showed higher CP concentration than the rangeland biomass sampled (Table 19.3) or the CP concentration of 4% obtained for maize residues in a study by Mudzengi et al. (2020). It is likely that crop residues are a mixture containing at least parts of C_3 plants such as legumes with higher CP concentration (~ 10%). Low protein concentration during a feed gap may be associated with low digestibility and, hence, poor livestock performance (Mudzengi et al. 2020). Despite disagreements presented by the utilization of crop residues on smallholder farms, i.e., "mulching or no mulching" (Valbuena et al. 2012), a mixture of crop residues may serve as a good source of additional feed. However, the quality and quantity of the residues should be more in balance with animals' demand especially in periods of pasture scarcity (winter, spring), to significantly contribute to feed gap mitigation.

Supplementary feed plays a crucial part in livestock production as it can greatly improve the productivity of the livestock (Bell 2009; Bell et al. 2017). In South Africa, different conventional supplements and agroindustrial by-products are available for purchase (Marandure et al. 2020). However, such feed purchase depends on the socioeconomic status of a farm, but also on the intensity of the livestock production. For smallholder livestock farming that is often financially constrained, first-choice supplementary feeds constitute crop residues and agricultural or household waste. However, our results of CP concentration show that feed supplements are more valuable than anticipated particularly when compared to rangeland biomass, which should be beneficial for the livestock enterprise during feed gaps overall.

However, since the quantity of supplementary feed may depend on herd size, resource-constrained farmers may fail to purchase enough to sustain production. In this case, a farmer will strategically feed animals that are too weak to search for herbage intake on rangelands. On the other hand, focus could be given to high-performing livestock such as lactating cows. Additionally, browse trees can also provide supplementary feed during the dry season (Mudzengi et al. 2020). Here, we found that indigenous species such as *Colophospermum mopane* (common on rangelands) are rich in crude protein (Table 19.3) and likely other nutrients.

19.3.3.2 Soil Resources

In relation to soil fertility, evidence from the literature demonstrated that the majority of smallholder farmers in the southern African region face land degradation (Rufino et al. 2011; Zingore et al. 2007), and this phenomenon is particularly true among smallholder farmers in South Africa (Kolawole 2013). We collected soil samples across land use (rangelands and cropping lands) to get an insight into the fertility status (Table 19.4). We are aware that site-specific nutrient allocation in soils, for instance, around home gardens, or fields close to homesteads have caused soil fertility gradients, problematic in terms of sustainable land use (Mtambanengwe and Mapfumo 2005; Rowe et al. 2006; Zingore et al. 2007).

Basing on Kotzé et al. (2013) who evaluated basic soil properties across different land use types and management situations, all the soil nutrients may be limiting plant production. Under communal set up, Kotzé et al. (2013) discussed low nutrient content (e.g., <2% C, < 0.2% N, < 10 mg/kg P). We found similar results for our study (Table 19.4) which demonstrates poor land use conditions. The C/N ratio of c. 10 points to organic matter quality which potentially readily supplies nitrogen to crops. However, both the N and C contents are very low demonstrating issues with soil quality. Soil degradation is also a reflection of grazing effects on rangelands as previously discussed by Descheemaeker et al. (2010) and Linstädter

Table 19.4 Selected soil chemical properties across different land use types in the studied Limpopo region. $n = 18 \pm 5.3$ soil samples (0–10 cm) per site

Site	Land use	pH	N total (%)	C total (%)	P (mg kg^{-1})	K (mg kg^{-1})	Mg (mg kg^{-1})
Site1	Cropland	6.5 ± 0.8	0.06 ± 0.01	0.66 ± 0.09	<1.00	17.45 ± 3.97	17.53 ± 2.25
Site1	Rangeland	6.3 ± 0.6	0.07 ± 0.03	0.80 ± 0.35	2.40 ± 1.6	20.88 ± 6.02	20.56 ± 9.16
Site 2	Cropland	5.4 ± 0.7	0.10 ± 0.02	1.18 ± 0.16	<1.00	6.56 ± 6.20	24.82 ± 3.42
Site 2	Rangeland	5.3 ± 0.5	0.12 ± 0.04	1.55 ± 0.83	<1.00	5.53 ± 3.77	31.43 ± 12.8
Site 3	Rangeland	5.0 ± 0.7	0.06 ± 0.03	0.62 ± 0.31	<1.00	9.98 ± 8.60	8.58 ± 13.26

et al. (2014), thus an issue of stocking intensity (Kotzé et al. 2013). Additionally, in an aerial cover study conducted under similar conditions in South Africa, Dlamini et al. (2014) showed from initially nondegraded soils, that grazing decreased soil organic carbon by 94% while nitrogen decreased by 40% on communal rangelands managed by smallholder livestock farmers. Such degradation was found under fine sandy loamy soils in the semiarid zones in South Africa. Most soils in the region where the present study was conducted refer to such soil textures (Swanepoel et al. 2015). Carbon is important for soil nutrient cycling and water storage. Nutrient limitation is generally potentially restricting herbage production. Therefore, soil fertility initiatives with an emphasis on C, N, and P through future research may be essential for improving pasture forage supply.

19.4 Discussion

19.4.1 Dealing with Feed Gaps

Though the variability in the supply of feed to livestock is linked to the variability in the rainfall patterns that restricts rangeland productivity, the vulnerability of communal livestock farmers to feed gaps may also depend on the adaptive capacities of rural communities. Therefore, the effects of feed gaps can highly be site-specific (Godde et al. 2021). Having this in mind, any strategies designed to deal with either a regular or an irregular feed gap must be context-specific with direct and indirect effects on livestock production. Moore et al. (2009) proposed two main approaches to deal with the occurrence of feed gaps: tactical and strategical approaches. According to these authors, a tactical response is implemented when needs arise. For instance, a farmer could buy or sell livestock depending on the balance between the number of herds and the available feed. A tactical response could also involve the application of fertilizers to pastures to boost seasonal production in the rainy season. This approach is usually preferable for irregular feed gaps where the supply of feed is less predictable in terms of its magnitude and timing (Bell 2009). Such management aims at the provisioning of conserves obtained during times of excess feed supply. The advantages of tactical responses are that these can easily be implemented without changing the existing land-use or farming patterns and that opportunity costs are generally low in years when the tactical response is not executed. On the other hand, a strategic approach can be deployed for situations with regular feed gaps and requires structural adjustments to the livestock farming system. A strategic response involves the introduction of multiyear permanently available forage shrubs as a feed base.

In a communal setup, a more efficient approach in alleviating feed gaps among resource-constrained livestock keepers in Limpopo should have benefits for the natural resources (e.g., rangelands). However, many approaches to improve the common grazing resources among livestock farmers through improved management have failed as demonstrated in other semiarid and arid areas (e.g., Tavirimirwa et al. 2019). Nevertheless, insights from systems evaluation emphasize farming system flexibility as a prerequisite for risk adaptation (Thornton and Herrero 2015). Particularly in the smallholder South African context, in the light of the absence of effective rangeland governance, clear tenure policies and entrenching inequalities in access to land and resources; smallholders' current drought responses are likely to continue (Müller et al. 2015; Vetter 2013). This echoes Atkinson's (2013) call for flexibility in both tenure and smallholder-oriented commercialization policies. We concur with Vetter (2013) that the involvement of livestock keepers in any solution-oriented debate is mandatory and critical in developing a contextual understanding of locally nuanced challenges. For this to happen, policy makers need to have a sense of accountability and interest in co-framing the needs of and with smallholding livestock keepers. Managing the political framework, thus, begins with understanding and recognizing the concerns and importance of communal livestock for local food security, cultural value, and livelihood asset (Ainslie 2013).

19.4.2 Managing Rangeland Stocking Density: Destocking to Reduce Pressure on Natural Resources

In their understanding of "better" rangeland management, stakeholders from our group discussion maintained that communal rangelands were unquestionably over-grazed. Thus, destocking or resting periods may be the only reasonable options to restore productivity and close dry-season feed gaps. The role of stocking densities and overgrazing in debates about the management of southern Africa's rangelands remains a very controversial topic. Despite its persistent promotion to ameliorate Africa's rangelands from degradation, the technocratic approach to destocking the rangelands is not a universal panacea that fits every social-ecological context (Godde et al. 2020; Tavirimirwa et al. 2019). Farmers persistently resisted to comply with such top-down approaches that were far from addressing their realities (Tavirimirwa et al. 2019). This is because farmers mainly seek to maximize herd size; hence, destocking initiatives fail to be implemented. Furthermore, grazing schemes or resting periods should not be recommended in this context as they reduce the flexibility of the common grazing resource (Tavirimirwa et al. 2019). However, as argued by Lamega et al. (2021), destocking can be attained if it is subsidized to be in balance with the seasonal feed budget. The longstanding debate still appears to be grounded on different understandings between top-down-oriented policies and stakeholders.

19.4.3 On-Farm Feed Production

Maize stover is particularly an important feed resource on smallholder farms. To improve livestock productivity using maize stover Dejene et al. (2021) demonstrated that upper maize stover fractions had higher total N concentrations and lower fiber content, and varied among different genotypes. The production of dry season (winter) forages, such as protein-rich legumes as cover crops, is a traditional practice across southern Africa, for example, Bennett et al. (2010) have reported that C₃ species such as oats (*Avena sativa L.*) and barley (*Hordeum vulgare L.*) can be intercropped with maize during the dry season. Such species can do well under the South African winter climate (cool season with low temperature), but with limited water during the winter period, irrigation schemes are crucial for high and effective production. Also, legumes have always been of interest to rural development agendas but their implementation also met with skepticism among smallholder farmers (Sumberg 2002, 2004). For instance, dual-purpose winter forage crops may provide higher feed availability during feed gaps, which can maintain livestock or accommodate higher stocking density. While the "sustainable intensification" narrative promotes cover crop legumes to close yield and thus feed gaps, the upscaling and practical implementation has been of limited success among smallholding mixed-crop livestock farmers (Tittonell and Giller 2013). It is important that feed improvement interventions fully address the quality and quantity of forage (Balehegn et al. 2020). From an agronomic point of view, however, recent field trials prove the underutilized and drought-tolerant legume lablab (*Lablab purpureus*) promising when grown in Limpopo under rain-fed conditions (Rapholo et al. 2019). Additionally, forage brassicas have the potential to alleviate regular feed gaps due to high productivity (Bell et al. 2020) if integrated as feed-base strategies in drier or mixed farming systems. However, feeding *Brassica rapa* has been associated with liver disease in Holstein cows in South Africa (Davis et al. 2021). Therefore, more research is needed in the context of feeding brassicas to local cattle breeds.

19.4.4 Feed Aid Schemes

Drought emergency support programs subsidize farmers during severe drought with supplementary feed obtained from commercial forage growers according to the farmers. A smallholder farmer commented on the present design of supplementary feeding support:

> I think the other challenge is, if we can get supplements from the government, that will help us a lot. But now they do sometimes, just as I said, I got 20 cattle and then they gave me 5 bags.

According to the farmers, feed aid comes rarely in periods of severe feed deficit. The program follows no specific criterion for acquiring such feed aids. Hence, farmers with very small herds (e.g., 5) may receive a one-time and free of charge supply, the same amount of feed (usually two to five bags of 25 kg) as a farmer that owns 20 plus cattle. In effect, such an approach to feed gap alleviation on smallholder farms is considered among farmers as not responding to the actual issue. A regular reception of such aids may help the livestock enterprise, but the question arises whether such programs can serve as a long-term sustainable adaptation strategy for smallholder farmers.

The need of strengthening the nutritional status of animals during seasonal feed gaps, through feed quality enhancement, may be achieved using combinations of different options. To make sure that farmers adopt strategies to reduce the impact of feed gaps, combined options should be considered, taking into account sociocultural factors associated with the smallholder livestock systems.

Farmers in Limpopo may learn from the pastoralists in the dry areas of Burkina Faso that deal with feed gaps by employing conservation methods such as building up fodder bundles from mowing grasses or plants when they are plentiful (Ouédraogo et al. 2021). Anyway, the success of feed interventions or any interventions to alleviate feed gaps on smallholder farms is highly dependent on specific local conditions (Balehegn et al. 2020), which cannot be overstated. Moreover, as argued by Balehegn et al. (2020), we need to also consider other related challenges that face smallholder farmers such as market access for selling stock, improved water or irrigation schemes, improved livestock breeding techniques, and diseases, all of which could reduce the effects of feed gaps and improved farm profitability. For scientists, there is also the need to develop proper research objectives, and set up necessary experiments (surveys, field trials, modeling exercises) as suggested by Garrett et al. (2017) that are site and context-specific to the subject of seasonal feed gaps.

19.5   Conclusions

As presented in this chapter, feed gaps are generally governed by the environmental conditions that regulate the demand for, and the supply of energy but also the capacity of livestock managers to utilize diverse feed sources. Feed gaps will remain a key issue for livestock farmers in the dry areas of Limpopo amid climate variability. Therefore, developing multiple options for farmers may be beneficial in sustaining livestock throughout the year. The success, however, of any given recommendations must consider location and farm type specificity but also include sociocultural values associated with livestock keeping. To support rural policies in the face of climate uncertainties, there is a need to reconfigure and restructure the livestock systems in a way that feed sources become more in balance with smallholder livestock and their demand on communal rangelands throughout the year. For instance, if the farmer engages in cropping, with access to irrigation, dual-purpose C_3 crops may serve as an option for alleviating winter feed gaps or may be used for trading. A cost-benefit analysis in relation to feed production and utilization may be helpful in evaluating adequate feeding strategies. However, the use of modeling to integrate different components of the system and management options as stated by Rötter et al. (2021) will become critical to determine ideal solutions for management issues against feed gaps.

Acknowledgments This work was financially supported by the German Federal Ministry of Education and Research (BMBF, "SALLnet-project" grant number 01LL1802A). The authors acknowledge the participation of the farmers of the selected communities and the help of extension officers from the Limpopo Department of Agriculture and Rural Development. The views expressed here are solely those of the authors.

References

Ainslie A (2013) The sociocultural contexts and meanings associated with livestock keeping in rural South Africa. Afr J Range Forage Sci 30(1–2):35–38. https://doi.org/10.2989/10220119.2013.770066

Archer E, Landman W, Malherbe J et al (2019) South Africa's winter rainfall region drought: a region in transition? Clim Risk Manag 25(May):100188. https://doi.org/10.1016/j.crm.2019.100188

Atkinson D (2013) Municipal commonage in South Africa: a critique of artificial dichotomies in policy debates on agriculture. Afr J Range Forage Sci 30(1–2):29–34. https://doi.org/10.2989/10220119.2013.785021

Avenant P (2019) Availability of rangeland in South Africa for livestock grazing availability of rangeland in South Africa for livestock grazing. Department of Agriculture, Forestry and Fisheries, April. https://www.researchgate.net/publication/332465336_Availability_of_rangeland_in_South_Africa_for_livestock_grazing

Balehegn M, Duncan A, Tolera et al (2020) Improving adoption of technologies and interventions for increasing supply of quality livestock feed in low- and middle-income countries. Glob Food Sec, 26 (June), 100372. doi:https://doi.org/10.1016/j.gfs.2020.100372

Batisani N, Pule-meulenberg F, Batlang U et al (2021) African handbook of climate change adaptation. Springer, Cham, pp 339–362. https://doi.org/10.1007/978-3-030-45106-6

Bell LW (2009) Building better feed systems. Trop Grassl 43(4):199

Bell LW, Moore AD, Thomas DT (2017) Integrating diverse forage sources reduces feed gaps on mixed crop-livestock farms. Animal 12:1–14. https://doi.org/10.1017/S1751731117003196

Bell LW, Watt LJ, Stutz RS (2020) Forage brassicas have potential for wider use in drier, mixed crop-livestock farming systems across Australia. Crop Pasture Sci 71(10):924–943. https://doi.org/10.1071/CP20271

Bennett J, Ainslie A, Davis J (2010) Fenced in: common property struggles in the management of communal rangelands in central eastern cape province, South Africa. Land Use Policy 27(2):340–350. https://doi.org/10.1016/j.landusepol.2009.04.006

Chikowo R, Zingore S, Snapp S et al (2014) Farm typologies, soil fertility variability and nutrient management in smallholder farming in Sub-Saharan Africa. Nutr Cycling Agroecosyst 100(1):1–18. https://doi.org/10.1007/s10705-014-9632-y

DAFF (2021) Newsletter: national livestock statistics. Department of Agriculture, Land Reform and Rural Development. August, 2

Davis AJ, Collet MG, Steyl JCA et al (2021) Hepatogenous photosensitisation in cows grazing turnips (Brassica rapa) in South Africa. J S Afr Vet Assoc 92:1–6. https://doi.org/10.4102/JSAVA.V92I0.2106

Dejene M, Dixon RM, Walsh KB et al (2021) High-cut harvesting of maize stover and genotype choice can provide improved feed for ruminants and stubble for conservation agriculture. Agron J 114:187. https://doi.org/10.1002/agj2.20874

Deniau C, Gaillard T, Ambagogo AMB et al (2017) 2017 EFITA WCCA CONGRESS conference proceedings. 2017 Efita Wcca Congress, 254. http://www.efita2017.org/wp-content/uploads/2017/09/EFITA_WCCA_2017_proceedings.pdf%0A

Descheemaeker K, Amede T, Haileslassie A (2010) Improving water productivity in mixed crop-livestock farming systems of sub-Saharan Africa. Agric Water Manag 97(5):579–586. https://doi.org/10.1016/j.agwat.2009.11.012

Descheemaeker K, Oosting SJ, Homann-Kee Tui S et al (2016) Climate change adaptation and mitigation in smallholder crop–livestock systems in sub-Saharan Africa: a call for integrated impact assessments. Reg Environ Chang 16(8):2331–2343. https://doi.org/10.1007/s10113-016-0957-8

Descheemaeker K, Zijlstra M, Masikati P et al (2018) Effects of climate change and adaptation on the livestock component of mixed farming systems: a modelling study from semi-arid Zimbabwe. Agric Syst 159. (May 2017):282–295. https://doi.org/10.1016/j.agsy.2017.05.004

Dlamini P, Chivenge P, Manson A et al (2014) Land degradation impact on soil organic carbon and nitrogen stocks of sub-tropical humid grasslands in South Africa. Geoderma 235:372–381. https://doi.org/10.1016/j.geoderma.2014.07.016

Garrett RD, Niles MT, Gil JDB et al (2017) Social and ecological analysis of commercial integrated crop livestock systems: current knowledge and remaining uncertainty. Agric Syst 155(May):136–146. https://doi.org/10.1016/j.agsy.2017.05.003

Godde CM, Boone RB, Ash AJ et al (2020) Global rangeland production systems and livelihoods at threat under climate change and variability. Environ Res Lett 15(4). https://doi.org/10.1088/1748-9326/ab7395

Godde CM, Mason-D'Croz D, Mayberry DE et al (2021) Impacts of climate change on the livestock food supply chain; a review of the evidence. Glob Food Sec 28(January):100488. https://doi.org/10.1016/j.gfs.2020.100488

HarvestChoice and IFPRI (2015) Agro-ecological zones for Africa south of the Sahara. Harvard Dataverse, https://doi.org/10.7910/DVN/M7XIUB/PW7APO

Homann-Kee Tui S, Valbuena D, Masikati P et al (2015) Economic trade-offs of biomass use in crop-livestock systems: exploring more sustainable options in semi-arid Zimbabwe. Agric Syst 134:48–60. https://doi.org/10.1016/j.agsy.2014.06.009

Klinck L, Ayisi KK, Isselstein J (2022) Drought-induced challenges and different responses by smallholder and semicommercial livestock farmers in semiarid limpopo, South Africa-an indicator-based assessment. Sustainability, 14(14):8796

Köhler-Rollefson I (2004) Farm animal genetic resources: safeguarding national assets for food security and trade. Summary publication about four workshops on animal genetic resources held in the SADC Region: GTZ, FAO, CTA, 346

Kolawole OD (2013) Soils, science and the politics of knowledge: how African smallholder farmers are framed and situated in the global debates on integrated soil fertility management. Land Use Policy 30:470. https://doi.org/10.1016/j.landusepol.2012.04.006

Kotzé E, Sandhage-Hofmann A, Meinel JA et al (2013) Rangeland management impacts on the properties of clayey soils along grazing gradients in the semi-arid grassland biome of South Africa. J Arid Environ 97:220–229. https://doi.org/10.1016/j.jaridenv.2013.07.004

Kutu F (2012) Effect of conservation agriculture management practices on maize productivity and selected soil quality indices under South Africa dryland conditions. Afr J Agric Res 7(26):3839–3846. https://doi.org/10.5897/ajar11.1227

Lamega SA, Komainda M, Hoffmann MP et al (2021) It depends on the rain: smallholder farmers' perceptions on the seasonality of feed gaps and how it affects livestock in semi-arid and arid regions. Clim Risk Manag 34(September):100362. https://doi.org/10.1016/j.crm.2021.100362

Linstädter A, Schellberg J, Brüser K et al (2014) Are there consistent grazing indicators in drylands? Testing plant functional types of various complexity in South Africa's grassland and savanna biomes. PLoS One 9(8):e104672. https://doi.org/10.1371/journal.pone.0104672

Makuvaro V, Murewi CTF, Dimes J et al (2018) Are smallholder farmers' perceptions of climate variability and change supported by climate records? A case study of Lower Gweru in Semiarid Central Zimbabwe. Weather Clim Soc 10(1):35–49. https://doi.org/10.1175/WCAS-D-16-0029.1

Maleko D, Msalya G, Mwilawa A et al (2018) Smallholder dairy cattle feeding technologies and practices in Tanzania: failures, successes, challenges and prospects for sustainability. Int J Agric Sustain 16(2):201–213. https://doi.org/10.1080/14735903.2018.1440474

Mapiye C, Chimonyo M, Dzama K et al (2009) Opportunities for improving Nguni cattle production in the smallholder farming systems of South Africa. Livest Sci 124(1–3):196–204. https://doi.org/10.1016/j.livsci.2009.01.013

Mapiye C, Chikwanha OC, Chimonyo M et al (2019) Strategies for sustainable use of indigenous cattle genetic resources in southern Africa. Diversity 11(11):1–14. https://doi.org/10.3390/d11110214

Marandure T, Bennett J, Dzama K et al (2020) Advancing a holistic systems approach for sustainable cattle development programmes in South Africa: insights from sustainability assessments. Agroecol Sustain Food Syst 44(7):827–858. https://doi.org/10.1080/21683565.2020.1716130

Martens C, Hickler T, Davis-Reddy C et al (2020) Large uncertainties in future biome changes in Africa call for flexible climate adaptation strategies. Glob Chang Biol February:1–19. https://doi.org/10.1111/gcb.15390

Masikati P, Tui SHK, Descheemaeker K et al (2015) Crop–livestock intensification in the face of climate change: exploring opportunities to reduce risk and increase resilience in southern Africa by using an integrated multi-modeling approach. In: Rosenzweig C, Hillel D (eds) Handbook of climate change and agroecosystems. Imperial College Press, London, pp 159–198. https://doi.org/10.1142/9781783265640_0017

Meza I, Eyshi Rezaei E, Siebert S et al (2021) Drought risk for agricultural systems in South Africa: drivers, spatial patterns, and implications for drought risk management. Sci Total Environ 799:149505. https://doi.org/10.1016/j.scitotenv.2021.149505

Miles MB, Huberman AM, Saldana J (2014) Qualitative data analysis: a methods sourcebook. Sage, London

Mogashoa R, Dlamini P, Gxasheka M (2021) Grass species richness decreases along a woody plant encroachment gradient in a semi-arid savanna grassland, South Africa. Landsc Ecol 36(2):617–636. https://doi.org/10.1007/s10980-020-01150-1

Moore AD, Bell LW, Revell DK et al (2009) Feed gaps in mixed-farming systems: insights from the grain & Graze program. Anim Prod Sci 49(10):736. https://doi.org/10.1071/AN09010

Moyo B, Sikhalazo D, Mota LPM (2012) Behavioural patterns of cattle in the communal areas of the Eastern Cape Province, South Africa. Afr J Agric Res 7(18):2824–2834. https://doi.org/10.5897/ajar11.930

Mpandeli S, Nesamvuni E, Maponya P (2015) Adapting to the impacts of drought by smallholder farmers in Sekhukhune District in Limpopo Province, South Africa. J Agric Sci 7(2). https://doi.org/10.5539/jas.v7n2p115

Mpofu TJ, Ginindza MM, Siwendu NA, Nephawe KA, Mtileni BJ (2017) Effect of agro-ecological zone, season of birth and sex on pre-weaning performance of Nguni calves in Limpopo Province, South Africa. Trop Anim Health Prod 49(1):187–194. https://doi.org/10.1007/s11250-016-1179-2

Mtambanengwe F, Mapfumo P (2005) Organic matter management as an underlying cause for soil fertility gradients on smallholder farms in Zimbabwe. Nutr Cycl Agroecosyst 73:227. https://doi.org/10.1007/s10705-005-2652-x

Mudzengi CP, Dahwa E, Simbarashe Kapembeza C (2020) Livestock feeds and feeding in semi-arid areas of Southern Africa. In: Abubakar M (ed) Livestock health and farming. IntechOpen, London. https://doi.org/10.5772/intechopen.90109

Müller B, Schulze J, Kreuer D et al (2015) How to avoid unsustainable side effects of managing climate risk in drylands — the supplementary feeding controversy. Agric Syst 139:153–165. https://doi.org/10.1016/j.agsy.2015.07.001

Munjonji L, Ayisi KK, Mudongo EI et al (2020) Disentangling drought and grazing effects on soil carbon stocks and CO2 fluxes in a semi-arid African savanna. Front Environ Sci 8(October):1–14. https://doi.org/10.3389/fenvs.2020.590665

Nardone A, Ronchi B, Lacetera N et al (2010) Effects of climate changes on animal production and sustainability of livestock systems. Livest Sci 130(1–3):57–69. https://doi.org/10.1016/j.livsci.2010.02.011

Nyamushamba GB, Mapiye C, Tada O et al (2017) Conservation of indigenous cattle genetic resources in southern Africa's smallholder areas: turning threats into opportunities - a review. Asian Australas J Anim Sci 30(5):603–621. https://doi.org/10.5713/ajas.16.0024

Ouédraogo K, Zaré A, Korbéogo G et al (2021) Resilience strategies of West African pastoralists in response to scarce forage resources. Pastoralism 8:11

Rapholo E, Odhiambo JJO, Nelson WCD et al (2019) Maize-lablab intercropping is promising in supporting the sustainable intensification of smallholder cropping systems under high climate risk in southern Africa. Exp Agric 56:1–14. https://doi.org/10.1017/S0014479719000206

Rootman GT, Stevens JB, Mollel NM (2015) Policy opportunities to enhance the role of smallholder. S Afr Jnl Agric Ext 43(2):91–104. https://doi.org/10.17159/2413-3221/2015/v43n2a360

Rötter RP, Scheiter S, Hoffmann MP et al (2021) Modeling the multi-functionality of African savanna landscapes under global change. Land Degrad Dev 32(6):2077–2081. https://doi.org/10.1002/ldr.3925

Rowe EC, van Wijk MT, de Ridder N et al (2006) Nutrient allocation strategies across a simplified heterogeneous African smallholder farm. Agric Ecosyst Environ 116(1–2):60–71. https://doi.org/10.1016/j.agee.2006.03.019

Rufino MC, Dury J, Tittonell P et al (2011) Competing use of organic resources, village-level interactions between farm types and climate variability in a communal area of NE Zimbabwe. Agric Syst 104(2):175–190. https://doi.org/10.1016/j.agsy.2010.06.001

Rusinamhodzi L, Corbeels M, Giller KE (2016) Diversity in crop residue management across an intensification gradient in southern Africa: system dynamics and crop productivity. Field Crop Res 185:79–88. https://doi.org/10.1016/j.fcr.2015.10.007

Schlecht E, Blümmel M, Becker K (1999) The influence of the environment on feed intake of cattle in semi-arid Africa. In: Van der Heide D, Huisman EA, Kanis E, Osse JWM, Verstegen MWA (eds) Regulation of feed intake. CABI, London, pp 167–186

Schüller H (1969) Die CAL- Methode, eine neue Methode zur Bestimmung des pflanzenverfügbaren Phosphates in Boden. Zeitschrift für Pflanzenernährung und Bodenkunde 123:48

Simelton E, Quinn CH, Batisani N et al (2013) Is rainfall really changing? Farmers' perceptions, meteorological data, and policy implications. Clim Dev 5(2):123–138. https://doi.org/10.1080/17565529.2012.751893

Statista (2021) Maize production in South Africa. https://www.statista.com/statistics/1135488/maize-production-in-south-africa-by-province/. Accessed 25 Feb 2022

Stats SA (2018) Economic growth better than what many expected. In Statistics South Africa. http://www.statssa.gov.za/?p=10985

Stroebel A, Swanepoel FJC, Pell AN (2011) Sustainable smallholder livestock systems: a case study of Limpopo Province, South Africa. Livestock Science 139(1–2):186–190. https://doi.org/10.1016/j.livsci.2011.03.004

Sumberg J (2002) The logic of fodder legumes in Africa. Food Policy 27(3):285–300. https://doi.org/10.1016/S0306-9192(02)00019-2

Sumberg J (2004) The logic of fodder legumes in Africa: a response to Ienné and wood. Food Policy 29(5):587–591. https://doi.org/10.1016/j.foodpol.2004.07.011

Swanepoel PA, du Preez CC, Botha PR et al (2015) A critical view on the soil fertility status of minimum-till kikuyu–ryegrass pastures in South Africa. Afr J Range Forage Sci 32(2):113–124. https://doi.org/10.2989/10220119.2015.1008043

Tavirimirwa B, Manzungu E, Washaya S et al (2019) Efforts to improve Zimbabwe communal grazing areas: a review. Afr J Range Forage Sci 36(2):73–83. https://doi.org/10.2989/10220119.2019.1602566

Terler G, Gruber L, Knaus WF (2019) Nutritive value of ensiled maize stover from nine different varieties harvested at three different stages of maturity. Grass Forage Sci 74(1):53–64. https://doi.org/10.1111/gfs.12390

Thornton PK, Herrero M (2015) Adapting to climate change in the mixed crop and livestock farming systems in sub-Saharan Africa. In Nature climate change 5, 9, pp. 830–836). Nature Publishing Group. https://doi.org/10.1038/nclimate2754

Thundathil JC, Dance AL, Kastelic JP (2016) Fertility management of bulls to improve beef cattle productivity. Theriogenology 86(1):397–405. https://doi.org/10.1016/j.theriogenology.2016.04.054

Tittonell P, Giller KE (2013) When yield gaps are poverty traps: the paradigm of ecological intensification in african smallholder agriculture. Field Crop Res 143:76–90. https://doi.org/10.1016/j.fcr.2012.10.007

Trollope WSW, Trollope LA, Bosch OJH (1990) Veld and pasture management terminology in southern Africa. J Grassland Soc Southern Africa 7(1):52–61. https://doi.org/10.1080/02566702.1990.9648205

Valbuena D, Erenstein O, Homann-Kee Tui S et al (2012) Conservation agriculture in mixed crop-livestock systems: scoping crop residue trade-offs in sub-Saharan Africa and South Asia. Field Crop Res 132:175–184. https://doi.org/10.1016/j.fcr.2012.02.022

Vetter S (2013) Development and sustainable management of rangeland commons – aligning policy with the realities of South Africa's rural landscape. Afr J Range Forage Sci 30(1–2):1–9

Vetter S, Goodall VL, Alcock R (2020) Effect of drought on communal livestock farmers in KwaZulu-Natal, South Africa. Afr J Range Forage Sci 37(1):93–106. https://doi.org/10.2989/10220119.2020.1738552

Ziervogel G, New M, Archer van Garderen E et al (2014) Climate change impacts and adaptation in South Africa. Wiley Interdiscip Rev Clim Chang 5(5):605–620. https://doi.org/10.1002/wcc.295

Zingore S, Murwira HK, Delve RJ et al (2007) Soil type, management history and current resource allocation: three dimensions regulating variability in crop productivity on African smallholder farms. Field Crop Res 101:296. https://doi.org/10.1016/j.fcr.2006.12.006

Agricultural Land-Use Systems and Management Challenges

20

Reimund P. Rötter ⓘ, Mandla Nkomo, Johannes Meyer zu Drewer, and Maik Veste ⓘ

Abstract ✉

This chapter aims at providing an overview of the diversity of agroecological conditions, features of main farming systems, agricultural land use, its dynamics and drivers during the last two decades as well as major threats in ten countries of southern Africa (SA10). Based on this, we attempt to identify the resultant challenges for sustainable land management and outline potential interventions with a focus on smallholder farmers. By analyzing cropland dynamics during 2000–2019, we show how land use has been shaped by climate, demographic development, economic imperatives and policy realities. Concrete examples of these complex interactions illustrate both considerable shrinkage in South Africa and Zimbabwe or expansion of cropland in Mozambique and Zambia.

R. P. Rötter (✉)
Tropical Plant Production and Agricultural Systems Modelling (TROPAGS), and Centre of Biodiversity and Sustainable Land Use (CBL)
e-mail: reimund.roetter@uni-goettingen.de

M. Nkomo
Solidaridad Southern Africa, Johannesburg, South Africa

CGIAR Excellence in Agronomy 2030 Initiative, IITA, Kasarani, Nairobi, Kenya
e-mail: M.Nkomo@cgiar.org

J. M. zu Drewer
Tropical Plant Production and Agricultural Systems Modelling (TROPAGS), University of Göttingen, Göttingen, Germany

Ithaka Institute for Carbon Strategies, Arbaz, Switzerland
e-mail: mzd@ithaka-institut.org

M. Veste
CEBra – Centrum für Energietechnologie Brandenburg e.V., Cottbus, Germany
e-mail: veste@cebra-cottbus.de; maik.veste@b-tu.de

© The Author(s) 2024
G. P. von Maltitz et al. (eds.), *Sustainability of Southern African Ecosystems under Global Change*, Ecological Studies 248,
https://doi.org/10.1007/978-3-031-10948-5_20

During the past 20 years, cropland increased by 37% on average across SA10 mainly at the expense of forestland—showing huge spatiotemporal heterogeneity among countries. Most smallholders face shrinking farm size and other resource limitations that have resulted in soil nutrient mining and low agricultural productivity—a highly unsustainable situation. We conclude with an outlook on potential transformation pathways ("TechnoGarden" and "AdaptiveMosaic") for the near future and thereby provide a frame for further studies on sustainable land management options under given local settings.

20.1   Overview of Agricultural Land-Use and Related Management Challenges

20.1.1 Introduction

Agricultural land use and management in southern Africa has always followed a pathway driven and shaped by climate and its variability, policy realities, economic imperatives and demographic development. The data gathered on land under production and agricultural productivity in the last two decades seems to confirm these realities. There is also huge variation regarding the dependence of livelihoods and the economic importance of the agricultural sector among the various countries in southern Africa. Although the Republic of South Africa's economy is not as dependent on agriculture (sector contributes less than 3% to GDP), in value and volume terms, *South Africa* has the largest agricultural economy in the region and therefore provides a good case study of a postagrarian economy, that, however, still depends to some extent on its agricultural sector. In the rest of the Southern African Development Community (SADC), agriculture's role in the economy is much more pronounced, contributing to the bulk of employment for citizens (on average 70%), as well as being a major contributor to GDP (on average 25%). Prominent agricultural economies in the region, such as *Zambia, Zimbabwe* and *Mozambique,* provide good, but different examples of the complex combination of factors that continue to influence land use and management within the agricultural sector in developing, largely agrarian-based economies. Whereas Zambia and Mozambique have been investing in agricultural development, in different ways and for different reasons, and have seen expansion of agricultural land use during the last one to two decades, Zimbabwe has paid much less attention to agricultural development and seen reductions in agriculturally used land.

Other countries considered here for an overall analysis of the agroecological conditions and agricultural developments of southern Africa are *Angola, Botswana, Eswatini, Lesotho, Malawi* and *Namibia.* Cutoff points, i.e., not included were the Democratic Republic of the Congo and Tanzania, commonly perceived as belonging to central, respective eastern Africa. Due to its geographical particularities, also Madagascar is not included in the analysis. Mainland southern Africa represents one interconnected agroecological region with often common, but distinct socioeconomic characteristics and varied biophysical constraints in terms of water availability and soil conditions across this geographical region. For easier reference, from here the *10 countries of southern Africa* listed above will be referred to as *SA10*.

The objectives of this chapter are:

1. To introduce agroecological features, agricultural land use and recent land-use dynamics in southern Africa (SA10).
2. To describe major farming systems including current production levels, resource use and constraints.
3. To present global change threats and discuss the quest for sustainable intensification and key constraints to achieving this.
4. To identify promising options that respond to key management challenges and sketch alternative future agricultural transformation pathways.

20.1.2 The Agroecological Conditions of Southern Africa

According to the original agroecological zone (AEZ) definition by FAO, an AEZ is mainly defined by its temperature regime and moisture availability conditions. Temperature regime is characterized by temperature belts with specific ranges chosen for mean, minima and maxima to coincide with temperature thresholds demarcating thermal suitabilities for major crops, whereas moisture availability is at first level characterized by the ratio of annual precipitation to potential evapotranspiration (expressed as aridity or humidity index) (see Fig. 20.1) (Fischer et al. 2012). There is a very strong gradient in annual rainfall spanning from well above 1500 mm in the northeastern parts of the southern African region to less than 50 mm along the Namibian coast (Hijmans et al. 2005) (Figs. 20.2 and 20.4). Rainfall seasonality and associated atmospheric circulation processes are other important factors for agriculture (Richard et al. 2001). Overall, most areas in southern Africa receive predominantly convective rainfall from October to March or all-season rainfall. On the contrary, the important agricultural region of the southwestern Cape receives predominantly frontal winter rainfall from April to September, driven by westerly derived mid-latitude cyclones. In South Africa a transition zone receives rainfall from both summer and winter rainfall systems along the southern coast. Characteristic "Walter-climate diagrams" illustrate rainfall seasonality in the different "eco-climatic zones" (Fig. 20.2).

An important additional characteristic of AEZs is the length of the growing period (LGP) expressed in days during which (on average) moisture availability conditions are considered "agrohumid," that is, with sufficient water supply to allow crop growth. For detailed agricultural and farm management planning at local scales, different authors (e.g., Jätzold and Kutsch 1982) have introduced the consideration of rainfall seasonality and the probability of rainfall received per growing season instead of only using mean values, for specifying the LGP. These authors also emphasized the role of soil conditions, especially soil depth and soil

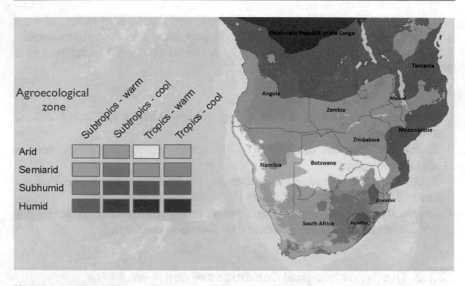

Fig. 20.1 Agroecological zones of southern Africa (AEZ16 System) [*Adapted from* Kate (2009). Permission to use granted as per open access status under creative common license: Attribution 4.0 International (CC BY 4.0)]

water retention to accurately describe the potential of AEZs including LGP. Since the 1970s, there have been numerous local to global AEZ classification systems (see Rötter et al. 2016). Most used is the Global Agroecological Zones (GAEZ) as defined by IIASA (Fischer et al. 2012). In Fig. 20.1, we present the main AEZs of southern Africa as compiled in the version of IFPRI (Kate 2009). We find that in most of the 10 countries (SA10) considered in our analysis the tropical warm, semiarid (dry savanna) zone is prevalent, followed by: the semiarid to semihumid tropical highland climates of Angola, Malawi, Namibia, South Africa, Zambia and Zimbabwe; the arid to subhumid subtopics of Namibia, South Africa and Lesotho, and the arid tropical warm zone covering large parts of Botswana and Namibia, and finally, the humid tropical lowland zone that is mainly found in Mozambique. Most of these zones except for the arid ones have a moderate to high agricultural potential. This is reflected clearly in the LGP map (Fig. 20.3), where LGPs with durations of less than 60 days are restricted to the arid zones of Botswana, Namibia and South Africa—with the marginal agricultural zones (LGPs from 60–119 days) wrapped around these (in pink).

Another serious constraint to agriculture is the fairly high variability of rainfall, both interannual/-seasonal and intraseasonal (e.g., Davis-Reddy and Vincent 2017). A detailed account of current and recent past climate variability is given in Chap. 5. The bad news is that ongoing climate change has already amplified the severity of weather phenomena causing high rainfall variability, such as El Nino Southern Oscillation (ENSO). Especially the strong El Nino events have led to extended drought and also yield loss (Verschuur et al. 2021). Figure 20.4 shows the long-term annual precipitation pattern averaged over the 118 years period (1901–2018),

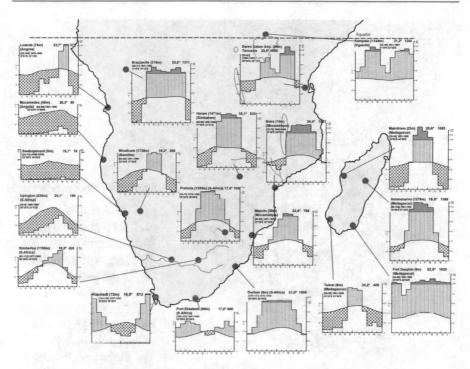

Fig. 20.2 Climate diagrams after Heinrich Walter showing different climate types and seasonal precipitation patterns in southern Africa [*Adapted from* Breckle and Rafiqpoor (2019). Permission to use granted as per written communication by S.W. Breckle (Author), 2022]

and Fig. 20.5 illustrates the difference between cumulative seasonal precipitation (in mm) over the months November to February (i.e., the main rainy season in most of our target region) averaged over the 30 year period 1981–2010 (Fig. 20.5a, *left*) *vis a vis* the cumulative precipitation averaged only over the years with strong El Nino events (Fig 20.5b, *right*). In strong El Nino years the area with low to very low rainfall (<150 mm) is considerably extended (especially in Namibia, South Africa and Botswana) compared to average conditions, and, on the other hand, the area with high to very high rainfall (>950 mm) is also expanded—in particular in Zambia (Fig. 20.5).

As for soil conditions (see Figs. 20.6 and 20.7), the region shows a very complex pattern due to the high spatial heterogeneity of the major soil building and forming factors and processes such as geology, including tectonic stability (factor time), topography, hydrological conditions, vegetation types and cultivation history (Sikora et al. 2020) next to the factor climate. Among the most predominant soil groups, we find the deeply weathered Ferralsols from the old land surfaces from Angola and Zambia, constituting low natural fertility mainly due to poor soil chemical properties. Also quite prevalent are the Arenosols poor in water holding capacity and low in nutrients, stretching from Angola in the north via Namibia

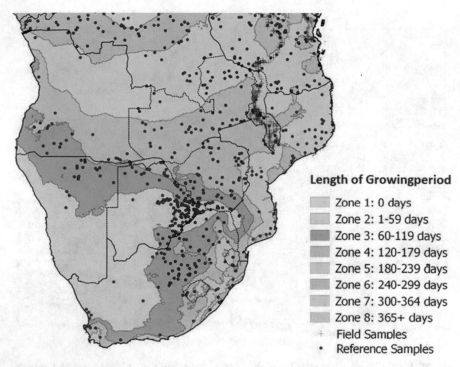

Length of Growingperiod

Zone 1: 0 days
Zone 2: 1-59 days
Zone 3: 60-119 days
Zone 4: 120-179 days
Zone 5: 180-239 days
Zone 6: 240-299 days
Zone 7: 300-364 days
Zone 8: 365+ days
+ Field Samples
• Reference Samples

Fig. 20.3 The length of growing period in southern Africa [*Adapted from* Xiong et al. (2017). Permission to use granted as per open access status under creative common license: Attribution Non-Commercial, No Derivatives 4.0 International (CC BY-NC-ND 4.0)]

and Botswana further south to the northern tips of South Africa. Widespread are the fairly fertile Cambisols of Zimbabwe and South Africa, and the Phaeozems of northern Mozambique, South Africa and central Zimbabwe.

Leptosols show a broad band that mainly stretches from the arid zones of Namibia to South Africa. The spatial pattern of selected soil properties is shown in Fig. 20.7 extracted from the high resolution iSDA digital soil map shows (from left to right): total soil carbon, extractable P and total soil N. A more detailed overview of soil conditions and soil fertility issues in the region is provided by Vlek et al. (2020).

20.1.3 Major Farming Systems in Southern Africa: Their Characteristics and Dynamics

The agroecological conditions along with the influence of socioeconomic factors such as market access generate a distinct geographical pattern of generic farming systems (see Fig. 20.8). Dominating in terms of land use is the group of maize-based cropping systems, widespread in the northern and eastern realms of southern Africa,

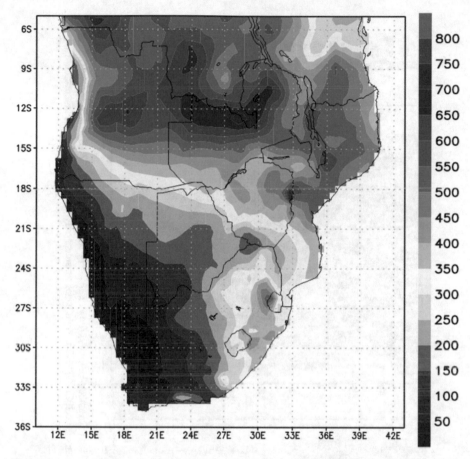

Fig. 20.4 Annual precipitation (mm) in southern Africa (1901–2019) [*Data: Climatology CRU TS4 (1901–2019) (extraction and mapping by NRC Ferreira, TROPAGS/University of Göttingen)*]

and a pocket of agroecological suitability stretching south through Zimbabwe and South Africa. Maize-mixed farming systems represent the livelihood basis for 100 million rural people in Sub-Saharan Africa (Auricht et al. 2014). The central and western parts of southern Africa, experiencing a semiarid to arid climate, are dominated by agropastoral systems, with Namibia and central South Africa being able to only sustain pastoralism on a large scale. Only the eastern coast and Western Cape region of South Africa, as well as the state of Eswatini, can naturally sustain perennial cropping systems on a large scale.

To a large extent, these cropping systems rely on rainfed water exclusively, with irrigation being common only in some limited areas. Future climate projections point toward reduced rainfall and increased variability for most of southern Africa, with severe reductions in the already marginal, western part (Nhemachena et al. 2020) (Fig. 20.9).

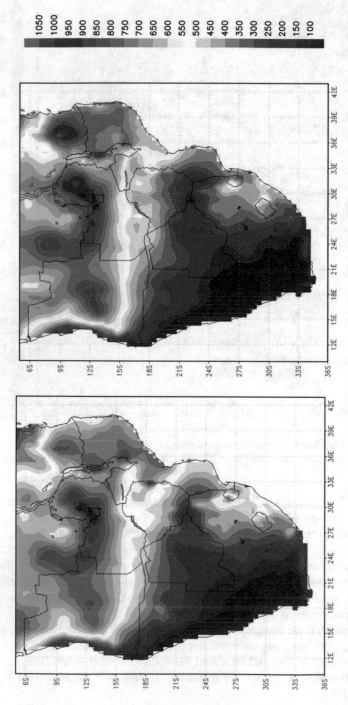

Fig. 20.5 Cumulative, seasonal precipitation in southern Africa (November–February) *(own analyses by NRC Ferreira, TROPAGS/University of Göttingen)*. The left panel indicates baseline conditions, considering all years from 1981–2010. The right panel considers only years with strong El Niño occurrence [*Data source: Climatology CRU TS4 (1981–2010)*]

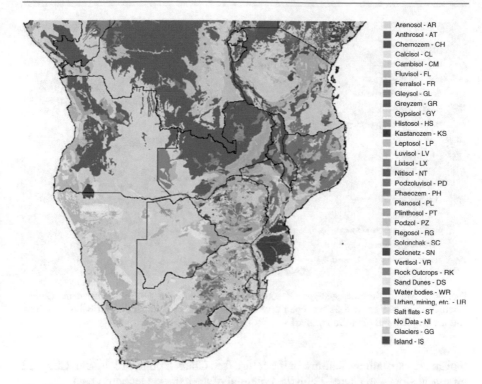

Fig. 20.6 FAO classification of southern African soils [*Adapted from* Fischer et al. (2008). Permission to use granted by written communication with FAO-GSP secretariat, chief publication branch, 2022]

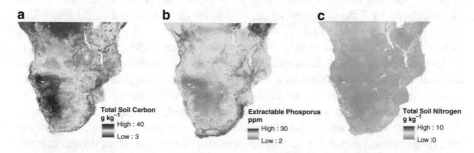

Fig. 20.7 Key soil characteristics in southern Africa, indicating total soil carbon (**a**), extractable phosphorus (**b**) and total soil nitrogen (**c**) [*Adapted from* iSDA Africa (2021). Permission to use granted as per open access status of iSDA database, attribution is given, and the original authors were notified]

These projections of a diminishing resource basis are alarming. That is, in first place, water resources for agriculture (Meza et al. 2021; Chap. 22 on macadamia) and fertile soils (Vlek et al. 2020). Especially improved water management and adaptation measures to drought and rainfall variability become imperative if the

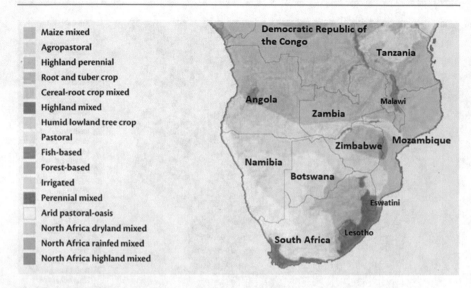

Fig. 20.8 Major farming systems of southern Africa [*Adapted from* Auricht et al. (2014). Permission to use granted as per open access status of IFPRI publication, attribution is given, and the original authors were notified]

region is to sustain agriculture in the future (see Chap. 5 on hydroclimate, Chap. 22 on macadamia and Chap. 23 on the potential of agricultural technologies).

Besides climate change, the other factor exerting considerable pressure on the natural resource quality and environment is the continuous expansion of agricultural land. In just two decades, from 2000 to 2019, the cropland in southern Africa has expanded by 37% to a total of 28 million ha in 2019 (see Fig. 20.10). While maize has clearly remained to be the dominating crop, novel industrial crops such as soybean have emerged rapidly and found their place within the major cropping systems (FAOstat 2021). Climate change and population growth and increasing food demand are directly mirrored in a shift and expansion of agricultural production, provoking conflicts with other land-use objectives (conservation, forestry). In the last decades, an unsustainable trend of deforestation has emerged. Drastic deforestation is observed in the Miombo woodlands across southern Africa where strong agricultural expansion and charcoal production are the major drivers of land-use changes (Ribeiro et al. 2020).

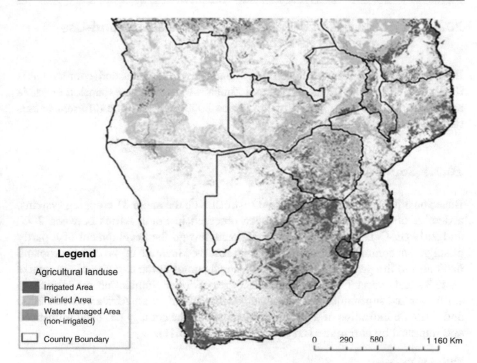

Fig. 20.9 Agricultural land-use patterns by type of water source [*Adopted from* Nhemachena et al. (2020). Permission to use granted as per open access status under creative common license: Attribution 4.0 International (CC BY 4.0)]

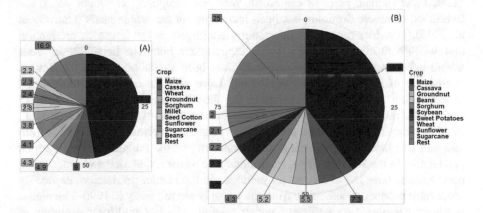

Fig. 20.10 Cropland composition and expansion in southern Africa (SA10) 2000–2019. (**a**) Composition of cropland in SA10 in 2000 (20 million ha). (**b**) Composition of cropland in SA10 in 2019 (28 million ha) [own analyses, J Meyer-zu-Drewer, based on *FAOstat (2021)*]

20.2 Selected Case Country Studies Illustrating Land-Use Dynamics and Its Drivers

In the following we have chosen four contrasting examples and country cases illustrating shrinkage (South Africa and Zimbabwe) as well as expansion (Zambia and Mozambique) of cropland for the period 2000–2019 and the different causes that led to such developments.

20.2.1 South Africa

Based on official national data and FAO estimates on the major 31 cropping systems, a decline of 22% of cropland has been observed in South Africa between 2000 and 2019 (FAOstat 2021). This change points toward the development of a partly postagrarian community. Remarkable is an observed increase of +679% in cropland dedicated to the production of soybean, which comes at the expense of o.a. maize (−43%) and wheat (−42%) production area. Further dominating crops include sunflower and sugarcane. The land-use patterns within South Africa between 2000 and 2019 are indicative of a number of factors that the country has been influenced and impacted by during the last two decades (Fig. 20.11).

Climate Factors
South Africa being largely a country with limited water resources has limited options for irrigation. Rainfed agriculture prevails with reliable crop production typically found in the higher rainfall areas (Eastern Cape, KwaZulu Natal, Free State, Mpumalanga, parts of the North West and Limpopo). In recent years, the likelihood of severe droughts has been increasing for the arable lands (Conway et al. 2015). The drier regions of the country are largely put to livestock production and wildlife farming, and high value crops under intensive irrigation in areas where infrastructure and water resources have been available. Due to the limited availability of irrigation water, one of the major areas of conflict in those areas in the period under review is related to the allocation and use of water rights for agricultural purposes. In both rainfed crop and rangeland-based livestock farming, trends indicate that the last two decades have been about increasing crop yields from less and less land, and increasing the productivity of the rangelands for livestock production. In the Western Cape, traditionally the wheat production area, successive poor seasons have led to a constant reduction in land under production, especially since most production subsidies from the wheat board fell away in 1996. The region is highly dependent on sufficient winter rainfall. The last multiyear droughts of 2015–2018 have led to drastic crop losses in the Western Cape region, likely to be exacerbated under future climate change (Theron et al. 2021).

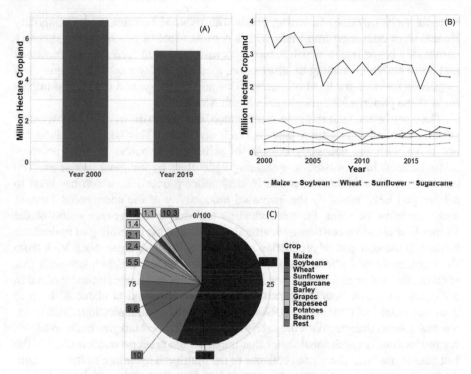

Fig. 20.11 Cropland dynamics: South Africa. (**a**) Total cropland area as of 2000 and 2019 in million hectares. (**b**) Area dynamics of five dominating cropping systems (2000–2019) in million hectares. (**c**) Composition of cropland as of 2019 [*FAOstat (2021)*]

Policy and Institutional Factors

The trend-line on land use for agriculture points to a steady decline in land use for agricultural purposes during 2000–2019, which to a large extent has been influenced by fundamental agropolicy changes that preceded the 2000s. The main influence has come from deregulation, land reform, land redistribution and changes in the agricultural finance and insurance environment. Prior to 1996, South Africa's agriculture was managed through the activities of statutory commodity boards, which were constituted under the Marketing Act 1968 to oversee the agricultural activities of various commodities such as the Wheat Board, the Maize Board, the Wool Board, etc. Their functions were to regulate the production and marketing of various commodities, by ensuring access to inputs, mechanization and favorable pricing through single market channels. For that reason from the 1930s up to the late 1980s, there was rapid expansion of land use for agricultural production. When the Marketing Act was replaced by the Marketing of Agricultural Products Act No 47 in 1996, the price protection afforded to farmers fell away, and only the market could determine price. As a result, marginal producers left the various sectors, as they could not compete, and the producers in favorable agroclimatic zones improved their productivity. Access to improved technology such as GMO maize also contributed to rapid yield increases in maize, albeit with reduced hectares. The Commodity Boards were converted into trusts with a narrower role of administering statutory commodity based levies, which in turn were mainly used to fund industry functions and board activities relating to information, grading, quality standards, training and inspection services for local producers. This policy change had the biggest impact on land-use patterns for agriculture in South Africa.

The other factor, land reform and restitution, resulted in the transfer of previously white controlled agricultural land to black occupants. This transfer of land did not come with the necessary transfer of skills and resources that the previous administration had invested in sustaining agriculture by marginal producers. The result has been a further reduction in land under production, which has been to a large part been offset by the increased productivity of the commercial farming sector in terms of yield. Further reduction was due to the reorganization of the former Homeland/Bantustans agricultural systems and the defunding of agriculture leading to the collapse of production in large parts of these areas. Nick Vink from the Department of Agricultural, University of Stellenbosch makes a key point that state-driven farmer assistance grew to a peak of 25% of all agricultural income in 1984, and although steadily decreasing thereafter, remained at about 20% up to commencement of democracy (1990/1991) (personal communication). Thereafter we saw a steep decline. With the safety net removed, less land was made available for production. An additional factor that impacted the grain production areas in the last decade has been the virtual collapse of the multiperil insurance system in South Africa. Successive poor seasons have made insuring crops such as maize and wheat unviable for the insurance industry and the multiperil insurance product was largely withdrawn from the market.

Economic and Demographic Factors

The economic and demographic impact on land use for agriculture in the review period, was driven by the rapid economic growth in the late 1990s and early 2000s which saw GDP growth of about 5% on average, and a massive postdemocracy increase in the middle class from 1.7 million individuals in 2004, to over 4.2 million individuals by 2012. The sheer spending power within this group grew the poultry industry (i.e., the largest part of the SA agricultural complex by revenue at over 15% of gross value generated by the agricultural sector) so much that South Africa currently needs to import up to 30% of its poultry feed requirements annually. This is because poultry demand has grown faster than supply, pushing consumption of maize and soy for feed, and resulting in imports to close the gap. Per capita poultry consumption between 2000 and 2017 has increased from 18.5 kg to 40.0 kg per capita. Since poultry relies on maize and soybeans, these grain sectors have shown massive increase in productivity (maize) and increase in plantings (soybean) in response to greater local demand.

20.2.2 Zimbabwe

Considering official FAO data on 71 dominant cropping systems in Zimbabwe, a decrease of −19% in total cropland was identified. A decrease can be observed both in staple crops such as maize (−36%), and commercial crops such as seed-cotton (−45%).

Climate Factors
Zimbabwe has experienced increasingly erratic rains over the past two decades and has been impacted by the El Nino phenomena seemingly more than neighboring countries (Setimela et al. 2018). The rainfall pattern is a major factor in influencing land use for rain-fed crops, livestock and irrigated crops. The variability between seasons and periods in land put under crop production has largely tracked rainfall amount and its variability. Extremely dry seasons recently, such as 2015/2016 and 2019 have led to drops in the area planted to maize—the main staple crop in Zimbabwe. Wheat is exclusively produced under irrigation and the variability in production has been linked to availability of irrigation water, and stability of electricity supply. Zimbabwe's primary source of power is hydroelectric generation from the Kariba dam. Successive drought has resulted in reduced inflows into the dam, and this has ultimately rendered the hydroelectric scheme inoperable due to low water levels. In turn, power supply has been erratic in the last decade, with load shedding reaching 18 h of planned power outages per day. Wheat production under irrigation was near impossible under these conditions. However, when considering the area put under cash crops such as tobacco and soybeans (Fig. 20.12), counterintuitively, the production of these cash crops was not subject to this distortion. After a land reform inspired collapse in the early 2000s, tobacco production has shown a rapid upward trend (Government of Zimbabwe 2018) driven largely by the financial incentive the crop offers to all sizes of producers where the producer price is in US dollars. A crop like soybean has stabilized due to the economic importance of the crop especially to the livestock sector in Zimbabwe. The reason, therefore, lies in a combination of policy and economic factors.

Policy Factors
The Zimbabwean agropolicy environment in the last two decades, after climate, has had the largest influence on agricultural land-use patterns in Zimbabwe (e.g., ZAIP 2013). The major policy was the start in 2000 of the Fast-Track land reform process that summarily stopped production in most acquired farms, as well as disruption to the agro inputs sector. The impact was a drop in area planted in the first few years of land reform, although there was a recovery midway into the first decade (2000–2010). In addition to land acquisition and transfer, state policies related to marketing of agricultural produce, in particular maize and wheat, created a disincentive to produce the crops, as government controlled prices were lower than regional benchmarks, and were paid in an unstable and failing local currency. Maize and wheat's loss seemed to have been tobacco and soybean's gain. These crops were

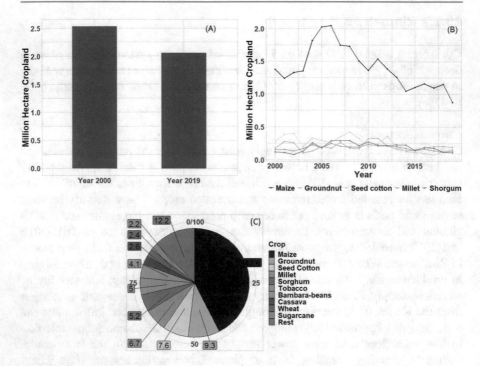

Fig. 20.12 Cropland dynamics: Zimbabwe. (**a**) Total cropland area as of 2000 and 2019 in million hectares. (**b**) Area dynamics of five dominating cropping systems (2000–2019) in million hectares. (**c**) Composition of cropland as of 2019 [*FAOstat (2021)*]

governed by a free market pricing regime and could be sold in foreign currency, and farmers could realize real value and returns. The explosive growth in tobacco and soybean is a reflection of the policy impacts of how they are marketed in Zimbabwe. The high growth in tobacco has come at the expense of the environment, due to the use of greenwood as the fuel source for curing the tobacco. The environmental damage associated with tobacco, will in the future become an existential challenge in the high rainfall areas of the country (Tatenda 2019).

Economic and Demographic Factors

Zimbabwe's economy has generally experienced erratic growth from the mid-1960s as a result of armed conflict and postindependence economic management issues. Agriculture has been a stabilizing factor since it impacts the majority of the population. However, a mismanaged post-2000 land reform pushed even this sector over the brink, with additional negative impacts by frequent droughts. In a country where 70% of the population derive their livelihood from agriculture, 20% of the GDP, 40% of all exports and 60% of manufacturing raw materials come from agriculture, economics and land use are intricately connected.

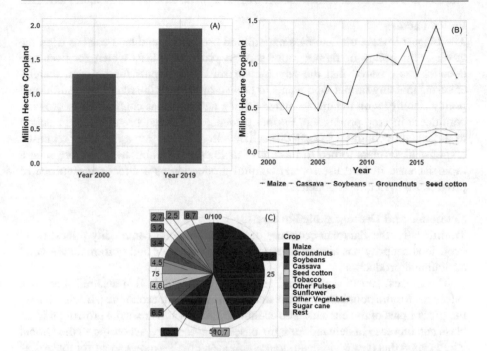

Fig. 20.13 Cropland dynamics: Zambia. (**a**) Total cropland area as of 2000 and 2019 in million hectares. (**b**) Area dynamics of five dominating cropping systems (2000–2019) in million hectares. (**c**) Composition of cropland as of 2019 [*FAOstat (2021)*]

20.2.3 Zambia

Considering the 28 dominating crops, an increase of cropland by 51% was observed, totaling 1.9 million ha in 2019 (FAOstat 2021). While the traditional crops maize and groundnut are still dominating, a strong development toward industrial and export-oriented crops can be observed. The soybean production area experienced a growth of +1083% since 2000, turning it into the top 3 crops (Fig. 20.13).

Climate Factors

Zambia's tropical climate and low population density are advantages that the agricultural sector has benefitted from. The country's agroclimatic conditions have remained favorable for crop production in most cultivation areas in the last two decades, and that has created a level of predictability within the agricultural sector. The main crop production areas of Central, Eastern and Southern Provinces lie along the fertile belt of so called agroecological zones I and IIa which combine high rainfall and good soils for crop production.

Policy Factors

The biggest impact on Zambia's agricultural production and land use has been the government policy of farmer input support program (FISP) which for over two decades has ensured that the government subsidizes inputs for farmers, thereby ensuring that any farmer who wants to grow crops has all the requisites. Additional policy stability with regards to maize pricing and trade has contributed to growing confidence in crop production. Further evidence of this has been farmers' reaction to unfavorable contracts in cash crops like cotton, which saw a rapid market based response as farmers turned to more profitable crops. Zambia has therefore seen a rapid increase in land use for agricultural production as a direct consequence of progressive agricultural policies.

Economic and Demographic Factors

Traditionally, the Zambian economy depends on mineral commodity prices; however, food security has always depended on the country's ability to maintain good agricultural production.

The longest run of agricultural surpluses (ReNAPRI 2014) ensured there is a safety net for the population and created opportunities for economic participation by the greater part of the citizens. The economic stability has created a growing middle class and increase in demand for agro-based commodities. According to the Global Yield Gap Atlas (www.yieldgap.org/zambia), Zambia's average yield for maize has averaged 1.1 million t ha^{-1}, against an achieved average 6.5 million t ha^{-1} in the much drier South Africa. Considering these current yields, there is still a long way for Zambia to achieve high productivity. Growth in agriculture, that has driven rising incomes and rapid urbanization has also created unintended consequences in that there is a serious energy deficit that has created unsustainable wood harvesting for of charcoal for energy. The Centre for Forestry Research (CIFOR) estimates that 30,000 ha of forest cover are lost annually (Day et al. 2014) The main drivers for forest cover loss are listed as agricultural expansion, urban infrastructure development, wood extraction (e.g., for charcoal and wood fuel) and uncontrolled fires. The impact of this rapid and large-scale deforestation has the potential to have negative and serious environmental impacts for Zambia in the near future and calls for corrective measures.

20.2.4 Mozambique

Considering the 40 dominant cropping systems, Mozambique experienced a strong expansion of cropland of +87% during 2000–2019. Remarkable is the strong increase in production area for staple crops such as maize (+109%), paddy rice (+293%) and sorghum (+109%) (FAOstat 2021) (Fig. 20.14).

Climate Factors

The tropical to subtropical climatic conditions of the region are largely influenced by the monsoons from the Indian Ocean and Mozambique current with warm

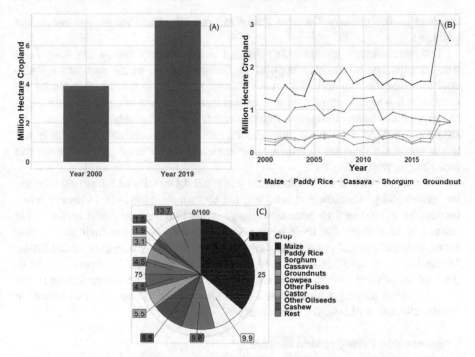

Fig. 20.14 Cropland dynamics: Mozambique. (**a**) Total cropland area as of 2000 and 2019 in million hectares. (**b**) Area dynamics of five dominating cropping systems (2000–2019) in million hectares. (**c**) Composition of cropland as of 2019 [*FAOstat (2021)*]

surface waters flowing south along the African east coast, while the southern area of Zambezi River is influenced by the subtropical anticyclonic zone. The south of Mozambique is generally drier with an average rainfall lower than 800 mm, decreasing to as low as 300 mm. Mozambique is already highly susceptible to climate variability and extreme weather events. Periods with floods are followed by droughts. Meanwhile, climate change has raised the frequency of extreme weather such as tropical cyclones with destructive effects on agriculture. In 2020 for example, Mozambique had two such cyclones making landfall on the country. Manuel et al. (2021) emphasize regional differences in climate change impacts due to differences in agroecological conditions. Higher negative impacts of climate change are expected on the agricultural outcomes in the central and northern regions, which are currently characterized by more favorable agroecological conditions than the drier southern regions (Swain et al. 2011).

Policy Factors

While Mozambique probably has the best agroclimatic conditions in southern Africa, it yet is one of the poorest countries in the region. A combination of decades of internal conflict since independence has held back the country's agricultural potential. However, after the 1992 Rome Peace Agreement, sufficient stability returned to the country for agriculture to take advantage of its agroecological potential.

The Mozambican government's limited fiscal capacity meant that they have limited capacity to directly support agriculture in the same way other SA10 countries like Zambia and Malawi are able to. In trying to manage this reality the government adopted a policy of concession agriculture wherein private companies are given concessions to operate outgrower schemes exclusively in a district, for a single commodity. Farmers in that district, growing that commodity, can only sell to the concession holder. In return, the concession holder must provide inputs and technical support to farmers.

This model worked well in the early years and drove a lot of the strong increase in agricultural production and land use for agriculture. However, in recent years, farmers have switched to nonconcessional crops like soybeans and sesame, that reward farmers fairly for their labour. Cotton and tobacco, the main concession crops have been under pressure as farmers turn to more open market traded crops. Mozambique has also been subject to controversial "land grab" issues as a result of the land concession system. It remains to be seen if and when the government will be able to start playing a bigger role in agricultural support and if this will result in greater utilization of land for agriculture.

Economic and Demographic Factors
The end of the civil war gave the economy a chance to grow almost exponentially, for slightly over a decade (World Bank 2006). Demand for basic commodities such as poultry, previously all imported, created the impetus for local production. The economic situation has deteriorated in the last decade, but local consumer demand remains (World Bank 2021). It is anticipated that land use for agriculture will grow multiple times as market systems take root and stabilize across a number of commodities.

20.3   Global Change Threats and the Quest for Sustainable Intensification and Diversification

20.3.1 Changes in Demography, Food Demand and Food Insecurity

Changes in Demography One of the big challenges for Africa in the twenty-first century is its rapid population growth. Looking at the medium variant of the United Nations projections for the continent as a whole, its population will nearly double between 2020 and 2050 to an estimated 2.6 billion people. Globally, the population is expected to grow by just 30% (UN DESA 2017), Africa accounting for half of this growth in that period. When we look at the SA10 treated in this chapter, the population of these countries together amounted to about 45 million in 1960. The population count increased to about 175 million in 2017 (UN DESA 2017) and projections suggest that by 2050 approx. 350 million people (Klingholz 2020) will live in the region—a doubling in just 33 years. Most of the population growth is still happening in rural areas and this increase in rural population is very unlikely to be absorbed by employment in the primary agricultural sector (Sikora et al. 2020).

Changes in Food Demand and Food Insecurity Changes in food demand is not just a matter of more people needing more calories, but depends on various factors such as demographic structure, changes in diet, economic development, etc. (Rötter et al. 2007). Changes in diets due to more wealth and associated changes in lifestyles and food consumption patterns (toward more meat) possibly have the strongest influence on increased per capita calorie demand (Tilman et al. 2011). There is a large food demand-supply gap for southern Africa that may even widen in the future decades as a consequence of rapid population and income growth. The World Food Summit (1996) defined: "food security represents a state when all people at all times have physical and economic access to safe and nutritious food to meet their dietary needs and food preferences for an active and healthy life." Among the rural population in southern African countries about 16% have consistently been classified as "food insecure" (SADC 2018). Geo-referenced data on the current status of food insecurity and related indicators at (sub-)national scale can be found in WFP. The ongoing COVID-19 pandemic has again since 2020 increased the total number and relative share of people experiencing chronic hunger—globally by about 120 million people, with a considerable share of those in Sub-Saharan Africa (FAO, IFAD, UNICEF, WFP and WHO 2021).

20.3.2 Climate Variability and Change, Natural Resource Limitations and Low Agricultural Productivity

Climate Variability Southern Africa is one of the world regions characterized by high rainfall variability (Davis-Reddy and Vincent 2017). There is evidence that inter-annual rainfall variability over southern Africa has increased since the late 1960s and that droughts have become more intense and widespread in the region (e.g., MacKellar et al. 2014). Among the many factors influencing rainfall variability, the El Nino Southern Oscillation (ENSO) phenomenon has possibly the strongest impact over large parts of southern African regions. Here, El Niño conditions are generally associated with below-average rainfall years over the summer rainfall regions (see Fig. 20.5, above), while La Niña conditions are associated with above-average rainfall. The 1982/1983 and 2015/2016 droughts in many parts of SA10 coincided with strong El Niño events. Chapter 5 gives more details on current and recent past climate variability.

Observed Impacts: In 2015/2016 South Africa experienced the worst drought since 1930. Large parts of maize (83%) and wheat (53%) are produced under rainfed conditions, making them especially vulnerable. In 2015, The Free State, KwaZulu-Natal, Limpopo, Mpumalanga, Northern Cape and North-West provinces were declared drought disaster areas. Also other countries (Lesotho, Swaziland,

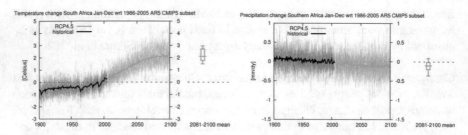

Fig. 20.15 Climate change scenarios: temperature change for South Africa, and precipitation change for southern Africa, 1900–2100 plotted from KNMI Climate Explorer website based on CMIP 5 multimodel ensemble—42 models, using one ensemble member per model (Source: http://climexp.knmi.nl/plot_atlas_form.py)

Zambia, Zimbabwe) experienced yield reductions and associated increases in maize prices (WFP 2016). Verschuur et al. (2021) showed that drought in South Africa and Lesotho in 2007 resulted in severe food insecurity in Lesotho.

Climate Change For southern Africa, Engelbrecht et al. (2015) report drastic increases in surface temperature for the region—about twice as high as the global rate of warming. A decrease in late summer rainfall (JF, i.e., January and February) has been reported over the western regions including Namibia and Angola. Long-term records have shown significant increases in average rainfall intensity and the length of the dry season (New et al. 2006). Trends in flood occurrences have been decreasing prior to 1980 and increasing afterward. Mean annual temperatures have increased in the last five decades and have reached 0.2–0.5°C/decade in some regions such as in south-western Africa. Under the highest emission scenarios (RCP8.5 or SSP5–8.5), almost all African regions will very likely experience a warming larger than 3°C, while under a low emission scenario (RCP2.6 or SSP1–2.6), the warming probably remains below 2°C (IPCC 2021). Some projections of annual temperature and precipitation changes (as anomalies referring to 1986–2005) are presented in Fig. 20.15. Chapter 7 gives information about the latest climate change projections for the region.

Consequences: Accelerated climate change will put additional pressure on the multifunctionality of southern African savanna ecosystems and the Western Cape winter rain area (Midgley and Bond 2015). Ecosystem services such as provision of food, feed, fuel, carbon sequestration, nutrient cycling, habitat quality, pollination and natural pest control are under threat (Rötter et al. 2021). Both agricultural and hydrological drought are projected to increase in southern Africa—most severe for agriculture will be the projected significant increase in the probability of extremes, especially heat waves and severe droughts (IPCC 2021). The number of days with maximum temperature exceeding 35°C is projected to increase in the range of 50–100 days by 2050 under high emission scenario SSP5–8.5 for most regions in Africa. Some adaptations are possible, e.g., through judicious choice of more suited crop cultivars (see Sect. 20.4.1.2). Global warming will very likely increase the frequency of extreme El Niño events.

Natural Resource Limitations, Land Degradation and Low Agricultural Productivity

Vlek et al. (2020) described the natural resource situation in southern Africa as "land rich but water poor," at the same time stressing the need for agricultural intensification and emphasizing options to stop soil nutrient mining and land degradation by integrated nutrient management practices (Vanlauwe et al. 2010) with special attention to soil organic matter.

Sub-Saharan Africa is dominated by low input agriculture with associated low farmer's yield which may be well below 20% of the climatic potential (Van Ittersum et al. 2016). This has often been compensated by additional land clearing and "over-cropping" of already marginalized land (Nkonya et al. 2016). Yield levels for major staples such as maize remain low (at 1–2 t ha^{-1}) due to low inputs. Average fertilizer application rates in 2017 in sub-Saharan Africa were still below 17 kg ha^{-1} (NPK together) (Vanlauwe and Dobermann 2020). The huge nutrient gaps, i.e., the gaps between the nutrients actually applied and those required to replenish the nutrients removed by harvested products (Ten Berge et al. 2019), resulting in soil nutrient depletion, are a main cause of stagnating low agricultural productivity, land degradation and poverty of smallholders (Vlek et al. 2020). Land clearing and deforestation to expand agricultural land use has led to rapid degradation of more than 95 million ha of land in SSA (Nkonya et al. 2016). The loss of vegetative cover, depletion of soil organic matter, lack of management skills and appropriate technologies are recognized factors codetermining soil degradation (Kuyah et al. 2021). Little fertilizer, no irrigation is what we may call the "status quo management practice" of smallholder farmers in southern Africa (Chap. 23). On the other hand, it has been demonstrated at many on-station and on-farm field experiments and by yield statistics of commercial farms that cereals yields of 6–7 ha^{-1} are achievable at high nutrient and water use efficiencies. Such yields can be sustained if applying appropriate technologies and good management, such as site-specific nutrient management, smart crop rotations and deficit irrigation (e.g., Swanepoel et al. 2018). Yet, the considerable increases in cropland and harvested area for the main crops are largely responsible for the recent increases in crop productivity in southern Africa (FAOSTAT 2021) (Sects. 20.1–20.2).

20.3.3 The Quest for Sustainable Intensification and Diversification

Southern Africa has also been identified as a hotspot for biodiversity, whereby agricultural expansion is a key driving force for the declining species diversity (Midgley and Bond 2015). The demographic and climate change projections underline the urgency of science-informed identification of sustainable land management options that, on the one hand, lead to sustainable increase of crop yields per unit area so as to meet increasing food demand and, on the other hand, protect biodiversity in forests and natural vegetation by saving these areas from agricultural expansion (IPCC 2019; Sikora et al. 2020).

From Definitions to Implementation Sustainable agricultural intensification (SI) means "to produce more with less land, water and labour to meet growing food demands and save space for biodiversity." This definition has been phrased during the 1990s in the context of market liberalization and economic growth in countries with densely populated areas in S, SE and E Asia (Rötter et al. 2007). Godfray et al. (2010) emphasizes that SI is the logical response to the threefold challenge of a further strong increase in food demand, increasing competition for natural resources and decline in resource availability and quality, and climate change threats. While under some conditions SI can be achieved through specialization on one/few crops, diversification of crop production, horticulture and livestock toward mixed farming systems can increase the economic viability and resilience of farms and farm households (Kuyah et al. 2021), especially climate variability and change. Tibesigwa et al. (2017), among others, found that such mixed farming systems are less vulnerable compared to specialist crop farms. SI is commonly defined as an increase of agricultural production and improvement of ecosystem services from the same area of land—with constant or reduced inputs and reduced negative externalities such as the agricultural carbon footprint (Garnett et al. 2013). In order to minimize greenhouse gas emissions from the African land-use sector, SI must be favored over an expansion scenario, as the latter leads to higher greenhouse gas emissions and jeopardizes climate protection more than intensification (Tilman et al. 2011; Van Loon et al. 2019). But so far, the contrary has been practiced.

Multiple techniques for SI in Africa, both traditional and novel, are already at hand (Rötter et al. 2007; Jeffery et al. 2017; Kuyah et al. 2021) Yet, still ongoing is the generation of knowledge on appropriate, site-specific measures that take the various sustainability dimensions (environmental, economic and social) into account. Likewise, the associated knowledge diffusion for wider adoption for a broad spectrum of crop and livestock systems is receiving increased attention. Besides efficiency and environmental impacts also cultural and financial limitations affecting adoption rate of sustainable management practices need to be considered. Market access, education, land rights and availability of inputs will finally steer the direction of the implementation and the magnitude of impact. In Sect. 20.4, a brief synthesis of recent review publications on SI with technologies suitable for southern Africa is made.

20.4   Agricultural Management Challenges and Transformation Pathways for a Sustainable Future

20.4.1 Most Pressing Agricultural Management Challenges

Key to a sustainable transformation of agriculture in SA10 will be to convert low productivity and food insecure subsistence farms to productive and economically viable commercially oriented systems applying sustainable land management practices (Sikora et al. 2020). Current smallholder systems suffer from vulnerability to climate variability and change, rapid soil nutrient depletion, shrinking farm size, lack of agricultural knowledge and technology and poor access to input and output markets. Hence, the most pressing management challenges include improving soil health, germplasm and water management from field to watershed in conjunction with climate change adaptation and mitigation, natural pest control and protection of biodiversity through its integration into farm management. Moreover, existing commercial farms have to become more resource-efficient, climate-smart and environment-friendly (e.g., Vanlauwe and Dobermann 2020; Kuyah et al. 2021). A few examples are given below.

20.4.1.1 The Need of Improving Soil Health in the Face of Climate Change

Soil fertility is broadly defined as the suitability of soil physical, chemical and biological characteristics, at a given site to match the site specific production and management objective. Measures counteracting nutrient-mining, as well as for preventing soil erosion (Chap. 13) and carbon-stock degradation must be implemented. It has often been reported that soil carbon losses are associated with a decline in soil quality and crop yield (Lal 2004). Given the many nutrient poor soils, year round high temperatures and the (semi-) humid to semiarid conditions, soil organic carbon (SOC) concentrations in southern Africa are generally low. Swanepoel et al. (2016) found that 58% of the top soils have SOC concentrations of <0.5% or less and that conventional farming has further depleted native SOC stocks, on average by 46%. The protection of the remaining SOC is therefore imperative. Conservation agriculture (CA) represents a possible avenue to enhance climate change adaptation and mitigation in conjunction with soil fertility improvement and increased yields (Thierfelder et al. 2017). The central challenge is to develop and implement measures enabling soil fertility improvements by smallholders, such as shown for Integrated Nutrient Management (INM) techniques in Africa (Vanlauwe et al. 2010).

20.4.1.2 Water Management from the Crop via Farm to the Watershed

According to Vlek et al. (2020), agriculture in SA10 consumes about 85% of the water withdrawn from nature (rivers, streams, aquifers, etc.). Nhemachena et al. (2020) provided a comprehensive analysis of the projected climate change impacts on the interrelated agriculture and water sectors of the Southern African Development Community (SADC); the largest share of this geo-region (75%) is characterized as marginal with arid to semiarid climatic conditions (<650 mm precipitation year^{-1}). The share of irrigation in SA10 is somewhere between 10% and 15%, whereby The Republic of South Africa keeps the lion's share with about 1.5 million ha in 2010 (Vlek et al. 2020). The overall picture is that of a fragile region with high livelihood dependence on variable rainfall regimes and prevalence of largely inefficient irrigation technologies—with climate change likely to even worsen that picture. Consequently, the SADC region may face losses in agricultural productivity ranging from 15% to 50%. This baseline situation and the projected outlook require rapid and widespread adaptations in water management from crop/field to watershed level (Chap. 22).

Watersheds and River Basins Major rivers such as the Okavango, Sambesi, Limpopo or Oranje have a transboundary character. Unsustainable management of these resources also has geo-political implications. From a hydrological perspective only in Mozambique and Zambia, and, to a lesser extent in Zimbabwe, viable options for increasing the share of irrigated crops exist (Vlek et al. 2020). Increasing water demands from urban areas is also lowering the agricultural use of the water resources.

Farm Level On farm water-harvesting structures must be mainstreamed and efficient irrigation schemes and technologies developed and deployed. Efficient drip irrigation systems can save water and extend watering times. Further, soil moisture conserving agronomic practices and rainwater harvesting must be adopted (e.g., Kuyah et al. 2021).

Field and Crop Level Varietal choice regarding water consumption and water-use complementarity, and adopted planting dates can improve the water use and utilization at plant level. Additionally, development of new breeds with reduced transpiration, increased water use-efficiency and deeper rooting must be developed. Breeding of climate-smart plants is a key-stone for the adaptation of the agricultural sector to future climates (Chap. 23).

20.4.1.3 Integration of Biodiversity at Farm and Landscape Level

Southern Africa landscapes harbor a significant part of global biodiversity. The Cape Floristic Region, the Succulent Karoo and the Maputaland-Pondoland-Albany biodiversity hotspot are recognized as a global priority for nature conservation in the context of the world's 34 biodiversity hotspots. Meanwhile, habitat loss has been accelerated by the ongoing transformation and fragmentation of landscapes. Unique biomes like fynbos, renosterveld and strandveld have been converted for fruit and cereal production. Nowadays, only 5% of the original renosterveld biome remains in the agricultural lowlands. According to the South African "Threatened Plant Species Program" (South African National Biodiversity Institute, SANBI), 67% of all threatened plant species occur in the fynbos biome of the Cape region. Many of these species have a very limited distribution range and only persist in small areas or even in a single location. Remaining patches are situated on private lands: implying that the integration of natural habitats into agricultural landscapes is a priority issue. In general, a functioning mosaic of agricultural fields, orchards, conservation areas and landscape-scale ecological networks can increase ecological resilience at watershed/landscape level. The direct interactions between crops and natural vegetation and fauna can have positive effects on crop production (Chap. 22). The increased introduction of natural and seminatural vegetation into the agricultural landscape promotes the settlement of animals. This is particularly important for predator-prey relationships for natural pest control and the abundance and diversity of pollinators. On the other hand, with the decrease of habitats and the simultaneous increase in population, the animals are driven toward closer contact with humans, which leads to a significant increase in human–wildlife conflicts with crop raiding by wildlife having become an important negative commercial factor for farmers (Seoraj-Pillai and Pillay 2017). Development of large-scale ecological networks serving as corridors that connect the remaining natural habitats in fragmented landscapes can improve structural and functional connectivity for the exchange of biodiversity, and increase the effective size of local protective areas. Redesign of integrated landscapes can result in a win-win situation for agriculture and conservation, but needs further strategic research and practical implementation.

20.4.2 Outlook on Sustainable Transformation Pathways

Southern Africa is in need of doubling its food production within the next two decades. Facing the contemporary challenges of climate change, land degradation, biodiversity loss and other interferences with the planetary boundaries, it is clear that staying within "safe operation space" (Rockström et al. 2020) will be a challenge but must be achieved without compromise.

Key to a sustainable transformation of agriculture in southern Africa will be to convert low productivity and food insecure smallholder farming systems, suffering from vulnerability to climate variability and change, rapid soil nutrient depletion, shrinking farm size, lack of agricultural knowledge and technology and poor access to input and output markets, to more productive and economically viable systems. This calls for sustainable land management practices (Sikora et al. 2020) and continued policy support of the ongoing structural transformation of African farming systems (Barrett et al. 2018). Transformation of agricultural land-use systems requires a systemic, integrative multiscale and multidisciplinary approach (HLPE 2020). The challenges that evolve—e.g., to develop climate-smart and resilient farming systems—are often studied with a reductionist approach (e.g., investigating single plants or animal breeds on their drought or heat tolerance) without subsequent integration of its findings with required adjustments of other system components of the farm or landscape. However, if the whole system is transformed, it is essential to study the mutual interactions of crop and livestock production systems jointly with other major land uses in a region and their interrelations with the natural resource base. Moreover, all agents (agricultural producers, extension services, other resource managers, etc.) have to be involved in the process, and apart from evaluating the systems for efficiency gains, also the impacts on economic, ecological and social aspects have to be taken into account. There are only a few projects that look at the multifunctionality of agricultural landscapes in view of possible transformation pathways, using such a systemic integrative approach. Exceptions include SPACES2-SALLnet that develops such scenarios for Limpopo (Rötter et al. 2021; Chaps. 22 and 23) and SPACES2-ASAP that integrates agroforestry systems into land management in southern Africa (Sheppard et al. 2020). There are strong arguments for the development and implementation of SI strategies tailored to the local biophysical and socioeconomic settings (Cassman and Grassini 2020). Plenty of choices exist on how to implement these strategies on the ground in terms of cropping systems, agronomic practices, breeding and other enabling technologies and support systems/infrastructure. A common goal is to move from the current situation, which most often is unsustainable (economically and/or ecologically) to desirable sustainable farming systems. Such transformations require SI and diversification strategies (Vanlauwe and Dobermann 2020). While the availability of many options for intervention may create the need to prioritize the means (i.e., the interventions/technology options), we think that it is equally important that solutions are codesigned by multiple stakeholders including scientists as otherwise they will not turn out to be sustainable. The best strategy will not only depend on the prevailing agroecological conditions but also on several other factors such as human capabilities (e.g., education, practical skills; entrepreneurship), as discussed by Gatzweiler and von Braun (2016).

There are many possible or conceivable transformation pathways for agriculture globally and for world regions such as southern Africa. Construction of agricultural development scenarios have been an integral part of future-oriented assessments for many years (IAASTD 2009). Also more recent agricultural development scenarios (e.g., Antle et al. 2017) go back to four archetype scenarios: Global orchestration; Fortress; TechnoGarden; AdaptiveMosaic (Du-Lattre-Gasquet et al. 2009). Here, we only consider the two archetype scenarios "TechnoGarden" and "Adaptive Mosaic" since both aim at environment-friendly sustainable management practices, although with slightly different foci (Fig. 20.16).

According to Du-Lattre-Gasquet et al. (2009), (1) TechnoGarden combines new technologies with focus on high resource use efficiency as globally developed/exchanged with site-specific knowledge, whereas (2) AdaptiveMosaic tailors diverse low cost management practices to the local specificities, continuously adapting them to changes and largely utilizing local resources & knowledge focusing on soil health and biodiversity conservation. A prerequisite for any sustainable

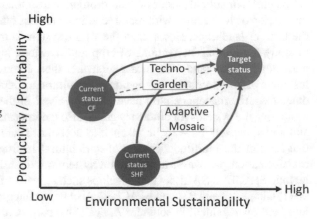

Fig. 20.16 Anticipated (schematic) transformation trajectories of TechnoGarden (upper arrows) and AdaptiveMosaic (lower arrows) from the current status of commercial farming (CF) and smallholder farming systems (SHF) to a sustainable future status

transformation pathway is that it is: (1) economically viable (2) environmentally sound, (3) resource-use efficient, (4) climate-resilient with (5) a low or negative carbon footprint and (6) based on equity among the various actors.

Here, we sketch some features of two potential future transformation pathways (Fig. 20.16), but refrain from prescribing where and under what conditions exactly these should be developed. Principally, we also do not claim that a "TechnoGarden" pathway would best fit to high or medium agroecological potential areas (e.g., where maize-mixed or root and tuber-based systems dominate), and the "AdaptiveMosaic" suits more to the marginal semiarid ecozones (e.g., Karoo) and savanna zones (where mixed crop-livestock/agropastoral systems dominate). Yet, there may be some argument that, initially, the "High Tech" would often be found rather closer to urban centers or in well-connected rural areas, while the "Adaptive local" pathway would initially rather be found in remote, less accessible rural areas. In the longer term we will likely see that the more commercially oriented, "TechnoGarden" with High Tech and relatively capital-intensive input will converge with the smallholder low input local "AdaptiveMosaic" systems that are based on diversity and agroecological principles. Both need to lead to systems that are profitable, highly productive and environmentally sound (Fig. 20.16).

TechnoGarden This will lead smallholders to become more commercial farmers, and current commercial farmers to apply environmentally sound practices using the best available technologies (e.g., precision farming). Apart from being capital-intensive, TechnoGarden recycles resources (water, nutrients) whether land-based or decoupled (e.g., vertical farming; cultivating insects for protein, etc.). Furthermore, it integrates renewable energy networks like photovoltaic, wind power, biogas and energy storage. This requires a high level of technical and managerial skills, and tailors technologies to local conditions by utilizing agroecological principles, complemented by local knowledge & resources. For southern Africa, this pathway could comprise climate-smart crop rotations or legume-based intercropping systems (e.g., Hoffmann et al. 2020), efficient irrigation or season-specific management of input use (based on weather forecasts & crop monitoring)—whatever is technically feasible/reasonable under the given local settings. Practices can include (climate- and pest-) resilient and new (food and fodder) crops/cultivars (Chaps. 19 and 23), integrated and site-specific nutrient management (INM and SSNM) (Vanlauwe and Dobermann 2020; VanLauwe et al. 2010) and natural pest control/integrated pest management (Chap. 22). Wherever feasible, such technology packages should be combined with mechanization (e.g., shared machinery at community level). A few studies have explored the impacts on yield, environment and/or farm economics of some elements of such technology packages in Africa (see, e.g., Rötter et al. 2016; Swanepoel et al. 2018; Hoffmann et al. 2020 and Chap. 23). Crop yield increases in the range of 100% to 300% compared to status quo have been reported—narrowing the yield gap from the usual 0.2 of the climatic potential yield to 0.5 or more (e.g., Van Ittersum et al. 2016). The TechnoGarden will aim to reduce different agricultural risks by newest technologies and adequate, if necessary capital-intensive, resource use and recycling technologies.

Fig. 20.17 Diversification and integrated farming concepts: (**a, b**) diversified cropping enabled by water-saving mulching and drip irrigation (Photo: Frank's smallholder at Ndengeza Village), Limpopo Province, South Africa; and Integrated Farming supporting sustainability, biodiversity and crop productivity: (**c**) in the Winelands of the Western Cape and (**d**) in the rangelands at the Bokkeveld Mountains in the Northern Cape, South Africa (Photo: Farm Papkuilsfontein, Niewouldtville)

AdaptiveMosaic will allow smallholders that are currently mainly subsistence-oriented to become more commercially oriented. The pathway will largely build on local resources and adaptive management (e.g., Kuyah et al. 2021). The term "adaptive" indicates that risk reduction takes place by continuously adapting/adjusting to changing conditions. It may comprise different local means of integrated soil nutrient and residue management, combining use of available organic materials with (little) industrial fertilizer (Vlek et al. 2020; Thierfelder et al. 2017). Furthermore, it applies the principles of conservation agriculture (CA) and natural pest control. Often irrigation will not be possible, but rainwater harvesting may be applied. Crop diversification and intercropping options need to be tailored to (the often limited) water availability (Fig. 20.17a, b). In more humid areas, diversification through diverse agroforestry (Chap. 21) and crop rotations consisting of cereals, legumes and root crops might be introduced. In their comprehensive review, Kuyah et al. (2021) identified fertilizer micro-dosing, planting basins, push and pull technologies (pest control), conservation agriculture, agroforestry and double-up legume cropping as appropriate SI measures for Sub-Saharan Africa.

Whenever possible, management practices/use of material inputs such as fertilizers should be season-specific based on weather forecasts (Phillips et al. 1998). The agronomic application of pyrogenic carbon, i.e., biochar has been recognized by IPCC (2019) as an appropriate and scalable negative emission technology with high impact potential. Biochar application as a means to amend soil fertility and sequester carbon should be an integral management component, especially in agroforestry systems. The cultivation of new indigenous cash crops can contribute to an integrated and sustainable farming system. Medical plants (e.g., Devil's claw *Harpagophytum procumbens*) and Rooibos tea (*Aspalatus linearis*) and honeybush (*Cyclopia* spec.) are good examples for increased use of indigenous crops. Integrated farming in combination with biodiversity and nature conservation can be observed in marginal regions of southern Africa (Fig. 20.17c, d). Nature-based tourism can generate income, which contributes significantly to the revenues obtained from traditional sheep and crop farming. This requires a good understanding of ecosystem services and their uses in the heterogeneous landscapes of the farms.

20.5   Conclusions

The high diversity of agroecologcial conditions in southern Africa in conjunction with the different economic and sociopolitical settings creates a multitude of agricultural management challenges. In most of the region, high to medium potential agricultural land is amply available, but water resources are scarce. While commercial farmers are usually well-endowed with resources, the many smallholders increasingly face serious resource limitations that have resulted in negative environmental impacts and persistently low productivity. It is very likely that climate change will further reduce water security and food security. This situation is unsustainable and, in conjunction with rapid increase in population and food demand, is likely to lead to social unrest and ecological disaster for the region, if no major transformation of agricultural systems will take place. Key future management challenges, required policy interventions and research needs include:

- agronomic means for smallholder farms to restore soil health, implement climate-resilient cropping systems, integrate biodiversity for pest control and increase productivity must be complemented by policy measures
- policy interventions must be tailored to smallholder needs so they get access to required inputs, technologies and knowledge, markets, etc. so that farming becomes more profitable and environmentally sound in the long term
- investments into training of extension services and farmers on sustainable management of soil, water, crops and livestock to increase resource use efficiencies, productivity and reduce undesired outputs
- investments into the design of local solutions for new technologies including precision farming, digitalization (e.g., weather and market information via mobile-phone), production of renewable energy, techniques for the recycling of

nutrients and water, GMO and advanced breeding tools for breeds resilient to climate extremes, pests, etc.

- prioritize research into climate-neutral and adaptive farm management
- stimulate systemic, multiscale and multidisciplinary research approaches to explore and evaluate options that support the multifunctionality of the diverse agricultural landscapes
- support research on the redesign of agricultural landscapes and integration of biodiversity to maintain ecosystem services and for conservation

The two agricultural transformation pathways sketched, each with somewhat different means, (1) aim to boost productivity by overcoming key constraints to agricultural production, (2) lead to economically viable farming systems, (3) restore/maintain ecosystem services and (4) reduce the environmental/carbon-footprint agricultural production.

References

Antle J, Mu JE, Zhang H et al (2017) Design and use of representative agricultural pathways for integrated assessment of climate change in U.S. Pacific Northwest cereal-based systems. Front Ecol Evol 5:99. https://doi.org/10.3389/fevo.2017.00099

Auricht C, Dixon J, Boffa JM, Garrity DP (2014) Farming systems of Africa. In: Sebastian K (ed) Atlas of African agriculture research and development: revealing agriculture's place in Africa. International Food Policy Research Institute (IFPRI), Washington, D.C., pp 14–15

Barrett CB, Christian P, Shimeles A (2018) The processes of structural transformation of African agriculture and rural spaces. World Dev 105:283–285. https://doi.org/10.1016/j.worlddev.2018.02.019

Breckle S-W, Rafiqpoor MD (2019) Vegetation und Klima. Springer Spektrum, Heidelberg

Cassman KG, Grassini P (2020) A global perspective on sustainable intensification research. Nat Sustain 3:262–268. https://doi.org/10.1038/s41893-020-0507-8

Conway D, van Gaderen EA, Derying D, Dorling S, Krueger T, Landman W et al (2015) Climate and southern Africa's water-energy-food nexus. Nat Clim Chang 5:837–846. https://doi.org/10.1038/nclimate2735

Davis-Reddy CL, Vincent K (2017) Climate risk and vulnerability: a handbook for Southern Africa, 2nd edn. CSIR, Pretoria

Day M, Gumbo D, Moombe KB, Wijaya A, Sunderland T (2014) Zambia country profile: monitoring, reporting and verification for REDD+. Occasional Paper 113. Centre for International Forestry Research. CIFOR, Bogor Barat

Du-Lattre-Gasquet M, Rötter RP, Kahiluoto H (2009) Looking into the future of agricultural knowledge, science and technology (Chap. 5). In: Watson R, Herren H, Wakungu J, MacIntyre I (eds) IAASTD, Washington

Engelbrecht F, Adegoke J, Bopape M, Naidoo M, Garland R, Thatcher M, McGregor J, Katzfey J, Werner M, Ichoku C (2015) Projections of rapidly rising surface temperatures over Africa under low mitigation. Environ Res Lett 10:085004

FAO, IFAD, UNICEF, WFP and WHO (2021) The State of Food Security and Nutrition in the World 2021. Transforming food systems for food security, improved nutrition and affordable healthy diets for all. FAO, Rome. https://doi.org/10.4060/cb4474en

FAOSTAT (2021) Food and Agriculture Organisation of the United Nations Statistical Database; Statistical Division; FAO: Rome, Italy, 2021; Available online: http://www.fao.org/statistics/en/ (accessed on 15 February 2021).

Fischer G, Nachtergaele F, Prieler S, Van Velthuizen HT, Verelst L, Wiberg D (2008) Global agro-ecological zones assessment for agriculture (GAEZ 2008). IIASA/FAO, Laxenburg/Rome, p 10

Fischer G, Nachtergaele FO, Prieler S, Teixeira E, Tóth G, Van Velthuizen H, Verelst L, Wiberg D (2012) Global Agro-Ecological Zones – Model Documentation GAEZ v. 3.0. IIASA/FAO, Laxenburg/Rome, p 179

Garnett T, Appleby MC, Balmford A, Bateman IJ, Benton TG, Bloomer P, ... Godfray HCJ (2013) Sustainable intensification in agriculture: premises and policies. Science 341(6141):33–34

Gatzweiler FW, von Braun J (eds) (2016) Technological and institutional innovations for marginal-ized smallholders in agricultural development. Springer, Cham. ISBN 978-3-319-25718-1 (eBook). https://doi.org/10.1007/978-3-319-25718-1

Godfray HCF, Beddington JR, Crute IR, Haddad L, Lawrence D, Muir JF, Pretty J, Robinson S, Thomas SM, Toulmin C (2010) Food security: the challenge of feeding 9 billion people. Science 327(5967):812–818. https://doi.org/10.1126/science.1185383

Government of Zimbabwe (2018) National Agricultural Policy Framework (2018–2030). Ministry of Lands, Agriculture and Rural Resettlement, Harare

Hijmans RJ, Cameron SE, Parra JL, Jones PG, Jarvis A (2005) Very high resolution interpolated climate surfaces for global land areas. Int J Climatol 25:1965–1978. https://doi.org/10.1002/joc.1276

HLPE (2020) Food Security & Nutrition. Report 15 report by the High Level Panel of Experts on Food Security and Nutrition of the Committee on World Food Security. FAO, Rome

Hoffmann MP, Swanepoel CM, Nelson WCD, Beukes DJ, van der Laan M, Hargreaves JNG, Rötter RP (2020) Simulating medium-term effects of cropping system diversification on soil fertility and crop productivity in southern Africa. Eur J Agron 119:126089. https://doi.org/10.1016/j.eja.2020.126089

IAASTD (2009) Agriculture at a cross-roads. North America-Europe report. In: Watson R, Herren H, Wakungu J, MacIntyre I (eds) International Assessment of Agriculture, Agricultural Science and Technology (IAASTD), Washington

IPCC (2019) Climate Change and Land: an IPCC special report on climate change, desertification, land degradation, sustainable land management, food security, and greenhouse gas fluxes in terrestrial ecosystems [P.R. Shukla, J. Skea, E. Calvo Buendia, V. Masson-Delmotte, H.-O. Pörtner, D. C. Roberts, P. Zhai, R. Slade, S. Connors, R. van Diemen, M. Ferrat, E. Haughey, S. Luz, S. Neogi, M. Pathak, J. Petzold, J. Portugal Pereira, P. Vyas, E. Huntley, K. Kissick, M. Belkacemi, J. Malley, (eds)]

IPCC (2021) Summary for policymakers. In: Climate Change 2021: the physical science basis. Contribution of Working Group I to the Sixth Assessment Report of the Intergovernmental Panel on Climate Change [Masson-Delmotte, V., P. Zhai, A. Pirani, S. L. Connors, C. Péan, S. Berger, N. Caud, Y. Chen, L. Goldfarb, M. I. Gomis, M. Huang, K. Leitzell, E. Lonnoy, J.B.R. Matthews, T. K. Maycock, T. Waterfield, O. Yelekçi, R. Yu and B. Zhou (eds)]. Cambridge University Press, UK and New York. pp. 3–32. https://doi.org/10.1017/9781009157896.001

iSDA Africa (2021) Innovative Solutions for Decision Agriculture Ltd

Jätzold R, Kutsch H (1982) Agro-ecological zones of the tropics with a sample from Kenya. Der Tropenlandwirt 83:15–34

Jeffery S, Abalos D, Prodana M, Bastos AC, Van Groenigen JW, Hungate BA, Verheijen F (2017) Biochar boosts tropical but not temperate crop yields. Environ Res Lett 12(5):053001

Kate S (2009) Agro-ecological zones of Africa. Washington, DC, International Food Policy Research Institute

Klingholz R (2020) Twice as many people in 2050: the need for agricultural transformation in Southern Africa (Chap. 3). In: Sikora AS, Terry ER, Vlek PLG, Chitjja J (eds) Transforming agriculture in southern Africa: constraints, technologies, policies and processes. Routledge, pp 17–26

Kuyah S, Sileshi GW, Nkurunziza L, Chirinda N, Ndayisaba PC, Dimobe K, Öborn I (2021) Innovative agronomic practices for sustainable intensification in sub-Saharan Africa. A review. Agron Sustain Dev 41(2):1–21

Lal R (2004) Soil carbon sequestration to mitigate climate change. Geoderma 123(1–2):1–22

MacKellar N, New M, Jack C (2014) Observed and modelled trends in rainfall and temperature for South Africa: 1960–2010. S Afr J Sci 110(7 & 8):51–63

Manuel L, Chiziane O, Mandhlate G, Hartley F, Tostão E (2021) Impact of climate change on the agriculture sector and household welfare in Mozambique: an analysis based on a dynamic computable general equilibrium model. Clim Chang 167:6. https://doi.org/10.1007/s10584-021-03139-4

Meza I, Rezzaei EE, Siebert S et al (2021) Drought risk for agricultural systems in South Africa: drivers, spatial patterns, and implications for drought risk management. Science of the Total Environment 799:149505

Midgley G, Bond W (2015) Future of African terrestrial biodiversity and ecosystems under anthropogenic climate change. Nat Clim Chang 5:823–829. https://doi.org/10.1038/nclimate2753

New M, Hewitson B, Stephenson DB, Tsiga A, Kruger A, Manhique A, Gomez B, Coelho CA, Masisi DN, Kululanga E (2006) Evidence of trends in daily climate extremes over southern and West Africa. J Geophys Res 111(7):14102

Nhemachena C, Nhamo L, Matchaya G, Nhemachena CR, Muchara B, Karuaihe ST, Mpandeli S (2020) Climate change impacts on water and agriculture sectors in Southern Africa: threats and opportunities for sustainable development. Water 12(10):2673

Nkonya E, Mirzabaev A, Von Braun J (2016) Economics of land degradation and improvement–a global assessment for sustainable development. Springer Nature, p 686

Phillips JG, Cane MA, Rosenzweig C (1998) ENSO, seasonal rainfall patterns and simulated maize yield variability in Zimbabwe. Agric For Meteorol 90:39–50. https://doi.org/10.1016/S0168-1923(97)00095-6

ReNAPRI, Regional Network of Agricultural Policy Research Institutes (2014) Presentation at annual ReNAPRI symposium, Lusaka, Zambia

Ribeiro NS, Katerere Y, Chirwa PW, Grundy IM (eds) (2020) Miombo woodlands in a changing environment: securing the resilience and sustainability of people and woodlands. Springer International Publishing, Cham. https://doi.org/10.1007/978-3-030-50104-4

Richard Y, Fauchereau N, Poccard I, Rouault M, Trzaska S (2001) 20th century droughts in southern Africa: spatial and temporal variability, teleconnections with oceanic and atmospheric conditions. Int J Climatol 21(7):873–885

Rockström J, Edenhofer O, Gaertner J, DeClerck F (2020) Planet-proofing the global food system. Nat Food 1:3–5

Rötter RP, Van Keulen H, Kuiper M, Verhagen J, Van Laar HH (eds) (2007) Science for agriculture and rural development in low-income countries. Springer, Dordrecht

Rötter RP, Sehomi FL, Höhn JG, Niemi JK, van den Berg M (2016) On the use of agricultural system models for exploring technological innovations across scales in Africa: a critical review ZEF. Discussion papers on Development Policy 223, Center for Development Research, Bonn, July 2016, p 85

Rötter RP, Scheiter S, Hoffman MP, Pfeiffer M, Nelson WCD, Ayisi K, Taylor P, Feil J-H, Bakhsh SY, Isselstein J et al (2021) Modeling the multi-functionality of African savanna landscapes under global change. Land Degrad Dev 32:2077–2081. https://doi.org/10.1002/ldr.3925SADC

SADC (2018) SADC Regional Vulnerability Assessment & Analysis (RVAA) synthesis report on the state of food anad nutrition security and vulnerability in southern Africa. SADC, Gaborone

Seoraj-Pillai N, Pillay N (2017) A meta-analysis of human–wildlife conflict: South African and global perspectives. Sustainability 9:34

Setimela P, Gasura E, Thierfelder C, Zaman-Allah M, Cairns JE, Boddupalli PM (2018) When the going gets tough: performance of stress tolerant maize during the 2015/16 (El Niño) and 2016/17 (La Niña) season in southern Africa. Agric Ecosyst Environ 268:79–89

Sheppard JP, Bohn Reckziegel R, Borrass L, Chirwa PW, Cuaranhua CJ, Hassler SK, Hoffmeister S, Kestel F, Maier R, Mälicke M, Morhart C, Ndlovu NP, Veste M, Funk R, Lang Seifert F, du Toit TB, Kahle HP (2020) Agroforestry: an appropriate and sustainable response to a changing climate in Southern Africa? Sustainability 12(17):6796

Sikora AS, Terry ER, Vlek PLG, Chitjja J (eds) (2020) Transforming agriculture in southern Africa: constraints, technologies, policies and processes, 1st edn. Routledge, New York. https://doi.org/10.4324/9780429401701

Swain A, Bali Swain R, Themnér A, Krampe F (2011) Climate change and the risk of violent conflicts in Southern Africa. Global crisis solutions. Menlo Park, Pretoria

Swanepoel CM, van der Laan M, Weepener HL, Du Preez CC, Annandale JG (2016) Review and meta-analysis of organic matter in cultivated soils in southern Africa. Nutr Cycl Agroecosyst 104(2):107–123

Swanepoel CM, Rötter RP, Van der Laan M, Annandale JG, Beukes DJ, du Preez CC, Hoffmann MP (2018) The benefits of conservation agriculture on soil organic carbon and yield in southern Africa are site-specific. Soil Tillage Res 183:72–82

Tatenda GN (2019) Natural resource degradation through tobacco farming in Zimbabwe: CSR implications and the role of the government. Communicatio 45(3):23–39. https://doi.org/10.1080/02500167.2019.1569541

Ten Berge HFM, Hijbeek R, van Loon M, Rurinda J, Tesfaye K, Zingore S, Craufurd P, van Heerwaarden J, Brentrup F, Schröder JJ, Boogaard HL, de Groot HLE, van Ittersum MK (2019) Maize crop nutrient input requirements for food security in sub-Saharan Africa. Glob Food Sec 23:9–21

Theron SN, Archer E, Midgley SJE, Walker S (2021) Agricultural perspectives on the 2015–2018 Western Cape drought, South Africa: characteristics and spatial variability in the core wheat growing regions. Agric For Meteorol 304–305:108405. https://doi.org/10.1016/j.agrformet.2021.108405

Thierfelder C, Chivenge P, Mupangwa W, Rosenstock TS, Lamanna C, Eyre JX (2017) How climate-smart is conservation agriculture (CA)?–its potential to deliver on adaptation, mitigation and productivity on smallholder farms in southern Africa. Food Secur 9(3):537–560

Tibesigwa B, Visser M, Turpe J (2017) Climate change and South Africa's commercial farms: an assessment of impacts on specialised horticulture, crop, livestock and mixed farming systems. Environ Dev Sustain 19:607–636. https://doi.org/10.1007/s10668-015-9755-6

Tilman D, Balzer C, Hill J, Befort BL (2011) Global food demand and the sustainable intensification of agriculture. Proc Natl Acad Sci U S A 108(50):20260–20264

UN DESA (2017) World population prospects: the 2017 revision. United Nations Department of Economic and Social Affairs, New York

Van Ittersum MK, van Bussel LGJ, Wolf J, Grassini P, van Wart J, Guilpart N, Claessens L, de Groot H, Wiebe K, Mason-D'Croz D, Yang H, Boogaard H, van Oort PA, van Loon MP, Saito K, Adimo O, Adjei-Nsiah S, Agali A, Bala A, Chikowo R, Kaizzi K, Kouressy M, Makoi JH, Ouattara K, Tesfaye K, Cassman KG (2016) Can sub-Saharan Africa feed itself? Proc Natl Acad Sci U S A 113(52):14964–14969. https://doi.org/10.1073/pnas.1610359113

Van Loon MP et al (2019) Impacts of intensifying or expanding cereal cropping in sub-Saharan Africa on greenhouse gas emissions and food security. Glob Chang Biol 25:3720–3730. https://doi.org/10.1111/gcb.14783

Vanlauwe B, Dobermann A (2020) Sustainable intensification of agriculture in sub-Saharan Africa: first things first! Front Agric Sci Eng 7(4):376–382. https://doi.org/10.15302/J-FASE-2020351

Vanlauwe B, Bationo A, Chianu J, Giller KE, Merckx R, Mokwunye U, Ohiokpehai O, Pypers P, Tabo R, Shepherd KD, Smaling EMA, Woomer PL, Sanginga N (2010) Integrated soil fertility management: operational definition and consequences for implementation and dissemination. Outlook Agric 39(1):17–24. https://doi.org/10.5367/000000010791169998

Verschuur J, Li S, Wolski P et al (2021) Climate change as a driver of food insecurity in the 2007 Lesotho-South Africa drought. Sci Rep 11:3852. https://doi.org/10.1038/s41598-021-83375-x

Vlek PLG, Tamene L, Bogardi J (2020) Land rich but water poor: the prospects of agricultural intensification in southern Africa (Chap. 5). In: Sikora AS, Terry ER, Vlek PLG, Chitjja J (eds) Transforming agriculture in southern Africa: constraints, technologies, policies and processes. Routledge, pp 36–44

WFP (2016) El Niño: undermining resilience - implications of El Niño in Southern Africa from a food and nutrition security perspective. World Food Programme

World Bank (2006) Post-conflict Mozambique's reconstruction: a transferable strategy in Africa. Multisectoral Report Number 35527

World Bank (2021) Mozambique economic update, setting the stage for recovery. The World Bank, IBRD, IDA

Xiong J, Thenkabail PS, Gumma MK, Teluguntla P, Poehnelt J, Congalton RG, Thau D (2017) Automated cropland mapping of continental Africa using Google Earth Engine cloud computing. ISPRS J Photogramm Remote Sens 126:225–244

ZAIP (2013) Zimbabwe Agricultural Investment Plan (2013–2017). A comprehensive framework for the development of Zimbabwe's Agricultural sector

The Need for Sustainable Agricultural Land-Use Systems: Benefits from Integrated Agroforestry Systems

21

Maik Veste [iD], Jonathan P. Sheppard, Issaka Abdulai,
Kwabena K. Ayisi, Lars Borrass, Paxie W. Chirwa, Roger Funk [iD],
Kondwani Kapinga, Christopher Morhart, Saul E. Mwale,
Nicholas P. Ndlovu, George Nyamadzaw, Betserai I. Nyoka,
Patricia Sebola, Thomas Seifert, Mmapatla P. Senyolo,
Gudeta W. Sileshi, Stephen Syampungani, and Hans-Peter Kahle

Abstract

This chapter introduces the different agroforestry systems (AFSs) as part of the diversification of agricultural landscapes and gives examples of their use in different related crop production systems in southern Africa. The introduction of trees into agriculture has several benefits and can mitigate the effects of climate change. For example nitrogen-fixing trees and shrubs contribute significantly to nutrient recycling and benefit soil conservation, which is particularly important for smallholder farms. In addition, shelterbelts play an important role in reducing

M. Veste (✉)
CEBra – Centrum für Energietechnologie Brandenburg e.V., Cottbus, Germany
e-mail: veste@cebra-cottbus.de

J. P. Sheppard · L. Borrass · C. Morhart · N. P. Ndlovu · G. Nyamadzaw · H.-P. Kahle
University of Freiburg, Faculty of Environment and Natural Resources, Freiburg, Germany

I. Abdulai
Georg-August-University of Göttingen, Department of Crop Sciences, Göttingen, Germany

K. K. Ayisi · P. Sebola
University of Limpopo, Risk and Vulnerability Science Centre, Sovenga, South Africa

P. W. Chirwa
University of Pretoria, Department of Plant and Soil Sciences, Pretoria, South Africa

R. Funk
ZALF - Leibnitz Institute for Agricultural Landscapes, Müncheberg, Germany

K. Kapinga
Mzuzu University, Biological Sciences Department, Mzuzu, Malawi

© The Author(s) 2024
G. P. von Maltitz et al. (eds.), *Sustainability of Southern African Ecosystems under Global Change*, Ecological Studies 248,
https://doi.org/10.1007/978-3-031-10948-5_21

wind speeds, and thus, evapotranspiration, and modifying the microclimatic conditions, which is an important factor for the adaptation of cropping systems to climate change. These integrated AFS landscapes provide important ecosystem services for soil protection, food security and for biodiversity. However, deficiencies in the institutional and policy frameworks that underlie the adoption and stimulus of AFS in the southern African region were identified. Furthermore, the following factors must be considered to optimise AFS: (1) selection of tree species that ensure maximum residual soil fertility beyond 3 years, (2) size of land owned by the farmer, (3) integrated nutrition management, where organic resources are combined with synthetic inorganic fertilisers and (4) tree-crop competition in the root zone for water.

21.1 Introduction

21.1.1 Land-Use Pressure

Agricultural production in sub-Saharan Africa (SSA) has been widely affected by the use of unimproved seed varieties, declining soil fertility, expensive inorganic fertilisers and, in some cases, poor pricing and marketing systems (Kuyah et al. 2021). In addition, continuous cropping with low inputs has resulted in devastating soil and land degradation effects. Amongst the major manifestation of land degradation are loss of soil organic matter (SOM), decline in fertility, elemental imbalances, deterioration of soil structure, as well as acidification and salinisation

S. E. Mwale
Copperbelt University, Dag Hammarskjöld Institute for Peace and Conflict Studies, Kitwe, Zambia

B. I. Nyoka
Bindura University of Science Education, Department of Environmental Science, Bindura, Zimbabwe

World Agroforestry Centre (ICRAF), Southern Africa Node, Lilongwe, Malawi

T. Seifert
University of Freiburg, Faculty of Environment and Natural Resources, Freiburg, Germany

Stellenbosch University, Department of Forestry and Wood Sciences, Stellenbosch, South Africa

M. P. Senyolo
University of Limpopo, Department of Agricultural Economics and Animal Production, Polokwane, South Africa

G. W. Sileshi
Addis Ababa University, College of Natural and Computational Sciences, Addis Ababa, Ethiopia

University of KwaZulu-Natal, Earth and Environmental Sciences, Pietermaritzburg, South Africa

S. Syampungani
Copperbelt University, ORTARCHI - Oliver R Tambo African Research Initiative, Kitwe, Zambia

(FAO and ITPS 2015). Reports have shown that 24% of the global land area has suffered degradation within the last 25 years, with the cultivated land area directly contributing approximately 19% (Henao and Baanante 2006; Nkonya et al. 2016). Due to an increasing human population, the luxury of traditional fallowing consistent with former farming practices has been curtailed, leading to other land uses being exploited for agricultural expansion. For instance, in southern Africa, large forested areas have been converted to agriculture (Gondwe et al. 2020; Dziba et al. 2020). This is overwhelmingly the main cause of deforestation (Fisher 2010). In the Miombo region of southern Africa, FAOSTAT reported an increase in cropped area from 100,000 km^2 to 272,000 km^2 between 1961 and 2014 (Dziba et al. 2020). It is clear that agriculture is the main cause of woodland conversion in the ecosystem. The drivers of both small- and large-scale cropland expansion in the region vary greatly between countries, with widely varying degrees of land-use intensification and expansion (Ryan et al. 2016). Overall, however, cropped area per rural person has remained around 0.3 ha per head, whilst the rural population has increased from 31 to 111 million (1961–2020; data from FAOSTAT).

Whilst a small human population allowed land to lay fallow in order to rebuild and sustain the soil physical and chemical properties, this has not been possible in southern Africa due to the immense pressure to provide food for a rapidly growing population. Increasing productivity within small pieces of land has been at the mercy of continuous application of synthetic inorganic fertilisers by smallholder farmers, which are mostly costly and inaccessible. Consequently, several soil-improving interventions were promoted with a farming systems approach in agriculture including crop rotation with leguminous crops. In later years, a sustainable investment in soil fertility management programmes through the adoption of low-cost agroforestry (AF) technologies or practices that increase the resilience of agricultural production was promoted in different agroecological regions of the world, including in southern Africa (Kuyah et al. 2021; Muchane et al. 2020). Such soil-fertility-improving interventions are intended to make Africa achieve food and nutritional security (Chap. 20). Indeed, this addresses a wide range of Sustainable Development Goals (SDGs) of the United Nations including Zero Hunger (SDG 2), Health (SDG 3), Climate Action (SDG 13) and Life and Land (SDG 15).

21.1.2 Agroecosystems of Southern Africa

Most parts of southern African vegetation are generally referred to as the Zambezian phytoregion. The region covers ten countries in central and southern Africa between latitudes 3° and 26° south with a total area of 377 million ha (White 1983). The region falls within the tropical summer-rainfall zone with a single rainy season (November–April) and two dry seasons, a cool season from May to August and a hot season from September to November (Geldenhuys and Golding 2008). Annual rainfall is 500–1500 mm, with a decreasing gradient from north to south (Chidumayo 1997). Within the SSA region, savanna constitutes the largest ecoregion (Eriksen 2007). These are ecosystems that have been heavily influenced by both natural and anthropogenic factors such as fire, cultivation practices and wood extraction for charcoal production. Degradation of the agroecosystems in the region has been associated with not only a massive loss of soil material, but also a loss of fauna and flora. Additionally, anthropogenic influences have had an impact on the distribution of the woodland ecosystems in the region. For example the current distribution of Miombo woodland, the principal vegetation type in the region, is the result of fire regimes and anthropogenic practices (Tarimo et al. 2015).

Winter rainfall occurs predominantly in the Western Cape. The Cape Floristic Region, for example, is one of the world's 34 biodiversity hotspots and is recognised as a global priority area for nature conservation. Habitat loss has been accelerated by the ongoing transformation and fragmentation of landscapes. In large areas of SSA, soil structural degradation, low SOC concentrations and nutrient limitations are widespread in both natural and man-made ecosystems (Tamene et al. 2019). Agricultural land for crop production and rangelands takes up more of the land surface of southern Africa than any other type of land-use. Cereals and grains are southern Africa's most important crops, occupying a large area of cultivatable land (Chap. 20). Maize is the most common crop and a dietary staple, a source of livestock feed and an export crop in some countries. Other crops include sorghum, millet, wheat and rice grown for subsistence use and income generation. A larger number of small-scale farmers and commercial farmers also produce cassava, peanuts, sunflower seeds, beans, potatoes, pumpkins and soybeans. The Western Cape is traditionally the second largest wheat producer in South Africa, but also fruits, grapes and vegetables and oilseeds are important agricultural products. An overview on the agroecological regions in SSA is given by Roetter et al. in Chap. 20.

21.1.3 Impact of Land Use on African Savannas

In the African savanna, the most significant land-use practices include arable and pastoral systems as well as the harvesting of timber products. Agriculture is normally practiced and traditionally takes a form of shifting cultivation, which comprises interchanging between a short phase of cultivation and a period of fallow. In this way, shifting cultivation transmutes savanna into a mosaic landscape with croplands, fallows of different ages and non-arable savanna sites that are not used for cultivation due to unfavourable soil and habitat conditions. Characteristic for these mosaic landscapes is the preservation of some highly valued tree species such as *Adansonia digitata* (baobab), *Parkia biglobosa* and *Vitellaria paradoxa* on croplands. Besides natural fires, people set fires for various reasons such as to clear ground for agriculture, to achieve higher visibility and to stimulate an off-season re-growth of perennial herbs (Krohmer 2004). During the last decades, the African savannas were subject to high climatic variability and land-use changes (Hickler et al. 2005; Wezel and Lykke 2006; Brink and Eva 2009). Land-use changes account for 70–80% of the biodiversity changes in the African savannas (de Chazal and Rounsevell 2009). The percentage of land intensively used for agriculture has increased in Africa, and agricultural systems have been intensified due to the growing use of fertilisers and pesticides. Land-use changes are driving the loss of natural habitats, biodiversity and stored carbon and the loss of other ecosystem services (Brink and Eva 2009). The reduction of natural resource capital leads to an increased risk of soil erosion, land degradation and of natural hazards such as floods.

21.2 Developing Sustainable Land Management Strategies for the Savannas

21.2.1 Current Land Management Strategies

Agriculture remains an important engine for the growth of the southern African economy due to its backward and forward linkages to the economy. A changing climate is widely acknowledged as a threat to the agricultural sector; however, the sector holds a great potential in contributing towards the greening of the southern African economy. One approach advocated to support a transition to an all-inclusive green economy is climate smart agriculture (CSA). CSA is defined as agriculture that sustainably increases crop productivity, enhances resilience (adaptation), reduces or removes greenhouse gases (mitigation) and is leading to the achievement of national food security and development goals. A widely promoted CSA in South Africa is conservation agriculture (CA) which is defined as a farming system that promotes the maintenance of minimum soil disturbance, permanent soil cover and diversifies crops per unit area or time. Crop diversification includes practices such as intercropping, crop rotation, cover cropping and AF, which are key to the sustenance of CA. The practice of conservation agriculture with trees (CAWT) is a term recently used to describe the combined CA practices and AF, and it is believed to be an important CSA technique, but its benefits are not well documented. The worldwide acknowledgement of AF as an integrated approach to sustainable land use owing to its production and environmental benefits spans over several decades (Nair et al. 2021). In both CSA and agroforestry systems (AFSs), an on-field assessment plays an important role in the evaluation of access modalities and provides an understanding of characteristics that have a bearing on the beneficiaries' choice and preferences regarding adoption and the use of feasible technologies and management practices.

Box 21.1 Case Study: Limpopo Climate Smart Agriculture
This study was initiated to address three objectives relating to CSA, namely: (1) to establish climate-smart (CSA) techniques and practices introduced and advocated with an understanding of factors that hinder farmer adoption, (2) smallholder maize farmers' perceptions and preference of specific CSA techniques and (3) document some dominant traditional AF practices for

(continued)

Box 21.3 (continued)

viable CSA interventions in the province. The study was carried out in Limpopo Province of South Africa. Limpopo Province was chosen as the study area due to its diverse farming activities, high climatic variability and largely arid to semi-arid nature, suggesting that CSA techniques and practices that reduce the effects of droughts, moisture stress and water scarcity are necessary. The province spans a total area of 20,011 km^2 and a population of 1,092,507, inclusive of a portion of Kruger National Park. In general, the bulk of precipitation in Limpopo Province occurs in summer with rainfall ranging between 400 mm and 600 mm.

Data Collection Methods

To achieve the research objectives, the study employed a combination of qualitative and quantitative methods which usually complement each other, as none of these methods are better than the other. Accordingly, in this study, literature review and semi-structured interviews with several groups of relevant stakeholders in the area of climate change, water management and agriculture were conducted using non-probability purposive sampling to identify factors impacted by water availability and climate change. Consequently, semi-structured interviews were held with farmers, NGOs and other stakeholders through both key informants and semi-structured interviews. The qualitative data was first transcribed by making memos and noting of main and key initial observations regarding the contextual information.

The Best-Worst Scaling (BWS) model was used to document farmers' perception and preferred CSA farming practices that are perceived as best and worst in sustaining crop productivity under climate change. This technique measures the relative importance that respondents attach to certain attributes. In developing the survey instrument for this objective, 15 farming practices suitable for dryland maize production based on literature were used (Table 21.1). The third objective on documentation of prevalent traditional AF practices in farmers' fields and home gardens was achieved by first reviewing a study on indigenous AF practices in the Limpopo Province carried out about 20 years ago (Ayisi et al. 2018) in the Mopane district. This was followed by site visits to farmers' fields across different rainfall regimes to assess dominant practices. Descriptive statistics was used to identify dominant systems and associated pros and cons of the practice.

CSA Technologies

Several technologies and practices consistent with CSA were noted, which included CA, DTSVs, infield rainwater harvesting (IRHW) and AF. For

(continued)

Box 21.3 (continued)

instance the adoption of seed varieties is anticipated to permit harvest even under adverse conditions, whilst helping farmers to deal with dry spells and mitigate against rain shortfall. Rainwater harvesting was also noted to have the potential to increase the rainwater productivity and yields with prospects to mitigate against the risk of crop failure associated with erratic and declining rainfall. AF was found to have prospects of improving soil fertility, whereas CA was an option for soil fertility improvement, whilst contributing to mitigation through limited tractor use and safeguarding soil carbon sequestration. To uncover the context within which the CSATIs are used, respondents revealed some key factors for adoption, which include proof of technology benefits, need for immediate benefits, involvement of end-users of the technologies and provision of support and complementary programmes, amongst others.

Farmers' Preference for Specific CSA Techniques

Report on the ranking farmers' preference for different CSA interventions in the Mopane District is presented in Table 21.1.

Traditional Agroforestry Practices

In general, AF in the Mopane district occurs in diverse forms in homesteads and farmlands. Fruit trees dominate the home gardens, whereas indigenous trees occur on the farmlands. However, planned or externally driven AF initiatives were found to be limiting, though few location-specific testing of species and systems had been carried out in the past. Leaving trees on farmlands as AF was prevalent in most agricultural production systems. AF in this sense is passive and has become a land management decision by which farmers choose not to remove specific trees when clearing land for farming. Farmers maintain trees with subsistence crops for several reasons amongst others (Tables 21.2 and 21.3).

Farmers within very high rainfall zones tend to focus on exotic fruit trees, grown in pure stands rather than in an intimate mixture with annual crops. However, AFS involving fruit trees such pawpaw, banana, mango and avocado planted with maize and vegetables can be found. The medium and drier localities are dominated by sparsely populated indigenous woody species mainly marula, Jackalsberry and acacia in association with maize. Interest in fruit tree production is largely encouraged by the favourable rainfall and availability of the local market for the fruits.

Conclusions

Whilst results have indicated some CSATIs with high prospects for the promotion of CSA in South Africa, high initial investment costs and additional labour required as well as management intensiveness associated with some CSATIs may render them unfavourable in the southern African context, particularly within smallholder agriculture. It is likely that a combination of

(continued)

Box 21.3 (continued)

technologies and practices will be necessary to achieve enhanced results with CSA attempts, so future research could unpack how this happens in practice. Diverse AFS occur in the study area, but the practice is more passive than planned interventions primarily and lowly ranked due to lack of information on the benefits of the practice.

Table 21.1 Respondents perceived the following attributes from best to worst

Ranking of practices	Description of practice
P1	Intercrop maize with legumes as nitrogen source.
P2	Apply maize residue as a mulch to bare soil.
P3	Changing planting date.
P4	Adopt drought-tolerant and fast-maturing maize cultivars.
P5	Changing maize plant density.
P6	Apply fertilisers according to maize fertiliser recommendations.
P7	Feed maize residues to livestock.
P8	Adopt ripper tillage for maize production.
P9	Apply fertiliser that releases nutrients slowly for maize production.
P10	Changing from maize to crops that require less nitrogen fertilisation.
P11	Intercrop maize with trees as the source fertilisers.
P12	Adopt no-till for maize production.
P13	Changing from maize production to livestock and dairy production.
P14	Changing from maize to sorghum production.
P15	Shift from farming to non-farming activities.

Table 21.2 Farmers' reasons for practicing agroforestry in the Mopane District

1	Food production for household consumption
2	Medicinal
3	Fodder
4	Material for building
5	Fuelwood
6	Fruit for sale and consumption

Table 21.3 Major limitations to the adoption of intensive agroforestry by farmers

Item	Constraints to adoption
1	Limited land area per household which cannot accommodate trees
2	Lack of land ownership for long-term investment in the woody perennial species
3	Lack of knowledge on agroforestry system
4	Inadequate water in drier areas for successful tree production

21.2.2 Low Input, No-Tillage Agriculture

Sustainable agriculture is an essential requirement to satisfy the needs of human beings, enhancement of natural resource base as well as environmental quality over a long period of time. The overarching purpose of sustainable agriculture is the conservation of the natural resource base, particularly soil and water by depending on the minimal utilisation of artificial inputs from outside the farming system. It ensures that land recovers from the disturbances caused by cultivation and the harvest of crops (Wezel and Lykke 2006; Francis and Porter 2011). Sustainable agriculture promotes the adoption of conservation practices such as crop rotation, integrated pest management, natural fertilisation methods, minimum tillage and biological control. Sustainable land management also requires an utilisation of techniques that reduces nitrogen loss (Küstermann et al. 2010). Sustainable agricultural practices can be effective in improving water use efficiency specifically in poor developing countries affected by water scarcity (Pretty et al. 2006). The use of agricultural practices such as no tillage or minimum tillage as some of the strategies to ensure sustainable land management has proven to be valuable in the reduction of soil loss and soil fertility restoration (Altieri 2002; Pretty et al. 2006; Lal 2007). These agricultural practices improve soil fertility by implementing farming practices such as using cover crops, leaving residues in the field, avoiding soil compaction, reducing the use of agrochemicals and unnecessary system inputs (e.g. World Bank 2008).

21.2.3 Perennial Crops

The cultivation of perennial crops has proven to reduce the detrimental effect of soil tillage, thereby promoting a sustainable management of land. Perennial crops have been reported to bring a valuable number of benefits. This is owing to the fact that their roots go beyond the depths of 2 m and can significantly improve the functioning of the ecological system such as conservation of water resources, nitrogen cycling as well as carbon sequestration. Compared to annual crops, perennial crops are reported to be more effective in the maintenance of the topsoil, that is to be 30–50 times more effective in the reduction of nitrogen losses, and to sequester between 300 and 1100 kg C ha^{-1} a^{-1}, compared to the 0 to 300 – 400 kg C ha^{-1} a^{-1} sequestered by annual crops (Cox et al. 2005). It is also believed that perennial crops could help restrain the impacts of climate change, reduce management costs, as they do not need to be replanted every year; hence, they require fewer passes of farm machinery and fewer inputs of pesticides and fertilisers. Perennial crops also require less harmful inputs such as the application of herbicides.

21.2.4 Usage of Crop Varieties

Sustainable land management requires an improvement of crop varieties as it becomes increasingly difficult to adjust the environment to the requirements of the plant. High yield plant varieties that are adapted to specific production environments and sustainable agricultural practices and that are resistant to specific pests and diseases will become increasingly important in the future. Livestock improvement will increase productivity and make more efficient use of scarce land and water. Biotechnology's potential as a tool for sustainable production systems should be evaluated and supported on a case-by-case basis (World Bank 2008).

21.2.5 Organic Farming

Organic farming has proven to be another approach for sustainable land management in the region. Conservation and enhancement of soil health is at the epicentre of organic farming. However, in order to conserve soil fertility, a number of farming practices that take full advantage of ecological cycles must be employed. This can be carried out by implanting practices such as crop rotation, intercropping, polyculture, cover crops and mulching. Long-term crop yield stability and the ability to buffer variations in yield against climatic adversity is critical in agriculture's capability to support society in the future. Sullivan (2009) estimates that for every 1% of soil organic matter (SOM) content, the soil can hold 10,000–11,000 L of plant-available water per ha of soil down to a depth of about 30 cm. Many studies have shown that, under drought conditions, crops within organically managed systems produce higher yields than comparable crops managed conventionally. This advantage can result in organic crops out-yielding conventional crops by 80% on average under severe drought conditions (Pimentel et al. 2005; Smolik et al. 1995). The primary reason for higher yield in organic crops is thought to be due to the higher water-holding capacity of the soils under organic management (Sullivan 2009). Nevertheless, other studies in the past have shown that organically managed crop systems have lower long-term yield variability and higher cropping system stability (Smolik et al. 1995).

21.2.6 Integrated Pest Management Systems

Integrated pest management (IPM) systems have been developed for many crops to control pests, weeds and diseases whilst reducing potential environmental damage from excessive use of chemicals. Scaling up IPM technologies is a challenge, as these management systems rely on farmers' understanding of complex pest ecologies and crop–pest relationships. Thus, although IPM messages need to be simplified, IPM systems require continuous research and technical support and intensive farmer education and training along with policy-level support (World Bank 2008).

21.2.7 Precision Agriculture

Precision agriculture improves productivity by better matching management practices to local crop and soil conditions. Relatively sophisticated technologies are used to vary input applications and production practices, according to seasonal conditions, soil and land characteristics and production potential (see Chap. 20).

21.3   Agroforestry Systems

21.3.1 Integration of Agroforestry into Sustainable Land-Use Systems

Under the conditions of global changes, there is an urgent need for alternative land-use systems and changes to current management to provide food security and resilient and climate-smart agricultural systems, as well as to combat desertification and the loss of biodiversity. In this context, the integration of AF is often discussed as a strategy that can be used both for the adaptation to, and for the mitigation of, climate change effects (e.g. Nair 2012; Zomer et al. 2016; Makate et al. 2019; Sheppard et al. 2020a). To effectively present AFS as a solution, we must present evidence of how AFS can be utilised as a means of buffering and mitigating the predicted climate change effects on agricultural production systems, rural livelihoods, food security and local microclimates.

21.3.2 What Is Agroforestry?

AFS can be defined as dynamic, ecologically based, natural resource management systems that, through the integration of trees on farms and in the agricultural landscape, diversify and sustain production for increased social, economic and environmental benefits for land users at all levels. The definition of AFS has evolved over the years and is now considered as a collective name for land-use systems and technologies where woody perennials (trees, shrubs, palms, bamboos, etc.) are deliberately used on the same land-management units as agricultural crops and/or animals, in some form of spatial arrangement or temporal sequence. Trees in AFS provide a range of goods (fruits, timber, fodder, leaf litter and green manure, medicines, firewood) and ecosystem services (carbon sequestration, windbreak, improvement of microclimate, soil protection, habitat structure, food for animals etc.), thereby enhancing food and nutrition security and resilience to climate change. AF is already practiced by both small and large-scale farmers in the southern Africa region and there is evidence that the wider practice has been prevalent for many decades in different parts of the world (Nair et al. 2021). In Malawi for example, the prevalence on farms of AF tree species was already observed nearly 90 years ago by Hornby (1934, cited by Dewees (1995)). Today, regeneration (by planting or natural) and management of tree species on farmland (croplands and on rangelands) is now widespread in all the ecological regions of southern Africa. Broadly, there are three main types of AFS namely (1) agri-silvicultural (crops and trees), (2) silvo-pastoral (trees and livestock) and (3) agro-silvopastoral (crops, trees and livestock) (Nair et al. 2021). On most farms in southern Africa, trees are either established through (1) retention during land clearing for crops and pastures, (2) natural regeneration from stumps and roots in places where trees had been cleared (farmer managed natural regeneration) and (3) planted from seeds and seedlings (planted agroforestry systems). Each of these three methods of tree establishment has its own advantages and disadvantages. For example planting trees in drylands is a challenge due to low survival rates, and the high costs associated with accessing germplasm, nursery and out planting (Reij and Garrity 2016; Brancalion and Holl 2020). In this case (2) should be recommended.

21.3.3 Origin of Systems

Retention of Trees Retention of selected tree species during land clearing is a common method of establishing trees on farms. The method is cheap and effective. Tree species retained depend on farmers' preference and the ecological zone. In Malawi, for example, tree species retained on crop fields and pasture lands include *F. albida*, *Vachellia* spp., *Erythrina abyssinica*, *Markhamia obtusifolia*, indigenous fruit trees (*Uapaca kirkiana*, *Azanza gackeana*, *Parinari curatellifolia*, *Strychnos* spp., *Sclerocarya birrea*, *Ziziphus mauritiana*) and fodder trees (*Kigelia africana*, *Piliostigma thonningii*), depending on the ecological region (Dewees 1995). In northern Namibia, tree retained on farms include indigenous fruit tree species such as *S. birrea*, *Berchemia discolor*, *Diospyros mespiliformis*, *Strychnos* spp. and *Hyphaene petersiana*. Trees are retained on contour bunds, farm boundaries or in the field where they are intercropped with field crops or combined with pasture.

Farmer Managed Natural Regeneration (FMNR) FMNR is a low-cost method of establishing desired tree species on farms where trees had originally been cleared. Trees are established by natural regeneration from stump and root stock sprouting whilst keeping the land under the primary function of agricultural production, whether crops or livestock (Lohbeck et al. 2020; Weston et al. 2015). The FMNR practice is effective on landscapes where propagules (stumps, roots, seeds) can still be found. In the case of seeds, these are either deposited by wind or through animal dung. With FMNR, farmers select preferred tree species as they regenerate, removing undesirable ones whilst tending those preferred. Tending includes the pruning of branches and canopy and the thinning out of some trees and stems to achieve the desired tree density and protecting the seedlings and saplings from animal damage. Documented evidence shows that FMNR practice is widespread in southern Africa (Reij and Garrity 2016; Moore et al. 2020). In Tanzania, a study found as many as 69 tree species being managed on farms, although the average number of species selected and retained by farmers on crop lands was three, with umbrella thorn (*Vachellia tortilis* syn. *Acacia tortilis*) being selected most often by the farmers (Moore et al. 2020). A survey of trees on farms in the central plains of Malawi showed that mango (*Mangifera indica*) is the dominant tree species on agricultural land, accounting for one-third of the tree population. Other tree species with significant numbers are *Piliostigma thonningii* and *Erythrina abyssinica*, both indigenous trees (Dewees 1995). A recent study also found that more than 50% of indigenous trees regenerated on the farms are *Piliostigma thonningii* and *Combretum* spp. in the mid altitude sub-humid ecological zone. In the semi-arid lakeshore ecological zone, the dominant tree species regenerated by farmers are *F. albida* and *Vachellia polyacantha* (syn. *Acacia polyacantha*).

Planted Agroforestry Systems These AFS are established from either seedlings or by direct seeding. Some tree species such as *S. birrea* and *Gliricidia sepium* are also established from truncheons. There are many types of planted AFS which include systematic and dispersed intercropping with either coppicing or full canopy tree species, improved relay fallows (utilising, e.g., *Tephrosia vogelii*, *Sesbania sesban* and *Cajanus cajan*), protein fodder banks and windbreaks. If not intercropped, these trees can also be planted along contour bunds, farm boundaries and fallowed fields. If intercropped with coppicing tree species, trees are cut back repeatedly to prevent shading of crops. Generally, management of planted AFS depends on the species and the system objectives. In Malawi, the shrubs, *C. cajan*, are estimated to cover about 113,000 ha (Simtowe et al. 2010).

21.3.4 Typical Types of Agroforestry Systems in Southern Africa

The last decades have witnessed an increase in the promotion and a corresponding increase in the uptake of AFS in southern Africa. Several agri-silvicultural AFS (crops and trees), with different spatial and temporal arrangements, have been promoted and these include:

1. *Intercropping* systems can be described as those which combine multiple crops at different spatial and temporal scales, for example, relay intercropping which is considered as a cropping arrangement where the lifecycle of one crop overlaps with that of another crop. Fertiliser trees such as *G. sepium* or *S. sesban* can be established between the crops (Kwesiga et al. 2003);
2. *Improved fallows* or fallow rotations, where planted tree fallows are left for a short period (2 years) and are followed by 2–3 years of maize crop (Fig. 21.1a, b). Short-duration fallows with herbaceous legumes have been examined widely and were found to increase yields of subsequent crops compared to traditional grass fallows or continuous cropping systems (Nyamadzawo et al. 2012). The trees are left growing on residual moisture once the maize crop has been harvested. Improved fallow is a practice whereby a piece of land is dedicated to

Fig. 21.1 *Tephrosia candida* (**a**) and *Gliricidia sepium* (**b**) are used in AFS to improve fallows in Malawi, (**c**) AF as part of a home garden in Caprivi, Namibia and (**d**) shelterbelts in the Western Cape Province, South Africa (Photos Rebekka Maier **a**, **b** and Maik Veste **c**, **d**)

fallowing with fast-growing nitrogen-fixing trees or shrubs. Improved fallows are an improvement over natural fallows, with the capability to attain the objective for using natural or traditional fallow systems more quickly, through careful choice of species, management of tree density, spatial arrangement, and pruning. From ecosystem perspective, the main function of the fallow is the transfer of mineral nutrients from the soil back to the woody biomass, which is then made available through burning, decomposition, and nutrient turnover from the organic biomass. These fallows also come in different forms depending on the size of the land holding: They can be non-coppicing fallows/rotational fallows or coppicing fallows (Akinnifesi et al. 2007). Several tree species have been used in these systems including *Sesbania sesban, Tephrosia candida, T. vogelii*, and *Crotalaria* spp. (Fabaceae), for rotational fallows. For coppicing fallows, species include: *Leucaena* spp., *Calliandra calothyrsus, Gliricidia sepium, Senna siamea, Flemingia macrophylla* and *Vachellia* spp. (Kwesiga et al. 2003; Mafongoya et al. 2006). Furthermore, woody biomass and nutrients can be provided also from intercropped pigeon pea (*Cajanus cajan*). The nutrient levels such as nitrogen are influenced by the species, their coppicing ability and the biomass production.

3. *Parkland systems*, for example, where *F. albida* is intercropped with crops. With its reverse phenology the tree-crop competition for resources is reduced, whilst enhancing crop yields and soil health (Barnes and Fagg 2003); The term fertiliser trees is premised on the impact of the various soil fertility improvement practices on key ecological functions including nitrogen fixation, soil fertility improvement and soil conservation (Sileshi et al. 2014). Essentially, these practices are modifications of the natural fallow and traditional shifting cultivation systems, which have become unsustainable in southern Africa (Akinnifesi et al. 2008). As indicated earlier, trees have the potential to improve soil fertility through nutrients contributed from decomposition of biomass or leaf residues, nutrient flow, atmospheric nitrogen fixation (legumes only), root turnover and nutrient cycling processes, as well as the influence on soil microclimate and associated faunal activities.

4. *Biomass transfer* is essentially moving green leaves and twigs of fertiliser trees or shrubs from one part of a farm to another to be used as mulch or green manure (Kwesiga et al. 2003; Sileshi et al. 2020a, b). The effect of biomass transfer is also dependent on the amount and quality of leaf manure. To improve the system, appropriate nutrient-rich tree species have been selected. Amongst the legume species tested for biomass transfer, so far *G. sepium* has shown superior performance in southern Africa. *Leucaena Leucocephala* and *T. vogelii* have also been used in biomass transfer technologies (Place et al. 2003; Kuntashula et al. 2004).

5. *Fodder banks*, which are concentrated units of forage legumes established and managed to provide additional protein for selected cattle during the dry season. They involve the establishment of high-quality fast-growing leguminous trees or shrubs, and often leguminous species with an objective of providing supplements to livestock to achieve high productivity and are mostly used during the dry season to bridge periods of forage shortage.

6. *Alley cropping* or hedgerow intercropping is an AF practice in which perennial, usually leguminous trees or shrubs are grown simultaneously with an arable crop. Alley cropping involves growing crops in alleys formed between planted hedgerows of widely spaced woody species that are regularly coppiced to reduce shading and below ground competition with companion crops, and to provide green manure and mulch (Kang and Wilson 1987). Tree species that have been tested in southern Africa include *L. leucocephala* and *G. sepium* (Kwesiga et al. 2003). In general, alley cropping is more promising in the humid tropics than in the drier areas, mainly due to below- and above-ground interactions between trees and companion crops, and the climatic conditions. The literature on the effect of alley cropping on crop yields in southern Africa is generally contradictory. In northern Zambia, alley cropping with *L. leucocephala* increased the yield by 90% compared to limed control after 6 years whilst *G. sepium* had no effect in the same trial (Matthews et al. 1992).

7. *Multi-story plantations*, which are characteristic AFS that involve growing several (tree) crops in different layers of a shaded perennial cropping system. Multi-story cropping will alter the light and radiation environment of understory species more than their nutrient relations.
8. *Tree/Home gardens*, where perennial agricultural crops and livestock are grown in association with seasonal multipurpose AF trees and shrubs within the compound of individual houses, under the management of family labour (Fig. 21.1c).
9. *Shelterbelts*, which are barriers that are erected to break down or slow down the ravages of wind which are placed on the windward side (Fig. 21.1d). Wind breaks consist of trees or shrubs maintained and arranged in such a way that they work as a protective measure against destructive winds and cold fronts.

Several of these AFS have been adopted in southern Africa ranging from improved fallows, alley intercroppings, parkland systems, biomass transfer systems and shelterbelts amongst others. These have resulted in increased crop productivity through improved soil organic matter and soil physical properties, water storage, soil fertility and soil biodiversity at farm level and landscape scale (Akinnifesi et al. 2010; Sileshi et al. 2014, 2020b).

21.3.5 Benefits and Limits of Agroforestry

AFS present the potential capacity to contribute to climate change mitigation and adaptation by enhancing agricultural landscape resilience, improving the microclimate, sequestering carbon, and reducing greenhouse gas emissions. AFS are one of those few land-use systems that provide adaptation and mitigation services in an integrated and synergistic manner (Duguma et al. 2014). AFS provide the potential to adapt and modify existing land-use management strategies to external pressures providing a stable long-term solution that is able to meet environmental and socio-economic needs as a replacement for unsustainable agricultural activities. AFS contribute to a wide range of important ecosystem services for protection of soils, optimise agricultural production systems, and provide additional income by forest and non-forest products (Sheppard et al. 2020b; Nair et al. 2021).

Mechanisms for Soil Improvement in AFS
AF practices have been demonstrated to increase soil fertility through benefits from fallowing using annual, biannual or perennial nitrogen fixing trees or 'leguminous fertiliser trees' which are either planted in rotation (e.g. improved fallows) or intercropped with crops (Kwesiga et al. 2003; Mafongoya et al. 2006; Sileshi et al. 2014). Leguminous trees such as *G. sepium* and *Acacia angustissima* and others such as *S. sesban and C. cajan* can fix nitrogen that can be of use to the crops that are grown after the fallow period (Sileshi et al. 2014, 2020a, b). Chikowo (2004) estimated that the total annual fixed nitrogen in *A. angustissima* (non-woody components + leaf litter) was 122 kg N ha^{-1} during the 2-year fallow period, whilst

C. cajan, S. sesban and cowpea (*Vigna unguiculata*) fixed 97, 84 and 28 kg N ha^{-1}, respectively.

AF practices also sequester more carbon compared to other agricultural land-use systems (Kumar and Nair 2011; Sileshi et al. 2014). However, the amount of biomass and SOC added is not the same between different systems and varies with tree species, soil type, rainfall and environmental conditions. Several studies have estimated biomass buildup in AFS. Nyamadzawo et al. (2008a) reported that improved fallows of *A. angustissima* and *S. sesban* accumulated 26.3 and 25.4 Mg ha^{-1} in leaf litter and twigs after 2 years of fallowing and resulted in 3.7–9.1 Mg ha^{-1} more SOC compared to continuous maize cropping. Fallowing also improves soil structure, build-up of soil organic matter and its carbon stocks, thus contributing to carbon sequestration. Build-up of SOM is critical to soil productivity and generally corresponds to nutrient build-up. The increase in SOM increases the cation exchange capacity (CEC) of the surface soil, which is especially important in kaolinitic soils and other light textured soils with low CEC. The associated benefits of high SOM include reduced phosphorus fixation in soils with high iron and aluminium oxide contents, buffering of soil against pH changes, improved water retention and nutrient retention against leaching and reduced mineralisation rates (Nyamadzawo et al. 2009).

In southern Africa and most of the tropical regions, the soils are acidic and deficient in phosphorus; hence, there is a need for inorganic P supplements. However, in most smallholder areas, mineral phosphorous is available but very expensive. In addition, in some soils, P may be present in the soil, but it is not available for plant uptake because of low bioavailability due to the high binding capacity of P to Al and Fe minerals in acidic soils. The use of AFS can be an option as some trees enhance P bioavailability to subsequent crops (Chikowo 2004; Mweta et al. 2007). Trees improve the P availability through secretion of organic acids and an increased mycorrhizal fungi population in the soil.

Impact of Fertiliser Trees on Soil Improvement and Crop Yield
Research has shown that the use of organic amendments may be a better and more sustainable option to improve soil health amongst resource-constrained smallholder farmers in SSA. However, the challenge of using organic amendments is that the range of the organic resources available to smallholder farmers is narrow, and in most cases, there are just animal manures and a few plant residues left after grazing and leaf litter collected from woodlands. The major challenge is to widen the range of organic nutrient resources in farming systems and increase quantities of those already in existence. Systems such as AF fertiliser tree systems, which mimic natural processes and make effective use of soil nutrients, rainfall, sunlight and natural resources are possible sustainable options. AF fertiliser tree systems encompass practices such as crop rotations, intercropping, no or low use of chemical fertilisers, composting, little or no tillage and direct seeding, maintenance of soil cover, maximisation of water infiltration, monitoring crop and water status (Garrity et al. 2010a, b; Sileshi et al. 2014). The application of these methods aims at

using water, land, nutrients and other natural resources in a manner that prevents deterioration of the land and provides examples of sustainable farming systems that can be utilised in the smallholder farming sector. AF fertiliser tree system could potentially serve as a reliable and cost-effective alternative to increase soil carbon and nutrient stocks in southern Africa soils (Sileshi et al. 2014; Bayala et al. 2018).

In southern Africa, traditionally farmers grow crops under scattered trees, and thus, the region has both, traditional fallow and mixed intercropping systems as well as improved AFS. These include parkland systems, improved fertiliser tree systems, and green leaf biomass transfer systems (Akinnifesi et al. 2008, 2010).

Trees in the parkland are retained in order to improve the yield of understory crops. The most common species in the landscape in the drylands is *F. albida* (Box 21.2). According to a recent meta-analysis by Sileshi (2016), soil organic carbon (SOC) was increased by 46%, total nitrogen by 50%, available phosphorus by 21%, exchangeable potassium by 32%, and grain yields of maize and sorghum by 150% and 73% respectively, under the tree canopy compared to the open area. Larger increases in SOC and nutrients were observed on inherently nutrient-poor sites than on nutrient-rich sites (Sileshi 2016). The improved crop growth under tree canopies can be explained in terms of a combination of different factors: (1) increased nutrient inputs including those from biological nitrogen fixation, manure and urine from livestock grazing or resting under the tree, and birds that take shelter under or perch in search for food; (2) increased nutrient availability through enhanced soil biological activities and rates of nutrient turnover; and (3) improved microclimate and soil physio-chemical properties (Akinnifesi et al. 2008). The *F. albida* was promoted in Malawi (Amadu et al. 2020) Other traditionally systems include shifting cultivation such as "chitemene" in northern Zambia (Kwesiga et al. 2003).

Soil Biodiversity

AF also increases the diversity and population of soil biota, thus ensuring a healthy ecosystem (Barrios et al. 2012; Muchane et al. 2020). Under improved fallow systems, the microbial biomass is higher (Nyamadzawo et al. 2009), the microbial community is much more diverse and the rate of plant material decomposition is much faster (Sileshi and Mafongoya 2006a, b), thus ensuring nutrient recycling and timely release of N and other nutrients as pointed out before. The fungi that are associated with increased P availability in agricultural soils are the arbuscular mycorrhizal (AM) *(phylum: Glomeromycota)*. Reported that N-fixing legumes resulted in better colonisation of cereal roots and an increase in AM fungal populations in the soil in addition to alleviating P-deficiency whilst enhancing N-fixation at the same time.

Fig. 21.2 Anatrees (*Faidherbia albida*) embedded in a parkland AFS with maize fields in Malawi (Photo: Rebekka Maier)

Box 21.2 The Anatree: A Key Species for Agroforestry in Africa

As leguminous nitrogen-fixing anatree (Faidherbia albida syn. *Acacia albida*, Fabaceae) is common in the Sudano-Sahelian region of sub-Saharan Africa, forming "parklands" (Fig. 21.2; van Wyk and van Wyk 2013) and grows in a wide range of ecological conditions either scattered or gregarious, in closed canopy woodlands or open savanna It grows on the banks of seasonal and perennial rivers and streams on sandy alluvial soils or on flat lands. *The tree species* is the most promising utilised 19 tree species in southern Africa and is one of the most recognised trees utilised for intercropping. The species is widespread within millions of farmers' fields throughout the eastern, western, and southern regions of Africa especially amongst low-lying areas (Barnes and Fagg 2003). It is highly compatible for cropping with food crops unlike other indigenous trees because it sheds its nitrogen-rich leaves during the early rainy season and remains dormant throughout the crop growing period, a phenomenon known as reverse phenology.

These leaves will start growing again at the beginning of the dry season. This reduces tree crop competition for resources, whilst enhancing crop yields

(continued)

Box 21.4 (continued)

and soil health (Barnes and Fagg 2003). *F. albida* creates a unique opportunity for increasing smallholder productivity by input of high-quality leaf residue for increased soil fertility (Garrity et al. 2010a, b; Sileshi 2016), thus reducing the amount of inorganic N fertiliser needed. The coincidence of litterfall and rainfall season ensures the timely decomposition of the tree leaves which releases nutrients, particularly N, one of the most deficient nutrients in the smallholder farming sector. However, when promoting this system, there is need for targeting certain areas, especially those places where the trees are naturally adapted.

F. albida also increases livestock production through supplying high-quality fodder and nutritious pods. The trees also produce seeds that can be used as food by humans during periods of food shortages (Barnes and Fagg 2003). In addition, it enhances carbon storage in farmed landscapes through increased carbon sequestration (both above and below ground). It is drought tolerant; hence, it can be considered a keystone species for climate-smart agriculture in much of Africa (Garrity et al. 2010a, b). In Malawi, maize yields under *F. albida* trees increased by 50% compared with maize alone (Saka et al. 1994). *F. albida* trees also resulted in a yield increase of between 10% and 100% for various other crops (Hadgu et al. 2009; Sileshi 2016). To show the importance of *F. albida*, for example the government of Ethiopia has launched an initiative to plant 100 million *F. albida* trees (Beedy et al. 2014).

Soil Physical Properties and Soil Water Availability

In most smallholder farming areas of southern Africa, conventional tillage is the most common method of land preparation before planting crops. The challenge is that conventional tillage has resulted in increased runoff losses and soil erosion. However, the use of fast-growing AF trees that fix nitrogen has been reported to increase soil organic matter, improving soil physical conditions. Improved soil physical conditions can result in better soil aggregation, lower bulk density, lower resistance to penetration (Lal 1989), improved soil porosity and reduced surface sealing. Improved soil structure also increases hydraulic conductivity, infiltration rates and water holding capacity (Lal 1989). Trees also break up plough layers and increase infiltration rates since they have deeper rooting systems (Nyamadzawo et al. 2008b). Nyamadzawo et al. (2003) reported that plots under *A. angustissima* maintained high infiltration rates of over 35 mm h^{-1} 2 years after fallow termination, because of the addition of biomass from the re-growth of cut stumps in the second cropping season and the presence of an active tree root system. In addition, AF trees also reduced the raindrop impact on the soil and, hence, reduced structural degradation. Trees may affect soil water content by reducing its due to high water consumption or competition with another tree (Bayala et al. 2008). However, trees act as water "pumps" and "safety nets" through hydraulic lift mechanisms. Hydraulic lift means that trees with access to deeper soil layers lift water through their roots or capillary forces to higher soil layers where crop roots can access it (Sakuratani et al. 1999; Liste and White 2008). Hydraulic lift from trees ensures water availability and, thus, enhances productivity of crops in AFS (Sileshi et al. 2014, 2020a). In general, available water can be used more efficiently in a tree–crop system than a sole crop system owing to favourable microclimate and improved water use efficiency (Beedy et al. 2014). Although trees can increase the potential soil-water-holding capacity, they can also have negative effects on the actual water volume available in the tree–crop–soil system. Tree roots can use water accumulated deeper in the soil profile, which can benefit crop growth, resulting in water deficit for shallow rooted crops and can use residual available water outside the crop growing season (García-Barrios and Ong 2004). Highly soluble nutrients such as N, K^+ and Ca^{2+}, which are leached into deep soil layers, can be brought to the surface through the deeper rooting habits of AF trees (Sileshi et al. 2020a). Beside the manifold positive aspects, root competition needs to be considered, for example in windbreaks where roots of *Casuarina* ssp. grow into the adjacent irrigated orchards and root pruning is often applied.

Modification of Microclimate by Shelterbelts
The improvement of the microclimatic growth conditions for crops is important especially in times of a changing climate. For that purpose, tree shelterbelts and AFS can be a suitable tool to mitigate climate change effects in agriculture. The coastal regions of the West Coast and the Overberg regions in South Africa as well as the Winelands of the Western Cape Province are characterised by high mean annual wind speeds of 5–8 m s^{-1} at 10 m above ground. High wind speeds are a threat for cultivated crops. Plantations of tree shelterbelts and hedges are traditional eco-engineering measures to reduce lee-side wind speed near the ground or near the crop canopy. In the Western Cape mainly, fast-growing tree species, including *Alnus cordata, Casuarina cunninghamiana, Pinus radiata, Populus simonii* and various *Eucalyptus* species, are used for the wind protection of fields, vineyards and fruit orchards (Fig. 21.3). The design and orientation of the windbreaks are arranged perpendicular to the prevalent wind directions and are modified by local topography.

Fig. 21.3 Windbreak with (**a**) *Casuarina cunninghamiana* for the protection of a citrus orchard and (**b**) with *Populus simonii* in a vineyard, Western Cape, South Africa (Photo Maik Veste)

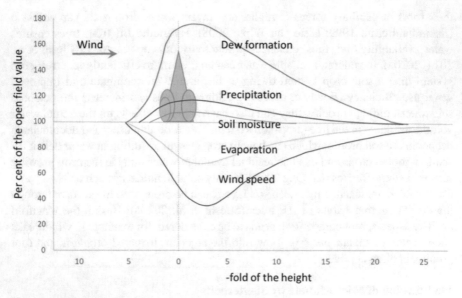

Fig. 21.4 Influence of a shelterbelt on microclimatic conditions

Trees can improve microclimatic conditions by reducing air temperature and wind speed and reducing evaporation from soils by shading crops, thereby increasing the availability of water in the soil. The microclimatic effects of linear windbreaks are summarised in Fig. 21.4.

The reduction of wind speed on the downwind side of a shelterbelt is a function of distance, aerodynamic porosity and tree height. Since in wind-prone areas, wind disturbs laminar layer of crop plants and leads to a significant increase in transpiration, wind shelter from trees is able to reduce transpiration and, consequently, soil water losses significantly (Veste et al. 2020). A poplar windbreak (see Fig. 21.3b) was demonstrated to reduce the mean wind speed at an 18 m distance from the hedgerow at 2 m canopy level (Fig. 21.5a) by 27.6% over the entire year and by 39.2% over the summer growing season compared to a reference in the open field. This effect leads to a parallel reduction of evapotranspiration of 15.5% during the whole year and of 18.4% over the growing season (Fig. 21.5b).

Furthermore, in the fruit growing regions of the Western Cape, shelterbelts are essential to minimise fruit damages of citrus and other wind-sensitive fruits. In a recent study, Geldenhuys et al. (2022) could show that the fruit quality was significantly affected by the presence of a windbreak, whilst it had no significant effect on citrus fruit yield. The increase of peel wind scar damage with increasing the distances from the windbreak resulted in a reduced export quality by 17.7% and the associated economic losses. In this case, the citrus orchard was protected by a windbreak built up by evergreen beefwood (*C. cunninghamiana*). Beside the wind effects, trees also reduce exposure to heat stress, which minimises tissue temperature to optimise the phenology and productivity of understory crops (Monteith et al.

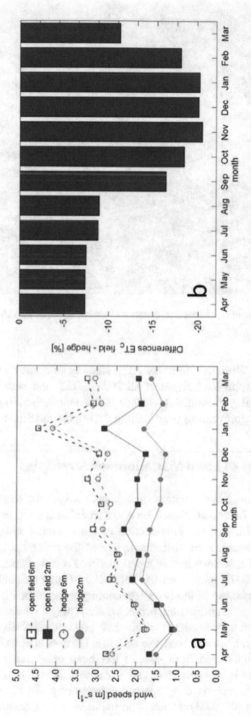

Fig. 21.5 (**a**) Monthly average of wind speed at 2 m and 6 m above ground near a poplar hedge (height 5 m) in 18 m north and as a reference in approximately 100 m distance and (**b**) monthly differences in crop-specific evapotranspiration by the hedge in the open vineyard, Western Cape (after Veste et al. 2020)

Fig. 21.6 Single trees providing shades for livestock (Western Cape, South Africa, Photo: Elbé du Toit)

1991; Vandenbelt and Williams 1992). In AFS, shading of crops by tree crowns is an essential feature (Bohn Reckziegel et al. 2021, 2022) and beneficial for crop productivity due to delayed stomatal closure under shade. Shading can be also beneficial for livestock, preventing over-heating during the daytime (Fig. 21.6).

21.4   Innovations of Land Management Strategies

Sustainable utilisation and conservation of savanna ecosystems requires an urgent intervention. This can be accomplished by encompassing human land use via the formation of protected areas, the introduction of management systems in human land-use areas that guarantee the sustainable use of the natural resources and by improving agricultural efficiency in forest peripheries. Protected areas, according to Adams and Hutton (2007), have been the backbone of international conservation strategies since the beginning of the twentieth century, even if their history is much older. In spite of their spatial limitation, protected areas play a vital role, specifically in the tropics, in protecting ecosystems within their borders, precisely by preventing land clearing arising from various land-use activities (Bruner et al. 2001;). Evidence of high diversity of fauna and flora species has been observed in a number of regions. Such examples have been observed in Zambia (Banda et al. 2006) where communal areas are characterised by a high heterogeneity, the ultimate source of biodiversity (Pickett et al. 2003). Hence, the maintenance of traditional land-use

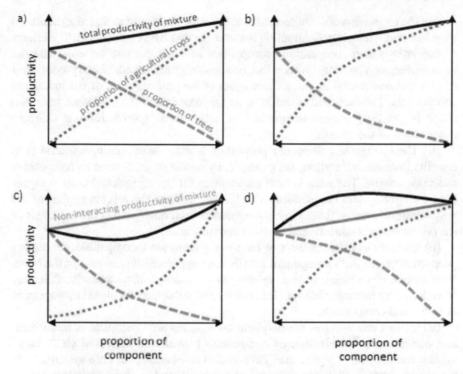

Fig. 21.7 Different effects of mixing agricultural crops and trees in agroforestry systems on the total productivity of the land-use system (solid line) and the individual productivities of the participating agricultural crops and the trees (dotted/orange and dashed/green lines, respectively). The figure shows four scenarios (**a-d**) where one system component is gradually replaced by the other towards full forest cover or pure agricultural cropping (after Sheppard et al. 2020a)

practices resulting in a mosaic-like distribution of various land units is the key to the maintenance of biodiversity in communal areas of the African Savannas (Augusseau et al. 2006).

Figure 21.7 depicts a land-use system replacement series applied to a conceptual and vastly simplified two-component AFS. This applies the conceptual ideas of production ecology to AFS exploring the idea of plant community mixtures as presented by Harper (1977) and nowadays applied to different forestry systems (Pretzsch et al. 2017). Within this conceptual example, the density of the AFS tree component is the same as in the monoculture cropping system and always totals 100%. Figure 21.7 describes four scenarios where one system component is gradually replaced by the other towards either full forest cover by increasing the proportion of trees or pure agricultural cropping with an increase in proportion of agricultural crops. In the given example, it is assumed that the agricultural crop is more productive than the tree culture and productivity is independent of external variables such as climate and site characteristics.

(a) The proportion of trees decreases at the same linear rate as that of agricultural crop increase. There is no interaction between the two AFS components. The effects of the inter-system competition (competition between the two systems) and the intra-system competition (within the two systems) are equal. Total productivity of this scenario results in an *additive effect* of the productivities of the individual components. This scenario is unlikely, as the interaction effect between trees and crops is generally proven to provide an influence on growth for one or more components of the system.

(b) The change in component proportion is non-linear. The agricultural crop benefits from the interaction, for example, by means of facilitation or competitive reduction factors. The intra-system competition for the agricultural crop is higher than the inter-system competition with the tree culture; the reverse applies to the tree culture. However, these effects compensate each other so that the net effect of the combination is *additive* and equal to scenario (a).

(c) Interactions between the two land-use systems are incompatible, decreasing proportion of one AFS component results in an opportunistic increase in the other. Intra-system competition is high, leading to an *under-yielding* scenario. This may be reflected by incompatible species choice or an influence of a biased management of individual components.

(d) Interactions between the two land-use systems are synergistic or mutualistic and non-linear, a combination of components provides an increased yield. Intra-system competition is higher than inter-system competition for both systems. This may result from facilitation, competitive reduction, and/or niche complementarity of both agricultural crops and trees (agricultural crops and trees utilising different soil resources). This leads to *over-yielding* at the level of the mixture and is the scenario that is most often touted as a benefit of AFS (i.e., increased land equivalent ratio (LER)).

Nevertheless, applying a simplified concept does not fully reflect the complexity of the interactions that occur within functioning AFS. Figure 21.8 is based on the work by Van Ittersum and Rabbinge (1997) presenting both the yield potentials and yield gaps between agricultural production systems and AFS. This further conceptual description highlights the actual, achievable and experimental yields when compared to a potential yield which is limited by growth-defining factors including temperature, CO_2, incoming direct solar insolation, individual plant physiology and phenology. This potential is further modified by site-based growth-limiting factors such as water and nutrient availability, growth-reducing factors such as biotic (e.g. competition from weeds, diseases, pests) and abiotic (e.g. drought, storm) influences, and also highlights an experimental yield gap which accounts for yield differences between field trials and practice.

In real life, such conceptual models must be tested and modified to provide elevated productivity over simple agricultural production methods accounting for species mixture and for limiting or reducing factors that prevent the full potential of AFS being realised. This is especially important within the southern African region where the effects of predicted climate change are multifaceted and far-reaching and are suggested to hit southern African communities hardest. The predicted instances of decreased rainfall can lead to loss of crops and land degradation and represent

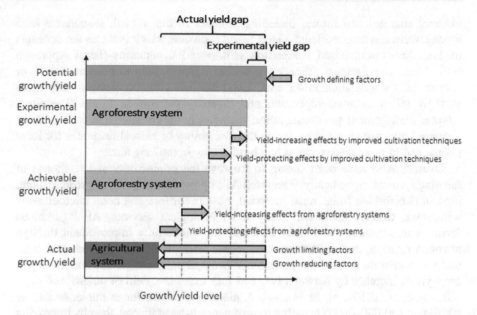

Fig. 21.8 Yield potentials, yield gaps and relationships amongst yield levels and growth-defining, growth-limiting and growth-reducing factors, as well as yield-increasing and yield-protecting measures (after Sheppard et al. 2020a)

a real and serious growth reduction factor. Increased frequency and severity of extreme weather events can also affect the viability of crops and can bring disruption and loss of profitability widening the gap among actual, achievable and potential yield (Fig. 21.8). As discussed in the sections above, the increased support and employment of AFS within southern Africa can help increase sustainability and resilience of smallholder farmers, brought about by integrating the benefits of suitable multipurpose tree and shrub species and adequate AFS practices to existing subsistence farming systems. It is not just subsistence farms either; the integration of trees within general agricultural practices can boost the productivity of the land and thus the economy of an area, providing employment, security and prosperity, laterally reducing investment risks supplying supplementary food and a variety of raw materials to trade a benefit that can also filter down and benefit individuals within the community.

21.5   Implications for Land Management Systems on the African Savannas

Sustainable land management is commonly considered as the main approach to prevent, mitigate and reverse land degradation, but it can also serve as an integral climate change adaptation strategy, being based on the fact that the healthier and more resilient the system is, the less vulnerable and more adaptive it will be to external changes and forces, including climate. In that regard, sustainable land management can be considered a land-based approach, which includes the concepts of both Ecosystem-Based Adaptation (EBA) and Community-Based Approach (CBA) land management practices, if widely adopted, help to prevent, reduce or reverse land degradation in an area. Land-use and climatic changes may more strongly affect savanna vegetation and diversity patterns in future. Therefore, adapted management and conservation strategies in the communal as well as in the protected area are required to ensure the availability of natural resources for local people and to protect ecosystems and biodiversity in the long term.

Overall, AFS have been shown to improve the productivity and resilience of farming systems. Specifically, integrated AFSs provide nitrogen-rich green manure, protein-rich fodder, fruits, nuts, firewood, flowers for foraging bees, microclimate, windbreak, timber, shade and many other ecological services. AF leguminous fertiliser tree systems are mostly managed for soil fertility improvement through nitrogen fixation, and production of copious amounts of nitrogen-rich leaf litter and green manure that is incorporation. When optimally established and managed, crop yields increase by between two and four times the yield of unfertilised plots (Garrity et al. 2010a, b). In Malawi, *F. albida* parklands, for example, enable an additional 150,000–300,000 metric tons of maize to be produced, thereby improving the food security of families farming under the systems and generating surpluses for sale. Besides direct benefits of increased crop yields, soil of AF plots shows a high diversity of soil biota (Sileshi and Mafongoya 2005), a highly desirable attribute of good soil. AF fertiliser trees also provide firewood which indirectly contributes to reduced deforestation. Other AFSs, for example coffee, can be used integrated with bee keeping and results in increased coffee yields. Fruit and nuts AFS provide nutrition and income generation from sale of fruits and nuts. The fruits and nuts contribute to family nutrition security and diversify farm income streams. In drylands, trees provide fodder, which is critical during the long dry season, whilst in smallholder dairy farming, trees provide cheaper but high-quality protein-rich fodder, enhancing milk production at lower costs.

21.6 Agroforestry in Policy Implementation

21.6.1 Challenges in Policy Coordination

In general, policies play an essential role in human–environment interactions as they define priorities, remove barriers, create capacities and potentially ensure the availability of key resources for the implementation of different programmes. Within AF, clear policies are also a necessary precondition in ensuring its wide-scale adoption and consequent harnessing of the proclaimed benefits. Over the last decade, there has been a growing interest in AF from a policy perspective. In a number of countries, national agencies are developing objectives and strategies that integrate AF into their policies and programmes. However, it is particularly important to note that despite a high-level policy recognition of AF, there is little knowledge on how policy aspects of AF are actually being integrated and implemented in different contexts. Policy and institutional factors with their connections have implications on how AF is approached. The cross-sectoral nature as well as the existing institutional dispositions can aggravate the difficulties for proper design, coordination and implementation of AF projects. These problems become more apparent and challenging to overcome when linked with complex issues such as land-use planning and administration, in particular issues such as land ownership and rights of use (including rights of possession, inheritance, use, usufruct and disposition). The cross-sectoral nature of AF also means that it is impossible for just one single institution or agency to implement proposed AF programmes without collaborating and coordinating with other sectors. Although coordination and collaboration are important ingredients for effective policy implementation, their potency is however fraught with challenges as they depend on contextual factors, such as the policy environment, existing policies, administration institutions, international pressure, the economy and other actors. Very few studies have detailed how these factors play out in AF implementation. Given that AF as a concept sits squarely between a number of complex policy fields, such as agriculture, forests and climate change, where coordination and collaboration play a huge part in its success, it is also worthy to focus on this strand of knowledge.

In pursuit of this knowledge, 15 interviews with different actors who are in the forefront in implementing AF and related technologies in Malawi were conducted. A policy document analysis was carried out to establish the prominence of AF. In the following section, four key challenges that they have encountered whilst trying to implement AF are reported. These include the lack of a clear framework for AF, lack of trust amongst actors, lack of resources and political interference.

1. Lack of harmonisation/no clear framework on agroforestry

Malawi has different policies and strategies that incorporate AF, and to a greater extent most of these policies emphasise that policy coordination and collaboration is vital for policy implementation. For instance this is mentioned in 71% (10/14) of the policies that we reviewed. The Food Security Policy of 2006 states, "If we are to guarantee the implementation of the policies and programmes of food security, it is necessary to guarantee the coordination, not only of government institutions, but also of all actors involved in the food economy."

Despite the existence of these policy documents, most interviewees expressed that in relation to the coordination of activities, the documents are vague and difficult to interpret. The interviewees mentioned that some of the policies are not sufficiently connected (integrated) across sectors and lack supporting instruments and resources to implement different activities. Additionally, there are no plans or measures to overcome these siloed coordination challenges.

2. Lack of mutual trust amongst organisations

Effective coordination and collaboration also depend on the level of trust amongst actors. Essentially, it improves relations, generates mutual understanding, legitimacy and commitment for a particular activity. One major reason for the lack of trust emanates from different philosophical and work approaches. These different approaches are usually related to donor organisations who exert influence in ideologies, power and resources. Although donors provide resources to supplement work efforts in AF, their influence consequently determines how each organisation engages with others. This eventually causes some projects to collapse, since they bring in new elements that might be different or contrary to what other actors are pursuing.

Although there are existing platforms and systems that have been created by both state and non-state actors to overcome these challenges and coordinate activities, some organisations still bypass these platforms. This has also led to different challenges: for example the introduction of black wattle (*Acacia mearnsii*), a tree species that is considered an alien invasive species in Malawi and other parts of southern Africa.

3. Lack of Resources/capacity for joint action

There is a lack of an effective and sustainable financing mechanism for the implementation of AF activities. AF is rarely a priority in national or sectoral budgets and it competes with other activities for the same resources. Although the agriculture sector receives more budgetary support from the government and over half of this allocation goes towards subsidy programmes, particularly maize seeds and inorganic fertilisers for smallholder farmers. Consequently, other programmes have to share the remainder—this includes AF activities. The remaining budget is usually only sufficient to pay staff salaries with very little resources left for other projects. Without project resources, no one is willing to take up AF activities. Most of the resources that support AF usually come from bilateral and multilateral donor arrangements. However, because of the low uptake of AF innovations by farmers, it has become difficult to get funding that is solely directed to AF as most donors seem to prefer other strategies and ideas.

4. Politicisation in agroforestry

According to the respondents, politicians attempt to gain political mileage by ignoring sustainable and long-term projects in favour of those that offer immediate benefits to the populace. Usually, these politically motivated projects are masked as pro-poor development programmes and very appealing to the farmers who cannot wait to witness the benefits of AF over a long time. These political projects present significant barriers in attempts to scale up as they discourage farmers from implementing AF activities. One of the causes of this challenge is that AF does not get enough political support. This scenario can be contrasted with the European Union's (EU) Common Agriculture Policy where AF enjoys EU-level recognition and support.

21.6.2 Policy Research in Agroforestry

Whilst supportive institutions and targeted policies are lauded as important towards upscaling and the wider adoption of AF, there is also a need to acknowledge that policies are not always implemented as envisioned and do not necessarily achieve intended results. It is therefore important to appreciate the role of policy research in policy implementation. Research can significantly contribute towards the development and implementation of effective policies for the adoption of AF technologies. Between August 2019 and June 2020, Ndlovu and Borrass (2021) conducted a literature review to assess the status of policy research in AF with a focus on the SADC region. Key to their findings was that most of the research has a strong bias towards the biophysical aspects and technical attributes of AF. However, in the last two decades, there is also a clear increase in studies that have a socio-economic orientation: mostly those with the intention to address the challenges of upscaling and adoption of AF in different contexts.

Whilst much literature is available, on the different barriers associated with adoption of AF, there is little research addressing policy and institutional aspects. There are few articles that have pursued to engage in understanding how different national and local policies influence the advancement of AF. In addition, the research community with a focus on policy issues is rather narrow and most articles are published by authors from specific institutions with a very direct interest in the propagation of AF. Critical perspectives are generally missing, and the variation of theoretical and conceptual approaches to the study of AF in the policy arena is very limited. Interestingly, none of these shortcomings have deterred scientific articles from presenting bold social scientific claims or defining institutional pre-requisites for a "successful" implementation or adoption of AF.

21.7 Conclusions

AF can make an important contribution to the diversification of agricultural landscapes and to increase resilience against a changing climate in southern Africa. The introduction of trees can also provide additional products, offering multiple ecosystem services, influencing crop production, and generates additional incomes for smallholder farmers. Protection against erosion and conservation of soil fertility are important arguments for the introduction of AFS. To optimise the benefits of AFS in terms of soil protection, the following critical factors must be considered: (1) selection of tree species that ensure maximum residual soil fertility beyond 3 years, (2) size of land owned by the farmer, (3) integrated nutrition management, where organic resources are combined with synthetic inorganic fertilisers and (4) tree–crop root competition for soil water. This is particularly important for the nitrogen and phosphate cycle, as it has a high savings potential and can contribute to sustainable soil development. The development of catch crop strategies in combination with the inclusion of N-fixing trees is important for closing the nitrogen cycle in AFS and enables an optimised nutrient cycle. This is an important aspect for the future development of sustainable agriculture in southern Africa. Furthermore, research can contribute to the adoption of AF by focusing on understanding the processes of policy interventions. Additionally, policy recommendations that actually reflect on the policy conditions of a particular context are likely to be accepted and actioned

(Sikora et al. 2020). Research and analysis on AF should tackle this assertion more frequently with the aim of effectively communicating with policy actors. Consequently, this calls for a shift towards a context-specific policy research agenda on AF.

In general, shelterbelts and alley-cropping systems are major eco-engineering measures to reduce water demands and influence directly soil evaporation and crop transpiration in the neighbouring fields. The redesign of the agricultural landscape by the introduction of specially designed obstacles to airflow will significantly influence the near-ground wind field. Further detailed information about tree water use is needed to optimise the water use efficiency and ecohydrological implications of the combined tree–crop interactions under climate change conditions. The integration of managed AFS, tree shelterbelt and hedges into climate-smart agriculture can mitigate the effects of climate changes to a certain extent and improve the growth conditions of crops and contribute to a resilient livelihood. Still an open scientific gap is the importance of AF for biodiversity and conservation. Not in all cases can the introduction of trees be seen as positive for the development and conservation of ecosystems in southern Africa. Invasive trees are of major concern for natural ecosystems, due to their drastic impacts on water resources and biodiversity. Further research and development of integrated landscapes combining different land uses and natural ecosystems are needed.

References

Adams WM, Hutton J (2007) People, parks and poverty: political ecology and biodiversity conservation. Conserv Soc 5:147–183

Akinnifesi FK, Makumba W, Sileshi G, Ajayi OC, Mweta D (2007) Synergistic effect of inorganic N and P fertilizers and organic inputs from *Gliricidia sepium* on productivity of intercropped maize in southern Malawi. Plant Soil 294:203–217

Akinnifesi FK, Chirwa P, Ajayi OC, Sileshi G, Matakala P, Kwesiga F, Harawa R, Makumba W (2008) Contributions of agroforestry research to livelihood of smallholder farmers in southern Africa: part 1. Taking stock of the adaptation, adoption and impacts of fertilizer tree options. Agric J 3:58–75

Akinnifesi FK, Ajayi OC, Sileshi G, Chirwa PW, Chianu J (2010) Fertilizer trees for sustainable food security in the maize-based production systems of east and southern Africa. A review. Agron Sustain Dev 30:615–629

Altieri MA (2002) Agroecology: the science of natural resource management for poor farmers in marginal environments. Agric Ecosyst Environ 93:1–24

Amadu FO, Miller DC, Mcnamara PE (2020) Agroforestry as a pathway to agricultural yield impacts in climate-smart agriculture investments: evidence from southern Malawi. Ecol Econ 167:106443

Augusseau X, Nikiéma P, Torquebiau E (2006) Tree biodiversity, land dynamics and farmers' strategies on the agricultural frontier of southwestern Burkina Faso. Biodivers Conserv 15:613–630

Banda T, Schwartz MW, Caro T (2006) Woody vegetation structure and composition along a protection gradient in a miombo ecosystem of western Tanzania. For Ecol Manag 230:179–185

Barnes RD, Fagg CW (2003) *Faidherbia albida*: monograph and annotated bibliography. Tropical forestry papers 41. Oxford Forestry Institute

Barrios E, Sileshi G, Shepherd K, Sinclair FL (2012) Agroforestry and soil health: linking trees, soil biota and ecosystem services. In: Wall DH (ed) The Oxford handbook of soil ecology and ecosystem services. Oxford University Press, Oxford, pp 315–330

Bayala J, Ouedraogo SJ, Teklehaimanot Z (2008) Rejuvenating indigenous trees in agroforestry parkland systems for better fruit production using crown pruning. Agrofor Syst 72:187–194. https://doi.org/10.1007/s10457-007-9099-9

Bayala J, Kalinganire A, Sileshi GW, Tondoh JE (2018) Soil organic carbon and nitrogen in agroforestry systems in sub-Saharan Africa: a review. In: Bationo A (ed) Improving the profitability, sustainability and efficiency of nutrients through site specific fertilizer recommendations in West Africa agro-ecosystems. Springer, pp 51–61. https://doi.org/10.1007/978-3-319-58789-9_4

Beedy TL, Nyamadzawo G, Luedeling E, Kim D-G, Place F, Hadgu K (2014) Soil fertility replenishment and carbon sequestration by agroforestry technologies for small landholders of eastern and southern Africa. In: Lal R, Stewart BA (eds) Soil management of smallholder agriculture. Advances in soil science. CRC Press, pp 237–277

Bohn Reckziegel R, Larysch E, Sheppard JP, Kahle H-P, Morhart C (2021) Comparing shading effects of 3DTree structures with virtual leaves. Remote Sens 13(3):532. https://doi.org/10.3390/rs13030532

Bohn Reckziegel R, Sheppard JP, Kahle HP, Larysch E, Spiecker H, Seifert T, Morhart C (2022) Virtual pruning of 3D trees as a tool for managing shading effects in agroforestry systems. Agrofor Syst 96:89–104. https://doi.org/10.1007/s10457-021-00697-5

Brancalion PHS, Holl KD (2020) Guidance for successful tree planting initiatives. J Appl Ecol 57:2349–2361. https://doi.org/10.1111/1365-2664.13725

Brink AB, Eva HD (2009) Monitoring 25 years of land cover change dynamics in Africa: a sample based remote sensing approach. Appl Geogr 29(4):501–512

Bruner AG, Gullison RE, Rice RE, da Fonseca GAB (2001) Effectiveness of parks in protecting tropical biodiversity. Science 291:125–128

Chidumayo EN (1997) Miombo ecology and management: an introduction. Intermediate Technology Publications, London. 166 pp

Chikowo R (2004) Nitrogen cycling in subhumid Zimbabwe: closing the loop. PhD thesis, University of Wageningen, Wageningen. ISBN: 90-5808-968-X

Cox TS, Glover JD, van Tassel DL, Cox CM, Dehaan LR (2005) Prospects for developing perennial grain crops. Bioscience 59:649–659

de Chazal J, Rounsevell M (2009) Land-use and climate change within assessments of biodiversity change: a review. Glob Environ Chang 19(2):306–315

Dewees PA (1995) Trees on farms in Malawi: private investment, public policy, and farmer choice. World Dev 23(7):1085–1102

Duguma LA, Minang PA, van Noordwijk M (2014) Climate change mitigation and adaptation in the land use sector: from complementarity to synergy. Environ Manag 54(3):420–432

Dziba L, Ramoelo A, Ryan C, Harrison S, Pritchard R, Tripathi H, Sitas N, Selomane O, Engelbrecht F, Pereira L et al (2020) Scenarios for just and sustainable futures in the miombo woodlands. In: Ribeiro NS, Katerere Y, Chirwa PW, Grundy IM (eds) Miombo woodlands in a changing environment: securing the resilience and sustainability of people and woodlands. Springer Nature Switzerland AG, Cham, pp 191–234. https://doi.org/10.1007/978-3-030-50104-4_6

Eriksen C (2007) Why do they burn the 'bush'? Fire, rural livelihoods, and conservation in Zambia. Geogr J 173(3):242–256

FAO and ITPS (2015) Status of the world's soil resources (SWSR) – main report. Food and Agriculture Organization of the United Nations and Intergovernmental Technical Panel on soils. FAO, Rome

Fisher B (2010) African exception to drivers of deforestation. Nat Geosci 3:375–376

Francis CA, Porter P (2011) Ecology in sustainable agriculture practices and systems. Crit Rev Plant Sci 30:64–73

Garcia-Barrios L, Ong CK (2004) Ecological interactions, management lessons and design tools in tropical agroforestry systems. In: Nair PKR (ed) Advances in agroforestry, vol 1, pp 221–236

Garrity DP, Akinnifesi FK, Ajayi OC, Weldesemayat SG, Mowo JG, Kalinganire A, Larwanou M, Bayala J (2010a) Evergreen agriculture: a robust approach to sustainable food security in Africa. Food Sec 2:197–214. https://doi.org/10.1007/s12571-010-0070-7

Garrity DP, Akinnifesi FK, Ajayi OC et al (2010b) Evergreen agriculture: a robust approach to sustainable food security in Africa. Food Sec 2:197–214. https://doi.org/10.1007/s12571-010-0070-7

Geldenhuys CJ, Golding JS (2008) Resource use activities, conservation and management of natural resources of African savannahs. In: Faleiro FG, Lopes A, Neto D (eds) Savannahs: Desafios e estastegiaspara o equilibrio entre sociedade, agronegocio e recursosnaturais. EmbrapaCerrados, Planatina, pp 225–260

Geldenhuys H, Lötze E, Veste M (2022) Fruit quality and yield of mandarin (*Citrus reticulata*) in orchards with different windbreaks in the Western Cape, South Africa. Erwerbs-Obstbau. https://doi.org/10.1007/s10341-022-00725-3

Gondwe MF, Cho MA, Chirwa PW, Geldenhuys CJ (2020) Land use land cover change and the comparative impact of co-management and government-management on the forest cover in Malawi (1999–2018). J Land Use Sci 14:281–305

Hadgu KM, Kooistra L, Rossing WA, van Bruggen AHC (2009) Assessing the effect of Faidherbia albida based land use systems on barley yield at field and regional scale in the highlands of Tigray, Northern Ethiopia. Food Sec 1:337–350

Harper JL (1977) Population biology of plants. Academic Press, London. ISBN: 9780123258502

Henao J, Baanante C (2006) Agricultural production and soil nutrient mining in Africa implications for resource conservation and policy development: summary. An International Center for Soil Fertility and Agricultural Development, IFDC

Hickler T, Eklundh L, Seaquist JW, Smith B, Ardö J, Olsson L, Sykes MT, Sjöström M (2005) Precipitation controls Sahel greening trend. Geophys Res Lett 32(21):L21415

Kang BT, Wilson GF (1987) The development of alley cropping as a promising agroforestry technology. In: Steppler HA, Nair PKR (eds) Agroforestry a decade of development. ICRAF, Nairobi, pp 227–243

Krohmer RW (2004) The male red-sided garter snake (*Thamnophis sirtalis parietalis*): reproductive pattern and behavior. ILAR J 45(1):65–74

Kumar BM, Nair PKR (eds) (2011) Carbon sequestration potential of agroforestry systems. Opportunities and challenges. Springer, Dordrecht. https://doi.org/10.1007/978-94-007-1630-8

Kuntashula E, Mafongoya PL, Sileshi G, Lungu S (2004) Potential of biomass transfer technologies in sustaining vegetable production in the wetlands (dambos) of eastern Zambia. Exp Agric 40:37–51

Küstermann B, Christen O, Hülsbergen KJ (2010) Modelling nitrogen cycles of farming systems as basis of site- and farm-specific nitrogen management. Agric Ecosyst Environ 135:70–80

Kuyah S, Sileshi GW, Nkurunziza L, Chirinda N, Ndayisaba PC, Dimobe K, Öborn I (2021) Innovative agronomic practices for sustainable intensification in sub-Saharan Africa. A review. Agron Sustain Dev 41(16). https://doi.org/10.10007/s13593-021-00673-4

Kwesiga F, Akinnifesi FK, Mafongoya PL, McDermott MH, Agumya A (2003) Agroforestry research and development in southern Africa during the 1990s: review and challenges ahead. Agrofor Syst 59:173–186

Lal R (1989) Agroforestry systems and soil management of a tropical alfisol. IV. Effects of soil physical and soil mechanical properties. Agrofor Syst 8:197–215

Lal R (2007) Evolution of the plow over 10,000 years and the rationale for no-till farming. Soil Tillage Res 93:1–12

Liste H-H, White JC (2008) Plant hydraulic lift of soil water — implications for crop production and land restoration. Plant Soil 313:1–17

Lohbeck M, Albers A, Boels L, Bongers F, Morel F, Sinclair F et al (2020) Drivers of farmer-managed natural regeneration in the Sahel. Lessons for restoration. Sci Rep 10:15038. https://doi.org/10.1038/s41598-020-70746-z

Mafongoya PL, Kuntashula E, Sileshi G (2006) Managing soil fertility and nutrient cycles through fertiliser trees in southern Africa. In: Uphoff N et al (eds) Biological approaches to sustainable soil systems. CRC Press, Boca Raton, pp 274–289

Makate C, Makate M, Mango N, Siziba S (2019) Increasing resilience of smallholder farmers to climate change through multiple adoption of proven climate-smart agriculture innovations. Lessons from southern Africa. J Environ Manag 231:858–868

Matthews RB, Lungu S, Volk J, Holden ST, Solberg K (1992) The potential of alley cropping in improvement of cultivation systems in the high rainfall areas of Zambia. II. Maize production. Agrofor Syst 17:241–261

Monteith JL, Ong CK, Corlett JE (1991) Microclimate interactions in agroforestry systems. For Ecol Manag 45:31–44

Moore E, van Dijk T, Asenga A, Bongers F, Sambalino F, Veenendaal E, Lohbeck M (2020) Species selection and management under farmer managed natural regeneration in Dodoma, Tanzania. Front For Glob Change 3:563364. https://doi.org/10.3389/ffgc.2020.563364

Muchane MN, Sileshi GW, Gripenberg S, Jonsson M, Pumarino L, Barrios E (2020) Agroforestry boosts soil health in the humid and sub-humid tropics: a meta-analysis. Agric Ecosyst Environ 295:106899. https://doi.org/10.1016/j.agee.2020.106899

Mweta DE, Akinnifesi FK, Saka JDK et al (2007) Green manure from prunings and mineral fertilizer affect phosphorus adsorption and uptake by maize crop in a Gliricidia-maize intercropping. Sci Res Essay 2:446–453

Nair PKR (2012) Carbon sequestration studies in agroforestry systems: a reality-check. Agrofor Syst 86:243–253

Nair PKR, Kumar BM, Nair VD (2021) An introduction to agroforestry – four decades of scientific developments, 2nd edn. Springer Nature, Cham

Ndlovu NP, Borrass L (2021) Promises and potentials do not grow trees and crops. A review of institutional and policy research in agroforestry for the southern African region. Land Use Policy 103:105298. https://doi.org/10.1016/j.landusepol.2021.105298

Nkonya E, Anderson W, Kato E, Koo J, Mirzabaev A, von Braun J, Meyer S (2016) Global cost of land degradation. In: Nkonya E, Mirzabaev A, von Braun J (eds) Economics of land degradation and improvement. Springer, Cham, pp 117–165

Nyamadzawo G, Nyamugafata P, Chikowo R, Giller KE (2003) Partitioning of simulated rainfall under maize-improved fallow rotations in kaolinitic soils. Agrofor Syst 59:207–214

Nyamadzawo G, Chikowo R, Nyamugafata P, Nyamangara J, Giller KE (2008a) Soil organic carbon dynamics of improved fallow–maize rotation systems under conventional and no-tillage in Central Zimbabwe. Nutr Cycl Agroecosyst 81:85–93. https://doi.org/10.1007/s10705-007-9154-y

Nyamadzawo G, Nyamugafata P, Chikowo P, Giller KE (2008b) Residual effects of fallows on infiltration rates and hydraulic conductivities in a kaolinitic soil subjected to conventional tillage (CT) and no-tillage (NT). Agrofor Syst 72:161–168. https://doi.org/10.1007/s10457-007-9057-6

Nyamadzawo G, Nyamangara J, Mzulu A, Nyamugafata P (2009) Soil microbial biomass and organic matter pools in fallow systems under conventional and no-tillage practices from Central Zimbabwe. Soil Tillage Res J 102:151–157. https://doi.org/10.1016/j.still.2008.08.007

Nyamadzawo G, Nyamangara J, Wuta M, Nyamugafata P (2012) Effects of pruning regimes on soil water and maize yields in coppicing fallow–maize intercropping systems in Central Zimbabwe. Agrofor Syst 84:273–286. https://doi.org/10.1007/s10457-011-9453-9

Pickett STA, Cadenasso ML, Benning TL (2003) Biotic and abiotic variability as key determinants of savannah heterogeneity at multiple spatio-temporal scales. In: du Toit JT, Rogers KH, Biggs HC (eds) The Kruger experience. Ecology and management of savannah heterogeneity. Island Press, New York, pp 22–40

Pimentel D, Hepperly P, Hanson J, Douds D, Seidel R (2005) Environmental, energetic, and economic comparisons of organic and conventional farming systems. Bioscience 55:573–582

Place F, Adato M, Hebinck P, Omosa M (2003) The impact of agroforestry-based soil fertility replenishment practices on the poor in Western Kenya. FCND discussion paper no. 160. International Food and Policy Research Institute, Washington, D.C.

Pretty JN, Noble AD, Bossio D, Dixon J, Hine RE, Penning der Vires FVT, Morrison JIL (2006) Resource-conserving agriculture increases yields in developing countries. Environ Sci Technol 40:1114–1119

Pretzsch H, Forrester DI, Bauhus J (2017) Mixed-species forests. Springer, Heidelberg. https://doi.org/10.1007/978-3-662-54553-9

Reij C, Garrity D (2016) Scaling up farmer-managed natural regeneration in Africa to restore degraded landscapes. Biotropica 48:834–843. https://doi.org/10.1111/btp.12390

Ryan CM, Pritchard R, McNicol I, Owen M, Fisher JA, Lehmann C (2016) Ecosystem services from southern African woodlands and their future under global change. Philos Trans R Soc B 371:20150312

Saka AR, Bunderson WT, Itimu OA, Phombeya HSK, Mbekeani Y (1994) The effects of Acacia albida on soils and maize grain yields under smallholder farm conditions in Malawi. For Ecol Manag 64:217–230

Sakuratani E, Aoe T, Higuchi H (1999) Reverse flow in roots of *Sesbania rostrata* measured using the constant power heat balance method. Plant Cell Environ 22:1153–1160

Sheppard JP, Bohn Reckziegel R, Borrass L, Chirwa PW, Cuaranhua CJ, Hassler SK, Hoffmeister S, Kestel F, Maier R, Mälicke M, Morhart C, Ndlovu NP, Veste M, Funk R, Lang F, Seifert T, du Toit B, Kahle H-P (2020a) Agroforestry: an appropriate and sustainable response to a changing climate in southern Africa? Sustainability 12(17):6796. https://doi.org/10.3390/su12176796

Sheppard JP, Chamberlain J, Agúndez D, Bhattacharya P, Chirwa PW, Gontcharov A, Sagona WCJ, Shen HL, Tadesse W, Mutke S (2020b) Sustainable forest management beyond the timber-oriented status quo: transitioning to co-production of timber and non-wood forest products—a global perspective. Curr Forestry Rep 6:26–40. https://doi.org/10.1007/s40725-019-00107-1

Sikora RA, Eugene RT, Paul LGV, Joyce C (eds) (2020) Transforming agriculture in southern Africa: constraints, technologies, policies and processes. Routledge, London. https://www.routledge.com/Transforming-Agriculture-in-Southern-Africa-Constraints-Technologies/Sikora-Terry-Vlek-Chitja/p/book/9781138393530#toc

Sileshi GW (2016) The magnitude and spatial extent of Faidherbia albida influence on soil properties and primary productivity in drylands. J Arid Environ 132:1–14. https://doi.org/10.1016/j.jaridenv.2016.03.002

Sileshi G, Mafongoya PL (2005) Variation in macrofauna communities and functional groups under contrasting land-use systems in eastern Zambia. Appl Soil Ecol 33(1):49–60

Sileshi G, Mafongoya PL (2006a) Long-term effect of legume-improved fallows on soil invertebrates and maize yield in eastern Zambia. Agric Ecosyst Environ 115:69–78

Sileshi G, Mafongoya PL (2006b) Variation in macrofaunal communities under contrasting land-use systems in eastern Zambia. Appl Soil Ecol 33:49–60

Sileshi GW, Mafongoya PL, Akinnifesi FK, Phiri E, Chirwa P, Beedy T, Makumba W, Nyamadzawo G, Njoloma J, Wuta M, Nyamugafata P, Jiri O (2014) Fertilizer trees. Encyclopedia of agriculture and food systems, vol 1. Elsevier, San Diego, pp 222–234

Sileshi GW, Mafongoya PL, Nath AJ (2020a) Agroforestry systems for improving nutrient recycling and soil fertility on degraded lands. In: Dagar JC et al (eds) Agroforestry for degraded landscapes: recent advances and emerging challenges, vol 1. Springer Nature, Singapore, pp 225–254. https://doi.org/10.1007/978-981-15-4136-0_8

Sileshi GW, Akinnifesi FK, Mafongoya PL, Kuntashula E, Ajayi OC (2020b) Potential of *Gliricidia*-based agroforestry systems for resource limited agro-ecosystems. In: Dagar JC et al (eds) Agroforestry for degraded landscapes: recent advances and emerging challenges, vol 1. Springer Nature, Singapore, pp 255–281. https://doi.org/10.1007/978-981-15-4136-0_9

Simtowe F, Shiferaw B, Kassie M, Abate T, Silim S, Siambi M, Kananji G (2010) Assessment of the current situation and future outlooks for the Pigeon pea sub-sector in Malawi. ICRISAT, Nairobi

Smolik JD, Dobbs TL, Rickerl DH (1995) The relative sustainability of alternative, conventional and reduced-till farming system. Am J Altern Agric 10:25–35

Sullivan P (2009) Drought resistant soil; ATTRA, National Center for Appropriate Technology. USDA, Washington, DC

Africa. Nutr Cycl Agroecosyst 104:107–123. https://doi.org/10.1007/s10705-016-9763-4

Tamene L, Sileshi GW, Ndengu G, Mponela P, Kihara J, Sila A, Tondoh J (2019) Soil structural degradation and nutrient limitations across land use and climatic zones in southern Africa. Land Degrad Dev 30:1288–1299. https://doi.org/10.1002/ldr.3302

Tarimo B, Dick ØB, Gobakken T et al (2015) Spatial distribution of temporal dynamics in anthropogenic fires in miombo savannah woodlands of Tanzania. Carbon Balance Manag 10:18. https://doi.org/10.1186/s13021-015-0029-2

Van Ittersum MK, Rabbinge R (1997) Concepts in production ecology for analysis and quantification of agricultural input-output combinations. Field Crop Res 52:197–208

Van Wyk B, van Wyk P (2013) Field guide to tress of southern Africa. Struik Nature, Cape Town

Vandenbelt RJ, Williams JH (1992) The effect of soil surface temperature on the growth of millet in relation to the effect of Faidherbia albida trees. Agric For Meteorol 60:93–100

Veste M, Littmann T, Kunneke A, Du Toit B, Seifert T (2020) Windbreaks as part of climate-smart landscapes reduce evapotranspiration in vineyards, Western Cape Province, South Africa. Plant Soil Environ 66:119–127. https://doi.org/10.17221/616/2019-pse

Weston P, Hong R, Kaboré C, Kull CA (2015) Farmer-managed natural regeneration enhances rural livelihoods in dryland West Africa. Environ Manag 55:1402–1417. https://doi.org/10.1007/s00267-015-0469-1

Wezel A, Lykke AM (2006) Woody vegetation change in Sahelian West Africa: evidence from local knowledge. Environ Dev Sustain 8:553–567

White F (1983) The vegetation of Africa. A descriptive memoir to accompany the UNESCO/AETFAT/UNSO vegetation map of Africa. UNESCO, Paris

World Bank (2008) Sustainable land management sourcebook. The World Bank, Washington D.C.. 178 pp

Zomer RJ, Neufeldt H, Xu J, Ahrends A, Bossio D, Trabucco A, van Noordwijk M, Wang M (2016) Global tree cover and biomass carbon on agricultural land: the contribution of agroforestry to global and national carbon budgets. Sci Rep 6:29987

Management Options for Macadamia Orchards with Special Focus on Water Management and Ecosystem Services

22

Sina M. Weier, Thomas Bringhenti, Mina Anders, Issaka Abdulai, Stefan Foord, Ingo Grass, Quang D. Lam, Valerie M. G. Linden, Reimund P. Rötter ⓘ, Catrin Westphal, and Peter J. Taylor

Abstract

South Africa is the World's largest producers of macadamia nuts, with about 51,000 ha of land covered by macadamia. This leads to major farming challenges, as the expansion of orchards is associated with the loss of habitat and biodiversity, the excessive use of and resistance to insecticides, and an increased pressure on water resources. More frequent and severe droughts and heat waves are projected to worsen the situation and have already negatively affected harvests. Here we review current literature and recent work conducted

S. M. Weier (✉)
University of Venda, SARChI Chair on Biodiversity Value & Change, School of Mathematical & Natural Science, Thohoyandou, South Africa
e-mail: Weier.SM@ufs.ac.za

P. J. Taylor
University of the Free State, Department of Zoology & Entomology & Afromontane Research Unit, Phuthaditjhaba, South Africa

T. Bringhenti · I. Abdulai · Q. D. Lam · R. P. Rötter
Tropical Plant Production and Agricultural Systems Modelling (TROPAGS), University of Göttingen, Göttingen, Germany

M. Anders · C. Westphal
Georg-August-University Göttingen, Department of Crop Sciences, Functional Agrobiodiversity, Göttingen, Germany

S. Foord · V. M. Linden
University of Venda, SARChI Chair on Biodiversity Value & Change, School of Mathematical & Natural Science, Thohoyandou, South Africa

I. Grass
University of Hohenheim, Department of Ecology of Tropical Agricultural Systems (490f), Stuttgart, Germany

© The Author(s) 2024
G. P. von Maltitz et al. (eds.), *Sustainability of Southern African Ecosystems under Global Change*, Ecological Studies 248,
https://doi.org/10.1007/978-3-031-10948-5_22

in the subtropical fruit growing area of Levubu, South Africa, which include
catchment-scale assessments of ground water, landscape-scale studies on pest
control and pollination services, through to evaluations of tree-level water use.
Several biological control options are being developed to replace pesticides.
Results suggest that bats and birds provide large and financially measurable
pest control services, and interventions should therefore focus on maintaining
functional landscapes that would be resilient in the face of global climate change.
This would include a landscape matrix that includes natural vegetation and
minimize water consumption by optimizing irrigation schedules.

22.1   Introduction

Macadamias are currently one of the most expensive nuts in the world, at nearly
twice the price of almonds. The industry is projected to have a compound annual
growth rate of 6.8% between 2020 and 2025 (APNEWS 2020). The macadamia
nut's 'healthy whole food' image has created a growing demand for macadamias
as ingredients and processed products in the food and beverage but also cosmetic
industry (Green and Gold Macadamias 2018).

While the trees originated in north-eastern Australia, the first commercial
cultivation of *Macadamia integrifolia* Maiden & Bechte started in Hawaii in 1931
(De Villiers and Joubert 2003). Macadamia trees require a tropical to subtropical
climate with high annual rainfall above 1000 mm. The trees grow well in soil with a
high organic content but not on saline or sodic soils (De Villiers and Joubert 2003).

According to SAMAC (2021), the largest producers of macadamia nuts in
2020 were South Africa (48,925 tons) and Kenya (42,530 tons) followed by
Australia (42,000 tons). However, China is projected to grow its production from ca.
18,000 tons in 2018 to ca. 450,000 tons by 2025 (AGTAG 2018). Other macadamia
growing countries include Zimbabwe, USA, Israel, New Zealand, Vietnam, and
Brazil (DAFF 2019).

In South Africa, macadamias have been one of the fastest growing tree crop
industries in the last decade, providing seasonal and permanent employment for
over 20,000 people. The value of the annual production was 4.8 billion ZAR or
about 330 million USD in 2019 (SAMAC 2020). Main growing areas in South
Africa are the provinces of Limpopo, Mpumalanga, coastal KwaZulu-Natal, and the
Eastern Cape. South African macadamia orchards cover approximately 51,000 ha,
of which around 6000 were planted in 2019 alone (DAFF 2019). This expansion
is set to continue until 2030, with 1000 additional seasonal and permanent farm
workers employed yearly (SAMAC 2020). However, the ongoing expansion of
macadamia orchards also leads to farm and landscape management challenges.
Monocultures, the excessive use of insecticides, and habitat loss at a local and
landscape scale are associated with the loss of natural enemies of crop pests such
as bats, birds, predatory insects as well as the loss of wild pollinators (Foley et al.
2005; Tilman et al. 2001; Tscharntke et al. 2012; Weier et al. 2021). Furthermore,

the increased and incorrect use of pesticides has led to resistance in pests such as stinkbugs (Hemiptera: Pentatomidae) (Schoeman 2018). Additionally, South Africa has experienced some severe droughts in recent years, with the combined effects of an El Niño event (Baudoin et al. 2017). Already one of the driest countries in the world, these droughts are predicted to worsen under future climate change and regularly impact harvests negatively throughout South Africa (Mogoatlhe 2020).

This chapter is based on work conducted in the subtropical fruit growing area of Levubu in the Luvuvhu river valley, situated in the northernmost South African province Limpopo. *Macadamia integrifolia* has been cultivated here for over 60 years (Ahrens 1991). Levubu is one of the two main growing areas in the province, with about 10,000 ha of macadamia planted thus far. This sub-Saharan African region receives its main rain in the summer months between November and April with around 1000 mm of annual rainfall. Apart from macadamia, the main agricultural products farmed are banana (*Musaceae*), avocado (*Persea*), timber (*Eucalyptus* and *Pinaceae*) and to a lesser extent mango (*Mangifera*), pecan (*Carya*), lychee (*Litchi*), or maize (*Zea*).

The Levubu area is on the south-eastern slopes of the Soutpansberg mountain range, part of the UNESCO Vhembe Biosphere Reserve (VBR). The Soutpansberg is a centre for plant endemism with a remarkably high animal biodiversity (Hahn 2017; Joseph et al. 2019; UNESCO 2010; Taylor et al. 2013, 2015; Van Wyk and Smith 2001). Also see Chap. 23.

Given the rapidly increasing demands on biodiversity and water-related ecosystem services due to agricultural intensification and climate change in South Africa, the aim of this chapter is to review the current literature and synthesize a decade of research in the Levubu area and provide recommendations on the mitigation of agricultural intensification and climate change effects for more sustainable agricultural practices.

22.2 Water Management

22.2.1 Water Availability and Macadamia Irrigation

The climatic conditions of the main macadamia cultivation area in Limpopo, are characterized by unevenly distributed annual rainfall that rarely exceed 1000 mm. Macadamia trees in the region therefore require supplementary irrigation for good yields and optimal quality (Carr 2012; Murovhi 2003). The South African Macadamia Growers Association (SAMAC) latest census in 2012 suggests that 80% of the macadamia growing area in Limpopo is irrigated, this figure has subsequently increased even further. The resultant growing demand for irrigation water increases pressure on the limited water resources of the province.

Long-term flow monitoring (80 years) of the Luvuvhu river (the main tributary in the catchment), where it leaves the commercially irrigated agricultural area, points to significant decreases in stream flow (Ramulifho et al. 2021). These decreases are highly seasonal with significant reductions from November to February, periods

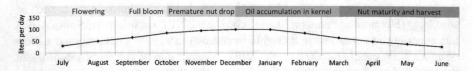

Fig. 22.1 Example of typical daily macadamia tree water requirements during the different phenological stages of trees, planted at a density of 312 trees per hectare and with a canopy coverage of 60%, as recommended by the South African Macadamia Growers Association (SAMAC)

that coincide with peak water demands of macadamia (Fig. 22.1), and have already resulted in the cessation of flow of the Luvuvhu during certain parts of the year (Ramulifho et al. 2019). Regionally, climate change is predicted to result in 5–10% decrease in rainfall (Hewitson and Crane 2006; Conway et al. 2015; IPCC 2021).

Groundwater levels measured from 2007–2013 in the Luvuvhu catchment show that the lowest groundwater level occurs between October and November and has decreased by around 2 m (23–25 m) (Makungo and Odiyo 2017). Although other land-use systems, such as gum (*Eucalyptus* ssp.) plantations, are less water efficient than macadamias per unit of land area (Botha 2018), commercial irrigated farming is one of the main sources of water consumption in the province (Shabalala et al. 2022). Finally, newly established macadamia orchards in the region (the fastest growing crop in the Soutpansberg in terms of its expansion) are increasingly located in more arid areas of the mountain and its surrounds, which is furthermore increasing the demand for ground water.

Historically, the Limpopo growers strongly relied on surface water but a combination of politics and poor maintenance of the water infrastructures leads to a major shift towards the use of groundwater in the early 2000s (Stephan Schoeman, personal communication). This is unsustainable given the slow recharge of the water table, which is around 4% of the mean annual precipitation and further evidenced by boreholes running dry in recent years.

Very little is known about the status of groundwater in the province. This is particularly concerning within the context of climate change and the relevance to monitoring the allocation of water licences by Water boards (National Water Act 1998). Licences for specific water volumes generally depend on the size of the production area and are legally allocated and verified through formal processes. Macadamia growers purchase water licences from local authorities, which also monitor on-farm water use.

22.2.2 Sustainable Water Management Practices

Given the circumstances, it is paramount for macadamia growers to increase the efficiency and sustainability of water use. This can be done by: (1) choosing the appropriate irrigation system, (2) meeting tree water requirements better with irrigation water supply, (3) using advanced technology for a better understanding

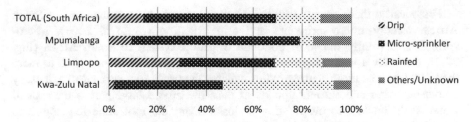

Fig. 22.2 Percentage of adoption of different irrigation systems in the three main macadamia production areas of South Africa (Data from the 2012 SUBTROP Census)

and monitoring of water dynamics in the orchards, and (4) adopting water-saving agricultural practices in the orchards.

A reflection on the best water management options for macadamia is bound to start with the question on whether such trees need to be irrigated at all. About 20% of macadamia producers in South Africa rely solely on rainfall as a source of water for their orchards (Fig. 22.2). Although irrigation is considered desirable, especially in areas where the average annual rainfall is less than 1000 mm (Carr 2012), there is no experimentally sound quantification of yield reductions for rainfed macadamia production (compared to that under irrigation). Different studies show contradictory results and strongly depend on the specific climatic conditions of the growing areas and seasons considered (Trochoulias and Johns 1992; Searle and Lu 2002). Moreover, macadamia yields are cyclical and highly variable, which makes it somewhat difficult to establish cause and effect (Carr 2012; Huett 2004).

Some critical phenological stages, i.e., periods during which lack of sufficient water supply can strongly affect macadamia production, are the periods between flowering and nut set (August–October), and the premature nut drop period (November–December). Trees experiencing water stress during such periods produce less flowers, with an overall reduced nut set (Murovhi 2003) and increased nut drop (Carr 2012).

Furthermore, water stress during the nut maturation stages decreases photosynthesis rates at a time when energy demands for oil accumulation are highest. This consequently reduces yield and nut quality (Stephenson et al. 2003). Another major disadvantage of water deficits in the roots zone is that the tree cannot take up nutrients (including those supplied by fertilizers). Therefore, according to a local independent macadamia consultant (S. Schoeman, personal communication) rainfed macadamia production in Limpopo is not considered a generally viable option for the future, although some niche microclimates allow for it. An option which might receive more attention by macadamia growers in the future is deficit irrigation, in the form of nearly rain-fed production with supplementary irrigation during specific critical phenological stages or in case of prolonged droughts. At present, such an approach is still rather uncommon and would require thorough scientific investigation as well as commercial evaluation.

Historically, there are two main irrigation systems for macadamias in South Africa, namely micro-sprinklers and drip irrigation (Murovhi 2003). Micro-sprinklers, currently the most common irrigation system in South Africa (Fig. 22.2), allow for a wide range of emitter flow rates, from as low as 15 L to more than 100 L/h. The large wetting radius matches the tree's root surface area, thereby increasing water and nutrient uptake. At the same time, a large part of the applied water could be lost to evaporation or losses from the root zone (e.g. by deep percolation), or be taken up by grass and weeds growing around the trees, thus strongly reducing the system's efficiency as well as water productivity. Due to the large amounts of water applied, this system is generally used at a low frequency of one to three times per week.

Drip irrigation requires more frequent applications (up to 300 days/year) of smaller water amounts (emitter delivery rates commonly vary between 0.7 L and 4.0 L/h). Dripper lines are placed close to the tree stems, with spacing's of 0.6 m between drippers. Compared to the micro-sprinkler irrigation, this system requires less water, despite maintaining a continuous wetted strip along the tree line, thus having a higher system efficiency and water productivity. Therefore, the adoption of drip irrigation has increased in recent years (and since the 2012 Census) to about 40% of the new installations, mainly at the expense of micro-sprinkler irrigation (S. Schoeman, personal communication).

Recently, a new system is emerging, following the principle of applying small amounts of water at high frequency, with the aim of better matching irrigation water supply with the rate of plant water uptake. This is the centralized low-flow drip fertigation concept (or the ultra-low flow drip in its most extreme version). In comparison to the regular drip irrigation, these irrigation systems have lower emitter delivery rates of 0.6–0.7 L/h (0.4 L/h for ultra-low flow drip) and wider spacing between drippers of 0.9–1.0 m, leading to low system delivery rates, which require higher, almost daily, irrigation frequencies. The concept is to apply the daily irrigation requirement of the trees almost at the same rate that the tree uses the water (i.e. with system flow rates of typically 0.15–0.3 mm/h) and mostly rely on capillary water movement rather than mass flow in the soil, thus leading to significantly reduced risk of soil saturation and run-off. With 1–2 mm of water applied per day over several hours in a very efficient way, the irrigation of the entire farm at the same time becomes possible with a centralized and labour-friendly application management. This also allows for the coupling of irrigation and fertilization (the so-called fertigation), which can both be targeted to the tree's daily needs year-round. The limitations of this system include its requirement of a more complicated design with multiple dedicated mainlines to each field valve, the higher installation costs, mainly related to the fertigation injection systems, and the need of rearranging the scheduling of irrigation events. Furthermore, due to the low delivery rates, it is important to monitor and fully understand the soil water dynamics (e.g. the required filling time for the soil reservoir) and manage it accordingly. Nevertheless, it is currently considered as the most promising irrigation system by sector experts (S. Schoeman and Barry Christie—technical manager of Green Farms Nut Company, personal communication), allowing the achievement of the highest system efficiency and reducing water consumption to a minimum. Therefore, low-flow drip irrigation systems are expected to increase in the future, with large investments forecasted for their establishment, especially in large farms. Another less common irrigation system is the so-called floppy sprinkler irrigation, however, its very low efficiency, and large water volumes applied, preclude it from receiving considerable attention in the South African context.

Despite the good intentions of most macadamia growers and consultants to improve water management there is still surprisingly little knowledge on the exact water requirements of macadamias. So far, growers have mostly relied on management guide charts based on accumulated empirical evidence on daily or weekly recommended water requirements for the different phenological stages of trees with different ages, planting densities, and canopy coverages (Fig. 22.1). However, these are merely used as guidelines, for example, for the planning of the irrigation system given a certain water allocation, and sometimes regarded as excessively high (Lee 2020).

In one of the few well-known attempts to experimentally quantify macadamia tree water use amounts for Australian conditions, Stephenson et al. (2003) reported estimates of daily evapotranspiration ranging between 52 L (winter) and 80 L per tree (summer) for 'HAES 246' cultivars growing on sandy soils. In South Africa, Ibraimo et al. (2014) measured average daily water uses ranging between 27 L and 51 L/day in 6-year-old (intermediate bearing) macadamia trees ('Beaumont' cultivar). In a follow-up study, Taylor et al. (2021) attempted to distinguish between the water use of intermediate bearing and full-bearing 'Beaumont' macadamia trees, reporting comparatively lower average daily water use values of 22–35 L for the full-bearing trees, about 60% higher than for the younger trees. In general, they claim that such values are strictly depending on local environmental conditions, tree canopy size, and management factors, thus making it very difficult to provide precise estimates of macadamia tree water use without having additional on-field measurements of tree transpiration in a range of different orchards (Taylor et al. 2021). Furthermore, different macadamia cultivars show different water requirements. For example, the widely popular variety 'Beaumont' ('HAES 695') is known to cope poorly with low water availability (S. Schoeman, personal communication). On the other hand, daily transpiration measurements in an Australian macadamia study (Searle and Lu 2003) showed almost double the water use by cultivar 'HAES 741' compared to that of 'HAES 344'. It would be therefore of great importance to get a better knowledge and understanding of the differences in transpiration between different cultivars, as well as of their specific performance in relation to the growing environment (Taylor et al. 2021). Common on-field strategies used by macadamia growers to determine the soil water status and to schedule irrigation accordingly include monitoring of the weather conditions (i.e. the variables that influence tree evapotranspiration) and using devices such as tensiometers and capacitance probes, which, respectively, measure soil water tension and soil moisture at different depths. However, oversimplified empirical norms are often followed to determine when it is necessary to irrigate. Yet an increased attention and investment by South African macadamia growers into monitoring the water status of their orchards has been observed in recent years (S. Schoeman, personal communication).

One of the risks growers are increasingly aware of is that of over-irrigating. According to recent studies, this is often the case (Ibraimo et al. 2014; Botha 2020). In experiments conducted on Beaumont macadamia orchards in Mpumalanga, a conservative water use behaviour of macadamias was observed, with a halt of tree transpiration when a certain level of evaporative demand (typically at a leaf-to-air vapour pressure deficit above 2 kilopascal) is reached (Smit et al. 2020). This climate-induced control is exerted through the closure of stomata (Lloyd 1991; Smit et al. 2020). This indicates that under hot and dry conditions, the trees will not necessarily use more water and the application of large irrigation amounts under these conditions would not lead to the desired outcome. On the contrary, excessive irrigation might lead to a reduction in soil aeration, especially in saturated soils, thereby further restricting water uptake and affecting tree health, growth, and nut yield (Botha 2020). Other negative effects of over-irrigation include iron deficiency, increased susceptibility to Phytophthora, and the loss of fertilizer by leaching, with the related economic and environmental impacts.

A number of promising technological innovations could further help to improve the sustainability of water management in macadamia orchards. For instance, better weather forecasts through improved climatic models would help to plan irrigation accordingly. Similarly, the increasing availability of more affordable weather stations and soil moisture probes shall facilitate the on-field monitoring of climatic conditions and soil water dynamics. Remote sensing and especially aerial photography are likely to play an increasingly important role, since they are proving to be very useful and labour-friendly tools to detect problems in the orchard. Farm management apps and portals will facilitate sharing information between macadamia growers and consultants. Nevertheless, farming from remote is far from being a feasible reality. In fact, according to macadamia expert Barry Christie (personal communication), although increasing, the adoption by macadamia growers of most of the above-mentioned innovations is still low. Other sustainable management options to reduce orchard water consumption include the adoption of water-saving agricultural practices like mulching. The presence of organic matter (leaves, husks, or compost) on the soil surface is crucial for tree health and it reduces evaporation and conserves water within the soil (Botha 2020; Steyn 2019). Inorganic options (e.g. weed mat) also exist and are sometimes used mainly in young orchards, where soil evaporation is especially high due to the greater area exposed to solar radiation between the tree rows. Careful management of weeds and grass cover in the orchards can also help reduce water losses caused by their transpiration (Botha 2020).

Recent droughts between 2015 and 2019 have increased the awareness of limits to water resources among South African macadamia growers, who are trying to decrease their water consumption, for example by switching to more efficient irrigation systems or by improving the monitoring of water use to avoid the risk of over-irrigating. Nevertheless, such efforts cannot fully counter-balance the overall increased water consumption, due to the continuous expansion of irrigated macadamia production areas. However, water availability and the impact of climate change on the local water resources are not yet perceived as major risks by macadamia growers. According to S. Schoeman (personal communication) and findings from a macadamia growers workshop held at Levubu in February 2019, major concerns include issues related to pest control, future political scenarios, energy prices, and industry developments. The general perception is that reduced water availability will increase production costs but growers will still prefer growing macadamias because of their high market value. However, under the projected drought scenarios and the stricter policing of water allocations, this might not be possible for much longer (Botha 2020; Shabalala et al. 2022). Therefore, the only solution lies in increasing the water use efficiency of macadamia orchards, making use of the best available knowledge, technologies, and practices to reduce non-beneficial water losses to a minimum.

22.2.3 Suggestions Towards More Sustainable Water Management in Macadamia Orchards by SALLnet

Science has to play a distinct role in improving water management in times of a changing climate. More science-informed decision-making can be provided, for example, by delivering experimental evidence on the performance of different systems and management options. This should be based on a deep understanding of the relevant ecophysiological processes, which influence tree water requirements, the effects of management and by more robust climate projections at the local scale.

In the framework of the SPACES II—SALLnet joint research project, an ongoing study of macadamia water use aims to increase our understanding and gain new insights into the processes determining the interactions of genotype, environment, and management in macadamia orchards, represented by different macadamia management systems in Levubu, with a focus on water dynamics.

Fig. 22.3 Scheme of recorded weather and tree physiological parameters in the selected macadamia orchards (picture credits: Thomas Bringhenti)

To this end, intensive field experiments monitoring the hourly tree transpiration rates, daily water use, and leaf water potential (an indicator of water stress) of two different macadamia cultivars ('Beaumont' and 'HAES 849') were set up and run for two consecutive seasons (Fig. 22.3). Additional measurements were made on tree phenological development (number of racemes and nuts per tree), tree morphology (e.g. tree height, canopy volume, leaf area density, etc.), and nut production. Moreover, microclimate, soil water dynamics, and orchard management (especially irrigation) were also recorded.

That dataset contributes to a better understanding of the water use behaviour of macadamia trees in response to different water supply and environmental conditions, as well as the quantification of macadamia water use efficiency for contrasting management intensities. A related objective is the development of a macadamia growth and water use model to, among others, assist in setting the upper and lower limits of required water inputs for macadamia trees in different environmental conditions, and thus improving the precision of current empirical approaches to compute fruit tree water requirements (Orgaz et al. 2007; Villalobos et al. 2013) and to avoid wasteful over-irrigation. Eventually this will allow for upscaling of results from field experimentation across the whole region and for different climatic scenarios (e.g. by simulating the impact of future projections of long-term climate change on macadamia water use).

22.3 Pollination

In addition to abiotic and management factors, biotic factors such as pollination and biological control also determine macadamia production (Grass et al. 2018; Linden et al. 2019). Here we present management strategies that facilitate pollination in macadamia orchards in order to increase nut set and hence yield in a sustainable way.

Pollination Requirements of Macadamia
Macadamia is a mass-flowering crop of which one mature tree can produce up to 2500 inflorescences in one season (Moncur et al. 1985). The inflorescences are arranged in racemes of 10–35 cm length and each one bears 100–300 flowers (Fig. 22.4) depending on the variety (Trueman 2013). The small white flowers (the Beaumont variety produces pink flowers) develop from several whirls on the stalk and form one conspicuous inflorescence. The flowers are open for 1 week (Ito and Hamilton 1980; Sedgley 1983) and given that the pollination was successful, they develop into initial nuts 3 weeks after anthesis (Trueman and Turnbull 1994a; Wallace et al. 1996). Many immature nuts abscise during the first 7–15 weeks after anthesis, whereas the time and extent of the drop depend on the site, cultivar, time since canopy pruning (McFadyen

(continued)

Fig. 22.4 Honeybee (*Apis mellifera*) sitting on a raceme with open macadamia flowers (picture credit: Mina Anders)

et al. 2011, 2012; Sakai and Nagao 1985; Trueman 2010; Trueman and Turnbull 1994b) as well as pest damage (see Sect. 22.4.1). However, the nuts that remain on the raceme for around 15–20 weeks (final nut set) are likely to remain until maturation. The nuts are harvested from the orchard floor after they drop maturely off the tree, although the variety Beaumont must be treated with ethylene-generating compound (2-chloroethyl) phosphonic acid to induce the nut drop (Richardson and Dawson 1993).

The flowers of macadamia show features that indicate dependence on insect pollination, namely the bright colour of the petals, a strong scent as well as resources like pollen or nectar. The most observed agents for pollen transfer are honeybees (*Apis mellifera* L.) and stingless bees (*Tetragonula* spp.), but beetles, flies and even birds have also visited flowers and been considered potential pollinators (Heard and Exley 1994; Howlett et al. 2015). Although wind pollination might be possible (Urata 1954), several pollinator exclusion experiments indicate a strong pollinator dependency (Grass et al. 2018; Tavares et al. 2015; Wallace et al. 1996). Grass et al.

(2018) demonstrated that where insects were prevented from visiting flowers, initial and final nut set was reduced by 80% and by 54%, respectively. Further, Heard (1993) showed initial and final nut set in macadamia to be correlated with increased insect visitation to flowers. As macadamia is partially self-incompatible (Hardner et al. 2009; Urata 1954), self-fertilization is possible, but minimized through flower morphology (Sedgley 1983; Urata 1954). This underpins the dependency of macadamia on animal-pollination.

22.3.1 Potential Pollinators of Macadamia Crops

In their Australian native range, macadamias have two main pollinators, endemic stingless bees (*Tetragonula* spp.) and introduced honeybees (*Apis mellifera*) (Howlett et al. 2015; Vithanage and Ironside 1986). Both are commonly used for pollination in commercial macadamia orchards.

A study by Heard and Exley (1994) in Australian macadamia orchards observed honeybees (60.5%), stingless bees (35.8%), while the remaining 4% were butterflies (*Lepidoptera*), hoverflies (*Syrphidae*), other Hymenopterans and even birds. Stingless bees mainly collect pollen and thus have intimate contact to the stigma; this is why they are considered to be very efficient. In contrast, honeybees first collect nectar and are considered less efficient, but compensate through high visitation rates (Heard 1994).

In Hawaii, where macadamia has been cultivated since the 1920s (Shigeura and Ooka 1984), honeybees are considered to be the most important pollinators, although other pollinator taxa such as hoverflies have also been observed to visit macadamia flowers (Tavares et al. 2015). In Brazil, butterflies accounted for 50% of flower visits, ensuring initial nut set of inflorescences in the same magnitude as hand cross-pollination (Santos et al. 2020).

In South African orchards, visual observations revealed that 90–99% of the flower visitors were honeybees, the remainder comprised of complemented by hoverflies, wasps, stingless bees, wild bees, and butterflies (Grass et al. 2018; Anders et al. unpublished data). Another study, in the same region, observed a lower ratio of honeybees (65%) associated and a higher frequencies of *Diptera* spp. (33%) (Ramotjiki 2020). In the macadamia region in Levubu, wild honeybee colonies are commonly found in natural or semi-natural habitat around the orchards, where they colonize suitable nesting sites. This means that besides managed honeybees, wild honeybees also provide pollination service in the orchards, as long as the landscape includes appropriate patches of natural habitat. Hence, both managed and wild honeybees must be taken into consideration as important pollinators for macadamia in this region.

Although thrips (*Thripidae*) are considered a pest on macadamia (see Sect. 22.4.1), their contribution to pollination remains unclear. They are found in vast numbers in flowers of a large range of plants. Whereas some species are pollinators (Mound 2005), individual thrips have been recorded consuming more than 1500 pollen grains per day, depending on the pollen grain size (Kirk 1987). Because they move between the flowers of a macadamia raceme, they are very likely to transport pollen between flowers of the same raceme. However, a recent pollinator exclusion study has shown that thrips might have been largely overlooked as an important pollinator of macadamias, as the final nut set of macadamia was positively correlated with the number of adult thrips on flowers (Meyer 2016).

22.3.2 Pollination Limitation

Although macadamia is dependent on insect pollination, there is evidence for pollination limitation despite honeybee management. High nutrient demand of the nuts results in high abscission rates and only a small proportion of flowers (3%) develops into mature nuts (Evans et al. 2021; Grass et al. 2018). Even lower proportions are not unusual, with only 0.3% (Ito and Hamilton 1980) and 0.6% observed (Sakai and Nagao 1985).

On the other hand, supplemental hand-pollination resulted in a significantly higher initial (66%) and final (44%) nut set than natural pollination in the study of Grass et al. (2018), corresponding with other studies (Howlett et al. 2019; Wallace et al. 1996). Further, recent studies showed that macadamia is much more self-incompatible than previously thought. Genetic analyses demonstrated that depending on the cultivar 80–90% of the harvested nuts were cross-pollinated while only up to 8% were self-pollinated (Richards et al. 2020; Kämper et al. 2021). Grass et al. (2018) concluded that honeybees fail to deliver adequate pollination services, especially as increasing their colony density could even result in reduced final nut set. Higher visitation rates were neither related to higher bee density nor nut set. Intraspecific competition at high colony densities may have led honeybees to repeatedly exploit the same resources, reducing cross-pollination between macadamia trees and varieties. This means, efficient pollination is not simply determined by a high number of pollinators, but also by other factors, for example their foraging behaviour on the flower or their movement between the trees. Also, the landscape configuration, i.e. the cover of natural habitat is likely to affect pollinator behaviour and pollination services.

In order to get a broader understanding of pollination limitation and services, we established another macadamia pollination experiment in the Levubu region in 2019 and 2020, where we simultaneously observed different potential influences on pollination and yield, incorporating irrigation or rain-fed production as well as landscape factors such as altitude and amount of semi-natural vegetation in the landscape. Grass et al. (2018) did not detect higher pollinator visitation rates or nut set on trees close to natural habitat and hence no spillover effect from these to macadamia orchards. However, their study did not consider landscape effects. The surrounding landscape of macadamia orchards can provide additional nesting and foraging resources for wild bees and thus can influence pollinator diversity in macadamia orchards and consequently crop pollination services (Bänsch et al. 2021; Beyer et al. 2021a). The objective of the project is to gain a deeper understanding of the interaction between different management and landscape contexts, and

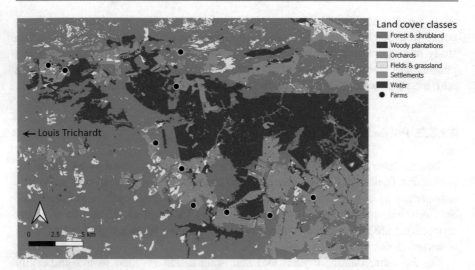

Fig. 22.5 Map of the current study area of the SPACES project including study sites (farms) used for pollination experiments (black dots) and broad land cover classes (map credit: Mina Anders)

pollination in order to improve pollination services in macadamia orchards (see Fig. 22.5).

22.3.3 Management Strategies to Facilitate Pollination Services in Macadamia Orchards

Pollination of macadamia is important for nut production, but the provision of optimal pollination services is not attained by simply increasing managed honeybee colonies. One option is the enhancement of cross-pollination. For commercial nut production, a large number of different varieties are cultivated. The role of cross-pollination between varieties has been explored in a couple of studies. Supplemental hand cross-pollination enhances not only fruit set (Herbert et al. 2019; Howlett et al. 2019; Trueman and Turnbull 1994a; Wallace et al. 1996) but also nut weight (Herbert et al. 2019). By manually cross-pollinating almost an entire tree, Trueman et al. (2022) achieved an increase of up to 109% of kernel yield. Empirical studies confirmed these results, identifying higher yield and nut mass in blocks where several varieties are grown than in single-variety blocks and a decrease in harvested nuts with distance to the cross-variety pollen source (Ito and Hamilton 1980; Kämper et al. 2020). However for individual varieties, recent genetic examination indicated unexpected high degrees of self-fertilization of up to 20–40% (Langdon et al. 2019).

The planting of different and ideally well matching varieties in close distances is still a promising management strategy to increase cross-pollination.

Another option to increase pollination service is the promotion of semi-natural pollinator habitat. Managed as well as wild pollinators profit from natural or semi-natural habitat, which provides resources throughout the year (Beyer et al. 2021a, b; Dainese et al. 2019). The access to continuous and diverse food resources is essential for general pollinator community health (Alaux et al. 2017) and thus pollination performance. Wild pollinators additionally depend on nesting sites (Kremen et al. 2007). For many crops, wild insects play an even bigger role for pollination than honeybees (Garibaldi et al. 2013). It has been shown that distance to natural habitat leads to a decline in pollinator abundance and visitation rate of native pollinators (Carvalheiro et al. 2010; Ricketts et al. 2008). A high proportion of semi-natural habitat, on the other hand, improves pollinator richness and abundance (Ecraerts et al. 2019; Beyer et al. 2021b). For example, in almond fields the percentage of natural area in the 2 km buffer zones increased both wild pollinator-species richness and honeybee visits (Alomar et al. 2018). To enhance provisioning of pollination service, access to natural or semi-natural habitat plays an important role for both, honeybees and wild pollinators.

Pollinator distributions in orchards can be optimized by the spatial arrangement of beehives. The pollinators should be distributed evenly in the orchards and be able to transmit pollen between trees and varieties. Cunningham et al. (2016) showed that the pollination service was improved by changing the spatial arrangement of honeybee colonies in almond orchards. At any given colony density, fruit set outcomes were better when smaller placements (<100 colonies) were used which were more closely spaced (<700 m apart) than was standard (Cunningham et al. 2016).

Similarly, a study in macadamia orchards in Australia showed both honeybees and managed stingless bees did not distribute evenly in the orchard, but rather occurred in higher densities close to their colonies. This applied particularly for stingless bees, as >96% of the recordings were within 100 m of the stingless bee hives (Evans et al. 2021).

An observational study during the first SPACES project, revealed a drastic effect of insecticide applications on honey bees (Linden 2019). Bee numbers observed in the macadamia orchards increased significantly with time after each chemical application. This indicates a negative effect of pesticide usage on honeybee activities inside the orchard, despite efforts of the farmers to minimize impacts on pollinators. Recovery of bee activities occurred faster at orchard edges next to natural vegetation. At these natural orchard edges bee numbers were in general significantly higher than at human-modified (e.g. continuous farmland, roads) orchard edges. Natural vegetation in and around orchards therefore seems to play a key role in the rehabilitation of pollinators in macadamia orchards and serves as source for wild bees as pollinators.

22.4 Natural Pest Control with a Special Focus on Insectivorous Bats

22.4.1 South African Macadamia Insect Pests

The main insect pests in the South African macadamia industry are several Heteropteran and Lepidopteran species. Some species of thrips (Thripidae) are considered a minor pest, which can cause damages to flowers, while other thrips species are possibly beneficial in predating on other arthropods and in aiding pollination. Major thrips infestations in orchards have been cause for concern in several South African growing regions including Levubu in recent times (Hepburn 2015; Schoeman 2009).

The indigenous two-spotted stink bug *Bathycoelia distincta* Distant (Hemiptera: Pentatomidae) is by far the major pest on macadamias in South Africa and economically the most significant Heteropteran species (Schoeman 2018). However, damage is also caused by several Tortricidae (Lepidoptera) species (Schoeman and De Villiers 2015; Schoeman 2018). According to Schoeman (2009), over 10% of immature nut drop in macadamia is linked to the tortricid complex, making them economically important pest species.

The competitive advantage of the two-spotted stinkbug over other Heteropterans is its extremely long mouthpart (±16 mm) compared to other species, enabling them to feed on all varieties of macadamia even after nuts have matured (Schoeman 2018). This damage to the macadamia is called 'late' stinkbug damage, referring to damage occurring late in the season when the macadamia shell is penetrated while the mature kernel is undergoing oil accumulation (Schoeman 2018).

The losses through direct insect damage to the macadamia kernel by early and late stinkbug damage, were estimated at 96 and 84 million ZAR, respectively, for the growing season of 2019 alone (SAMAC 2020). Additionally, there are also indirect effects of insect damage such as promoting immature nut drop, kernel germination, and fungus infestation, which were estimated losses of 52, 17, and 32 million ZAR for 2019, respectively (La Croix and Thindwa 1986; Nagao et al. 1992; Schoeman and de Villiers 2015; SAMAC 2020).

22.4.2 Avoided Cost Models and Exclusion Studies of Vertebrate Predators

The concept of 'ecosystem services', defined as the benefits that humans derive from biodiversity and ecosystems such as regulating, supporting, and provisioning processes (Wangai et al. 2016), has been increasingly appreciated and understood by the global community in the last decades (Millennium Assessment Board 2005). Although crucial in providing many of these services, bats have always suffered from unfounded negative public perceptions and have only received scientific attention as important ecosystem service providers in recent years (Voigt and Kingston 2016). Probably the most significant early contribution to our understanding of the

Fig. 22.6 Vertebrate exclusion cages covering two macadamia trees each in Levubu, Limpopo (picture credit: Dr. Valerie Linden)

economic value of bats to the agricultural industry was an avoided-cost model by Boyles et al. (2011), estimating that pest suppression by insectivorous bats has an annual value of about 22.9$ billion to the agricultural industry of the United States. Following other studies (López-Hoffman et al. 2014; Puig-Montscrrat et al. 2015; Wanger et al. 2014) using this modelling approach, a study conducted as part of SPACES by Taylor et al. (2018) estimated the value of bats to the South African macadamia industry in suppressing stinkbugs alone at 57–139$ per hectare per year.

However, a later exclusion study, also conducted as part of the SPACES programme, by Linden et al. (2019) shows that the values provided by the avoided-cost model were likely an underestimation and that the combined value of ecosystem services provided by insectivorous bats and birds through pest predation even outweighs the disservice by crop raiding vervet monkeys (*Chlorocebus pygerythrus*).

Using exclusion cages, the effect of the absence of birds and or bats as well as crop raiding mammals was tested, distinguishing between diurnal, nocturnal, or constant exclusions and comparing the yield, quality, and economic value of the exclusions at either natural or human-modified orchard edges (Fig. 22.6). The cages were erected in between macadamia crops, after the previous nuts had been harvested and before new flowers had started to develop, experiments were then running over three consecutive years.

At the natural orchard edge, where crop raiding by monkeys occurs, the avoided cost by bats and birds suppressing insect pests was about $5000 per hectare per year. Whereas, crop loss through crop raiding was about $1600 per hectare per year (Linden et al. 2019).

However, estimates based on exclusion studies cannot account for the total ecosystem service of pest suppression including open-air foraging bats (Monadjem et al. 2020), which feed in open spaces on certain pests such as moth before they descend into orchards. The open-air feeding guild of bats (families Molossidae and Emballonuridae in South Africa) generally hunt above the canopy of vegetation. McCracken et al. (2008, 2012) showed on the example of the open-air feeding Brazilian free-tailed bat (*Tadarida brasiliensis*) that these bats are not only able to exploit local pest infestations of the corn earworm but also hunt at altitudes up to 900 m above ground level. Most importantly, McCracken et al. (2008) suggest that the high foraging activity levels of this species at 400–500 m above ground level are linked to the migration of insects such as certain moths.

Similar to the diet analyses of McCracken et al. (2012), a study under SPACES conducted in the Levubu macadamia orchards showed that local bat population is presumably much more generalist and opportunistic in their foraging behaviour than previously assumed (Weier et al. 2019a). Testing for four pest insect species (*B. distincta, N. viridula, T. batrachopa, and C. peltastica*), the study showed that nearly all faecal samples analysed from four families of bats (Molossidae, Nycteridae, Rhinolophidae, and Vespertilionidae) contained genetic sequences of at least one stinkbug and one moth pest insect.

22.4.3 Habitat Use of Bats in Macadamia Orchards

Having established the importance of insectivorous bats in macadamia pest control, further research within the SPACES programme investigated habitat use of bats in order to guide agro-environmental management. By means of acoustic monitoring during active drive transects in Levubu orchards, Weier et al. (2018) found that bat activity increases with Hemiptera abundance but also with the amount of natural and semi-natural vegetation near orchards. Generally, the ecosystem service of pest suppression was higher at natural orchard edges in Levubu (Linden et al. 2019; Weier et al. 2021).

Crisol-Martínez et al. (2016) found that the activity of the clutter and clutter-edge guilds of bats decreased going away from natural orchards into macadamia monoculture in eastern Australia, while the most common species preferred the least fragmented and therefore the least isolated areas. As found by Weier et al. (2018), the abundance of insects and water availability had an influence on the abundance of species (Crisol-Martínez et al. 2016). Water availability, for both foraging and drinking, through artificial water sources such as dams can also increase the activity and diversity of bats in other agroecosystems (Shapiro et al. 2020; Sirami et al. 2013). Bats seem to prefer polyculture or organic agroecosystems (Kelly et al. 2016; Wickramasinghe et al. 2003; Wordley et al. 2017).

However, most of the agroecological studies on bats are currently focusing on common insectivorous species and it is worth mentioning that rare clutter feeding species such as the Rhinolophidae might also have a key role in suppressing certain pest insect species (Russo et al. 2018). From studies conducted in southern African agroecosystems these species seem to be already affected considerably by the ongoing land-use change and possibly also the competition and displacement by more generalist species as they have been recorded in very low numbers in more intensive agroecosystems (Linden et al. 2019; Shapiro et al. 2020; Weier et al. 2018, 2021). While, many (but not all) species of the open air and clutter edge feeding guilds of bats do use anthropogenic structures (such as tunnels, bridges, and roofs) for roosting the rhinolophids most commonly require their habitat to provide caves or old hollow trees. Generally, bat species benefit from natural vegetation which provides a variety of roosting sites such as loose bark, large curled leaves, tree hollows, woodpecker holes, and more.

In conclusion, a heterogeneous landscape in and around orchards, which provides connectivity, foraging, and roosting sites through natural and semi-natural vegetation promotes the activity and diversity of bat species and their ecosystem service provision. The same can be assumed for the ecosystem services provided by birds, therefore the diversity and richness of bird species in Levubu macadamias is currently investigated under the SPACES programme. The installation of bat houses is considered a way to buffer decreasing natural roost sites in many countries at the moment. However, it is unclear whether this has a positive effect on the overall bat communities and pest control service provision in general (Griffiths et al. 2017). Building on a previous study looking into the occupancy of bat houses in Levubu macadamia orchards under the SPACES programme (Weier et al. 2019b), a currently ongoing study conducted in the same area is investigating the effect of occupied bat houses on the surrounding bat species composition and activity in more detail.

22.4.4 The Effect of Pesticide Application on Ecosystem Services

The approach generally recommended for stinkbug pest control in southern African macadamia orchards is to base pesticide application on scouting for nymphs and adults, monitoring numbers using a knockdown method (Schoeman 2012). Scouting should focus on the edges of the orchards, where stinkbugs immigrate into orchards in the early season to ensure that the first application significantly reduces the first generation of nymphs, while another minimum of four applications of pesticides throughout the season are applied according to the life history of two-spotted stinkbugs (Nortje and Schoeman 2016). The use of pesticides in the winter months and over the flowering period is generally not recommended, as pest numbers are low, no crop on the tree, and pollination could be impacted by sprays affecting bee and other pollinator populations. While some stinkbugs overwinter in or near the macadamia orchards, others migrate into the orchards when food becomes available. Natural vegetation is seen both as a source of stinkbugs and a deterrent as it is serving as an alternative food source. Most recently it has been recommended and practised by many farmers to leave grasses and weeds to grow around the orchards and within the tree lines. Experience shows that this reduces the activity of stinkbugs on the macadamia trees, as they stay within these weed beds. Once these sections are mowed, stinkbug numbers were observed to increase on macadamia trees. While this is based on anecdotal evidence, many farmers are applying it in an attempt to minimize damages in a natural manner.

The common threshold at which spraying is recommended is four stink bugs found per 10 trees. However, according to a sector expert an estimated 10–15% of farmers still rely on scheduled or the so-called calendar sprays against stinkbugs, meaning that pesticides are applied in regular intervals independent from the confirmed presence of pest insects or their abundance on trees. A particular concern with this approach, apart from ecosystem (service) degradation, is that it increases the likelihood of stinkbugs developing resistances to pesticides (Schoeman 2018), which can also be aided by tree height and shape. Stinkbugs prefer the dark and dense areas of the orchards for foraging and stinkbug damage increases with tree density (Schoeman 2014). Conventional sprays applied with tractors become less efficient if the macadamia tree height exceeds 6 m (Drew 2003). It is recommended that trees should not be taller than 80% of the width of rows between trees (Schoeman 2018).

Another promising future alternative for the pest management of two-spotted stinkbugs is the use of semiochemicals especially alarm pheromones (Pal et al. 2020). While trap crops such as *Crotalaria juncea* might help to decrease the kernel damage caused by other stinkbug species, no suitable trap crop has been found to attract the two-spotted stinkbug as it seems to be highly monophagous (Steyn 2019). The other main pest for macadamias are lepidopteran species of the nutborer complex, namely the macadamia nutborer and the false codling moth. Monitoring of the nutborer complex is facilitated by means of species-specific pheromone traps, which can also be used to control them. Additionally, young nuts can be monitored for oviposition by the nutborer moths. Apart from pesticides there are several biological control agents including fungi, viruses, and bacteria registered for the use of these pests.

In 2018, the worldwide average use of pesticides per hectare of cropland was 2.63 kg (FAO 2018). While the average for African countries was much lower (0.3 kg/ha; FAO 2018), the bioaccumulation of pesticides in non-target species is generally of great concern and has been reported to negatively affect the behaviour and life history of invertebrates as well as vertebrates (Oliveira et al. 2021).

Overall, the effect of pesticides on bat species has been vastly understudied and represents a large scientific research gap (Oliveira et al. 2021; Torquetti et al. 2021). In a review of studies published in English between 1964 and 2019, Oliveira et al. (2021) found only 28 studies on the effect of pesticides on bats worldwide. These studies showed that the ingestion of pesticides by bats through insects, fruits, or water can have serious negative consequences including impaired torpor and echolocation, liver pathologies, oxidative stress, and endocrine disruption as well as decreased energy reserves. In a review of declines in bird populations in agroecosystems, nearly half of the reviewed studies ($N = 122$) found pesticide use had a negative effect on local species (Stanton et al. 2018). Recent studies on the effect of neonicotinoids on birds, particularly insectivorous birds, in the USA and the Netherlands have linked its usage to a decline in bird diversities and populations of an average annual 3% and 3.5%, respectively (Hallmann et al. 2014; Li et al. 2020). Given the high longevity of bats, it seems likely that the annual declines of bat populations due to the effects of pesticide usage are higher than those reported for birds, making it an urgent field for future research.

Mostly, farmers tend to spray their macadamia orchards in the early morning or late evening hours. The recommendation is to spray while the maximum ambient temperatures are below <18 °C, at which stinkbugs are immobile and cannot fly out of the orchards. While insectivorous bats are active throughout the night from sunset to sunrise, their peak activity is for about 3 h after sunset. It is much harder to determine peak activities for birds in the area as there are both diurnal and nocturnal bird species active in the Levubu orchards (Linden et al. 2019).

Linden et al. (2019) observed that hymenopterans took the longest to recover after a pesticide application event. Several beneficial insect species fall within this order, chief among which are parasitoids that are specialist predators of pest species. Spiders are the dominant invertebrate predator in these orchards and are some of the first taxa to recolonize trees after a spray event. Assemblages in macadamia orchards are dominated by wandering spiders (>90%) and mainly belong to family Salticidae (73%) (Dippenaar-Schoeman et al. 2001). Haddad and Dippenaar-Schoeman (2004) observed that a salticid species that dominates Pistachio orchard assemblages ate at least one lygaeid bug a day. However, calendar spraying over the long-term results in an almost complete collapse in spider assemblages, particularly if the surrounding vegetation is highly transformed.

22.5 Conclusions

- To increase the efficiency and sustainability of macadamia water use, growers should adopt water-saving irrigation systems and reduce their irrigation water supply to small and targeted applications, aiming at meeting the specific tree water requirements. Research can help to determine such amounts for different environmental conditions.
- Technological innovations (e.g. soil moisture probes and remote sensing) allow for a better understanding and monitoring of water dynamics in macadamia orchards, which, combined with the adoption of water-saving agricultural practices (e.g. mulching), can considerably reduce the orchards' water footprint in view of future water limitations.
- Recommendations for growers to maximize biodiversity services in macadamia orchards include retaining natural and semi-natural habitats in the landscape and enhancing agrobiodiversity, increasing wild pollinator abundances and optimizing the spatial arrangement of beehives. The installation of bat houses might be another option to improve natural pest control services. There is a range of alternative, ecologically friendly recommendations for natural pest control.

Any successful long-term control of pest insect damage in macadamia requires an integrated pest management (IPM) approach.

• The timing and application of pesticide sprays should be modified based on ecological and biological principles, such as a day-degree models of stinkbug development, or based on scouting for threshold pest stinkbug numbers in orchards and taking peak activity times of bats and birds into account, to mitigate the ecological impact of pesticides.

Generally, the research focus of industry bodies needs to shift from short-term economic benefits for farmers, to focus more on the long-term security of the industry, identifying the threats deriving from a changing climate and developing corresponding risk management strategies to mitigate their impact (e.g. water availability). A priority should be to maintain sustainable agroecosystems which provide resilient biodiversity services under the predicted decrease in annual rainfall.

References

AGTAG (2018) 63% of global macadamia crop forecast to come from China by 2025. https://www.agtag.co.za/category/3/post/21247. Accessed 21 July 2021

Ahrens P (1991) History and statistics of macadamias in Levubu. In: Proceedings of the macadamia mini symposium, Levubu, 18–19 September 1991, pp 3–4

Alaux C, Allier F, Decourtye A et al (2017) A 'landscape physiology' approach for assessing bee health highlights the benefits of floral landscape enrichment and semi-natural habitats. Sci Rep. https://doi.org/10.1038/srep40568

Alomar D, González-Estévez MA, Traveset A et al (2018) The intertwined effects of natural vegetation, local flower community, and pollinator diversity on the production of almond trees. Agric Ecosyst Environ 264:34–43

APNEWS (2020) Worldwide Macadamia Market (2020 to 2025) - growth, trends and forecasts. https://apnews.com/press-release/business-wire/. Accessed 15 May 2021

Bänsch S, Tscharntke T, Gabriel D et al (2021) Crop pollination services: complementary resource use by social vs solitary bees facing crops with contrasting flower supply. J Appl Ecol 58:476–485

Baudoin MA, Vogel C, Nortje K et al (2017) Living with drought in South Africa: lessons learnt from the recent El Niño drought period. Int J Disaster Risk Reduct 23:128–137

Beyer N, Gabriel D, Westphal C (2021a) Contrasting effects of past and present mass-flowering crop cultivation on bee pollinators shaping yield components in oilseed rape. Agric Ecosyst Environ. https://doi.org/10.1016/j.agee.2021.107537

Beyer N, Kirsch F, Gabriel D et al (2021b) Identity of mass-flowering crops moderates functional trait composition of pollinator communities. Landsc Ecol 36:2657–2671

Botha L (2018) Do we have enough water for all our macs? The Macadamia, Winter 2018 edition

Botha L (2020) Research shows South African macadamia orchards are over-irrigated. The Macadamia, Autumn 2020 edition

Boyles JG, Cryan PM, McCracken GF et al (2011) Economic importance of bats in agriculture. Science 332:41–42

Carr MKV (2012) The water relations and irrigation requirements of Macadamia (*Macadamia* spp.): a review. Exp Agric 49(1):74–94

Carvalheiro LG, Seymour CL, Veldtman R et al (2010) Pollination services decline with distance from natural habitat even in biodiversity-rich areas. J Appl Ecol 47:810–820

Conway D, van Garderen EA, Deryng D et al (2015) Climate and southern Africa's water-energy-food nexus. Nat Clim Chang 5:837–846

Crisol-Martínez E, Moreno-Moyano LT, Wormington KR et al (2016) Using next-generation sequencing to contrast the diet and explore pest-reduction services of sympatric bird species in macadamia orchards in Australia. PLoS One. https://doi.org/10.1371/journal.pone.0150159

Cunningham SA, Fournier A, Neave MJ et al (2016) Improving spatial arrangement of honeybee colonies to avoid pollination shortfall and depressed fruit set. J Appl Ecol 53:350–359

DAFF (2019) A profile of the South African macadamia nut market value chain. https://www.daff.gov.za. Accessed 22 June 2021

Dainese M, Martin EA, Aizen MA et al (2019) A global synthesis reveals biodiversity-mediated benefits for crop production. Sci Adv. https://doi.org/10.1126/sciadv.aax0121

De Villiers E, Joubert P (2003) The cultivation of Macadamia. ARC-Institute for Tropical and Subtropical Crops, Nelspruit

Dippenaar-Schoeman AS, Van den Berg MA, Van den Berg AM et al (2001) Spiders in macadamia orchards in the Mpumalanga Lowveld of South Africa: species diversity and abundance (Arachnida: Araneae). Afr Plant Prot 7(1):39–46

Drew H (2003) Critical issues in spray application in macadamias using ground-based air-assisted sprayers. Proceedings of the Second International Macadamia Symposium, Tweed Heads, New South Wales, pp 120–125

Eeraerts M, Smagghe G, Meeus I (2019) Pollinator diversity, floral resources and semi-natural habitat, instead of honey bees and intensive agriculture, enhance pollination service to sweet cherry. Agric Ecosyst Environ. https://doi.org/10.1016/j.agee.2019.106586

Evans LJ, Jesson L, Read SFJ et al (2021) Key factors influencing forager distribution across macadamia orchards differ among species of managed bees. Basic Appl Ecol. https://doi.org/10.1016/j.baae.2021.03.001

FAO (2018) Pesticides - average use per area of cropland 1990–2018. Food and Agriculture Organization of the United Nations (FAO), Statistics Division (ESS), Environment Statistics team. http://www.fao.org/faostat/en/#data/EP/visualize. Accessed 15 June 2021

Foley JA, DeFries R, Asner GP et al (2005) Global consequences of land use. Science 309:570–574

Garibaldi LA, Steffan-Dewenter I, Winfree R et al (2013) Wild pollinators enhance fruit set of crops regardless of honey bee abundance. Science 339:1608–1611

Grass I, Meyer S, Taylor PJ et al (2018) Pollination limitation despite managed honeybees in South African macadamia orchards. Agric Ecosyst Environ 260:11–18

Green & Gold Macadamias (2018) International Macadamia Symposium wrap up. https://www.greenandgoldmacadamias.com/. Accessed 21 May 2021

Griffiths SR, Bender R, Godinho LN et al (2017) Bat boxes are not a silver bullet conservation tool. Mammal Rev 47(4):261–265

Haddad CR, Dippenaar-Schoeman AS (2004) An assessment of the biological control potential of Heliophanus pistaciae (Araneae: Salticidae) on Nysius natalensis (Hemiptera: Lygaeidae), a pest of pistachio nuts. Biol Control 31(1):83–90

Hallmann CA, Foppen RP, Van Turnhout CA, De Kroon H, Jongejans E (2014) Declines in insectivorous birds are associated with high neonicotinoid concentrations. Nature 511(7509):341–343

Hahn N (2017) Endemic flora of the Soutpansberg, Blouberg and Makgabeng. S Afr J Bot 113:324–336

Hardner CM, Peace C, Lowe AJ et al (2009) Genetic resources and domestication of Macadamia. In: Janick J (ed) Horticultural reviews. Wiley, Hoboken, pp 1–125

Heard TA (1993) Pollinator requirements and flowering patterns of Macadamia integrifolia. Aust J Bot 41:491–497

Heard TA (1994) Behaviour and pollinator efficiency of stingless bees and honey bees on macadamia flowers. J Apic Res 33:191–198

Heard TA, Exley EM (1994) Diversity, abundance, and distribution of insect visitors to macadamia flowers. Environ Entomol 23:91–100

Hepburn C (2015) The phenologies of Macadamia (Proteaceae) and thrips (Insecta: Thysanoptera) communities in Mpumalanga province, South Africa. Dissertation, Rhodes University

Herbert SW, Walton DA, Wallace HM (2019) Pollen-parent affects fruit, nut and kernel development of Macadamia. Sci Hortic 244:406–412. https://doi.org/10.1016/j.scienta.2018.09.027

Hewitson BC, Crane RG (2006) Consensus between GCM climate change projections with empirical downscaling: precipitation downscaling over South Africa. Int J Climatol 26:1315–1337

Howlett BG, Nelson WR, Pattemore DE et al (2015) Pollination of macadamia: review and opportunities for improving yields. Sci Hortic 197:411–419

Howlett BG, Read SFJ, Alavi M et al (2019) Cross-pollination enhances Macadamia yields, even with branch-level resource limitation. HortScience 54:609–615

Huett DO (2004) Macadamia physiology review: a canopy light response study and literature review. Aust J Agric Res 55:609–624

Ibraimo N, Taylor N, Ghezehei S, Gush M, Annandale J (2014) Water use of macadamia orchards. In: Gush M, Taylor N (eds) The water use of selected fruit tree orchards, vol 2: Technical report on measurements and modelling. WRC report no. 1770/2/14, Water Research Commission, Pretoria

IPCC (2021) Climate change 2021: the physical science basis. In: Masson-Delmotte V, Zhai P, Pirani A et al (eds) Contribution of working group I to the sixth assessment report of the Intergovernmental Panel on Climate Change. Cambridge University Press, Cambridge. https://doi:10.1017/9781009157896

Ito PJ, Hamilton RA (1980) Quality and yield of "Keauhou" macadamia nuts from mixed and pure block plantings. HortScience 15:307

Joseph GS, Muluvhahothe MM, Seymour CL et al (2019) Stability of Afromontane ant diversity decreases across an elevation gradient. GECCO 17. https://doi.org/10.1016/j.gecco.2019.e00596

Kämper W, Wallace HM, Ogbourne SM et al (2020) Dependence on cross-pollination in macadamia and challenges for orchard management. Proceedings 36. https://doi.org/10.3390/proceedings2019036076

Kämper W, Trueman SJ, Ogbourne SM, Wallace HM (2021) Pollination services in a macadamia cultivar depend on across-orchard transport of cross pollen. J Appl Ecol. https://doi.org/10.1111/1365-2664.14002

Kelly RM, Kitzes J, Wilson H, Merenlender A (2016) Habitat diversity promotes bat activity in a vineyard landscape. Agric Ecosyst Environ 223:175–181. https://doi.org/10.1016/j.agee.2016.03.010

Kirk WDJ (1987) How much pollen can thrips destroy? Ecol Entomol 12:31–40

Kremen C, Williams NM, Aizen MA et al (2007) Pollination and other ecosystem services produced by mobile organisms: a conceptual framework for the effects of land-use change. Ecol Lett 10:299–314

La Croix EAS, Thindwa HZ (1986) Macadamia pests in Malawi. III. The major pests. The biology of bugs and borers. Trop Pest Manag 32(1):11–20

Langdon KS, King GJ, Nock CJ (2019) DNA paternity testing indicates unexpectedly high levels of self-fertilisation in macadamia. Tree Genet Genomes 15:29

Lee P (2020) Why irrigate macadamias? The Macadamia, Winter 2020 edition

Li Y, Miao R, Khanna M (2020) Neonicotinoids and decline in bird biodiversity in the United States. Nat Sustain 3(12):1027–1035

Linden VMG (2019) How vertebrate communities affect quality and yield of macadamia farms in L Levubu, South Africa. Dissertation, University of Venda

Linden VMG, Grass I, Joubert E et al (2019) Ecosystem services and disservices by birds, bats and monkeys change with macadamia landscape heterogeneity. J Appl Ecol. https://doi.org/10.1111/1365-2664.13424

Lloyd J (1991) Modelling stomatal responses to environment in macadamia integrifolia. Funct Plant Biol 18(6):661–671

López-Hoffman L, Wiederholt R, Sansone C et al (2014) Market forces and technological substitutes cause fluctuations in the value of bat pest-control services for cotton. PLoS One. https://doi.org/10.1371/journal.pone.0087912

Makungo R, Odiyo JO (2017) Estimating groundwater levels using system identification models in Nzhelele and Luvuvhu areas, Limpopo province, South Africa. Phys Chem Earth 100:44–50

McCracken GF, Gillam EH, Westbrook JK et al (2008) Brazilian free-tailed bats (*Tadarida brasiliensis*: Molossidae, Chiroptera) at high altitude: links to migratory insect populations. Integr Comp Biol 48:107–118

McCracken GF, Westbrook JK, Brown VA et al (2012) Bats track and exploit changes in insect pest populations. PLoS One 7(8). https://doi.org/10.1371/journal.pone.0043839

McFadyen LM, Robertson D, Sedgley M et al (2011) Post-pruning shoot growth increases fruit abscission and reduces stem carbohydrates and yield in macadamia. Ann Bot 107:993–1001

McFadyen L, Robertson D, Sedgley M et al (2012) Time of pruning affects fruit abscission, stem carbohydrates and yield of macadamia. Funct Plant Biol 39:481

Meyer S (2016) Effects of spatial arrangement of bee hives and landscape context on pollination of Macadamia in South Africa. Master's thesis, Georg-August-Universität Göttingen, Göttingen

Millennium Assessment Board (2005) Millennium ecosystem assessment. New Island Press, Washington DC

Mogoatlhe L (2020) South Africa repeals state of disaster for drought. Here's why it's a 'grave concern' for farmers https://wwwglobalcitizenorg/en/content/south-africa-drought-national-crisis-farmers/. Accessed 23 May 2021

Monadjem A, Taylor PJ, Cotterill FDP et al (2020) Bats of southern and Central Africa: a biogeographic and taxonomic synthesis. Wits University Press, Johannesburg

Moncur MW, Stephenson RA, Trochoulias T (1985) Floral development of macadamia integrifolia Maiden & Betche under Australian conditions. Sci Hortic 27:87–96

Mound LA (2005) THYSANOPTERA: diversity and interactions. Annu Rev Entomol 50:247–269

Murovhi N (2003) Irrigation. In: De Villiers EA, Joubert PH (eds) The cultivation of Macadamia. ARC-Institute for Tropical and Subtropical Crops, Nelspruit

Nagao MA, Hirae HH, Stephenson RA (1992) Macadamia: cultivation and physiology. Crit Rev Plant Sci 10(5):441–470

Nortje GP, Schoeman S (2016) Biology and management of stink bugs in Southern African Macadamia orchards - current knowledge and recommendations. Working paper. https://www.researchgate.net/publication/319351182. Accessed 2 May 2021

Oliveira JM, Destro ALF, Freitas MB et al (2021) How do pesticides affect bats?–a brief review of recent publications. Braz J Biol 81(2):499–507

Orgaz F, Villalobos FJ, Testi L, Fereres E (2007) A model of daily mean canopy conductance for calculating transpiration of olive canopies. Funct Plant Biol 34(3):178–188

Pal E, Hurley B, Slippers B, Fourie G (2020) Progress towards the characterisation of pheromones of the two-spotted stinkbug (Bathycoelia distincta). SAMAC, September 2020. https://www.fabinet.up.ac.za/publication/pdfs/4059-pal.2020.pdf

Puig-Montserrat X, Torre I, Lopez-Baucells A et al (2015) Pest control service provided by bats in Mediterranean rice paddies: linking agroecosystems structure to ecological functions. Mamm Biol 80:237e245

Ramotjiki ML (2020) Does observational methods affect the observed impacts of exotic plants on flower visitors in around macadamia orchards. http://hdl.handle.net/11602/1661

Ramulifho P, Ndou E, Thifhulufhelwi R et al (2019) Challenges to implementing an environmental flow regime in the Luvuvhu river catchment, South Africa. Int J Environ Res Public Health 16(19):3694

Ramulifho PA, Rivers-Moore NA, Foord SH (2021) Loss of intermediate flow states only evident when considering sub-daily flow metrics in a major tributary in the Limpopo basin. Ecohydrology. https://doi.org/10.1002/eco.238

Richards TE, Kämper W, Trueman SJ, Wallace HM, Ogbourne SM, Brooks PR, Nichols J, Hosseini Bai S (2020) Relationships between nut size, kernel quality, nutritional composition and levels of outcrossing in three macadamia cultivars. Plan Theory 9:228. https://doi.org/10.3390/plants9020228

Richardson AC, Dawson TE (1993) Enhancing abscission of mature macadamia nuts with ethephon. N Z J Crop Hortic Sci 21:325–329

Ricketts TH, Regetz J, Steffan-Dewenter I et al (2008) Landscape effects on crop pollination services: are there general patterns? Ecol Lett 11:499–515. https://doi.org/10.1111/j.1461-0248.2008.01157.x

Russo D, Bosso L, Ancillotto L (2018) Novel perspectives on bat insectivory highlight the value of this ecosystem service in farmland: research frontiers and management implications. Agric Ecosyst Environ 266:31–38. https://doi.org/10.1016/j.agee.2018.07.024

Sakai WS, Nagao MA (1985) Fruit growth and abscission in macadamia integrifolia. Physiol Plant 64:455–460

SAMAC (2020) Industry statistics. https://www.samac.org.za/industry-statistics. Accessed 12 Mar 2021

SAMAC (2021) Crop forecast 2021. https://www.samac.org.za/industry-statistics/. Accessed 25 May 2021

da Santos RS, de Milfont MO, Silva MM, Carneiro LT, Castro CC (2020) Butterflies provide pollination services to macadamia in northeastern Brazil. Sci Hortic 259:108818. https://doi.org/10.1016/j.scienta.2019.108818

Schoeman PS (2009) Key biotic components of the indigenous Tortricidae and Heteroptera complexes occurring on Macadamia in South Africa. Dissertation, North West University, Potchefstroom

Schoeman PS (2012) Macadamia scouting. https://www.samac.org.za/wp-content/uploads/2016/08/macadamia-scouting_prelim.pdf. Accessed 16 Oct 2016

Schoeman PS (2014) Aspects affecting distribution and dispersal of the indigenous Heteroptera complex (Heteroptera: Pentatomidae & Coreidae) in South African macadamia orchards. Afr Entomol 22(1):191–196

Schoeman PS (2018) Relative seasonal occurrence of economically significant heteropterans (Pentatomidae and Coreidae) on macadamias in South Africa: implications for management. Afr Entomol 26(2):543–549

Schoeman PS, De Villiers EA (2015) Macadamia. In: Prinsloo GL, Uys VM (eds) Insects of cultivated plants and natural pastures in southern Africa. Entomological Society of Southern Africa, Hatfield

Searle C, Lu P (2002) Optimising irrigation scheduling for the production of high quality Macadamia nuts. Horticulture Australia Ltd, Australia. Report No. MC98019

Searle C, Lu P (2003) Whole–tree water use and irrigation scheduling in macadamias. Proceedings of the Second International Macadamia Symposium, Tweed Heads, Queensland

Sedgley M (1983) Pollen tube growth in macadamia. Sci Hortic 18:333–341

Shabalala M, Toucher M, Clulow A (2022) The macadamia bloom – what are the hydrological implications? Sci Hortic 292

Shapiro JT, Monadjem A, Röder T et al (2020) Response of bat activity to land cover and land use in savannas is scale-, season-, and guild-specific. Biol Conserv 241. https://doi.org/10.1016/j.biocon.2019.108245

Shigeura GT, Ooka H (1984) Macadamia nuts in Hawaii: history and production. Res Ext Ser 39:95

Sirami C, Jacobs DS, Cumming GS (2013) Artificial wetlands and surrounding habitats provide important foraging habitat for bats in agricultural landscapes in the Western Cape, South Africa. Biol Conserv 164:30–38

Smit TG, Taylor NJ, Midgley SJE (2020) The seasonal regulation of gas exchange and water relations of field grown macadamia. Sci Hortic 267. https://doi.org/10.1016/j.scienta.2020.109346

Stanton RL, Morrissey CA, Clark RG (2018) Analysis of trends and agricultural drivers of farmland bird declines in North America: a review. Agric Ecosyst Environ 254:244–254

Stephenson RA, Gallagher EC, Doogan VJ (2003) Macadamia responses to mild water stress at different phenological stages. Aust J Agric Res 54:67–75

Steyn JN (2019) Alternative practices for optimising soil quality and crop protection for macadamia orchards, Limpopo Province, South Africa. Dissertation, University of Venda

Tavares JM, Villalobos EM, Wright MG (2015) Contribution of insect pollination to macadamia integrifolia production in Hawaii. Proc Hawaii Entomol Soc 47:35–49

Taylor PJ, Sowler S, Schoeman MC et al (2013) Diversity of bats in the Soutpansberg and Blouberg mountains of northern South Africa: complementarity of acoustic and nonacoustic survey methods. S Afr J Wildl Res 43:12–26

Taylor PJ, Munyai A, Gaigher I et al (2015) Afromontane small mammals do not follow the hump-shaped rule: elevational variation in a tropical biodiversity hotspot (Soutpansberg Mountains, South Africa). J Trop Ecol 31:37–48

Taylor PJ, Grass I, Alberts AJ et al (2018) Economic value of bat predation services – a review and new estimates from macadamia orchards. Ecosyst Serv 30:372–381

Taylor NJ, Smit T, Smit A, Midgley SJE, Clulow A, Annandale JG, Dlamini K, Roets N (2021) Water use of Macadamia orchards, vol 2. Report to the Water Research Commission and macadamias South Africa NPC, WRC report no. 2552/2/21, Pretoria, South Africa

Tilman D, Fargione J, Wolff B et al (2001) Forecasting agriculturally driven global environmental change. Science 292:281–284

Torquetti CG, Guimarães ATB, Soto-Blanco B (2021) Exposure to pesticides in bats. Science of the Total Environment 755:142509

Trochoulias T, Johns G (1992) Poor response of macadamia (macadamia integrifolia Maiden and Betche) to irrigation in a high rainfall area of subtropical Australia. Aust J Exp Agric 32(4):507–512

Trueman SJ (2010) Benzyladenine delays immature fruit abscission but does not affect final fruit set or kernel size of Macadamia. Afr J Agric Res 5:1523–1530

Trueman SJ (2013) The reproductive biology of macadamia. Sci Hortic 150:354–359

Trueman SJ, Turnbull CGN (1994a) Effects of cross-pollination and flower removal on fruit set in macadamia. Ann Bot 73.23–32

Trueman SJ, Turnbull CGN (1994b) Fruit set, abscission and dry matter accumulation on girdled branches of macadamia. Ann Bot 74:667–674

Trueman SJ, Kämper W, Nichols J, Ogbourne SM, Hawkes D, Peters T, Hosseini Bai S, Wallace HM (2022) Pollen limitation and xenia effects in a cultivated mass-flowering tree, macadamia integrifolia (Proteaceae). Ann Bot 129(2):135–146

Tscharntke T, Clough Y, Wanger TC et al (2012) Global food security, biodiversity conservation and the future of agricultural intensification. Biol Conserv 151(1):53–59

UNESCO (2010) MAB Biosphere Reserves Directory, Biosphere Reserve Information, South Africa, Vhembe. http://www.unesco.org. Accessed 12 May 2021

Urata U (1954) Pollination requirements of macadamia. Hawaii Agricultural Experiment Station, University of Hawaii, Technical Bulletin 22

Van Wyk AE, Smith GF (2001) Regions of floristic endemism in southern Africa: a review with emphasis on succulents. Umdaus Press, Pretoria

Villalobos FJ, Testi L, Orgaz F et al (2013) Modelling canopy conductance and transpiration of fruit trees in Mediterranean areas: a simplified approach. Agric For Meteorol 171–172:93–103

Vithanage V, Ironside DA (1986) The insect pollinators of macadamia and their relative importance. J Aust Inst Agric Sci 52:155–160

Voigt C, Kingston T (2016) Bats in the Anthropocene: conservation of bats in a changing world. Springer International Publishing, Cham

Wallace HM, Vithanage V, Exley EM (1996) The effect of supplementary pollination on nut set of macadamia (Proteaceae). Ann Bot 78:765–773

Wangai PW, Burkhard B, Müller F (2016) A review of studies on ecosystem services in Africa. J Sustain Built Environ 5(2):225–245

Wanger TC, Darras K, Bumrungsri S et al (2014) Bat pest control contributes to food security in Thailand. Biol Conserv 171:220–223

Weier SM, Grass I, Linden VMG et al (2018) Natural vegetation availability and bugs promote insectivorous bat activity in macadamia orchards, South Africa. Biol Conserv. https://doi.org/10.1016/j.biocon.2018.07.017

Weier SM, Moodley Y, Fraser MF et al (2019a) Insect pest consumption by bats in macadamia orchards established by molecular diet analyses. GECCO. https://doi.org/10.1016/j.gecco.2019.e00626

Weier SM, Linden VMG, Grass I et al (2019b) The use of bat houses as day roosts in macadamia orchards, South Africa. Peer J. https://doi.org/10.7717/peerj.6954

Weier SM, Linden VMG, Hammer A et al (2021) Bat guilds respond differently to habitat loss and fragmentation at different scales in macadamia orchards in South Africa. Agric Ecosyst Environ 320. https://doi.org/10.1016/j.agee.2021.107588

Wickramasinghe LP, Harris S, Jones G, Vaughan N (2003) Bat activity and species richness on organic and conventional farms: impact of agricultural intensification. J Appl Ecol 40:984–993. https://doi.org/10.1111/j.1365-2664.2003.00856.x

Wordley CFR, Sankaran M, Mudappa D, Altringham JD (2017) Bats in the Ghats: agricultural intensification reduces functional diversity and increases trait filtering in a biodiversity hotspot in India. Biol Conserv 210:48–55. https://doi.org/10.1016/j.biocon.2017.03.026

Potential of Improved Technologies to Enhance Land Management Practices of Small-Scale Farmers in Limpopo Province, South Africa

23

Jan-Henning Feil, Reimund P. Rötter ⓘ, Sara Yazdan Bakhsh,
William C. D. Nelson, Bernhard Dalheimer, Quang Dung Lam,
Nicole Costa Resende Ferreira, Jude Odhiambo,
Gennady Bracho-Mujica, Issaka Abdulai, Munir Hoffmann,
Bernhard Bruemmer, and Kingsley Kwabena Ayisi ⓘ

J.-H. Feil (✉)
Department of Agriculture, South Westphalia University of Applied Sciences, Soest, Germany
e-mail: feil.jan-henning@fh-swf.de

R. P. Rötter · W. C. D. Nelson · Q. D. Lam · N. C. R. Ferreira · G. Bracho-Mujica · I. Abdulai
Tropical Plant Production and Agricultural Systems Modelling (TROPAGS), University of
Göttingen, Göttingen, Germany

S. Y. Bakhsh · B. Dalheimer
Department of Agricultural Economics and Rural Development, University of Göttingen,
Göttingen, Germany

J. Odhiambo
University of Venda, School of Agriculture, Thohoyandou, South Africa

M. Hoffmann
AGVOLUTION GmbH, Göttingen, Germany

B. Bruemmer
Department of Agricultural Economics and Rural Development, University of Göttingen,
Göttingen, Germany

Centre of Biodiversity and Sustainable Land Use (CBL), University of Göttingen, Göttingen,
Germany

K. K. Ayisi
University of Limpopo, Risk and Vulnerability Science Center, Polokwane, South Africa

© The Author(s) 2024
G. P. von Maltitz et al. (eds.), *Sustainability of Southern African Ecosystems
under Global Change*, Ecological Studies 248,
https://doi.org/10.1007/978-3-031-10948-5_23

Abstract

In this chapter, we explore how, in the face of increasing climatic risks and resource limitations, improved agro-technologies can support sustainable intensification (SI) in small-scale farming systems in Limpopo province, South Africa. Limpopo exhibits high agro-ecological diversity and, at the same time, is one of the regions with the highest degree of poverty and food insecurity in South Africa. In this setting, we analyze the effects of different technology changes on both food security dimensions (i.e., supply, stability, and access) and quality of ecosystem service provision. This is conducted by applying a mixed-method approach combining small-scale farmer survey data, on-farm agronomic sampling, crop growth simulations, and socioeconomic modeling. Results for a few simple technology changes show that both food security and ecosystem service provision can be considerably improved when combining specific technologies in a proper way. Furthermore, such new "technology packages" tailored to local conditions are economically beneficial at farm level as compared to the status quo. One example is the combination of judicious fertilizer application with deficit or full irrigation in small-scale maize-based farming systems. Provided comparable conditions, the results could be also beneficial for decision-makers in other southern African countries.

23.1 Introduction

23.1.1 Background and Motivation

Southern Africa has been identified as a hotspot for global change processes and biodiversity, whereby agricultural expansion is regarded as a key driving force for the declining species diversity (IPBES 2018). The projected doubling of the African human population by 2050 (as compared to 2010) and the climate change-induced increased frequency of extreme droughts underline the urgency of science-informed assessments in support of identifying sustainable land management options (IPCC 2019; Rötter et al. 2021). About 70% of the population of southern Africa relies on agriculture. Most of them are smallholders, of which about 94% depend on rainfed agriculture. Around 16% of the rural population has been characterized as "food insecure" during the last 5 years (Sikora et al. 2020).

In recent years, there has been growing attention and support for innovation for supporting sustainable agricultural production (e.g., Herrero et al. 2020) such as agro-ecology, sustainable intensification (SI), and climate-smart agriculture (Cassman and Grassini 2020; FAO 2010; Kuyah et al. 2021; Wilkus et al. 2021). Yet, there is some debate about which approaches should be applied in which contexts and to whose benefits. Site- and season-specific, knowledge-intensive agricultural management practices combined with advanced breeding tools hold promise to increase resource use efficiencies and crop performance considerably (e.g., Hoffmann et al. 2018; Hammer et al. 2020) and can be supported by digital technologies (Herrero et al. 2020; Von Braun et al. 2021).

There exist a considerable number of socioeconomic constraints that need to be overcome to create a fertile ground so that technological innovations can unfold (Gatzweiler and von Braun 2016). Economic weaknesses have been sharpened since the start of the COVID-19 pandemic with considerable negative consequences for food security (Savary et al. 2020). Limited access to land, water, other resources and markets for smallholder farmers has negatively affected rural livelihoods. Moreover, recent shifts of a considerable proportion of agricultural production and land use away from human food-related activities toward animal feed, timber, and biofuels in some regions have presented trade-offs between food security and energy needs. In other regions, land use change from agriculture toward mining, nature conservation, or settlements has reduced the agricultural production area.

Southern African savanna landscapes are composed of arable land, rangelands, and orchards/homegardens (Rötter et al. 2021). In this chapter, we focus on the potential of technological improvements on crop productivity and rural livelihoods of small-scale farmers who largely depend on the ecosystem services (ES) these three major land use types provide. Small-scale farmers in the region are highly diverse in terms of resource endowments such as land and water. The generally huge yield gaps (with yield levels at 20% of the potential), food insecurity, and shrinking land holdings call for radical changes in land use policies and management to avoid societal unrest growing in the future. In national plans on sustainable development, sustainable intensification (Cassman and Grassini 2020) of these systems, not surprisingly, has the highest policy priority (Sikora et al. 2020). It is seen as an important means to provide incentives to the younger farmer generation, boost agricultural development, and to set land aside for nature conservation.

A broad range of management interventions has been suggested for promoting sustainable intensification (e.g., Kuyah et al. 2021; Vanlauwe and Dobermann 2020; Wilkus et al. 2021), including cereal intercropping with legumes, conservation agriculture, agroforestry, site-specific fertilizer application and irrigation. Most experimental studies on testing such interventions have just looked at impacts on dry matter production and yield, but a few also looked at other ecosystem functions such as carbon sequestration and water and nutrient use efficiency. Yet, to date, no study has looked in an integrated manner at the complexity of smallholder systems with a broad range of interacting ES at the landscape level. The SALLnet project has that ambition (Rötter et al. 2021), and here we present a few of the results of such integrated analyses across different scale levels, from field via farm to landscape level.

23.1.2 Problem Statement and Objectives of the Chapter

A key question for many smallholder-dominated agricultural landscapes in southern Africa is: "how can the multiple Ecosystem Services (ES) be enhanced through sustainable land management interventions and enabling policies?" (Rötter et al. 2021).

Limpopo province in the northern-most corner of South Africa combines most of the global change threats, and is also featuring several of the typical land use changes that have been observed across southern Africa over the last few decades (Chap. 20). High population growth, severe land degradation, and high climate variability in conjunction with low agricultural productivity and poverty have already led to a decline of essential ES in Limpopo province such as provision of food and feed, nutrient cycling, and habitat quality (Hoffmann et al. 2020; Pfeiffer et al. 2019).

Against this background, the present chapter aims at investigating how agro-technology improvements could support SI in small-scale farming systems in Limpopo province, South Africa. To do this, we suggest an integrated crop model APSIM (for applications in Africa, see, e.g., Whitbread et al. 2010) and downstream socioeconomic modeling by means of agent-based modeling. These models are calibrated to Limpopo province by using data from small-scale farmer surveys, on-farm agronomic sampling, and long-term crop experiments. To demonstrate the respective impact analysis of agro-technology improvements within our modeling framework, we use the example of improved soil nutrient and irrigation practices in combination. We also discuss further agricultural technology improvements and innovations that could be likewise assessed going forward.

The remainder of the chapter is structured as follows: To provide a basis for our analysis, an overview of the small-scale farming sector in Limpopo province is presented in Sect. 23.2 by using the results of a large-scale survey conducted within the course of the SALLnet project. Section 23.3 then analyzes the current yield gaps and resource use efficiencies in small-scale farming systems in Limpopo province, on which basis improvements shall be worked out in the following. Accordingly, Sect. 23.4 first provides an overview of the methods and tools to be used to analyze the impacts of different technologies and innovations on small-scale farming systems. Subsequently, these are calibrated to the study region in Sect. 23.5. Finally, Sect. 23.6 provides a summary of the main findings and draws some implications.

23.2 Farm Household Characteristics: Small-Scale Subsistence Versus Emerging Farmers

Limpopo is one of the least developed provinces in South Africa and currently experiencing both strong population growth and a high poverty rate. A large share of the population (89%) is living in rural areas, and farming is the main occupation (Gyekye and Akinboade 2003; LDARD 2012). In order to understand the structure of the smallholder farming sector in Limpopo province of South Africa, five study areas were selected from Limpopo based on differences in climatic aridity, demography, and socioeconomic factors. The selected sites are located in rural areas of the Mopani district: Mafarana, Gavaza, Ga-Selwana, Makushane, and Ndengeza. A comprehensive small-scale farmer survey was conducted between April and July 2019 after pretesting in selected villages; the interviews were conducted in person with farm household heads or individuals who are responsible for farm management. Permission to access farmers was obtained from tribal authorities of each village. The purpose of the survey was to collect information on socioeconomic, demographic, farm, and household characteristics as well as information on resource endowment and agricultural activities during the 2018/2019 crop season. Using a purposive random sampling procedure, data were collected from 215 farm households across the five villages in Limpopo, of which three households had to be excluded afterwards due to incomplete information. Therefore, the final data set for the following analysis covered 212 households.

Table 23.1 presents a summary of selected descriptive statistics regarding crucial farmer and farm characteristics, including farm performance, resource management, socioeconomics, as well as external incentives (e.g., agricultural extension services, access to credits and markets). Accordingly, we found that the average farm household in the survey sample has a household head who is on average 66 years old. The share of female-headed households was the same as the national general household survey in 2019 with 48.8% (Statistics South Africa 2019). The average household in the survey owned 4.4 ha land, of which 70% is left fallow during the winter (dry season). We found considerable variation in farm size, especially regarding their cultivated area. In terms of production systems, the small-scale farms in the sample were mainly characterized by mixed crop-livestock production. Our survey showed that maize (*Zea mays* L.) is most important to ensure household food security and cultivated by nearly all farms. The secondary major crops are legumes such as peanuts (*Arachis hypogaea*), Bambara nuts (*Vigna subterranea* L.), and cowpea (*Vigna unguiculata*) which are produced by 59% of farms. Horticultural crops such as fruits (e.g., mango, banana) are grown by 32%, and vegetables (e.g., tomato, onion, cabbage, paprika) by 15% of the surveyed farms. Maize and legumes were mainly grown for household consumption but vegetables contributed to both household consumption and income generation. With regard to livestock farming, cattle provided the main source of livestock income while farms also kept goats, pigs as well as chickens. On average, 41% of agricultural income stemmed from crop sales and 25% from livestock sales. Moreover, the degree of commercialization for crops was 39% and for livestock was 6%, indicating the proportion of selling value of the total value of the production, based on market prices in 2019.

Agricultural products were mainly traded on informal on-farm markets (58%). Only 17% of the local farmers had access to formal off-farm markets. Social grants including old age and child support grants played an important role on farm household incomes for most smallholders. According to Statistics South Africa (2019), around 59% of the households received grants as their main sources of income in Limpopo. Direct support from the government as well as extension services mainly occurred in the form of input supplies, mechanization, livestock health services, and training. The number of visits of extension services on average was 1.32 times in a year. The field preparation was usually carried out by a rented tractor or donkey. Nevertheless, among these farmers, only 6% had their own private

Table 23.1 Descriptive statistics of the small-scale farmer survey conducted in Limpopo in 2019

Variable (Scale/measurement)	Mean (SD)
Farmer	
Age (number of years)	66.45 (11.19)
Gender (1 = male; 0 = female)	0.52 (0.50)
Education (number of years)	4.76 (5.04)
Off-farm job of the farmer (1 = yes; 0 = no)	0.22 (0.41)
Social grant income (in Rand)	26,689.8 (15,308.7)
Remittance income (in Rand)	4168.3 (12,026.5)
Risk attitude (Likert scale: 1: Highly risk averse–10: Highly risk seeking)	4.29 (2.85)
Farm	
Total area of the farm (hectares)	4.44 (6.13)
Total area under cultivation (hectares)	3.02 (3.33)
Share of fallow area in winter (share: 0–1)	0.70 (0.43)
Number of crops cultivated in winter (numbers)	0.25 (0.72)
Cultivating vegetables (1 = yes; 0 = no)	0.15 (0.36)
Cultivating fruits (1 = yes; 0 = no)	0.32 (0.47)
Cultivating legumes (1 = yes; 0 = no)	0.59 (0.49)
Share of sale value crops to total value crops cultivated (share: 0–1)	0.40 (0.41)
Share of sale value animals to total value of animals (share: 0–1)	0.06 (0.13)
Having animal (1 = yes; 0 = no)	0.58 (0.49)
Number of cattle (number)	4.6 (9.4)
Income of selling crops and animals (in Rand)	25,137.9 (121,098)
Crop share of total on-farm income (in Rand)	0.41 (0.46)
Animal share of total on-farm income (in Rand)	0.25 (0.40)

Variable (Scale/measurement)	Mean (SD)
Resource management and external incentives	
Having tractor (1 = yes; 0 = no)	0.06 (0.23)
Water source (1 = yes; 0 = no)	
Depends on rain	0.34 (0.47)
Tap water	0.41 (0.49)
Public dam, lake	0.09 (0.29)
Private borehole	0.16 (0.36)
Hours of irrigation in year (hours)	91.56 (310.50)
Methods of irrigation(1 = yes; 0 = no)	
No irrigation	0.34 (0.47)
Primitive irrigation method	0.49 (0.50)
Advanced irrigation method	0.16 (0.36)
Applying pesticide on farm (1 = yes; 0 = no)	0.14 (0.34)
Applying fertilizer on farm (1 = yes; 0 = no)	0.31 (0.46)
Number of hired permanent worker in year (man-day numbers)	48.50 (255.60)
Number of hired seasonal worker in year (man-day numbers)	17.33 (59.36)
Selling at farm (1 = yes; 0 = no)	0.58 (0.49)
Selling at market (1 = yes; 0 = no)	0.17 (0.38)
Access to credits (1 = yes; 0 = no)	0.10 (0.30)
Investment in the past 5 years (1 = yes; 0 = no)	0.37 (0.48)
Number of visits of extension services (number of visits)	1.32 (4.35)

tractor. In this respect, merely 10% of the respondents had access to formal credits but 37% of the farmers invested in the last 5 years mainly in equipment for irrigation, fences, and machinery. Besides household members working on their own farms, the permanent and seasonal employed labor worked amounted on average to 48.5 and 17.33 man-days per year (1 man-day = 8 h/person). Regarding irrigation, the most common source of water was tap water (41%) which was usually only available in the home garden next to their residential building. 34% of the sample was purely rain-dependent, while on average 9% and 16% of farmers had access to public water sources and private boreholes. Hence, 49% of the sample used primitive irrigation methods (e.g., buckets, furrows).

According to the collected information described above, the smallholder farmers in the sample were found to be highly heterogeneous in terms of farm and farmer characteristics, resource management as well as external incentives such as agricultural extension services, access to credits, and markets. Moreover, the heterogeneous groups of smallholder farmers were reliant on different forms of government interventions and agricultural policies, depending on the objective and characteristics of each group.

23.3 Yield Gaps and Current Resource Use Efficiencies in Small-Scale Farming Systems

In Sect. 23.3.1, we give a brief account of the different yield gaps (see Kassie et al. 2014) as well as of current water and nitrogen efficiencies for maize cultivated by smallholders. Based on literature and simulation results, we show the scope for closing or narrowing down the yield gap between actual farmer's yields and potential yields that could be attained under irrigated or rainfed conditions with best management. Next, we look at the efficiency gains that might be obtained by distinct management interventions—restricting ourselves to water and nutrient management, whereby in the latter with focus on the macro-nutrients nitrogen (N) and phosphorus (P). Furthermore, Sect. 23.3.2 presents the results of efficiency analysis of current maize-based small-scale farming systems in five villages in Mopani district, Limpopo province. As a consequence of this, we discuss the potentials of a number of alternative management interventions in enhancing farm income and other ecosystem services in Sect. 23.3.3.

23.3.1 Current Resource Use Efficiencies for Different Small-Scale Farming Systems in the Region

Regarding crop production in South Africa, maize is the major staple crop and mostly grown by smallholders under rainfed conditions. The yield of maize in the study regions within Limpopo province is fairly low—for small-scale farmers ranging between 1 t ha^{-1} and 2 t ha^{-1}. This is mainly due to manual farming techniques together with low input provision such as no or little fertilization, lack of

Table 23.2 Average water use efficiency under different treatments (in kg grain DM yield/m^3 water)

Treatments	Rainfed cultivation	Deficit irrigation (100 mm)	Full irrigation (200 mm)
Applied amount (kg/ha)			
0N: 0P	0.35	0.39	0.31
10N: 5P	0.41	0.47	0.35
40N: 30P	0.71	0.79	0.62
120N: 60P	1.09	1.15	0.95

quality seeds, and no irrigation (FAO 2010). The observed increase in water scarcity and land degradation, and particularly poor soil fertility through nutrient mining in most smallholder farming systems poses a serious threat to crop production in southern Africa (Vlek et al. 2020). Therefore, a logical agricultural intervention measure is to produce more grain yield per volume of water used ("more crop per drop") and replenish the nutrient-depleted soils with mineral and organic fertilizer. Water use efficiency (WUE) needs to be increased through appropriate water, nutrient, and crop management interventions measures to sustainably raise agricultural productivity. In a water management context, the term WUE refers to crop production per unit of water used, with units such as kg grain ha^{-1} mm^{-1} or kg m^{-3} (Sadras et al. 2012).

The simulated WUE under different treatments was derived from simulated maize results for 35 years (1985–2020) at the experimental Syferkuil site within the Limpopo province (Table 23.2). The WUE varied between 0.1 kg m^{-3} and 1.39 kg m^{-3} among treatments and years. Those different, predefined fertilizer treatments in maize production could also be mirrored by the farmer behavior observed within the small-scale farmer survey as presented in the second section. Accordingly, the two lower levels of 0N:0P and 10N:5P were approximately applied by the vast majority of the small-scale farmers within the survey, who were primarily producing for self-subsistence purposes. 40N:30P represents the level used by those small-scale farmers, who already emerged to a more market-oriented role. The highest intensity treatment of 120N:60P again approximates the level of intensity commonly used by commercial farmers in maize production in this region.

The results showed that combined deficit irrigation and the 120N: 60P kg ha^{-1} fertilizer application gave the highest WUE value of 1.15 kg m^{-3} (average over 35 years). This finding is in agreement with the results of Kurwakumire et al. (2014) who found that WUE under rainfed conditions ranged from 0.038 kg m^{-3} to 0.113 kg m^{-3} (control), while it improved from 0.3 kg m^{-3} to 0.8 kg m^{-3} for crops receiving 120N:40P:60K fertilization.

NUE is here defined as the incremental maize yield per applied nutrient. The average NUE ranged from 23.67 to 45.51 kg grain yield/kg nutrient (Table 23.3). The highest NUE was obtained in the case of 40N: 30P (kg ha^{-1}) fertilizer application under rainfed cultivation. The NUE values presented in Table 23.3 are

Table 23.3 Average nutrient use efficiency (NUE) under rainfed cultivation (in kg grain DM yield/kg nutrient)

Nutrient application rate (kg ha⁻¹)	Average grain yield (kg ha⁻¹)	NUE (kg/kg)
0N: 0P (control)	1307.4	–
10N: 5P	1544.1	23.67
40N: 30P	3127.6	45.51
120N: 60P	4751.5	28.70

Fig. 23.1 Box plots of simulated yield among the treatments at the experimental Syferkuil location. Yields are simulated from 1985–2020 (R, D, and F stand for rainfed, deficit, and full irrigation, respectively; N0, N10, N40, and N120 stand for the following N:P combinations: 0N: 0P; 10N: 5P; 40N: 30P; and 120N: 60P, kg ha⁻¹)

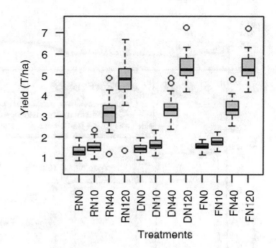

within the range of values for a similar study reported by Ngome et al. (2013) for smallholder farmers on three dominant soil types of Western Kenya. The values are also largely in agreement with findings from a national soil fertility program in Kenya tailored to the needs of smallholder farmers in agro-ecological zones with medium to high agricultural potential (Smaling et al. 1992). In an early review, van Duivenboden (1992) found for West Africa similar NUE values as presented in Table 23.3.

The simulated maize yields presented in Fig. 23.1 show that maize yield increased with increases in the rate of fertilizer and irrigation applications at the experimental Syferkuil site. However, the increase in fertilizer rates had a stronger effect on maize yield improvement than the irrigation treatment. Increasing the N:P fertilization rate alone increased average annual maize yield of RN40 to 3.12 t ha⁻¹ (i.e., 144%) and FN120 to 4.75 t ha⁻¹ (i.e., 271%) compared to that of RN0 which was 1.3 t ha⁻¹, respectively. Considering irrigation alone, treatments of DN0 and FN0 increased maize yields only by 10.5% and 21.9%, respectively, compared to RN0. The combination of fertilization and irrigation application gave the highest average annual yields for the DN120 and FN120 treatments, achieving 5.4 t ha⁻¹ and 5.7 t ha⁻¹, respectively. Note, that in these treatments it is further assumed that other nutrients are always in ample supply and the crop is well protected from pests and diseases.

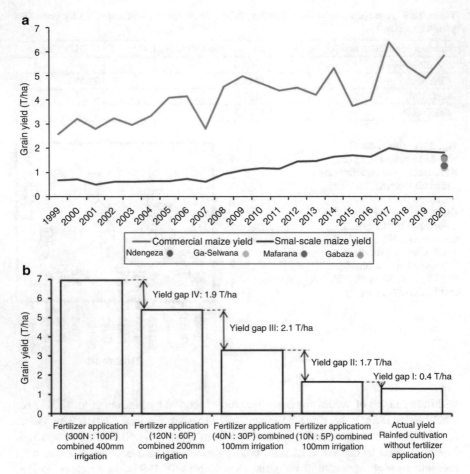

Fig. 23.2 Maize yield gap between commercial and small-scale farmers—averaged for South Africa—colored dots show farmers yields in four villages of Mopani district in 2020 (**a**); Simulated average maize yields under different treatments in the Syferkuil, SA for period 1999–2020 (**b**) data source maize yield series: https://www.sagis.org.za/historic%20hectares%20and%20production%20info.html

Figure 23.2a illustrates the maize yield gap over the last 30 years between yields actually obtained by large-scale commercial farmers and those of small-scale farmers in South Africa. It shows that although maize yields fluctuated over time for both groups of farmers, the yield trend for commercial farmers significantly increased over the last 30 years. However, the maize yields for small-scale farmers did not exceed the level of 1.5 t ha^{-1}. Based on the simulated maize yields at Syferkuil (Fig. 23.2b), yields for small-scale farmers can be increased considerably by applying different treatments combining fertilizer and irrigation. Yet, the highest yield increases that could be obtained come at very poor water and nutrient use efficiencies—and thus high losses to the environment. The different yield gaps illustrated in Fig. 23.2a, b indicate the most effective interventions for raising yields.

23.3.2 Results of Efficiency Analysis of Current Maize-Based Small-Scale Farming Systems

Maize dominates the South African food system, being both the vital dietary staple crop and feed grain. Therefore, it is the most prevalent agricultural crop for small-scale farmers in rural areas (Obi and Ayodeji 2020). As the majority of small-scale farmers cultivate maize mainly for subsistence purposes, the levels of production and supply of maize from small-scale farming are important indications of food security. Over the past century, South African maize production experienced some significant changes (Greyling and Pardey 2019). Small-scale farmers played an important role in producing maize in the country in previous years.

In addition to environmental stress, agricultural production in general and maize cultivation in particular in southern Africa are both confronted with several macro- and micro structural constraints (Mpandeli and Maponya 2014). Some of these constraints are inefficient policies and extension system support, limited access to agricultural credits which results in reduced investment in the agricultural sector, inadequate infrastructure facilities such as transportation and communication, along with the lack of proper economic incentives (e.g., Baloyi et al. 2012). These constraints result in yield reduction and harvest failures in recent years (Hove and Kambanje 2019), which exacerbates food insecurity and poverty, especially among smallholder farmers who often practice subsistence farming.

Despite government support and various strategies implemented to improve the productivity of the agricultural sector in South Africa in recent years (e.g., FAO 2017), small-scale farmers still perform considerably below their potential production capacity (Baloyi et al. 2012). Improving agricultural productivity and, thus, crop yields can generally be achieved by more efficient use of available farm resource endowments and by adopting new technologies (e.g., Ali et al. 2019).

For agricultural productivity growth, the concept of technical efficiency (TE) plays a major role and is widely discussed in the literature. It provides information on the performance of the farmers and their potential to improve productivity and efficiency by utilizing existing farm resources and technologies. In addition, technical efficiency analysis allows identifying the main factors affecting the efficiency level of farms. According to various studies in the literature, improvement in the efficiency levels of agricultural production, which is the main component of total factor productivity (TFP) growth, can be regarded as key in alleviating food insecurity in developing economies (Ogundari 2014).

To date, to our best knowledge, less attention has been paid to the analysis of efficiencies of small-scale maize farmers in South Africa and in particular the Limpopo province. To address this research gap, the aim of the present analysis is to evaluate the TE of smallholder maize farmers and the potential factors that lead to the deviations from the common production frontier. Accordingly, a parametric efficiency analysis [a single-step Stochastic Frontier Model (SFA)] considering a dual heteroskedastic production frontier is applied to 190 households cross-sectional survey data in 2019 from five selected study areas in the Limpopo province in South Africa. The model is designed to estimate TE of small-scale maize farmers by adopting a holistic approach of analysis that considers production inputs, perceived risks, and socioeconomic/socio-demographic characteristics to examine their joint influences on productivity and efficiency of the maize production among smallholders in the Limpopo province of South Africa. Moreover, by identifying the determinants of TE, the results of this analysis will offer policy implications to improve maize production, as well as to tackle food insecurity and poverty.

Following the model specification tests,[1] a single-step SF model with a Cobb–Douglas production function is implemented to estimate the stochastic frontier production function and the inefficiency function simultaneously using maximum likelihood estimation. Moreover, we considered heteroscedasticity in both error components (u_i and v_i), following the twofold heteroskedastic model of Hadri (1999). This implies that both variances of the inefficiency and idiosyncratic error terms ($\sigma^2_{u_i}$ and $\sigma^2_{v_i}$) are depending on the explanatory variables with the half-normal distribution of inefficiency term $\sigma^2_{u_i}$ and independently normal distribution of $\sigma^2_{v_i}$. For a more detailed description of the methodology, see Yazdan Bakhsh et al. (2022).

Table 23.4 represents the summary of the results of the analysis. The dependent variable is maize output in kilograms. Considered production inputs include area under maize cultivation in ha, quantity of seeds and fertilizers in kg, pesticide in number of applications per year, labor used for maize cultivation in man-days unit, preparation costs (machinery and animal capital) in currency unit (ZAR), as well as the dummy of irrigating the maize fields. Furthermore, for the inefficiency component, the explanatory variables include: age, gender, and education level of household head, off-farm income, social grants, extension service support, credit access, member of the agricultural organizations, access to off-farm markets, having agricultural training, owning cattle, and total cultivated land. There is a wide variation in output and input use between farmers among selected villages. In this regard, the heterogeneity in production technologies is addressed by considering the geographical location of each village in the selected province, with the assumption that technology is homogeneous within the same geographical location but different across them. Therefore, dummies of location were included in the production frontier as a control variable. Since drought and pest were the two key uncertainties that the selected farmers perceived as the main reasons of the maize harvest failure in this year compared to previous years, we considered the dummies of farmers' perceived risks (failure in crop production due to drought and pest) in both

---

[1] These tests were implemented based on the likelihood ratio (LR) tests to specify the appropriate functional form for the production frontier model (Cobb–Douglas vs. transcendental logarithmic (Translog)), as well, to test the technical efficiency and production risks effects.

Table 23.4 Maximum likelihood estimates of the stochastic frontier analysis for maize production of small-scale farmers in Limpopo, including the stochastic frontier function, the technical inefficiency function, and the production variability function

Stochastic frontier		Technical inefficiency		
Variables	Coefficient (robust std. err.)	Variables	Coefficient (robust std. err.)	Marginal effects
Constant	2.91*** (1.09)	Constant	2.28*** (0.91)	
ln_Land	0.34* (0.08)	Age (<60)	−1.81*** (0.72)	−0.41
		Age (>60 and <74)	−0.79*** (0.30)	−0.18
ln_Seed	0.17** (0.10)	Age (>75)	0.25 (−)	0.06
ln_Fertilizer	0.19*** (0.04)	Gender	−0.74*** (0.19)	−0.17
		Education level	0.10 (0.10)	0.02
Pesticide	0.26*** (0.05)	Off-farm income	−0.32 (0.64)	−0.07
ln_Labor	−0.13 (0.10)	Social grants	−1.56*** (0.41)	−0.35
ln_Preparation costs	0.11 (0.08)	Access to credits	−2.94** (1.47)	−0.66
D_Irrigation	0.54* (0.33)	Extension service support	−0.26*** (0.10)	−0.06
		Organization member	−2.51*** (0.80)	−0.56
D_risk_drought	−0.76*** (0.23)	Access to market	−0.92** (0.43)	−0.21
D_risk_pest	−0.81*** (0.21)	Own cattle	−1.03*** (0.40)	−0.23
		Training in agriculture	−3.07* (1.75)	−0.69
D_risk_drought and pest	−1.62*** (0.27)	ln_Cultivated land	1.07** (0.51)	0.24
D_Battese_Fertilizer	0.52*(0.31)	*Production variability (risk)*		
Region (base:Gabaza)		Variables	Coefficient (robust Std. Err.)	Marginal effects
Ga-Selwana	−0.48*** (0.08)	Constant	−0.81***(0.23)	
Mafarana	−0.35*** (0.05)	D_risk_drought	0.75 (0.48)	0.60
Makushaneh	−0.12 (0.08)	D_risk_pest	0.78 (0.60)	0.65
Ndengeza	0.17** (0.08)	D_risk_drought and pest	1.03*** (0.23)	0.90
Number of observations	190			
Wald chi2 (6)	179.24			
Prob > chi2	0.000			
Log (likelihood)	−246.02			

Note: Statistically significant at levels of *0.1, **0.05, and ***0.01

production frontier and idiosyncratic error, investigating their influences on both level and variability of the maize production.

In the following, the results of the three functional components of the analysis, as can be seen in Table 23.4, are briefly described:

Production Function

The results of the stochastic frontier model indicate the main determinants of the productivity level of the respective small-scale farmers in maize production. Since the output and input variables in the Cobb–Douglas production frontier model are in logarithmic form, the estimated coefficients can be directly interpreted as the partial production elasticities. The positive signs of the coefficients for almost all included variables follow the monotonicity property of the production function, whereas labor shows a negative sign, which is however not statistically significant. Irrigation exhibits the greatest coefficient among the all production factors (0.62), indicating that smallholder farmers with access to irrigation were found to have substantially higher maize yields compared to those cultivating under rainfed conditions. Moreover, pesticide application (0.52), maize area (0.34), fertilizer (0.19), and seed (0.17) are also important technological factors. Furthermore, farmers' perceived risks with regard to drought and pest have significantly negative effects on maize productivity.

Technical Inefficiency

The results of the technical inefficiency model indicate the main drivers of inefficiency of small-scale maize farmers. Since the dependent variable in the technical inefficiency part of the model is the technical inefficiency, a negative sign of the coefficients indicates the negative effect on the technical inefficiency, or in other words, a positive effect on technical efficiency (Kumbhakar et al. 2015). Accordingly, the results indicate that gender (male farmer), age of less than 74 years, off-farm income, access to credits, social grants, member of agricultural organizations, extension services supports, access to markets, owning cattle, and agricultural training have positive effects on TE of the small-scale maize farmers. Education shows a negative but not significant effect on efficiency. On the contrary, the overall cultivated acreage of a farmer has a significantly negative effect on TE of smallholder maize farmers.

The estimated coefficients of the technical inefficiency cannot be directly interpreted from the model, due to the non-linear relationship between $E(u_i)$ and each of the explanatory variables. Therefore, the marginal effects are calculated to investigate the magnitude of the exogenous factors on inefficiency (Kumbhakar et al. 2015). Based on the results, having access to credits, being a member of an agricultural organization, and receiving training in agricultural practices significantly reduce production inefficiency within the investigated sample of small-scale farmers. The effect translates into achieving higher and more stable output. Following training in agricultural practices with 69%, having access to credits has a statistically significant effect on technical inefficiency, so that the level of technical inefficiency can be reduced, on average, by 66% with 1% increase in accessibility of

Table 23.5 Estimated technical efficiency (TE) in the selected regions

TE scores	Obs.	Mean	Std. dev.	Min	Max
Technical efficiency (all regions)	190	0.67	0.23	0.07	0.99
TE_Gavaza	24	0.75	0.17	0.41	0.96
TE_Ga-Selwana	44	0.68	0.25	0.07	0.99
TE_Mafarana	27	0.78	0.13	0.43	0.99
TE_Makushane	49	0.58	0.25	0.07	0.97
TE_Ndengeza	46	0.64	0.25	0.13	0.97

credits. This is followed by the factors organization members (56%), social grants (35%) and market access (21%).

Production Variability

The results of the production variability model indicate the determinants of the variability in maize production. Here, the perceived risks for pests and droughts, as well as an interaction term out of both variables were included in the analysis. The results indicate that drought and pests can be seen as the main drivers for production risk in small-scale farming in Limpopo, especially in regard to maize.

Following the estimation of production frontier and technical inefficiency model and identifying the main determinants of technical inefficiency, we additionally investigated the technical efficiency scores of the smallholder maize farmers at both farm and regional level (Table 23.5). The results reveal that the mean estimated technical efficiency is 0.67, indicating that, on average, the smallholder maize farms produce 67% of potential output at given input levels and technology. The mean scores are relatively similar across the selected regions. This result suggests that there is opportunity to improve maize production by using current inputs and technology.

Individual efficiency levels range from 0.07 to 0.99 and vary due to farm-specific characteristics (e.g., financial and agricultural supports, management practices, production specialization, etc.). According to the comparison of the main characteristics of the farmers within the 10% upper and lower efficiency levels, efficiency is strongly influenced by agricultural training, membership in agricultural organization, and number of visits by extension officers. In terms of farm characteristics, the farmers within the 10% upper levels of efficiency have access to more land and cultivated area than the farmers with the 10% lowest efficiency level. Besides maize as their staple crop, the farmers with the 10% highest level of efficiency focus more on vegetables (e.g., chili, green pepper, okra, cabbage, spinach, tomato, and onion) and legumes (e.g., green bean, peanut, cowpeas, and Bambara nuts) and more cattle than the lowest levels of efficiency. Moreover, these farmers apply more endowments such as fertilizer, pesticide, and irrigation on their field.

23.3.3 Potential of Different Types of Technological Improvements

In spite of the fact that climate risks have been shown to be among the main sources of variability and uncertainty in crop production, quantified information is still scarce for sub-Saharan Africa and far from perfect. This concerns information on climate-induced yield variability (Rötter and Van Keulen 1997), climatic yield potential and its variability for a given location, and the magnitude of different types of yield gaps (e.g., Kassie et al. 2014). There is a need for localized yield gap analysis that can indicate the scope for yield increases through well-targeted management interventions and/or introduction of new technologies such as new breeds, micro-dosing of fertilizer, or gravitational drip irrigation.

Yield gap analysis in the first place involves quantifying the differences between simulated potential yield and actual farmers' yield level and identifying those factors most responsible for the yield differences (Van Ittersum et al. 2013). Factors that are most common in causing yield gaps include water stress, nutrient deficiency, pests, and diseases through sub-optimal agronomic management (Hoffmann et al. 2018; Kassie et al. 2014). The benchmarks for calculating yield gaps are yields under optimum management which are potential yield attainable under irrigation (Yp) or yield attainable under water-limited or rainfed conditions (Yw). These benchmarks are most commonly quantified by using crop simulation models (Van Ittersum et al. 2013; Hoffmann et al. 2018). The potential yield is limited by climate conditions (temperature, solar radiation, CO_2 concentration) and plant genetic characteristics. The water-limited yield, which is also known as water-limited yield potential, is additionally influenced by precipitation and soil water characteristics.

Here we illustrate the current low agricultural productivity with a few examples from Limpopo, one of the least developed provinces in South Africa compounded by a high population growth rate and a relatively high poverty level of the rural population. About 89% of Limpopo's population (5.8 million) lives in rural areas with farming as their main occupation. Most smallholder farmers have limited resource endowments and are still producing for subsistence purposes (see Sect. 23.2). These farmers are particularly vulnerable to climate variability and other environmental stresses.

While there is some debate surrounding the meaning of Sustainable Intensification (SI) (Garnett et al. 2013), few would refute the necessity of yield increases for SI under typical smallholder conditions. Regardless of the socioeconomic scenario, few farmers would adopt technological innovations without thorough explanations or demonstration, and most will expect immediate benefits (Vanlauwe and Dobermann 2020). The spatial and time-wise variability, which is characteristic of the systems in question, mean that uniform "best practice" is usually not a solution (Rurinda et al. 2020). Crop simulation studies can be used to explore the potential benefits of different SI options and assess the ecosystem services of different management interventions (e.g., Hoffmann et al. 2020). Reviews, such as the recent one performed by Kuyah et al. (2021) can help to select promising options from technology packages that have shown to be successful in different smallholder environments of Africa.

Table 23.6 Technology packages (or treatments) defined for the simulation runs

Treatment name	Fertilizer (urea N kg ha^{-1})	Water
Status quo		Rainfed
TMT 23N	25 at sowing	Rainfed
TMT 50N	50 at sowing	Rainfed
TMT 25N deficit	25 at sowing	Deficit irrigation
TMT 50N deficit	50 at sowing	Deficit irrigation

Deficit irrigation: 20 mm when fraction of alternative water supply (AWS) is half of capacity (0.5 from 0–1.0) down to 600 mm in soil depth

Two undisputed problems for many smallholders in SSA are: (1) the general degradation of the soil (Vanlauwe et al. 2014) and (2) climate variability and change (Rötter and Van Keulen 1997; IPCC 2019). The latter include increased frequency of extreme weather events, such as heat-waves and prolonged droughts (IPCC 2019) and within-season dry spells (Rurinda et al. 2020). Such conditions are particularly treacherous for low input systems, as they exacerbate pest and disease problems (see, e.g., Rötter et al. 2018), such as the fall armyworm in crops like maize, which recently has been common in the study region. Poor soil and residue management also leads to low soil organic carbon (SOC) levels and nutrient mining, which is often aggravated in drier and more marginal soil regions (Lal 2004). Vanlauwe and Dobermann (2020) propose for sub-Saharan Africa that SI cannot happen without the use of fertilizer, and that its use needs to be combined with other agronomic improvements. In addition, the low-income nature of smallholder systems means that many cannot afford to waste fertilizer in seasons with unfavorable weather.

Using crop model APSIM (for applications in Africa, see, e.g., Whitbread et al. 2010), we set-up simulation experiments for different management interventions meant to show agronomic pathways of how to help subsistence farmers to move onto the next step of the "ladder" of farming (see. Table 23.6). Two sites were chosen which are contrasting in terms of soil parameters as well as climate conditions— Gabaza is the more fertile and more humid site compared to the semi-arid Selwane site. Planting density and cultivar choice was as recommended by extensions officers for the selected villages in Mopani district. Simulation results for the main rainy season (with weather data from 1998 to 2019, i.e. 20 seasons: October–May) for each site are presented in Fig. 23.3. The treatments (thermo mechanical treatment (TMT)), as detailed in Table 23.6, were based on the current management (status quo) as observed throughout ground truthing field trips to the villages studied in 2019. The so-called status quo technology consists of no input in terms of fertilizer or irrigation. Further TMTs explore urea N application at sowing under rainfed conditions (25 and 50 kg ha^{-1}), as well as the implementation of deficit irrigation (based on rainwater harvesting).

Results of this simulation experiments showed that overall, maize yields were, as expected, estimated to be distinctly higher at the more humid and fertile site, Gabaza, where, on average, maize yields were reaching well above 3 t ha^{-1} with the best available combination of technologies, whereas at Selwana the maize yield

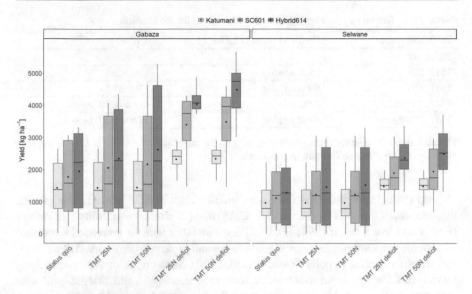

Fig. 23.3 Simulated maize yields for Gabaza and Selwane over 20 seasons [1998–2019; source: NASA POWER Data (https://power.larc.nasa.gov/)]. For scenario details, see Table 7. Maize cultivars are represented by shades of green and mean values by solid black points

ceiling has not been exceeding 2 t ha^{-1}. The deficit irrigation treatment reduced yield variability. Cultivar trends were similar in both sites, with the lowest yields from landrace Katumani, followed by SC601, and the hybrid H 614.

23.4 Tools Required to Model Impacts of Different Technological Innovations on Sustainable Agricultural Land-Use (at Multiple Scales)

In this section, the challenges of developing a framework for analyzing sustainable land management scenarios are firstly described in Sect. 23.4.1. Based on this, the used modeling techniques of agro-ecosystems (crop) modeling as well as agent-based modeling are described in general in Sects. 23.4.2 and 23.4.3.

23.4.1 The Challenge of Developing a Framework for Analyzing Sustainable Land Management Scenarios

Establishing a framework that helps analyze sustainable land use and management options at multiple scales is multifaceted and complex as discussed by Rötter et al. (2021). In regions with a high climate variability and diverse agro-ecological conditions, site- and context-specific interventions are increasingly needed, but these must consider other connected land uses and the resources they share. A suitable strategy demands a multidisciplinary approach. This is particularly the case for modeling multiple ecosystem functions and services, results of which require careful interpretation by a multidisciplinary team of scientists together with local experts (see Rötter et al. 2016).

A typical case that can exemplify a complex land management scenario framework for dry savannas in southern Africa is the crop-livestock system as found in the villages studied in the Mopani district of Limpopo province. From an arable crop production perspective, once the main maize crop has been harvested, leaving the maize biomass and stubble in the field would be one way to ensure the soil is not left bare throughout the dry season and soil integrity is maintained through biomass decomposition. Depending on the water balance of the soil and access to seed, there may even be an opportunity to implement the cultivation of a legume cover crop during this part of the season to add fresh biomass and fix nitrogen. It is only when looking into the livestock aspect of the broader landscape system that key limitations become clear. Post maize harvest, livestock herders graze their cattle on the remaining maize biomass for dry-season feed, removing any soil cover and chance of crop residue nutrient cycling and soil organic carbon development. This is one of the use cases examined by the SALLnet project (see, e.g., Rötter et al. 2021) where a well-designed, combined use of various ecosystem modeling approaches can play an important role in the development of sustainable land use management interventions. In our example of the villages in Mopani district, the crop model APSIM has been combined with the rangeland vegetation model A Dynamic Global Vegetation Model (aDGVM2) (e.g., Hoffmann et al. 2018 and Pfeiffer et al. 2019, respectively).

While developing a usable land management framework is certainly a challenge, there are a number of suitable modeling components, existing data and tools upon which it can be based. With the digitalization and open-access nature of data, such new datasets as digital soil map (e.g., iSDA) and weather information (e.g., NASA Power, CHIRPS) are becoming increasingly accessible and reliable, as are the techniques and technologies to utilize these. For example, via the Internet of Things (IoT) and growth of smartphone ownership the world over. But, it needs pulling together, be managed by multidisciplinary teams and networks of scientists and other key stakeholders if it is going to be used to help implement sustainable land management options for current and future landscapes. This type of novel, multi-scale, and integrated approach promoted by the SALLnet project for southern African savanna landscapes is further described by Rötter et al. (2021).

23.4.2 Agro-Ecosystems Modeling

Agro-ecosystems models or crop simulation models (CSMs) are powerful tools to assess the risk of producing a given crop in a particular soil–climate regime and to assist in management decisions that minimize the risks to crop production. CSMs integrate knowledge from different disciplines and provide researchers with capabilities for conducting simulation experiments to either complement actual field experiments with additional outputs, link ecophysiological understanding to genetics for target crop improvement (Cooper et al. 2021), or to extrapolate experimental results in time and space (e.g., Rötter and van Keulen 1997). A precondition is that models have been well calibrated and validated using pluriannual and multi-locational field data. For sub-Saharan Africa, there are only very few generic crop simulation models (for the main cereals and legumes) that have been evaluated thoroughly under different agro-ecological conditions and with comprehensive and complete soil-weather-crop data sets (Rötter and Van Keulen 1997; Whitbread et al. 2010); The most widely applied and tested among these crop models are APSIM, Decision Support System for Agrotechnology Transfer (DSSAT), and World Food Studies (WOFOST) (see Rötter et al. 2018).

The essence of these three dynamic, process-based crop simulation models (CSMs) is that they can take interactions between genotype (G), environment (E), and management practices (M) into account (Rötter et al. 2015; Cooper et al. 2021). Working on a daily time step, these models are driven by daily values of the most important meteorological variables for plant growth and yield formation such as temperature, global radiation and precipitation and are supplied by a range of soil and crop parameters (and initial system conditions) supplying information on soil physical and chemical properties and major characteristics/traits regarding physiological properties of different crops/crop cultivars and their canopy architecture. Last, but not least, usually management data on material inputs (water, fertilizer; biocides) and their timing can be specified (see Fig. 23.4). The three models APSIM, DSSAT, and WOFOST differ in the original purpose of their development that is reflected in the type/detail and number of input and output data (see, e.g., Pirttioja et al. 2015).

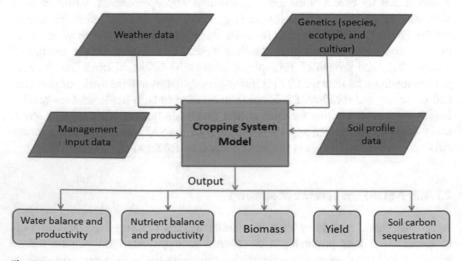

Fig. 23.4 Schematic of a crop simulation modeling platform: input and output

CSMs have been applied in various ways to support plant breeding, e.g. to design crop ideotypes for different environments aimed at minimizing resource use per unit of dry matter produced (Rötter et al. 2015), or to estimate yield potential of crop ideotypes designed for current and projected future climates (e.g., Senapati and Semenov 2020).

The strength of process-based models (see Fig. 23.4) in ex ante evaluation of new technologies and in aiding ideotype breeding is because of their capability to describe causal relationships between crop growth and environmental and management factors driving them, and hence, to quantify the interactions between genotype (G), environment (E), and management (M). The models' limitations relate to the accuracy of the process descriptions and uncertainties related to their parameters. Together these affect their usability for ideotype design and assessment of the effects of improved or new agro-technologies on biomass, final yield, water use, and emissions.

23.4.3 Agent-Based Economic Modeling

To integrate all biophysical and socioeconomic descriptive data and results gathered so far and link them with behavioral economics aspects of the respective decision-makers, i.e. the small-scale farmers, an agent-based farm household model (ABFHM) has been developed and is presented in this section. The model determines improved decisions under different future scenarios at farm level and for policy impact assessment at regional level.

The ABFHM presented in this section build upon the model in Feil et al. (2013). They develop and validate a generic numerical agent-based real options model that describes the simultaneous investment decisions of all firms, including their respective interactions, within a market under uncertainty. The uncertainty enters the model as one or more exogenous stochastic variables that represent uncertain exogenous demand, for instance. Based on this, the firms make investment and production decisions in each production period, which again form the sectoral supply of the respective product at the aggregate level. Accordingly, the endogenous equilibrium price process for the produced commodity can be derived directly. Hence, their model does not rely on some restrictive preconditions of other real options applications, which merely focus on one myopic firm and take market prices as exogenously given and deterministic. The numerical model is solved by a combination of stochastic simulations and genetic algorithms (GA) (for the detailed description of the numerical optimization procedure, cf. Feil et al. 2013). We adjust and enhance this generic model for application to the object of research at hand, i.e. small-scale farming in Limpopo. The basic structure of the ABFHM is illustrated in Fig. 23.5 and will be briefly explained in the following.

Consider a number of N homogenous and risk-neutral competing small-scale farms, each having repeatedly the opportunity to adapt predefined technology packages up to their maximum acreage X_{cap}, on which they cultivate maize in this example, either now or at a later point with in the period under consideration T.

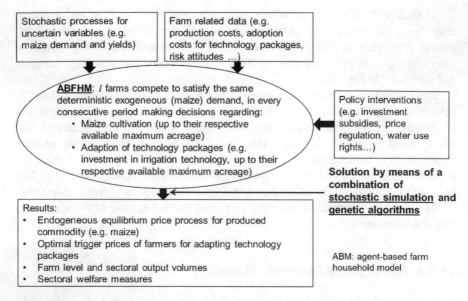

Fig. 23.5 Basic structure of the agent-based farm household model (ABFHM)

This technology package could, for instance, be the application of full irrigation combined with fertilizer application of 120:60 kg N:P per ha, which would cause additional investment and operational costs, on the one hand, but increasing maize yields and hence income on the other hand. Size, adaptation costs, and production are assumed to be proportional, i.e. there are no economies of scale. The production capacity of a farm n in t, resulting in a production output X_t^n, can be adjusted via technology adoption just once in a period, resulting in an additional production output $\Delta X_{t+\Delta t}^n$ in the following period. If it is assumed that the adoption costs are sunk in total, there are no possibilities to reverse the technology adoption, i.e. the associated investment costs are perfectly irreversible. However, in every period the production output declines corresponding to a geometric depreciation rate. Then production follows:

$$X_{t+\Delta t}^n = X_t^n \cdot (1 - \lambda) + \Delta X_{t+\Delta t}^n \tag{23.1}$$

The stochastic demand process μ_t and the price elasticity η (e.g., for maize) are assumed to be known. Prices result from the reactions of *all* market participants on the exogenous stochastic demand process and, hence, need to be determined endogenously within the model. Without loss of generality, the relationship between market supply X_t and price P_t is defined by an isoelastic demand function according to Eq. (23.2):

$$P_t = D(X_t, \mu_t) = \left(\frac{\mu_t}{X_t}\right)^{\Pi} \text{ with } \Pi = -\frac{1}{\eta} \tag{23.2}$$

Within the model, perfect competition is assumed. Accordingly, the farms are assumed to have rational expectations and complete information regarding the development of demand and the technology adoption behavior of all competitors. Because of this, it should be expected that in equilibrium all farms have the same critical price which triggers the adoption, in the following called the trigger price. However to derive this equilibrium by means of the GA approach, the competing firms need to interact, which they do by defining their (at first different) trigger prices. This interaction of the firms equals a second price sealed bid auction in which each farm can sell its product if it asks less or equal the market price. To derive the adoption decisions of the farms, it is assumed that firms with lower trigger prices have a stronger tendency to adopt. Thus, all farms can be ranked by their trigger prices, starting with the lowest, i.e. $\acute{P}^n \leq \acute{P}^{n+1}$. Consequently, firm $n + 1$ does not adopt if firm n has not already completely adopted the respective technology package on all of their available acreage. Likewise, if firm n has fully adopted the TP, firm $n - 1$ has fully adopted it also. Moreover, in every period t, a marginal (or last) firm exists which adopts to the extent that its trigger price equals the expected product price of the next period. For the size of investment of a firm $\tilde{n}$ in t, corresponding to its additional production output in $t + \Delta t$, follows:

$$\Delta X_{t+\Delta t}^{\tilde{n}} \left(\overline{P}^{\tilde{n}} \right) = \max \tag{23.3}$$

The goal of the model is to identify the optimal trigger prices of the farms, which can be expected to be (nearly) identical in equilibrium according to the above assumptions. For this, an objective function needs to be established which determines the investment behavior of the agents in the model. Each farm's adoption decisions aim to maximize the expected net present value (NPV) of the future cash flows F_0^n, in the real options terminology also called option value, by choosing its farm-specific trigger price $\acute{P}^n$:

$$\max_{\overline{p}n} \left\{ F_0^n \left(\overline{p}n \right) \right\} = \max_{\overline{p}n} \left\{ \sum_{t=0}^{T} (-k + P_t) \cdot X_t^n \left(\overline{P}_t^n \right) \cdot e^{-r \cdot t} \right\} \tag{23.4}$$

k denotes the total costs of technology adoption per output unit and period, which are composed of the capital cost of the initial investment outlay I (e.g., for irrigation machinery) and all other relevant costs c (e.g., material costs for fertilizer, labor costs for running the irrigation system):

$$k_t' = I \cdot \left\{ e^{r \cdot \Delta t} - (1 - \lambda) \right\} + c \tag{23.5}$$

The numerical model is solved by a combination of stochastic simulations and genetic algorithms (GA). Furthermore, policy interventions like investment subsidies, price regulations, or water use rights can be flexibly integrated into the model as needed. For the detailed description of the numerical optimization procedure, cf. Feil et al. (2013).

23.5 Assessing Effects of Selected Technology Improvements on Ecosystem Services

A number of promising, alternative agricultural management interventions in support of SI have been identified during the course of the SALLnet project. Apart from improved soil nutrient and water management practices, a whole range of new technologies could be very useful to increase agricultural productivity in a sustainable manner, including new plant seeds and breeds more resilient to climate variability and anticipated changes, mechanization and utilization of digital tools. However, while a few of such agricultural innovations were found among the most advanced farms in Limpopo, the vast majority of farms surveyed in the villages during the 2019 campaign showed only very basic management interventions— mostly relating to soil fertility and water management.

Given that the status quo in terms of crop management can generally be considered as minimal, in the following section, we concentrate on improving those technologies that, according to Vanlauwe and Dobermann (2020) are considered as the most crucial building blocks for lifting subsistence farmers to the first step out of poverty and of increasing productivity, i.e. fertilizer application and water management.

23.5.1 Example of Integrated Analysis of the Effects of Selected Improved Technologies and Agricultural Innovations on Rural Livelihoods and Other Ecosystem Services at Community Level

In Table 23.7, the assumptions used for the simulations by means of the ABFHM as described in Sect. 23.4.3 are listed. Three technology packages (TP) for maize cultivation are investigated following Sect. 23.3.3. TP 0 represents the status quo treatment of no fertilizer application and rainfed irrigation, TP 1 combines deficit irrigation with a fertilizer application of 40 kg N and 30 kg P per ha, and TP 2 combines full irrigation with a fertilizer application of 120 kg N and 60 kg P per ha. Therefore, irrigation technology has to be invested in and implemented for TP 1 and 2, which causes capital costs.

In the ABFHM, both TP 1 and TP 2 are each applied to 100 small-scale farmers (i.e., the agents), who are assumed to be in close proximity to each other in one region. According to the classification conducted in Sect. 23.2, at the start of the ABFHM simulations these small-scale farmers represent subsistence-oriented farmers with a sole focus on maize farming and, so far, no use of irrigation technology and fertilizer (i.e., they all start with TP 0). Without loss of generality, these farmers are assumed to have a standardized acreage of one ha for maize cultivation. Based on this, they make decisions in every consecutive production period starting with year one, whether or not they adopt TP 1 or TP 2, depending on which TP of both is exogenously available. On the one hand, this would imply

Table 23.7 Parameters used in the agent-based farm household model

Basic model parameters	
Total number of farms N	100
Farm size	1 ha
Period under consideration T	Infinite, approximated by 30 years
Time step length Δt	1.00 (i.e., one period equals one year)
Simulation runs	10,000
Considered technology packages (TP)	
TP 0	Rainfed maize cultivation with no fertilizer application
TP 1	Deficit irrigation (100 mm) combined with 40:30 kg N:P per ha
TP 2	Full irrigation (200 mm) combined with 120:60 kg N:P per ha
Maize yield related parameters (simulated from 1985 to 2019) [a]	
Mean maize corn yield	TP 0 = 1.31, TP 1 = 3.35, TP 2 = 5.39 t per ha
Standard deviation	TP 0 = 0.28, TP 1 = 0.60, TP 2 = 0.64 t per ha
Maize demand related parameters [b]	
Estimated stochastic process for the process of the demand parameter μ_t	Geometric Brownian motion
Price elasticity of demand η	0.99
Drift rate α	−0.42%
Standard deviation σ	5.30%
Cost related parameters [c]	
Fertilizer costs	TP 1 = 1708, TP 2 = 5124 ZAR per ha and a
Irrigation investment capital costs	TP 1 = TP 2 = 11,616 ZAR per ha and a
Irrigation running costs	TP 1 = 860, TP 2 = 1232 ZAR per ha and a

[a] Own simulations with APSIM
[b] Source: South African Government (2021)
[c] Source: own small-scale farmer survey conducted in SALLnet from February to May 2019

higher expected yields and therefore additional expected revenues, as farmers would sell their excess maize corn on a local market at equilibrium market prices as derived within the ABFHM (cf. Sect. 23.4.3). On the other hand, this would also entail additional costs, in fact the investment and running costs of the irrigation machinery as well as the costs for purchasing and applying fertilizer.

The results of the ABFHM simulations are presented in Table 23.8. Accordingly, for each TP 1 and TP 2, separate model runs are conducted over infinite time, approximated by a discrete time period in the numerical model of 30 years. For both these runs, the average resulting maize corn trigger prices of the farms for adopting the respective TP are provided in the third column. Furthermore, the share of farms out of the sample of 100 farms that adopt the respective TP over time within the respective ABFHM run (column four) is shown. Moreover, the resulting average maize corn yield within the overall sample of 100 farms (column five) as well as

Table 23.8 Results of the agent-based farm household model

Technology package (TP)	Description of TP	Trigger price for TP adoption (in ZAR per kg maize corn)	Share of farms adopting TP 1 or TP 2 over time (in %)	Average maize corn yield within overall farm sample (in t per ha and year)	Increase in maize corn production at village level (in %)
TP 0	Rainfed, no fertilizer	–	–	1.31	–
TP 1	Deficit irrigation combined with 40:30 kg P:N/ha	4.30	28.1	1.88	42.8
TP 2	Full irrigation combined with 120:60 kg P:N/ha	3.10	55.9	3.60	174.7

the resulting overall increase in maize corn production at community, respectively, village level (column six), is also displayed.

Accordingly, the average maize corn trigger price of the farms for adopting TP 1 is 4.30 and for TP 2 is 3.10 ZAR per kg, respectively. Comparing this to a considerably higher observed actual average maize corn price of about 9.00 ZAR per kg in the villages at the time of the survey in early 2019 (cf. Sect. 23.2), it becomes obvious that it could be economically worthwhile for the small-scale farmers to adopt the described TP's already now. This, of course, would require the technology to be available to small-scale farmers in the region in general, that is water access, irrigation machinery, and fertilizer, at the assumed conditions, which in reality is often not the case in the investigated villages (cf. section two). The lower average trigger price for TP 2 compared to TP 1 indicates that it is economically worthwhile for the farmers to adopt TP 2 sooner. Based on this, 28% of small-scale farmers would adopt TP 1 and even 56% would adopt TP 2 over time. The resulting increases in regional maize corn supply would be 43% in the case of TP 1 and as much as 175% in the case of TP 2 in the long run.

23.5.2 Discussion of Other Promising Land Management Interventions

The SALLnet farm survey campaign in five villages during 2019 revealed for some farmers a few other management interventions. Some farmers had started diversifying crops with the relatively drought-tolerant sorghum and the bambara groundnut legume—as had been promoted by local extension service providers. Farm machinery consisted of poorly maintained moldboard plows that were few in number and therefore shared by many. Instead of site-and season-specific crop cultivar selection, farmers were generally unknowledgeable about the cultivars they planted and tended to simply make do with what was available. This was largely dependent on what was saved from the previous harvest, what was available in local agricultural supply stores, and in some instances what was given to them by the University of Limpopo. Hence, clearly, there is a wide range of promising options that can be used to go beyond those currently found at the study sites and beyond moderate fertilizer application in combination with deficit irrigation.

For many regions in the semi-arid tropics, including the Limpopo province, serious water-resource constraints are increasing and set to continue. Agricultural management needs to buffer such constraints, but not all interventions can be implemented in every farming context. In terms of integrated soil-nutrient management, it must be recognized that soils in the region are poor and low in soil organic matter (SOM) and therefore water retention capacity. This in turn makes soils and their resources vulnerable to erosion. In South Africa, more than 60% of the soils are low in SOM, highly degraded and, furthermore, low in terms of productivity due to nutrient mining and erosion (Vlek et al. 2020). In addition, climate change effects such as warmer and drier conditions throughout most of South Africa are expected to increase the rates at which soil C mineralization occurs, along with associated losses in soil ecosystem functions. As taken into account through our modeling experiments, the application of mineral fertilizers (especially N and P fertilizer) is the major soil-nutrient/fertility replenishing option, which can lead to substantial yield gains—as shown by our analyses. However, in order to sustain any yield gains, the application of mineral fertilizer must be complemented with conservation agriculture (CA) techniques, organic inputs, and good agronomic practices (see Thierfelder et al. 2017). Moreover, meaningful soil management must focus on restoring soil health through stopping and reversing nutrient depletion and organic matter decline (Swanepoel et al. 2018). A few considerations when working toward such goals:

- Conservation agriculture (CA) is knowledge intensive. It not only does it require specialist skills, but it often necessitates special seeds (e.g., cover crops) and machinery (e.g., seed drills). While certainly key to sustainable farming systems design, CA is challenging in the smallholder context (see Baudron et al. 2015).
- Constraints to soil amendments arise from the scarcity of organic material due to multiple claims on biomass, i.e. crop residues, wood litter, cattle manure. For example, crop residues in the field, including stubble, can be used as animal feed or left to decompose and partially replenish the soil.
- In marginal, food insecure, high-risk environments, farmer advisory services and new technologies are often lacking. This makes it difficult to employ appropriate risk-management strategies that enable site- and season-specific input adjustments and the design of efficient cropping systems. This brings us to the reality that the benefits of digitalization have not yet reached the smallholder context, despite already benefiting larger scale producers and agribusiness (see Baumüller and Kah 2020).

Kuyah et al. (2021) and Wilkus et al. (2021) described a whole range of promising options, i.e., innovative agronomic practices for SI in Sub-Saharan Africa (SSA), including doubled-up legume intercropping, CA, agroforestry practices, push-pull technology, etc. While these all can boost production, there are additional options for pushing small-scale production levels toward those of commercial operations and increasing the efficiency of the farming systems in the medium to long term (Baudron et al. 2015). Here are a just a few suggestions:

Crop Improvement Programs Acceleration of crop improvement is increasingly difficult in highly variable climates of already risk-prone systems—as found in Limpopo. Basis for any targeted breeding program taking shifts in future environmental conditions into account, is ex ante evaluation of genotype-by-environment-by-management (GxExM) interactions, as supported by CSMs. To make progress, crop improvement strategies need to develop crop varieties (G) and agronomic practices (M) specifically adapted to current and future local conditions (E) (Cooper et al. 2021). In a review on the bottlenecks of developing climate-resilient maize for the stress-prone (sub-) tropics, Cairns and Prasanna (2018) present the benefits of tailor-made genotypes for well-defined target environments. One major hurdle in this context has been data availability (e.g., lack of high-resolution soil and climate data). However, progress is being made, exemplified through the release of online and open-access databases such as the iSDAsoil (see: https://www.isda-africa.com), a mapping system for SSA at a resolution of 30 m, including 24 billion locations across Africa. Such resources could contribute toward site-specific recommendations for farmers in the near future.

Mechanization of Farming Practices Farmers of Limpopo region shared tractors and plows and were therefore limited in terms of when they could prepare the fields and sow, often leading to poorly-timed cultivation with yield penalties. The region is also impacted by El Nino-Southern Oscilliation (ENSO) events, with marked inter-annual differences in soil moisture supply (Moeletsi et al. 2011). Future improvement for these systems is therefore likely to be centered around timing, i.e. the ability of farmers to adapt quickly to climate conditions. The use of scale-relevant, affordable machinery in general could enable farmers to react in a more agile manner, for example through implementing seedbed management independently without having to wait for the availability of shared tractors and plows.

Digital Tools While digital tools can help connect agricultural stakeholders and spread information, also the digitalization of the agricultural systems themselves is also high on the research and development agenda. The German development organization "Welthungerhilfe," for example, shows how a mobile phone-based app "AgriShare" helps connect farmers in Zimbabwe without assets with those who can supply it, such as commercial or private hiring services (Welthungerhilfe 2018). This includes services for production, processing, and transportation.

23.6   Synthesis and Outlook

In this chapter, we explored how, in the face of increasing climatic risks and resource limitations, improved agro-technologies could support SI in smallholder farming systems in Limpopo province, South Africa. We used the example of improved soil nutrient and irrigation practices in combination to perform an integrated bio-economic analysis of the potential benefits for smallholder farmers and some essential ESs they rely on. From the perspective of technical feasibility, economic viability, and environmental benefits the selected technology packages clearly show that they can support SI pathways for small-scale farmers. However, for implementing such technological improvements successfully at a larger (e.g., landscape and regional) scale in the near (short- to medium term) future, establishment of the required mandatory foundations for this will be necessary. That is, enabling policies as well as dissemination and training of farmers in new applying technologies.

Policy Recommendations Solid foundations can be laid by adjusting national and local policies and resource management regulations at higher aggregation levels—in accordance with the scale targeted at. To achieve this, concerted actions for SI of small-scale farming systems will be required involving politicians, extension services, up- and downstream agribusinesses, and other supporting institutions. This includes in the first instance the gradual establishment of an appropriate infrastructure in the respective regions regarding transport routes to allow the access of input and output markets, specific packaging needs (for fertilizer) of small-scale farmers as well as effective and knowledge-oriented and competent extension services.

Research Challenges Regarding crop sciences and agro-ecosystems modeling there are a number of future research challenges. A very urgent one is to accelerate breeding of climate-resilient crop cultivars as needed under progressive climate change. For developing these, next-generation breeding technologies such as high-density genotyping coupled to high throughput (precision) phenotyping and the use of crop simulation models is required. These methods can help untangle GxExM interactions and provide farmers with suitable genotypes, along with tailor-made management recommendations, informed by in silico experimentation (Hammer et al. 2020). Yet, an essential pre-requisite will be to improve agro-ecosystems for this task. That can be realized in various manners, e.g. modifying modeling routines for better capturing relevant processes at sub-daily scale, e.g. transpiration under low air humidity, collecting new phenotypic data to incorporate accurate information about genetic diversity, or linking genomic to ecophysiological information to improve model calibration (Rötter et al. 2015).

References

Ali I, Huo X, Khan I, Ali H, Khan B, Khan SU (2019) Technical efficiency of hybrid maize growers: A stochastic frontier model approach. J Integr Agric 18(10):2408–2421. https://doi.org/10.1016/S2095-3119(19)62743-7

Baloyi RT, Belete A, Hlongwane JJ, Masuku MB (2012) Technical efficiency in maize production by small-scale farmers in Ga-Mothiba of Limpopo province, South Africa. Afr J Agric Res 7(40):5478–5482. https://doi.org/10.5897/AJAR11.2516

Baudron F, Sims B, Justice S et al (2015) Re-examining appropriate mechanization in eastern and southern Africa: two-wheel tractors, conservation agriculture, and private sector involvement. Food Sec 7:889–904. https://doi.org/10.1007/s12571-015-0476-3

Baumüller H, Kah M (2020) Going digital: harnessing the power of emerging technologies for the transformation of Southern African agriculture. https://doi.org/10.4324/9780429401701-23

Cairns JE, Prasanna BM (2018) Developing and deploying climate-resilient maize varieties in the developing world. Curr Opin Plant Biol 45:226–230. https://doi.org/10.1016/j.pbi.2018.05.004

Cassman KG, Grassini P (2020) A global perspective on sustainable intensification research. Nat Sustain 3:262–268. https://doi.org/10.1038/s41893-020-0507-8

Cooper M, Voss-Fels KP, Messina CD, Tang T, Hammer GL (2021) Tackling G × E × M interactions to close on-farm yield-gaps: creating novel pathways for crop improvement by predicting contributions of genetics and management to crop productivity. Theor Appl Genet 134:1625–1644. https://doi.org/10.1007/s00122-021-03812-3

FAO (2017) Defining small-scale food producers to monitor target 2.3 of the 2030 agenda for sustainable development. FAO statistics working paper series, vol 17, issue 12. www.fao.org/3/a-i6858e.pdf

FAO (Food and Agriculture Organization of the United Nations) (2010) 'Climate–smart' agriculture. Policies, practices and financing for food security, adaptation and mitigation. Conference on agriculture, food security and climate change, The Hague, 2010

Feil JH, Musshoff O, Balmann A (2013) Policy impact analysis in competitive agricultural markets: a real options approach. Eur Rev Agric Econ 40(4):633–658

Garnett T, Appleby MC, Balmford A, Bateman IJ, Benton TG, Bloomer P, Burlingame B, Dawkins M, Dolan L, Fraser D, Herrero M, Hoffmann I, Smith P, Thornton PK, Toulmin C, Vermeulen SJ, Godfray HCJ (2013) Sustainable intensification in agriculture: premises and policies. Science 341(6141):33–34. https://doi.org/10.1126/science.1234485

Gatzweiler FW, von Braun J (eds) (2016) Technological and institutional innovations for marginalized smallholders in agricultural development. Springer Cham, Heidelberg. ISBN 978-3-319-25718-1 (eBook). https://doi.org/10.1007/978-3-319-25718-1

Greyling JC, Pardey PG (2019) Measuring maize in South Africa: the shifting structure of production during the twentieth century, 1904–2015. Agrekon 58(1):21–41. https://doi.org/10.1080/03031853.2018.1523017

Gyekye A, Akinboade OA (2003) A profile of poverty in the Limpopo Province of South Africa. East Afr Soc Sci Res Rev 19(2):89–109. https://doi.org/10.1353/eas.2003.0005

Hadri K (1999) Estimation of a doubly heteroscedastic stochastic frontier cost function. J Bus Econ Stat 17(3):359–363. https://doi.org/10.2307/1392293

Hammer GL, McClean C, Van Oosterom E et al (2020) Designing crops for adaptation to the drought and high temperature risks anticipated in future climates. Crop Sci 60:605–621. https://doi.org/10.1002/csc2.20110

Herrero M, Thornton PK, Mason-D'Croz D et al (2020) Innovation can accelerate the transition towards a sustainable food system. Nat Food 1:266–272. https://doi.org/10.1038/s43016-020-0074-1

Hoffmann MP, Odhiambo JJO, Koch M, Ayisi K, Zhao G, Soler AS, Rötter RP (2018) Exploring adaptations of groundnut cropping to prevailing climate variability and extremes in Limpopo Province, South Africa. Field Crop Res 219:1–13. https://doi.org/10.1016/j.fcr.2018.01.019

Hoffmann MP, Swanepoel CM, Nelson WCD, Beukes DJ, van der Laan M, Hargreaves JNG, Rötter RP (2020) Simulating medium-term effects of cropping system diversification on soil fertility and crop productivity in southern Africa. Eur J Agron 119. https://doi.org/10.1016/j.eja.2020.126089

Hove L, Kambanje C (2019) Lessons from the El Nino e induced 2015/16 drought in the southern Africa region. Curr Dir Water Scarcity Res 2:33–54. https://doi.org/10.1016/B978-0-12-814820-4.00003-1

IPBES (2018) The IPBES regional assessment report on biodiversity and ecosystem services for Africa. In: Archer E, Dzoba E, Mulongoy KJ, Maoela M, Walters M (eds) Secretariat of the intergovernmental science-policy platform on biodiversity and ecosystems services. IPBES, Bonn

IPCC (2019) Climate change and land: an IPCC special report on climate change, desertification, land degradation, sustainable land management, food security, and greenhouse gas fluxes in terrestrial ecosystems [P.R. Shukla, J. Skea, E. Calvo Buendia, V. Masson-Delmotte, H.-O. Pörtner, D. C. Roberts, P. Zhai, R. Slade, S. Connors, R. van Diemen, M. Ferrat, E. Haughey, S. Luz, S. Neogi, M. Pathak, J. Petzold, J. Portugal Pereira, P. Vyas, E. Huntley, K. Kissick, M. Belkacemi, J. Malley, (eds.)]

Kassie BT, Van Ittersum MK, Hengsdijk H, Asseng S, Wolf J, Rötter RP (2014) Climate-induced yield variability and yield gaps of maize (Zea mays L.) in the central Rift Valley of Ethiopia. Field Crop Res 160:41–53. https://doi.org/10.1016/j.fcr.2014.02.010

Kumbhakar SC, Wang H-J, Horncastle P, A. (2015) A practitioner's guide to stochastic frontier analysis using STATA. Cambridge University Press

Kurwakumire N, Chikowo R, Mtambanengwe F, Mapfumo P, Snapp S, Johnston A, Zingore S (2014) Maize productivity and nutrient and water use efficiencies across soil fertility domains on smallholder farms in Zimbabwe. Field Crops Res 164(1):136–147

Kuyah S, Sileshi GW, Nkurunziza L, Chirinda N, Ndayisaba PC, Dimobe K, Öborn I (2021) Innovative agronomic practices for sustainable intensification in sub-Saharan Africa. A review. Agron Sustain Dev 41. https://doi.org/10.1007/s13593-021-00673-4

Lal R (2004) Soil carbon sequestration impacts on global climate change and food security. Science 304:1623–1627. https://doi.org/10.1126/science.1097396

LDARD (2012) Limpopo Department of Agriculture and Rural Development. Webpage of the Limpopo Provincial Government, Republic of South Africa

Moeletsi ME, Walker S, Landman WA (2011) ENSO and implications on rainfall characteristics with reference to maize production in the Free State Province of South Africa. Phys Chem Earth 36:715–726. https://doi.org/10.1016/j.pce.2011.07.043

Mpandeli S, Maponya P (2014) Constraints and challenges facing the small scale farmers in Limpopo Province, South Africa. J Agric Sci 6(4), 135. https://pdfs.semanticscholar.org/ffef/f498d2fb153db4a96b1a07c351186c6d3b36.pdf

Ngome AF, Becker M, Mtei MK, Mussgnug F (2013) Maize productivity and nutrient use efficiency in Western Kenya as affected by soil type and crop management. Int J Plant Prod 7(3). https://www.researchgate.net/publication/287352032

Obi A, Ayodeji BT (2020) Determinants of economic farm-size–efficiency relationship in smallholder maize farms in the eastern Cape Province of South Africa. Agriculture 10(4):98. https://doi.org/10.3390/agriculture10040098

Ogundari K (2014) The paradigm of agricultural efficiency and its implication on food security in Africa: what does meta-analysis reveal? World Dev 64(1920):690–702. https://doi.org/10.1016/j.worlddev.2014.07.005

Pfeiffer M, Langan L, Linstädter A, Martens C, Gaillard C, Ruppert JC et al (2019) Grazing and aridity reduce perennial grass abundance in semi-arid rangelands - insights from a trait- based dynamic vegetation model. Ecol Model 395:11–22. https://doi.org/10.1016/j.ecolmodel.2018.12.013

Pirttioja N, Carter TR, Fronzek S, Bindi M, Hoffmann H, Palosuo T, Ruiz-Ramos M, Tao F, Trnka M, Acutis M, Asseng S, Baranowski P, Basso B, Bodin P, Buis S, Cammarano D, Deligios P, Destain MF, Dumont B, Ewert F, Ferrise R, François L, Gaiser T, Hlavinka P, Jacquemin

I, Kersebaum KC, Kollas C, Krzyszczak J, Lorite IJ, Minet J, Minguez MI, Montesino M, Moriondo M, Müller C, Nendel C, Öztürk I, Perego A, Rodríguez A, Ruane AC, Ruget F, Sanna M, Semenov MA, Slawinski C, Stratonovitch P, Supit I, Waha K, Wang E, Wu L, Zhao Z, Rötter RP (2015) A crop model ensemble analysis of temperature and precipitation effects on wheat yield across a European transect using impact response surfaces. Clim Res 65:87–105. https://doi.org/10.3354/cr01322

Rötter RP, Van Keulen H (1997) Variations in yield response to fertilizer application in the tropics: II. Risks and opportunities for smallholders cultivating maize on Kenya's arable land. Agric Syst 53:69–95. https://doi.org/10.1016/S0308-521X(96)00037-6

Rötter RP, Tao F, Höhn JG, Palosuo T (2015) Use of crop simulation modelling to aid ideotype design of future cereal cultivars. J Exp Bot 66(12):3463–3476. https://doi.org/10.1093/jxb/erv098erv098

Rötter RP, Sehomi FL, Höhn JG, Niemi JK, Van den Berg M (2016) On the use of agricultural system models for exploring technological innovations across scales in Africa: a critical review. ZEF - discussion papers on development policy no. 223, University of Bonn, Bonn, 85 pp, ISSN: 1436-9931

Rötter RP, Hoffmann MP, Koch M, Müller C (2018) Progress in modelling agricultural impacts of and adaptations to climate change. Curr Opin Plant Biol 45(B):255–261. https://doi.org/10.1016/j.pbi.2018.05.009

Rötter RP, Scheiter S, Hoffmann M, Pfeiffer M, Nelson WCD et al (2021) Letter-to-the-editor: modelling the multi-functionality of African savanna landscapes under global change. Land Degrad Dev. https://doi.org/10.22541/au.161425259.98733826/v1

Rurinda J, Zingore S, Jibrin JM, Balemi T, Masuki K, Andersson JA, Pampolino MF, Mohammen I, Mutegi J, Kamara AY, Vanlauwe B, Craufurd PQ (2020) Science-based decision support for formulating crop fertilizer recommendations in sub-Saharan Africa. Agric Syst 180. https://doi.org/10.1016/j.agsy.2020.102790

Sadras VO, Grassini P, Steduto P (2012) Status of water use efficiency of main crops. SOLAW background thematic report–TR07

Savary S, Akter S, Almekinders C, Harris J, Korsten L, Rötter R, Waddington S, Watson D (2020) Mapping disruption and resilience mechanisms in food systems. Food Secur 12:695–717. https://doi.org/10.1007/s12571-020-01093-0

Senapati N, Semenov MA (2020) Large genetic yield potential and genetic yield gap estimated for wheat in Europe. Glob Food Sec 24:100340. https://doi.org/10.1016/j.gfs.2019.100340

Sikora AS, Terry ER, Vlek PLG, Chitjja J (eds) (2020) Transforming agriculture in southern Africa: constraints, technologies, policies and processes, 1st edn. Routledge, New York. https://www.taylorfrancis.com/books/transforming-agriculture-southern-africa-richard-sikora-eugene-terry-paul-vlek-joyce-chitja/e/10.4324/9780429401701

Smaling EMA, Nandwa SM, Prestele H, Rötter RP, Muchena FN (1992) Yield response of maize to fertilizers and manure under different agro-ecological conditions in Kenya. Agric Ecosyst Environ 41:241–252. https://doi.org/10.1016/0167-8809(92)90113-P

South African Government (2021) Agriculture, land reform and rural development. South African Government. https://www.gov.za/about-sa/agriculture

Statistics South Africa (2019) General household survey. http://www.ncbi.nlm.nih.gov/pubmed/11469378

Swanepoel C, Rötter RP, Laan M, Annandale J, Beukes DJ, Preez CC, Swanepoel L, Van der Merwe A, Hoffmann M (2018) The benefits of conservation agriculture on soil organic carbon and yield in southern Africa are site-specific. Soil Tillage Res 183. https://doi.org/10.1016/j.still.2018.05.016

Thierfelder C, Chivenge P, Mupangwa W et al (2017) How climate-smart is conservation agriculture (CA)? – its potential to deliver on adaptation, mitigation and productivity on smallholder farms in southern Africa. Food Sec 9:537–560. https://doi.org/10.1007/s12571-017-0665-3

van Duivenboden N (1992) Sustainability in terms of nutrient elements with special reference to West Africa. Report 160, CABO-DLO, Wageningen

Van Ittersum MK, Cassman KG, Grassini P, Wolf J, Tittonell P, Hochman Z (2013) Yield gap analysis with local to global relevance—a review. Field Crop Res 143:4–17. https://doi.org/10.1016/j.fcr.2012.09.009

Vanlauwe B, Dobermann A (2020) Sustainable intensification of agriculture in sub-Saharan Africa: first things first! Front Agric Sci Eng 7(4):376–382. https://doi.org/10.15302/J-FASE-2020351

Vanlauwe B, Coyne D, Gockowski J, Hauser S, Huising J, Masso C, Nziguheba G, Schut M, van Asten P (2014) Sustainable intensification and the African smallholder farmer. Curr Opin Environ Sustain 8:15–22. https://doi.org/10.1016/j.cosust.2014.06.001

Vlek PLG, Tamene L, Bogardi J (2020) Land rich but water poor: the prospects of agricultural intensification in southern Africa (Chapter 5). In: Sikora AS, Terry ER, Vlek PLG, Chitjja J (eds) Transforming agriculture in southern Africa: constraints, technologies, policies and processes, 1st edn. Routledge, pp 36–44. https://doi.org/10.4324/9780429401701

Von Braun J, Afsana K, Fresco LO et al (2021) Food system concepts and definitions for science and political action. Nat Food. https://doi.org/10.1038/s43016-021-00361-2

Welthungerhilfe (2018) AgriShare, helping farmers to produce and earn more. Concept paper: https://www.welthungerhilfe.org/news/publications/detail/agrishare/

Whitbread AM, Robertson MJ, Carberry PS, Dimes JP (2010) How farming systems simulation can aid the development of more sustainable smallholder farming systems in southern Africa. Eur J Agron 32:51–58. https://doi.org/10.1016/j.eja.2009.05.004

Wilkus E, Mekuria M, Rodriguez D, Dixon J (2021) Sustainable intensification of maize–legume systems for food security in eastern and southern Africa (SIMLESA): lessons and way forward. ACIAR monograph no. 211, Australian Centre for International Agricultural Research, Canberra, 503 pp

Yazdan Bakhsh S, Brümmer B, Rötter RP, Ayisi KK, Twine WT, Feil JH (2022) Analyzing maize production efficiency of small-scale farmers in South Africa using a stochastic frontier approach. DARE discussion paper series 2201

Part IV

Monitoring and Modelling Tools

A New Era of Earth Observation for the Environment: Spatio-Temporal Monitoring Capabilities for Land Degradation

24

Christiane Schmullius ⓘ, Ursula Gessner ⓘ, Insa Otte ⓘ,
Marcel Urban ⓘ, George Chirima, Moses Cho, Kai Heckel,
Steven Hill, Andreas Hirner ⓘ, Pawel Kluter, Nosiseko Mashiyi,
Onisimo Mutanga, Carsten Pathe, Abel Ramoelo, Andrew Skowno,
Jasper Slingsby, and Jussi Baade ⓘ

C. Schmullius (✉) · M. Urban · K. Heckel · C. Pathe
Department for Earth Observation, Institute of Geography, Friedrich-Schiller-University Jena,
Jena, Germany
e-mail: c.schmullius@uni-jena.de; kai.heckel@mail.de; carsten.pathe@uni-jena.de

U. Gessner · A. Hirner
German Aerospace Center (DLR), German Remote Sensing Data Center (DFD), Wessling,
Germany
e-mail: ursula.gessner@dlr.de; andreas.hirner@dlr.de

I. Otte · S. Hill · P. Kluter
Department of Remote Sensing, Institute of Geography and Geology, University of Wuerzburg,
Wuerzburg, Germany
e-mail: insa.otte@uni-wuerzburg.de; steven.hill@uni-wuerzburg.de;
pawel.kluter@uni-wuerzburg.de

G. Chirima
ARC-Soil, Climate, Water and Agricultural Engineering, Arcadia, South Africa
e-mail: chirimaj@arc.agric.za

M. Cho
Department of Geography, Geoinformatics and Meteorology, University of Pretoria, Pretoria,
South Africa
e-mail: mcho@csir.co.za

N. Mashiyi
South African National Space Agency, Directorate Earth Observation, Pretoria, South Africa
e-mail: nmashiyi@sansa.org.za

Abstract

Land degradation can be defined as a persistent reduction or loss of the biological and economic productivity resulting from climatic variations and human activities. To quantify relevant surface changes with Earth observation sensors requires a rigorous definition of the observables and an understanding of their seasonal and inter-annual temporal dynamics as well as of the respective spatial characteristics. This chapter starts with brief overviews of suitable remote sensing sources and a short history of degradation mapping. Focus is on arising possibilities with the new European Sentinel satellite fleet, which ensures unprecedented spatial, spectral, and temporal monitoring capabilities. Synergistic retrieval of innovative degradation indices is illustrated with mapping examples from the SPACES II (Science Partnerships for the Adaptation/Adjustment to Complex Earth System Processes) SALDi (South Africa Land Degradation Monitor) and EMSAfrica projects plus South African contributions. Big data approaches require adapted exploration techniques and infrastructures—both aspects conclude this chapter.

24.1   Introduction

Land degradation (LD) is a global problem affecting approximately 70% of drylands with 73% of Africa's agricultural lands already degraded (DFFE 2018). The narrative of land degradation, its location and causes are evolving over time (Scholes

O. Mutanga
Department of Geography, University of KwaZulu-Natal, Pietermaritzburg, South Africa
e-mail: Mutangao@ukzn.ac.za

A. Ramoelo
Centre for Environmental Studies (CFES), Department of Geography, Geoinformatics and Meteorology, University of Pretoria, Pretoria, South Africa
e-mail: abel.ramoelo@up.ac.za

A. Skowno
National Biodiversity Assessment Unit, South African National Biodiversity Institute (SANBI), Silverton, South Africa

Department of Biological Sciences and Centre for Statistics in Ecology, Environment and Conservation, University of Cape Town, Rondebosch, South Africa
e-mail: a.skowno@sanbi.org.za

J. Slingsby
Department of Biological Sciences and Centre for Statistics in Ecology, Environment and Conservation, University of Cape Town, Rondebosch, South Africa

Fynbos Node, South African Environmental Observation Network (SAEON), Cape Town, South Africa
e-mail: jasper.slingsby@uct.ac.za

J. Baade
Department of Geography, Friedrich Schiller University Jena, Jena, Germany
e-mail: jussi.baade@uni-jena.de

2009; von Maltitz et al. 2019). From a government perspective, concerns such as soil erosion are being viewed as less important in the immediate time frame, compared to issues such as bush encroachment and the invasion of alien plants, which are becoming the most prominent current degradation concerns (von Maltitz et al. 2019; O'Connor et al. 2014). Since the middle of the twentieth century, the term land degradation in South Africa had been linked to veld and soil degradation and has been addressed by numerous measures (Hoffman and Todd 2000; Hoffman and Ashwell 2001).

Target 15.3 of the Sustainable Development Goals (SDG) aims to achieve Land Degradation Neutrality (LDN) worldwide by 2030. Three global indicators for assessing land degradation are suggested in the LDN Scientific Conceptual Framework: land cover (physical land cover class), land productivity (net primary productivity, NPP), and carbon stocks [soil organic carbon (SOC) stocks] (Orr et al. 2017; Cowie et al. 2018). The South African LDN target setting process during 2017/2018 showed that these global indicators are not appropriate to fully describe the nature, extent, and location of degradation in South Africa (von Maltitz et al. 2019, see also Chap. 3). Global satellite-based NPP time series, for example, indicate areas of negative trends that clearly differ from perception-based assessments of land degradation. Reasons might be that some degradation aspects such as bush encroachment and invasive alien species can result in increased plant cover and NPP; also, high inter-season variability of rainfall and its impact on plant productivity impedes identifying management-related land degradation (Wessels et al. 2012). Thus, it is rather suggested to locate land degradation hotspots and respective target intervention areas based on the results of the perception-based assessments such as the Land Degradation Assessment in Drylands (LADA) Project (Lindeque and Koegelenberg 2015), in combination with issue-specific maps, even though currently such maps are hardly available (von Maltitz et al. 2019).

Land degradation processes in South Africa are as complex as the country's ecosystems and are intricately linked to food security, poverty, urbanization, climate change, and biodiversity. Therefore, South African authors contribute their experience and emerging tools. The chapter starts with a brief overview of Earth observation (EO) sensors suitable for degradation monitoring, followed by historic and emerging EO strategies for the following six LD topics: vegetation development and cover change, woody cover trends, bush encroachment, invasive species, drought and soil moisture assessments, and overall degradation. A summarizing table to highlight the achievements and perspectives from the Copernicus programme and other international emerging technologies concludes the chapter. The SPACES II SALDi-project's methodological implementations are described for representative regions in Chap. 29.

24.2 Overview of Satellite Earth Observation Data Sources Suitable for Degradation Monitoring

Until the mid-2010s, the most common satellite datasets used for degradation mapping were time series of Moderate Resolution Imaging Spectroradiometer (MODIS) $(250–500 \text{ m}^{-1} \text{ km})$, Satellite Pour l'Observation de la Terre-VEGETATION (SPOT-VGT) (1 km), and Advanced Very High Resolution Radiometer (AVHRR) (1–8 km), as well as Landsat and aerial images which were usually not available in the form of time series but rather as mono- or irregular multi-temporal acquisitions. Even though important and highly valuable Earth-observation based analyses have been conducted based on these datasets, one major challenge remained: The relatively coarse spatial resolution of these time-series data and the irregular temporal availability of higher spatial resolution imagery resulted in general difficulties to assess degradation processes where small-scale landscape elements and patterns show distinct seasonal or multi-annual dynamic behaviour.

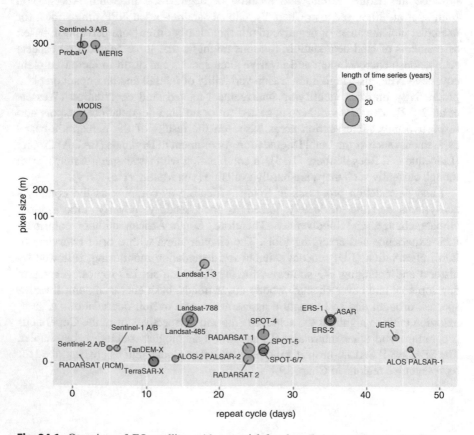

Fig. 24.1 Overview of EO-satellites with potential for degradation assessment. Satellites are shown which allow time-series assessments at medium to high spatial resolution (<=300 m) and where data can (partly) be accessed free of charge

With the launch of the Copernicus Sentinel satellite fleet in 2015 freely available radar and since 2017 optical Earth observation data can be used (Aschbacher 2017), which fulfil the requirements of a reliable observation system as proposed by Main et al. (2016), e.g. Woody Cover mapping based on the data's high geometric (10 m and 20 m) and temporal (5–12 days) resolutions, plus the guarantee of decades of data consistency due to the commitment of the European Commission to the Copernicus Programme (Article 4(1) of Regulation (EU) No 377/2014). As can be seen in Fig. 24.1, the increasing availability of free Earth observation data during recent years introduces new challenges and opportunities in the development of synergistic approaches combining optical and microwave information, e.g. Sentinel-1/-2 and NASA's Landsat-8.

New lidar sensors, specifically NASA's Global Ecosystem Dynamics Investigation (GEDI), complement surface monitoring capabilities by adding a vertical component, thus making 3-dimensional monitoring of vegetation structure feasible. Figure 24.1 also illustrates the rich heritage of radar sensors in space, which are being continued and extended by an increasing number of space agencies. The Sentinel-1 fleet constitutes the break-through to operationalizing microwave remote sensing, known to deliver information about vegetation volume and soil moisture estimates. New hyperspectral sensors in space (the German EnMAP was launched on 1 April 2022) deliver a wealth of spectral signatures to support degradation monitoring as described, e.g., by Oldeland et al. (2010) or assembled by Cawse-Nicholson et al. (2021).

The need for analysis-ready data (ARD) in the optical and especially radar domain has been recognized and formerly complex information is increasingly easy to be used and applied (e.g. through companies such as SINERGISE and Google Earth Engine). Various processing tools are accessible without cost for large datasets (e.g. PyroSAR, SNAP).

24.3 History, Opportunities, and Challenges for Degradation Monitoring in South Africa

Von Maltitz et al. (2019) developed five major assessment goals for land degradation monitoring strategies: rangeland degradation, bush encroachment, degradation of croplands, vegetation cover change, and alien species. Accordingly, the SPACES II project SALDi defined a workflow to generate the necessary indicators by exploiting Sentinel-1 and -2 time series as well as additional products, e.g. rainfall estimates from rain gauge and satellite observations [i.e. Climate Hazards Group InfraRed Precipitation with Station data (CHIRPS)] (Fig. 24.2).

The following sections explain the retrieval of EO products with further contributions from South African authors leading to (1) vegetational development, (2) woody cover trends, (3) bush encroachment, (4) invasive species, (5) drought and soil moisture assessments, and (6) overall degradation, thus covering the degradation aspects exemplified in Sect. 24.1 and summarized in Sect. 24.5.

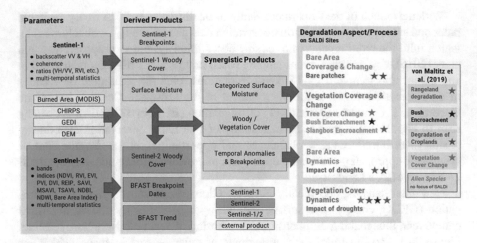

Fig. 24.2 Workflow of deriving land degradation indicators from Sentinel-1 and Sentinel-2 time series with additional products to deliver maps of four major degradation aspects. How these maps contribute to von Maltitz's assessment goals is indicated by coloured asterisks

24.3.1 Vegetational Development

Degradation processes related to vegetation development and productivity have been analysed in different studies based on EO data. A common way is the multi-temporal analysis of vegetation index data as derived from multispectral sensors which can measure aspects of ecosystem health and changes (e.g. Higginbottom and Symeonakis 2014; Wessels et al. 2004). Multi-annual analyses usually include the identification of trends and/or of abrupt changes in a given temporal behaviour. Up to the release of NASA's Landsat archive in 2008 and the computational advancements required to process it, the key optical sensor used for such analyses was AVHRR, due to its long temporal coverage and short repeat cycles, but also MODIS, SPOT-VGT, and Proba-V were used. These datasets allow for a large spatial and temporal coverage, but have clear limitations with respect to their spatial resolution of 250–1000 m. Early analyses though with 1 km AVHRR data from South Africa showed that Normalized Difference Vegetation Index (NDVI) growing season sums are in many cases lower than in non-degraded areas (Wessels et al. 2004). For Kruger National Park, these growing season sums were related to herbaceous biomass and its inter-annual variations, but sub-pixel heterogeneity of the coarse data as well as considerable scale differences to in situ data hindered the production of reliable biomass estimates (Wessels et al. 2006).

In South Africa, large portions of rangelands have seen extensive modifications from centuries of livestock farming (Hoffman and Ashwell 2001), alien plant species have invaded extensively in mountainous, riparian, and coastal regions (Van Wilgen and Wilson 2018), and natural fire regimes have seen major disruptions in grassy biomes and the Fynbos (Slingsby et al. 2020a, b). The ecological condition in these modified areas ranges from near natural to heavily modified depending on the degree to which ecosystem structure, function, and composition have been altered. Unfortunately, mapping these degrees of degradation is not easy, because one is dealing with continuous variation in the degree of modification rather than relatively distinct classes. These difficulties have led to some distinction between land cover mapping and degradation mapping in southern Africa.

A critical aspect when interpreting temporal changes of remotely sensed vegetation development with regard to human-induced degradation in arid and semi-arid areas is considering the influence of precipitation on observed vegetation trends and changes. This is particularly relevant for those large areas where rainfall is both highly variable and significantly influences vegetation productivity. The effect of precipitation can, for example, be considered in multiple regression analyses (e.g. Wessels et al. 2007a) or by comparing trends in time series of vegetation data and climate variables (e.g. Niklaus et al. 2015). Further prominent approaches are RUE (Rain-Use Efficiency) or RESTREND (Residual Trends). Using 1 km AVHRR-NDVI data and modelled 8 km NPP data, Wessels et al. (2007b) found RESTREND being better suited than RUE for the assessment of degradation in South Africa. However, several years later, Wessels et al. (2012) conducted a detailed study on the sensitivity of AVHRR-based trend and RESTREND analyses for degradation assessment and found these methods not capable of indicating land degradation with 1 km resolution NDVI data for a north-eastern study region in South Africa. Higginbottom and Symeonakis (2020) analysed changes in NASA's Goddard Space Flight Center (NASA/GSFC) Global Inventory and Modelling Studies (GIMMS) NDVI and RUE time series for break points and trends using Breaks for Additive Season and Trend (BFAST). They concluded that in southern Africa, constant positive trends in RUE combined with varying trend types of NDVI may be indicative of shrub encroachment, but they likewise highlighted difficulties in correct interpretation of drivers and processes. High-resolution trend analyses for South Africa were conducted using a 30 m Landsat EVI (Enhanced Vegetation Index) time series (Venter et al. 2020). The authors revealed patterns of degradation (e.g. bush encroachment) and restoration for some landscapes and thus demonstrated the high potential of such trend analyses with higher spatial detail. But at the same time, they underlined that Landsat data scarcity in the 1980s is a potential source of error.

Many approaches to time series analysis were developed in economics, social sciences, and engineering, and later found their way into phenology and thus remote sensing. While most models in the aforementioned disciplines try to come up with a prediction, many applications in remote sensing look for times when a break and change in a cycle (breakpoints) has occurred. The basic assumption of many methods is the notion that time series can be decomposed into three components: a seasonal, trend, and residual component. For land degradation issues, it is the trend component that provides information on whether fluctuations or even changes in the seasonal component are associated with a long-term change, or are just a one-time event that settles back to the old cycle after a certain time. Such analyses can be performed using the BFAST (Breaks for Additive Season and Trend) algorithm (Verbesselt et al. 2010), a well-established method for characterizing break points

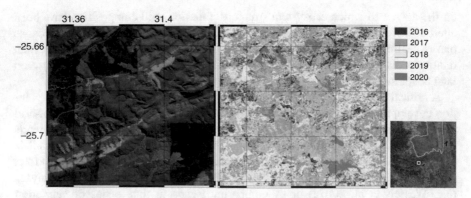

Fig. 24.3 BEAST-based breakpoint determination of Sentinel-2 NDVI time series. The area displayed is characterized by intensive forestry and located near the southern border of Kruger National Park. Left: Google Maps Overview. Centre: BEAST classification with year of major breakpoints. Right: Location map, white square indicating position of sub-scene

and associated trends. Considerations to reduce the computational effort led to the evaluation of a method capable of processing no-data values (reduction of pre-processing) and a faster determination of breakpoints using a different approach. For the map displayed in Fig. 24.3, the Bayesian ensemble algorithm of Zhao et al. (2019) called Bayesian Estimator of Abrupt change, Seasonality, and Trend (BEAST) was chosen to establish break points in a Sentinel-2 NDVI and Bare Soil Index (BSI) time series. Evaluation of the BEAST products show that the accuracy of the results varies with land cover type. Changes in forestry or fire scars are picked up as homogenous areas, whereas patterns over open grassland with shrubs are irregular. This kind of analysis also requires very good data preparation, because results are sensitive to outliers (e.g. undetected clouds) leading to negative spikes in time series, which might be mistaken for real events.

To evaluate Sentinel-1 time series to detect surface changes, irregularities in the radar backscatter and coherence time series were analysed. The processing procedure is based on the recurrence plot analysis (Marwan et al. 2007) followed by the detection of breakpoints using a Sobel filter. The aim is to identify regions where possible degradation processes take place, such as land-use changes susceptible for erosion (e.g. clearings for macadamia plantations), fallow farmland or shrub encroachment (compare Chap. 29 for slangbos mapping). The method for detecting breakpoints is initially based on a pixel-based smoothing procedure using a median filter over the period under investigation, here March 2015 to March 2020, followed by the identification of breakpoints in the time series using the Sobel filter. Figure 24.4 shows examples of Sentinel-1 breakpoint maps for the SALDi project regions Ehlanzeni and Sol Plaatje illustrating the detection of surface changes.

In addition to spatial and temporal resolution issues, several authors underlined, that degradation identified by EO-based vegetation products—even though corrected for rainfall influences—can likewise result from other processes. Southern Africa is comprised of largely open vegetation with low tree cover, often a high

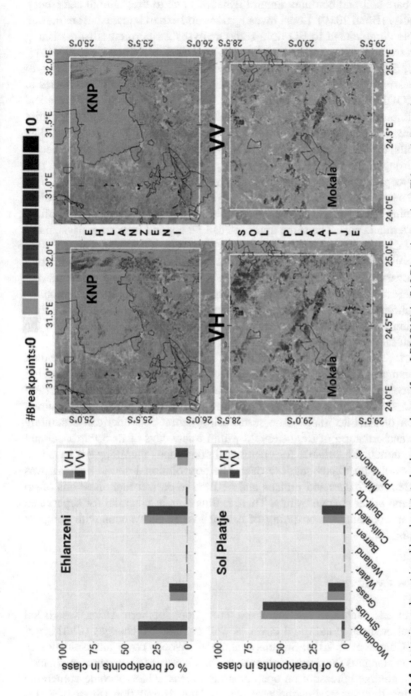

Fig. 24.4 Assigning radar-retrieved breakpoints to land cover classes (from LRI 2018) in SALDi project regions Ehlanzeni (upper images) and Sol Plaatje (lower images). Kruger and Mokala National Park boundaries are shown. Left: Sentinel-1 VV-, like-polarization and VH-, cross-polarization derived breakpoints per land cover class in % reflecting the different land usage systems and that both polarizations are suitable. Right: number of breakpoints in the Sentinel-1 time series 2015–2020 featuring areas and number of abrupt changes indicating regions of possible degradation processes (contains modified Copernicus Sentinel data [2015–2020])

fraction of bare soil and complex natural dynamics due to fire, rainfall sensitivity, and seasonality (Bond 2019). Local investigations and expert knowledge are thus an indispensable complement to EO time-series analyses for correctly distinguishing degradation from other processes (e.g. Wessels et al. 2007b; Prince et al. 2009).

Recently, Slingsby et al. (2020b) developed an approach that identifies degradation processes in the Fynbos biome of South Africa by identifying anomalies in observed MODIS NDVI relative to the expected NDVI produced by a hierarchical Bayesian time-series model. The model predicts the natural dynamics (postfire recovery rate, seasonality, maximum NDVI) based on abiotic environmental data (climate, soils, topography) and fire history, allowing identification of alien species invasions, drought or pathogen driven mortality, vegetation cover loss or fire. Their proof of concept including Landsat and high-resolution satellite data leading to an operational near-real-time change detection system for land managers and policy makers is being implemented as "Ecosystem Monitoring for Management Application" (http://emma.eco/). The same approach can be applied to other ecosystems if their natural dynamics can be suitably characterized with a time-series model. It is of great importance, that the dynamical features are based on factors such as fire and postfire recovery, a greater contribution of bare soil to observed vegetation indices, as well as high sensitivity to rainfall and a strong seasonality. This allows to monitor and detect abrupt or gradual changes in the state of an ecosystem in near-real time by identifying areas where the observed vegetation signal has deviated from the expected natural variation.

Moncrieff (2022) focusses on a different degradation process: the complex landscape changes of the Renosterveld. This is a hyperdiverse, critically endangered shrubland ecosystem in South Africa with less than 5–10% of its original extent remaining in small, highly fragmented patches. His work demonstrates that direct classification of satellite image time series using neural networks can accurately detect the transformation of Renosterveld within a few days of its occurrence, and that trained models are suitable for operational continuous monitoring if based on daily, high-resolution Planet satellite data. The convolutional neural network was applied to Sentinel 2 data and indices and resulted in correct identifications of up to 89% of land cover change events. There is thus a great potential for supervised approaches to continuous monitoring of habitat loss in ecosystems with complex natural dynamics.

24.3.2 Woody Cover

Woody cover encroachment has increased throughout southern Africa, which led to substantial environmental, land cover as well as land-use changes (Eldridge et al. 2011; O'Connor et al. 2014; Stevens et al. 2016). Woody cover intensification in rangelands by slangbos (*Seriphium plumosum*), black wattle (*Acacia mearnsii*), etc., will result in enlarged pressure on open grassland areas, which become vulnerable to overgrazing thus increasing the potential of land degradation (Snyman 2012; Oelofse et al. 2016). In protected areas and national parks, the knowledge of woody plants abundance and change are essential information for park management and conservation efforts. An intensification of woody plants will likely cause a reduction in grass and herbaceous biomass (Berger et al. 2019), which has direct influence on grazing animals, their territories, and migration as well as predators seeking herbivores (Munyati and Sinthumule 2016).

The high potential of combining multispectral and radar data for woody cover assessments has been illustrated in several studies. As an often-referenced example, Bucini et al. (2010) utilized Landsat Enhanced Thematic Mapper (ETM)+ (2000–2001) and radar data Japanese Earth Resources Satellite (JERS-1, 1995–1996) jointly with field measurements to map woody cover for Kruger National Park at 90 m spatial resolution. Skowno et al. (2017) combined Advanced Land Observing Satellite (ALOS) Phased Array L-band Synthetic Aperture Radar (PALSAR) data with national, Landsat-derived land cover maps to quantify changes in woodlands and grasslands of South Africa between 1990 and 2013.

Another example of synergistic radar/multispectral analyses is the work of Higginbottom et al. (2018) who predicted woody cover at 30–120 m spatial resolution for the South African province Limpopo by fusing Landsat Thematic Mapper (TM)/ETM+ and ALOS-PALSAR data and aerial imagery. Urban et al. (2020) generated a woody vegetation cover map at four different spatial resolutions (10 m, 30 m, 50 m, 100 m) for Kruger National Park based on Sentinel-1 data in combination with airborne lidar measurements. Holden et al. (2021) mapped invasive alien trees in the Boland mountains of the Fynbos biome, a key driver of biodiversity loss and run-off reduction in the region, at 10 m using a combination of Sentinel-1 and Sentinel-2 data. Multispectral time series from multiple resolution sensors were used, e.g., in a multi-scale analyses of woody vegetation cover in Namibian savannas by Gessner et al. (2013), including MODIS, Landsat, and very high-resolution satellite data. As woody cover is considered as an essential biodiversity variable (EBV) (Pettorelli et al. 2016), EO time series are key to develop wall-to-wall monitoring strategies (Urban et al. 2020). However, current remote sensing techniques are not likely to replace field measurements completely, as sustainable validation strategies for EO-derived woody vegetation composition with in situ data is still of very high importance (Kiker et al. 2014).

SALDI's woody cover retrieval for Kruger National Park used an airborne Light Detection and Ranging (LiDAR) strip from 2014 with a 2 m spatial resolution, made available through SANParks Scientific Services. These LIDAR data were converted to a woody cover percentage map with 10 m resolution to match Sentinel-1 and -2 pixel sizes and were then used in a random forest, machine learning (RF-ML) approach as training input. The resulting map products are shown in Figs. 24.5 and 24.6, illustrating the advantage of joint radar-optical analyses and additionally—with the new Copernicus satellites—achieving at the same time greatly improved spatio-temporal resolutions. In order to compare different sites in the future and over time, a uniform training data set is necessary, as now being available with NASA's Global Ecosystem Dynamics Investigation Lidar (GEDI), launched in 2018. This latter procedure has been applied to map woody cover changes illustrated in Fig. 24.7.

The woody cover estimates based on Sentinel-1 (radar) and Sentinel-2 (optical) data show generally similar patterns and overall agreement in Fig. 24.5. The radar

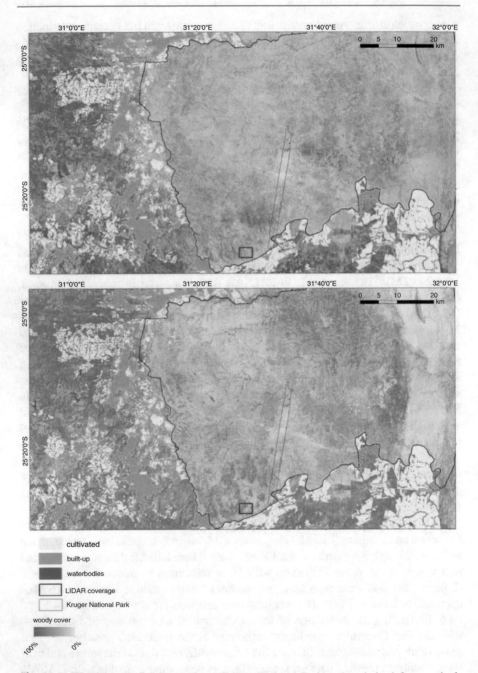

Fig. 24.5 Woody cover for the southern Kruger National Park region derived from a single airborne Lidar strip (small stripe in centre) and (upper map) Sentinel-1 time series 2016–2019 and (lower map) Sentinel-2 time series 2016–2017. The two products show complementarities and thus the need for radar-optical synergy. Waterbodies, built-up areas, and cultivated areas were masked using the National Land Cover 2017/2018 product (LRI 2018)

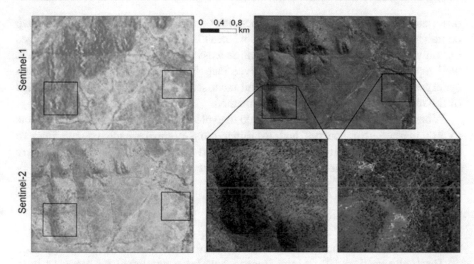

Fig. 24.6 Detailed view of Sentinel-1 vs. -2 woody cover estimates (green maps on left) and very high-resolution images (Google Earth Pro [GU1]) for the red-framed subset in Fig. 24.5, exemplifying advantages and disadvantages of either method. See text for further information

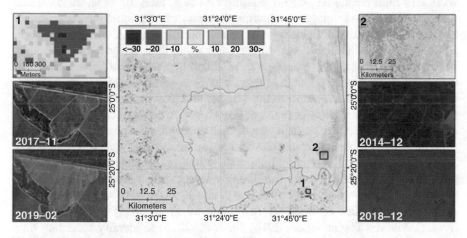

Fig. 24.7 Deriving woody cover change between 2016 and 2019 of the southern Kruger National Park and surrounding areas using Copernicus Sentinel-1 data and NASA's GEDI LiDAR at 50 m spatial resolution. Left side: box 1—woody cover decline due to harvested forest plantations, subimages: Google Earth (Maxar Technologies), right side: box 2—woody cover regrowth after fire, subimages: NASA Landsat-8, RGB = Bands 5-4-3 (EO Browser/Sinergise Ltd.)

product, though, exhibits more contrast between areas of high and low woody cover in flat terrain, but is still more affected by topography despite radiometric corrections. This finding is exemplified in Fig. 24.6: Sentinel-1 shows more contrast in flat areas, where areas of high and low woody cover are better discernable than in the optical dataset (framed box on the right). Sentinel-2 better detects differences in mountainous areas (framed box on the left). Here, the Sentinel-1 map contains uniformly high woody cover, whereas in reality denser woody coverages are only found on slopes oriented towards west and south. This spatial pattern is well represented in the Sentinel-2-derived map. This example underlines the high synergistic potential of using optical and radar sensors jointly for taking advantage of the respective strengths of each EO method.

When looking at change, it is feasible to explore either radar or optical products to not merge error sources and to rather cross-compare each change map to distinguish between consistent change areas and sensor-specific detections. Figure 24.7 illustrates a Sentinel-1 SALDi change product.

24.3.3 Bush Encroachment

There have been a number of studies on bush encroachment in southern Africa based on Earth observation data. Aerial images have been employed for mapping bush encroachment in several projects (e.g. Hudak and Wessman 1998, 2001; Stevens et al. 2016; Ward et al. 2014; Wigley et al. 2010), sometimes also in combination with very high-resolution (<2 m) satellite data (e.g. Shekede et al. 2015). These datasets have the advantage of allowing detection of individual bushes (compare also Sect. 24.3.7), and aerial images often date back to decades where space-borne EO data were not available yet. At the same time, their analysis is hindered by the enormous efforts for data pre-processing and because quantitative inter-image comparisons are hardly possible or extremely time-consuming for large areas due to missing radiometric comparability (Hudak and Wessman 2001). Furthermore, the low temporal frequency of multiple years between airborne campaigns does not allow the needed temporal monitoring of bush encroachment processes.

With respect to multispectral remote sensing analyses at spatial resolutions of 30 m or less, usually selected radiometrically optimal acquisitions were analysed prior to the availability of dense satellite time series such as now being recorded by the two Copernicus Sentinel-2 satellites A and B since 2016 resp. 2018. Earlier examples are studies using SPOT (e.g. Munyati et al. 2011) and Landsat (e.g. Symeonakis and Higginbottom 2014; Ng et al. 2016) to identify spreading bush areas or for mapping the distribution of encroaching bush species in southern Africa. Cho and Ramoelo (2019) have developed a methodology for detecting increasing tree and bush cover in the grassland and savanna biomes of South Africa using MODIS data despite its coarser geometric resolution. The methodology is based on asynchronous NDVI phenologies of grasses and trees in these semi-arid systems. Using a 16-day NDVI time series generated from MODIS NDVI data between 2001 and 2018, the authors first determined the best time for mapping tree cover in the region, which turned out to be a narrow period from Julian day 161–177 (June 10–26). This is the period of maximum contrast between grasses and trees. Eight tree-cover maps (2001–2018) were generated using linear regression models derived from Julian day 161 for each year.

In rarer cases, radar satellite data were applied: Main et al. (2016) combined time series of the European Space Agency's (ESA's) Envisat Advanced Synthetic

Aperture Radar (ASAR) and airborne LiDAR data for 2006–2009 to analyse woody cover in Kruger National Park and surrounding areas with a spatial resolution of 75 m, Urban et al. (2021) synergistically combined Sentinel-1 radar and optical Sentinel-2 time-series data to monitor slangbos encroachment, a woody shrub, on arable land (see Chap. 29).

24.3.4 Invasive Species (*Acacia mearnsii*)

The invasion of productive lands by alien plants is an important contribution to land degradation. Large monocultures of alien species have a negative impact on water resources, pasture and crop production and biodiversity. This is the case with Black Wattle, *Acacia mearnsii*, a prominent invasive alien species in South Africa. Remote sensing of invasive species is mainly aimed at mapping the extent of the invasion, quantifying their biomass, and highlighting invasion hotspots. Masemola et al. (2019, 2020a, b) conducted numerous studies to: (1) explore the utility of hyperspectral data to discriminate *A. mearnsii* from native species, (2) determine the optimal period to spectrally distinguish *A. mearnsii* from native plants, (3) determine if vegetation indices related to the unique biological traits of *A. mearnsii* have the potential to distinguish it from native species, (4) assess the potential of multispectral Sentinel-2 data to map the distribution at the landscape level, and (5) explore the applicability of radiative transfer model (RTM) simulations to characterize the differences between *A. mearnsii* and native species. Through this multi-step approach, *Acacia mearnsii* could be distinguished from native species with overall accuracies ranging from 75% to 90%.

24.3.5 Drought and Soil Moisture

Southern African biomes are particularly prone to drought because more than 65% of the area is semi-arid and thus environmental conditions alternate strongly in space and time. Monitoring and spatio-temporal assessment of drought impacts is a challenge due to a limited number of weather stations. In addition, drought monitoring is difficult, since effects can accumulate over time. Droughts have major impact on ecosystems, such as fire severity (Mukheibir and Ziervogel 2007), biodiversity and ecosystem functioning (Masih et al. 2014; Graw et al. 2017), and economy, e.g. food production (Verschuur et al. 2021). Therefore, analysing surface moisture dynamics is of high importance, as it is also highly correlated to vegetation and soil respiration, which represents both root and microbial respiration, and is one of the main ecosystem fluxes of carbon (Makhado and Scholes 2011).

Various studies focus on the development of drought monitoring concepts using EO data from different sources (AghaKouchak et al. 2015) and for different applications—agriculture (Bijaber et al. 2018; Zeng et al. 2014; Winkler et al. 2017), grasslands (He et al. 2015; Villarreal et al. 2016), savanna ecosystems (Graw et al. 2017; Western et al. 2015). The majority of these applications utilized optical information with coarse spatial resolution (MODIS and AVHRR). Surface moisture parameters and drought conditions were analysed using different ratios, e.g. NDVI (Normalized Difference Vegetation Index), EVI (Enhanced Vegetation Index), VCI (Vegetation Condition Index) as well as SPI (Standard Precipitation Index) (Graw et al. 2017).

Marumbwa et al. (2019, 2020, 2021) conducted several studies to assess the impact of meteorological drought on southern African biomes. To achieve this objective, the authors first analysed spatio-temporal rainfall trends to establish trends at pixel-level and then assessed drought impact on biomes using VCI products. Further, they analysed drought and land cover interactions using land cover data and a novel land cover "village pixel" developed from livestock density data as a proxy for the type of rural community. Based on the 2015–2016 season analysis, the Kruskal–Wallis test showed significant difference in drought impact (mean VCI) among different land cover types. This study provided information to adapt usage to different ecosystems to enable communities to better cope with droughts.

Recent microwave satellite missions for operational soil moisture retrieval are ESA's SMOS (Soil Moisture and Ocean Salinity) and NASA's SMAP (Soil Moisture Active Passive) sensors, which utilize L-Band passive radar data to estimate global soil moisture with coarse spatial resolution (10–25 km) (Entekhabi et al. 2010; Kerr et al. 2012). The most recent global dataset is ESA's Climate Change Initiative (ESA-CCI) soil moisture product at 25 km pixel spacing (Dorigo et al. 2017), for which Khosa et al. (2020) found a correlation greater than 0.6 with in situ measurements for two sites in Kruger National Park. Soil moisture applications with higher spatial resolution have been carried out since the launches of ESA's radar sensors on-board the European Remote Sensing (ERS-1/-2) satellites (e.g. Bourgeau-Chavez et al. 2007; Haider et al. 2004) and ENVISAT ASAR (e.g. Paloscia et al. 2008; Zribi et al. 2005), Canada's commercial RADARSAT (e.g. Glenn and Carr 2004; Leconte et al. 2004), and are continued with the recent Copernicus Sentinel-1 data (e.g. Alexakis et al. 2017; Lievens et al. 2017). The potential of a synergetic combination of Sentinel-1 and Sentinel-2 has been addressed by only a few studies so far (e.g. Gangat et al. 2020; Gao et al. 2017; Urban et al. 2018), indicating that the high repetition rates of the Sentinel-scenes offer good determination rates for surface moisture estimates.

The open SPACES II SALDi-project data cube offers—amongst many other data sets and products (see Chap. 29)—analysis-ready Sentinel-1 radar time series, which allow application of multi-temporal change detection methods for surface moisture retrieval. A sample sequence of six such surface moisture maps derived for December 2020 are shown in Fig. 24.8. The maps begin with dry surface conditions indicated by reddish tones. After December 14, surface moisture generally increases, as indicated by yellow to blue tones. These model results illustrate the heterogeneity of surface conditions (vegetation structure, soil features) and precipitation effects over time and space. The model is briefly explained in Chap. 29.

To validate Sentinel-1 radar-retrieved products, it is essential to consider type and status of vegetation and relate to moisture conditions either through information about precipitation or in situ soil moisture measurements. Therefore, the SALDi-

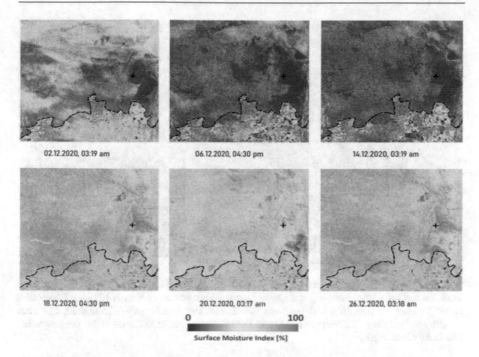

Fig. 24.8 Relative surface moisture derived from Copernicus Sentinel-1 radar data for December 2020 for the south-east corner of Kruger National Park (KNP) (black line indicates park-border). The Surface Moisture Index is sensitive to water content of vegetation and the underlying soil surface. It therefore highlights structural differences (e.g. fire scars with in KNP), land-use practices (agricultural patterns in southern territories), and effects of precipitation. The cross indicates the location of SALDi's surface moisture in situ instrument. Times are in UCT

project deployed one SMT100 Time Domain Transmission instrument, which measures volumetric soil moisture, in each of the six project areas. At each site, a total of eight sensors take continuously every 5 min measurements in a depth of 6–10 cm. Figure 24.9 gives geographic information and illustrates the harsh surface conditions.

Figure 24.10 illustrates the temporal behaviour of Sentinel-2 NDVI and the two complex retrievals ESA CCI Soil Moisture and SALDi Sentinel-1 Surface Moisture (SurfMi) product. Precipitation data combined with local knowledge and in situ soil moisture is the key for interpretation: the NDVI-lag in October and November 2020 was caused by a severe fire; thus the smooth surface (see photo in Fig. 24.9) results in specular scattering of the Sentinel-1 signals, keeping SurfMi to Zero (except during rain) although the in situ moisture is increasing. These important insights are needed for a synergistic savanna monitoring tool to map simultaneously vegetation status, realistic surface moisture conditions and even time and location of local precipitation events.

Fig. 24.9 Left: Map of SALDi's project area Ehlanzeni stretching 100 × 100 km² south from Skukuza and west from Komatipoort. Green line: southern border of Kruger National Park, black cross: location of SALDi's soil moisture instrument near Lower Sabie. Instrument site after flooding in February 2022 (upper right) and fire in September 2020 (lower right) (Photographs: Tercia Strydom, KNP)

24.3.6 Nation-Wide Assessments

The first national survey, conducted by Hoffman et al. (1999) and based on expert interviews at the district level, concluded that about 26% of the surface of South Africa was heavily degraded (own calculations). Fairbanks et al. (2000) calculated on the basis of Landsat satellite images for the years 1994 to 1996, that 80% of the country is covered by "natural" tree and grassland vegetation and only 20% were transformed by humans. Using a similar approach, Schoeman et al. (2013) identified for the year 2005, 16% as being transformed and corrected the value for the year 1994 to 14.5%. For the period from 1981 to 2003, Bai and Dent (2007) identified a degradation process for nearly 30% of the country using NASA/GSFC GIMMS NDVI remote sensing time series.

The first national-scale, remote-sensing based land cover map was produced in the late 1990s using 1994–1995 Landsat 5 data (Fairbanks et al. 2000). Thirty-one classes were mapped manually on hardcopy space maps at approximately 1:250,000 scale. In the early 2000s, a semi-automated land cover classification was produced using Landsat 5/7 imagery based on 2000–2001 data (van den Berg et al. 2008). Both datasets were widely used in research and conservation planning. In 2013, the national government began a programme to regularly produce national-scale land cover maps using a standardized classification scheme.

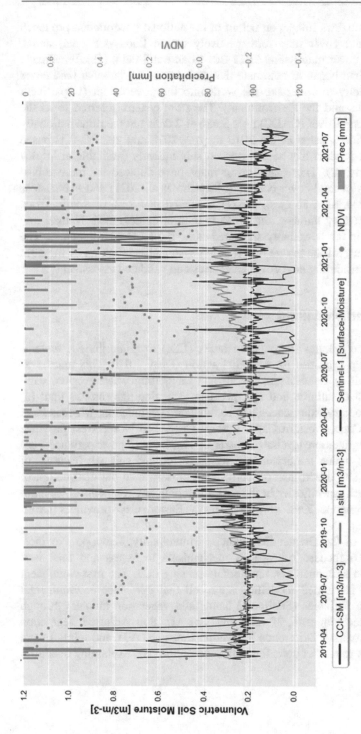

Fig. 24.10 Lower Sabie time series plot since installation of SALDi in situ instrument in March 2019 (in situ gaps occurred when instrument was flooded) until August 2021 with SANParks Lower Sabie precipitation data, SALDi Sentinel-2 NDVI time series at 20 m pixel size, SALDi Sentinel-1 Surface Moisture (SurfMi) product as 30 × 30 m²-mean over nine pixel every 4–6 days, and daily ESA CCI Soil Moisture (SM) of 1 pixel covering 25 × 25 km². The ESA- and Sentinel-1 products rather agree with vegetation growth, where the latter has a significantly larger dynamic range and shows fine sensitivity to precipitation any time during the year. During the rainy season many Sentinel-2 scenes are cloud-covered and NDVI-values drop to 0 (but are shown to illustrate missing optical information). Lowest CCI SM and SALDi SurfMi values from October to December 2020—despite increasing in situ moisture—are a result of wrong model assumptions, which disregard possible backscatter mechanisms, such as specular reflection from a severely burnt surface as in this case (see Fig. 24.9)

The company GeoTerra Image, on behalf of the national government, produced a 1990 national land cover map retrospectively (using Landsat 5 data) and a 2014 national land cover map (using 2013–2014 Landsat 8 data) (GeoTerraImage 2015). A process then began to automate the production of a national land cover and bring the capacity to undertake the work into the government (Department of Forestry Fisheries and the Environment). The new system, referred to as the Computer Aided Land Cover (CALC) uses Sentinel-2 data and the same nationally accepted classification scheme as the 2013 map. A 2018 and 2020 CALC based national land cover product has been produced subsequently (available at https://egis.environment.gov.za/). These land cover maps have allowed for more robust time series analysis of land cover change (Musetsho et al. 2021) and habitat loss (Skowno et al. 2021). Skowno et al. (2021) estimated that 22% of South Africa has been transformed by humans, and 78% is natural or degraded. Natural and degraded areas could not be reliably separated using the NLC 2018 methods. The uncertainties are thus still large due to a missing systematic definition of the term land degradation considering land cover categories and validated assessment.

24.3.7 Integrated Modelling

The ARC developed a Land Degradation Index (LDI) to guide the SA government with setting up priorities for remedial action (DEA 2016). This index was generated by exploiting Landsat time-series and integrating water erosion, wind erosion, aridity index, salinity, and soil pH data. The findings showed that: (a) most parts of the country experiences low to moderate degradation, whereas large parts of Northern Cape, Western Cape, Eastern Cape, and North West province experience high degradation levels; (b) parts of Northern Cape experience very high degradation of which the drivers are wind erosion and soil pH; (c) areas of severe degradation and desertification correspond closely with the distribution of communal rangelands, specifically in the steeply sloping environments adjacent to the escarpment in Limpopo, KwaZulu-Natal, and the Eastern Cape provinces (DFFE 2018).

In a case study of the Greater Sekhukhune Municipality, Limpopo Province, Nzuza et al. (2020) focused on various mechanisms to assess and map land degradation using in situ, ancillary, and remote sensing data. The first component was the use of an integrated modelling approach that combines environmental and remote sensing variables (Sentinel-2 bands and vegetation indices) through machine learning techniques. Leaf Area Index (LAI), Soil Adjusted Vegetation Index (SAVI), Normalized Difference Vegetation Index (NDVI), and rainfall were significant variables to predict potential land degradation risk, explaining over 80% of the variation.

The second component (Nzuza et al. 2021) was to evaluate a triangulation approach consisting of remote sensing products, Sustainable Land Management (SLM) practices [using the World Overview Conservation Approaches and Technology (WOCAT) mapping questionnaire (Gonzalez-Roglich et al. 2019)], and a participatory expert assessment. The climatic variability, overgrazing, poor governance, and unsustainable land management practices were cited as major causes of land degradation. Perceived types of land degradation were soil erosion and loss of vegetation cover. The study demonstrated that complementary information from various sources is crucial for monitoring and assessing land degradation (see Fig. 24.11).

24.3.8 High-Resolution Validation

Land degradation monitoring relies on precise in situ information. Reference data to be used to estimate parameters linked to degradation status need to follow specific requirements in terms of temporal and spatial precision. In order to map the extent of, e.g., erosion patterns or woody plants, greater spatial resolution is essential. Here, a product is presented, that has been generated in the context of the SPACES II Ecosystem Management Support for Climate Change in Southern Africa (EMSAfrica) project to support research and management in SANParks' Kruger National Park and serves as a reference data set for validation of EO-retrieved surface products (Heckel et al. 2021). The necessary aerial imagery is freely available from the archives of the Chief Directorate, National Geospatial Information (CDNGI), Department of Agriculture, Land Reform and Rural Development (DALRRD). The derivation of the surface models was accomplished by (a) metadata preparation (definition of flight parameters and input of camera specific information), (b) ingestion of data to the bulk processor, (c) selection of tie points (epipolar pairing of stereoscopic imagery), (d) semi-global matching to derive the Digital Surface Model (DSM), (e) surface height removal to retrieve the Digital Terrain Model (DTM), and (f) orthorectification of the initial aerial imagery using the derived DSM. Processing of the extensive data set was carried out using the CATALYST Enterprise software environment. The data sets were modelled with a ground sampling distance of 0.25 m, 1 m, and 5 m and tiled according to CDNGI standards. Free download of the wall-to-wall terrain data coverage (large map in centre of Fig. 24.12) is possible through the Centre for Environmental Data Analysis (CEDA) platform of the British National Centre for Atmospheric Science. For challenges and limitations, see Table 24.1.

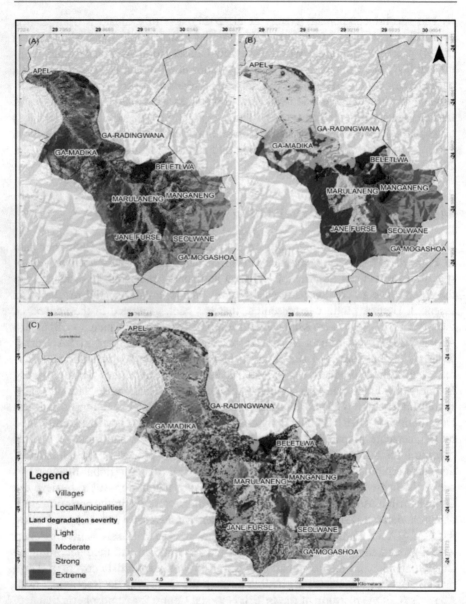

Fig. 24.11 Spatial distribution of land degradation severity based on (**a**) a remote sensing derived map, (**b**) participatory assessment workshop, and (**c**) triangulation approach severity map (Nzuza et al. 2021). Map **c** illustrates how observations and social aspects diversify degradation classification and thus remedial needs

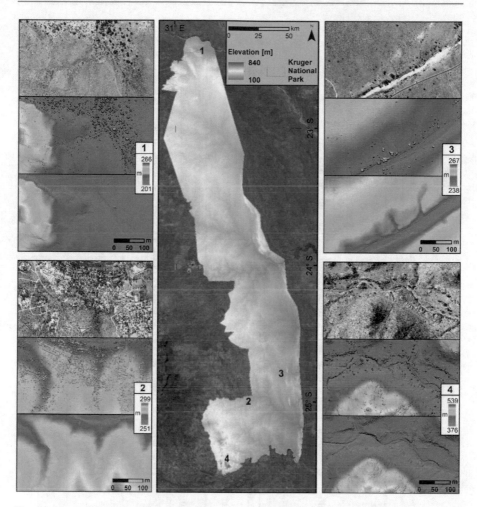

Fig. 24.12 Centre: First wall-to-wall, 25 cm posting terrain model of Kruger National Park. Numbers indicate the locations of the four subsets 1 (upper left), 2 (lower left), 3 (upper right), 4 (lower right). Each subset contains the orthomosaicked airphotos (top), digital surface model (middle), and digital terrain model (bottom). The subsets illustrate the great detail of cover types and subtle topographies: single trees and larger bushes can clearly be recognized, dry riverbeds and channels are reconstructed including run-off features (Background data: Environmental Systems Research Institute (ESRI), Maxar, GeoEye, Earthstar Geographics, Centre National d'Etudes Spatiales (CNES)/Airbus Defence and Space, United States Department of Agriculture (USDA), United States Geological Survey (USGS), AeroGRID, Institut Geographique National, and the Geo-Information Systems user community)

Table 24.1 Summary of this chapter findings for new spatio-temporal EO monitoring capabilities for land degradation in southern Africa

Observable	Key achievements	Key constraints	Emerging technologies
Vegetation development	Increasingly improved spatial and temporal resolution Easy to access data cubes with analysis-ready data Multiple time series techniques and available RESTREND, RUE	Length of available time series Poor compatibility between different sensors Inter-seasonal differences greater than between area differences Overarching effect of rainfall—Stronger signal than the degradation signal CO2 fertilization effect vs. degradation effects Missing in situ information (phenological observations, weather stations, soil moisture, etc.)	Exploitation of full spatio-temporal radar-optical synergies Hybrid modelling techniques merging physical understanding with machine learning approaches Citizen science support
Woody cover trends	Well mapped for expansion of woody plants into grassland Constantly improving woody cover products	High range of uncertainty Possible interference from rainfall effects Radar expertise missing	New spaceborne LiDAR data availability Upcoming radar EO systems (e.g. NASA/ISRO) with longer wavelengths (L- and P-band) and polarimetric capabilities
Bush encroachment	Aerial images, sometimes in combination with very high-resolution satellite imagery, allow discrimination of individual bushes	Enormous efforts for data pre-processing, extremely time-consuming and lack of radiometric comparability Low temporal frequency of aerial imagery, often only available for smaller areas	Upcoming constellations of hundreds of smallsats (miniaturized satellites under 1200 kg) allowing daily coverages at high spatial resolutions around 2–5 m Operational processing opportunities with cloud-based, bulk processing
Invasive species	Reliable discrimination from native species	Mapping accuracy—Spatially, i.e. minimum mapping areas, and thematically, i.e. needs very good spectral library	Hyperspectral data: increasing airborne and drone availabilities and EnMAP satellite—launched April 2022 Radiative transfer model simulations

Drought and soil moisture estimates	Different states of surface moisture can be mapped over large areas at medium spatial resolutions several times per month, additionally provision of uncertainty flags for the user	Influence of vegetation on the ground (do we map moisture of the surface or the canopy) Only relative measurements More in situ references needed Radar expertise missing	Synergistic use of active/passive microwave data plus optical data to better constrain radar products Big data—If all governing parameters can be sampled and included in modelling Fully polarimetric L-band sensors allow better differentiation of backscattering mechanisms (NASA's and the Indian Space Research Organisation's (ISRO) Synthetic Aperture Radar NISAR
Overall degradation	Development of nation-wide indices Availability of EO data and products Spatial methods (online) courses	Many degradation processes give false positives or negatives Degradation is a multi-phased problem Conflating degradation potential and actual degradation	Nation-wide coordinative actions of process understanding (e.g. Expanded Freshwater and Terrestrial Environmental Observation Network EFTEON) and SANSA's digital earth South Africa (DESA) initiative

24.4   Big Data Challenges and Insight into SALDi Process Flow

Remote sensing data as provided by various satellites has proven as a useful tool to monitor the environment and its changing land use and cover through time (e.g. Wulder et al. 2008). It provides the potential of a synoptic view of an explicit spatial extent.

Since the launch of the Sentinel-1 and Sentinel-2 satellites in 2015, the temporal and spatial resolution of freely and open available earth observation data has significantly increased (Gómez et al. 2016). The volume of data is quickly growing especially in the earth observation domain. Multiple scientific projects have produced an exorbitant amount of data in recent years. In the era of big data, information retrieval is increasingly gaining importance. There is an emerging need for developing tools that are able to face the challenge of big data. The methods are expected to be precise and tolerant to noise. The results are expected to be interpretable in order to provide a better understanding of data structures (Giuliani et al. 2019).

One of the areas of research in which great progress has been made in recent years to address the aforementioned big data challenges are earth observation data cubes. Earth Observation Data Cubes (EODC) are known as a promising solution to efficiently and effectively handle big Earth Observation (EO) data generated by satellites and made freely and openly available from different data repositories (Giuliani et al. 2019). So far various EODC implementations throughout the globe are currently operational: (1) Digital Earth Africa and Digital Earth Australia (Dhu et al. 2017), (2) the Swiss Data Cube (Giuliani et al. 2017a, b), (3) the EarthServer (Baumann et al. 2016), (4) the E-sensing platform (Camara et al. 2017), (5) the Copernicus Data and Information Access Services (DIAS) (European Commission 2018), or (6) the Google Earth Engine (Gorelick et al. 2017). These initiatives are paving the way for broadening the use of EO data to larger communities of users, thereby supporting decision-makers with timely information converted into meaningful geophysical variables and ultimately unlocking the information power of EO data (Giuliani et al. 2019).

24.4.1 General Big Data Situation

Since the start of continuous satellite observations in 1972, EO data exceeded the petabyte-scale and has been made available to a broad audience by increasingly open access options from different platforms. Big data analysis challenges besides data storage include searching for the data, downloading it, data pre-processing, conducting data analysis, and finally decision-making based on the data analysis. Nevertheless, the full potential of EO data is still not exploited due to the lack of (1) scientific knowledge, (2) difficult to access, (3) lack of expertise, (4) particular structure, and the (5) effort and storage costs. As Laney stated (2001), in the future, there will be no greater barrier to effective data management than the variety of incompatible data formats, non-aligned data structures, and inconsistent data semantics. Surely, one of the most demanding aspects is the need to develop cross and multidisciplinary applications and integrating heterogeneous data sources (Nativi et al. 2015). Therefore, the *big five* of the big data challenges could be named as (1) volume, (2) variety, (3) velocity, (4) veracity, value, and validity as well as (5) visualization. To address these challenges within the SALDi project, the SALDi data cube was established. Digital Earth Africa and Digital Earth South Africa were established by national entities to address the above challenges on a national level.

24.4.2 Necessary Big Data Exploration Methodologies

Conventional technologies in the EO context have limited storage capacity, rigid management tools and are cost expensive. Besides, they often lack scalability, flexibility, and performance which are urgently needed in the context of big data. Therefore, big data management requires new methods and powerful technologies. Local processing of big data will be limited due to the steadily growing amount of data. The solution to overcome these problems is cloud computing, especially in terms of data processing and usage. It can provide the possibility to make large volume of EO data available to a wide range of users, as it provides an environment which is designed for easy EO data handling and visualizing without a need of downloading and pre-processing them beforehand, but with a strong focus on data analyses and decision-making based on large spatio-temporal datasets in an analysis-ready format.

This is especially handy when it comes to time series analysis. As remote sensing satellites revisit a given location on the earth's surface in regular time steps, image sequences of the same areas over time are generated (Ferreira et al. 2020). Time series derived from these sequences can be utilized for analysing land cover change and soil degradation.

To simplify the workflow of time series analysis which have been derived from satellite images, analysis-ready data (ARD) can be produced and organized in multidimensional earth observation data cubes (EODC). ARD can be defined as "satellite data processed to a minimum set of requirements and organized into a form that allows immediate analysis with a minimum of additional user effort and interoperability both through time and with other data sets" (Siqueira et al. 2019). Thus, ARD implicates processing satellite imagery from data acquisition to radiometric calibration, and through additional conversions, to top-of-atmosphere (TOA) reflectance, and finally surface reflectance (Giuliani et al. 2017a, b).

The term data cube refers to a set of image time series associated with spatially aligned pixels (Appel and Pebesma 2019). Each element of an Earth observation (EO) data cube has two spatial dimensions and one temporal dimension, and is associated with a set of values (Giuliani et al. 2019). The SALDi data cube from optical and radar satellite data includes all necessary pre-processing steps and is generated to monitor vegetation dynamics of 5 years for six focus areas within South Africa. Intra- and inter-annual variability in both, a high spatial and temporal resolution will be accounted for to monitor land degradation. Therefore, spatial high-resolution Earth observation data from 2016 to 2021 from Sentinel-1 (C-Band radar) and Sentinel-2 (optical) are integrated in the SALDi data cube. Additionally, a number of vegetation indices as well as the Bare Soil Index (BSI) are implemented to account for explicit land degradation and vegetation monitoring. A national land cover classification (South African Department of Forestry, Fisheries and Environment) with 72 various land cover classes as well as a digital elevation model in a spatial resolution of 30 m (based on Copernicus DEM with global 30 m resolution (GLO-30) is available. Thus, the SALDi data cube builds a platform which can be utilized for an efficient data analysis of various multi-temporal land surface products (cf. Fig. 24.13).

All current developments in the context of big (EO) data would not have been possible without Free and Open Data policies to facilitate access to data and open source code to efficiently develop software solutions (Ferrari et al. 2018). Open Science is not only a new approach to research but also to educational processes, which seeks to make scientific research more collaborative and transparent. It makes knowledge accessible by using digital technologies and new collaborative tools (European Commission 2016). The open data practice enables scientific research to be reused, redistributed, and reproducible.

24.4.3 Available Infrastructures

To ensure the long-lasting availability and permanent access to EO data and results including the ability to continuously develop and adapt data processing chains a so-called EO-Data-Repository was established for the six research sites within the SALDi project. It allows flexible data management and furthermore provides an analysis environment for earth observation data. Through an interactive user interface all partners are empowered to actively participate in data handling.

Therefore, an Earth Observation Data Cube (ODC) was set up for the six research sites within the SALDi project (SALDi data cube). It allows the handling of large data amounts from various data sources and different data types. The SALDi data cube is used for data download, storage, and pre-processing of the Sentinel-1 and Sentinel-2 satellite data which can be used for further remote sensing products and methods. The users can access the data cube through an interactive user interface. The SALDi data cube serves as a central infrastructure for geospatial data and thus forms the interface between remote sensing data-based data provision and method development.

The added value of the SALDi data cube is making earth observation data easily accessible to end-users who do not have in-depth expertise in the evaluation of remote sensing data. The data cube considerably simplifies the access and use of satellite data, since complex and time-consuming steps such as (1) the download, (2) the storage, and (3) the pre-processing are already implemented in the SALDi

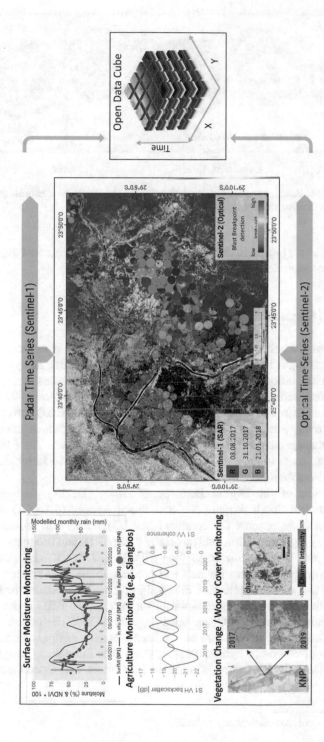

Fig. 24.13 The SALDi data cube is a platform for an efficient and user-oriented analysis of land surface dynamics based on multi-temporal earth-observation data and products

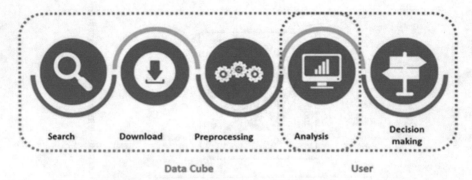

Fig. 24.14 Simplified basic remote sensing workflow. The SALDi data cube is capable of easing this workflow by automating the first three to four processing steps

data cube[1] and the user (via an interactive user interface) has direct access to gridded data sets ready for analysis and decision-making (Fig. 24.14).

24.4.4 Digital Earth Africa

There are various initiatives in Africa and South Africa that have been established. The Digital Earth Platform provides access to near real-time analysis-ready medium- to high-resolution data and products for various applications derived from the USGS datasets such as Landsat and Copernicus Sentinel-1 and -2 for the entire African continent. The production of analysis-ready data (ARD) derived from Sentinel-1 is currently underway. The South African National Space Agency hosts the programme management office (PMO) for Digital Earth Africa. The PMO will ensure that various users in the continent have access to earth observation data and products that address their needs.

The development of the Digital Earth South Africa (DESA) platform is a collaboration between South African National Space Agency (SANSA) and the South African Radio Astronomy Observatory (SARAO). DESA will provide users with access to very high-resolution analysis-ready data (ARD) and products derived from Satellite Pour l'Observation de la Terre (SPOT) 1–7. It will also provide a common and consistent platform for data and product access, and sharing which will enable users to focus on application-driven algorithms. This reduces the burden of downloading and pre-processing data for the end-users. The ARD is developed according to Committee on Earth Observation Satellites (CEOS) definition. DESA uses the Open Data Cube architecture and utilizes a variety of open source tools such as the Jupyter Notebooks, Python libraries, Open Data Cube Stats, PostgreSQL database, Open Data Cube Explorer, Command Line Tools, and Open Geospatial

---

[1] https://datacube.remote-sensing.org/projects/saldi/.

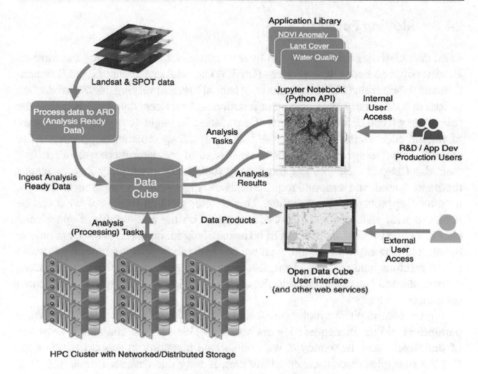

Fig. 24.15 Overview of the Digital Earth South Africa Platform

Consortium (OGC) web services (Mhangara and Mapurisa 2019). The tools allow users to access and analyse big datasets and products in a cloud environment. The overview of the DESA platform is shown in Fig. 24.15.

24.4.5 International Cooperation and Knowledge Exchange

Regional-scale data cubes like the SALDI data cube and the data within this cube are easier to handle—compared to conventional earth observation data— especially for those users who are not explicit remote sensing specialists. To ensure both, a successful international cooperation and efficient knowledge exchange, it is essential that data and processes within the data cube are consistent and well documented.

In addition, to realize the full potential of the ODC products to address local and regional decision-making and policies, it is important to increase research and gather in situ ground data for proper algorithm and product validation. Over time, it is expected that open data products will increase, their accuracy will improve, and data access and usage will become easier and more efficient for everyone.

24.5 Moving Forward

Land degradation, as defined by the Intergovernmental Science-Policy Platform on Biodiversity and Ecosystem Services (IPBES) Global Assessment of Land Degradation and Restoration (2018) includes both habitat loss and varying degrees of decline or loss of biodiversity and ecosystem function and services, thus encompassing the full range of ecological conditions. Degradation is slight to severe modification of natural ecosystems due to factors like overgrazing, erosion, inappropriate fire regimes, invasive species, etc., but some vestige of the natural ecosystem remains (see also Chap. 3). Hence, EO monitoring tools have to suit very heterogeneous thematic, spatial, and temporal requirements. A single sensor, a single methodology, a "mono"-approach will not suffice. This chapter gave an insight of what can be achieved with state-of-the-art procedures based on the new wealth of space- and airborne observational data. It has to be acknowledged, however, that we are only at the start of data exploration and what we can learn about spatio-temporal dynamics of our precious land surfaces. Table 24.1 is an attempt to summarize achievements, constraints, and emerging technologies as a quick reference for further scientific and programmatic actions.

Figure 24.2 depicts which optical and radar EO products can lead to surface parameters aiding in degradation monitoring. This chapter contains a selection of derivables, such as woody cover and surface moisture, to support monitoring. But the examples also illustrate limitations, if only one data set or one approach is applied or if in situ data is missing. The presented key EO indicators, which are treated as correspondent to relevant degradation processes, consist of poorly validated surface products with respect to savanna vegetation state, structure, dynamics, and surface moisture conditions. They are, e.g., called "woody cover", but they strictly rather represent a "spectral product", not an "information product" (analogue to unsupervised and supervised classification). The most promising methods therefore are based on a better physical understanding of spectral responses, thus enabling a knowledge-based interpretation of annual and inter-annual variations. Understanding spatio-temporal patterns lead to meaningful machine learning (ML) approaches—or vice versa, ML-retrieved results can then be associated with either experience or with discoveries. Having the Copernicus Sentinel-fleet and Landsat time-series for the future decade(s) available and thus an unprecedented wealth of spectral and spatial data sets, multi-temporal characteristics on the pixel-level (statistics, trends, break points, etc.) are assets, which still need to be associated with relevant surface features. Possibly, a new product nomenclature can be drafted, which is closer to spectral characteristics, including radar, and thus to the true nature of EO observations: e.g. radar water clouds and their relevance for structural changes, spectral indices and their significance for pixel-unmixing strategies for complex savanna biomes.

To accomplish break-throughs in the exploitation of EO big data sets, it needs the respective technical infrastructure as described in Sect. 24.4, and it needs experienced natural scientists from the regions. The methods and EO products developed during the SPACES 2 projects were only accomplished based on the strong cooperation that grew between the South African and German team partners and colleagues. Regular conferences, such as SANParks' Savanna Science Network Meeting, where interaction is yearly greatly fostered, have tremendously improved mutual methodological understanding beyond project concepts. Dedicated Summer Schools are a further important asset to develop strong scientific grounds for implementation of the achieved procedures and to educate the next generation of responsible Earth observation experts.

Acknowledgement   The authors would like to thank all colleagues who contributed to the ideas developed in this chapter during field work or in numerous discussions in South Africa. This contribution benefited from funding by the Federal Ministry of Education and Research (BMBF) for the SPACES 2 Joint Projects: South Africa Land Degradation Monitor (SALDi) (BMBF grant 01LL1701 A–D) and EMSAfrica—Ecosystem Management Support for Climate Change in Southern Africa—Subproject 4: Remote Sensing Based Ecosystem Monitoring (01LL 1801 D) within the framework of the Strategy "Research for Sustainability" (FONA) www.fona.de/en. The authors thank the anonymous reviewer and the editor, G. von Maltitz, for their critical and helpful comments to an earlier draft of this manuscript.

References

AghaKouchak A, Farahmand A, Melton FS, Teixeira J, Anderson MC, Wardlow BD, Hain CR (2015) Remote sensing of drought: progress, challenges and opportunities. Rev Geophys 53:452–480

Alexakis DD, Mexis FDK, Vozinaki AEK, Daliakopoulos IN, Tsanis IK (2017) Soil moisture content estimation based on Sentinel-1 and auxiliary earth observation products. A hydrological approach. Sensors 17(6):1–16

Appel M, Pebesma E (2019) On-demand processing of data cubes from satellite image collections with the Gdalcubes library. Data 4(3):1–16. https://doi.org/10.3390/data4030092

Aschbacher J (2017) ESA's earth observation strategy and Copernicus. In: Onoda M, Young OR (eds) Satellite earth observations and their impact on society and policy. Springer, Singapore, pp 81–86

Bai ZG, Dent DL (2007) Land degradation and improvement in South Africa 1. Identification by remote sensing. Report 2007/03, ISRIC World Soil Information, Wageningen, 58 pp. https://www.isric.org/sites/default/files/isric_report_2007_03.pdf

Baumann P, Mazzetti P, Ungar J, Barbera R, Barboni D, Beccati A, Bigagli L, Boldrini E, Bruno R, Calanducci A, Campalani P, Clements O, Dumitru A, Grant M, Herzig P, Kakaletris G, Laxton J, Koltsida P, Lipskoch K, Mahdiraji AR, Mantovani S, Merticariu V, Messina A, Misev D, Natali S, Nativi S, Oosthoek J, Pappalardo M, Passmore J, Rossi AP, Rundo F, Sen M, Sorbera V, Sullivan D, Torrisi M, Trovato L, Veratelli MG, Wagner S (2016) Big data analytics for earth sciences: the EarthServer approach. Int J Digital Earth 9(1):3–29. https://doi.org/10.1080/17538947.2014.1003106

Berger C, Bieri M, Bradshaw K, Brümmer C, Clemen T, Hickler T, Kutsch WL, Lenfers UA, Martens C, Midgley GF, Mukwashi K, Odipo V, Scheiter S, Schmullius C, Baade J, du Toit JCO, Scholes RJ, Smit IPJ, Stevens N, Twine W (2019) Linking scales and disciplines: an interdisciplinary cross-scale approach to supporting climate-relevant ecosystem management. Climate Change 156(1–2):139–150. https://doi.org/10.1007/s10584-019-02544-0

Bijaber N, El Hadani D, Saidi M, Svoboda M, Wardlow B, Hain C, Poulsen C, Yessef M, Rochdi A (2018) Developing a remotely sensed drought monitoring indicator for Morocco. Geosciences 8(2):55. https://doi.org/10.3390/geosciences8020055

Bond WJ (2019) Open ecosystems: ecology and evolution beyond the forest edge. Oxford University Press, Oxford. https://doi.org/10.1093/oso/9780198812456.001.0001

Bourgeau-Chavez LL, Kasischke ES, Riordan K, Brunzell S, Nolan M, Hyer E, Slawski J, Medvecz M, Walters T, Ames S (2007) Remote monitoring of spatial and temporal surface soil moisture in fire disturbed boreal forest ecosystems with ERS SAR imagery. Int J Remote Sens 28(10):2133–2162. https://doi.org/10.1080/01431160600976061

Bucini G, Hanan N, Boone R, Smit I, Saatchi S, Lefsky M, Asner G (2010) Woody Fractional Cover in Kruger National Park, South Africa: remote sensing–based maps and ecological insights. In: Hill MJ, Hanan NP (eds) Ecosystem function in savannas. CRC Press, Boca Raton, pp 219–237

Camara G, Queiroz G, Vinhas L, Ferreira K, Cartaxo R, Simoes R, Llapa E, Assis L, Sanchez A (2017) The E-sensing architecture for big earth observation data analysis. In: Proc Conf Big Data from Space BIDS, November, pp 402–405. https://doi.org/10.2760/383579

Cawse-Nicholson K, Townsend PA, Schimel D, Assiri AM, Blake PL, Buongiorno MF, Campbell P et al (2021) NASA's surface biology and geology designated observable: a perspective on surface imaging algorithms. Remote Sens Environ 257:112349. https://doi.org/10.1016/j.rse.2021.112349

Cho MA, Ramoelo A (2019) Optimal dates for assessing long-term changes in tree-cover in the semi-arid biomes of South Africa using MODIS NDVI time series (2001–2018). Int J Appl Earth Obs Geoinf 81:27–36. https://doi.org/10.1016/j.jag.2019.05.014

Cowie AL, Orr BJ, Sanchez VMC, Chasek P, Crossman ND, Erlewein A, Louwagie G, Maron M, Metternicht GI, Minelli S, Tengberg AE, Walter S, Welton S (2018) Land in balance: the scientific conceptual framework for land degradation neutrality. Environ Sci Policy. https://doi.org/10.1016/j.envsci.2017.10.011

DEA - Department of Environmental Affairs (2016) Report: phase 1 of Desertification, Land Degradation and Drought (DLDD) land cover mapping impact indicator of the United Nations Convention to Combat Desertification (UNCCD). Pretoria

DFFE - Department of Forestry, Fisheries and the Environment (2018) The second National Action Programme for South Africa to combat desertification, land degradation and the effects of drought (2018–2030), Pretoria, pp 1–35

Dhu T, Dunn B, Lewis B, Lymburner L, Mueller N, Telfer E, Lewis A, McIntyre A, Minchin S, Phillips C (2017) Digital earth Australia–unlocking new value from earth observation data. Big Earth Data 1(1–2):64–74. https://doi.org/10.1080/20964471.2017.1402490

Dorigo W, Wagner W, Albergel C, Albrecht F, Balsamo G, Brocca L, Chung D, Ertl M, Forkel M, Gruber A, Haas E, Hamer PD, Hirschi M, Ikonen J, de Jeu R, Kidd R, Lahoz W, Liu YY, Miralles D, Mistelbauer T, Nicolai-Shaw N, Parinussa R, Pratola C, Reimer C, van der Schalie R, Seneviratne SI, Smolander T, Lecomte P (2017) ESA CCI soil moisture for improved Earth system understanding: State-of-the art and future directions. Remote Sens Environ 203:185–215. https://doi.org/10.1016/j.rse.2017.07.001

Eldridge DJ, Bowker MA, Maestre FT, Roger E, Reynolds JF, Whitford WG (2011) Impacts of shrub encroachment on ecosystem structure and functioning: towards a global synthesis. Ecol Lett 14:709–722

Entekhabi D, Njoku EG, O'Neill PE, Kellogg KH, Crow WT, Edelstein WN, Entin JK, Goodman SD, Jackson TJ, Johnson J, Kimball J, Piepmeier JR, Koster RD, Martin N, McDonald KC, Moghaddam M, Moran S, Reichle R, Shi JC, Spencer MW, Thurman SW, Tsang L, Van Zyl J (2010) The soil moisture active passive (SMAP) mission. Proc IEEE 98(5):704–716. https://doi.org/10.1109/JPROC.2010.2043918

European Commission (2016) Open innovation, open science, open to the world - publications office of the EU. https://op.europa.eu/de/publication-detail/-/publication/3213b335-1cbc-11e6-ba9a-01aa75ed71a1. Accessed 14 Oct 2021

European Commission (2018) The DIAS: user-friendly access to Copernicus data and information. https://www.copernicus.eu/sites/default/files/Copernicus_DIAS_Factsheet_June2018.pdf. Accessed 07 Oct 2021

Fairbanks DHK, Thompson MW, Vink DE, Newby TS, Van den Berg HM, Everard DA (2000) The South African land-cover characteristics database: a synopsis of the landscape. SA J Sci 96(2):69–82. http://hdl.handle.net/10204/1087

Ferrari T, Scardaci D, Andreozzi S (2018) The open science commons for the European research area. Earth Obs Open Sci Innov 43–67. https://doi.org/10.1007/978-3-319-65633-5_3

Ferreira KR, Queiroz GR, Vinhas L, Marujo RFB, Simoes REO, Picoli MCA, Camara G, Cartaxo R, Gomes VCF, Santos LA, Sanchez AH, Arcanjo JS, Fronza JG, Noronha CA, Costa RW, Zaglia MC, Zioti F, Korting TS, Soares AR, Chaves MED, Fonseca LMG (2020) Earth observation data cubes for Brazil: requirements, methodology and products. Remote Sens 12(24):1–19. https://doi.org/10.3390/rs12244033

Gangat R, van Deventer H, Naidoo L, Adam E (2020) Estimating soil moisture using Sentinel-1 and Sentinel-2 sensors for dryland and palustrine wetland areas. S Afr J Sci 116(7/8) https://sajs.co.za/article/view/6535

Gao Q, Zribi M, Escorihuela MJ, Baghdadi N (2017) Synergetic use of Sentinel-1 and Sentinel-2 data for soil moisture mapping at 100 m resolution. Sensors 17(9):1966. https://doi.org/10.3390/s17091966

GeoTerraImage (2015) Technical report: 2013/2014 South African National Land Cover Dataset version 5, Pretoria

Gessner U, Machwitz M, Conrad C, Dech S (2013) Estimating the fractional cover of growth forms and bare surface in savannas. A multi-resolution approach based on regression tree ensembles. Remote Sens Environ 129:90–102. https://doi.org/10.1016/j.rse.2012.10.026

Giuliani G, Chatenoux B, De Bono A, Rodila D, Richard JP, Allenbach K, Dao H, Peduzzi P (2017a) Building an Earth Observations Data Cube: lessons learned from the Swiss Data Cube (SDC) on generating Analysis Ready Data (ARD). Big Earth Data 1(1–2):100–117. https://doi.org/10.1080/20964471.2017.1398903

Giuliani G, Nativi S, Obregon A, Beniston M, Lehmann A (2017b) Spatially enabling the global framework for climate services: reviewing geospatial solutions to efficiently share and integrate climate data & information. Clim Serv 8:44–58

Giuliani G, Camara G, Killough B, Minchin S (2019) Earth observation open science: enhancing reproducible science using data cubes. Data 4:147

Glenn NF, Carr JR (2004) Establishing a relationship between soil moisture and RADARSAT-1 SAR data obtained over the Great Basin, Nevada, U.S.A. Can J Remote Sens 30(2):176–181. https://doi.org/10.5589/m03-057

Gómez C, White JC, Wulder MA (2016) Optical remotely sensed time series data for land cover classification: a review. ISPRS J Photogramm Remote Sens 116:55–72. https://doi.org/10.1016/j.isprsjprs.2016.03.008

Gonzalez-Roglich M, Zvoleff A, Noon M, Liniger H, Fleiner R, Harari N, Garcia C (2019) Synergizing global tools to monitor progress towards land degradation neutrality: trends. Earth and the world overview of conservation approaches and technologies sustainable land management database. Environ Sci Pol 93:34–42. https://doi.org/10.1016/j.envsci.2018.12.019

Gorelick N, Hancher M, Dixon M, Ilyushchenko S, Thau D, Moore R (2017) Google earth engine: planetary-scale geospatial analysis for everyone. Remote Sens Environ 202:18–27. https://doi.org/10.1016/j.rse.2017.06.031

Graw V, Ghazaryan G, Dall K, Gómez AD, Abdel-Hamid A, Jordaan A, Piroska R, Post J, Szarzynski J, Walz Y, Dubovyk O (2017) Drought dynamics and vegetation productivity in different land management systems of Eastern Cape, South Africa - a remote sensing perspective. Sustainability 9(10):1728. https://doi.org/10.3390/su9101728

Haider SS, Said S, Kothyari UC, Arora MK (2004) Soil moisture estimation using ERS 2 SAR data: a case study in the Solani River catchment. Hydrol Sci J 49(2):323–334. https://doi.org/10.1623/hysj.49.2.323.34832

He B, Liao Z, Quan X, Li X, Hu J (2015) A global Grassland Drought Index (GDI) product: algorithm and validation. Remote Sens 7(10):12704–12736. https://doi.org/10.3390/rs71012704

Heckel K, Urban M, Bouffard J-S, Baade J, Boucher P, Davies A, Hockridge EG, Lück W, Smit I, Jacobs B, Norris-Rogers M, Schmullius C (2021) The first sub-meter resolution digital elevation model of the Kruger National Park, South Africa. Koedoe 63(1):1–13. https://doi.org/10.4102/koedoe.v63i1.1679

Higginbottom TP, Symeonakis E (2014) Assessing land degradation and desertification using vegetation index data: current frameworks and future directions. Remote Sens 6:9552–9575. https://doi.org/10.3390/rs6109552

Higginbottom TP, Symeonakis E (2020) Identifying ecosystem function shifts in Africa using breakpoint analysis of long-term NDVI and RUE data. Remote Sens 12:1894. https://doi.org/10.3390/rs12111894

Higginbottom TP, Symeonakis E, Meyer H, van der Linden S (2018) Mapping fractional woody cover in semi-arid savannahs using multi-seasonal composites from Landsat data. ISPRS J Photogramm Remote Sens 139:88–102. https://doi.org/10.1016/j.isprsjprs.2018.02.010

Hoffman MT, Todd S, Ntshona Z and Turner S (1999) Land degradation in South Africa. Final report to the Department of Environmental Affairs and Tourism, South Africa. http://hdl.handle.net/11427/7507

Hoffman T, Todd S (2000) A national review of land degradation in South Africa: the influence of biophysical and socio-economic factors. J South Afr Stud 26(4):743–758. https://doi.org/10.1080/713683611

Hoffman T, Ashwell A (2001) Nature divided: land degradation in South Africa. University of Cape Town Press, Cape Town. 168 pp

Holden PB, Rebelo AJ, New MG (2021) Mapping invasive alien trees in water towers: a combined approach using satellite data fusion, drone technology and expert engagement. Remote Sens Appl Soc Environ 21(January):100448. https://doi.org/10.1016/j.rsase.2020.100448

Hudak AT, Wessman CA (1998) Textural analysis of historical aerial photography to characterize woody plant encroachment in South African savanna. Remote Sens Environ 66:317–330

Hudak AT, Wessman CA (2001) Textural analysis of high resolution imagery to quantify bush encroachment in Madikwe Game Reserve, South Africa, 1955–1996. Int J Remote Sens 22(14):2731–2740. https://doi.org/10.1080/01431160119030

Kerr YH, Waldteufel P, Richaume P, Wigneron JP, Ferrazzoli P, Mahmoodi A, Al Bitar A, Cabot F, Gruhier C, Juglea SE, Leroux D, Mialon A, Delwart S (2012) The SMOS soil moisture retrieval algorithm. IEEE Trans Geosci Remote Sens 50(5 Part 1):1384–1403. https://doi.org/10.1109/TGRS.2012.2184548

Khosa FV, Mateyisi MJ, van der Merwe MR, Feig GT, Engelbrecht FA, Savage MJ (2020) Evaluation of soil moisture from CCAM-CABLE simulation, satellite-based models estimates and satellite observations: a case study of Skukuza and Malopeni flux towers. Hydrol Earth Syst Sci 24(4):1587–1609. https://hess.copernicus.org/articles/24/1587/2020/

Kiker GA, Scholtz R, Smit IPJ, Venter FJ (2014) Exploring an extensive dataset to establish woody vegetation cover and composition in Kruger National Park for the late 1980s. Koedoe 56(1):1–10. https://doi.org/10.4102/koedoe.v56i1.1200

Laney D (2001) 3D data management: controlling data volume, velocity, and variety. META Group. https://pdfcoffee.com/ad949-3d-data-management-controlling-data-volume-velocity-and-varietypdf-pdf-free.html

Leconte R, Brissette F, Galarneau M, Rousselle J (2004) Mapping near-surface soil moisture with RADARSAT-1 synthetic aperture radar data. Water Resour Res 40(1). https://doi.org/10.1029/2003WR002312

Lievens H, Reichle RH, Liu Q, De Lannoy GJM, Dunbar RS, Kim SB, Das NN, Cosh M, Walker JP, Wagner W (2017) Joint Sentinel-1 and SMAP data assimilation to improve soil moisture estimates. Geophys Res Lett 44(12):6145–6153. https://doi.org/10.1002/2017GL073904

LRI (Land Resources International) (2018) Automated land cover classification South Africa. Final Report - SSC WC 03(2017/2018) DRDLR. Land Resources International, Pietermaritzburg

Lindeque GHL, Koegelenberg FA (2015) Perceptions on land degradation and current responses to land degradation problems in South Africa: local municipality fact sheet series. Department of Agriculture, Forestry and Fisheries, Pretoria. http://media.dirisa.org/inventory/archive/spatial/carbon-atlas/metadata-sheets/lada_south_africa_loss_of_cover_daff_apr2016_metadata.pdf

Main R, Mathieu R, Kleynhans W, Wessels K, Naidoo L, Asner GP (2016) Hyper-temporal C-band SAR for baseline woody structural assessments in deciduous savannas. Remote Sens 8(8):1–19. https://doi.org/10.3390/rs8080661

Makhado RA, Scholes RJ (2011) Determinants of soil respiration in a semi-arid savanna ecosystem, Kruger National Park, South Africa. Koedoe 53(1):1–8. https://doi.org/10.4102/koedoe

Marumbwa FM, Cho MA, Chirwa P (2019) Analysis of spatio-temporal rainfall trends across southern African biomes between 1981 and 2016. Phys Chem Earth 114:102808. https://doi.org/10.1016/j.pce.2019.10.004

Marumbwa FM, Cho MA, Chirwa P (2020) An assessment of remote sensing-based drought index over different land cover types in southern Africa. Int J Remote Sens 41(19):1–15. https://doi.org/10.1080/01431161.2020.1757783

Marumbwa FM, Cho MA, Chirwa P (2021) Geospatial analysis of meteorological drought impact on southern Africa biomes. Int J Remote Sens 42(06):2155–2173. https://doi.org/10.1080/01431161.2020.1851799

Marwan N, Romano MC, Thiel M, Kurths J (2007) Recurrence plots for the analysis of complex systems, Phys Rep 438(5-6):237–329. https://doi.org/10.1016/j.physrep.2006.11.001. https://www.sciencedirect.com/science/article/pii/S0370157306004066

Masemola C, Cho MA, Ramoelo A (2019) Assessing the effect of seasonality on leaf and canopy spectra for the discrimination of an alien tree species, Acacia mearnsii, from co-occurring native species using parametric and nonparametric classifiers. IEEE Trans Geosci Remote Sens 57(8):5853–5867. https://doi.org/10.1109/TGRS.2019.2902774

Masemola CM, Cho MA, Ramoelo A (2020a) Towards a semi-automated mapping of Australia native invasive alien Acacia trees using a radiative model and Sentinel-2 in South Africa. ISPRS J Photogramm Remote Sens 166:153–168. https://doi.org/10.1016/j.isprsjprs.2020.04.009

Masemola C, Cho MA, Ramoelo A (2020b) Sentinel-2 time series based optimal features and time window for mapping invasive Australian native Acacia species in KwaZulu Natal, South Africa. Int J Appl Earth Obs Geoinf 93:102207. https://doi.org/10.1016/j.jag.2020.102207

Masih I, Maskey S, Mussá FEF, Trambauer P (2014) A review of droughts on the African continent: a geospatial and long-term perspective. Hydrol Earth Syst Sci 18(9):3635–3649. https://doi.org/10.5194/hess-18-3635-2014

Mhangara P, Mapurisa W (2019) Multi-mission earth observation data processing system. Sensors 19(18):3831. https://doi.org/10.3390/s19183831

Moncrieff GR (2022) Continuous land cover change detection in a critically endangered shrubland ecosystem using neural networks. Remote Sens 14:2766. https://doi.org/10.3390/rs14122766

Mukheibir P, Ziervogel G (2007) Developing a municipal adaptation plan (MAP) for climate change: the city of Cape Town. Environ Urban 19(1):143–158. https://doi.org/10.1177/0956247807076912

Munyati C, Sinthumule NI (2016) Change in woody cover at representative sites in the Kruger National Park, South Africa, based on historical imagery. Springerplus 5(1):1417. https://doi.org/10.1186/s40064-016-3036-1

Munyati C, Shaker P, Phasha MG (2011) Using remotely sensed imagery to monitor savanna rangeland deterioration through woody plant proliferation: a case study from communal and biodiversity conservation rangeland sites in Mokopane, South Africa. Environ Monit Assess 176(1–4):293–311. https://doi.org/10.1007/s10661-010-1583-4

Musetsho KD, Chitakira M, Nel W (2021) Mapping land-use/land-cover change in a critical biodiversity area of South Africa. Int J Environ Res Public Health 18. https://doi.org/10.3390/ijerph181910164

Nativi S, Mazzetti P, Santoro M, Papeschi F, Craglia M, Ochiai O (2015) Big data challenges in building the global earth observation system of systems. Environ Model Softw 68:1–26. https://doi.org/10.1016/j.envsoft.2015.01.017

Ng WT, Meroni M, Immitzer M, Böck S, Leonardi U, Rembold F, Gadain H, Atzberger C (2016) Mapping Prosopis spp. with Landsat 8 data in arid environments: evaluating effectiveness of different methods and temporal imagery selection for Hargeisa, Somaliland. Int J Appl Earth Obs Geoinf 53:76–89. https://doi.org/10.1016/j.jag.2016.07.019

Niklaus M, Eisfelder C, Gessner U, Dech S (2015) Land degradation in South Africa - a degradation index derived from 10 years of net primary production data. In: Remote sensing and digital image processing, pp 247–267. https://doi.org/10.1007/978-3-319-15967-6_12

Nzuza P, Ramoelo A, Odindi J, Mwenge-Kahinda J, Madonsela S (2020) Predicting land degradation using Sentinel-2 and environmental variables in the Lepellane catchment of the Greater Sekhukhune District, South Africa. Phys Chem Earth 124(1). https://doi.org/10.1016/j.pce.2020.102931

Nzuza P, Ramoelo A, Odindi J, Mwenge Kahinda J, Lindeque L (2021) A triangulation approach for assessing and mapping land degradation in Lepellane Catchment of the Sekhukhune District. S Afr Geogr J. https://doi.org/10.1080/03736245.2021.2000481

O'Connor TG, Puttick JR, Hoffman MT (2014) Bush encroachment in southern Africa: changes and causes. Afr J Range Forage Sci 31(2):67–88. https://doi.org/10.2989/10220119.2014.939996

Oelofse M, Birch-Thomsen T, Magid J, de Neergaard A, van Deventer R, Bruun S, Hill T (2016) The impact of black wattle encroachment of indigenous grasslands on soil carbon, eastern cape, South Africa. Biol Invasions 18(2):445–456. https://doi.org/10.1007/s10530-015-1017-x

Oldeland J, Dorigo W, Wesuls D, Jürgens N (2010) Mapping bush encroaching species by seasonal differences in hyperspectral imagery. Remote Sens 2(6):1416–1438. https://doi.org/10.3390/rs2061416

Orr BJ, Cowie AL, Sanchez VMC, Chasek P, Crossman ND, Erlewein A, Louwagie G, Maron M, Metternicht GI, Minelli S, Tengberg AE, Walter S, Welton S (2017) Scientific conceptual framework for land degradation neutrality. A report of the science-policy interface. United Nations Convention to Combat Desertification (UNCCD), Bonn, Germany, pp 1–98, ISBN: 978-92-95110-42-7 (hard copy), 978-92-95110-41-0 (electronic copy)

Paloscia S, Pampaloni P, Pettinato S, Santi E (2008) Comparison of algorithms for retrieving soil moisture from ENVISAT/ASAR images. IEEE Trans Geosci Remote Sens 46:3274–3284. https://doi.org/10.1109/TGRS.2008.920370

Pettorelli N, Wegmann M, Skidmore A, Mücher S, Dawson TP, Fernandez M, Lucas R, Schaepman ME, Wang T, O'Connor B, Jongman RHG, Kempeneers P, Sonnenschein R, Leidner AK, Böhm M, He KS, Nagendra H, Dubois G, Fatoyinbo T, Hansen MC, Paganini M, de Klerk HM, Asner GP, Kerr JT, Estes AB, Schmeller DS, Heiden U, Rocchini D, Pereira HM, Turak E, Fernandez N, Lausch A, Cho MA, Alcaraz-Segura D, McGeoch MA, Turner W, Mueller A, St-Louis V, Penner J, Vihervaara P, Belward A, Reyers B and Geller GN (2016) Framing the concept of satellite remote sensing essential biodiversity variables: challenges and future directions. Remote Sens Ecol Conserv 2:122–131. https://doi.org/10.1002/rse2.15

Prince SD, Becker-Reshef I, Rishmawi K (2009) Detection and mapping of long-term land degradation using local net production scaling: application to Zimbabwe. Remote Sens Environ 113(5):1046–1057. https://doi.org/10.1016/j.rse.2009.01.016

Schoeman F, Newby TS, Thompson MW and Van den Berg EC (2013) South African National Land-Cover Change Map. South African J Geom 2(2):94–105

Scholes RJ (2009) Syndromes of dryland degradation in southern Africa. Afr J Range Forage Sci 26(3):113–125. https://doi.org/10.2989/AJRF.2009.26.3.2.947

Shekede MD, Murwira A, Masocha M (2015) Wavelet-based detection of bush encroachment in a savanna using multi-temporal aerial photographs and satellite imagery. Int J Appl Earth Obs Geoinf 35(PB):209–216. https://doi.org/10.1016/j.jag.2014.08.019

Siqueira A, Lewis A, Thankappan M, Szantoi Z, Goryl P, Labahn S, Ross J, Hosford S, Mecklenburg S, Tadono T, Rosenqvist A, Lacey J (2019) CEOS analysis ready data for land - an overview on the current and future work. In: International geoscience and remote sensing symposium (IGARSS), Institute of Electrical and Electronics Engineers Inc., pp 5536–5537

Skowno AL, Thompson MW, Hiestermann J, Ripley B, West AG, Bond WJ (2017) Woodland expansion in South African grassy biomes based on satellite observations (1990–2013): general patterns and potential drivers. Glob Chang Biol 23(6):2358–2369. https://doi.org/10.1111/gcb.13529

Skowno AL, Jewitt D, Slingsby JA (2021) Rates and patterns of habitat loss across South Africa's vegetation biomes. S Afr J Sci 117(1/2). https://doi.org/10.17159/sajs.2021/8182

Slingsby JA, Moncrieff GR, Rogers AJ, February EC (2020a) Altered ignition catchments threaten a hyperdiverse fire-dependent ecosystem. Glob Chang Biol 26(2):616–628. https://doi.org/10.1111/gcb.14861

Slingsby JA, Moncrieff GR, Wilson AM (2020b) Near-real time forecasting and change detection for an open ecosystem with complex natural dynamics. ISPRS J Photogramm Remote Sens 166(August):15–25. https://doi.org/10.1016/j.isprsjprs.2020.05.017

Snyman HA (2012) Habitat preferences of the encroacher shrub, Seriphium plumosum. S Afr J Bot 81:34–39. https://doi.org/10.1016/j.sajb.2012.05.001

Stevens N, Erasmus BFN, Archibald S, Bond WJ (2016) Woody encroachment over 70 years in South African savannahs: overgrazing, global change or extinction aftershock? Philos Trans R Soc B Biol Sci 371(1703):20150437. https://doi.org/10.1098/rstb.2015.0437

Symeonakis E, Higginbottom T (2014) Bush encroachment monitoring using multi-temporal landsat data and random forests. In: International Archives of the Photogrammetry, Remote Sensing and Spatial Information Sciences - ISPRS Archives. International Society for Photogrammetry and Remote Sensing, pp 29–35

Urban M, Berger C, Mudau TE, Heckel K, Truckenbrodt J, Odipo VO, Smit IPJ, Schmullius C (2018) Surface moisture and vegetation cover analysis for drought monitoring in the southern Kruger National Park using Sentinel-1, Sentinel-2, and Landsat-8. Remote Sens 10(9):1482. https://doi.org/10.3390/rs10091482

Urban M, Heckel K, Berger C, Schratz P, Smit IPJ, Strydom T, Baade J, Schmullius C (2020) Woody cover mapping in the savanna ecosystem of the Kruger National Park using sentinel-1 C-band time series data. Koedoe 62(1):1–6. https://doi.org/10.4102/koedoe.v62i1.1621

Urban M, Schellenberg K, Morgenthal T, Dubois C, Hirner A, Gessner U, Mogonong B, Zhan Z, Baade J, Collett A, Schmullius C (2021) Using Sentinel-1 and Sentinel-2 time series for Slangbos mapping in the Free State Province, South Africa. Remote Sens 13(3342):1–20. https://doi.org/10.3390/rs13173342

Van den Berg EC, Plarre C, Van den Berg HM, Thompson MW (2008) The South African National Land-Cover 2000. Agricultural Research Council-Institute for Soil, Climate and Water, Unpublished report no. GW/A/2008/86

Van Wilgen BW, Wilson JR (eds) (2018) The status of biological invasions and their Management in South Africa in 2017, vol 204. South African National Biodiversity Institute, Kirstenbosch and DST-NRF Centre of Excellence for Invasion Biology, Stellenbosch

Venter ZS, Scott SL, Desmet PG, Hoffman MT (2020) Application of Landsat-derived vegetation trends over South Africa: potential for monitoring land degradation and restoration. Ecol Indic 113:106206. https://doi.org/10.1016/j.ecolind.2020.106206

Verbesselt J, Hyndman R, Newnham G, Culvenor D (2010) Detecting trend and seasonal changes in satellite image time series, Remote Sens Environ 114(1):106–115. https://doi.org/10.1016/j.rse.2009.08.014

Verschuur J, Li S, Wolski P, Otto FEL (2021) Climate change as a driver of food insecurity in the 2007 Lesotho-South Africa drought. Sci Rep 11(1):3852. https://doi.org/10.1038/s41598-021-83375-x

Villarreal ML, Norman LM, Buckley S, Wallace CSA, Coe MA (2016) Multi-index time series monitoring of drought and fire effects on desert grasslands. Remote Sens Environ 183:186–197. https://doi.org/10.1016/j.rse.2016.05.026

von Maltitz GP, Gambiza J, Kellner K, Rambau T, Lindeque L, Kgope B (2019) Experiences from the South African land degradation neutrality target setting process. Environ Sci Pol 101:54–62. https://doi.org/10.1016/j.envsci.2019.07.003

Ward D, Hoffman MT, Collocott SJ (2014) A century of woody plant encroachment in the dry Kimberley savanna of South Africa. Afr J Range Forage Sci 31(2):107–121. https://doi.org/10.2989/10220119.2014.914974

Wessels KJ, Prince SD, Frost PE, Van Zyl D (2004) Assessing the effects of human-induced land degradation in the former homelands of northern South Africa with a 1 km AVHRR NDVI time-series. Remote Sens Environ 91(1):7–67. https://doi.org/10.1016/j.rse.2004.02.005

Wessels KJ, Prince SD, Zambatis N, MacFadyen S, Frost PE, Van Zyl D (2006) Relationship between herbaceous biomass and 1-km2 Advanced Very High Resolution Radiometer (AVHRR) NDVI in Kruger National Park, South Africa. Int J Remote Sens 27(5):951–973. https://doi.org/10.1080/01431160500169098

Wessels KJ, Prince SD, Carroll M, Malherbe J (2007a) Relevance of rangeland degradation in semiarid Northeastern South Africa to the nonequilibrium theory. Ecol Appl 17(3):815–827. https://doi.org/10.1890/06-1109

Wessels KJ, Prince SD, Malherbe J, Small J, Frost PE, VanZyl D (2007b) Can human-induced land degradation be distinguished from the effects of rainfall variability? A case study in South Africa. J Arid Environ 68(2):271–297. https://doi.org/10.1016/j.jaridenv.2006.05.015

Wessels KJ, van den Bergh F, Scholes RJ (2012) Limits to detectability of land degradation by trend analysis of vegetation index data. Remote Sens Environ 125:10–22. https://doi.org/10.1016/j.rse.2012.06.022

Western D, Mose VN, Worden J, Maitumo D (2015) Predicting extreme droughts in savannah Africa: a comparison of proxy and direct measures in detecting biomass fluctuations, trends and their causes. PLoS One 10(8):e0136516. https://doi.org/10.1371/journal.pone.0136516

Wigley BJ, Bond WJ, Hoffman MT (2010) Thicket expansion in a South African savanna under divergent land use: local vs. global drivers? Glob Chang Biol 16(3):964–976. https://doi.org/10.1111/j.1365-2486.2009.02030.x

Winkler K, Gessner U, Hochschild V (2017) Identifying droughts affecting agriculture in Africa based on remote sensing time series between 2000–2016: rainfall anomalies and vegetation condition in the context of ENSO. Remote Sens 9(8):831. https://doi.org/10.3390/rs9080831

Wulder MA, White JC, Goward SM, Masek JG, Irons JR, Herold M, Cohen WB, Loveland TR, Woodcock CE (2008) Landsat continuity: issues and opportunities for land cover monitoring. Remote Sens Environ 112:955–969. https://doi.org/10.1016/j.rse.2007.07.004

Zeng L, Shan J, Xiang D (2014) Monitoring drought using multi-sensor remote sensing data in cropland of Gansu Province. In: IOP conference series: earth and environmental science

Zhao K, Wulder MA, Hu T, Bright R, Wu Q, Qin H, Li Y, Toman E, Mallick B, Zhang X (2019) Detecting change-point, trend, and seasonality in satellite time series data to track abrupt changes and nonlinear dynamics: a Bayesian ensemble algorithm. Remote Sens Environ 232:111–181. https://doi.org/10.1016/j.rse.2019.04.034

Zribi M, Baghdadi N, Holah N, Fafin O (2005) New methodology for soil surface moisture estimation and its application to ENVISAT-ASAR multi-incidence data inversion. Remote Sens Environ 96(3–4):485–496. https://doi.org/10.1016/j.rse.2005.04.005

The Marine Carbon Footprint: Challenges in the Quantification of the CO₂ Uptake by the Biological Carbon Pump in the Benguela Upwelling System

25

Tim Rixen, Niko Lahajnar, Tarron Lamont, Rolf Koppelmann,
Bettina Martin, Luisa Meiritz, Claire Siddiqui,
and Anja K. Van derPlas

Abstract

Quantifying greenhouse gas (GHG) emissions is essential for mitigating global warming, and has become the task of individual countries assigned to the Paris agreement in the form of National Greenhouse Gas Inventory Reports (NIR). The NIR informs on GHG emissions and removals over national territory encompassing the 200-mile Exclusive Economic Zone (EEZ). However, apart from only a few countries, who have begun to report on coastal ecosystems, mostly mangroves, salt marshes, and seagrass meadows, the NIR does not cover

T. Rixen (✉)
Leibniz Centre for Tropical Marine Research - ZMT, Bremen, Germany

Institute of Geology, Universität Hamburg, Hamburg, Germany
e-mail: Tim.Rixen@leibniz-zmt.de

N. Lahajnar · L. Meiritz
Institute of Geology, Universität Hamburg, Hamburg, Germany

T. Lamont
Oceans & Coasts Research Branch, Department of Forestry, Fisheries, and the Environment,
Cape Town, South Africa

Department of Oceanography, University of Cape Town, Rondebosch, South Africa

Bayworld Centre for Research & Education, Cape Town, South Africa

R. Koppelmann · B. Martin
Institute of Marine Ecosystem and Fishery Science, Universität Hamburg, Hamburg, Germany

C. Siddiqui
Leibniz Centre for Tropical Marine Research - ZMT, Bremen, Germany

A. K. Van der Plas
National Marine Information and Research Centre, Ministry of Fisheries and Marine Resources,
Swakopmund, Namibia

© The Author(s) 2024
G. P. von Maltitz et al. (eds.), *Sustainability of Southern African Ecosystems
under Global Change*, Ecological Studies 248,
https://doi.org/10.1007/978-3-031-10948-5_25

or report on GHG sources and sinks of the 200-mile exclusive economic zone which, for Namibia and South Africa includes the Benguela Upwelling System (BUS). Based on our results, we estimated a CO_2 uptake by the biological carbon pump of 18.5 ± 3.3 Tg C year^{-1} and 6.0 ± 5.0 Tg C year^{-1} for the Namibian and South African parts of the BUS, respectively. Even though it is assumed that the biological carbon pump already responds to global change and fisheries, uncertainties associated with estimates of the CO_2 uptake by the biological carbon pump are still large and hamper a thorough quantification of human impacts on the biological carbon pump. Despite these uncertainties, it is suggested to include parameters such as preformed nutrient supply, carbon export rates, Redfield ratios, and CO_2 concentrations measured at specific key sites into the NIR to stay focussed on the biological carbon pump and to support research addressing open questions, as well as to improve methods and observing concepts.

25.1 Introduction

The implementation of the Paris agreement to keep global warming below 1.5–2.0 °C is accompanied by a variety of measures such as the compilation of the NIR to monitor the emission of CO_2 and other greenhouse gases in order to review the realization of climate pledges at national levels (e.g. UNEP 2019). The National Greenhouse Gas Inventory Report (NIR) splits CO_2 emissions into four sectors: (I) energy, (II) industrial process and product use (IPPU), (III) waste, as well as (IV) agriculture, forestry, and other land use (AFOLU). In order to achieve a comparability of greenhouse gas emissions among different countries, the IPCC provides guidelines to quantify greenhouse gas emissions within these sectors (IPCC 2006). According to the South African, Namibian, and German NIR's the first three sectors are net sources of CO_2, while AFOLU acts as a CO_2 sink to the atmosphere in all three countries (Table 25.1, German Environment Agency 2020; Department of Environmental Affairs 2019; Government of Namibia 2021).

AFOLU quantifies CO_2 emissions caused by land use and land cover changes and is based on the quantification of net changes in carbon stocks. It considers six land-use categories. Among them are wetlands including also coastal ecosystems such as

Table 25.1 CO_2 Emissions from South Africa, Namibia, and Germany according to the NIRs of these countries

Country	South Africa	Namibia	Germany
Reference year	2015	2015	2017
	(Tg C year^{-1})	(Tg C year^{-1})	(Tg C year^{-1})
Total CO_2 emissions	117.6	−29.6	213.0
Sector (I–III)	125.2	1.0	221.4
AFOLU	−7.5	−30.6	−8.4

mangrove forests, tidal marshes, and seagrass meadows (IPCC 2014). Carbon stocks of these coastal wetland reservoirs are named 'blue carbon' due to the colour of the ocean (e.g. Nellemann et al. 2009). Despite being linked by name to the marine environment, blue carbon ignores the CO_2 uptake by phytoplankton and its storage within the ocean's biological carbon pump.

In terms of carbon fixation, phytoplankton in the ocean are as productive as terrestrial plants with a global rate of about 50 Pg C year^{-1} (Field et al. 1998), but the storage of carbon differs in systems dominated by terrestrial plants and marine phytoplankton. Terrestrial plants are the world's largest reservoir of living biomass with a reservoir size of 450 Pg C, whereas phytoplankton represent, with 1–2 Pg C, a relatively small carbon stock (Bar-On et al. 2018; Falkowski et al. 1998). Instead of building-up a huge biomass, carbon fixed by phytoplankton is exported into the deep ocean where the vast majority of the exported biomass is respired and stored as dissolved inorganic carbon (DIC). This results in a strong gradient between low DIC concentrations in surface waters where carbon is fixed via photosynthesis into biomass, and high DIC concentrations at greater depth where the exported organic matter is remineralized. The transfer of DIC into the deep sea favours the CO_2 uptake by the ocean by decreasing the CO_2 concentration and therewith the partial pressure of CO_2 (pCO_2) in surface waters. A hypothetical collapse of the biological carbon pump is assumed to increase the atmospheric CO_2 concentrations by 200–300 ppm, representing a doubling of the pre-industrial atmospheric CO_2 concentration (Heinze et al. 2015).

The CO_2 uptake efficiency of the biological carbon pump is linked to the global thermohaline circulation of the ocean (Heinze et al. 1991; Tschumi et al. 2011) and believed to have played a significant role in controlling atmospheric CO_2 concentrations during glacial/interglacial transitions (Broecker and Barker 2007; Bauska et al. 2016; Schmitt et al. 2012). Even though it is widely assumed that the biological carbon pump also responds to the current climate change (e.g. Laufkötter et al. 2017; Duce et al. 2008; DeVries and Deutsch 2014; Riebesell et al. 2007) and fisheries (Bianchi et al. 2021), neither the magnitude nor the direction of change is predictable (Passow and Carlson 2012; Laufkötter et al. 2017; Laufkötter and Gruber 2018). Considering their potential impact on the CO_2 concentration in the atmosphere and the Paris goals, it appears as a necessity to strongly reduce these uncertainties. This means that variations of the CO_2 uptake by the biological carbon pump should be quantified and respective drivers should be identified, which, in turn, provides background to the discussion on responsibilities (national versus the international community). Here we use data obtained within the BMBF (Federal German Ministry for Education and Research) funded project TRAFFIC (Trophic tRAnsfer eFFICiency) to develop concepts that help to describe the status of the biological carbon pump and to quantify changes of the CO_2 uptake within the BUS.

25.2 Study Area: The Benguela Upwelling System

The BUS stretches from the Angola Benguela Frontal Zone (ABFZ) at ~15°S to Cape Agulhas (~35°S, Fig. 25.1). The south-easterly trade winds emanating from the interplay between the South Atlantic Anticyclone and the continental low-pressure trough, cause the emergence of distinct upwelling cells along the coast (e.g. Kämpf and Chapman 2016; Veitch et al. 2009; Shannon and Nelson 1996). The Lüderitz Upwelling Cell at 26°40′S is the strongest of these cells and divides the BUS into a northern (NBUS) and southern (SBUS) subsystem (e.g. Hutchings et al. 2009; Duncombe Rae 2005; Shannon 1985).

Sub-Antarctic Mode Water mainly feeds upwelling (Marinov et al. 2006; Sarmiento et al. 2004), but its properties vary due to regional differences at sites of its formation and mixing with other water masses. The South Atlantic Central Water (SACW) is comprised of Sub-Antarctic Mode Water that originates as mixture of mainly Antarctic Intermediate Water and Subtropical Mode Water (Souza et al. 2018; Karstensen and Quadfasel 2002; McCartney 1977). It is subducted beneath subtropical surface waters north of the Sub-Antarctic Front at about 36°S–54°S and is fed via the South Atlantic Current into the Benguela Current (e.g. Gordon 1981; Donners et al. 2005). In the southern part of the SBUS, the Benguela Current follows the shelf break towards the equator, while the Benguela Jet, also known as the Cape Jet, flushes the SBUS shelf and carries Indian Ocean Central Water into the SBUS (Durgadoo et al. 2017; Tim et al. 2015). The mixture between South Atlantic Central Water and Indian Ocean water creates a new water mass, which is referred to as the Eastern South Atlantic Water (ESACW, e.g. Liu and Tanhua 2021). The ESACW as well as the SACW which largely encompasses the SBUS along with the Benguela Current, feed the complex equatorial current system. It comprises a number of east- and westward flowing currents and undercurrents (e.g. Pitcher et al. 2021) and finally feed the Angola Current from where modified SACW enters the NBUS as a poleward undercurrent at the ABFZ (Fig. 25.1). During its voyage through the south and equatorial Atlantic Ocean the SACW gets enriched in nutrients and depleted in oxygen (Mohrholz et al. 2008).

South of Lüderitz, the interplay between the Benguela Current, the Benguela Jet, and coastal upwelling creates a complex and temporally varying frontal system starting with the Oceanic Front (e.g. Veitch et al. 2010; Hardman-Mountford et al. 2003). This front is considered as the outer boundary of the SBUS (Barange and Pillar 1992; Shannon and Nelson 1996) whereas the Shelf Break Front and Upwelling Front develop in response to the variable nature of upwelling across the wide shelf regions (Shannon 1985).

During the last two decades, satellite data indicate no significant trends in productivity (Demarcq 2009; Lamont et al. 2019; Verheye et al. 2016), although

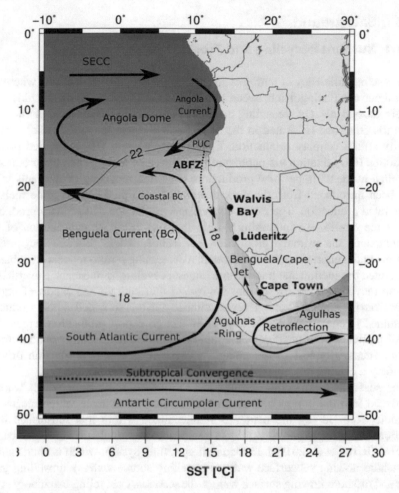

Fig. 25.1 Southeast Atlantic Ocean with its main currents as adopted and simplified from Verheye et al. (2016) and Hardman-Mountford et al. (2003) as well as annual mean sea surface temperatures (Smith et al. 2008). ABFZ (Angola Benguela Frontal Zone) and 'Subtropical Convergence' (dotted line) mark major oceanographic fronts. SECC South Equatorial Counter Current, PUC Poleward Under Current

there is a tendency towards an intensification of upwelling in the SBUS (Lamont et al. 2018, 2019; Tim et al. 2015; Wang et al. 2015; Sydeman et al. 2014). Associated with this trend are shifts in the ecosystem structure such as an overall decrease in zooplankton abundance and a shift from larger to smaller zooplankton species (Lamont et al. 2019; Verheye et al. 2016; Jarre et al. 2015; Hutchings et al. 2012).

25.3 Background

25.3.1 Nutrient Recycling and Productivity

The recycling efficiency of nutrients strongly influences the productivity whereas in general one can distinguish between two different nutrient recycling machineries in pelagic ecosystems. One recycling machinery operates in the seasonal thermocline, while the other one is located in the sunlit surface ocean (euphotic zone) where it directly affects primary production. Eppley and Peterson (1979) divided primary production (CO_2 fixation via photosynthesis) into new and regenerated production. Upwelled nutrients impel new production, while the recycling of nutrients which have been introduced from the dark deep ocean into the euphotic zone fuels the regenerated production. The ratio between new production and primary production defines the f-ratio. An increasing f-ratio indicates a higher contribution of new production to the primary production and hence, a less efficient recycling of nutrients in the euphotic zone. Vice versa, a decreasing f-ratio reflects an enhanced regenerated production and thus a more intense recycling of nutrients. Nevertheless, nutrient recycling does not prevent the loss of nutrients via the export of organic matter from the euphotic zone. Over an annual cycle this so-called export production is assumed to equal new production. Even though export production represents a loss of nutrients from the euphotic zone, the exported organically bound nutrients are not necessarily lost to the pelagic ecosystem as export production drives a secondary recycling machinery in the seasonal thermocline.

The seasonal thermocline defines the subsurface layer from which water is introduced into the euphotic zone on the seasonal times scale which means that the more exported organic matter is remineralized within this subsurface layer, the more formerly exported nutrients return into the euphotic zone and can be recycled (Rixen et al. 2019a). In a coastal upwelling system which is characterized by onshore flowing subsurface waters (upwelling source waters), upwelling at the coast and offshore flowing surface waters, the secondary recycling machinery is part of the nutrient trapping system provoking new production by increasing nutrient concentrations in the upwelling source waters (Dittmar and Birkicht 2001; Tyrrell and Lucas 2002; e.g. Barange and Pillar 1992; Flynn et al. 2020; Rixen et al. 2021). Hereby, the utilization of upwelled nutrients and the associated development of plankton blooms in the offshore flowing upwelled waters initiate the nutrient trapping by exporting the formerly upwelled and now organically bound nutrients into the subsurface waters. This reduces their advection into the open ocean along with offshore flowing surface waters and keeps them within the upwelling system. Although these nutrients increase new production they come from the remineralization of organic matter as those nutrients which drive the regenerated production in the euphotic zone. Hence, these subsurface nutrients are also called regenerated nutrients. Since in addition to nutrients, CO_2 is also released during the remineralization of organic matter, primary production driven by the utilization of regenerated nutrients consumes the associated regenerated CO_2 and creates no need for additional CO_2 as long as the Redfield carbon to nutrient uptake and remineralization ratio is constant. Hence, even though recycling of nutrients in the euphotic zone and the seasonal thermocline contributes to an elevated productivity of upwelling systems through their impact on regenerated and new production, respectively, it hardly affects the CO_2 storage of the biological carbon pump. This differs with the utilization of preformed nutrients.

25.3.2 Preformed Nutrients and the CO_2 Uptake Efficiency of Biological Carbon Pump

In contrast to regenerated nutrients which are released during the remineralization of exported organic matter, physical processes carry preformed nutrients into the deep ocean. This also occurs during the Sub-Antarctic Mode Water formation in winter when the lack of light prevents the utilization of regenerated nutrients in surface waters. During the subduction of mode water, the biologically unused preformed nutrients are transported into the deep ocean (Knox and McElroy 1984; Sarmiento and Toggweiler 1984; Siegenthaler and Wenk 1984; Duteil et al. 2012; Ito and Follows 2005). The regenerated CO_2 formerly associated with the regenerated nutrients, is, in turn, released from the biological carbon pump. Vice versa, the biological carbon pump takes up CO_2 by the utilization of preformed nutrients and their retransformation into regenerated nutrients at sites such as the BUS where upwelling introduces mode waters into the euphotic zone. To which extent the CO_2 release and uptake affects the CO_2 flux across the air–sea interface depends on its influence on the solubility pump. It could absorb CO_2 released from the biological carbon pump or favour its emission into the atmosphere.

25.3.3 Solubility Pump and Upwelling from a Carbon Cycling Perspective

The solubility pump is an abiotically driven carbon pump which responds to the partial pressure difference of CO_2 (ΔpCO_2) between the surface ocean and the atmosphere. The net flux of CO_2 follows the pressure gradient, which means that CO_2 invades the ocean if the pCO_2 in the atmosphere exceeds the oceanic pCO_2 and vice versa the ocean emits CO_2 into the atmosphere if the pCO_2 in the ocean is higher than it is in the atmosphere. The CO_2 concentration and its temperature- and salinity-controlled solubility determines the pCO_2 in the ocean (Weiss 1974). Hence, CO_2 concentrations link the biological carbon pump and the solubility pump, which gains strength in cold waters due to an increased solubility of CO_2 at low temperatures. If, for example, ocean currents carry warm tropical waters with a pCO_2 exceeding those in the atmosphere into polar regions, the pCO_2 could fall below those in the atmosphere. Consequently, the water takes up CO_2, simply due to decreasing sea surface temperatures. Vice versa, if water masses which are formed in the Southern Ocean such as the SACW, upwell in the BUS, they warm up and release CO_2 that was previously taken up in the Southern Ocean. Hence, the opposing impacts of the biological carbon pump and the solubility pump on the CO_2 concentrations during the Sub-Antarctic Mode Water formation and upwelling of mode waters in the BUS are parts of one system.

25.4   Data and Methods

In order to study the biological carbon pump within the framework of TRAFFIC, two cruises were carried out with the German research vessels *Meteor* (M153: February, 19th–March 31st 2019) and *Sonne* (SO285: August 20th–November 2nd 2021, Fig. 25.2). Mooring operations were additionally carried out during the *RV Sonne* cruise SO283, the *RS Algoa* cruise ALG 269, and other pre-TRAFFIC cruises with the German RVs *Meteor* and *M.S. Merian*. In addition to sediment trap samples, we analysed water samples to characterize source water masses and the chemical composition of plankton. This allows us to quantify the supply and utilization of preformed nutrients as well as Redfield ratios, which are crucial to translate nutrient consumption into carbon uptake. Furthermore, satellite data were evaluated and linked to sediment trap data in order to estimate the reliability of export production rates. Satellite-derived sea surface temperature (SST) and primary production rates (PP) were downloaded from the MODIS-Aqua website in August 2021 (https://oceandata.sci.gsfc.nasa.gov/MODIS-Aqua/Mapped/Monthly/4km/sst/) and the ocean primary production website in August 2020 (http://www.science.oregonstate.edu/ocean.productivity/).

25.4.1 Particulate Matter

Particulate matter was collected by moored and drifting sediment traps, as well as through the filtration of water samples and by net catches.

A sediment trap mooring was deployed off Walvis Bay (14.04°E/23.02°S) and equipped with a Hydrobios MST-12 sediment trap at water-depths between 65 m and 75 m (~74 m above the sea floor). Since the moored TRAFFIC traps have yet not fully been recovered, we focus on the pre-TRAFFIC deployment periods. This includes seven deployment periods, which are in part not time-coherent. The first deployment started in December 2009 and the last deployment ended in November 2017. Furthermore, four short-term drifting sediment trap systems were deployed in the NBUS and SBUS for about 41–83 h during the cruise M153 (Fig. 25.2). These systems were equipped with four to five single sediment traps and a Hydrobios MST-12 trap at the bottom of the drifter at water-depths of down to 500 m. In total, 24 drifter samples were collected and analysed in Germany.

According to common sediment trap processing procedures, sediment trap samples were split into fractions of >1 mm and <1 mm (Haake et al. 1993; Honjo et al. 2008). The >1 mm fraction contained larger swimmers and the <1 mm fraction was assumed to represent the gravitationally driven export of particles (= particle flux). In addition, the 24 drifter samples were additionally macroscopically

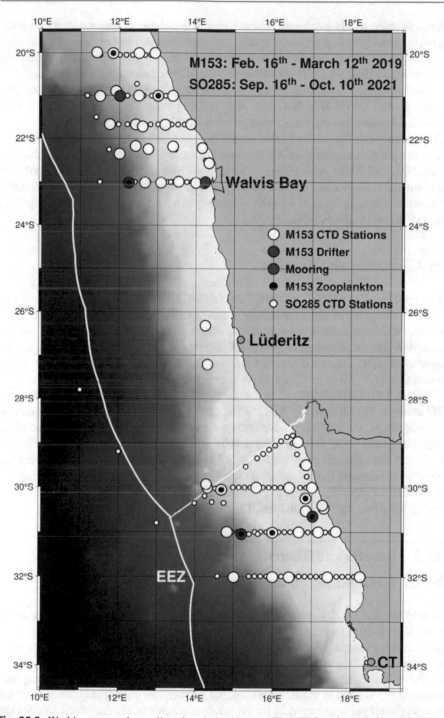

Fig. 25.2 Working area and sampling sites during the two TRAFFIC cruises M153 and SO285. EEZ = 200-mile Exclusive Economic Zones, CT indicates Cape Town. Please note that the timing of the cruises indicates the time within the working area

and microscopically examined and the picked zooplankton was divided into three genera, namely copepods, amphipods, and euphausiids. Furthermore, we obtained zooplankton and fish samples from net catches at the deployment stations during the cruise M153 (Fig. 25.2). These samples were fully homogenized and analysed.

Water samples (5 L and 30 L) obtained from the CTD rosette water sampler were filtered (WHATMAN GF/F; ~0.7 μm; 47 mm diameter) and rinsed with deionized water. Subsequently, filters were dried, shipped to Germany, and analysed in accordance to the sediment trap samples. A detailed description of the sediment trap sample procedure including the biogeochemical analysis is given by Haake et al. (1993) and Rixen et al. (1996). The analysis included the determination of total carbon and nitrogen and organic carbon. The phosphorus content was additionally determined in zooplankton samples according to a method modified from Aspila et al. (1976) and Grasshoff et al. (1999).

25.4.2 Water Sampling

The profiling SEABIRD ELECTRONICS (SBE) 911PLUS CTD system was equipped with a DIGIQUARTZ pressure sensor, a SBE3 temperature and SBE4 conductivity sensor within a double sensor setup, and a SBE43 dissolved oxygen sensor. Additionally, water samples were taken from the CTD rosette water sampler to analyse DIC concentrations as well as the total alkalinity (TA) and nutrient concentrations. Therefore, we used a cavity ringdown spectrometer (PICARRO G2201-I, 1510CFIDS2047_v1.0) attached to a Liaison A0301 (DIC), a VINDTA 2C analyser (MARIANDA, Kiel, Germany, TA), and a continuous flow injection system (SKALAR SAN PLUS SYSTEM). Detection limits for nitrate/nitrite (NO_x) and phosphate were 0.08 μM and 0.07 μM, respectively (Flohr et al. 2014). Both the PICARRO and the VINDTA 2C analyser were calibrated using Certified Reference Material (CRM, batch #177) provided by A. Dickson (Scripps Institution of Oceanography, La Jolla, CA, USA). The accuracy was ±12 μmol kg^{-1} and ± 4.3 μmol kg^{-1} for DIC and TA, respectively.

25.5   Results and Discussion

25.5.1 CO_2 Concentrations

Within the BUS, CO_2 concentrations were high in a narrow belt along the coast (Emeis et al. 2018) where ESACW and SACW enriched in nutrients and CO_2 reach the surface. Plankton blooms developing in response to the nutrient input consume nutrients and CO_2. Thereby they decrease the concentration of both components in the offshore flowing surface water until nitrate is consumed and phosphate concentrations drop to values of ~0.2 μmol kg^{-1} (Flohr et al. 2014). This so-called excess phosphate is assumed to be the result of a nutrient uptake which follows the Redfield ratio and a relative enrichment of phosphate over nitrate due to anaerobic processes within the Oxygen Minimum Zone (OMZ) on the shelf and surface sediments (Flohr et al. 2014; Mashifane 2021; Nagel et al. 2013; Goldhammer et al. 2010). However, after the consumption of nitrate within offshore flowing surface waters, a balance between nitrate consumption and release during the respiration of organic matter prevents nitrate accumulation in the euphotic zone. Since such a regenerated nutrient cycle hardly affects CO_2 concentrations, the variability of pCO_2 is strongly reduced at approximately 340 km from the coast. This can be seen in the BUS (Siddiqui et al. 2023) as well as in the California Upwelling System (Chavez and Messié 2009) and marks in addition to nitrate concentrations below the detection limit, the diminishing influence of upwelling on the pelagic ecosystem. At these outskirts of the upwelling driven ecosystem, CO_2 concentrations reflect the balance between the CO_2 uptake via the utilization of preformed nutrients (biological carbon pump) and the warming of the upwelling water (solubility pump). Vice versa, changing CO_2 concentrations at such sites indicate varying intensities of these two pumps within the upwelling system. Hence, moored CO_2 observing buoys, including oxygen, nitrate, and pH sensors could help to detect long-term changes in the balance between the biological carbon pump and solubility pump. Since the region influenced by upwelling varies and is fragmented by mesoscale eddies and filaments (e.g. Rubio et al. 2009), it is suggested to deploy such CO_2 observing buoys along transects perpendicular to the coast encompassing the region in which nitrate depletion and low variability of pCO_2 indicate the diminishing effect of upwelling on pelagic ecosystems.

25.5.2 Export Production

Export/new production can be used in conjunction with satellite-derived primary production rates to characterize the recycling efficiency in the euphotic zone as indicated by the f-ratio. Combined with the supply of preformed nutrients along with upwelling source water masses, it is also a parameter that allows us to estimate the CO_2 uptake by the biological carbon pump as we will discuss in the following sections. However, so far there are no reliable methods to measure export production directly. Due to uncertainties involved in methods commonly applied to determine export production, estimates of global mean export production rates reveal a wide range with values between 1.8 and 27.5 Pg C $year^{-1}$ (Lutz et al. 2007; del Giorgio and Duarte 2002; Honjo et al. 2008). A widely accepted global mean export production is centred around 10 Pg C $year^{-1}$ (Emerson 2014: and references therein) resulting in a mean area normalized export production rate of 28 g C m^{-2} $year^{-1}$ (area of the ocean 360 10^{12} m^2) and a global mean f-ratio of 0.2.

25.5.2.1 Export Production: A Top-Down Approach

In contrast to the bottom-up approach in which nutrient concentrations are multiplied by upwelling velocities to quantify new/export production (e.g. Chavez and Messié 2009; Messié et al. 2009; Waldron et al. 2009), other methods address the issue of export production from a top-down approach by looking at primary production. Thereby, export production is described as a simple function of primary production. Eppley and Peterson (1979) introduced one of the first functions which assumes that export/new production (POC_{Export}) contributes 50% (f-ratio 0.5) to the total primary production (PP) in high productive systems (see Eq. 25.1). Laws et al. (2000) and Henson et al. (2011) additionally considered impacts of sea water temperatures (SST) on the link between export and primary production as shown in Eqs. (25.2) and (25.3).

$$POC_{Export} = \begin{matrix} 0.0025 \cdot PP^2, & \text{if } PP < 200 \\ 0.5 \cdot PP, & \text{if } PP > 200 \end{matrix} \qquad (25.1)$$

$$POC_{Export} = (-0.02 \cdot SST + 0.63) \cdot PP \qquad (25.2)$$

$$POC_{Export} = 0.23 \cdot \exp^{(-0.08 \cdot SST)} \cdot PP \qquad (25.3)$$

In the BUS, SSTs vary in general between <20 °C at the ABFZ and about 10 °C at sites where upwelling source waters reach the surface. Within this temperature range, f-ratios (= POC_{Export}/PP) as derived from Eqs. (25.2) and (25.3) are far below 0.5, with means of 0.3 ± 0.06 and 0.07 ± 0.02, respectively (Fig. 25.3). This implies a much stronger nutrient recycling in the euphotic zone and a lower export production at the same primary production. We used these equations in addition to satellite-derived primary production rates and SST to calculate export production rates, which were compared to organic carbon fluxes measured by a sediment trap deployed in the NBUS (Figs. 25.2 and 25.4).

The sediment trap was deployed in a water-depth of 60–75 m close to the Namibian coast within the narrow belt along the coast where CO_2 concentrations and primary production are high. According to satellite data, the mean primary production rate over the period of the sediment trap experiment from 2009 to 2018 was about 3000 mg m^{-2} day^{-1}. This is high, but within the upper range of primary production rates measured in situ during expeditions (Barlow et al. 2009; Wasmund et al. 2005). Export production rates derived from the three equations vary (Eq. 25.1 = 1794 ± 426 mg m^{-2} day^{-1}; Eq. 25.2 = 1101 ± 295 mg m^{-2} day^{-1}) due to the different f-ratios, whereas results obtained from Eq. (25.3) (229 ± 64 mg m^{-2} day^{-1}) are similar to organic carbon fluxes measured by sediments traps (136 ± 90 mg m^{-2} day^{-1}). However, since the sediment trap was deployed at a water-depth of 60–75 m within the thermocline where exported organic matter is respired as indicated by low oxygen concentrations, trap results do not reflect export production rates but only the export of organic matter at water-depth 60–75 m. Hence, the difference between the export production and the measured organic carbon flux indicates the organic matter remineralization between the base of the euphotic zone and the trap-depth. Data obtained in the SBUS allowed us to study organic matter decomposition at such low water-depths in more detail.

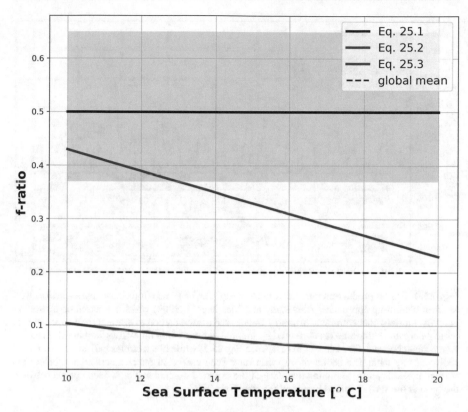

Fig. 25.3 f-ratio versus sea surface temperatures according to Eqs. (25.1), (25.2), and (25.3) as well as in comparison to the global mean f-ratio (dotted line) and those determined by our SBUS drifter study (grey shaded area)

During the cruise M153 (February, 19th–March 31st 2019) a drifter was deployed close to the South African coast, with sediment traps attached at four different water-depths (20, 30, 50, and 75 m) at one of our 24 hour-stations. At this station in situ primary production was also determined by using measured photosynthesis-light-curves (PE curves) in combination with incubation experiments at in situ temperatures and different light intensities. Furthermore, CTD casts were conducted four times a day in order to capture a diurnal cycle and during the cruise SO285 (August 20th–November 2nd 2021) this site was revisited (Fig. 25.5). The drifter traps showed that the total flux of organic carbon was highest just below the surface mixed layer at a water-depth of 20 m (1588 mg m^{-2} day^{-1}) and decreased step-wise with increasing water-depth. At a water-depth of 75 m, the '20 m-flux' had already decreased by 78% to 358 mg m^{-2} day^{-1}. Measured in situ primary production reached values of about 2430 mg m^{-2} day^{-1} and were similar to monthly mean satellite-derived primary production rates of 3428 mg m^{-2} day^{-1} (March, 2019). Primary production of 2430–3428 mg m^{-2} day^{-1} and an export production of

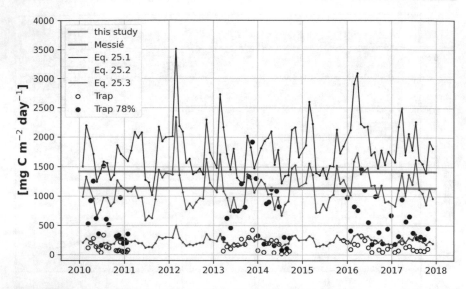

Fig. 25.4 Export production rates derived from Eqs. (25.1)–(25.3) (top-down approaches) as well as from bottom-up approaches (this study and Messié et al. 2009) which are based on upwelling velocity, nutrient concentrations in source waters, and a fixed carbon to nutrient ratio. Open circles indicate organic carbon fluxes measured by moored sediment trap off Walvis Bay in the NBUS at water-depth between 60 m and 74 m (Trap, see Fig. 25.2) while black circles indicate the corrected sediment trap data. The sediment trap data have been corrected for remineralization below the surface mixed layer by assuming that 78% of the exported organic matter is decomposed between the base of the surface mixed layer and the trap (Trap 78%)

1588 mg m^{-2} day^{-1} result in a f-ratio of 0.65 and 0.37, respectively. f-ratios used in Eq. (25.1) and obtained by Eq. (25.2) partly fall in this range whereas f-ratios derived from Eq. (25.3) seem to be too low (Fig. 25.3).

Assuming that 78% of the export production is decomposed between the base of the surface mixed layer and a water-depth of about 75 m implies that the measured mean organic flux of 136 ± 90 mg m^{-2} day^{-1} at our NBUS trap site represents only 78% of the export production which, in turn, suggests an export production of 618 ± 410 mg m^{-2} day^{-1}. If one applies this 78%-correction to all measured sampling intervals, it shows that there are several periods at which the corrected organic carbon export rates are similar to those derived from Eq. (25.1) and partly even Eq. (25.2) (Fig. 25.4). This by far does not solve the dilemma of determining export production rates nor the accuracy of sediment trap measurements, but merely increases the reliability of Eq. (25.1) and (25.2) as well as our drifter experiment.

The revisit of the SBUS drifter site during the cruise SO285 showed that the seasonally varying mixed layer depth is another factor that needs to be considered when comparing shallow water sediment trap results and satellite data. Even though a similar situation was observed during the cruises M153 and SO285 at this site (Fig. 25.5), the comparison of data obtained during these cruises also shows some pronounced differences. For instance, the surface mixed layer was about

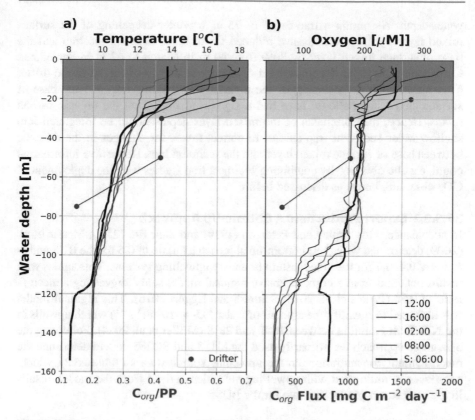

Fig. 25.5 Temperature and oxygen profiles obtained at the near shore drifter site in the SBUS (see Fig. 25.2) at different times of the day during the cruise M153 (March 2019, station 7) and during the cruise SO285 (September/October 2021, station 43) at 06:00 in the morning (S: 6:00). The red circles indicate the ratio between exported organic matter and primary production (C_{org}/PP, a) and organic carbon fluxes determined by the drifter study in four different water depth. The dark and light grey shades indicate the depth of the surface mixed layer during the cruises M153 and SO285

14 m deeper, while bottom water oxygen concentrations were higher in September 2021 (SO285) than in March 2019 (M153). The intensity of the bottom water OMZ is known to reveal a pronounced seasonal cycle with decreasing oxygen concentrations during the main upwelling season in austral summer and increasing oxygen concentrations during the austral winter (Pitcher et al. 2014). Hence, the bottom water OMZ was more intense in March at the end of the upwelling season than in September at the beginning of the upwelling season. Vertical mixing is, in turn, assumed to supply oxygen to the intermediate water layer so that this layer maintains nearly constant oxygen concentrations over the years. Nevertheless, spatial variations of the Upwelling Front cause temporally limited events during which the oxygen concentration drop by up to ~150 µM (Rixen et al. 2021).

However, the 14 m winter deepening of the surface mixed layer could have strongly influenced results obtained by sediment traps which are deployed at a fixed water-depth. Assuming a trap-depth of 75 m, a winter deepening of the surface mixed layer by 14 m could have reduced the distance between the trap and the base of surface mixed layer by 30% from 60 m in March to 42 m in September. Considering the rapid decomposition of organic matter as observed in the drifter experiment, such a shortening of the distance between the trap and the base of surface mixed layer should have had a significant impact on the organic carbon fluxes. Hence, measurements of the mixed layer depth should be integrated into shallow water sediment trap studies to correct fluxes for changes in the distance between base of surface mixed layer and the sediment trap. Respective information could, e.g., be obtained by combining sediment trap studies with the deployment of CO_2 observing buoys as discussed before.

25.5.2.2 Export Production: A Bottom-Up Approach

In accordance with Eppley and Peterson (1979) and their Eq. (25.1), Messié et al. (2009) determined with their bottom-up approach a f-ratio of 0.5 for the BUS and f-ratios of 0.4–0.7 for the other eastern boundary upwelling systems. This agrees with results obtained from a comprehensive regional model study suggesting a mean f-ratio of 0.58 (Emeis et al. 2018; Schmidt and Eggert 2016). This regional model was also used to quantify the amount of water (0.9×10^6 m^3 s^{-1}) which upwells in the NBUS off Namibia between 16°S and 28°S (Müller et al. 2014). Following the bottom-up approach we primarily used the M153 and SO285 data to determine the mean nutrient concentrations in the upwelling source waters (Table 25.2), which, subsequently multiplied with upwelling velocities derived from the model, results in the new/export production rate for the BUS.

Table 25.2 Mean properties of the upwelling source water between 18–28°S during the cruises M153 and SO285. Please note that nutrient concentrations have been measured in discrete samples during the cruise M153 and nitrate concentrations by the *SEASCAN SUNA Deep Nitrate Sensor* during the cruise SO285. Other nutrient data from cruise SO285 are not yet available since this cruise was still on-going while this paper was written, n = number of samples and Std = standard deviation

	M153			SO285		
	March, 2019			September, 2021		
	Mean	Std	N	Mean	Std	N
Density (g cm^{-3})	1.028	0.153	8249	1.028	0.156	9454
Sigma-theta (sθ)	26.69	0.15	8249	25.95	0.16	9454
θ (T$_{pot}$) (°C)	10.36	1.63	8249	10.13	1.77	9454
Salinity	34.93	0.17	8249	33.97	0.20	9454
O$_2$ (μmol kg^{-1})	64.84	33.18	8249	86.29	43.23	9454
NO$_x$ (μmol kg^{-1})	22.59	6.14	141	24.39	4.95	8214
PO$_4$ (μmol kg^{-1})	1.75	0.38	141			
PO$_4$-r (μmol kg^{-1})	1.21	0.36	141			
PO$_4$-p (μmol kg^{-1})	0.53	0.14	141			

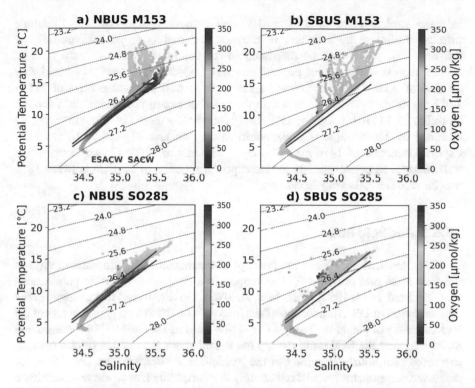

Fig. 25.6 θ-S-diagrams including concentrations of dissolved oxygen derived from CTD casts in NBUS and SBUS during the cruises M153 (February/March 2019) and SO285 (September/October 2021). The red and blue straight lines indicate the θ-S characteristics of the ESACW and SACW according to the equation given by Flohr et al. (2014)

SACW and ESACW, the principal source waters in the BUS, can be identified in θ-S-diagrams as they fall on distinct straight lines within a density range of 27.3 to 26.4 (Fig. 25.6). The comparison between cruises M153 and SO285 shows pronounced differences in the surface water properties. Summer warming decreased the density of surface water and increased the density gradient between surface and upwelling source waters in March, whereas winter cooling had an opposing effect (Fig. 25.6). It decreased the density gradient between surface waters and upwelling source water masses. In contrast to the surface water, the physical properties of upwelling source waters with this density horizon showed hardly any variations except in concentrations of dissolved oxygen (Table 25.2). In the NBUS, they were lower in March than in September due to an enhanced inflow of oxygen-poor SACW between March and May (Mohrholz et al. 2008).

In comparison to oxygen the seasonal variability of nitrate was low (Table 25.2). A mean nitrate concentration of 23.5 μmol kg^{-1} (= (22.59 + 24.39)/2) and a phosphate concentration of 1.75 μmol kg^{-1} multiplied by the mean upwelling velocity of 0.9 × 10^6 m^3 s^{-1} (~0.93 × 10^6 kg s^{-1}) suggest nutrient inputs into the surface layer of 52.5×10^9 mol phosphate and 718.5×10^9 mole nitrate. Multiplied by a carbon to nitrogen and carbon to phosphorus ratio of 7.6 and 106, which will be discussed in the following sections, these nutrient inputs suggest a new/export production rate of 65–67 Tg C year^{-1}. Considering an area of 0.16×10^{12} m^2 these values imply export production rates of 1113–1147 mg m^{-2} day^{-1} which is within the range of export production rates derived from Eq. 25.2 (1101 ± 295 mg m^{-2} day^{-1}) and below those (1416 mg m^{-2} day^{-1}) obtained by Messié et al. (2009). Multiplied by an area of 0.16×10^{12} m^2 the export production of 1416 mg m^{-2} day^{-1} amounts to 83 Tg C year^{-1} which we will consider as upper estimate for the export production in the BUS between 18°S and 28°S in the following discussion.

25.5.3 Redfield Ratio

In 1934, Redfield introduced the concept of a constant carbon to nutrient ratio in the year 1934 and provided further support in a couple of follow-up publications (e.g. Redfield 1934, 1958; Redfield et al. 1963). He discovered that changes in the concentration of DIC, nitrate, phosphate, and oxygen in the water column reflect the stoichiometry of plankton. Today a C/N/P/-O$_2$ ratio of $106/16/1/-138$ is considered as the classical Redfield ratio. During the last 60 years, the number of observations increased enormously showing that the Redfield ratios derived from correlations of dissolved components reveal a remarkably low variability (Anderson and Sarmiento 1994; Takahashi et al. 1985) in comparison to those derived from the elementary composition of plankton (Boyd and Trull 2007; Planavsky 2014; Martiny et al. 2013). Hence, we are again confronted with a discrepancy between bottom-up (dissolved components) and top-down (plankton stoichiometry) approaches.

25.5.3.1 Redfield Ratio: A Top-Down Approach

A main factor controlling the Redfield ratio is the ratio between cyanobacteria and eukaryotic plankton as cyanobacteria often show enhanced and highly variable Redfield ratios (Karl et al. 1997; Bertilsson et al. 2003; Sanudo-Wilhelmy et al. 2004). Hence, C/N/P ratios in oligotrophic subtropical gyres dominated by cyanobacteria are high, with C/N/P values of 195:28:1 (Martiny et al. 2013; Teng et al. 2014). However, in addition to variations in plankton community structure, environmental changes such as temperature and pH variations as well as nutrient stress could influence the carbon to nutrient ratios in individual plankton clades (Boyd and Trull 2007; Planavsky 2014; Martiny et al. 2013; Geider and La Roche 2002; Riebesell et al. 2007). Hence, organic matter in nutrient-rich environments and at high latitudes often reveal lower Redfield ratios (78/13/1, C/N = 6) than in nutrient-depleted regions (138/18/1, C/N = 7.7) at lower latitudes (Martiny et al. 2013; Teng et al. 2014).

This compilation also shows that in comparison to C/P ratios, the variability of C/N ratios is quite low (6–7.7) and hardly deviates from the classical Redfield ratio

Table 25.3 Mean C/N and C/P ratios with corresponding standard deviations (std) and n number of samples measured. The samples were obtained from net catches at locations marked as 'M153 Zooplankton' in Fig. 25.2

	C/N		C/P		n
	Mean	Std	Mean	Std	
Fish	4.6	0.5	61.6	15.6	13
Squid	4.5	0.5	106.8	10.5	5
Zooplankton	5.1	0.7	82.0	11.5	15
Jelly fish	4.0	0.2	110.0	16.3	4

of 6.6. We obtained a similar result from plankton samples which were collected within the chlorophyll maximum during the cruise M153. The chlorophyll maximum is the zone within surface waters where the highest chlorophyll concentration occurs. C/N ratios determined from these samples vary between 5.6 and 8.6, with a mean of about 6.7 in the NBUS and SBUS. Zooplankton, fish, squid, and jelly fish samples collected by net catches during the cruise M153 reveal, in turn, lower C/N ratios of about 4.7 ± 0.7 ($n = 37$), suggesting an enrichment of nitrogen within higher trophic levels of the food web (Table 25.3).

Our zooplankton samples include euphausiids, amphipods, and decapods, but excluded copepods. Copepods are abundant in the BUS as well as in our drifter samples and are known to reveal highly variable C/N ratios of 3.6–10.4 (Bode et al. 2015; Schukat et al. 2014). In order to study the role of zooplankton in more detail, we also deployed drifting sediment traps during the cruise M153 as mentioned before. The results show that the zooplankton carbon contributes on average $65 \pm 21\%$ to the total organic carbon collected by the traps with a mean C/N ratio of 7.8 ± 1.1. Considering the total organic carbon flux (zooplankton and the <1 mm fractions) results in a mean C/N ratio of 8.1 ± 1.8. This is similar to the mean C/N ratios of 8.7 determined by other sediment trap experiments in the region including traps deployed at water-depths down to 2500 m (Vorrath et al. 2018) and those C/N ratios measured in the moored NBUS traps (8.4 ± 1.3). Compared to phytoplankton (5.6–8.6, mean: 6.7), these slightly enhanced C/N ratios could be caused by a preferential decomposition of organically bound nitrogen in the water column or inputs of organic matter from land and resuspended sediments. On the other hand, we discovered in line with an earlier study along the European continental margin (Antia 2005) a preferential leaching of phosphorus containing organic components, which raised C/P ratios to values of >400. Hence, C/P ratios measured in trap samples have been ignored in this discussion. Along the European continental margin where traps had been deployed at water-depths between 600 and 3200 m, leaching of organically bound nitrogen raised the C/N ratio on average from 8.1 to 11.3 (Antia 2005). Even though a mean C/N ratio of 8.1 ± 1.8 as measured by our traps at a water-depth of 60–75 m implies a comparably low leaching of organically bound nitrogen, it could suffice to explain the slightly enhanced C/N ratio in the trapped exported organic matter. To which extent C/N ratios derived from concentrations of DIC, phosphate, and nitrate reflect those of particulate matter will be discussed in the next section.

25.5.3.2 Redfield Ratio: A Bottom-Up Approach

In the NBUS and SBUS, nitrate and phosphate concentrations are correlated, while the slope derived from the regression equation suggests mean N/P ratios of 13.7–14 (Fig. 25.7). The slightly lower N/P ratio in the NBUS (Fig. 25.7b) is most likely a consequence of nitrate reduction (Tyrrell and Lucas 2002). This occurs in the NBUS OMZ (e.g. Nagel et al. 2013), which is more intense than the SBUS OMZ. However, C/N and C/P ratios are more difficult to determine since the solubility pump accounts for the majority of the carbon dissolved in ocean waters. The y-axis intercept of the regression line derived from the correlation between phosphate and DIC concentrations indicates the DIC background concentrations of 2144 μmol kg^{-1} in the NBUS and 2085 μmol kg^{-1} in SBUS. These differences could be caused by the dissolution of carbonates, as well as varying pCO_2 disequilibria between surface waters and the atmosphere and the uptake of anthropogenic

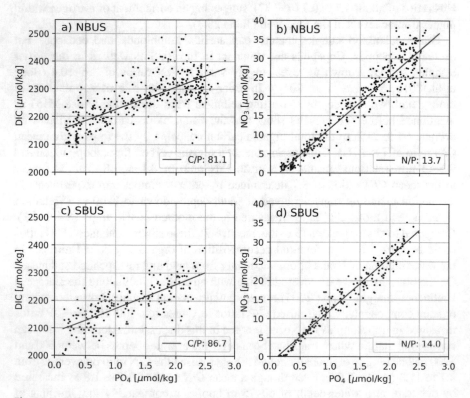

Fig. 25.7 DIC versus phosphate concentrations as well as phosphate versus nitrate concentrations measured in samples obtained in NBUS and SBUS during the cruise M153

CO_2 during the formation of ESACW and SACW. The slopes, in turn indicate C/P ratios of 81.1 and 86.7 in the NBUS and SBUS, respectively.

Well-accepted and often used methods (Sabine et al. 1999; Gruber et al. 2019) associated with the quasi-conservative tracer ΔC^* can be used to quantify the influence of the pCO_2 disequilibria between ocean and atmosphere, the dissolution of carbonate and the uptake of anthropogenic CO_2 on the DIC concentration (Gruber et al. 1996). But since these methods already rely on the assumption of a constant Redfield ratio as derived from Anderson and Sarmiento (1994) of 117/16/1/170 (C/N/P/$-O_2$), they are not applicable here where we aim to determine Redfield ratios in the BUS. Nevertheless, deviation from the minimum TA was considered as a consequence of carbonate dissolution and the corresponding release of DIC ($DIC_{dissolution} = (TA_{measured} - TA_{minimum})/2$) as well as the DIC background concentrations were subtracted from the measured DIC concentrations ($DIC_{corrected} = DIC_{measured} - DIC_{dissolution} - DIC_{background}$). These corrections reduced the differences between NBUS and SBUS, and the slope of the resulting regression equation implied a C/P ratio of 93 (Fig. 25.8). C/P ratios varying between 81.1 and 93.0 combined with N/P ratios of 13.7–14.0 suggest a mean Redfield ratio of $87 \pm 6/14 \pm 0.2/1$. The resulting C/N of about 6.3 falls slightly below the classical Redfield C/N ratio of 6.6 and C/N ratios determined by us in the chlorophyll maximum with a mean of about 6.7. Considering the chlorophyll maximum C/N ratios instead of the N/P ratio derived from nitrate and phosphate concentrations results in a mean Redfield ratio of $87 \pm 6/13 \pm 3/1$. The differences between these two approaches are low and the overall relatively low Redfield ratios are similar to those derived from plankton in nutrient-rich environments and high latitudes

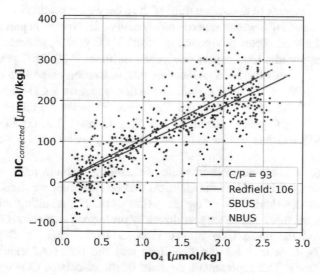

Fig. 25.8 Corrected DIC versus phosphate concentrations from the NBUS and SBUS. The red line marks the regression equation ($DIC_{corrected} = 92.9 \times PO_4 + 2.129$, $r = 0.782$, $n = 617$) and the blue line shows show the Redfield ratio

(78/13/1). However, in February 2011, we determined with a bottom-up approach a Redfield ratio of about $100 \pm 1/16 \pm 2/1$ off Namibia and also located a region on the Namibian shelf where anaerobic processes in association with nutrient fluxes across the sediment water interface reduced the Redfield ratio to 66/10/1 (Flohr et al. 2014). Hence, Redfield ratios seem to vary interannually and could be influenced by sediment water interactions within pronounced bottom water OMZ. Nevertheless, regarding the CO_2 uptake by the biological carbon pump, our bottom-up and top-down approach based on the M153 data support Redfield's findings and imply that the consumption of upwelled nitrate and phosphate adsorbs as much CO_2 as is released during the respiration of exported organic matter. This, in turn, poses the question as to how the carbon export could influence the CO_2 uptake by the biological carbon pump.

25.5.4 CO_2 Sequestration

In order to estimate carbon sequestration by the biological carbon pump, we applied the bottom-up approach to calculate export production and used the Redfield ratio to convert phosphate supply into surface waters into carbon export production rates. In principle, this approach is similar to those used by Waldron et al. (2009) for the SBUS but in contrast to Waldron et al. (2009), we used only preformed and not the total nutrient concentrations of the source waters. As discussed before, preformed nutrients have been detached from CO_2 so that their utilization acts as a CO_2 sink. Since the preformed phosphate concentrations contribute about 30% to the total phosphate concentration (Table 25.2), the CO_2 uptake by the biological carbon pump amounts to about one third of the total export production. In the BUS between 18°S and 28°S which approximately covers the Namibian part of the BUS, we estimated a total export production of 65–83 Tg C year^{-1}, whereas Waldron et al. (2009) suggested a total export production of about 7–39 Tg C year^{-1} for the South African part of the BUS. Assuming that utilization of preformed nutrients accounts for 30% of the total export production suggests a CO_2 uptake of 20–25 Tg C year^{-1} (mean $= 22.5 \pm 2.5$ Tg C year^{-1}) in the Namibian part of the BUS and 2–12 Tg C year^{-1} (mean $= 7 \pm 5$ Tg C year^{-1}) in the South African part of the BUS which is similar to the results of our previous study (Siddiqui et al. 2023).

In order to estimate possible impacts of changing Redfield ratio, we repeated the calculations but instead of the classical C/P ratio of 106, we used a C/P ratio of 87 ± 6 as determined by using the M153 data. The resulting effect can be calculated by the rule of three (export driven by preformed nutrients/$106 \times 87 \pm 6$) which lowered the mean CO_2 uptake rates to 18.5 ± 3.3 Tg C year^{-1} and 6.0 ± 5.0 Tg C year^{-1} for the Namibian and the South African part of the BUS, respectively. This represents a decrease of the calculated CO_2 uptake by the biological carbon pump of 17% (4 Tg C year^{-1}) and 13% (1 Tg C year^{-1}) off Namibia and South Africa.

The CO_2 uptake rates by the biological carbon pump of 18.5 ± 3.3 Tg C year^{-1} and 6.0 ± 5.0 Tg C year^{-1} are in the same order of magnitude as the CO_2 uptake by AFOLU of 30.6 Tg C year^{-1} (Namibia) and 7.5 Tg C year^{-1} (South Africa, see Table 25.1), but these carbon fluxes cannot be compared to each other since the latter represents human impacts on the CO_2 uptake by terrestrial ecosystems. Even though marine ecosystems respond to global change (Lamont et al. 2019; Verheye et al. 2016; Jarre et al. 2015; Hutchings et al. 2012), fisheries (Bianchi et al. 2021) and in particular bottom trawling which will be discussed in the following section, effects on the CO_2 uptake by the biological carbon pump via influences on the Redfield ratio have not yet been quantified. Nevertheless, the magnitude of the CO_2 uptake rates by the biological carbon pump is on levels at which small changes due to, e.g., decreasing C/P ratios of 17% or 13%, respectively, could influence the CO_2 uptake by AFOLU significantly.

25.5.5 Benthos

So far, we discussed the CO_2 uptake by the biological carbon pump in the water column, but it also influences the burial of carbon in sediments. This represents a removal of carbon from the climate active carbon cycle and humans could favour and reduce it. An enhanced supply of clay and other minerals via dust inputs and river discharges due to soil as well as coastal erosion favours the organic carbon sedimentation by reducing the remineralization of organic matter in the water column and surface sediments mainly by two processes. Firstly, minerals incorporated into particles increase their sinking speed and the resulting accelerated transport of organic matter from the euphotic zone onto the sediments lowers the residence time of organic matter in the water columns and therewith its remineralization (e.g. Rixen et al. 2019b). Secondly, the adsorption of organic matter to mineral surfaces reduces its remineralization in the water column and surface sediments by protecting organic matter against bacterial attacks (e.g. Armstrong et al. 2002; Hedges 1977). Vice versa, bottom water trawling remobilizes the sedimentary organic matter and favours its return into the climate active carbon cycle. Sala et al. (2021) introduced an approach to quantify impacts of bottom trawling on the remobilization of carbon from marine sediments on a global scale. Assuming a continuous bottom trawling, Sala et al. (2021) suggested, for instance, a global long-term sedimentary organic carbon loss of about 158 Tg C year^{-1}, and also broke it down to the BUS where it amounts to about 5 Tg C year^{-1}. In comparison to AFOLU and our estimates of the CO_2 uptake by the biological Compared to AFOLU and our estimates of the CO_2 uptake by the biological carbon pump of about 24 Tg C/year-1 in the entire BUS (NBUS and SBUS), this is a substantial amount. It shows that humans can significantly affect the CO2 uptake of the biological carbon pump which still needs to be quantified.

25.6　Conclusion

In contrast to the often assumed and used constant global mean Redfield ratio, our results indicate varying Redfield ratios whereas the one determined by us during the TRAFFIC cruise M153 ($87 \pm 6/13 \pm 3/1$) is similar to those derived by other studies in nutrient-rich environments and high latitudes (78/13/1). CO_2 uptake by the biological carbon pump based on the supply of preformed nutrients and the Redfield ratio amounts to 18.5 ± 3.3 Tg C year^{-1} and 6.0 ± 5.0 Tg C year^{-1} for the Namibian and South African parts of the BUS, respectively. These uptake rates are on a level at which small changes could influence the CO_2 uptake by AFOLU significantly. Considering the release of sedimentary carbon by bottom trawling and that ecosystems in the BUS respond to global change and fisheries, it is quite likely that humans affect the CO_2 uptake by the biological carbon pump already but uncertainties are still too large to quantify such impacts. Hence, we suggest to improve methods to estimate the supply of preformed nutrients, carbon export rates, and Redfield ratios and expand monitoring strategies by linking established observing methods such as remote sensing and sediment trap studies, and focus on CO_2 observation at the outer rim of the upwelling system. This appears to be a key site to detect long-term changes in the balance between the biological carbon pump and solubility pump. Despite large uncertainties regarding the CO_2 uptake by the biological carbon pump and its influence on the solubility pump, we propose to include parameters such as preformed nutrient supply, carbon export rates, Redfield ratios, and CO_2 concentrations measured at specific key sites into the NIR.

Acknowledgments The authors would like to thank all scientists, technicians, captains, and crew members of the German and South African research vessels Meteor, Sonne, and Algoa for their support during the cruises M153, SO283, SO285, and ALG 269. In particular, the authors are very grateful to F. Hüge and M. Birkicht for analysing the nutrient samples and the German Federal Ministry of Education and Research (BMBF) for funding our research under the grant no. 03F0797A (ZMT) and 03F0797C (Universität Hamburg).

References

Anderson LA, Sarmiento JL (1994) Redfield ratios of remineralization determined by nutrient data analysis. Glob Biogeochem Cycles 8(1):65–80

Antia AN (2005) Solubilization of particles in sediment traps: revising the stoichiometry of mixed layer export. Biogeosciences 2(2):189–204

Armstrong RA, Lee C, Hedges JI et al (2002) A new, mechanistic model for organic carbon fluxes in the ocean: based on the quantitative association of POC with ballast minerals. Deep-Sea Res 49(II):219–236

Aspila KI, Haig A, Chau ASY (1976) A semi-automated method for the determination of inorganic, organic and total phosphate in sediments. Analyst 101:187–97

Barange M, Pillar SC (1992) Cross-shelf circulation, zonation and maintenance mechanisms of Nyctiphanes capensis and Euphausia hanseni (Euphausiacea) in the northern Benguela upwelling system. Cont Shelf Res 12(9):1027–1042

Barlow R, Lamont T, Mitchell-Innes B et al (2009) Primary production in the Benguela ecosystem, 1999 - 2002. Afr J Mar Sci 31(1):97–101

Bar-On YM, Phillips R, Milo R (2018) The biomass distribution on earth. Proc Natl Acad Sci 115(25):6506–6511

Bauska TK, Baggenstos D, Brook EJ et al (2016) Carbon isotopes characterize rapid changes in atmospheric carbon dioxide during the last deglaciation. Proc Natl Acad Sci 113(13):3465

Bertilsson S, Berglund O, Karl DM et al (2003) Elemental composition of marine Prochlorococcus and Synechococcus: implications for the ecological stoichiometry of the sea. Limnol Oceanogr 48(5):1721–1731

Bianchi D, Carozza David A, Galbraith Eric D et al (2021) Estimating global biomass and biogeochemical cycling of marine fish with and without fishing. Sci Adv 7(41):eabd7554

Bode M, Hagen W, Schukat A et al (2015) Feeding strategies of tropical and subtropical calanoid copepods throughout the eastern Atlantic Ocean – latitudinal and bathymetric aspects. Prog Oceanogr 138:268–282

Boyd PW, Trull TW (2007) Understanding the export of biogenic particles in oceanic waters: is there consensus? Prog Oceanogr 72(4):276–312

Broecker W, Barker S (2007) A 190‰ drop in atmosphere's $\Delta14C$ during the "mystery interval" (17.5 to 14.5 kyr). Earth Planet Sci Lett 256(1–2):90–99

Chavez FP, Messié M (2009) A comparison of eastern boundary upwelling ecosystems. Prog Oceanogr 83(1-4):80–96

del Giorgio PA, Duarte CM (2002) Respiration in the open ocean. Nature 420(6914):379–384

Demarcq H (2009) Trends in primary production, sea surface temperature and wind in upwelling systems (1998-2007). Prog Oceanogr 83(1-4):376–385

Department of Environmental Affairs, 2019. South Africa's 3rd Biennial Update Report to the United Nations Framework Convention on Climate Change. Department of Environmental Affairs, South Africa, Pretoria. South Africa., p. 234.

DeVries T, Deutsch C (2014) Large-scale variations in the stoichiometry of marine organic matter respiration. Nat Geosci 7(12):890–894

Dittmar T, Birkicht M (2001) Regeneration of nutrients in the northern Benguela upwelling and the Angola-Benguela front areas. S Afr J Sci 97:239–246

Donners J, Drijfhout SS, Hazeleger W (2005) Water mass transformation and subduction in the South Atlantic. J Phys Oceanogr 35(10):1841–1860

Duce RA, LaRoche J, Altieri K et al (2008) Impacts of atmospheric anthropogenic nitrogen on the open ocean. Science 320(5878):893–897

Duncombe Rae CM (2005) A demonstration of the hydrographic partition of the Benguela upwelling ecosystem at 26° 40′S. Afr J Mar Sci 27(3):617–628

Durgadoo JV, Rühs S, Biastoch A et al (2017) Indian Ocean sources of Agulhas leakage. J Geophys Res Oceans. https://doi.org/10.1002/2016JC012676

Duteil O, Koeve W, Oschlies A et al (2012) Preformed and regenerated phosphate in ocean general circulation models: can right total concentrations be wrong? Biogeosciences 9(5):1797–1807

Emeis K, Eggert A, Flohr A et al (2018) Biogeochemical processes and turnover rates in the northern Benguela upwelling system. J Mar Syst 188:63–80

Emerson S (2014) Annual net community production and the biological carbon flux in the ocean. Glob Biogeochem Cycles 28(1):14–28

Eppley RW, Peterson BJ (1979) Particulate organic matter flux and planktonic new production in the deep ocean. Nature 282:677–680

Falkowski PG, Barber RT, Smetacek V (1998) Biogeochemical controls and feedbacks on ocean primary production. Science 281:200–206

Field CB, Behrenfeld MJ, Randerson J et al (1998) Primary productivity of the biosphere: an integration of terrestrial and oceanic components. Science 281:237–240

Flohr A, van der Plas AK, Emeis K-C et al (2014) Spatio-temporal patterns of C: N: P ratios in the northern Benguela upwelling regime. Biogeosciences 11:885–897

Flynn RF, Granger J, Veitch JA et al (2020) On-shelf nutrient trapping enhances the fertility of the southern Benguela upwelling system. J Geophys Res Oceans 125(6):e2019JC015948

Geider RJ, La Roche J (2002) Redfield revisited: variability of C:N:P in marine microalgae and its biochemical basis. Eur J Phycol 37:1–17

German Environment Agency (2020) Submission under the United Nations Framework Convention on Climate Change and the Kyoto Protocol 2020 National Inventory Report for the German Greenhouse Gas Inventory 1990 – 2018, in: Strogies, M., Gniffke, P. (Eds.). Umweltbundesamt, Germany, Dessau-Roßlau, Germany, p. 997.

Goldhammer T, Bruchert V, Ferdelman TG et al (2010) Microbial sequestration of phosphorus in anoxic upwelling sediments. Nat Geosci 3(8):557–561

Gordon AL (1981) South Atlantic thermocline ventilation. Deep Sea Res A 28(11):1239–1264

Government of Namibia, 2021. Fourth Biennial Update Report (BUR4) to the United Nations Framework Convention on Climate Change, Windhoek, Namibia, p. 120.

Grasshoff, Klaus, Klaus Kremling, and Manfred Ehrhardt. 1999. *Methods of seawater analysis* (Wiley-VCH: Weinheim, Germany).

Gruber N, Sarmiento JL, Stocker TF (1996) An improved method for detecting anthropogenic CO2 in the oceans. Global Biogeochem Cycles 10(4):809–837

Gruber N, Clement D, Carter BR et al (2019) The oceanic sink for anthropogenic CO2 from 1994 to 2007. Science 363(6432):1193

Haake B, Ittekkot V, Rixen T et al (1993) Seasonality and interannual variability of particle fluxes to the deep Arabian Sea. Deep Sea Res I 40(7):1323–1344

Hardman-Mountford NJ, Richardson AJ, Agenbag JJ et al (2003) Ocean climate of the South East Atlantic observed from satellite data and wind models. Prog Oceanogr 59(2):181–221

Hedges JI (1977) The association of organic molecules with clay minerals in aqueous solutions. Geochim Cosmochim Acta 41(8):1119–1123

Heinze C, Maier-Reimer E, Winn K (1991) Glacial pCO_2 reduction by the world ocean: experiments with the Hamburg Carbon Cycle Model. Paleoceanography 6(4):395–430

Heinze C, Meyer S, Goris N et al (2015) The ocean carbon sink – impacts, vulnerabilities and challenges. Earth Syst Dynam 6(1):327–358

Henson SA, Sanders R, Madsen E et al (2011) A reduced estimate of the strength of the ocean's biological carbon pump. Geophys Res Lett 38(4):L04606

Honjo S, Manganini SJ, Krishfield RA et al (2008) Particulate organic carbon fluxes to the ocean interior and factors controlling the biological pump: a synthesis of global sediment trap programs since 1983. Prog Oceanogr 76(3):217–285

Hutchings L, van der Lingen CD, Shannon LJ et al (2009) The Benguela current: an ecosystem of four components. Prog Oceanogr 83(1-4):15–32

Hutchings L, Jarre A, Lamont T et al (2012) St Helena Bay (southern Benguela) then and now: muted climate signals, large human impact. Afr J Mar Sci 34(4):559–583

IPCC, 2006. 2006 IPCC Guidelines for National Greenhouse Gas Inventories, Prepared by the National Greenhouse Gas Inventories Programme, in: Eggleston H.S., Buendia L., Miwa K., T., N., K., T. (Eds.). IPCC National Greenhouse Gas Inventories Programme Technical Support Unit, IGES, Japan

IPCC, 2014. 2013 Supplement to the 2006 IPCC Guidelines for National Greenhouse Gas Inventories: Wetlands Methodological Guidance on Lands with Wet and Drained Soils, and Constructed Wetlands for Wastewater Treatment, in: Hiraishi, T., Krug, T., Tanabe, K., Srivastava, N., Baasansuren, J., Fukuda, M., Troxler, T.G. (Eds.). IPCC, Switzerland

Ito T, Follows MJ (2005) Preformed phosphate, soft tissue pump and atmospheric CO_2. J Mar Res 63:813–839

Jarre A, Hutchings L, Kirkman SP et al (2015) Synthesis: climate effects on biodiversity, abundance and distribution of marine organisms in the Benguela. Fish Oceanogr 24(S1):122–149

Kämpf J, Chapman P (2016) The Benguela current upwelling system. In: Kämpf J, Chapman P (eds) Upwelling systems of the world. Springer, Cham. https://doi.org/10.1007/978-3-319-42524-5_7

Karl D, Letelier R, Tupas L et al (1997) The role of nitrogen fixation in biogeochemical cycling in the subtropical North Pacific Ocean. Nature 388:533–538

Karstensen J, Quadfasel D (2002) Formation of southern hemisphere thermocline waters: water mass conversion and subduction. J Phys Oceanogr 32(11):3020–3038

Knox F, McElroy MB (1984) Changes in atmospheric CO2: influence of the marine biota at high latitude. J Geophys Res Atmos 89(D3):4629–4637

Lamont T, García-Reyes M, Bograd SJ et al (2018) Upwelling indices for comparative ecosystem studies: variability in the Benguela upwelling system. J Mar Syst 188:3–16

Lamont T, Barlow RG, Brewin RJW (2019) Long-term trends in phytoplankton chlorophyll a and size structure in the Benguela upwelling system. J Geophys Res Oceans 124:1170–1195

Laufkötter C, Gruber N (2018) Will marine productivity wane? Science 359(6380):1103

Laufkötter C, John JG, Stock CA et al (2017) Temperature and oxygen dependence of the remineralization of organic matter. Glob Biogeochem Cycles 31(7):1038–1050

Laws EA, Falkowski PG, Smith WO et al (2000) Temperature effects on export production in the open ocean. Glob Biogeochem Cycles 14(4):1231–1246

Liu M, Tanhua T (2021) Water masses in the Atlantic Ocean: characteristics and distributions. Ocean Sci 17(2):463–486

Lutz MJ, Caldeira K, Dunbar RB et al (2007) Seasonal rhythms of net primary production and particulate organic carbon flux to depth describe the efficiency of biological pump in the global ocean. J Geophys Res Oceans 112(C10):C10011

Marinov I, Gnanadesikan A, Toggweiler JR et al (2006) The Southern Ocean biogeochemical divide. Nature 441(7096):964–967

Martiny AC, Pham CTA, Primeau FW et al (2013) Strong latitudinal patterns in the elemental ratios of marine plankton and organic matter. Nat Geosci 6:279

Mashifane TB (2021) Denitrification and anammox shift nutrient stoichiometry and the phytoplankton community structure in the Benguela upwelling system. J Geophys Res Oceans 126(8):e2021JC017816

McCartney MS (1977) Subantarctic mode water. In: Angel MV (ed) A voyage of discovery: George Deacon 70th anniversary volume, Supplement to deep-sea research. Pergamon Press, Oxford, pp 103–119

Messié M, Ledesma J, Kolber DD et al (2009) Potential new production estimates in four eastern boundary upwelling ecosystems. Prog Oceanogr 83(1-4):151–158

Mohrholz V, Bartholomaeb CH, van der Plas AK (2008) The seasonal variability of the northern Benguela undercurrent and its relation to the oxygen budget on the shelf. Cont Shelf Res 28(3):424–441

Müller AA, Reason CJ, Schmidt M et al (2014) Computing transport budgets along the shelf and across the shelf edge in the northern Benguela during summer (DJF) and winter (JJA). J Mar Syst 140(Part B(0)):82–91

Nagel B, Emeis K-C, Flohr A et al (2013) N-cycling and balancing of the N-deficit generated in the oxygen minimum zone over the Namibian shelf—an isotope-based approach. J Geophys Res Biogeo 118:361–371

Nellemann, C., Corcoran, E., Duarte, C.M., Valdés, L., De Young, C., Fonseca, L., Grimsditch, G., 2009. Blue Carbon. A Rapid Response Assessment. United Nations Environment Programme, Arendal, Norway, p. 79.

Passow U, Carlson CA (2012) The biological pump in a high CO2 world. Mar Ecol Prog Ser 470:249–271

Pitcher GC, Probyn TA, du Randt A et al (2014) Dynamics of oxygen depletion in the nearshore of a coastal embayment of the southern Benguela upwelling system. J Geophys Res Oceans 119(4):2183–2200

Pitcher GC, Aguirre-Velarde A, Breitburg D et al (2021) System controls of coastal and open ocean oxygen depletion. Prog Oceanogr 197:102613

Planavsky NJ (2014) The elements of marine life. Nat Geosci 7(12):855–856

Redfield AC (1934) On the proportions of organic derivations in sea water and their relation to the composition of plankton. In: Daniel RJ (ed) James Johnstone memorial volume. University Press of Liverpool, Liverpool, pp 176–192

Redfield AC (1958) The biological control of chemical factors in the environment. Am Sci:205–222

Redfield AC, Ketchum BH, Richards FA (1963) The influence of organisms on the composition of sea-water. In: Hitt MN (ed) The sea. Wiley, New York, pp 26–77

Riebesell U, Schulz KG, Bellerby RGJ et al (2007) Enhanced biological carbon consumption in a high CO_2 ocean. Nature 450(7169):545–548

Rixen T, Haake B, Ittekkot V et al (1996) Coupling between SW monsoon-related surface and deep ocean processes as discerned from continuous particle flux measurements and correlated satellite data. J Geophys Res 101(C12):28569–28582

Rixen T, Gaye B, Emeis K-C (2019a) The monsoon, carbon fluxes, and the organic carbon pump in the northern Indian Ocean. Prog Oceanogr 175:24–39

Rixen T, Gaye B, Emeis KC et al (2019b) The ballast effect of lithogenic matter and its influences on the carbon fluxes in the Indian Ocean. Biogeosciences 16(2):485–503

Rixen T, Lahajnar N, Lamont T et al (2021) Oxygen and nutrient trapping in the southern Benguela upwelling system. Front Mar Sci 8:1367

Rubio A, Blanke B, Speich S et al (2009) Mesoscale eddy activity in the southern Benguela upwelling system from satellite altimetry and model data. Prog Oceanogr 83(1):288–295

Sabine CL, Key RM, Johnson KM et al (1999) Anthropogenic CO_2 inventory of the Indian Ocean. Glob Biogeochem Cycles 13(1):179–198

Sala E, Mayorga J, Bradley D et al (2021) Protecting the global ocean for biodiversity, food and climate. Nature 592(7854):397–402

Sanudo-Wilhelmy SA, Tovar-Sanchez A, Fu F-X et al (2004) The impact of surface-absorbed phosphorus on phytoplankton Redfield stoichiometry. Nature 432:897–901

Sarmiento JL, Toggweiler JR (1984) A new model for the role of the oceans in determining atmospheric pCO_2. Nature 308:621–624

Sarmiento JL, Gruber N, Brzezinski MA et al (2004) High-latitude controls of thermocline nutrients and low latitude biological productivity. Nature 427(6969):56–60

Schmidt M, Eggert A (2016) Oxygen cycling in the northern Benguela upwelling system: modelling oxygen sources and sinks. Prog Oceanogr 149:145–173

Schmitt J, Schneider R, Elsig J et al (2012) Carbon isotope constraints on the deglacial CO2 rise from ice cores. Science 336(6082):711–714

Schukat A, Auel H, Teuber L et al (2014) Complex trophic interactions of calanoid copepods in the Benguela upwelling system. J Sea Res 85:186–196

Shannon LV (1985) The Benguela ecosystem 1. Evolution of the Benguela, physical features and processes. In: Barnes M (ed) Oceanography and marine biology. University Press, Aberdeen, pp 105–182

Shannon LV, Nelson G (1996) The Benguela: large scale features and processes and system variability. In: Wefer G, Berger WH, Siedler G et al (eds) The South Atlantic: present and past circulation. Springer, Berlin, pp 163–210

Siddiqui C, Rixen T, Lahajnar N et al (2023) Regional and global impact of CO2 uptake in the Benguela upwelling system through preformed nutrients. Nat Commun 14:2582

Siegenthaler U, Wenk T (1984) Rapid atmospheric CO_2 variations and ocean circulation. Nature 308:624–627

Smith TM, Reynolds RW, Peterson TC et al (2008) Improvements to NOAA's historival merged land-ocean surface temperature analysis (1880 - 2006). J Clim 21:2283–2296

Souza AGQ, Kerr R, Azevedo JLL (2018) On the influence of subtropical mode water on the South Atlantic Ocean. J Mar Syst 185:13–24

Sydeman WJ, García-Reyes M, Schoeman DS et al (2014) Climate change and wind intensification in coastal upwelling ecosystems. Science 345(6192):77

Takahashi T, Broecker WS, Langer S (1985) Redfield ratio based on chemical data from Isopycnal surfaces. J Geophys Res 90(C4):6907–6924

Teng Y-C, Primeau FW, Moore JK et al (2014) Global-scale variations of the ratios of carbon to phosphorus in exported marine organic matter. Nat Geosci 7(12):895–898

Tim N, Zorita E, Hünicke B (2015) Decadal variability and trends of the Benguela upwelling system as simulated in a high-resolution ocean simulation. Ocean Sci 11(3):483–502

Tschumi T, Joos F, Gehlen M et al (2011) Deep ocean ventilation, carbon isotopes, marine sedimentation and the deglacial CO_2 rise. Clim Past 7(3):771–800

Tyrrell T, Lucas MI (2002) Geochemical evidence of denitrification in the Benguela upwelling system. Cont Shelf Res 22(17):2497–2511

UNEP, 2019. Emissions Gap Report 2019. United Nations Environment Programme (UNEP), Nairobi, Kenya, p. 80.

Veitch J, Penven P, Shillington F (2009) The Benguela: a laboratory for comparative modeling studies. Prog Oceanogr 83(1):296–302

Veitch J, Penven P, Shillington F (2010) Modeling equilibrium dynamics of the Benguela Current System. J Phys Oceanogr 40(9):1942–1964

Verheye HM, Lamont T, Huggett JA et al (2016) Plankton productivity of the Benguela Current Large Marine Ecosystem (BCLME). Environ Dev 17:75–92

Vorrath M-E, Lahajnar N, Fischer G et al (2018) Spatiotemporal variation of vertical particle fluxes and modelled chlorophyll a standing stocks in the Benguela upwelling system. J Mar Syst 180:59–75

Waldron HN, Monteiro PMS, Swart NC (2009) Carbon export and sequestration in the southern Benguela upwelling system: lower and upper estimates. Ocean Sci 5(4):711–718

Wang D, Gouhier TC, Menge BA et al (2015) Intensification and spatial homogenization of coastal upwelling under climate change. Nature 518:390

Wasmund N, Lass HU, Nausch G (2005) Distribution of nutrients, chlorophyll and phytoplankton primary production in relation to hydrographic structures bordering the Benguela-Angolan frontal region. Afr J Mar Sci 27(1):177–190

Weiss RF (1974) Carbon dioxide in water and seawater: the solubility of a non-ideal gas. Mar Chem 2:203–215

Dynamics and Drivers of Net Primary Production (NPP) in Southern Africa Based on Estimates from Earth Observation and Process-Based Dynamic Vegetation Modelling

26

Mulalo P. Thavhana, Thomas Hickler, Marcel Urban ⓘ, Kai Heckel, and Matthew Forrest

Abstract

Terrestrial net primary production (NPP) is a fundamental Earth system variable that also underpins resource supply for all animals and fungi on Earth. We analysed recent past NPP dynamics and its drivers across southern Africa. Results from the Dynamic Global Vegetation Model (DGVM) LPJ-GUESS correspond well with estimates from the Moderate Resolution Imaging Spectroradiometer (MODIS) satellite sensor as they show similar spatial patterns, temporal trends, and inter-annual variability (IAV). This lends confidence to using LPJ-GUESS for future climate impact research in the region. Temporal trends for both datasets between 2002 and 2015 are weak and much smaller than inter-annual variability both for the region as a whole and for individual biomes. An increasing NPP trend due to CO_2 fertilisation is seen over the twentieth century in the LPJ-GUESS simulations, confirming atmospheric CO_2 as a long-term driver of NPP. Precipitation was identified as the key driver of spatial patterns and inter-annual variability. Understanding and disentangling the effects of these changing drivers on ecosystems in the coming decades will present challenges pertinent to both climate change mitigation and adaptation. Earth observation and process-

M. P. Thavhana · T. Hickler
Institute of Physical Geography, Goethe University Frankfurt am Main, Frankfurt am Main, Germany

Senckenberg Biodiversity and Climate Research Centre (SBiK-F), Frankfurt am Main, Germany
e-mail: Mulalo.Thavhana@senckenberg.de

M. Urban · K. Heckel
Department for Earth Observation, Friedrich Schiller University, Jena, Germany

M. Forrest
Senckenberg Biodiversity and Climate Research Centre (SBiK-F), Frankfurt am Main, Germany

© The Author(s) 2024
G. P. von Maltitz et al. (eds.), *Sustainability of Southern African Ecosystems under Global Change*, Ecological Studies 248,
https://doi.org/10.1007/978-3-031-10948-5_26

based models such as DGVMs have an important role to play in meeting these challenges.

26.1 Introduction

Plant photosynthesis on land takes up about 120 billion tons of carbon (C) per year, which is equivalent to 440 billion tons of CO_2 (Friedlingstein et al. 2020). This is about 10 times more than the global annual CO_2 emissions of 43 billion tons of CO_2 (Friedlingstein et al. 2020). About half of this uptake is used by plants as respiration to maintain their metabolism and for nutrient uptake (Gonzalez-Meler et al. 2004). The rest is available as net primary production (NPP) to grow new biomass, replace leaves and fine roots, transfer sugars to mycorrhizal fungi in the soil, and produce root exudates and biogenic volatile organic compounds (Chapin et al. 2011).

NPP is the carbon gained by plants using photosynthesis at the ecosystem level after subtracting the respiration costs, and can thus be calculated as the gross primary production (GPP) minus plant autotrophic respiration (R_a) (Chapin et al. 2011). NPP is important for providing fundamental resources for all animals and fungi on Earth including ecosystem services for people such as food, fibre, and timber production (Melillo et al. 1993; Abdi et al. 2014; Ardö 2015; Pan et al. 2015). The importance for human society can also be illustrated through the large fraction of NPP used by humans, which has been estimated as human appropriation of net primary production (HANPP) (Fetzel et al. 2012). HANPP is currently estimated to be about 25% of global NPP (Haberl et al. 2007; Niedertscheider 2011; Abdi et al. 2014; Andersen and Quinn 2020).

The main direct drivers of NPP include: temperature, precipitation, solar radiation, the CO_2 concentration in the atmosphere, nutrient availability, and vegetation structure such as the amount of leaves per ground area (Heisler-White et al. 2008; Reeves et al. 2014; Gao et al. 2016; Feng et al. 2019; Ji et al. 2020; Zhang et al. 2021). Atmospheric CO_2 influences NPP both directly as the photosynthesis of plants with the common C_3 photosynthesis partly CO_2-limited, and indirectly, as many plants reduce stomatal conductance under enhanced CO_2 levels, which can lead to more conservative water use (Archibald et al. 2009; Reeves et al. 2014; Xu et al. 2016). In savannas, these plant-physiological CO_2 effects can lead to complex changes in vegetation dynamics and fire as plants with C_4 photosynthesis, which are most savanna grasses, benefit much less from increasing CO_2 than woody plants, which can lead to woody encroachment (Midgley and Bond 2015). Terrestrial NPP patterns are expected to change in the future in response to these drivers and human population dynamics, thus necessitating assessment of NPP sensitivity to climate and other environmental change (Mohamed et al. 2004; Reeves et al. 2014). The overall vegetation NPP in warm dry regions, such as most of southern Africa is commonly mostly limited by precipitation or the amount of available moisture (Nemani et al. 2003; Hickler et al. 2005; Ji et al. 2020).

Given that NPP is one of the most variable components of the terrestrial C cycle, ecologists aim to make accurate estimates of this component when conducting research on terrestrial ecosystems, C cycles, and climate change (Sala and Austin 2000; Yu et al. 2018). Answering important questions concerning the global C balance, and predictions of the effects of global climate change rely on estimates of this fundamental quantity (Sala and Austin 2000). However, directly measuring GPP and NPP in the field is close to impossible and hence researchers resort to estimating the vegetation production components (Clark et al. 2001; Chapin et al. 2011; Peng et al. 2017).

There are various ways to estimate NPP across large extents and this includes: (1) field surveys (and subsequently extrapolating field measurements for local NPP to larger regions, using a vegetation map), (2) Earth Observation-based products, which are also informed by field survey, and (3) process-based ecophysiological modelling (Ruimy et al. 1994; Zhao et al. 2005). In this chapter, we focus on the latter two. Temporal trends, particularly if shown by both of these two approaches, will highlight whether there is increase (greening) or decrease (browning) of NPP in the study region (Zhu et al. 2016).

Southern Africa is one of the regions identified as most vulnerable to climate change (Pan et al. 2015; Ranasinghe et al. 2021; Chap. 3). Precipitation is expected to decrease over the summer rainfall region of southern Africa, and with the drying effect of increased temperature will thus lead to a robust and pronounced decrease in soil moisture over the region (IPCC 2021; Chaps. 6 and 7). In the summer rainfall areas, this is due to the El Niño/Southern Oscillation phenomenon (ENSO) which is negatively correlated with the amount of rainfall during the summer season in southern Africa (Malherbe et al. 2015; Chap. 6). Furthermore, there is high confidence in projected mean precipitation decreases in west southern Africa and medium confidence in east southern Africa by the end of the twenty-first century (Ranasinghe et al. 2021; Chap. 7).

Climate change will challenge agriculture, forestry, water systems, health, and the adaptive capacity of the natural ecosystems of the region (Pan et al. 2015). Robust projections of NPP will be highly relevant to meeting these coming challenges. However, in order to produce such projections, a solid understanding of the current NPP dynamics and drivers must be established. To this end, we seek to shed light on the following research questions: (1) what are the current NPP spatial distribution and temporal trends in southern Africa? (2) how consistent are the estimates of NPP from different methods? (3) what are the drivers of the dynamics that are producing these spatial patterns and temporal trends? and (4) can process-based models give robust estimates of future NPP by capturing these dynamics? Therefore, in this chapter, we: (1) examine and compare spatial and temporal NPP patterns for southern Africa derived from a Dynamic Global Vegetation Model (DGVM) and Earth observation-based estimates and (2) investigate some possible drivers (climate variables and atmospheric CO_2 concentration) of NPP in the region. (3) Furthermore, we subset our data using a well-known biome map to determine the NPP patterns and drivers for different ecosystems.

26.2   Materials and Methods

26.2.1 Study Region

The southern African region (here defined as -35.0 S to -20.0 S and 13.5 E to 35.0 E) is located on the southernmost part of the African continent consisting of several countries, namely: Angola, Botswana, Lesotho, Malawi, Mozambique, Namibia, South Africa, Swaziland, Zambia, and Zimbabwe. Southern Africa has both low-lying coastal areas, and mountains with varied terrain, ranging from forest and grasslands to deserts. Furthermore, the region has diverse ecoregions that includes grassland, bushveld, Karoo, savanna, and shrublands (Rutherford et al. 2006; Schoeman and Monadjem 2018). The climate across southern Africa varies from arid conditions in the west to humid subtropical conditions in the north and east, while much of the central part of southern Africa is classified as semi-arid (Cooper et al. 2004; Daron 2015). Despite the wide range of climate types, agriculture is a critical sector for all of the economies of southern African countries, so the effects of climate change on NPP and the knock-on effects on agricultural productivity, ecosystem service development, and food security are highly relevant across the study region (Gornall et al. 2010).

26.2.2 Data Sources

26.2.2.1 Earth Observation Data

MODIS and other Earth observation missions are essential tools for the development and evaluation of Earth system models predicting global ecosystem changes, which are an important information source for political decision-makers (Simmons et al. 2016; Chaps. 24 and 29). The use of Earth observation data allows the monitoring of different ecosystem variables (e.g. vegetation/biomass changes, surface moisture dynamics, etc.) with high spatial resolution and short temporal intervals (Gao et al. 2013). Earth observation data from various sources has become a valuable tool for analysing vegetation productivity in combination with *in situ* NPP estimates (Zhao et al. 2005; Fukano et al. 2021). A variety of light use efficiency (LUE) models have been developed (Running et al. 2004; Zhang et al. 2015) to calculate GPP from measured absorbed photosynthetically active radiation (APAR) (Xiao et al. 2019). Earth observation data have also been integrated with machine learning approaches (Xiao et al. 2008; Jung et al. 2009) and process-based models (Hazarika et al. 2005; Liu et al. 2019) for quantifying C fluxes (e.g. GPP and NPP). Autotrophic respiration can be estimated using modelling approaches that use daily climate variables and estimated biomass, and this can then be subtracted from GPP to derive NPP (Clark et al. 2001; Ardö 2015). In addition to estimates of these fluxes, the so-called vegetation indices (formed by combining two or more spectral bands) have been developed to characterise different aspects of vegetation from Earth observation data (Masoudi et al. 2018). Studies have shown the normalised difference vegetation index (NDVI) to be a good proxy for NPP at high spatial resolution (Zhao et al. 2005; Pachavo and Murwira 2014; Cui et al. 2016). NDVI is expressed as:

$$NDVI = \frac{(NIR - RED)}{(NIR + RED)} \qquad (26.1)$$

where *NIR* and *RED* are reflectance values in the near-infrared and red wavelengths, respectively (Tucker 1979). NDVI values range from $+1.0$ to -1.0, where negative values may be representative of cloudy conditions or areas over water bodies. Areas of barren rock, sand, or snow show very low NDVI values of 0.1–0; sparse vegetation such as shrubs and grasslands or senescing crops may show moderate NDVI values of 0.2–0.5; and high NDVI values of 0.6–0.9 correspond to dense vegetation found in temperate and tropical forests or crops at their peak growth stage (Higginbottom and Symeonakis 2014).

26.2.2.1.1 Earth Observation Platforms and Products
This study primarily utilised time series information from MODIS sensors onboard the satellites TERRA and AQUA (Minnett 2001; Yang et al. 2006; Cao 2020). Both platforms have sun-synchronous orbits with a revisit time of 1–2 days (Savtchenko et al. 2004). Data is acquired in 36 spectral bands with wavelengths ranging from 0.4 to 14.385 μm (Cao 2020).

For this study, NPP was taken from the MOD17A3 (UM Collection 5) annual totals. The MOD17 products are the first MODIS operational data sets to regularly monitor global vegetation productivity (Zhao et al. 2005; Yu et al. 2018). Details of the MODIS NPP derivations are provided in the section below.

NDVI estimates were taken from the MOD13C2 Collection 6, 16-day product (at monthly intervals) which is at 1 km spatial resolution and is provided in 0.05 degree geographic climate modelling tiles (Solano et al. 2010). The 16-day product was processed to mean annual values in R. This dataset has been used for modelling global biogeochemical and hydrologic processes in both global and regional climates (Didan 2015). Furthermore, the data have been utilised in studies to characterise land surface biophysical properties and processes, including primary production and land cover conversion (Solano et al. 2010; Didan 2015).

In addition, NDVI time series from the Advanced Very-High-Resolution Radiometer (AVHRR) sensor on board of the National Oceanic and Atmospheric Administration's (NOAA) polar-orbiting satellites was used to investigate NDVI long-term trends. Mounted on a polar-orbiting satellite it acquires images of the visible, near-infrared, and thermal infrared parts of the electromagnetic spectrum (Kidwell 1995; Sus et al. 2018). The sensor has a spatial resolution of approximately 1.1 km at satellite nadir and covers the time period from 1981 to 2015 (Trishchenko et al. 2002). NOAA AVHRR is a widely used sensor due to its long-term monitoring period and retrieval of various land surface parameters such as land cover/use dynamics, NDVI, Land Surface Temperature (LST), and Albedo (Forkel et al. 2013; Wang et al. 2020; Gulev et al. 2021; Urban et al. 2013).

26.2.2.1.2 MODIS NPP Derivation

The MOD17 algorithm is based on the original LUE logic of Monteith (1972). Input data for the model include climatic variables such as temperature, solar radiation, vapour pressure deficit (VPD) from meteorology dataset from NASA global modelling and assimilation office (Running et al. 2004). MOD15 leaf area index and fraction of absorbed photosynthetically active radiation (FAPAR) products are also utilised (Running et al. 2004; Zhao et al. 2005). Land cover classification from MODIS MCD12Q1 data product is used (Running et al. 2004). A Biome Parameter Lookup Table (BPLUT) containing values of ε_max (Eq. 26.2) was derived and later updated (Running et al. 2004; Zhao et al. 2005). The table contains different vegetation types, temperature, and VPD limits and other biome-specific physiological parameters for respiration calculations (Running et al. 2004). The different vegetation types obtained from the land cover type 2 classification include: evergreen needleleaf forest, evergreen broadleaf forest, deciduous needleleaf forest, deciduous broadleaf forest, mixed forests, closed shrublands, open shrublands, woody savannas, savannas, grasslands, and croplands. Environmental multipliers represent limitations by low temperature and high VPD, and autotrophic respiration is estimated with a Q_{10} relationship (Zhao et al. 2005; Ardö 2015). The MOD17 algorithm calculates daily GPP as:

$$\text{GPP} = \varepsilon_{\max} \times 0.45 \times \text{SW}_{\text{rad}} \times \text{FPAR} \times f\,(\text{VPD}) \times f\,(T_{\min}) \tag{26.2}$$

where $\varepsilon_{\max}$ is the maximal, biome-specific light use efficiency (g C MJ^{-1}), SW_{rad} is incoming short-wave radiation [assuming 45% to be photosynthetic active radiation (PAR)], FPAR is the fraction of photosynthetically active radiation, $f(\text{VPD})$ and $f(T_{\min})$ are linear scalars reducing GPP due to water and temperature stress (Ardö 2015). The model estimation utilises the fact that GPP is closely related to the APAR and that APAR can be measured continuously using Earth observation sensors (e.g. MODIS) (Cui et al. 2016; Xiao et al. 2019). The FPAR is estimated as a function of NDVI, derived from the standard MODIS land product (MOD15) (Eqs. 26.3 and 26.4) (Running et al. 2004; Zhao et al. 2005; Gonsamo and Chen 2017).

$$\frac{\text{APAR}}{\text{PAR}} = \text{NDVI} \tag{26.3}$$

$$\text{FPAR} = \frac{\text{APAR}}{\text{PAR}} = \text{NDVI} \tag{26.4}$$

$$\text{GPP} = \varepsilon_{\max} \times \text{PAR} \times \text{NDVI} \times f\,(\text{VPD}) \times f\,(T_{\min}) \tag{26.5}$$

$$\text{PsnNet} = \text{GPP} - R_{\text{ml}} - R_{\text{mr}} \tag{26.6}$$

NPP is calculated annually as:

$$\text{NPP} = \sum_{i=1}^{365} \text{PsnNet} - \left(R_{\text{mo}} + R_{\text{g}} \right) \tag{26.7}$$

where PsnNet is the maintenance respiration by leaves (R_{ml}) and fine roots (R_{mr}) and is calculated daily. R_{mo} is the annual maintenance respiration by all other living parts except leaves and fine roots, R_{g} is the annual growth respiration (Zhao et al. 2005).

The products were projected to a geographic grid while resampling to 0.5 degree using "average" resampling type in order to match the resolution of the climate input data used to drive our DGVM. Grid cells without valid MOD17 NPP (MOD12Q1 land cover barren, water, or urban) were masked out from the LPJ-GUESS data in order to make the data sets comparable with identical spatial extent, land cover classes, and number of grid cells. The datasets were aggregated (annual sums for NPP, annual means for NDVI) to a 14-year annual time series from 2002 to 2015.

26.2.2.2 Dynamic Vegetation Models

A class of ecosystem models known as dynamic global vegetation models (DGVMs) have been developed to simulate vegetation dynamics and biogeochemical cycling either at regional or global scales (Prentice et al. 2007; Sitch et al. 2008; Smith et al. 2014). DGVMs represent basic ecophysiological processes, such as photosynthesis, plant and soil respiration, C allocation, and plant growth, competition between plant types for resources (commonly light and water, increasingly also nutrients) and disturbances such as fire (Ardö 2015; Hantson et al. 2016). Simulating the impacts of climate change on ecosystems and feedback from ecosystems on climate, in particular via the terrestrial carbon cycle, has been a research priority (Prentice et al. 2007; Kelley et al. 2013). Subsequently, the representation of land-use has also received attention and has been integrated into DGVMs (Bondeau et al. 2007; Lindeskog et al. 2013; Pugh et al. 2019; Drüke et al. 2021).

DGVM's largest potential lies in process-understanding (Hickler et al. 2005) rather than short-term predictions, which can be even more accurate with empirical approaches. DGVMs are powerful tools to quantify spatial and temporal variations in ecosystem C fluxes and to analyse the underlying mechanisms of NPP at large scales (Tao et al. 2003; Yu et al. 2018). DGVMs have the potential to accurately explain how ecosystem processes will interact in future climatic conditions, CO_2 concentration, nitrogen deposition, land-use changes, and soil conditions (Melillo et al. 1993; Luo et al. 2004; Ardö 2015). DGVMs are driven with climate and other environmental data (e.g. soil properties and nutrient deposition) which can be either historical (observed) data sets or projections of past/future environmental conditions.

26.2.2.2.1 LPJ-GUESS Model and Setup

The Lund–Potsdam–Jena (LPJ) model has been developed as a process-based DGVM which can efficiently represent the land–atmosphere interaction and potentially be applied for broader global problems (Gerten et al. 2004; Sitch et al. 2003). The Lund–Potsdam–Jena General Ecosystem Simulator (LPJ-GUESS) framework was originally developed to add a more detailed representation of vegetation dynamics through a "forest-gap model" to the LPJ DGVM (Smith et al. 2001). Thus, LPJ-GUESS is an individual (or cohort) based model which combines biogeography and biogeochemistry typical of a DGVM with a comparatively more detailed individual and patch-based plant functional type (PFT) representation of vegetation structure, demography, growth, mortality, reproduction, carbon allocation, and resource competition (Smith 2001; Sitch et al. 2003). The model now includes an interactive nitrogen cycle (Smith et al. 2014), which can limit photosynthesis and is so important to constrain future potential CO_2 fertilisation effects (Hickler et al. 2015), and a representation of agricultural land and management (Lindeskog et al. 2013).

In the framework, productivity is simulated as the emergent outcome of growth and competition for light, space, and soil resources among woody plant individuals and a herbaceous understory in a number of a replicate patches (typically 15–50) representing "random samples" of each simulated locality or grid cell (Smith 2001). Natural, cropland, and pasture land cover types are distinguished and their fractional covers are prescribed from the dataset by Hurtt et al. (2011). Within the cropland land cover type, crop fractions from the MIRCA database (Portmann et al. 2010) are used and nitrogen fertiliser application rates from Zaehle et al. (2010) are prescribed.

In this study, we used the standard version of the cohort-based LPJ-GUESS model using 20 replicate patches at a spatial resolution of $0.5° \times 0.5°$. Climate forcing data from the CRU JRA v2.0 dataset (details below) were used. Land-use dataset by Hurtt et al. (2011) was included. The global plant functional types (PFTs) analysed in our model were: boreal needleleaf evergreen (BNE), boreal shade-intolerant needleleaf evergreen (BINE), boreal needleleaf summergreen (BNS), temperate needleleaf evergreen (TeNE), temperate broadleaf summergreen (TeBS), shade-intolerant broadleaf summergreen (IBS), temperate broadleaf evergreen (TeBE), tropical broadleaf evergreen (TrBE), tropical shade-intolerant broadleaf evergreen (TrIBE), tropical broadleaf raingreen (TrBR), C3 grasses (C3G), C4 grasses (C4G). The model output includes GPP, NPP (kg C m^{-2} year^{-1}), respiration, carbon pools, burnt area fraction, and potential vegetation among other outputs. The model was run on a daily time step. All simulations were initialised with a 500 years spinup to allow vegetation, soil carbon and nitrogen pools to build up from "bare ground" to a "steady state" and then the full transient period (1901–2018) was simulated. Fire was enabled through the SIMFIRE-BLAZE fire model (Knorr et al. 2016; Nieradzik et al. 2015).

26.2.2.3 Meteorological Data

Precipitation, temperature, and solar radiation data from the CRU JRA v2.0 dataset were used for both driving the LPJ-GUESS simulations and for investigating the correlation between NPP and its potential drivers. As a basis, this dataset starts with the Japanese reanalysis (JRA) (Harada et al. 2016; Kobayashi et al. 2015) data produced by the Japanese Meteorological Agency (JMA). This is then adjusted, where possible, to align with the monthly values of the CRU TS 3.26 dataset (Harris and Jones 2019), a gridded land surface dataset based on meteorological station data produced by the Climatic Research Unit (CRU). The data availability spans from January 1901 to December 2017. The dataset is a 6-hourly, gridded time series of ten meteorological variables and is intended to be used to drive models of the global land surface and biosphere such as DGVMs. The variables are provided on a 0.5 × 0.5 degree grid.

26.2.3 Data Analysis

NDVI time series from MODIS and AVHRR products along with MODIS NPP products were used to assess the vegetation productivity. NPP was simulated to assess whether the LPJ-GUESS model agrees with MODIS and AVHRR estimates by following the inter-annual variation of remotely estimated NPP and NDVI. This was to evaluate the model's capability to reproduce past data and ultimately adopted for future predictions in southern Africa.

The potential driving factors and trend analysis of NPP and NDVI were conducted per biome, according to the South African National Biodiversity Institute (SANBI) 2006 biome map (Fig. 26.1) (Rutherford et al. 2006). The biome map is made up of nine well established biomes in South Africa which includes: Savanna, Grassland, Nama Karoo, Succulent Karoo, Fynbos, Albany Thicket, Forest, Indian Ocean Coastal Belt (IOCB), and Desert (Rutherford et al. 2006). This was to show the ecological and climatic variability experienced across the South African region. Although the study area covers southern Africa and not just South Africa, the SANBI biome was used because it has been studied extensively. Furthermore, the

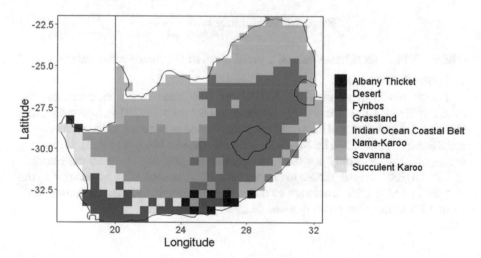

Fig. 26.1 South African National Biodiversity Institute (SANBI) 2006 biome map by Rutherford et al. (2006)

study focuses on the dynamics and drivers of NPP per biome where actually the South African biomes extend into the upper regions of the study area.

26.2.3.1 Analysis Software

The R statistical programming language was used for processing and for statistical analysis of the data. MODIS R package (Mattiuzzi et al. 2017) was used to download and process MODIS data. We used the DGVMTools R package[1] to perform comparisons, analysis, and plotting of the spatially explicit simulated and remotely sensed NPP distribution across the study region. DGVMTools is a high-level framework for processing, analysing, and visualising DGVM data output which easily interfaces with both the raster package and base R functionality. The *ggplot2* package (Wickham 2016) was used for additional plotting and linear trend analysis.

26.2.3.2 Time Series Analysis

The NDVI and NPP time series were analysed along with the key climate variables (precipitation, temperature, and solar radiation) over the period 2002–2015. The region experienced extreme rainfall events in the year 2000 which were far outside its normal variability (Dyson and van Heerden 2001; Smithers et al. 2001). Accordingly, the years 2000 and 2001 were excluded from the analysis since including them was found to produce spurious trends and thus produced misleading analysis and conclusions. The AVHRR NDVI time series were analysed from 1982 to 2015 in order to gain some perspective of the longer NDVI trend. As these variables have different units and magnitudes, we derived aggregated standardised anomalies following Seaquist et al. (2008) and this is expressed as:

$$\text{Standardised anomaly} = \frac{\left(x - \bar{x}\right)}{\text{sd}} \tag{26.8}$$

where x is the annual mean values, $\bar{x}$ is the mean of the annual mean values, and sd is the standard deviation.

Linear trends were fitted to NDVI and NPP time series to determine the productivity long-term trend (Higginbottom and Symeonakis 2014). Due to the large inter-annual variation and relatively short time series of our data, most trends are not statistically significant. The Mann-Kendall test (Mann 1945) was used to quantify the significance and only $p < 0.05$ was considered. We also analysed the response of LPJ-GUESS simulated NPP to atmospheric CO_2 concentration by comparing the standard LPJ-GUESS simulation to one with atmospheric CO_2 concentration fixed from 1901 at its corresponding value of 296.4 ppm.

---

[1] https://github.com/MagicForrest/DGVMTools

26.2.3.3 Statistical Correlation Analyses

We used Pearson's product moment correlation coefficient (r) to quantify the level of agreement between LPJ-GUESS and MODIS NPP. To examine the relationship between different variables (NPP, NDVI, CO_2, and different climate factors) Spearman's rank correlation coefficient was used. This is because we are interested in monotonic relationships between these variables which do not necessarily need to be linear. Spearman's coefficient is insensitive to any nonlinearity that could undermine the detection of a monotonic relationship between the variables. The coefficient ranges from -1 to $+1$ and was examined using the "PerformanceAnalytics" R package (Peterson et al. 2020). Large positive values indicate strong agreement, large negative values indicate strong disagreement, and values near 0 indicate random agreement (Seaquist et al. 2008; Bon-Gang 2018).

26.3   Results

26.3.1 NPP Geographical Patterns

The broad spatial patterns of NPP are as expected, with higher NPP in regions of higher rainfall and lower NPP in areas that experience extreme heat and receive less rainfall. In general, NPP simulated by LPJ-GUESS and estimated by MODIS showed similar spatial patterns, with Pearson's $r = 0.85$ and Root Mean Squared Error (RMSE = 0.1446 kg C m^{-2} year^{-1}). The main difference is lower LPJ-GUESS simulated NPP values along the coastal regions (Fig. 26.2). Contrastingly, LPJ-GUESS showed higher NPP than MODIS as one moves in from the coast. The inter-annual dynamics of NPP on a gridcell level correspond reasonably well. This is indicated by Pearson's correlation coefficients of the time series of the individual

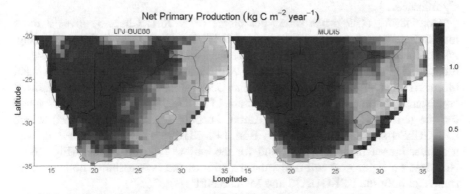

Fig. 26.2 Spatial distribution of modelled NPP and remotely sensed NPP averaged over time (2002–2015) to evaluate spatial patterns of NPP over the southern African region. The plot presents a strong linear relationship between LPJ-GUESS and MODIS of $r = 0.85$ and RMSE = 0.1446 (kg C m^{-2} year^{-1})

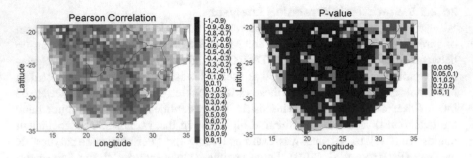

Fig. 26.3 Pearson correlation and significance level of modelled NPP and Earth observation-based NPP per gridcell, averaged over time (2002–2015). The plot presents a high correlation of LPJ-GUESS and MODIS NPP over most parts of the region with great significance level ($p < 0.05$) within the correlated areas

grid cells with high statistical significance ($p < 0.05$) over most parts of the study region (Fig. 26.3).

26.3.2 NPP Temporal Development

When averaged over the whole study region, LPJ-GUESS simulated and MODIS-estimated NPP showed good agreement in terms of the overall magnitude and inter-annual variation (Fig. 26.4). The inter-annual variation was found to be large, annual values ranged from a maximum of over 0.5 kg C m^{-2} year^{-1} in 2006, to less than 0.35 kg C m^{-2} year^{-1} (minima in 2003 and 2015). The time series also revealed a small tendency of the LPJ-GUESS model to simulate higher values than estimated with MODIS. However, when applying the Mann-Kendall test the trends showed no significance.

Considering NPP times series per biome (Fig. 26.1; Chap. 3) gives a more nuanced view of the regional disparities between LPJ-GUESS and MODIS (Fig. 26.5). The grass-dominated biomes (Grassland and Savanna) showed good agreement in NPP magnitude between LPJ-GUESS and MODIS, the tree-dominated biomes (Forest, Albany Thicket, and Indian Ocean Coastal Belt) and Fynbos showed consistently higher NPP in MODIS than LPJ-GUESS. This pattern is reversed for the most arid biomes (Nama Karoo, Succulent Karoo, and Desert) where LPJ-GUESS simulates higher NPP. Both LPJ-GUESS and MODIS NPP showed a similar increasing but weak trend for the majority of the biomes (Fig. 26.5). However, on a per grid cell basis, the majority of the region shows no significant trends for both the LPJ-GUESS and MODIS NPP (Fig. 26.6b).

Examination of simulated LPJ-GUESS NPP over a longer period (1901–2015) showed high inter-annual variation against a backdrop of increasing NPP (Fig. 26.7b) caused by rising atmospheric CO$_2$ (Fig. 26.7a). According to the model, NPP has increased by 18% since 1901. This is in agreement to the increasing trend for

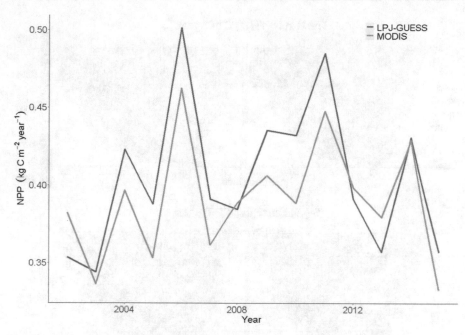

Fig. 26.4 Temporal distribution of LPJ-GUESS modelled and MODIS-estimated NPP (kg C m^{-2} year^{-1}) averaged over the whole southern Africa study region

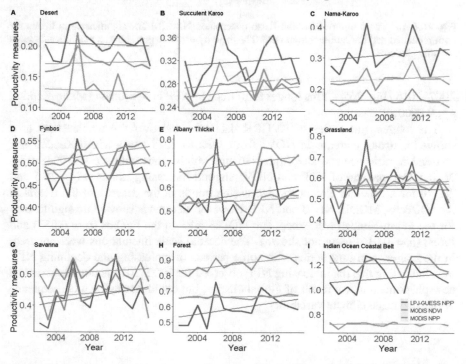

Fig. 26.5 Absolute values and linear trends of NPP (kg C m^{-2} year^{-1}) and NDVI (unitless) per biome

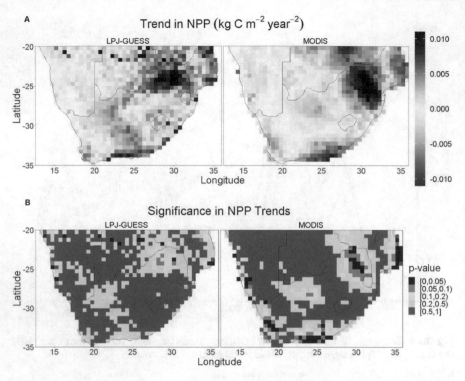

Fig. 26.6 (a) Trend in modelled and Earth observation NPP. (b) The significance in trends for both modelled and Earth observation NPP. The majority of the region shows no significant trends for both products

2002–2015 (Fig. 26.4). This longer term trend in NPP is significant (Mann-Kendall $p < 0.00005$).

The longer time series of AVHRR data averaged over the whole study area showed a strong increase in NDVI from 1982 to 2015 and a Mann-Kendall test of trend significance revealed high statistical significance with a $p < 0.0048$ (Fig. 26.7b). Comparison of NPP and NDVI showed increasing trends (Figs. 26.5 and 26.7b) for all biomes except for decreasing trends in the desert and flat trends in the IOCB for MODIS NDVI and NPP. However, the trends showed no significance for the Mann-Kendall test except for MODIS NPP in the Fynbos ($p = 0.037$) and forest ($p = 0.016$) (data not shown). The decadal scale fluctuations were apparent in the longer term time series with some periods of increasing and declining NPP, thus indicating that the increasing NPP observed for 2002–2015 is not necessarily a new phenomenon or a result of global change, but could simply be a consequence of decadal scale climate variability.

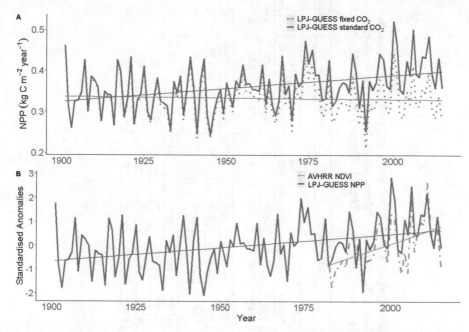

Fig. 26.7 (a) Annual average NPP (kg C m^{-2} year^{-1}) across the whole study region simulated by LPJ-GUESS with historically varying atmospheric CO_2 concentration and CO_2 fixed at values from the year 1901 (296.3785 ppm) and (**b**) Aggregated standardised anomalies of historical LPJ-GUESS NPP (1901–2015) and AVHRR NDVI (1982–2015). This plot evaluates the trends and Inter-annual variability (IAV) of NPP over a longer time span

26.3.3 NPP and NDVI Correlations

The LPJ-GUESS and MODIS NPP showed high Pearson correlation coefficient (ranges from 0.68 to 0.82) throughout the biomes except for Fynbos (0.37) (Table 26.1). The correlations between the two products showed significance of $p < 0.001$ while Fynbos showed no significance. Succulent Karoo (0.49), Albany Thicket (0.50), and IOCB (0.16) showed no significance when comparing LPJ-GUESS NPP and MODIS NDVI while high and significant ($p < 0.001$) correlation was observed for the rest of the biomes (ranges from 0.64 to 0.91). The IOCB biome (0.36) showed low and not significant correlation when comparing MODIS NPP and NDVI, while the rest of the biomes showed significantly high correlations (ranges from 0.61 to 0.93).

26.3.4 Potential Driving Factors of NPP

Precipitation showed similar inter-annual variation to that of LPJ-GUESS and MODIS NPP per biome, as demonstrated by significantly high correlations with

Table 26.1 Spearman rank correlation coefficient for aggregated standardised anomalies (MODIS NDVI, MODIS NPP, LPJ-GUESS NPP, precipitation, temperature, and solar radiation) per biome, from 2002 to 2015. The Pearson correlation was applied for the LPJ-GUESS and MODIS NPP comparisons. The table shows the value of the correlation plus the significance level as stars. Each significance level is associated to a symbol: *p*-values (0, 0.001, 0.01, 0.05, 0.1, 1) refers to symbols (***, **, *, †, ' ') respectively

Biome	NPP_LPJ-GUESS Variables					NPP_MODIS				NDVI_MODIS			Precipitation		Temperature
	Precipitation	Temperature	Solar radiation	NDVI_MODIS	NPP_MODIS[a]	Precipitation	Temperature	Solar radiation	NDVI_MODIS	Precipitation	Temperature	Solar radiation	Temperature	Solar radiation	Solar radiation
Desert	0.81***	−0.44	−0.31	0.82***	0.69**	0.42	−0.064	−0.48†	0.89***	0.58*	−0.20	−0.45	−0.51†	−0.10	0.42
Succulent Karoo	0.84***	−0.39	−0.32	0.49†	0.68***	0.56*	−0.077	−0.30	0.86***	0.43	−0.033	−0.30	−0.37	0.051	0.073
Nama Karoo	0.84***	−0.015	−0.73**	0.68**	0.81***	0.70**	−0.56*	−0.65*	0.73***	0.75***	−0.14	−0.79**	−0.13	−0.85***	0.21
Fynbos	0.89***	−0.40	−0.27	0.64*	0.37	0.42	−0.086	0.0022	0.64**	0.56*	−0.24	0.046	−0.45	−0.059	−0.20
Albany Thicket	0.81***	−0.27	−0.58*	0.50†	0.70**	0.70**	−0.40	−0.25	0.62*	0.55*	−0.16	0.0022	−0.40	−0.29	−0.064
Grassland	0.92***	−0.30	−0.76**	0.91***	0.78***	0.79**	−0.50†	−0.65*	0.76**	0.86***	−0.27	−0.78**	−0.42	−0.74**	0.44
Savanna	0.84***	−0.13	−0.61*	0.81***	0.74**	0.95***	−0.55*	−0.73**	0.93***	0.93***	−0.39	−0.81***	−0.38	−0.71**	0.35
Forest	0.85***	−0.38	−0.37	0.70**	0.82***	0.61*	−0.22	−0.25	0.61*	0.80***	−0.17	−0.13	−0.33	−0.35	0.015
IOCB	0.95***	−0.48†	−0.13	0.16	0.72***	0.58*	−0.67*	−0.11	0.36	0.0022	−0.35	−0.43	−0.35	−0.099	0.36

[a]Pearson correlation

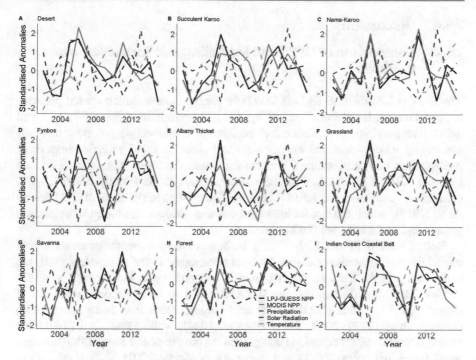

Fig. 26.8 Aggregated standardised anomalies of precipitation, temperature, and solar radiation in comparison with both the modelled and Earth observed NPP per biome

both MODIS (0.56–0.95) and LPJ-GUESS NPP (0.81–0.95) (Fig. 26.8 and Table 26.1). However, there were two notable exceptions to this: the correlations between MODIS NPP and precipitation in the Desert (0.42) and Fynbos (0.42) biomes were low and not significant. In contrast, solar radiation and temperature showed negative correlations with MODIS and LPJ-GUESS NPP across all biomes, although they were not significant in most cases (Table 26.1). Of all the climate drivers, temperature showed a significantly increasing trend with the Mann-Kendall $p < 0.049$ for Succulent Karoo and 0.009 for Desert while the rest were not significant. This is consistent with global warming in areas of high inter-annual variability and relatively short time series (as is the case here).

The model simulation in which atmospheric CO_2 concentration was held constant at 1901 levels (Fig. 26.7a) showed that increasing CO_2 concentration is responsible for a large increase in NPP. This effect was built to be around +18% (+0.064 kg C m^{-2} year^{-1}) by 2015. LPJ-GUESS NPP showed a slight decline without significant trend when the effects of increasing atmospheric CO_2 were removed.

26.4 Discussion

26.4.1 Comparison of NPP from Modelling and Earth Observation Products

The DGVM LPJ-GUESS and MODIS NPP products show similar spatial patterns, temporal trends, inter-annual variability, and good correlation (Figs. 26.2, 26.3, and 26.4). The good correspondence and correlation is encouraging as both methods are largely independent and only share some common input variables (e.g. daily precipitation and temperature), but from different data sets. The correspondence between MODIS17 and LPJ-GUESS here appears to be slightly better than in comparison with an earlier LPJ-GUESS version applied to simulate all of Africa by Ardö (2015), which did not include land-use and nitrogen limitation to vegetation growth and had a different fire module.

Both the MODIS and LPJ-GUESS estimates are to a greater or lesser extent model derived estimates and so subject to uncertainty in the parameters used, the input data and, in the case of LPJ-GUESS, uncertainty in the process representations. The MODIS estimate might be expected to be the more reliable as it uses Earth observation FPAR as an input (which means it is more constrained) (Myneni et al. 1999, 2002; Running et al. 2004), while FPAR in LPJ-GUESS emerges from complex equations that govern the growth of and competition between plant functional types (PFTs) and disturbances (Smith 2001; Sitch et al. 2003). It should also be noted that the MODIS NPP algorithm has been criticised, e.g., for its assumed temperature effects (Medlyn 2011), and the representation of environmental controls of NPP is not very sophisticated, see above. However, given that MODIS NPP uses Earth observation FPAR and so is in principle better constrained, we can interpret the comparison of LPJ-GUESS with MODIS as an evaluation of LPJ-GUESS to some extent. As we find good correspondence between the two, we can conclude that past and future projections for southern Africa are also reliable.

LPJ-GUESS is commonly used to derive future projections to guide climate adaptation and mitigation measures, so this result is promising and indicates that LPJ-GUESS can be fruitfully applied to the study area. As a small caveat, it should be noted that the per biome comparisons between MODIS and LPJ-GUESS did show some disparities in the more arid and the more tree-dominated biomes. This is not surprising as we applied the standard global LPJ-GUESS configuration here and a global model cannot be expected to reproduce all the features of a diversely vegetated region such as southern Africa. Future studies would benefit from using a regionally parameterised version of LPJ-GUESS, in particular the inclusion of shrub PFTs to better represent the arid regions.

All analysed time series products (LPJ-GUESS and MODIS NPP, and MODIS and AVHRR NDVI) show increasing trends for all the biomes except for the desert which showed a decreasing trend and grasslands which showed no trend (Figs. 26.5 and 26.7b). One should note that the inter-annual variability for our study region is large compared to the magnitudes of these trends and so, when tested, these trends did not show statistical significance. This is consistent with other results that have pointed out that vegetation trends are mostly not statistically significant for our study region and the MODIS operational period (Samanta et al. 2011; Cortés et al. 2021). However, an examination of the trends plotted spatially on a per gridcell basis (Fig. 26.6a) shows regional hotspots of statistical significant greening and browning. Furthermore it should be noted that even within a particular biome, different regions can show opposite trends (consider, for example, LPJ-GUESS in the western and eastern parts of the Fynbos biome in Fig. 26.6a) which is not apparent from the biome averaged results (Fig. 26.4). The LPJ-GUESS derived NPP over the whole southern Africa region shows a statistically significant rise since 1901 but with high IAV and decadal scale fluctuations due to internal climate variability or oscillations. Therefore, we can conclude the future NPP trajectory will be governed by both climate variability and long-term trends resulting from changing drivers due to global change.

26.4.2 Drivers of NPP Patterns and Trends

The spatial pattern of NPP, shorter-term temporal trends, and inter-annual variability appear to be mainly driven by gradients or changes in precipitation (Figs. 26.2, 26.4, and 26.8). This can be seen, for example, with the high NPP occurring along the eastern coastal parts of the region (Fig. 26.2). This region receives the highest rainfall while the western and central part of the region experiences high temperatures and less rainfall (Botai et al. 2018). As outlined in the introduction, we expected this dominant role of precipitation as a driver of NPP variations. This is in agreement with the study conducted by Zhu et al. 2016, where they concluded that the greening in South Africa over the period from 1982 to 2009 is primarily driven by increasing precipitation. Recent climate projections show decreasing precipitation trends in southern Africa both for low and high warming scenarios, with corresponding increases in aridity and drought (IPCC 2021). Higher levels of warming result in stronger precipitation decreases, with the highest percentage decreases projected in the west of the region, which is already arid (IPCC 2021, Fig. SPM.5c). Our results indicate that, up to 2015, precipitation decreases have not yet caused widespread negative impacts on NPP as we can see slightly increasing trends in Figs. 26.4 and 26.7a, although there are some regional exceptions (Fig. 26.6a). The potential effects of future drying on NPP and the consequences for ecosystems, ecosystem services, agriculture, and forestry remain an open question. To tackle this question, predictive methods such as DGVMs, with more detailed or regionally parameterized representations of land-use, or more empirical modelling approaches must be applied in combination with future climate projections and, ideally, informed by experimental work.

The observed high negative correlation with solar radiation is likely due to the fact that precipitation is associated with cloud cover and therefore decreased solar radiation. This indicates that whilst solar radiation is essential for photosynthesis and therefore NPP, it is not a limiting factor to the same degree as precipitation and so is in fact not a driver of NPP in the study region (Nemani et al. 2003). The same was found by Hickler et al. (2005) (Fig. 3) for the Sahel region.

In our study region, NPP shows a negative correlation with temperature. This is consistent with the fact that most of the study region is in the subtropical to warm temperate climate zones, so photosynthesis is unlikely to be temperature limited (not applicable for winter rain regions in the south west) and no positive correlation is expected (Nemani et al. 2003). Instead, this negative correlation may occur due to a combination of increased autotrophic respiration rates that accompany higher temperatures, a net reduction of photosynthesis due to decreased efficiency and heat stress at higher temperatures, and decreased soil water availability due to stronger atmospheric moisture demand. Among all the climatic variables analysed, only temperature in the Succulent Karoo and Desert biome showed statistically significant trends (increasing) from 2002 to 2015. Taking these findings together and putting them in the context of global change, it is likely that temperature will have an escalating and negative effect on NPP in these biomes in the coming decades, at least under moderate warming without unprecedented heat stress for the vegetation, which is not represented in the applied model version.

According to LPJ-GUESS, the plant-physiological effects of increasing atmospheric CO_2 have been large and positive between 1901 and 2015 (Fig. 26.7a). However, the simulated CO_2 effects over the last decades are only minor in similar arid environments (Hickler et al. 2005). The strong effects since the beginning of the last century are consistent with other global estimates. Ciais et al. (2012) estimated a pre-industrial GPP of 80 Pg year^{-1}, which is substantially lower than the present estimate of about 120 Pg (Ciais et al., 2013; Friedlingstein et al., 2020). Using another independent method, Campbell et al. (2017) estimated a GPP increase of about 31% since the early last century. GPP and NPP are strongly related, NPP being about half of GPP (Ciais et al., 2013; Yu et al., 2018), which also applies to LPJ-GUESS (Ardö 2015). The CO_2 effect is also in line with a global greening trend attribution study using different leaf area index (LAI) estimates and ten DGVMs (Zhu et al. 2016). These authors concluded that CO_2 fertilisation explains 70% of the observed greening trend between 1982 and 2009, particularly in the tropics. Their analyses, however, suggested other major drivers for most of southern Africa, namely climate change in the north-eastern part of our study region and land cover change in central South Africa (Zhu et al. 2016; Fig. 26.3c). In our study region, CO_2 effects have been found to be minor under extreme drought stress when meristem growth is strongly limited by leaf turgor (Xu et al., 2016). Moreover, future CO_2 fertilisation effects might be smaller as the CO_2 limitation of photosynthesis increasingly Saturates and because of progressive nutrient limitation (Hickler et al. 2015; Wang et al. 2020).

26.5 Conclusion

The two rather independent approaches to estimate spatial patterns and recent past trends in NPP we used (Earth-observation-based estimations from MODIS and the LPJ-GUESS DGVM) provided similar results. This suggests that the spatio-temporal results presented here are robust. As the MODIS estimate is more constrained by observed data, this result may also be considered as an evaluation of the LPJ-GUESS model. Thus, the model should be suitable for future projections in southern Africa in order to inform climate adaptation and mitigation measures.

Precipitation change is the most important driver of NPP dynamics in the study region, particularly of the inter-annual variations and spatial distribution. However, increasing atmospheric CO_2 has had a large effect since the early twentieth century and might continue to play a significant role as anthropogenic CO_2 emissions continue. Solar radiation and temperature were not found to be significant drivers of NPP in the study region.

Given the importance of precipitation in southern Africa and its projected decline, it is likely that ecosystems and agriculture will be under increasing pressure as their foundational building block, NPP, might decrease in areas where precipitation strongly decreases. The findings presented here alert us of the importance of mitigation and adaptation strategies going into the future and the challenges global change will bring. Both simulation models and Earth observation data can play an important role in meeting these challenges.

Acknowledgements The authors would like to thank Prof. Dr. Lukas Lehnert, from Ludwig-Maximilians University in Munich, Germany, for his comments regarding trends on remote sensing data. The authors further like to extend their gratitude to Dr. Jasper Slingsby from the University of Cape Town, South Africa and their anonymous reviewer for their constructive and insightful comments which significantly improved their chapter.

References

Abdi AM, Seaquist J, Tenenbaum DE, Eklundh L, Ardö J (2014) The supply and demand of net primary production in the Sahel. Environ Res Lett 9(9). https://doi.org/10.1088/1748-9326/9/9/094003

Andersen CB, Quinn J (2020) Human appropriation of net primary production. Encycl World's Biomes:22–28. https://doi.org/10.1016/B978-0-12-409548-9.12434-0

Archibald SA, Kirton A, Van Der Merwe MR, Scholes RJ, Williams CA, Hanan N (2009) Drivers of inter-annual variability in Net Ecosystem Exchange in a semi-arid savanna ecosystem, South Africa. Biogeosciences 6(2):251–266. https://doi.org/10.5194/bg-6-251-2009

Ardö J (2015) Comparison between remote sensing and a dynamic vegetation model for estimating terrestrial primary production of Africa. Carbon Balance Manag 10(1). https://doi.org/10.1186/s13021-015-0018-5

Bondeau A, Smith P, Zaehle S, Schaphoff S, Lucht W, Cramer W, Gerten D, Lotze-Campen H, Müller C, Reichstein M, Smith B (2007) Modelling the role of agriculture for the 20th century global terrestrial carbon balance. Glob Change Biol 13:679–706

Bon-Gang H (2018) Methodology. In: Performance and improvement of green construction projects. Elsevier, pp 15–22

Botai CM, Botai JO, Adeola AM (2018) Spatial distribution of temporal precipitation contrasts in South Africa. S Afr J Sci 114(7–8):1–9. https://doi.org/10.17159/sajs.2018/20170391

Campbell JE, Berry JA, Seibt U, Smith SJ, Montzka SA, Launois T, Belviso S, Bopp L, Laine M (2017) Large historical growth in global terrestrial gross primary production. Nature 544(7648):84–87. https://doi.org/10.1038/NATURE22030

Cao Z (2020) Chapter 9 – Assessment methods for air pollution exposure. In: Li L, Zhou X, Tong W (eds) Spatiotemporal analysis of air pollution and its application in public health. Elsevier, pp 197–206., ISBN 9780128158227. https://doi.org/10.1016/B978-0-12-815822-7.00009-1

Chapin FS, Matson PA, Vitousek PM (2011) Principles of terrestrial ecosystem ecology. https://doi.org/10.1007/978-1-4419-9504-9

Ciais P, Tagliabue A, Cuntz M, Bopp L, Scholze M, Hoffmann G, Lourantou A, Harrison SP, Prentice IC, Kelley DI, Koven C, Piao SL (2012) Large inert carbon pool in the terrestrial biosphere during the Last Glacial Maximum. Nat Geosci 5(1):74–79. https://doi.org/10.1038/ngeo1324

Ciais P, Sabine C, Bala G, Bopp L, Brovkin V, Canadell J, Chhabra A, DeFries R, Galloway J, Heimann M, Jones C, Le Quéré C, Myneni RB, Piao S, Thornton P (2013) Carbon and other biogeochemical cycles. In: Stocker TF, Qin D, Plattner G-K, Tignor M, Allen SK, Boschung J, Nauels A, Xia Y, Bex V, Midgley PM (eds) *Climate Change 2013* The physical science basis. Contribution of working group I to the fifth assessment report of the intergovernmental panel on climate change. Cambridge University Press, Cambridge, pp 465–570. https://doi.org/10.1017/CBO9781107415324.015

Clark DA, Brown S, Kicklighter DW, Chambers JQ, Thomlinson JR, Ni J (2001) Measuring net primary production in forests: concepts and field methods. Ecol Appl 11(2):356–370

Cooper JJ, Fleming GJ, Malungani TP, Misselhorn AA (2004) Ecosystem services in Southern Africa: a regional assessment

Cortes J, Mahecha MD, Reichstein M, Myneni RB, Chen C, Brenning A (2021) Where are global vegetation greening and browning trends significant? Geophys Res Lett 48:9

Cui T, Wang Y, Sun R, Qiao C, Fan W, Jiang G, Hao L, Zhang L (2016) Estimating vegetation primary production in the Heihe River Basin of China with multi-source and multi-scale data. PLoS One 11(4):153971. https://doi.org/10.1371/journal.pone.0153971

Daron J (2015) Challenges in using a Robust Decision Making approach to guide climate change adaptation in South Africa. Clim Change 132(3):459–473. https://doi.org/10.1007/s10584-014-1242-9

Didan K (2015) MOD13C2 MODIS/Terra Vegetation Indices Monthly L3 Global 0.05Deg CMG. NASA LP DAAC. https://doi.org/10.5067/MODIS/MOD13C2.006

Dyson LL, van Heerden J (2001) The heavy rainfall and floods over the northeastern interior of South Africa during February 2000. South African Journal of Science 97(3):80–86

Drüke M, Von Bloh W, Petri S, Sakschewski B, Schaphoff S, Forkel M, Huiskamp W, Feulner G, Thonicke K (2021) CM2Mc-LPJmL v1.0: Biophysical coupling of a process-based dynamic vegetation model with managed land to a general circulation model. Geosci Model Dev 14(6):4117–4141. https://doi.org/10.5194/GMD-14-4117-2021

Feng Y, Zhu J, Zhao X, Tang Z, Zhu J, Fang J (2019) Changes in the trends of vegetation net primary productivity in China between 1982 and 2015. Environ Res Lett 14(12):124009. https://doi.org/10.1088/1748-9326/AB4CD8

Fetzel T, Niedertscheider M, Erb K-H, Gaube V, Gingrich S, Haberl H, Krausmann F, Lauk C, Plutzar C (2012) Human appropriation of net primary production in Africa: patterns, trajectories, processes and policy implications. Soc Ecol Work Pap 37(August):725

Forkel M, Carvalhais N, Verbesselt J, Mahecha MD, Neigh CSR, Reichstein M (2013) Trend change detection in NDVI time series: effects of inter-annual variability and methodology. Remote Sens 5(5):2113–2144. https://doi.org/10.3390/rs5052113

Friedlingstein P, O'Sullivan M, Jones MW, Andrew RM, Hauck J, Olsen A, Peters GP, Peters W, Pongratz J, Sitch S, Le Quéré C, Canadell JG, Ciais P, Jackson RB, Alin S, Aragão LEOC, Arneth A, Arora V, Bates NR, Becker M, Benoit-Cattin A, Bittig HC, Bopp L, Bultan S, Chandra N, Chevallier F, Chini LP, Evans W, Florentie L, Forster PM, Gasser T, Gehlen M,

Gilfillan D, Gkritzalis T, Gregor L, Gruber N, Harris I, Hartung K, Haverd V, Houghton RA, Ilyina T, Jain AK, Joetzjer E, Kadono K, Kato E, Kitidis V, Korsbakken JI, Landschützer P, Lefèvre N, Lenton A, Lienert S, Liu Z, Lombardozzi D, Marland G, Metzl N, Munro DR, Nabel JEMS, Nakaoka SI, Niwa Y, O'Brien K, Ono T, Palmer PI, Pierrot D, Poulter B, Resplandy L, Robertson E, Rödenbeck C, Schwinger J, Séférian R, Skjelvan I, Smith AJP, Sutton AJ, Tanhua T, Tans PP, Tian H, Tilbrook B, Van Der Werf G, Vuichard N, Walker AP, Wanninkhof R, Watson AJ, Willis D, Wiltshire AJ, Yuan W, Yue X, Zaehle S (2020) Global Carbon Budget (2020). Earth Syst Sci Data 12(4):3269–3340. https://doi.org/10.5194/essd-12-3269-2020

Fukano Y, Guo W, Aoki N, Ootsuka S, Noshita K, Uchida K, Kato Y, Sasaki K, Kamikawa S, Kubota H (2021) GIS-based analysis for UAV-supported field experiments reveals soybean traits associated with rotational benefit. Front Plant Sci 0:1003. https://doi.org/10.3389/FPLS.2021.637694

Gao Y, Zhou X, Wang Q, Wang C, Zhan Z, Chen L, Yan J, Qu R (2013) Vegetation net primary productivity and its response to climate change during 2001–2008 in the Tibetan Plateau. Sci Total Environ 444:356–362. https://doi.org/10.1016/j.scitotenv.2012.12.014

Gao Q, Guo Y, Xu H, Ganjurjav H, Li Y, Wan Y, Qin X, Ma X, Liu S (2016) Climate change and its impacts on vegetation distribution and net primary productivity of the alpine ecosystem in the Qinghai-Tibetan Plateau. https://doi.org/10.1016/j.scitotenv.2016.02.131

Gerten D, Schaphoff S, Haberlandt U, Lucht W, Sitch S (2004) Terrestrial vegetation and water balance-hydrological evaluation of a dynamic global vegetation model. https://doi.org/10.1016/j.jhydrol.2003.09.029

Gonsamo A, Chen JM (2017) 3.11 – Vegetation primary productivity. In: Liang S (ed) Comprehensive remote sensing. Elsevier, pp 163–189. https://doi.org/10.1016/B978-0-12-409548-9.10535-4. ISBN 9780128032213

Gonzalez-Meler MA, Taneva L, Trueman RJ (2004) Plant respiration and elevated atmospheric CO2 concentration: cellular responses and global significance. Ann Bot 94:647–656. https://doi.org/10.1093/aob/mch189

Gornall J, Betts R, Burke E, Clark R, Camp J, Willett K, Wiltshire A (2010) Implications of climate change for agricultural productivity in the early twenty-first century. Philos Trans R Soc B Biol Sci 365:2973–2989. https://doi.org/10.1098/rstb.2010.0158

Gulev SK, Thorne PW, Ahn J, Dentener FJ, Domingues CM, Gerland S, Gong D, Kaufman DS, Nnamchi HC, Quaas J, Rivera JA, Sathyendranath S, Smith SL, Trewin B, von Shuckmann K, Vose RS (2021) Changing state of the climate system. In: Masson Delmotte V, Zhai P, Pirani A, Connors SL, Péan C, Berger S, Caud N, Chen Y, Goldfarb L, Gomis MI, Huang M, Leitzell K, Lonnoy E, Matthews JBR, Maycock TK, Waterfield T, Yelekçi O, Yu R, Zhou B (eds) Climate Change 2021: The Physical Science Basis. Contribution of Working Group I to the Sixth Assessment Report of the Intergovernmental Panel on Climate Change. Cambridge University Press. In Press

Haberl H, Erb KH, Krausmann F, Gaube V, Bondeau A, Plutzar C, Gingrich S, Lucht W, Fischer-Kowalski M (2007) Quantifying and mapping the human appropriation of net primary production in earth's terrestrial ecosystems. Proc Natl Acad Sci 104(31):12942–12947. https://doi.org/10.1073/PNAS.0704243104

Hantson S, Arneth A, Harrison SP, Kelley DI, Prentice IC, Rabin SS, Archibald S, Mouillot F, Arnold SR, Artaxo P, Bachelet D, Ciais P, Forrest M, Friedlingstein P, Hickler T, Kaplan JO, Kloster S, Knorr W, Lasslop G, Li F, Mangeon S, Melton JR, Meyn A, Sitch S, Spessa A, van der Werf GR, Voulgarakis A, Yue C (2016) The status and challenge of global fire modelling. Biogeosciences 13:3359–3375. https://doi.org/10.5194/bg-13-3359-2016

Harada Y, Kamahori H, Kobayashi C, Endo H, Kobayashi S, Ota Y, Onoda H, Onogi K, Miyaoka K, Takahashi K (2016) The JRA-55 reanalysis: representation of atmospheric circulation and climate variability. J Meteorol Soc Jpn Ser II 94:269–302. https://doi.org/10.2151/jmsj.2016-015

Harris IC, Jones PD (2019) CRU TS3.26: Climatic Research Unit (CRU) Time-Series (TS) Version 3.26 of High-resolution gridded data of month-by-month variation in climate (Jan. 1901–Dec.

2017). Centre for Environmental Data Analysis, 01 March 2019. https://dx.doi.org/10.5285/7ad889f2cc1647efba7e6a356098e4f3

Hazarika MK, Yasuoka Y, Ito A, Dye D (2005) Estimation of net primary productivity by integrating remote sensing data with an ecosystem model. Remote Sens Environ 94(3):298–310. https://doi.org/10.1016/j.rse.2004.10.004

Heisler-White JL, Knapp AK, Kelly EF (2008) Increasing precipitation event size increases aboveground net primary productivity in a semi-arid grassland. Oecologia 158(1):129–140. https://doi.org/10.1007/s00442-008-1116-9

Hickler T, Eklundh L, Seaquist JW, Smith B, Ardo J, Olsson L, Sykes MT, Sjo M (2005) Precipitation controls Sahel greening trend. Geophys Res Lett 32:2–5. https://doi.org/10.1029/2005GL024370

Hickler T, Rammig A, Werner C (2015) Modelling CO2 impacts on forest productivity. Curr For Rep 1(2):69–80. https://doi.org/10.1007/s40725-015-0014-8

Higginbottom TP, Symeonakis E (2014) Assessing land degradation and desertification using vegetation index data: current frameworks and future directions. Remote Sens 6(10):9552–9575. https://doi.org/10.3390/rs6109552

Hurtt GC, Chini LP, Frolking S, Betts RA, Feddema J, Fischer G, Fisk JP, Hibbard K, Houghton RA, Janetos A, Jones CD, Kindermann G, Kinoshita T, Klein Goldewijk K, Riahi K, Shevliakova E, Smith S, Stehfest E, Thomson A, Thornton P, van Vuuren DP, Wang YP (2011) Harmonization of land-use scenarios for the period 1500–2100: 600 years of global gridded annual land-use transitions, wood harvest, and resulting secondary lands. Clim Change 109:117. https://doi.org/10.1007/s10584-011-0153-2

IPCC (2021) Summary for policymakers. In: Masson Delmotte V, Zhai P, Pirani A, Connors SL, Péan C, Berger S, Caud N, Chen Y, Goldfarb L, Gomis MI, Huang M, Leitzell K, Lonnoy E, Matthews JBR, Maycock TK, Waterfield T, Yelekçi O, Yu R, Zhou B (eds) Climate Change 2021: The Physical Science Basis. Contribution of Working Group I to the Sixth Assessment Report of the Intergovernmental Panel on Climate Change. Cambridge University Press. In Press

Ji Y, Zhou G, Luo T, Dan Y, Zhou L, Lv X (2020) Variation of net primary productivity and its drivers in China's forests during 2000–2018. For Ecosyst 7(1):1–11. https://doi.org/10.1186/S40663-020-00229-0

Jung M, Reichstein M, Bondeau A (2009) Towards global empirical upscaling of FLUXNET eddy covariance observations: validation of a model tree ensemble approach using a biosphere model. Biogeosciences 6:2001–2013

Kelley DI, Prentice I, Harrison S, Wang H, Simard M, Fisher JB, Willis K (2013) A comprehensive benchmarking system for evaluating global vegetation models. Biogeosciences 10:3313–3340. https://doi.org/10.5194/bg-10-3313-2013

Knorr W, Arneth A, Jiang L (2016) Demographic controls of future global fire risk. Nature Clim Change 6:781–785. https://doi.org/10.1038/nclimate2999

Kidwell KB (comp, ed) (1995) NOAA Polar Orbiter Data (TIROS-N, NOAA-6, NOAA-7, NOAA-8, NOAA-9, NOAA-10, NOAA-11, NOAA-12, and NOAA-14, NOAA-15, NOAA-16, NOAA-17, NOAA-18, NOAA-19) Users Guide NOAA/NESDIS, Washington, D.C.

Kobayashi S, Ota Y, Harada Y, Ebita A, Moriya M, Onoda H, Onogi K, Kamahori H, Kobayashi C, Endo H, Miyaoka K, Kiyotoshi T (2015) The JRA-55 reanalysis: general specifications and basic characteristics. J Meteorol Soc Jpn 93(1):5–48. https://doi.org/10.2151/jmsj.2015-001

Lindeskog M, Arneth A, Bondeau A, Waha K, Seaquist J, Olin S, Smith B (2013) Implications of accounting for land use in simulations of ecosystem carbon cycling in Africa. Earth Syst Dyn 4(2):385–407. https://doi.org/10.5194/esd-4-385-2013

Liu Y, Kumar M, Katul GG, Porporato A (2019) Reduced resilience as an early warning signal of forest mortality. Nat Clim Chang 9:880–885. https://doi.org/10.1038/s41558-019-0583-9

Luo T, Pan Y, Ouyang H, Shi P, Luo J, Yu Z, Lu Q (2004) Leaf area index and net primary productivity along subtropical to alpine gradients in the Tibetan Plateau. Glob Ecol Biogeogr 13(4):345–358. https://doi.org/10.1111/j.1466-822X.2004.00094.x

Malherbe J, Dieppois B, Maluleke P, Van Staden M, Pillay DL (2015) South African droughts and decadal variability. Nat Hazards 80(1):657–681. https://doi.org/10.1007/S11069-015-1989-Y

Mann HB (1945) Non-parametric tests against trend. Econometrica 13:245–259

Masoudi M, Jokar P, Pradhan B (2018) A new approach for land degradation and desertification assessment using geospatial techniques. Hazards Earth Syst Sci 18:1133–1140. https://doi.org/10.5194/nhess-18-1133-2018

Mattiuzzi M, Verbesselt J, Hengl T, Klisch A, Stevens F, Mosher S, Evans B, Lobo A, Hufkens K, Detsch F (2017) MODIS – acquisition and processing of MODIS products. https://github.com/MatMatt/MODIS

Medlyn BE (2011) Comment on "Drought-induced reduction in global terrestrial net primary production from 2000 through 2009". Science 333:1

Melillo JM, McGuire AD, Kicklighter DW, Moore B, Vorosmarty CJ, Schloss AL (1993) Global climate change and terrestrial net primary production. Nature 363(6426):234–240. https://doi.org/10.1038/363234a0

Midgley GF, Bond WJ (2015) Future of African terrestrial biodiversity and ecosystems under anthropogenic climate change. Nat Clim Change 5:823–829. https://doi.org/10.1038/nclimate2753

Minnett PJ (2001) Satellite remote sensing of sea surface temperatures. Encycl Ocean Sci:91–102. https://doi.org/10.1016/b978-012374473-9.00343-x

Mohamed MAA, Babiker IS, Chen ZM, Ikeda K, Ohta K, Kato K (2004) The role of climate variability in the inter-annual variation of terrestrial net primary production (NPP). Sci Total Environ 332:123–137. https://doi.org/10.1016/j.scitotenv.2004.03.009

Monteith J (1972) Solar radiation and productivity in tropical ecosystems. J Appl Ecol 9(3):747–766

Myneni RB, Knyazikhin Y, Privette JL, Running SW, Nemani R, Zhang Y, Tian Y, Wang Y, Morissette JT, Glassy J, Votava P (1999) MODIS Leaf Area Index (LAI) and fraction of photosynthetically active radiation absorbed by vegetation (FPAR) product. Modis Atbd Version 4.(4.0):130

Myneni RB, Hoffman S, Knyazikhin Y, Privette JL, Glassy J, Tian Y, Wang Y, Song X, Zhang Y, Smith GR, Lotsch A, Friedl M, Morisette JT, Votava P, Nemani RR, Running SW (2002) Global products of vegetation leaf area and fraction absorbed PAR from year one of MODIS data. Remote Sens Environ 83(1–2):214–231. https://doi.org/10.1016/S0034-4257(02)00074-3

Nemani RR, Keeling CD, Hashimoto H, Jolly WM, Piper SC, Tucker CJ, Myneni RB, Running SW (2003) Climate-driven increases in global terrestrial net primary production from 1982 to 1999. Science 300(5625):1560–1563. https://doi.org/10.1126/science.1082750

Niedertscheider M (2011) Human appropriation of net primary production in South Africa, 1961–2006. A socio-ecological analysis. Master thesis, Vienna University, Vienna

Nieradzik LP, Haverd VE, Briggs P, Meyer CP, Canadell J (2015) BLAZE, a novel Fire-model for the CABLE Land-surface model applied to a Re-assessment of the australian continental carbon budget. AGU Fall Meeting Abstracts December 14–18

Pachavo G, Murwira A (2014) Remote sensing net primary productivity (NPP) estimation with the aid of GIS modelled shortwave radiation (SWR) in a Southern African Savanna. Int J Appl Earth Obs Geoinf 30(1):217–226. https://doi.org/10.1016/J.JAG.2014.02.007

Pan S, Dangal SRS, Tao B, Yang J, Tian H (2015) Recent patterns of terrestrial net primary production in Africa influenced by multiple environmental changes. Ecosyst Heal Sustain 1(5):1–15. https://doi.org/10.1890/EHS14-0027.1

Peng D, Zhang B, Wu C, Huete AR, Gonsamo A, Lei L, Ponce-Campos GE, Liu X, Wu Y (2017) Country-level net primary production distribution and response to drought and land cover change. Sci Total Environ 574:65–77. https://doi.org/10.1016/j.scitotenv.2016.09.033

Peterson BG, Carl P, Boudt K, Bennett R, Ulrich J, Zivot E, Cornilly D (2020) "Performance-Analytics" Econometric tools for performance and risk analysis. https://github.com/braverock/PerformanceAnalytics

Portmann FT, Siebert S, Döll P (2010) MIRCA2000—Global monthly irrigated and rainfed crop areas around the year 2000: a new high-resolution data set for agricultural and hydrological modeling. Glob Biogeochem Cycle 24. https://doi.org/10.1029/2008GB003435

Prentice IC, Bondeau A, Cramer W, Harrison SP, Hickler T, Lucht W, Sitch S, Smith B, Sykes MT (2007) Dynamic global vegetation modelling: quantifying terrestrial ecosystem responses to large-scale environmental change. In: Canadell JD, Pataki E, Pitelka LF (eds) Terrestrial ecosystems in a changing world. Springer, Berlin, pp 175–192

Pugh TAM, Lindeskog M, Smith B, Poulter B, Arneth A, Haverd V, Calle L (2019) Role of forest regrowth in global carbon sink dynamics. Proc Natl Acad Sci U S A 116:4382–4387

Ranasinghe R, Ruane AC, Vautard R, Arnell N, Coppola E, Cruz FA, Dessai S, Islam AS, Rahimi M, Carrascal DR, Sillmann J, Sylla MB, Tebaldi C, Wang W, Zaaboul R (2021) Climate change information for regional impact and for risk assessment. In: Masson Delmotte V, Zhai P, Pirani A, Connors SL, Péan C, Berger S, Caud N, Chen Y, Goldfarb L, Gomis MI, Huang M, Leitzell K, Lonnoy E, Matthews JBR, Maycock TK, Waterfield T, Yelekçi O, Yu R, Zhou B (eds) Climate Change 2021: The Physical Science Basis. Contribution of Working Group I to the Sixth Assessment Report of the Intergovernmental Panel on Climate Change. Cambridge University Press. In Press

Reeves MC, Moreno AL, Bagne KE, Running SW (2014) Estimating climate change effects on net primary production of rangelands in the United States. Clim Change 126(3–4):429–442. https://doi.org/10.1007/s10584-014-1235-8

Ruimy A, Saugier B, Dedieu G (1994) Methodology for the estimation of terrestrial net primary production from remotely sensed data. J Geophys Res 99(D3):5263–5283. https://doi.org/10.1029/93JD03221

Running SW, Nemani RR, Heinsch FA, Zhao M, Reeves M, Hashimoto H (2004) A continuous satellite-derived measure of global terrestrial primary production. BioScience 54(6):547–560. https://doi.org/10.1641/0006-3568(2004)054[0547:ACSMOG]2.0.CO;2

Rutherford MC, Mucina L, Powrie LW (2006) Biomes and bioregions of southern Africa: the vegetation of South Africa, Lesotho and Swaziland. Strelitzia 19:31–51

Sala OE, Austin AT (2000) Methods of estimating aboveground net primary productivity. In: Sala OE, Jackson RB, Mooney HA, Howarth RW (eds) Methods in ecosystem science. Springer, New York, pp 31–43

Samanta A, Costa MH, Nunes EL, Vieira SA, Xu L, Myneni RB (2011) Comment on "Drought-induced reduction in global terrestrial net primary production from 2000 through 2009". Science 333:2

Savtchenko A, Ouzounov D, Ahmad S, Acker J, Leptoukh G, Koziana J, Nickless D (2004) Terra and Aqua MODIS products available from NASA GES DAAC. Adv Sp Res 34(4):710–714. https://doi.org/10.1016/j.asr.2004.03.012

Schoeman CM, Monadjem A (2018) Community structure of bats in the savannas of southern Africa: influence of scale and human land use. Hystrix, Ital J Mammal 29(1):3–10. https://doi.org/10.4404/hystrix-00038-2017

Seaquist JW, Hickler T, Eklundh L, Ardö J, Heumann BW (2008) Disentangling the effects of climate and people on Sahel vegetation dynamics. Biogeosci Discuss 5(4):3045–3067. https://doi.org/10.5194/bgd-5-3045-2008

Simmons A, Fellous JL, Ramaswamy V, Trenberth K, Asrar G, Burrows JP, Ciais P, Drinkwater M, Friedlingstein P, Gobron N, Guilyardi E, Halpern D, Heimann M, Johannessen J, Levelt PF, Lopez-Baeza E, Penner J, Scholes R, Shepherd T (2016) Observation and integrated Earth system science: a roadmap for 2016–2025. Adv Space Res 57:2037–2103. https://doi.org/10.1016/j.asr.2016.03.008

Sitch S, Smith B, Prentice IC, Arneth A, Bondeau A, Cramer W, Kaplan JO, Levis S, Lucht W, Sykes MT, Thonicke K, Venevsky S (2003) Evaluation of ecosystem dynamics, plant geography and terrestrial carbon cycling in the LPJ dynamic global vegetation model. Glob Chang Biol 9(2):161–185. https://doi.org/10.1046/j.1365-2486.2003.00569.x

Sitch S, Huntingford C, Gedney N, Levy PE, Lomas M, Piao SL, Betts R, Ciais P, Cox P, Friedlingstein P, Jones CD, Prentice IC, Woodward FI (2008) Evaluation of the terrestrial carbon cycle, future plant geography and climate-carbon cycle feedbacks using five dynamic global vegetation models (DGVMs). Glob Change Biol 14(9):2015–2039. https://doi.org/10.1111/j.1365-2486.2008.01626.x

Smith B (2001) LPJ-GUESS – an ecosystem modelling framework. Dep Phys Geogr Ecosyst Anal INES, Sölvegatan 12:22362

Smith B, Wärlind D, Arneth A, Hickler T, Leadley P, Siltberg J, Zaehle S (2014) Implications of incorporating N cycling and N limitations on primary production in an individual-based dynamic vegetation model. Biogeosciences 11(7):2027–2054. https://doi.org/10.5194/bg-11-2027-2014

Smith B, Prentice IC, Sykes MT (2001) Representation of vegetation dynamics in the modelling of terrestrial ecosystems: comparing two contrasting approaches within European climate space. Glob Ecol Biogeogr 10:621–637

Smithers JC, Schulze RE, Pike A, Jewitt GPW (2001) A hydrological perspective of the February 2000 floods : a case study in the Sabie River catchment. Water SA, 27(3):325–332. https://doi.org/doi:10.4314/wsa.v27i3.4975

Solano R, Didan K, Jacobson A, Huete A (2010) MODIS Vegetation Index User's Guide (MOD13 Series). Univ Arizona 2010(May):38

Sus O, Stengel M, Stapelberg S, McGarragh G, Poulsen C, Povey AC, SchlundtC TG, Christensen M, Proud S, Jerg M, Grainger R, Hollmann R (2018) The Community Cloud retrieval for CLimate (CC4CL) – Part 1: A framework applied to multiple satellite imaging sensors. Atmos Measur Tech 11:3373–3396

Tao B, Li K, Shao X, Cao M (2003) The temporal and spatial patterns of terrestrial net primary productivity in China. J Geogr Sci 13(2):163–171. https://doi.org/10.1007/bf02837454

Trishchenko AP, Fedosejevs G, Li Z, Cihlar J (2002) Trends and uncertainties in thermal calibration of AVHRR radiometers onboard NOAA-9 to NOAA-16. J Geophys Res Atmos 107(24):ACL 17-1–ACL 17-13. https://doi.org/10.1029/2002JD002353

Tucker CJ (1979) Red and photographic infrared linear combinations for monitoring vegetation. Remote Sens Environ 8(2):127–150. https://doi.org/10.1016/0034-4257(79)90013-0

Urban M, Forkel M, Schmullius C, Hese S, Hüttich C, Herold M (2013) Identification of land surface temperature and albedo trends in AVHRR Pathfinder data from 1982 to 2005 for northern Siberia. Int J Remote Sens 34(12):4491–4507. https://doi.org/10.1080/01431161.2013.779760

Wang S, Zhang Y, Ju W, Chen JM, Ciais P, Cescatti A, Sardans J, Janssens IA, Wu M, Berry JA, Campbell E, Fernández-Martínez M, Alkama R, Sitch S, Friedlingstein P, Smith WK, Yuan W, He W, Lombardozzi D, Kautz M, Zhu D, Lienert S, Kato E, Poulter B, Sanders TGM, Krüger I, Wang R, Zeng N, Tian H, Vuichard N, Jain AK, Wiltshire A, Haverd V, Goll DS, Peñuelas J (2020) Recent global decline of CO2 fertilization effects on vegetation photosynthesis. Science 370(6522):1295–1300. https://doi.org/10.1126/science.abb7772

Wickham H (2016) ggplot2: Elegant graphics for data analysis. Springer, New York. ISBN 978-3-319-24277-4. https://ggplot2.tidyverse.org

Xiao J, Zhuang Q, Baldocchi DD, Law BE, Richardson AD, Chen J, Oren R, Starr G, Noormets A, Ma S, Verma SB, Wharton S, Wofsy SC, Bolstad PV, Burns SP, Cook DR, Curtis PS, Drake BG, Falk M, Fischer ML, Foster DR, Gu L, Hadley JL, Hollinger DY, Katul GG, Litvak M, Martin TA, Matamala R, McNulty S, Meyers TP, Monson RK, Munger JW, Oechel WC, Tha Paw U K, Schmid HP, Scott RL, Sun G, Suyker AE, Torn MS (2008) Estimation of net ecosystem carbon exchange for the conterminous United States by combining MODIS and AmeriFlux data. 148(11):1827–1847. https://doi.org/10.1016/j.agrformet.2008.06.015

Xiao X, Doughty R, Wu X, Zhang Y, Chang Q, Qin Y, Wang J, Bajgain R (2019) Spatial-temporal dynamics of global terrestrial gross primary production during 2000-2018: an update on vegetation photosynthesis model and it simulations with Terra/MODIS images. American Geophysical Union, Fall Meeting 2019

Xu Z, Jiang Y, Jia B, Zhou G (2016) Elevated-CO2 response of stomata and its dependence on environmental factors. Front Plant Sci 7(657). https://doi.org/10.3389/fpls.2016.00657

Yang W, Shabanov NV, Huang D, Wang W, Dickinson RE, Nemani RR, Knyazikhin Y, Myneni RB (2006) Analysis of leaf area index products from combination of MODIS Terra and Aqua data. Remote Sens Environ 104:297–312. https://doi.org/10.1016/j.rse.2006.04.016

Yu T, Sun R, Xiao Z, Zhang Q, Liu G, Cui T, Wang J (2018) Estimation of global vegetation productivity from Global LAnd Surface Satellite data. Remote Sens 10(2). https://doi.org/10.3390/rs10020327

Zaehle S, Friedlingstein P, Friend AD (2010) Terrestrial nitrogen feedbacks may accelerate future climate change. Geophys Res Lett 37. https://doi.org/10.1029/2009GL041345

Zhang L-X, Zhou D-C, Fan J-W, Hu Z-M (2015) Comparison of four light use efficiency models for estimating terrestrial gross primary production. Ecol Model 300:30–39. https://doi.org/10.1016/j.ecolmodel.2015.01.001

Zhang Y, Hu Q, Zou F (2021) Spatio-temporal changes of vegetation net primary productivity and its driving factors on the Qinghai-Tibetan Plateau from 2001 to 2017. Remote Sens 13:1566. https://doi.org/10.3390/rs13081566

Zhao M, Heinsch FA, Nemani RR, Running SW (2005) Improvements of the MODIS terrestrial gross and net primary production global data set. Remote Sens Environ 95(2):164–176.http://dx.doi.org/10.1016/j.rse.2004.12.011

Zhao F, Xu B, Yang X, Jin Y, Li J, Xia L, Chen S, Ma H (2005) Remote sensing estimates of grassland aboveground biomass based on MODIS net primary productivity (NPP): a case study in the Xilingol Grassland of Northern China. Remote Sens 6:5368–5386. https://doi.org/10.3390/rs6065368

Zhu Z, Piao S, Myneni RB, Huang M, Zeng Z, Canadell JG, Ciais P, Sitch S, Friedlingstein P, Arneth A, Cao C, Cheng L, Kato E, Koven C, Li Y, Lian X, Liu Y, Liu R, Mao J, Pan Y, Peng S, Peuelas J, Poulter B, Pugh TAM, Stocker BD, Viovy N, Wang X, Wang Y, Xiao Z, Yang H, Zaehle S, Zeng N (2016) Greening of the Earth and its drivers. Nat Clim Chang 6(8):791–795. https://doi.org/10.1038/nclimate3004

Comparison of Different Normalisers for Identifying Metal Enrichment of Sediment: A Case Study from Richards Bay Harbour, South Africa

27

Paul Mehlhorn, Brent Newman, and Torsten Haberzettl

Abstract

South Africa's ecosystems are challenged in various ways by anthropogenic effects, such as land-use change, leading to soil erosion in concert with industrial or agricultural pollution, leading to an increase in pollutants in final depositional systems. Here we focus on metals in the marine environment of Richards Bay Harbour. The use for Al, Fe, Rb, Ti and the silt fraction of the sediment as normalisers of Cr, Cu, Co and Pb concentrations in sediment is compared to determine if they provide the same understanding on the enrichment. Baseline metal concentration models were defined and Enrichment Factors calculated to quantify the magnitude of enrichment.

Exceedingly high Cr and Cu concentrations in defined parts of the harbour lead to similar trends rather than a similar effectiveness of the normalisers. Probable biogeochemical processes hinder the effectiveness of Fe and geological background or hydrodynamic properties hinder the effectiveness of Ti as normaliser. Differences in the spatial extent of sediment identified as enriched and the area where metal concentrations exceed guidelines detracts from fully appreciating the extent of metal contamination of sediment using guidelines, with management implications. Furthermore, in the case of Cu, the guidelines for this metal might be underproductive.

P. Mehlhorn · T. Haberzettl (✉)
Institute for Geography and Geology, University of Greifswald, Greifswald, Germany
e-mail: paul.mehlhorn@uni-greifswald.de; torsten.haberzettl@uni-greifswald.de

B. Newman
Coastal Systems Research Group, CSIR, Durban, South Africa

Nelson Mandela University, Port Elizabeth, South Africa

© The Author(s) 2024
G. P. von Maltitz et al. (eds.), *Sustainability of Southern African Ecosystems under Global Change*, Ecological Studies 248,
https://doi.org/10.1007/978-3-031-10948-5_27

27.1   Introduction

Metals are common, and often significant contaminants of sediment in ports (Birch et al. 2020; Mehlhorn et al. 2021) and are the subject of considerable attention in the scientific literature. The focus is founded on valid concerns, including that sediment is a major fate for, and through remobilisation a potential source of metals in aquatic ecosystems (Newman and Watling 2007), and metals are known to present ecological risks when present at elevated concentrations (Chapman and Wang 2001). A major challenge, however, is identifying metal concentrations that reflect the natural state and those that are enhanced through an anthropogenic contribution (i.e. contamination). This is complicated for several reasons. First, metals are a ubiquitous, naturally occurring component of sediment. The mere presence of metals in sediment does not thus infer contamination. Second, metal concentrations in uncontaminated sediment can vary by orders of magnitude over small spatial scales depending on the sediment's mineralogy, granulometry and organic content amongst other factors (Schropp and Windom 1988; Windom et al. 1989; Loring 1991; Krumgalz et al. 1992; Balls et al. 1997; Grant and Middleton 1998; Thomas and Bendell-Young 1999; Rubio et al. 2000; Kersten and Smedes 2002; Amorosi et al. 2007; Du Laing et al. 2007; Woods et al. 2012). Third, despite input and transport dissimilarities, naturally occurring and anthropogenically introduced metals tend to accumulate in sediment in the same areas (Loring 1991; Hanson et al. 1993). Due to these complexities, similar metal concentrations in two sediment samples from the same system may reflect contamination in one sample but not the other, due to a difference in the sediments granulometry. Similarly, very different metal concentrations in two sediment samples from the same aquatic system might in both cases reflect the natural condition, for the same reasons. Therefore, "high" metal concentrations do not necessarily reflect increased levels of contamination and vice versa (Newman and Watling 2007). The direct comparison of metal concentrations amongst sediment samples is thus important in terms of ecological and human health, if certain thresholds are exceeded, but not in the context of identifying contamination itself.

To properly interpret metal concentrations in sediment, it is necessary to compensate for the factors that control their natural variation before background or baseline concentrations can be distinguished from enriched (higher than "expected") concentrations. There are two approaches to normalisation, namely: a) using a metal that acts conservatively—geochemical normalisation, or b) using a grain size fraction—granulometric normalisation (Birch and Snowdon 2004; Newman and Watling 2007). In this way, metal concentrations that are atypical of the bulk of the data can be identified. The investigator must then decide if atypically high metal concentrations reflect contamination or can be explained by natural biogeochemical or hydrodynamic processes.

Geochemical normalisation makes use of a metal that acts as a proxy for the grain size variation of sediment, and more specifically for the silt and clay (mud) fraction (Birch and Snowdon 2004). A metal normaliser should (a) be highly refractory,

(b) be structurally combined to one or more of the major metal-bearing phases of sediment, (c) co-vary in proportion to the naturally occurring concentrations of the metals of interest, (d) be insensitive to inputs from anthropogenic sources and (e) be stable and not subject to environmental influences such as reduction/oxidation, adsorption/desorption and other diagenetic processes that may alter sediment concentrations (Luoma 1990). Commonly, Al, Fe, Li, Rb or Ti are used as geochemical normalisers (Daskalakis and O'Connor 1995; Santos et al. 2005; Tůmová et al. 2019).

In granulometric normalisation, metal concentrations are either normalised to a specific grain size fraction of sediment, or metal concentrations are analysed in a defined grain size fraction of the sediment after sieving. Metals are predominantly incorporated in and preferentially bind to fine sediment rather than to coarse material, meaning that silt and clay are the most effective normalisers in this approach (Newman and Watling 2007; Szava-Kovats 2008; Koigoora et al. 2013). In addition, Ti is often associated with fine sand to silt or clay, whereas K and Al are often associated with clay type minerals (Watling 1977; Haberzettl et al. 2019), while Zr has been reported as enriched in the silt fraction (Cuven et al. 2010; Kylander et al. 2011; Ohlendorf et al. 2014). Therefore, it is necessary to identify the best-fit grain size range for granulometric normalisation. However, caution is required when a grain size fraction is used as a normaliser, since two separate samples are processed and analysed, one for grain size and one for metals. Errors in sample splitting can lead to errors in normalisation. Similarly, isolating a sediment fraction for metal analysis is usually done by wet sieving and this can lead to the loss of metals weakly adsorbed to the surface of sediment grains.

Site-specific circumstances might alter the usefulness of normalisers. In this study, we present data from Richards Bay Harbour on the northeast coast of South Africa, where different factors need to be considered in the choice of the normaliser. Industries near Richards Bay Harbour include two aluminium smelters and a ferrochrome smelter. These industries import alumina and export aluminium and ferroalloys through the port. The effect of open air bulk handling (e.g. of ferrochrome) can introduce metals to the harbour, and point sources of Cr or Cu at bulk handling terminals were recognised by Mehlhorn et al. (2021). Contamination arising from the spillage during import or export and from the smelters (e.g. through atmospheric deposition) could thus influence the utility of Al and Fe as normalisers. In a previous study (Mehlhorn et al. 2021), we used Al to normalise metal concentrations in sediment sampled in Richards Bay Harbour. However, considering the industrial impact outlined above, there are potential limitations to the use of Al and Fe as normalisers of metal concentrations in sediment in the harbour. Similarly, the catchment geology must be considered. For example Ti-bearing heavy mineral dune sands of the Maputaland Group Sibayi Formation surrounding Richards Bay Harbour (Kelbe 2010; Botha 2018) question the usefulness of certain geochemical normalisers, such as Ti, as heavy minerals might be differently influenced by prevailing hydrodynamic conditions compared to other normalisers. Furthermore, the basin topography of Richards Bay Harbour is optimised to serve marine port operations. To keep the port operational, the basin shape of the harbour is largely controlled by dredging (Greenfield et al. 2011; Dladla et al. 2021). Remobilisation of metals from sediment by seabed disturbances can occur via bioturbation, increased current velocities (storms, tides) or dredging (Förstner 1989; Daskalakis and O'Connor 1995; Saulnier and Mucci 2000). Ship operations, especially propeller wash currents, impact the depositional environment (Mehlhorn et al. 2021) and disturb natural sedimentation processes.

To understand changes in an environment, it is necessary to identify a system's reference state using various proxies, in this case, the release of (heavy) metals by surrounding industries and their potential health risks. This enables their comparability to past and future changes (cf. Mehlhorn et al. 2021), but also shifts in the spatial extent of contamination, indicating sink and source relationships. However, each basin or catchment is subject to site-specific environmental influences that need special consideration when evaluating (normalising) parameters. In this study, we evaluate the efficacy of several potential normalisers for concentrations of Cr, Cu, Co and Pb in Richards Bay Harbour sediment, by defining baseline models using different normalisers and comparing trends in the enrichment of sediment identified by the models.

27.2   Materials and Methods

27.2.1 Site Description

Richards Bay Harbour (Fig. 27.1) is situated in the province of KwaZulu-Natal, on the subtropical northeast coast of South Africa. Prior to port construction, Richards Bay was a large, shallow estuary of about 30 km^2 fed by five rivers (Begg 1978). Harbour development started in the early 1970s and involved the construction of

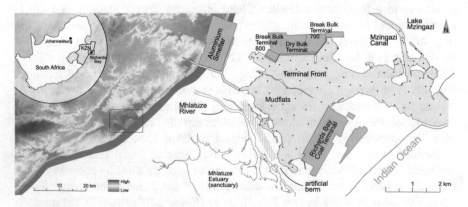

Fig. 27.1 The study area of Richards Bay Harbour. Indicated is the harbour area, which is separated from the Mhlatuze Estuary (sanctuary) by an artificial berm. Also shown are potential contamination sources (red) and sampling locations (black dots). A low altitude dune ridge extending from WNW to ESE divides the harbour and separates the shallow water Mudflats from the deep water Terminal Front

an artificial berm that divided the estuary into two parts. The northern part was developed into Richards Bay Harbour. A new mouth was dredged for the southern part, which was designated a nature sanctuary (Mhlathuze Estuary), and part of the Mhlathuze River was canalised and diverted into this part. A major reason for the construction of Richards Bay Harbour was to facilitate the export of coal through the Richards Bay Coal Terminal, which is now one of the largest coal export terminals in the world (Nel et al. 2007). Other industries that were established in the area to take advantage of the import and export opportunities provided by the harbour are the following: aluminium smelters, a phosphoric acid and fertiliser plant, a ferrochrome smelter and heavy minerals mining and refining operations. A wide range of bulk materials are imported and exported through the harbour in addition to coal, including ferroalloys, sulphur, phosphoric acid, alumina, aluminium, heavy minerals and woodchips.

27.2.2 Sampling

Sampling was conducted in August and September of 2018. Eighty surface sediment samples were collected (Fig. 27.1) using an Ekman-Birge bottom sampler (HYDROBIOS, Kiel, Germany). Only the topmost 1 cm was sampled with a plastic spoon. Samples were stored in sterile polyethylene *Nasco* Whirl Pak's and cooled until further processing.

27.2.3 Granulometric and Geochemical Analysis

The sediment was prepared for granulometric analyses by soaking 0.5–2 g aliquots of wet sediment in 2 ml of hydrochloric acid (HCl, 10%) and 5 ml of hydrogen peroxide (H_2O_2, 10%). The residues were dispersed overnight in 5 ml tetrasodium pyrophosphate ($Na_4P_2O_7$ 10 H_2O, 0.1 M) in an overhead shaker. The samples were then analysed using a Laser Diffraction Particle Size Analyser (Fritsch Analysette 22; FRITSCH GmbH, Germany).

Freeze-dried aliquots of sediment were ground to a particle size <60 µm. Subsamples of the sediment were digested at the University of Greifswald using a modified aqua regia treatment. The procedure involved the addition of 1.25 ml of HCl (37%, suprapur) and 1.25 ml of HNO_3 (65%, suprapur) to 100 mg of sediment, which was then digested in PTFE crucible pressure bombs in an oven at 160 °C for 3 h. Each batch included a laboratory blank as well as a reference sediment sample of certified estuarine sediment (BCR-667) and indicated a good recovery range of 76.9–98.6% for the measured metals (Co = 86%; Cr = 96%; Cu = 96%; Fe = 99%; Pb = 77%). Elemental concentrations were measured at Friedrich Schiller University Jena using an Agilent 725 ES ICP-OES (Al, Ca Fe, K, Mg, Mn, Na, P, S, Sr, Ti) and a Thermo Fischer Scientific X-Series II ICP-MS (As, Co, Cr, Cu, Ni, Pb, Zn, Rb), as described in Mehlhorn et al. (2021).

27.2.4 Data Analysis

Scatterplots of the relationship between potential normalisers (Al, Rb, Ti, Fe, Silt) and co-occurring Cu, Cr, Co and Pb concentrations in sediment showed that, apart from Ti, there was a linear relationship between the bulk of the concentrations and the normalisers. Baseline models were thus defined by fitting a linear regression and 95% prediction limits to scatter plots of Cu, Cr, Co and Pb concentrations and the potential normalisers. Cu, Cr, Co and Pb concentrations falling outside the prediction limits were deemed outliers and sequentially trimmed, starting with the concentration with the largest residual, reiterating the regression and proceeding in this manner until all concentrations fell on or within the prediction limits. The resultant regression and associated prediction limits define the baseline model. Apart from Ti, error terms for the regressions approximated normality, but their variance was usually not homogenous. The Cu, Cr, Co and Pb concentrations were not transformed to approximate this assumption. In general, the lack of error term homogeneity does not result in biased estimates of regression parameters, but does result in an increase in variance about these estimates (Hanson et al. 1993). Schropp et al. (1990), Weisberg et al. (2000) and Woods et al. (2012) used a similar approach for defining baseline models, but continued to trim metal concentrations until the error terms were normally distributed and homogenous. This approach was not followed in this study, since it required the trimming of a larger number of Cu, Cr, Co and Pb concentrations than the approach described above, including concentrations that were subjectively considered to be part of the baseline range.

Sediment with Cu, Cr, Co and Pb at a concentration above a baseline model upper prediction limit was interpreted as enriched (i.e. the metal concentration is in excess of the baseline). The upper prediction limit was thus used to discriminate baseline from enriched metal concentrations. The magnitude of enrichment for each metal concentration was quantified by computing an Enrichment Factor (EF), which is calculated by dividing the observed metal/normaliser ratio by the reference (upper prediction limit) metal/normaliser ratio. EF's >1 represent enriched concentrations, noting that this does not imply an enhancement through anthropogenic contribution, but rather that the concentration is atypical of the data that define the model (Horowitz 1991). Spatial trends for EF's were plotted using ArcMap v. 10.8.1., using inverse distance weighting with barrier function.

Reaching a conclusion on whether metal enrichment reflects contamination thus requires consideration of ancillary factors, including possible (bio)geochemical processes that can lead to natural enrichment (e.g. diagenetic enhancement), the absolute difference between a metal concentration and the baseline model upper prediction limit, the number of metals in a particular sediment sample at an enriched concentration and the proximity of metal-enriched sediment to known or strongly suspected anthropogenic sources of metals. The larger the difference between a metal concentration and the baseline model upper prediction limit, the closer the enriched sediment sampling site is to known or strongly suspected anthropogenic sources of metals, and the greater the number of metals enriched in sediment at a particular sampling site the more likely the excess concentrations reflect contamination.

27.3 Results and Discussion

27.3.1 Evaluation of Normaliser Suitability

In a mineralogically homogenous area, the absolute concentrations of metals in sediment are largely controlled by the sediments grain size (Taylor and McLennan 1981; Horowitz and Elrick 1987; Förstner 1989; Horowitz 1991; Loring 1991; Larrose et al. 2010; Matys Grygar and Popelka 2016; Liang et al. 2019). Aluminosilicates, the dominant natural metal-bearing phase of sediment, predominate in silt and clay (mud). Sand, in contrast, is comprised largely of metal-deficient quartz. Metals naturally adsorb onto Fe/Mn oxides and organic matter in sediment in quantities that are usually proportional to grain size (Kersten and Smedes 2002). As a result, there is usually a strong positive correlation between the concentration of metals and the mud fraction, and between the concentration of different metals in uncontaminated sediment (Rubio et al. 2000; Sabadini-Santos et al. 2009; Coynel et al. 2016; but see Matys Grygar et al. 2013, 2014; Jung et al. 2014, 2016 for evidence of non-linear relationships). These relationships provide the basis for geochemical normalisation, by modelling the linear relationship between metal and co-occurring element (Hanson et al. 1993; Weisberg et al. 2000; Kersten and Smedes 2002). Geochemical normalisation through linear regression is premised on a two-component linear mixing model, one end member representing metal deficient quartz (sand) and the other metal rich aluminosilicates (mud; Hanson et al. 1993). A fundamental requirement for the use of a metal as a geochemical normaliser, therefore, is that the metal must be strongly correlated to the fine-grained fraction of the sediment for which it acts as a proxy.

Richards Bay Harbour includes multiple heterogeneous sedimentary environments (Mehlhorn et al. 2021). To identify the most effective granulometric normaliser we plotted correlation coefficients of individual elements to different grain size classes (Fig. 27.2). Several elements show a high positive correlation with parts of the clay and silt fraction. The curve shape of Al is highlighted as an example in Fig. 27.2. Therefore, this grain size fraction appears to be a good normaliser. As the grain size increases the correlation coefficient decreases and changes to strongly negative at the onset of 250 μm (medium sand). Elements showing a visually similar pattern to Al, but at a lower correlation coefficient of $r = 0.5$, include Cu, Cr and Mn. Other elements, like Sr, Ti and Ca, show no distinct correlation to any grain size class. The correlation between the Silt and Mud (Clay + Silt, which appear to be good normalisers) fractions of sediment in Richards Bay Harbour is strong ($r = 0.993, p < 0.01$). Consequently, only Silt is hereafter considered as a normaliser of metal concentrations in sediment in the harbour.

Al and Rb concentrations in sediment in Richards Bay Harbour are highly correlated to the silt fraction ($r = 0.954, p < 0.001$), but less so to the clay

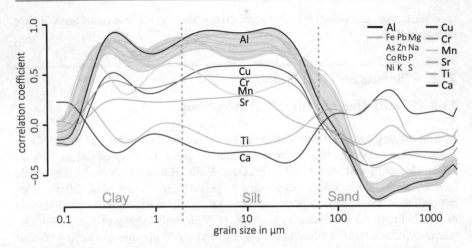

Fig. 27.2 Correlation coefficient versus grain size. The unique distribution curves indicate the varying correlation coefficients of individual elements against grain size classes. Elements displayed by grey lines show a similar behaviour to Al. Coloured curves indicate a differing and individual behaviour when plotted against grain size

fraction ($r = 0.782 - 0.792$, $p < 0.001$). The implication is that Al and Rb are closely associated with or incorporated in the silt fraction of the sediment and are thus effective proxies for this fraction. Despite the import of alumina, export of aluminium and presence of two large aluminium smelters near Richards Bay Harbour, sediment in the harbour does not indicate contamination by Al. This conclusion is based on the strong correlation of Al with most of the elements, and by the absence of pronounced outliers in the relationship between Al and the silt fraction of the sediment and elements that are unlikely to be influenced by anthropogenic activities, such as Rb.

Fe concentrations are also highly correlated to the silt fraction of the sediment ($r = 0.834$, $p < 0.001$), but less so to the clay fraction ($r = 0.798$, $p < 0.001$). The weaker correlation of Fe concentrations to the silt fraction compared to Al and Rb is a result of anomalously high Fe concentrations in sediment at five stations in a shallow area opposite the Richards Bay Coal Terminal, colloquially known as the Mudflats (Fig. 27.1). If the data for sediment sampled on the Mudflats is excluded, the correlation between Fe concentrations and the silt fraction increases ($r = 0.928$, $p < 0.001$), but the correlation to the clay fraction remains moderate ($r = 0.777$, $p < 0.001$). The implication is that Fe is closely associated with or incorporated into the silt fraction of the sediment and is thus an effective proxy for this fraction across most of the harbour, but not on the Mudflats.

Ti concentrations in sediment of Richards Bay Harbour are not correlated to the silt ($r = -0.094$, $p = 0.405$) or clay ($r = 0.061$, $p = 0.589$) fractions. Ti is thus not an effective proxy for these fractions of sediment in the harbour, and by implication also not for the concentrations of other metals.

Table 27.1 Coefficients of determination (r^2) for copper, chromium, cobalt and lead baseline models defined using different normalisers

		Metal			
		Cu	Cr	Co	Pb
Normaliser	Al	0.935	0.813	0.962	0.966
	Fe	0.95	0.931	0.988	0.897
	Rb	0.929	0.768	0.959	0.967
	Ti	0.002	0.031	0.007	0.003
	Silt	0.932	0.898	0.923	0.877

The concentrations/values of normalisers should be strongly positively correlated to the concentrations of metals in uncontaminated sediment. However, in some parts of Richards Bay Harbour, the sediment is metal contaminated (Mehlhorn et al. 2021). The baseline models for Cu, Cr, Co and Pb do not thus represent background concentrations, since it is possible, and for Cu and Cr likely, that certain concentrations included in the baseline models reflect low magnitude contamination of the sediment but are not identified as outliers through the approach used to define the models. Nevertheless, the stronger the relationship between the normalisers and metals, the more effective the baseline model should theoretically be in identifying metal enrichment of sediment.

There is not much difference in the coefficients of determination for baseline models apart from Ti, for which coefficients are consistently very low (Table 27.1). The Fe normalised baseline models provide the highest coefficients for Cu, Cr and Co, and the Rb model for Pb. The baseline model coefficients of determination suggest, therefore, that Fe is the best normaliser for Cu, Cr and Co, and Rb for Pb. However, the Fe normalised baseline models are strongly influenced by the exclusion of Cu and Cr and inclusion of Co and Pb concentrations in sediment on parts of the Mudflats due to the anomalous concentrations of Fe, Co and Pb in this part of the harbour. The anomalous concentrations of Fe might be caused by a local change in redox processes (Wündsch et al. 2014; Zolitschka et al. 2019). The critical Eh for the reduction of Fe^{3+} to more soluble Fe^{2+} is 100 mV (Sigg and Stumm 1996). If the redox potential is lowered, Fe ions start to migrate in pore water. When an Eh above 100 mV is encountered, the ions will oxidise and precipitate, leading to the enrichment of sediment that is not necessarily linked to an anthropogenic impact (Haberzettl et al. 2007; Brunschön et al. 2010).

27.3.2 Baseline Model Comparison

If normalisers are to be considered similarly effective, then (a) they should identify a similar number of sediment samples as enriched by any metal, (b) the spatial extent of the enrichment should be similar, (c) the magnitude of the enrichment should be similar and (d) the enrichment should be logical in the context of known or potential anthropogenic sources of the metal. The baseline models defined for Cu, Cr, Co and Pb using different normalisers are provided in Figs. 27.3, 27.4, 27.5 and 27.6,

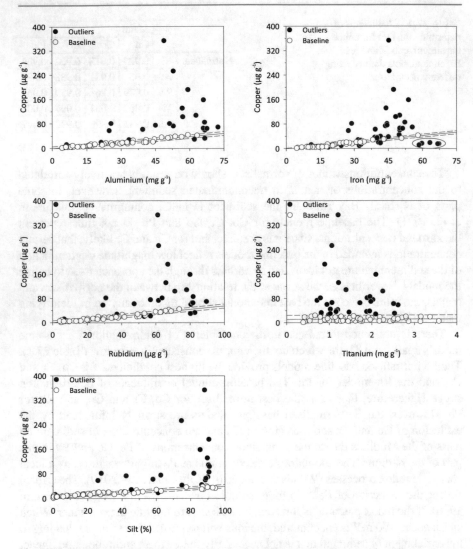

Fig. 27.3 Baseline models for copper defined using different normalisers, with outlier concentrations superimposed (black filled dots). Samples of the mudflat region are highlighted (red circle) in the iron-normalised model

with outlier concentrations superimposed. Parameters for the baseline models are provided in Table 27.2.

As stated previously, sediment with metals at a concentration above the upper prediction limit was deemed to be enriched by the metal. The number of sediment samples identified as enriched by individual baseline models is provided in Fig. 27.7.

The number of sediment samples identified as enriched by Cu is similar for the Al, Rb and Silt normalised baseline models, but the Fe and Ti models identify a

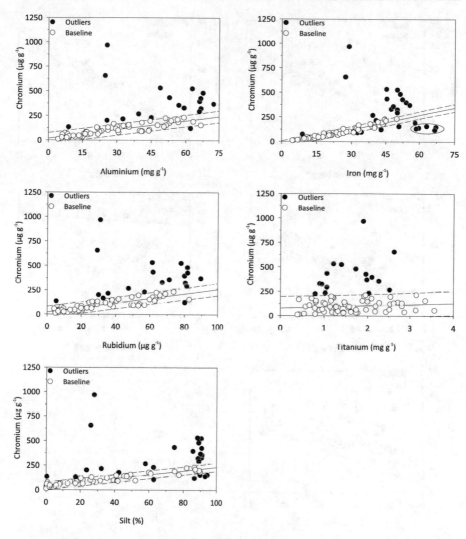

Fig. 27.4 Baseline models for chromium defined using different normalisers, with outlier concentrations superimposed (black filled dots). Samples of the mudflat region are highlighted (red circle) in the iron-normalised model

higher number. The Al, Fe, Rb and Ti normalised baseline models identify a similar number of sediment samples as enriched with Cr, but the Silt model identifies a higher number. The number of samples identified as Co and Pb enriched varies widely amongst the normalisers and is most similar for Al, Rb, and Silt and the lowest for Ti. The number of metals identified as enriched by baseline models defined using different normalisers is not particularly convincing as to which of the normalisers is more effective than others. However, we consider those normalisers

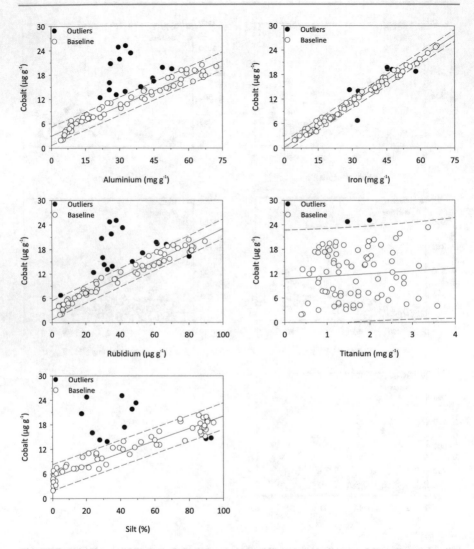

Fig. 27.5 Baseline models for cobalt defined using different normalisers, with outlier concentrations superimposed (black filled dots)

that identify a comparable number of sediment samples as enriched by a metal are likely more effective than others.

27.3.3 Spatial Enrichment Trends

The spatial trend for Cu and Cr EFs that are >1 is comparable when computed using Al, Rb or Silt normalised baseline models (Figs. 27.8, 27.9, 27.10 and

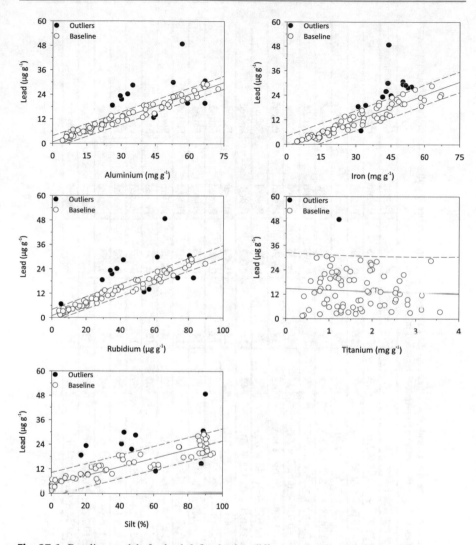

Fig. 27.6 Baseline models for lead defined using different normalisers, with outlier concentrations superimposed (black filled dots)

27.11). However, the trend for EFs computed using Fe and Ti normalised models is different, especially for Co and Pb (Figs. 27.10 and 27.11). A larger area of sediment is identified as Cu enriched by the Fe and Ti normalised baseline models, the additional area extending from the 700 series berth basin into the northern part of the Richards Bay Coal Terminal basin and also including one or more isolated areas in the latter basin (Fig. 27.8). Fe and Ti identify a similar area of the 600 and 700 series berth basins (part of the Dry Bulk Terminal, Fig. 27.1) as Cr enriched compared to other normalisers but identify little (Fe) or no (Ti) Cr enrichment of sediment on the Mudflats.

Table 27.2 Baseline model regression parameters, number of samples included in the baseline model (n), number of outliers, coefficient of determination (r^2) and assessment of normality and constant variance for metal concentrations included in the models

Metal	Normaliser	Model regression formula	n	Outliers	r^2	Normality	Constant variance
Chromium	Al	Cr = 15.273 + (3.102 × Al)	59	21	0.813	Passed	Failed
	Fe	Cr = −28.061 + (4.935 × Fe)	53	27	0.931	Passed	Failed
	Rb	Cr = 21.723 + (2.363 × Rb)	60	20	0.768	Passed	Failed
	Ti	Cr = 77.428 + (13.009 × Ti)	63	17	0.031	Failed	Passed
	Silt	Cr = 38.500 + (1.949 × SILT)	55	25	0.898	Passed	Failed
Copper	Al	Cu = −0.625 + (0.660 × Al)	57	23	0.935	Passed	Failed
	Fe	Cu = −3.439 + (0.764 × Fe)	42	38	0.950	Passed	Passed
	Rb	Cu = 0.0983 + (0.546 × Rb)	56	24	0.929	Passed	Failed
	Ti	Cu = 11.071 + (0.345 × Ti)	49	31	0.002	Failed	Passed
	Silt	Cu = 6.004 + (0.332 × SILT)	57	23	0.932	Passed	Failed
Lead	Al	Pb = 0.715 + (0.388 × Al)	68	12	0.966	Passed	Failed
	Fe	Pb = −0.869 + (0.417 × Fe)	66	14	0.897	Passed	Failed
	Rb	Pb = 0.984 + (0.315 × Rb)	67	13	0.967	Passed	Failed
	Ti	Pb = 14.769 − (0.647 × Ti)	79	1	0.003	Failed	Passed
	Silt	Pb = 4.334 + (0.215 × SILT)	70	10	0.877	Passed	Failed
Cobalt	Al	Co = 2.918 + (0.245 × Al)	65	15	0.962	Passed	Passed
	Fe	Co = 0.198 + (0.366 × Fe)	73	7	0.988	Passed	Failed
	Rb	Co = 3.066 + (0.200 × Rb)	63	17	0.959	Passed	Failed
	Ti	Co = 10.741 + (0.644 × Ti)	78	2	0.007	Failed	Passed
	Silt	Co = 5.053 + (0.151 × SILT)	69	11	0.923	Passed	Passed

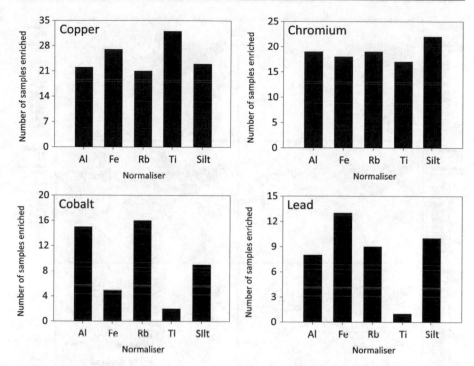

Fig. 27.7 Number of sediment samples with copper, chromium, cobalt and lead concentrations identified as enriched by baseline models defined using different normalisers

There is clear evidence for significant anthropogenic inputs of Cu and Cr in Richards Bay Harbour. Chromium ore and ferrochrome are exported through the harbour. During export, chromium ore and ferrochrome particles are spilled into the harbour and account for the Cr contamination (Mehlhorn et al. 2021). The source of the excess Cu in sediment is less certain. Cu concentrate was historically exported through the harbour and the contamination might reflect the spillage of Cu concentrate particles during export, noting that exports ceased in 2012. The Cu and Cr enrichment of sediment sampled across much of the 600 and 700 series berth basins (dry bulk terminals) is of such a high magnitude that comparable areas of sediment are identified as enriched by baseline models defined using different normalisers, regardless of the effectiveness of the normalisers. The sediment in part of the Small Craft Harbour is identified as Cu enriched by all normalisers (Fig. 27.8), possibly reflecting inputs from antifouling coatings on vessel hulls amongst other possible sources. Some of the baseline models also identify Cu enrichment of sediment in canals joining the north-eastern part of the port, but the source of the Cu is uncertain.

All baseline models identify Cu and Cr enrichment near the 600 and 700 series berths. An extension of the Cu enrichment from the 700 series berth basin to the northern part of the Richards Bay Coal Terminal basin is identified using the Fe

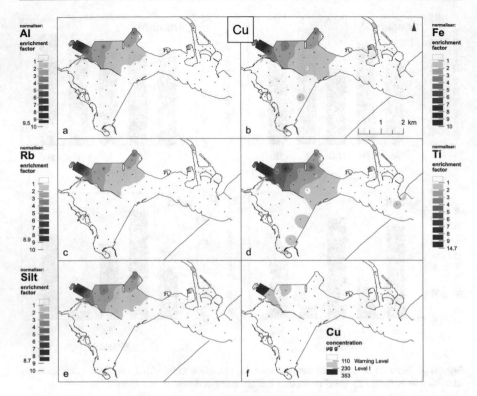

Fig. 27.8 Spatial distributions of the copper (Cu) baseline models normalised to (**a**) Al, (**b**) Fe, (**c**) Rb, (**d**) Ti and (**e**) Silt. The enrichment factors are plotted to the same scale to achieve visual comparability, note the extended legend of Ti ($EF_{max} = 14.7$), maximum values are shown next to the classification. For additional comparison, the distribution map of Cu (**f**) in surface sediments of Richards Bay Harbour is plotted, making use of guideline concentrations as presented by the Department of Environmental Affairs (2012)

and Ti normalised baseline models (Fig. 27.8). This is contradictory as it seems unlikely that Cu but not Cr (Fig. 27.9) would be dispersed to and accumulated in this area considering the strong similarity in parts of the 600 and 700 series berth basins where the sediment is identified as Cu and Cr enriched by Al, Rb and Silt normalised baseline models. The wider area of Cu enrichment identified by the Fe normalised baseline model is probably related to the wider scatter of Cu versus Fe concentrations compared to other normalisers, which led to the trimming of a larger number of concentrations compared to other normalisers (Table 27.2). The Cr enrichment of sediment identified on the Mudflats using the Al, Rb and Silt normalised baseline models, but absence of Cu enrichment also seems illogical for the same reason discussed above for Cu enrichment extending to the Richards Bay Coal Terminal basin. As stated above, the magnitude of Cu and Cr enrichment of sediment at many stations in the 600 and 700 series berth basins is large and it

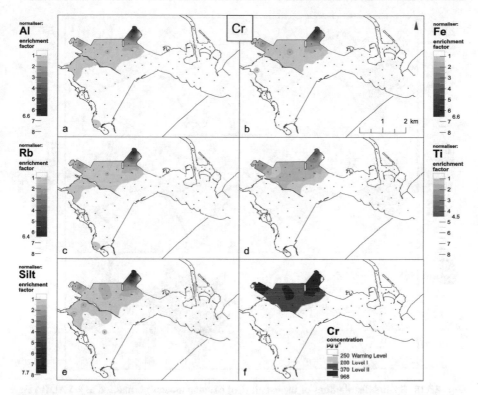

Fig. 27.9 Spatial distributions of the chromium (Cr) baseline models normalised to (**a**) Al, (**b**) Fe, (**c**) Rb, (**d**) Ti and (**e**) Silt. The enrichment factors are plotted to the same scale; maximum values are shown next to the classification. For additional comparison, the distribution map of Cr (**f**) in surface sediments of Richards Bay Harbour is plotted, making use of guideline concentrations as presented by the Department of Environmental Affairs (2012)

is almost inevitable the sediment will be identified as enriched regardless of the effectiveness of the baseline models defined using different normalisers.

Trends in Co and Pb enrichment of sediment are more revealing on the effectiveness of different normalisers. Sediment on the Mudflats of Richards Bay Harbour is identified as enriched by Co and Pb using Al, Rb and Silt normalised baseline models, but not by Fe and/or Ti normalised baseline models (Figs. 27.10 and 27.11). These elements might share a similar accumulation mechanism or be affected similarly by sediment biogeochemical processes.

Examples of Fe enrichment identified using Al and Silt as the normaliser are provided in Fig. 27.12. The consequence is that Fe normalised Co and Pb concentrations are shifted to the right in scatterplots, to the extent they fall within the prediction limits of baseline models defined using the data (Fig. 27.13). In contrast, the Cu and Cr concentrations in sediment at these stations were trimmed during baseline model definition and fall below the lower prediction limit (Figs. 27.3 and

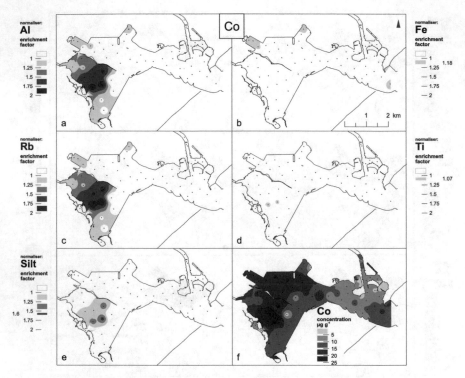

Fig. 27.10 Spatial distributions of the cobalt (Co) baseline models normalised to (**a**) Al, (**b**) Fe, (**c**) Rb, (**d**) Ti and (**e**) Silt. The enrichment factors are plotted to the same scale; maximum values are shown next to the classification. For additional comparison, the concentration of Co (**f**) in surface sediments of Richards Bay Harbour is plotted

27.4). The anomalous concentrations do not thus have the same influence on the baseline models as the Co and Pb concentrations.

The differences between Al, Rb and Silt normalised baseline models and the Fe baseline model for Co and Pb suggest there is either an anthropogenic source of Fe, Co and Pb to the Mudflats, or there is a biogeochemical or hydrodynamic process leading to the enrichment of these metals in sediment in this part of the harbour. The Bhizolo Canal connects to the Mudflats and is a potential source of effluents from adjacent aluminium smelters, but it seems unlikely this will be limited largely to Fe, Co and Pb, but rather to the introduction of Al. A more likely explanation is that redox processes resulted in an inhomogeneous Fe distribution (Demory et al. 2005; Brunschön et al. 2010). Support for this hypothesis is provided by Mn, which is also a redox-sensitive metal (Haberzettl et al. 2007) and which is also identified as enriched in sediment on the Mudflats (data not reported here) using Al, Rb and Silt normalised baseline models.

The EF plots presented in Figs. 27.8–27.11 make it easy to visually compare the spatial extent of enrichment of sediment by various metals in Richards Bay Harbour

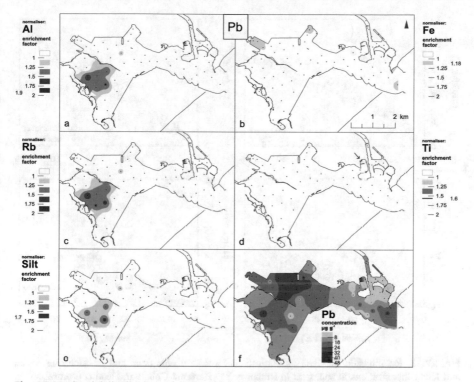

Fig. 27.11 Spatial distributions of the Lead (Pb) Baseline models normalised to (**a**) Al, (**b**) Fe, (**c**) Rb, (**d**) Ti and (**e**) Silt. The enrichment factors are plotted to the same scale; maximum values are shown next to the classification. For additional comparison, the concentration of Pb (**f**) in surface sediments of Richards Bay Harbour is plotted

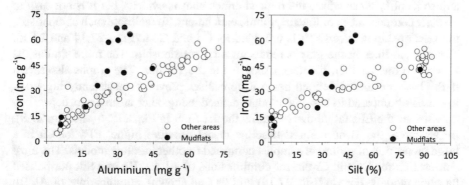

Fig. 27.12 Relationship between the aluminium concentration and silt fraction and co-occurring iron concentrations in sediment in Richards Bay Harbour. Iron concentrations in sediment sampled on the Mudflats are highlighted (black filled dots)

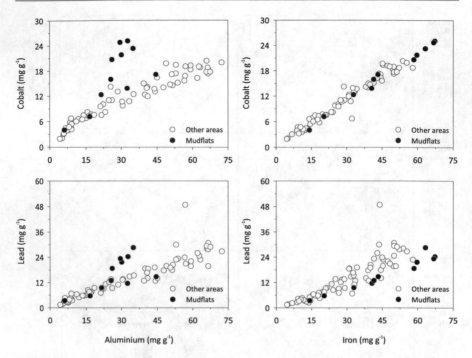

Fig. 27.13 Relationship between aluminium and iron concentrations and co-occurring cobalt and lead concentrations in sediment in Richards Bay Harbour. Cobalt and lead concentrations in sediment sampled on the Mudflats are highlighted (black filled dots). Note how the anomalously high cobalt and lead concentrations in the aluminium normalised plots are shifted to the right in the iron normalised plots

and to identify areas where the highest enrichment is evident, but it is not easy to visually compare areas of low magnitude enrichment. To facilitate such comparison, the relationship between EFs is provided for Cu and Co in Figs. 27.14 and 27.15. The oblique lines in the graphs represent a 1:1 relationship. The more similar the EFs are to one another, the nearer they fall to the 1:1 line. The graphs also reveal if the baseline model defined using a normaliser provides consistently higher or lower EFs compared to baseline models defined using other normalisers depending on whether the EFs fall above or below the 1:1 line. In Fig. 27.14 (and Fig. 27.8), for example, the Ti normalised baseline model provides higher EFs for Cu in a large proportion of sediment samples compared to other baseline models. The most similar EFs for Cu and Cr are for combinations of Al, Fe, Rb and Silt normalised baseline models. For Co (Fig. 27.15) and Pb (not shown), combinations of Al, Rb and Silt normalised baseline models provide similar EFs, but these differ from EFs computed using Fe and Ti normalised models. The differences for Co and Pb reflect the influence of the high Fe, Co and Pb concentrations in sediment on the Mudflats on baseline models defined using Al, Rb and Silt (Figs. 27.12 and 27.13). Thus, while these sediment samples are included in the baseline models for Co and Pb, the

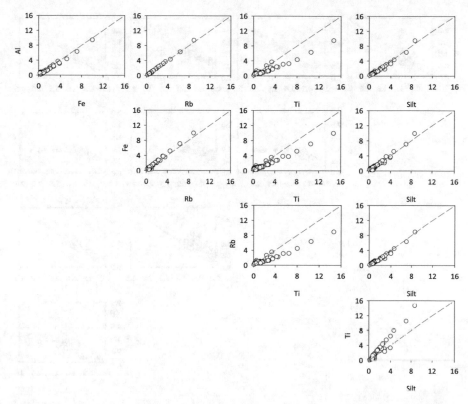

Fig. 27.14 Relationship between copper enrichment factors computed using baseline models defined using different normalisers. The diagonal dashed line represents a 1:1 line

Cu and Cr concentrations fall below the lower prediction limit of the Fe normalised models. They do not thus exert a similarly pronounced influence on the baseline model parameters.

A random choice of any of the normalisers considered in this study will thus provide a different understanding on the spatial extent and magnitude of enrichment. The differences will vary depending on the chosen normaliser, but are small when Al, Rb and Silt are used as normalisers. If only Fe was considered for normalisation, then this would have provided a different understanding on metal enrichment of sediment in some parts of the harbour compared to that provided by Al, Rb and Silt normalised baseline models. Ti provides a very different understanding of enrichment for most metals compared to other normalisers, reflecting its unsuitability as a normaliser in the case of Richards Bay Harbour.

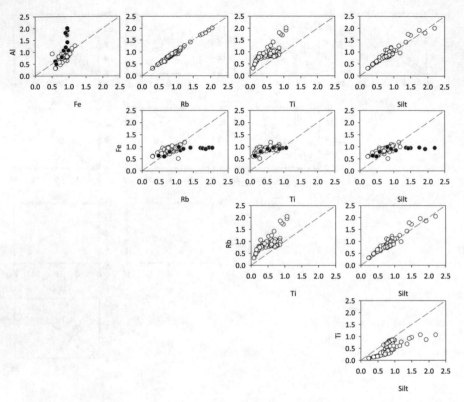

Fig. 27.15 Relationship between cobalt enrichment factors computed using baseline models defined using different normalisers. The diagonal dashed line represents a 1:1 line. Black symbols represent enrichment factors for sediment sampled on the Mudflats

27.3.4 Comparison to Guidelines

A common approach to estimating the toxicological significance of metal concentrations in sediment is to compare them to sediment quality guidelines. A misconception is that sediment quality guidelines indicate the onset of metal contamination of sediment. Although this is usually true for metal concentrations that exceed a guideline, depending on how conservative the guideline is, the sediment might be contaminated by a metal at concentrations well below the guideline. As examples, spatial trends for Cu and Cr concentrations in sediment in Richards Bay Harbour that exceed sediment quality guidelines used by the Department of Environmental Affairs (2012) to manage dredged material in South Africa are included in Figs. 27.8 and 27.9. There are three guidelines, known as the Warning Level, Level I and Level II. The Warning Level provides a warning of incipient metal contamination but is not used for decision-making. Sediment with metals at a concentration below the Level I is considered to pose a low risk and is suitable for open water disposal. Sediment with metals at a concentration between the Level I and Level II is considered cause for concern, with the degree of concern increasing as the concentrations approach the Level II. Sediment with metals at a concentration exceeding the Level II is considered to pose a high risk and is unsuitable for open water disposal unless other evidence (e.g. toxicity testing) shows the metals are not toxic to sediment-dwelling organisms.

There are large differences in the spatial extent of sediment identified as Cu enriched by baseline models defined using different normalisers and the area of sediment with Cu at a concentration exceeding the guidelines (Fig. 27.8). The area of maximum intensity (either concentration or EF) is confined to the 600 series berth basin. However, the area defined by EFs >1 extends into the 700 berth basin (Terminal Front) and to other areas, such as the Small Craft Harbour.

The difference in the spatial extent of sediment identified as metal enriched using the baseline models and the spatial extent of sediment with metals at a concentration exceeding Action List guidelines is often pronounced. This alludes to the guidelines perhaps being under- or over-protective, noting that the guidelines were not derived using empirical data from South African coastal waters but rather were adopted from guidelines used in North America. The guidelines might not, therefore, be appropriate to (all) parts of the South African coastline. Taking Cu as an example, large areas of sediment in the harbour are identified as enriched by this metal, yet the area where Cu concentrations exceed sediment quality guidelines is quite small. The use of the guidelines thus fails to identify sediment that is quite significantly enriched by Cu as potentially problematic, and alludes to management challenges larger than those alluded to by using the guidelines alone. This may mean the Cu guidelines are under-protective, that is the guideline concentrations are too high for sediment in Richards Bay Harbour.

27.4   Conclusion

This study compares the effectiveness of different normalisers on assorted metal in sediment in Richards Bay Harbour. The aim was to define metal concentration baselines and to identify areas of the harbour that are currently metal enriched, to aid in the tracking of changes in metal enrichment of sediment over time. The baseline models defined using some normalisers provide a powerful tool for this purpose, compensating for grain size influences on natural metal concentrations and reducing subjectivity in deciding if sediment is metal enriched. The models are equally effective in identifying changes in metal concentrations that may result from anthropogenic metal inputs and those that might arise from future climate changes, such as the introduction of increased metal and sediment loads to coastal waters associated with changes in precipitation patterns.

The use of different normalisers provides a different understanding on metal enrichment of sediment in Richards Bay Harbour. Despite the import of alumina and the presence of two large aluminium smelters near Richards Bay harbour, this does not appear to result in a significant Al contamination of sediment in the harbour affecting its use as a normaliser. The EF trends for metals in Richards Bay

Harbour, including those not discussed here, are most similar when Al, Rb or Silt are used as normalisers. The EF trends using Fe as the normaliser are similar to the trends for most other metals normalised by Al, Rb and Silt. However, the (potential hydrodynamic or redox-induced) enhancement of Co, Pb and Mn concentrations in sediment across part of the Mudflats influences the Fe normalised baseline model parameters and results in differences in EF trends for these metals in this part of the harbour. Ti is not considered an effective normaliser of metal concentrations in sediment of Richards Bay Harbour, as it is not correlated with the fine-grained fraction of sediment or other metal concentrations. However, if contamination is of a very high magnitude, contaminated areas will be identified using different normalisers, regardless of the effectiveness of the normalisers. The comparison of baseline models to Action List guidelines shows that baseline models reveal a wider area of contamination (e.g. Cu), thus leading to a better understanding of the system's state.

The findings of this study highlight the need to investigate the utility of different normalisers before a decision is made on the most effective one. Different normalisers should not be expected to provide precisely the same understanding on metal enrichment of sediment, even if this might be an aspirational outcome. Small differences in sample processing and analysis, and biogeochemical processes in sediment, will lead to small differences in the identification of enrichment, especially if it is of a low magnitude. Further, baseline models should not be used as precise boundaries, to clearly delineate enriched from unenriched sediment. Certain normalisers might prove more effective for certain metals than others. This does not imply different normalisers should be used for different metals, but by investigating the utility of different normalisers, anomalies in relationships for metals that allude to important features in the environment that need to be considered may be identified. The implication is that more than one, and preferably several potential normalisers should be analysed in studies focussing on metals in sediment if the aim is to identify enrichment that reflects contamination. Although the inclusion of several potential normalisers has financial implications from an analytical perspective, the ecological or management implications based on conclusions reached using a single (faulty) normaliser that might over- or underestimate enrichment may be significant and warrants such investment.

Acknowledgements This work was supported by the German Federal Ministry of Education and Research [grant number: 03F0798C] and is part of project TRACES (Tracing Human and Climate impact in South Africa) within the SPACES II Program (Science Partnerships for the Assessment of Complex Earth System Processes). The field work support by Andrew Green (University of KwaZulu-Natal), Peter Frenzel, Olga Gildeeva (both Friedrich-Schiller-University Jena) and George Best is gratefully acknowledged. We also thank Tammo Meyer (University of Greifswald) for providing help for sample digestion and Dirk Merten (Friedrich-Schiller-University Jena) for ICP-MS and -OES analyses.

References

Amorosi A, Sammartino I, Tateo F (2007) Evolution patterns of glaucony maturity: a mineralogical and geochemical approach. Deep-Sea Res II Top Stud Oceanogr 54:1364–1374. https://doi.org/10.1016/j.dsr2.2007.04.006

Balls PW, Hull S, Miller BS, Pirie JM, Proctor W (1997) Trace metal in Scottish estuarine and coastal sediments. Mar Pollut Bull 34:42–50. https://doi.org/10.1016/S0025-326X(96)00056-2

Begg G (1978) The estuaries of Natal: a resource inventory report to the Natal Town and Regional Planning Commission conducted under the auspices of the Oceanographic Research Institute, Durban. Natal Town and Regional Planning Report, pp 1–657

Birch GF, Snowdon RT (2004) The use of size-normalisation techniques in interpretation of soil contaminant distributions. Water Air Soil Pollut 157:1–12. https://doi.org/10.1023/B:WATE.0000038854.02927.1f

Birch GF, Lee J-H, Tanner E, Fortune J, Munksgaard N, Whitehead J, Coughanowr C, Agius J, Chrispijn J, Taylor U, Wells F, Bellas J, Besada V, Viñas L, Soares-Gomes A, Cordeiro RC, Machado W, Santelli RE, Vaughan M, Cameron M, Brooks P, Crowe T, Ponti M, Airoldi L, Guerra R, Puente A, Gómez AG, Zhou GJ, Leung KMY, Steinberg P (2020) Sediment metal enrichment and ecological risk assessment of ten ports and estuaries in the World Harbours Project. Mar Pollut Bull 155:111129. https://doi.org/10.1016/j.marpolbul.2020.111129

Botha GA (2018) Lithostratigraphy of the late Cenozoic Maputaland Group. S Afr J Geol 121:95–108. https://doi.org/10.25131/sajg.121.0007

Brunschön C, Haberzettl T, Behling H (2010) High-resolution studies on vegetation succession, hydrological variations, anthropogenic impact and genesis of a subrecent lake in southern Ecuador. Veg Hist Archaeobotany 19:191–206. https://doi.org/10.1007/s00334-010-0236-4

Chapman PM, Wang F (2001) Assessing sediment contamination in estuaries. Environ Toxicol Chem 20:3–22. https://doi.org/10.1002/etc.5620200102

Coynel A, Gorse L, Curti C, Schafer J, Grosbois C, Morelli G, Ducassou E, Blanc G, Maillet GM, Mojtahid M (2016) Spatial distribution of trace elements in the surface sediments of a major European estuary (Loire Estuary, France): source identification and evaluation of anthropogenic contribution. J Sea Res 118:77–91. https://doi.org/10.1016/j.seares.2016.08.005

Cuven S, Francus P, Lamoureux SF (2010) Estimation of grain size variability with micro X-ray fluorescence in laminated lacustrine sediments, Cape Bounty, Canadian High Arctic. J Paleolimnol 44:803–817. https://doi.org/10.1007/s10933-010-9453-1

Daskalakis KD, O'Connor TP (1995) Normalization and elemental sediment contamination in the coastal United States. Environ Sci Technol 29:470–477

Demory F, Oberhänsli H, Nowaczyk NR, Gottschalk M, Wirth R, Naumann R (2005) Detrital input and early diagenesis in sediments from Lake Baikal revealed by rock magnetism. Glob Planet Chang 46(1–4):145–166. https://doi.org/10.1016/j.gloplacha.2004.11.010

Department of Environmental Affairs (2012) National Environmental Management: Integrated Coastal management Act, 2008 (Act No. 24 of 2008). National Action List for the screening of dredged material proposed for marine disposal in term of section 73 of the National Environmental Management: Integrated Coastal management Act, 2008 (Act No. 24 of 2008). Government Gazette, pp 6–9

Dladla NN, Green AN, Cooper J, Mehlhorn P, Haberzettl T (2021) Bayhead delta evolution in the context of late Quaternary and Holocene sea-level change, Richards Bay, South Africa. Mar Geol 441:106608. https://doi.org/10.1016/j.margeo.2021.106608

Du Laing G, Vandecasteele B, de Grauwe P, Moors W, Lesage E, Meers E, Tack FMG, Verloo MG (2007) Factors affecting metal concentrations in the upper sediment layer of intertidal reedbeds along the river Scheldt. J Environ Monit 9:449–455. https://doi.org/10.1039/B618772B

Förstner U (1989) Contaminated sediments. Springer, Berlin

Grant A, Middleton R (1998) Contaminants in sediments: using robust regression for grain-size normalization. Estuaries 21:197. https://doi.org/10.2307/1352468

Greenfield R, Wepener V, Degger N, Brink K (2011) Richards Bay Harbour: metal exposure monitoring over the last 34 years. Mar Pollut Bull 62:1926–1931. https://doi.org/10.1016/j.marpolbul.2011.04.026

Haberzettl T, Corbella H, Fey M, Janssen S, Lücke A, Mayr C, Ohlendorf C, Schäbitz F, Schleser GH, Wille M, Wulf S, Zolitschka B (2007) Lateglacial and Holocene wet—dry cycles in southern Patagonia: chronology, sedimentology and geochemistry of a lacustrine record from Laguna Potrok Aike, Argentina. The Holocene 17:297–310. https://doi.org/10.1177/0959683607076437

Haberzettl T, Kirsten KL, Kasper T, Franz S, Reinwarth B, Baade J, Daut G, Meadows ME, Su Y, Mäusbacher R (2019) Using 210Pb-data and paleomagnetic secular variations to date anthropogenic impact on a lake system in the Western Cape, South Africa. Quat Geochronol 51:53–63. https://doi.org/10.1016/j.quageo.2018.12.004

Hanson PJ, Evans DW, Colby DR, Zdanowicz VS (1993) Assessment of elemental contamination in estuarine and coastal environments based on geochemical and statistical modeling of sediments. Mar Environ Res 36:237–266. https://doi.org/10.1016/0141-1136(93)90091-D

Horowitz AJ (1991) A primer on sediment-trace element chemistry. Lewis Publishers, Chelsea

Horowitz AJ, Elrick KA (1987) The relation of stream sediment surface area, grain size and composition to trace element chemistry. Appl Geochem 2:437–451. https://doi.org/10.1016/0883-2927(87)90027-8

Jung H-S, Lim D, Xu Z, Kang J-H (2014) Quantitative compensation of grain-size effects in elemental concentration: a Korean coastal sediments case study. Estuar Coast Shelf Sci 151:69–77. https://doi.org/10.1016/j.ecss.2014.09.024

Jung H, Lim D, Xu Z, Jeong K (2016) Secondary grain-size effects on Li and Cs concentrations and appropriate normalization procedures for coastal sediments. Estuar Coast Shelf Sci 175:57–61. https://doi.org/10.1016/j.ecss.2016.03.028

Kelbe B (2010) Hydrology and water resources of the Richards Bay EMF area

Kersten M, Smedes F (2002) Normalization procedures for sediment contaminants in spatial and temporal trend monitoring. J Environ Monit 4:109–115. https://doi.org/10.1039/b108102k

Koigoora S, Ahmad I, Pallela R, Janapala VR (2013) Spatial variation of potentially toxic elements in different grain size fractions of marine sediments from Gulf of Mannar, India. Environ Monit Assess 185:7581–7589. https://doi.org/10.1007/s10661-013-3120-8

Krumgalz BS, Fainshtein G, Cohen A (1992) Grain size effect on anthropogenic trace metal and organic matter distribution in marine sediments. Sci Total Environ 116:15–30. https://doi.org/10.1016/0048-9697(92)90362-V

Kylander ME, Ampel L, Wohlfarth B, Veres D (2011) High-resolution X-ray fluorescence core scanning analysis of Les Echets (France) sedimentary sequence: new insights from chemical proxies. J Quat Sci 26:109–117. https://doi.org/10.1002/jqs.1438

Larrose A, Coynel A, Schäfer J, Blanc G, Massé L, Maneux E (2010) Assessing the current state of the Gironde Estuary by mapping priority contaminant distribution and risk potential in surface sediment. Appl Geochem 25:1912–1923. https://doi.org/10.1016/j.apgeochem.2010.10.007

Liang J, Liu J, Xu G, Chen B (2019) Distribution and transport of heavy metals in surface sediments of the Zhejiang nearshore area, East China Sea: sedimentary environmental effects. Mar Pollut Bull 146:542–551. https://doi.org/10.1016/j.marpolbul.2019.07.001

Loring DH (1991) Normalization of heavy-metal data from estuarine and coastal sediments. ICES J Mar Sci 48:101–115. https://doi.org/10.1093/icesjms/48.1.101

Luoma SN (1990) Processes affecting metal concentrations in estuarine and coastal marine sediments. In: Furness RW, Rainbow PS (eds) Heavy metals in the marine environment. CRC Press, Boca Raton

Matys Grygar T, Popelka J (2016) Revisiting geochemical methods of distinguishing natural concentrations and pollution by risk elements in fluvial sediments. J Geochem Explor 170:39–57. https://doi.org/10.1016/j.gexplo.2016.08.003

Matys Grygar T, Nováková T, Bábek O, Elznicová J, Vadinová N (2013) Robust assessment of moderate heavy metal contamination levels in floodplain sediments: a case study on the

Jizera River, Czech Republic. Sci Total Environ 452–453:233–245. https://doi.org/10.1016/j.scitotenv.2013.02.085

Matys Grygar T, Elznicová J, Bábek O, Hošek M, Engel Z, Kiss T (2014) Obtaining isochrones from pollution signals in a fluvial sediment record: a case study in a uranium-polluted floodplain of the Ploučnice River, Czech Republic. Appl Geochem 48:1–15. https://doi.org/10.1016/j.apgeochem.2014.06.021

Mehlhorn P, Viehberg F, Kirsten K, Newman B, Frenzel P, Gildeeva O, Green A, Hahn A, Haberzettl T (2021) Spatial distribution and consequences of contaminants in harbour sediments – a case study from Richards Bay Harbour, South Africa. Mar Pollut Bull 172:112764. https://doi.org/10.1016/j.marpolbul.2021.112764

Nel EL, Hill TR, Goodenough C (2007) Multi-stakeholder driven local economic development: reflections on the experience of Richards Bay and the uMhlathuze municipality. Urban Forum 18:31–47. https://doi.org/10.1007/s12132-007-9004-7

Newman BK, Watling RJ (2007) Definition of baseline metal concentrations for assessing metal enrichment of sediment from the south-eastern Cape coastline of South Africa. Water SA 33. https://doi.org/10.4314/wsa.v33i5.184089

Ohlendorf C, Fey M, Massaferro J, Haberzettl T, Laprida C, Lücke A, Maidana NI, Mayr C, Oehlerich M, Ramón Mercau J, Wille M, Corbella H, St-Onge G, Schäbitz F, Zolitschka B (2014) Late Holocene hydrology inferred from lacustrine sediments of Laguna Cháltel (southeastern Argentina). Palaeogeogr Palaeoclimatol Palaeoecol 411:229–248. https://doi.org/10.1016/j.palaeo.2014.06.030

Rubio B, Nombela M, Vilas F (2000) Geochemistry of major and trace elements in sediments of the Ria de Vigo (NW Spain): an assessment of metal pollution. Mar Pollut Bull 40:968–980. https://doi.org/10.1016/S0025-326X(00)00039-4

Sabadini-Santos E, Knoppers BA, Oliveira EP, Leipe T, Santelli RE (2009) Regional geochemical baselines for sedimentary metals of the tropical São Francisco estuary, NE-Brazil. Mar Pollut Bull 58:601–606

Santos IR, Silva-Filho EV, Schaefer CEGR, Albuquerque-Filho MR, Campos LS (2005) Heavy metal contamination in coastal sediments and soils near the Brazilian Antarctic Station, King George Island. Mar Pollut Bull 50(2):185–194. https://doi.org/10.1016/j.marpolbul.2004.10.009

Saulnier I, Mucci A (2000) Trace metal remobilization following the resuspension of estuarine sediments: Saguenay Fjord, Canada. Appl Geochem 15:191–210. https://doi.org/10.1016/S0883-2927(99)00034-7

Schropp SJ, Windom HL (1988) A guide to the interpretation of metal concentrations in estuarine sediments coastal zone management section. Florida Department of Environmental Regulation, Florida

Schropp SJ, Lewis FG, Windom HL, Ryan JD, Calder FD, Burney LC (1990) Interpretation of metal concentrations in estuarine sediments of Florida using aluminum as a reference element. Estuaries 13:227. https://doi.org/10.2307/1351913

Sigg L, Stumm W (1996) Aquatische Chemie: eine Einführung in die Chemie wässriger Lösungen und natürlicher Gewässer. vdf, Hochschulverl.-AG an d. ETH Zürich, Zürich

Szava-Kovats RC (2008) Grain-size normalization as a tool to assess contamination in marine sediments: is the <63 micron fraction fine enough? Mar Pollut Bull 56:629–632. https://doi.org/10.1016/j.marpolbul.2008.01.017

Taylor SR, McLennan SM (1981) The composition and evolution of the continental crust: rare earth element evidence from sedimentary rocks. Phil Trans R Soc A 301:381–399. https://doi.org/10.1098/rsta.1981.0119

Thomas C, Bendell-Young L (1999) The significance of diagenesis versus riverine input in contributing to the sediment geochemical matrix of iron and manganese in an intertidal region. Estuar Coast Shelf Sci 48:635–647. https://doi.org/10.1006/ecss.1998.0473

Tůmová Š, Hrubešová D, Vorm P, Hošek M, Grygar TM (2019) Common flaws in the analysis of river sediments polluted by risk elements and how to avoid them: case study in the Ploučnice

River system, Czech Republic. J Soils Sediments 19(4):2020–2033. https://doi.org/10.1007/s11368-018-2215-9

Watling RJ (1977) Trace metal distribution in the Wilderness-lakes. CSIR Special Report FIS 147

Weisberg SB, Wilson HT, Heimbuch DG, Windom HL, Summers JK (2000) Comparison of sediment metal:aluminum relationships between the eastern and gulf coasts of the United States. Environ Monit Assess 61:373–385. https://doi.org/10.1023/A:1006113631027

Wepener V, Vermeulen LA (2005) A note on the concentrations and bioavailability of selected metals in sediments of Richards Bay Harbour, South Africa. Water SA 31(4):589–596. https://doi.org/10.4314/wsa.v31i4.5149

Windom HL, Schropp SJ, Calder FD, Ryan JD, Smith RG Jr, Burney LC, Lewis FG, Rawlinson CH (1989) Natural trace metal concentrations in estuarine and coastal marine sediments of the southeastern United States. Environ Sci Technol 23:314–320

Woods AM, Lloyd JM, Zong Y, Brodie CR (2012) Spatial mapping of Pearl River Estuary surface sediment geochemistry: influence of data analysis on environmental interpretation. Estuar Coast Shelf Sci 115:218–233. https://doi.org/10.1016/j.ecss.2012.09.005

Wündsch M, Biagioni S, Behling H, Reinwarth B, Franz S, Bierbaß P, Daut G, Mäusbacher R, Haberzettl T (2014) ENSO and monsoon variability during the past 1.5 kyr as reflected in sediments from Lake Kalimpaa, Central Sulawesi (Indonesia). The Holocene 24:1743–1756. https://doi.org/10.1177/0959683614551217

Zolitschka B, Fey M, Janssen S, Maidana NI, Mayr C, Wulf S, Haberzettl T, Corbella H, Lücke A, Ohlendorf C, Schäbitz F (2019) Southern Hemispheric Westerlies control sedimentary processes of Laguna Azul (south-eastern Patagonia, Argentina). The Holocene 29:403–420. https://doi.org/10.1177/0959683618816446

Catchment and Depositional Studies for the Reconstruction of Past Environmental Change in Southern Africa

28

Annette Hahn, Enno Schefuß, Nicole Burdanowitz, Hayley C. Cawthra, Jemma Finch, Tarryn Frankland, Andrew Green, Frank H. Neumann, and Matthias Zabel

Abstract

Terrestrial signals in marine sedimentary archives are often used for reconstructing past environments, vegetation and climate, as well as for determining sediment fluxes, pathways, and depositional sites and changes in erosional runoff. It is therefore important to understand the origin, transport, and depositional processes of the various terrestrial sedimentary components in a depositional system. In this chapter, we use examples from southern Africa to show how source-to-sink studies have led to a clearer interpretation of downcore proxy records. Twelve rivers in four river catchment areas of various scales and in distinct climatic settings and geological formations are included in this

A. Hahn (✉) · E. Schefuß · M. Zabel
MARUM – Center for Marine Environmental Sciences, University of Bremen, Bremen, Germany
e-mail: ahahn@marum.de

N. Burdanowitz
Institute for Geology, Universität Hamburg, Hamburg, Germany

H. C. Cawthra
Geophysics and Remote Sensing Unit, Council for Geoscience, Bellville, South Africa

African Centre for Coastal Palaeoscience, Nelson Mandela University, Port Elizabeth, South Africa

J. Finch · T. Frankland
Discipline of Geography, School of Agricultural, Earth and Environmental Sciences, University of KwaZulu-Natal, Pietermaritzburg, South Africa

A. Green
Discipline of Geology, University of KwaZulu-Natal, Durban, South Africa

F. H. Neumann
Evolutionary Studies Institute, University of the Witwatersrand, Johannesburg, South Africa

© The Author(s) 2024
G. P. von Maltitz et al. (eds.), *Sustainability of Southern African Ecosystems under Global Change*, Ecological Studies 248,
https://doi.org/10.1007/978-3-031-10948-5_28

compilation. We also discuss studies from the current-swept South African east coast, the broad western and southern margins, and investigations from protected marine embayment settings. We consider a large suite of commonly used proxies (plant wax isotopes, elemental composition, and fossil pollen) as well as hydroacoustic surveying techniques (PARASOUND and multibeam bathymetric profiling). Sampling strategies and sample types that may be used in catchment analyses are discussed. Challenges and limitations of the above-mentioned approaches are outlined. In conclusion, we underline the importance of a thorough source-to-sink approach to paleo-environmental reconstructions using terrigenous proxies.

28.1 Introduction

Knowledge of natural, preindustrial climate change and associated processes is essential for assessing current and projected environmental changes. Only when the interrelationships of all essential influencing factors are sufficiently understood will we be able to recognize environmental changes at an early stage, to assess them, and to react to them with sustainable (utilization) concepts. To achieve this goal, sediments as climate archives are of inestimable importance. The application of different methodological and analytical approaches is a prerequisite for the reliable interpretation of the multitude of proxies parameters and their information. The case studies presented here certainly do not cover the entire spectrum of comparable studies in southern Africa, but they represent successful model examples of innovative paleoclimate research. The reconstruction of past environments is critical to the objectives of several projects within the SPACES program. Strong links exist, for example, between the science presented in Chaps. 1 and 2 where a sound understanding of past environments is considered necessary to anticipate present and future response to climate change. Additionally, a detailed understanding of vegetation and hydrologic change during the Holocene can be considered as critical reference information for conservation and land use management practices (Chap. 12). Marine and terrestrial sediment cores are used to understand past environmental conditions from proxy data sources, including a large variety of physical, chemical, and biological indicators. These proxies are used to infer past fluctuations in vegetation, precipitation and weathering regimes, sediment fluxes, pathways, and depositional sites as well as changes in erosional runoff. Some of these indicators are measured on allochthonous sediment components, that is, components that were transported to the site by water or wind. At offshore coring sites located close to a river mouth or leeward of areas with high winds and sparse vegetation cover, sediments may be almost entirely of allochthonous nature. The interpretation of terrigenous (terrestrial) climatic signals measured on these components requires a thorough understanding of their source, or provenance, as well as of their transport pathways to final deposition. A complex factor in downcore analysis of proxies measured on terrestrial-derived material is that different components (such as pollen and spores, charred fragments, plant waxes, minerals, and rock fragments) may have different source regions, as well as transport pathways. Within a particular catchment, the provenance of inorganic and organic material is likely to be distinct, especially if there is a large variability in relief and vegetation cover. This has been shown in overview studies of riverine transport (Leithold et al. 2016) as well as in individual catchment studies (e.g., Bouchez et al. 2014; Häggi et al. 2016; Galy et al. 2011; Hahn et al. 2016; Hemingway et al. 2016; Herrmann et al. 2016; Zhao et al. 2015). Furthermore, the transport pathways and deposition of the various terrigenous fractions will be a function of the characteristics of source areas that govern the size, shape, density, and stability of sedimentary particles as shown by several on-shelf sediment distribution studies (e.g., Petschick et al. 1996; Rogers and Rau 2006). Sediment depocenters on the shelf are also in many ways controlled by sea-level history and ocean currents as they play a large role in fractionating grain size, grain shape, and composition (Green and Mackay 2016; Flemming and Hay 1988).

Source-to-sink studies shed light on the catchment processes and depositional conditions that influence proxy signatures. They facilitate reliable paleoclimatic interpretation of proxies measured on allochthonous sediment components. The diversity of the climatic settings, types of rivers, accompanying wetland systems, and depositional environments in southern Africa are immense and the work presented here has attempted to target different climatic and environmental zones within this (Fig. 28.1 shows simplified climatic and oceanic circulation systems in southern Africa). The case studies presented from southern Africa illustrate how source-to-sink studies can be applied in a variety of paleo-environmental and paleo-climatic contexts.

28.2   Depositional Studies/Stratigraphy of Marine Environments at the Sink

28.2.1 Description of Mapping Strategies

A core or grab sample provides a window of insight into that specific site location, but context is necessary to construct robust interpretations. Hydroacoustic surveys of the seafloor and coastal water bodies have been carried out in both ocean and coastal water body settings. The purpose of this work was to (1) provide suitable geological context around the deposition of sediments that were cored by allowing for a holistic "picture" of the sedimentary environment and (2) mapping was carried out before coring to inform an appropriate site for sampling. In all cases, these water bodies were mapped with subbottom profilers, and specifically, offshore voyage M123 traversed the continental margin from Walvis Bay to the Delagoa Bight and continuous coverage of multibeam echosounder data was acquired for depth, and PARASOUND for sub-seafloor context. In the coastal water bodies, boomer subbottom profiles were collected. Around each proposed site, numerous

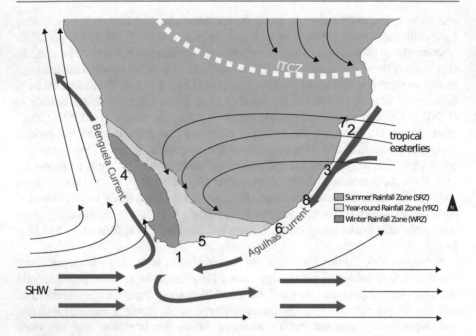

Fig. 28.1 Schematic overview of the main components of the oceanic (thick arrows) and atmospheric (thin arrows) circulation over southern Africa (modified from Truc et al. 2013). Southern Hemispheric Westerlies (SHW) and Intertropical Convergence Zone (ITC) are shown in their summer position. The numbers indicate the approximate location of the case studies in this chapter, more detailed maps are provided in the following paragraphs. Figure modified from Hahn et al. (2017)

profiles were specifically collected to contextualize appropriate coring sites prior to sampling.

Boomers are low-frequency seismic profilers, characterized by a broadband frequency spectrum and high peak intensity. Seismic energy is derived from a bank of capacitors, discharged into an electromagnetic coil and providing a clear pulse and resolvability of seismic return. Boomer subbottom profilers are considered medium-penetrating seismic systems and allow an insight into upper continental shelf stratigraphy.

The PARASOUND method of subbottom profiling works on two frequencies (18 kHz and a secondary variable frequency of 20.5–23.5 kHz). The superimposition of these variable frequencies results in a "nonlinear parametric effect" that reduces the area of reflecting the surface compared to conventional subbottom profilers (Hempel et al. 1994).

Echosounders determine the travel time of an acoustic pulse by detecting the sharp leading edge of the return echo, providing a depth to seafloor (Mayer and Hughes-Clarke 1995). Acquisition of sonar swath bathymetry using a multibeam echosounder encompasses the principle of a three-dimensional fan shape of acoustic energy, subjected to pitch, roll, yaw, and vessel oscillation. The array transmits pulses triggered at known intervals to insonify an area of seafloor normal to the ship's track. High-resolution motion reference units compensate for the inherent motion in the data acquired. The angular coverage sector of beam angles varies with depth, allowing accurate bathymetric determination from a relatively small number of passes across a deeper area.

28.2.2 Challenges and Limitations

For voyage M123 (Zabel 2016), areas on the seafloor (the continental shelf and shelf-edge environments) were located that preserve terrestrial deposits, such as fluvial, lacustrine, and mudbelt depo-centers. Reconnaissance sites were broadly defined using legacy data of Dingle et al. (1987) and Birch et al. (1986) and past and ongoing work of South African marine geologists. These maps and existing data allowed the sites to be constrained according to surficial textures and sediment characteristics, and proved particularly beneficial in identifying the loci of muddy sediments on the shelf.

28.2.3 Examples of Southern African Hydroacoustic Studies That Have Been Used for Source-to-Sink Understanding

28.2.3.1 Case Study 1: Southernmost South African Shelf

The inner to mid continental shelf of the Agulhas Bank, which forms part of the Paleo-Agulhas Plain when it was exposed as a terrestrial landscape in the past, is covered with Pleistocene deposits. The wide lateral extension of these remnant deposits is the expression of a flat underlying substrate, which differs markedly to the adjacent onshore area (Cawthra et al. 2018, 2020). Six sediment vibrocores were obtained across the Paleo-Agulhas Plain in submerged environments of estuaries, lakes, and river floodplains (GeoB18305-3, GeoB18306-1, GeoB18307-2, GeoB18308-1, GeoB20628-1 and GeoB20629-1) that revealed sediments preserved since the Last Glacial Maximum. A seismic stratigraphic framework was developed from the inner to mid-shelf between the Breede River in the West and Plettenberg Bay in the East, for the RAiN project, and this work elucidated 20 different Quaternary units within 2 depositional sequences (Fig. 28.2). Incised paleo-river channels were mapped and cored deposits also mapped seismically from estuarine, lacustrine, and fluvial systems are grouped to represent the lower floodplain (Fig. 28.3). The most pervasive stratigraphic pattern in these shelf deposits was left behind during the sea-level drop from 125,000 years ago to 20,000 years ago, or the penultimate cycle of warming into the Last Glacial Maximum.

28.2.3.2 Case Study 2: Limpopo Shelf

The Limpopo shelf of the Delagoa Bight comprises the extensive submerged and asymmetrical wave-dominated delta of the combined Matola, Incomati, Lusutfu, and Limpopo rivers. Hydroacoustic studies in the region (e.g., Dyer et al. 2021) gave

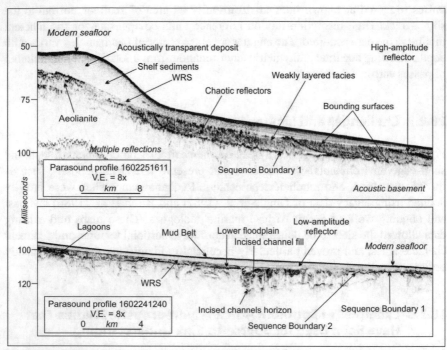

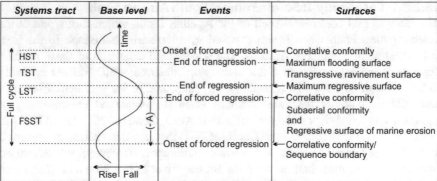

Fig. 28.2 (Top) Representative subbottom profiles to demonstrate how the bounding horizons, e.g., wave ravinement surface ("WRS") and sequence boundaries, as well as seismic units and facies were delineated in this study. (Bottom) Schematic showing the basis for definition of systems tracts (–A): negative accommodation space

further background information on sediment depositional processes to correctly interpret core stratigraphy. The seismic stratigraphy and architecture of the shelf was revealed by PARASOUND subbottom profiling to comprise two submerged, wave-dominated deltas sandwiched between three associated paleo-shorelines, all draped by sediment that had accumulated after the postglacial transgression. Given the complexity of the underlying submerged delta stratigraphy, especially the

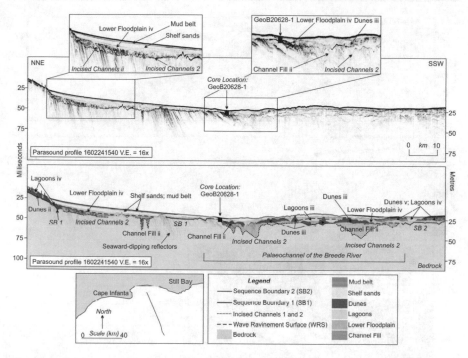

Fig. 28.3 Subbottom profile intercepting the paleochannel of the Breede River near Cape Infanta. The site of core GeoB20628-1 is shown

convoluted delta top deposits that comprise most of the proximal areas of the shelf, the marine sediment cores were located in the most stratigraphically consistent areas of the postglacial cover sediment. These included at the shelf edge, amid high frequency and continuous well-layered reflections, and behind aeolianite pinnacles on the mid shelf, where post glacial transgressive erosion was minimized and the contemporary mud clinoform of the rivers could be preserved (Fig. 28.4).

28.2.3.3 Case Study 3: Durban Shelf

Unlike the above-mentioned examples, the Durban shelf is a current-swept and mostly bedrock-controlled feature where Holocene sediments are typically only preserved as the infill of old river courses (Green et al. 2013a). Despite the scarcity of sediment, previous multibeam bathymetric surveying revealed a series of submerged lagoonal deposits on the seabed that appeared to comprise muddy infilling material in line with the contemporary lagoons of the southeast African margin (Green et al. 2013b, 2014). The detailed hydroacoustic surveying and seismic stratigraphy of these lagoonal features was highlighted by Pretorius et al. (2016), who identified multiple aeolianite and beachrock shorelines preserved on the seabed, behind which extensive Holocene sediment had accumulated (Fig. 28.5). Much of the deeper Holocene sediments on the shelf were considered to be relict portions of a detached shoreface, driven by abrupt rises in sea level (Pretorius et

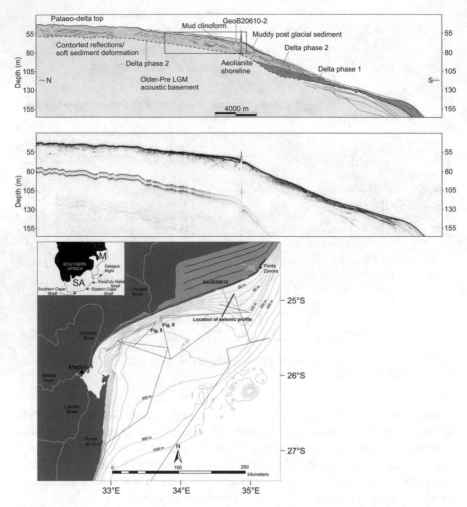

Fig. 28.4 Seismic stratigraphy of the transgressive deposits of the Limpopo shelf. Two phases of submerged deltas are evident, covered to landward by the prograding delta top. This is in turn draped by the modern mud clinoform of the area, which has dammed behind an aeolianite pinnacle (after Dyer et al. 2021)

al. 2016). The core sites were chosen on the basis of sediment thickness and their position within a large incised valley that had back-flooded to form the lagoons during the Younger Dryas period; the aim was to retrieve muddy material suitable as a climate archive spanning that period.

The cores retrieved comprised a series of alternating gravels and sands; however, with the aid of the seismic stratigraphy and stratigraphic architecture of the subbottom, it became apparent that these reflected a series of sediments that had been deposited during periods of intense marine storminess. The longer core, GeoB18303-2, was associated with wash-over into the back-barrier lagoon system

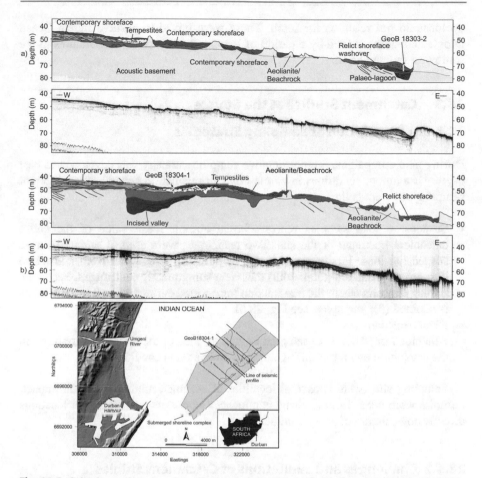

Fig. 28.5 Seismic stratigraphy of the Durban continental shelf. (**a**) Core location GeoB18303-2, which retrieved a mixture of shoreface and wash-over deposits, sheltered behind an aeolianite barrier. These overlie the paleo-lagoon complex imaged by Green et al. (2013a). (**b**) Core location of GeoB18304-1, which intersected a series of tempestites (hummocky reflectors) that adjoin the modern shoreface

during a period of storminess that spanned ca. 11,400–11,200 years ago (Green et al. 2022), consistent with the post Younger Dryas submergence of the lagoon system (Pretorius et al. 2016). The shallower core was located in the more proximal incised valley of the system, the Holocene cover deposits of which comprised a series of intercalated marine- and terrestrial-derived sediments (Green et al. 2022). In the subbottom profiles, these had a characteristic hummocky seismic appearance and were consequently interpreted as classic shelf tempestites, their ages spanning the period ca. 7000–4800 years ago. Green et al. (2022) linked this period of storminess to an unprecedented series of southward tracking tropical cyclones that made landfall in Durban, in contrast to the contemporary situation where tropical cyclones do not reach as far south. These were linked to positive Indian Ocean Dipole anomalies, driven by a period of increased sea surface temperatures in the southwest Indian Ocean.

28.3 Catchment Studies at the Source

28.3.1 A Description of Sampling Strategies

During field campaigns, a representative sampling strategy was developed to best capture catchment conditions in a time-efficient manner. In general, three different sample types were collected:

- Suspension load samples representing a current mean value of the entire catchment upstream of the site. Two techniques were used to sample riverine suspension load. Firstly, an approach of pumping river water from midstream into several canisters (total: 100l) that were subsequently centrifuged. Secondly, finding deposits where the fine fraction transported during higher flow conditions is retained (for examples, see Fig. 28.6).
- River bank samples.
- Paleoflood deposits were sampled to gain an idea of the temporal variability in the catchment and representing nonmodern analogue conditions.

Sampling sites were chosen at localities of assumed minimum human impact. Samples were taken from as many confluences as possible, as well as at locations directly downstream of every confluence.

28.3.2 Challenges and Limitations of Catchment Studies

The catchments of most rivers are diverse. Large river systems may traverse several climatic and vegetation zones, as well as a suite of geological formations (examples from southern Africa: Orange and Limpopo River systems). Small catchments can be equally diverse due to, for example, microclimates, the effects of large altitudinal differences, or localized wetland occurrences near the river channels. The variability throughout a catchment increases further when seasonal changes are considered. Capturing the total diversity of entire catchment areas with the number of samples that can be collected by a research team within a limited time frame of a research project is close to impossible. Catchment areas situated in more rural areas (e.g., parts of the Eastern Cape, South Africa) are often inaccessible via the road network, limiting the availability of sampling sites. The above-mentioned sampling strategies (Sect. 28.3.1) are employed to obtain the best representation of the temporal and spatial diversity in a catchment. Although sampling sites with obvious signs of human or animal (e.g., bioturbation due to hippopotamus or cattle trampling) disturbance are avoided during sampling, contamination of

Fig. 28.6 Examples of sediments collected from dried out puddles (**a,c**) and flood deposits (**b,d**) along the Orange River. Copyright Ralph Kreutz, MARUM

samples by human activity can never be excluded, with some catchments being heavily polluted especially near urban areas. Sites that are easily accessible to sampling are often used for common recreational activities such as fishing or boating. Depositional processes might therefore be disturbed and contaminants (oils, plastics, sewage) may be found in the samples. The downcore application of proxy interpretations based on catchment studies may be further distorted by situations when nonanalogue conditions occurred. The interpretations of proxy indicators that can be deduced from modern catchment studies is only valid if the conditions in the catchment remain comparable with the modern environment throughout time. However, vegetation and erosion processes within a catchment may change with large climatic shifts, for example, during glacial-interglacial transitions. During no-modern-analogue scenarios, changes in sediment source area have to be accounted for, and downcore proxy interpretation may have to be adjusted.

Info boxes on sediment proxies:

Palynology is the analysis of organic-walled microfossils, which comprise pollen and spores as well as algal cysts, fungal remains, dinoflagellate cysts, scolecodonts, and even the inner linings of foraminifera. Fossil pollen in particular offers one of the most widely used tools for understanding past environmental conditions, because pollen grains are morphologically diverse, abundant and generally well dispersed, and extremely resistant to decay. By counting and identifying pollen and other microfossils preserved in sediments, it becomes possible to reconstruct past environments including vegetation composition, climate shifts, changes in soil properties and hydrology, and sometimes also human disturbances through farming and pastoralism.

The *elemental composition* of sediment is used to gain insights into geological mechanisms and earth processes such as climate change, local and regional events (e.g., floods, landslides, storms), and anthropogenic changes (e.g., changing land use, pollution). A large number of element proxies have been recognized as important indicators of climate, weathering, and erosion conditions as well as sediment provenance.

Plant-waxes as vegetation and hydrological indicators. All terrestrial plants protect their leaves but also other plant surfaces with a surficial layer of wax (Eglinton and Hamilton 1967). These waxy compounds are made to resist, so can persist in the environment (e.g., soils, sediments) for long times after plant decay. In particular, n-alkanes as nonfunctionalized lipids are very refractory, preserving their distributional and isotope signatures up to millions of years (Schimmelmann et al. 2006).The concentrations of waxes in sedimentary archives depend on the terrestrial input, either due to higher plant coverage in the source area or stronger (fluvial, aeolian) transport. Different plants, however, produce differential amounts of waxes, complicating a quantitative approach (Garcin et al. 2014). As resistant compounds, they get relatively enriched during organic matter degradation, providing information on the degradation status of total organic material. Other degradation information can be obtained from their internal distribution. Plant synthesize preferentially even- (fatty acids, n-alcohols) or odd-numbered (n-

alkanes) waxes, a signal that diminishes with proceeding wax degradation (Bray and Evans 1961).Some additional information on the contributing vegetation might be obtained from the internal distributions of the waxes. Generally, grasses tend to produce longer-chain waxes than woody vegetation (Vogts et al. 2009). However, this information might be misleading as plants from specific environments can contribute very specific chain-lengths, such as sedges, for example, Cyperaceae in swamps and floodplains producing preferentially very long-chain waxes (Schefuß et al. 2011). In sedimentary archives, the terrestrial wax signal has to be disentangled from other, microbial and/or algal, contributions. Overall, wax distributions alone should thus only be interpreted with caution and such interpretations preferentially be based on calibration studies on wax distributions of regional plant types (Carr et al. 2014; Herrmann et al. 2016). On the other hand, the analysis of multiple wax components is able to provide information on the diversity of plant changes and environmental responses within an ecosystem (Hoetzel et al. 2013). More specific information on contributing photosynthetic plant types can be obtained from their compound-specific stable carbon (δ^{13}C) isotope compositions (Diefendorf and Freimuth 2017). A broad distinction can be made between waxes from C_3 or C_4 plants with waxes from C_4 vegetation being isotopically enriched in ^{13}C due to a more effective photosynthetic carbon (CO_2) fixation (Collins et al. 2011). Most tropical grasses are of C_4 type, while all trees and shrubs are C_3 plants. In this sense, the occurrence of C_4, that is, ^{13}C-enriched, waxes has been attributed to contributions from drier, grassy environments. However, also plants from specific environments, such as the Cyperaceae from swamps or salt-tolerant plants in saline settings, can be of C_4 type, complicating a straightforward interpretation of ^{13}C compositions alone (Schefuß et al. 2011). An additional complication are CAM plants, being able to switch carbon fixation mechanisms and thus producing variable and intermediate ^{13}C compositions, such as, for instance, in biomes along the west coast of south Africa (Boom et al. 2014). Plants also adjust their ^{13}C compositions due to environmental stress, such as water shortage. Under drought conditions, plants increase their water use efficiency, leading to a ^{13}C enrichment (Hou et al. 2007). Thus, ^{13}C compositions of plant waxes should not be interpreted alone but in conjunction with other parameters. Direct hydrologic information can be obtained through compound-specific hydrogen isotope (δD) analyses of waxes. The δD composition of waxes depends on the isotope composition of their meteoric water source, ultimately related to the isotopic composition of rainfall with an overprint of isotopic enrichment due to evaporation and/or plant transpiration in dry settings. Such isotopic enrichment due to evapo-transpiration has been detected in dry environments of South Africa (Herrmann et al. 2017). Additionally, a secondary dependency of δD composition of waxes is observed between different plant types (Sachse et al. 2012). Nonetheless, the δD composition of waxes provides a direct link to atmospheric isotope hydrology. Rainfall isotope effects such as the amount-, temperature-, altitude- and source-effects are all observed in specific settings in South Africa (Burdanowitz et al. 2018; Hahn et al. 2017, 2018, 2021; Miller et al. 2019, 2020). On longer time-scales (i.e., pre-Holocene) the ice volume effect additionally leads to an enrichment of isotopes in the hydrological cycle. For some of these effects, their contribution to the δD_{wax} signal can be accounted for (e.g., temperature, ice volume) or can be assumed to be constant (e.g., altitude) in specific settings. Others, that is, amount and source effect, are often complicated to disentangle, in particular, in southern Africa where several moisture sources overlay. A careful multiwax parameter and multiproxy (incl. pollen, inorganic geochemistry) analysis, however, can lead to meaningful interpretations and far-reaching insights into (paleo-)hydrological changes compared to single parameter analyses. Plant-wax analyses thus yield integrated information on broad-scale vegetation changes, organic matter degradation, and the eco-hydrological status of the entire ecosystem. Combined in a multiproxy approach with palynology, inorganic geochemistry, etc., plant-wax analyses are able to provide very specific and detailed insights on the (vegetation and hydrologic) status of modern ecosystems, transport pathways of terrestrial organic matter, climatic-driven vegetation changes, and atmospheric-driven hydrologic changes.

28.3.3 Examples of Southern African Source-to-Sink Studies That Have Been Used for Paleoclimatic Work

28.3.3.1 Case Study 4: Western South African Coast

The Orange River, the largest river system in southern Africa (catchment area of almost 10^6 km^2, that is, equivalent to about 77% of the land area of South Africa), drains large parts of South Africa from the headwaters in the Maloti-Drakensberg Mountains in the east to the mouth in Namaqualand where it enters the Atlantic Ocean (Fig. 28.7). It transports about 106×10^6 m^3 of sediment annually into the Atlantic Ocean (Birch 1977; Compton et al. 2010) and formed the ~70–120 m below sea level coastal-parallel Namaqualand mudbelt, which contains sedimentary records of the discharged material of the past ca. 11,000 years (Compton et al. 2010). For a thorough interpretation of the stored paleo-climatic and paleo-environmental signals, it is important to know where the material originates from and how it is altered along the way from the source to the Namaqualand mudbelt. Therefore, inorganic and organic geochemical analyses of soil samples, river suspended material, riverbed samples, marine surface sediments, and marine downcore sediments were carried out.

Hydrogen isotopic analyses of plant-wax derived n-alkanes (δD_{wax}) in soils show a diverse picture across the different rainfall zones in South Africa (Herrmann et al. 2017) (Fig. 28.8). In the summer rainfall zone (SRZ), δD_{wax} reflects the annual δD of precipitation (δD_p), although affected by evapotranspiration, which is indicated by the apparent fractionation factor (εapp, Fig. 28.8a). Thus, δD_{wax} is a suitable qualitative paleo-hydrological recorder in the SRZ. In the winter rainfall zone (WRZ), the situation is less straightforward. There is no relationship between δD_{wax} and annual δD_p, probably due to wide microclimatic variability, distinct vegetation communities, and diversity as well as potential influence of summer rain, especially in the eastern parts. In the WRZ, the processes overprinting the δD_p signal in the δD_{wax} remain unclear and require further research. The n-alkane distribution

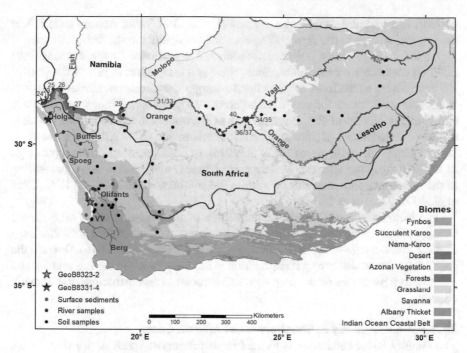

Fig. 28.7 Map of all sampling sites relevant for case study 1 (soils, river, surface sediments, cores), biomes (after Mucina and Rutherford 2006). Figure after Burdanowitz et al. (2018)

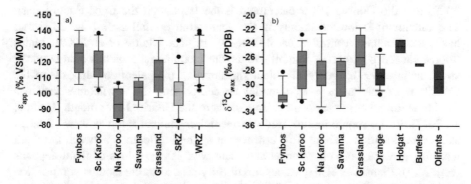

Fig. 28.8 Box and whisker plots for (**a**) the apparent hydrogen isotope fractionation (εapp) of plant-wax-derived n-alkanes in soils from different biomes (dark grey) and rainfall zones (light grey) and (**b**) $\delta 13C_{wax}$ of soils from different biomes (dark grey) and river samples (red). Boxes comprise the middle 50% of samples and the horizontal black line within the box represents the median. Black dots outside the whisker plots indicate the uppermost and lowermost 10%. Na Karoo and Sc Karoo indicate Nama Karoo and Succulent Karoo, respectively. Note that εapp is shown on an inverse axis. Figure after Herrmann et al. (2016)

patterns and carbon isotopes of n-alkanes ($\delta^{13}C_{wax}$) of South African soils reflect the vegetation type in the surrounding area and are distinct in the distinct biomes.

As the Orange River is the main sediment contributor to the Namaqualand mudbelt offshore western South Africa, there is a temptation to regard the Orange River sediment as the basis of past climate change interpretation. However, several small ephemeral rivers located in the adjacent Succulent Karoo biome and the Berg and Olifants Rivers in the southern part of the mudbelt are additional possible sediment contributors (Benito et al. 2011; Burdanowitz et al. 2018; Granger et al. 2018; Hahn et al. 2016; Herrmann et al. 2016). In addition, wind-driven input (e.g., of material from the adjacent western coast biomes) may be also important adding to the sediment contribution in the middle and southern mudbelt (Birch 1977; Gray et al. 2000; Zhao et al. 2015).

The isotopic and geochemical composition of the sediment cores offshore the Orange River mouth varies with shifting rainfall patterns throughout the large Orange River catchment area (Herrmann et al. 2017; Hahn et al. 2016). Overall, the case study shows that proxy interpretation is not simple. Knowing the sources and overprinting processes of the used proxies is crucial to reconstructing past climatic changes.

28.3.3.2 Case Study 5: Southernmost African Coast

The Gouritz River catchment is located in southernmost South Africa (Fig. 28.9). Despite the relatively small size of the catchment (53,139 km^2) compared to the Orange (973,000 km^2), the altitudinal gradient is steep as the Swartberg Mountains rise abruptly above 2000 m a.s.l. within 100 km of the coast (Le Maitre et al. 2009) and the Gouritz River catchment is the largest on the Cape South Coast. The catchment is located mainly in the year-round rainfall zone (Fig. 28.1) and has a mean annual runoff of ca. 488 $\times$ 10^6 m^3 (Le Maitre et al. 2009). Major floods, caused by extreme rainfall events, are characteristic of this area (Desmet and Cowling 1999). The geochemical signatures of the paleoflood deposits of the Gouritz river catchment were essential for deciphering the downcore signal in a marine sediment core near Mossel Bay, offshore the Gouritz River mouth (Hahn et al. 2017). In the source-to-sink study, terrestrial catchment samples were all taken at lowland locations. There was, however, a distinct difference between the plant wax δD_{C31} values of soil samples and plant wax δD_{C31} values of flood deposits (Fig. 28.9). Plant wax δD_{C31} values are directly related to the isotope composition of precipitation (Sessions et al. 1999). Therefore, the conditions under which plant waxes contained in soils and plant waxes derived from paleoflood deposits were synthesized must have been different. Rainfall δD signatures become deuterium-depleted with altitude (ca. 10–15‰ per 1000 m; Gonfiantini et al. 2001). In view of the extreme elevation difference in the Gouritz River catchment, it is most likely that the relatively deuterium-depleted paleoflood deposits contain a considerable amount of upper-catchment material. Southern Hemispheric Westerlies-related precipitation events in the otherwise arid winters have been described as the main source of precipitation in the upper Gouritz River catchment (Chase et al. 2013).

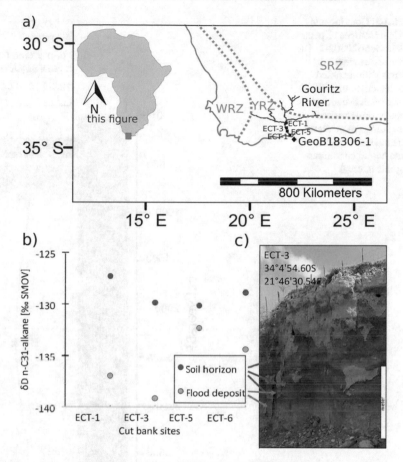

Fig. 28.9 (**a**) Map of the Gouritz River catchment and sampling sites. The winter rainfall zone (WRZ), summer rainfall zone (SRZ), and year-round rainfall zone (YRZ) are indicated. (**b**) Variations in δD of the n-C31-alkane (‰ VSMOW) in the distinct horizons of paleoflood vs. soil formation horizons. (**c**) Gouritz River tributary cut bank is depicted as an example illustrating these alternating horizons. Figure modified after Hahn et al. (2017)

For the interpretation of layers with deuterium-depleted plant waxes (Fig. 28.10), in the marine sediment core GeoB18308-1 in Mossel Bay offshore the Gouritz River mouth (Hahn et al. 2017), this information was particularly valuable. These layers could be linked to an increase in Southern Hemispheric Westerlies-related precipitation events in the upper Gouritz River catchment. In this manner, the Gouritz River catchment study led to a better understanding of variations in past regional climatic conditions.

28.3.3.3 Case Study 6: South-Eastern African Coast
A comparative study of catchment samples (riverbank sediments, flood deposits, suspension loads, and soils) and offshore deposits along the south-east African

Fig. 28.10 Core log and deuterium isotopes of plant waxes in GeoB18308-1. The core log shows internal structures with recorded colors demonstrating the presence of an event deposit. The deuterium-enriched values during the medieval climate anomaly (MCA) indicate humid conditions during this interval

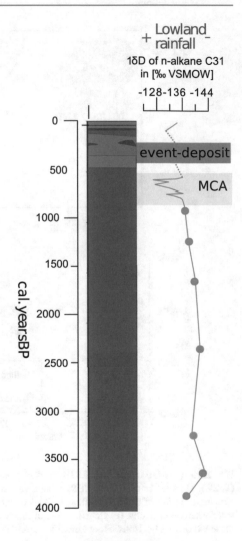

coast was used to investigate the source of various terrigenous sediment fractions and to delimit the movement of sediments along the shelf initiated by the Agulhas Current (Hahn et al. 2018). In the river catchments along the coast, subtle differences in sampling locations and associated environments (e.g., riverbank sediments, flood deposits, suspension loads, and soils) were reflected in the various proxy-indicators (Fig. 28.11). Large-scale trends, such as the climatic and environmental shift from the temperate winter rainfall zone in the southwest to the tropical summer rainfall zone in the northeast (Fig. 28.1), were also reflected in the data (Fig. 28.12). Plant wax δD signatures are indicative of increasing rainfall amounts toward the north concurring with a change from subtropical to tropical climate. Pollen assemblages and plant-wax $\delta^{13}C$ signatures document a shift from a mixed C_4/C_3 signature farther north in the subtropical grasslands/savanna to Mediterranean

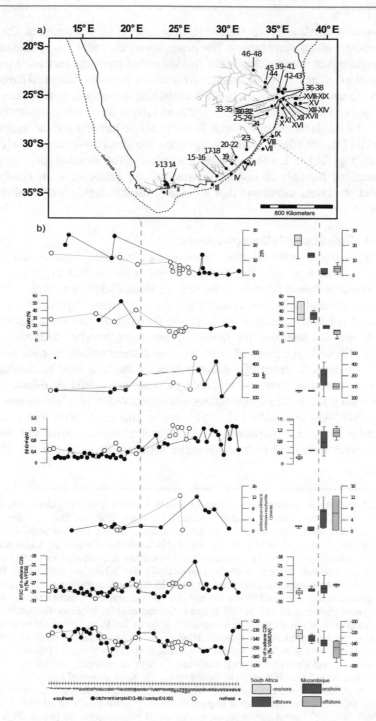

Fig. 28.11 (a) Study area of case study 3 in south-eastern Africa showing relevant drainage systems plotted with the sampling locations. The river catchments depicted on the map are (from

and mountain shrublands vegetation (C_3 dominated) in the southern Cape and Drakensberg regions, respectively. The petrographic, chemical, heavy-mineral, and bulk mineralogical composition of the riverine material reflects catchment geology, which consists of metamorphic and igneous rocks in the north (Kaapvaal Craton and Bushveld Igneous Complex), and mainly sandstones in the south (sedimentary rocks of the Cape and Karoo Supergroups). Offshore sample signatures were comparable to those of the adjacent river systems. It was therefore inferred that the influence of the Agulhas Current affects sediment deposition and distribution only seaward of the mid-shelf (Fig. 28.11). It was concluded that sediments on the inner shelf, especially from locations leeward of coastal protrusions, are protected from erosion and redistribution. These sediments thus constitute valuable archives of local climatic change.

28.3.3.4 Case Study 7: Delagoa Bight

Catchment samples from the river systems (e.g., Incomati, Matola, and Lusutfu rivers) that discharge into the Delagoa Bight were studied to determine the provenance of sediment deposits in the Delagoa Bight (Schüürman et al. 2019). The terrigenous material in sediments on the Delagoa Bight near Maputo, Mozambique, has several potential source areas including the coastal river catchments of the Incomati, Matola, and Lusutfu rivers and the large, interior catchment of the Limpopo River. A selection of trace element concentrations in river sediment samples was used to determine end-members for the four river catchments. All element concentrations were normalized by aluminum to avoid the dilution effects by the marine fraction. The entire dataset was scrutinized by multivariate analysis of variance (MANOVA) in order to identify the elements most useful for the regional end-member analysis. Relative end-member contributions to core top samples from the shelf were then used to trace inorganic sediment deposition in the Delagoa

Fig. 28.11 (continued) southwest to northeast): Breede, Gouritz, Great Fish, Great Kei, Mbashe, Tsitsa, Roadant, Tugela, Lusutfu, Matula, Incomati, Limpopo. The grey line separates samples grouped as Mozambique from the samples grouped as South Africa. (**b**) Southwest to northeast transect of multiple proxy-indicator trends. The x axis labels and symbols are coded according to terrestrial sampling site (Arabic numbers and full circles, respectively) or marine core top location (Latin numbers and empty circles), respectively. See (**a**) for numbered sites. Box and whisker plots representing the statistical relevance of (**b**). The bottom and top of the box represent the interquartile range, the band inside the box the median, and the whiskers of the minimum and maximum. Please note that for all proxies, the interquartile range of the South African offshore samples does not overlap the interquartile range of the Mozambique offshore samples. The proxies can be interpreted as follows: High zircon + tourmaline + rutile (ZTR) values indicate more extensive recycling of sedimentary source rocks. High quartz% indicates input from Cape Supergroup and Karoo Supergroup sandstones. A high metamorphic detritus (MI*) index is indicative of detritus from high-grade basement rocks. The main source of high (Fe+Cr+Ni)/Al values is the Bushveld Igneous Complex. Podocarpus, Clematis, Mimosa, and Euphorbia versus Poaceae (grass) pollen is a ratio between tropical grasslands and montane shrubland/Mediterranean forest. High $\delta13CC29$ values reflect predominance of C4 plants versus C3 plants. High $\delta DC29$ values indicate low rainfall amounts. Figure modified from Hahn et al. (2018)

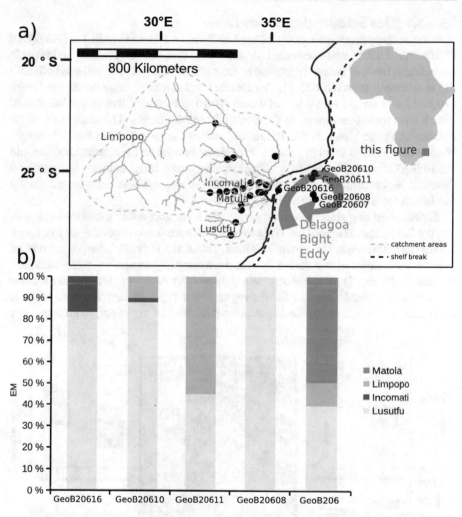

Fig. 28.12 (**a**) Overview map of the Incomati, Limpopo, Lusutfu, and Matola catchments including sample locations from terrestrial sampling campaigns as well as the Delagoa Bight with GeoB surface sediment sites. The dashed line marks the shelf break. (**b**) Contribution of empirical end-members of the Limpopo, Incomati, Matola, and Lusutfu catchments to sea-floor GeoB surface samples from the Delagoa Bight. The pathway of the sediment drift in the Delagoa Bight is indicated by the red arrow marking the Delagoa Bight Eddy. Figure after Schüürman et al. (2019)

Bight (Fig. 28.12). The results show that the local cyclonic circulation induces a strong eastward sediment drift in part preventing sedimentation in the bight. This was essential for the paleoclimatic interpretation of several marine sediment cores located in the bight (Miller et al. 2020; Hahn et al. 2021). In particular, the low relative contribution of the Limpopo River was unexpected.

28.3.3.5 Case Study 8: Umzimbuvu River

A multiseason provenance study of the Umzimbuvu River (Fig. 28.13; (Frankland 2020; Sect. 28.3.1) was conducted to understand catchment dynamics and provide taphonomic basis to the interpretation of the marine sediment core collected offshore the river mouth (GeoB20623-1). The Umzimvubu is a large, undammed catchment (19,852 km²) on the east coast of South Africa comprising five major tributaries, which have their headwaters in the Drakensberg Mountains. The main stem of the catchment is the Umzimvubu River, which flows over ~400 km through deeply incised river valleys from the source through the coastal belt before discharging into the Indian Ocean at Port St. Johns. The catchment transitions between the grassland biome at higher altitudes to the savanna biome at intermediate altitudes and finally the Indian Ocean Coastal Belt biome at the coast.

Eleven sampling sites were identified at the most accessible locations representing the full spatial extent of the Umzimvubu Catchment and capturing all five major tributaries (Mzintlava, Umzimvubu, Kinira, Thina, and iTsitsa). To capture temporal sediment dynamics within the catchment, three sampling campaigns were conducted to sample the dry (June), intermediate (September), and wet (November) months according to monthly rainfall for the region. Four types of samples were taken at each location to determine the provenance dynamics of sediment, hydrologic, and

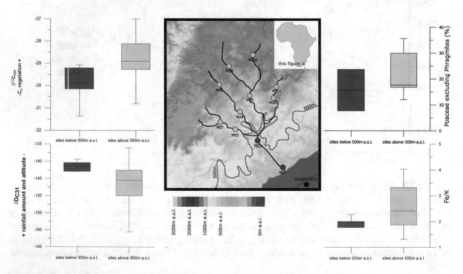

Fig. 28.13 Box and whisker plots illustrating the differences in sediment geochemistry and palynology between the upper and lower catchment areas of the Umzimvubu River. The minimum, maximum, median, upper, and lower quartile δ13C, δD, and Fe/K values of upper and lower catchment samples are shown. The geochemistry plots include river bed sediment sampled from each site during the dry (July), intermediate (September), and wet (November) seasons of 2017. Phragmites/Cyperaceae pollen ratio is based on 18 viable catchment samples across the three seasons. The upper and lower catchments are defined in the accompanying elevation map by the dotted line marking 500 m a.s.l. δ13C values of the C31 n-alkane are presented in ‰ VPDB and δD of the C31 n-alkane in ‰ VSMOW. Figure after Hahn et al. (2021)

vegetation signals within the catchment, viz., suspension load, riverbed sediment, dried puddle sediments, and water isotope samples. Suspension load samples were collected using a portable water pump deployed off of a bridge at the center of the river. Riverbed sediment was collected using grab samples. Dried puddle sediments were sampled from dried flood deposits where there was evidence of potential past flood events. Water isotope samples were collected downstream of each sample site for $\delta^{18}O$ and δ^2H analysis. These were ultimately excluded from the overall dataset as they did not show a major relation to the inorganic and organic signal provenance, nor did they have a strong correlation to the other proxies. Collectively, the catchment-wide sampling of river bed and suspension load sample was used to distinguish upper from lower catchment sediment signatures, which could be applied downcore on a sediment archive offshore the river system (GeoB20623-1, Hahn et al. 2021).

In terms of geochemistry, the catchment study showed that there was a distinct difference in the sediment signatures of the upper and lower catchment. Upper catchment sediments are characterized by high $\delta^{13}C$ values (due to the prevailing C_4 grassland vegetation), low δD values (due to the altitude effect; Gonfiantini et al. 2001), and higher Fe/K value (due to the high chemical weathering in the arid upper catchment) (Fig. 28.13). This is relevant for the interpretation of proxy indicators in the sediment core, since a greater proportion of upper catchment material in the sediments transported into the ocean occurs during high rainfall events.

For the palynological aspects of the study, suspension load samples yielded poor pollen concentration; thus, river bed samples were selected as a more appropriate sample type for pollen analysis. These surface river bed samples are not indicative of the time of sampling, but rather reflect pollen accumulation over an extended period, likely spanning several months or even years depending on the site. Moreover, samples taken in the downstream catchment represent a greater allochthonous signal, incorporating pollen grains transported from higher up the catchment. A further caveat to catchment pollen interpretation is the fact that seasonal variations in the pollen signal, such as those observed for *Zea mays* (maize), are indicative of flowering season of the parent vegetation, rather than of any seasonal climatic variations influencing vegetation type. This leads to the broader consideration of pollen as a direct indicator for parent vegetation, but an indirect indicator of prevailing climate, with inherent lags in the timing of vegetation response to climatic change. The catchment pollen analysis samples were therefore grouped across seasons, since seasonal flowering patterns do not inform paleo-vegetation or paleo-climatic interpretation. The aim of the catchment pollen research was to determine whether there is a coherent pollen signal indicative of the dominant biomes (Mucina and Rutherford 2006) across the Umzimvubu River catchment, viz. grassland in the upper catchment area (n = 14), savanna at intermediate altitudes (n = 12), and Indian Ocean Coastal Belt biome along the coastline (n = 6). Perhaps unsurprisingly, the pollen signature was dominated by a few ubiquitous taxa, including grasses (Poaceae excluding *Phragmites*-type) and sedges (Cyperaceae), which tend to be overrepresented in South African pollen records from wetlands. This overrepresentation obscured any clear biome-related signals in the vegetation.

The pollen signal also revealed the presence of pollen from exotic pine trees (*Pinus* spp.) in the upper catchment areas, reflecting the transformed nature of the catchment area, with parts of the natural grassland vegetation having been afforested by commercial plantations. A multiproxy comparison of upper and lower catchment samples showed clear altitudinal differences in organic and inorganic geochemical proxies, but also in the ratio of *Phragmites*/Cyperaceae pollen (Fig. 28.13; Hahn et al. 2021). In the upper (lower) catchment areas, *Phragmites*-type pollen is on average less (more) abundant than Cyperaceae. *Phragmites australis* is an aquatic/semiaquatic reed that grows in shallow water, whereas Cyperaceae (sedges) is a more ubiquitous taxon, with sedges growing in a range of habitats from wetlands to grasslands and forest fringes (Gaigher 1990; Archer 2000). The higher *Phragmites*/Cyperaceae pollen ratio in the lower altitude parts of the catchment likely reflects the prevalence of *P. australis* in the floodplain areas near the river mouth, and more humid climatic conditions than the higher altitude sites.

The paper by Hahn et al. (2021) was able to use these insights from modern catchment geochemistry and pollen sampling to apply the *Phragmites*/Cyperaceae pollen ratio, the isotope geochemistry, and the elemental composition downcore. The *Phragmites*/Cyperaceae pollen ratio is used as a moisture indicator to interpret the offshore marine fossil pollen record of core GeoB20623-1 recovered offshore the Umzimvubu River mouth, with ratio increases (decreases) suggesting more (less) humid conditions. Likewise, as a result of the catchment study, high $\delta^{13}C$ values, low δD values, and higher Fe/K values in core GeoB20623-1 are associated with an increased input of upper catchment sediments. Since a greater proportion of upper catchment material in the sediments transported into the ocean occurs during high rainfall periods, intervals with high $\delta^{13}C$ values, low δD values, and higher Fe/K values in core GeoB20623-1 are interpreted as humid phases.

28.4   Synthesis and Practical Recommendations

Summing up, we state that the suite of source-to-sink and depositional studies described in this chapter has been essential for the accurate interpretation of environmental conditions from sediment cores taken in and around southern Africa. Exploring depositional systems and the stratigraphy of marine environments at the sediment sink has proven indispensable for locating sediment bodies suitable for coring and therefore consequent analysis of past changes. This is particularly true for current swept shelves, such as the South African shelf, where great expanses of the seafloor are devoid of sediment. Furthermore, seismic stratigraphy can give insights into changing sedimentary conditions that will entirely change the interpretation of downcore proxies. Eminent examples of this are the lake sediment sequences dating from lower sea level conditions found in several Agulhas Bank/Paleo-Agulhas Plain sediment cores. In more practical terms, seismic stratigraphy studies can enable the choice of suitable coring equipment (gravity corer, vibrocorer, box corer, and multicorer) depending on the sediment grain size and texture. Possible penetration depth and thus barrel lengths for coring can also be estimated using seismic surveying. For the interpretation of the proxies measured on the retrieved sediment cores, we consider catchment studies a crucial. The examples outlined in these studies show how the suite of organic and inorganic indicators can vary with changes in the source area that can only be understood, and thus correctly interpreted, in a framework of a catchment area study. Considering source-to-sink study conduct, after this work, we recommend choosing sampling locations before and after tributaries and/or confluences and analyzing a suite of different samples (suspension load, river bed, flood deposits) from each site. Organic parameters can be excellent indicators of vegetation present at the site (river bank samples) as well as upstream of the confluence or tributary (suspension load samples). Compound-specific hydrogen isotope analyses can inform on present and past eco-hydrologic status. Inorganic parameters can give insights into changes in sediment provenance (geological formations) and/or shifts in weathering regimes at or upstream of the site. As best practice for future paleo-environmental studies, it is recommended that sediment cores are interpreted in the context of their depositional conditions explored using seismic surveying techniques and that catchment samples are analyzed in order to shed light on the processes that influence proxy parameters measured on allochthonous sedimentary components.

28.5   Conclusions

Using examples from southern Africa, we illustrate that catchment studies and hydroacoustic/stratigraphic studies are essential for reliable paleo-climatic and paleo-environmental interpretations of paleo-archives:

- Paleoclimate studies are important for anticipating present and future responses to climate and climate change as they provide reference information on past climate changes and variability.
- Often, several rivers discharge into the same embayment and/or strong oceanic currents may displace sediments. In both cases, it is vital to determine from which river systems sediments originate from prior to commencing environmental interpretations.
- Rainfall is not always evenly distributed over a catchment area. By identifying source areas within a catchment, a more locally constrained climate reconstruction can be made.
- Hydroacoustic surveying is vital for identifying suitable coring locations and contextualizing depositional processes.
- The present is not always an analogue of past conditions. This work has shown the value in comparison/verification at different temporal scales.

The overall messages for land use management and environmental protection policies emerging from this chapter are as follows:

- Reliable records of past climate and ecologic changes are indispensable for understanding present and future responses to climate and climate changes.
- Vegetation patterns during the Holocene differed across southern Africa, as did the past climate. Management strategies thus need to be regionally specific and consider region-specific variabilities and sensitivities.

Acknowledgments This work was financially supported by Bundesministerium für Bildung und Forschung (BMBF, Bonn, Germany) within the framework of the SPACES program under the projects "Regional Archives for Integrated Investigation (RaiN) phase 1 and 2," and "Tracing Human and Climate impacts in South Africa (TRACES)." This study would not have been possible without the MARUM—Center for Marine Environmental Sciences, University of Bremen, Germany. In particular, we thank the GeoB Core Repository at the MARUM and Pangaea (www.pangaea.de) for archiving the sediments and the data used in this chapter. The captain, crew, and scientists of the Meteor M123, M102, and M57/1 cruises are acknowledged for facilitating the recovery of the studied material.

References

Archer C (2000) Cyperaceae. In: Leistner OA (ed) Seed plants of southern Africa families and genera, vol 10. Strelitzia, pp 594–605

Benito G, Thorndycraft VR, Rico MT, Sánchez-Moya Y, Sopeña A, Botero B, Machado MJ, Davis M, Pérez-González A (2011) Hydrological response of a dryland ephemeral river to southern African climatic variability during the last millennium. Quat Res 75(3):471–482. https://doi.org/10.1016/j.yqres.2011.01.004

Birch GF (1977) Surficial sediments on the continental margin off the west coast of South Africa. Mar Geol 23(4):305–337. https://doi.org/10.1016/0025-3227(77)90037-8

Birch GF, Rogers J, Bremner JM (1986) Marine Geoscience Map Series: texture and composition of surficial sediments of the continental margin of the Republics of South Africa. Transkei and Ciskei, 1 p

Boom A, Carr AS, Chase BM, Grimes HL, Meadows ME (2014) Leaf wax n-alkanes and δ13C values of CAM plants from arid southwest Africa. Org Geochem 67:99–102. https://doi.org/10.1016/j.orggeochem.2013.12.005

Bouchez J, Galy V, Hilton RG, Gaillardet J, Moreira-Turcq P, Pérez MA et al (2014) Source, transport and fluxes of Amazon River particulate organic carbon: insights from river sediment depth-profiles. Geochim Cosmochim Acta 133:280–298. https://doi.org/10.1016/j.gca.2014.02.032

Bray EE, Evans ED (1961) Distribution of n-paraffins as a clue to recognition of source beds. Geochim Cosmochim Acta 22:2–15

Burdanowitz N, Dupont L, Zabel M, Schefuß E (2018) Holocene hydrologic and vegetation developments in the Orange River catchment (South Africa) and their controls. The Holocene 28(8):1288–1300. https://doi.org/10.1177/0959683618771484

Carr AS, Boom A, Grimes HL, Chase BM, Meadows ME, Harris A (2014) Leaf wax n-alkane distributions in arid zone South African flora: environmental controls, chemotaxonomy and palaeoecological implications. Org Geochem 67:72–84. https://doi.org/10.1016/j.orggeochem.2013.12.004

Cawthra HC, Jacobs Z, Compton JS, Fisher EC, Karkanas P, Marean CW (2018) Depositional and sea-level history from MIS 6 (Termination II) to MIS 3 on the southern continental shelf of South Africa. Quat Sci Rev 181:156–172

Cawthra HC, Frenzel P, Gander L, Hahn A, Compton JS, Zabel M (2020) Seismic stratigraphy of the inner to mid Agulhas Bank, South Africa. Quat Sci Revs 235:105979

Chase BM, Boom A, Carr AS, Meadows ME, Reimer PJ (2013) Holocene climate change in southernmost South Africa: rock hyrax middens record shifts in the southern westerlies. Quat Sci Rev 82:199–205. https://doi.org/10.1016/j.quascirev.2013.10.018

Collins JA, Schefuss E, Heslop D, Mulitza S, Prange M, Zabel M, Tjallingii R, Dokken TM, Huang EQ, Mackensen A, Schulz M, Tian J, Zarriess M, Wefer G (2011) Interhemispheric symmetry of the tropical African rainbelt over the past 23,000 years. Nat Geosci 4:42–45

Compton JS, Herbert CT, Hoffman MT, Schneider RR, Stuut J-B (2010) A tenfold increase in the Orange River mean Holocene mud flux: implications for soil erosion in South Africa. The Holocene 20(1):115–122. https://doi.org/10.1177/0959683609348860

Desmet P, Cowling R (1999) The climate of the Karoo–a functional approach. The Karoo: Ecol Patterns Process:3–16

Diefendorf AF, Freimuth EJ (2017) Extracting the most from terrestrial plant-derived n-alkyl lipids and their carbon isotopes from the sedimentary record. a review. Org Geochem 103.1–21

Dingle RV, Birch GF, Bremner JM, De Decker RH, Du Plessis A, Engelbrecht JC, Fincham MJ, Fitton T, Flemming BW, Gentle RI, Goodlad SW, Martin AK, Mills EG, Moir GJ, Parker RJ, Robson SH, Rogers J, Salmon DA, Siesser WG, Simpson ESW, Summerhayes CP, Westall F, Winter A, Woodborne MW (1987) Deep-sea sedimentary environments around southern Africa (South-East Atlantic and South-West Indian Oceans). Ann S Afr Mus 98:1–27

Dyer SE, Green AN, Cooper JAG, Hahn A, Zabel M (2021) Response of a wave-dominated coastline and delta to antecedent conditioning and fluctuating rates of postglacial sea-level rise. Mar Geol 434:106435. https://doi.org/10.1016/j.margeo.2021.106435

Eglinton G, Hamilton RJ (1967) Leaf epicuticular waxes. Science 156:1322–1335

Flemming B, Hay R (1988) Sediment distribution and dynamics on the Natal continental shelf. Coast Ocean Stud off Natal, South Africa 26:47–80

Frankland T (2020) Sediment dynamics and provenance of hydrologic and vegetation signals of the Mzimvubu Catchment, Eastern Cape, South Africa. Unpublished MSc thesis, University of KwaZulu-Natal

Gaigher C (1990) Wetlands. Cape Conserv Ser 6:1–21

Galy V, Eglinton T, France-Lanord C, Sylva S (2011) The provenance of vegetation and environmental signatures encoded in vascular plant biomarkers carried by the Ganges-Brahmaputra rivers. Earth Planet Sci Lett 304(1–2):1–12. https://doi.org/10.1016/j.epsl.2011.02.003

Garcin Y, Schefuß E, Schwab VF, Garreta V, Gleixner G, Vincens A, Todou G, Séné O, Onana J-M, Achoundong G, Sachse D (2014) Reconstructing C3 and C4 vegetation cover using n-alkane carbon isotope ratios in recent lake sediments from Cameroon, Western Central Africa. Geochim Cosmochim Acta 142:482–500

Gonfiantini R, Roche M-A, Olivry J-C, Fontes J-C, Zuppi GM (2001) The altitude effect on the isotopic composition of tropical rains. Chem Geol 181(1–4):147–167. https://doi.org/10.1016/S0009-2541(01)00279-0

Granger R, Meadows ME, Hahn A, Zabel M, Stuut J-BW, Herrmann N, Schefuß E (2018) Late-Holocene dynamics of sea-surface temperature and terrestrial hydrology in southwestern Africa. The Holocene 28(5):0959683617744259. https://doi.org/10.1177/0959683617744259

Gray CED, Meadows ME, Lee-Thorp JA, Rogers J (2000) Characterising the Namaqualand mudbelt of southern Africa: chronlogy, palynology and palaeoenvironments. S Afr Geogr J 82(3):137–142. https://doi.org/10.1080/03736245.2000.9713705

Green AN, MacKay CF (2016) Unconsolidated sediment distribution patterns in the KwaZulu-Natal Bight, South Africa: the role of wave ravinement in separating relict versus active sediment populations. Afr J Mar Sci 38(Suppl 1):S65–S74

Green AN, Cooper JAG, Leuci R, Thackeray Z (2013a) Formation and preservation of an overstepped segmented lagoon complex on a high-energy continental shelf. Sedimentology 60:1755–1768. https://doi.org/10.1111/sed.12054

Green AN, Dladla N, Garlick GL (2013b) Spatial and temporal variations in incised valley systems from the Durban continental shelf, KwaZulu-Natal, South Africa. Mar Geol 335:148–161

Green AN, Cooper JAG, Salzmann L (2014) Geomorphic and stratigraphic signals of postglacial meltwater pulses on continental shelves. Geology 42(2):151–154. https://doi.org/10.1130/g35052.1

Green AN, Cooper JAG, Loureiro C et al (2022) Stormier mid-Holocene southwest Indian Ocean due to poleward trending tropical cyclones. Nat Geosci 15:60–66. https://doi.org/10.1038/s41561-021-00842-w

Häggi C, Sawakuchi AO, Chiessi CM, Mulitza S, Mollenhauer G, Sawakuchi HO et al (2016) Origin, transport and deposition ofleaf-wax biomarkers in the Amazon Basin and the adjacent Atlantic. Geochim Cosmochim Acta 192:149–165. https://doi.org/10.1016/j.gca.2016.07.002

Hahn A, Compton JS, Meyer-Jacob C, Kirsten KL, Lucasssen F, Pérez Mayo M, Schefuß E, Zabel M (2016) Holocene paleo-climatic record from the South African Namaqualand mudbelt: a source-to-sink approach. Quat Int 404 Part B:121–135. https://doi.org/10.1016/j.quaint.2015.10.017

Hahn A, Schefuß E, Andò S, Cawthra HC, Frenzel P, Kugel M, Zabel M (2017) Southern Hemisphere anticyclonic circulation drives oceanic and climatic conditions in late Holocene southernmost Africa. Clim Past 13(6):649–665. https://doi.org/10.5194/cp-13-649-2017

Hahn A, Miller C, Andó S, Bouimetarhan I, Cawthra HC, Garzanti E, Zabel M (2018) The provenance of terrigenous components in marine sediments along the east coast of southern Africa. Geochem Geophys Geosyst

Hahn A, Neumann FH, Miller C, Finch J, Frankland T, Cawthra HC, Zabel M (2021) Mid-to Late Holocene climatic and anthropogenic influences in Mpondoland, South Africa. Quat Sci Rev 261:106938. https://doi.org/10.1016/j.quascirev.2021.106938

Hemingway JD, Schefuß E, Dinga BJ, Pryer H, Galy VV (2016) Multiple plant-wax compounds record differential sources and ecosystem structure in large river catchments. Geochim Cosmochim Acta 184:20–40. https://doi.org/10.1016/j.gca.2016.04.003

Hempel P, Spiess V, Schreiber R (1994) Expulsion of shallow gas in the Skagerrak—evidence from sub-bottom profiling, seismic, hydroacoustical and geochemical data. Estuar Coast Shelf Sci 38(6):583–601

Herrmann N, Boom A, Carr AS, Chase BM, Granger R, Hahn A, Zabel M, Schefuß E (2016) Sources, transport and deposition of terrestrial organic material: a case study from southwestern Africa. Quat Sci Rev 149:215–229. https://doi.org/10.1016/j.quascirev.2016.07.028

Herrmann N, Boom A, Carr AS, Chase BM, West AG, Zabel M, Schefuß E (2017) Hydrogen isotope fractionation of leaf wax n-alkanes in southern African soils. Org Geochem 109:1–13. https://doi.org/10.1016/j.orggeochem.2017.03.008

Hoetzel S, Dupont L, Schefuß E, Rommerskirchen F, Wefer G (2013) The role of fire in Miocene to Pliocene C-4 grassland and ecosystem evolution. Nat Geosci 6:1027–1030

Hou JZ, D'Andrea WJ, MacDonald D, Huang YS (2007) Hydrogen isotopic variability in leaf waxes among terrestrial and aquatic plants around Blood Pond, Massachusetts (USA). Org Geochem 38:977–984

Le Maitre D, Colvin C, Maherry A (2009) Water resources in the Klein Karoo: the challenge of sustainable development in a water-scarce area. S Afr J Sci 105:39–48

Leithold EL, Blair NE, Wegmann KW (2016) Source-to-sink sedimentary systems and global carbon burial: a river runs through it. Earth Sci Rev 153:30–42. https://doi.org/10.1016/j.earscirev.2015.10.011

Mayer L, Hughes-Clarke JE (1995) STRATAFORM Cruise Report: R/V Pacific Hunter. Multibeam Survey, July 14–28

Miller C, Finch J, Hill T, Peterse F, Humphries M, Zabel M, Schefuss E (2019) Late Quaternary climate variability at Mfabeni peatland, eastern South Africa. Clim Past 15:1153–1170

Miller C, Hahn A, Liebrand D, Zabel M, Schefuß E (2020) Mid- and low latitude effects on eastern South African rainfall over the Holocene. Quat Sci Rev 229:106088. https://doi.org/10.1016/j.quascirev.2019.106088

Mucina L, Rutherford M (2006) The vegetation of South Africa, Lesotho and Swaziland, pp 749–790

Petschick R, Kuhn G, Gingele F (1996) Clay mineral distribution in surface sediments of the South Atlantic: sources, transport, and relation to oceanography. Mar Geol 130(3):203–229. https://doi.org/10.1016/0025-3227(95)00148-4

Pretorius L, Green A, Cooper A (2016) Submerged shoreline preservation and ravinement during rapid postglacial sea-level rise and subsequent "slowstand". Geol Soc Am Bull 128(7–8):1059–1069. https://doi.org/10.1130/b31381.1

Rogers J, Rau AJ (2006) Surficial sediments of the wave-dominated Orange River Delta and the adjacent continental margin off south-western Africa. Afr J Mar Sci 28(3–4):511–524. https://doi.org/10.2989/18142320609504202

Sachse D, Billault I, Bowen GJ, Chikaraishi Y, Dawson TE, Feakins SJ, Freeman KH, Magill CR, McInerney FA, van der Meer MTJ, Polissar PJ, Robins RJ, Sachs JP, Schmidt H-L, Sessions AL, White JWC, West JB, Kahmen A (2012) Molecular paleohydrology: interpreting the hydrogen-isotopic composition of lipid biomarkers from photosynthesizing organisms. Annu Rev Earth Planet Sci 40:221–249

Schefuß E, Kuhlmann H, Mollenhauer G, Prange M, Patzold J (2011) Forcing of wet phases in southeast Africa over the past 17,000 years. Nature 480:509–512

Schimmelmann A, Sessions AL, Mastalerz M (2006) Hydrogen isotopic (D/H) composition of organic matter during diagenesis and thermal maturation. Annu Rev Earth Planet Sci 34:501–533

Schüürman J, Hahn A, Zabel M (2019) In search of sediment deposits from the Limpopo (Delagoa Bight, southern Africa): deciphering the catchment provenance of coastal sediments. Sediment Geol 380:94–104. https://doi.org/10.1016/j.sedgeo.2018.11.012

Sessions AL, Burgoyne TW, Schimmelmann A, Hayes JM (1999) Fractionation of hydrogen isotopes in lipid biosynthesis. Org Geochem 30(9):1193–1200. https://doi.org/10.1016/S0146-6380(99)00094-7

Truc L, Chevalier M, Favier C, Cheddadi R, Meadows ME, Scott L, Chase BM (2013) Quantification of climate change for the last 20,000 years from Wonderkrater, South Africa: implications for the long-term dynamics of the Intertropical Convergence Zone. Palaeogeogr Palaeoclimatol Palaeoecol 386:575–587. https://doi.org/10.1016/j.palaeo.2013.06.024

Vogts A, Moossen H, Rommerskirchen F, Rullkötter J (2009) Distribution patterns and stable carbon isotopic composition of alkanes and alkan-1-ols from plant waxes of African rain forest and savanna C-3 species. Org Geochem 40:1037–1054

Zhao X, Dupont L, Meadows ME, Wefer G (2015) Pollen distribution in the marine surface sediments of the mudbelt along the west coast of South Africa. Quat Int. https://doi.org/10.1016/j.quaint.2015.09.032

Zabel M (2016) Climate archives in coastal waters of southern Africa–Cruise No. M123–February 3–February 27, 2016–Walvis Bay (Namibia)–Cape Town (Rep. of South Africa). METEOR-Berichte M, 123, 50

Observational Support for Regional Policy Implementation: Land Surface Change Under Anthropogenic and Climate Pressure in SALDi Study Sites

29

Jussi Baade ⓘ, Ursula Gessner ⓘ, Eugene Hahndiek,
Christiaan Harmse ⓘ, Steven Hill, Andreas Hirner ⓘ,
Nkabeng Maruping-Mzileni ⓘ, Insa Otte ⓘ, Carsten Pathe,
Paul Renner ⓘ, Konstantin Schellenberg ⓘ,
Shanmugapriya Selvaraj ⓘ, Chris Smith, Tercia Strydom ⓘ,
Annette Swanepol, Frank Thonfeld ⓘ, Marcel Urban ⓘ,
Zhenyu Zhang ⓘ, and Christiane Schmullius ⓘ

Abstract

South Africa is a vast, very diverse and dynamic country experiencing rapidly changing demands for the utilization of its natural resources. At the same time, global climate change and related processes affect the land's agricultural utilization. The considerable expansion of high-resolution Earth observation

J. Baade (✉)
Department of Geography, Friedrich Schiller University Jena, Jena, Germany

U. Gessner · A. Hirner · F. Thonfeld
German Aerospace Center (DLR), Wessling, Germany
e-mail: ursula.gessner@dlr.de; andreas.hirner@dlr.de; frank.thonfeld@dlr.de

E. Hahndiek
Nuwejaars Wetlands SMA, Bredasdorp, Republic of South Africa
e-mail: eugene@nuwejaars.com

C. Harmse · A. Swanepol
Northern Cape Department of Agriculture, Environmental Affairs, Land Reform and Rural Development, Eiland Research Station, Upington, South Africa
e-mail: charmse@ncpg.gov.za

S. Hill · I. Otte
Department of Remote Sensing, Institute of Geography and Geology, University of Wuerzburg, Wuerzburg, Germany
e-mail: steven.hill@uni-wuerzburg.de; insa.otte@uni-wuerzburg.de

© The Author(s) 2024
G. P. von Maltitz et al. (eds.), *Sustainability of Southern African Ecosystems under Global Change*, Ecological Studies 248,
https://doi.org/10.1007/978-3-031-10948-5_29

systems providing high temporal resolution acquisitions of the land surface free of charge provides the opportunity to monitor land surface dynamics at an unprecedented temporal and spatial resolution. Based on the work in the SPACES II project South Africa Land Degradation Monitor (SALDi), we present examples highlighting the new observational opportunities potentially supporting regional policy implementation. Thus, the main objective of this chapter is to present applied examples from the six SALDi study sites spread across the country illustrating some of the new capabilities providing simultaneously a regional overview of land surface dynamics as well as high-resolution information on specific areas, for example a district, a municipality, a farm or a specific field plot.

29.1 Introduction

Land use and land cover reflect at any given time the utilization of the resources of a country. Utilization patterns and trends are dependent on underlying natural resources (e.g., the geology, soils, vegetation, climate and hydrology conditions)

N. Maruping-Mzileni
Scientific Services, South African National Parks (SANParks), Kimberley, South Africa
e-mail: nkabeng.mzileni@sanparks.org

C. Pathe · P. Renner · K. Schellenberg · C. Schmullius
Department for Earth Observation, Institute of Geography, Friedrich Schiller University Jena, Jena, Germany
e-mail: carsten.pathe@uni-jena.de; paul.renner@uni-jena.de; konstantin.schellenberg@uni-jena.de; c.schmullius@uni-jena.de

S. Selvaraj
Research Centre for Agricultural Remote Sensing, Institute for Crop and Soil Science, Julius Kuehn Institute, Braunschweig, Germany
e-mail: shanmugapriya.selvaraj@julius-kuehn.de

C. Smith
Soil Conservation/LandCare, Department of Agriculture and Rural Development, Free State Province, Thaba Nchu, South Africa
e-mail: csmith@dard.gov.za

T. Strydom
Scientific Services, South African National Parks (SANParks), Skukuza, South Africa
e-mail: tercia.strydom@sanparks.org

M. Urban
ESN (EnergiesystemeNord) GmbH, Jena, Germany

Z. Zhang
Institute of Geography, University of Augsburg, Augsburg, Germany

Institute of Meteorology and Climate Research (IMK-IFU), Karlsruhe Institute of Technology, Campus Alpin, Garmisch-Partenkirchen, Germany
e-mail: zhenyu.zhang@partner.kit.edu

and the opportunities, conditions and conventions to exploit them. This leads to a specific land use pattern in specific regions which might change due to anthropogenic and/or climate pressure over time. Often, the changes seemingly introduced by external forcing like climate change are considered land degradation, while anthropogenic changes are often considered development (e.g. the transformation of rangeland to settlement areas or cultivated land into mining areas). This kind of obvious land use and land cover change is regularly monitored within the framework of repeated national land cover and land use assessments. Compared to this, the SPACES II project South Africa Land Degradation Monitor (SALDi) targets more subtle land cover changes which eventually might cause land degradation. Thus, the project addresses the dynamics and functioning of multi-use landscapes with respect to land use and land cover change, water fluxes and implications for habitats and ecosystem services.

An overarching goal of the project is to implement new and sustainable adaptive land degradation assessment tools. The project aims to advance current methodologies for multi-use landscapes by innovatively incorporating inter-annual and seasonal variability in a spatially explicit approach and takes advantage of the emerging availability of high spatio-temporal resolution Earth observation data (e.g. ESA Sentinels, DLR TanDEM-X, NASA Landsat), growing sources of in-situ data and advancements in modelling approaches.

In this chapter, we will provide observational support and discoveries on subtle land cover changes in the six SALDi study sites in support of regional policy implementation to identify areas under the threat of land degradation. The aim is to thus support the remedial actions. Following this introduction, we will first provide an overview of the six SALDi study sites and discuss the land degradation challenges at these sites. Subsequently, we will introduce the SALDi product catalog developed based on the high spatio-temporal resolution Earth observation data mainly acquired by the ESA Sentinels since 2017. Finally, we will provide examples of the observational support and discoveries found at each site. The examples have been chosen based on land surface processes (those perceived as most important) to highlight the opportunities to deal with observations on a regional and local scale at the same time. In addition, we try to highlight the advantages of time series data and analysis when dealing with land surface processes.

29.2 SALDi Study Sites

In order to develop a broadly applicable set of products to support detection of land surface dynamics and eventually land degradation processes, we identified six study areas covering approximately 100 km × 100 km each (Fig. 29.1). The study sites cover the main climatic gradient from the semi-arid winter-rainfall region in the southwest across the central semi-arid year-round-rainfall region to the semi-humid summer-rainfall region in the northeast. The climate diagrams according to Walter and Lieth (1967) (Fig. 29.2) clearly identify average dry and wet months and highlight the great aridity in the SALDi study area No. 2: Kai !Garib, located

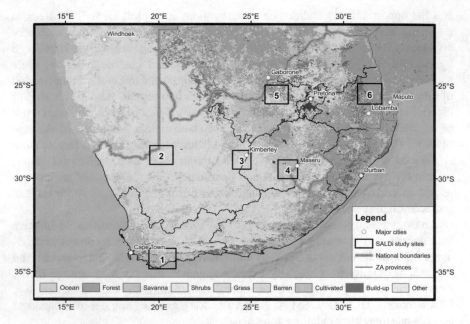

Fig. 29.1 Land cover in southern Africa, that is the SALDi SP2 WRF Hydro modelling domain, and location of SALDi study areas in South Africa. 1: Overberg, 2: Kai !Garib, 3: Sol Plaatje, 4: Mantsopa, 5: Bojanala Platinum, 6: Ehlanzeni (data source: ©ESRI 2013, simplified MODIS Land Cover from Friedl et al. 2010, compilation: J. Baade)

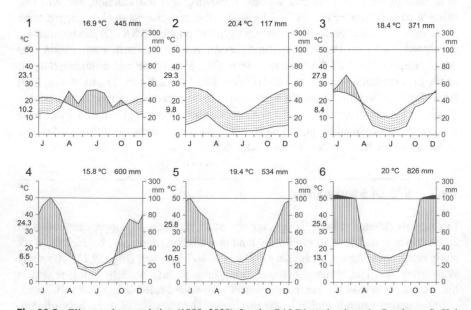

Fig. 29.2 Climate characteristics (1980–2020) for the SALDi study sites. 1: Overberg, 2: Kai !Garib, 3: Sol Plaatje, 4: Mantsopa, 5: Bojanala Platinum, 6: Ehlanzeni (data source: Funk et al. 2015, Harris et al. 2020, compilation: Z. Zhang)

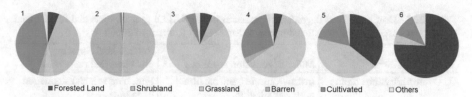

Fig. 29.3 Land cover characteristics in the six SALDi study areas. 1: Overberg, 2: Kai !Garib, 3: Sol Plaatje, 4: Mantsopa, 5: Bojanala Platinum, 6: Ehlanzeni (data source: simplified land cover from DFFE 2021, compilation: J. Baade)

at the boundary to Namibia. In addition, the selection of study areas considers physical (e.g. biomes, relief, soils and land types) and socio-economic (e.g. land tenure, farming practices) differences within South Africa. Within our study sites, protected areas (often National Parks) represent benchmark sites (Wiese et al. 2011), providing a foundation for Baseline Trend Scenarios (Diaz et al. 2015), against which climate-driven ecosystem service dynamics of multi-used landscape (cropland, rangeland, forests) are evaluated.

The analysis and comparison of the land cover in the six study areas (Fig. 29.3) provides evidence for the strong impact of climate conditions on the utilization of land. In the Overberg region in the Southwest (SALDi site 1), 45% of the area is cultivated and 40% covered by shrubland (i.e. mostly fynbos), while forested land[1] cover more than 75% of the area in Ehlanzeni in the Northeast (SALDi site 6). In Kai !Garib (SALDi site 2), shrubs (50%) and barren land (48%) characterize the landscape, while grassland (75%) represent the dominant land cover in Sol Plaatje (SALDi site 3). Compared to this, Mantsopa (SALDi site 4) and Bojanala Platinum (SALDi site 5) are characterized by a more diverse land cover. It is this diversity of land utilization in the different study sites which supports the aspiration to eventually provide overall results applicable to the whole of South Africa and beyond.

In accordance with the different climatic conditions as well as the different land cover and land use conditions, which also reflect other basic conditions, such as relief, geological conditions, soils and climate, it is possible to identify dominant degradation processes and related land surface processes (Table 29.1) for the study areas in cooperation with local stakeholders. Within the framework of SALDi, these major degradation processes and related land surface processes are to be addressed based on the SALDi remote sensing product catalog (see below and Chap. 24, Fig. 24.2). Comparing the six sites shows that desertification, bush encroachment or invasive species and soil erosion are concerns in all six sites. Reservoir siltation as an off-site damage from soil erosion (Chap. 13) is a concern in all sites with sufficient rainfall to support reservoirs, that is Overberg, Mantsopa, Bojanala Platinum and

---

[1] We follow the terminology used in the South African National Land-Cover classification (SANLC, DFFE 2021). Often savanna would be the ecologically more appropriate term for forested land, but that is included in the SANLC terminology as it follows international guidelines.

Table 29.1 Dominant land degradation processes in the SALDi project areas and related land surface processes (Biome classification acc. to Mucina and Rutherford 2006)

Study area/biome	Major degradation processes	Related land surface processes and properties
1. Overberg/Fynbos biome	Invasive species, soil erosion, reservoir siltation,	Fire, bare soil and vegetation dynamics
2. Kai!Garib/Nama-Karoo biome	Bush encroachment, soil & gully erosion (?), salinization	Fire, bare soil and vegetation dynamics (bush encroachment), gully formation
3. Sol Plaatje/Nama-Karoo and Savanna biome	Bush encroachment, soil & gully erosion, salinization	Fire, bare soil and vegetation dynamics (specif. on game farms), gully formation
4. Mantsopa/Grassland biome	Bush encroachment, soil & gully erosion, reservoir siltation	Fire, bare soil and vegetation dynamics (Slangbos encroachment), gully formation
5. Bojanala Platinum/Savanna biome	Bush encroachment, soil erosion, reservoir siltation	Fire, bare soil and vegetation dynamics (specif. on game farms)
6. Ehlanzeni/Savanna biome	Bush encroachment, soil & gully erosion, reservoir siltation, land use change	Fire, bare soil and vegetation dynamics (bush encroachment vs. loss of trees), land use and cover change

Ehlanzeni. Salinization is mentioned in areas where irrigation is important, while gully erosion is of concern in areas with high relief and steep slopes. The related land surface processes and properties observable with a remote sensing approach include general vegetation dynamics, fire, bush encroachment, the temporal and spatial proportion of bare soil and possibly the formation of gullies (see Chap. 13, Fig. 13.5).

One difficulty in distinguishing climatic-driven land surface dynamics from anthropogenic land degradation in the sense of a continuous declining process is the high inter-seasonal variation of climate on a decadal scale in many parts of South Africa. This can be clearly illustrated by the more than 100-year long precipitation record from Bredasdorp in Overberg, Western Cape Province (Fig. 29.4). Based on this, the 5-year moving average of annual precipitation varies regularly and partly for several decades between 400 and over 500 mm. Under these conditions, it is rather difficult to attribute observations from a few years (~5 years) to either 'normal' land surface dynamics or land degradation.

29.3 Description of SALDi Product Catalog

In the SALDi project, several products are developed from satellite data to be used as a basis for assessing land degradation in the project areas (Fig. 29.1). The product

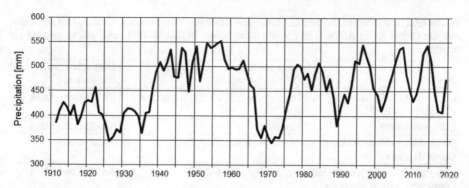

Fig. 29.4 Precipitation variability (5 year moving average) 1910–2020, Zeekoevlei Farm, Bredasdorp, Overberg (data courtesy of P. Albertyn and E. Hahndiek, Nuwejaars SMA, compilation: J. Baade)

catalog comprises data derived mainly from Sentinel-1 radar and Sentinel-2 optical satellite time series, airborne and spaceborne lidar data and further earth observation sensors. It includes information products on temporal anomalies and breakpoints, on woody and vegetation cover, on surface moisture and a number of indices that can serve as proxies for land surface properties, for example bare area abundance, vegetation activity or moisture. A detailed overview over the SALDi product catalog is given in Table 29.2. At the top of the table, a number of basic data sets providing background information on the study sites are listed. The next section lists the analysis ready data and products related to the Sentinel-1 radar satellites followed by the Sentinel-2 optical data and products. To avoid reiterations within this book, products already described in Chap. 24 are not discussed in greater detail here.

The spectral indices mentioned in Table 29.2 can be used as proxies for biophysical variables describing the status of the land surface concerning, for example, vegetation cover or moisture. They have proven to have a wide and growing range of applications (Tucker 1979, Qi et al. 1994; Diek et al. 2017). As they are mostly used to monitor land surface and vegetation conditions, they can provide early warning on droughts and famines. The performance of these indices differs in accordance with land cover (e.g. sparse vegetation vs. very dense vegetation) and various environmental conditions. Therefore, we provide in Table 29.3 some details on the most utilized spectral indices provided in the SALDi data cube.

29.4 Regional Examples of Land Surface Change

Based on the SALDi Earth observation product catalog (Sect. 29.3), we present in the following examples of times series analysis of Sentinel-1 and Sentinel-2 data aimed at observations on relevant land surface dynamics, eventually on processes which might be or are considered land degradation. The examples cover all six

Table 29.2 Overview of the SALDi Earth observation products provided in the SALDi data cube and the Digital Earth Africa (DEA[a]) data cube, access to SALDi data cube will be granted on request

Product	Spatial and temporal resolution, additional information	Data cubes providing the product
Copernicus GLO-30 Digital Elevation Model (DEM)	30 m, acquired: 2010–2015, released 2019	SALDiCube, DEA
ESA worldcover	10 m, 2020	SALDiCube, DEA
South African National Land Cover data sets	30 m, 1990, 2013, 2017, 2020	SALDiCube
Historic Flood Mapping Water Observations from Space (ls wofs)	30 m, 16 days, 2014–2022	SALDiCube, DEA
Sentinel-1 Gamma normalized radar backscatter	20 m, every 6–12 days, 2018–2022	SALDiCube, DEA
Sentinel-1 Coherence Product (ascending)	45 m, every 6–12 days, 2017–2021	SALDiCube
Sentinel-1 Breakpoints (see Chap. 24)	10 m, one 5-year raster for each SALDi site with a number of breakpoints, 2015–2020	SALDiCube
Sentinel-1 Surface moisture indicator (see Chap. 24)	50 m, every 6–12 days, 2017–2021	SALDiCube
Sentinel-1 Woody cover (see Chap. 24)	10 m, one raster for each SALDi site, 2019–2020, Information content: percentage of woody cover (0–100%)	SALDiCube
Sentinel-2 imagery, Level 2A (surface reflectance) converted to GeoTIFFs	10, 20, 60 m (band dependent), every 5 days, 2017–2022	SALDiCube, DEA
Sentinel-2A level 1C—cloud and shadow masks from Fmask	10, 20, 60 m (band dependent), every 5 days, 2017–2022	SALDiCube
Sentinel-2 spectral indices	10, 20 m (band dependent), every 5 days, 2017–2022, Indices: BSI, EVI, MSAVI, NDBI, NDVI, NDWI, PVI, SAVI, TSAVI, REIP	SALDiCube
Sentinel-2 Canopy Cover (see Chap. 24)	10 m, one raster for each SALDi site, 2019–2020, Information content: percentage of canopy cover (0–100%) and canopy height	SALDiCube
Sentinel-2 Breakpoints (see Chap. 24)	10 m, one 3-layer raster for each SALDi site, 2017–2021, Information content: date of breakpoint; probability of breakpoint occurrence; probability of exact timing of breakpoint	SALDiCube

[a]https://www.digitalearthafrica.org/

SALDi study sites picking specific issues for each site. The presented examples and the interpretation are the outcome of project works including dedicated personnel

Table 29.3 Overview of the spectral indices derived from Sentinel-2 data provided in the SALDi data cube

Spectral indices	Explanation
BSI	The Bare soil index is a spectral indicator that enhances the identification of bare soil areas/patches and vegetated areas. One should note that the BSI scales from -1 to 1 and cannot easily be transformed to percentage of bare soil.
EVI	The enhanced vegetation index can be used to quantify vegetation greenness. In contrast to NDVI, EVI corrects for certain atmospheric conditions and canopy background noise and is known to be more sensitive in areas with dense vegetation.
MSAVI	The modified soil adjusted vegetation index has been rated as the superior index to detect leaf area index for certain crop types and is therefore useful for information linked to agriculture.
NDBI	The normalized difference built-up index uses the NIR and SWIR bands to emphasize manufactured built-up areas. It can be utilized to mitigate the effects of terrain illumination differences as well as atmospheric effects.
NDVI	The normalized difference vegetation index is associated with vegetation characteristics and is probably one of the most used indices. It uses the red and the near-infrared spectral bands, and high values correspond to denser and healthier vegetation.
NDWI	The normalized difference water index has proven its usefulness for water body monitoring as it enhances water efficiently in most of the cases. However, it is sensitive to built-up land and might result in over-estimation of water bodies.
PVI	The perpendicular vegetation index is known as a more complex index that takes the soil emissivity into account. That is of special importance as this is one of the major limitations of NDVI. The PVI measures the changes from the bare soil reflectances caused by the vegetation. In this way, it gives an indication of vegetative cover independent of the effects of the soil.
SAVI	The soil adjusted vegetation index is mostly utilized to correct the NDVI for the influence of soil brightness in areas where vegetative cover is low.
TSAVI	The transformed soil-adjusted vegetation index is a vegetation index that attempts to minimize soil brightness influences by assuming the soil line has an arbitrary slope and intercept.
REIP	The red-edge inflection point is an indicator allowing to draw conclusions about chlorophyll content and plant vitality.

and the exchange with local experts. It is hoped that the examples are not only scientifically interesting, but useful in solving everyday problems on the ground. One should note that although most presented analyses are easy to automate, it often needs an experienced person to evaluate the results and assess the implications of observed land surface dynamics.

29.4.1 Overberg

The SALDi study site 1, Overberg, is located in the winter rainfall region in the Southwest of the country. With about 450 mm of mean annual rainfall over the whole area, the land is largely used for mixed farming, consisting of winter grains and livestock (beef cattle and Merino sheep) and to a lesser degree for livestock farming (in particular dairy cattle). Especially, the NW around Robertson is well known for winegrowing which is expanding into the Southern Overberg area. According to an analysis of the South African National Land Cover (SANLC) data (DFFE 2021), about 43% of the land surface (5175 km^2 of 13,080 km^2) is cultivated. Only about 1% of the cultivated area (50 km^2) is under pivot irrigation, highlighting the importance of rainfed arable farming in the Overberg region. Most of the arable land is located on the undulating Southern Coastal Platform characterized by mudrock, shales and siltstone. These basic conditions make soil erosion by water a potentially relevant soil degradation process in the area (Chap. 13, Fig. 13.4). In addition to this, veld fires occurring during the summer fire season and flooding of valley bottoms and lowland wetlands during the winter rainfall season are important ecological process in this region shaping land surface dynamics.

Concerning the soil erosion issue, we are investigating the applicability of the Bare Soil Index (adapted BI acc. to Rikimaru et al. 2002) to assess the soil erosion hazard caused by sparse vegetation cover and the exposure of bare soil. For this, we derived the BSI for each year from 2017 through 2020 based on the available (cloud-free) Sentinel 2 optical acquisitions (10 m resolution) and calculated basic statistics, that is the minimum, maximum, mean, median, standard deviation and variance for every year. To advance the understanding of the signal, we distinguished between pivot irrigated and rainfed fields. As expected, irrigated fields often have a higher vegetation cover and thus a lower BSI. The same is true for not cultivated land, like shrubland and grassland, which shows higher vegetation cover and a lower BSI. However, the mean annual BSI or a mean BSI over several years (not shown) is not very selective. Instead, we provide in Fig. 29.5, showing an 85 km^2 section of the intensively utilized hilly Southern Coastal Lowlands (Partridge et al. 2010) northwest of Elim, the mean intra-annual range of the BSI, which is a proxy for the management intensity of the areas. Shrublands and wooded valley bottoms are characterized by rather continuous vegetation cover and a low range of BSI while harvest operations on cropland induce large changes in vegetation cover and the BSI. Furthermore, on cultivated areas where tillage operations and the growth of different crops in different years supports a high range of BSI, a time series analysis of the annual range might indicate areas which were not worked on, i.e. fallow in a particular year. Thus, a field-specific analysis of the BSI over a number of years might provide a tool to identify the abandonment of farmland being a potential indicator of land degradation, if land abandonment is not due to socio-economic developments as e.g. recently documented for sites in the Eastern Cape (de la Hay and Beinhart 2017).

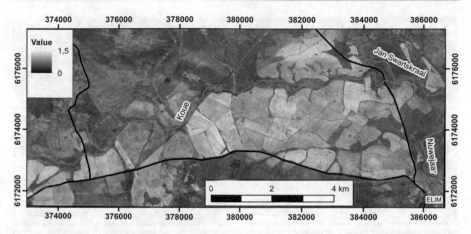

Fig. 29.5 Multi-annual (2017–2020) mean annual range of the bare soil index (BSI) in a section of the SALDi Overberg study site along the Koue River close to Elim. The eastern half of the section is part of the Nuwejaar Special Management Area (SMA). Cultivated areas are characterized by high intra-annual dynamics due to tillage operations, while neighbouring shrubland areas and valley bottoms are characterized by low intra-annual change, i.e. continuous vegetation cover. A closer look depicts vegetated contour bunds representing effective soil conservation measures on most fields (data source: Contains modified Copernicus Sentinel data 2017–2020, compilation: A. Hirner, J. Baade)

We further analysed time series of a Sentinel-2 based vegetation index (Normalized Difference Vegetation Index, NDVI) for the entire Overberg Study area (13,000 km^2) for abrupt land surface changes (breakpoints) between April 2018 and March 2020. Using the BEAST-breakpoint detection method (Zhao et al. 2019) allows not only to delineate the position of abrupt changes in space but also in time. Most of the identified abrupt changes can be related to fire events during the summer and are located in areas of (semi-)natural vegetation such as shrublands, grasslands and naturally wooded land (Fig. 29.6a–d). The largest veld fire scars identified were associated with the Riviersonderend and Bonnivale fire in December 2018 (Overberg FPA 2019) in the eastern part of the Riviersonderend Nature Reserve (Fig. 29.6c, e) and a veld fire east of Bredasdorp in the Overberg Test Range in September 2019. The presented Sentinel-2 based results at 10 m spatial resolution show good agreement with the established coarse resolution (500 m) MODIS burned area product (Fig. 29.6e, f). At the same time, they indicate that the higher resolution data is able to reduce the minimum mapping unit from 25 ha to less than 1 ha supporting post-fire management activities.

During the winter rainfall season, culminating in June through August with on average >50 mm of rainfall per month (personal comm. E. Hahndiek), valley bottom areas and specially the coastal lowland wetlands are prone to regular flooding. Here, flooding is not only a hazard to infrastructure, but a very important ecological process sustaining valley bottom and other wetlands (Ellery et al. 2016) in the area. As flooding often coincides with overcast conditions, synthetic aperture radar

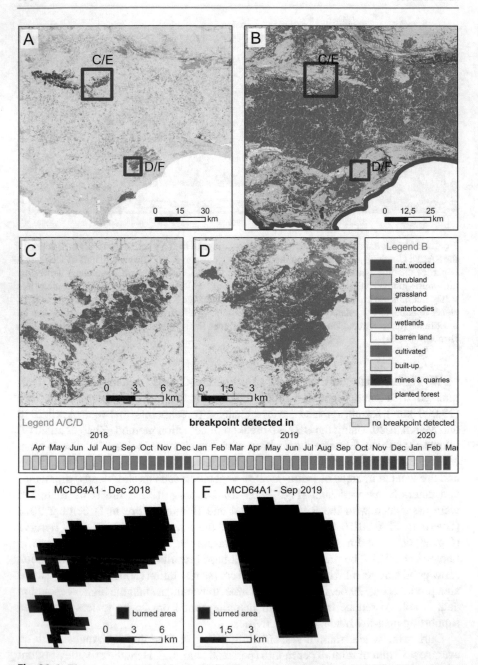

Fig. 29.6 Fire scars and their timing in the Overberg study area as detected by BEAST breakpoint analysis based on Sentinel-2 data for Apr/2018–Mar/2020. (**a**) Overview of breakpoint results for the entire Overberg site with zoom-ins (**c** & **d**) (data source: Contains modified Copernicus Sentinel data 2018–2020), (**b**) land cover of the Overberg site (LRI 2018), and (**e** & **f**) Burned areas as detected by MODIS Product MCD64A1 (Giglio et al. 2015, compilation: A. Hirner, U. Gessner)

(SAR) satellites, like Sentinel-1, penetrating clouds are a most useful instrument to map flooding extent and flooding frequency. This is based on the fact that (smooth) water surfaces are characterized by very low backscatter. To illustrate the capability and some limitations of this approach, we focus on an 815 km^2 area (excluding the ocean surface) located between Cape Agulhas and Bredasdorp (Fig. 29.7) and including parts of the Agulhas National Park and the Nuwejaars Wetlands SMA. Here we analysed the time series of Sentinel-1 VV backscatter images from July 2018 through November 2021, providing a total of over 100 raster images. Firstly, VV backscatter was subjected to adaptive speckle filtering using the Lee filter with a 5 by 5 window size, which preserves the local spatial variations and minimize classification error. Then, the speckle filtered linear intensity backscatter was converted into decibel (dB) for easier statistical evaluation. Using the method of grey-level thresholding (White et al. 2015) with a fixed threshold of −18.0 dB, we mapped surface water occurrence in every single image and finally added up the observations to provide a flooding frequency map showing areas covered by water, varying from rarely to constantly (Fig. 29.7). Comparison with higher spatial resolution optical imagery indicates, that the minimum mapping unit of water surfaces is about 2 ha, despite of a 20 m spatial resolution of the raster data. Another limitation results from the fact that wind causes surface roughness on water surfaces providing backscatter values which might be as high as from rough land surfaces. While still indicating the spatial extent of flooding, this clearly impedes a rigorous temporal flooding frequency analysis. Taking the ocean surface as a reference, pixels from the Ocean are classified as water on average in 40 out of 100 observations using the −18.0 dB threshold. The seemingly very wet coastal dune field west of Agulhas, highlights another textbook ready issue when analyzing SAR backscatter data. Keeping these uncertainties in mind, the map clearly shows the major pans and vleis in the region and their seasonal dynamics. Furthermore, a number of arable fields, especially in the Kars river valley bottom and prone to seasonal flooding and water logging are clearly visible. These fields have been claimed early last century in the framework of the channelling of the Kars river and are used for arable farming of wheat and canola or for fodder production and grazing.

29.4.2   Kai !Garib

The SALDi study site 2, Kai !Garib, centered around the Augrabies Falls National Park (AFNP) in the NW at the border with Namibia represents one of the driest parts of South Africa. With about 120 mm of mean annual rainfall and a similar amount of mean annual evapotranspiration the climate is arid around the year (Fig. 29.2). The area belongs to the Nama-Karoo biome and is characterized by sparse, grassy vegetation (Mucina and Rutherford 2006). Episodically water-filled, non-perennial valleys are characterized by rather dense bushes and trees. According to the DFFE (2021) land cover map 48% of the site is barren land, 50% is shrubland, and only slightly more than 1% is cultivated. The most prominent feature characterized by abundant riverine vegetation is the Orange River Valley, traversing the study

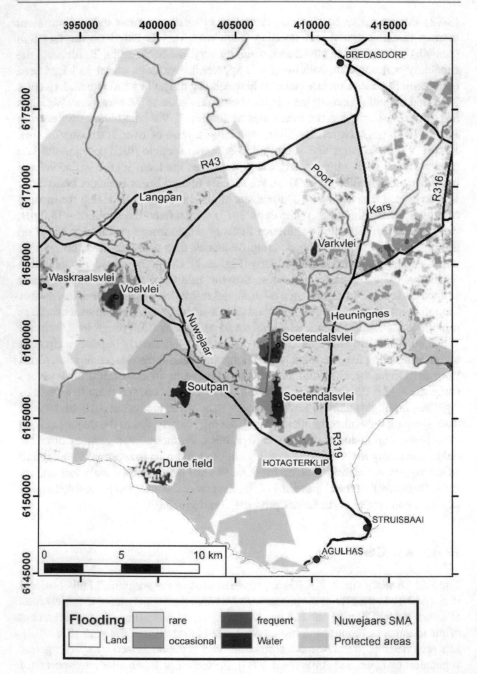

Fig. 29.7 Multi-annual (2018–2021) flooding frequency map of the Cape Agulhas region, a section of the SALDi Overberg study site. A considerable part of the study area belongs either to the Agulhas National Park or is part of the Nuwejaar Special Management (SMA). The analysis provides evidence of the seasonal dynamics of the coastal pans and vleis as well as the valley bottom wetlands in the region (data source: contains modified Copernicus Sentinel data 2018–2021, compilation: J. Baade, S. Selvaraj)

Table 29.4 South African National Land Cover data available for analysing the development of irrigated permanent vines and orchards in Kai !Garib, Overview (sources: GEOTERRAIMAGE 2015, 2016; LRI 2018; DEFF 2021)

Year	Spatial resolution [m]	No of classes	Class name(s) (abbr.)	Area [km^2]
1990	30	1 of 72	Permanent vines	20.98
1995	N.A.	1	Permanent irrigated	34.27
2014	30	2×3 of 72	Commercial permanent (Orchards/Vines)	35.111.47
2018	30	1 of 10	Cultivated	47.59
2020	20	2 of 47	Commercial permanent (Orchards/Vines)	48.790.64

area from East to West. Here, the perennial waters are the source for widespread irrigation agriculture, especially vineyards and orchards. While large parts of the Kai !Garib area have been rather stable over the past years, an ongoing expansion of irrigated areas along the Orange River can be observed.

Here we focus on the suitability of the digital South African National Land Cover data sets, available since 1990, to characterize land surface change by focusing on the development of irrigated areas in the Orange River valley in the vicinity of Augrabies. Over a period of 30 years, one needs to consider that mapping procedures changed due to technical developments, new methods and requirements. This concerns, for example, the recently increased spatial and spectral resolution. In brief, Table 29.4 provides an overview of the available data and provides insight into the changing land cover class definitions complicating the comparison over the years. A detailed quality check provides evidence for the 1995 land cover data set deviations from previous and following mapping principles. Hence, it was not considered for further analysis. Apart from this limitation, Fig. 29.8 indicates a noteworthy persistence of irrigated vines and orchards in the centre of the Orange River valley since 1990 (dark green colour) and the ongoing expansion of irrigated permanent cultures onto the higher slopes of the valley along the road from Kakamas (not shown) to Augrabies and further to the West as well as north of the river. Overall, the area of permanent orchards and vines has more than doubled from 21 km^2 to 49 km^2 in the analysed section over the past 30 years.

29.4.3 Sol Plaatje

The SALDi study site 3, Sol Plaatje, is located in the centre of the country at the border between the Northern Cape and the Free State provinces with Kimberley being located in the NE corner of the study site. Climate wise, the area is located in the summer rainfall zone (SRZ) and is characterized by a semi-arid climate based on an average annual rainfall amount of about 370 mm. Concerning vegetation, the study area is split into a northern part belonging to the Savanna Biome (Kimberley Thornveld) and the southern part belonging to Nama-Karoo Biome (Northern Upper

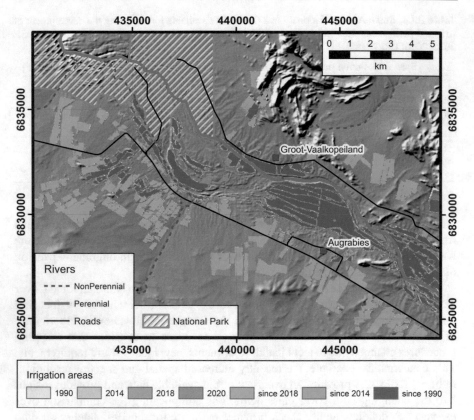

Fig. 29.8 Persistence and expansion of irrigated areas in the Orange River Valley upstream of Augrabies Falls National Park between 1990 and 2020. The brownish colours indicate the year of establishment of irrigated areas and the green colours the persistence of the areas for the given period, e.g. since 1990, since 2014, etc. (projection: UTM34S, data source: ESA 2021, GEOTERRAIMAGE 2015, 2016; LRI 2018; DFFE 2021, compilation: J. Baade)

Karoo) while along the rivers azonal, alluvial vegetation is present (Mucina and Rutherford 2006). According to the DFFE (2021) land cover map the study site is mainly characterized by grassland (77%); shrubland (8.7%), forested land (6.3%) and cultivated land (4.7%) are the other main land cover classes present in the region.

Here we focus on the role of land use and land tenure on the status of the land surface and vegetation cover, respectively. Figure 29.9 provides a detail of the Sentinel-1 radar-based Earth observation time series analysis for this area. It shows a mean backscatter image of the northern part of Mokala National Park (MONP), located about 50 km southwest of Kimberley and west of the N12 at Ritchie (Motswedimosa). The bright area in the lower center of the figure represents the steep rocky outcrops of dolerite forming vast mesas in the landscape. South of Ritchie a number of pivot irrigation fields are clearly discernible and to the

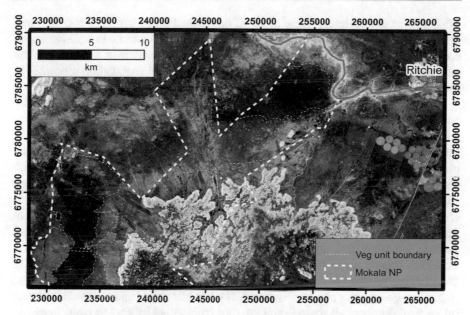

Fig. 29.9 Sentinel-1 mean polarized (VV) backscatter for the period July 2018 to November 2021 ($N = 208$) for Mokala National Park and surroundings (projection: UTM35S, data source: contains modified Copernicus Sentinel data 2018–2021, vegetation units from SANParks (2017), compilation: C. Pathe, J. Baade)

West of Ritchie the incised valley of the Riet River is marked by rather light hues. Apart from this one can observed considerable variation in the grayscale reflecting differences in the density of the vegetation. Where straight lines, angular features or circles dominate the pattern, these differences can clearly be attributed to human activities, that is agriculture. Some of the angular features are located even within the boundaries of MONP representing the legacy of prior agricultural activities in this area, which was proclaimed a protected site in 2007 (SANParks 2017). Besides these feature, clear differences and patterns are visible in many areas. Within the MONP boundaries, these patterns correspond very well with the vegetation units identified by SANParks (2017, p. 138) based on optical imagery. At the same time, one can distinguish in places differences in hue on either side of the National Park boundary indicating slight differences in vegetation density due to slightly different management. This clearly highlights the potential of radar-based Earth Observation to depict differences in vegetation in a savanna setting.

29.4.4 Mantsopa

Located in the Highveld grassland ecosystem (1300–1700 m asl., Mucina and Rutherford 2006), the SALDi study site 4, Mantsopa, is characterized by plateau mountains with extensive rangeland, pastures and arable land at the lower altitudes.

With a mean annual temperature of 16 °C and annual precipitation of 500–700 mm (cf. Fig. 29.2), the region is situated in the summer rainfall region. In addition to soil erosion (Le Roux et al. 2008), the loss of pristine vegetation cover, changes in plant composition and bush encroachment represent major land degradation processes in the area (Mucina and Rutherford 2006).

Here we present an example for the monitoring of bush encroachment by analyzing the dynamics of slangbos (*Seriphium plumosum*), also known as bankrupt bush, over the recent years. Slangbos has been documented to increasingly spread into pastures and fallow land (Snyman 2012; Avenant 2015) diminishing the productivity of agricultural land and is considered the main encroaching shrub species in the area. The study focuses on a 1375 km^2 section of the SALDi Mantsopa study area between Maseru in the SE and Sandspruit in the NW. This area is characterized by strong human induced land surface dynamics as is evident from the Sentinel-1 radar backscatter RGB composite demonstrating the inter-annual surface and vegetation dynamics in the area (Fig. 29.10). Arable land is clearly visible in mostly green hues due to high backscatter of predominantly maize in the growing season (fall). This seasonal effect is less pronounced for grassland (e.g. olive-green areas close to Sandspruit) while hill slopes and settlements scatter intensely throughout all seasons and are thus represented by gray to white hues.

Against this background, slangbos sprawl is analysed using a continuous time series of (1) the backscatter and (2) the interferometric phase coherence from the Sentinel-1 C-band radar system for the 6-year period from 2015 to 2020 (Fig. 29.11). To validate the findings based on high-resolution optical imagery, 45 sampling sites characterized by slangbos encroachment were selected in the NE part of the SALDi Mantsopa study site (Fig. 29.10). The sampled area covers in total about 122 ha and equals 1357 pixels of 30 × 30 m.

In the time series analysis (Fig. 29.11), the smoothed S-1 VH backscatter shows an undulating progress over the course of a year peaking every summer. In addition, one can identify an overall strong inter-annual increasing trend for the sites with slangbos sprawl as evident from the increasing annual minima, mean and maxima. This trend can be attributed to the spreading of water-filled vegetation, that is slangbos. The seasonal amplitude of S-1 VH backscatter reflects the strong alternating wet and dry seasons, that is an increase of soil moisture and vegetation (structure and water content) coinciding with the beginning of the growing season. The rather low backscatter in spring 2015 and summer 2016 is a result of the poor vegetation development due to the strong droughts of the years 2014–2016 (Archer et al. 2017) clearly visible as well in optical data from that time (Urban et al. 2021). Featuring a rather similar undulating but lagged signal compared to the S-1 VH backscatter, the smoothed 14-day repeat-pass S1 VV coherence decreases with leaf development and increases with senescence of vegetation. Since coherence is related to vegetation properties, it is an additional and independent parameter for land surface structure.

Based on the observed trends of the two named radar signals, a slangbos occurrence map was prepared and discussed in detail for a number of sites in the Mantsopa study area (Urban et al. 2021). In addition to this, a detailed 65 km^2

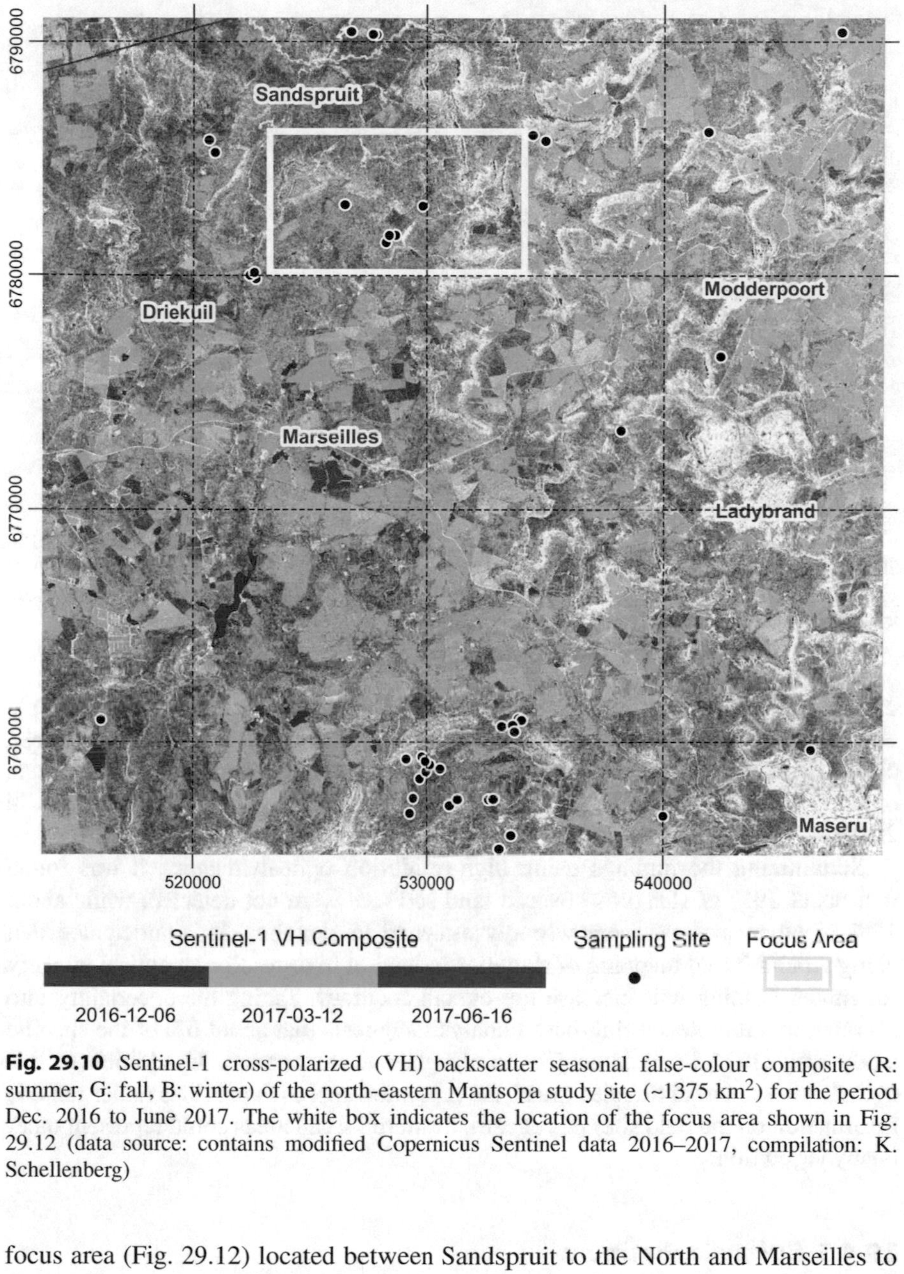

Fig. 29.10 Sentinel-1 cross-polarized (VH) backscatter seasonal false-colour composite (R: summer, G: fall, B: winter) of the north-eastern Mantsopa study site (~1375 km^2) for the period Dec. 2016 to June 2017. The white box indicates the location of the focus area shown in Fig. 29.12 (data source: contains modified Copernicus Sentinel data 2016–2017, compilation: K. Schellenberg)

focus area (Fig. 29.12) located between Sandspruit to the North and Marseilles to the South (see Fig. 29.10 for the location of the focus area) is presented here. The colours show surface areas (ground pixel size: 30 × 30 m) where the Sentinel-1 based slangbos detection model indicated the dominant presence of slangbos (>50% cover) in a given year. Areas with persistent slangbos cover are presented in white. Apart from some stable slangbos areas distributed throughout the focus area and

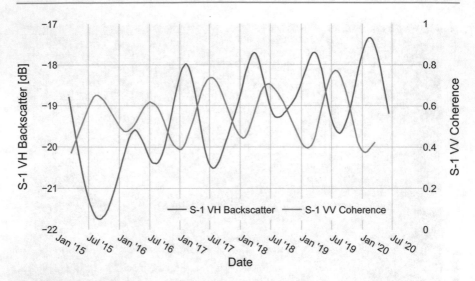

Fig. 29.11 Sentinel-1 (S-1) radar backscatter (blue) and coherence time series (orange) 2015–2019 for areas with observed slangbos encroachment. Both the aggregation to a 30 × 30 m pixel spacing and time series smoothing helped leveraging the high signal uncertainty inherent to imaging radar (compilation: K. Schellenberg)

often associated with escarpments, the mapping provides evidence of considerable land surface dynamics. While some areas like in the SW have obviously been cleared from slangbos at the beginning of the observation period, i.e. in 2016, others reached the status of being slangbos encroached only at the end of the period, i.e. in 2019.

Scrutinizing the findings using high-resolution optical imagery, it was found that about 19% of slangbos-covered land surfaces were not detected, while about 15% of other surfaces were wrongly assigned to slangbos. It is anticipated that using ground-based mapping of slangbos instead of high-resolution optical imagery for model training will increase the overall accuracy. Taking the uncertainty into account, this remote sensing-based analysis suggests that about 6% of the specific study area (1375 km^2) is subject to slangbos encroachment. In addition to this overall assessment of affected areas, the presented approach provides rather reliable information on the field sites of concern. Scattered small areas could represent other bushy vegetation.

29.4.5 Bojanala Platinum

The SALDi study site 5, Bojanala Platinum, is located in the summer rainfall region in the North of the country close to the border with Botswana. With about 550 mm of mean annual rainfall over the whole area, the land is largely used for livestock farming as indicated by the high proportion of grassland (42%) and

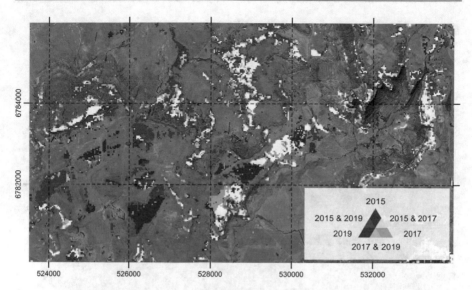

Fig. 29.12 Sentinel-1 based slangbos occurrence map for the years 2015–2019 (projection: UTM35S). Colours indicate detection of slangbos coverage (>50%) in the given years. White hues indicate persistent occurrence of slangbos (data source: contains modified Copernicus Sentinel data 2015–2019, compilation: K. Schellenberg)

forested land (36%) (DFFE 2021) representing often savannas, more specifically the Central Bushveld Bioregion (Mucina and Rutherford 2006). About 18% of the area is cultivated and less than 1% of the cultivated area is under pivot irrigation, highlighting the importance of rainfed arable farming in the Bojanala Platinum region. As visible in the South African National Land Cover map (LRI 2018) reproduced in Fig. 29.13, one can identify three major regions in the study area: the southern part is dominated by grassland and arable land, partly intensively used through pivot irrigation schemes. The central, rather mountainous part is cleanly dominated by forests and grasslands. In the north, the landscape is increasingly flat and dry and not suitable for arable farming and mostly covered by shrublands characteristic of a savanna type landscape.

Here, we analyzed the dynamics of the relatively stable, natural vegetation types of Bojanala Platinum at interannual scale based on the Sentinel-2-derived BSI (Rikimaru et al. 2002). We computed the maximum BSI per year for 2017–2021 as an indicator of the maximum fraction of exposed soil for each year under investigation. In addition, we calculated mean values of maxBSI for the land cover types woodland, grassland and shrubland, according to the South African National Land Cover (LRI 2018) (Fig. 29.13) and compared them to annual CHIRPS precipitation sums (Funk et al. 2015).

Figure 29.13 shows relatively little variations of maxBSI in a relatively stable study site that is only affected by minor land use change. The most conspicuous differences are related to striking patterns of patches with high bare soil fractions

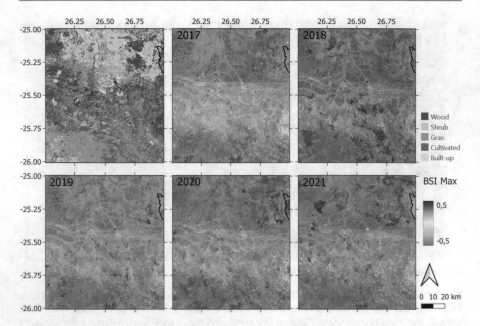

Fig. 29.13 Maximum bare soil index (maxBSI) per year, based on Sentinel-2 time series for the years 2017–2021. Low BSI values indicate low and high BSI values high bare soil fractions. Westernmost part of the Pilanesberg National Park borders indicated by black Polygon in the NE of the maps (data source: land cover: LRI 2018, contains modified Copernicus Sentinel data 2017–2021, compilation: A. Hirner)

(i.e. high maximum BSI values) which are discussed below (Fig. 29.15). Apart from these patches, there are general trends that demonstrate that even in this relatively stable landscape, the Sentinel-2-based BSI allows to reveal general temporal dynamics in the vegetation cover and hence changes in the condition of land cover types. The BSI values of 2018, which were affected by drought, are generally slightly higher and therefore, the corresponding image is darker red than in the other years. These subtle variations become particularly obvious when regarding the land-cover-based statistics of mean annual BSI (Fig. 29.14).

Figure 29.14 shows the annual mean BSI for three major land cover classes of the Bojanala Platinum study site. For all years, the diagram depicts—as expected—lowest bare soil fractions in woodlands, followed by shrublands, and grasslands with the highest proportion of open soil. The 2018 and 2019 droughts lead to lower vegetation cover and thus to increased bare soil fractions for all three land cover types in 2018 and 2019, while the effect for woodland is less pronounced than for shrubland and grassland. Large fire scars can be observed as patches of high BSI values in all years. Their abundance, location, size and extent, however, vary from year to year. The drivers of fire are manifold and their interaction complex (Archibald et al. 2009). Therefore, the fire activity depicted in Fig. 29.15 is not necessarily associated with annual rainfall variation.

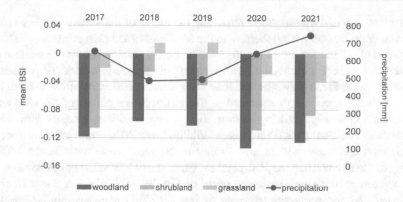

Fig. 29.14 Mean bare soil index (BSI) per land cover based on Sentinel-2 time series for the years 2017–2021. Land cover types were extracted from Land Resource International (LRI 2018). Lower BSI values show lower bare soil and higher vegetation cover fractions. In addition, annual precipitation sums (CHIRPS data, Funk et al. 2015) for the study site are shown in blue (compilation: U. Gessner)

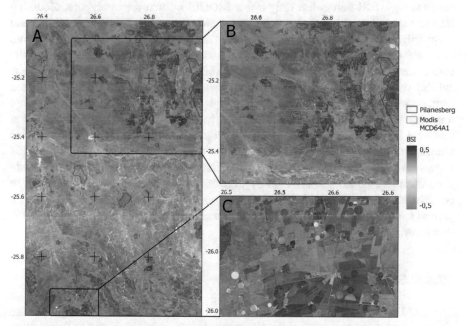

Fig. 29.15 Details of Sentinel-2 bare soil index (BSI) 2020 time series. The two main causes of distinctive patches of high BSI values are either fires or agricultural activities like tilling. All images show the BSI together with polygons of the MODIS burned area product (MDC64A1, Giglio et al. 2015) of 2020. Almost all dark red patches in inset **b** can be correlated to burned areas, whereas the patterns in inset **c** are mostly associated with fields of various types (data source: contains modified Copernicus Sentinel data 2017–2021, compilation: A. Hirner, F. Thonfeld)

A more detailed investigation regarding the cause of distinctive patches of high BSI was done by comparing them with the MODIS burned area product MCD64A1 (Giglio et al. 2015) for the years 2017–2021. Figure 29.15a shows the western part of the site together with MODIS burned area polygons. Most dark red patches are associated with these polygons and thus show that the BSI is sensitive to fresh burn scars. This is especially pronounced in the northwestern corner of the site, which is highlighted in Fig. 29.15b. During 2020 strong wildfires, roughly 50% of the Pilanesberg National Park (Pilanesberg Wildlife Trust 2020) were severely affected, and burned areas are evident also further to the west. Here most of the MODIS polygons clearly overlap the BSI patches. Particularly low values of maximum BSI can be seen in the south-western corner of Fig. 29.15b, where a water reservoir is located. At the north-eastern edge of Fig. 29.15b, open-cast mining areas are visible that show comparatively low maximum BSI values. There is hardly any vegetation cover there. However, the spectral signature of the exposed rock and the unweathered tailings differs from that of bare soil, resulting in lower BSI values.

A different situation is presented in the southern centre of Fig. 29.15a, which exhibits high BSI values but only a few MODIS burned area polygons. Zooming into this area (Fig. 29.15c) shows a dense pattern of arable fields, both rainfed and pivot irrigation. Some of the irrigated fields are seemingly almost always covered by vegetation and hence show some of the lowest BSI values encountered in the entire scene. However most of the rainfed fields and some pivot irrigation areas exhibit very high BSI values, indicating that the soil is exposed at least once a year, most probably during tilling. The high BSI values in this area are visible not only in 2020 but in all years covered in our investigation and confirm that this area is one of the few areas in the Bojanala Platinum site where intensive arable agriculture is practiced. Compared to this, the central part of Fig. 29.15a shows lower BSI maximum values. This region is predominantly covered by woody vegetation and hence less affected by intra-annual changes in vegetation cover. In contrast to cultivated land with large variation due to management (e.g. tillage), natural and semi-natural woody vegetation is characterized by less pronounced BSI dynamics.

29.4.6 Ehlanzeni

The SALDi study site 6, Ehlanzeni, is located in the summer rainfall region in the Northeast of the country. Covering the transition from the highveld in the western part of the study area to the lowveld in the central and eastern part, it is characterized by considerable diversity. Climate wise the difference in altitude results not only in considerable differences in mean air temperature, but also in precipitation amounts. Indigenous forests and plantation forest, as well as grassland is characteristic for the highveld areas, while plantations, mixed farming and extensive settlement areas are to be found in the lowveld with the exception of the part belonging to the Kruger National Park (KNP) and neighbouring private game reserves. Specially, the KNP, which has never been affected by European style agriculture and set aside as a protected area more than 100 years ago, can be considered a pristine benchmark site providing the opportunity to study land cover dynamics in semi-arid areas under near-natural conditions.

In this section, we present an example highlighting the differences of surface moisture conditions in the pristine benchmark site on the one hand and the neighbouring areas under human impact on the other. As an approach to support land management on a local and regional level, an example of multidimensional land surface change analysis in one of the most thriving local municipalities of South Africa, that is Bushbuckridge, is provided.

During 2015 and 2016, South Africa faced one of the most severe meteorological droughts due to exceptional ENSO (El Niño Southern Oscillation) conditions (Di Liberto 2016), causing a delay in the start and a strong reduction in the amount of rainfall during the summer rainfall season. Despite the Cape Town region dominating international headlines, according to Di Liberto (2016), the Mpumalanga province in northeastern South Africa was as severely impacted by this drought, despite it being less publicized.

Figure 29.16 shows a comparison of the mean bi-monthly surface moisture conditions expressed as Surface Moisture Index (SurfMI, Urban et al. 2018) at the end of the dry seasons 2016 and 2017 for the SALDi Ehlanzeni site. The approach is based on Sentinel-1 backscatter time series and a change detection algorithm developed by Wagner et al. (1999) and Naeimi et al. (2009). This radar-retrieved surface moisture data reflects the water content of either the vegetation or the soil surface depending on the surface cover dominating the Sentinel-1 backscatter. In combination with optical products on canopy cover and phenology information, it is possible to separate biophysical versus geophysical spatio-temporal patterns. The Bare Soil Index (BSI) (Rikimaru et al. 2002) derived from optical Sentinel-2 data was used to mask out areas with dense vegetation cover, that is areas with a BSI < 0.02 according to Diek et al. (2017). To avoid radiometric effects, slopes steeper than 15% (based on the Copernicus GLO-30 Digital Elevation Model, ESA 2021) were also masked out. Thus, the coloured areas in Fig. 29.16 represent pixels with only moderate slope and either no or sparse vegetation.

The surface moisture conditions at the end of the dry season 2016 and 2017 in the SALDi Ehlanzeni study region, dominated by the southern Kruger National Park (KNP), is used to (1) show the variation in surface moisture conditions at the end of the severe drought and (2) to document, 1 year later, the recovery of the area from the drought. At the end of the winter 2016, most areas inside and outside the KNP are characterized by exceptionally dry conditions. Hilly areas, for example around Pretoriuskop in the SW of the KNP) and the footslopes of the escarpment in the west, are characterized by slightly more moist conditions. Outside the Park, moist areas are observed most probably due to irrigation. Inside the Park, one can distinguish variations in the surface moisture conditions which coincide with geological and soil patterns. A good example is the North-South striking wide strip indicating very dry conditions close to the eastern boundary of the Park, which corresponds to the occurrence of basalts and the corresponding clayey soils as compared to the granitic rocks and sandy soils to the East and West of

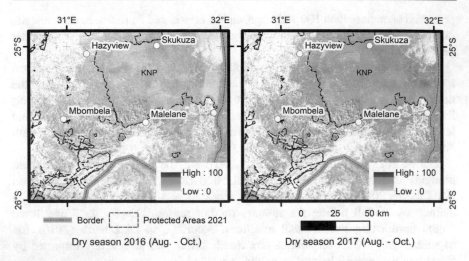

Fig. 29.16 Bi-monthly average of Sentinel-1 retrieved surface moisture index SurfMI at the end of the dry seasons 2016 and 2017 for SALDi project region Ehlanzeni illustrating the impact of the severe 2015/2016 drought and the recovery. White regions represent masked pixels due to either dense vegetation or sloped relief. Black polygons show protected areas, southern KNP is covering the upper right corner (data source: contains modified Copernicus Sentinel data 2016–2017, compilation: M. Urban, C. Pathe, J. Baade)

this feature (CGS 2019). The pattern visible in the western part of the Park strongly represents the distribution of thickets and woodlands in this area (Gertenbach 1983). Most interestingly, a year later, at the end of the 2017 dry season, these patterns are obliterated and it is postulated that the spatial variation of surface moisture conditions reflects spatial variations in rainfall amounts during the previous summer.

Located at the NW corner of the SALDi Ehlanzeni study area is the Bushbuckridge focus area of the EMSAfrica (Ecosystem Management Support for Climate Change in Southern Africa) project. This covers the Bushbuckridge Local Municipality as well as adjacent areas of the escarpment and highveld to the West and the lowveld to the East, that is the private game reserves like MalaMala and Sabie Sands as well as parts of the KNP (Fig. 29.17). This area is very diverse concerning natural resources and conditions as well as the utilization of the landscape making it a textbook example for a multi-use landscape.

For this area, we provide an example of a land-use and land-cover (LULC) change and trend analysis based on Sentinel-2 optical earth observation data time series analysis for a period of 7 years starting in 2015. The targeted changes are savanna browning and greening trends, afforestation and deforestation, fire events, settlement expansion, as well as water level changes in reservoirs. To identify LULC change, all Sentinel-2 data acquired from 2015 to 2021 at a temporal resolution of 2–3 days and a spatial resolution of up to 10 m were analysed. To improve change detection, firstly, the annual peak of the phenological season (i.e. the growing season) acquisitions was identified.

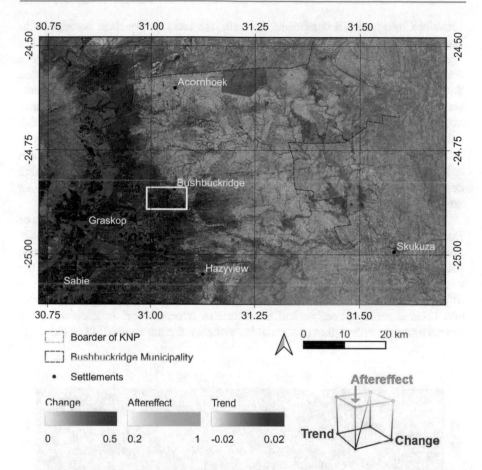

Fig. 29.17 Change-Aftereffect-Trend (CAT) Analysis 2015–2021 derived from annual peak of the phenological season Tasseled Cap Disturbance (TC D) analysis at 10 m resolution. The white frame indicates the large-scale section shown in Fig. 29.18 (data source: contains modified Copernicus Sentinel data 2015–2021, compilation: P. Renner)

Based on these, the Change-Aftereffect-Trend (CAT) analysis (Hird et al. 2016) developed in FORCE (Framework for Operational Radiometric Correction for Environmental monitoring) (Frantz 2019) on the Tasseled Cap-Disturbance index (TC-D, Healey et al. 2005) was performed. The TC-D is an index created to highlight relevant vegetation changes, but due to its complex combination of multiple bands, other LULC changes can be accentuated as well. The bands used are combined into three sub-indices TC-Wetness, TC-Brightness, and TC-Greenness from which the TC-D is derived [TC-Disturbance = TC-Brightness − (TC-Greenness + TC-Wetness)]. Details of the workflow are documented in Renner (2022).

In Fig. 29.17, the results of the CAT analysis, that is the most significant change from the time series, the average value after the change and the overall trend, are combined into an RGB composite. Overall, one can identify three broad, North-South striking zones with dominant change and trend patterns: The highveld in the West and including most part of the escarpment and the foothills, where blueish hues dominate, is characterized by a clear browning trend of the vegetation. Between Sabie, Graskop, Bushbuckridge and Hazyview forestry activities, that is clear-cuts and afforestation, are clearly detected by the dominant reddish hues. The centre of the area is characterized by bluish hues indicating browning and light greenish colours indicating stable conditions, for example on grassland, for roads and in settlements. In the East, that is the protected areas including KNP, yellow to greenish colours indicate a combination of sudden change and a trend towards greening despite rather sparse vegetation. This greening trend could however be mainly an artefact of the period of observation which started in the 2015/2016 drought.

The large-scale section focusing on the surroundings of the Inyaka Dam south of Bushbuckridge (Fig. 29.18) clearly shows the high spatial resolution of the analysis, providing the opportunity to identify spatial features in great detail. These include the road (R533) and the aisle of the power line crossing the dam, forest and other roads and settlement patterns in Bushbuckridge. In particular, boundaries between different land cover and land use areas are characterized by sharp delineations and become more noticeable and accessible to interpretation. In addition to the observations described above, it provides evidence for the application of the CAT-

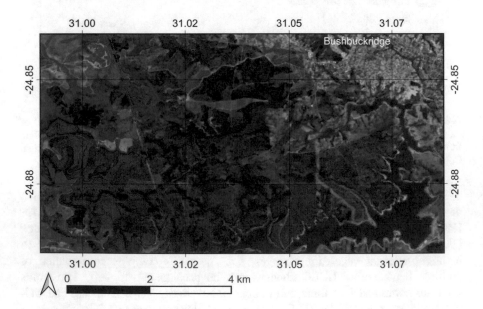

Fig. 29.18 Change-Aftereffect-Trend (CAT) Analysis 2015–2021 derived from annual peak of the phenological season Tasseled Cap-Disturbance (TC-D) analysis at 10-m resolution. Section focusing at the Inyaka Dam south of Bushbuckridge. For the color legend see Fig. 29.17 (data source: contains modified Copernicus Sentinel data 2015–2021, compilation: P. Renner)

analysis to highly dynamic processes like the change of water levels in reservoirs. These are evident from the pinkish frame around the blue water body of the dam indicating a mixture of sudden significant change and a continuous drying trend. In the West, the CAT times series analysis of the boundary of the municipality, clearly marked in the field by a fire break, indicates marked differences in the management of forested areas on either side of the boundary. This can be attributed to more homogeneous forests plots and plot-wise management (clear-cuts and afforestation) west of the boundary. In Bushbuckridge, one can distinguish a pattern of light greenish colours indicating rather stable conditions and turquoise colours indicating change due to vegetation removal and exposure of bare soil in the framework of building activities.

29.5 Summary of Implementation Opportunities and Conclusions

This chapter showcased examples for Earth-observation-based support for regional policy implementations against the background of land surface change under anthropogenic and climate pressure. Based on long Sentinel-1 and Sentinel-2 time series assembled in the SALDi data cube, we present a wide range of application approaches to analyse land surface and vegetation dynamics. Table 29.5 provides an overview and summary of the surface processes and properties targeted, the methods implemented and an evaluation of possible obstacles. The table shows that all data and procedures needed for an analysis of the addressed processes are available. The necessary satellites exist and the data streams are guaranteed for decades to come. However, an operational retrieval of land surface dynamics and degradation indicators need hard- and software implementations as described in Sect. 24.4 (Chap. 24). Furthermore, expertise of analysts in the field of optical and radar remote sensing, support from local experts and in-situ data is essential. In addition, data on land properties as well as temporal changes on the surface (including, e.g. precipitation) facilitates the interpretation of the observations. Furthermore, to evaluate observed land surface trends, consistent long duration time series are needed against the background of the high inter-annual and decadal variation of precipitation characterizing semi-arid climates (e. g., Fig. 29.4).

As a conclusion, further research is indeed needed on validating and re-fining the existing procedures and products with respect to spatial and temporal detail and accuracy. For example investigations of radar-optical synergies have only started, and longer time-series will result in better pattern recognition. National requirements on thematic contents of the remote sensing retrieved maps have to be considered, too. A close and continuous dialogue between Earth observation experts and stakeholders is considered most beneficial for both scientific excellence and programmatic implementation for sustainable land management. The results presented showcase local to regional investigations and cannot—at this stage— be extrapolated to national assessments. Since the presented results are based on physical interactions, regularities exist (e.g. for slangbos monitoring). However,

Table 29.5 Summary of examples for observational support for regional policy implementations with respect to land surface processes and properties

Study sites	Processes and properties	Implementation	Obstacles
1 Overberg	Land management (bare soil)	Easy implementation: Sentinel-2 Bare soil index (BSI) multi-temporal statistics maps based on existing products and methodologies.	None, but BSI not easy to convert to proportion of bare soil (see Table 29.3)
	Land surface and vegetation dynamics (general and fire)	Feasible implementation: Sentinel-2 NDVI multi-temporal statistic and breakpoint analyses methodologies available.	None, but a multi-year time series is required
	Land surface dynamics (flooding)	Easy implementation: Sentinel-1 backscatter is sensible to water surfaces.	None, but water surface roughness impacts detection of water bodies
2 Kai!Garib	Land surface and vegetation dynamics (land use change)	Easy implementation: usage of South African National Land Cover (SANLC) map products.	None, but classification and terminology changed over time
3 Sol Plaatje	Land surface and vegetation dynamics (vegetation mapping)	Feasible implementation: Sentinel-1 multi-temporal statistics supports subtle grassland and savanna structural mapping, comparable to optical approaches.	None, but requires radar backscatter understanding
4 Mantsopa	Land surface and vegetation dynamics (bush encroachment)	Challenging implementation: Sentinel-1 multi-temporal backscatter and coherence analysis for bush encroachment mapping needs understanding of radar scattering mechanisms.	None, but Sentinel-1 coherence analysis requires radar expertise and in-situ observations for model building
5 Bojanala Platinum	Vegetation dynamics (inter-annual change)	Easy implementation: combination of Sentinel-2 BSI and MODIS burnt area products are readily available.	None
6 Ehlanzeni	Land surface and vegetation dynamics (surface moisture)	Feasible implementation: Sentinel-1 Surface Moisture (SurfMi-model) is available, but constraints regarding vegetation cover and topography need to be understood.	None, but Sentinel-1 time series need to be modelled, which requires radar and local expertise
	Land cover and land use change	Feasible implementation: Sentinel-2 CAT trend analysis and FORCE modelling framework is available.	None, but Tasseled Cap-procedure needs to be understood thoroughly

these need to be consolidated, as suggested above and adapted to specific regional concerns and requirements, especially in a diverse country like South Africa.

Acknowledgement We like to thank all people who contributed to the ideas developed in this chapter during field work or in numerous discussions. This contribution benefited from funding by the Federal Ministry of Education and Research (BMBF) for the SPACES2 Joint Projects: South Africa Land Degradation Monitor (SALDi) (BMBF grant 01LL1701 A-D) and EMSAfrica—Ecosystem Management Support for Climate Change in Southern Africa—Subproject 4: Remote Sensing Based Ecosystem Monitoring (01LL 1801 D) within the framework of the Strategy "Research for Sustainability" (FONA) www.fona.de/en. We are very grateful to four Generations of PK Albertyns for collecting rainfall data over a century on their farm and Pieter Albertyn who made that data available to us. We thank the reviewer, A. Skowno and the editor, G. von Maltitz, for their critical and helpful comments to an earlier draft of this manuscript.

References

Archer ERM, Landman WA, Tadross MA, Malherbe J, Weepener H, Maluleke P, Marumbwa FM (2017) Understanding the evolution of the 2014–2016 summer rainfall seasons in southern Africa: key lessons. Clim Risk Manag 16:22–28. https://doi.org/10.1016/j.crm.2017.03.006

Archibald S, Roy DP, Van Wilgen BW, Scholes RJ (2009) What limits fire? An examination of drivers of burnt area in Southern Africa. Glob Chang Biol 15:613–630. https://doi.org/10.1111/j.1365 2486.2008.01754.x

Avenant P (2015) Report on the National Bankrupt Bush (Seriphium plumosum) Survey (2010-2012). https://doi.org/10.13140/RG.2.2.27655.50088

CGS (Council for Geoscience) (2019) Geological map of the Republic of South Africa and the Kingdoms of Lesotho and Swaziland 1:1.000.000. Council of Geoscience, Pretoria

de la Hey M, Beinart W (2017) Why have South African smallholders largely abandoned arable production in fields? A case study. J South Afr Stud 43(4):753–770. https://doi.org/10.1080/03057070.2016.1265336

DFFE (Department of Forestry, Fisheries and the Environment) (2021) South African National Land-Cover 2020 accuracy assessment report. Vers. V1.0.4. DEFF, Pretoria. https://egis.environment.gov.za/data_egis/

Di Liberto T (2016) A not so rainy season: drought in Southern Africa in January 2016. https://www.climate.gov/news-features/event-tracker/not-so-rainy-season-drought-southern-africa-january-2016. Accessed 25 Apr 2018

Díaz S, Demissew S, Carabias J, Joly C, Lonsdale M, Ash N, Larigauderie A, Adhikari JR, Arico S, Báldi A, Bartuska A, Baste IA, Bilgin A, Brondizio E, Chan KMA, Figueroa VE, Duraiappah A, Fischer M, Hill R, Koetz T, Leadley P, Lyver P, Mace GM, Martin-Lopez B, Okumura M, Pacheco D, Pascual U, Pérez ES, Reyers B, Roth E, Saito O, Scholes RJ, Sharma N, Thaman R, Watson R, Yahara T, Hamid ZA, Akosim C, Al-Hafedh Y, Allahverdiyev R, Amankwah E, Asah ST, Asfaw Z, Bartus G, Brooks LA, Caillaux J, Dalle G, Darnaedi D, Driver A, Erpul G, Escobar-Eyzaguirre P, Failler P, Mokhtar Fouda AM, Fu B, Gundimeda H, Hashimoto S, Homer F, Lavorel S, Lichtenstein G, Mala WA, Mandivenyi W, Matczak P, Mbizvo C, Mehrdadi M, Metzger JP, Mikissa JB, Moller H, Mooney HA, Mumby P, Nagendra H, Nesshover C, Oteng-Yeboah AA, Pataki G, Roué M, Rubis J, Schultz M, Smith P, Sumaila R, Takeuchi K, Thomas S, Verma M, Yeo-Chang Y, Zlatanova D (2015) The IPBES conceptual framework – connecting nature and people. Curr Opin Environ Sustain 14:1–16. https://doi.org/10.1016/j.cosust.2014.11.002

Diek S, Fornallaz F, Schaepman ME, de Jong R (2017) Barest Pixel Composite for agricultural areas using landsat time series. Remote Sens 9:1245. https://doi.org/10.3390/rs9121245

Ellery WN, Grenfell SE, Grenfell MC, Powell R, Kotze DC, Marren PM, Knight J (2016) Wetlands in southern Africa. In: Knight J, Grab SW (eds) Quaternary environmental change in Southern Africa: physical and human dimensions. Cambridge University Press, Cambridge, pp 188–202. https://doi.org/10.1017/CBO9781107295483.012

ESA (European Space Agency) (2021) Copernicus Global Digital Elevation Model, COP-DEM-GLO-30. Paris. https://doi.org/10.5270/ESA-c5d3d65

Frantz D (2019) FORCE – Landsat + Sentinel-2 analysis ready data and beyond. Remote Sens 11(9):1–21. https://doi.org/10.3390/rs11091124

Friedl MA, Sulla-Menashe D, Tan B, Schneider A, Ramankutty N, Sibley A, Huang X (2010) MODIS Collection 5 global land cover: algorithm refinements and characterization of new datasets. Remote Sens Environ 114:168–182. https://doi.org/10.1016/j.rse.2009.08.016

Funk C, Peterson P, Landsfeld M, Pedreros D, Verdin J, Shukla S, Husak G, Rowland J, Harrison L, Hoell A, Michaelsen J (2015) The climate hazards infrared precipitation with stations—a new environmental record for monitoring extremes. Scientific Data 2(1):150066. https://doi.org/10.1038/sdata.2015.66

GEOTERRAIMAGE (South Africa) (2015) 2013–2014 South African National Land-Cover Dataset. Data User Report and MetaData. Vers. 05. Department of Environmental Affairs, Pretoria

GEOTERRAIMAGE (South Africa) (2016) 1990 South African National Land-Cover Dataset. Data User Report and Metadata. Vers. 05#2. Department of Environmental Affairs, Pretoria

Gertenbach W (1983) Landscapes of the Kruger National Park. Koedoe 26(1):9–121. https://doi.org/10.4102/koedoe.v26i1.591

Giglio L, Justice C, Boschetti L, Roy D (2015) MCD64A1 MODIS/Terra+Aqua Burned Area Monthly L3 Global 500m SIN Grid V006 [Data set]. NASA EOSDIS Land Processes DAAC. https://doi.org/10.5067/MODIS/MCD64A1.006. Accessed 22 Feb 2022

Harris I, Osborn TJ, Jones P, Lister D (2020) Version 4 of the CRU TS monthly high-resolution gridded multivariate climate dataset. Scientific Data 7(1):109. https://doi.org/10.1038/s41597-020-0453-3

Healey SP, Cohen WB, Zhiqiang Y, Krankina ON (2005) Comparison of Tasseled Cap-based Landsat data structures for use in forest disturbance detection. Remote Sens Environ 97(3):301–310. https://doi.org/10.1016/j.rse.2005.05.009

Hird JN, Castilla G, McDermid GJ, Bueno IT (2016) A simple transformation for visualizing non-seasonal landscape change from dense time series of satellite data. IEEE J Select Topics Appl Earth Observ Remote Sens 9(8):3372–3383. https://doi.org/10.1109/JSTARS.2015.2419594

Le Roux JJ, Morgenthal TL, Malherbe J, Pretorius DJ, Sumner PD (2008) Water erosion prediction at a national scale for South Africa. Water SA 34(3):305–314. https://doi.org/10.4314/wsa.v34i3.180623

LRI (Land Resources International) (2018) Automated Land Cover Classification South Africa. Final Report – SSC WC 03(2017/2018) DRDLR. Land Resources International, Pietermaritzburg

Mucina L, Rutherford MC (eds) (2006) The vegetation of South Africa, Lesotho and Swaziland. Strelitzia 19. South African National Biodiversity Institute, Pretoria

Naeimi V, Scipal K, Bartalis Z, Hasenauer S, Wagner W (2009) An improved soil moisture retrieval algorithm for ERS and METOP scatterometer observations. IEEE Trans Geosci Remote Sens 47(7):1999–2013. https://doi.org/10.1109/TGRS.2008.2011617

Overberg FPA (Greater Overberg Fire Protection Association) (2019) Fire season 2018/19. Conting the costs. https://overbergfpacoza/a-devastating-fire-season-the-gofpa-counts-the-costs/. Accessed 10 Apr 2022. Last updated: 2019-06-12

Partridge TC, Dollar ESJ, Moolman J, Dollar LH (2010) The geomorphic provinces of South Africa, Lesotho and Swaziland: a physiographic subdivision for earth and environmental scientists. Trans R Soc South Africa 65(1):1–47. https://doi.org/10.1080/00359191003652033

Pilanesberg Wildlife Trust (2020) Wild fire update. https://pilanesbergwildlifetrustcoza/wild-fire-update/. Accessed 10 Jul 2022. Last updated: 2020-07-17

Qi J, Chehbouni A, Huete AR, Kerr YH, Sorooshian S (1994) A modified soil adjusted vegetation index. Remote Sens Environ 48(2):119–126. https://doi.org/10.1016/0034-4257(94)90134-1

Renner P (2022) Optical earth observation time series for analysis of land surface dynamics utilising the framework for operational radiometric correction for environmental monitoring (FORCE) in the Bushbuckridge Region, South Africa. Unpubl. M.Sc. Thesis. Department of Earth Observation, Friedrich Schiller University Jena

Rikimaru A, Roy PS, Miyatake S (2002) Tropical forest cover density mapping. Trop Ecol 43(1):39–47

SANParks (South African National Parks) (2017) Mokala National Park Management Plan 2017–2027. SANParks, Pretoria. https://www.sanparks.org/assets/docs/conservation/park_man/mokala-plan.pdf. Accessed 15 Aug 2022

Snyman HA (2012) Habitat preferences of the encroacher shrub, Seriphium plumosum. S Afr J Bot 81:34–39. https://doi.org/10.1016/j.sajb.2012.05.001

Tucker CJ (1979) Red and photographic infrared linear combinations for monitoring vegetation. Remote Sens Environ 8:127–150

Urban M, Berger C, Mudau TE, Heckel K, Truckenbrodt J, Onyango Odipo V, Smit IPJ, Schmullius C (2018) Surface moisture and vegetation cover analysis for drought monitoring in the Southern Kruger National Park Using Sentinel-1, Sentinel-2, and Landsat-8. Remote Sens 10(9):1482. https://doi.org/10.3390/rs10091482

Urban M, Schellenberg K, Morgenthal T, Dubois C, Hirner A, Gessner U, Mogonong B, Zhang Z, Baade J, Collett A, Schmullius C (2021) Using Sentinel-1 and Sentinel-2 time series for Slangbos mapping in the Free State Province, South Africa. Remote Sens 13(17):3342. https://doi.org/10.3390/rs13173342

Wagner W, Lemoine G, Rott H (1999) A method for estimating soil moisture from ERS Scatterometer and soil data. Remote Sens Environ 70(2):191–207. https://doi.org/10.1016/S0034-4257(99)00036-X

Walter H, Lieth H (1967) Klimadiagramm-Weltatlas. Fischer, Jena

White L, Brisco B, Dabboor M, Schmitt A, Pratt A (2015) A collection of SAR methodologies for monitoring wetlands. Remote Sens 7(6). https://doi.org/10.3390/rs70607615

Wiese L, Lindeque L, De Villiers M (2011) Land Degradation Assessment in Drylands Project Policy Report. ARC-ISCW Report Nr. GW/A/2011/52 & GW/56/17. ARC-ISCW, Pretoria

Zhao K, Wulder MA, Hu T, Bright R, Wu Q, Qin H, Li Y, Toman E, Mallick B, Zhang X (2019) Detecting change-point, trend, and seasonality in satellite time series data to track abrupt changes and nonlinear dynamics: a Bayesian ensemble algorithm. Remote Sens Environ 232:111–181. https://doi.org/10.1016/j.rse.2019.04.034

Part V
Synthesis and Outlook

Research Infrastructures as Anchor Points for Long-Term Environmental Observation

30

Gregor Feig, Christian Brümmer ⓘ, Amukelani Maluleke, and Guy F. Midgley ⓘ

Abstract

In this chapter, we highlight the importance and value of key Environmental Research Infrastructures, and how these can act as anchor points for long-term environmental observations and facilitate interdisciplinary environmental research. We briefly summarize the development of these efforts in South and southern Africa over the last three decades and from this perspective discuss how their successful maintenance and further implementation may turn such RIs into important anchor points for long-term environmental scientific work in support of environmental sustainability, national commitments under selected international policy discussions, and societal well-being. The fundamental role of Environmental Research Infrastructures is multifold and includes the provision of data that enable reporting and policy development, the provision of validation sites in the development of new observational sensors, measurement techniques and models, and the provision facilities for training of scientists and technicians. Humanity currently faces a number of global crises, including the impact of

G. Feig (✉)
South African Environmental Observation Network (SAEON), Pretoria, South Africa

Department of Geography, Geoinformatics and Meteorology, University of Pretoria, Pretoria, South Africa

C. Brümmer
Thünen Institute of Climate-Smart Agriculture, Braunschweig, Germany

A. Maluleke
Department of Botany and Zoology, University of Stellenbosch, South Africa

G. F. Midgley
Department of Botany and Zoology, School for Climate Studies and Global Change Biology Group, Stellenbosch University, Stellenbosch, South Africa

© The Author(s) 2024
G. P. von Maltitz et al. (eds.), *Sustainability of Southern African Ecosystems under Global Change*, Ecological Studies 248,
https://doi.org/10.1007/978-3-031-10948-5_30

changes in the climate, resulting in droughts, floods, fires, storms, and other extreme events. These crises are significantly stressing and transforming the lives and livelihoods of the vast majority of humanity. The societal response to these events is dependent on the availability of scientific knowledge and its effective transfer to governance structures, industry, and the broader society. In order to effectively address these challenges, large amounts of long-term social-ecological data are required across a broad range of intersecting disciplines that are available for analysis by the scientific community. Research Infrastructures have the ability to act as anchor points in the provision and utilization of this data, and the development of indigenous capacity to develop the observations and technical skills.

30.1 Introduction

The well-being of modern human societies is deeply dependent on natural resources (Angelstam 2018; Mirtl et al. 2018; Loescher et al. 2022), and recent global assessments have strongly advanced the predictive understanding of multiple, interacting dependencies (e.g., Díaz et al. 2015; Diaz 2019; Pörtner, Hans-Otto et al. 2021). Ecosystem processes and structures (both geophysical and biological) themselves interact in complex ways across various temporal and spatial scales, and it is important to understand these processes and how they are responding in a rapidly changing environment. This is especially important when trying to understand some of the grand challenges facing human societies, such as climate change, loss of biodiversity, land-use change, pollution, and eutrophication (Mirtl et al. 2018; Loescher et al. 2022). Many regions worldwide are experiencing the impacts of increasingly frequent and damaging climatic events such as heat waves, extended droughts, storms, or changes in rainfall distribution and intensity with increasingly adverse consequences for ecosystem functioning, biodiversity, and ecosystem services that support livelihoods (Masson-Delmotte et al. 2021).

A number of research activities deployed since the 1990s have been enhancing knowledge of the functioning of biophysical components of the Earth system and their interactions. These enhanced understandings of fundamental processes now inform the implementation of mitigation and/or adaption measures for climate change impacts through guidelines, regulations, or policy briefs. Among the best developed, and most vital for global climate stability, is the science behind the global carbon cycle (e.g., Friedlingstein et al. 2022). However, significant regional gaps in the coverage of such research limit a fuller understanding, resulting in a reliance on the use of extrapolation and assumptions that have not been tested in these underrepresented regions. This weakens important global level insights into optimal policy development and implementation of planned responses. This is particularly true in southern Africa (López-Ballesteros et al. 2018; Nickless et al. 2020), which has historically had limited amounts of data collection and remains inadequately integrated with respect to key research infrastructure, human capacity, and networks to support this regionally and globally important work.

The contribution of Africa to the global carbon cycle is characterized by its low fossil fuel emissions (with the exception of South Africa), and its rapidly increasing and urbanizing population, which is expected to change the fuel use patterns. However, due to the limited number of long-term measurements conducted in Africa, this region contributes significantly to the uncertainty in the global CO_2 budget (Ciais et al. 2011). Indeed, it is still not known if Africa is a net carbon source or a sink of carbon to the atmosphere, nor how it is likely to change in the future (Merbold et al. 2009; Ciais et al. 2011). Current risks include an expansion in cropland and increased rates of degradation and deforestation in the extensive dryland and savanna systems and the tropical forests of central Africa (Ciais et al. 2011).

Key questions that are increasingly highlighted for study in an African context include the ways in which ecosystems and biological communities are changing and potentially adapting as a result of both local and global drivers of change. These studies include analyses of factors that result in ecosystems approaching or crossing tipping points beyond which irreversible change may occur (Taylor and Rising 2021) and the related determination of ecosystem resilience vital for reducing environmental risks. These questions are also highly relevant for broader global and national policy commitments by African nations, providing evidence-based support for national positions including international UN conventions under which increasingly stringent requirements for credible data are needed, such as the Paris Accord of the UNFCCC, targets under the CBD, and the Sustainable Development Goals (United Nations 2021).

Research Infrastructures include facilities, resources, and related services used by the scientific community to conduct cutting-edge research, knowledge transmission, knowledge exchange, and knowledge preservation (European Strategy Forum on Research Infrastructures 2018). In this chapter, we aim to highlight the importance and value of key Environmental Research Infrastructures focused on land surface–atmosphere interactions, with relevance for the carbon cycle and associated biogeochemical functioning, and challenges inherent in building and maintaining these efforts. Our contribution is based on experience gained in South Africa, and more broadly across the subcontinent, with regard to the installation, maintenance, and long-term sustainability of such infrastructures, including capacity building (e.g., Bieri et al. 2022; Chap. 31). We briefly summarize the development of these efforts over the last three decades and, from this perspective, discuss how their successful maintenance and further implementation may turn such RIs into important anchor points for the positioning and long-term development of environmental scientific work in support of environmental sustainability, national commitments under selected international policy discussions, and societal well-being.

30.2   Rationale for Coordinated Terrestrial Research Infrastructure in Southern Africa

To obtain an Earth systems' view of environmental processes, large amounts of diverse data are required that are often measured over the long term in a consistent manner. This is beyond the capacity of individual scientists or research sites to maintain and synthesize. Therefore, collective efforts have been needed to create Environmental Research Infrastructures (ERIs) at a large enough scale to provide data to answer the types of large ecosystem-scale questions being asked (Mirtl et al. 2018; Loescher et al. 2022).

Research Infrastructures that focus on biogeochemical cycles must confront the challenge of measuring relevant aspects of systems with a high degree of temporal and spatial complexity. For example, understanding changes in atmospheric composition requires information about the sources and sinks of terrestrial and marine ecosystems as well as the processes governing the surface–atmosphere exchange.

Biogeochemically focused research infrastructure in a region like southern Africa cannot focus merely on biophysical aspects, but must also consider the complexity of the region's biological diversity and ecosystems, and the vital activities of people in these landscapes. The status of multiple drivers of ecosystem structure and functioning are particularly relevant, including vegetation, soil, land-use and disturbance regimes, hydrological flows, and the omnipresent role of human activities in all aspects. For this reason, independent research projects focusing on subelements of the greater Earth system may not capture important linkages to factors that are beyond the scope of the specific project aims, even when coupled to larger-scale models of frameworks. By contrast, well-coordinated efforts supported by RIs designed to provide comprehensive platforms of deliberately monitored variables can enhance the potential for improved systems' understanding. These are to be supported by RI staff to provide and operate the platform, while external researchers have access to the infrastructure for undertaking additional research, on a project basis.

These concepts are being addressed by a number of national or regional science programs such as the Integrated Carbon Observation System (ICOS) in Europe (Heiskanen et al. 2021), the National Ecological Observatory Network (NEON) of the United States (Keller et al. 2008; Metzger et al. 2019), the Terrestrial Ecosystem Research Network (TERN) of Australia (Cleverly et al. 2019), the Chinese Ecosystem Research Network (CERN) (Li Shenggong et al. 2015), and the international Long-Term Environmental Observation Network (ILTER) (Haberl et al. 2006; Mirtl et al. 2018). Within Africa, the establishment of such continental scale networks is not as advanced and there is a marked shortage of observations that cover the variety of natural and human-altered biomes that occur in Africa. This is detrimental in the assessment of the drivers of global change of feedback interactions. This is also of consequence to understanding the contribution of the African continent to global processes, such as its contribution to the global carbon cycle (López-Ballesteros et al. 2018; Nickless et al. 2020).

Long-term Environmental RIs have been noted to have four characteristics or "Conceptual pillars" (Loescher et al. 2022), including long-term time horizon of decades to centuries, the need for in situ observations at different spatial scales across ecosystem compartments of in natura sites, zones, and socio-ecological regions, strong process orientation on the study of ecosystem processes as they respond to both internal and external drivers related to ecosystem and social processes, and the use of a systems approach where abiotic and biotic components interact at different scales and human use of the systems is highlighted. For such a challenging set of characteristics to be met in a southern African setting, strong networking within and between research and academic institutions will be needed, and this would need to be supported by commitments to funding support and investment in human capacity on a time scale of at least a decade.

The fundamental position of Environmental RIs in the scientific value chain is the provision of reliable long-term observational data. This data is then available to support research into the process-level understanding of ecosystem interactions, the development of remote sensing and modeling products, data applications, and the support of national and global policies (Fig. 30.1).

One of the core roles of Environmental RIs is to drive and facilitate further research, through a number of processes including the provision of data for use by national and international researchers, and the provision of a research platform on which local and international researchers can conduct studies and train students, a set of sites and infrastructure to train environmental observation technicians, and a focal area to establish citizen science projects and engagement (Ramoutar-Prieschl and Hachigonta 2020). The research platform nature of many RIs allows for

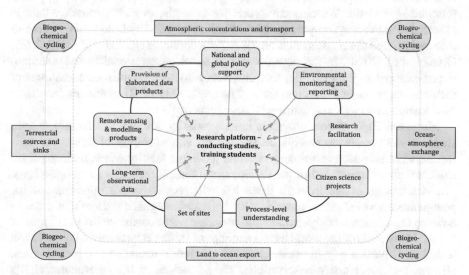

Fig. 30.1 Conceptual diagram illustrating the foundational components of an environmental Research Infrastructure and their interaction embedded in natural processes of global biogeochemical cycling

collaboration with universities and other research organizations to conduct research at the higher levels of the knowledge generation pyramid.

The provision of long-term large-scale data is one of the core functions of environmental RIs. These data can be used in a myriad of ways, including providing essential background and contextual data for research at the RI sites and supplementary observations for national and international routine environmental monitoring and reporting. For example, specific output in terms of the carbon fluxes and ecosystem carbon storage will support efforts to comply with the United Nations Framework Convention of Climate Change (UNFCCC) Paris agreement reporting regulations (Edenhofer et al. 2014). Data from the observations of C exchange and environmental carbon stocks will provide an independent observational-based estimate of the state of annual CO_2 inventories in understudied sections of the Agriculture, Forestry and Other Land Use (AFOLU) sectors, complementary to the traditional activity-based emissions estimates, and it could help in assessing the efficacy of CO_2 mitigation strategies.

The large suite of environmental observations will provide much of the necessary background information and will allow researchers to build on the data being produced in order to develop ecological theory and delve into process-based studies. This in turn will drive theory development, with a stronger emphasis on incorporating processes related to the functioning of the ecosystem in which the RIs operate into the global knowledge base.

By design, RIs offer innovation platforms for the development and validation of novel sensing and data acquisition technologies, instruments, and methods. Examples of this may be the provision of essential validation datasets for researchers to use in the development of remote sensing products for vegetation and ecosystem functioning and the hydrological cycle: for example, evapotranspiration data is essential in validating components of the Earth system and hydrological models and there are limited observation sites on the African continent (Khosa et al. 2019, 2020; Gokool et al. 2020). In the development of novel observational instrumentation, it is necessary to compare the instrumentation to well-established measurement methodologies under field conditions. The availability of RI platforms facilitates such intercomparisons and instrument development.

Moreover, RI Platforms provide a human capacity development facility to train students and young researchers on the use of, and operation of advanced environmental observation instrumentation and provide a facility where undergraduate students can be introduced to the various measurement techniques and operations. The datasets generated through these RIs will provide ample opportunities for postgraduate level students to work with large and integrated data sets in order to develop Data Science competencies and provide scientific value to the operations of the RI. Many parts of the world have a shortage of skilled environmental observation technicians, particularly in areas related to air quality management, hydrological, climate, and biodiversity observations. The operations at the environmental RIs will need to train and develop the skills of junior technicians in maintaining these observation networks. This effectively functions as a pipeline for the development of trained and experienced environmental observation technicians who will be able to move into roles in other spheres of government or industry.

30.3   Status of a Coordinated Terrestrial Environmental Research Infrastructure in Southern Africa

Over the past three decades, a number of land surface atmosphere flux-related infrastructures have developed across the African continent. While these have provided some early indications of the relevant carbon cycle functioning, the density of these installations across the continent is well below the level required to derive globally credible insights, especially given the diversity of ecosystems, land uses, and soil and climatic gradients within this vast region. The first of these were developed through the SAFARI 2000 project (Scholes and Andreae 2000; Scholes et al. 2001; Scholes 2006), which established the Skukuza (South Africa) and Mongu (Zambia) towers (Gatebe et al. 2003). Further observations in southern Africa were established by Veenendal et al. (2004) in Botswana and Brümmer et al. (2008) for Burkina Faso. This individual work was eventually consolidated through the CarboAfrica project (Ciais et al. 2011); however, this has not continued as a coherent integrated network of observation platforms, thereby limiting the interoperability between these measurements, nor has it allowed for the development of a cohort of skilled technicians and researchers. Besides the need for highly skilled staff for installing and maintaining an eddy flux tower as well as large investment costs for instruments, limited implementation is likely due to funding limitations and conflicting priorities for scarce funds. The limited funding is illustrated in that africa receives less than 5% if the global-climate-related funding (IPPC) of which less than half goes to the maintenance of institutions.

At present, there are only seven flux measurement sites on the continent that are reporting to Fluxnet (https://fluxnet.org/), and many of these are out of date (Table 30.1). FluxNet is an international "network of networks," tying together regional networks of Earth system scientists who use the eddy covariance technique to measure the cycling of carbon, water, and energy between the biosphere and atmosphere.

The challenge of maintaining and growing this capability has constrained the further elaboration of these African RIs, but in 2016, South Africa selected an ecosystem flux RI called the Expanded Freshwater and Terrestrial Environmental Observation Network (EFTEON) as one of its key national investments in environmental monitoring. The roll-out of this program has been a vindication of SPACES and SPACES II investments in similar components, and the landscape scale approaches taken in the SPACES II program, in particular.

With the rising need for developing environmental observation and research capacity in Africa, there have recently been a number of projects initiated. Particular emphasis has been placed on how the in-house skills are developed and how the operational transfer of the infrastructure to local institutions can be accomplished (Bieri et al. 2022). Within the context of highlighting several research activities in

Table 30.1 List of African sites on Fluxnet (https://fluxnet.org/)

Country	Site name	Lat	Long	Ecosystem (IGBP)[M2]	Elevation (masl)	Reference	Time period
Ghana	Ankasa	5.2685	−2.6942	Evergreen broadleaf forest	124		2011–2014
Senegal	Dahra	15.4028	−15.4322	Savanna	40	Tagesson et al. (2015a, b)	2010–2013
South Africa	Skukuza	−25.0197	31.4969	Savanna	359	Scholes et al. (2001), Kutsch et al. (2008), Archibald and Kirton (2009), Williams et al. (2009), Fan et al. (2015)	2001–2016
	Malopeni	−23.8366	31.2137	Savanna	389	Khosa et al. (2020)	2008–2016
	Welgegund	−26.5698	26.9393	Grassland	1480	Räsänen et al. (2017)	2010–2013
Zambia	Mongu	−15.4378	23.2589	Deciduous broad leaf	1053	Gatebe et al. (2003)	2000–2009
Republic of Congo	Tchizalamou	−4.2892	11.6564	Savanna	82		2006–2009

this book, the BMBF program Science Partnerships for the Adaption to Complex Earth System Processes in Southern Africa (SPACES) offered the opportunity to coestablish both equipment and capacity for scientific monitoring that is intended to operate beyond typical project lifetimes, thereby eventually turning into an RI or becoming a part of an already existing network coordinated by a national science institution. With regard to greenhouse gas (GHG) flux measurements, the projects ARS AfricaE (Adaptive Resilience of Southern African Ecosystems, https://ars-africae.org/) and EMS Africa (Ecosystem Management Support for Climate Change in Southern Africa, (https://www.emsafrica.org/) were designed to set up flux towers for continuous observation of CO_2 and energy exchange between the land surface and the atmosphere in managed and (semi-)natural South African ecosystems. An example of the variety of research options provided by operating flux towers over the long term is given by Rybchak et al. (in Chap. 17). The authors demonstrate the effect of grazing intensity and weather on the CO_2 sequestration potential and biodiversity in typical Nama-Karoo ecosystems. In that chapter, they offer useful suggestions and a roadmap for a transfer of project infrastructure and capacity into a longer-term initiative, such as the described RIs.

Recent experience gained during the initial establishment of an RI in the EFTEON network (Benfontein Nature Reserve) revealed the potential to capture the impact of rare extreme events. An extreme wildfire event occurred in the footprint of the flux tower sites, following thorough vegetation sampling and a lead-up period of 2 years of flux measurements. The impact of the fire on the magnitude of the CO_2 Fluxes can clearly be seen in Fig. 30.2 where the flux response before and after the fire can be seen (Fig. 30.3).

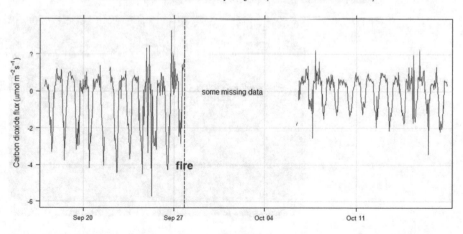

Fig. 30.2 Eddy-covariance flux data for the Benfontein Site before and following the rare fire event of the 28 September 2021

Fig. 30.3 Repeat
photographs for the
Benfontein Tower, top taken
on 24 August 2021 (1 month
prior to the fire), middle 6
October 2021 (7 days
post-fire) (courtesy
Amukelani Maluleke), and
bottom 9 March 2022 (at the
end of the following growing
season)

30.4 Design and Observational Aims of an Environmental RI

Environmental RIs are designed to implement a broad set of observations and allow for the deployment of additional research through projects. This provides the opportunity to study processes through identifying and quantifying the drivers of environmental change and the ecosystem response to those drivers.

In the recent paper on the development of a Global Ecosystem Research Infrastructure (GERI), a number of key features of environmental RIs include that they estimate and provide essential environmental observations (including GHG flux), they adopt a cause and effect paradigm, they implement a focus on understanding spatial and temporal variability in ecological drivers and processes, they have implemented a scaling strategy and a focus on reporting observational uncertainty (Loescher et al. 2022). The initial questions of an environmental RI are important and influence the scope and design, however much of the greatest value in a RI may be derived from opportunistic studies that build on the baseline of the infrastructure that has been set up and operated. Therefore, it is imperative that environmental RIs accommodate researchers and research questions outside the original scope of the design.

In the context of the South African Environmental Observation Network (SAEON), the drivers of terrestrial environmental change that have been considered include

- Weather and climatic conditions, such as long-term climatic change, the impacts of ultraviolet radiation, and hydrological functioning and sediments.
- Changes in atmospheric composition as drivers of environmental change include issues such as an increase in atmospheric CO_2 concentration, changes in the concentration of other atmospheric gases and particulates, the deposition of acidic species to the land and water surface, and changes in nutrient loading (eutrophication) through atmospheric deposition processes.
- Land-use change is an important management option through alteration in the way land is used and valued, and the activities (economic and otherwise) that occur on the land.
- Biotic changes can drive ecosystem processes; classic examples include events such as disease epidemics or pests and the introduction and spread of invasive alien organisms.
- Finally, disturbance events can drive changes in ecosystem structure and function; these might include issues such as fires (see Fig. 30.2), droughts or floods, or other large infrequent events.

The establishment of long-term environmental RIs in the landscape with a detailed record of baseline characteristics to benchmark impact of changes is essential to document and elucidate the magnitude of the drivers. At the same time, the geographic distribution of terrestrial South African RIs allows one to monitor and quantify large-scale ecosystem responses through various thematic studies including inter alia:

- *Biodiversity:* Observations of changes in biodiversity are conducted at four (interrelated) hierarchical levels, with the aim to quantify changes and understand the drivers responsible for observed changes, such as shift in extent and position of biomes, shifts in the extent and position of ecosystems within biomes, changes in biodiversity integrity (richness, composition, and structure) across trophic levels within ecosystems, and changes in the distribution and abundance of species.
- *Biogeochemical cycling and productivity:* Biogeochemical cycling plays a central role in the fate of greenhouse gases and the supply of provisioning ecosystem services; observational foci relate to carbon cycling and storage, primary and secondary production and other biogeochemical cycles, such as (the N and P) cycles.
- *Hydrological functioning and sediments:* These processes play a crucial role in the provisioning and quality of water; observational foci relate to the hydrological flow regime, the quality of the water in the various components of the system, and other impacts such as redistribution of sediments.
- *Fire Regime:* Fire is a key determinant and management tool of the structure and function of many terrestrial biomes, and changes in the fire regime (type, intensity season, and frequency of burning) may have widespread consequences for biodiversity, biogeochemical cycling, carbon sequestration, and hydrological functioning.
- *Social response:* How do societies drive and respond to a changing environment?

These themes highlight, for example, a number of overarching research questions that would be appropriately underpinned by terrestrial RIs:

- *Provisioning ecosystem services:* How do different land use, disturbance regimes, soil fertility, and climate constrain the capacity of South African ecosystems to deliver human needs such as clean water, clean air, nutrition, energy, and a safe, productive, and attractive environment?
- *Biogeochemistry and productivity:* What is the potential for South African ecosystems to sequester CO_2? What is the likely size of the change in carbon pools and fluxes in South Africa as a result of changes in land cover and land use, and what trends are observable? What is their resilience under changing climatic and land-use conditions?
- *Biodiversity:* What are the biodiversity and ecosystem services implications of using South African landscapes in different ways?
- *System variability:* What are the spatial patterns in South Africa of diurnal, seasonal, annual, and interannual ecosystem pools and fluxes of C, water, nutrients, and energy and how are they changing?
- *Ecosystem resilience:* What are the implications of changing biodiversity for the resilience of the ecosystem functioning and ecosystem service delivery, under ongoing climate and land-use change? How do various land management approaches affect ecosystem productivity, efficiency, and sustainability? Which strategies maximize societal resilience to climate extremes and other shocks?

With these considerations in mind, when South Africa embarked on developing an environmental RI for freshwater and terrestrial processes (EFTEON) (Feig 2018), the following principles were considered essential in the infrastructure planning:

Long-term environmental research (LTER) The EFTEON is intended for long-term continuous operation and the primary purpose of the network is to provide long-term environmental data for the national and global research community. Site operations need to undertake measurements and observations that are of value at both the short and long term. As a result, the selected landscapes need to be available for multidecade operation and allow regular (daily) access to the core site by the EFTEON staff and researchers using the platform.

Research Platform EFTEON is intended as a research platform with an open data and open platform use policy. This is in order to facilitate the use of the infrastructure and data by other researchers, both nationally and internationally. The selected landscapes must allow for the use of the facilities by multiple researchers or partners from multiple research organizations (managed by rules of site usage and selected via evaluation of submitted project proposals).

Spatial diversity coverage Landscapes selected for the network are to represent South African biomes and human-transformed ecosystems and their embedded aquatic systems, this design concept places a focus on lived-in landscapes and landscapes in transition (driven by climate change or land-use change). The selected landscapes shall enable the long-term observation of the coupled terrestrial/aquatic systems in the face of change and shall include a number of relevant land uses, such as conservation, urbanization, agriculture, post-mining rehabilitation, etc.

Historical observations and experimental datasets Incorporation of existing research and linking to existing socio-ecological datasets is a strong focus of the EFTEON design concept. The availability of long-term existing social-ecological and Earth system data sets would be considered an advantage in the selection of the landscapes. There is a strong emphasis on data archiving and data archaeology to ensure long-term availability and continuity of datasets.

Experiments and manipulations The use of experiments and manipulations provides considerable insight into environmental processes. The landscapes 2*must* offer the opportunity to implement and sustain appropriate experiments, at a scale matched to the scales of key processes, to help elucidate process-level understanding of ecosystem changes. They could include things like disturbances, irrigation or fertilization, withholding or adding herbivores, excluding fire or increasing fire frequency. These may be experiments undertaken and managed by the EFTEON staff or those experiments operated by landscape users from one or many institutions (managed by rules of site usage and selected via evaluation of submitted project proposals). Landscapes that *only* consist of strictly protected land covers failed this principle.

The general situational characteristics of the landscape These include the opportunity to observe the coupled terrestrial and aquatic (fluvial and groundwater) systems, the opportunity to observe social-ecological systems in the South-African developmental context, the opportunity to act as a National RI allowing and encouraging the use of these landscapes by multiple research organizations (both nationally and internationally), and spatial coverage across important biomes.

The landscape location in the face of Global Change This focused on the presence of representative near-natural land cover (i.e., land uses where most key ecological processes are autonomous, rather than imposed by human agents) and modified land uses (e.g., cultivation, plantation, urbanization, mining, etc.), the climatic impacts (e.g., anticipated climate change hotspots), or gradients, which can be optimally observed (altitudinal gradients/projected climate change hotspots), transition zones between biomes, which occur within the proposed landscape, and the expected development pathways for the landscape (including any evidence or published plans for regional developments or other evidence as may be appropriate).

Logistical and operational suitability of the core and associated sites This includes, inter alia, security of tenure for the operations, particularly for the core site, existing facilities for hydrological observations such as gauging weirs, dams, testing boreholes et cetera, suitability for the deployment of micrometeorological observations (i.e., the assumptions of horizontal heterogeneity and steady-state conditions are met), any existing long-term observations or experiments, including details of the research, availability of the data, data users and key findings, and the availability of office facilities for staff and guest researchers.

Stakeholder analysis This includes an in-depth analysis of the relevant stakeholder communities within the nominated landscapes, including land owners and land custodians, communities and residential areas, engagement with relevant authorities or resident groups, and assessments of current land uses.

30.5 Toward the Regional and Multidisciplinary Integration of Terrestrial Biogeochemical Research Infrastructures

In order to meet the long-term environmental research and monitoring needs, a number of authors have suggested the concept of essential variables. Essential Climate Variables (ECVs) were first defined by GCOS as "Physical, chemical or biological variables or groups of linked variables that critically contribute to the characterization of the Earth's climate" (Reyers et al. 2017; López-Ballesteros et al. 2018). A number of organizations have published lists of essential variables, including the Global Climate Observation System (GCOS) for ECVs (WMO 2015), the Group on Earth Observations Biodiversity Observation Network (GEOBON) for Essential Biodiversity Variables (EBVs) (Guerra et al. 2019), the Ecosystem and Socio-Ecosystem Functional Types project (ESEFT) for essential social-ecological functional variables (ESEFT 2019) and the EU funded SEACRIFOG project (Supporting EU-African Cooperation on Research Infrastructures for Food Security and Greenhouse Gas Observations, reference) has published essential variables for GHG observations in Africa. The Global Climate Observation System (GCOS) stipulates a suite of 54 essential climate variables: these are divided into the components for the land and atmosphere (reference). GEOBON lists 6 classes of essential biodiversity variables with 21 variables identified, the classes include genetic composition, species and populations, species traits, community composition, ecosystem function, and ecosystem structure (Guerra et al. 2019). ESEFT lists the Essential Social-Ecological variables in classes (Components and Functional Dimensions): the social system, the ecosystem, and interactions (ESEFT 2019).

The availability of long-term observations aligned to the applicable essential variables mentioned above is of crucial importance in the development of, and the validation of, ecosystem models and remote sensing products. A prime example of a closely linked relationship between the flux observations infrastructure and the modeling community can be found in the link between Oz Flux and the CABLE model development team (Kowalczyk et al. 2006; Haverd et al. 2013, 2016a, b; De Kauwe et al. 2015). In South Africa, this same dynamic has been demonstrated with the flux measurements at Skukuza and the CSIR CABLE development team (Khosa et al. 2019, 2020).

A wider fitting frame for the SPACES projects and the above-mentioned activities form the BMBF-initiated and funded science centers, that is, SASSCAL, and its West African sister organization WASCAL. SASSCAL is the Southern African Science Service Centre for Climate Change and Adaptive Land Management, a joint initiative of Angola, Botswana, Namibia, South Africa, Zambia, and Germany in response to the challenges of Global Change. The center understands the role of science as a service to societies that are most severely affected by climate change and to provide decision-makers with evidence-based results and advice. Hence, SASSCAL may act as a facilitator for supporting research infrastructures as it is conceptualized and operationalized to complement the existing research and capacity development and research initiatives in the region. Very similar objectives, although a bit more tailored to the regional conditions, are pursued by WASCAL, the West African Science Service Centre for Climate Change and Adaptive Land Management, thereby likewise offering opportunities for supporting RIs in the long term. Cooperation within all these activities and among the several groups involved is a key aspect for the success of the work and for achieving the specific aims of the infrastructure. Critical issues like unbalanced resource distribution, paternalism, or misuse of good scientific practices have been observed in the past and must be combatted in all present and future endeavors. A review of North–South relationships and recommendations on how to avoid mistakes of the past are given in Chap. 31 by Lütkemeier et al.

In South Africa, the South African Environmental Observation Network (SAEON) has been tasked with the development of long-term environmental in-situ research infrastructure and the data management facilities that support such research. Recently three Research Infrastructures have been awarded to SAEON:

these include the Expanded Freshwater and Terrestrial Environmental Observation Network (EFTEON), the Shallow Marine and Coastal Research Infrastructure (SMCRI), and the South African Polar Research Infrastructure (SAPRI). These RIs all focus on facilitating interdisciplinary environmental research. In the terrestrial sphere, the EFTEON RI is establishing six research landscapes, with a thematic focus on (1) *Biogeochemistry*, including eddy covariance measurements and atmospheric deposition in order to quantify the exchange of CO_2, nutrients, and energy between the land surface and the atmosphere, (2) *Biodiversity*, to quantify the abundance, diversity of a diverse range of biological communities, including vegetation, avifauna, invertebrates, and others, (3) *Hydrology*, with a focus on quantifying water quantity and quality and the movement through the landscape, (4) *Climatology and Atmospheric Processes*, to establish a detailed climatological record and understanding of atmospheric chemical exchanges, and (5) *Social Ecological Systems*, to gain an understanding of how humans interact with and make decisions relating to the ecosystems in which they operate (Feig 2018).

Similarly, the SMCRI is developing infrastructure across four Sentinel sites; these include (1) the Algoa Bay Site, located around Gqeberha (Formally Port Elizabeth), (2) the Two Oceans Sentinel Site, around Cape Town, (3) the Natal Bight Sentinel Site, located north of Durban, and (4) the Marion Island Sentinel site in the South Indian Ocean. SMCRI is developing a number of research platforms that include airborne remote sensing, a Coastal Biogeochemistry Laboratory, acoustic telemetry arrays, and marine remote imaging. In addition to the science platforms, they operate a number of platforms that facilitate the work, including a Coastal Craft Platform, a Hyperbaric Chamber, and a Science Engagement Platform.

The SAPRI is in the early stages of development and will facilitate research in the Southern Ocean and the Antarctic.

30.6   Impact of a Research Infrastructure and Its Assessment

Setting up an environmental RI like outlined above requires comprehensive coordinated effort and resources including a variety of communities. Expectations from operators, scientists, the private and the public sector, and other stakeholders may vary with individual perspectives and aims. Therefore, it is important to understand and assess what impact on the environment, science, and society a fully functional RI is able to cause.

Broad types of impact usually fall under one of the following categories:

- Science and technology
- Social impact
- Human capital impact
- Economic and innovation impact

For environmental RIs, these can be more specifically tailored to the main strategic objectives of the groups involved. As an example, ICOS has undergone an impact assessment in 2018 where key indicators were defined along these strategic lines (Heiskanen et al. 2021):

- Producing standardized high-precision long-term observational data
- Stimulating scientific studies and modeling efforts and providing a platform for data analysis and synthesis
- Communicating science-based knowledge toward society and contributing timely information relevant to the greenhouse gas policy and decision making
- Promoting technical developments
- Ensuring high visibility of the RI

It is then possible to fully grasp the RI impact at various levels by evaluating specifically designed indicators such as, but not limited to, lengths of the acquired data sets, degree of harmonization of the data sets, number of related articles published, media appearances, provision of policy-relevant data, publications used outside the scientific domain, new knowledge generated on carbon sources and sinks, investments mobilized by the RI, and application of data in globally leading models. Two further, highly important indicators are the improvement of long-term decisions through enhanced political discourse based on evidence and a reduction of damage by extreme weather events and through more effective climate mitigation policy.

30.7   Conclusion

In this chapter, we aimed to highlight the importance and value of key Environmental RIs focused on land surface–atmosphere interactions, with relevance for the carbon cycle and associated biogeochemical functioning, and challenges inherent in building and maintaining these efforts.

It is clear that the presence of long-term environmental RI provides a strong opportunity for furthering a wide range of environmental research. In some cases, these observations can be undertaken by the RI itself; however, the nature of the questions to be asked and the techniques that may be employed in these RIs advocate for the role of the RIs acting as a research platform where they provide the long-term high-quality baseline observations, while other research entities such as universities, science councils, and others build off of the base that the RIs create and build the knowledge base through short-term projects that either elucidate process level questions, the link between fields in a multi- or interdisciplinary manner, or utilize different or novel observation techniques to fill in the observational gaps left in the initial RI design.

Studies presenting a synthesis of available methods and data for estimating the African carbon budget stressed the associated large uncertainties due to a very limited number of long-term observations (Ciais et al. 2011). The contribution of Africa to the global carbon cycle is characterized by its low fossil fuel emissions, a rapidly increasing population concomitant with cropland expansion, potential degradation, and deforestation. Published estimates are in the range of -0.6 to -0.2 $PgCyr^{-1}$ associated with uncertainties in the same order of magnitude indicating a small net sink of carbon for the whole African continent (Valentini et al. 2014). Coordinated endeavors such as environmental RIs will help reduce the large uncertainties in the continental-scale carbon budget. In this chapter, we briefly summarized the development of these efforts over the last three decades and, from this perspective, discussed how their successful maintenance and further implementation may turn such RIs into important anchor points for long-term environmental scientific work in support of environmental sustainability, national commitments under selected international policy discussions, and societal well-being.

Humanity currently faces a number of global crises, including the impact of changes in the climate, resulting in droughts, floods, fires, and other extreme events. These crises are significantly stressing the lives and livelihoods of the vast majority of humanity. The societal response to these events is dependent on the availability of scientific knowledge and its effective transfer to governance structures, industry, and the broader society. In order to effectively address these challenges, large amounts of data are required across a broad range of intersecting disciplines that are available for analysis by the scientific community. Research Infrastructures have the ability to act as anchor points in the provision and utilization of this data.

References

Angelstam P (2018) LTSER platforms as a place-based transdisciplinary research infrastructure: learning landscape approach through evaluation. Landsc Ecol 24. https://doi.org/10.1007/s10980-018-0737-6

Archibald SA, Kirton A (2009) Drivers of inter-annual variability in Net Ecosystem Exchange in a semi-arid savanna ecosystem, South Africa. 16

Bieri M, Du Toit J, Feig G, Maluta NE, Mantlana B, Mateyisi MJ, Midgley GF, Mutanga S, von Maltitz G, Bruemmer C (2022) Integrating project-based infrastructures with long-term greenhouse gas observations in Africa. Clean Air J 32:9. https://doi.org/10.17159/caj/2022/32/1.13081

Brümmer C, Brüggemann N, Butterbach-Bahl K, Falk U, Szarzynski J, Vielhauer K, Wassmann R, Papen H (2008) Soil-atmosphere exchange of N2O and NO in near-natural Savanna and agricultural land in Burkina Faso (W. Africa). Ecosystems 11:582–600. https://doi.org/10.1007/s10021-008-9144-1

Ciais P, Bombelli A, Williams M, Piao S, Chave J, Ryan CM, Henry M, Brender P, Valentini R (2011) The carbon balance of Africa: synthesis of recent research studies. Philos Trans R Soc A Math Phys Eng Sci 369:2039–2057. https://doi.org/10.1098/rsta.2010.0328

Cleverly J, Eamus D, Edwards W, Grant M, Grundy MJ, Held A, Karan M, Lowe AJ, Prober SM, Sparrow B, Morris B (2019) TERN, Australia's land observatory: addressing the global challenge of forecasting ecosystem responses to climate variability and change. Environ Res Lett 14:095004. https://doi.org/10.1088/1748-9326/ab33cb

De Kauwe MG, Kala J, Lin Y-S, Pitman AJ, Medlyn BE, Duursma RA, Abramowitz G, Wang Y-P, Miralles DG (2015) A test of an optimal stomatal conductance scheme within the CABLE land surface model. Geosci Model Dev 8:431–452. https://doi.org/10.5194/gmd-8-431-2015

Diaz S (2019) Summary for policymakers of the global assessment report on biodiversity and ecosystem services of the Intergovernmental Science-Policy Platform on Biodiversity and Ecosystem Services. ipbes

Díaz S, Demissew S, Carabias J, Joly C, Lonsdale M, Ash N, Larigauderie A, Adhikari JR, Arico S, Báldi A, Bartuska A, Baste IA, Bilgin A, Brondizio E, Chan KM, Figueroa VE, Duraiappah A, Fischer M, Hill R, Koetz T, Leadley P, Lyver P, Mace GM, Martin-Lopez B, Okumura M, Pacheco D, Pascual U, Pérez ES, Reyers B, Roth E, Saito O, Scholes RJ, Sharma N, Tallis H, Thaman R, Watson R, Yahara T, Hamid ZA, Akosim C, Al-Hafedh Y, Allahverdiyev R, Amankwah E, Asah ST, Asfaw Z, Bartus G, Brooks LA, Caillaux J, Dalle G, Darnaedi D, Driver A, Erpul G, Escobar-Eyzaguirre P, Failler P, Fouda AMM, Fu B, Gundimeda H, Hashimoto S, Homer F, Lavorel S, Lichtenstein G, Mala WA, Mandivenyi W, Matczak P, Mbizvo C, Mehrdadi M, Metzger JP, Mikissa JB, Moller H, Mooney HA, Mumby P, Nagendra H, Nesshover C, Oteng-Yeboah AA, Pataki G, Roué M, Rubis J, Schultz M, Smith P, Sumaila R, Takeuchi K, Thomas S, Verma M, Yeo-Chang Y, Zlatanova D (2015) The IPBES Conceptual Framework — connecting nature and people. Curr Opin Environ Sustain 14:1–16. https://doi.org/10.1016/j.cosust.2014.11.002

Edenhofer O, Pichs-Madruga R, Sokona Y, Farahani E, Kadner S, Sayboth K, Adler A, Baum I, Brunner S, Eickemeier P, Kriemann B, Savolainen J, Schloemer S, von Stechow C, Zwickel T, Minx JC (2014) IPCC, 2014: Summary for Policymakers. In: Climate Change 2014: Mitigation of Climate Change. Contribution of Working Group III to the Fifth Assessment Report of the Intergovernmental Panel on Climate Change. Cambridge University Press

ESEFT (2019) Essential social-ecological functional variables (ESEFVs). Ecosyst Socio-ecosyst Funct Types. http://functionaltypes.caescg.org/esefvs/. Accessed 27 Aug 2022

European Strategy Forum on Research Infrastructures (2018) Strategy Report on Research Infrastructures Roadmap 2018. Milan

Fan Z, Neff JC, Hanan NP (2015) Modeling pulsed soil respiration in an African savanna ecosystem. Agric For Meteorol 200:282–292. https://doi.org/10.1016/j.agrformet.2014.10.009

Feig G (2018) The expanded freshwater and terrestrial environmental observation network (EFTEON). Clean Air J 28. https://doi.org/10.17159/2410-972x/2018/v28n2a14

Friedlingstein P, Jones MW, O'Sullivan M, Andrew RM, Bakker DCE, Hauck J, Le Quéré C, Peters GP, Peters W, Pongratz J, Sitch S, Canadell JG, Ciais P, Jackson RB, Alin SR, Anthoni P, Bates NR, Becker M, Bellouin N, Bopp L, Chau TTT, Chevallier F, Chini LP, Cronin M, Currie KI, Decharme B, Djeutchouang LM, Dou X, Evans W, Feely RA, Feng L, Gasser T, Gilfillan D, Gkritzalis T, Grassi G, Gregor L, Gruber N, Gürses Ö, Harris I, Houghton RA, Hurtt GC, Iida Y, Ilyina T, Luijkx IT, Jain A, Jones SD, Kato E, Kennedy D, Klein Goldewijk K, Knauer J, Korsbakken JI, Körtzinger A, Landschützer P, Lauvset SK, Lefèvre N, Lienert S, Liu J, Marland G, McGuire PC, Melton JR, Munro DR, Nabel JEMS, Nakaoka S-I, Niwa Y, Ono T, Pierrot D, Poulter B, Rehder G, Resplandy L, Robertson E, Rödenbeck C, Rosan TM, Schwinger J, Schwingshackl C, Séférian R, Sutton AJ, Sweeney C, Tanhua T, Tans PP, Tian H, Tilbrook B, Tubiello F, van der Werf GR, Vuichard N, Wada C, Wanninkhof R, Watson AJ, Willis D, Wiltshire AJ, Yuan W, Yue C, Yue X, Zaehle S, Zeng J (2022) Global Carbon Budget 2021. Earth Syst Sci Data 14:1917–2005. https://doi.org/10.5194/essd-14-1917-2022

Gatebe CK, King MD, Platnick S, Arnold GT, Vermote EF, Schmid B (2003) Airborne spectral measurements of surface-atmosphere anisotropy for several surfaces and ecosystems over southern Africa: REMOTE SENSING OF SURFACE REFLECTANCE DURING SAFARI 2000. J Geophys Res-Atmos 108:n/a–n/a. https://doi.org/10.1029/2002JD002397

Gokool S, Riddell E, Jarmain C, Chetty K, Feig G, Thenga H (2020) Evaluating the accuracy of satellite-derived evapotranspiration estimates acquired during conditions of water stress. Int J Remote Sens 41:704–724. https://doi.org/10.1080/01431161.2019.1646940

Guerra CA, Pendleton L, Drakou EG, Proença V, Appeltans W, Domingos T, Geller G, Giamberini S, Gill M, Hummel H, Imperio S, McGeoch M, Provenzale A, Serral I, Stritih A, Turak E, Vihervaara P, Ziemba A, Pereira HM (2019) Finding the essential: improving conservation monitoring across scales. Glob Ecol Conserv 18:e00601. https://doi.org/10.1016/j.gecco.2019.e00601

Haberl H, Winiwarter V, Andersson K, Ayres RU, Boone C, Castillo A, Cunfer G, Fischer-Kowalski M, Freudenburg WR, Furman E, Kaufmann R, Krausmann F, Langthaler E, Lotze-Campen H, Mirtl M, Redman CL, Reenberg A, Wardell A, Warr B, Zechmeister H (2006) From LTER to LTSER: conceptualizing the socioeconomic dimension of long-term socioecological research. Ecol Soc 11:art13. https://doi.org/10.5751/ES-01786-110213

Haverd V, Raupach MR, Briggs PR, Canadell JG, Isaac P, Pickett-Heaps C, Roxburgh SH, van Gorsel E, Viscarra Rossel RA, Wang Z (2013) Multiple observation types reduce uncertainty in Australia's terrestrial carbon and water cycles. Biogeosciences 10:2011–2040. https://doi.org/10.5194/bg-10-2011-2013

Haverd V, Smith B, Raupach M, Briggs P, Nieradzik L, Beringer J, Hutley L, Trudinger CM, Cleverly J (2016a) Coupling carbon allocation with leaf and root phenology predicts tree–grass partitioning along a savanna rainfall gradient. Biogeosciences 13:761–779. https://doi.org/10.5194/bg-13-761-2016

Haverd V, Smith B, Trudinger C (2016b) Process contributions of Australian ecosystems to interannual variations in the carbon cycle. Environ Res Lett 11:054013. https://doi.org/10.1088/1748-9326/11/5/054013

Heiskanen J, Brümmer C, Buchmann N, Calfapietra C, Chen H, Gielen B, Gkritzalis T, Hammer S, Hartman S, Herbst M, Janssens IA, Jordan A, Juurola E, Karstens U, Kasurinen V, Kruijt B, Lankreijer H, Levin I, Linderson M-L, Loustau D, Merbold L, Myhre CL, Papale D, Pavelka M, Pilegaard K, Ramonet M, Rebmann C, Rinne J, Rivier L, Saltikoff E, Sanders R, Steinbacher M, Steinhoff T, Watson A, Vermeulen AT, Vesala T, Vítková G, Kutsch W (2021) The integrated carbon observation system in Europe. Bull Am Meteorol Soc:1–54. https://doi.org/10.1175/BAMS-D-19-0364.1

Keller M, Schimel DS, Hargrove WW, Hoffman FM (2008) A continental strategy for the National Ecological Observatory Network. Front Ecol Environ 6:282–284

Khosa FV, Feig GT, van der Merwe MR, Mateyisi MJ, Mudau AE, Savage MJ (2019) Evaluation of modeled actual evapotranspiration estimates from a land surface, empirical and satellite-based models using in situ observations from a South African semi-arid savanna ecosystem. Agric For Meteorol 279:107706. https://doi.org/10.1016/j.agrformet.2019.107706

Khosa FV, Mateyisi MJ, van der Merwe MR, Feig GT, Engelbrecht FA, Savage MJ (2020) Evaluation of soil moisture from CCAM-CABLE simulation, satellite-based models estimates and satellite observations: a case study of Skukuza and Malopeni flux towers. Hydrol Earth Syst Sci 24:1587–1609. https://doi.org/10.5194/hess-24-1587-2020

Kowalczyk EA, Wang YP, Law RM, Davies HL, McGregor JL, Abramowitz G (2006) The CSIRO Atmosphere Biosphere Land Exchange (CABLE) model for use in climate models and as an offline model. p. 42

Kutsch WL, Hanan N, Scholes B, McHugh I, Kubheka W, Eckhardt H, Williams C (2008) Response of carbon fluxes to water relations in a savanna ecosystem in South Africa. Biogeosciences 5:1797–1808. https://doi.org/10.5194/bg-5-1797-2008

Li S, Yu G, Yu X, He H, Guo X (2015) A brief introduction to Chinese Ecosystem Research Network (CERN). J Resour Ecol 6:192–196. https://doi.org/10.5814/j.issn.1674-764x.2015.03.009

Loescher HW, Vargas R, Mirtl M, Morris B, Pauw J, Yu X, Kutsch W, Mabee P, Tang J, Ruddell BL, Pulsifer P, Bäck J, Zacharias S, Grant M, Feig G, Zheng L, Waldmann C, Genazzio MA (2022) Building a global ecosystem research infrastructure to address global grand challenges for macrosystem ecology. Earth's Future 10. https://doi.org/10.1029/2020EF001696

López-Ballesteros A, Beck J, Bombelli A, Grieco E, Lorencová EK, Merbold L, Brümmer C, Hugo W, Scholes R, Vačkář D, Vermeulen A, Acosta M, Butterbach-Bahl K, Helmschrot J, Kim D-G, Jones M, Jorch V, Pavelka M, Skjelvan I, Saunders M (2018) Towards a feasible and representative pan-African research infrastructure network for GHG observations. Environ Res Lett 13:085003. https://doi.org/10.1088/1748-9326/aad66c

Masson-Delmotte V, Zhai P, Pirani A, Connors SL, Péan C, Berger S, Caud N, Chen Y, Goldfarb L, Gomis MI, Huang M, Leitzell K, Lonnoy E, Matthews JBR, Maycock TK, Waterfield T, Yelekçi Ö, Yu R, Zhou B (eds) (2021) Climate Change 2021: the physical science basis. Contribution of Working Group I to the Sixth Assessment Report of the Intergovernmental Panel on Climate Change. Cambridge University Press

Merbold L, Ardo J, Arneth A, Scholes RJ, Nouvellon Y, de Grandcourt A, Archibald S, Bonnefond JM, Boulain N, Brueggemann N, Bruemmer C, Cappelaere B, Ceschia E, El-Khidir HAM, El-Tahir BA, Falk U, Lloyd J, Kergoat L, Dantec VL, Mougin E, Muchinda M, Mukelabai MM, Ramier D, Roupsard O, Timouk F, Veenendaal EM, Kutsch WL (2009) Precipitation as driver of carbon fluxes in 11 African ecosystems. Biogeosciences 6:1027–1041. https://doi.org/10.5194/bg-6-1027-2009

Metzger S, Ayres E, Durden D, Florian C, Lee R, Lunch C, Luo H, Pingintha-Durden N, Roberti JA, SanClements M, Sturtevant C, Xu K, Zulueta RC (2019) From NEON field sites to data portal: a community resource for surface–atmosphere research comes online. Bull Am Meteorol Soc 100:2305–2325. https://doi.org/10.1175/BAMS-D-17-0307.1

Mirtl M, Borer E, Djukic I, Forsius M, Haubold H, Hugo W, Jourdan J, Lindenmayer D, McDowell WH, Muraoka H, Orenstein DE, Pauw JC, Peterseil J, Shibata H, Wohner C, Yu X, Haase P (2018) Genesis, goals and achievements of Long-Term Ecological Research at the global scale: a critical review of ILTER and future directions. Sci Total Environ 626:1439–1462. https://doi.org/10.1016/j.scitotenv.2017.12.001

Nickless A, Scholes RJ, Vermeulen A, Beck J, López-Ballesteros A, Ardö J, Karstens U, Rigby M, Kasurinen V, Pantazatou K, Jorch V, Kutsch W (2020) Greenhouse gas observation network design for Africa. Tellus Ser B Chem Phys Meteorol 72:1–30. https://doi.org/10.1080/16000889.2020.1824486

Pörtner H-O, Scholes RJ, Agard J, Archer E, Arneth A, Bai X, Barnes D, Burrows M, Chan L, Cheung WL, Diamond S, Donatti C, Duarte C, Eisenhauer N, Foden W, Gasalla MA, Handa C, Hickler T, Hoegh-Guldberg O, Ichii K, Jacob U, Insarov G, Kiessling W, Leadley P, Leemans R, Levin L, Lim M, Maharaj S, Managi S, Marquet PA, McElwee P, Midgley G, Oberdorff T, Obura D, Osman Elasha B, Pandit R, Pascual U, APF P, Popp A, Reyes-García V, Sankaran M, Settele J, Shin Y-J, Sintayehu DW, Smith P, Steiner N, Strassburg B, Sukumar R, Trisos C, Val AL, Wu J, Aldrian E, Parmesan C, Pichs-Madruga R, Roberts DC, Rogers AD, Díaz S, Fischer M, Hashimoto S, Lavorel S, Wu N, Ngo H (2021) Scientific outcome of the IPBES-IPCC co-sponsored workshop on biodiversity and climate change. Zenodo. https://doi.org/10.5281/ZENODO.4659158

Ramoutar-Prieschl R, Hachigonta S (2020) Management of research infrastructures: a South African funding perspective. Springer International, Cham. https://doi.org/10.1007/978-3-030-37281-1

Räsänen M, Aurela M, Vakkari V, Beukes JP, Tuovinen J-P, Van Zyl PG, Josipovic M, Venter AD, Jaars K, Siebert SJ, Laurila T, Rinne J, Laakso L (2017) Carbon balance of a grazed savanna grassland ecosystem in South Africa. Biogeosciences 14:1039–1054. https://doi.org/10.5194/bg-14-1039-2017

Reyers B, Stafford-Smith M, Erb K-H, Scholes RJ, Selomane O (2017) Essential variables help to focus sustainable development goals monitoring. Curr Opin Environ Sustain 26–27:97–105. https://doi.org/10.1016/j.cosust.2017.05.003

Scholes M (2006) A holistic and integrated approach to the understanding of biogeo-chemistry: Library Letter. Glob Ecol Biogeogr 15:431–431. https://doi.org/10.1111/j.1466-822X.2006.00249.x

Scholes M, Andreae MO (2000) Biogenic and pyrogenic emissions from Africa and their impact on the global atmosphere. Ambio 29:23–29

Scholes RJ, Gureja N, Giannecchinni M, Dovie D, Wilson B, Davidson PK, McLoughlin C, van der Velde K, Freeman A, Bradley S, Smart DR, Ndala S (2001) The environment and vegetation of the flux measurement site near Skukuza, Kruger National Park. Koedoe 44

Tagesson T, Fensholt R, Cropley F, Guiro I, Horion S, Ehammer A, Ardö J (2015a) Dynamics in carbon exchange fluxes for a grazed semi-arid savanna ecosystem in West Africa. Agric Ecosyst Environ 205:15–24. https://doi.org/10.1016/j.agee.2015.02.017

Tagesson T, Fensholt R, Guiro I, Rasmussen MO, Huber S, Mbow C, Garcia M, Horion S, Sandholt I, Holm-Rasmussen B, Göttsche FM, Ridler M-E, Olén N, Lundegard Olsen J, Ehammer A,

Madsen M, Olesen FS, Ardö J (2015b) Ecosystem properties of semiarid savanna grassland in West Africa and its relationship with environmental variability. Glob Chang Biol 21:250–264. https://doi.org/10.1111/gcb.12734

Taylor CA, Rising J (2021) Tipping point dynamics in global land use. Environ Res Lett 16:125012. https://doi.org/10.1088/1748-9326/ac3c6d

United Nations (2021) THE 17 GOALS | Sustainable Development. https://sdgs.un.org/goals. Accessed 4 Jun 2021

Valentini R, Arneth A, Bombelli A, Castaldi S, Cazzolla Gatti R, Chevallier F, Ciais P, Grieco E, Hartmann J, Henry M, Houghton RA, Jung M, Kutsch WL, Malhi Y, Mayorga E, Merbold L, Murray-Tortarolo G, Papale D, Peylin P, Poulter B, Raymond PA, Santini M, Sitch S, Vaglio Laurin G, van der Werf GR, Williams CA, Scholes RJ (2014) A full greenhouse gases budget of Africa: synthesis, uncertainties, and vulnerabilities. Biogeosciences 11:381–407. https://doi.org/10.5194/bg-11-381-2014

Veenendaal EM, Kolle O, Lloyd J (2004) Seasonal variation in energy fluxes and carbon dioxide exchange for a broad-leaved semi-arid savanna (Mopane woodland) in Southern Africa. Glob Chang Biol 10. https://doi.org/10.1111/j.1365-2486.2003.00699.x

Williams CA, Hanan N, Scholes RJ, Kutsch W (2009) Complexity in water and carbon dioxide fluxes following rain pulses in an African savanna. Oecologia 161:469–480. https://doi.org/10.1007/s00442-009-1405-y

WMO (2015) Status of the global observing system for climate GCOS-195. WMO

Lessons Learned from a North-South Science Partnership for Sustainable Development

31

Robert Luetkemeier, Mari Bieri, Ronja Kraus, Meed Mbidzo, and Guy F. Midgley ⓘ

Abstract

SDG goal 17 seeks to strengthen global partnerships, especially between Global North and South. However, in research and development, experiences indicate a mismatch in expectations with perceived power, funding and workload imbalances, a situation derogated as 'parachute science' or 'helicopter research'. The research programme SPACES seeks to enhance North-South collaborations. As an inter- and transdisciplinary research programme focusing on the interactions between land, sea, atmosphere, biosphere and society, it aims to be a forum for fruitful partnerships. In this chapter, we carve out lessons learned from the nine projects involved in the programme's second phase. Based on a survey amongst 66 SPACES II scientists, we explored their motivations for collaboration, their involvement in decision-making, the assigned resources and workloads as well as conflicts between Northern and Southern teams. Furthermore, we conducted

R. Luetkemeier (✉)
Institute for Social-Ecological Research (ISOE), Frankfurt, Germany

Goethe University Frankfurt, Frankfurt, Germany
e-mail: luetkemeier@isoe.de

M. Bieri
Thünen Institute of Climate-Smart Agriculture, Braunschweig, Germany

R. Kraus
Institute for Social-Ecological Research (ISOE), Frankfurt, Germany

Stockholm University, Stockholm, Sweden

M. Mbidzo
Namibia University of Science and Technology, Windhoek, Namibia

G. F. Midgley
Stellenbosch University, Stellenbosch, South Africa

© The Author(s) 2024
G. P. von Maltitz et al. (eds.), *Sustainability of Southern African Ecosystems under Global Change*, Ecological Studies 248,
https://doi.org/10.1007/978-3-031-10948-5_31

bibliometric analyses and observed an intensification of the North-South co-authorship network over time. We conclude that SPACES can be considered a success as researchers acted largely as peers on an equal footing. Nevertheless, our insights show that (1) the asynchrony in funding is a threat for effective collaborations, (2) continuous project evaluation should incorporate a North-South component and (3) collaborative publications should be formalised as a tool for integration.

31.1   Introduction

A rapidly growing human population and global environmental change, particularly the impacts of climate change, are expected to put pressure on the natural resource base of the African continent (Gasparatos et al. 2017), on which the majority of the population depends for their livelihoods and which are important for economic growth and inclusive development (Gupta and Vegelin 2016). In order to meet the Sustainable Development Goals (SDGs), there is a need to deal with the multi-faceted and interlinked sustainability challenges faced by countries in Africa. This needs to be done from an inter- and transdisciplinary research approach with multiple perspectives to produce knowledge relevant to real-world problems (Pohl and Hirsch Hadorn 2007). In addition to working on pressing environmental challenges, respective research projects are now increasingly expected to consider the needs of society (Tress et al. 2005), whilst also having policy implications. Fulfilling these expectations requires collaboration of teams from the natural and social sciences, including non-academic stakeholders (e.g. institutional actors and local communities) to work on sustainability challenges (Patel et al. 2017). One key component for sustainable development is described in SDG 17, which calls for strengthening global partnerships (Waage et al. 2015). Respective partnerships can take the shape of South-South research collaborations, which are important for regional scientific and political integration (Boshoff 2010). Since the mid-twentieth century, however, North-South collaborations (NSCs) have received more attention particularly in the discourses around post-colonialism and the diverse strands of development aid (Gaillard 1994). In 1979, the United Nations adopted the 'Vienna Program of Action' stating key criteria to strengthen international research collaborations such as accounting for development priorities of the South, joint participation and control as well as capacity development (Gaillard 1994; UN 1979). Unfortunately, respective aspirations have still not been fully met as current critical perspectives on NSC confirm. Bradley (2017), for instance, explores the agenda-setting processes and power imbalances between Northern and Southern research partners and finds prevailing deficits (Bradley 2017). The term 'parachute science' or 'helicopter research' evolved in this regard to critically describe the way Northern researchers conduct research in the South without adding any benefit to the Southern science sphere (Giller 2020; Stefanoudis et al. 2021). Approaches to account for the deficits in NSC often refer to the necessity to follow a transdisciplinary mode of research (Schmidt and Pröpper 2017). More specifically, recent investigations particularly highlight the need to acknowledge the double-role of scientists from the South (acting as researchers and field facilitators), the need for joint publications to foster capacity building, as well as methodologically guided knowledge integration and co-production (Luetkemeier et al. 2021).

One such NSC initiative is the research programme Science Partnerships for the Adaptation to Complex Earth System Processes in Southern Africa (SPACES), funded by the German Federal Ministry for Education and Research (BMBF). The programme's first phase was launched in 2012 (SPACES I), whilst the current second phase was initiated in 2018 (SPACES II). The programme can be classified as a type-I NSC (Bradley 2007:13) as it is characterizsed by research teams that are explicitly constructed to carry out a certain project. The BMBF funds collaborative German-African research projects in South Africa and Namibia that contribute to the formulation of policy recommendations for Earth system management. The programme is aligned with the national and international trends and initiatives on international collaboration, including SDG 17 and the BMBF's Africa strategy of 2014. The programme was from the start defined as a joint endeavour with the Department of Science and Innovation (DSI) of South Africa and was intended to contribute to intensified cooperation with the Namibian Ministry of Education (MET). A core aim of the programme is to promote scientific networking between African and German research institutions. The programme's research projects were intended to focus on jointly defined areas, consider both African and German interests, emphasise partnership and ownership and should enable continuity and reliability of collaborations. Scientific exchange between partner countries and international networking were stressed as expectations from BMBF. In this regard, the success indicators put strong emphasis on NSC in the form of, for example, the number of joint German-African publications, and jointly supervised students (BMBF 2017).

Against this background, this chapter examines the structure and quality of NSC within SPACES II to carve out both successful ways of collaboration, but also deficits in how Northern and Southern researchers worked together. We draw lessons learned to contribute to more efficient and equitable future research and development projects between Northern and Southern partners. Section 31.2 provides an overview on the methods we applied to explore the North-South collaborations. Section 31.3 presents the results of our investigations and Sect. 31.4 discusses our findings against the current state of knowledge in the scientific literature. Section 31.5 draws conclusions for future collaborative research projects.

31.2 Material and Methods

For exploring the structure and quality of NSC in SPACES II, we applied three distinct methods. First, we screened the research proposals of the projects involved to qualitatively assess how the research teams initially designed the North-South collaboration technically and conceptually. Second, we conducted a structured online survey amongst researchers involved in SPACES II to obtain their perspectives on how the collaboration actually manifested. Third, we quantitatively explored the evolution of co-authorships amongst Northern and Southern project partners over time. In general, we separate between Northern and Southern researchers based on their actual institutional affiliation and not based on their personal nationality. This means that a researcher from South Africa who is currently affiliated to a German institution is considered a Northern researcher in our investigations and vice versa.

31.2.1 Structured Online Survey

Most studies that seek to evaluate North-South collaborations follow a qualitative research design (Luetkemeier et al. 2021; Schmidt and Pröpper 2017; Casale et al. 2011), as the topic itself is highly diverse and case-specific as well as often subject to a small number of researchers involved in respective projects. As SPACES is a research programme with nine individual research projects in its second phase that partly build upon prior collaborations, a reasonable number of researchers were involved. Taking advantage of this large number of researchers, a quantitative empirical research approach was considered as a valuable way forward to representatively identify similarities and differences on how researchers consider NSC amongst German, Namibian, and South African partners.

We initially compiled a list of researchers who were involved in ongoing research activities of SPACES II at the time of writing this chapter, based on the feedback of the respective project leads. Though the projects apply varying definitions in terms of who qualifies as a 'project researcher' (e.g. Southern researchers are not formally funded [cf. Sect. 31.3.1], so their actual affiliation to SPACES may be hidden), we consider the compiled contact list as reasonably comprehensive. All projects listed both Northern (124) and Southern researchers (87) with an average of 10 on the Southern and 14 on the Northern side. This results in a total statistical population of $N = 211$. The online survey was pre-tested amongst four researchers (two from the North, two from the South) and adapted somewhat to increase clarity in the language used in some questions. It was implemented and rolled out using the software LimeSurvey (2021) in July 2021.

In agreement with the project leads, we initially screened the nine project proposals for statements on how the collaboration between Northern and Southern researchers was designed. We specifically looked into the structural project setup in terms of (1) how the lead team was composed, (2) if prior collaborations existed, (3) what tasks Northern and Southern partners were assigned to and (4) which tools were planned to foster collaboration. This helped us to gain a basic understanding of how the nine projects were developed and what importance NSC received in initial considerations. Against this background, we built a structured questionnaire to gain quantitative insights into the researchers' viewpoints on selected NSC issues. In order to receive as many responses as possible, we considered a time span of about 10 min for answering the questions as a suitable compromise between scope and applicability of the survey. The survey received an ethical clearance from the ethics board of Institute for Social-Ecological Research (ISOE). All respondents were guaranteed that their participation was fully anonymous and the results were solely used for the analysis of NSC as presented in this chapter.

As a basic structure of the survey, we made use of the 'SDG Partnership Guidebook' that suggests four pillars for effective partnerships in international collaborations for sustainable development (Stibbe and Prescott 2020). Though this framework is not specifically targeted towards research projects, we consider it as a comprehensive list of aspects to analyse the quality of NSC. The structured part of the survey was set up along the following topics:

- *Fundamentals*: What is the benefit for project partners to collaborate and how were the projects developed?
- *Partnership relationship*: Do researchers work transparently and trustfully with adequate involvement in decision-making processes?
- *Structure and setup*: Were resources, tasks and workload assigned fairly to all project partners?
- *Management and leadership*: Did tensions and conflicts occur between project partners and how was the communication implemented?

In addition to the structured questions, open questions on the impact of COVID-19 on project work and crucial issues to consider when developing future NSC projects were posed at the end of the survey. The structured questions were designed in a way to apply a four-point Likert-scale (Joshi et al. 2015). The respondents were hence asked to reveal either a tendency for approval or rejection of the questions and statements presented. As a fifth category, respondents were able to select 'no answer', which we did not treat as a neutral category in between approval and rejection, but rather as a non-response or missing value. The ordinal scale responses can be statistically evaluated, especially with respect to the differences between Northern and Southern partners. For doing this, the non-parametric Mann-Whitney U test (also known as Wilcoxon rank-sum test) for two independent samples was applied (MacFarland and Yates 2016). The null hypothesis that the two samples' means do not significantly deviate from each other was rejected, if $p < 0.05$.

For analysing the open questions posed in the survey, we used the software MaxQDA (VERBI Software 2019) to code the respondents' statements. The coding scheme evolved whilst working through the statements with a final grouping of similar codes.

31.2.2 Co-authorship Network Analyses

The purpose of the bibliometric analyses was to back up and verify statements and literature insights on the role of collaborative publications. The analyses served to visualise the evolution of the cooperative publication network, with a focus on Northern and Southern researchers, prior and during the SPACES I and II programmes. Publication data of SPACES II researchers (211 researchers as well as other academic staff in the nine research projects) was collected from the Scopus database. Scopus was selected due to the availability of full author affiliation data. The searches were limited to articles, conference papers, reviews, book chapters, books and data papers in English language. The scientific field was limited to Earth and Environmental sciences, Geosciences, Ecology, Marine sciences, computer sciences and related disciplines.

To study the evolution of collaborations, we analysed three 4-year periods: (1) immediately prior to SPACES (2009–2012); (2) during the first phase of SPACES (2013–2016); (3) immediately before and during SPACES II (2017–2020). These time brackets are indicative, as there were large differences in the start and end dates of the different projects (i.e. the first SPACES II project started in June 2018 and the last in February 2019). The final dataset for 2009–2012 consisted of 852 publications; 2013–2016 consisted of 1175 publications; and 2017–2020 of 1436 publications.

For visualising co-authorship networks by individual researchers and by institutes, we used the network visualisation function of VOSviewer (van Eck and Waltman 2010). For the analyses, different forms of author names were merged by building a dictionary (e.g. G.F. Midgley would be synonymous with G. Midgley). Similarly, different forms of spelling institute names were combined for the analyses, and furthermore, different departments of the same institute were merged to represent the parent institute as unit of analysis (e.g. 'Stellenbosch University'). Only collaborating institutes of SPACES I or II were selected in the visualisations.

For the network visualisations, we included all individuals and institutes who had co-authored at least one publication with another SPACES II researcher or another SPACES II institute. We used fractional counting to control the disproportionate impact of multi-author documents (see Perianes-Rodriguez et al. 2016 for discussion on benefits). In fractional counting, the number of co-authors affects the weighting: for example a paper co-authored by 20 authors gets a weight of 1/20. We also excluded publications with more than 50 authors from the analyses, with the expectation that these outputs are less representative of true collaborative relationships. To construct the mapping, VOSviewer uses the 'association strength' similarity measure, which was normalised to control the large differences between author publishing activities. Clustering resolution was set to 1.0, as by default in VOSviewer; by changing the resolution, clusters can be made larger or smaller to explore the data (van Eck and Waltman 2009). Network visualisation was used to show the development of co-authorship links amongst individual researchers, and overlay visualisation was used to identify the main institutes contributing to NSC publications.

31.3   Results

For presenting the results of our study, we begin with a brief overview on how the proposals of the nine SPACES II projects addressed NSC. Subsequently, we present selected results of the online survey with a particular focus on common grounds and significant differences between the viewpoints of Northern and Southern researchers. Afterwards, we particularly shed light on the evolution of the collaboration between Northern and Southern researchers, based on co-authorships in scientific publications.

31.3.1 Designing Effective Collaborations in Project Proposals

Screening the proposals confirmed that the projects cover a broad range of topics from Earth system sciences and likewise differ strongly in their scientific setup ranging from clearly focused interdisciplinary to extensive transdisciplinary approaches. In this sense, the projects also deviated in terms of (1) formal assignment of responsibilities to African partners, (2) allocation of grants and resources for African institutions as well as (3) tools planned to enhance collaboration between Northern and Southern researchers.

The responsibilities of African partners varied strongly across the projects. Whilst the work packages of some projects were solely led by German partners, most projects explicitly implemented a tandem structure in which both a Northern and a Southern partner were formally considered as work package leads. For some projects, this even held true for the structure of the overall project lead. Though the lead structures were rather heterogeneous, most proposals clearly outlined the responsibilities and tasks of Northern and Southern partners.

In SPACES II, the BMBF funded the German partners, whereas the African partners were expected to apply for funding from their country or provide evidence of counter-financing on their own, according to prior agreements with the South African and Namibian ministries. Therefore, African partners were not formally allowed to be funded in the same way as the German partners. Some projects subcontracted their African collaborators and hence made financial means available for researchers to carry out particular tasks. However, funds provided through subcontracts are limited to a certain proportion of the total project budget and to service-based rather than research contracts.

In terms of collaborative tools, the projects formally referred to the SPACES-associated programme of the German Academic Exchange Service (DAAD) to fund mutual visits of researchers. Some projects went further and explicitly mentioned publications as a key tool for collaboration as well as data-sharing protocols. One project even introduced a steering committee composed of Northern and Southern researchers to guarantee joint decision-making.

31.3.2 Researchers' Perspectives on North-South Collaboration

In total, 66 researchers of the SPACES II programme participated in the online survey (response rate of 31%) during the month of July in 2021. About 36% of them came from a South African or Namibian institution, whilst the remaining 64% were affiliated to a German institution. These shares correspond to the distribution in the statistical population of Southern (87) and Northern (124) researchers. About 38% of the respondents were female, 59% were male and about 3% preferred to not answer this question. Most of the respondents (12 Southern and 23 Northern researchers) described themselves as holding a leading position in their respective projects, either as overall project leads, work package leads or principal investigators. Only about one-fourth of the respondents declared themselves as PhD students. About 48% of the respondents (11 Southern and 21 Northern partners) stated that they did not have any prior collaborations with their current Northern or Southern project partners, respectively.

31.3.2.1 Fundamentals

When considering the motivations of project partners to join the SPACES II programme, the differences between Northern and Southern researchers were not statistically significant (question 5). They expected benefits (1) in expanding their professional network, (2) in sharing knowledge and mutual learning as well as (3) from gaining novel research insights, which conventional projects might not provide (Fig. 31.1). 'Access to a new study area', 'tapping into new research funds' and 'gaining access to resources to carry out research' were less important than the aforementioned ones but still more than 50% of the respondents consider

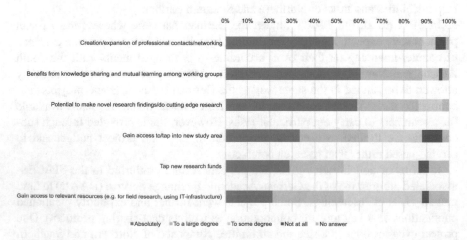

Fig. 31.1 Motivations for researchers and research groups to join the North-South collaboration. Results are based on question 5 of the online survey with a sample size of $n = 66$. According to the Mann-Whitney-U-test, no significant differences could be found between Northern and Southern researchers

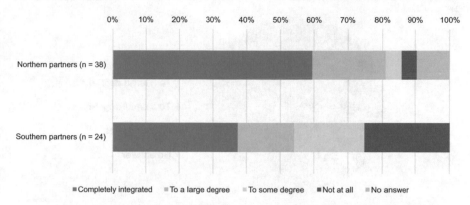

Fig. 31.2 Researchers' evaluation regarding their integration into the initial process of defining/drafting the overall research question(s). Results are based on question seven of the online survey with a sample size of $n = 66$. The responses of Northern and Southern partners are significantly different with $p < 0.05$

them relevant. Looking at the fulfilment of these motivations and expectations, the majority of the Southern researchers indicated that they were 'absolutely satisfied' or at least 'to a large degree' (question 6). The expectations for improved networking scored highest with a consent of about 78%.

In terms of designing the overall research question, the participants' responses revealed that this process was organised in a collaborative manner in most cases. Often, SPACES II projects could build on prior collaborations to develop the project proposal, and researchers carried out an integrating process, for example, via workshops (question 8). As Fig. 31.2 presents, a high proportion of Northern partners (81%) compared to Southern partners (54%) were either 'completely' or 'to a larger degree' involved in the initial drafting of the overall research questions. This difference is statistically significant.

Aside from the initiating process, a partnering mindset amongst the researchers is a key attribute of successful North-South collaborations (Stibbe and Prescott 2020:70). Our respondents used terms like 'communication', 'transparency', 'knowledge sharing' and 'respect' to illustrate what they consider a partnering mindset. In this regard, about 69% had a positive view on the partnering mindset without a significant difference between Northern and Southern researchers (question 10).

31.3.2.2 Partnership Relationship

To look at how the researchers evaluated their partnership in more detail, we asked the respondents if they consider their respective Northern or Southern counterparts as acting transparently (question 13) and as committed in the project (question 12). The responses showed an appreciation for each side with no significant differences between the North and the South. Both groups evaluated the commitment and the transparent way of acting as largely positive (>90%). However, when looking deeper

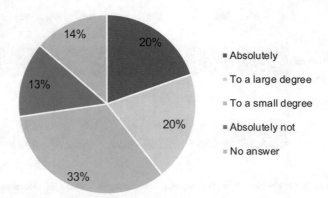

Fig. 31.3 Researchers' evaluation on whether the fact that official project reports to the funder had to be written exclusively in German language was an obstacle to collaboration. Results are based on question 14 of the online survey with a total sample size of $n = 57$. Differences in responses from Northern and Southern partners are not statistically significant at $p < 0.05$

into specific issues of project collaboration, some critical aspects appeared to be of concern. One issue was that according to formal funding requirements, official project reports had to be written exclusively in German. This may pose a challenge to a collaborative project, in which working and publication language is English. Figure 31.3 shows that about 40% of the researchers considered this requirement as a serious obstacle for project collaboration, whilst no significant difference could be found between the two groups (question 14).

A discrepancy between Northern and Southern researchers became obvious in terms of involvement in decision-making processes with respect to project administration and management (question 16) as well as practical research conduction (question 15). In both cases, the Southern researchers felt that they had been less involved in decision-making processes than their Northern counterparts had. For both questions, the differences between the groups were statistically significant (Fig. 31.4).

Moving on to even more practical and tangible issues in research collaboration, the question of authorship was assessed. We asked the researchers if they considered the way they have been involved in publishing collaborative research results was adequate or not (question 17). Figure 31.5 shows that overall, 84% of the respondents had a rather positive view on their involvement in publishing collaborative research results. None of the researchers chose to answer that 'absolutely no' involvement took place. However, in contrast to all other answer categories more Southern researchers answered 'somewhat involved', though this difference is not statistically significant due to the small sample size. Section 31.3.4 looks deeper into the evolution of co-authorship amongst Northern and Southern partners throughout the course of the programme.

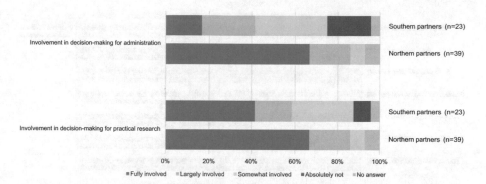

Fig. 31.4 Perceived degree of involvement in (**a**) decision-making process with respect to project administration (e.g. funding formalities, reporting, representation) and (**b**) decision-making processes with respect to practical research conduction (e.g. field work planning, case study selection, method selection). Results are based on questions 15 and 16 of the online survey with a total sample size of $n = 62$. The responses of Northern and Southern partners are significantly different with $p < 0.05$

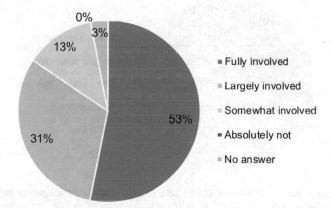

Fig. 31.5 Evaluation of researchers from the North and the South, whether they have been adequately involved in the publication of joint research. Results are based on question 17 of the online survey with a total sample size of $n = 64$. The responses from Northern and Southern researchers do not differ statistically significantly ($p < 0.05$)

31.3.2.3 Structure and Setup

From a structural perspective, the survey results revealed insights into the researchers' focus of work (question 18). Figure 31.6 shows to which activities the respondents allocated their personal workforce. More than 60% of all respondents indicated that they spent most of their work on empirical research, followed by student supervision and field logistics. Overall, the differences between Northern and Southern researchers were not statistically significant, except for the communication with the funding agency, for which the Northern partners allocated more work than their Southern partners did.

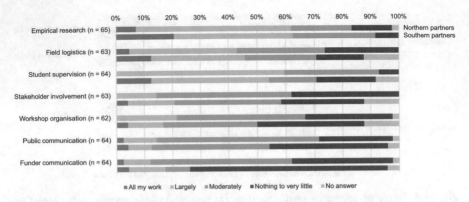

Fig. 31.6 Workload assigned to certain project activities by researchers. Results are based on question 18 of the online survey with a total sample size ranging between $n = 62$ and $n = 65$. The responses from Northern and Southern researchers do not differ statistically significantly ($p < 0.05$), except for 'funder communication'

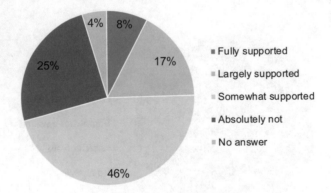

Fig. 31.7 Researchers' evaluation of if Southern partners have been adequately supported with resources (e.g. funds, staff, equipment). Results are based on question 20 of the online survey with a total sample size of $n = 63$

In terms of funding and resources endowment, the online survey assessed respective viewpoints. With respect to the Southern partners' in-kind or non-monetary contributions, such as the provision of working hours and equipment (question 19), the results show that 58% of the respondents consider these as adequately acknowledged. Here, no significant differences could be found between the North and the South.

Considering the endowment of Southern partners with adequate resources for conducting project activities, Fig. 31.7 shows the perception of Northern and Southern researchers (question 20). The responses showed a critical perspective as about 71% of the respondents considered the Southern partners as only 'somewhat supported' or worse. The views between Northern and Southern researchers did not vary. Interestingly, though not statistically significant, a larger share of the Southern

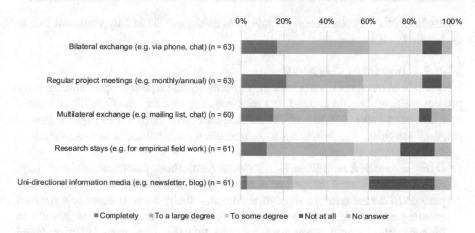

Fig. 31.8 Evaluation of researchers, which communication channels/tools helped them during the course of the project to stay informed about the activities of their Northern/Southern partners. Results are based on question 22 of the online survey with a sample size ranging from $n = 61$ to $n = 63$. The responses from Northern and Southern researchers do not differ significantly

partners considered their own resources endowment as being adequate, whilst more Northern partners considered it the other way round.

31.3.2.4 Management and Leadership

As a key component of formal project collaboration and integration, sharing data and results was highlighted by most projects in their proposals. In this regard, 86% of our respondents confirmed respective processes were implemented as intended (question 23). Similarly, the researchers had a positive perception of how responsibilities and expectations were shared amongst the involved researchers. The majority declared that only minor tensions and conflicts were registered during the project period (question 21). In both cases, no significant differences could be found between the two groups.

Figure 31.8 shows how the researchers evaluated different tools in their potential to keep Southern or Northern colleagues, respectively, up to date about ongoing research activities (question 22). Regular project meetings and bilateral exchange were considered as the primary means for mutual updates. However, a considerable number of respondents revealed that respective tools have either not been used in the projects, or have been implemented poorly so that no continuous mutual update was possible.

31.3.3 Qualitative Insights from the Survey

Alongside the structured survey part, the respondents were invited to comment on two issues in an open, unrestricted way. First, we assessed how the COVID-19 pandemic affected NSC and second, the researchers were asked to mention the key issues in which they see a need for action to enhance NSC in a hypothetical follow up funding programme.

31.3.3.1 Impact of COVID-19

About 91% of our respondents took the opportunity and reported about outcomes for project collaboration. In general, the researchers agreed that the COVID-19 related restrictions of public life like travel restrictions (Devi 2020) led to impairments of research activities. The following consequences were highlighted in more detail.

- *Delay of research activities:* Due to travel restrictions, empirical fieldwork (e.g. data acquisition) was delayed or even cancelled. Some projects missed essential periods to assess relevant data to accomplish their research goals (e.g. missed growing season). Experiments had to be cancelled or restarted, which led to higher costs. Overall, researchers assume that this may have led to reduced scientific output.
- *Exchange between researchers suffered:* Travel restrictions prohibited joint fieldwork sessions, which are usually seen as a key element for productive exchange between Northern and Southern researchers. No conferences and project meetings could be held in person and thus, important forums for exchange and networking could not be implemented. However, some respondents observed an improved communication between Northern and Southern partners via the increased use of digital communication technologies (cf. Fig. 31.8). Long-established research collaborations were considered as more effective in coping with the COVID-19-induced research restrictions.
- *Higher workload for Southern partners:* Though the quantitative survey results did not indicate a significant difference between the workload of Northern and Southern researchers, some respondents indicated that COVID-19 put more pressure on the South. As travel restrictions prohibited German researchers to travel, Southern partners had to take over essential research tasks (e.g. data acquisition, stakeholder communication).
- *Negative psychological impacts:* The delay of research activities was considered by some of our respondents as having negative impacts on the motivation of researchers to conduct project tasks. This may have been particularly true for student researchers who only had a limited period available to conduct their research tasks and who may have cancelled their activities.

31.3.3.2 Fields of Action to Improve North-South Collaborations

We asked our respondents to state crucial aspects that should be considered for future research projects to enhance NSC (question 24). In this regard, 55 out of 66 shared their opinions on this issue and we condensed these into the following dimensions.

- *Improved funding opportunities for Southern partners:* Most respondents from both the North and the South indicated that they see the necessity to improve the funding opportunities of Southern partners. The recommendations covered a wide range of topics, but frequently touched upon the points of enhancing financial endowment for fieldwork and laboratory analyses, more options for Southern researchers and students to visit their Northern counterparts (e.g. for mutual learning and utilising research infrastructure) and improved funding modalities for Southern students, in general. One female African scientist boiled it down to *'funding directed to Southern partners reduces the "parachute science" feeling and empowers the Southern partners'*.

- *More equitable project design and management:* Respondents often highlighted the necessity to better integrate Southern partners into overall project management. This includes both the design phase for proposal writing (aligning research activities to local needs) as well as continuous project management and decision-making processes. One key issue raised was the request to more explicitly declare certain research objectives and clearly assign tasks and responsibilities to certain actors of the project team.

- *Enhanced exchange between researchers and students:* The respondents saw the need to improve the exchange between Northern and Southern researchers on both the Post-Doc and the student level. Personal exchange with one another was seen as a key element to enhance NSC, which could, for instance, be fostered in the field of empirical research, collaborative teaching, student supervision and exchange programmes.

31.3.4 Role of SPACES in Fostering Collaborative Publications

In addition to the insights into the role of co-authorships between Northern and Southern researchers, we explored this issue in a quantitative way to reveal if SPACES was able to enhance collaboration as measured by this indicator. For this purpose, the following subsections first look into the co-authorship networks of individual researchers, then zoom out onto the institutional level and finally assess the relative proportion of collaborative papers against the total number of papers published by SPACES researchers.

31.3.4.1 Co-authorship Networks Between Individual Researchers

During the years 2009–2012, one-third (69 out of 211) of SPACES II network researchers were involved in publications co-authored with another SPACES researcher, irrespective of his/her regional affiliation. During 2013–2016, the proportion had increased to 45% (95 researchers), and during the last time bracket, 2017–2020, the majority 82% (147 researchers) had co-authored publications with another SPACES II researcher. All authors contributing most to the total co-authorships were German: during 2009–2012 F Jeltsch (total link strength, i.e. the total strength of the co-authorship links of the researcher with others in the network 14.0) and N Blaum (11.0); during 2013–2016 S Higgins (17.0), T Haberzettl and R Mäusbacher (14.0), and during 2017–2020, H Kunstmann (27.0) and P Laux (22.0). The strongest co-authorship links (i.e. those with most co-authored publications; by fractional counting, publications with higher number of co-authors are given less weight) were between two German authors (during 2009–2012 Jeltsch and Blaum, 9.3; during 2013–2016 Higgins and Scheiter 8.5; and during 2017–2020 Kunstmann and Laux 21.5) or two South African authors (during 2009–2012 Chirwa and Syampungani, 7.0).

Similarly, an expanding network is seen when limiting the focus to North-South co-authored publications (i.e. links between German and southern African authors). Figure 31.9 shows the co-authorship network constructed from the publications of SPACES II researchers during the three selected time brackets. Each circle represents a researcher, and the size of the circle indicates 'total link strength'. Lines amongst researchers represent co-authorship links: researchers connected with a line have at least one co-authored publication, and those with thicker lines have a stronger link. In addition, authors are clustered together according to collaboration strength, i.e. those authors that publish more with each other are located closer together. To highlight the evolution of NSC, only respective collaborative linkages are shown in the visualisations. Note that the vertical and horizontal coordinate locations of the authors in the visualisation reflect the full collaboration network: this way, natural collaboration clusters (marine-terrestrial clusters; remote sensing cluster etc.) were maintained.

During the years 2009–2012, only 9% (20 authors) were involved in North-South co-authored publications with other SPACES II researchers. Within 2013–2016, the network had broadened to 41 researchers (19%), and in 2017–2020, nearly half (103) of the researchers had published with a Northern or Southern SPACES II co-author. Nearly all authors contributing most to NSC co-authorships were Namibian or South African: during 2009–2012 B Strohbach (3.0) and H Verheye (3.0); during 2013–2016 M Meadows (3.8), A van der Plas (3.6) and K Kirsten (3.5); and during 2017–2020, I Smit (6.5) and I Grass (5.0). The strongest co-authorship links were between H Verheye and W Ekau (1.5) and H Verheye and W Hagen (1.5) (2009–2012); B Bookhagen and T Smith (2.0) and V Morholz and A van der Plas (1.5) (2013–2016) and B Bookhagen and T Smith (4.0) and I Grass and S Weier (2.5) (2017–2020) (Fig. 31.9). Notably, most of these are senior researchers and none are early-career researchers. When considering all NSC collaborative publications, German collaborators dominated lead authorship during all three time brackets. During the years 2009–2012, 52% of collaborative papers were led by a German author, whereas 26% had a southern African lead author. During the years 2013–2016, the difference was proportionally slightly smaller, with 51% German-led, and 34% southern African led publications. During the last bracket 2017–2020, 47% of the collaborative publications were German-led and 27% by a southern African lead author.

Specific observations can be made when focusing on the clusters that indicate topical collaborations of SPACES II researchers (based on broad definition of the researchers' main field). The oceanography researchers cluster publishing partic-ularly on the Benguela upwelling system, including A Van der Plas and D Louw (Namibia), W Ekau (Germany), H Verheye (South Africa) and others, exists before SPACES and stays relatively stable throughout the research programme. In the last time bracket (2017–2020), this cluster is larger and linked to others, with P Brandt

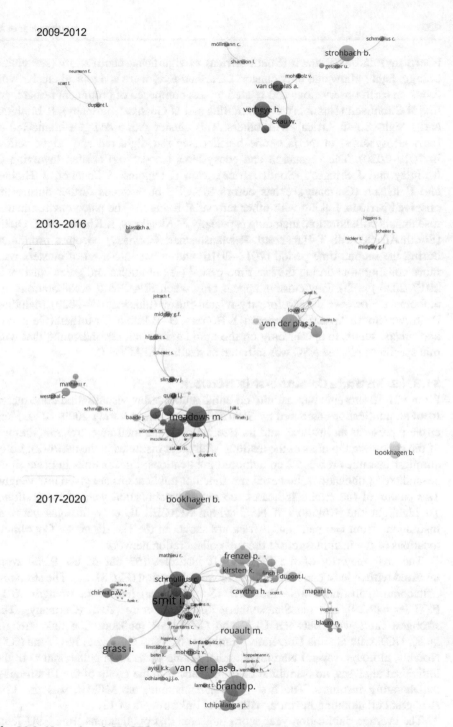

Fig. 31.9 Networks of SPACES (II) researchers with North-South co-authorship links (i.e. publishing with a Northern or Southern SPACES (II) research partner) on the 2009–2012 (above); 2013–2016 (middle) and 2017–2020 (below) publications. Size of circles indicates total link strength (i.e. authors with more co-authored publications show larger). South-South and North-North collaboration links have been removed from the visualisation

(Germany) and M Rouault (South Africa) as significant contributors (see violet, orange, light yellow and pink clusters). Furthermore, there is a topical cluster with focus on earth observations and related topics, composed of prominent researchers like B Strohbach (Namibia), C Schmullius and U Gessner (Germany), R Mathieu, and I Smit (South Africa), with others. This cluster grows and gets connected to (agri-)ecosystems, in the last time bracket (see red, light red and bright yellow in 2017–2020). The vegetation and ecosystems (modelling) cluster, involving G Midgley and J Slingsby (South Africa), with S Higgins, S Scheiter, T Hickler and F Jeltsch (Germany), exists before SPACES but extends further during the observed periods, linking with other terrestrial research. The paleo environmental researchers collaboration, including especially M Meadows, K Kirsten and L Quick (South Africa) with T Haberzettl, R Mäusbacher (Germany) becomes prominent during the second time period (2013–2016) and grows into a set of clusters with more collaborators during the last time period (see turquoise and green clusters in 2017–2020 graph). Few clusters appear only when SPACES II collaborations are elaborated. However, the agroforestry-related cluster (blue in 2017–2020), including P Chirwa (South Africa), J Sheppard, L Borrass, H.-P. Kahle, C Morhart (Germany) and others, seems to appear only on the third studied bracket, indicating that with this specific topic, the NSC was initiated as result of SPACES II.

31.3.4.2 Institute Co-authorship Network

Figure 31.10 presents the institute co-authorship overlay visualisation constructed from all publications authored by SPACES II researchers during 2009–2020. Each circle represents an institute, and its size indicates the total link strength (strength of the co-authorship links of the institute with other institutes in the network). Lines amongst institutes represent co-authored publications; thicker lines indicate more co-authored publications, however, multi-author publications are given less weight. The colour of the circles indicates the average publication year of the institute. To highlight the evolution of NSC within SPACES II, only linkages between institutions from Germany and Africa are shown in the visualisations. Coordinate locations of the institutes reflect the full collaboration network.

The vast majority of the SPACES II institutes (63 out of 66, 95%) were involved with at least one NSC collaborative publication (Fig. 31.10). The strongest collaborating institutes were the UCT (South Africa) (total link strength 56.2), FSU Jena (26.9), and the Senckenberg Research Centre (20.0) (Germany). The strongest links were between UCT and GEOMAR (collaboration link strength 14.8), UCT with Goethe University Frankfurt (6.5) and UCT with FSU Jena (6.5). Notably, although several Namibian authors came up as main collaborators in the individual analyses, no Namibian institute made it in the group of the 10 strongest collaborating institutes. The first Namibian institute, the MFMR, was the 12th strongest collaborating institute, with a total link strength of 12.

The average publication year score indicates that publications from FSU Jena, CSIR, and the University of Hamburg came out on average a bit earlier than

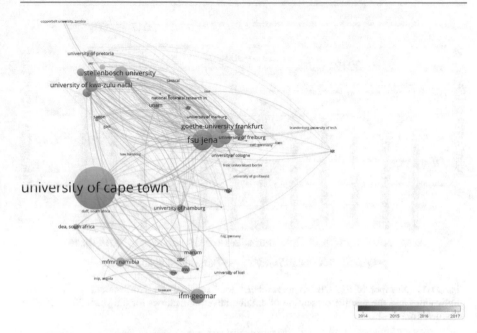

Fig. 31.10 Network of SPACES institutes (both SPACES and SPACES II) with North-South co-authorship links (i.e. publishing with a Northern or Southern SPACES II research partner) during 2009–2020. Size of circles indicates total link strength (i.e. institutes with more co-authored publications show larger, however fractional counting reduces the influence of multi-author publications). The colour of the circles indicates the average publishing year. South-South and North-North collaboration links have been removed from the visualisation

those from e.g. the UCT, Goethe-University, University of Göttingen, Senckenberg Research Institute, Stellenbosch University, and the University of Kwa-Zulu Natal.

31.3.4.3 Relative Proportion of Collaborative Publications

We further expected that the impact of SPACES I and II would show as an increased proportion of North-South co-authored publications over time. Using the same set of SPACES I and II authors publications and counting the total number of publications per country, those publications where at least one author's affiliation was a German university or institution were considered 'German'. Those publications where at least one author's affiliation was an African university or institution (including South Africa, Namibia, Angola, Botswana, Lesotho, Mozambique, Swaziland/Eswatini, Zambia, Zimbabwe, and SASSCAL) were considered 'African'. Figure 31.11 shows that the proportion of collaborative publications peaked in 2018, probably indicating the higher number of project publications that came out at the end of the SPACES I projects.

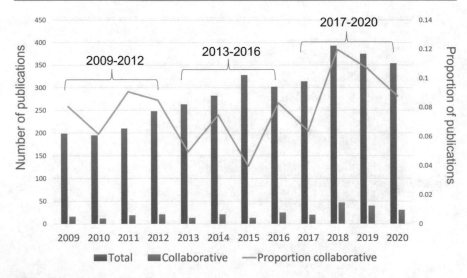

Fig. 31.11 Number of SPACES II researchers' total publications, North-South 'collaborative' publications and the relative proportion of collaborative publications for the years 2009–2020 as obtained from Scopus

31.4   Discussion

Having presented a wide spectrum of results above, in the following subsections, we take a closer look at certain issues uncovered or raised that we consider as essential to address when designing future NSC. In this respect, we first reflect upon the (financial) resources endowment of the Southern partners, as the majority of our respondents indicated the need for improved funding schemes. Second, we discuss the way Southern partners were involved in project design and continuous management as a significant difference in viewpoints between the North and the South could be uncovered. Third, we explore the results of the co-authorship analysis to discuss the potential of collaborative publications as a key tool for effective NSC. Finally, we critically reflect upon the applied methods and the representativeness of our results.

31.4.1 Resources Endowment of Southern Partners

Both the quantitative survey results as well as the qualitative statements of our respondents are critical of the resources endowment of the Southern partners, especially with respect to monetary means (cf. Sects. 31.3.2.2 and 31.3.3.2). The observation of an imbalance in funding is shared by researchers from the North and the South. Comparable results were found in other meta-studies that looked into

NSC programmes elsewhere (Giller 2020; Skupien and Rüffin 2020; Schneider et al. 2019).

Formally, the SPACES II programme was set up as a cooperative funding scheme together with the South African Department of Science and Innovation (DSI) and the Namibian Ministry of Education (MET) (BMBF 2017). With both countries, Germany had developed pre-existing arrangements to enhance collaborative research. According to the funding regulations, which are legally bound by the German federal budget law, only German institutions were entitled to apply for research grants awarded by BMBF. African partners had to apply for respective linked grants at their home country's collaborating partner institution (DSI or MET), had to be subcontracted by their German partners for particular service-based activities or had to contribute in-kind or non-monetary means (BMBF 2017). This model applied to all project research activities, whilst support for student exchange and mutual research visits by partners was enhanced through an exchange and researcher mobility programme established by the German Academic Exchange Service (DAAD) for which both African and German students and researchers could apply.

These funding formalities do not appear to have been sufficient to establish an adequate resources endowment of the Southern partners to conduct research and ensure sufficient researcher mobility, as the experiences of our respondents indicate. This imbalance became exacerbated during the COVID-19 pandemic, when travel restrictions (Devi 2020) inhibited German researchers from carrying out activities in South Africa and Namibia. According to the respondents, African partners stood in to replace these functions, without being fully compensated for their efforts, though the situation was somewhat alleviated by flexible responses in funding arrangements. Overall, the COVID-19 pandemic can be considered a caesura for project collaboration, but it might also provide a window of opportunity to enhance NSC (Jeppesen and Miklian 2020). Despite the drawbacks that our respondents state with respect to the limitations of meetings in person, we particularly see the potential of digital communication tools to enhance the quality of collaboration in a post-pandemic world. Respective tools like video conferencing, data sharing and project management can act complementary to standard project tools such as annual meetings, phone conferences and research visits. In this regard, Sowe et al. (2021) found an increased utilisation of digital communication and collaboration tools amongst Northern and Southern research partners during lockdown periods with certain challenges in particular for Southern researchers like the availability of reliable power and internet connections (Sowe et al. 2021).

In order to improve the funding situation of Southern researchers in NSC programmes, potential improvements can be explored in (1) more concrete multilateral agreements for securing research funds prior to programme start, (2) equal treatment of Southern and Northern researchers in terms of (staff, material, travel) funding to conduct research as well as (3) the reduction of structural institutional barriers for Southern researchers which prohibit southern partners from getting the same incentives as their northern counterparts from third-party grants (e.g. lack of research incentivising policies in some Southern universities). Furthermore, we consider multilateral independent institutions such as the Southern African Science Service Centre for Climate Change and Adaptation (SASSCAL, Helmschrot and Jürgens 2015) as a blueprint. It is intended to be capable of funding research projects based on its own agendas and with its own monetary means. If such an institution is successfully set up and operated, efforts in this direction could provide an effective model to overcome respective funding imbalances.

31.4.2 Involvement in Project Design and Management

Our survey revealed significant differences amongst the views from Northern and Southern researchers with respect to their involvement in project design and ongoing management, as well as fieldwork decisions (cf. Sects. 31.3.2.1 and 31.3.2.2). As these findings might indicate a top-down approach from the Northern to the Southern partners—possibly caused by funding imbalance—we consider the following insights as crucial in order to enable effective, balanced and equal NSC.

The initial steps in project design are known to be particularly crucial for a fruitful and collaborative atmosphere and thus an overall successful project. Referring to transdisciplinary research in this regard, the process of 'problem framing' is essential in which perceptions of scientific disciplines and non-scientific stakeholders are brought together (Jahn et al. 2012). Almost half of the respondents from the South indicated that they were not adequately involved in these processes, which may relate to the long time lag of more than a year between the participatory process that involved workshops with scientists and funding representatives, and the announcement for grant applications and their award. On the contrary, these perceptions are also likely due to ongoing staff turnover and the subsequent lack of awareness of these original discussions by those who joined the projects later or were not in a position to participate in respective processes (e.g. PhD students). Nonetheless, this result confirms to a degree previous findings on the difficulties of Southern researchers to participate in the processes of agenda setting (Bradley 2017). Ideas to overcome an inadequate involvement of Southern researchers in the agenda-setting process could be found in an explicit initial project phase for collaborative problem framing via 'quick-initiation-funding' schemes (Luthe 2017).

In terms of involvement in overall project administration and management, the discrepancy between Northern and Southern researchers continues. In both fields, Southern partners indicated lower levels of involvement, as confirmed by other studies previously (Luetkemeier et al. 2021; Schmidt and Pröpper 2017), which might be a direct consequence from being less involved in project design. Though we could assume that heavy involvement in project administration and management is somewhat time-consuming and even burdensome, it affords power to influence overarching strategies. Hence, being not involved at these levels limits the subjective feeling of ownership. Likewise, Southern researchers indicated that they had not been involved in fieldwork planning or practical project decisions, largely. This is again crucial, because it could even be counterproductive, as important Southern knowledge stocks remain untapped (i.e. environmental and institutional setting). Excluding them from decision-making processes therein may hamper the effectiveness of actual research tasks.

Overall, the evaluated projects within SPACES II show a heterogeneous degree of involvement. Whilst some projects explicitly applied a 'tandem structure' in lead positions, in other projects, the role of African collaborators was less obvious. Tandem-structuring the leadership of work packages within a project can be a tool to balance opportunities and responsibilities for Northern and Southern partners. However, having two leaders might also complicate the decision-making process. Furthermore, especially if funding comes from the North, the actual leadership might remain in the hands of the Northern partner. One option to circumvent these challenges could be to keep the leadership of each work package to one person only, either coming from the South or North and in any case equally balanced across all work packages. Alternatively or in addition, a Steering Committee can be an adequate tool to foster joint decision-making within the collaborative research project.

The aspect of under-valuation may also be reflected in the answers of our respondents to the question, if in-kind or non-monetary contributions were adequately acknowledged (cf. Sect. 31.3.2.3). Here, 42% of the Southern respondents indicated that their contributions were not adequately appreciated. This could possibly lead to dissatisfaction and even frustration amongst Southern partners, as found in a qualitative evaluation of two German-Namibian collaborative projects by Luetkemeier et al. (2021). The imbalance in involvement and appreciation of contributions could likely result in unclear responsibilities and research tasks as well as skewed workloads.

Interestingly, our survey does not support a qualitative finding from Luetkemeier et al. (2021) that Southern researchers were less involved in actual empirical research, but rather in activities such as field logistics, stakeholder communication or student supervision. Our results do not show a significant difference between the tasks and workloads of Northern and Southern researchers, indicating that Southern partners did not implicitly have to act as 'field facilitators' of Northern researchers. Both Northern and Southern researchers have been equally involved in empirical research, which we consider a positive sign—apart from the considerations above—for a truly cooperative research project.

31.4.3 Co-authorship as a Key Tool for Effective Collaborations

When observing all SPACES II researchers' collaborative publications via bibliometric networks, the main contributors were German (senior) researchers, and strongest links were formed within organisations. These relationships remained stable, probably indicating that the SPACES II projects were largely established on existing collaborations within Germany. The total network of collaborating authors grew rapidly with time, as researchers were taking up collaborative publication activities with their SPACES II colleagues. Similar growth was observed when limiting focus to NSC co-authored publications. Prior to SPACES, less than 10%

of authors were publishing together with their Northern or Southern partners, whilst immediately before and during SPACES II (2017–2020), this had increased to a half. This indicates that SPACES II was successful in meeting its objective of improving the networks of NSC, when using co-authored publications as a measure. On the other hand, whilst during the last time bracket, the vast majority of researchers (82%) co-published papers with SPACES II-colleagues, only a half published with their Northern or Southern partners.

In the light of earlier studies, the proportion of NSC publications amongst SPACES II researchers (12% during the 'best' year of 2018) appears relatively low. Pouris and Ho (2014) found that in 2011, 58% of papers published on the African continent were co-authored with international partners; Germany was the fourth most important international collaboration partner. Access to funding and equipment were assumed to be the major reasons behind the high proportion of South-North co-authored papers in Africa, and also main factors explaining the comparatively low numbers of collaborative publications between southern African countries (Pouris and Ho 2014; Zravkovic et al. 2016). In the case of SPACES II, we can assume that a higher proportion of NSC papers would have been recorded especially from the German side if only those publications officially published 'under SPACES II' were counted. In particular, the senior researchers were undoubtedly involved in various other (non-NSC) projects. Furthermore, the COVID-19 pandemic and the cancellations of the associated researcher mobility programme may have decreased the overall number of co-authored NSC publications from SPACES II. For example Zravkovic et al. (2016) found that southern African researchers, who left their home country temporarily for scientific education and training (e.g. master/PhD studies), also published more NSC papers.

In contrast to the overall network, the NSC co-authored publications network was dominated by Southern researchers as the most productive, best-linked researchers. It appeared that the NSC papers were more focussed on fewer, very active collaborators particularly on the Southern side. Similarly, a few South African institutes contributed disproportionately to the NSC co-authorship network. The strongest collaborator, UCT, is both the highest-ranked South African university, and the highest-ranked African university on the main global university rankings (rank 269 on the Center of World University Rankings 2021–2022 Edition; rank 155 on the Times Higher Education University Rankings 2021; and rank 226 on the QS Top Universities Ranking of 2022). This trend may have two distinct reasons: First, most fully funded project researchers (i.e. PhD and post-doctoral researchers) are German-affiliated, including southern African PhD students funded with 4-year PhD grants, each publishing some NSC papers. Second, the largest and wealthiest institutes on the Southern side may have longer history and better preparedness for international collaborations and thus outperform smaller institutions in the South. The question of how this may be a result of or may even consolidate power asymmetries in research amongst Southern partners goes beyond the scope of this study but can be a valid entry point for further investigations of South-South collaborations.

Besides bibliometrics of published articles, it is relevant to look at the process leading to co-authored papers. The question of authorship is often a difficult and disputed topic in research collaborations (see, e.g., Luetkemeier et al. 2021). It would be an unwanted outcome if Southern collaborators felt themselves not invited to co-author in publications where they would have had a clear role. Previous studies have raised concern over North-South collaborations where Southern partners were confronted with implicit expectations of acting as facilitators of communication to stakeholders and policy-makers, or even field managers (Luetkemeier et al. 2021).

31.4.4 Critical Reflection on Methodology

Against the background of our experiences in carrying out the survey, we consider a structured online survey as a suitable method to collect perspectives from researchers in the field of Earth system sciences. We consider them as being open to such a survey format. Deliberately, we assembled an author team of Northern and Southern researchers to keep a balanced view on relevant NSC issues that should be part of the assessment. Furthermore, we designed the survey in a way to make the barriers for participation as small as possible. In this regard, we consider the available period of about a month and the required time for completion of only 10 min as a good compromise. In addition, we assume that the inclusion of both quantitative and qualitative questions offered the researchers a suitable tool to share their views on NSC in a comfortable and anonymous way.

Although we consider the response rate of about 31% as being adequate, the critical question is: why did not more researchers take the opportunity to participate? We can only make assumptions but reasons could be found in an incomplete or a non–up-to-date list of contact details, holiday season in the North in July, low interest in the subject, low involvement in practical NSC (e.g. on a student level) or frustration and resignation about the topic itself or current framing conditions (e.g. COVID-19 situation).

To better evaluate the representativeness of the survey, it should be noted that the respondents evaluated their collaboration with all their Southern or Northern partners, respectively. However, it could likely be the case that they actually had different experiences with different partners. Hence, their actual responses either may be an averaged assumption about all their partners, or may be biased by a certain single case. Furthermore, the results we presented are averaged over all SPACES II projects. It is likely the case that the expectations towards NSC differed amongst the projects, depending on the actual research subject, the required methodology and, hence, (dis)satisfaction might differ.

With regard to the co-authorship analysis, we must note that due to COVID-19, the publishing of project results was, in many cases, delayed, and due to the normal publication time lag, the proportion of NSC authors may still significantly increase after the publishing of this chapter. In addition, the Scopus database does not necessarily cover all publications, but was selected here because of the availability of institute affiliations data.

31.5 Policy Message

Handling Global Change phenomena such as climate change, biodiversity loss and environmental degradation, on the one hand, and approaching the challenges of poverty reduction and enhancing water and food security in the developing world are amongst the most crucial challenges today. SDG Goal 17 highlights the important role of international collaborations to develop applied solutions for these problems. Effective NSC in science is one key component therein.

In this study, we assessed the quality and structure of NSC in the research programme SPACES II based on a structured online survey that was carried out amongst the involved researchers from South African, Namibian and German institutions as well as an extensive co-authorship network analysis. In synopsis of our results, we consider the SPACES programme as a good step forward in carrying out NSC, because Northern and Southern researchers acted largely as peers on an equal footing. Despite this success, room for improvement exists, in particular, with respect to the integration of Southern researchers in the processes of project design and continuous management as well as the equal availability of funds for research. In particular, the latter point turned out to be crucial for successful NSC as most challenges, we identified, likely, have their origin in unequal resources endowment. Against this background, we recommend the following key points to enhance NSC for future research programmes:

- *Asynchrony of funding:* Research requires funds to cover costs for staff, equipment and mobility for all actors involved in a project. The persisting asynchrony of funds available to Northern and Southern partners is, however, a fundamental area of conflict and frustration and hence a potential threat to effective NSC. If collaborative research initiatives are an explicit intention by funders and research consortia, appropriate and sufficiently timed monetary means are required for all parties involved to carry out the intended research. We see four options to getting closer to this desirable state: First, multilateral negotiations between funders prior to an NSC programme should put emphasis on legally securing sufficient funds from all countries involved. Second, multilateral funding agencies based 'in the South' should be (financially) supported to enhance the opportunities for independent research funding. Third, funding regulations should acknowledge the financial requirements of Southern researchers, for example, via direct provision of funds or explicit subcontracting opportunities. Fourth, remaining institutional barriers in Southern institutions that prohibit researchers from benefitting equally from third-party grants should be identified and reduced.
- *Continuous NSC evaluation:* We consider an investigation like ours not just of interest for final project evaluation. If a large number of scientists are involved in a collaborative research project, especially comprehensive quantitative metrics could well be used to periodically monitor the quality of NSC within a research programme or project. Hence, we suggest evaluating partnership expectations and perceptions in the project beginning as well as in the mid-

term to immediately respond to dissatisfaction amongst partners that might negatively influence the achievement of project goals and lead to the abortion of existing or rejection of future collaborations. Therefore, we suggest monitoring the number of involved researchers as an important metric, especially from the South, where partners are often not formally funded. In addition, conducting surveys on the researchers' changing perceptions throughout the project concerning levels of involvement and opportunities to engage can quickly uncover negative developments and enable countermeasures. Furthermore, monitoring of continued co-publishing activities of the involved researchers (even after formal NSC programmes ended) may provide an effective indication of the quality and long-term impact of collaborative projects. Although we did not investigate the ratio between female and male project members within this study, we strongly recommend including gender balance in the evaluation of collaborative research projects. This includes the share of women in the overall project as well as their position (e.g. as leaders) and differences or similarities between Northern and Southern partners. Furthermore, knowledge gained over the last decades on transdisciplinary science with respect to formally setting up interdisciplinary projects in a methodologically guided way can be fruitful to structure future activities.

- *Collaborative publications:* Though alternative outputs and outcomes of research, such as societal impacts, are increasingly considered relevant when evaluating the quality of research, scientific publications remain the key currency, due to their quantifiability and associated metrics of researchers' performance. In this regard, publications are a central tool to enhance collaboration between Northern and Southern researchers and our analysis shows that an intensification did emerge over time. However, we see a need to enhance this tool, as in particular on the Southern side, co-authored publications tend to be disproportionately focused on a few established, senior partners. Here, respective successes are strongly bound to available funding for Southern researchers to be able to invest time and money in respective authorship processes.

References

BMBF (2017) Zweite Bekanntmachung der Richtlinie zur Förderung von 'SPACES – Forschungspartnerschaften zur Anpassung komplexer Prozesse im System Erde in der Region Südliches Afrika' im BMBF-Rahmenprogramm 'Forschung für nachhaltige Entwicklung' (FONA). ENGLISH: Second call of the directive for funding of 'SPACES – Science Partnerships for the Adaptation to Complex Earth System Processes in Southern Africa' in the BMBF framework programme 'Research for sustainable development' (FONA). German Federal Ministry of Education and Research (BMBF). https://www.bmbf.de/bmbf/shareddocs/bekanntmachungen/de/2017/03/1336_bekanntmachung.html. Accessed 01 Sept 2021

Boshoff N (2010) South–South research collaboration of countries in the Southern African Development Community (SADC). Scientometrics 84(2):481–503. https://doi.org/10.1007/s11192-009-0120-0

Bradley M (2007) North-South research partnerships: challenges, responses and trends: a literature review and annotated bibliography. Working Paper 1, IDRC Canadian Partnerships Working Paper Series

Bradley M (2017) Whose agenda? Power, policies, and priorities in North-South research partnerships. In: Mougeot LJA (ed) Putting knowledge to work – collaborating, influencing and learning for international development, pp 37–70

Casale MAJ, Flicker S, Nixon SA (2011) Fieldwork challenges: lessons learned from a north-south public health research partnership. Health Promot Pract 12(5):734–743. https://doi.org/10.1177/1524839910369201

Devi S (2020) Travel restrictions hampering COVID-19 response. Lancet 395(10233):1331–1332. https://doi.org/10.1016/S0140-6736(20)30967-3

Gaillard JF (1994) North-South research partnership: is collaboration possible between unequal partners? Knowl Policy 7(2):31–63. https://doi.org/10.1007/BF02692761

Gasparatos A, Takeuchi K, Elmqvist T et al (2017) Sustainability science for meeting Africa's challenges: setting the stage. Sustain Sci 12(5):635–640. https://doi.org/10.1007/s11625-017-0485-6

Giller KE (2020) Grounding the helicopters. Geoderma 373(114302). https://doi.org/10.1016/j.geoderma.2020.114302

Gupta J, Vegelin C (2016) Sustainable development goals and inclusive development. Int Environ Agreements Polit Law Econ 16(3):433–448. https://doi.org/10.1007/s10784-016-9323-z

Helmschrot J, Jürgens N (2015) Integrated SASSCAL research to assess and secure current and future water resources in Southern Africa. Proc Int Assoc Hydrol Sci 366:168–169. https://doi.org/10.5194/piahs-366-168-2015

Jahn T, Bergmann M, Keil F (2012) Transdisciplinarity: Between mainstreaming and marginalization. Ecol Econ 79:1–10. https://doi.org/10.1016/j.ecolecon.2012.04.017

Jeppesen S, Miklian J (2020) Introduction: Research in the time of Covid-19. Forum Dev Stud 47(2):207–217. https://doi.org/10.1080/08039410.2020.1780714

Joshi A, Saket K, Satish C et al (2015) Likert scale: explored and explained. Br J Appl Sci Technol 7(4):396–403. https://doi.org/10.9734/BJAST/2015/14975

LimeSurvey (2021) LimeSurvey manual. https://manual.limesurvey.org/LimeSurvey_Manual. Accessed 14 Sept 2021

Luetkemeier R, Mbidzo M, Liehr S (2021) Water security and rangeland sustainability: transdisciplinary research insights from Namibian–German collaborations. S Afr J Sci 117(1/2). https://doi.org/10.17159/sajs.2021/7773

Luthe T (2017) Success in transdisciplinary sustainability research. Sustainability 9(1):71. https://doi.org/10.3390/su9010071

MacFarland TW, Yates JM (2016) Mann–Whitney U test. Introduction to nonparametric statistics for the biological sciences using R. Springer International, pp 103–132. https://doi.org/10.1007/978-3-319-30634-6_4

Patel Z, Greyling S, Simon D et al (2017) Local responses to global sustainability agendas: learning from experimenting with the urban sustainable development goal in Cape Town. Sustain Sci 12(5):785–797. https://doi.org/10.1007/s11625-017-0500-y

Perianes-Rodriguez A, Waltman L, van Eck NJ (2016) Constructing bibliometric networks: a comparison between full and fractional counting. J Informet 10:1178–1195. https://doi.org/10.1016/j.joi.2016.10.006

Pohl C, Hirsch Hadorn G (2007) Principles for designing transdisciplinary research: proposed by the Swiss Academies of Arts and Sciences. Oekom Verlag. https://www.research-collection.ethz.ch/handle/20.500.11850/158306

Pouris A, Ho Y (2014) Research emphasis and collaboration in Africa. Scientometrics 98:2169–2184. https://doi.org/10.1007/s11192-013-1156-8

Schmidt L, Pröpper M (2017) Transdisciplinarity as a real-world challenge: a case study on a North–South collaboration. Sustain Sci 12(3):365–379. https://doi.org/10.1007/s11625-017-0430-8

Schneider F, Buser T, Keller R et al (2019) Research funding programmes aiming for societal transformations: ten key stages. Sci Public Policy 46(3):463–478. https://doi.org/10.1093/scipol/scy074

Skupien S, Rüffin N (2020) The geography of research funding: semantics and beyond. J Stud Int Educ 24(1):24–38. https://doi.org/10.1177/1028315319889896

Sowe SK, Schönfeld M, Samimi C et al (2021) Impact of the COVID-19 pandemic for North-South research collaboration: an experience report. In: Companion Volume to the Proceedings, Madrid, pp 18–21

Stefanoudis PV, Licuanan WY, Morrison TH et al (2021) Turning the tide of parachute science. Curr Biol 31(4):R184–R185. https://doi.org/10.1016/j.cub.2021.01.029

Stibbe D, Prescott D (2020) The SDG partnership guidebook. A practical guide to building high impact multi-stakeholder partnerships for the Sustainable Development Goals, 1st edn. The Partnering Initiative and UNDESA

Tress B, Tress G, Fry G (2005) Integrative studies on rural landscapes: policy expectations and research practice. Landsc Urban Plan 70(1):177–191. https://doi.org/10.1016/j.landurbplan.2003.10.013

UN (1979) United Nations conference on science and technology for development: Vienna program of action. Int Leg Mater 18(6):1608–1643

van Eck NJ, Waltman L (2010) Software survey: VOSviewer, a computer program for bibliometric mapping. Scientometrics 84:523–538. https://doi.org/10.1007/s11192-009-0146-3

VERBI Software (2019) MAXQDA 2020, Berlin

Waage J, Yap C, Bell S et al (2015) Governing the UN sustainable development goals: interactions, infrastructures, and institutions. Lancet Glob Health 3(5):e251–e252. https://doi.org/10.1016/S2214-109X(15)70112-9

Zdravkovic M, Chiwona-Karltun L, Zink E (2016) Experiences and perceptions of South–South and North–South scientific collaboration of mathematicians, physicists and chemists from five southern African universities. Scientometrics 108:717–743. https://doi.org/10.1007/s11192-016-1989-z

Synthesis and Outlook on Future Research and Scientific Education in Southern Africa

32

Graham P. von Maltitz ⓘ, Guy F. Midgley ⓘ, Jennifer Veitch ⓘ,
Christian Brümmer ⓘ, Reimund P. Rötter ⓘ, Tim Rixen,
Peter Brandt, and Maik Veste ⓘ

Abstract

The sustainability of southern Africa's natural and managed marine and terrestrial ecosystems is threatened by overuse, mismanagement, population pressures, degradation, and climate change. Counteracting unsustainable development requires a deep understanding of earth system processes and how these are affected by ongoing and anticipated global changes. This information must be translated into practical policy and management interventions. Climate models project that the rate of terrestrial warming in southern Africa is above the global terrestrial average. Moreover, most of the region will become drier. Already

G. P. von Maltitz (✉)
School for Climate Studies, Stellenbosch University, Stellenbosch, South Africa

South African National Biodiversity Institute, Cape Town, South Africa

G. F. Midgley
School for Climate Studies, Stellenbosch University, Stellenbosch, South Africa
e-mail: gfmidgley@sun.ac.za

J. Veitch
South African Environmental Observation Network, Marine Offshore Node, Cape Town, South Africa
e-mail: ja.veitch@saeon.nrf.ac.za

C. Brümmer
Thünen Institute of Climate-Smart Agriculture, Braunschweig, Germany
e-mail: christian.bruemmer@thuenen.de

R. P. Rötter
University of Göttingen, Tropical Plant Production and Agricultural Systems Modelling, (TROPAGS) and Centre of Biodiversity and Sustainable Land Use (CBL), Göttingen, Germany
e-mail: reimund.roetter@uni-goettingen.de

there is evidence that climate change is disrupting ecosystem functioning and the provision of ecosystem services. This is likely to continue in the foreseeable future, but impacts can be partly mitigated through urgent implementation of appropriate policy and management interventions to enhance resilience and sustainability of the ecosystems. The recommendations presented in the previous chapters are informed by a deepened scientific understanding of the relevant earth system processes, but also identify research and knowledge gaps. Ongoing disciplinary research remains critical, but needs to be complemented with cross-disciplinary and transdisciplinary research that can integrate across temporal and spatial scales to give a fuller understanding of not only individual components of the complex earth-system, but how they interact.

32.1 Introduction

"We are not living in a stable world. Science needs to unravel the nature, causes and consequences of this change in a way that can lead to transformative change with positive impact on society." This was the closing challenge from the South African Department of Science and Innovation's (DSI) Prof Yonah Seleti at the SPACES II Programme Synthesis Meeting in Pretoria June 2022. This view strongly supports the idea that environmentally focused policy-related research, especially in developing regions such as southern Africa, be motivated by and embedded in the need to enhance the social-ecological basis to advance human well-being. While environmentally focused research addresses important unknowns about the functioning of the natural (ecological) world, such research thus cannot be decoupled from pressing social issues, such as the security and sustainability of food production, water supply and its quality, and how these issues link with poverty and developmental challenges. However, strengthening the relevant links between explanatory scientific research and its policy implications remains a difficult and complex challenge (Stringer and Dougill 2013; von Maltitz 2020).

T. Rixen
Department of Biogeochemistry and Geology, Leibniz Centre for Tropical Marine Research (ZMT), Bremen, Germany
e-mail: tim.rixen@leibniz-zmt.de

P. Brandt
GEOMAR Helmholtz Centre for Ocean Research Kiel and Experimental Oceanography, Christian-Albrechts-Universität zu Kiel, Kiel, Germany
e-mail: pbrandt@geomar.de

M. Veste
CEBra – Centrum für Energietechnologie Brandenburg e.V., Cottbus, Germany
e-mail: veste@cebra-cottbus.de

Institute for Environmental Sciences, Brandenburg University of Technology Cottbus-Senftenberg, Cottbus, Germany
e-mail: veste@cebra-cottbus.de

It is now almost certain that the southern African social-ecological system, like several other regions of the world, is approaching potentially consequential "tipping points," sets of conditions that entrain self-perpetuating changes that have adverse impacts (Chap. 7; IPCC 2022). These could include, for example, prolonged and intense drought leading to major cities and their surrounding regions running out of water, the collapse of food production systems and related food insecurity, novel intense weather events such as category 4 cyclones making landfall on southern Africa's east coast ever further southward, and toward major cities and centers of human settlement, or unprecedented heatwaves (Mbokodo et al. 2020; Chaps. 6 and 7). The impacts of such events would have extensive health and social consequences including loss of human life, disease, human displacement, human migration, infrastructure damage and related food and water insecurity, and increases in poverty and deprivation. These impacts would adversely impact on the region achieving the United Nations Sustainable Development Goals (SDG) (Chap. 3). A well-developed predictive understanding of such climate change-driven impacts enables society to develop strategies, approaches, and policies that can anticipate and avoid, adapt to, or mitigate against impacts. Working toward this goal underpins the Science Partnerships for the Adaptation to Complex Earth System Processes (SPACES II) program, with key outputs from this research and important gaps in knowledge highlighted across the chapters of this book.

The southern African region is highly complex in terms of its environmental and socioeconomic setting and its diverse flora and fauna, and this is a fundamental condition that challenges explanatory scientific work in its efforts to tease out a predictive understanding at a level of credibility sufficient to inform appropriate societal responses. The region is an area of convergence of several global scale geophysical drivers of climate, ecosystems, and ecosystem processes. The mid-latitude southerly extent of this region gives rise to a globally unique meeting of warm, high salinity, and nutrient-poor waters carried polewards by the Agulhas Current down the east coast and the cold, nutrient-rich waters of the Benguela Current originating from the Antarctic Circulation and feeding the Benguela Upwelling System along the west coast. These currents have major impacts on precipitation, with the warm Agulhas Current bringing rainfall to the east, while the cold Benguela Current inhibits rain on the west. Two distinct climatic systems, partly linked to these ocean currents, exert seasonally variable impacts on the region. Midlatitude cyclones bring rainfall to the western Cape region with high predictability, and sometimes further northward and inland with lower dependability, during the austral winter months. The seasonal southward migration of the Inter-Tropical Convergence Zone (ITCZ) as summer months approach tends to enhance summer convective rainfall, as well as displacing the path of the midlatitude cyclones from their winter landfall track to the southern ocean. This West-East winter-summer seasonality pattern is overlaid by a roughly orthogonal trend in rainfall amount, with arid and semiarid conditions in the southwest transitioning over semiarid grasslands and savannas to subtropical woodlands and tropical forests in the north. High interannual variance in rainfall is associated with the lower influence of predictable winter rainfall regime, and annual rainfall total throughout the region. These climatic gradients, together with related disturbances like wildfire, have had major impacts on the biodiversity that has evolved in the region over millennia. The spatial variation in interannual variability in rainfall also poses multiple challenges for both agriculture and the provisioning of water for human and industrial use (Chaps. 5, 6 and 12).

Southern Africa's diversity in ocean currents, topography, including the ocean shelf, climate, geology, and disturbance regimes, has resulted in southern Africa having unique and high levels of biodiversity, with a disproportionate number of endemic species in both the marine and terrestrial environments (Chap. 2). This diversity is under increasing pressure from global change drivers (Chap. 3) including anthropogenic land cover and land-use change (Chaps. 3, 4, 13, 15, 17, 20 and 22), spread of invasive alien plant and animal species (Chaps. 3 and 24), and overexploitation of terrestrial and marine biological resources (Chaps. 3, 11, 16, 19, 25 and 26). Critical feedbacks of these impacts on the global climate system derive from resulting trends such as land degradation (Chaps. 13, 20 and 23), which is a CO_2 source to the atmosphere (Chaps. 14, 17, 23, 25 and 30). Loss of biodiversity and ecosystems goods and services has profound impacts on local livelihoods (Chaps. 3, 4, 20, 22 and 23). Sustainable agriculture, ecotourism, and other nondepleting uses of natural resources are strategic imperatives for the region and would ensure the persistence of large contributors to individual livelihoods and countries' Gross Domestic Products (GDPs), with varying priority in different countries (Chaps. 4 and 20). Both agriculture and tourism are disproportionately large contributors to job creation and livelihoods opportunities, in any region where unemployment is a key concern (Chap. 4). Direct access to natural resources is important to rural communities and is especially a safety net for the poor (Chaps. 19, 20, 21, 22, 23 and 25). Despite this, the region has a distinct tension between developmental needs versus biodiversity protection and nature conservation (Chaps. 3, 4, 20, 22 and 23). While biodiversity is the mainstay of a large number of both formal and informal livelihood opportunities in sub-Saharan Africa, for example, Shumsky et al. (2014), from a policy perspective, environmental concerns are often perceived as constraints to economic development (Chaps. 4 and 31).

From a social-ecological perspective, a merging of vastly different resource management systems has occurred over four millennia in sub-Saharan Africa (Bjornlund et al. 2020). Colonialist practices from Europe converged with or dominated traditional Khoisan and Bantu tenure, land use, and land management systems, resulting in complex and often contested land management regimes in the region. Ancient traditional land management systems have been replaced or modified along with the conflicted history of land ownership in the region. Land tenure in sub-Saharan African countries is a postcolonial mix of private land holdings with their history in the colonial occupation of the region, mixed with a number of different forms of customary tenure with roots in traditions of the Bantu tribes that were resident in the area at the time of colonization. Some of these traditional approaches have themselves been modified through colonial and postcolonial processes.

These differences in tenure and management legacies link to vastly different land-use practices, which range from large-scale high intensity farming through to small-scale subsistence types of farming (Chaps. 3, 15, 20–23). By far the single largest land use is rangelands for livestock or game production (Chaps. 15–20). Rangeland management systems differ substantially with rangeland being either commercially managed, often using livestock rotational techniques reliant on fencing and with relatively low stocking densities, or being communally managed where livestock stocking rates can be close to the ecological carrying capacity (Chaps. 16, 19, 20, and 23). Increasingly, rangeland is also being managed for wildlife-based agri- and nature-based tourism industries (Chaps. 15 and 18). This diversity of agricultural practices poses multiple research priorities and challenges to support management decision-making in these varied land-based subsectors. The urgency for solutions is growing with the need to reconcile demands for increasing agricultural productivity with resource use constraints and environmental concerns (Chaps. 19–23).

There is clear evidence that near-shore marine resources have been exploited by humans dating back at least to the late Pleistocene as evidenced by analysis at Pinnacle Point (Marean 2010; Wren et al. 2020). The advent of the commercial fisheries industry in the late seventeenth century greatly increased the levels of marine exploitation and gave rise to thriving fisheries industry, which can be divided into four main sectors: subsistence fishing, small-scale (artisanal) fishing, large-scale commercial fishing, and recreational fishing. The fishing industry is important in terms of supporting local food security, and the provision of job opportunities (Baust et al. 2015). However, it remains a politically contentious industry with ongoing disputes around licensing among main sectors and the associated access to resources, with clear historical evidence of discrimination against the artisanal sector (van Sittert 2017). Deep-sea, mostly commercial, and inshore (mixed commercial and artisanal) trawling target hake is a key component of the commercial fishing industry, while pelagic purse-seine fisheries target sardine, anchovy, and herring. Other important, largely artisanal sectors include shrimps, chokka squid, and numerous line fish species. The high value of crayfish and abalone creates a disproportionate revenue and amount of jobs for the biomass harvested (Baust et al. 2015), and pressure from poaching. Harvest levels have fluctuated greatly over time: between the different sectors and between the west and east coasts. Due to upwelling, the west coast has a globally high rate of net primary production and has been the main contributor to the tonnage of fish caught. Initially, this was mostly based on hake, with catches peaking in the late 1960s, but declined substantially thereafter. In Namibia, the small pelagic (mostly sardine) fisheries industry had largely collapsed by 1977 (Paterson et al. 2013). On a tonnage basis, the east coast contributes far less to the fishing catch, but it has far greater involvement of small-scale and subsistence fisheries, which play a proportionately more important role than on the west coast (Baust et al. 2015). Stricter management and quota systems reduced fishing pressure but have not resulted in a complete recovery of stocks (Paterson et al. 2013).

32.2 Overview

32.2.1 Climate Change

Climate change is a global challenge that manifests differently across and between regions and subregions. The southern Africa region is anticipated to be subjected to severe impacts (Chaps. 3, 7, 20–23), being largely warm and dry. The interior, especially to the west, has historically warmed at twice the global average, a trend that is projected to continue for years to decades (Chaps. 6 and 7). Rainfall projections are less certain, but there is growing consensus that almost all of the region is likely to become more arid, with this effect being strongest in the already arid west. The combined impact of warming and reduced rainfall will greatly increase drought stress, an impact that will be felt both within the natural environment as well as in crop agriculture and rangelands (Chaps. 6, 7, 14, 19–23 and 26). The climate of the region already has high interseasonal variance, and this is likely to increase. Maximum, rather than mean, impacts are likely to have the greatest short-term social and ecological consequences, with more prolonged and intense droughts likely to occur. Despite overall projections of greater aridity, the warming of the Indian Ocean is likely to result in tropical cyclones of greater strength that move further south, bringing episodic flooding and related impacts on built infrastructure and the natural environment (Chap. 7).

Southern Africa has two distinct rainfall regimes: the winter rainfall, Mediterranean-like climate of the Fynbos, and semiarid Succulent Karoo of the southwestern tip of the continent, where rainfall is dependent on midlatitude cyclones that sweep over the region in winter, and the summer rainfall interior region that is dominated by convection rainfall (Chaps. 5–7). These regions are separated by a zone with all-season rain. The winter rainfall region's rainfall is projected to be reduced, because the path of the midlatitude cyclones will be displaced southward. This circulation pattern has occurred in years when the Western Cape has experienced severe drought, such as during the 1920s, and more recently during the 2015–2018 drought that almost resulted in a "day-zero" scenario, that is, the depletion of an urban water supply to the Cape Town metropolis. Climate change projections indicate a high likelihood of decreased rainfall and even more severe drought over the next few decades (Chap. 7).

A key feature of the summer rainfall region is a high interannual variance in rainfall. There is good evidence that this is linked, in part, to El Niño Southern Oscillation (ENSO) events, with dry periods associated with the El Niño phase, and wet periods associated with La Niña phase. This relationship is not perfectly dependable, however, and can only be used as a partial predictor of annual rainfall in the summer rainfall region (Chaps. 5, 6 and 12). Additional influences of rainfall variability include oceanic conditions in the Angola-Benguela region, the Agulhas Current and its behavior within the Mozambique Channel, and the Indian Ocean (Chaps. 5–9).

Reasons for the low correlation between the strength of ENSO events and realized rainfall remain unresolved. A better understanding of the drivers of climate is therefore imperative to fully understand climate change impacts, or to be able to accurately conduct seasonal weather forecasts. Chapter 8 considered changes in what is termed the Agulhas leakage and how this has a relatively small but important impact on local climate. The extent of Agulhas leakage is likely to increase as a result of climate change. The need to regionalize climate drivers that are often at scales below those of large-scale climate models is highlighted by this work. Feedbacks from vegetation cover to the climate system were explored in Chap. 10 where models were used to show that wide-scale land cover change (in this case, deforestation) would have direct and negative feedbacks on precipitation patterns. This preliminary evidence provides some justification for action to reduce deforestation and to promote reforestation in Angola and Zambia, and could add a novel interpretation to the impacts of exotic commercial plantations in South Africa if confirmed by credible modeling of the processes involved.

Understanding past climate (Chaps. 5, 27 and 28) is important for calibrating future climate models and can also be directly used to better understand both current vegetation patterns and how vegetation has responded to climate variance in the past. Proxy climate data shows that there has been a succession of cooler and warmer periods over the past millennium with rainfall in the winter rainfall region negatively correlated with temperature, and the opposite in the summer rainfall region. This trend could not be replicated by modeling efforts, suggesting further research is required to explain this discrepancy (Chap. 5).

32.2.2 Climate Change Impacts

The combined impact of a hotter and drier future on an already arid region will have adverse impacts on net primary production (Chap. 26), biodiversity (Chaps. 15–17), agricultural systems (Chaps. 17–23), water provision, tourism, and many other natural resources (Chap. 3). Severe extremes, such as heat waves, droughts, and floods, which are assumed to be associated with this long-term trend, are expected to further increase these devastating negative impacts on the terrestrial biosphere (Chaps. 6, 7, 20, 22 and 23).

Climate change also influences ocean currents, upwelling systems, and marine biogeochemical cycles. These changes become evident in oceanic warming with increased stratification within the upper ocean and a lowered supply of oxygen from the atmosphere. Since the marine fauna responds to temperature variations and are sensitive to oxygen levels, such changes can directly affect exploitable fish stocks (Chaps. 2, 11 and 25). Decreasing oxygen concentrations have, for instance, been linked to the southward shift of the rock lobster distribution and an increased frequency of rock lobster walkouts in response to associated events during which dissolved oxygen is completely consumed in the water column (e.g., Hutchings et al. 2009).

Additionally, changes of upwelling systems are expected to influence primary production but due to the complex interplay of upwelling, mesoscale oceanographic features, and frontal zones, the link between upwelling intensity and primary production may not be linear (e.g., Rixen et al. 2021). Ocean changes are also likely to have direct feedbacks into the local climatic systems, but these feedbacks are far from fully resolved (Chaps. 8 and 9).

32.2.3 Carbon Cycles

The fate of atmospheric CO_2 is critical in understanding climate futures. The vast extent of the southern African land surface and the surrounding oceans means that small changes in overall trends in emitting or sequestering carbon are important for understanding global carbon dynamics. Thus, both oceanic and terrestrial realms are potentially important global carbon sinks (Chaps. 2, 15 and 25).

The global ocean is a critical sink of carbon, through what is termed the solubility pump and the biological carbon pump, and while the solubility pump is well studied, the response of the biological carbon pump to global change remains poorly understood. Currently, neither the magnitude nor the direction of its expected change is predictable. However, the Benguela Upwelling System along the west coast is known to act as CO_2 source to atmosphere in the north and as a CO_2 sink in the south due to varying relative strengths of these two marine carbon pumps (Chaps. 2 and 25; Siddiqui et al. 2023). While in both systems, the solubility pump increases CO_2 concentrations in the surface water due to the warming of upwelled water, the biological carbon pump lowers the CO_2 concentration, because it fixes CO_2 into biomass through primary production. Carbon is exported via the produced organic matter into deeper parts of the ocean and the underlying sediments. The CO_2 uptake by the biological carbon pump was estimated to be 18.5 ± 3.3 Tg C year^{-1} and 6.0 ± 5.0 Tg C year^{-1} for the Namibian and South African part of the Benguela Upwelling System, sizeable numbers that confirm the accumulated effect of small changes of CO_2 exchange that could be of great relevance for national carbon budgets (Chap. 25).

Bottom trawling for fish and the associated remobilization of sedimentary organic carbon could potentially impact CO_2 uptake both by the biological carbon pump and through changes in the pelagic food web structure (Chap. 25). A near-collapse of the fishing industry due to overharvesting has resulted in the system having currently a relatively low secondary production of fish relative to primary production, with a significant change in fish species dominance (Chaps. 2, 11 and 25). There is also an order of magnitude increase in neritic copepods, though it is unclear if this is due to climate change or changes in fish stocks (Chap. 2).

With respect to terrestrial trends in carbon sequestration, it was found that in the semiarid Karoo, grass-dominated, degraded, and less diverse sites had higher carbon sequestration rates than the more typical dwarf-shrub-dominated sites. This raises important questions about the preferred state of the landscape and how it might be managed. It also appears that slight changes to the management regimes can tip such systems from being a net carbon emitter to being a net carbon sink. Annual precipitation patterns both in terms of magnitude and seasonal distribution are also critical in this regard, with wet years resulting in net CO_2 sinks, while arid years tend to show a net emission (Chap. 17).

In savanna systems, both empirical data and modeling found that the vegetating productivity was able to recover rapidly from artificially created prolonged drought periods (Chap. 26). Landscape level changes in standing biomass due to tree encroachment in savanna are seen as a degradation of the natural vegetation and have major adverse impacts on ranching industries. It does however lead to large, but poorly quantified amounts of carbon sequestration, and Namibia has quantified this trend in its greenhouse gas inventory, reporting that the country is carbon neutral largely due to the wide-scale occurrence of bush encroachment. This raises important questions around trade-offs among carbon sequestration, biodiversity, livelihood opportunities, and other environmental goods and services (Chaps. 14–16 and 26).

32.2.4 Marine Systems

In contrast to the terrestrial system where the ecology is largely determined by precipitation, temperature, disturbance regimes, and the nature of the soil, within the marine environment, the physical and biogeochemical properties of the sea water are the key determining variables affecting the marine ecosystem. Climate change impacts the thermal structure of the ocean with enhanced temperature increase near the surface, increasing stratification and reducing oxygen levels. Climate-warming driven changes to global and local wind patterns will impact both ocean currents and the strength of ocean upwellings. The nature of the southern African marine environment is largely determined by the contrasting ocean currents, the Agulhas Current transporting warm, salty, and nutrient-depleted tropical waters along the east coast, while the Benguela current system on the west coast is associated with cold and nutrient-rich water that upwells to the surface impacting the ecology of the region, particularly as it relates to regionally enhanced primary production (Chap. 25). The southern tip of Africa is where these transition zones between Atlantic and Indo-Pacific water bodies occur, giving unique habitats with high levels of both speciation and endemism (Chap. 2). The marine habitat also differs from the terrestrial habitat in that it can be zoned by depth of the water column and depth of the seabed. Obviously, the nature of the seabed at each depth is also of importance. As discussed above, there is a level of mixing of the converging ocean currents in the form of rings of the Agulhas water "leaking" into the Benguela, which result in 300-km wide swirling masses of Indian Ocean water moving northwestward within the Atlantic and that may persist for 3 years (Chap. 8).

As in the terrestrial environment, southern Africa has above-average endemism within its marine waters (Chap. 2). This makes the region biologically important from a conservation perspective. As in the terrestrial biomes, it is the combined impacts of past and current human disturbances, in this instance pollution, bottom trawling, and overharvesting, combined with the impacts of climate change, that pose major threats to the region's marine biodiversity in the longer term (Chap.

25). Already there are clear impacts in changes to trophic chains due to the intensity of fish harvesting, with possible feedbacks into the biogeochemistry cycles as discussed before. Since the 1950s–1960s, there have been substantial, long-term changes in abundance, biomass, production as well as species and size composition of neritic (on the shelf) mesozooplankton communities in both the Northern Benguela Upwelling System (nBUS) and Southern Benguela Upwelling System (sBUS) subsystems (Chaps. 2 and 25). Long-term changes in zooplankton communities from the mid-1990s to mid-2000s, are likely to have fundamental effects on biogeochemical processes, food web structure, and ecosystem functioning of the BUS as well as on the ecosystem services, for example, carbon sequestration and fisheries, supported by the plankton (Verheye et al. 2016).

Decoupling the climate change impacts on zooplankton and fish stocks from the management impacts on fish stock depletion remains an important question for the long-term sustainable harvesting and management of these globally important fish resources. This will also require a better understanding of what appears to be a jellyfish dominated dead-end food chain (Ekau et al. 2018).

These changes have been described for the nBUS by Bode et al. (2014) and Verheye et al. (2016), for the sBUS by Huggett et al. (2009), Blamey et al. (2015) and Verheye et al. (2016). Abundances of neritic copepods have increased by at least one order of magnitude in both subsystems, with turning points reached around the mid-1990s in the south and around the mid-2000s in the north, but declining afterward. At the same time, there were marked changes in copepod community structure, with a gradual shift in dominance from larger to smaller species in both subsystems.

32.2.5 Terrestrial Environment

At the regional scale, a combination of climate change and enhanced CO_2 concentrations was modeled to have profound biome level shifts in vegetation structure (Chap. 14).

Loss of soil through soil erosion has long been identified as a threat to South Africa as soil loss in the region can be one to two orders of magnitude higher than soil genesis (Chap. 13). Conserving soil is fundamental to long-term sustainable agriculture, including adapting to climate change impacts (Chaps. 20 and 23). This critical and basic aspect of land management, though not new, must not be overlooked when focusing on new climate smart land management solutions (Chap. 20). However, a reanalysis of potential erosion rates suggests that previous estimates may have been an order of magnitude too high (Chap. 13). Land transformation and degradation impact on the natural environment, potentially reducing net primary production (NPP) and causing a loss of biodiversity. The combined impact from land degradation, when coupled with climate change, may be more severe than when considering these impacts in isolation (Chap. 3).

The process of increasing density of indigenous woody vegetation, locally referred to as bush encroachment, was identified as a major threat to current rangeland livelihood activities. Its occurrence is widespread over most savanna areas, and increased woodiness within grasslands is also widely reported. Although SPACES II projects did not consider bush encroachment specifically, it is a background theme to many chapters considering rangeland management (e.g., Chaps. 15–19).

Drought is a natural aspect of the savanna environment (Chaps. 5 and 6), and predictions are that droughts are likely to intensify with climate change (Chap. 7). Savanna vegetation responses to drought were investigated in Chap. 26. Drought causes substantive production and biomass loss, and a shift from perennial to annual species of grasses, but no signs of tipping points to alternative states were observed. Recovery from drought occurred rapidly, even after an artificial 6-year prolonged drought period. Resting periods from grazing were, however, identified as important. The methodology used could help identify undesirable vegetation shifts and recommend management interventions.

32.2.5.1 Natural Rangelands and Agriculture

Use of the terrestrial landscape ranges from total resource conservation through to land transformation for crop-based agricultural production. All countries in the region have complex land tenure arrangements where there is a dualism of tenure with some farmers having freehold or leasehold on vast individually owned farms, whereas other land users exist on areas of customary tenure where the rangeland resource is communally used, though small crop plots are privately used and managed, at least during the growing season (Chaps. 2–4). In addition, there are vast areas of state-owned and managed national game reserves, especially within the savanna biome (Chaps. 15 and 18). Within the private and communal sector, there has been a move from almost all land being used for livestock, to much of the dryer areas now being used for wildlife (Chaps. 15, 16, 18 and 19). It is suggested that wildlife is the economically optimal use for areas of low rainfall (Chap. 18).

32.2.5.2 Primary and Secondary Production and Use in the Terrestrial Environment

Given that the savanna biome is by far the largest biome within southern Africa, it received a disproportionately large focus in the SPACES II program. Use of the savanna ranges from pure conservation through wildlife and livestock management to food crop production. Much of the savanna is used by local communities who have traditional tenure to the land and use it for livestock, game, and a multitude of other natural products including woodfuels. However, large commercial farms, including game farms and private reserves, are also common. Chapter 15 gives an overview of rangeland use within the savanna biome and the threats and degradation challenge it faces. Bush encroachment is identified as one of the biggest threats to livestock and tourism dependent linked livelihood activities. Drivers of bush encroachment are complex and traditionally are attributed to livestock and fire regime management. The direct impact of raised CO_2 levels in promoting bush encroachment is being increasingly recognised as one of the contributing factors to bush encroachment (Bond and Midgley 2012; Chap. 15).

32.2.6 Understanding Rates of Change

Monitoring and the critical importance of long term datasets was a common theme running through almost all chapters of this book, with monitoring data either being the basis of the studies, an identified shortcoming in resolving long term impacts or developed as part of the methodologies. Although our understanding of the degree and rate of environmental change as a consequence of climate change has advanced tremendously over the past three decades, many of the current studies highlight the degree to which uncertainty still remains. Fortunately the southern African region has extensive long term climatic and natural history data creating a sound baseline for some variables from which to track further change. The southern African region in general, and especially the Cape Floristic region, has attracted extensive botanical research which is mostly well archived (Chap. 2). The local avian research community, using citizen science approaches in addition to more traditional approaches has long term, extensive and replicated bird distribution data resolved to a quarter degree spatial scale and covering most of the subregion (Hugo and Altwegg 2017). In common with marine data, sampling intensity varies greatly and tends to bias to the more accessible areas. Most other taxa also have reasonable data sets due to a large and active population of environmental scientists in the region, linked to almost all of the main universities and additional research institutions.

The need for ongoing and increased levels of monitoring in both the terrestrial and marine environments is critical for tracking and better understanding the long-term impacts that the region is experiencing from climate change and human interventions. This needs to include both abiotic monitoring of climate, the atmosphere and oceans, as well as monitoring of key biological variables such as net primary production and the associated sequestration of atmospheric CO_2 (Chap. 25). Earth based or ocean based, location specific monitoring is needed to calibrate satellite based observations. In this context, SPACES contributed with long-term greenhouse gas (GHG) measurement infrastructures to overcome the limits of our understanding of the temporal dynamics of the biosphere-atmosphere exchange of carbon (Bieri et al. 2022). This requires highly sophisticated monitoring equipment and well-educated scientists and technicians for data-analysis and maintenance. However, in the ocean, we have not yet reached this stage and further efforts are needed to understand the exchange processes.

In the South African context the newly approved Expanded Freshwater and Terrestrial Environmental Observation Network (EFTEON) and the South African Polar Research Infrastructure (SAPRI) within the South African Earth Observation Network (SAEON, van Jaarsveld et al. 2007) is a major commitment by the South African government to further its monitoring and research network for the terrestrial, freshwater and marine environments.

Despite this, across the subregion there is still insufficient long term commitment to long term environmental and climatic monitoring given the high probability of adverse livelihood and environmental impacts that will be experienced from climate change. This shortfall is probably most pronounced in the vast marine environment, partly due the cost of installing and servicing permanent monitoring sites being orders of magnitude more expensive than for terrestrial based monitoring sites.

Global change drivers, including climate change, are impacting and changing natural environments in novel and fairly unpredictable ways. Understanding the nature of the change and the rates at which this change is taking place is critical for long-term sustainable management of these systems (e.g., Challinor et al. 2018).

32.3   Emerging Issues for Integrated Ecosystems Research

As is common in attempting to understand complex coupled human-environmental (or socio-ecological) systems, much uncertainty remains. Despite the considerable advances in knowledge from the previous studies, most of these studies have tended to be stand-alone investigations on single or few components of the system, and have not attempted integration across the entire system. Future studies should not only increase the depth of knowledge within systems components, but also develop an understanding of the linkages between (sub-)systems. Understanding of the linked ocean–climate–land interactions in southern Africa is not fully resolved, despite this triggering the climate change responses of the entire system. For a better understanding of climate change impacts and feedbacks and development of land management and climate-smart agriculture in the region, an integrated research approach supported by due involvement of, and interaction with, different stakeholder groups will be the next step.

32.3.1 Ocean–Atmosphere–Land Interactions and Feedbacks

The interactions between ocean, atmosphere and land are complex, with many feedbacks between the three components (Fig. 32.1). The ocean surrounding southern Africa has direct but still purely quantified feedbacks into the climate of the subregion as they influence air temperatures, rainfall, and the west coast fog banks. The seasonality of precipitation is also of high ecological relevance and controls the development of the vegetation and the biome. Climate change impacts on ocean currents could therefore have profound and poorly understood impacts on local weather and climate.

Transports from the terrestrial landscape into the oceans via river discharges and eolian dust deposits strongly influence the ocean currents as, for example, seen in front of Congo river mouth (Chap. 9) as well as seawater chemistry and the sedimentation (Chap. 2). As illustrated in Chaps. 12, 27, and 28, sediment studies can indicate the major changes to biodiversity, land use, and pollution. Anthropogenic changes in the terrestrial environment from land use (Chaps. 20–22), increased erosion (Chap. 13), and other forms of pollution are transported by rivers and as eolian dust into the ocean (Chaps. 28 and 29). Eventually, these

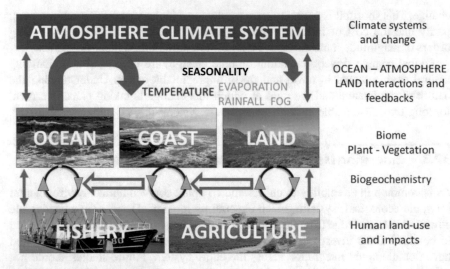

Fig. 32.1 Ocean–atmosphere–land interactions and feedbacks

components are deposited at the continental shelf, potentially with multiple impacts on the marine environment and its biodiversity (Eckardt et al. 2020).

Climate change and other human activities will directly impact on the intensity and nature of land–atmosphere–ocean linkages. For example, climate change is likely to increase drought conditions, which lead to dust storms where large quantities of dust are transported off-shore into the oceans (UNEP 2016; Eckardt et al. 2020). Land management practices, including a move to nontill agriculture, together with large-scale adoption of windbreak technologies, could greatly reduce wind speed, soil erosion, and have positive effects on the microclimate (Chap. 21; Veste et al. 2020).

A feature of both savanna and fynbos vegetation is that the vegetation burns naturally at intervals ranging from twice a year to only once in 20–30 years, depending on vegetation type and rainfall. Timing and frequency of fires has direct anthropogenic links. The fires produce vast smoke plumes, which have direct and indirect climate feedbacks (Lu et al. 2018; Ichoku 2020).

Wind-driven pollutants, soil, and smoke also enter the oceans, with both positive and negative effects. Extensive rangeland fires are the norm over much of the savanna during the late winter. During this period, a high-pressure system forms over the interior, with the smoke typically circulating out over the Atlantic and back again in what is termed the gyre effect, but also leading to extensive deposition over the Atlantic. At odd periods, especially during neutral El Niño Southern Oscillation (ENSO) events, there is also an observed fast flow of what has been termed a "river of smoke" that exits the east coast, travelling in a southeast direction over the Indian Ocean toward Australia (Kanyanga 2009). The actual smoke clouds have direct albedo effects, particles act as nuclei for cloud formation, and in addition there are nutrients, including soluble forms of iron, that are deposited in the oceans and leading to extensive algal blooms (Piketh et al. 2000).

32.3.2 Mesoscale Effects and Across Scale Effects

Many drivers of local climatic impacts (Fig. 32.2), and particularly precipitation, are driven by, or occur, at a spatial scale far smaller than can be determined from global circulation models. Examples include, but are not limited to, localized convectional thunderstorms, which are the mainstay of the summer rainfalls precipitation; deep mixing of surface and subsurface waters in the southern oceans due to localized events (Nicholson et al. 2016, 2022); the Agulhas leakage, which brings warm and salty waters as well as multiple organisms into the Benguela Current (Chap. 8); and the Benguela Upwelling System, which are critical to driving the high primary production along the west coast (Chap. 9). Resolving the impacts of these mesoscale events on both climate and ecological function is critical for a full understanding of the risks posed from climate change.\

Complex interactions occur between processes at different scales and from different drivers (Fig. 32.2). Clearly, global circulatory patterns impact on local drivers of climate such as the formation of convectional storms. Equally prevailing winds and their strength along the west coast of southern Africa have impacts on the location and intensity of upwelling systems (Chap. 8). A multitude of small, localized fires cause smoke plumes to extend far into the Atlantic and Indian oceans, These localized impacts feed directly into ecological processes, effecting for instance net primary production (NPP) at the local scale. Feedbacks can be in

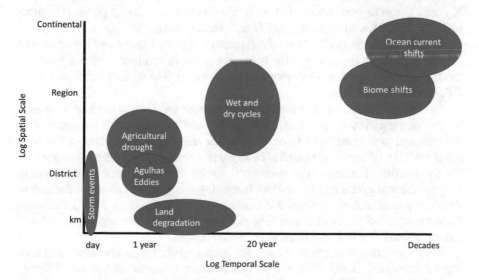

Fig. 32.2 The temporal and spatial scales of important climate change related processes

both directions, and very direct, for instance increased evapotranspiration being a trigger for local thunderstorm development. Equally feedbacks can be slow and very weakly coupled, for instance increased NPP resulting in CO_2 uptake from the atmosphere and hence feeding back into global CO_2 dynamics and climate.

Identifying the key mesoscale drivers of climatic and environmental processes as well as the way these are driven by and impact on processes at both the local and global scale remains poorly resolved within the southern African context.

32.3.3 Biogeochemical Processes in Hydrosphere–Biosphere–Lithosphere

Climate change leads to an increase in precipitation variability with changes of dry-wet cycles and thus controls vegetation development and compositions in biomes. Previous studies have mainly focused on the carbon cycle (Chaps. 17 and 25), which is also linked to the increased atmospheric CO_2 concentration, and the hydrological cycle (Chap. 6). Less attention has been paid to the interactions and feedbacks between carbon and other biogeochemical processes (particularly those including nitrogen, N, and phosphorus, P), water availability, and changes in vegetation in southern African biomes. Generally, southern African soils are characterized by a low nitrogen and phosphate content. The P-bioavailability is largely affected by adsorption and desorption processes of Fe-and Al-hydroxides in acid soils and P immobilization by Ca in alkaline soils. The mineral crystallinity is one important aspect, affecting P adsorption and desorption processes. Biological nitrogen fixation (BNF) with symbiotic bacteria plays an important biological role to overcoming the low soil nitrogen levels in nutrient-poor soils, and allows plants to build up their own N pool even in dry conditions, but under the conditions of larger portions of plant carbon transported toward the symbiotic microorganisms to maintain their BNF. Little information is available on the interconnected biological and soil chemical processes in the soil-plant system for most biomes in southern Africa. Studies in the fynbos show specific adaptations of plants depending on soil conditions (e.g., Griebenow et al. 2022).

The biogeochemical processes on the ecosystem level are externally controlled by the precipitation (Fig. 32.3a). Beside the rainfall amount, the frequency of drought and wet conditions is crucial for the ecological processes in the soil-plant systems. Therefore, changes in dry-wet cycles have drastic consequences for the uptake of macro- and micronutrients by the plants, but also the entire biogeochemical cycles and vegetation. Increasing droughts will alternate the carbon fluxes in plants and ecosystems and reducing productivity (Fig. 32.3b); however, for the coupled changes of P and N uptake and cycling by the vegetation and soil microbes, only few investigations were carried out.

The implication of climate change for hydrological, biogeochemical, and ecological processes in the different biomes is more complex due to the different rainfall amounts and the seasonality in the summer, winter, and year-round rainfall zones. In general, spatiotemporal characteristics of the various weather systems and

Fig. 32.3 (**a**) Water is the most critical factor for the interactions among carbon (C), nitrogen (N), and phosphate in soil-plant systems. (**b**) Schematic illustration of the interactions and transport of N and C with good water supply and drought stress (modified after Gypser and Veste 2019)

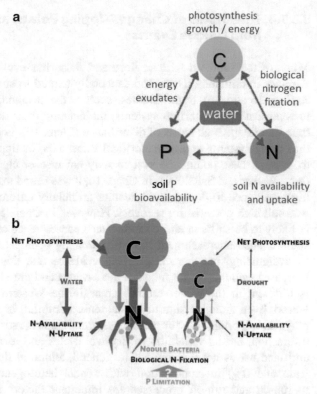

the biomes are broadly understood. However, the transition between these rainfall zones under climate change in southern Africa remains an open question and the transition zones are complex (Conradie et al. 2022). Changes in the rain seasonality and the length of dry-wet cycles were observed in the recent decades (Roffe et al. 2021, 2022), which are coupled to the complex ocean–climate–land interactions in the different seasons (Sect. 32.3.2). Furthermore, there is clear evidence that with climate change, the frequency of extreme events will increase, like the extreme droughts in the Western Cape in 2015–2017 (Theron et al. 2021; Wolski et al. 2021; Chap. 7) or heavy rainfalls and floodings in the Karoo recently in summer 2022 and the KwaZulu-Natal floods in April 2022.

Therefore, a better understanding of these changes in relation to the ocean and their impacts on terrestrial ecosystem functioning is needed. These requires a transcontinental study approach across the summer, winter, and year-round rainfall zones in southern Africa on the influence of changes of rainfall seasonality and their feedback on biogeochemical processes, vegetation composition and dynamics, biome development, and land-use systems. Such information is also important for the development of resilient and climate-smart agriculture in the region.

32.3.4 Thresholds of Change, Tipping Points, and Impacts from Unique Events

Much of the southern African flora and fauna has evolved in a situation of high variance in climatic drivers, and especially related to annual precipitation (Chaps. 2, 5, 6, 16 and 26). Some biomes, such as the savanna, are considered by many to represent disequilibrium systems, maintaining their structure because of, rather than despite the high levels of disturbance (Chap. 15). Ecological theory speculates that many systems experience critical thresholds, or tipping points, and if pushed to beyond these points, the system may collapse or change to a new stable state (e.g., Walker and Salt 2006). In Chap. 16, it was found that savanna systems exhibit high resilience to drought and despite artificially enforced intense drought, there was still high potential for recovery. However, in Chap. 14, it was shown that there is likely to be shifts in entire biomes as a consequence of climate change. Aspects such as bush encroachment have traditionally been seen as vegetation responses to overgrazing, but there is growing evidence that this might be a precursor to increased woodiness over much of the savanna and grassland biomes. The situation is different in the winter rainfall areas in the Western Cape and the Succulent Karoo. Even though a significant decrease in rainfall is expected here, seasonality plays an important role for the ecosystem. This is clearly linked to the ocean–land interaction, but its variability of dry-wet cycles and its importance for vegetation and land use is not understood. In general, temporal fluctuations in precipitation (rain and fog) in combination with spatial heterogeneity of water supply due to run-on and run-off processes are important factors for plant communities. A model developed by Reineking et al. (2006) shows that spatiotemporal variation of resource supply can maintain diversity over a long time. Furthermore, the importance of fire disturbances (Hebbelmann et al. 2022) and invasive plants (van Wilgen et al. 2020) for changes in natural ecosystems as well in agri-systems will be increased under climate change. These dramatic changes to thresholds in several stable ecosystems can be irreversible if caused by humans. The characteristic return time to equilibrium increases as a threshold is approached (Wissel 1984). Thus, in addition to information on the effects of stress on systems, we also need studies on the length of recovery phases, whose ecological importance has been underestimated.

The environment is not in isolation of humans, and operates in a complex coupled human environmental system. Human population density has increased by between one and two orders of magnitude over the past 200 years. This has been accompanied by large scale replacement of indigenous mammals with domestic livestock, extensive deforestation, clearing of natural vegetation for crop fields, impoundment of streams, extensive fencing and provision of artificial water which change natural animal migration patterns and many other impacts (Chaps. 3, 4, 18 and 29). Systems wide consequences of these changes are hard to model. Both climate change and anthropogenic activity are resulting in new and unique pressures on the environment. Responses of the environment to unique events that fall outside of the envelope in which the environment evolved may result in unanticipated environmental change. Breaching thresholds and tipping points in the climate and ocean systems could have even greater consequences for the region.

32.3.5 Distinguishing Signals from Complex Drivers

As discussed above, there are many complex changes taking place in the subregion driven by a mix of climate change and other anthropogenic impacts. We also are slowly developing better monitoring tools and building long term sequences of environmental data, including a better understanding of natural cycles of change. Disentangling the causes of observed impacts remains hugely difficult and problematic, especially since there may be synergistic, or dampening effects from alternate drivers of change. For instance, the CO_2 fertilization effect from anthropogenic emissions can to some extent offset impacts from poor land management. In the west-coast fisheries, the impacts from overharvesting may be difficult to distinguish from changes in the frequencies and intensities from near-shore upwellings (Blamey et al. 2015). To overcome this, there is a need for a combination of enhanced systems' understanding of the drivers of systems dynamics, use of experimental manipulation, improved and enhanced long-term monitoring, and the use of modeling to understand probable consequences from specific impacts.

32.3.6 Multifunctional Land Use and Integrated Landscapes

It is expected that due to climate change, high rainfall variability and water scarcity will increase in the future, and this will affect the water use patterns in agricultural production (Chap. 7). Water resources in the region are limited and a more water-efficient landscape is needed to mitigate negative effects of climate change. An integrated approach to water and landscape management needs to combine agricultural, forestry, and natural ecosystems to increase ecosystem resilience, to optimize water use efficiency, to ensure a sustainable bioeconomy, and to strengthen conservation of biodiversity during the next decades. Land-use systems of the future have to be holistic to support different needs of the society. Systems must be able to balance ecosystem services such as water and food provision, carbon sequestration, and biodiversity (O'Farrell et al. 2010; Mastrangelo et al. 2014; Grass et al. 2020). Complex trade-offs are inevitable and so-called win-win solutions are not guaranteed, though some choices have less societal and ecological consequences than others. Coordinated actions between different stakeholders are needed to improve recent farming practices, while reducing the pressure on land conversion and improving food security in a sustainable way (Rötter et al. 2016). An improved understanding of the water–land-use relations is essential to develop strategies for integrating resource-efficient land use, agricultural productivity, sustainable water management and conservation of the natural ecosystem and ecosystem functioning (Quinn et al. 2011; Rötter et al. 2021; Pfeiffer et al. 2022). As the region urbanizes,

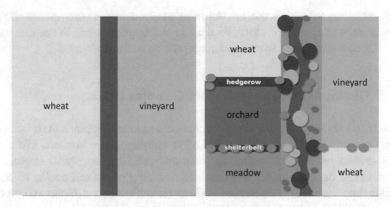

Fig. 32.4 Scheme of the concept of differentiated land use on a farm/landscape level: (**a**) state of development with large-scale agriculture and (**b**) ideal image of differentiated land use in integrated landscapes with the integration of natural habitats (modified after Veste and Böhm 2018)

complex trade-offs between urban and rural needs intensify, with water allocation already identified as a critical resource to the region (Matchaya et al. 2019). The recent increase of renewable energy from photovoltaic and wind parks needs to be integrated in a sustainable way as part of a multifunctional land-use system (Murombo 2022).

To develop more sustainable and resilient farming systems, there is a need to redesign agricultural landscapes and to integrate different land-use types including natural and seminatural areas from the farm to the landscape level (Chaps. 20, 22 and 23; Landis 2017; Pfeiffer et al. 2022). A differentiated land use with complex landscape structure (Fig. 32.4) has been previously proposed for Central European agricultural landscapes (Haber 2007) but has rarely been realized in practice. A key challenge for implementing sustainable and resilient agricultural systems is to develop appropriate governance structures and set appropriate incentives for land users (Leventon et al. 2017).

A particular challenge is the integration of habitats for the conservation of biodiversity within the farmed landscapes (see, Chap. 23). Depending on the species, different scales and demands must be considered. The combination of utilized agricultural landscapes and natural habitats has various advantages for the use of the landscape (Chaps. 21–23; Rötter et al. 2021). In the last decades, various initiatives have been developed in southern Africa, but more integration of biodiversity into the landscape and management is needed. There is clear evidence that transformation of farmlands and the integration of wildflower vegetation, old fields, or patches of natural vegetation within crop fields or orchards supports a wide range of insects and spiders and enhances pollinator activity and pollination services to crops and fruit trees (Chap. 22; Theron et al. 2020; Ratto et al. 2021). A recent study by Eckert et al. (2022) emphasizes the importance of small-scale spatial heterogeneity for soil arthropod biodiversity in two different regions in South Africa. For many African big game species, on the other hand, larger habitats and routes are needed to enable migration. Restrictions and direct confrontations often lead to conflicts between humans and wildlife, which must be regulated accordingly. However, smaller mammals, insects, birds, and even plants will need to migrate as a consequence of climate change and a more integrated landscape will aid in facilitating this movement (von Maltitz et al. 2007).

In addition to the integration of biodiversity, the interactions and feedbacks of biogeochemical processes between the subcomponents must also be increasingly taken into account in the analysis of integrative landscapes (Chaps. 20–23; Rötter et al. 2021). In this context, agroforestry (Chap. 21; Sheppard et al. 2020) or adjacent riparian forests/woodland are combined land use that also has positive impacts on soil and hydrology. Furthermore, intercropping systems with diversified crop cultivation can enhance yield, environmental quality, production stability, and ecosystem services.

To meet these challenges, more transdisciplinary research is needed to better understand the land-water-climate nexus and linking different ecological processes, including nature conservation research. Moreover, innovative methodologies and decision-support tools to manage land for a balanced provision of food, feed, timber, fiber, and fuel, a sustainable bioeconomy and conservation of biodiversity and development of natural ecosystems are urgently needed. A recent study by Chrysafi et al. (2022) emphasized the importance of understanding the Earth system interactions for sustainable food production.

32.3.7 Connection Between Different Research Topics and Approaches: The Need for Integration Across Disciplines and for Interdisciplinary and Transdisciplinary Collaboration

The complexity of climate change and environmental sustainability forces researchers to draw on interdisciplinary knowledge that spans social and natural sciences (Schipper et al. 2021). Climate change and its impacts on the socio-environmental system is often referred to as a "wicked problem" (Rittel and Webber 1973), and there is a growing branch of science around how to best deal with wicked problems (e.g., Wohlgezogen et al. 2020; Daviter 2017; de Abreu and de Andrade 2019). Solutions to resolving climate change impacts to achieve environmental and social sustainability require a system-based approach integrating across multiple environmental and social disciplines (Stock and Burton 2011). As illustrated throughout the book chapters, there are complex bidirectional feedbacks between the climate and the environment, as well as between different environmental systems, and subsystems. Clearly, there are also strong linkages between the oceans and the terrestrial environment, adding additional complexity. Further, in both the terrestrial and marine environments, there is a long history of human-led environmental change, much of this likely to interact in synergy with climate change to amplify the climate change impacts on the environment (Chap. 3).

An example of how transdisciplinary approaches have been implemented in agricultural science is given in the following section. Ecological science is influencing agricultural management choices—for example, climate smart agriculture/soil health/carbon.

In agricultural systems science, for quite some time, integrated system modeling for ex-ante evaluation of agro-technologies involving multiple disciplines has been an important research topic. For instance, the approach has been applied to support explorations of alternative agricultural development scenarios, and constituted an important building block of generating information to feed the development cycle of policies on land use and natural resource management (Van Ittersum et al. 2004; Rötter et al. 2016). Such an inter- and transdisciplinary collaboration approach—supported by modeling—is described in the following section and illustrated in Fig. 32.5.

To identify technically feasible land management options, first, biophysical modeling (step 3, left) is needed as shown in Fig. 32.5. Integrated agro-economic modeling (step 3, right) can then deliver valuable inputs for identifying technically feasible as well as socially acceptable and economically viable options in close interaction with (key) stakeholders. Modeling frameworks for such kinds of anal-

Model-based identification of options

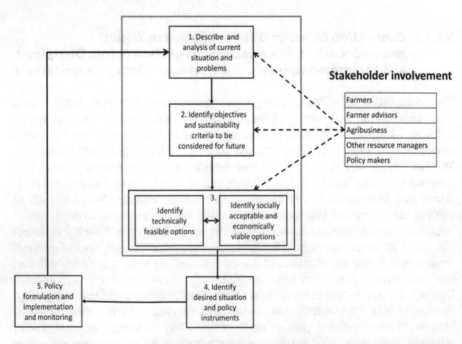

Fig. 32.5 Schematic development cycle of policies for natural resource management and land use (steps 1–3 in the green box) supported by agricultural system modeling studies and stakeholder interaction (Rötter et al. 2016, modified from van Ittersum et al. 2004)

yses have been developed in different contexts (see Van Ittersum et al. 2004); in the framework of climate change and land-use dynamics in southern Africa, such a framework has been designed and partly realized in the SALLnet project (Chap. 23; Hoffmann et al. 2020; Rötter et al. 2021; Nelson et al. 2022; Pfeiffer et al. 2022).

32.3.8 Societal Impacts: Visions for the Future

Although there is an intrinsic benefit from conserving ecosystems and understanding climate risks, it is the direct impacts on human well-being that is one of the key justifications of environmental research. Large parts of southern Africa can be considered developing, with countries varying in their level of development. Despite this, all the countries share many key developmental needs including job creation, ensuring food security, reducing inequalities, and the provisioning of basic services such as water. Although the level of agricultural dependency varies between the countries (see Chap. 20), all of the southern African countries have a high reliance on the agricultural sector for support of the economy and job creation. Throughout the region, the agriculture section is predicted to have negative impacts from climate change unless substantive adaptation measures are implemented (Chap. 20). Increasingly, the role of nature based tourism is also becoming an important component of most countries' economies, and is seen as an important component of rural job creation. Reducing climate risk through adaptation interventions is critical. These interventions can take many forms and have been given multiple names but, in most instances, relate to enhancing the resilience of the landscape or farming activity through improved management. This often involves reducing land degradation, enhancing water and soil conservation, and restoring biodiversity. In the biodiversity sector, this is referred to as ecosystem-based adaptation (EBA), while in the agriculture sector, sustainable agriculture practices such as conservation tillage, agroforestry, or climate smart agriculture may be used. Ensuring land based adaptation practices are implements and increasingly important for the sustainable future of livelihoods in the region. Two recent examples illustrate how critical this is. The 3 years of drought culminating in the 2018 water crisis in the city of Cape Town, which almost culminated in Cape Town becoming the first major city globally to totally run out of water (Sousa et al. 2018; Burls et al. 2019). This same problem is now occurring in a number of other towns and cities within the region, and predictions are that the situation is likely to become worse in the future. Secondly, the 2022 floods in KwaZulu Natal have illustrated the devastating impacts that can be anticipated from increased frequencies and intensities of storm events. In both cases, integrated management of the catchments would have helped mitigate the severity of the impacts.

Impacts of bad management are less obvious in the marine environment and might only manifest through secondary measures such as fish yields per harvest effort. Sustainability of the marine resources is critical to many coastal industries as well as providing a critical source of protein. The marine resources need to be managed so that they can provide sustained social benefit in the future. In areas where the marine resource is already severely degraded, management strategies to allow for the recovery of the resource are needed, which will require a sound understanding of the ecological dynamics.

There is an ongoing need to link results coming out from theoretical environmental research to the actual livelihood impacts of individual people and communities. This can take place in a number of different ways. Short-term weather forecasting can assist in avoiding impacts from floods, dry spells, and heatwaves. Medium-term predictions can assist farmers in determining planting patterns for the season. Apart from these tactical decisions, strategic decisions need to be supported by transdisciplinary approaches. For example, projections on long-term climate change impacts will guide in the types of land-use systems that are biophysically and technically feasible for the future. In addition negotiation with various stakeholders (different interest groups) must also be considered to ensure social acceptability. (see Fig. 32.4), may even impact on spatial planning for settlement and investments.

32.3.9 Governance

Achieving sound environmental governance is a critical component for effective environmental management and conservation (Bennett and Satterfield 2018). Achieving sound governance is complex and although it involves many processes, at its core should be sound scientifically based decision making. Governance is generally considered to require institutions, structures, and processes that determine how, by whom, and for whom decisions are made (Bennett and Satterfield 2018). Clearly, for any decision making to be effective, it is critical that appropriately skilled and capacitated human resources are in place. Many aspects of governance, such as law enforcement, fall outside of the realms of environmental science; however, where possible, all aspects of governance from the initial formation of policy, rules, and norms through to on-the-ground implementation of policies and plans should be based on the best available scientific evidence. This requires both that policy relevant environmental research is being undertaken, and that there are appropriately skilled human resources within relevant government departments to assimilate this knowledge into governance processes.

The fact that environmental issues are split across what are typically multiple government departments (e.g., Environment, Agriculture, Water, Forestry) and have impacts on many other departments (e.g., Health, Rural Development, Tourism, Industrial Development, Economics) means that sound environmental management requires integration across departments and cannot be achieved in isolation. Effective integration requires special skills as well as appropriate research to understand these cross sector linkages. Further, it requires political will and appropriate policies and structures to facilitate intergovernmental department collaboration.

Given that our scientific understanding is not perfect, and that the environment is constantly changing, an adaptive management process is recommended. Over-arching environmental objectives are likely to be based on national and global policy consideration. These need to be translated into operational and verifiable goals, which guide action that is taken. A monitoring system is then required to understand if the verifiable goals are being achieved as anticipated. If not, the action and potentially the goals (if they are unachievable) need to be reassessed. This entire system needs to be based on the most up-to-date scientific understanding, with the science underpinning the management assumptions being reassessed based on new data and the outcomes from the monitoring and evaluation process. Capacity building of both state and academic staff within the southern African region remains critical in ensuring the above process (see Chap. 31).

32.4   Mutual Learning, Capacity Development, and Citizen Science

Interdisciplinary programs such as SPACES play an important role in developing skills within young emerging scientists in the region. Southern Africa has world-class expertise relating to its fauna and flora, and has also established itself as a continental leader in terms of climate change research. This makes it an ideal location for collaborative research with European partners in joint endeavors to understand the new challenges that global change is placing on the environment. European partners bring access to the most modern research ideas as well as access to advanced monitoring and modeling equipment and skills. The long-term data records from the region, together with good local infrastructure, make the region ideal for ongoing climate change research in a southern hemisphere developing world context.

Developing and retaining new scientific capacity in southern Africa has its unique challenges. High turnover in national departments is a reality of the region, driven in part by the transitional nature of expertise put in place during the colonial past, which is being replaced by new appointees aligned to national transformation priorities. Currently, there is high demand for the relatively limited pool of top-class graduates from what are termed previously disadvantaged racial groups. One unfortunate consequence is that many emerging young scientists are promoted into managerial positions at a relatively young age, and the capacities they have gained around the science need replacing. The benefit is that they are bringing their scientific background into the management arena. This, coupled with decreasing local funding for scientific research, means that partnering with European institutions can greatly facilitate this local capacity generation.

Climate change in particular is leading to new constellations and new ecological systems that cannot be understood in this form with previous knowledge alone. Mutual learning can better link previous experiences and point to new common paths for research and application. This also applies to knowledge transfer to Europe, where increased heat waves and drought will change living conditions. Joint summer schools and the exchange of students and scientists also offer the opportunity to get to know the other perspective between the continents. The newer IT-based technologies allow for shared classrooms and (blended) e-learning for certain courses (e.g., socio-ecological modeling). However, these learning methods cannot replace classical field work and experience in nature, but should be used in a complementary way in future projects.

A high upcoming potential for scientific projects has the integration of citizen science. These formats can mobilize specific long-term local knowledge and have additional benefits for the interactions among science, stakeholders, and policy (Peter et al. 2021; Vohland et al. 2021). Particularly, in biodiversity research, the involvement of individuals or natural history societies is a growing source for long-term monitoring and science. In the future, the direct involvement of people with expertise as citizen scientists should be supported in scientific projects in the context of land-use, landscape development, and climate change. These groups provide a link between research and education and, furthermore, thus are also multipliers of scientific results into society and for policy makers.

32.5   Lessons Learned and Recommendations for the Future

This book summarizes a very wide range of primary experimental, synthesis, and analytical work that has been generated over 5–10 years by a large group of authors. The composition and relationships between these authors has changed, in some cases dramatically (Chap. 1; Bieri et al. 2022), careers have been launched, new capacity has been created, and some strong long-term collaborative relationships have been built. Many new insights have been gained, new scientific findings have been made, and many important scientific and policy-relevant questions can now be asked from a new perspective and with a deeper background understanding.

At a higher level, self-analysis (Chap. 31) provides many guidelines on how future multi- and transdisciplinary cross-national collaborative programs could be further enhanced. In an era when capacity building and skills transfer are paramount in international negotiations like the UNFCCC, UNCCD, and UNCBD, this learning is valuable indeed. While the conception behind this collaborative work lies more than a decade in the past, the fruits of this work show that much of that conception was sound, and despite the many changes in regional and global political and environmental imperatives, the results presented here add great value to the fundamental understanding of regional functioning, risks, and potential solutions to the essential challenges facing the region.

Quoting Butts et al. 2016: "as we enter a world where extreme events, from natural disasters to deliberate attacks, become ever more common, sustainability and resilience become essential elements of our national security strategy. We ignore these values at our peril". The concept of environmental security is not new, but with an increased certainty of the adverse impacts of climate change on both the environment and livelihoods in the southern African region (IPCC 6th report, Chap. 7), it is clear that the work summarized here provides increased support for addressing the environmental impacts that pose major risks to the security of the region. Substantive social disruption and displacements due to extreme weather events, and longer-term risks such as from a biodiversity perspective confirm that the products of this work are of increasing value.

Climate-related risk has often been expressed in terms of the nature of the hazard, the vulnerability of the environment (or human population) being impacted, and the level of exposure. The work described here is particularly motivated by this framing. However, a fourth component to understanding risk was added recently, among others by WGII of IPCC (2014), and further elaborated by Simpson et al. (2021). This relates to the ability of society to respond to or manage the risks. Under this framing, there are complex interactions within and between the risk components and the responses to them that can amplify or reduce the final outcomes. Simpson et al. (2021) state "Indeed, recent evidence indicates how some of the most severe climate change impacts, such as those from deadly heat or sudden ecosystem collapse, are strongly influenced by interactions across multiple sectoral, regional, and response-option boundaries." It is with these complexities that the chapters in this book have only, in some instances, started to grapple.

Understanding risk and determining response options is now seen as increasingly requiring a more integrated understanding of the dynamics of the complex systems involved (see also Challinor et al. 2018). The model of collaboration that the work described here followed could be applied to better understand the interlinkages between components that have been addressed here, and to answer the more complex questions that arise from a strong theoretical basis of understanding the components themselves.

The SPACES I and SPACES II programs focused on key elements of the complex coupled earth system processes in the southern African region and have advanced our understanding on both the likely nature of future impacts as well as management and policy interventions that may assist communities in mitigating the effects of global environmental change. In this regard, these and previous international and national programs (e.g., BMBF-BIOTA, SAEON) have made substantive progress in understanding key and prioritized aspects of the environment and its climatic drivers. Looking further into the future, it is clear that that while an understanding of ecological processes remains the basis for robust policy recommendations, the critical feedbacks and influences of human responses at multiple spatial and temporal scales must become much more seamlessly integrated into such work.

Taking the above into account, it becomes clear that the southern African region lacks a well-integrated view on the implications of the wide range of policy options now proposed and being implemented to address biodiversity loss, climate change, and development imperatives. There is no credible set of projections, for example, on how these will interact at the regional level to influence land-use change and ecological sustainability. Not only is this view lacking, the region does not have the predictive means to inform such a view, beyond some partial analysis. In the northern Hemisphere, many integrated modeling approaches have been developed along these lines, and this is now a critical skills gap that needs to be filled to allow the work presented here to be leveraged for better decision making.

This synthetic view of the results of SPACES thus emphasizes more than ever the urgent need for integrated approaches to predictive modeling of coupled processes between the ocean, atmosphere and land, their feedback to the biosphere, and the outcomes as extrapolated spatially and taking into account a range of social-ecological responses. It is on this basis that concepts and plans for sustainable use of ecosystems on land and in the ocean, and related issues of climate protection, can be credibly developed in the future while taking into account social and economic development imperatives. Such approaches can provide guidance for implementation in practice without oversimplifying the scientific insights and findings that more properly represent a complex and dynamic world .

References

Baust S, Teh L, Harper S, Zeller D (2015) South Africa's marine fisheries catches (1950–2010). In: Le Manach F, Pauly D (eds) Fisheries catch reconstructions in the Western Indian Ocean, 1950–2010. Fisheries Centre Research Reports 23(2). Fisheries Centre, University of British Columbia, pp 129–150 [ISSN 1198–6727].

Bennett NJ, Satterfield T (2018) Environmental governance: a practical framework to guide design, evaluation and analysis. Conserv Lett 11:e12600. https://doi.org/10.1111/conl.12600

Bieri M, du Toit J, Feig G, Maluta NE, Mantlana B, Mateyisi M, Midgley GF, Mutanga S, von Maltitz G, Brümmer C (2022) Integrating project-based infrastructures with long-term greenhouse gas observations in Africa. Clean Air J 32(1):1–9

Bjornlund V, Bjornlund H, Van Rooyen AF (2020) Why agricultural production in sub-Saharan Africa remains low compared to the rest of the world – a historical perspective. Int J Water Resour Dev 36:S20–S53. https://doi.org/10.1080/07900627.2020.1739512

Blamey L, Shannon L, Bolton J, Crawford R, Dufois F, Evers-King H, Griffiths C, Hutchings L, Jarre A, Rouault M, Watermeyer K, Winker H (2015) Ecosystem change in the southern Benguela and the underlying processes. J Mar Syst 144:9–29. https://doi.org/10.1016/j.jmarsys.2014.11.006

Bode M, Kreiner A, van der Plas AK, Louw DC, Horaeb R, Auel H et al (2014) Spatio-temporal variability of copepod abundance along the 20°S monitoring transect in the Northern Benguela Upwelling System from 2005 to 2011. PLoS One 9(5):e97738. https://doi.org/10.1371/journal.pone.0097738

Bond W, Midgley G (2012) Carbon dioxide and the uneasy interactions of trees and savannah grasses. Philos Trans R Soc B 367:601–612. https://doi.org/10.1098/rstb.2011.0182

Burls NJ, Blamey RC, Cash BA, Swenson ET, Fahad A, Bopape M-JM, Straus DM, Reasin CJC (2019) The Cape Town "Day Zero" drought and Hadley cell expansion. npj Clim Atmos Sci 2:27. https://doi.org/10.1038/s41612-019-0084-6

Butts K, Goodman S, Nugent N (2016) The concept of environmental security. Solutions J. https://thesolutionsjournal.com/2016/02/22/the-concept-of-environmental-security-2/

Challinor AJ, Adger WN, Benton TG, Conway D, Joshi M, Frame D (2018) Transmission of climate risks across sectors and borders. Phil Trans R Soc A376:20170301. https://doi.org/10.1098/rsta.2017.0301

Chrysafi A, Virkki V, Jalava M et al (2022) Quantifying Earth system interactions for sustainable food production via expert elicitation. Nat Sustain 2022. https://doi.org/10.1038/s41893-022-00940-6

Conradie WS, Wolski P, Hewitson BC (2022) Spatial heterogeneity in rain-bearing winds, seasonality and rainfall variability in southern Africa's winter rainfall zone. Adv Stat Clim Meteorol Oceanogr 8:31–62. https://doi.org/10.5194/ascmo-8-31-2022

Daviter F (2017) Coping, taming or solving: alternative approaches to the governance of wicked problems. Policy Stud 38:571–588. https://doi.org/10.1080/01442872.2017.1384543

de Abreu MCS, de Andrade R (2019) Dealing with wicked problems in socio-ecological systems affected by industrial disasters: a framework for collaborative and adaptive governance. Sci Total Environ 694:133700. https://doi.org/10.1016/j.scitotenv.2019.133700

Eckardt FD, Bekiswa S, Von Holdt JR et al (2020) South Africa's agricultural dust sources and events from MSG SEVIRI. Aeolian Res 47:100637. https://doi.org/10.1016/j.aeolia.2020.100637

Eckert M, Gaigher R, Pryke JS, Samways MJ (2022) Conservation of complementary habitat types and small-scale spatial heterogeneity enhance soil arthropod diversity. J Environ Manag 317:115482. https://doi.org/10.1016/j.jenvman.2022.115482

Ekau W, Auel H, Hagen W, Koppelmann R, Wasmund N, Bohata K, Buchholz F, Geist F, Martin B, Schukat A, Verheye HM, Werner T (2018) Pelagic key species and mechanisms driving energy flows in the northern Benguela upwelling ecosystem and their feedback into biogeochemical cycles. J Mar Syst 88:49–62. https://doi.org/10.1016/j.jmarsys.2018.03.001

Grass I, Kubitza C, Krishna VV et al (2020) Trade-offs between multifunctionality and profit in tropical smallholder landscapes. Nat Commun 11:1186. https://doi.org/10.1038/s41467-020-15013-5

Griebenow S, Makunga NP, Privett S, Strauss P, Veste M, Kleinert A, Valentine A (2022) Soil pH influences the organic acid metabolism and exudation in cluster roots of Protea species from the Mediterranean-type Fynbos ecosystem, Western Cape, South Africa. Rhizosphere 21:100486. https://doi.org/10.1016/j.rhisph.2022.100486

Gypser S, Veste M (2019) Importance of P limitation for the C:N:P ratio in Robinia plantations on marginal sites. Verhandlungen GfÖ 49:82. https://doi.org/10.13140/RG.2.2.21245.31206

Haber W (2007) Naturschutz und Kulturlandschaften – Widersprüche und Gemeinsamkeiten. Anliegen Natur 31(2):3–11

Hebbelmann L, O'Connor TGO, du Toit JCO (2022) Fire as a novel disturbance and driver of vegetation change in Nama-Karoo rangelands, South Africa. J Arid Environm 203:104777. https://doi.org/10.1016/j.jaridenv.2022.104777

Hoffmann MP, Swanepoel CM, Nelson WCD, Beukes DJ, van der Laan M, Hargreaves JNG, Rötter RP (2020) Simulating medium-term effects of cropping system diversification on soil fertility and crop productivity in southern Africa. Eur J Agron 119. https://doi.org/10.1016/j.eja.2020.126089

Huggett J, Verheye H, Escribano R, Fairweather T (2009) Copepod biomass, size composition and production in the Southern Benguela: spatio-temporal patterns of variation, and comparison with other eastern boundary upwelling systems. Prog Oceanogr 83:197–207

Hugo S, Altwegg R (2017) The second Southern African Bird Atlas Project: causes and consequences of geographical sampling bias. Ecol Evol 7:6839–6849. https://doi.org/10.1002/ece3.3228

Hutchings L, Augustyn CJ, Cockcroft A et al (2009) Marine fisheries monitoring programmes in South Africa: review article. S Afr J Sci 105:182–192. https://doi.org/10.10520/EJC96925

Ichoku C (2020) African biomass burning and its atmospheric impacts. In: Oxford research encyclopedia of climate science. Oxford University Press

IPCC (2014) Climate Change 2014: impacts, adaptation and vulnerability: Part A: Global and Sectoral Aspects: Working Group II Contribution to the IPCC Fifth Assessment Report (pp. I-Ii). Cambridge: Cambridge University Press

IPCC (2022) Climate Change 2022: impacts, adaptation, and vulnerability. In: Pörtner H-O, Roberts DC, Tignor M, Poloczanska ES, Mintenbeck K, Alegría A, Craig M, Langsdorf S, Löschke S, Möller V, Okem A, Rama B (eds) Contribution of Working Group II to the Sixth Assessment Report of the Intergovernmental Panel on Climate Change. Cambridge University Press. In Press

Kanyanga JK (2009) El Niño Southern Oscillation (ENSO) and atmospheric transport over Southern Africa. PhD Thesis. University of Johannesburg, Johannesburg

Landis DA (2017) Designing agricultural landscapes for biodiversity-based ecosystem services. Basic Appl Ecol 18:1–12. https://doi.org/10.1016/j.baae.2016.07.005

Leventon J, Schaal T, Velten S, Dänhardt J, Fischer J, Abson DJ, Newig J (2017) Collaboration or fragmentation? Biodiversity management through the common agricultural policy. Land Use Policy 64:1–12. https://doi.org/10.1016/j.landusepol.2017.02.009

Lu Z, Liu X, Zhang Z et al (2018) Biomass smoke from southern Africa can significantly enhance the brightness of stratocumulus over the southeastern Atlantic Ocean. Proc Natl Acad Sci U S A 115:2924–2929. https://doi.org/10.1073/pnas.1713703115

Marean CW (2010) Pinnacle Point Cave 13B (Western Cape Province, South Africa) in context: the Cape Floral kingdom, shellfish, and modern human origins. J Hum Evol 59:425–443. https://doi.org/10.1016/j.jhevol.2010.07.011

Mastrangelo ME, Weyland F, Villarino SH, Barral MP, Nahuelhual L, Laterra P (2014) Concepts and methods for landscape multifunctionality and a unifying framework based on ecosystem services. Landsc Ecol 29(2):345–358

Matchaya G, Nhamo L, Nhlengethwa S, Nhemachena C (2019) An overview of water markets in southern Africa: an option for water management in times of scarcity. Water 11:1006. https://doi.org/10.3390/w11051006

Mbokodo I, Bopape M-J, Chikoore H, Engelbrecht F, Nethengwe N (2020) Heatwaves in the future warmer climate of South Africa. Atmosphere 11(7):712. https://doi.org/10.3390/atmos11070712

Murombo T (2022) Regulatory imperatives for renewable energy: South African perspectives. J Afr Law 66:97–122. https://doi.org/10.1017/S0021855321000206

Nelson WCD, Hoffmann MP, May C, Mashao F, Ayisi K, Odhiambo J, Bringhenti T, Feil J-H, Yazdan Bakhsh S, Abdulai I, Rötter RP (2022) Tackling climate risk to sustainably intensify smallholder maize farming systems in southern Africa. Environ Res Lett 17:075005. https://doi.org/10.1088/1748-9326/ac77a3

Nicholson S-A, Lévy M, Llort J et al (2016) Investigation into the impact of storms on sustaining summer primary productivity in the Sub-Antarctic Ocean. Geophys Res Lett 43:9192–9199. https://doi.org/10.1002/2016GL069973

Nicholson S-A, Whitt DB, Fer I et al (2022) Storms drive outgassing of CO2 in the subpolar Southern Ocean. Nat Commun 13:158. https://doi.org/10.1038/s41467-021-27780-w

O'Farrell PJ, Reyers B, Le Maitre DC, Milton SJ, Egoh B, Maherry A, Colvin C, Atkinson D, De lange W, Blignaut JN, Cowling RM (2010) Multi-functional landscapes in semi-arid environments: implications for biodiversity and ecosystem services. Landsc Ecol 25(8):1231–1246

Paterson B, Kirchner C, Ommer RE (2013) A short history of the Namibian Hake fishery – a social-ecological analysis. Ecol Soc 18:art66. https://doi.org/10.5751/ES-05919-180466

Peter M, Diekötter T, Höffler T, Kremer K (2021) Biodiversity citizen science: outcomes for the participating citizens. People Nat 3:294–311. https://doi.org/10.1002/pan3.10193

Pfeiffer M, Hoffmann MP, Scheiter S, Nelson W, Isselstein J, Ayisi KK, Odhiambo J, Rötter RP (2022) Effects of alternative crop-livestock management scenarios on selected ecosystem services in smallholder farming – a landscape perspective. Biogeosci Discuss [preprint]. https://doi.org/10.5194/bg-2022-61

Piketh SJ, Tyson PD, Steffen W (2000) Aeolian transport from southern Africa and iron fertilization of marine biota in the South Indian Ocean. S Afr J Sci 96:244–246. https://doi.org/10.10520/AJA00382353_8986

Quinn C, Ziervogel G, Taylor A, Takama T, Thomalla F (2011) Coping with multiple stresses in rural South Africa. Ecol Soc 16(3):2. https://doi.org/10.5751/ES-04216-160302

Ratto F, Steward P, Sait S, Pryke JS, Gaigher R, Samways MJ, Kunin W (2021) Proximity to natural habitat and flower plantings increases insect populations and pollination services in South African apple orchards. J Appl Ecol 58:2540–2551. https://doi.org/10.1111/1365-2664.13984

Reineking B, Veste M, Wissel C, Huth A (2006) Environmental variability and allocation trade-offs maintain species diversity in a process-based model of succulent plant communities. Ecol Model 199:486–504

Rittel HWJ, Webber MM (1973) Dilemmas in a general theory of planning. Policy Sci 4:155–169. https://doi.org/10.1007/BF01405730

Rixen T, Lahajnar N, Lamont T, Koppelmann R, Martin B, van Beusekom JEE, Siddiqui C, Pillay K, Meiritz L (2021) Oxygen and nutrient trapping in the Southern Benguela upwelling system. Front Mar Sci 8:1367

Roffe SJ, Fitchett JM, Curtis CJ (2021) Investigating changes in rainfall seasonality across South Africa: 1987–2016. Int J Climatol 41(Suppl 1):E2031–E2050. https://doi.org/10.1002/joc.6830

Roffe SJ, Steinkopf J, Fitchett M (2022) South African winter rainfall zone shifts: a comparison of seasonality metrics for Cape Town from 1841–1899 and 1933–2020. Theor Appl Climatol 147:1229–1247. https://doi.org/10.1007/s00704-021-03911-7

Rötter RP, Sehomi FL, Höhn JG, Niemi JK, van den Berg M (2016) On the use of agricultural system models for exploring technological innovations across scales in Africa: a critical review. ZEF – Discussion Papers on Development Policy 223, University of Bonn, Germany, 85 pp. https://doi.org/10.2139/ssrn.2818934

Rötter RP, Scheiter S, Hoffman MP, Pfeiffer M, Nelson WCD, Ayisi K, Taylor P, Feil J-H, Bakhsh SY, Isselstein J, Linstaedter A, Behn K, Westphal C, Odhiambo J, Twine W, Grass I, Merante P, Bracho-Mujica G, Bringhenti T, Lamega S, Abdulai I, Lam QD, Anders M, Linden V, Weier S, Foord S, Erasmus B (2021) Modelling the multi-functionality of African savanna landscapes under global change. Land Degrad Dev 32:2077–2081. https://doi.org/10.1002/ldr.3925

Schipper ELF, Dubash NK, Mulugetta Y (2021) Climate change research and the search for solutions: rethinking interdisciplinarity. Clim Chang 168:18. https://doi.org/10.1007/s10584-021-03237-3

Sheppard JP, Bohn Reckziegel R, Borrass L, Chirwa PW, Cuaranhua CJ, Hassler SK, Hoffmeister S, Kestel F, Maier R, Mälicke M, Morhart C, Ndlovu NP, Veste M, Funk R, Lang F, Seifert T, du Toit B, Kahle H-P (2020) Agroforestry: an appropriate and sustainable response to a changing climate in southern Africa? Sustainability 12:6796. https://doi.org/10.3390/su12176796

Shumsky S, Hickey G, Pelletier B, Johns T (2014) Understanding the contribution of wild edible plants to rural Socialecological resilience in semi-arid Kenya. Ecol Soc 19:34. https://doi.org/10.5751/ES-06924-190434

Siddiqui C et al (2023) Regional and global impact of CO2 uptake in the Benguela Upwelling System through preformed nutrients. Nat Commun 14(1):2582

Simpson NP, Mach KJ, Constable A et al (2021) A framework for complex climate change risk assessment. One Earth 4(4):489–501. https://doi.org/10.1016/j.oneear.2021.03.005

Sousa PM, Blamey R, Reason C, Ramos AM, Trigo RM (2018) The "Day Zero" Cape Town drought and the poleward migration of moisture corridors. Environ Res Lett 13:124025. https://doi.org/10.1088/1748-9326/aaebc7

Stock P, Burton RJF (2011) Defining terms for integrated (multi-inter-trans-disciplinary) sustainability research. Sustainability 3:1090–1113. https://doi.org/10.3390/su3081090

Stringer LC, Dougill AJ (2013) Channelling science into policy: enabling best practices from research on land degradation and sustainable land management in dryland Africa. J Environ Manag 114:328–335. https://doi.org/10.1016/j.jenvman.2012.10.025

Theron KJ, Gaigher R, Pryke JS et al (2020) Abandoned fields and high plant diversity support high spider diversity within an agricultural mosaic in a biodiversity hotspot. Biodivers Conserv 29:3757–3782. https://doi.org/10.1007/s10531-020-02048-9

Theron SN, Archer ERM, Midgley SJE, Walker S (2021) Agricultural perspectives on the 2015–2018 Western Cape drought, South Africa: characteristics and spatial variability in the core wheat growing regions. Agric For Meteorol 304–305:108405. https://doi.org/10.1016/j.agrformet.2021.108405

UNEP, WMO, UNCCD (2016) Global assessment of sand and dust storms. United Nations Environment Programme, Nairobi

Van Ittersum MK, Rötter RP, Van Keulen H, De Ridder N, Hoanh CT, Laborte AG, Aggarwal PK, Ismail AB, Tawang A (2004) A systems network (SysNet) approach for interactively evaluating strategic land use options at sub-national scale in South and South-east Asia. Land Use Policy 21:101–113. https://doi.org/10.1016/j.landusepol.2004.02.001

Van Jaarsveld AS, Pauw JC, Mundree S, Mecenero S, Coetzee BWT, Alard GF (2007) South African Environmental Observation Network: vision, design and status: SAEON reviews. S Afr J Sci 103:289–294

van Sittert L (2017) The marine fisheries of South Africa. In: Oxford research encyclopedia of African history. Oxford University Press

Van Wilgen NJ, van Wilgen BW, Midgley GF (2020) Biological invasions as a component of South Africa's global change research effort. In: van Wilgen B, Measey J, Richardson D, Wilson J, Zengeya T (eds) Biological invasions in South Africa, Invading Nature – Springer Series in Invasion Ecology, vol 14. Springer, Cham. https://doi.org/10.1007/978-3-030-32394-3_29

Verheye HM, Lamont T, Huggett JA, Kreiner A, Hampton I (2016) Plankton productivity of the Benguela Current Large Marine Ecosystem (BCLME). J Environ Dev 17:75–92. https://doi.org/10.1016/j.envdev.2015.07.011

Veste M, Böhm C (2018) Nachhaltige Holzproduktion in der Agrarlandschaft. In: Veste M, Böhm C (eds) Agrarholz - Schnellwachsende Bäume in der Landwirtschaft. Springer Spektrum, Heidelberg, pp 1–16. https://doi.org/10.1007/978-3-662-49931-3_1

Veste M, Littmann T, Kunneke A, Du Toit B, Seifert T (2020) Windbreaks as part of climate-smart landscapes reduce evapotranspiration in vineyards, Western Cape Province, South Africa. Plant Soil Environ 66:119–127. https://doi.org/10.17221/616/2019-PSE

Vohland K, Land-Zandstra A, Ceccaroni L, Lemmens R, Perelló J, Ponti N, Samson R, Wagenknecht K (2021) The science of citizen science. Springer Cham. https://doi.org/10.1007/978-3-030-58278-4

von Maltitz GP (2020) Harnessing science policy-interface processes to inform sustainable development policies in Sub-Saharan Africa. In: Gasparatos A, Ahmed A, Naidoo M, Karanja A, Fukushi K, Saito O, Takeuchi K (eds) Sustainability challenges in sub-Saharan Africa II: insights from Eastern and Southern Africa

von Maltitz GP, Scholes RJ, Erasmus B, Letsoalo A (2007) Adapting conservation strategies to accommodate climate change impacts in southern Africa. In: Leary N, Adejuwon J, Barros V, Burton I, Kulkarni J, Lasco R (eds) Climate change and adaptation. Earthscan, London, pp 1–27. ISBN 978-1-84407-470-9

Walker B, Salt D (2006) Resilience thinking. Sustaining ecosystems and people in a changing world. Island Press, Washington

Wissel C (1984) A universal law of the characteristic return time near thresholds. Oecologia 65:101–107. https://doi.org/10.1007/BF00384470

Wohlgezogen F, McCabe A, Osegowitsch T, Mol J (2020) The wicked problem of climate change and interdisciplinary research: tracking management scholarship's contribution. J Manag Org 26:1048–1072. https://doi.org/10.1017/jmo.2020.14

Wolski P, Conradie S, Jack C, Tadross M (2021) Spatio-temporal patterns of rainfall trends and the 2015–2017 drought over the winter rainfall region of South Africa. Int J Climatol 41(Suppl 1):E1303–E1319. https://doi.org/10.1002/joc.6768

Wren CD, Botha S, De Vynck J et al (2020) The foraging potential of the Holocene Cape south coast of South Africa without the Palaeo-Agulhas Plain. Quat Sci Rev 235:105789. https://doi.org/10.1016/j.quascirev.2019.06.012

Index

Printed in the United States
by Baker & Taylor Publisher Services